Statics
for Engineers

Springer

New York
Berlin
Heidelberg
Barcelona
Budapest
Hong Kong
London
Milan
Paris
Santa Clara
Singapore
Tokyo

B.B. Muvdi
A.W. Al-Khafaji
J.W. McNabb

Bradley University

Statics
for Engineers

With 1354 Illustrations

Springer

Bichara B. Muvdi
Amir W. Al-Khafaji
J.W. McNabb
Civil Engineering and Construction
Bradley University
Peoria, IL 61625
USA

Library of Congress Cataloging-in-Publication Data
Muvdi, B.B.
    Statics for engineers / B.B. Muvdi, A.W. Al-Khafaji, J.W. McNabb.
      p.   cm.
    Includes bibliographical references and index.
    ISBN 0-387-94779-5 (hc : alk. paper)
    1. Statics. I. Al-Khafaji, Amir Wadi. II. McNabb, J. W.
  III. Title.
  TA351.M87 1996
  620.1′03—dc20   96-17825

Printed on acid-free paper.

Production managed by Steven Pisano; manufacturing supervised by Jeffrey Taub.
Typeset by Asco Trade Typesetting Ltd., Hong Kong.
Printed and bound by R.R. Donnelley and Sons, Harrisonburg, VA.
Printed in the United States of America.

9 8 7 6 5 4 3 2 1

ISBN 0-387-94779-5 Springer-Verlag New York Berlin Heidelberg   SPIN 10538089

To my wife, Gladys, children, B. Charles, Diane, Katherine, Patti, and George, and grandchildren, Valerie, Christopher, and Richard, as well as my close friends, for their patience and continued support and encouragement during the preparation of the manuscripts.

B.B. Muvdi

To my children, Ali, Laith, and Elise, for keeping me young and happy.

A.W. Al-Khafaji

To Ada Spring McNabb, our children, their spouses, and young writers: Amy Chloe, Nicholas, and Phoebe.

J.W. McNabb

Preface

"Mechanics is one of the branches of physics in which the number of principles is at once very few and very rich in useful consequences. On the other hand, there are few sciences which have required so much thought—the conquest of a few axioms has taken more than 2000 years."—Rene Dugas, *A History of Mechanics*

Introductory courses in *engineering mechanics* (statics and dynamics) are generally found very early in engineering curricula. As such, they should provide the student with a thorough background in the basic fundamentals that form the foundation for subsequent work in engineering analysis and design. Consequently, our primary goal in writing *Statics for Engineers* and *Dynamics for Engineers* has been to develop the fundamental principles of *engineering mechanics* in a manner that the student can readily comprehend. With this comprehension, the student thus acquires the tools that would enable him/her to think through the solution of many types of engineering problems using logic and sound judgment based upon fundamental principles.

Approach

We have made every effort to present the material in a concise but clear manner. Each subject is presented in one or more sections followed by one or more examples, the solutions for which are presented in a detailed fashion with frequent reference to the basic underlying principles. A set of problems is provided for use in homework assignments. Great care was taken in the selection of these problems to ensure that all of the basic fundamentals discussed in the preceding section(s) have been given adequate coverage. The problems in a given set are organized so that the set begins with the simplest and ends with the most difficult ones. A sufficient number of problems is provided in each set to allow the use of both books for several semesters without having to assign the same problem more than once. Also, each and every one of the twenty three chapters in both books contains a set of review problems which, in general, are a little more challenging than the homework set of problems. This feature allows the teacher the freedom to choose homework assignments from either or from both sets. The two books have a total number of problems in excess of 2,600. All of the examples and problems were selected to reflect realistic situations. However, because these two books were written for beginning courses in statics and dynamics, the principal objective of these examples and problems remains to demonstrate the subject matter and to illustrate how the fundamental principles of mechanics may be used in the solution of practical problems.

Math Background

The prerequisite mathematical background needed for mastery of the material in both books consists primarily of high school courses in algebra and trigonometry and a beginning course in differential and integral calculus. We have made occasional use of vector analysis in the development of some concepts and basic principles and in the solution of some problems. The needed background in vector operations, however, is introduced and developed as needed throughout the books, particularly, in Chapters 3, 5, and 13. It should be emphasized, however, *that the vector approach is used only when it is judged to offer distinct advantages over the scalar method.* Such is the case, for example, in the solution of three-dimensional problems.

Free-Body Diagram

The very important concept of the free-body diagram is introduced in Chapters 2 and 4 but is used extensively throughout. It is our firm belief that the free-body diagram greatly enhances the understanding of the fundamental principles of mechanics. Thus, the free-body diagram is used not only in the solution of statics problems but also in the solution of dynamics problems whether Newton's second law, the energy method, or the impulse-momentum technique is used in their solution.

Organization

We have organized the material in both books to enable us to present the simple concepts before embarking on more difficult ones. Thus, the treatment of the mechanics of particles is dealt with before considering the mechanics of rigid bodies. Also, two-dimensional mechanics is presented separately, and before, three-dimensional mechanics. This approach allows the instructor to focus on simple concepts early in the semester and to postpone the more complex concepts until a later date when the student has had a chance to develop some maturity in the principles of mechanics. Furthermore, the separation between two- and three-dimensional mechanics in both *Statics for Engineers* and *Dynamics for Engineers*, provides the instructor flexibility in selecting topics to teach during a given semester.

Nontraditional Topics

In addition to the traditional topics found in existing books, we have included several new topics that we felt may be of interest to some teachers. These include: Axial Force and Torque Diagrams, General Theorem for Cables, a brief treatment of the Six Fundamental Machines and the use of Lagrange's Equations in the formulation of the equations of motion. It should be stated, however, that among the topics covered (both traditional and new) are some that are judged to be *not* essential for an understanding of the basic concepts of mechanics of rigid bodies. These topics are identified by asterisks in the Table of Contents.

Units

In view of the fact that the international system of units, referred to as SI (System International), is now beginning to gain acceptance in this country, it was decided to use it in this book. However, it is realized that a complete transition from the U.S. Customary to SI units will be a slow and costly process that may last as long as 20 years and possibly longer. Some factors that will play a significant role in slowing down the transformation process are the existing literature of engineering research and development, plans, and calculations, as well as structures and production machinery, that have been conceived and built using largely the U.S. Customary system of units. Thus, the decision was made to use both systems of units in this book. Approximately one-half of the examples and one-half of the problems are stated in terms of the U.S. Customary system whereas the remainder are given in terms of the emerging SI system of units.

Special Features

In addition to the features described under APPROACH, these two textbooks contain some special features that may be summarized as follows:

1. Many of the example and homework problems are designed to obtain *general symbolic solutions* that allow the student to view engineering problems from a broad point of view before assigning specific numerical data.
2. Each of the twenty-three chapters in both books is prefaced by a carefully written vignette designed to motivate the student prior to undertaking the study of the chapter.
3. The two books contain over 350 examples carefully worked out in sufficient detail to make the solutions easily understood by the student.
4. The two textbooks are characterized by the extensive use of free-body diagrams as well as impulse-momenta and inertia-force diagrams. Each of these diagrams is accompanied by a right-handed coordinate system that establishes the sign convention being used. With a few exceptions, three-dimensional, right-handed, x-y-z coordinate systems are shown with the x axis coming out of the paper. However, all of the two-dimensional, right-handed, x-y coordinate systems are shown with the x axis pointing to the right.
5. Each of the two volumes, *Statics for Engineers* and *Dynamics for Engineers*, has a companion *Solutions Manual*.

In addition to providing complete solutions to all of the homework problems, each *Solutions Manual* contains suggested outlines for courses in statics and in dynamics.

Appendices

Six Appendices containing information useful in the solution of many problems have been included at the end of each of the two books. Appendix A contains information about a selected set of areas, Appendix B, information about a selected set of masses, Appendix C, useful mathematical relations, Appendix D, selected derivatives, Appendix E, selected integrals, and Appendix F, information about supports and connections.

Acknowledgments

We acknowledge with much gratitude the assistance we received in the typing of the manuscript by Ms. Sharon McBride, Ms. Janet Maclean, and Wilma Al-Khafaji. We also acknowledge with thanks the help given by Dr. Farzad Shahbodaghlou and Dr. Akthem Al-Manaseer in the typing of the Solutions Manuals. The authors are grateful to the many colleagues and students who have contributed significantly and often indirectly to their understanding of statics and dynamics. Contributions by many individuals are given credit by reference to their published work and by quotations. The source of photographs is indicated in each case. These books have been written on the proposition *that good judgment comes from experience and that experience comes from poor judgment.* We certainly feel that the books are a real contribution to our profession, but we have miscalculated the enormous sacrifices they required. The quality of these books has been and will continue to be judged by our students and colleagues whose comments and suggestions have contributed greatly to the successful completion of the final manuscript. We would not have been able to complete this project without the help and support of our families. We apologize for any omissions.

The authors appreciate the efforts of the reviewers, who, by their criticisms and helpful comments, have encouraged us in the preparation and completion of the manuscript.

Our thanks to our editor Mr. Thomas von Foerster. All the people at Springer-Verlag who were involved with the production of this book deserve special acknowledgment for their dedication and hard work.

B.B.M.
A.W.A.
J.W.M.

Contents: Statics for Engineers

5   Equilibrium of Rigid Bodies in Three Dimensions   228

6   Truss Analysis   308

7   Frames and Machines   381

8   Internal Forces in Members   437

Contents: Dynamics for Engineers

1

Introductory Principles

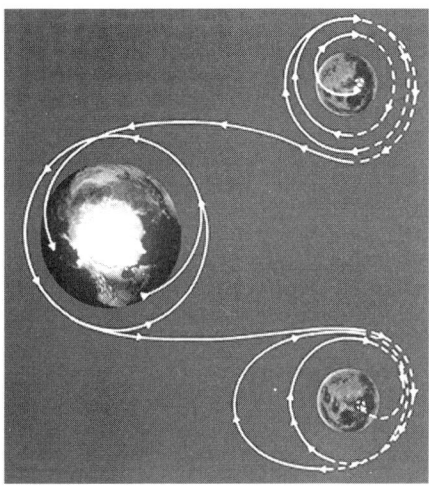

Newton's laws played a significant role in humankind's first landing on the Moon. This picture shows the path that Apollo 11 took.

Unlike art, which is dependent on a given person's gift, science relies on the collective contributions and discoveries made by many people throughout history. Additionally, the advancement of engineering science is rooted in the development of sound scientific principles upon which further development is made possible.

Three of the most important principles ever formulated were introduced by Isaac Newton more than three centuries ago and are covered in this chapter. These basic laws have greatly contributed to engineering science and practice. In describing nature, Newton's laws unite the heaven and Earth. Thus, while Galileo focused on earthly motion and Kepler had obtained his three laws on the motion of the heavenly bodies, Newton perfected these theories by unifying both. As Newton stated: "If I have seen a little farther than others, it is because I have stood on the shoulders of giants."

Newton's philosophy of science and his work on gravitation are most relevant to our present subject. His philosophy stated that laws are to be framed from verifiable phenomena that state nature's behavior in the precise language of mathematics, that is, mathematical models are developed to describe the physical world with measured precision and accuracy. However, engineering models and mathematical principles are inherently inexact because they often represent approximations of physical processes and may involve simplifying assumptions. Consequently, engineers use safety factors to help reduce the probability of failure, not necessarily to prevent it from happening.

In this chapter, we will demonstrate some basic concepts pertaining to systems of measurements and physical principles. Errors, models, and solutions techniques are summarized. Accordingly, the foundations upon which statics is based are introduced.

1.1
Review of Mechanics

Applied mechanics can be subdivided into *rigid body mechanics, mechanics of deformable bodies,* and *fluid mechanics.* Rigid body mechanics is further divided into *statics* and *dynamics.* In this text we deal with statics, the study of rigid bodies at rest or in motion with constant velocity and the forces acting on them. A companion text deals with dynamics, the study of rigid bodies in motion, the forces acting on them, and their corresponding velocities and accelerations.

More than 20 centuries ago, Archimedes began the development of statics but it was not until the work of Galileo, about four centuries ago, that a clear understanding of dynamics began to emerge. Table 1.1 "Historical Developments in Mechanics" is a brief presentation of historic milestones in the emergence of man's knowledge of mechanics. It

TABLE 1.1. Historical developments in mechanics

300 BC	Euclid developed geometry. Logical deductive method.
285–215 B.C.	Archimedes begins development of statics: law of the lever, location of center of gravity and concept of buoyancy. Greatest mathematician of antiquity.
1214–1294 AD	Roger Bacon—Beginnings of scientific method.
1452–1519	Leonardo da Vinci—Creative engineering genius. Understood statical moments and had begun to understand work of forces and perpetual motion.
1473–1543	Copernicus—Revolutionized man's place in the universe by showing that planets move around the Sun and that Earth is not fixed at the center of the universe.
1564–1642	Galileo—Began the development of dynamics and mechanics of materials.
1596–1650	Descartes—Combined algebra and geometry to create analytic geometry.
1642–1727	Newton—Wrote *Principia.* Discovered laws of dynamics, gravitational attraction, and calculus.
1646–1716	Leibnitz—Discovered calculus independently of Newton.
1654–1807	The Bernoullis—Deflections of beams, catenary equation, fundamental fluid mechanics, and additions to calculus.
1707–1783	Leonhard Euler—Most prolific writer of all time on mathematics. Improved mechanics and calculus.
1717–1783	Jean D'Alembert—Published *Traite de Dynamique.*
1736–1806	Charles Coulomb—Laws of dry friction.
1736–1813	Joseph L. Lagrange—*Mechanique Analytique* published in 1788.
1777–1859	Lewis Poinsot—*Elements de Statique* published in 1803.
1850–present	Refinements of classical mechanics, discovery of relativistic mechanics by Dr. Albert Einstein and an enormous number of applications. Specialization in branches and sub-branches of applied mechanics became a necessity.

is meant to arouse student interest in this exciting past and to illustrate that our present day knowledge was developed slowly over centuries.

Classical applied mechanics is a logical deductive science devoted to engineering applications. This engineering science is founded on undefined concepts and six fundamental principles. In geometry, we use terms such as *point* and *line*, but they remain undefined. Similarly, in mechanics, we accept the terms *space, time, force,* and *mass* as undefined concepts. We all have intuitive notions of what we mean by these terms, but we are unable to define them satisfactorily in simpler terms. Six fundamental principles, verifiable by experiment, are analogous to the axioms of geometry. We accept these principles because they can readily be shown to agree closely with test results. These principles are

Newton's Three Laws of Motion (Sec. 1.3)
Newton's Law of Universal Gravitation (Sec. 1.4)
The Parallelogram Law or the Law of Vector Addition (Sec. 2.2)
The Force Transmissibility Principle (Sec. 4.3)

These foundations of applied mechanics will be discussed in the Sections noted. It is obvious why this subject is also termed *Newtonian mechanics.*

1.2 Idealizations and Mathematical Models

Concentrated forces, particles, and *rigid bodies* are idealizations repeatedly used in statics and dynamics. Other idealizations will be discussed as required.

Concentrated Force

A concentrated force (a push or a pull) acts along a straight line termed its *line of action,* and this line intersects a body at a point. A force exerted on a body at a single point is termed a *concentrated force.* In reality, any applied force is distributed over a finite area. For example, if we were to apply a force by pushing a pencil point into the surface of a sheet of paper resting on a desk, then, this force is applied over a small area but, for convenience, we may choose to assume that the force is applied at a point. It is much simpler to deal with this concentrated force than to deal with a force distributed in some manner over a finite area. Local force intensities and deformations will depend upon the area of distribution, but such problems are considered in another branch of applied mechanics termed *The mechanics of deformable bodies.* We restrict the use of concentrated forces to cases where the area of distribution would be small compared to the overall size of the body.

Particle

A particle is a mass concentrated at a geometric point. Of course, all bodies with mass occupy a certain volume in space rather than being concentrated at a point. It is often convenient to idealize a body of finite dimensions as a *point mass*, and, in doing this, we have used the concept of a particle.

Rigid Body

To define what we mean by a rigid body, we consider two states of the body. The first state refers to the unloaded body, that is, a body not subjected to forces, and the second state refers to the loaded body, that is, a body subjected to forces. In the first or unloaded state, we consider two points, A and B, of the body and measure the straight line distance between them. For any real body in the second or loaded state, if we measure this distance AB, it will, in general, differ from the measurement taken in the first or unloaded state. *A perfectly rigid body is an idealized body such that the distance between two points, A and B chosen randomly, will be the same in the unloaded and loaded states.* In other words, all sets of points remain in their same relative positions as forces are applied. *Briefly, we may state that a perfectly rigid body is one which does not deform under the action of forces.* All real bodies do deform when subjected to forces, but, in many cases, these deformations are negligible.

If we use these and other idealizations and neglect certain other effects, then, we have created a mathematical model of a physical system. The particular model to be used depends upon the problem under investigation. For example, if we wish to study the motion of the planet Mars in its orbit about the Sun, we would idealize the planet as a particle. It is massive and occupies a considerable volume but it is very small in size compared to the extent of its path around the Sun. However, if we were to study the propagation of earthquake waves in this planet, we might initially idealize the planet as an elastic sphere. Comparison of theoretical predictions with experimental results may require modifying our mathematical model or rejecting it. *In general, the validity of a given mathematical model is assessed by comparing theoretical predictions with experimental results.*

1.3 Newton's Laws

The beginning of the scientific study of dynamics and the mechanics of deformable bodies can be traced back to the publication in 1638 of *Dialogue concerning Two New Sciences by Galileo.* Understanding both of these branches of applied mechanics is of primary importance in the scientific and technological revolution which is now about three and a half centuries old. This is a very short time compared to the time since the beginning of humankind.

Isaac Newton was born on Christmas Day, 1642, the year Galileo died. Newton went to Cambridge University in 1661 and graduated early in 1665. In autumn of that year, the great plague was raging in London and the University was closed. Newton returned to the Woolsthorpe house where he was born. He stayed until the spring of 1667 and during this time, in his early twenties, he discovered the law of universal gravitation, the calculus, and the composition of white light. In the Portsmouth Collection, written when he was about 73 years old, Newton stated: ... "All this was in the two plague years of 1665 and 1666, for in those days I was in the prime of my age for invention, and minded mathematics and philosophy more than at any time since." The *Principia* was published in 1687 when Newton was forty-five years old. It contained statements of the three laws of motion and the law of universal gravitation which are the foundations of classical engineering mechanics. The *Principia* has been called the greatest scientific work ever published. Newton spent about 18 months assembling and writing this revolutionary book. Sir Isaac Newton died in 1727 at the age of 85.

A Newtonian Reference Frame is a set of coordinate axes in terms of which Newton's three laws of motion enable us to predict motions satisfactorily when compared to experimental results. The coordinate axes are assumed to be attached to a rigid body which is readily identified. For earthbound engineering problems, we usually choose a reference frame attached to Earth as Newtonian.

Isaac Newton described a frame of reference which is fixed in the universe, but we have learned that there are no such frames of reference. Earth rotates about its polar axis and orbits the Sun. The Sun is one of billions of stars which move about the center of the Milky Way galaxy. Billions of galaxies comprise the expanding universe, and we know that these galaxies are in motion with respect to each other. A particle moving with respect to a reference frame attached to Earth participates in these motions of Earth, the Sun and the Milky Way galaxy. Nevertheless, our predictions using Newton's laws and a reference frame attached to the Earth are satisfactory for engineering applications. Any reference frame whose origin moves at constant velocity with respect to a known Newtonian Reference Frame and whose axes remain parallel to those of the known frame is also a Newtonian Reference Frame.

Newton's First Law: *A particle will remain at rest or move with constant velocity unless it is subjected to an unbalanced force.*

The phrase *remain at rest* means that the particle is not moving with respect to a Newtonian Reference Frame. If such a frame is attached to the Earth, then *remain at rest* means that the particle does not move with respect to Earth. The phrase *move with constant velocity* means that the particle moves along a straight line path with constant speed. Velocity has both magnitude and direction and is termed a *vector quantity*. If the velocity vector remains constant, it cannot change in magnitude or direction. The phrase *unless it is subjected to an unbal-anced force* means that the particle which remains at rest or moves at constant velocity has a zero force acting on it and, only if the force acting on it is nonzero, will it begin to move with nonconstant velocity. These ideas are summarized in Figure 1.1.

In Figure 1.1(a), a particle rests on a horizontal surface and the only forces acting on it are its weight W and the upward push P = W from the surface. These two forces are perfectly balanced and, as long as no unbalanced force acts on it, the particle will remain at rest (i.e. its velocity $v = 0$) indefinitely. In Figure 1.1(b), a particle moves along a smooth horizontal surface at constant velocity v. It is inconsequential how the particle acquired this constant velocity. But, as long as the only forces acting on it are its weight W and the upward push P, which are in perfect balance, the particle will continue to move indefinitely at constant velocity along the horizontal surface.

Newton's Second Law: A particle of mass m subjected to an unbalanced force will be accelerated in the direction of the force, and the magni-tude of this acceleration will be proportional to the magnitude of the unbalanced force. (This law may also be stated by saying that the unbalanced force acting on the particle is proportional to the time derivative of its linear momentum vector.)

To understand Newton's second law, we need to understand the meaning of acceleration. Acceleration of a particle is defined as the time rate of change of the velocity vector. Acceleration is a vector quantity which reflects *changes in magnitude and direction* of the velocity vector. Mathematically we express the second law as $\mathbf{F} = km\mathbf{a}$

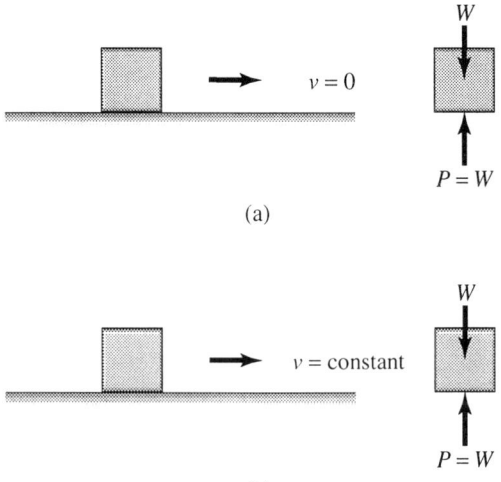

FIGURE 1.1.

where $\mathbf{F}$ is the unbalanced force acting on the particle, m is the mass of the particle, $\mathbf{a}$ is the acceleration of the particle and k is a proportionality constant which depends upon the system of units chosen. SI and U.S. Customary units, to be introduced later, are used in this text and, for both of these units, the proportionality constant k has a unit value. Thus,

$$\mathbf{F} = m\mathbf{a}. \tag{1.1}$$

Note that this equation expresses the equivalence of two vectors. The vector $\mathbf{F}$ is the unbalanced force acting on the particle. The vector $m\mathbf{a}$ is termed the *inertial force* which is the product of the particle mass m, a scalar, and its acceleration $\mathbf{a}$, a vector. The inertial force vector has the same direction as the acceleration vector. The inertial force has a magnitude equal to the product of the mass m and the magnitude a of the acceleration vector $\mathbf{a}$. Newton's second law is shown pictorially in Figure 1.2.

Newton's first law is a special case of the second law. If we start with $\mathbf{F} = m\mathbf{a}$ and consider the special case of the acceleration vector vanishing, or $\mathbf{a} = \mathbf{0}$ where $\mathbf{0}$ symbolizes a null vector, then $\mathbf{F} = \mathbf{0}$. Because the acceleration vector is a null vector, the velocity vector $\mathbf{v}$ remains constant. A constant vector is either a null vector or a vector which does not change in magnitude or direction with time. This leads to the conclusion that the particle either remains at rest (i.e., a null velocity vector) or moves with constant velocity (i.e., the acceleration vector is null) unless subjected to an unbalanced force (i.e., $\mathbf{F} \neq \mathbf{0}$).

Newton's Third Law: *If bodies A and B are in contact, then, the force exerted by body B on body A is equal and opposite in sense to the force exerted by body A on body B.*

Another way of expressing Newton's third law is to say that for every action there is an equal and opposite reaction. Forces, in general,

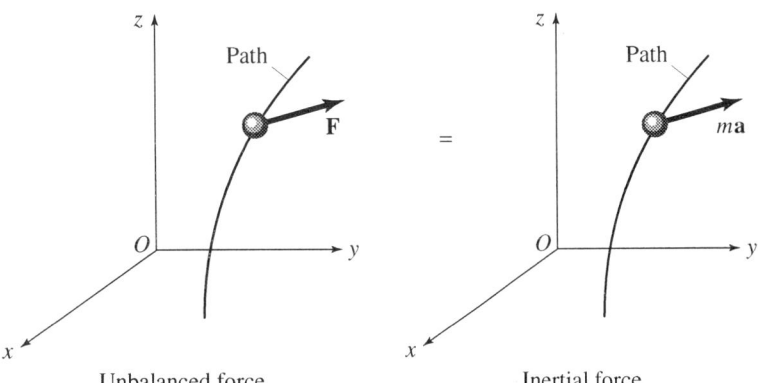

FIGURE 1.2.   Unbalanced force   Inertial force

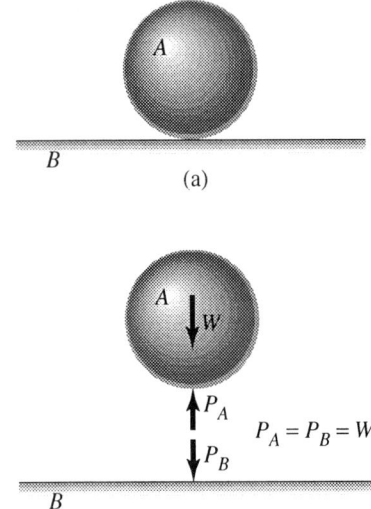

$P_A = P_B = W$

FIGURE 1.3.

(a)

(b)

may be described as contact, gravitational, electrostatic, electromagnetic, and nuclear. In our study of engineering mechanics, we will deal primarily with contact and gravitational forces. Contact forces arise between bodies which are in contact with each other whereas gravitational forces act at a distance through space.

These forces of action and reaction are vectors with a common line of action. This law applies to forces acting at points of contact and to the resultant of distributed forces which act over areas in contact. Newton's third law is shown pictorially in Figure 1.3. Figure 1.3(a) shows a sphere A of weight W at rest on a horizontal surface B. At their point of contact, equal forces of action and reaction exist according to Newton's third law. To expose these forces, the two bodies are shown separated in Figure 1.3(b). Here P_A represents the force that the horizontal surface exerts on the sphere A and P_B is the force that the sphere exerts on the horizontal surface. According to Newton's third law, $P_A = P_B$. Note that because $P_A = W$, it follows that $P_A = P_B = W$.

1.4
Newton's
Law of
Universal
Gravitation

An understanding of the Universal Law of Gravitation is essential for astronomical predictions and space exploration. This law has been termed the greatest generalization of all time because it applies to any two particles in the universe. It is given by

$$F = \frac{Gm_1 m_2}{r^2} \tag{1.2}$$

where F = the gravitational attractive force,
G = the universal gravitational constant,

$m_1, m_2 =$ the masses of the two particles which attract each other, and $r =$ the straight line distance between the particles.

Equation (1.2) states that the gravitational attractive force between two particles varies directly with the product of their masses and inversely with the square of the distance between them. This law is termed the *inverse square law of gravity*. Figure 1.4 shows the relatively large reductions in F with increasing values of r. Doubling r reduces F to one fourth its original value whereas tripling r reduces F to one-ninth its original value. As r becomes small, this attractive force, which tends to pull the bodies together, becomes relatively large, and, as r becomes very large, this attractive force becomes practically negligible. SI and U.S. Customary units for the universal law of gravitation are shown in Table 1.2 together with numerical values for the universal gravitational constant G.

As an illustration of this law, consider the motion of a spacecraft in elliptic orbit about Earth, shown in Figure 1.5(a). The spacecraft is shown at any position in its orbit at a distance r measured from the mass center of the Earth, point G. In this case, $m_1 = M_E$ where M_E is the mass of Earth and m_2 is the mass of the spacecraft.

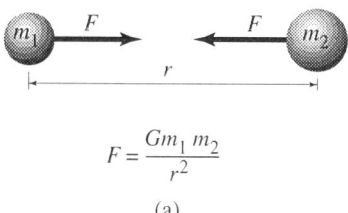

$$F = \frac{Gm_1 m_2}{r^2}$$

(a)

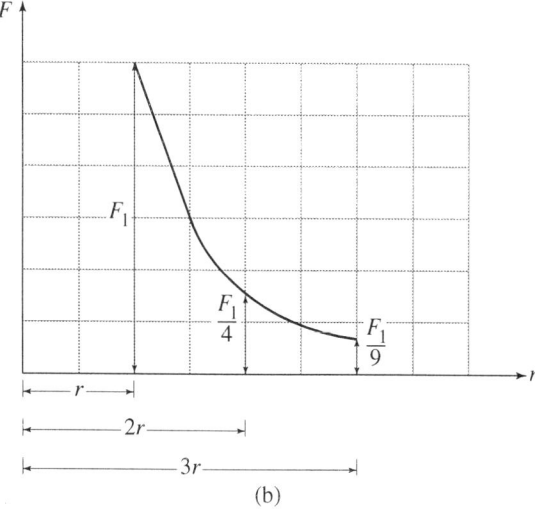

(b)

FIGURE 1.4.

TABLE 1.2. Units and values of G for the universal law of gravitation

Quantity	Symbol	SI Units	U.S. Customary Units
Force of attraction	F	N (Newton)	lb (pounds)
Gravitational constant	G	66.73×10^{-12} m^3/(kg·s^2)	34.4×10^{-9} ft^4/(lb·s^4)
Masses	m_1, m_2	kg (kilograms)	slugs
Distance	r	m (meters)	ft (feet)

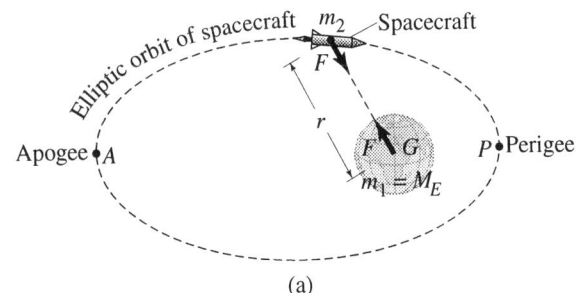

(a)

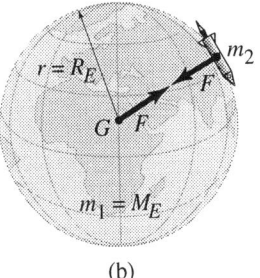

(b)

FIGURE 1.5.

As the spacecraft moves along its elliptic path, only r changes, and, as r changes, the magnitude and direction of the attractive force F changes. When the spacecraft reaches the point P nearest Earth's mass center, termed the *perigee*, the force F reaches its largest value. At point A furthest from the Earth's mass center, termed the *apogee*, the force F reaches its smallest value. Because the spacecraft is above Earth's atmosphere and we assume no meteorite collisions, then, the gravitational force F is the only force acting on the spacecraft.

When the spacecraft of mass m_2 is at the surface of Earth as shown in Figure 1.5(b), m_1 is the mass of the Earth M_E, and r is the radius of Earth R_E. The attractive force F becomes

$$F = \frac{Gm_2 M_E}{R_E^2}. \tag{1.3}$$

This gravitational attractive force F at the surface of Earth is known as the weight W of the spacecraft. By Newton's second law, $F = ma$. Thus,

$$F = W = m_2 g \qquad (1.4a)$$

where g is the acceleration due to gravity at the surface of Earth. Equation (1.4a) may be solved to obtain the mass of the spacecraft. Thus,

$$m_2 = \frac{W}{g}. \qquad (1.4b)$$

Equation (1.4b) shows that the mass of an object, such as the spacecraft in question, is a quantity that can be obtained by dividing the weight of the object by the acceleration of gravity. It should be observed, however, that both the weight of the object W and the acceleration of gravity g vary with the position of the object relative to the center of Earth. Nevertheless, the ratio W/g (i.e., the mass of the object) remains constant regardless of where in the universe the object is located. Thus, for example, while the weight of the spacecraft is different on Earth from its weight on the Moon, the mass of the spacecraft is the *same* in both places.

Equating the values of F of Eqs. (1.3) and (1.4b) yields,

$$m_2 g = \frac{Gm_2 M_{\mathrm{E}}}{R_{\mathrm{E}}^2},$$

$$GM_{\mathrm{E}} = gR_{\mathrm{E}}^2. \qquad (1.5)$$

Equation (1.5) provides a convenient means of computing the product GM_{E} which arises in the study of space mechanisms.

1.5 Systems of Units and Conversion Factors

Two systems of units are in common use today, identified as the *International System of Units (SI)* and the *U.S. Customary system of units*. Both systems are divided into three classes of units as follows:

1. *Base units*
2. *Supplementary units*
3. *Derived units*

The basic distinction between the two systems lies in the choice of the *base units*. In the SI system, the base units are those of *length, mass,* and *time,* as shown in Table 1.3, and all other units including force, are either supplementary or derived, as shown in Table 1.4. This system of units is referred to as an *absolute* system because the three base units remain constant regardless of the location in the universe where measurements are made. Thus, for example, as stated earlier, the mass m of an object does not change from location to location. Table 1.5 shows

TABLE 1.3. Base SI units for applied mechanics

Quantity	Name	Symbol
Length	meter	m
Mass	kilogram	kg
Time	second	s

TABLE 1.4. Derived SI units for applied mechanics

Quantity	Units	*Convenient Units
Acceleration	m/s^2	
Angle	rad	
Angular acceleration	rad/s^2	
Angular impulse	$kg \cdot m^2/s$	$N \cdot s \cdot m$
Angular momentum	$kg \cdot m^2/s$	$N \cdot s \cdot m$
Angular velocity	rad/s	
Area	m^2	
Density	kg/m^3	
Energy	$kg \cdot m^2/s^2$	$J = N \cdot m$ (Joules)
Force	$kg \cdot m/s^2$	N (Newton)
Frequency	s^{-1}	Hz (Hertz)
Impulse	$kg \cdot m/s$	$N \cdot s$
Length	m	
Mass	kg	
Moment of force	$kg \cdot m^2/s^2$	$N \cdot m$
Momentum	$kg \cdot m/s$	$N \cdot s$
Period	s	
Power	$kg \cdot m^2/s^3$	$W = J/s$
Pressure	$kg/(m \cdot s^2)$	$Pa = N/m^2$
Stress	$kg/(m \cdot s^2)$	$Pa = N/m^2$
Time	s	
Velocity	m/s	
Volume	m^3	
Weight	$kg \cdot m/s^2$	N
Work	$kg \cdot m^2/s^2$	$J = N \cdot m$

* If no entry, then base units are convenient.

some of the most commonly used derived SI quantities in applied mechanics along with their names and abbreviated symbols. On the other hand, in the U.S. Customary system, the base units are those of length, force, and time as shown in Table 1.6. All other units, including mass, are either supplementary or derived, as shown in Table 1.7 which also includes some convenient conversion factors that may be used in converting from the U.S. Customary to the SI system. The U.S. Cus-

TABLE 1.5. Names of SI symbols for applied mechanics

Quantity	Symbol	Name	Base Units
Energy	J	Joule	$kg \cdot m^2/s^2$
Force	N	Newton	$kg \cdot m/s^2$
Frequency	Hz	Hertz	s^{-1}
Pressure	Pa	Pascal	$kg/(s^2 \cdot m)$
Stress	Pa	Pascal	$kg/(s^2 \cdot m)$
Power	W	Watt	$kg \cdot m^2/s^3$

TABLE 1.6. Base U.S. Customary units for applied mechanics

Quantity	Name	Symbol
Length	foot	ft
Force	pound	lb
Time	second	s

tomary system is referred to as a *gravitational* system of units because the mass of an object in this system depends upon the gravitational attraction of the planet (e.g., Earth) on which the object is located.

The SI base units have been defined by the International Conference Generale des Poids et Mesures (CGPM) as follows:

1. *Length: The meter is the length equal to the distance traveled by light in a vacuum during 1/299 792 458 s (CGPM-1983).*
2. *Time: The second is the duration of 9 192 631 770 periods of radiation corresponding to the transition between the two hyperfine levels of the ground state of the ^{133}cesium atom (CGPM-1967).*
3. *Mass: The kilogram is the mass of a cylinder of platinum-iridium alloy kept in France by the International Bureau of Weights and Measures. The United States has a duplicate of this cylinder (CGPM-1889 and 1901).*

Only a single *supplementary unit* is widely used in applied mechanics. It is the radian for plane angles signified by the symbol *rad*. The radian is defined as the central angle of a circular arc such that the arc length intercepted between the two rays of the angle equals the radius of the circle. The radian is used in both the International System of Units (SI) and the U.S. Customary System of Units.

Derived units such as those of Tables 1.4 and 1.7 are obtained by combining base, supplementary, or other derived units. Note that all the conversion factors given in Table 1.7 are based upon the following two equalities: 1 ft = 0.3048 m and 1 lb = 4.448 N.

TABLE 1.7. Derived U.S. Customary units and conversion factors

Quantity	Units	Conversion Factors
Acceleration	ft/s^2	0.3048 m/s^2
Angle	rad	rad
Angular acceleration	rad/s^2	rad/s^2
Angular impulse	lb·ft·s	1.356 N·m·s
Angular momentum	lb·ft·s	1.356 N·m·s
Angular velocity	rad/s	rad/s
Area	ft^2	0.0929 m^2
Density	lb·s^2/ft^4 (slugs/ft^3)	515.2 kg/m^3
Energy	ft·lb	1.356 J
Force	lb	4.448 N
Frequency	s^{-1}	1.000 Hz
Impulse	lb·s	4.448 N·s
Length	ft	0.3048 m
Mass	lb·s^2/ft (slug)	14.59 kg
Moment of force	lb·ft	1.356 N·m
Momentum	lb·s	4.448 N·s
Period	s	s
Power	ft·lb/s	1.356 W
Pressure	lb/ft^2	47.88 Pa
Stress	lb/ft^2	47.88 Pa
Time	s	s
Velocity	ft/s	0.3048 m/s
Volume	ft^3	0.02832 m^3
Weight	lb	4.448 N
Work	ft·lb	1.356 J

Conversion of units from one system to another is based upon multiplication by unit ratios and cancellation of units not desired in the final answer. For example, to convert 1225 N to lb, we make use of Table 1.7 to write

$$1225 \text{ N} \times \frac{1 \text{ lb}}{4.448 \text{ N}} = 275.4 \text{ lb}.$$

Note that the N units cancel and the lb unit remains. Now, to convert 75.50 lb·ft to N·m we write

$$75.50 \text{ lb·ft} \times \frac{4.448 \text{ N}}{1 \text{ lb}} \times \frac{0.3048 \text{ m}}{1 \text{ ft}} = 102.4 \text{ N·m}.$$

Note that the lb and ft units cancel and the N and m units remain. Examples 1.1 and 1.2 present further unit conversions using the same method introduced above.

Let us use the SI system of units and Newton's second law to investigate some of the characteristics of the 1.0 kg mass shown in Figure 1.6. In Figure 1.6(a), the force F needed to produce an acceleration of 1.0 m/s^2 may be found from Newton's second law. Thus,

$$F = ma = (1.0 \text{ kg})(1.0 \text{ m/s}^2) = 1.0 \text{ kg·m/s}^2 = 1.0 \text{ N}.$$

Thus, as stated earlier, in the SI system, the force is a derived quantity whose unit is the kg·m/s^2 which has been named the Newton (symbol N) in honor of Sir Isaac Newton who discovered the three basic laws of mechanics as well as the law of gravitation. Thus, by definition, *a force of 1.0 N applied to a body of 1.0 kg mass produces an acceleration of 1.0 m/s^2*. Figure 1.6(b) shows a mass of 1.0 kg in free fall near the surface of Earth. This mass will have an acceleration due to gravity $a = g = 9.81$ m/s^2. Thus, by Newton's second law,

$$F = ma = (1.0 \text{ kg})(9.81 \text{ m/s}^2) = 9.81 \text{ kg·m/s}^2 = 9.81 \text{ N}.$$

Therefore, Earth attracts a 1.0 kg mass with a force of 9.81 N. In other words, *the weight of a 1.0 kg mass at the surface of the Earth is W = 9.81 N*.

Let us now use the U.S. Customary system along with Newton's second law to examine some of the characteristics of the mass m shown in Figure 1.7. In Figure 1.7(a), the mass m is subjected to a force of 1.01b which produces an acceleration of 1.0 ft/s^2. Solving for the mass m from Newton's second law,

$$m = F/a = (1.0 \text{ lb})/(1.0 \text{ ft/s}^2) = 1.0 \text{ lb·s}^2/\text{ft} = 1.0 \text{ slug}.$$

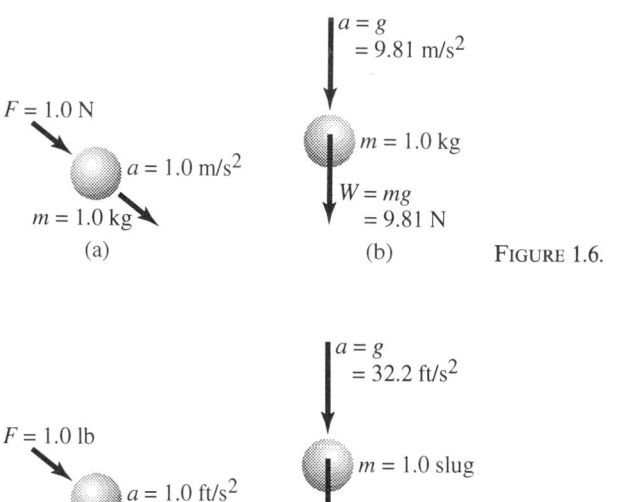

(a)   (b)   FIGURE 1.6.

(a)   (b)   FIGURE 1.7.

Thus, as stated earlier, in the U.S. Customary system, the mass m is a derived quantity whose unit is lb·s²/ft which has been named the *slug*. Therefore, by definition, *a force of 1.0 lb applied to a mass of 1.0 slug produces an acceleration of 1.0 ft/s²*. Figure 1.7(b) shows a mass of 1.0 slug in free fall near the surface of Earth. This mass will have an acceleration due to gravity of $g = 32.2$ ft/s². Thus, by Newton's second law,

$$F = ma = (1.0 \text{ lb·s}^2/\text{ft})(32.2 \text{ ft/s}^2) = 32.2 \text{ lb.}$$

Therefore, the Earth attracts the 1.0 slug mass with a force of 32.2 lb. In other words, *the weight of a 1.0 slug mass at the surface of the Earth is W = 32.2 lb.*

It should be emphasized that the value of g varies with altitude and latitude. However, for convenience, we will use a value of 9.81 m/s² in the SI system and 32.2 ft/s² in the U.S. Customary system.

It is useful to determine the product $GM_E = gR_E^2$ given by Eq. (1.5) and to find from it the mass of Earth M_E in both the SI and the U.S. Customary systems. Thus,

SI System

$$GM_E = gR_E^2 = (9.81 \text{ m/s}^2)(6.373 \times 10^6 \text{ m})^2 = 3.984 \times 10^{14} \text{ m}^3/\text{s}^2$$

From Table 1.2, $G = 66.73 \times 10^{-12} \text{ m}^3/\text{kg·s}^2$. Therefore,

$$M_E = GM_E/G = \frac{3.984 \times 10^{14} \text{ m}^3/\text{s}^2}{66.73 \times 10^{-12} \text{ m}^3/\text{kg·s}^2} = 5.970 \times 10^{24} \text{ kg;}$$

U.S. Customary System

$$GM_E = gR_E^2 = (32.2 \text{ ft/s}^2)(3960 \times 5280 \text{ ft})^2 = 1.408 \times 10^{16} \text{ ft}^3/\text{s}^2.$$

From Table 1.2, $G = 34.4 \times 10^{-9} \text{ ft}^4/(\text{lb·s}^4)$. Therefore,

$$M_E = GM_E/G = \frac{1.408 \times 10^{16} \text{ ft}^3/\text{s}^2}{34.4 \times 10^{-9} \text{ ft}^4/(\text{lb·s}^4)}$$

$$= 4.093 \times 10^{23} \text{ lb·s}^2/\text{ft} = 4.093 \times 10^{23} \text{ slug.}$$

SI Prefixes

In the SI system of units, prefixes are used to report very large and very small numbers. The SI prefixes in use are shown in Table 1.8, and they represent multiples or submultiples of a given unit. Thus, for example 5000 m may be conveniently written as 5 km (kilometer) and 0.030 m as 30 mm (millimeters). In our study of engineering mechanics, we will primarily have use for the following prefixes:

Length	1 m = 1000 mm
	1 km = 1000 m
Mass	1 kg = 1000 g
	1 Mg = 1000 kg
Force	1 kN = 1000 N
	1 MN = 1000 kN

TABLE 1.8. SI prefixes

Prefix	Symbol	Power of 10	Multiplication Factor
tetra	T	12	1 000 000 000 000
giga	G	9	1 000 000 000
mega	M	6	1 000 000
kilo	k	3	1 000
hecto*	h	2	100
deka*	da	1	10
deci*	d	−1	0.1
centi*	c	−2	0.01
milli	m	−3	0.001
micro	μ	−6	0.000 001
nano	n	−9	0.000 000 001
pico	p	−12	0.000 000 000 001
femto	f	−15	0.000 000 000 000 001
atto	a	−18	0.000 000 000 000 000 001

* Avoid the use of these prefixes except for areas and volumes

■ Example 1.1

Convert the following values expressed in U.S. Customary units into the stated SI units. (a) 60.0 mph to m/s, (b) 100.0 lb to N, (c) 10.0 slug to kg, and (d) 50.0 lb·ft to N·m. Refer to Table 1.7 as required for conversion factors.

Solution

In each case make use of the concept of multiplication by unit values and cancellation of units not required in the answer.

(a) $\dfrac{60 \text{ miles}}{1 \text{ hr}} \times \dfrac{1 \text{ hr}}{3600 \text{ s}} \times \dfrac{5280 \text{ ft}}{1 \text{ mile}} \times \dfrac{0.3048 \text{ m}}{1 \text{ ft}} = 26.8 \text{ m/s}$ ANS.

(b) $100 \text{ lbs.} \times \dfrac{4.448 \text{ N}}{1 \text{ lb}} = 445 \text{ N}$ ANS.

$10.0 \text{ slugs} \times \dfrac{14.59 \text{ kg}}{1 \text{ slug}} = 146 \text{ kg}$ ANS.

(d) $50.0 \text{ lb. ft} \times \dfrac{4.448 \text{ kg}}{1 \text{ lb}} \times \dfrac{0.3048 \text{ m}}{1 \text{ ft}} = 67.8 \text{ N·m}$ ANS.

■ Example 1.2

Convert the following values expressed in SI units into the stated U.S. Customary units. (a) 500.0 N·m to lb·in, (b) 125 Pa to psi, (c) 750 J to in·lb, and (d) 840 N·s to lb·s. Refer to Table 1.7, as required, for conversion factors.

Solution

In each case, make use of the concept of multiplication by unit values and cancellation of units not required in the answer.

(a) $500.0 \text{ N·m} \left(\dfrac{1 \text{ lb}}{4.448 \text{ N}} \right) \left(\dfrac{1 \text{ ft}}{0.3048 \text{ m}} \right) \left(\dfrac{12 \text{ in}}{1 \text{ ft}} \right) = 4430 \text{ lb·in}$ ANS.

(b) $125 \text{ Pa} \left(\dfrac{1 \text{ lb/ft}^2}{47.88 \text{ Pa}} \right) \left(\dfrac{1 \text{ ft}}{12 \text{ in}} \right)^2 = 0.0181 \text{ lb/in}^2$ ANS.

(c) $750 \text{ J} \left(\dfrac{1 \text{ ft·lb}}{1.356 \text{ J}} \right) \left(\dfrac{12 \text{ in}}{1 \text{ ft}} \right) = 6640 \text{ in·lb}$ ANS.

(d) $840 \text{ N·s} \left(\dfrac{1 \text{ lb·s}}{4.448 \text{ N·s}} \right) = 189.0 \text{ lb·s}$ ANS.

■

1.6
Dimensional
Analysis

Physical quantities are described by dimensions which are independent of units. The length of a football field, the wingspan of an airplane or the diameter of a pipe are all lengths. To simplify, we refer collectively to their dimensions as lengths and use the symbol L to denote them. Table 1.9 shows fundamental dimensions and their symbols. Those for mass M, length L, force F, and time T will suffice for our study of engineering mechanics but it is interesting to observe that all of physics needs only the eight fundamental dimensions shown in Table 1.9.

If we express the length of a football field in U.S. Customary units we would say that it is 300 ft long and in SI units we would say that it is 91.44 m long. The numerical value, 300 or 91.44, depends upon the systems of units used but the dimension is length L and does not depend upon the system of units.

Dimensions are either fundamental or derived. Derived dimensions

TABLE 1.9. Fundamental dimensions and symbols

Fundamental Dimensions	Symbol
Mass	M
Length	L
Force	F
Time	T
Temperature	t
Charge	Q
Molecular substance	n
Luminous intensity	I_L

are formed by combining fundamental dimensions. Kinetic energy is defined in terms of mass and speed squared so that dimensionally it is represented by ML^2T^{-2} which is a derived dimension. Linear momentum is defined in terms of mass and velocity and is represented by MLT^{-1} which is a derived dimension.

As stated earlier, systems of units are either absolute or gravitational. If the system of units is an absolute one (as SI), then, length L, time T, and mass M are considered fundamental, and all other dimensions are derived. The dimension of force F follows from Newton's second law and the dimensions of acceleration (LT^{-2}) by definition. $F = ma = ML^{-2}$ is the derived force dimension. If the system of units is a gravitational one (as the U.S. Customary) then, length L, time T, and force F are considered fundamental, and all other dimensions are derived. The dimension of mass M follows from Newton's second law and the dimensions of acceleration (LT^{-2}) by definition. From Newton's second law $M = F/a = FL^{-1}T^2$ is the derived mass dimension. All other derived dimensions are based on definitions of physical quantities.

Table 1.10 shows both absolute and gravitational systems of dimensions. Fundamental dimensions are shown together with selected derived dimensions for both systems. Once a quantity is defined, it is relatively easy to derive its dimensions.

TABLE 1.10. Systems of dimensions

Absolute System Fundamental Dimensions		Gravitational System Fundamental Dimensions	
Quantity	Symbol	Quantity	Symbol
Length	L	Length	L
Time	T	Time	T
Mass	M	Force	F
Derived Dimensions		Derived Dimensions	
Quantity	Symbol	Quantity	Symbol
Force	MLT^{-2}	Mass	$FL^{-1}T^2$
Acceleration	LT^{-2}	Acceleration	LT^{-2}
Area	L^2	Area	L^2
Energy	ML^2T^{-2}	Energy	FL
Impulse	MLT^{-1}	Impulse	FT
Moment of force	ML^2T^{-2}	Moment of force	FL
Momentum	MLT^{-1}	Momentum	FT
Power	ML^2T^{-3}	Power	FLT^{-1}
Velocity	LT^{-1}	Velocity	LT^{-1}

The following two examples present the method of dimensional analysis for several rational equations of engineering mechanics. This analysis is based upon the concept that rational equations must be dimensionally homogeneous. Each term of an equation must have the same dimensions or an error has been made in writing the equation.

■ Example 1.3

Perform a dimensional analysis for the following equations:

(a) Determine the dimensions of the constant G in *Newton's Universal Law of Gravitation*,

$$F = \frac{Gm_1 m_2}{r^2}.$$

Note that F is the magnitude of the attractive force which the masses exert on each other, m_1 and m_2 are the masses of the particles which attract each other, r is the straight line distance between the particles, and G is the universal gravitational constant.

(b) Determine the dimensions of the energy E in the *Mechanical Energy Equation*,

$$E = \tfrac{1}{2}mv^2 + mgh.$$

Note that E is the total energy of a particle, m is the particle mass, v is the particle speed, g is the acceleration due to gravity, and h is the elevation measured from an arbitrary datum.

(c) Determine the dimensions of the constants A and B in the *Particle Vibrations Equation*,

$$x = A \sin pt + B \cos pt$$

Note that x is a position coordinate, A and B are arbitrary constants which depend on the initial conditions of the motion, t is the time, and p is the circular frequency expressed as an angular velocity.

Solution

(a) Solve the equation for G to obtain

$$G = \frac{Fr^2}{m_1 m_2}.$$

Replace each symbol by its dimensional symbols using F for force, M for mass, L for length and T for time. Thus,

$$G = \frac{FL^2}{MM} = FL^2 M^{-2}.$$

Using Newton's second law, we replace F by $ML^{-2}T^2$ to give

$$G = M^{-1}T^2. \hspace{2cm} \text{ANS.}$$

(b) In this case, the constant $1/2$ is a pure number which is dimensionless and may be dropped from the dimensional analysis equations. Substituting the appropriate symbols for quantities on the right side of this equation gives

$$E = M\left(\frac{L}{T}\right)^2 + M\left(\frac{L}{T^2}\right)L$$

$$= ML^2T^{-2} + ML^2T^{-2}. \hspace{2cm} \text{ANS.}$$

We observe that the equation is dimensionally homogeneous and that the dimensions of E are $E = ML^2T^{-2}$. If we replace M by $FL^{-1}T^2$, then $E = FL$.

(c) Trigonometric functions may be defined as length ratios and are dimensionless quantities. Arguments of trigonometric functions are angles in radians and must also be dimensionless. Consider pt which is the argument in the sine and cosine functions of this equation. The circular frequency expressed as an angular velocity would be stated in units of radians per second, where the radian is a dimensionless ratio of an arc length to a radius. Thus, the dimensions of p are T^{-1} and $pt = T^{-1}T$ or dimensionless. Because we have shown that $\sin pt$ and $\cos pt$ are dimensionless, we may write

$$x = A + B.$$

The position coordinate x is a length and

$$L = A + B.$$

We conclude that $A = B = L$, that is, both constants have length dimensions. \hspace{1cm} ANS.

■ Example 1.4

Perform a dimensional analysis of the following equations:

(a) Determine the dimensions of the angular acceleration in the differential equation for simple pendulum motion,

$$\frac{d^2\theta}{dt^2} + \frac{g}{L}\sin\theta = 0.$$

The quantity $\frac{d^2\theta}{dt^2}$ is the angular acceleration of the simple pendulum, g is the acceleration due to gravity, L is the pendulum length, and θ is the angle which positions the pendulum with respect to a vertical line.

(b) Verify that the following equation for the horizontal cable tension component for a uniformly loaded cable is dimensionally homogeneous.

$$H = \frac{wL^2}{8h}.$$

Note that w is the uniform loading intensity per unit length applied over a horizontal projection of the cable, L is the span of the cable, h is the cable sag measured at the center of a symmetric cable, and H is the horizontal cable tension component.

(c) Determine the dimensions of the integral definition of the area moment of inertia,

$$I_x = \int_A y^2 \, dA.$$

Note that I_x is the area moment of inertia with respect to the x axis, y is the ordinate measured to the differential area dA, and dA is a differential area.

Solution

(a) Solve for the angular acceleration to obtain

$$\frac{d^2\theta}{dt^2} = -\frac{g}{L} \sin \theta.$$

Replace each symbol on the right hand side of this equation by its dimensional symbol. Thus,

$$\frac{d^2\theta}{dt^2} = -\frac{L}{T^2} \frac{1}{L} \frac{L}{L} = -\frac{1}{T^2}.$$

We may drop the negative sign because only the dimensions are required. Thus,

$$\frac{d^2\theta}{dt^2} = T^{-2}. \qquad \text{ANS.}$$

(b) Substitution of appropriate symbols for dimensions throughout this equation gives

$$H = \frac{\dfrac{F}{L} \cdot L^2}{L}.$$

The pure number 8 in the denominator on the right hand side of the equation has been omitted because it is dimensionless. Therefore,

$$H = F.$$

Because H is a force with dimensional units of F, it follows that the equation is dimensionally homogeneous. ANS.

(c) For purposes of dimensional analysis, we may drop the integral symbol. Note that dA represents a differential area which, dimensionally, is the product of two lengths of L^2. The equation becomes

$$I_x = L^2L^2 = L^4.$$ ANS.

Thus, the area moment of inertia has dimensions of length to the fourth power.

■

1.7 Problem Solving Techniques

An engineering education is of great value primarily because it is, in large measure, an integrated education of both sides of the human brain. Modern research on the brain has shown that the left side of the brain does the logical, step-by-step reasoning and that the right side of the brain is the intuitive, pictorial part. Successful problem solvers in applied mechanics deal with equations and with diagrams. Development of an integrated approach to the use of equations and diagrams is the basis of the problem solving techniques of this text.

The first step in problem solving is to read the problem statement carefully and relate this information to the associated figure which is usually provided. If not provided, a diagram should be constructed from the given information. The next step is to record all the given information. A sketch, approximately to scale, showing geometric and vector information is a necessary part of any solution. In statics, this sketch is termed a *free-body diagram*. It shows all external forces acting on the body as well as the given geometry. In dynamics, two sketches are normally made, one showing external real forces and another showing inertial forces. Given geometry may be shown on either or both of these sketches.

Given numerical values, not shown as part of these sketches, should be carefully recorded. At this stage, a complete record of known information has been stated in pictorial or written form.

Next, the unknown quantities which are to be determined should be listed. The word *Find* followed by mathematical symbols for these unknowns enables us to identify the quantities to be determined. If questions are asked, they should be written out in abbreviated form.

Equations available for a solution should be listed in symbolic form and a comparison made of the number of unknowns with the number of available equations. The equations of statics and dynamics may have to be supplemented by trigonometric, geometric, or other mathematical equations. Careful choice of equations to be written will enable us to eliminate unknowns which are not requested. Confidence in problem solving capabilities will be gained if we strive to compare unknowns to available equations and to outline a complete solution.

Writing the equations, which contain both known and unknown

quantities, is a vital step in the solution. A sign convention should be established and stated symbolically or by a coordinate system for equations which are to be written. Sign errors in terms of equations and incorrect trigonometric functions are frequent sources of incorrect solutions. Intense concentration is required for writing correct equations. Integration of visual and mathematical skills is required because reference to sketches is necessary to formulate the equations which describe the physical problem.

Once the equations are stated, they must be solved for the desired unknown quantities. In many cases they will be simultaneous, linear algebraic equations. Substitution and elimination methods learned in algebra courses will suffice for solving these equations. In many cases, it will be possible to write equations each of which contains only a single unknown. In this case, it will be very simple to solve for the unknown quantity. In some cases, the equations to be solved may be differential equations or equations involving integrals. Careful study of text discussions and detailed solutions of the examples will increase our capabilities for solving equations arising in engineering problems.

Solutions of equations yield magnitudes of physical quantities. Proper units and directional information, where appropriate, must be stated along with marked answers so that a reviewer can readily follow our solutions.

The reasonableness of your answers should always be appraised to discover possible errors. Back substitution of answers into the equations is also a valuable method for detection of possible errors. Supplementary equations can often be written, and the answers obtained from the original equations can be substituted in these supplementary equations for checking.

A word of caution is in order about writing and solving equations. It is possible to write equations for a given problem, solve them and back check the solutions and yet not have correctly solved the problem. If the equations formulated for the given physical system do not precisely describe that system, then, incorrect equations have been correctly solved and checked, but the answers do not correspond to the given physical situation. In other words, the mathematics is correct in all detail but it does not correspond to the physical system of interest.

Let us summarize these ideas in a series of critical questions relating to problem solving.

1. Have we carefully read the problem statement?
2. Do we understand the problem?
3. If a figure is given, do we visualize the problem? Have we sketched our own figure?
4. If a figure is not given, should we sketch one based upon the written information?

5. Have we carefully shown the vectors and used dimension lines for the geometry and have we established a sign convention?
6. Have we carefully recorded all given information?
7. Have we identified the unknowns? Have we recorded them symbolically?
8. Have we seen a similar problem? Does it resemble one of the examples?
9. What general concepts and equations are available to solve this problem? Have we listed them?
10. Do the number of unknowns and number of equations agree?
11. Are there geometric or trigonometric equations available?
12. Have we carefully written the governing equations?
13. Do the solutions satisfy these equations?
14. Are the answers reasonable? Do they satisfy my intuition?
15. Have we stated directional information for vectors? Have we stated the units of the solutions? Have we marked the answers for review by someone else?
16. Could someone else review my work and understand it? After six months, could we review our own work and understand it?

The reader should review these questions from time to time while solving problems. There is *no royal road* to mastery of applied mechanics. One must solve many problems and outline the solution of many additional problems to fully understand this fascinating subject.

1.8 Accuracy of Data and Solutions

Solutions to many engineering problems are achieved using mathematical models and/or testing. The accuracy of a given solution is related to several factors including properties inputted into a model. Generally, there are three sources of errors in engineering computations. These include the following:

1. *Errors in the data:* Errors in inputs or given values. Measurements are involved in many of these values and errors are unavoidable when measurements are taken.
2. *Errors in the calculations:* These arise because calculators and computers use only finite-digit mantissas.* Many calculators use 10-digit mantissas whereas 15-digit or 30-digit (double precision) mantissas are not unusual for computers. Other errors are introduced if an approximate method is used during the calculations. For example, the trapezoidal rule may be used to integrate a function approximately.

---

*If the number 195.62843 is written in scientific notation as 0.195262843×10^3, then 0.19562843 is known as an 8-digit mantissa.

3. *Errors in the mathematical model:* No matter how sophisticated a given computer program may be, it is based upon assumptions, and it has limitations. Comparisons with experimental results may be required to refine mathematical models. In our study of applied mechanics, we will be using some widely accepted modeling techniques and will not be comparing theoretical and experimental values. It is well to remember that the physical world cannot be represented exactly by mathematical models and that errors are inherent whenever we do such an analysis.

In general, these errors can be limited to acceptable ranges in engineering computations. It is important to be aware of these sources of errors to approximate their influence and to realistically report results of our calculations.

The following discussion concludes with a practical method for reporting results of computations.

A significant figure (or digit) is any digit used in writing a number except zeros used for decimal point location. A study of the following examples will clarify the definition.

Number	Significant Figures	Number	Significant Figures
3 204.5	5	0.204	3
362.7	4	0.008 7	2
20.8	3	0.004	1
1.850	4	0.0490 008	6
9.006	4	726.200	6
8.04	3	90.003	5
6.0	2	5.090	4

Ambiguity is possible when trailing zeros are involved. Consider 8000 and note that it could have 1, 2, 3 or 4 significant figures. In such cases it is best to use scientific notation to make clear the number of significant figures.

Number in Scientific Notation	Significant Figures
$8. \times 10^3$	1
8.0×10^3	2
8.00×10^3	3
8.000×10^3	4

Rules for finding the number of significant figures, after arithmetic operations have been performed, enable us to prevent loss of significant figures or to imply precision which does not exist. These rules are described as follows:

Addition and Subtraction

Report significant digits only as far to the right as shown in the *least precise* number. Consider the following examples:

64.1	(one digit to right of decimal)
4 826.753	(three digits to right of decimal)
8.2905	(four digits to right of decimal)
Sum 4 899.1435	Report as 4 899.1 (one digit to right of decimal)

4 524.6	(one digit to right of decimal)
−0.208	(three digits to right of decimal)
Sum 524.392	Report as 524.4 (one digit to right of decimal)

Multiplication and Division

Report significant figures of the number with the *fewest significant figures*. Consider the following examples:

$(4.86)(19.52) = 94.8672$	Report as 94.9
$(6.9)(275.33) = 1899.777$	Report as 1.9×10^3
$39.425/4.28 = 9.2114486$	Report as 9.21
$586.294/8.3 = 70.637831$	Report as 71

Combination of Operations

First, perform multiplication and division and determine the correct number of significant figures. Then, perform additions and subtraction and report the correct number of significant figures. Obviously this procedure is not practical for computers and is usually too time consuming for calculators. *Usually, we perform the entire calculation and use judgment in reporting the number of significant figures.*

The error in a quantity is defined as the difference between an approximate value and the true value of this quantity. If the error is large, we say that the approximate value is inaccurate. Accuracy refers to the nearness of a value to its true value.

Practical Method of Reporting Answers

If we use four significant figures for numbers beginning with *1* and three significant figures for all other numbers, we will show that the maximum error range will be about 0.5 percent. Consider the range from 1000 to 1999 and compute the largest percentage error range. Reporting the quantity 1000 means that the number lies between 999.5 and 1000.5 or an absolute error range of $1000.5 - 999.5 = 1.0$. The corresponding percentage error range is $(1/1000) \times 100 = 0.1\%$ Consider the range from 200 to 999 and compute the largest percentage error ranges. The quantity 200 means that the number lies between

199.5 and 200.5 or an absolute error range of $200.5 - 199.5 = 1.0$. The corresponding percentage error range is $(1/200) \times 100 = 0.5\%$. Because the maximum error range is 0.5%, the maximum error is 0.25%. Seldom in applied mechanics do we know data (i.e., inputs or given values) with greater accuracy than 0.25%. Computations with a computer or a calculator will practically always be more accurate than 0.25%.

Thus, as we solve problems, we assume, for data or results, the use of four significant figures for numbers beginning with 1 and three significant figures for all other numbers. In this text, example and problem answers are stated according to this rule but given information does not, in general, follow this rule. Except for unusual situations, this will be a very satisfactory engineering method.

Problems

Newton's Second Law

1.1 A particle of 4 kg mass has an acceleration of 6.00 m/s^2 along a straight line path. What resultant force acts on this particle? Express your answer in Newtons and pounds.

1.2 At a given instant during its motion, a particle of mass 10 kg is acted upon by a resultant force of 286 N. Determine the magnitude of its acceleration at this instant. Express your answer in m/s^2 and ft/s^2.

1.3 A particle of 5.25 slug mass has an acceleration along a straight line path of 6.20 ft/s^2. What resultant force acts on this particle? Express your answer in pounds and Newtons.

1.4 For a particle of unit mass (1.0 kg), determine the corresponding acceleration in magnitude and direction for each of the forces stated below. Show the direction with an arrow.

(a) 30 N horizontally to the right
(b) 6.35 N up to the right at 30° to the horizontal
(c) 20.5 N down to the left at 60° to the vertical

(d) 18.6 N up to the left at 40° to the vertical
(e) 30.9 N down to the right at 20° to the horizontal.

1.5 At a given instant during its motion a particle of mass $m = 2.5$ slug is accelerated at 10.0 ft/s^2 directed up to the right at 10° with the horizontal. What resultant force in magnitude and direction acts on this particle? Express your answer in pounds and Newtons.

Newton's Law of Universal Gravitation

1.6 If a spacecraft weighs 2400 lb at the Earth's surface, what attractive force is exerted on it by Earth at an altitude of 800 miles? (The radius of Earth is 3,960 miles.)

1.7 A body has a mass of 120 kg. What is its weight at Earth's surface in Newtons? What attractive force will be exerted on it by Earth at an altitude of 1200 km? (The radius of Earth is 3,960 miles.)

1.8   Consider a unit mass (1 kg) at the surface of Earth (i.e., $r = R_E$), and at $2R_E$, $3R_E$, $4R_E$ and $5R_E$. Calculate the attractive force at these distances exerted on this mass. Plot the variation of F versus r.

1.9   If we imagine a body of mass m at the surface of Earth, how does the attractive force F vary as m varies? Justify your answer and sketch the function F vs. m.

1.10   Given that the mass of the Moon is 7.35×10^{22} kg and the radius of the Moon is 1,079 miles, consider (a) a unit mass at the surface of the Moon and compare the force exerted on this mass by the Moon to the force exerted on it by Earth. (Assume that the Moon's orbit is circular with a radius of 238,840 miles.) (b) a unit mass at the surface of Earth and compare the force exerted on this mass by the Moon to the force exerted on it by Earth. (The mass of Earth is 5.976×10^2 kg.)

1.11   If Earth is 93.0 million miles from the Sun, determine the force of attraction exerted on Earth by the Sun. Express your answer in MN. Is an equal and opposite force exerted on the Sun? The mass of the Sun is 1.990×10^{30} kg and the mass of Earth is 5.976×10^{24} kg.

Dimensional Analysis

1.12   Perform a dimensional analysis of the following equation:

$$M_o = -P \cos \theta \, L \sin \phi + P \sin \theta \, L \cos \phi.$$

In this equation where M_o is the moment of P with respect to the origin O, expressed as the product of force and length dimensions, P is a force which makes an angle θ with the horizontal, L is the straight line distance from the origin to any point on the line of action of the force P. Note that this line makes an angle ϕ with the horizontal.

1.13   Perform a dimensional analysis of the following equation:

$$I_A = \tfrac{2}{5}mr^2 + md^2$$

Note that I_A is the mass moment of inertia of a sphere with respect to an axis through point A. It has dimensions of mass multiplied by a distance squared, m is the mass of the sphere, r is the radius of the sphere, and d is the distance from the center of the sphere to the axis through point A.

1.14   Perform a dimensional analysis of the following equations:

(a) $$x_G = \frac{A_1 x_1 + A_2 x_2 + A_3 x_3}{A_1 + A_2 + A_3}$$

where x_G, x_1, x_2, and x_3 are distances and A_1, A_2, and A_3 are areas.

(b) $$y_G = \frac{\displaystyle\int y \, dA}{\displaystyle\int dA}$$

where y, y_G are distances and A is an area.

(c) $$I_u = \frac{I_x + I_y}{2} + \sqrt{\left(\frac{I_x - I_y}{2}\right)^2 + P_{xy}^2}.$$

All quantities are area moments of inertia expressed as the product of an area and a distance squared.

Units Conversion

1.15   Convert the following quantities from the U.S. Customary system to SI units:
(a) 20 slug to kg
(b) 25 ft to m
(c) 400 lb to N
(d) 4800 lb·in. to J.

1.16   Convert the following quantities from SI units to the U.S. Customary system:
(a) 500 kg to slug
(b) 40 m to ft

(c) 50 kN to lb

(d) 650 J to lb·ft.

1.17 Convert the following quantities from the U.S. Customary system to SI units:

(a) 20000 W to ft·lb/s

(b) 75 MPa to psi

(c) 25 m/s to ft/s

(d) 20 km/s to mph.

Significant Figures

1.18 State the number of significant figures contained in each of the following quantities:

(a) 2 085.2

(b) 394

(c) 0.038 9

(d) 7.8460

(e) 5.820 × 10^6

(f) 8 000

(g) 0.002 95

(h) 9 643 000.

1.19 State the number of significant figures contained in each of the following quantities:

(a) 3.759 2 × 10^8

(b) 625.30

(c) 500

(d) 0.836

(e) 8 240

(f) 6 005

(g) 7.235 × 10^4

(h) 0.0400.

1.20 Use your calculator to perform the indicated operations and record the answer to the proper number of significant figures

(a) 82.4/9.875

(b) (42.35)(8.2)

(c) (0.072 × 0.558 4)/2.4

(d) 621.52/0.854

(e) (85.29 × 10^3) + (248.657 × 10^{-2}).

Review Problems

1.21 Let the weight of an object at the surface of the Earth be W_0. Show that the weight of the object W at any height h above the surface of Earth is given by the relationship $W = \left(\dfrac{R_E}{R_E + h}\right)^2 W_0$ where R_E is the radius of Earth. Assume that Earth is the *only* planet that exerts a force on the object. What is the weight of the object if $h = \frac{1}{2}R_E$ and $h = \frac{1}{4}R_E$?

1.22 (a) The displacement s measured in feet, as a function of time t measured in seconds, is given by $s = At^3 + Bt^2 + Ct + D$. Use dimensional analysis to find the units for the constants A, B, C, and D.

1.23 The mass of a body on Earth is $M_E = 20$ kg. Determine (a) its mass on Mars expressing the answer in kg and in slug and (b) its weight on Mars expressing the answer in N and in lb, if the acceleration of gravity on Mars is $g_M = 3.72$ m/s^2.

2

Equilibrium of a Particle in Two Dimensions

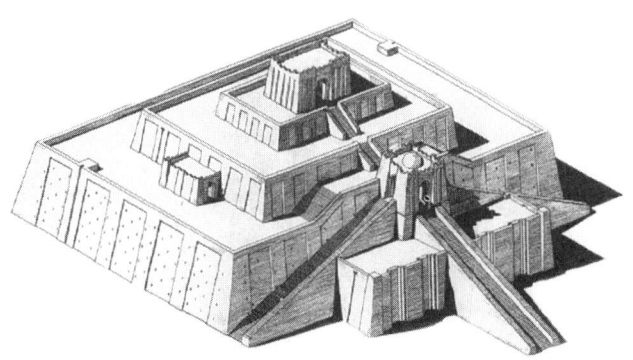

Reconstruction of the Sumerian temple tower, or ziggurat, at Ur from 2112 B.C. speaks eloquently to the genius of engineers from years past.

Throughout the ages, engineers have led the way as the makers of history. Their creative designs have impacted world civilizations like no other profession. In Mesopotamia, clay tablets have been uncovered which show that Babylonian engineers were familiar with basic engineering measurements, arithmetic, and algebra that are still in use today. Their number system, based on 60, has been handed down to us through the centuries in our measures of time and angle. It is still true today that engineers create designs to meet the needs of their society.

The world in which we live offers endless examples of magnificent symmetry and equilibrium. In developing the conditions of equilibrium, it is necessary to introduce the important concepts of vectors, forces *and the* free-body diagram. *The characteristics of vector quantities are described, and methods are developed for manipulating vectors and forces. It should be pointed out that the free-body diagram is one of the most basic and most significant ideas encountered in the study of mechanics. The student is urged to make every effort to obtain a clear understanding of this important concept.*

The successful conquest of engineering problems requires imagination. Albert Einstein once said that, "Imagination is more important than knowledge." As an engineer you must learn to examine, question, understand, then, solve. This chapter will help you begin this wonderful journey into engineering statics and the basic concepts and methods involved.

We begin by defining a particle *as a infinitesimal object that possesses mass but occupies no space. It follows that forces acting on a particle must necessarily intersect at the same point (*concurrent*). Furthermore, we will discuss forces that act in a single plane (*coplanar*). Thus, this chapter is concerned primarily with the necessary and sufficient conditions for the equilibrium of concurrent, coplanar force systems.*

2.1
Scalar and
Vector
Quantities

The physical quantities that we deal with in mechanics fall into two distinct types. The first type is known as a *scalar* quantity, and it is that type of quantity that can be fully defined by specifying its *magnitude*. Examples of scalar quantities include length, volume, time, mass and density. The second type is known as a *vector* quantity and is characterized by requiring a *magnitude* as well as a *direction* and a *sense* for complete definition. Furthermore, vector quantities cannot be added by the ordinary rules of algebra. They *must* be added by the law of the *parallelogram* which will be discussed in Section 2.2. Examples of vector quantities include force, velocity, momentum, and acceleration. All of these vector quantities will be encountered and discussed as we proceed with the study of statics and dynamics. Of immediate interest, however, is the vector quantity *force*, some characteristics of which are given below.

A *force* is the push or pull of one body on another through direct physical contact or through gravitational action. Other types of forces, such as electrostatic, electromagnetic, and nuclear do exist but will not be of primary concern in our study of mechanics. As a vector quantity, force possesses the following characteristics:

1. The *magnitude* of a force is a measure of its size. In the U.S. Customary system, the basic unit of force is the pound (lb). Other units commonly used include the kilo-pound (abbreviated as kip or k and equal to 1000 lb.) and the ton which is equal to 2000 lb. In the SI system, the basic unit of force is the Newton (N) although the kilo-newton (equal to 1000 N and abbreviated as kN) is widely used. As shown in Figure 2.1, a force **F** may be represented graphically by a directed line segment OB, also known as its line of action, whose length, according to some scale, gives the magnitude of the force. Thus, for example, if the scale in Figure 2.1 is 1 in. = 200 N, we would measure the distance in inches, from the *tail* of the force vector, point O, to its *head*, point B, and would, then, multiply this distance by 200 to obtain the magnitude of the force in Newtons.

2. The *direction* of a force is represented by the orientation of its *line of action* from some convenient reference line. Thus, for example, the direction of the force **F** in Figure 2.1 may be stated by giving the angle θ which gives the inclination of its line of action OB from the fixed *x* axis.

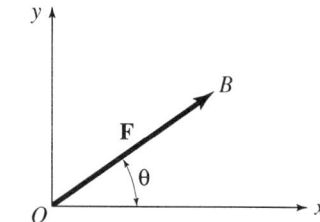

FIGURE 2.1.

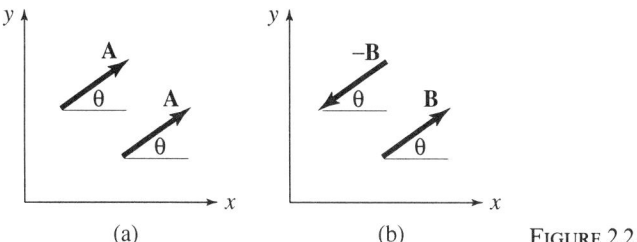

FIGURE 2.2.

3. The *sense* of a force is represented by its arrowhead. Thus, in Figure 2.1, the arrowhead indicates that the sense of the force is upwards and to the right, i.e., from O to B and not from B to O. Two forces may have the same magnitudes and the same directions, but, if they have different senses, they would produce different effects.

A vector quantity may be *fixed, sliding,* or *free.* A *fixed vector* is one that is required to occupy a unique position in space, and any change in this position would alter the nature of its action. A *sliding vector* is a vector that may occupy an infinite number of positions along its line of action. Thus, this vector may be moved (slid) along its line of action and still preserve the nature of its effect, provided that its magnitude and sense are not changed. Finally, a *free vector* is one that may occupy any position in space without altering the nature of its action. However, such a vector must preserve its magnitude, direction, and sense.

In this textbook, vector quantities are represented symbolically in boldface type and scalar quantities in italic type. Thus, for example, the symbol **V** represents a vector whereas the symbol *V* indicates only the magnitude of this vector. To distinguish between vector and scalar quantities in handwritten work, it is recommended that a vector quantity be represented by a letter with an arrow on top, e.g., $\vec{V}$.

Two vectors are said to be *equal* if they have the same magnitude, the same direction, and the same sense. The two vectors shown in Figure 2.2(a) are *equal vectors* and are labeled using the same symbol **A**. A *negative vector* −**B** is defined as a vector having the same magnitude and the same direction as vector **B** but the opposite sense, as shown in Figure 2.2(b).

2.2 Elementary Vector Operations

Two basic vector operations that are useful in the study of two-dimensional statics are introduced in this section. These operations are expanded in Chapter 3 to include the third dimension, and new operations are added in other chapters, as the need for them arises.

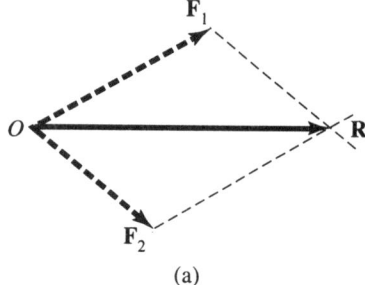

(a)

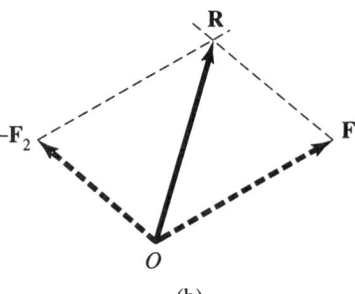

FIGURE 2.3.   (b)

Addition of Vectors

As stated earlier, vector quantities *cannot* be added by using the ordinary rules of algebra. For example, two forces of magnitudes 10 lb and 20 lb cannot be added to obtain 30 lb, unless they happen to act in the same direction and have the same sense. This is so because the ordinary rules of algebra consider only the magnitudes of the forces and ignore their directions and senses. A method that accounts for the magnitude, direction, and sense in the addition process, and, which, proven experimentally, yields the correct answer, is based upon the *law of the parallelogram.* This law states that two given vectors may be added by using them as two adjacent sides of a parallelogram. The diagonal of the parallelogram is the vector sum or the *resultant* of the two given vectors. Consider, for example, the two forces F_1 and F_2 acting at the same point O as shown in Figure 2.3(a). Forces that act at the same point are referred to as *concurrent* forces. The two vectors F_1 and F_2 form two adjacent sides of a parallelogram which may, then, be completed as shown in Figure 2.3(a). The diagonal of the parallelogram originating at point O at which the two forces act, is the vector sum or the resultant **R** of the forces F_1 and F_2. In other words, *the single force* **R** *is capable of producing the same effect at point O as the two forces* F_1 *and* F_2 *acting simultaneously.* The operation represented in Fig. 2.3(a) may be expressed symbolically by the vector relationship

$$\mathbf{R} = \mathbf{F}_1 + \mathbf{F}_2. \tag{2.1}$$

The process of subtracting one vector from another is accomplished in a similar manner. Thus, for example, if force $\mathbf{F}_2$ is to be subtracted from force $\mathbf{F}_1$, we may construct a parallelogram by using forces $\mathbf{F}_1$ and $-\mathbf{F}_2$ as two adjacent sides of the parallelogram and determining the diagonal that originates at point O, as shown in Figure 2.3(b). This diagonal represents the resultant $\mathbf{R}$ or the vector sum of forces $\mathbf{F}_1$ and $-\mathbf{F}_2$. The operation shown in Figure 2.3(b) may be expressed symbolically by the vector relationship

$$\mathbf{R} = \mathbf{F}_1 + (-\mathbf{F}_2). \tag{2.2}$$

The addition of forces $\mathbf{F}_1$ and $\mathbf{F}_2$ shown in Figure 2.3(a) may also be accomplished by constructing what is known as the vector triangle shown in Figure 2.4. The construction may be started with either of the two forces. Thus, in Figure 2.4(a) the force $\mathbf{F}_1$ is constructed as shown. The tail of force $\mathbf{F}_2$ is then placed at the head of force $\mathbf{F}_1$ and constructed as indicated. The vector from the tail of $\mathbf{F}_1$ to the head of $\mathbf{F}_2$ represents the vector sum or the resultant $\mathbf{R}$ of $\mathbf{F}_1$ and $\mathbf{F}_2$. Note that the vector triangle shown in Figure 2.4(a) is, in fact, the upper half of the parallelogram of Figure 2.3(a). A similar construction is shown in Figure 2.4(b) except that the initial force is $\mathbf{F}_2$ instead of $\mathbf{F}_1$. However, the resultant $\mathbf{R}$ is identical to that obtained in Figure 2.4(a). This type of construction is particularly convenient when dealing with more than two forces.

Consider, for example, the addition of forces $\mathbf{F}_1$, $\mathbf{F}_2$, $\mathbf{F}_3$ and $\mathbf{F}_4$ shown in Figure 2.5(a). The graphical determination of their resultant $\mathbf{R}$ is shown in Figure 2.5(b) where the sequence of head-to-tail construction, known as the *polygon method*, was accomplished in the order

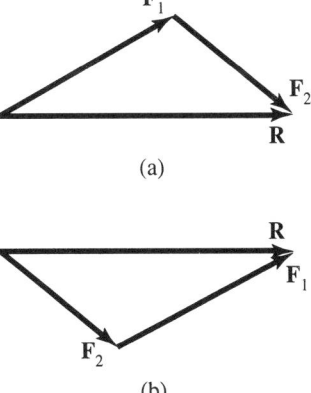

(a)

(b)   FIGURE 2.4.

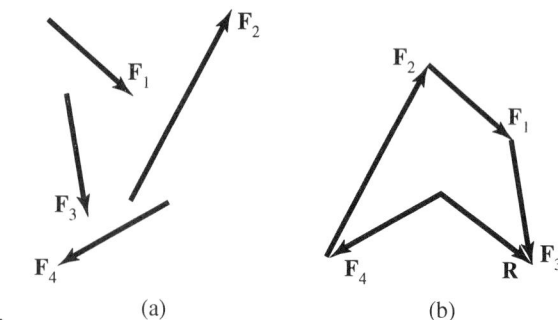

FIGURE 2.5.   (a)   (b)

$\mathbf{F}_4$, $\mathbf{F}_2$, $\mathbf{F}_1$ and $\mathbf{F}_3$ which led to a five-sided polygon. The vector from the tail of $\mathbf{F}_4$ (the first force) to the head of $\mathbf{F}_3$ (the last force) represents the resultant $\mathbf{R}$ of forces $\mathbf{F}_1$, $\mathbf{F}_2$, $\mathbf{F}_3$ and $\mathbf{F}_4$. Note that *any other* convenient order of head-to-tail construction would serve equally well. Note, also, that, for the special case of two forces, the polygon degenerates to a triangle (a three-sided polygon) as in Figure 2.4. The operation represented in Figure 2.5(b) may be expressed symbolically by the vector relationship

$$\mathbf{R} = \mathbf{F}_1 + \mathbf{F}_2 + \mathbf{F}_3 + \mathbf{F}_4. \tag{2.3a}$$

For the case where the summation includes n forces, Eq. (2.3a) may be written in the form

$$\mathbf{R} = \sum \mathbf{F} \tag{2.3b}$$

where $\sum$ is the summation symbol.

It should be pointed out that the construction of the parallelogram or the polygon need not be executed using instruments and according to a given scale. As a matter of fact, a neat freehand sketch would be sufficient because the resultant $\mathbf{R}$ may be determined using geometry and trigonometric analysis. These concepts are illustrated in examples at the end of this section.

Decomposition of Vectors

The reverse of vector addition is known as *vector decomposition* or *vector resolution*. Referring back to Figure 2.3(a), for example, the two forces $\mathbf{F}_1$ and $\mathbf{F}_2$ are known as the *components* of the resultant force $\mathbf{R}$. Actually, any force may be resolved (decomposed) into any two components by using the law of the parallelogram. Thus, consider the force $\mathbf{F}$ which is to be decomposed into two components along some arbitrary directions a and b as shown in Figure 2.6(a). The magnitudes and senses of the two components $\mathbf{F}_a$ and $\mathbf{F}_b$ are determined by completing the parallelogram, as shown. Generally speaking, however, the most commonly used components of a force are determined along two mutually perpendicular directions such as the x and y axes in Figure

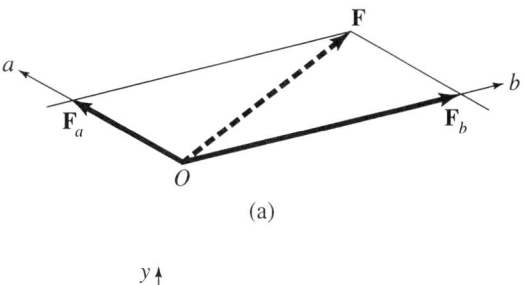

(a)

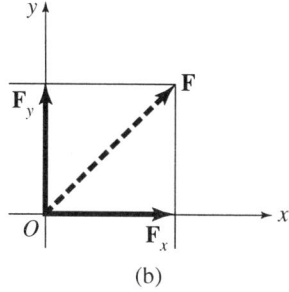

(b)

FIGURE 2.6.

2.6(b). The magnitudes and senses of F_x and F_y are determined by completing the parallelogram which, in this case, reduces to a simple rectangle. Here again, a neat freehand sketch is sufficient because determining the components may be accomplished by trigonometric analysis.

The following examples illustrate some of the concepts discussed above.

■ **Example 2.1**

Two eye bars are connected to a bracket and subjected to the forces shown in Figure E2.1(a). Determine the magnitude, direction, and sense of the resultant force acting on the bracket.

Solution

The 20 k and 15 k forces are used as two adjacent sides of a parallelogram to draw the diagram in Figure E2.1(b) according to the law of the parallelogram. Note that, in this special case, the parallelogram reduces to a simple rectangle. Alternatively, the force polygon method (i.e., the head-to-tail construction) may be used to obtain the right-angle triangle shown in Figure E2.1(c). Applying the *law of cosines* from trigonometry to triangle OAB,

$$(OB)^2 = (OA)^2 + (AB)^2 - 2(OA)(AB)\cos 90°. \qquad (a)$$

Because $\cos 90° = 0$, Eq. (a) reduces to

$$(OB)^2 = (OA)^2 + (AB)^2 \qquad (b)$$

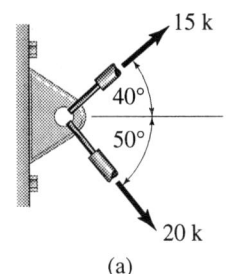

(a)

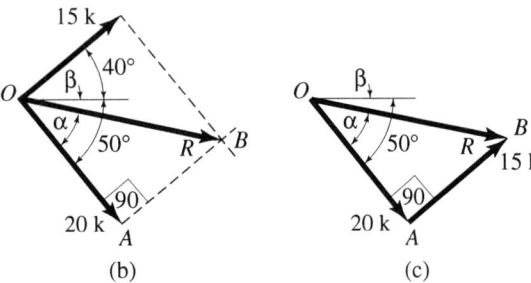

FIGURE E2.1.   (b)   (c)

which is an expression of the Pythagorean theorem. Thus, the law of cosines reduces to the Pythagorean theorem for the special case of a right-angle triangle. Therefore, using Eq. (b),

$$R = OB = \sqrt{(20)^2 + (15)^2} = 25.0 \text{ k}.$$

Also,

$$\alpha = \tan^{-1} \frac{15}{20} = 36.87°,$$

and

$$\beta = 50 - 36.87 = 13.13°.$$

Thus, the resultant has a magnitude of 25.0 k, a direction of 13.13° cw from the horizontal, and a sense downward and to the right. All of this information is summarized as follows:

$$R = 25.0 \text{ k} \quad 13.13°. \qquad \text{ANS.}$$

■ **Example 2.2**

The gusset plate of a roof truss is subjected to the two forces shown in Figure E2.2(a). Find the magnitude, direction, and sense of the resultant **R** of these two forces.

Solution

The law of the parallelogram is used to construct the diagram shown in Figure E2.2(b) where the 10 k and 20 k forces are used as two adjacent

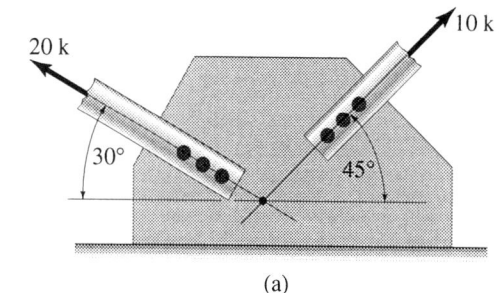

(a)

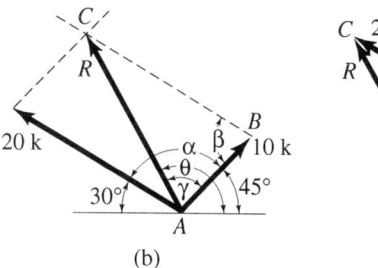

(b)

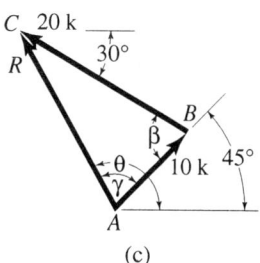

(c)   FIGURE E2.2.

sides of the parallelogram. Alternatively, the force polygon (triangle) shown in Figure E2.2(c) may be drawn using the head-to-tail construction method. From the geometry of the parallelogram in Figure E2.2(b),

$$\alpha = 180 - 45 - 30 = 105°$$

and

$$\beta = \frac{360 - 2(105)}{2} = 75°.$$

In the case of the triangle of Figure E2.2(c), the angle β may be found directly from $\beta = 45 + 30 = 75°$. By the *law of cosines* from trigonometry,

$$(AC)^2 = (AB)^2 + (BC)^2 - 2(AB)(BC)\cos\beta.$$

Therefore, because $R = AC$,

$$R^2 = 10^2 + 20^2 - 2(10)(20)\cos 75°,$$

$$R = 19.91 \text{ k}.$$

Using the *law of the sines* from trigonometry,

$$\frac{\sin\gamma}{20} = \frac{\sin 75°}{19.91}$$

from which

$$\gamma = 76.0°.$$

Therefore,

$$\theta = 45 + \gamma = 121.0°$$

Thus, the resultant has a magnitude of 19.91k, a direction defined by the angle θ from the horizontal, and a sense indicated by the arrowhead, i.e., upward and to the left. All of this information may be summarized as follows:

$$R = 19.91 \text{ k} \quad\underline{\hspace{0.5em}121.0°}.$$ ANS.

■ **Example 2.3**

Two tugboats A and B are towing a disabled ship S as shown in Figure E2.3(a). The resultant **R** of the pulling forces **F**$_A$ and **F**$_B$, produced by the two tugboats A and B, respectively must have a magnitude of 75 kN and must be directed along the axis CD. (a) If $\beta = 35°$, what are the magnitudes of **F**$_A$ and **F**$_B$. (b) If **F**$_B$ is to have the least possible magnitude, what should be the angle β and the magnitudes of **F**$_A$ and **F**$_B$.

Solution

(a) Using the law of the parallelogram, we construct the diagram shown in Figure E2.3(b) in the following manner. The resultant line OG, is drawn parallel to axis CD in Figure E2.3(a) and with a magnitude of 75 kN. At point O, the tail of the resultant, lines OE and OH are constructed parallel to the directions of forces **F**$_B$ and **F**$_A$, respectively. The parallelogram is now completed by starting at the head of the resultant, point G, and first drawing a line parallel to OH to intersect line OE at point E. The distance from O to E defines the magnitude of F_B. Secondly, starting at point G, a line is drawn parallel to OE to intersect line OH at point H. The distance from O to H defines the magnitude of **F**$_A$. These magnitudes are found by considering, for example, triangle OEG and the law of the sines. Thus,

$$\frac{F_B}{\sin 35°} = \frac{75}{\sin 120°}$$

which gives

$$F_B = 49.7 \text{ kN}.$$ ANS.

Also,

$$\frac{F_A}{\sin 25°} = \frac{75}{\sin 120°}$$

which gives

$$F_A = 36.6 \text{ kN}.$$ ANS.

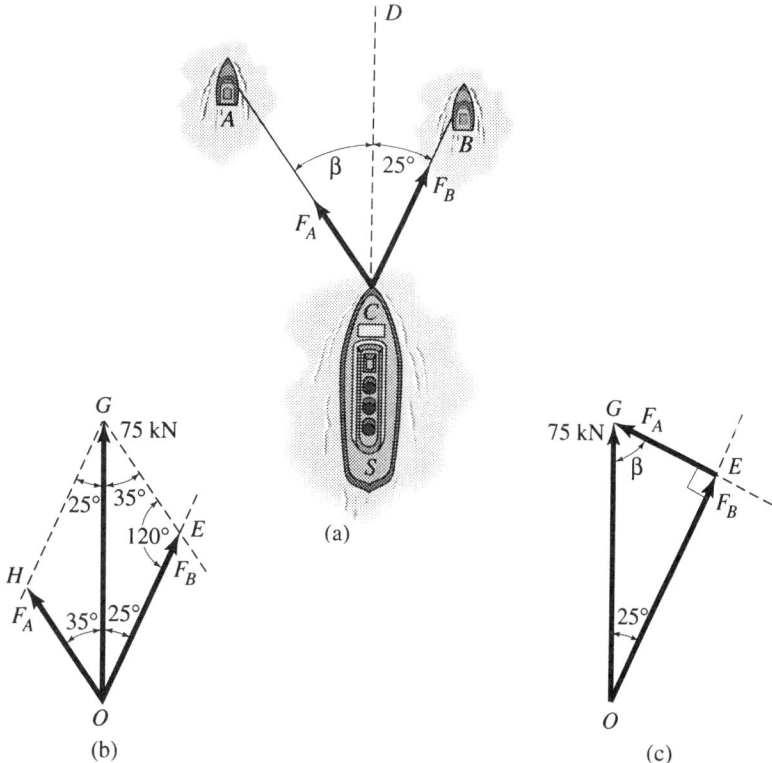

FIGURE E2.3.

(b) The polygon method, which yields a triangle in this case, is most convenient for the solution of this part. Because the resultant is known fully, it is represented by line OG as shown in Figure E2.3(c). From point O, line OE is constructed parallel to the given direction of F_B. According to the polygon method, the magnitude of F_A is defined by a line that originates at some point E on the line of action of F_B and terminates at fixed point G, the head of the resultant. For F_A to have the least possible magnitude, the distance from point G to line OE must be the shortest possible. This is obtained when line GE is perpendicular to line OE. Thus, the force polygon yields a right-angle triangle from which

$$\beta = 180° - 90° - 25° = 65.0°, \qquad \text{ANS.}$$

$$F_B = 75\cos 25° = 68.0 \text{ kN}, \qquad \text{ANS.}$$

and

$$F_A = 75\sin 25° = 31.7 \text{ kN}. \qquad \text{ANS.}$$

Problems

Use a trigonometric analysis, not a graphical method, to solve the following problems.

2.1 Two eye bars are attached to a bracket as shown in Figure P2.1. Let $F_1 = 2500$ lb, $F_2 = 6000$ lb, $\alpha = 70°$ and $\beta = 20°$. Find the magnitude, direction, and sense of the resultant force.

2.2 Two eye bars are attached to a bracket as shown in Figure P2.1. Let $\alpha = 40°$, $\beta = 25°$, $F_1 = 10.0$ kN and $F_2 = 4.5$ kN. Determine the magnitude, direction, and sense of the resultant force.

2.3 Two truss members are riveted to a gusset plate as shown in Figure P2.3. Let $\gamma = 30°$, $\theta = 60°$, $P = 5.4$ k, $Q = 3.0$ k. Determine the magnitude, direction, and sense of the resultant force.

2.4 Two truss members are riveted to a gusset plate as shown in Figure P2.3. Let $P = 10.0$ kN, $Q = 7.5$ kN, $\gamma = 45°$ and $\theta = 60°$. Compute the magnitude, direction, and sense of the resultant force.

2.5 A disabled automobile is being towed to the garage by a tow truck that applies a pull $P = 750$ lb as shown in Figure P2.5.

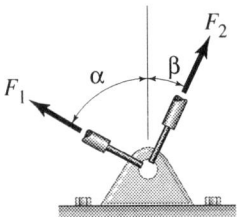

FIGURE P2.1.

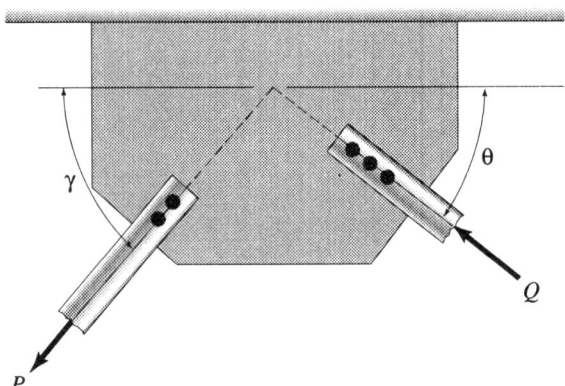

FIGURE P2.3.

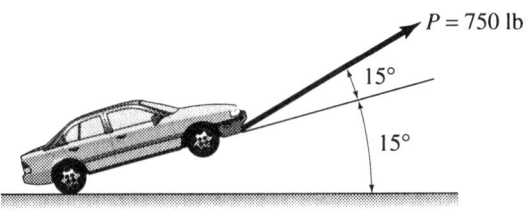

FIGURE P2.5.

Determine the horizontal and vertical components of the force *P*. State their magnitudes and senses.

2.6 A push of 30 N is applied to the handle of a lawnmower which moves along the horizontal surface as shown in Figure P2.6. Decompose the applied force into two perpendicular components, one along the horizontal and the second along the vertical.

FIGURE P2.7.

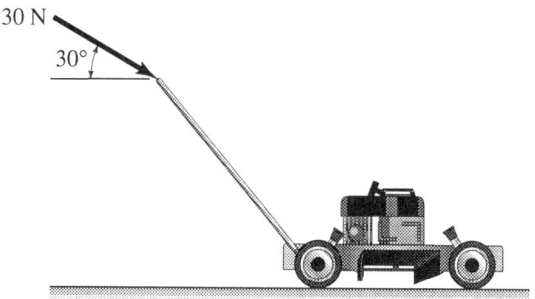

FIGURE P2.6.

2.7 A shopper needs to provide a horizontal force of 5 lb to move the shopping basket forward, If he/she pushes with a force *P* at an angle of 20° to the horizontal, as shown in Figure P2.7, determine the magnitude of the force *P*. Also determine, the magnitude and sense of the vertical component of *P*.

2.8 A stalled pick-up truck is being pulled by the forces $F_1 = 500$ N and $F_2 = 700$ N as shown in Figure P2.8. If $\alpha = 20°$ and $\beta = 40°$, find the magnitude, direction, and sense of the resultant force.

2.9 A stalled pick-up truck is being pulled by the two forces $\mathbf{F}_1$ and $\mathbf{F}_2$ as shown in Figure P2.8. If $\alpha = 35°$ and $\beta = 25°$ and the resultant is known to be a horizontal force of magnitude 250 lb directed from **B** to **C**, determine the magnitudes of $\mathbf{F}_1$ and $\mathbf{F}_2$.

2.10 A stalled pick-up truck is being pulled by the two forces $\mathbf{F}_1$ and $\mathbf{F}_2$ as shown in Figure P2.8. If $\alpha = 35°$ and the resultant is known to be a horizontal force of magnitude 900 N directed from **B** to **C**, determine the value of β so that the force $\mathbf{F}_2$ has the least possible magnitude. What is the magnitude of $\mathbf{F}_1$ under these conditions?

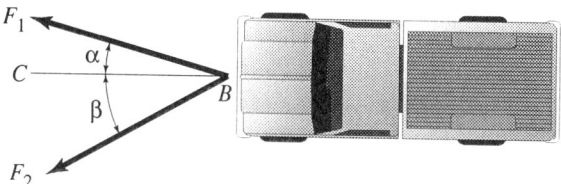

FIGURE P2.8.

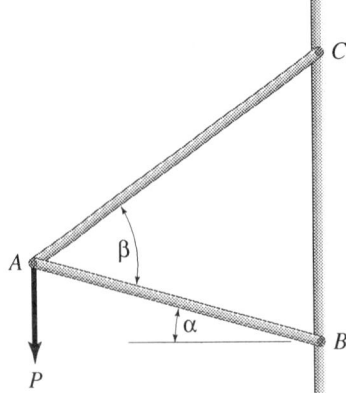

FIGURE P2.11.

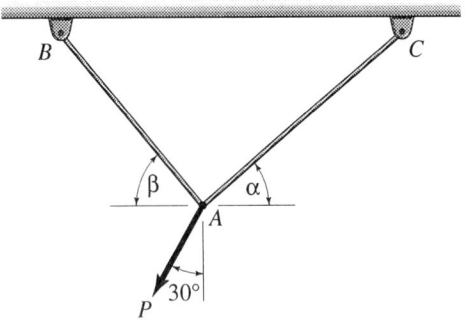

FIGURE P2.13.

2.11 Refer to Figure P2.11. When the force **P** is applied at A, it creates component forces $\mathbf{F}_{AB}$ and $\mathbf{F}_{AC}$ along members AB and AC, respectively. Let $P = 8.0$ k and, if $\mathbf{F}_{AC}$ is known to have a magnitude of 12.0 k and a sense from C to A and if $\alpha = 15°$, determine the angle β and the magnitude and sense of $\mathbf{F}_{AB}$.

2.12 Refer to Figure P2.11. When the force **P** is applied at A, it causes component forces $\mathbf{F}_{AB}$ and $\mathbf{F}_{AC}$ along members AB and AC, respectively. Let $P = 20$ kN, $F_{AC} = 15$ kN with a sense from C to A, and $F_{AB} = 7$ kN with a sense from A to B, then, determine the angles α and β.

2.13 Consider the system shown in Figure P2.13 which consists of two cables AB and AC supporting a load **P** as shown. Let $P = 5.0$ k, $\alpha = 20°$, and $\beta = 50°$. Determine the magnitudes, directions, and senses of the components of P along lines AC and AB.

2.14 Consider the system shown in Figure P2.13 which consists of two cables AB and AC supporting a load **P** as shown. Let $P = 250.0$ kN and $\alpha = 25°$. Find the value of β so that the component of **P** along line AB is minimum. What is the value of the component of **P** along line AC under these conditions?

2.3
Force Expressed in Vector Form

In the preceding section, we learned how to represent a force by a directed line segment whose length, according to some scale, signifies the magnitude, whose orientation from some reference axis indicates the direction, and whose arrowhead specifies the sense of the force. In many cases, however, it becomes desirable to express a given force symbolically using a vector formulation. Two equivalent vector formulations are used: one expresses the force in terms of its two rectangular components and the second expresses the force as a magnitude multiplying a vector quantity known as a *unit vector* defined in the direction of the force. These two vector formulations are developed and discussed in the following paragraphs.

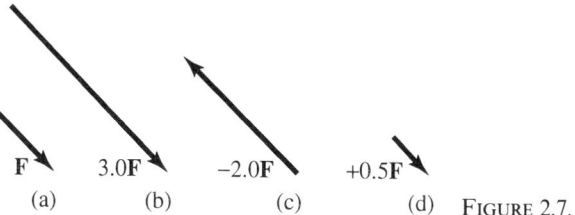

F   3.0F   −2.0F   +0.5F

(a)   (b)   (c)   (d)   FIGURE 2.7.

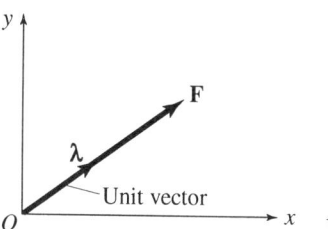

FIGURE 2.8.

Multiplication of a Vector by a Scalar

When a vector **V** is multiplied by a scalar q, the resulting quantity $q\mathbf{V}$ is a vector whose magnitude is qV and whose direction is the same as that of **V**. If q is positive, the sense of vector $q\mathbf{V}$ is the same as that of **V**. If, however, q is negative, the sense of vector $q\mathbf{V}$ is opposite to that of **V**, and the resulting vector is $-q\mathbf{V}$. Consider, for example, the force **F** shown in Figure 2.7(a). If this force is multiplied by $+3.0$, we would obtain the force shown in Figure 2.7(b), if multiplied by -2.0, the force shown in Figure 2.7(c) is obtained, and if multiplied by $+0.5$ the result would be that shown in Figure 2.7(d).

Unit Vector

By definition, a *unit vector* is a vector that has a specified direction and sense and a magnitude of *unity*. Thus, as shown in Figure 2.8, if the force **F** is fully defined, then, a unit vector λ in the direction of, and having the same sense as, the force **F** would be defined by the relationship

$$\lambda = \frac{\mathbf{F}}{F}. \tag{2.4}$$

Equation (2.4) may also be expressed in the form

$$\mathbf{F} = F\lambda \tag{2.5}$$

which expresses the fact that a force, or any other vector quantity, may be written as the product of its magnitude F (a scalar) and a unit vector λ in its direction. Equation (2.5) shows, once again, that the product of

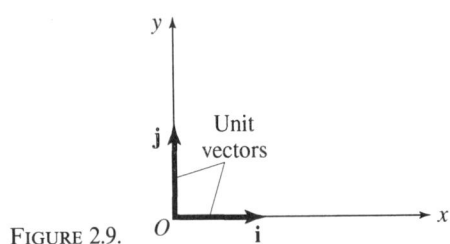

FIGURE 2.9.

a scalar and a vector is another vector having the same direction and sense as the original vector and a magnitude equal to the product of the scalar and the magnitude of the original vector.

Of particular interest are the unit vectors defined along rectangular coordinate axes. Thus, for example, in the case of the two-dimensional x-y coordinate system shown in Figure 2.9, unit vectors $\mathbf{i}$ and $\mathbf{j}$ are defined along the *positive* x and y axes, respectively. Note that, although the unit vectors λ, $\mathbf{i}$, and $\mathbf{j}$ are shown in Figures 2.8 and 2.9 originating, at the origins of the coordinate systems, they are sliding vectors that could be positioned anywhere along their respective lines of action.

Rectangular Components

As a consequence of Eq. (2.5), recall that any vector may be written as the product of its magnitude and a unit vector in its direction. Thus, for example, the force $\mathbf{F}_x$, having a magnitude of F_x and pointed in the positive x direction, may be written as $F_x\mathbf{i}$. Similarly, the vector $\mathbf{F}_y = F_y\mathbf{j}$ represents a force whose magnitude is F_y and is pointed in the positive y direction. These two forces are shown in relation to an x-y coordinate system in Figure 2.10(a). Using the polygon method (the head-to-tail construction), we may determine the resultant $\mathbf{F}$ of the two forces (components) $\mathbf{F}_x$ and $\mathbf{F}_y$ as shown in Figure 2.10(b). This was accomplished by constructing $\mathbf{F}_x$, first, followed by $\mathbf{F}_y$. Of course, the same resultant would have been obtained if we had first constructed $\mathbf{F}_y$ and, then, followed it with $\mathbf{F}_x$. The quantities $\mathbf{F}_x$ and $\mathbf{F}_y$ are known as the rectangular *vector components* of the force $\mathbf{F}$, and the quantities F_x and F_y are its rectangular *scalar components*. The vector addition depicted in Figure 2.10(b) may be stated symbolically by the vector equation

$$\mathbf{F} = \mathbf{F}_x + \mathbf{F}_y \tag{2.6a}$$

or

$$\mathbf{F} = F_x\mathbf{i} + F_y\mathbf{j} \tag{2.6b}$$

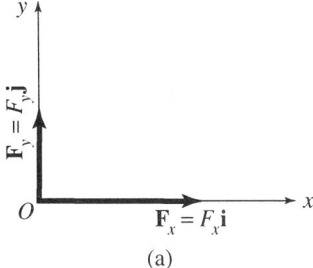

(a)

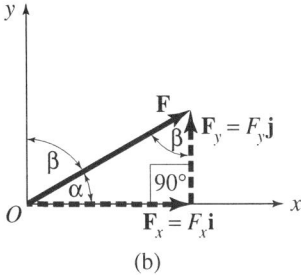

(b)

FIGURE 2.10.

which expresses the force **F** in terms of its x and y vector components. In Chapter 3, we will extend this vector formulation to include a vector component in the z direction.

The scalar components F_x and F_y may be related to the magnitude F of the resultant force **F** by the following relations which are obtained directly from the basic definitions of the trigonometric functions *sine* and *cosine* of an angle. Thus, referring to Figure 2.10(b),

$$\left.\begin{aligned} F_x &= F \cos\alpha, \\ F_y &= F \sin\alpha \end{aligned}\right\} \tag{2.7}$$

and

where the ccw angle α ranging between 0 and 360°, defines the direction of the force **F** from the positive x axis. It is generally more practical to deal with angles which are less than 90° and, therefore, when there is no chance of confusion, the angle α will occasionally be measured from either the positive or negative x axis.

Applying the Pythagorean theorem to the vectors in Figure 2.10(b), we obtain the magnitude of the force **F**. Thus,

$$F = \sqrt{F_x^2 + F_y^2} \tag{2.8}$$

Also, from Eqs. (2.7) and from Figure 2.10(b), we conclude that $\cos\alpha = F_x/F$ and $\cos\beta = F_y/F$. Using the definition of a unit vector, (Eq.

(2.4)), we may obtain a unit vector λ in the direction of the force $\mathbf{F}$. Thus,

$$\lambda = \frac{\mathbf{F}}{F}$$

$$= \left(\frac{F_x}{F}\right)\mathbf{i} + \left(\frac{F_y}{F}\right)\mathbf{j},$$

$$= (\cos\alpha)\mathbf{i} + (\cos\beta)\mathbf{j},$$

$$= \lambda_x\mathbf{i} + \lambda_y\mathbf{j}. \tag{2.9}$$

The quantities λ_x and λ_y are known as the direction cosines of vector (force) $\mathbf{F}$ and represent, respectively, the x and y components of the unit vector λ. Because a unit vector has a magnitude of 1, by definition, it follows from Eq. (2.8) that

$$\cos^2\alpha + \cos^2\beta = 1, \tag{2.10}$$

the familiar trigonometric equation relating two complementary angles.

2.4 Addition of Forces Using Rectangular Components

In Section 2.2, we learned how to add two or more vectors, such as forces, by the law of the parallelogram or by the polygon method. Another method, which is developed in this section, uses the rectangular components of a force.

In Section 2.3, we learned how to decompose a given force $\mathbf{F}$ into x and y (or $\mathbf{i}$ and $\mathbf{j}$) components and how to express it in the vector form $\mathbf{F} = F_x\mathbf{i} + F_y\mathbf{j}$. This ability to decompose a vector, such as a force, into rectangular components yields a very convenient and very practical method of adding together two or more vectors.

Consider, for example, the forces $\mathbf{F}_1$, $\mathbf{F}_2$ and $\mathbf{F}_3$, shown in Figure 2.11(a), which are to be added to obtain their resultant $\mathbf{R}$. Using the concept of vector addition defined in Section 2.2,

$$\mathbf{R} = \mathbf{F}_1 + \mathbf{F}_2 + \mathbf{F}_3 \tag{2.11}$$

Equation (2.11) may be generalized to include any number of forces by writing it in the form

$$\mathbf{R} = \sum \mathbf{F} \tag{2.12}$$

where the symbol $\sum \mathbf{F}$ signifies the vector sum of all forces in the system. If $\mathbf{F}_1$ to $\mathbf{F}_n$ are expressed in terms of their x and y components (see Figs. 2.11(b) to (d)), Eq. (2.12) becomes,

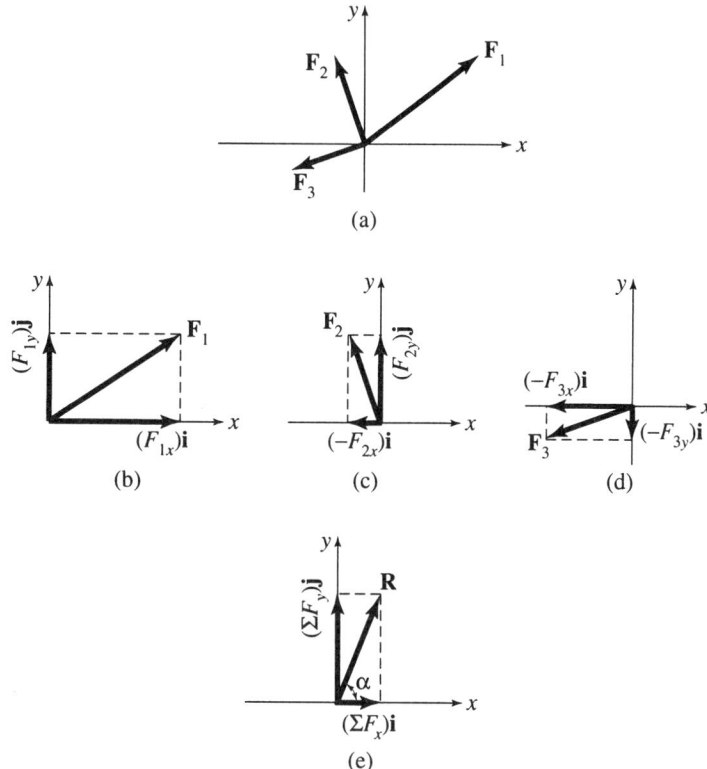

FIGURE 2.11.

$$\mathbf{R} = (\mathbf{F}_{1x})\mathbf{i} + (\mathbf{F}_{1y})\mathbf{j}$$

$$+ (F_{2x})\mathbf{i} + (F_{2y})\mathbf{j}$$

$$+ (\mathbf{F}_{3x})\mathbf{i} + (F_{3y})\mathbf{j}$$

$$\vdots$$

$$+ (F_{nx})\mathbf{i} + (F_{ny})\mathbf{j},$$

$$\mathbf{R} = (F_{1x} + F_{2x} + F_{3x} + \cdots + F_{nx})\mathbf{i} + (F_{1y} + F_{2y} + F_{3y} + \cdots + F_{ny})\mathbf{j}$$

$$\mathbf{R} = (\textstyle\sum F_x)\mathbf{i} + (\textstyle\sum F_y)\mathbf{j} = R_x\mathbf{i} + R_y\mathbf{j} \qquad (2.13)$$

where the symbol $R_x = \sum F_x$ is the component of $\mathbf{R}$ along the x axis, representing the algebraic sum of *all* force components in the x direction, and $R_y = \sum F_y$ is the component of $\mathbf{R}$ along the y axis, representing the algebraic sum of *all* force components in the y direction. These rectangular components are, then, added by the law of the parallelogram, as shown in Figure 2.11(e), to obtain the resultant $\mathbf{R}$. The magnitude of $\mathbf{R}$ is obtained by applying the Pythagorean theorem to one of the triangles in Figure 2.11(e). Thus,

$$R = \sqrt{\left(\sum F_x\right)^2 + \left(\sum F_y\right)^2}. \qquad (2.14)$$

The angle α which defines the direction of R from the positive x axis, may also be obtained from Figure 2.11(e). Thus,

$$\alpha = \tan^{-1}\left(\frac{\sum F_y}{\sum F_x}\right) = \tan^{-1}\frac{R_y}{R_x} \qquad (2.15)$$

The following examples illustrate some of the concepts discussed in Sections 2.3 and 2.4.

■ **Example 2.4**

A force $\mathbf{Q}$, applied to a bracket, has a magnitude of 5.0 k and is directed as shown in Figure E2.4. Express this force in vector form (a) in terms of its x and y components and (b) as a magnitude multiplying a unit vector.

Solution

(a) The components Q_x and Q_y are obtained by Eqs. (2.7). Thus

$$Q_x = Q\cos\alpha$$

$$= (5.0)\cos 60°$$

$$= 2.50 \text{ k},$$

and

$$Q_y = Q\sin\alpha$$

$$= (5.0)\sin 60°$$

$$= 4.33 \text{ k}.$$

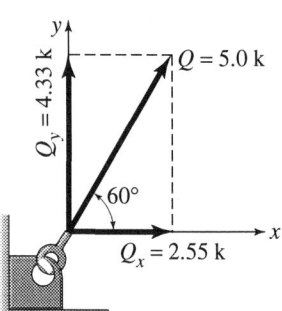

FIGURE E2.4.

Therefore, by Eq. (2.6),

$$\mathbf{Q} = (2.50\mathbf{i}) + (4.33\mathbf{j})\text{ k}.\qquad\text{(a)}\quad\text{ANS.}$$

(b) A unit vector along $\mathbf{Q}$ may be determined by Eq. (2.4). Thus,

$$\lambda = \frac{\mathbf{Q}}{Q}$$

$$= \frac{1}{5\text{ k}}[(2.50\text{ k})\mathbf{i} + (4.33\text{ k})\mathbf{j}]$$

$$= 0.500\mathbf{i} + 0.866\mathbf{j}.$$

Therefore, by Eq. (2.5),

$$\mathbf{Q} = [5.0(0.500\mathbf{i} + 0.866\mathbf{j})]\text{ k}\qquad\text{(b)}\quad\text{ANS.}$$

Equations (a) and (b) express the same vector in two different but *equivalent* forms. Whereas Eq. (a) gives the force $\mathbf{Q}$ in terms of its x and y components, Eq. (b) gives it in terms of its magnitude and its direction defined by the unit vector λ. Both formulations are useful, particularly, when dealing with three-dimensional force systems.

■ Example 2.5

A force is given in vector form by the expression

$$\mathbf{P} = (15\mathbf{i} - 20\mathbf{j})\text{ kN}$$

(a) State the rectangular components P_x and P_y.
(b) Find the magnitude of the force $\mathbf{P}$.
(c) Show a neat sketch of the force $\mathbf{P}$ along with its two vector components $\mathbf{P}_x$ and $\mathbf{P}_y$.
(d) Find the angle α which defines the direction of the force $\mathbf{P}$ from the positive x axis.
(e) Find a unit vector along the force $\mathbf{P}$.

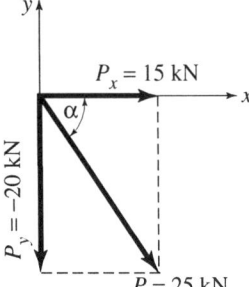

$P = 25$ kN   FIGURE E2.5.

Solution

(a) The magnitudes of the x component P_x and the y component P_y are given by the scalar multipliers of the $\mathbf{i}$ and $\mathbf{j}$ terms of the force $\mathbf{P}$. Thus,

$$P_x = 15 \text{ kN} \quad \text{and} \quad P_y = -20 \text{ kN}. \qquad \text{ANS.}$$

(b) The magnitude of the force P is given by Eq. (2.8). Thus,

$$P = \sqrt{P_x^2 + P_y^2} = \sqrt{15^2 + 20^2} = 25.0 \text{ kN}. \qquad \text{ANS.}$$

(c) The required sketch is shown in Figure E2.5. ANS.

(d) The angle α may be obtained by the second part of Eq. (2.7). Thus,

$$\alpha = \sin^{-1}\left(\frac{P_y}{P}\right) = \sin^{-1}\left(\frac{-20}{25}\right) = -53.1°. \qquad \text{ANS.}$$

The minus sign indicates that the angle α is cw from the positive x axis. Note that the angle α may also be obtained from the relationships $\alpha = \cos^{-1}(P_x/P)$ and $\alpha = \tan^{-1}(P_y/P_x)$. However, in the case of $\alpha = \cos^{-1}(P_x/P)$, the sign of the angle needs to be obtained by inspection.

(e) A unit vector λ in the direction of the force $\mathbf{P}$ may be found by Eq. (2.4). Thus,

$$\lambda = \frac{\mathbf{P}}{P} = \frac{1}{25 \text{ kN}}[(15\mathbf{i} - 20\mathbf{j}) \text{ kN}] = 0.600\mathbf{i} - 0.800\mathbf{j}. \qquad \text{ANS.}$$

■ **Example 2.6**

The four forces shown in Figure E2.6(a) are represented in four slightly different, but equivalent, forms. $\mathbf{F}_1$ is shown in terms of its $\mathbf{i}$ and $\mathbf{j}$ components, $\mathbf{F}_2$ in terms of a magnitude and a direction given by an angle from the positive x axis, $\mathbf{F}_3$ in terms of a magnitude multiplying a unit vector, and $\mathbf{F}_4$ in terms of a magnitude and a direction expressed in terms of a horizontal and a vertical distance. Determine the resultant of the four forces expressing it in terms of its $\mathbf{i}$ and $\mathbf{j}$ components. Determine its magnitude and the angle α that it makes with the x axis.

Solution

Each of the four forces is decomposed into x and y (or $\mathbf{i}$ and $\mathbf{j}$) components as follows:

$$\mathbf{F}_1 = (350\mathbf{i} + 500\mathbf{j}) \text{ lb},$$

$$\mathbf{F}_2 = (600 \cos 60°)\mathbf{i} - (600 \sin 60°)\mathbf{j}$$

$$= (300\mathbf{i} - 520\mathbf{j}) \text{ lb},$$

$$\mathbf{F}_3 = [-250(0.8)]\mathbf{i} + [250(0.6)]\mathbf{j}$$

$$= (-200\mathbf{i} + 150\mathbf{j}) \text{ lb},$$

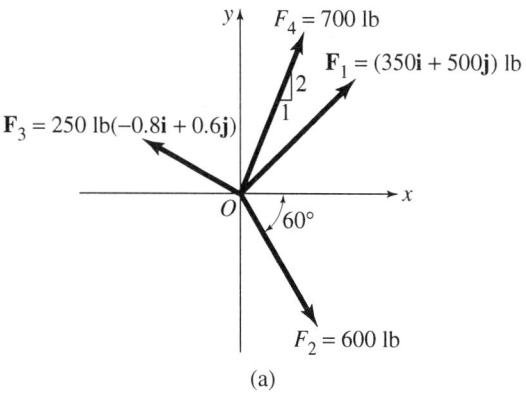

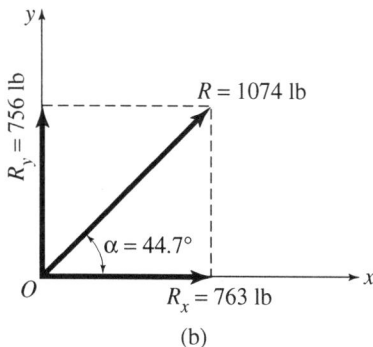

(b)   FIGURE E2.6.

$$\mathbf{F_4} = \left[700\left(\frac{1}{\sqrt{5}}\right) \right]\mathbf{i} + \left[700\left(\frac{2}{\sqrt{5}}\right) \right]\mathbf{j}$$

$$= (313\mathbf{i} + 626\mathbf{j}) \text{ lb.}$$

In the expression for $\mathbf{F_4}$, the quantities $1/\sqrt{5}$ and $2/\sqrt{5}$ represent the cosine and the sine of the angle, respectively, that $\mathbf{F_4}$ makes with the x axis.

The components R_x and R_y of the resultant R are obtained by adding, algebraically, all of the components in the x (or $\mathbf{i}$) and in the y (or $\mathbf{j}$) directions, respectively, Thus,

$$R_x = \sum F_x = 350 + 300 - 200 + 313 = 763 \text{ lb.}$$

and

$$R_y = \sum F_y = 500 - 520 + 150 + 626 = 756 \text{ lb.}$$

Therefore, by Eq. (2.13),

$$\mathbf{R} = (763\mathbf{i} + 756\mathbf{j}) \text{ lb.} \qquad\qquad \text{ANS.}$$

The magnitude of **R** may be found from Eq. (2.14). Thus,

$$R = \sqrt{(763)^2 + (756)^2} = 1074 \text{ lb.} \qquad \text{ANS.}$$

The angle α is obtained from Eq. (2.15). Thus,

$$\alpha = \tan^{-1}\frac{756}{763} = 44.7°. \qquad \text{ANS.}$$

The resultant **R** as well as its x and y components are shown in Figure E2.6(b).

■ Example 2.7

An eye bar shown in Figure E2.7 is subjected to the three forces **F**$_1$, **F**$_2$ and **F**$_3$ of which **F**$_1$ and **F**$_3$ are fully known. Determine the magnitude and direction of **F**$_2$ so that the resultant **R** of **F**$_1$, **F**$_2$, and **F**$_3$ has a magnitude of 4.00 kN, a direction along the axis of the eye bar, and a sense from A to O.

Solution

The resultant **R** may be written as

$$\mathbf{R} = -(4.00 \cos 30°)\mathbf{i} - (4.00 \sin 30°)\mathbf{j}$$

$$= (-3.46\mathbf{i} - 2.00\mathbf{j}) \text{ kN.}$$

The x component, $R_x = -4.00 \cos 30° = 3.46$ kN, of the resultant is obtained by adding, algebraically, the x components of the three forces **F**$_1$, **F**$_2$, and **F**$_3$. Thus,

$$-3.46 = -5.25 \cos 30° - F_2 \cos \alpha + 2.75 \cos 60°$$

from which

$$F_2 \cos \alpha = 0.29 \text{ kN} \qquad (a)$$

where α defines the direction of **F**$_2$ from the x axis, as shown in Figure E2.7. Also, the y component, $R_y = -4.00 \sin 30° = -2.00$ kN, of the

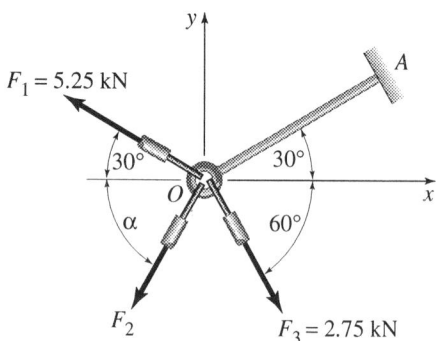

FIGURE E2.7.

resultant is obtained by adding, algebraically, the y components of the three forces $\mathbf{F}_1$, $\mathbf{F}_2$, and $\mathbf{F}_3$. Thus

$$-2.00 = 5.25 \sin 3.0° - F_2 \sin \alpha - 2.75 \cos 60°$$

from which

$$F_2 \sin \alpha = 2.25 \text{ kN.} \qquad \text{(b)}$$

Divide Eq. (b) by Eq. (a) to obtain

$$\frac{\sin \alpha}{\cos \alpha} = \tan \alpha = 7.7586$$

and

$$\alpha = 82.7°. \qquad \text{ANS.}$$

Also, from Eq. (b),

$$F_2 = \frac{2.25}{\sin 82.7°} = 2.27 \text{ kN.} \qquad \text{ANS.}$$

■

Problems

2.15 A cable anchored to a bracket is subjected to a force $\mathbf{F}$ as shown in Figure P2.15. Let $F = 8.0$ k and $\alpha = 25°$. Express the force F in vector form in terms of its x and y components.

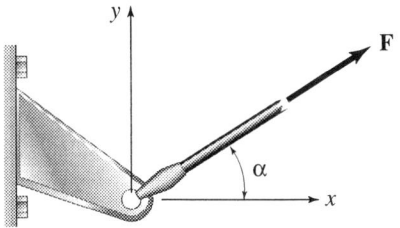

FIGURE P2.15.

2.16 Refer to Problem 2.15. Let $F = 15.0$ kN and $F_y = 12.0$ kN. Find F_x and α.

2.17 Refer to Problem 2.15. Let $F = 10.50$ k and $F_y = 7.75$ k. Express the force $\mathbf{F}$ in vector form in terms of a magnitude multiplying a unit vector in the direction of the force.

2.18 A workman pulls on a loaded cart with a force $\mathbf{P}$ as shown in Figure P2.18. If the magnitude of the horizontal component of $\mathbf{P}$ needed to move the cart is 150 N, determine the magnitude of $\mathbf{P}$ and the magnitude of its vertical component. Let $\alpha = 60°$.

2.19 Refer to Problem 2.18. Let $P_x = 50.0$ lb and $P_y = 40.0$ lb and express the force $\mathbf{P}$ in vector form in terms of a magnitude and a unit vector in the direction of $\mathbf{P}$.

2.20 Refer to Problem 2.18. Let $\mathbf{P} = (175\mathbf{i} + 200\mathbf{j})$ N. Find the magnitude of the force $\mathbf{P}$ and the value of the angle α that it makes with the x axis.

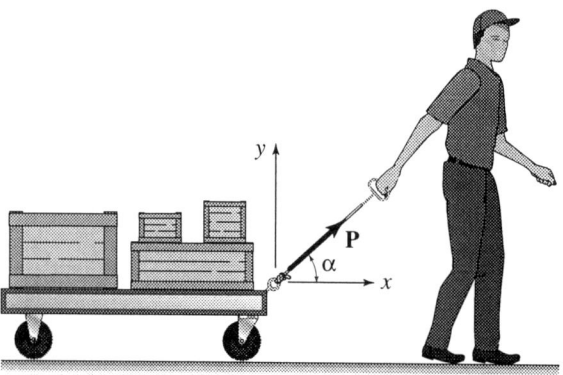

FIGURE P2.18.

2.21 A force is expressed in vector form by the equation $Q = (-5.3i + 3.5j)$ k. Show a neat sketch of the force Q along with its two vector components Q_x and Q_y. Express the force Q as a magnitude multiplying a unit vector in the direction of the force.

2.22 A force is expressed in vector form by the relationship $F = 750$ N $(-0.760i - 0.650j)$. Show a neat sketch of the force F along with its two vector components F_x and F_y. Find the angle β that the force F makes with the y axis.

2.23 The force F_1 shown in Figure P2.23 has a magnitude of 12.250 k. Find the x and y components of F_1.

2.24 The force F_2 shown in Figure P2.23 has a component in the y direction whose magnitude is 25.7 kN. Find the magnitude of the force F_2 and its component in the x direction.

2.25 Three eye bars are attached to a hook and subjected to forces, as shown in Figure P2.25. Let $P_1 = 8.65$ k, $\alpha_1 = 15°$, $P_2 = 4.20$ k, $\alpha_2 = 25°$, $P_3 = 12.25$ k, and $\alpha_3 = 30°$. Use rectangular components to find the magnitude of the resultant R and the angle α that it makes with the x axis. Show a sketch of this resultant.

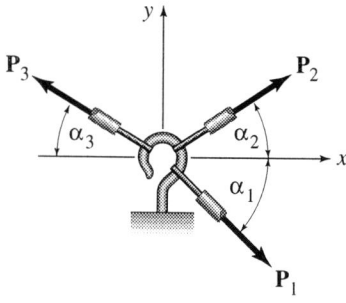

FIGURE P2.25.

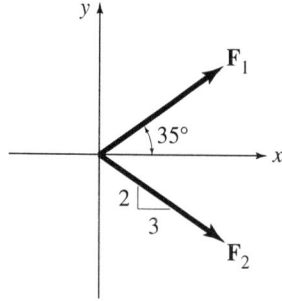

FIGURE P2.23.

2.26 Refer to Problem 2.25. Let $P_1 = 25.0$ kN, $\alpha_1 = 30°$, $P_2 = 31.5$ kN, $\alpha_2 = 40°$, $P_3 =$

43.2 kN, and $\alpha_3 = 25°$. Use rectangular components to determine the magnitude of the resultant **R** and the angle α that it makes with the x axis. Sketch this resultant.

2.27 Four forces are described as $F_1 = (750\mathbf{i} - 500\mathbf{j})$ lb, $F_2 = [1200(-0.55\mathbf{i} - 0.84\mathbf{j})]$ lb, F_3 has a magnitude of 950 lb and a direction given by $\alpha = -45°$ from the positive x axis, and F_4 has a magnitude of 700 lb and a direction and sense given by ⟋₂⟍¹. Use rectangular components to compute the magnitude of the resultant **R** and the angle α that it makes with the x axis.

2.28 Three truss members frame into a gusset plate as shown in Figure P2.28. Let $F_1 = 17.50$ kN, $\alpha_1 = 30°$, $F_3 = 25.75$ kN, and $\alpha_3 = 45°$. Find F_2 and α_2 so that the resultant **R** has a magnitude of 20.00 kN directed along the positive y axis.

2.29 Refer to Problem 2.28. Let $F_1 = 13.550$ k, $\alpha_1 = 20°$, $F_2 = 25.700$ k, $\alpha_2 = 35°$, and $F_3 = 18.375$ k. Find the angle α_3 so that the resultant **R** is directed along the positive x axis. What is the magnitude of this resultant?

2.30 The cart shown in Figure P2.30 is subjected to forces $\mathbf{Q}_1$, $\mathbf{Q}_2$, and $\mathbf{Q}_3$. The force $\mathbf{Q}_1$ is vertical and has a magnitude of 8.0 kN. The force $\mathbf{Q}_2$ is horizontal and has a magnitude of 15.0 kN. The force $\mathbf{Q}_3$ has a

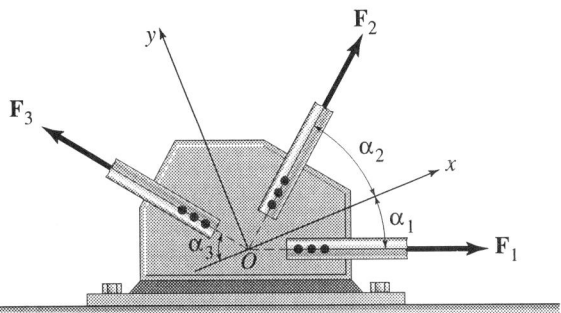

FIGURE P2.28.

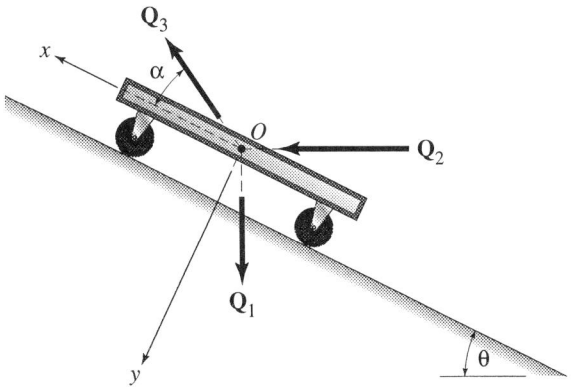

FIGURE P2.30.

magnitude of 25.0 kN and the angle $\alpha =$ 30°. Determine the angle ($0 < \theta < 90°$) of the inclined plane so that the resultant of the three forces is directed along the positive x axis. Find also the magnitude of the resultant.

2.31 Refer to Problem 2.30. Let $Q_1 = 10.0$ k, $Q_3 = 30.0$ k and $\theta = 25°$. If the resultant is to be directed along the positive x axis and have a magnitude of 27.0 k, determine the magnitude of the force $\mathbf{Q}_2$ and the value of the angle α.

2.5
Supports
and
Connections

It is necessary that we understand the nature of the forces produced at various types of supports and connections before we can solve problems dealing with the concept of equilibrium. Appendix F summarizes many types of supports and connections generally used with particles and rigid bodies and the unknown reactive force components associated with them. Five types of connections most commonly encountered in two-dimensional particle equilibrium (represented in Appendix F by cases 1 through 5) are discussed in the following paragraphs.

Springs

A spring is a mechanical device capable of stretching or contracting under the action of a force. Experimental evidence shows that when a linearly elastic spring is subjected to a force, the spring becomes longer if the force is a *pull* (tension) and shorter if the force is a *push* (compression). Furthermore, the applied force is directly proportional to the change in length of the spring according to the relationship

$$F = ks \qquad (2.16)$$

In Eq. (2.16), F is the applied force, s is the change in the length of the spring, known as its *deformation*, and k is the factor of proportionality between F and s called the spring *constant* or the spring *stiffness*. Examination of Eq. (2.16) shows that the units of k must be those of force divided by units of deformation which is a length. Thus, in the U.S. Customary system, such compound units as lb/in. or k/in. are common and, in the SI system, the compound units used are N/m or kN/m.

Consider, for example, the spring of spring constant k shown in Figure 2.12. In Figure 2.12(a) the spring is free of any forces and its length L_u is termed the *undeformed length* of the spring. In Figure 2.12(b), the spring is subjected to a pulling force F that produces an increase in length or an *extension*. Finally, in Figure 2.12(c), the spring is subjected to a pushing force that produces a decrease in length or a *contraction*. In either case, the magnitude of the extension or contraction is $s = F/k$ as obtained from Eq. (2.16). Note that the applied force F is carried by the spring to its supporting hook at the wall so that its magnitude, direction, and sense are unchanged. By the principle of action and reaction (Newton's third law), therefore, the supporting hook develops a resisting force B equal in magnitude and direction, but opposite in sense to F, as shown by the free-body diagrams of the hook in Figures 2.12(b) and (c). The concept of the free-body diagram will be discussed in detail in Section 2.6.

Short Cables and Short Links

Short flexible *cables* and short rigid (inflexible) members, referred to as *links*, are often used to transmit (carry) a force from one position to another. Consider, for example, the crate of weight W which is supported by a short flexible cable AB as shown in Figure 2.13(a) and by a short rigid link AB as shown in Figure 2.13(b). In either case, the weight (force) W is transmitted unchanged in magnitude, direction, and sense along the cable or link to the ceiling at the pin in the hinge. By

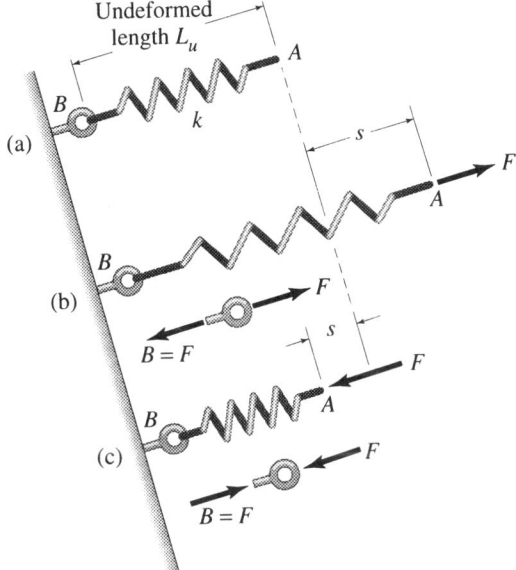

FIGURE 2.12.

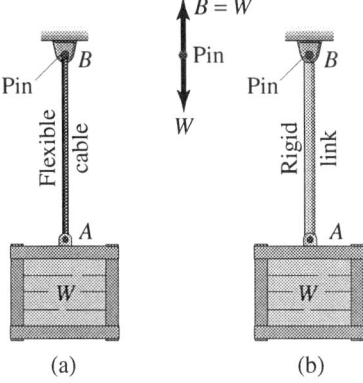

FIGURE 2.13.

the principle of action and reaction (Newton's third law), therefore, the support develops a resisting force B equal in magnitude and direction, but opposite in sense to W, as shown by the free-body diagram of the pin in Figure 2.13. The flexible cable and the rigid link are both subjected to a pulling force (tension) and perform exactly the same function in this particular case. However, there is a significant difference between these two structural members. A flexible cable can only be subjected to a pulling action; any pushing action (compression), regardless how small, will cause the cable to buckle or collapse and become ineffective in carrying the load. On the other hand, a short rigid link is capable of being subjected to either tension or compression.

Pulley-Cable Arrangements

Pulley-cable arrangements are versatile devices often used to generate a force at one location by the application of a much smaller force at another location. This characteristic of a pulley-cable arrangement will be demonstrated in Example 2.12 at the end of Section 2.7.

The basic function of the pulley in a pulley-cable arrangement is to enable the force in the cable to change its direction and sense without changing its magnitude. Thus, consider the flexible cable ABC which passes over a pulley hinged at point O as shown in Figure 2.14. If the pulley is assumed to be mounted on a frictionless (smooth) hinge at O, it can be shown (refer to Chapter 4) that the magnitude of the force in segment BC of the cable is equal to the force P applied at point A as shown in Figure 2.14. Consequently, regardless of the orientation of the cable on either side of the pulley, the force P is the same at any point along the cable.

Frictionless Hinge

The hinge is a connection or a support that is used in many structural and machine application. A *hinge connection* consists of a pin passing through two or more members (links) joining them together as shown in Figure 2.15(a). As discussed earlier, the force in each link is transmitted, unchanged, along the link's axis and would, thus, act on the pin. Therefore, the pin at A would be subjected to the three forces F_1, F_2, and F_3, as shown by the free-body diagram in Figure 2.15(b). A *hinge support* consists of a pin passing through a fixed support and through one or more members (links) framing into this support. A

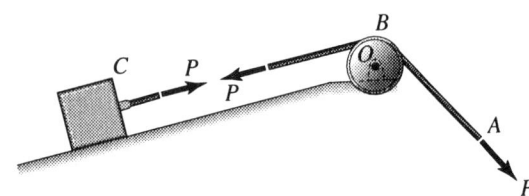

FIGURE 2.14.

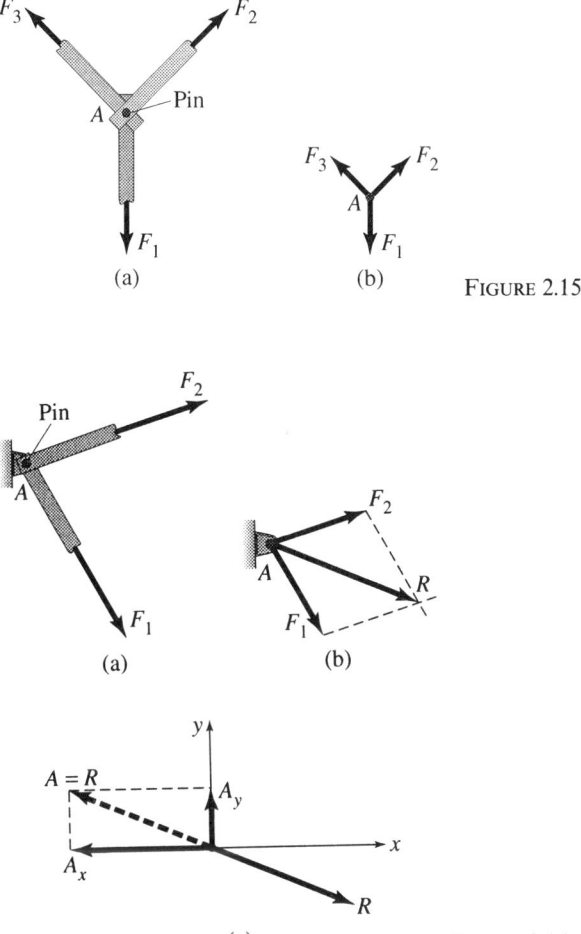

FIGURE 2.15.

FIGURE 2.16.

hinge support is shown in Figure 2.16(a) in which two members are connected to a fixed support by pin A. As in the case of the hinge connection, the force in each of the two members is transmitted, unchanged, to the pin, so that the pin at A will carry the forces $\mathbf{F}_1$ and $\mathbf{F}_2$, as shown in Figure 2.16(b). The resultant R of these two forces (or any number of forces acting on the pin) may be determined as shown. By the principle of action and reaction (Newton's third law), the support pin develops a resisting force $\mathbf{A}$ equal in magnitude and direction but opposite in sense to the resultant $\mathbf{R}$. Instead of the force $\mathbf{A}$, it is usually more convenient to deal with its x and y components, $\mathbf{A}_x$ and $\mathbf{A}_y$, respectively, as shown by the free-body diagram of the supporting pin in Figure 2.16(c).

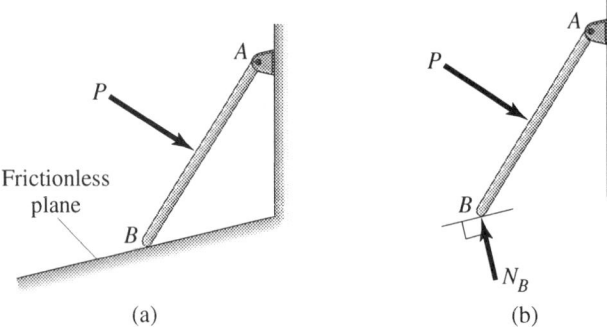

FIGURE 2.17.   (a)   (b)

Frictionless (ideal) conditions are generally assumed to eliminate the existence of moments due to frictional forces (discussed in later chapters) on the surfaces of contact between the pin and the holes in which it fits.

Frictionless Plane

A frictionless plane is an idealization used often to approximate conditions in a *real* plane. When used as a support, the frictionless plane makes it possible to simplify the solution of problems that, otherwise, would be complex if friction were to be included.

Consider, for example, member AB hinged at A and supported at B by a frictionless plane as shown in Figure 2.17(a). If the member is subjected to loads, such as *P*, reactive forces would develop at both the hinge at A and the frictionless plane at B. The reactive force at a hinge was discussed earlier and would not be considered here. The reactive force exerted by the frictionless plane on member AB would have only one component normal (perpendicular) to the frictionless plane as shown in Figure 2.17(b). The symbol N_B is used to signify the magnitude of this *normal* reactive force at point B. Note that, because of the absence of friction between the member and the plane, forces parallel to the surface of the plane *cannot* exist. Therefore, when a frictionless plane serves as a support, the direction of the reactive force is *always* normal to this frictionless plane.

2.6 The Free-Body Diagram

A *particle*, as defined in Chapter 1, is a minute object possessing mass but occupying no volume in space. A particle, therefore, may be viewed as a mathematically defined point which has mass. There are many physical systems, man-made and natural, that may be modeled mathematically as though they were particles. Such physical systems are, therefore, tractable using the principles developed for the analysis of a particle.

The concept of the *free-body diagram* of a particle is extremely important and should be well understood before embarking on a study of the free-body diagram of a rigid body discussed in later chapters. By definition, the *free-body diagram* of a particle is a drawing of the particle after isolating it from the rest of the system (structure, machine, etc.) of which it is a component part. The drawing should show, clearly, all of the forces that act on the particle. These forces should include the weight of the isolated particle which represents the gravitational pull of the earth on the particle, acts through the particle, and is directed toward Earth's center of gravity. Such a force is known as a *gravitational force* and usually given the symbol W to represent the weight. Other forces that should be included are those *externally applied* to the particle as well as *contact forces* which are exerted on the particle by its supports and other neighboring bodies attached to it. It is essential to ensure that the free-body diagram does *not* include forces that have no direct bearing on the isolated particle even though they act at other locations in the system of which the particle is one component. Known forces should be shown fully in magnitude, direction, and sense. Unknown forces are indicated using directed line segments with the unknown magnitude and/or sense represented by a suitable symbol. The sense of an unknown force may be assumed arbitrarily. A positive solution would indicate a correct assumption and a negative solution an incorrect one.

The concept of the free-body diagram and its significance in the solution of mechanics problems cannot be overemphasized. It is, therefore, essential that a systematic procedure be followed in constructing free-body diagrams. Such a procedure is described as follows:

1. A decision is made at the very outset as to what body (treatable as a particle) is to be isolated. This body is, then, separated from the rest of the system of which it is a part and its external outline shown neatly in a separate diagram.
2. *All* forces acting *on* the isolated body are, then, shown. These should include the weight of the body and any externally applied forces, as well as those representing the reactions at removed connections and supports. As indicated earlier, known forces should be clearly shown with proper magnitude, direction, and sense whereas an unknown force is shown by a directed line segment with an assumed sense and represented with a convenient symbol which may be subscripted, if necessary, using the chosen coordinate system or any other convenient symbolic device.
3. If a coordinate system has not been provided, one should be clearly indicated.

The following examples illustrate the construction of free-body diagrams of bodies or objects that may be treated as particles.

■ **Example 2.8**

The cable system shown in Figure E2.8(a) holds the automobile engine of weight $W = 350$ lb in the position shown by wedging a small and weightless sphere D between two frictionless planes, contacting these two planes at points E and H. Construct the free-body diagram of (a) the engine, (b) the connecting ring at B, and (c) the sphere D.

Solution

(a) The free-body diagram of the engine is shown in Figure E2.8(b) and was constructed by the three-step procedure outlined above. The following observations are made.

1. The two forces acting on the engine are its weight $W = 350$ lb and the force F_{EB} in the supporting cable BE. As stated earlier, the weight W is always directed toward the center of the earth, a direction which is generally portrayed as being vertically downward. Also, the weight W may be assumed to act through point G, the center of gravity of the engine (see Chapter 10).
2. Although the engine has a finite size, it is treated as a particle because the two forces acting on it are concurrent. The point of concurrency in this case is point G because, if extended, the line of action of the force F_{EB} will pass through this point.
3. The force $W = 350$ lb is known fully in magnitude, direction, and sense. As discussed in Section 2.5, the force F_{EB} in the supporting cable is directed along the cable and must be a pulling force (tension). Thus, both the direction and sense of the cable force are known, and the only unknown is its magnitude represented by the symbol F_{EB}.
4. It is good practice to show a coordinate system along with the free-body diagram. The origin of the coordinate system may be positioned at any location but, usually, it is convenient to place it at the point of concurrency of the forces acting on the particle. In this particular case the point of concurrency is point G. It is customary to place the x axis horizontally and the y axis vertically, but such placement is arbitrary.

(b) The free-body diagram of the connecting ring at B, constructed by the three-step procedure, is shown in Figure E2.8(c). The following comments are in order:

1. The three cables are connected to the ring at B. The force in each cable is tension directed along the axis of the cable. By the principle of action and reaction (Newton's third law), the force F_{BE} acting on the ring has the same magnitude as the force F_{EB} acting on the engine. The magnitude of the tension in cable BA is represented by the symbol F_{BA}. Similarly, the magnitude of the tension in cable BC is represented by the symbol F_{BC}.
2. Even though the connecting ring has a finite size, it is treated as a particle because the three forces acting on it are concurrent at the geometric center of the ring.

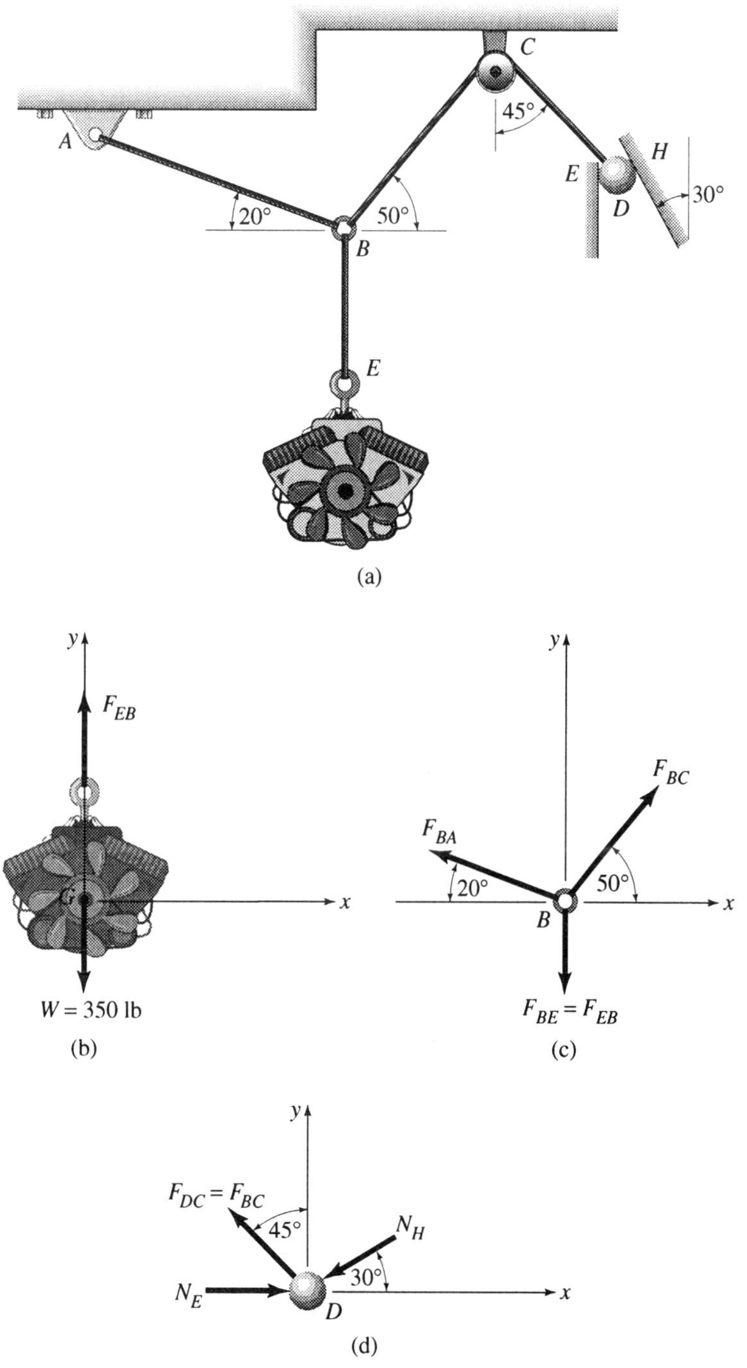

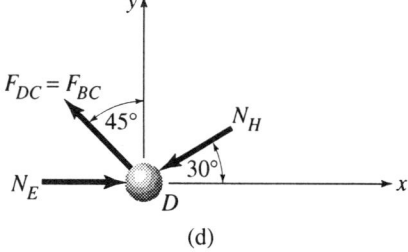

(a)

(b)

$W = 350$ lb

F_{EB}

(c)

F_{BA}

$20°$

B

$50°$

F_{BC}

$F_{BE} = F_{EB}$

(d)

$F_{DC} = F_{BC}$

$45°$

N_H

N_E

$30°$

D

FIGURE E2.8.

3. An x-y coordinate system was chosen such that its origin is at the geometric center of the ring.

(c) Using the three-step procedure, the free-body diagram for sphere D is constructed as shown in Figure E2.8(d). The following observations are made:

1. Because the sphere is assumed weightless, there are only three forces acting on it. The first is the tension in cable DC labeled F_{DC}. However, as discussed in Section 2.5, the tensions in the cable on both sides of the pulley are equal in magnitude. Consequently, F_{DC} has the same magnitude as F_{BC} which acts on the connecting ring at B. The second and third forces acting on the sphere are the reactive forces at the points of contact with the frictionless planes. Thus, as discussed in Section 2.5, the reactive force at F, N_F, is normal to the vertical frictionless plane and the reactive force at H, N_H, is normal to the inclined frictionless plane.
2. The sphere has a finite size. However, it is treated as a particle because the three forces acting on it are concurrent at the geometric center of the sphere.
3. An x-y coordinate system is shown with origin at the point of concurrency of the forces.

■ Example 2.9

Consider the structure shown in Figure E2.9(a) which is subjected to the force $Q = 5\,\mathrm{kN}$. Construct the free-body diagram of (a) the pin at A and (b) the pin at B.

Solution

(a) The free-body diagram of the pin at A is shown in Figure E2.9(b). It was obtained by the three-step procedure recommended for the construction of a free-body diagram. Note the following:

1. The forces in the two members (links) AB and AC are labeled F_{AB} and F_{AC}, respectively and act along the axes of their respective members. They are both shown as pulling on (acting away from) the pin. By action and reaction, this means that the pin pulls back on the two members creating tension in both. In general, it is good practice to assume that unknown *member* forces are tension. A positive solution of the equilibrium equations, as illustrated in the next section, would indicate a correct assumption (tension), whereas a negative solution would indicate compression.
2. The three forces acting on the pin at A are concurrent at the geometric center of the pin, justifying the treatment of a finite-sized pin as a particle.
3. An x-y coordinate system was chosen using the geometric center of the pin as the origin.

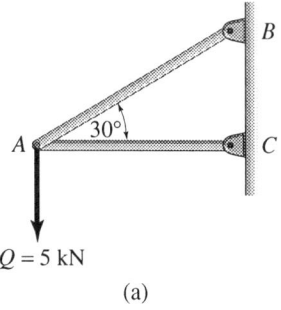

(a)

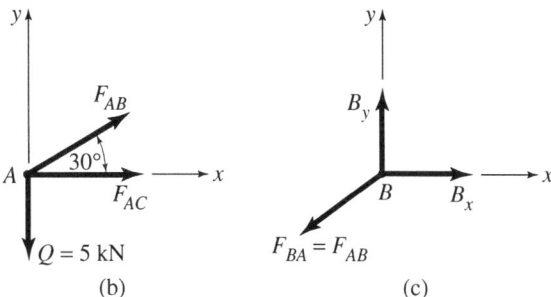

(b)   (c)   FIGURE E2.9.

(b) Using the three-step procedure, the free-body diagram of the pin at B was constructed as shown in Figure E2.9(c). The following observations are made:

1. By the principle of action and reaction, the force F_{BA} acting on pin B, must be equal in magnitude and direction but opposite in sense to the force F_{AB} acting on the pin at A. The unknown reactive force at the support where pin B is located is shown in terms of its x and y components B_x and B_y. This is good practice making it very convenient to solve the equilibrium equations discussed in the next section. It is also good practice to assume that the unknown reactive components are pointed in the positive directions on their respective coordinate axes. A positive solution of the equilibrium equations would indicate a correct assumption whereas a negative solution would indicate an incorrect one.
2. Although the pin has a finite size, it is treated as a particle because the three forces acting on it are concurrent at the geometric center of the pin.
3. The geometric center of the pin was again used as the origin of the chosen coordinate system.

Problems

2.32 Consider the system shown in Figure P2.32 and construct the free-body diagram of (a) the connecting ring at B and (b) the pin at C. Assume that the pulley at D is frictionless.

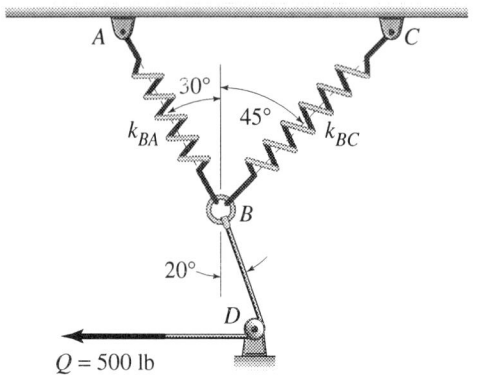

FIGURE P2.32.

2.33 Assume that the pulley at B and the inclined plane in Figure P2.33 are friction-

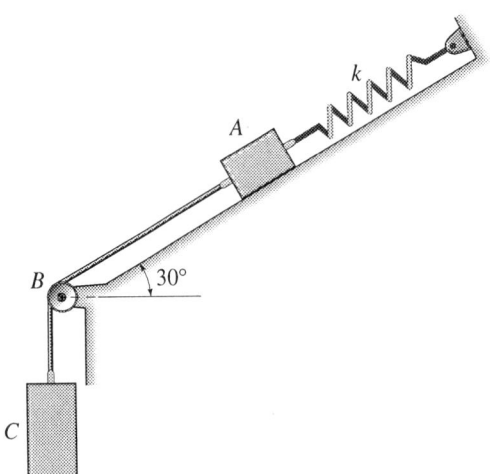

FIGURE P2.33.

less. Let $W_A = 500$ N, the weight of body A, and $W_C = 900$ N, the weight of body C. Construct the free-body diagram of (a) body A and (b) body C. Assume that both bodies are very small.

2.34 The weight $W = 650$ lb is held in position by the cable system shown in Figure P2.34. Construct the free-body diagram of (a) the weight W and (b) the connecting ring at B.

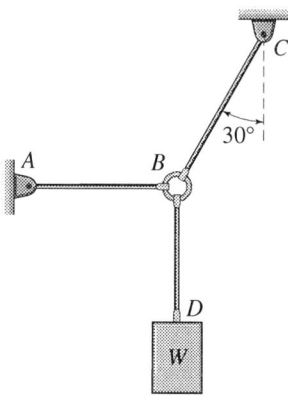

FIGURE P2.34.

2.35 Refer to Problem 2.34 and construct the free-body diagram for (a) the pin at A and (b) the pin at C.

2.36 The pulley-cable system shown in Figure P2.36 is used to move crate A up the frictionless inclined plane. Assume that the pulley at D is frictionless and crate A is very small, and construct the free-body diagram of (a) crate A and (b) the connecting ring at B.

2.37 Three short and weightless links are hinged together as shown in Figure P2.37. The compressive force in link DB has a magnitude $F_{DB} = 15$ kN. Construct the

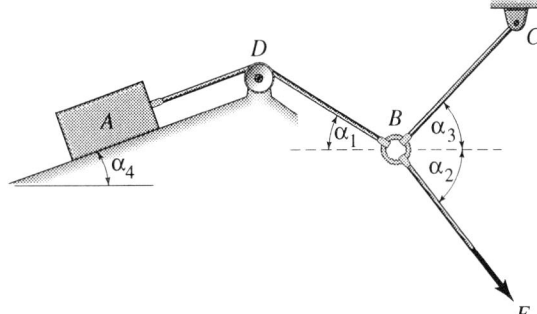

FIGURE P2.36.

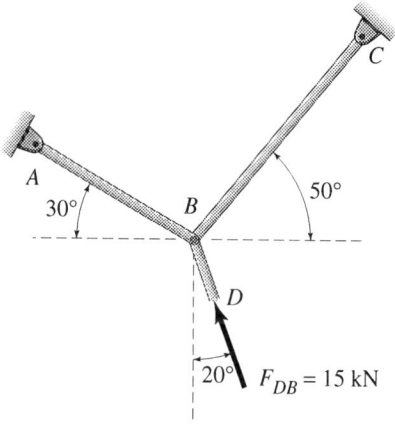

FIGURE P2.37.

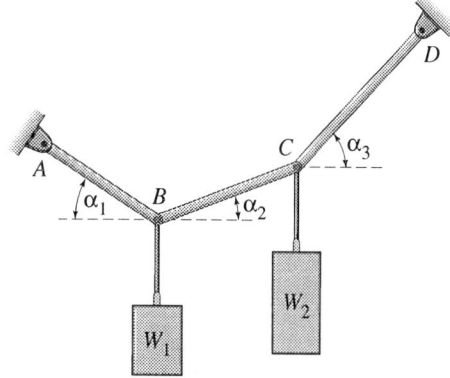

FIGURE P2.38.

free-body diagram of (a) the pin at B and (b) the pin at C.

2.38 Three short and weightless links AB, BC, and CD are hinged together at B and C and pinned to hinge supports at A and D. The system is used to support the two weights W_1 and W_2, as shown in Figure P2.38. Construct the free-body diagram of (a) the pin at B and (b) the pin at C.

2.39 A wall bracket is constructed as shown in Figure P2.39 to support a drum of weight $W = 750$ lb. Assume that all contacting surfaces are frictionless, and construct the free-body diagram of the drum.

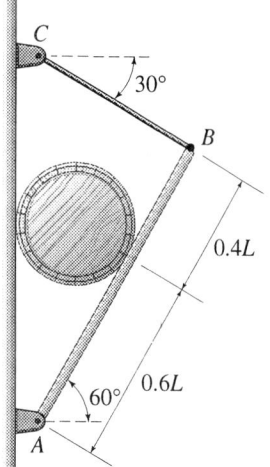

FIGURE P2.39.

2.40 A cylindrical pipe, which may be assumed weightless, is placed in the V-shaped support and subjected to a force $P = 1.2$ kN, as shown in Figure P2.40. Construct the free-body diagram of the pipe.

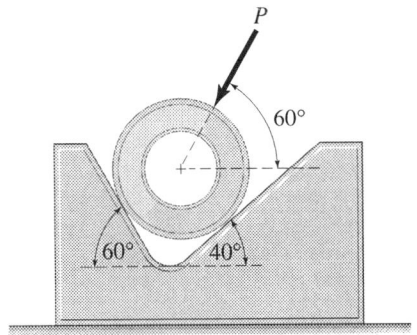

FIGURE P2.40.

2.7 Equilibrium Conditions and Applications

A particle is said to be in *equilibrium* if it is at rest or if it is *moving along a straight path at constant speed*. According to Newton's first law, a particle cannot change its state of either rest or constant-speed motion along a straight path, *unless* there is an unbalanced (resultant) force acting on it. Thus, equilibrium of a particle exists if, and only if, the resultant of *all* forces acting on it is equal to zero.

Because a force system acting on a particle must necessarily be concurrent and because our focus in this chapter is on two-dimensional force systems, it follows that a discussion of particle equilibrium must be limited to the equilibrium of concurrent two-dimensional force systems. The term *coplanar* is sometimes used to describe a two-dimensional force system or forces acting in the same plane. Therefore, the condition of particle equilibrium in two dimensions is achieved by assuring that the resultant of the coplanar and concurrent forces acting on it vanishes. By Eqs. (2.12) and (2.13), the resultant **R** of a coplanar and concurrent force system is expressed as

$$\mathbf{R} = \sum \mathbf{F} = (\sum F_x)\mathbf{i} + (\sum F_y)\mathbf{j}.$$

Because equilibrium requires that **R** be null, it follows that

$$\mathbf{R} = \sum \mathbf{F} = \mathbf{0} \qquad (2.17)$$

$$= (\sum F_x)\mathbf{i} + (\sum F_y)\mathbf{j} = \mathbf{0}. \qquad (2.18)$$

The only way that Eq. (2.18) can be satisfied is for each of its two terms to vanish independently. This, obviously, means that the coefficients of **i** and **j** must each be equal to zero. Thus,

and
$$\left.\begin{aligned} \sum F_x &= 0, \\ \sum F_y &= 0 \end{aligned}\right\} \qquad (2.19)$$

which are both *necessary* and *sufficient* conditions for the equilibrium of a particle in two dimensions. The first of Eqs. (2.19) shows that, if a particle is in equilibrium in the x direction, all of the force components in this direction must be in *perfect* balance. This means that, if there is a resultant force component acting in the positive x direction, another resultant force component of equal magnitude acting in the negative x direction must exist. The same statement applies to resultant force components in the y direction. Furthermore, the two conditions expressed in Eq. (2.19) are entirely independent of one another and either one may be satisfied even though the other may not be. Thus, for *complete* equilibrium of a particle in two dimensions, both of the independent conditions expressed in Eq. (2.19) must be satisfied.

The fact that we have available two independent equations of equilibrium in the case of a particle in two dimensions means that we can solve for a maximum of two unknown quantities in a given case. The two unknown quantities may be the magnitude and direction of a given force, the magnitudes of two different forces, the directions of two different forces or the magnitude of one force and the direction of another. In solving a given equilibrium problem, the beginner is faced with a choice between the vector equilibrium condition expressed in Eq. (2.17) and the scalar equilibrium conditions expressed in Eq. (2.19). Obviously, because the two sets of equilibrium conditions are equivalent, either set may be used. *However, under certain conditions, one set has distinct advantages over the other. In general, and as a rule of thumb, the vector approach is more elegant and has advantages only in solving three-dimensional force systems. The scalar approach is more direct and more suited to solving two-dimensional force systems. The procedure followed in this text is to use whichever approach is move convenient in a given situation.* Except for Example 2.10, all of the solutions presented in this chapter use the scalar method. The solution in Example 2.10 uses the scalar and the vector method for purposes of comparison.

Instead of using the scalar (algebraic) relations expressed in Eq (2.19), it is sometimes convenient to use the polygon method to determine the resultant **R** which must vanish if the particle is in equilibrium. This approach is especially useful in solving some two-dimensional equilibrium problems, particularly those in which the particle is subjected only to three forces. Because, the resultant **R** of the three forces must be zero for equilibrium, it follows that the force polygon, in such a case, is a closed triangle whose geometry may be used to facilitate solving the problem.

The following examples illustrate the solution of two-dimensional particle equilibrium problems.

■ Example 2.10

A crane is used to lift a crate whose mass is 2000 kg and a motorized pulley system provides the force needed to position the crate horizon-

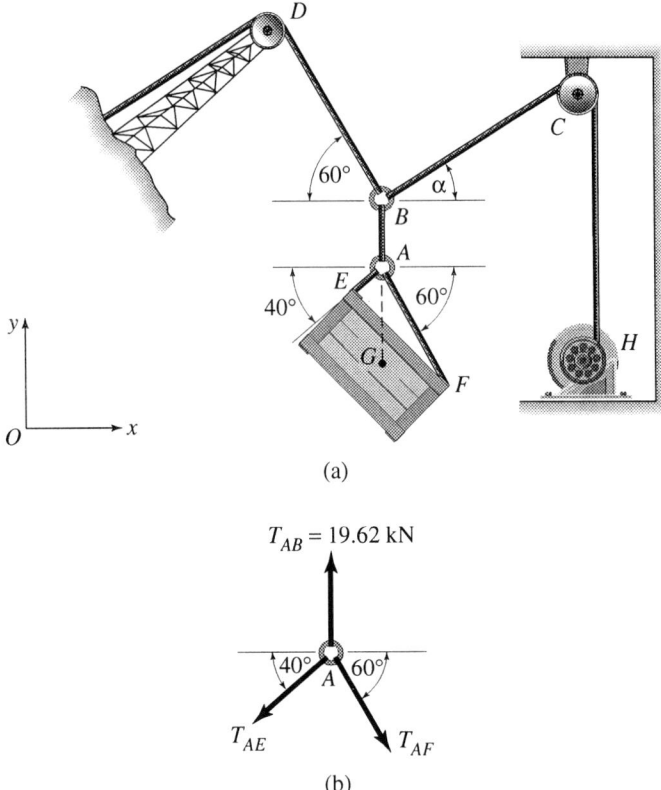

(a)

$T_{AB} = 19.62$ kN

(b)

FIGURE E2.10.

tally as shown in Figure E2.10(a). For the position shown, determine the tension in cables AE and AF.

Solution

The tensions in cables AE and AF may be determined by using the free-body diagram of connecting ring at A as shown in Figure E2.10(b). Note that the tension in cable AB must be equal to the weight of the crate which is found as follows:

$$W = mg$$

$$= 2000\ (9.81)$$

$$= 19620\ \text{N}$$

$$= 19.62\ \text{kN}.$$

The free-body diagram shown in Figure E2.10(b) represents a concurrent force system with the two unknown quantities T_{AE} and T_{AF}. Two approaches are used for this solution: the first is the scalar approach and the second is the vector approach. Thus,

Scalar Approach

Equations (2.19) are applied to obtain

$$\sum F_x = 0, \qquad T_{AF} \cos 60° - T_{AE} \cos 40° = 0,$$

from which

$$0.500 T_{AF} - 0.766 T_{AE} = 0. \tag{a}$$

$$\sum F_y = 0, \quad 19.62 - T_{AF} \sin 60° - T_{AE} \sin 40° = 0,$$

which reduces to

$$19.62 - 0.866 T_{AF} - 0.643 T_{AE} = 0. \tag{b}$$

A simultaneous solution of Eqs. (a) and (b) yields

$$T_{AF} = 15.30 \text{ kN} \qquad \text{and} \qquad T_{AE} = 10.00 \text{ kN.} \qquad \text{ANS.}$$

Vector Approach

Express all forces shown in Figure E2.10(b) in vector form. Thus

$$\mathbf{T}_{AB} = 19.62\mathbf{j},$$

$$\mathbf{T}_{AF} = (T_{AF} \cos 60°)\mathbf{i} - (T_{AF} \sin 60°)\mathbf{j}$$
$$= (0.500 T_{AF})\mathbf{i} - (0.866 T_{AF})\mathbf{j},$$

and

$$\mathbf{T}_{AE} = -(T_{AE} \cos 40°)\mathbf{i} - (T_{AE} \sin 40°)\mathbf{j}$$
$$= -(0.766 T_{AE})\mathbf{i} - (0.643 T_{AE})\mathbf{j}.$$

By Eq. (2.17),

$$\mathbf{R} = \sum \mathbf{F} = \mathbf{T}_{AB} + \mathbf{T}_{AF} + \mathbf{T}_{AE} = \mathbf{0},$$

$$(0.500\mathbf{T}_{AF} - 0.766\mathbf{T}_{AE})\mathbf{i} + (19.62 - 0.866\mathbf{T}_{AF} - 0.643\mathbf{T}_{AE})\mathbf{j} = \mathbf{0}.$$

Equating scalar multipliers of $\mathbf{i}$ and $\mathbf{j}$ on both sides of this vector equation gives two scalar equations. Thus,

$$0.500\mathbf{T}_{AF} - 0.766\mathbf{T}_{AE} = 0 \tag{c}$$

and

$$19.62 - 0.866\mathbf{T}_{AF} - 0.643\mathbf{T}_{AE} = 0. \tag{d}$$

Equations (c) and (d) are, of course, identical with Eqs. (a) and (b) obtained by the scalar approach.

■ **Example 2.11**

Refer to Example 2.10 and determine the angle α and the tension in cable BD if the motor produces a tension in cable HC equal to 10 kN.

Solution

The frictionless pulley at C serves to change only the direction of the force supplied by the motor but not its magnitude. Thus, the tension in

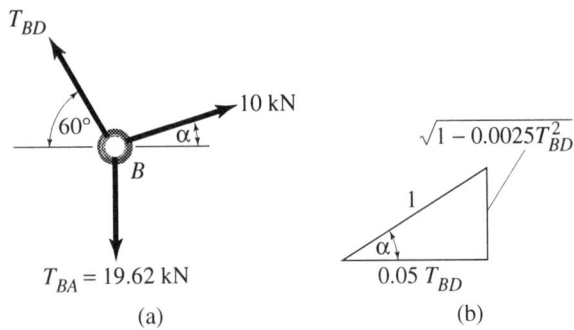

FIGURE E2.11.   (a)   (b)

cable BC is equal in magnitude to that in cable HC, namely, 10 kN. The unknown angle α and the unknown tension in cable BD may be found by considering the free-body diagram of the connecting ring at B as shown in Figure E2.11(a). This free-body diagram represents a concurrent force system with the unknown quantities T_{BD} and α. The conditions of equilibrium expressed in Eq. (2.19) are now applied to obtain the values of T_{BD} and α. Two types of solutions are presented here. The first is a trial-and-error solution whereas the second is a closed-form solution.

Trial-And-Error Solution

$$\sum F_x = 0, \quad 10 \cos \alpha - T_{BD} \cos 60° = 0, \tag{a}$$

form which

$$T_{BD} = 20.0 \cos \alpha. \tag{b}$$

$$\sum F_y = 0, \quad 10 \sin \alpha + T_{BD} \sin 60° - 19.62 = 0,$$

which yields

$$\sin \alpha + 0.0866 T_{BD} - 1.962 = 0. \tag{c}$$

Substituting Eq. (b) into Eq. (c) and simplifying,

$$\sin \alpha + 1.732 \cos \alpha - 1.962 = 0. \tag{d}$$

Equation (d) requires a trial-and-error solution which yields

$$\alpha = 18.8° \quad \text{and} \quad 41.2°. \qquad \text{ANS.}$$

Equation (b), then, provides the value of T_{BD}. Thus,

$$T_{BD} = 18.93 \text{ kN} \quad \text{and} \quad 15.05 \text{ kN}. \qquad \text{ANS.}$$

Closed-Form Solution

The trial-and-error solution above can be avoided by using an exact solution procedure. From Eq. (b),

$$\cos \alpha = 0.05 T_{BD} \tag{e}$$

which may be interpreted by the use of Figure E2.11(b) to yield

$$\sin \alpha = \sqrt{1 - 0.0025 T_{BD}^2} \qquad \text{(f)}$$

Substituting Eq. (f) in Eq. (c) yields

$$T_{BD}^2 - 33.98 T_{BD} + 284.9 = 0. \qquad \text{(g)}$$

This is a quadratic equation that may be solved in closed form to give $T_{BD} = 18.93$ kN and 15.05 kN which, when substituted into Eq. (e) yields $\alpha = 18.8°$ and $41.2°$, respectively, the same as the answers obtained previously.

■ **Example 2.12**

A package of weight $W = 200$ lb is held in position by a system of two short links and a spring, as shown in Figure E2.12(a). (a) What must be the angle α if the force Q is to have the least possible magnitude consis-

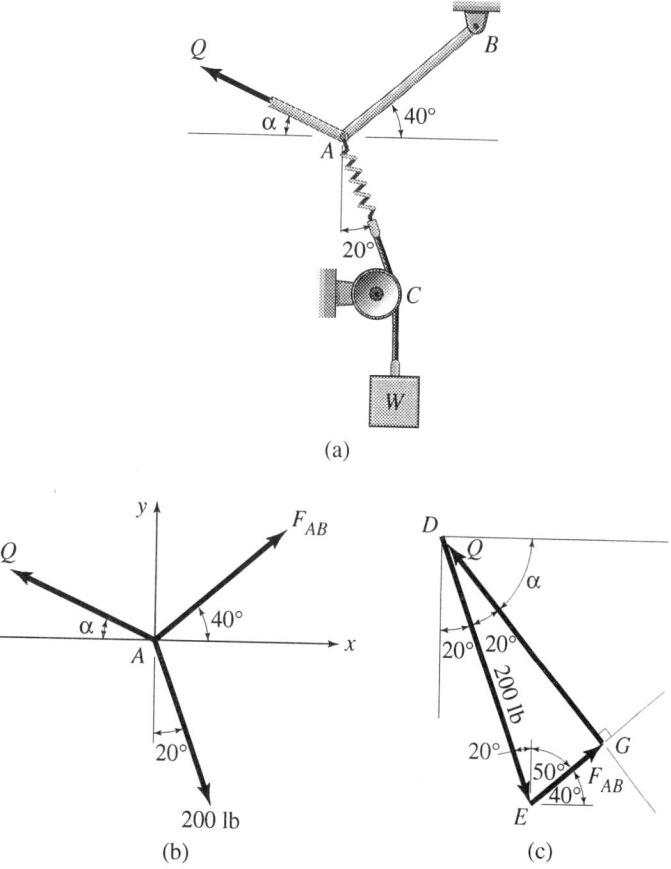

FIGURE E2.12.

tent with the equilibrium of the system. (b) What are the magnitudes of Q and the force in link AB. (c) What is the stretch in the spring if its spring constant $k = 2000$ lb/in. Assume that all hinges are frictionless.

Solution

(a) and (b): Two types of solutions are presented for parts (a) and (b) of the problem. The first is based on the equilibrium equations and the second makes use of the force-polygon method. The free-body diagram of the joint at A is shown in Figure E2.12(b) revealing that there are three unknown quantities, Q, α and F_{AB}. Because we have two equations of equilibrium in addition to the condition that Q must be a minimum, we have a sufficient number of conditions to solve for the three unknowns.

Equilibrium Equations Method

Apply the conditions of equilibrium expressed in Eq. (2.19). Thus,

$$\sum F_x = 0, \qquad 200 \sin 20° + F_{AB} \cos 40° - Q \cos \alpha = 0,$$

from which

$$F_{AB} = \frac{Q \cos \alpha - 68.4}{0.766}. \tag{a}$$

$$\sum F_y = 0, \qquad F_{AB} \sin 40° + Q \sin \alpha - 200 \cos 20° = 0$$

which reduces to

$$0.643 F_{AB} + Q \sin \alpha - 187.9 = 0. \tag{b}$$

Substituting Eq. (a) in Eq. (b) and solving for Q in terms of α

$$Q = \frac{245.36}{0.839 \cos \alpha + \sin \alpha}. \tag{c}$$

The condition that Q be a minimum is obtained by setting $dQ/d\alpha$ equal to zero. Thus, from Eq. (c),

$$\frac{dQ}{d\alpha} = \frac{-245.36(\cos \alpha - 0.839 \sin \alpha)}{(0.839 \cos \alpha + \sin \alpha)^2} = 0. \tag{d}$$

Therefore, if the denominator in Eq. (d) is not zero, it follows that

$$205.86 \sin \alpha - 245.36 \cos \alpha = 0. \tag{e}$$

Dividing through in Eq. (e) by $\cos \alpha$ yields

$$\tan \alpha = 1.1919$$

and

$$\alpha = 50.0° \qquad\qquad \text{ANS.}$$

which shows that the denominator in Eq. (d) is, in fact, not zero. Equations (c) and (a) now yield

$$Q = 188.0 \text{ lb} \qquad\qquad \text{ANS.}$$

and

$$F_{AB} = 68.4 \text{ lb.} \qquad \text{ANS.}$$

Force-Polygon Method

The force-polygon method which, in this case, reduces to a force triangle, possesses real advantages in determining the angle that makes the force Q a minimum.

The known force in the spring (200 lb) is drawn first starting at point D as shown in Figure E2.12(c). The direction of the force F_{AB} is, then, represented by line EG. Because the three forces acting on the pin at A are in equilibrium, the force triangle must be closed. This means that the force Q must have its tail some place on line EG and its head at point D. Because Q is to be a minimum, point G is selected so that the force Q is perpendicular to line EG as shown. Other points on line EG would yield a larger value for Q.

Analysis of the geometry of the force triangle yields the following:

$$\alpha = 90 - 40 = 50.0°,$$

$$Q = 200 \cos 20° = 187.9 \text{ lb}, \qquad \text{ANS.}$$

and

$$F_{AB} = 200.0 \sin 20° = 68.4 \text{ lb.}$$

Any differences between these answers and those obtained previously are insignificant and due to approximations.

(c) From Eq. (2.16) we obtain

$$s = \frac{F}{k} = \frac{200}{1000} = 200 \times 10^{-3} \text{ in.} \qquad \text{ANS.}$$

■ Example 2.13

Refer to the pulley arrangement shown in Figure E2.13(a). If a workman is able to apply a downward pull $P = 300$ N, as shown, determine the mass m of block A that he can maintain in equilibrium on the frictionless inclined plane. Also find the reaction between the block and the inclined plane. Assume that all hinges are frictionless and all dimensions (pulleys and block) are very small.

Solution

Based on the assumption that all sizes are very small, the force systems acting on all pulleys and on block A may be assumed concurrent. Thus, all of these bodies may be treated as particles. However, their free-body diagrams are shown as finite objects to indicate clearly the relationships that exist among the various forces. The free-body diagrams of the various pulleys, shown in Figure E2.13(b), were constructed on the basis that the cable tension is the same on either side of the pulley.

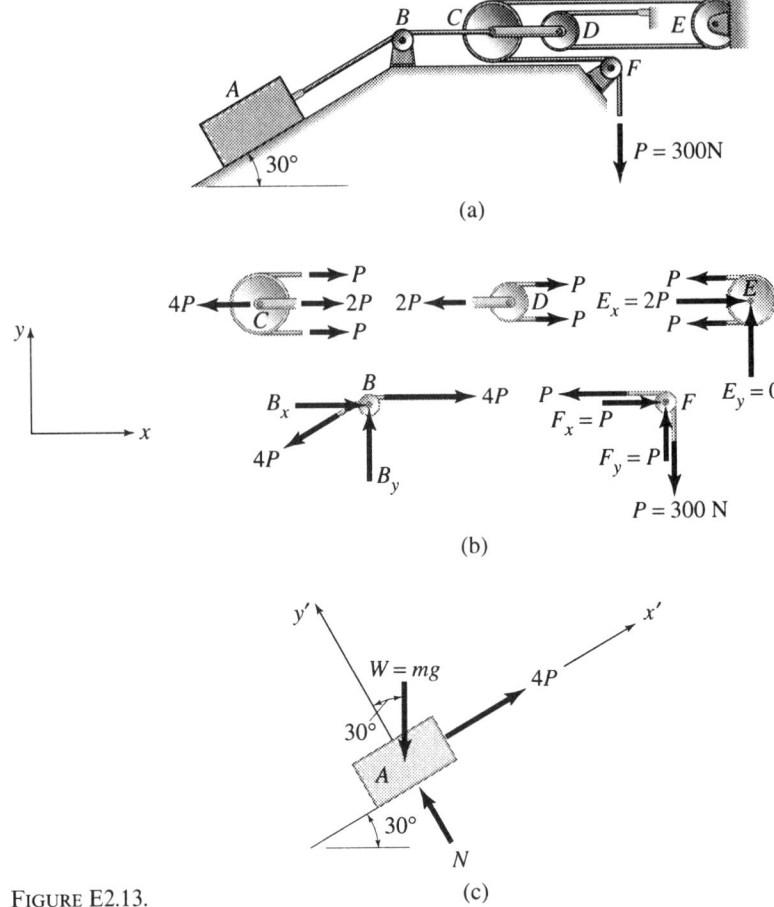

FIGURE E2.13.

Thus, starting with pulley F, we conclude that the tension in the horizontal cable is P. Therefore, the cable around pulley C has a tension equal to P on either side of this pulley. The same statement may be made about pulleys E and D. Now, applying the equilibrium equation $\sum F_x = 0$ to pulley D, we conclude that the link connecting pulleys D and C has a force in it equal to $2P$. Proceeding to pulley C and applying the equilibrium equation $\sum F_x = 0$, we determine that the tension in cable CB is equal to $4P$. Pulley B shows that the tension in cable BA is $4P$, also. Therefore, the free-body diagram of block A may be constructed as shown in Figure E2.13(c). Applying the equilibrium conditions Eqs. (2.19) to the free-body diagram of block A,

$$\sum F_{x'} = 0, \qquad 4P - mg \sin 30° = 0$$

from which

$$m = \frac{4P}{g \sin 30°} = \frac{4(300)}{9.81 \sin 30°} = 245 \text{ kg.} \qquad \text{ANS.}$$

$$\sum F_{y'} = 0, \qquad N - mg \cos 30° = 0$$

which yields

$$N = mg \cos 30° = 244.6(9.81) \cos 30° = 2.08 \times 10^3 \text{ N.} \qquad \text{ANS.}$$

■ Example 2.14

The system shown in Figure E2.14(a) consists of a very small sphere A of weight W attached to a flexible but inextensible cable. The cable passes over a small frictionless pulley and a force F is applied to the other end, as shown, to slide the sphere up the frictionless circular surface of radius R. In terms of W and θ, determine the magnitudes of F and the normal reaction N between the sphere and the circular surface, as the angle θ varies form 0 to 90°. Form the dimensionless ratios F/W and N/W, and plot them vs. the angle θ for $0 \leq \theta \leq 90°$.

Solution

The free-body diagram of the sphere is shown in Figure E2.14(b) along with an x-y coordinate system. This free-body diagram shows that the force system acting on the sphere is concurrent at the geometric center of the small sphere and that there are two unknown quantities F and N. These two quantities may be determined in terms of W and θ by using the two equilibrium conditions expressed in Eq. (2.19). Thus,

$$\sum F_x = 0, \qquad F \sin \phi - N \cos \theta = 0 \qquad \text{(a)}$$

where, from the geometry of Figure E2.14(b),

$$\sin \phi = AD/AB,$$

$$= \frac{\cos \theta}{\sqrt{5 - 4 \sin \theta}}, \qquad \text{(b)}$$

and

$$\cos \phi = DB/AB,$$

$$= \frac{2 - \sin \theta}{\sqrt{5 - 4 \sin \theta}}. \qquad \text{(c)}$$

Substituting Eq. (b) in Eq. (a) and solving for N,

$$N = \frac{F}{\sqrt{5 - 4 \sin \theta}}. \qquad \text{(d)}$$

$$\sum F_y = 0, \qquad F \cos \phi + N \sin \theta - W = 0. \qquad \text{(e)}$$

Substituting Eqs. (c) and (d) in Eq. (e) and solving for F,

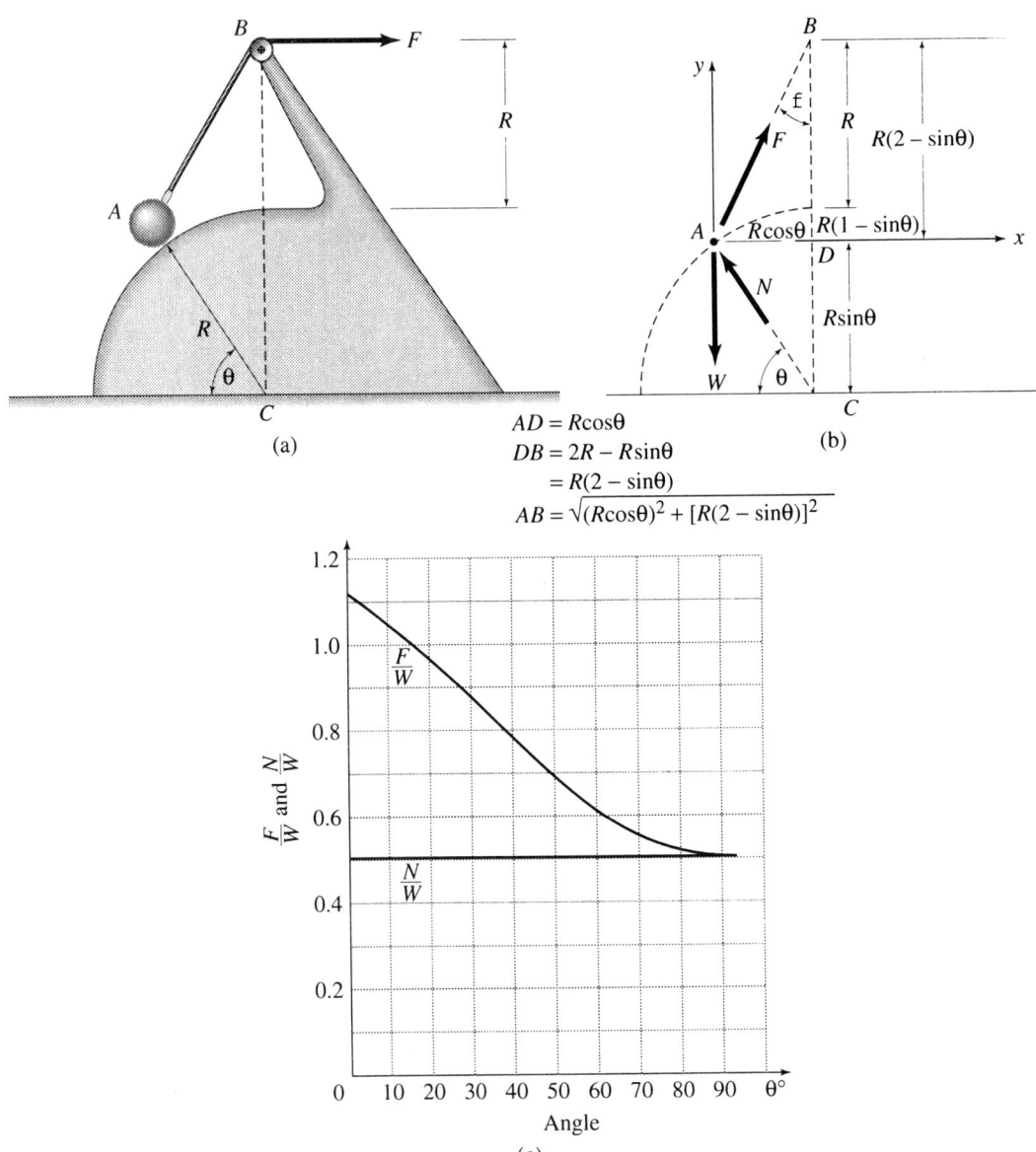

$AD = R\cos\theta$
$DB = 2R - R\sin\theta$
$= R(2 - \sin\theta)$
$AB = \sqrt{(R\cos\theta)^2 + [R(2 - \sin\theta)]^2}$

FIGURE E2.14.

$$F = \frac{W}{2}\sqrt{5 - 4\sin\theta}. \tag{f}$$

If Eq. (f) is substituted in Eq. (d),

$$N = \frac{W}{2} \tag{g}$$

which shows that the normal reaction N is constant at $W/2$ regardless of the position of the cylinder. The ratios F/W and N/W are obtained from Eqs. (f) and (g), respectively. Thus.

$$\frac{F}{W} = \frac{1}{2}\sqrt{5 - 4\sin\theta},$$

and

$$\frac{N}{W} = \frac{1}{2}.$$

These ratios versus the angle θ are plotted in Figure E2.14(c).

◼

Problems

2.41 The weight $W = 650$ lb is held in position by the cable system shown in Figure P2.41. Determine the tensions in cables BA and BC.

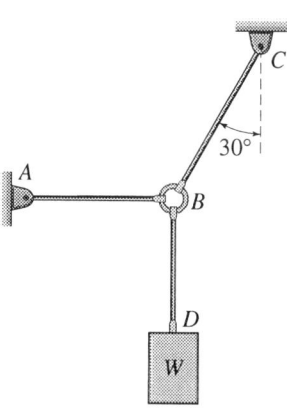

FIGURE P2.41.

2.42 Three short and weightless links are hinged together, as shown in Figure P2.42. The compressive force in link DB has a magnitude $F_{DB} = 15$ kN. Determine the forces in links BA and BC. Assume that all hinges are frictionless.

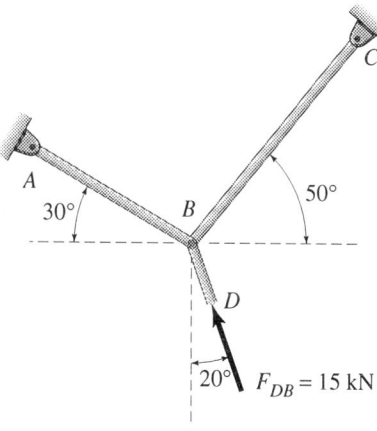

FIGURE P2.42.

2.43 Consider the system shown in Figure P2.43. Let $k_{BA} = 400$ lb/in. and $k_{BC} = 700$ lb/in. For the position shown, determine the amount of stretch (deformation) that each of the two springs has experienced.

2.44 The bracket shown in Figure P2.44 supports a cylinder of mass $m = 300$ kg. If the support reaction at A $N_A = 2000$ N, determine angle α and the support reaction N_B. Assume that all contacting surfaces are frictionless.

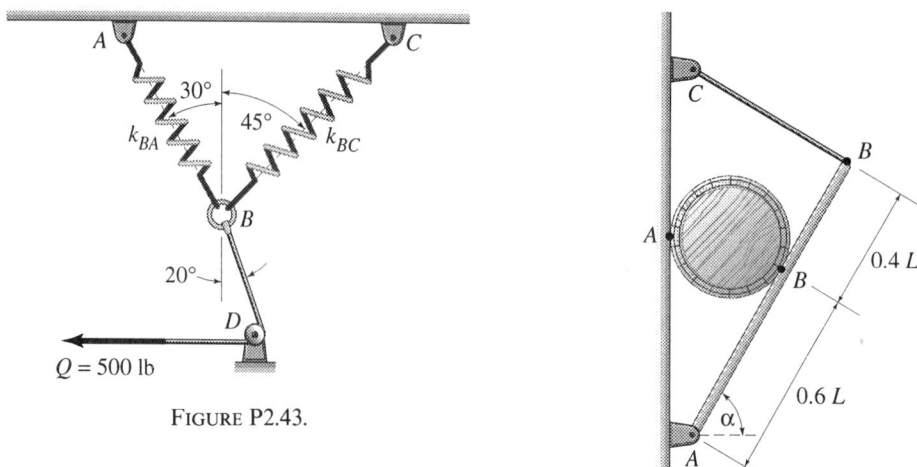

FIGURE P2.43.

FIGURE P2.44.

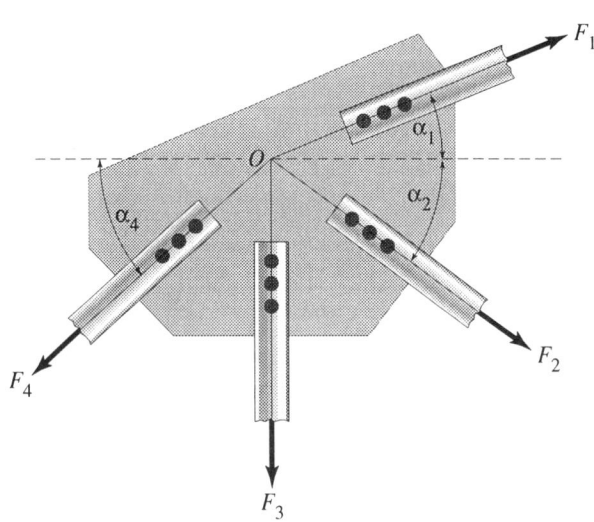

FIGURE P2.45.

2.45 A gusset plate is used to join members of a truss, as shown in Figure P2.45 so that the forces in the various members are concurrent at point O. Let $F_2 = 14.0$ k, $F_3 = 4.5$ k, $\alpha_1 = 25°$, $\alpha_2 = 35°$, and $\alpha_4 = 40°$. Determine the forces F_1 and F_4.

2.46 Refer to Problem 2.45. Let $F_1 = 35.5$ kN, $F_2 = 15.0$ kN, $F_3 = 5.0$ kN, $\alpha_1 = 30°$ and $\alpha_2 = 45°$. Find F_4 and α_4.

2.47 Refer to Problem 2.45. Let $F_2 = 7.4$ k, $F_3 = 3.5$ k, $F_4 = 11.7$ k, $\alpha_2 = 20°$, and $\alpha_4 = 30°$. Determine F_1 and α_1.

2.48 Block A of mass $m = 100$ kg slides on a frictionless inclined plane and is attached to a spring, as shown in Figure P2.48. In the position shown, the stretch of the spring is known to be 0.04 m. Determine the spring constant k and the normal re-

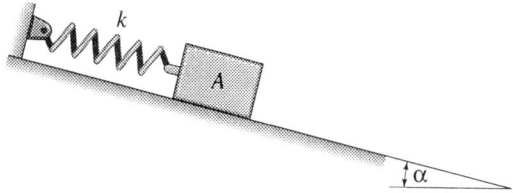

FIGURE P2.48.

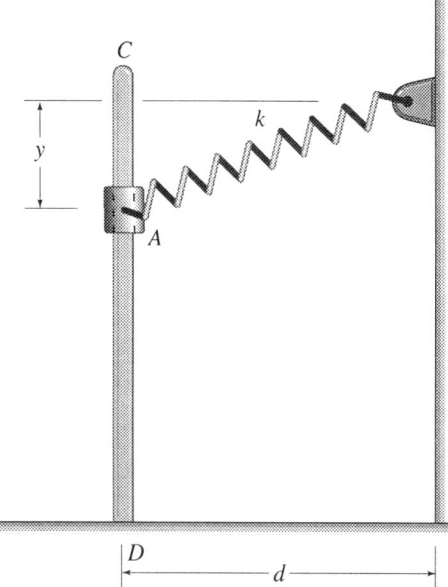

FIGURE P2.50.

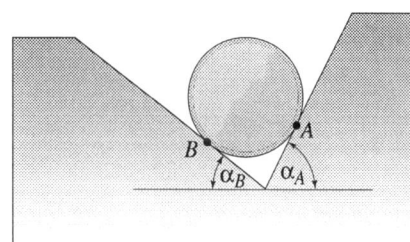

FIGURE P2.52.

action between the block and the inclined plane if $\alpha = 30°$. Assume that the block is very small.

2.49 Refer to Problem 2.48. Let the weight of block A $W = 500$ lb, the spring constant $k = 1000$ lb/in, and the stretch in the spring $s = 0.2$ in. Determine the angle α of the frictionless inclined plane and the normal reaction between this plane and block A.

2.50 Block A of weight W slides freely on the vertical frictionless rod CD as shown in Figure P2.50. The attached spring has a spring constant $k = 1000$ N/m and an undeformed length $L_u = 0.5$ m. At equilibrium, the stretched length of the spring was measured at 0.7 m. Find the weight W and the normal reaction between the block and the rod in this position. Assume that the block is very small, and let the distance d $= 0.5$ m.

2.51 Refer to Problem 2.50. Let $W = 30$ lb, $d = 2$ ft, and $y = 8$ in. Also, the undeformed length of the spring $L_u = 2$ ft. Determine the spring constant k and the normal reaction between the block and the vertical frictionless rod.

2.52 A cylinder of mass $m = 750$ kg is wedged between two frictionless inclined planes as shown in Figure P2.52. Let $\alpha_A = 50°$ and $\alpha_B = 30°$, and determine the normal reactions at points A and B where the cylinder contacts the two planes.

2.53 Refer to Problem 2.52. The weight of the cylinder $W = 300$ lb and the angle $\alpha_A = 35°$. Determine the angle α_B so that the normal reaction at B is the least possible (i.e. a minimum). Find this minimum reaction and the normal reaction at A.

2.54 Two cylinders are placed in a container as shown in Figure P2.54. The mass of the top cylinder $m_1 = 50$ kg and that of the bottom cylinder $m_2 = 300$ kg. Assume frictionless conditions at all contacting surfaces and determine the normal reactions at the points of contact A, B, C, and D.

2.55 Block A of weight $W = 55$ lb is attached

to a spring and can slide freely on an inclined frictionless plane, as shown in Figure P2.55. The spring constant $k = 72$ lb/in. and, at equilibrium, the stretch of the spring was measured to be $s = 0.5$ in. Let $\alpha = 40°$ and determine (a) the angle β defining the inclination of the spring from the vertical and (b) the normal reaction between the block and the inclined plane. Assume that the block is very small.

2.56 Block A of mass $m = 15$ kg is attached to a spring and can slide freely on an inclined frictionless plane as shown in Fig-

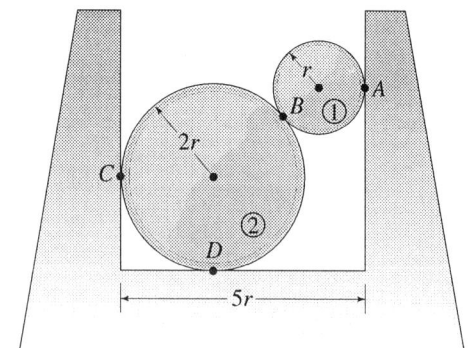

FIGURE P2.54.

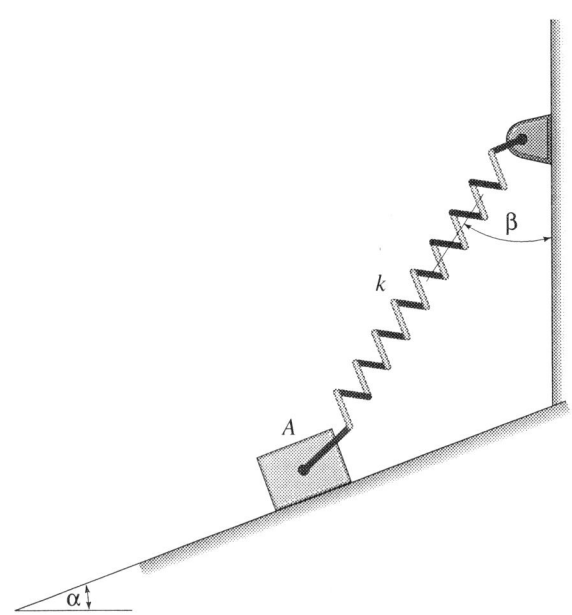

FIGURE P2.55.

ure P2.55. Let $\alpha = 30°$ and determine the angle β so that the force in the spring is a minimum. Under these conditions, what is the force in the spring and the normal reaction between block A and the inclined plane?

2.57 Three short links are hinged together at A as shown in Figure P2.57. Let $P = 20$ k

and $\alpha = 30°$. Determine the forces in members AB and AC.

2.58 Consider the frame shown in Figure P2.57. The force in link AB has a magnitude of 7.8 kN directed from B to A. Determine the angle α for equilibrium of point A so that the force P is minimum. Furthermore, determine the magnitude

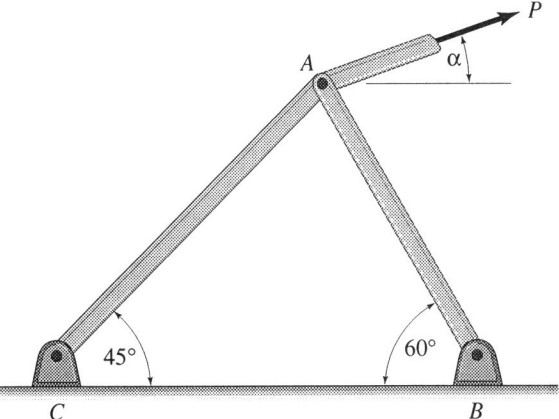

FIGURE P2.57.

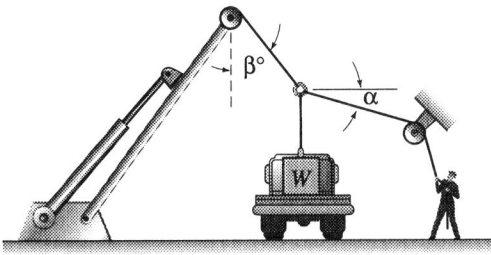

FIGURE P2.59.

of the minimum P and the force in link CA.

2.59 A crane is used to lift a weight $W = 600$ lb and a worker positions it properly on the truck as shown in Figure P2.59. The worker pulls on the rope with a force of 30 lb when the angle $\alpha = 20°$. Determine the tension in the crane cable and the angle β that it makes with the vertical.

2.60 The system shown in Figure P2.60 consists of a 2m long continuous cord that

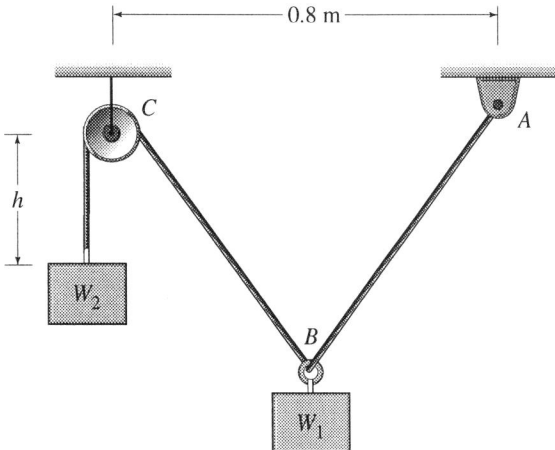

FIGURE P2.60.

is fixed at A and passes through a small frictionless ring at B and over a frictionless pulley at C. Develop the ratio W_1/W_2 as a function of h for the equilibrium position of the system. What is the value of h for the case when $W_1 = 1.5\ W_2$?

2.61 The pulley-cable arrangement shown in Figure P2.61 is used to lift a crate of mass m by applying a pull P at the end of the cable as shown. Determine the mass m that may be lifted when $P = 200$ N. Assume that all pulleys are very small and frictionless.

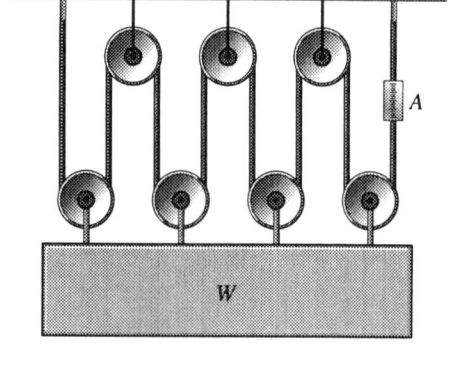

FIGURE P2.62.

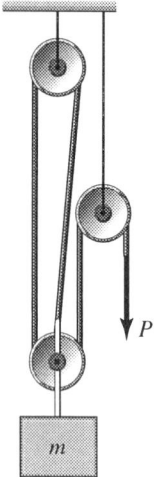

FIGURE P2.61.

2.62 Consider the pulley-cable arrangement shown in Figure P2.62. The spring scale at A indicates a reading 7.5 lb. Determine the weight W that is supported by the system. Assume that all pulleys are very small and frictionless.

2.63 A workman maintains himself in position by applying the necessary force to the end of the cable of the pulley-cable arrangement shown in Figure P2.63. If the spring scale at A indicates a reading of 150 N,

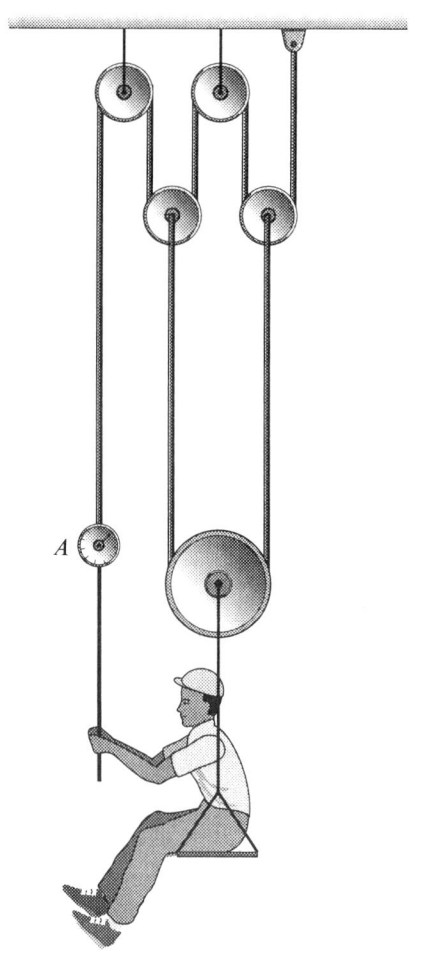

FIGURE P2.63.

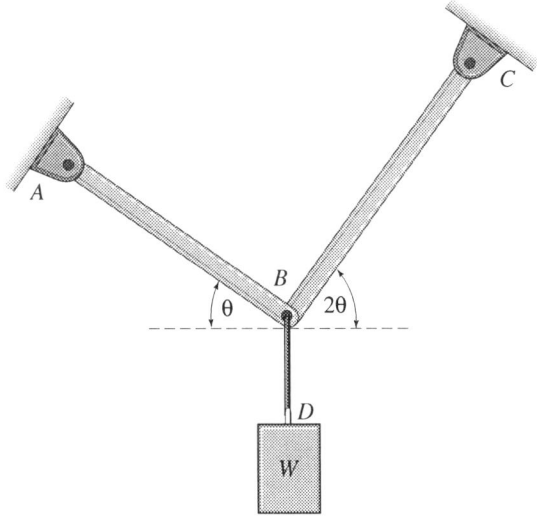

FIGURE P2.64.

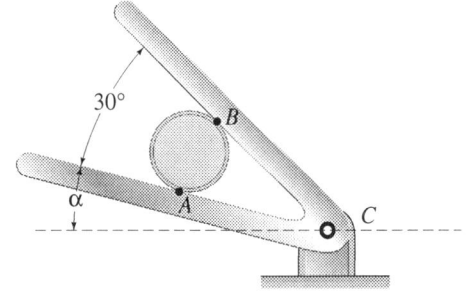

FIGURE P2.65.

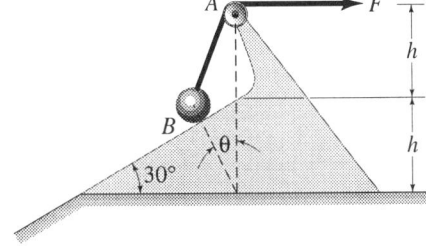

FIGURE P2.66.

determine the weight of the workman. Assume that all pulleys to be very small and frictionless.

2.64 Refer to the system shown in Figure P2.64. In terms of W and θ, determine the values of F_{BA} and F_{BC}, the forces in members BA and BC, respectively, and form the rations F_{BA}/W and F_{BC}/W.

2.65 The 30° bracket, shown in Figure 2.65, is hinged at C and may be rotated about this hinge to change the angle α at will. A cylinder of weight W is placed within the bracket as shown. Assume that all con-

tacting surfaces are frictionless and, in terms of W and α, determine the values of N_A and N_B, the normal reactions at points A and B, respectively, and form the ratios N_A/W and N_B/W.

2.66 A small cylinder of weight W rests on a frictionless inclined plane and is supported by a flexible cable that passes over a small frictionless pulley at A, as shown in Figure P2.66. In terms of W and θ, determine the values of F and of the normal reaction N_B between the cylinder and the inclined plane, and form the ratio F/W and N_B/W.

Review Problems

2.67 The force **F** shown in Figure P2.67 has a magnitude of 200 lb and its scalar components along lines *a* and *b* are 300 lb and 250 lb, respectively. Determine the angles α and β.

2.68 The frame, shown in Figure P2.68, is subject to the force *P* as indicated. If the component of *P* along AB is 15 kN directed from A to B and its component along BC is 8 kN directed from B to C, determine the magnitude of *P* and the angle α that defines its direction from the horizontal.

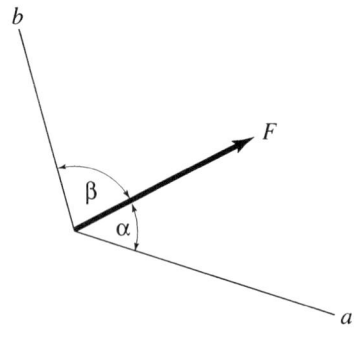

FIGURE P2.67.

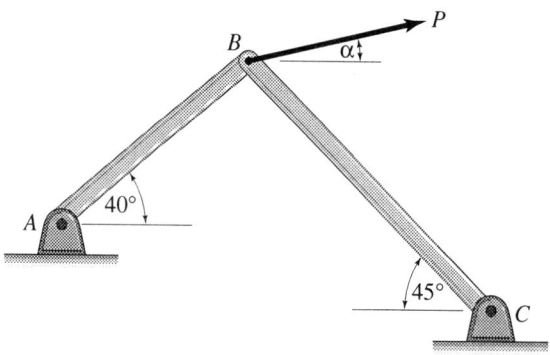

FIGURE P2.68.

2.69 Compute the resultant of the force system shown in Figure P2.69. (a) Express the answer in vector form in terms of two rectangular components. (b) Express the answer as a magnitude and an angle defining the direction of the resultant from the horizontal.

2.70 The force system shown in Figure P2.70 has a resultant of 75 N along the positive *x* axis. Determine the magnitude of *F* and the angle α defining its direction from horizontal.

2.71 The magnitude of the resultant of the force system shown in Figure P2.71 is to

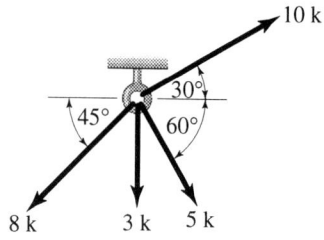

FIGURE P2.69.

be equal to or less than 15 kN. What range of values of the force *F* is consistent with this condition?

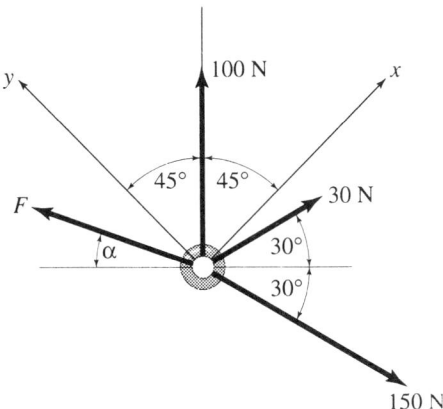

FIGURE P2.70.

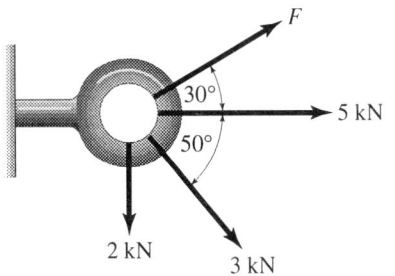

FIGURE P2.71.

2.72 A disabled car of weight W is to be re-
tained on a smooth inclined plane by the
force P as shown in Figure P2.72. Determine the least force P that will accom-

FIGURE P2.72.

plish the purpose and the angle β that it
makes with the horizontal. Treat the car
as a particle and express the answers in
terms of W and θ. Find P and β for the
case when $W = 4000$ lb and $\theta = 20°$.
What is the normal reaction between the
smooth plane and the car under these
conditions? Assume that the weight of the
car acts through point G.

2.73 The system shown in Figure P2.73 is in
equilibrium in the position indicated. If
the mass of block A is $m_A = 5$ kg and that
of block B is $m_B = 7$ kg, determine the
mass of block C and the height H consistent with equilibrium.

2.74 The weightless eye bar AB of Figure P2.74
is attached to a frictionless collar at B
and, therefore, remains horizontal for any
equilibrium position of the system shown.
The dimensions given in Figure P2.74
correspond to the equilibrium position
when $W = 500$ lb. Determine the spring

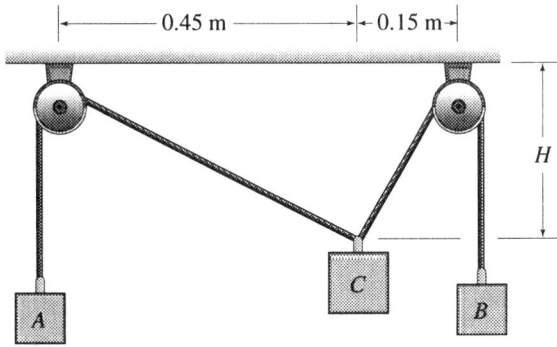

FIGURE P2.73.

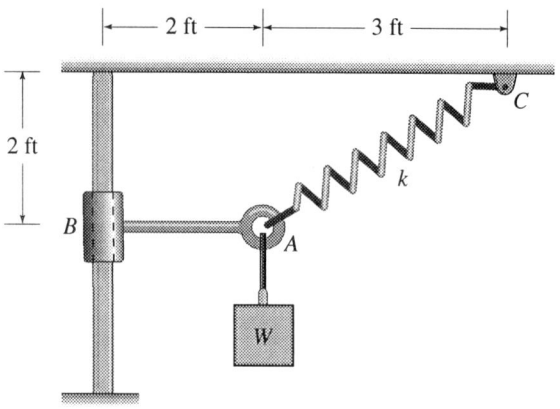

FIGURE P2.74.

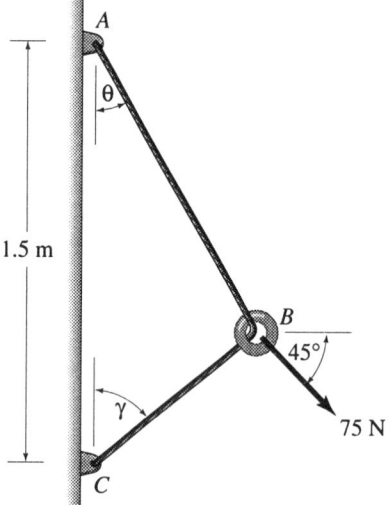

FIGURE P2.75.

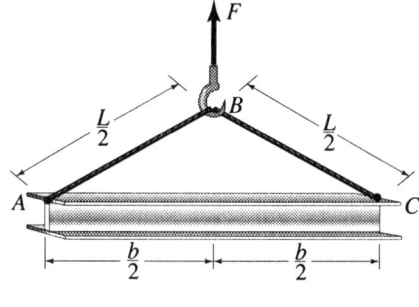

FIGURE P2.76.

equilibrium, determine (a) the angle γ, (b) the corresponding lengths of AB and BC, and (c) the force in the string ABC.

2.76 A beam of weight W and length b is being lifted into position at constant speed at a construction site, as shown in Figure P2.76. For safety reasons, the tension in the lifting cable ABC is not to exceed a certain fraction of the weight W expressed as kW, where $0.5 < k < 1$. (a) Show that the minimum length L of cable ABC is $L = (2k/\sqrt{4k^2 - 1})b$. (b) Find the minimum length L for the case of a beam of weight $W = 2000$ lb and length $b = 20$ ft and for a limiting tension in the lifting cable of 1500 lb.

constant k of the spring AC if the unstretched length of the spring $L_u = 40$ in.

2.75 The string ABC of Figure P2.75 has a total length of 2.2 m, is anchored at supports A and C, and passes through a smooth ring at B. When a force of 75 N magnitude inclined to the horizontal at $45°$ is applied to the ring as shown, the angle θ becomes $25°$ at equilibrium. At

3

Equilibrium of Particles in Three Dimensions

Earth and other planets in our solar system are a marvelous example of a three-dimensional system in equilibrium.

It is hard to look at planets in our solar system as particles! Well, this is not such a bad approximation if we are to consider our entire universe with its billions of stars and galaxies. The basic premise here is that everything is relative. As an engineer, you need to understand the implications of the assumptions you make.

One of the most important aspects of engineering is efficiency. The engineer must not resort to elaborate analysis when elementary analysis suffices. The test is not how good it looks, but how well it works. There is no value in compromising the integrity of a given engineering system by ignoring its real attributes. As Einstein once said, "Things should be made simple but not any simpler."

The analysis of a force system in three dimensions using vector algebra is introduced in this chapter. Several basic and efficient techniques are covered to facilitate the handling of such a system. The necessary and sufficient conditions for equilibrium of a particle in three dimensions are introduced and applied in solving engineering problems.

3.1 Force in Terms of Rectangular Components

In Section 2.3, dealing with two-dimensional force systems, we learned how to express a force in terms of two rectangular components using Eq. (2.7), namely, $\mathbf{F} = F_x\mathbf{i} + F_y\mathbf{j}$. In this equation, F_x and F_y are the scalar components of the force, and $\mathbf{i}$ and $\mathbf{j}$ the unit vectors, in the x and y directions, respectively. Equation (2.7) will be expanded in this section to represent properly a force in a three-dimensional space.

Consider the force $\mathbf{F}$ referred to an x-y-z coordinate system as shown in Figure 3.1. This force acts in the plane of rectangle OABC and may be decomposed into components $\mathbf{F}_z$, along the z axis, and $\mathbf{F}_{xy}$ in the plane of rectangle ODAE, i.e., the x-y plane. Therefore, $\mathbf{F}_{xy}$ may be decomposed into components $\mathbf{F}_x$, along the x axis, and $\mathbf{F}_y$, along the y axis. In effect, then, the force $\mathbf{F}$ was resolved into the three rectangular components $\mathbf{F}_x$, $\mathbf{F}_y$, and $\mathbf{F}_z$. Inversely, these three rectangular components may be added vectorially to obtain the force $\mathbf{F}$. Thus, by the law of the parallelogram, the two components $\mathbf{F}_x$ and $\mathbf{F}_y$ may be added to obtain $\mathbf{F}_{xy}$ which, in turn, may be added to $\mathbf{F}_z$ to obtain the force $\mathbf{F}$. Alternatively, the three components $\mathbf{F}_x$, $\mathbf{F}_y$, and $\mathbf{F}_z$ may be added by using the head-to-tail construction of the polygon method to obtain the resultant $\mathbf{F}$. This is done in Figure 3.2 in which the construction used the sequence $\mathbf{F}_x$ followed by $\mathbf{F}_y$ which was, then, followed by $\mathbf{F}_z$. Any other sequence, however, would achieve the same result. Either of these methods of addition may be expressed symbolically by the relationship

$$\mathbf{F} = \mathbf{F}_x + \mathbf{F}_y + \mathbf{F}_z. \tag{3.1}$$

If, as was done in Section 2.3, we define unit vectors $\mathbf{i}$ and $\mathbf{j}$ along the x and y axes, respectively, we may express $\mathbf{F}_x$ as $F_x\mathbf{i}$ and $\mathbf{F}_y$ as $F_y\mathbf{j}$. Also, if $\mathbf{k}$ represents a unit vector along the z direction we may write $\mathbf{F}_z$ in the form $F_z\mathbf{k}$. Therefore, Eq. (3.1) may be rewritten in the form

$$\mathbf{F} = F_x\mathbf{i} + F_y\mathbf{j} + F_x\mathbf{k} \tag{3.2}$$

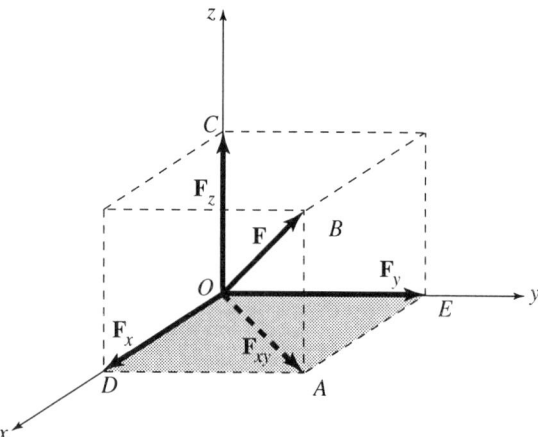

FIGURE 3.1.

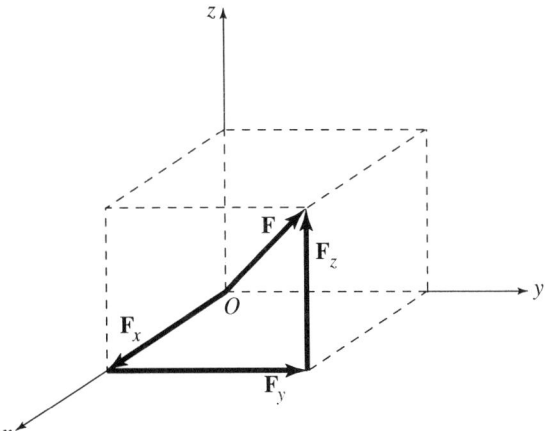

FIGURE 3.2.

where F_x, F_y, and F_z represent, respectively, the scalar components of **F** in the x, y, and z directions. Thus, Eq. (2.6) is but a special case of Eq. (3.2).

The scalar components F_x, F_y, and F_z, may be determined in terms of the magnitude F of the force **F** and trigonometric functions of the angles α, β, and γ that it makes with the x, y, and z axes, respectively. Figure 3.3(a) shows a force **F** along with the three angles α, β, and γ and the three rectangular components $F_x\mathbf{i}$, $F_y\mathbf{j}$ and $F_z\mathbf{k}$. The scalar component F_x, for example, may be obtained by trigonometry in terms of F by analyzing right-angle triangle OBA which, for simplicity of visualization, is redrawn in the plane of the page as shown in Figure 3.3(b). Similarly, by a trigonometric analysis of the right-angle triangles shown in Figures 3.3(c) and (d), we may obtain the scalar components F_y and F_z, respectively, in terms of F. The results of these analyses lead to the relationships

and

$$\left. \begin{aligned} F_x &= F \cos \alpha, \\ F_y &= F \cos \beta, \\ F_z &= F \cos \gamma. \end{aligned} \right\} \tag{3.3}$$

Alternatively, Eqs. (3.3) may be solved for the angles α, β, and γ to obtain

$$\left. \begin{aligned} \alpha &= \cos^{-1} \frac{F_x}{F} \\ \beta &= \cos^{-1} \frac{F_y}{F} \\ \gamma &= \cos^{-1} \frac{F_z}{F} \end{aligned} \right\} \tag{3.4}$$

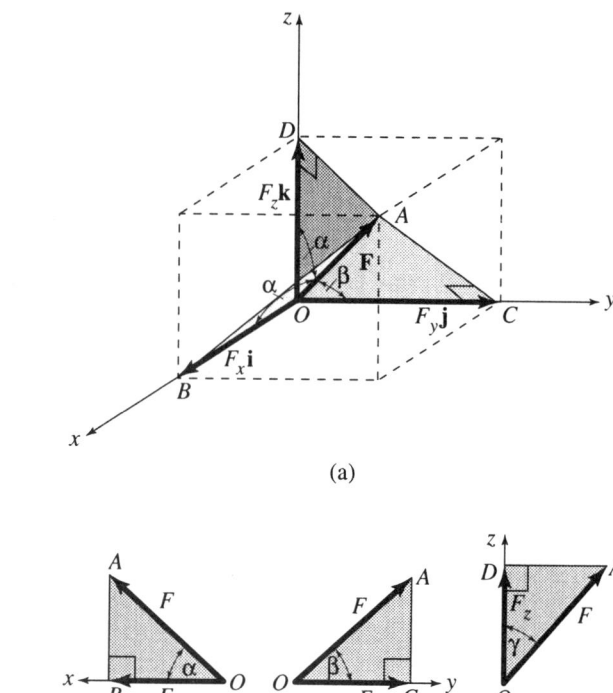

FIGURE 3.3.   (b)   (c)   (d)

The angles α, β, and γ are known as the direction angles for the force $\mathbf{F}$ and their cosines as the direction cosines. These angles are measured from the positive ends of their respective axes and have values that range between zero and 180°.

The magnitude F of the force $\mathbf{F}$ may be determined in terms of its scalar components F_x, F_y, and F_z by reference to Figure 3.1. Applying the Pythagorean theorem to right-angle triangle OAB, we conclude that $F = \sqrt{F_{xy}^2 + F_z^2}$. Now, if we apply the same theorem to right-angle triangle OEA, we conclude that $F_{xy}^2 = F_x^2 + F_y^2$ which, when substituted in the above equation for F, yields

$$F = \sqrt{F_x^2 + F_y^2 + F_z^2}. \qquad (3.5)$$

We observe that Eq. (2.9) is the two-dimensional special case of Eq. (3.5). Substituting from Eq. (3.3) in Eq. (3.5),

$$F = \sqrt{F^2 \cos^2 \alpha + F^2 \cos^2 \beta + F^2 \cos^2 \gamma}$$

$$= F\sqrt{\cos^2 \alpha + \cos^2 \beta + \cos^2 \gamma} \qquad (3.6)$$

from which

$$\cos^2 \alpha + \cos^2 \beta + \cos^2 \gamma = 1 \qquad (3.7)$$

Equation (3.7) represents a very useful relationship among the three direction cosines of a given vector and shows that these direction cosines are not independent, i.e., if two are known, the third is determined by Eq. (3.7). Note that Eq. (2.11) is the two-dimensional special case of Eq. (3.7). If Eqs. (3.3) are substituted in Eq. (3.2),

$$\mathbf{F} = (F \cos \alpha)\mathbf{i} + (F \cos \beta)\mathbf{j} + (F \cos \gamma)\mathbf{k}$$

$$= F[(\cos \alpha)\mathbf{i} + (\cos \beta)\mathbf{j} + (\cos \gamma)\mathbf{k}]$$

$$= F\lambda \qquad (3.8)$$

where $\lambda = \mathbf{F}/F$ defines a unit vector in the direction of the force $\mathbf{F}$ and is given by

$$\lambda = (\cos \alpha)\mathbf{i} + (\cos \beta)\mathbf{j} + (\cos \gamma)\mathbf{k}$$

$$= \lambda_x\mathbf{i} + \lambda_x\mathbf{j} + \lambda_x\mathbf{k}. \qquad (3.9)$$

The quantities λ_x, λ_y, and λ_z represent, respectively, the x, y and z scalar components of the unit vector λ and are given by

$$\left. \begin{array}{l} \lambda_x = \cos \alpha, \\ \lambda_y = \cos \beta, \\ \lambda_z = \cos \gamma. \end{array} \right\} \qquad (3.10)$$

and

Note that Eq. (3.8) expresses the same fact originally expressed in Chapter 2 by Eq. (2.6), that a force, or any other vector quantity, may be written as the product of its magnitude F (a scalar) and a unit vector λ in its direction.

The following examples illustrate some of the concepts discussed above.

■ **Example 3.1**

A force $\mathbf{F}$ has a magnitude of 1500.0 lb and makes angles of 50° and 120° with respect to the positive x and z coordinate axes, respectively. Express $\mathbf{F}$ in terms of its three rectangular components.

Solution

Because the angles α and γ are given as 50° and 120°, respectively, Eq. (3.7) may be used to find the angle β that the force $\mathbf{F}$ makes with the positive y axis. Thus, by Eq. (3.7),

$$\cos \beta = \sqrt{1 - \cos^2 \alpha - \cos^2 \gamma} = \pm 0.5804$$

The negative sign leads to $\beta = 125.5°$ which is inconsistent with the given data. Therefore, we use the positive sign which leads to

$$\beta = 54.5°.$$

Therefore, using Eqs. (3.3),

$$F_x = F \cos \alpha = 1500.0 \cos 50°$$

$$= 964.2 \text{ lb},$$

$$F_y = F \cos \beta = 1500.0 \cos 54.5°$$

$$= 870.6 \text{ lb},$$

and

$$F_z = F \cos \gamma = 1500.0 \cos 120°$$

$$= -750.0 \text{ lb}.$$

The force **F** is now expressed in terms of its rectangular components by Eq. (3.2). Thus,

$$\mathbf{F} = (964\mathbf{i} + 871\mathbf{j} - 750\mathbf{k}) \text{ lb}. \qquad \text{ANS.}$$

■ Example 3.2

A force P has components $P_x = -750$ N, $P_y = 1200$ N, and $P_z = -500$ N. Determine the angles α, β, and γ that the force P makes with the coordinate axes x, y, and z, respectively.

Solution

The magnitude of the force P is determined by Eq. (3.5). Thus,

$$P = \sqrt{P_x^2 + P_y^2 + P_z^2},$$

$$= \sqrt{(-750)^2 + (1200)^2 + (-500)^2},$$

$$= 1501 \text{ N}.$$

Equations (3.4) are now used to determine the direction angles. Thus,

$$\alpha = \cos^{-1} \frac{P_x}{P} \qquad\qquad \beta = \cos^{-1} \frac{P_y}{P}$$

$$= \cos^{-1} \frac{-750}{1501} = 120.0°, \qquad = \cos^{-1} \frac{1200}{1501} = 36.9°, \quad \text{ANS.}$$

and

$$\gamma = \cos^{-1} \frac{P_z}{P}$$

$$= \cos^{-1} \frac{-500}{1501} = 109.5°. \qquad \text{ANS.}$$

The calculations above may be checked, in part, by Eq. (3.7). Thus,

$$\cos^2(120.0°) + \cos^2(36.9°) + \cos^2(109.5°) = 1.001 \quad \text{(OK)}.$$

■

Problems

3.1 A force has a magnitude of 350 lb and makes the angles $\alpha = 45°$ and $\gamma = 112°$ with the x and z axes, respectively. Find the x, y, and z components of the force.

3.2 A force has a magnitude of 800 N and has components along the y and z directions equal to 500 N and -375 N, respectively. Find its component in the x direction and the angles α, β, and γ that it makes with the x, y, and z axes, respectively.

3.3 A force $\mathbf{Q}$ has components $Q_x = -1750$ lb, $Q_y = -2200$ lb, and $Q_z = 1670$ lb. Find the magnitude of $\mathbf{Q}$ and the direction angles α, β, and γ that it makes with the x, y, and z axes, respectively.

3.4 Consider the force $\mathbf{P}$ shown in Figure P3.4. Let $\alpha = 125°$, $\beta = 37°$, and $P_z = 2.55$ kN. Express the force $\mathbf{P}$ in vector form in terms of its x, y, and z components.

3.5 Refer to Figure P3.4. Let $P = 3.75$ k, $\beta = 75°$ and $P_x = 2.50$ k. Find α, γ, P_y, and P_z.

3.6 Refer to Figure P3.4. Let $\alpha = 55°$, $P_x = 700$ N, and $P_z = -400$ N. Express the force $\mathbf{P}$ in terms of $\mathbf{i}$, $\mathbf{j}$ and $\mathbf{k}$ components, and find the direction angles β and γ.

3.7 Find the magnitude and the direction angles α, β, and γ of the force $\mathbf{Q} = (-750\mathbf{i} + 500\mathbf{j} - 375\mathbf{k})$ lb.

3.8 Determine the direction angles α, β, and γ of the force $\mathbf{F} = (45\mathbf{i} - 300\mathbf{j} - 625\mathbf{k})$ N.

3.9 Consider the force $\mathbf{Q}$ shown in Figure P3.9. Let $\theta_1 = 65°$, $\theta_2 = 30°$ and $Q = 17.25$ k. Express $\mathbf{Q}$ in terms of its $\mathbf{i}$, $\mathbf{j}$, and $\mathbf{k}$ components.

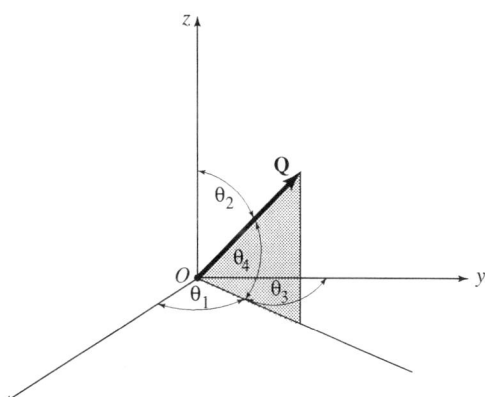

FIGURE P3.9.

3.10 Refer to Figure P3.9. Let the x component of the force $\mathbf{Q}$ be $Q_x = 550$ N, $\theta_3 = 40°$, and $\theta_4 = 50°$. Find the magnitude of $\mathbf{Q}$ and the direction angles α, β, and γ that it makes with the x, y, and z axes, respectively.

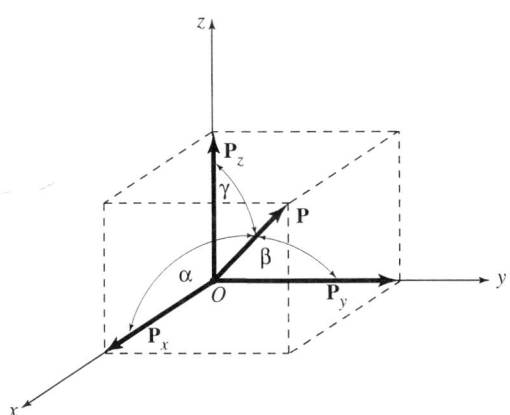

FIGURE P3.4.

3.2
Force in Terms of Magnitude and Unit Vector

Position Vector

The definition of a force in terms of its magnitude and a unit vector along its direction was introduced in Section 2.3 and expressed in Eq. (2.6). This definition requires determining a unit vector along the line of action of the force. It is possible to determine a unit vector along the line of action of the force if the coordinates of two points on its line of action are known. This determination is facilitated by the concept of the position vector.

By definition, a *position vector* $\mathbf{r}$ is a vector that defines the position of one point in relation to another point in space. The position vector is very useful in developing many concepts in both statics and dynamics. In this section, it will be used to define a force along any direction where two points are known.

The position of point Q in Figure 3.4 may be fully ascertained with respect to point O, the origin of the x-y-z coordinate system, by specifying the position vector $\mathbf{r}_Q$. Because the three rectangular components of $\mathbf{r}_Q$ in the x, y, and z directions are, respectively, $x_Q\mathbf{i}$, $y_Q\mathbf{j}$, and $z_Q\mathbf{k}$, as shown in Figure 3.4, it follows that

$$\mathbf{r}_Q = x_Q\mathbf{i} + y_Q\mathbf{j} + z_Q\mathbf{k}. \tag{3.11}$$

Now, consider the two points A and B shown in Figure 3.5. The positions of points A and B relative to point O, the origin of the x-y-z coordinate system, are specified by the position vectors $\mathbf{r}_A$ and $\mathbf{r}_B$, respectively. Using the same ideas that led to Eq. (3.11), we conclude that

and
$$\left.\begin{array}{l} \mathbf{r}_A = x_A\mathbf{i} + y_A\mathbf{j} + z_A\mathbf{k} \\ \mathbf{r}_B = x_B\mathbf{i} + y_B\mathbf{j} + z_B\mathbf{k}. \end{array}\right\} \tag{3.12}$$

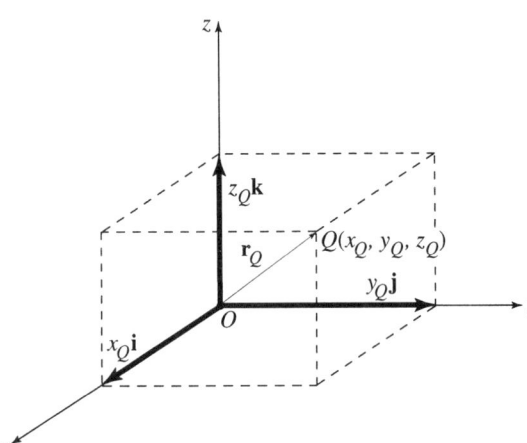

FIGURE 3.4.

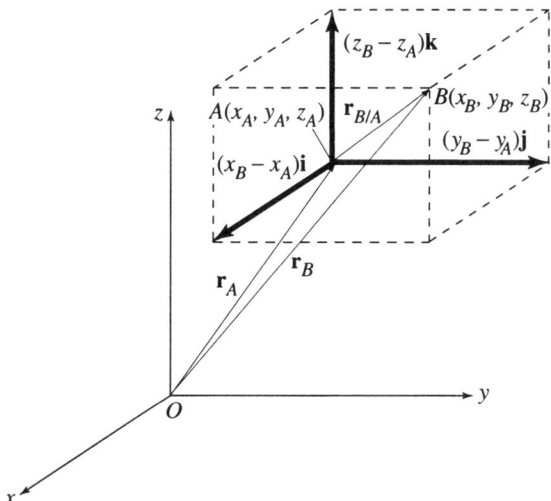

FIGURE 3.5.

Also, using vector addition and referring to Figure 3.5, we may write

$$\mathbf{r}_A + \mathbf{r}_{B/A} = \mathbf{r}_B \tag{3.13}$$

where $\mathbf{r}_{B/A}$ is the position vector of point B relative to point A. Solving Eq. (3.13) for $\mathbf{r}_{B/A}$ and substituting from Eqs. (3.12) and after simplifying,

$$\mathbf{r}_{B/A} = \mathbf{r}_B - \mathbf{r}_A = (x_B - x_A)\mathbf{i} + (y_B - y_A)\mathbf{j} + (z_B - z_A)\mathbf{k} \tag{3.14}$$

where the quantities $(x_B - x_A)\mathbf{i}$, $(y_B - y_A)\mathbf{i}$, and $(z_B - z_A)\mathbf{i}$ are, respectively, the x, y, and z vector components of $\mathbf{r}_{B/A}$. The magnitude $r_{B/A}$ of position vector $\mathbf{r}_{B/A}$, which represents the distance from point B to point A, is found in the same manner used to develop Eq. (3.5). Thus,

$$r_{B/A} = \sqrt{(x_B - x_A)^2 + (y_B - y_A)^2 (z_B - z_A)^2} \tag{3.15}$$

where the quantities $(x_B - x_A)$, $(y_B - y_A)$ and $(z_B - z_A)$ are, respectively, the x, y, and z scalar components of $\mathbf{r}_{B/A}$.

Unit Vector

The concept of a unit vector was introduced in Section 2.3 and defined in Eq. (2.5). This equation states that a unit vector in the direction of a given vector is obtained by dividing this vector by its magnitude. Thus, a unit vector λ_{AB} in a direction from point A to point B in Figure 3.5, is defined by the relationship

$$\lambda_{AB} = \frac{\mathbf{r}_{B/A}}{r_{B/A}} = \frac{(x_B - x_A)\mathbf{i} + (y_B - y_A)\mathbf{j} + (z_B - z_A)\mathbf{k}}{\sqrt{(x_B - x_A)^2 + (y_B - y_A)^2 + (z_B - z_A)^2}} \tag{3.16}$$

Definition of a Force

In Section 2.3, we learned how to express a vector, such as a force, in terms of its magnitude and a unit vector along its direction (see Eq. (2.6)). Frequently, in dealing with three-dimensional equilibrium problems, we encounter a force **F** whose direction is specified by the coordinates of two points along its line of action. This information, along with Eq. (3.16), makes it possible to define a unit vector λ along the line of action of the force. The force **F** may, then, be expressed in vector form by multiplying its magnitude F by the unit vector λ according to the relationship expressed in Eq. (3.8), namely, $\mathbf{F} = F\lambda$. These concepts are illustrated in the following example.

■ Example 3.3

A 10 ft × 10 ft hinged plate is supported in the position shown in Figure E3.3(a) by cable BA. The tension force F in the cable is known to be 300 lb and on the plate it is directed from B to A. Determine (a) the x, y, and z scalar components of the cable force acting on the plate at point B and (b) the direction angles α, β, and γ that the cable force makes, respectively, with the x, y, and z axes at point B.

Solution

(a) A unit vector λ_{BA} from B to A may be determined by Eq. (3.16) if the coordinates of points A and B are known. From the given geometry, the coordinates of point A are 5 ft, -7 ft and 15 ft, and those of point B are $10\cos 30° = 8.66$ ft, 10 ft, and $-10\sin 30° = -5$ ft. Once determined, it is good practice to indicate the coordinates on a sketch of the system as shown in Figure E3.3(b), which also shows the force F in cable BA. Thus, by Eqs. (3.14) and (3.15), respectively,

$$\mathbf{r}_{A/B} = (5 - 8.66)\mathbf{i} + (-7 - 10)\mathbf{j} + (15 + 5)\mathbf{k}$$

$$= (-3.66\mathbf{i} - 17\mathbf{j} + 20\mathbf{k}) \text{ ft,}$$

and

$$r_{A/B} = \sqrt{(-3.66)^2 + (-17)^2 + (20)^2} = 26.503 \text{ ft.}$$

Therefore, by Eq. (3.16),

$$\lambda_{BA} = \frac{\mathbf{r}_{A/B}}{r_{A/B}} = \frac{-3.66\mathbf{i} - 17\mathbf{j} + 20\mathbf{k}}{26.503}$$

$$= -0.138\mathbf{i} - 0.641\mathbf{j} + 0.755\mathbf{k}.$$

By Eq. (3.8), we may express the force in the cable from B to A in the form

$$\mathbf{F} = F\lambda_{BA} = [300(-0.138\mathbf{i} - 0.641\mathbf{j} + 0.755\mathbf{k})] \text{ lb.}$$

The above equation expresses the force F in vector form in terms of its

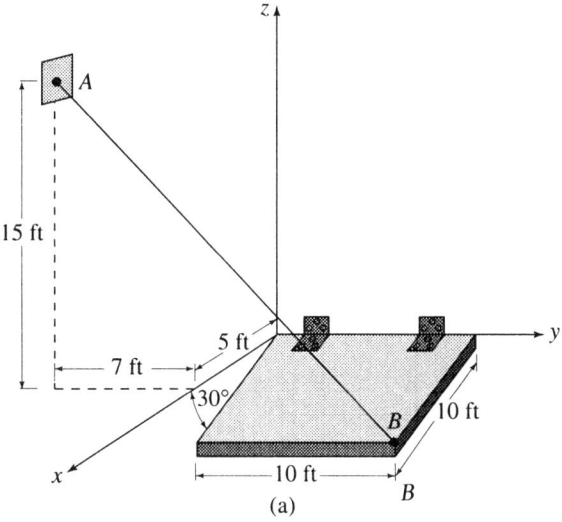

(a)

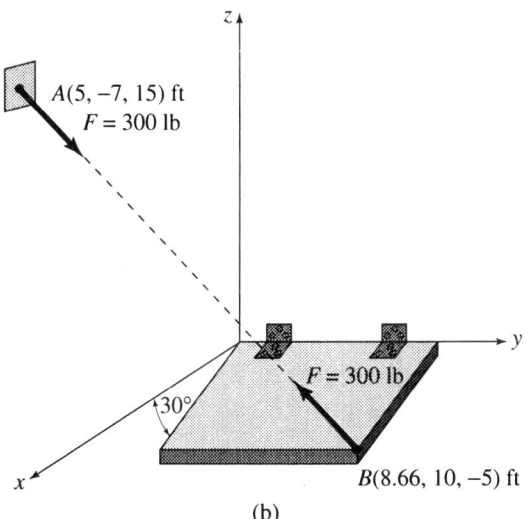

(b)   FIGURE E3.3.

magnitude (300 lb) and a unit vector $\lambda_{BA} = (-0.138\mathbf{i} - 0.641\mathbf{j} + 0.755\mathbf{k})$ along its line of action. To find the rectangular (i.e., the $\mathbf{i}$, $\mathbf{j}$ and $\mathbf{k}$) components, multiply the magnitude by the components of the unit vector to obtain

$$\mathbf{F} = (-41.4\mathbf{i} - 192.3\mathbf{j} + 227\mathbf{k}) \text{ lb.}$$

Thus, the components of the cable force at point B are

$$F_x = -41.4 \text{ lb, } F_y = -192.3 \text{ lb,} \quad \text{and} \quad F_z = 227.0 \text{ lb.} \quad \text{ANS.}$$

(b) The direction angles may be obtained by Eqs. (3.4) or by Eqs. (3.10). Using Eqs. (3.10)

$$\lambda_x = -0.138 = \cos \alpha$$

from which

$$\alpha = 97.9°. \qquad \text{ANS.}$$

Similarly,

$$\lambda_y = -0.641 = \cos \beta,$$

$$\beta = 129.9° \qquad \text{ANS.}$$

$$\lambda_z = 0.755 = \cos \gamma,$$

$$\gamma = 41.0°. \qquad \text{ANS.}$$

■

3.3
Dot (Scalar) Product

There are two fundamental types of products that may be defined with vector quantities. The first one is known as the *dot* or *scalar product* and the second as the *cross* or *vector product*. Each of these two vector products is defined to solve certain specific problems that arise in studying the subjects of Statics and Dynamics. The *dot (scalar) product* is discussed in this section whereas the *cross (vector) product* is introduced in Chapter 5.

In many three-dimensional problems in mechanics, it becomes necessary to determine the component of a vector, such as a force, in some specified direction, as well as the angle that one vector makes with another intersecting vector in space. These types of problems are easily handled by the use of the dot product of two vectors. The *dot product* of two intersecting vectors **V** and **W** is stated as **V·W** and defined by the relationship

$$\mathbf{V} \cdot \mathbf{W} = VW \cos \theta \qquad (3.17)$$

where V and W are the magnitudes, respectively, of vectors **V** and **W** and θ is the angle between them. Thus, by definition, the dot product of two vectors yields a scalar quantity whose magnitude, $VW \cos \theta$, may be viewed in one of two ways as illustrated in Figure 3.6. In Figure 3.6(a), the quantity $VW \cos \theta$ is viewed as W, multiplied by $V \cos \theta$ which is the scalar component of **V** in the direction of **W**. In Figure 3.6(b), the quantity $VW \cos \theta$ is viewed as V, multiplied by $W \cos \theta$ which is the scalar component of **W** in the direction of **V**. This property of the dot product is particularly useful in determining the component of a vector, such as a force, in any specified direction along which a unit vector is defined. Thus, consider the force **F** and the direction OA

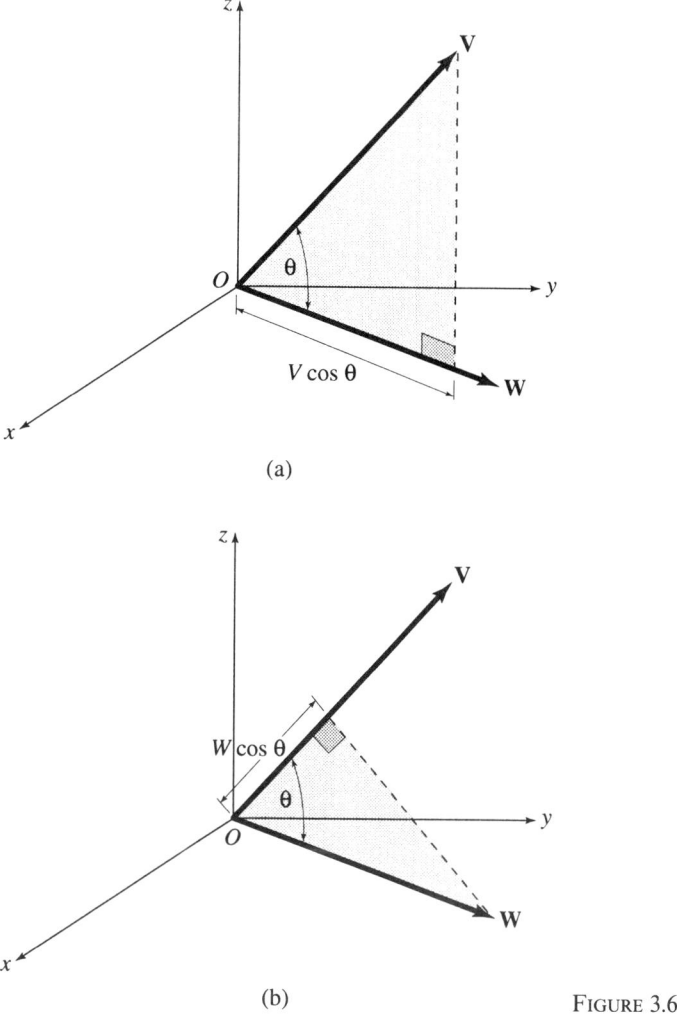

(a)

(b)   FIGURE 3.6.

along which the unit vector λ_{OA} has been defined as shown in Figure 3.7. The scalar component of force $\mathbf{F}$ in the direction OA, F_{OA}, is found by performing the dot product of λ_{OA} and $\mathbf{F}$. Thus

$$\mathbf{F}_{OA} = \lambda_{OA} \cdot \mathbf{F} \tag{3.18}$$

The justification for Eq. (3.18) is provided by the basic definition of the dot product given in Eq. (3.17). Using this definition, the dot product $\lambda_{OA} \cdot \mathbf{F}$ has a magnitude equal to $(\lambda_{OA})(F \cos \theta)$. Because $\lambda_{OA} = 1$, the magnitude of the dot product $\lambda_{OA} \cdot \mathbf{F}$ reduces to $F \cos \theta$ which, as seen in Figure 3.7, is the scalar component of force $\mathbf{F}$ in the direction OA.

Because $\cos 0° = 1$ and $\cos 90° = 0$, it follows from the basic definition of the dot product, Eq. (3.17), that

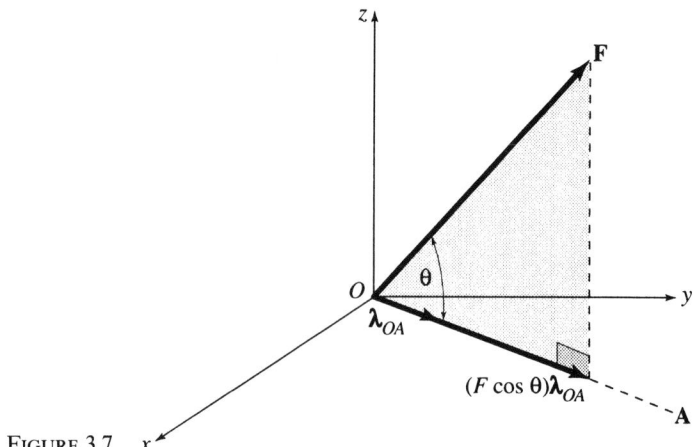

FIGURE 3.7.

and

$$i \cdot i = j \cdot j = k \cdot k = 1$$
$$i \cdot j = j \cdot k = k \cdot i = 0. \qquad (3.19)$$

Therefore, if two vectors V and W are expressed in terms of their rectangular components, their dot product may be written as

$$\mathbf{V} \cdot \mathbf{W} = (V_x \mathbf{i} + V_y \mathbf{j} + V_z \mathbf{k}) \cdot (W_x \mathbf{i} + W_y \mathbf{j} + W_z \mathbf{k})$$
$$= V_x W_x + V_y W_y + V_z W_z. \qquad (3.20)$$

Equation (3.20) states that the dot product of two vectors is obtained by forming the products of their respective x, y, and z components and, then, adding them together. In the special case when $V = W = F$, Eq. (3.17) yields $\mathbf{F} \cdot \mathbf{F} = F^2 \cos 0° = F^2$. Also, because $\mathbf{F} \cdot \mathbf{F} = F_x^2 + F_y^2 + F_z^2$ by Eq. (3.20), it follows that $F^2 = F_x^2 + F_y^2 + F_z^2$ which, of course, agrees with Eq. (3.5).

Equation (3.17) may be rewritten to obtain a convenient relationship for finding the angle θ between the vectors V and W. Thus,

$$\theta = \cos^{-1}\left(\frac{\mathbf{V} \cdot \mathbf{W}}{VW}\right) = \cos^{-1}\left(\frac{V_x W_x + V_y W_y + V_z W_z}{VW}\right) \qquad (3.21a)$$

Because $V/W = \lambda_v$ and $W/W = \lambda_w$, it follows that Eq. (3.21a) may also be expressed as

$$\theta = \cos^{-1}(\lambda_v \cdot \lambda_w). \qquad (3.21b)$$

It is stated here without proof that the dot product is both *commutative* and *distributive*. Thus, according to the commutative law.

$$\mathbf{V} \cdot \mathbf{W} = \mathbf{W} \cdot \mathbf{V}. \qquad (3.22)$$

Also, by the distributive law,

$$\mathbf{V} \cdot (\mathbf{W} + \mathbf{U}) = \mathbf{V} \cdot \mathbf{W} + \mathbf{V} \cdot \mathbf{U}. \qquad (3.23)$$

The following examples illustrate the concepts above.

■ Example 3.4   Member OA is fastened rigidly at O and is subjected to the force
$\mathbf{F} = (7.5\mathbf{i} - 10.0\mathbf{j} - 4.3\mathbf{k})$ kN at A as shown in Figure E3.4(a). Determine the scalar component of $\mathbf{F}$ in the direction from A to O by (a)
using Eq. (3.18) and (b) finding the angle θ between $\mathbf{F}$ and line AO and
forming the product $F \cos \theta$.

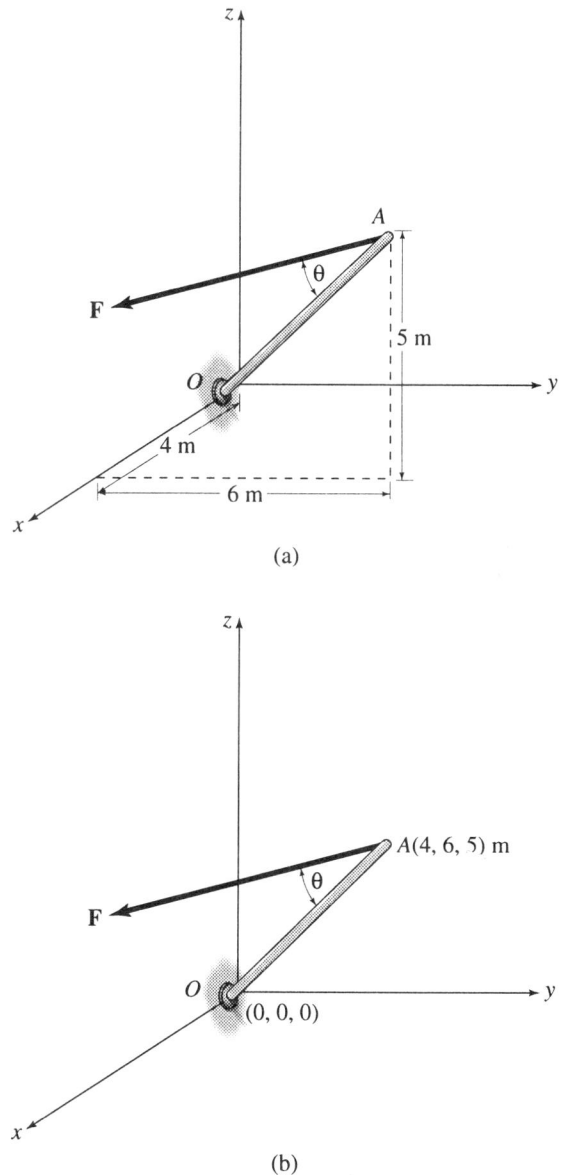

(a)

(b)   FIGURE E3.4.

Solution

(a) A position vector $\mathbf{r}_{O/A}$ and its magnitude $r_{O/A}$ are determined by using Eqs. (3.14) and (3.15), respectively. This requires knowledge of the coordinates of points A and O, which are obtained from the geometry given and shown in Figure E3.4(b). Thus,

$$\mathbf{r}_{O/A} = (0 - 4)\mathbf{i} + (0 - 6)\mathbf{j} + (0 - 5)\mathbf{k} = (-4\mathbf{i} - 6\mathbf{j} - 5\mathbf{k})m$$

and

$$r_{O/A} = \sqrt{(4)^2 + (6)^2 + (5)^2} = 8.775 \text{ m}.$$

Therefore, a unit vector λ_{AO} may be obtained by Eq. (3.16). Thus,

$$\lambda_{AO} = \frac{\mathbf{r}_{O/A}}{r_{O/A}} = \frac{-4\mathbf{i} - 6\mathbf{j} - 5\mathbf{k}}{8.775} = -0.456\mathbf{i} - 0.684\mathbf{j} - 0.570\mathbf{k}.$$

Therefore, by Eq. (3.18),

$$F_{AO} = [-0.456\mathbf{i} - 0.684\mathbf{i} - 0.570\mathbf{k}] \cdot [7.5\mathbf{i} - 10.0\mathbf{j} - 4.3\mathbf{k}]$$

$$= -(0.456)(7.5) + (0.684)(10.0) + (0.570)(4.3),$$

$$= 5.87 \text{ kN}. \qquad \text{ANS.}$$

The fact that the answer is positive shows that the component F_{AO} has the same sense as the unit vector λ_{AO}, namely, from A to O.

(b) Because we already have the unit vector λ_{AO} in the direction from A to O, we need only determine a unit vector λ_F in the direction of the force $\mathbf{F}$ and use Eq. (3.21) to find the angle θ. Thus, because $F = \sqrt{(7.5)^2 + (10.0)^2 + (4.3)^2} = 13.219 \text{ kN}$,

$$\lambda_F = \frac{\mathbf{F}}{F} = \frac{7.5\mathbf{i} - 10.0\mathbf{j} - 4.3\mathbf{k}}{13.219} = 0.567\mathbf{i} - 0.756\mathbf{j} - 0.325\mathbf{k}.$$

Therefore, by Eq. (3.21b),

$$\theta = \cos^{-1} \lambda_{AO} \cdot \lambda_F$$

$$= \cos^{-1}[-(0.456)(0.567) + (0.684)(0.756) + (0.570)(0.325)]$$

$$= 63.65°.$$

It follows, therefore, that

$$F_{AO} = F \cos \theta = 13.219 \cos 63.65°$$

$$= 5.87 \text{ kN} \qquad \text{ANS.}$$

which, of course, is the same answer as obtained in part (a).

■ **Example 3.5**

A 7 ft × 7 ft hinged plate is partially supported by cables BA and BC as shown in Figure E3.5(a). Determine (a) the angle θ between the two cables and (b) the scalar component along axis BO of the tension F_{AB} in cable BA if the magnitude of this tension is 500 lb and, at support B, it is directed from B to A.

Solution

(a) The coordinates of points A, B, and C, found from the geometry given, are shown in Figure E3.5(b) which also shows the force $\mathbf{F}_{AB}$.

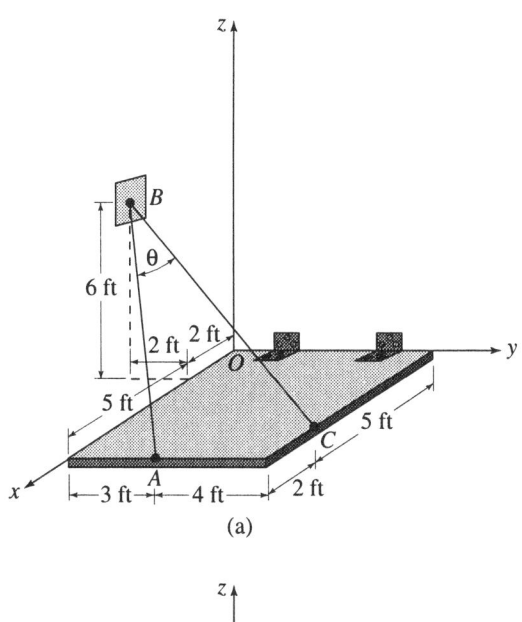

(a)

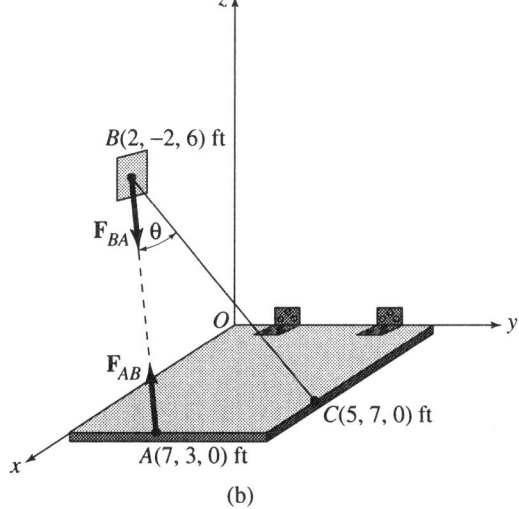

(b)

FIGURE E3.5.

Define the position vectors $\mathbf{r}_{A/B}$ and $\mathbf{r}_{C/B}$ from point B as follows:

$$\mathbf{r}_{A/B} = (5\mathbf{i} + 5\mathbf{j} - 6\mathbf{k}) \text{ ft}$$

and

$$\mathbf{r}_{C/B} = (3\mathbf{i} + 9\mathbf{j} - 6\mathbf{k}) \text{ ft.}$$

The magnitudes of $\mathbf{r}_{A/B}$ and $\mathbf{r}_{C/B}$ which represent, respectively, the lengths of cables BA and BC are found as follows:

$$r_{A/B} = \sqrt{(5)^2 + (5)^2 + (6)^2} = 9.274 \text{ ft,}$$

and

$$r_{C/B} = \sqrt{(3)^2 + (9)^2 + (6)^2} = 11.225 \text{ ft.}$$

Therefore, by Eq. (3.21),

$$\theta = \cos^{-1}\left(\frac{\mathbf{r}_{A/B}\cdot\mathbf{r}_{C/B}}{r_{A/B}r_{C/B}}\right) = \cos^{-1}\left[\frac{(5)(3) + (5)(9) + (6)(6)}{(9.274)(11.225)}\right]$$

$$= 22.8°. \hspace{4cm} \text{ANS.}$$

(b) A unit vector $\boldsymbol{\lambda}_{BA}$ from B to A is defined by

$$\boldsymbol{\lambda}_{BA} = \frac{\mathbf{r}_{A/B}}{r_{A/B}} = \frac{5\mathbf{i} + 5\mathbf{j} - 6\mathbf{k}}{9.274} = 0.539\mathbf{i} + 0.539\mathbf{j} - 0.647\mathbf{k}.$$

Therefore, the force $\mathbf{F}_{BA}$ in cable BA at B may be expressed in vector form by Eq. (3.8). Thus,

$$\mathbf{F}_{BA} = (500\boldsymbol{\lambda}_{BA}) \text{ lb} = (270\mathbf{i} + 270\mathbf{j} - 324\mathbf{k}) \text{ lb.}$$

Define a unit vector $\boldsymbol{\lambda}_{BO}$ from B to O by the relationship

$$\boldsymbol{\lambda}_{BO} = \frac{\mathbf{r}_{O/B}}{r_{O/B}} = \frac{-2\mathbf{i} + 2\mathbf{j} - 6\mathbf{k}}{\sqrt{(-2)^2 + (2)^2 + (-6)^2}} = -0.302\mathbf{i} + 0.302\mathbf{j} - 0.905\mathbf{k}.$$

Thus, by Eq. (3.18),

$$F_{BO} = \boldsymbol{\lambda}_{BA}\cdot\mathbf{F}_{BA} = -(0.302)(270) + (0.302)(270) + (0.905)(324)$$

$$= 293 \text{ lb.} \hspace{4cm} \text{ANS.}$$

The fact that the answer is positive shows that the scalar component F_{BO} has the same sense as the unit vector $\boldsymbol{\lambda}_{BO}$, namely, from B to O.

Problems

3.11 Refer to Figure P3.11 and express the position vector $\mathbf{r}_{A/O}$ in terms of its x, y, and z components. Find its direction angles with respect to the three coordinate axes. What is the distance from point O to A?

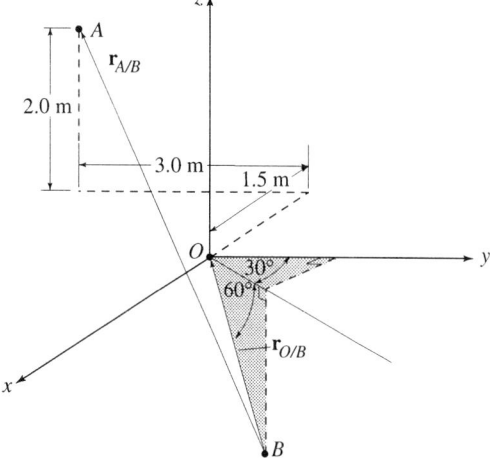

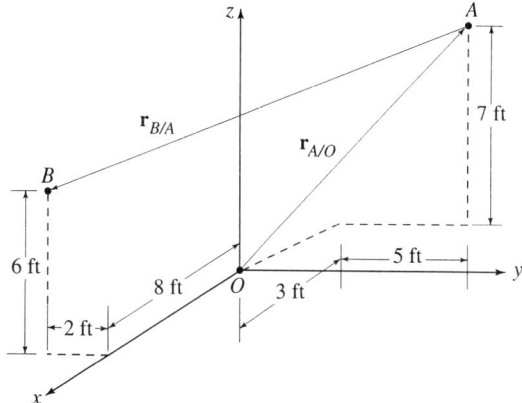

FIGURE P3.11.

FIGURE P3.13.

3.12 Refer to Figure P3.11 and express the position vector $\mathbf{r}_{B/A}$ in terms of its x, y, and z components. Find its direction angles with respect to the three coordinates axes. What is the distance from point A to point B?

3.13 Express the position vector $\mathbf{r}_{O/B}$ in Figure P3.13 in terms of its x, y, and z components. Find its direction angles with respect to the three coordinate axes. The distance from point B to point O is 4.5 m.

3.14 Express the position vector $\mathbf{r}_{A/B}$ in Figure P3.13 in terms of its x, y, and z components. Compute its direction angles with respect to the three coordinate axes and the distance from point B to point A.

3.15 The system shown in Figure P3.15 supports the weight W. If the compressive

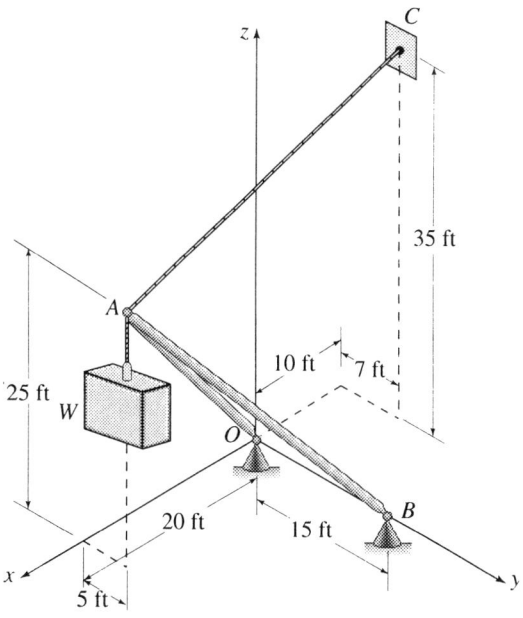

FIGURE P3.15.

force in member AB is 750 lb, determine (a) the x, y, and z components of this force on support B and (b) the direction angles that the force makes with the coordinate axes. Note that the compressive force on support B has a sense from A to B.

3.16 The system shown in Figure P3.15 supports the weight W. If the tensile force in cable AC is 2.55 k determine (a) the x, y, and z components of this force at point A and (b) the direction angles that the force makes with the coordinate axes. Note that the cable tensile force at A has a sense from A to C.

3.17 A tow truck is pulling a car out of a ravine, as shown in Figure P3.17. At the instant considered, the tension in cable OA is 4.5 kN. Determine (a) the x, y, and z components of this cable tension at attachment point A and (b) the direction angles that the cable tension makes with the coordinate axes. Note that the cable tension at point A has a sense from A to O.

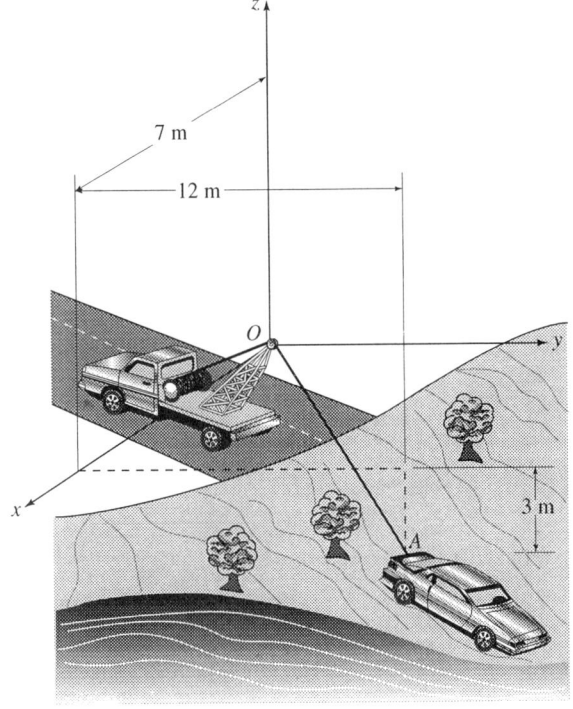

FIGURE P3.17.

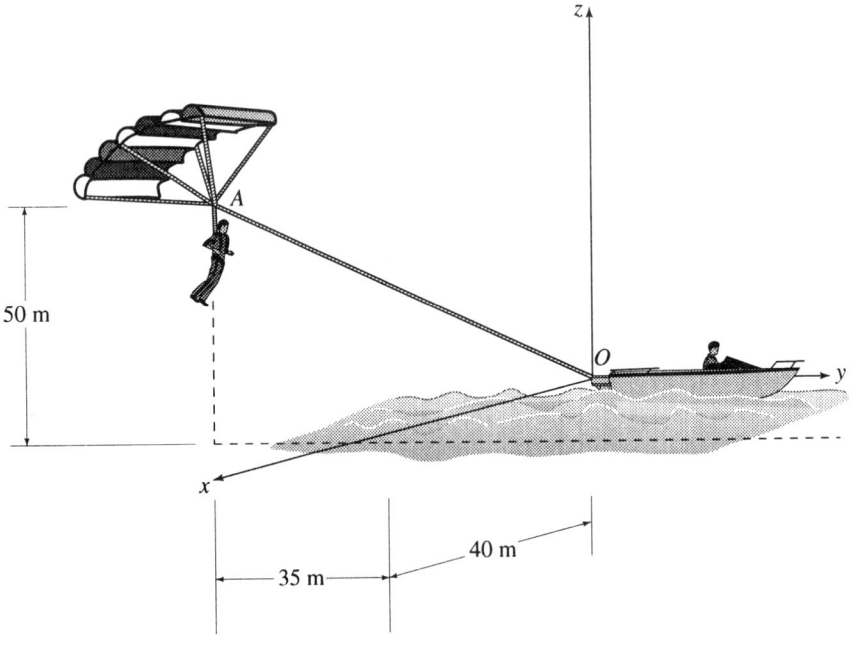

FIGURE P3.18.

3.18 A stunt parachutist is being lifted into the air by a boat as shown in Figure P3.18. For the position shown, the tension in line OA is 2500 N. Determine (a) the x, y, and z components of this tension acting on the boat at point O and (b) the direction angles that the cable makes with the coordinate axes. Note that the tension at point O has a sense from O to A.

3.19 A cable is stretched between points A and B as shown in Figure P3.19. If the force in the cable at point A is $\mathbf{F} = (-300\mathbf{i} + 450\mathbf{j} + 275\mathbf{k})$ lb and if the length of the cable from A to B is 25 ft, determine the x, y, and z coordinates of point A.

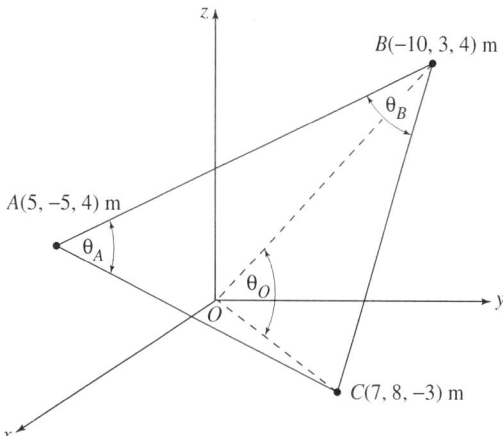

FIGURE P3.21.

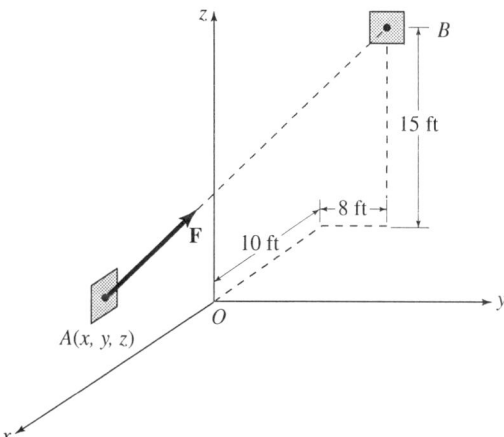

FIGURE P3.19.

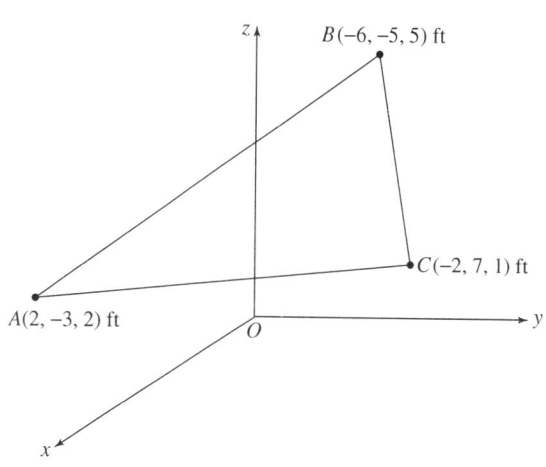

FIGURE P3.24.

3.20 Repeat Problem 3.19 if $\mathbf{F} = (150\mathbf{i} + 200\mathbf{j} - 325\mathbf{k})$ lb and the distance between A to B is 18 ft.

3.21 The coordinates of points A, B and C are given in Figure P3.21. Find the angle θ_O.

3.22 Refer to Figure P3.21, and determine the angle θ_A.

3.23 Refer to Figure P3.21, and determine the angle θ_B.

3.24 The coordinates of points A, B, and C are given in Figure P3.24. Construct a two-

dimensional sketch of triangle ABC showing the angles and the lengths of the sides.

3.25 The forces $\mathbf{F}_1$ and $\mathbf{F}_2$ represent the forces in two cables attached at point A as shown in Figure P3.25. Let $\mathbf{F}_1 = (-5\mathbf{i} + 3\mathbf{j} + 2\mathbf{k})$ kN and $F_2 = 7$ kN. Determine the angle θ between $\mathbf{F}_1$ and $\mathbf{F}_2$.

3.26 Refer to Problem 3.25. Determine the scalar component of $\mathbf{F}_1$ in the direction from A to B.

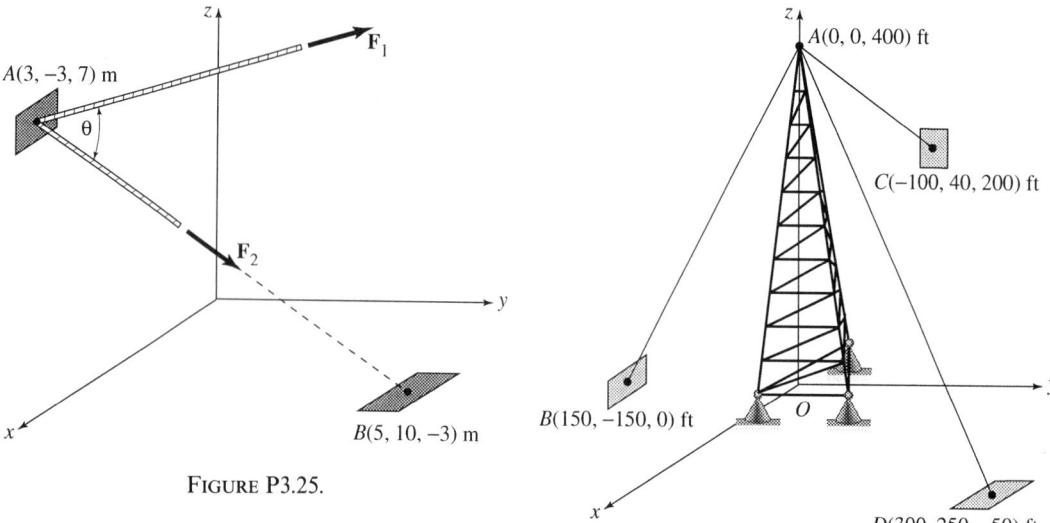

FIGURE P3.25.

FIGURE P3.27.

3.27 A transmission tower is stabilized by three guy cables as shown in Figure P3.27. If the tensile force in cable AB is 500 lb, determine, at support B, the scalar component of this force in a direction from B to D. Note that, at support B, the tensile force in the cable is directed from B to A.

3.28 Refer to the transmission tower in Figure P3.27. If the tensile force in cable AC is 750 lb, determine, at support C, its scalar component along a direction from C to

B. Note that, at support C, the tensile force is directed from C to A.

3.29 Refer to the transmission tower in Figure P3.27. If the tensile force in cable AD is 975 lb, determine, at support D, its scalar component in a direction from D to C. Note that, at support D, the tensile force is directed from D to A.

3.4
Addition of Forces Using Rectangular Components

The graphical and trigonometric techniques developed in Section 2.2 for the addition of concurrent two-dimensional forces are *not*, in general, *very practical* for the addition of two or more forces in space. In Section 2.4, however, we learned how to add two or more two-dimensional forces by summing their respective x and y components. This technique may be easily extended to include components in the z direction for a three-dimensional force system and represents a very convenient and practical method for adding two or more concurrent three-dimensional forces to determine their resultant in space.

Consider the case of several concurrent three-dimensional forces $\mathbf{F}_1$, $\mathbf{F}_2$, ..., $\mathbf{F}_n$ that need to be added to obtain their resultant $\mathbf{R}$ *which is a single force capable of producing the same effect as all of the other forces acting simultaneously*. The use of the head-to-tail construction developed in Section 2.2 leads to

$$R = F_1 + F_2 + \cdots + F_n = \sum F \tag{3.24}$$

where the symbol $\sum F$ represents the vector sum of all forces to be added. If F_1 to F_n are expressed in terms of their x, y, and z components, Eq. (3.24) becomes

$$
\begin{aligned}
R &= F_{1x}\mathbf{i} + F_{1y}\mathbf{j} + F_{1z}\mathbf{k} + F_{2x}\mathbf{i} + F_{2y}\mathbf{j} + F_{2z}\mathbf{k} + \cdots \\
&\quad + F_{nx}\mathbf{i} + F_{ny}\mathbf{j} + F_{nz}\mathbf{k} \\
&= (F_{1x} + F_{2x} + \cdots + F_{nx})\mathbf{i} + (F_{1y} + F_{2y} + \cdots + F_{ny})\mathbf{j} \\
&\quad + (F_{1z} + F_{2z} + \cdots + F_{nz})\mathbf{k} \\
&= (\textstyle\sum F_x)\mathbf{i} + (\textstyle\sum F_y)\mathbf{j} + (\textstyle\sum F_z)\mathbf{k} \\
&= R_x\mathbf{i} + R_x\mathbf{j} + R_x\mathbf{k}. \tag{3.25}
\end{aligned}
$$

The symbol $R_x = \sum F_x$ is the x scalar component of the resultant R and may be found by adding, algebraically, all force components in the x direction. Similarly, the symbols $R_y = \sum F_y$ and $R_z = \sum F_z$ represent, respectively, the y and z scalar components of the resultant R and are obtained by algebraic addition of all of their respective components. As shown in Figure 3.8, the vector components R_x, R_y, and R_z may be added to obtain their resultant R by the law of the parallelogram or by the head-to-tail construction. Obviously, Eq (3.25) represents both of these methods symbolically. The magnitude of R may be obtained by two successive applications of the Pythagorean theorem. The first application yields $R_{xy}^2 = R_x^2 + R_y^2$ and the second $R = \sqrt{R_{xy}^2 + R_z^2} = \sqrt{R_x^2 + R_y^2 + R_z^2}$. Substituting $\sum F_x$ for R_x, $\sum F_y$ for R_y, and $\sum F_z$ for R_z,

$$R = \sqrt{(\textstyle\sum F_x)^2 + (\textstyle\sum F_y)^2 + (\textstyle\sum F_z)^2}. \tag{3.26}$$

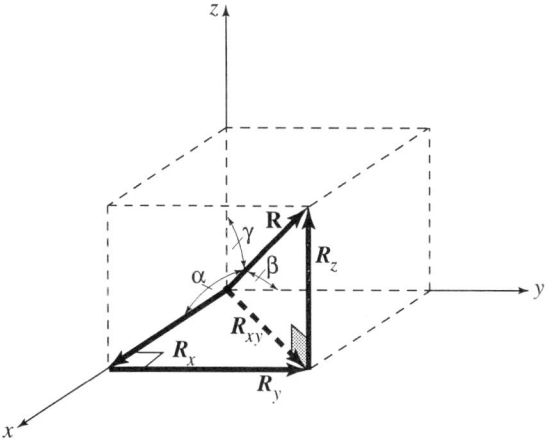

FIGURE 3.8.

The angles α, β, and γ made by the resultant **R** with the x, y, and z coordinate axes, respectively, are obtained using the same steps that led to Eqs. (3.4). Thus,

$$\left.\begin{aligned} \alpha &= \cos^{-1}\left(\frac{\sum F_x}{R}\right), \\[6pt] \beta &= \cos^{-1}\left(\frac{\sum F_y}{R}\right), \\[6pt] \gamma &= \cos^{-1}\left(\frac{\sum F_z}{R}\right). \end{aligned}\right\} \qquad (3.27)$$

and

Example 3.6 illustrates some of the above concepts.

■ **Example 3.6**

The system shown in Figure E3.6 is used to support a crate of weight W. The tension in cable AC $F_{AC} = 700$ N and that in cable AB $F_{AB} = 1350$ N. Determine the resultant of these two forces acting at point A. Express this resultant (a) in vector form in terms of its x, y, and z components and (b) as a magnitude along with its direction angles with respect to the three coordinate axes. Note that the tension at A in cable AC is directed from A to C and that in cable AB is directed from A to B.

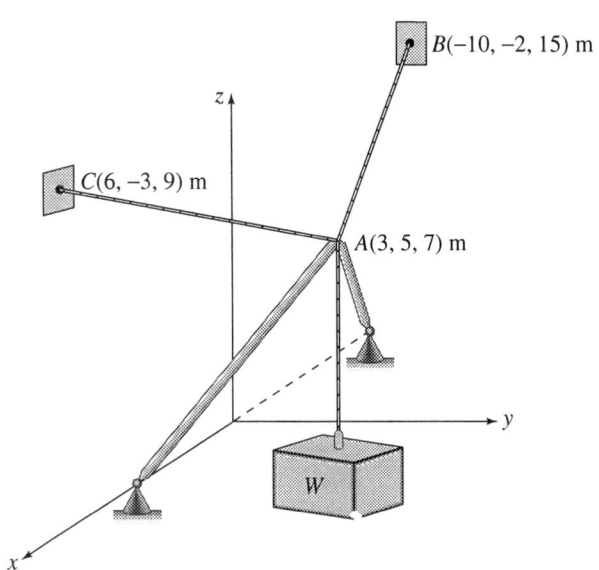

FIGURE E3.6. x

Solution

Note that the coordinates of points A, B, and C have been determined and are indicated in Figure E3.6. The position vectors and their magnitudes from A to B and from A to C are found as follows:

$$\mathbf{r}_{B/A} = (-13\mathbf{i} - 7\mathbf{j} + 8\mathbf{k})\,\text{m},$$

$$r_{B/A} = 16.793\,\text{m},$$

$$\mathbf{r}_{C/A} = (3\mathbf{i} - 8\mathbf{j} + 2\mathbf{k})\,\text{m},$$

and

$$r_{C/A} = 8.775\,\text{m}.$$

Unit vectors λ_{AB} and λ_{AC} are now determined by Eq. (3.16). Thus,

$$\lambda_{AB} = \frac{\mathbf{r}_{B/A}}{r_{B/A}} = -0.774\mathbf{i} - 0.417\mathbf{j} + 0.476\mathbf{k},$$

and

$$\lambda_{AC} = \frac{\mathbf{r}_{C/A}}{r_{C/A}} = 0.412\mathbf{i} - 0.912\mathbf{j} + 0.228\mathbf{k}.$$

Therefore, the forces $\mathbf{F}_{AB}$ and $\mathbf{F}_{AC}$ may be found by Eq. (3.8). Thus,

$$\mathbf{F}_{AB} = F_{AB}\lambda_{AB} = [1350(-0.774\mathbf{i} - 0.417\mathbf{j} + 0.476\mathbf{k})]\,\text{N}$$

$$= (-1045\mathbf{i} - 563\mathbf{j} + 643\mathbf{k})\,\text{N},$$

and

$$\mathbf{F}_{AC} = F_{AC}\lambda_{AC} = [700(0.412\mathbf{i} - 0.912\mathbf{j} + 0.228\mathbf{k})]\,\text{N}$$

$$= (288\mathbf{i} - 638\mathbf{j} + 160\mathbf{k})\,\text{N}.$$

(a) By Eq. (3.25),

$$\mathbf{R} = (\textstyle\sum F_x)\mathbf{i} + (\sum F_y)\mathbf{j} + (\sum F_z)\mathbf{k}$$

$$= (-1045 + 288)\mathbf{i} + (-563 - 638)\mathbf{j} + (643 + 160)\mathbf{k},$$

and

$$\mathbf{R} = (-757\mathbf{i} - 1201\mathbf{j} + 803\mathbf{k})\,\text{N}. \qquad \text{ANS.}$$

(b) By Eq. (3.26),

$$R = \sqrt{(\textstyle\sum F_x)^2 + (\sum F_y)^2 + (\sum F_z)^2}$$

$$= \sqrt{(757)^2 + (1201)^2 + (803)^2}$$

$$= 1631\,\text{N}. \qquad \text{ANS.}$$

The direction angles α, β, and γ are now found by Eqs. (3.27). Thus,

$$\alpha = \cos^{-1}\left(\frac{\sum F_x}{R}\right) = \cos^{-1}\left(\frac{-757}{1631}\right)$$

$$\alpha = 117.7°, \qquad \text{ANS.}$$

$$\beta = \cos^{-1}\left(\frac{\sum F_y}{R}\right) = \cos^{-1}\left(\frac{-1201}{1631}\right)$$

$$\beta = 137.4°, \qquad\qquad \text{ANS.}$$

and

$$\gamma = \cos^{-1}\left(\frac{\sum F_z}{R}\right) = \cos^{-1}\left(\frac{803}{1631}\right)$$

$$\gamma = 60.5°. \qquad\qquad \text{ANS.}$$

■

Problems

3.30 Let $F_1 = (1.5i - 2.7j - 3.2k)$ kN, $F_2 = (5.0i - 3.7j - 6.5k)$ kN, and $F_3 = (-2.0i + 7.7j + 4.3k)$ kN. Find the magnitude and direction angles α, β, and γ of the resultant force R.

3.31 Let $P_1 = (-550i + 700j - 357k)$ lb, $P_2 = (-235i - 150j + 500k)$ lb, and $P_3 = (300i - 275j - 625k)$ lb. Determine the magnitude and direction angles α, β, and γ of the resultant force R.

3.32 A plate is suspended in a horizontal position, partially, by two cables, as shown in Figure P3.32. The tensions in cables AB and AC are 675 N and 925 N, respectively. Find the resultant R of these two tensions acting at point A on the plate. Express this resultant (a) in vector form in terms of its x, y, and z components and (b) as a magnitude together with its direction angles α, β, and γ. Note that the tensions in cables AB and AC at point A are directed from A to B and from A to C, respectively.

3.33 A plate is suspended in a horizontal position, partially, by three cables, as shown in Figure P3.33. The tensions in cables DE, DF, and DO are 850 N, 775 N, and 550 N, respectively. Determine the resultant R of these three tensions acting at

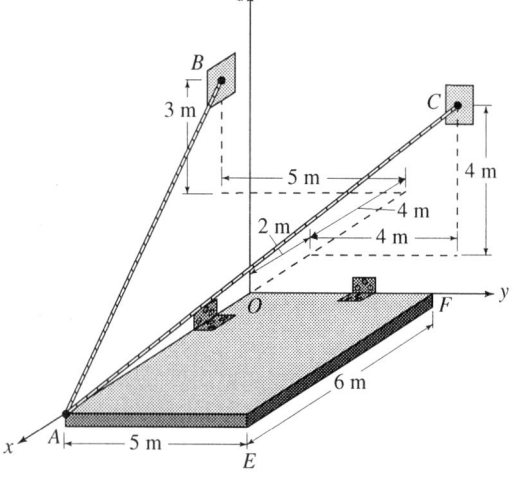

FIGURE P3.32.

support D. Express this resultant (a) in vector form in terms of its x, y, and z components and (b) as a magnitude together with its direction angles α, β, and γ. Note that the tensions in cables DE, DF, and DO at point D are directed from D to E, from D to F, and from D to O, respectively.

3.34 Let $P_1 = (200i + 175j - 300k)$ lb and $P_2 = (-50i - 300j + 150k)$ lb. The resul-

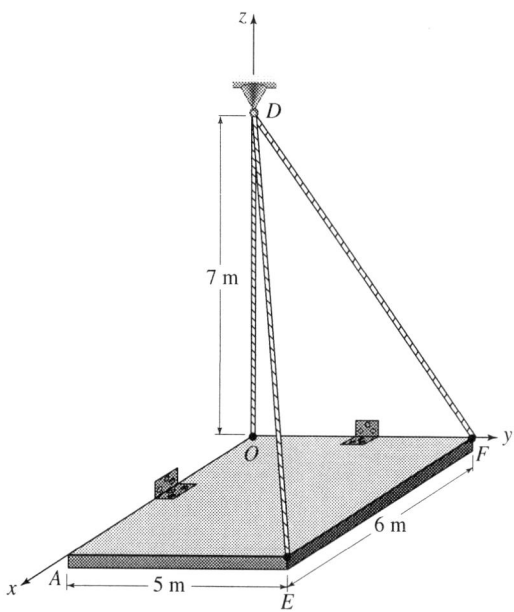

FIGURE P3.33.

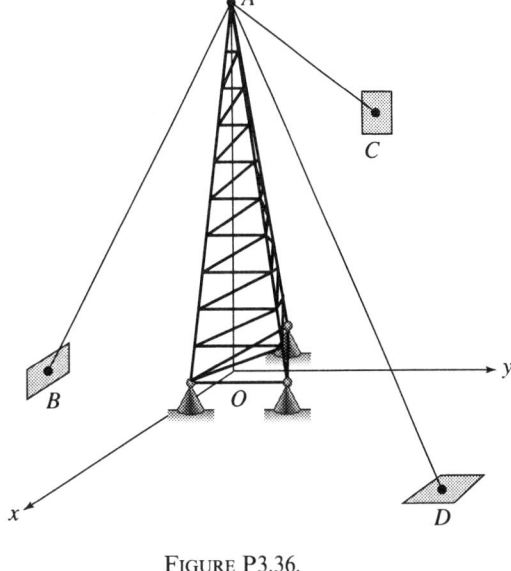

FIGURE P3.36.

tant **R** of **P**$_1$, **P**$_2$, and a third force **P**$_3$ has a magnitude 500 lb and is directed along the positive z axis. Determine the x, y, and z components of **P**$_3$.

3.35 Let **Q**$_1 = (-500\mathbf{i} - 300\mathbf{j} + 450\mathbf{k})$ N and **Q**$_2 = (275\mathbf{i} - 350\mathbf{j} + 75\mathbf{k})$ N. The resultant **R** of **Q**$_1$, **Q**$_2$, and a third force **Q**$_3$ has a magnitude of 300 N and is directed along the negative x axis. Find the x, y, and z components of **Q**$_3$.

3.36 A transmission tower is stabilized by three guy cables as shown in Figure P3.36. The resultant **R**, at point A, of the tensions in the three cables has a magnitude of 2400 lb directed along the negative z axis. Compute the tensions in the three cables. The coordinates of the support points are: A(0, 0, 400) ft, B(150, −150, 400) ft, C(−100, 40, 200) ft, and D(300, 250, −50) ft. Note that the tensions at A in cables AB, AC, and AD are directed from A to B, from A to C, and from A to D, respectively.

3.37 A transmission tower is supported by three guy cables as shown in Figure P3.36. The resultant **R**, at point A, of the tensions in the three cables has a magnitude of 9000 N directed along the negative z axis. Determine the tensions in the three cables. The coordinates of the support points are A(0, 0, 100) m, B(40, −40, 0) m, C(−30, 15, 60) m, and D(70, 60, −20) m. Note that the tensions at A in cables AB, AC, and AD are directed from A to B, from A to C, and from A to D, respectively.

3.38 A transmission tower is stabilized by three guy cables as shown in Figure P3.36. The tensions in cables AB, AC, and AD are, respectively, F_{AB}, 700 lb, and 650 lb. The length of cable AB is 500 ft and the resultant of the three cable forces at point A has a magnitude of 900 lb directed along the negative z axis. The coordinates of the three support points are A(0, 0, 600) ft,

C($-150, 70, 250$) ft, and D($300, 275, -75$) ft. Find the magnitude F_{AB} and the coordinates of point B. Note that the tensions at point A in cables AB, AC, and AD are directed from A to B, from A to C, and from A to D, respectively.

3.5 Equilibrium Conditions and Applications

As discussed in Chapter 2 for two-dimensional particle equilibrium, the condition of particle equilibrium in three dimensions is achieved by ensuring that the resultant of the concurrent force system acting on it vanishes. By Eqs. (3.24) and (3.25), the resultant R of a concurrent three-dimensional force system is expressed by $\mathbf{R} = \sum \mathbf{F} = (\sum F_x)\mathbf{i} + (\sum F_x)\mathbf{j} + (\sum F_x)\mathbf{k}$. Because equilibrium requires that $\mathbf{R}$ be zero, it follows that

$$\mathbf{R} = \sum \mathbf{F} = \mathbf{0} \tag{3.28}$$

or

$$\mathbf{R} = \sum \mathbf{F} = (\sum F_x)\mathbf{i} + (\sum F_x)\mathbf{j} + (\sum F_x)\mathbf{k} = \mathbf{0}. \tag{3.29}$$

The only way that Eq. (3.29) can be satisfied is for each of its three terms to vanish independently. This, obviously, means that the coefficients of $\mathbf{i}$, $\mathbf{j}$, and $\mathbf{k}$ must each be equal to zero. Thus, equilibrium of a particle in three dimensions requires that

and

$$\left.\begin{array}{l} \sum F_x = 0, \\ \sum F_y = 0, \\ \sum F_z = 0. \end{array}\right\} \tag{3.30}$$

Equations (3.30) are both *necessary* and *sufficient* conditions for the equilibrium of a particle in space. The first of Eqs. (3.30) shows that, if a particle is in equilibrium in the x direction, all of the force components in this direction must be in perfect balance. This means that, if there is a resultant force component acting in the positive x direction, another resultant force component of equal magnitude must exist acting in the negative x direction. The same statements apply to resultant force components in the y and z directions. Also, the three conditions of equilibrium expressed in Eq. (3.30) are entirely independent of one another and any one of them may be satisfied even though the other two may not be. Thus, for complete equilibrium of a particle in space, all three of the conditions expressed in Eq. (3.30) must be satisfied.

The fact that we have available three independent equations for the three-dimensional equilibrium of a particle implies that we can solve for no more than three unknown quantities in a given case. In solving for the three unknown quantities, we may proceed directly to scalar Eqs. (3.30) or we may begin with vector Eq. (3.28). In general, in dealing with three-dimensional force systems, it is much more practical to use the vector approach and begin the solution with Eq. (3.28), which will, then lead, to Eqs. (3.30), because it is usually difficult to visualize geometric relations in space. This, however, must be done if the x, y, and z components of the forces are to be determined using trigonometric relations to go directly to Eqs. (3.30). *For these reasons, the vector approach is generally used in solving three-dimensional equilibrium problems.*

As in the case of two dimensions, solving three-dimensional particle equilibrium problems requires the construction of an appropriate free-body diagram. This construction is accomplished using the same three-step procedure introduced in Section 2.6. Therefore, the student is urged to review Section 2.6 before proceeding with the study of three-dimensional particle equilibrium.

The construction of free-body diagrams requires knowledge of reactive forces produced at supports and connections. The supports and connections that were discussed in Section 2.5 for the two-dimensional case are equally usable in the three-dimensional case, and the student is again urged to review this material before proceeding. However, there is one additional support system known as the ball and socket, that is often used with three-dimensional systems. The characteristics of the ball and socket are needed when constructing three-dimensional, free-body diagrams and, therefore, they are discussed here and summarized as case 6 in Appendix F.

The Ball and Socket

The ball and socket is the three-dimensional counterpart of the hinge discussed in Section 2.5. As the name implies, it consists of a ball, attached to the end of a member, that fits inside a container known as the socket.

Consider the ball-and-socket support shown in Figure 3.9(a). The ball B is assumed to be able to rotate freely inside the socket. Thus, if member BA is subjected to any arbitrary force **F**, the ball in the socket would permit the member to rotate until it assumes the same direction as the force. The socket, however, does not allow the ball to translate in any direction. This, means that, if the applied force **F** has a component in the x direction, for example, a reaction component B_x is developed at the support to keep the ball from translating in the x direction. Similarly, if the applied force has components in the y and z direction, reaction components B_y and B_z are developed at the support to keep the ball from translating in the y and z directions, respectively. The three reaction components B_x, B_y, and B_z are shown in the free-body diagram of Figure 3.9(b) pointed in the positive direction of their respective coordinate axes. This is, obviously, only an assumption that is confirmed or denied after satisfying the equations of equilibrium.

The case of a member in contact with a rough surface is identical to

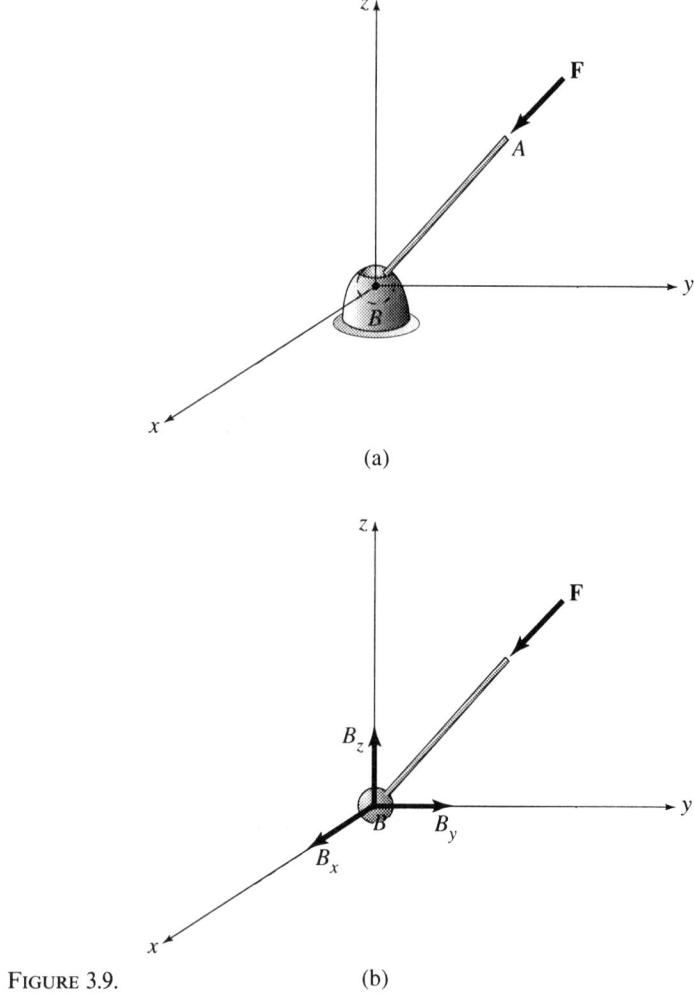

FIGURE 3.9.   (b)

that of a ball and socket in that three reaction components are developed at the point of contact. In view of this similarity, the rough
surface is not discussed any further here but is summarized, along with
the ball and socket, in Appendix F, case 6.

The following examples illustrate some of the above concepts.

■ **Example 3.7**

A crate of weight $W = 2500$ lb is supported, as shown in Figure E3.7(a),
by members AO and AB and cable AC. The supports at O and B are
balls and sockets and the forces in members AO and AB may be
assumed to act along their respective axes. Determine (a) the forces in

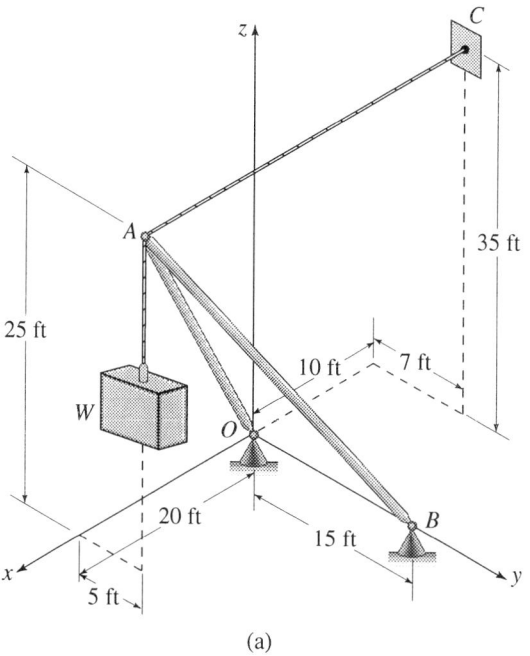

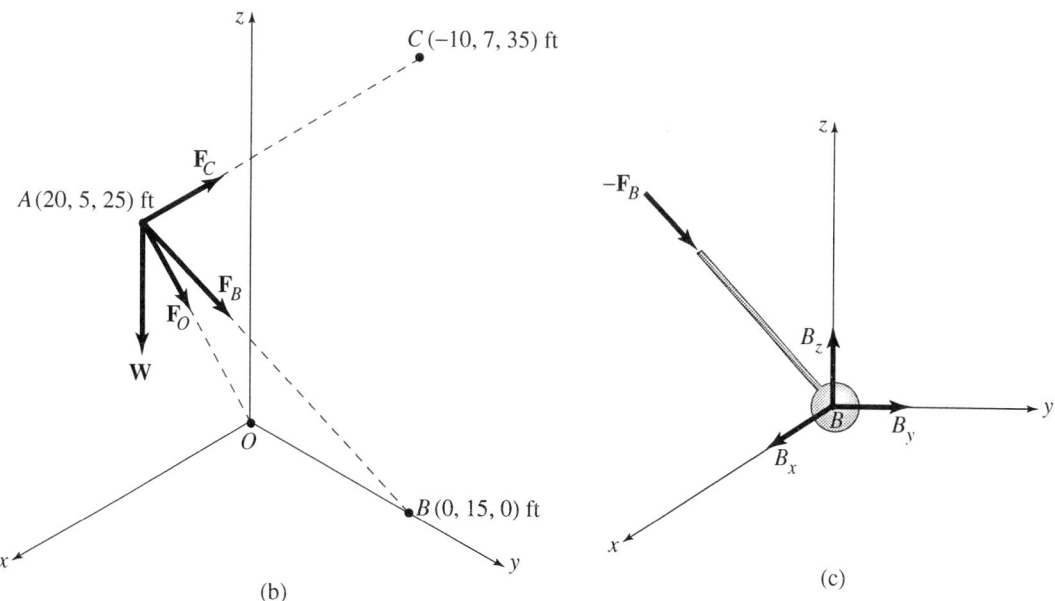

FIGURE E3.7.

cable AC and in members AO and AB and (b) the x, y, and z components of the reaction at support B.

Solution

(a) The free-body diagram of the joint at A is constructed as shown in Figure E3.7(b). This joint was chosen because it is the point of concurrency of the weight **W** as well as the force $\mathbf{F_C}$ in the cable and the forces and F_O and F_B in members AO and AB, respectively. Note that the free-body diagram shows the coordinates of all significant points as well as the given coordinate system. Note also that all unknown member forces have been assumed to be tension. This means that the unknown forces in all members framing into joint A are assumed to pull at this joint and, in turn, joint A pulls back on these members creating the assumed tensions.

Because the free-body diagram of Figure E3.7(b) represents a three-dimensional force system, we will solve the equilibrium problem using the vector method which requires the use of Eq. (3.29). The use of Eq. (3.29) necessitates expressing all forces in vector form. Thus,

$$\mathbf{W} = -(2500)\mathbf{k},$$

$$\mathbf{F_C} = F_C \left[\frac{-30\mathbf{i} + 2\mathbf{j} + 10\mathbf{k}}{\sqrt{(30)^2 + (2)^2 + (10)^2}} \right]$$

$$= -(0.947F_C)\mathbf{i} + (0.063F_C)\mathbf{j} + (0.316F_C)\mathbf{k},$$

$$\mathbf{F_O} = F_O \left[\frac{-20\mathbf{i} - 5\mathbf{j} - 25\mathbf{k}}{\sqrt{(20)^2 + (5)^2 + (25)^2}} \right]$$

$$= -(0.617F_O)\mathbf{i} - (0.154F_O)\mathbf{j} - (0.772F_O)\mathbf{k},$$

(a)

and

$$\mathbf{F_B} = F_B \left[\frac{-20\mathbf{i} + 10\mathbf{j} - 25\mathbf{k}}{\sqrt{(20)^2 + (10)^2 + (25)^2}} \right]$$

$$= -(0.596F_B)\mathbf{i} + (0.298F_B)\mathbf{j} - (0.745F_B)\mathbf{k}.$$

Now by Eq. (3.29)

$$\sum \mathbf{F} = \mathbf{W} + \mathbf{F_C} + \mathbf{F_O} + \mathbf{F_B} = \mathbf{0}. \tag{b}$$

Substituting Eqs. (a) in Eq. (b) and grouping the $\mathbf{i}$, $\mathbf{j}$ and $\mathbf{k}$ terms,

$$-[0.947F_C + 0.617F_O + 0.596F_B]\mathbf{i}$$

$$+[0.063F_C - 0.154F_O + 0.298F_B]\mathbf{j}$$

$$+[0.316F_C - 0.772F_O - 0.745F_B - 2500]\mathbf{k} = \mathbf{0} \tag{c}$$

It follows from Eq. (c), therefore, that

$$0.947F_C + 0.617F_O + 0.596F_B = 0,$$

$$0.063F_C - 0.154F_O + 0.298F_B = 0, \qquad \text{(d)}$$

and

$$0.316F_C - 0.772F_O - 0.745F_B = 2500$$

Note that the first of Eqs. (d) represents $\sum F_x = 0$, the second $\sum F_y = 0$ and the third $\sum F_z = 0$ which are expressions of Eq. (3.30) and are the three necessary and sufficient conditions for the equilibrium of joint A.

If the first of Eqs. (d) is solved for F_C,

$$F_C = -0.652F_O - 0.629F_B. \qquad \text{(e)}$$

When Eq. (e) is substituted in the last two of Eqs. (d),

$$-0.195F_O + 0.258F_B = 0,$$

and

$$-0.978F_O - 0.944F_B = 2500. \qquad \text{(f)}$$

Equations (f) are now solved simultaneously to obtain

$$F_O = -1479 \text{ lb},$$

and

$$F_B = -1117 \text{ lb}.$$

Substituting these values in Eq. (e) yields

$$F_C = 1666 \text{ lb}.$$

Therefore,

$$F_C = 1666 \text{ lb (T)},$$

$$F_O = 1479 \text{ lb (C)},$$

and

$$F_B = 1117 \text{ lb (C)} \qquad \text{ANS.}$$

The forces F_O and F_B are both compression (signified by the symbol (C)) because the solution indicated negative values. However, F_C is tension (signified by the symbol (T)) because the solution indicated a positive value.

(b) The force $\mathbf{F_B}$ acting at joint A may be expressed by the third of Eqs. (a) as

$$\mathbf{F_B} = (666\mathbf{i} - 333\mathbf{j} + 832\mathbf{k}) \text{ lb}.$$

By Newton's third law, it follows that the force acting on the ball and socket at B is $-\mathbf{F_B}$ as shown in the free-body diagram in Figure E3.7(c).

The reaction components, B_x, B_y, and B_z are assumed to act in the positive directions of the corresponding coordinate axes. Thus, by Eq. (3.25), the resultant reaction at the ball and socket is expressed by

$$\mathbf{B} = B_x\mathbf{i} + B_y\mathbf{j} + B_z\mathbf{k}$$

Therefore, by Eq. (3.29),

$$\sum \mathbf{F} = -\mathbf{F_B} + \mathbf{B} = (-666 + B_x)\mathbf{i} + (333 + B_y)\mathbf{j} + (-832 + B_z)\mathbf{k} = 0$$

from which

$$B_x = 666 \text{ lb,}$$

$$B_y = -333 \text{ lb,}$$

and

$$B_z = 832 \text{ lb.} \qquad \text{ANS.}$$

The minus sign on B_y indicates that this force component is opposite to the sense assumed in the free-body diagram of Figure E3.7(c).

■ Example 3.8

The system shown in Figure E3.8(a) consists of the four cables EA, EB, EC, and ED joined together at E. Cable ED passes over a small frictionless pulley at D and is used to apply the 5.00 kN force needed to place the container G in the position shown. If the tension in cable EA is 4.50 kN, determine the mass m of container G.

Solution

Because the frictionless pulley at D changes the direction but not the magnitude of the tensile force in the cable, the tension in cable ED, $T_{ED} = 5.00$ kN. Also, the tension in cable EA, $T_{EA} = 4.50$ kN, as given.

The free-body diagram of joint E is shown in Figure E3.8(b). Joint E was chosen because it is the point of concurrency of the unknown forces $\mathbf{T_{EB}}$, $\mathbf{T_{EC}}$ and $\mathbf{W_G}$ as well as the known forces $\mathbf{T_{EA}}$ and $\mathbf{T_{ED}}$. All of these forces are expressed in vector form as follows:

$$\mathbf{W_G} = -(mg)\mathbf{k} = -(9.81 \times 10^{-3} m)\mathbf{k}$$

where m is the mass of the container in kg,

$$\mathbf{T_{EB}} = T_{EB}\left[\frac{-3\mathbf{i} - \mathbf{j} + 3\mathbf{k}}{\sqrt{(3)^2 + (1)^2 + (3)^2}}\right]$$
$$= -(0.688T_{EB})\mathbf{i} - (0.229T_{EB})\mathbf{j} + (0.688T_{EB})\mathbf{k},$$

$$\mathbf{T_{EC}} = T_{EC}\left[\frac{5\mathbf{i} + 3\mathbf{j} + 2\mathbf{k}}{\sqrt{(5)^2 + (3)^2 + (2)^2}}\right]$$
$$= (0.811T_{EC})\mathbf{i} + (0.487T_{EC})\mathbf{j} + (0.324T_{EC})\mathbf{k},$$

$$\text{(a)}$$

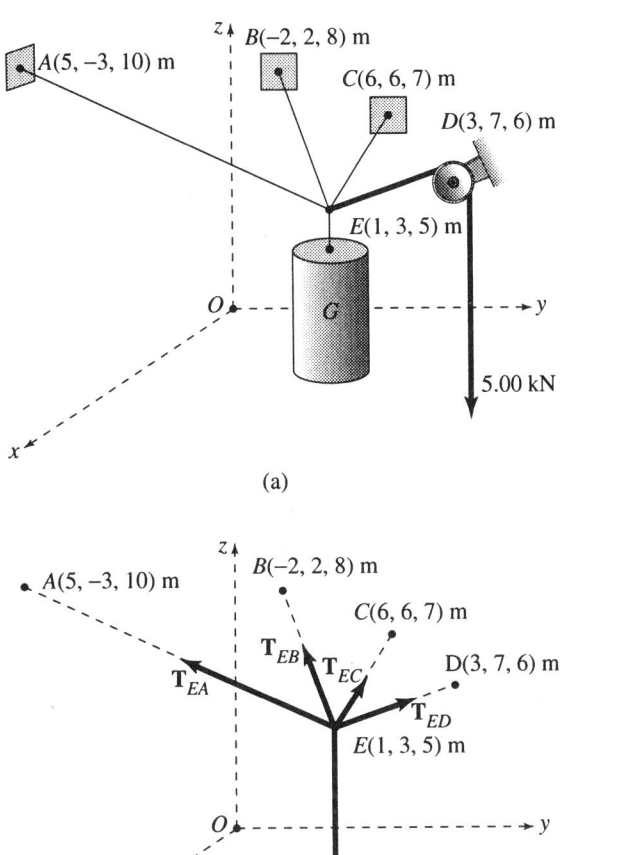

(a)

(b)   FIGURE E3.8.

$$\left.\begin{array}{l} \mathbf{T_{EA}} = 4.50\left[\dfrac{4\mathbf{i} - 6\mathbf{j} + 5\mathbf{k}}{\sqrt{(4)^2 + (6)^2 + (5)^2}}\right] \\[4mm] = (2.051)\mathbf{i} - (3.077)\mathbf{j} + (2.564)\mathbf{k}, \\[4mm] \mathbf{T_{ED}} = 5.00\left[\dfrac{2\mathbf{i} + 4\mathbf{j} + \mathbf{k}}{\sqrt{(2)^2 + (4)^2 + (1)^2}}\right] \\[4mm] = (2.182)\mathbf{i} + (4.364)\mathbf{j} + (1.091)\mathbf{k}. \end{array}\right\}$$ (a) Continued

and

By Eq. (3.29)

$$\sum \mathbf{F} = \mathbf{W_G} + \mathbf{T_{EB}} + \mathbf{T_{EC}} + \mathbf{T_{EA}} + \mathbf{T_{ED}} = \mathbf{0}.$$ (b)

Substituting Eqs. (a) in Eq. (b) and grouping the **i**, **j**, and **k** terms,

$$\left.\begin{aligned}
&[-0.688T_{EB} + 0.811T_{EC} + 2.051 + 2.182]\mathbf{i} \\
+\, &[-0.229T_{EB} + 0.487T_{EC} - 3.077 + 4.364]\mathbf{j} \\
+\, &[0.688T_{EB} + 0.324T_{EC} + 2.564 + 1.091 - 9.81 \times 10^{-3}\,m]\mathbf{k} = \mathbf{0}
\end{aligned}\right\} \quad (c)$$

We conclude, from Eq. (c), that

$$\left.\begin{aligned}
0.688T_{EB} - 0.811T_{EC} &= 4.233, \\
0.229T_{EB} - 0.487T_{EC} &= 1.287, \\
0.688T_{EB} + 0.324T_{EC} - (9.81 \times 10^{-3})m &= -3.655
\end{aligned}\right\} \quad (d)$$

and

Note that the first of Eqs. (d) represents $\sum F_x = 0$, the second $\sum F_y = 0$ and the third $\sum F_z = 0$ which are expressions of Eq. (3.30) and are the three necessary and sufficient conditions for the equilibrium of joint E.

Solve the first two of Eqs. (d) simultaneously to obtain

$$T_{EB} = 6.82 \text{ kN},$$

and

$$T_{EC} = 0.562 \text{ kN}.$$

Substituting these values in the third of Eqs. (d) yields

$$m = 869 \text{ kg}. \qquad\qquad \text{ANS.}$$

Problems

3.39 The particle at A is in equilibrium under the action of five concurrent forces as shown in Figure P3.39. Determine the magnitudes of $\mathbf{P}_1$, $\mathbf{P}_2$, and $\mathbf{P}_3$.

3.40 The particle at B is in equilibrium under the action of five concurrent forces as shown in Figure P3.40. Determine the magnitudes of $\mathbf{Q}_1$, $\mathbf{Q}_2$, and $\mathbf{Q}_3$.

3.41 If the particle at point A is in equilibrium under the action of the concurrent forces shown in Figure P3.41, find the magnitude and direction of the unknown force **F**.

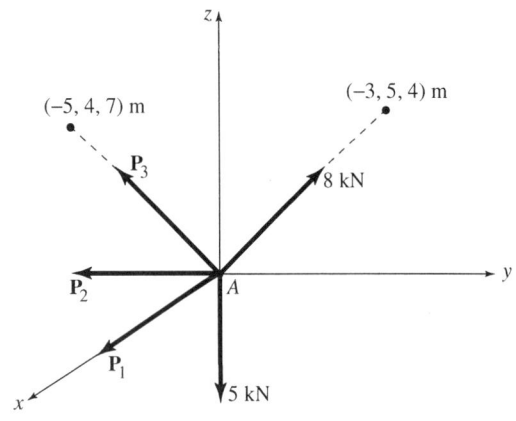

FIGURE P3.39.

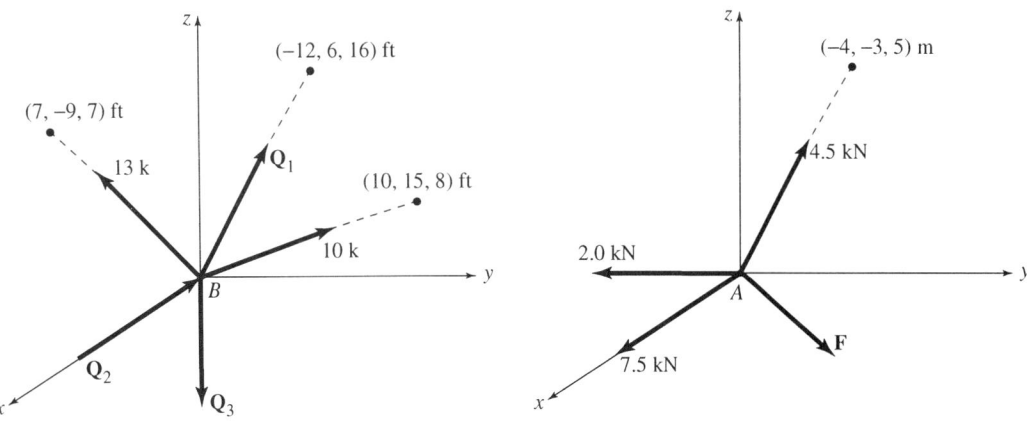

FIGURE P3.40.

FIGURE P3.41.

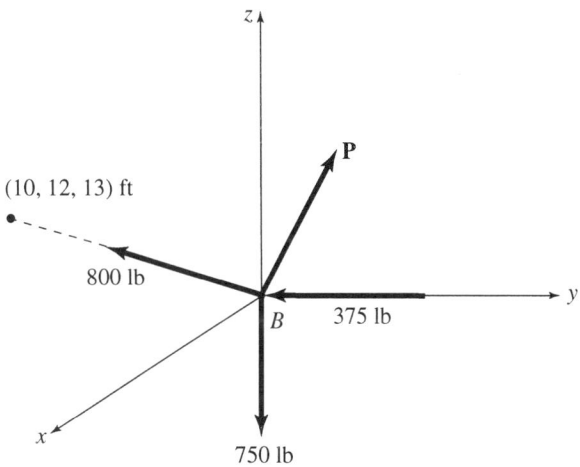

FIGURE P3.42.

3.42 If the particle at point B is in equilibrium under the action of the concurrent forces shown in Figure P3.42, find the magnitude and direction of the unknown force **P**.

3.43 The frame shown in Figure P3.43 consists of three short links capable of carrying tension or compression joined together by balls and sockets. If the force applied at joint D is $\mathbf{F} = (-5\mathbf{i} + 3\mathbf{j} + 7\mathbf{k})$ kN, determine the forces in the three links. Note that the force in a short link acts along its axis.

3.44 The frame shown in Figure P3.44 consists of three short links capable of carrying tension or compression joined together by balls and sockets. Determine (a) the forces in the three links knowing that the

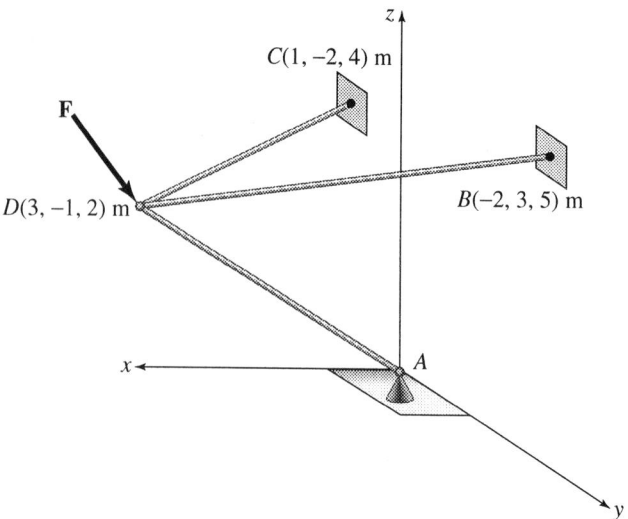

FIGURE P3.43.

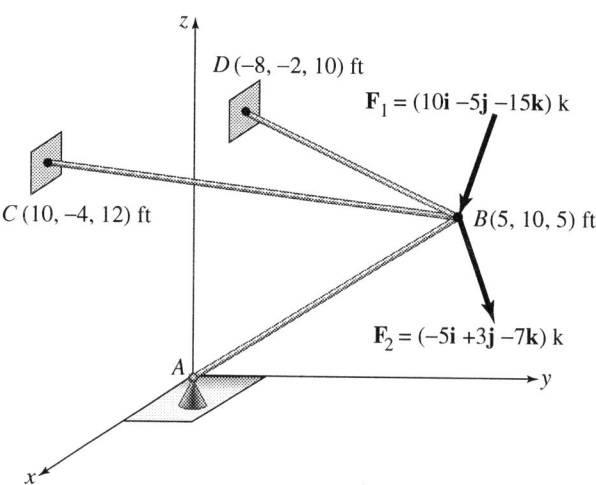

FIGURE P3.44.

force in a link acts along its axis and (b) the x, y, and z components of the reaction at support A.

3.45 A package of mass m is supported by two wires AB and AC and a spring AD as shown in Figure P3.45. In the position shown, the stretch of the spring was mea-

sured as 0.425 m. If the spring constant is $k = 900$ N/m, determine the mass m and the tension in the two wires.

3.46 The system shown in Figure P3.46 consists of four ropes attached together at A. Rope AE passes over a small frictionless pulley at E and a pull of 50 lb is applied

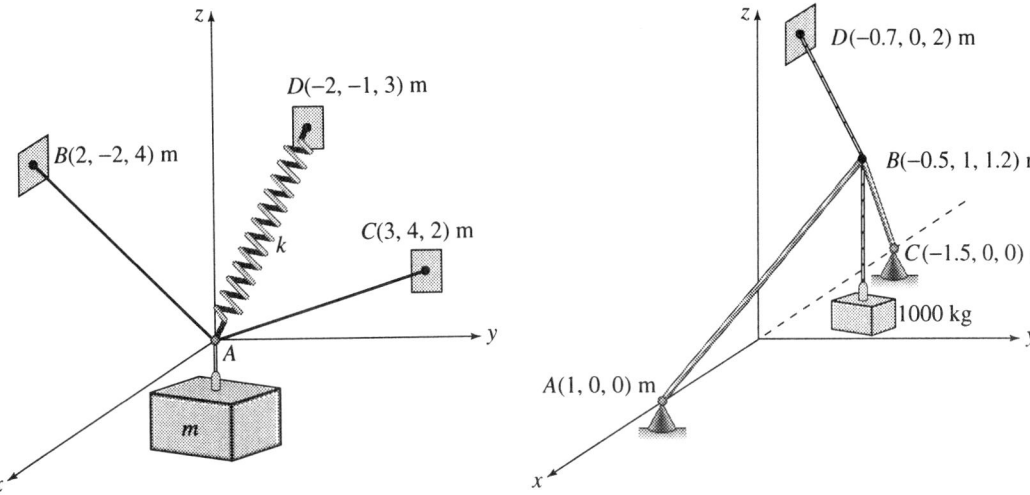

FIGURE P3.45.

FIGURE P3.47.

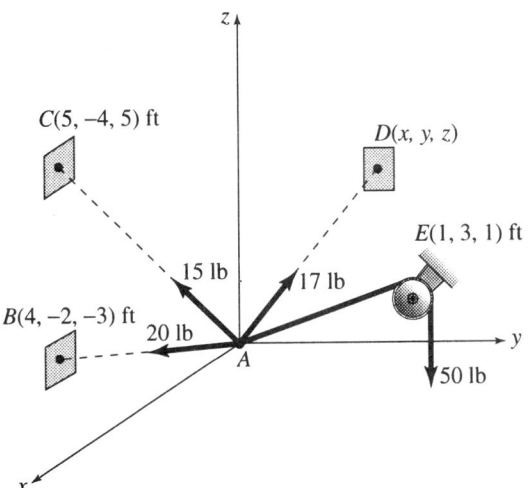

FIGURE P3.46.

at the end. The tensions in ropes AB, AC and AD are 20 lb, 15 lb, and 17 lb respectively. The length of rope AD is 5.0 ft. Find the coordinates of point D.

3.47 The structure shown in Figure P3.47 consists of two links AB and BC and a cable BD. A 1000 kg mass is suspended as shown. Assume that all joints and supports are balls and sockets and determine (a) the forces in links AB and BC and in cable BD. The loads in all the members may be assumed to act along their respective axes. Also, links AB and BC are capable of carrying tension or compression. (b) the x, y, and z components of the reaction at the ball-and-socket support at C.

3.48 A crate of weight W is suspended as shown in Figure P3.48. If the tension in cable BD is 700 lb, determine the weight W and the forces in members BA and BC. Note that forces in links and cables act along the axes of these members. Assume a ball and socket at all joints and supports.

3.49 The system shown in Figure P3.49 consisting of links AB and BD and cable BC is used to support a crate of mass m. (a) If the compression in cable BA is 5.0 kN, find the mass m and the forces in members BD and BC. Note that forces in links and cables act along the axes of these members. (b) Assume a ball and socket at all joints and supports, and determine the

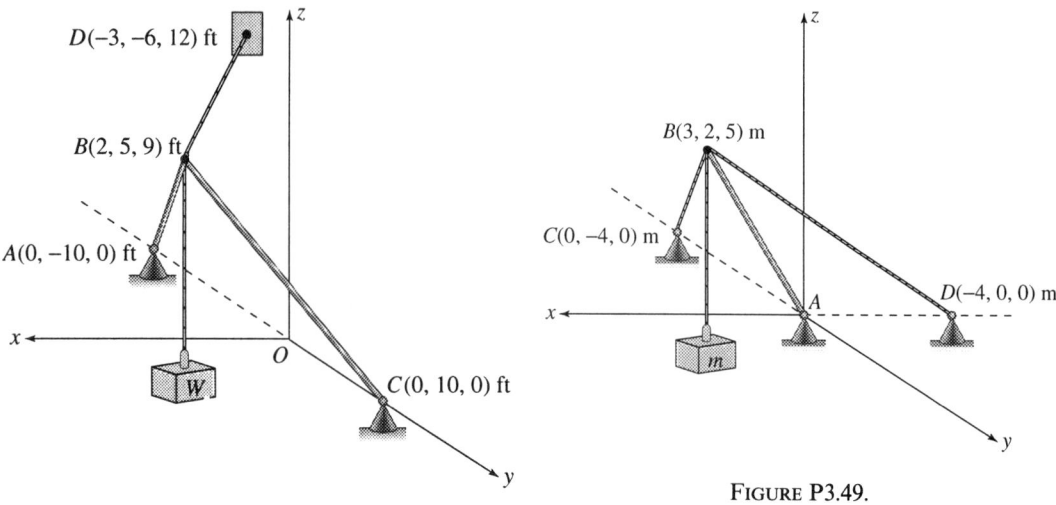

$D(-3, -6, 12)$ ft

$B(2, 5, 9)$ ft

$A(0, -10, 0)$ ft

x

O

W

$C(0, 10, 0)$ ft

y

FIGURE P3.48.

$B(3, 2, 5)$ m

$C(0, -4, 0)$ m

x

A

$D(-4, 0, 0)$ m

m

y

FIGURE P3.49.

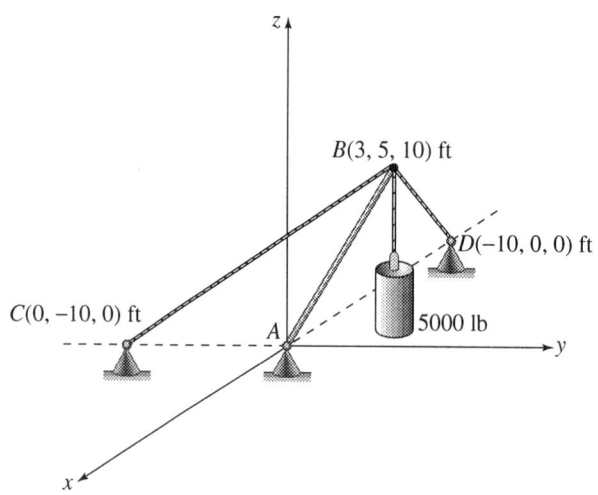

$B(3, 5, 10)$ ft

$D(-10, 0, 0)$ ft

$C(0, -10, 0)$ ft

A

5000 lb

y

x

FIGURE P3.50.

$x, y,$ and z components of the reaction at support A.

3.50 The structure shown in Figure P3.50, consisting of a link AB and two cables BC and BD, is used to support a load of 5000 lb. Assume a ball and socket at all joints and supports, and determine the forces in AB, BC, and BD. Note that

forces in links and cables act along the axes of these members.

3.51 In the position shown in Figure P3.51, the spring is stretched by 0.05 m. If the magnitude of the force **F** is 30 kN, determine the spring constant k and the forces in cables AB and AC. Assume that the pulley at D is very small and frictionless.

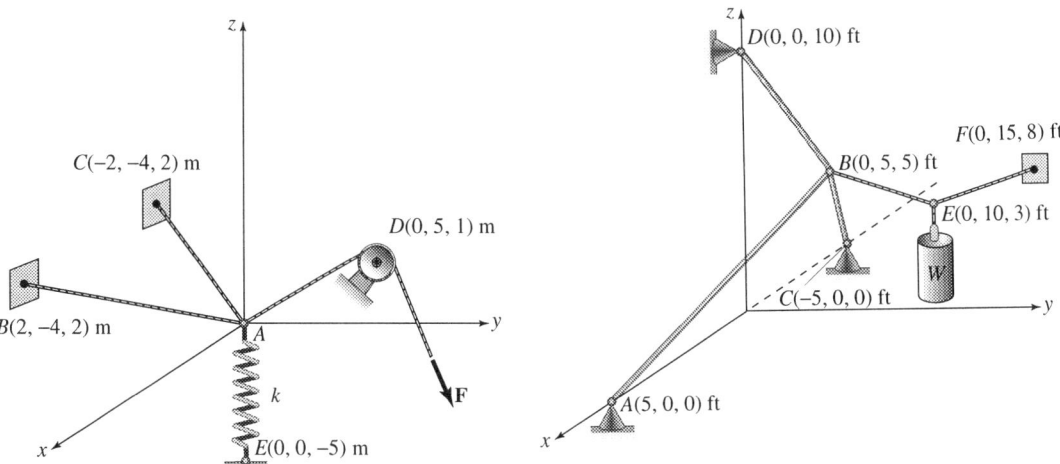

FIGURE P3.51.

FIGURE P3.52.

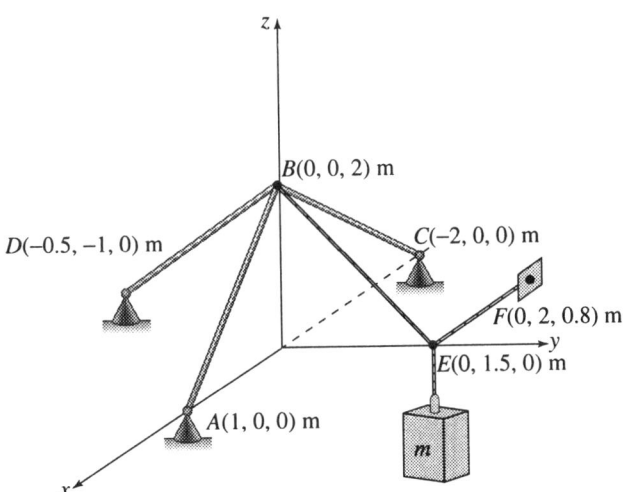

FIGURE P3.54.

3.52 The tripod, consisting of three links BA, BC, and BD, supports cables BE and EF which, in turn, support a load $W = 2000$ lb, as shown in Figure P3.52. Note that forces in links and cables act along their respective axes and that links can support either tension or compression. Determine (a) the forces in cables EF and EB and (b) the forces in the three links BA, BC, and BD.

3.53 Refer to the system shown in Figure P3.52. The force in link BD is 950 lb tension. Determine the weight W. Note that forces in links and cables act along their respective axes and that links can support either tension or compression.

3.54 The tripod, consisting of three links BA, BC, and BD, supports cables BE and EF which in turn support a mass $m = 1200$ kg, as shown in Figure P3.54. Note that forces in links and cables act along their respective axes and that links can support either tension or compression. Determine (a) the forces in cables EF and EB and (b) the forces in the three links BA, BC, and BD.

3.55 Refer to the system shown in Figure P3.52. The force in link BA is 1.75 kN tension. Determine the mass m of the weight W. Note that forces in links and cables act along their respective axes and that links can support either tension or compression.

Review Problems

3.56 The force $\mathbf{F}$ shown in Figure P3.56 has a magnitude of 150 lb. (a) Determine the scalar components of the force along the x, y, and z directions. (b) Find the angles α, β, and γ made by the force $\mathbf{F}$ with the x, y, and z axes, respectively.

3.57 A structural member OA is supported partially by cable AB, as shown in Figure P3.57. (a) If the tension in the cable is 1.75 kN, determine the component of this cable tension along the structural member. Note that the tension in the cable at A is directed from A to B. (b) Find the angle at B made by the cable with line BO.

3.58 The pipe assembly is supported partially by cable BC as shown in Figure P3.58. (a) The component of the cable tension along segment AB of the pipe is not to exceed 150 lb. Determine the maximum tension in the cable. Note that the tension in cable BC at B is directed from B to C. (b) Find the angle at C made by cable BC with line AC.

3.59 Let $\mathbf{F}_1 = (4\mathbf{i} + F_{1y}\mathbf{j} + 7\mathbf{k})$ kN, $\mathbf{F}_2 = (F_{2x}\mathbf{i} + 10\mathbf{j} - 12\mathbf{k})$ kN and $\mathbf{F}_3 = (-8\mathbf{i} - 6\mathbf{j} + F_{3z}\mathbf{k})$ kN. (a) If the resultant of the three forces $\mathbf{R} = 20\mathbf{i}$, find F_{1y}, F_{2x},

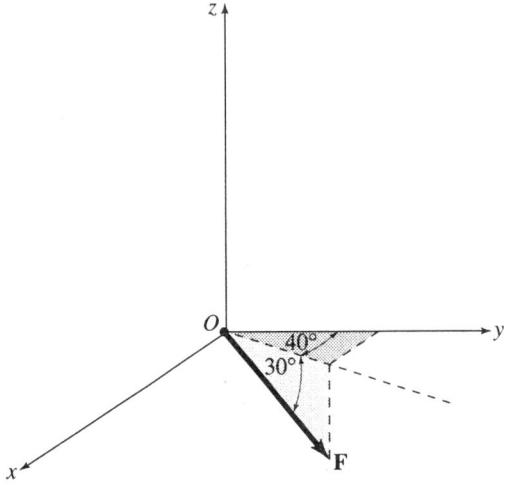

FIGURE P3.56.

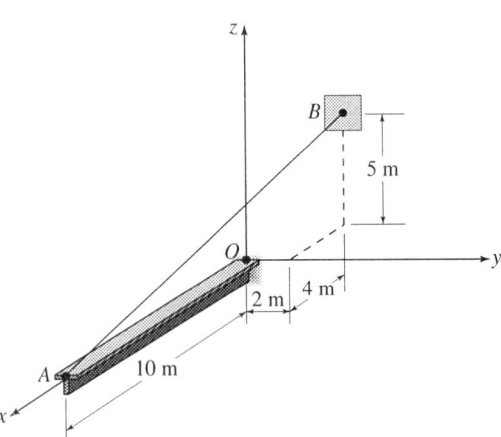

FIGURE P3.57.

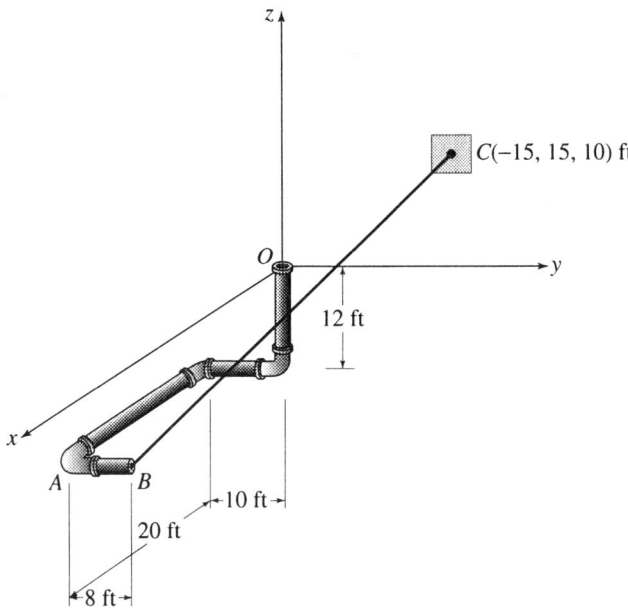

FIGURE P3.58.

and F_{3z}. (b) If the resultant of the three forces $\mathbf{R} = (-4\mathbf{i} + 3\mathbf{j} - 15\mathbf{k})$ kN, find F_{1y}, F_{2x}, and F_{3z}.

3.60 The three forces shown in Figure P3.60 have the following magnitudes: $F_1 = 10$ k, $F_2 = 15$ k, and $F_3 = 20$ k. If the resultant of the three forces $\mathbf{R} = (-8\mathbf{i} - 5\mathbf{j} + 30\mathbf{k})$ k, determine the angles α_3, β_3, and γ_3 defin-

ing the direction of $\mathbf{F}_3$ with respect to the x, y, and z axes, respectively.

3.61 An automobile with a mass of 1600 kg is lifted at constant speed by a crane as shown in Figure P3.61. The crane applies

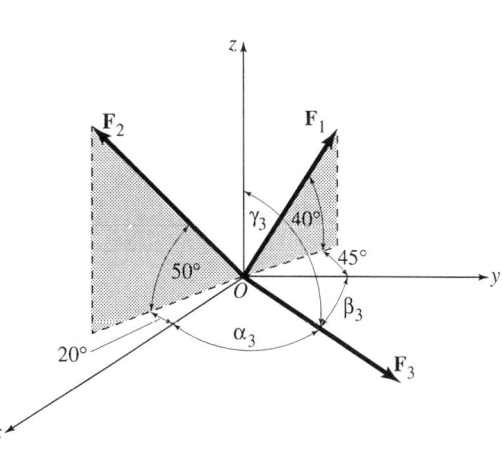

FIGURE P3.60.

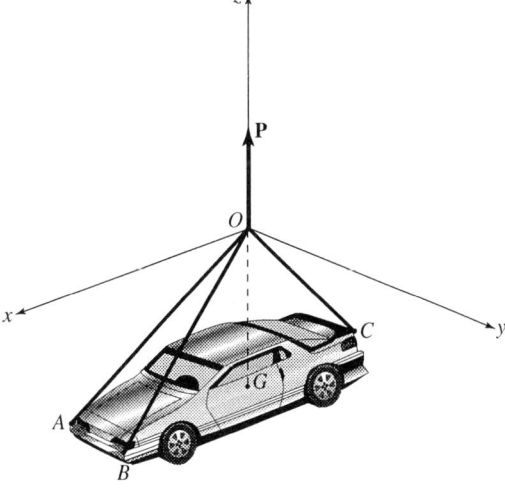

FIGURE P3.61.

a force P at the point of attachment of the three cables OA, OB, and OC, as shown. At the instant shown, the co-ordinates of points A, B, and C are A(3.0, −0.7, −5.5) m, B(3.0, 0.7, −5.5) m, and C(−2.5, 0.0, −3.0) m. Determine the tensions in cables OA, OB, and OC.

3.62 The cable system shown in Figure P3.62 is used to support the crate which has a weight $W = 500$ lb. Cable BAD is con-tinuous through the smooth ring at A whereas cables AC and AE are attached to the ring. Determine the tension in each of the four cables.

3.63 Collar A may slide on a smooth member AC that is parallel to the x axis, and col-lar B may slide on a smooth member OB that lies along the y axis, as shown in Figure P3.63. Rod AB of length $L = 3.00$ m is attached to both collars by ball-and-socket joints. Forces $P_A = 2.75$ kN and P_B maintain the system in equilib-rium in the position shown. Determine the force P_B and the force in rod AB. Also determine the forces that members AC and OB exert on their respective col-lars. Note that, because these members are smooth, the forces they exert on the collars must necessarily be perpendicular to the members.

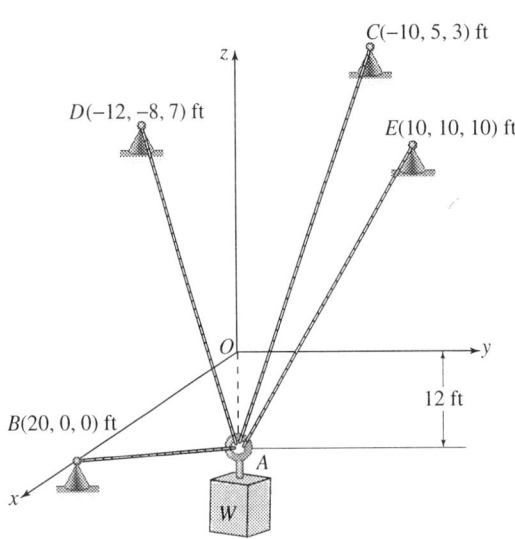

FIGURE P3.62.

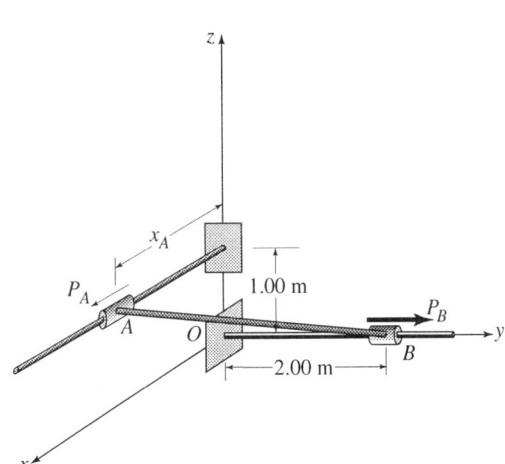

FIGURE P3.63.

4

Equilibrium of Rigid Bodies in Two Dimensions

This crane is one of many examples that can be analyzed as a rigid body in two dimensions.

A rigid body *may be viewed as an arrangement of an infinitely large number of particles whose positions in space are fixed relative to one another. In other words, a rigid body is one with finite dimensions that do not change under the action of applied forces. Consequently, when loaded, rigid bodies do not experience dimensional changes known as* deformations. *A real body does deform when subjected to forces. The degree and type of deformation depends upon the nature of the body and the magnitude and type of forces acting on it. However, for most bodies of engineering interest, such deformations are relatively small and may be ignored without much effect on the conditions of equilibrium or on the conditions relating to their motion. Thus, the assumption of rigidity of bodies is justified not only in statics but also in dynamics.*

In Chapter 2, the conditions of equilibrium were discussed for a particle in two dimensions. Of necessity, the forces acting on a particle are concurrent. In the case of a rigid body, however, the forces acting do not necessarily have to be concurrent because of the finite dimensions of rigid bodies. This condition of noncurrency of forces requires the introduction of the concept of the moment *of a force which is done in Section 4.1. The distinction between* internal *and* external *forces is given in Section 4.2 and, in Sections 4.3 to 4.6, methods are developed to replace complex force systems by a single force. Our knowledge of supports and connections is expanded in Section 4.7, and the concept of the free-body diagram for a rigid body is discussed in Section 4.8. Finally, the conditions of equilibrium are developed in Section 4.9 where they are applied in solving two-dimensional problems in equilibrium.*

4.1
Concept of the Moment-Scalar Approach

Moment of a Force

As a consequence of the application of a force to a rigid body, the rigid body experiences a tendency to rotate about axes which neither intersect nor are parallel to the line of action of the force. The rotating tendency produced by a force about a given axis is known as the *moment* of the force about this axis. The magnitude of this rotating tendency about an axis depends upon the position of the force with respect to the axis. Thus, consider the force **F** applied at A on a rigid body as shown in Figure 4.1(a). The axis *a-a* is any axis that does not intersect the line of action of the force **F** and is not parallel to it. The moment of the force **F** about axis *a-a* is defined as a vector **M** whose magnitude M is the product of the magnitude of the force F and *the perpendicular distance d* from the line of action of the force to axis *a-a*, as shown in Figure 4.1(b) by the curved arrow. Thus

$$M = Fd. \tag{4.1}$$

The distance d is known as the *moment arm* of the force **F**. The direction of the moment vector **M** is normal to the plane defined by the line

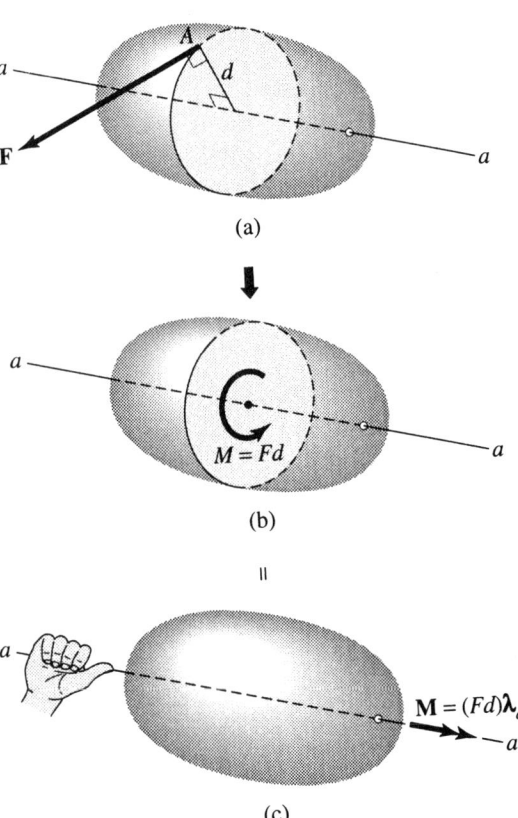

FIGURE 4.1.

of action of **F** and the perpendicular distance *d*. The sense of this moment is established by the *right-hand rule*. Application of this rule requires that the four fingers of the right hand be curled in the direction of the rotating tendency, in which case the thumb would point in the correct sense of the vector **M** as shown in Figure 4.1(c). *Note that a double-headed arrow is used for moments in this text to distinguish them from forces which are represented by single-headed arrows.* The sense of the moment represented by a double-headed arrow is obtained by the use of the right-hand rule. This is accomplished by placing the thumb of the right hand in the direction of the arrows in which case the curled four fingers point in the correct sense of the moment. This technique eliminates ambiguities particularly when dealing with three-dimensional diagrams of force systems.*

Because a vector may be written as the product of its magnitude and a unit vector in its direction, the vector **M** may be expressed in the form

$$\mathbf{M} = (Fd)\lambda_a \qquad (4.2)$$

where λ_a is a unit vector along the axis *a-a* with a sense determined by the right-hand rule.

The rotating tendency produced by the force **F** lies in the plane defined by its line of action and the perpendicular distance *d*. The rotating tendency, the moment **M**, may be depicted in this plane by a curved arrow representing both the sense and the magnitude of the moment **M** as shown in Figure 4.1(b). Note that the sense of the moment agrees with the direction in which the four fingers of the right hand are curled when establishing the sense of vector **M**. The magnitude of the moment *M*, given by Eq. (4.1), has units of force multiplied by units of length. Thus, such units as k·ft and lb·in. are used in the U.S. Customary system and units, such as kN·m and N·m, are common in the SI system.

The use of the curved arrow to represent the direction and sense of a moment is a very convenient technique when dealing with two-dimensional force systems acting on rigid bodies. Consider, for example, the wrench that is used to turn the hex nut as shown in Figure 4.2. The force has a magnitude of $F = 20$ lb applied to the wrench so that its line of action is at a distance $d = 5$ in. from point A, which represents the axis of the bolt as shown in Figure 4.2(a). Therefore, the moment of this force about the bolt axis with a magnitude $M = (20)(5) = 100$ lb·in. tends to rotate the hex nut in a ccw direction. This

---

* This same method is used in the companion book "Dynamics for Engineers" to distinguish between such quantities as linear velocity (single-headed arrow) and angular velocity (double-headed arrow).

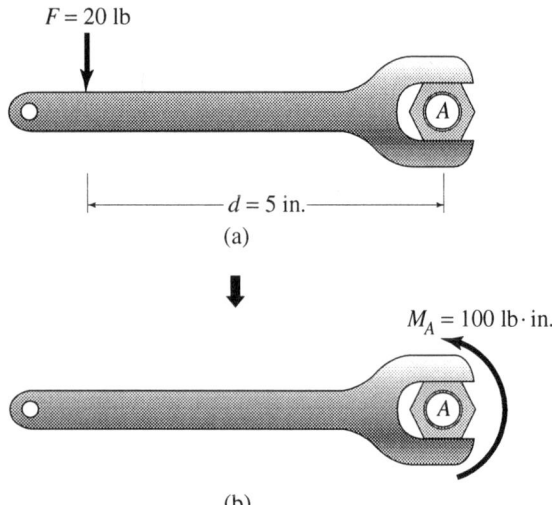

FIGURE 4.2.   (b)

fact may be represented symbolically by a curved arrow. Thus, as shown in Figure 4.2(b), the notation $M_A = 100$ lb·in. ↻ signifies that the moment acting on the hex nut about the bolt axis through point A has a magnitude of 100 lb·in. and a ccw sense. The right-hand rule applied to the curved arrow indicates that the vector representing this moment is perpendicular to the page and, in this case, points out of the page. Thus, the curved arrow, used in the two-dimensional case, signifies not only the sense of the moment but its direction as well.

Varignon's Theorem

A French mathematician named Varignon (1654–1722) introduced a theorem that carries his name and is also known as the *principle of moments. Varignon's theorem,* or *the principle of moments,* is a very useful concept in solving problems relating to moments of forces. It states that *the moment of a force about an axis is equal to the sum of the moments of the components of this force about the same axis.* The proof of this statement is given here for the two-dimensional case and in Chapter 5 for the three-dimensional case.

Consider the force F shown in Figure 4.3 in relation to an x-y coordinate system. The line of action of the force passes through point A (x, y). The perpendicular distance from the z axis through point O to the line of action of F is d. The magnitudes of the x and y components of the force F shown acting at point A are $F \sin \theta$ and $F \cos \theta$, respectively. The moment of F about the z axis through point O is expressed as

$$M_O = Fd \; \curvearrowleft.$$

Also, the sum of the moments of the two components of F about the z

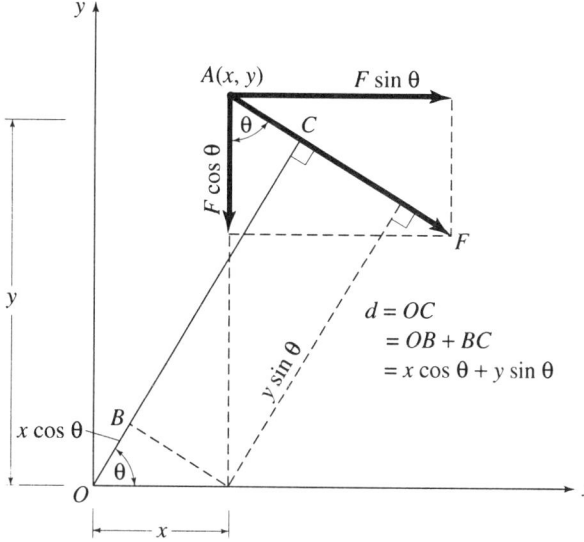

FIGURE 4.3.

axis through point O is given by

$$M_O = (F\cos\theta)x + (F\sin\theta)y = F(x\cos\theta + y\sin\theta) \quad \circlearrowleft$$

where the direction and sense of the moments were found by the right-hand rule. However,

$$x\cos\theta = OB,$$

and

$$y\sin\theta = BC.$$

Therefore,

$$x\cos\theta + y\sin\theta = OB + BC = OC = d$$

It follows, therefore, that the sum of the moments of the components of F about the z axis through point O is identical to the moment of F about the same axis which, of course, proves Varignon's theorem or the principle of moments. As a general rule, the solution of problems dealing with moments of forces about axes is simpler if this theorem is employed as opposed to the direct use of Eq. (4.1).

The following examples illustrate some of the above concepts.

■ Example 4.1

Determine the moment produced by the force $P = 20\,\text{kN}$ acting on the post as shown in Figure E4.1(a) about (a) an axis through point A and (b) an axis through point B. Note that both axes are perpendicular to the page.

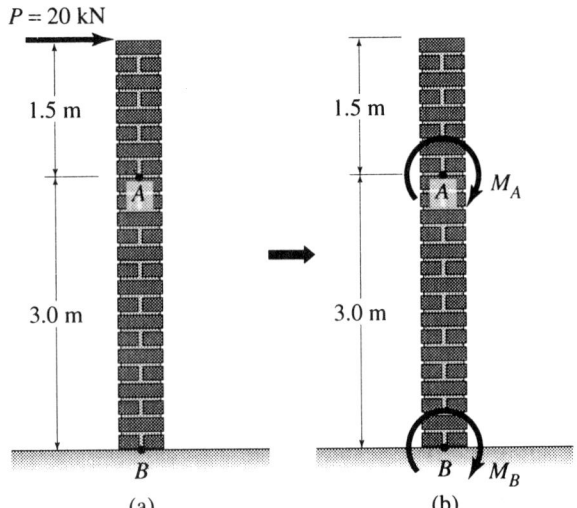

FIGURE E4.1.   (a)   (b)

Solution

(a) The magnitude of the moment of P about an axis through point A perpendicular to the page is given by Eq. (4.1) where the perpendicular distance from the axis through A to the line of action of the force (the moment arm) is $d = 1.5$ m. Thus,

$$M = Fd,$$

$$M_A = (20)(1.5) = 30.0 \text{ kN·m}.$$

The direction and sense of this moment are obtained by applying the right-hand rule. This leads to the conclusion that the moment is cw as shown in Figure E4.1(b). Therefore, the moment M_A may be completely specified by

$$M_A = 30.0 \text{ kN·m} \, \circlearrowright.$$ ANS.

(b) The magnitude of the moment of P about an axis through point B perpendicular to the page is given by Eq. (4.1) where the moment arm is $d = 4.5$ m. Thus,

$$M = Fd,$$

$$M_B = (20)(4.5) = 90.0 \text{ kN·m}.$$

The direction and sense of this moment are obtained by applying the right-hand rule as in part (a) above and the moment M_B may be fully defined by

$$M_B = 90.0 \text{ kN·m} \, \circlearrowright.$$ ANS.

This moment is also shown in Figure E4.1(b).

■ **Example 4.2**

Consider the truss in Figure E4.2(a) which is subjected to a force of 15 k as shown. Find the moment of this force about an axis through point A perpendicular to the page (a) using Eq. (4.1) and (b) using Varignon's theorem.

Solution

The magnitude of the moment produced by the 15-k force about an axis through point A may be found by Eq. (4.1), $M = Fd$, where $F = 15$ k and d is the perpendicular distance from point A to the line of action of the force F as shown in Figure E4.2(a). The distance d, the moment arm, is determined as follows:

$$\alpha = \tan^{-1}\frac{32}{24} = 53.13°$$

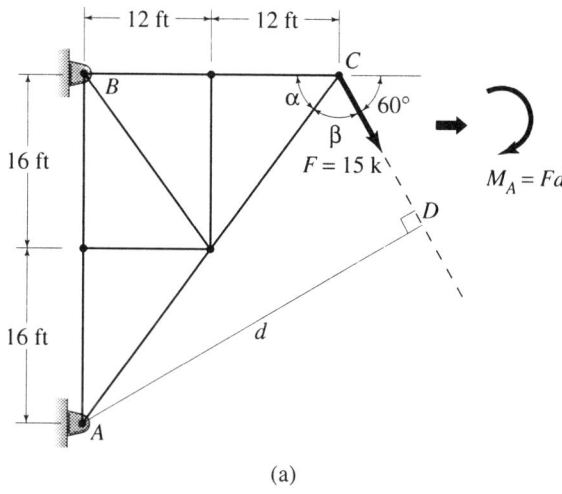

(a)

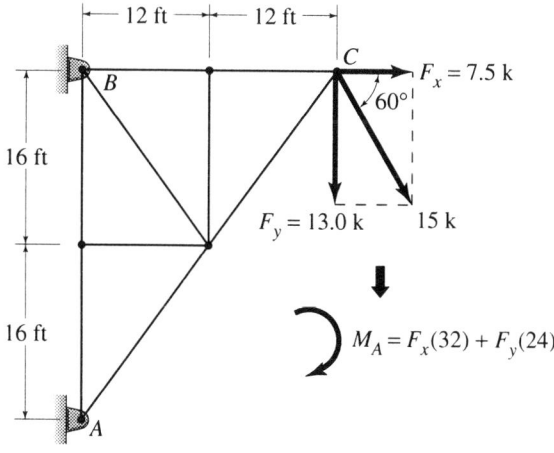

(b)

FIGURE E4.2.

$$\beta = 180 - 60 - 53.13 = 66.87°$$

Using triangle ACD, where $AC = \sqrt{24^2 + 32^2} = 40$ ft,

$$d = AD = 40 \sin 66.87° = 36.8 \text{ ft.}$$

Therefore,

$$M_A = (15)(36.8) = 552 \text{ k·ft.}$$

The direction and sense of this moment are obtained by the right-hand rule which leads to the conclusion that the moment about an axis at A is cw, as shown in Figure E4.2(a). Thus, the moment M_A may be completely defined by stating that

$$M_A = 552 \text{ k·ft } \circlearrowright.$$ ANS.

(b) The moment of the 15-k force about the perpendicular axis through point A is conveniently obtained by the use of Varignon's theorem. Thus, the 15-k force is decomposed into the two rectangular components $F_x = 15 \cos 60° = 7.5$ k and $F_y = 15 \sin 60° = 13.0$ k as shown in Figure E4.2(b). Therefore, by Varignon's theorem,

$$M_A = (7.5)(32) + (13.0)(24) = 240 + 312 = 552 \text{ k·ft } \circlearrowright$$

which, of course, is the same as obtained in part (a) of the solution. Note that the direction and sense of the moments above were obtained by the right-hand rule.

■

Moment of a Couple

A *couple* is a special type of moment that is created by the action of two parallel forces which are equal in magnitude, opposite in sense, and *noncollinear*. The word *noncollinear* signifies that the two parallel forces do not have the same line of action. These two parallel forces define a plane known as the *plane of the couple*. Thus, the two parallel forces **F** and $-$**F** in Figure 4.4 constitute a couple because they have the same magnitude F, are opposite in sense, and are noncollinear. Because the two forces of a couple **F** and $-$**F** yield a resultant force which is zero, it follows that a couple produces only a tendency to rotate without producing a tendency to translate. Like the moment of a force, the moment of a couple is a vector and, therefore, has magnitude, direction, and sense.

The direction and sense of the moment of a couple are determined by the right-hand rule in the same manner as for the moment of a force. Thus, for example, the right-hand rule leads us to conclude that the direction and sense of the couple in Figure 4.4 would be given by a cw curved arrow. In other words, the right-hand rule shows that the vector representing this couple would be perpendicular to the page and

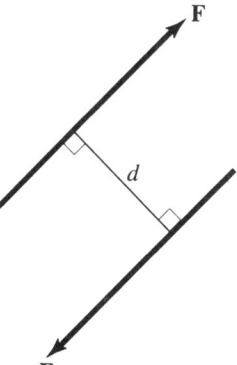

FIGURE 4.4.

would point into it. The magnitude of the moment of the couple may be obtained by algebraically adding the magnitudes of the moments produced by the two forces of the couple. This procedure shows that the magnitude of the moment of a couple is *the same about any axis perpendicular to its plane and is equal to the magnitude F of the two forces in the couple multiplied by the perpendicular distance between them.* Thus,

$$M = Fd \qquad (4.3)$$

where M represents the magnitude of the moment of the couple and d is the perpendicular distance between the two forces. To prove this statement, consider the couple shown in Figure 4.5. An axis perpendicular to the plane of the couple is chosen through an arbitrary point O. Because the two forces are parallel to one another, it follows that the perpendicular directions from point O to these two forces coincide. Thus, in Figure 4.5, direction OAB is perpendicular to both forces and, whereas OA represents the perpendicular distance to $-\mathbf{F}$, OB represents the perpendicular distance to $\mathbf{F}$. Assuming that a ccw moment is positive,* the magnitude of the moment of the couple is found as follows:

$$\text{For } \quad \mathbf{F}, \qquad (Mo)_1 = F(\text{OB}).$$

$$\text{For } \quad -\mathbf{F}, \qquad (Mo)_2 = -F(\text{OA}).$$

Therefore, the net moment M produced by the two forces of the couple is

$$M = (Mo)_1 + (Mo)_2 = F(\text{OB}) - F(\text{OA}) = F[(\text{OB}) - (\text{OA})] = Fd \; \circlearrowleft$$

---

* This convention is consistent with the right-hand rule.

The positive sign of the answer indicates that the moment of the couple is ccw (as represented by the ccw curved arrow) which, of course, is in agreement with the direction and sense obtained by the right-hand rule. Also, the above result shows that the *magnitude* of the moment of a couple is independent of the moment axis and is only a function of the magnitude of the forces of the couple and the perpendicular distance between them. Thus, a couple may be looked upon as a *free vector*. These important conclusions will be generalized in Chapter 5 where couples in three dimensions are discussed. However, in the two-dimensional case, a couple may be expressed in vector form by Eq. (4.2). Thus, using Eq. (4.2), the couple represented in Figure 4.5, for example, may be written in the form $\mathbf{M} = (Fd)\mathbf{k}$.

As free vectors, two or more couples may be added by the law of the parallelogram or by the use of rectangular components to obtain their resultant. Determining the resultant of three-dimensional couples will be discussed in Chapter 5. In the two-dimensional case, however, two or more couples that act in the same or in parallel planes are represented by vectors normal to these planes and, therefore, are parallel to one another. Thus, such two-dimensional couples may be represented by their magnitudes along with curved arrows which are either ccw (positive) or cw (negative), and their resultant may be obtained by a simple algebraic addition. Alternatively, the vector representation may be employed. For example, if all of the couples to be added act in the same or in parallel x-y planes, then, they may be represented by expressions of the form $\mathbf{M} = M\mathbf{k}$ or $\mathbf{M} = -M\mathbf{k}$ and the resultant $\mathbf{M_R}$ obtained by a vector addition. Thus,

$$\mathbf{M_R} = \sum \mathbf{M}. \qquad (4.4)$$

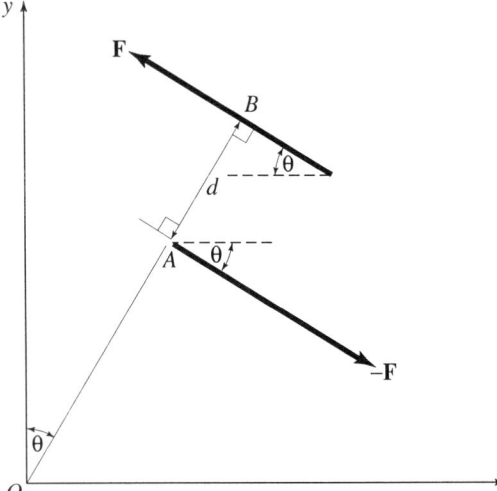

FIGURE 4.5.

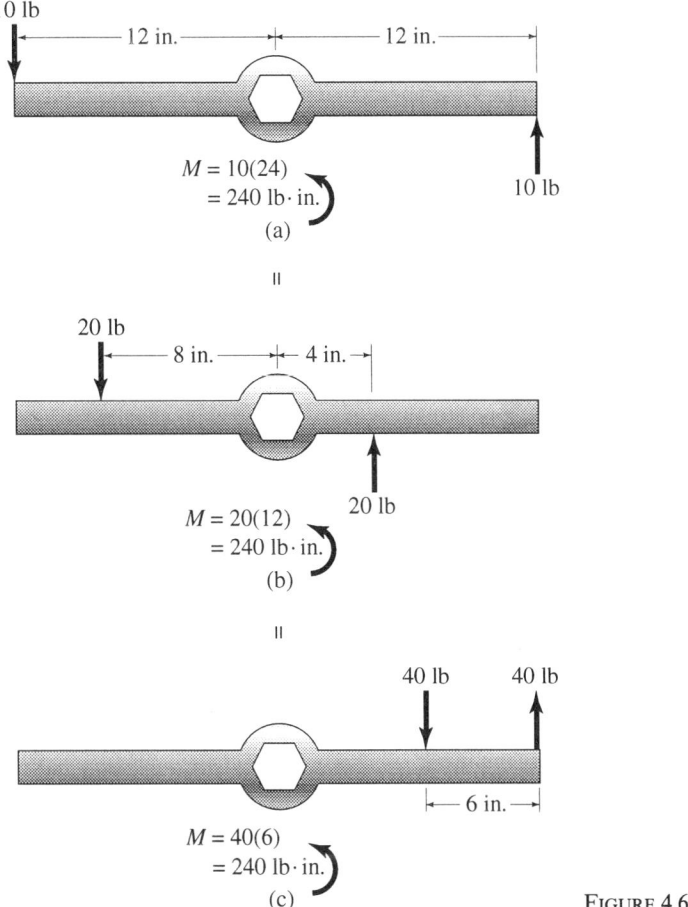

FIGURE 4.6.

Couples acting on a rigid body are said to be *equivalent* if they produce the same effect on the rigid body. Thus, equivalent couples must have the same magnitude, direction, and sense. Therefore, the three couples shown in Figure 4.6 are equivalent because each has a magnitude of 240 lb·in. and a direction and sense that may be represented by a ccw curved arrow. To be equivalent, however, couples do not have to act in the same plane. Couples acting in parallel planes are equivalent if they possess the same magnitude, direction, and sense. Thus, the two couples shown in Figure 4.7, acting in different but obviously parallel x-y planes, are equivalent, even though they do not act in the same plane. Each has a magnitude of 10 kN·m and a direction and sense that may be represented by a cw curved arrow or by a double-headed vector pointed in the negative z direction.

The following examples illustrate some of the concepts discussed

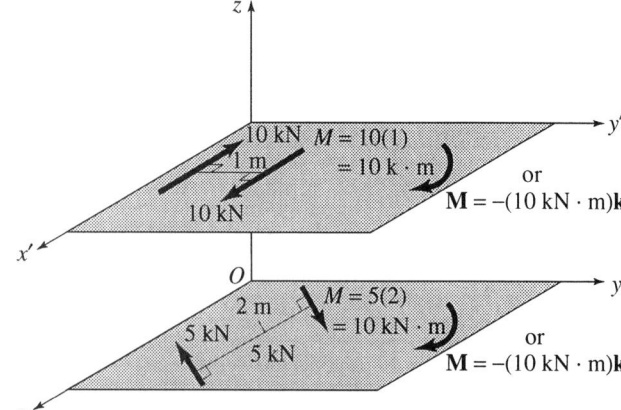

FIGURE 4.7.

above dealing with the two-dimensional treatment of the moment of a couple.

■ **Example 4.3**

Consider the couple shown in Figure E4.3, and determine its magnitude, direction, and sense. Express the answer as a magnitude along with a curved arrow.

Solution

This problem will be solved in two different ways as follows:

(a) The magnitude of the moment of the couple may be determined from its basic definition expressed by Eq. (4.3), $M = Fd$, where $F = 500$ N and d is determined with the aid of Figure E4.3. Thus,

$$GD = 3 \tan 20° = 1.092 \text{ m.}$$

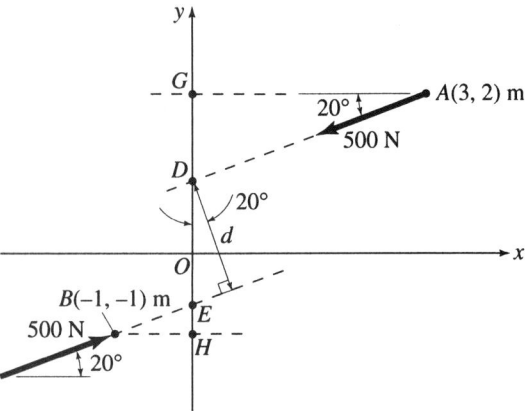

FIGURE E4.3.

Therefore,

$$OD = 2 - 1.092 = 0.908 \text{ m}.$$

Also,

$$HE = 1.0 \tan 20° = 0.364 \text{ m},$$

and

$$OE = 1.0 - 0.364 = 0.636 \text{ m}.$$

Thus, the perpendicular distance d between the lines of action of the two forces is given by

$$d = (OD + OE) \cos 20° = (0.908 + 0.636) \cos 20° = 1.451 \text{ m}$$

and

$$M = Fd = 500(1.451) = 726 \text{ N·m} \circlearrowleft \qquad \text{ANS.}$$

where the direction and sense of the couple, indicated by the ccw curved arrow, were obtained by applying the right-hand rule.

(b) The magnitude of the moment produced by the couple may be determined by selecting any convenient axis perpendicular to the plane of the couple and using the basic definition of the moment of a force. This is possible because of the conclusion reached earlier that the moment of a couple is constant and does not depend upon the axis of rotation. Thus, if an axis through point O is selected and if Varignon's theorem is used,

$$M = (500 \cos 20°)2 - (500 \sin 20°)3 + (500 \cos 20°)1 - (500 \sin 20°)1$$

$$= 939.7 - 513.0 + 469.8 - 171$$

$$= 726 \text{ N·m} \circlearrowleft. \qquad \text{ANS.}$$

which, of course, is identical to that obtained in part (a) of the solution. Again, the ccw curved arrow indicates the direction and sense of this couple which were obtained by the right-hand rule. Note that, instead of the two dimensional scalar notation of a magnitude with a curved arrow, the answer may be represented in vector form as

$$\mathbf{M} = (726\mathbf{k}) \text{ N·m} \qquad \text{ANS.}$$

which has the same physical significance when interpreted according to the right-hand rule.

Both of the methods illustrated in parts (a) and (b) above are acceptable and viable. However, under certain conditions, one may have advantages over the other.

■ **Example 4.4**

Two 50-lb forces are applied to a shaft as shown in Figure E4.4. Determine the magnitude of (a) the two forces F and (b) the two forces Q, that will produce the same moment as the two 50-lb forces.

Solution

The two applied 50-lb forces constitute a couple in the x-y plane whose magnitude $M = 50(20) = 1000$ lb·in. The direction and sense of this couple are obtained from the right-hand rule and may be represented by a cw curved arrow when looking toward the origin of the x-y-z coordinate system along the z axis. Alternatively, this couple may be represented in vector form by the expression

$$\mathbf{M} = (-1000\mathbf{k}) \text{ lb·in.} \tag{a}$$

(a) Similarly, the couple $\mathbf{M}^F$, produced by the two forces $\mathbf{F}$ acting in the same x-y plane as the 50-lb forces, may be written in the form

$$\mathbf{M}^F = (-5F\mathbf{k}) \text{ lb·in.}$$

where the perpendicular distance between the two forces $\mathbf{F}$ is the 5-in. diameter of the upper shaft. Because $\mathbf{M}^F$ must be the same as $\mathbf{M}$, it follows that

$$5F = 1000$$

from which

$$F = 200 \text{ lb.} \qquad \text{ANS.}$$

(b) The couple $\mathbf{M}^Q$, produced by the two forces $\mathbf{Q}$ acting in the x'-y' plane which is parallel to the plane of the 50-lb forces, may be expressed in the form

$$\mathbf{M}^Q = (-10Q\mathbf{k}) \text{ lb·in.}$$

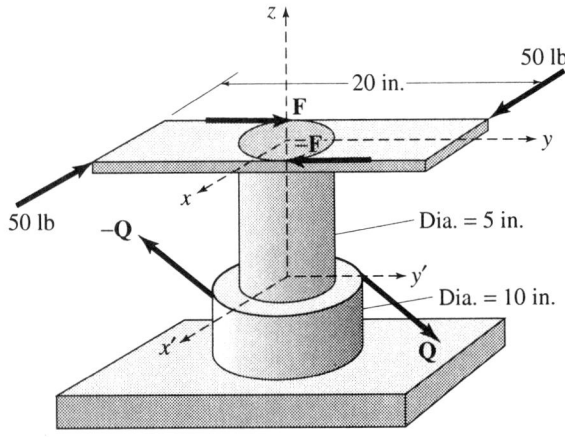

FIGURE E4.4.

where the perpendicular distance between the two forces **Q** is the 10 in. diameter of the lower shaft. Because $\mathbf{M}^Q$ must be equal to **M**, it follows that

$$10Q = 1000$$

from which

$$Q = 100 \text{ lb.} \qquad\qquad \text{ANS.}$$

In summary, the couple produced by the two **Q** forces and that produced by the two **F** forces is equivalent to the couple produced by the two 50 lb forces because they all have the same magnitude, direction, and sense even though they act in different planes.

Problems

4.1 (a) Use the basic definition of moment to find the moment produced by the 200-N force in Figure P4.1 about an axis perpendicular to the page at point O. (b) Use Varignon's theorem to find the moment described in part (a).

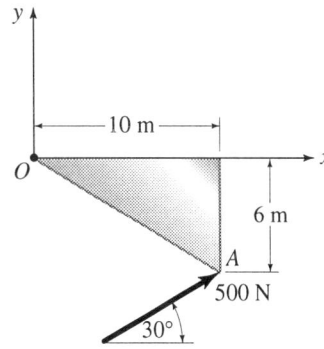

FIGURE P4.2.

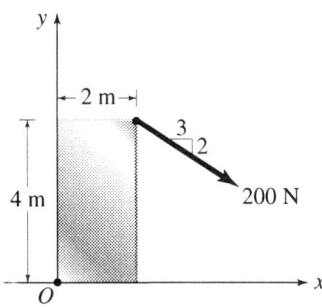

FIGURE P4.1.

4.2 (a) Use the basic definition of moment to find the moment produced by the 500-N force in Figure P4.2 about an axis perpendicular to the page at point O. (b) Use Varignon's theorem to find the moment described in part (a).

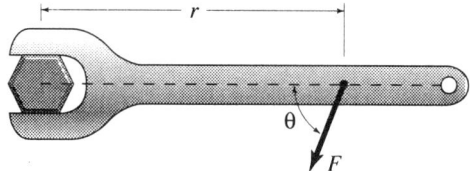

FIGURE P4.3.

4.3 Consider the wrench in Figure P4.3, and use the basic definition of moment (a) to find the moment of $F = 20$ lb about the axis of the hex nut if $\theta = 60°$ and $r = 20$ in. (b) find θ for the maximum moment about the axis of the hex nut if $F = 20$ lb and $r = 20$ in. (c) find r if the force $F = 40$ lb is to produce the same moment as in (a) when $\theta = 30°$.

4.4 Repeat Problem 4.3 using Varignon's theorem.

4.5 Consider the truss shown in Figure P4.5. (a) Determine the moment produced about an axis perpendicular to the page at point A by the 15-kN force. Use Varignon's theorem. (b) Find the magnitude and sense of a vertical force at E that would produce the same moment as in part (a).

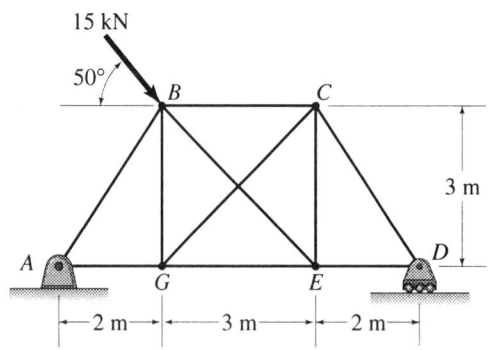

FIGURE P4.5.

4.6 Consider the truss shown in Figure P4.5. (a) Determine the moment produced about an axis perpendicular to the page at point D by the 15-kN force. Use Varignon's theorem. (b) Find the magnitude and sense of a horizontal force at C that would produce the same moment as in part (a).

4.7 Refer to the two forces acting on the beam shown in Figure P4.7. (a) Determine the moment of each of the two forces about an axis perpendicular to the

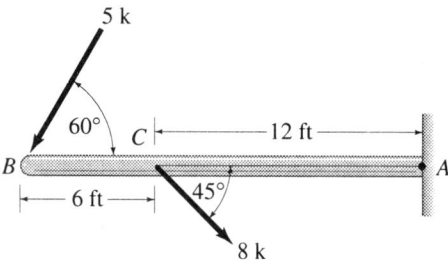

FIGURE P4.7.

page at point A. What is the resultant moment about this axis? (b) Determine the magnitude and sense of a vertical force at B that would produce the same moment as the resultant moment of the two forces in part (a).

4.8 Consider the two forces acting on the truss shown in Figure P4.8. (a) Find the moment of each force about an axis perpendicular to the page at point E. What is the resultant moment about this axis? (b) Find the magnitude, direction, and sense of the smallest force at C that would produce the same moment as the resultant moment of the two forces in part (a). (c) Find the magnitude and sense of a horizontal force applied at C that would produce the same moment as the resultant moment of the forces in part (a).

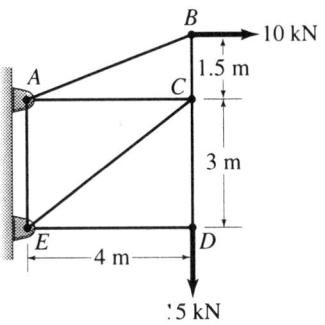

FIGURE P4.8.

4.9   Consider the mechanism shown in Figure P4.9. The pressure p in the cylinder produces a force F in push rod AB equal to 750 lb, as shown. Use the basic definition of moment to determine the moment of the force F about an axis perpendicular to the page at point O.

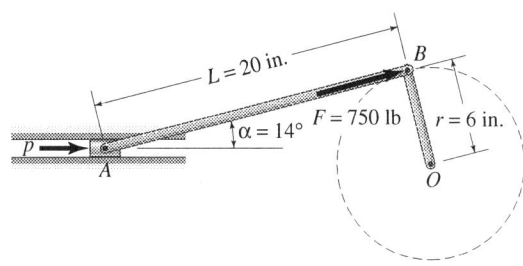

FIGURE P4.9.

4.10   Repeat Problem 4.9 using Varignon's theorem by decomposing the force F into (a) horizontal and vertical components at point B and (b) a component parallel and one perpendicular to the crank OB at point B.

4.11   Consider the pliers shown in Figure P4.11. Use the basic definition of moment to find the moment of the 200-N force about an axis perpendicular to the page at point O.

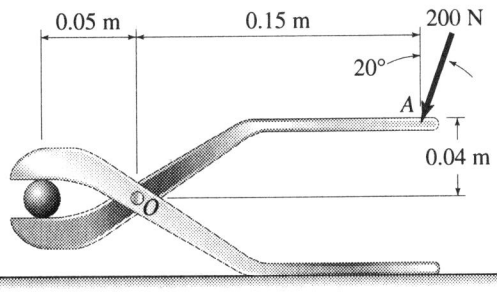

FIGURE P4.11.

4.12   Repeat Problem 4.11 using Varignon's theorem by resolving the 200-N force into (a) horizontal and vertical components at point A and (b) a component parallel and one perpendicular to line OA at point A.

4.13   Refer to the system shown in Figure P4.13. (a) Determine the moment of the 800-lb weight about an axis perpendicular to the page at point O. (b) What must be the weight W to produce a moment of the same magnitude and direction but opposite in sense to that in part (a)?

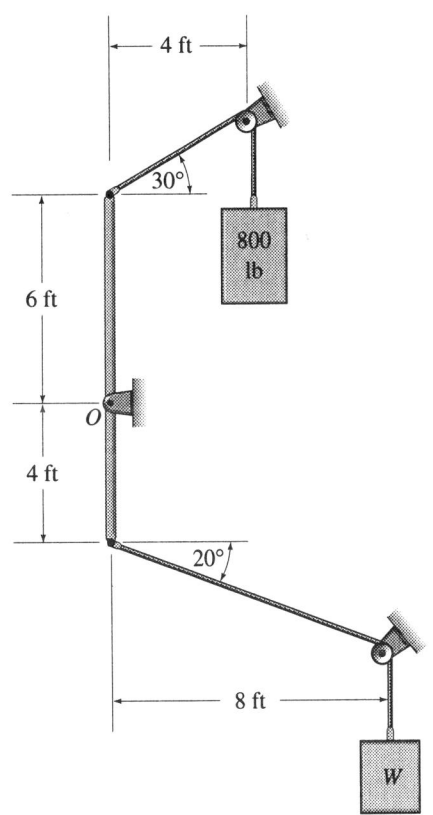

FIGURE P4.13.

4.14   (a) Use the basic definition of moment to find the moment of the 2-kN force about

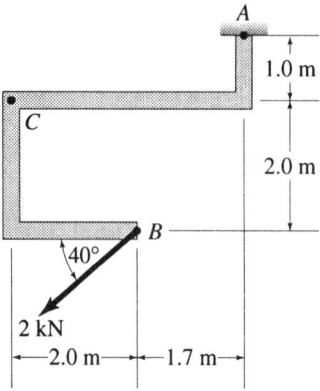

FIGURE P4.14.

an axis at point A perpendicular to the page in Figure P4.14. (b) What horizontal force at C would produce the same moment as in part (a).

4.15 Repeat Problem 4.14(a) using Varignon's theorem by resolving the 2-kN force into (a) horizontal and vertical components at point B and (b) components parallel and perpendicular to line AB at point B.

4.16 Find the magnitude, direction, and sense of the couple shown in Figure P4.16 using the basic definition of a couple expressed in Eq. (4.3). State your answer (a) as a magnitude along with a curved arrow and (b) in vector form.

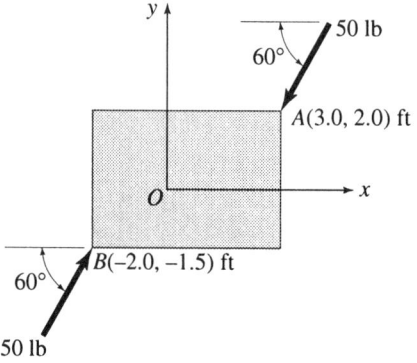

FIGURE P4.16.

4.17 Repeat Problem 4.16 by adding the moments of the two forces about an axis through point O perpendicular to the page.

4.18 Find the magnitude, direction, and sense of the couple shown in Figure P4.18 using the basic definition of a couple expressed in Eq. (4.3). State your answer (a) as a magnitude along with a curved arrow and (b) in vector form.

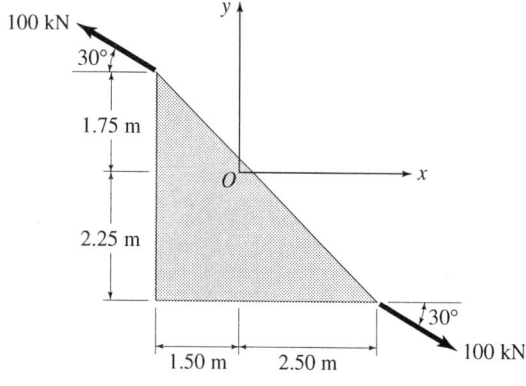

FIGURE P4.18.

4.19 Repeat Problem 4.18 by adding the moments of the two forces about an axis through point O perpendicular to the page.

4.20 A plate is shaped into the form of a 5 ft × 4 ft parallelogram and is subjected to the couples shown in Figure P4.20. Determine the resultant couple acting on the plate.

4.21 A plate is shaped into the form of a trapezoid and is subjected to the couples shown in Figure P4.21. Determine the resultant couple acting on the plate.

4.22 The axis of the shaft shown in Figure P4.22 coincides with the x axis. It is subjected to three couples (torques) that lie in planes parallel to the y-z plane. Determine a single couple in the y-z plane

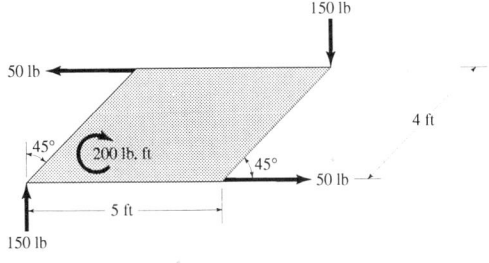

FIGURE P4.20.

that would produce the same effect as the three applied couples.

4.23 The socket wrench is subjected to a couple consisting of two 50-N forces as shown in Figure P4.23. What forces *F* applied as shown are needed to produce the same couple on the socket wrench?

4.24 The drill bit of a drilling machine creates a cw couple of 60 lb·ft on the triangular plate of Figure P4.24 as the hole is being drilled. Determine the magnitudes and senses of two equal and horizontal forces applied at A and B that would create a

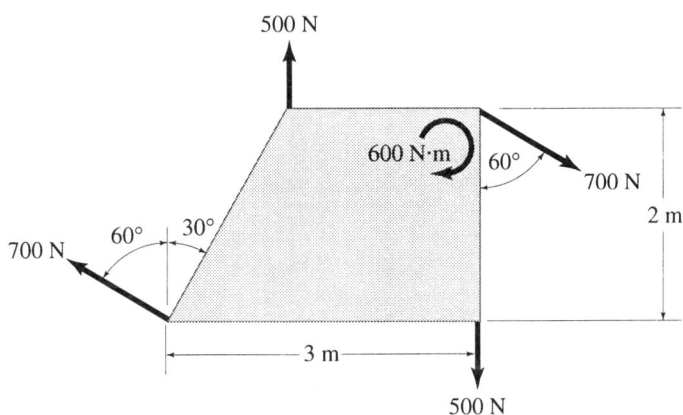

FIGURE P4.21.

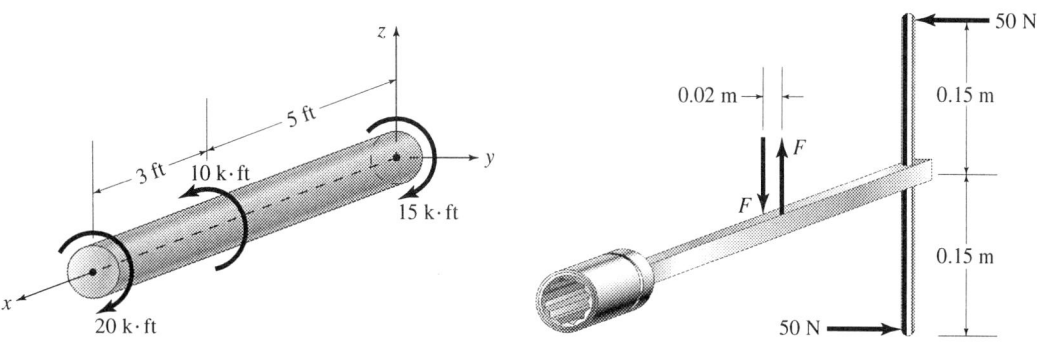

FIGURE P4.22.   FIGURE P4.23.

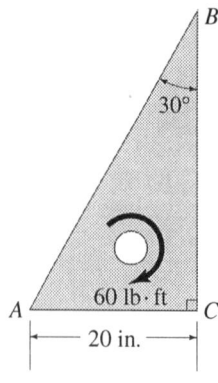

FIGURE P4.24.

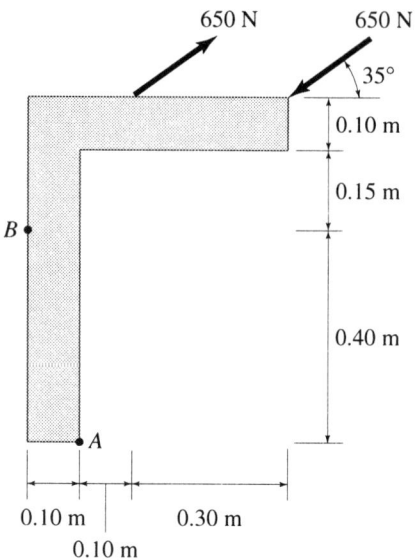

FIGURE P4.25.

couple of equal magnitude and direction but opposite in sense.

4.25 A couple consisting of two 650-N forces is applied to the angle section as shown in Figure P4.25. Determine the magnitudes and senses of two equal and vertical forces applied at A and B that would produce a couple of equal magnitude and direction but opposite in sense.

4.2
Internal and External Forces— Force Transmissibility Principle

As a push or a pull, a force may be viewed as *internal* or *external* with respect to the rigid body on which it acts. Briefly, a force *external* to a rigid body is one that represents the action on this body by another body through either direct contact or gravitational attraction. *Internal* forces, on the other hand, represent those forces that maintain together the various components of the rigid body. These components are the particles composing a single rigid body or several rigid bodies connected together in some fashion.

In constructing the free-body diagram of a rigid body (see Section 4.8), we must be very careful to distinguish between internal and external forces because *only* external forces appear on the free-body diagram. The external forces are the only forces that contribute to the external behavior of the rigid body and determine whether the rigid body remains at rest or moves in some manner.

Consider, for example, the case of the pulley shown in Figure 4.8(a) which is hinged at its geometric center, point A, assumed frictionless and coincident with its center of mass*. The belt around the pulley is

---

* The concept of the center of mass is discussed in Chapter 10.

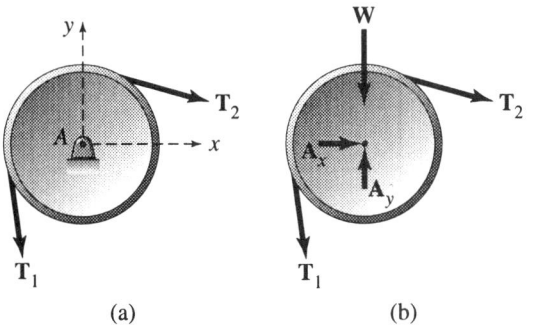

(a)   (b)   FIGURE 4.8.

subjected to the tensions T_1 and T_2. The free-body diagram of the
pulley is shown in Figure 4.8(b) which shows *only* the external forces
W, T_1, T_2, A_x, and A_y. The weight W represents the pull of the Earth's
gravity on the pulley. It is actually distributed throughout the body of
the pulley but for our purposes, only the resultant W is shown acting
through point A and pointed downward toward the center of Earth.
The forces T_1 and T_2 are those acting on the belt and are, therefore,
external to the pulley. Finally, the forces A_x and A_y represent the action
on the pulley of the smooth pin at the hinge. Note that, if $T_1 = T_2$, the
pulley remains at rest under the action of the forces shown in Figure
4.8(b). However, if $T_1 \neq T_2$, the pulley will rotate about the hinge at A.
Thus, as stated earlier, only the *external* forces acting on the pulley
determine its external behavior. It should be pointed out that the pulley is also subjected to *internal* forces of action and reaction throughout its body existing in equal and opposite pairs that cancel each other
and, therefore, do not contribute to its external behavior. Note that
internal forces are discussed in Chapter 8.

The *force transmissibility principle* is extremely useful in the study of
the mechanics of rigid bodies. This principle states that *the external
effects of a force on the equilibrium or motion of a rigid body is the same
regardless of the point of application of the force along its line of action.*
It should be pointed out that this principle is based upon experimental
evidence and is applicable only to the case of rigid bodies. Thus, consider, for example, the frame shown in Figure 4.9 which is subjected to
a force P. According to this principle, the external support reactions at
C and D will be the same regardless of whether the applied force
pushes at point A (Figure 4.9(a)) or pulls at point B (Figure 4.9(b)) as
long as the magnitude and sense of the force remain unchanged. As a
matter of fact, the external support reactions at C and D will remain
the same if the force P were applied at any other point along its line of
action defined by the direction AB. However, it should be made clear
that the internal forces induced in the frame (if assumed to be nonrigid)

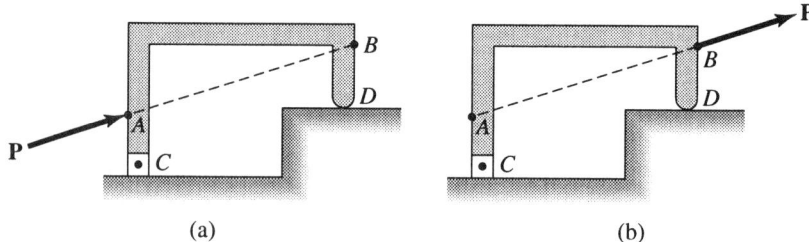

FIGURE 4.9.   (a)   (b)

would depend upon the exact location of the applied force *P*. Thus, if the force *P* were applied at A, the internal forces and resulting deformations in the neighborhood of point A will *not* be the same if the force *P* were applied at point B or at any other point along its line of action.

4.3 Replacement of a Single Force by a Force and a Couple

In the study of rigid body mechanics, it is very useful to be able to replace one force system by a second *equivalent* to the first. The word *equivalent* implies that the two force systems produce identical external effects on the rigid body on which they act. This process makes it possible to reduce a complex force system to an equivalent force system, in general, consisting of one single force and a single couple.

In this section, we will focus on the process of replacing a single force, applied at some point on a two-dimensional rigid body, by a force and a couple, where the force acts at a second point on the body. This process is facilitated through the concept of a *null couple* which is defined as a set of two forces equal in magnitude, opposite in sense, and collinear. In other words, it is a couple whose magnitude is zero because the perpendicular distance *d* between the two forces is zero. An example of a null couple acting on a rigid body is shown in Figure 4.10. Because the two forces cancel each other and produce no moment, such a couple will not have a tendency to translate or rotate the rigid

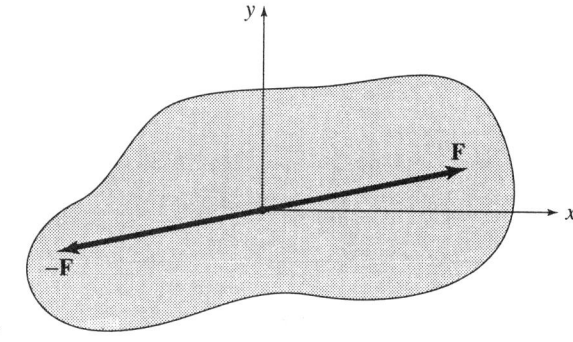

FIGURE 4.10.

body on which it acts. Therefore, such a couple may be added to a force system acting on a rigid body without changing the external effects that this force system has on the rigid body.

Consider, for example, the rigid plate supported at A and B, shown in Figure 4.11(a), and subjected to the force **F** which lies in the same plane (the x-y plane) as the plate. Let us suppose that the force **F** is to be replaced by a force-couple system at an arbitrary point D. This may

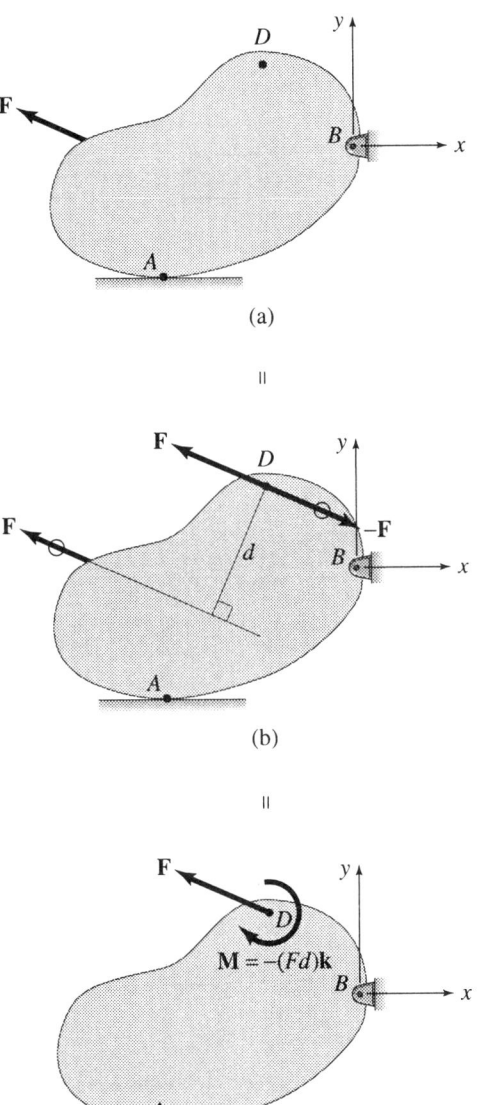

(a)

∥

(b)

∥

(c)

FIGURE 4.11.

be accomplished by adding a null couple at D, as shown in Figure 4.11(b). The forces in the null couple are equal in magnitude and direction to the force **F**. The force system in Figure 4.11(b) may be viewed as consisting of the couple, $\mathbf{M} = -(Fd)\,\mathbf{k}$, composed of the two forces **F** that have been marked with small circles, plus the original force now acting at point D. Because a couple is a free vector, it may be placed at any position on the rigid body. For convenience, it is generally shown at the new point where the force is now acting. Thus, the given force of Figure 4.11(a) may be replaced by the equivalent force system shown in Figure 4.11(c) consisting of a force **F** and a couple $\mathbf{M} = -(Fd)\mathbf{k}$. In other words, the force system of Figure 4.11(c) will produce the same reactions at supports A and B as the original force of Figure 4.11(a). The fact that the three force systems shown in Figure 4.11 are equivalent is indicated by the equal signs shown between Figures 4.11(a) and (b) and between Figures 4.11(b) and (c). The above process may be summarized by the following principle:

A force at one point on a rigid body may be replaced at any other point by a force-couple system in which the force is equal in magnitude, direction, and sense to the original force and the couple is equal in magnitude, direction, and sense to the moment produced by the original force about an axis perpendicular to the page at the second point.

Although the above concepts were developed and discussed using vector notation, it is usually more convenient to use the scalar notation in solving two-dimensional problems as will be illustrated in the following example.

■ **Example 4.5**

A 10-kN force is applied at point A on the structure as shown in Figure E4.5(a). Replace this force by a force-couple system (a) at point B and (b) at point D.

Solution

(a) The force-couple system at point B, shown in Figure E4.5(b), is equivalent to the given force and consists of a force equal in magnitude, direction, and sense to the given 10-kN force plus a couple **M** with magnitude, direction, and sense identical to those of the moment produced by the given force about point B. The magnitude M_B of the couple is equal to the moment of the force acting at A about point B and may be determined by Varignon's theorem. Thus, the given force at A is decomposed into horizontal and vertical components. The vertical component, $10\sin 20°$, passes through point B producing no moment about this point. However, the horizontal component, $10\cos 20°$, produces a ccw moment about this point. Therefore,

$$M_B = (10\sin 20°)(0) + (10\cos 20°)(0.5) = 4.70 \text{ kN·m } \circlearrowleft. \quad \text{ANS.}$$

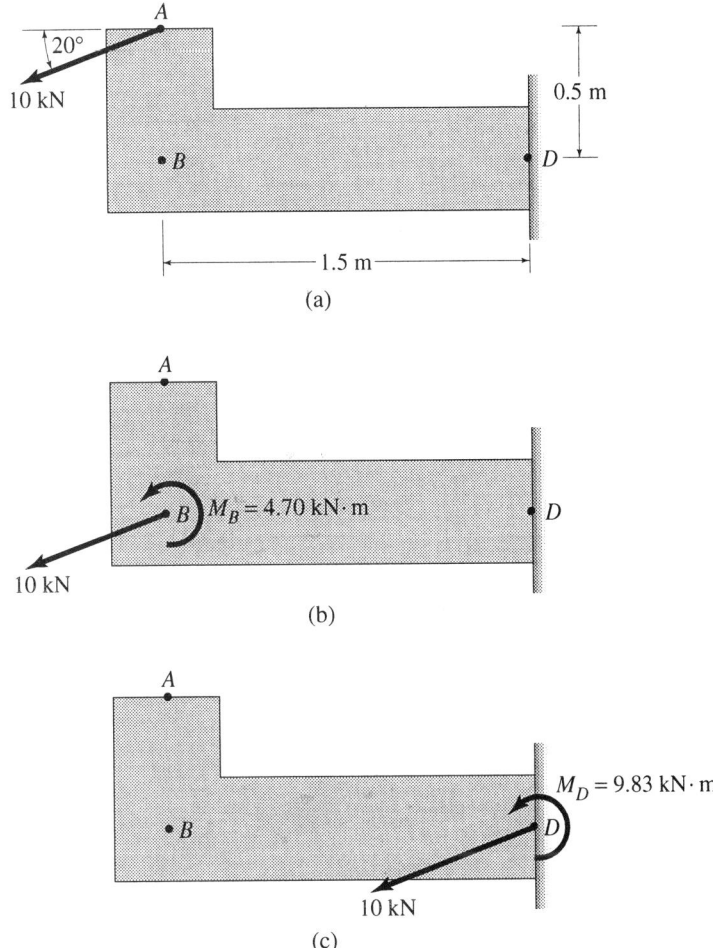

$$M_D = (10 \sin 20°)(1.5) + (10 \cos 20°)(0.5) = 9.83 \text{ kN·m} \; \text{\reflectbox{$\circlearrowleft$}}.$$

FIGURE E4.5.

(b) The force-couple system at point D, shown in Figure E4.5(c), is equivalent to the given force and consists of a force equal in magnitude, direction and sense to the given 10-kN force plus a couple **M** with magnitude, direction, and sense identical to those of the moment produced by the given force about point D. As in part (a), the magnitude M_D of the couple is found using Varignon's theorem.

$$M_D = (10 \sin 20°)(1.5) + (10 \cos 20°)(0.5) = 9.83 \text{ kN·m} \; \text{\reflectbox{$\circlearrowleft$}}. \quad \text{ANS.}$$

In summary, the force systems shown in Figures E4.5(a), (b), and (c) are all equivalent because they produce the same effect on the rigid body on which they act. In particular, each of the three force systems produces the same reactive forces at the fixed support at D.

4.4 Replacement of a Force System by a Force and a Couple

A force system, consisting of one or more forces and one or more couples, that acts on a two-dimensional rigid body may be replaced by a single resultant force at any arbitrary point and a single resultant couple using the method developed in Section 4.3. Consider, for example, the rigid plate shown in Figure 4.12(a) which is supported at A and B and subjected to the forces $\mathbf{F}_1$, $\mathbf{F}_2$, and $\mathbf{F}_3$ and to the couples $\mathbf{M}_4 =$

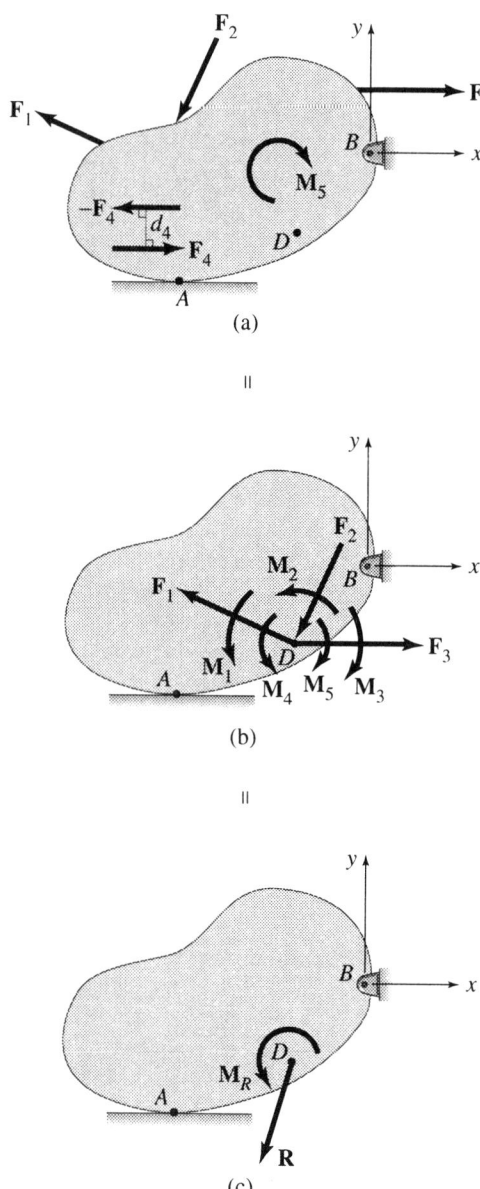

FIGURE 4.12.

$(F_4 d_4)\mathbf{k}$ and $\mathbf{M}_5$, all of which act in the plane of the plate, i.e., in the x-y plane. It is required to replace this force system by an equivalent force-couple system at any arbitrary point D. Using the principle developed in Section 4.3 for replacing a single force by a force-couple system, we proceed to replace each of the three forces $\mathbf{F}_1$, $\mathbf{F}_2$, and $\mathbf{F}_3$ separately by equivalent force-couple systems as shown in Figure 4.12(b). For example, the force $\mathbf{F}_1$ is replaced by $\mathbf{F}_1$ and $\mathbf{M}_1 = (F_1 d_1)\mathbf{k}$ at D, where d_1 (not shown in Figure 4.12) is the perpendicular distance from point D to the line of action of $\mathbf{F}_1$. Similarly, $\mathbf{F}_2$ is replaced by $\mathbf{F}_2$ and $\mathbf{M}_2 = (F_2 d_2)\mathbf{k}$ and $\mathbf{F}_3$ by $\mathbf{F}_3$ and $\mathbf{M}_3 = -(F_3 d_3)\mathbf{k}$, where d_2 and d_3 (not shown in Fig. 4.12) are the perpendicular distances from point D to the lines of action of $\mathbf{F}_2$ and $\mathbf{F}_3$, respectively. Also, the couples $\mathbf{M}_4 = (F_4 d_4)\mathbf{k}$ and $\mathbf{M}_5 = -(M_5)\mathbf{k}$ are simply shifted to point D because they are free vectors. The three forces $\mathbf{F}_1$, $\mathbf{F}_2$, and $\mathbf{F}_3$ are combined by the relationship $\mathbf{R} = \sum \mathbf{F}$ to obtain the resultant force $\mathbf{R}$. Similarly, the five couples $\mathbf{M}_1$ to $\mathbf{M}_5$ are added using the relationship $\mathbf{M}_R = \sum \mathbf{M}$ to obtain the resultant couple $\mathbf{M}_R$. These quantities are shown in Figure 4.12(c) which represents a force-couple system equivalent to the original force couple system of Figure 4.12(a). The equal signs between the force systems shown in Figures 4.12(a) and (b) and between Figures 4.12(b) and (c) signify that all of these force systems are equivalent implying that they all produce the same external effects on the rigid body. Thus, for example, the support reactions at A and B are the same regardless of which force system is applied to the rigid body.

As indicated earlier, although the vector notation was used in the development of the concepts, the scalar notation is generally more convenient in solving two-dimensional problems. Also, the replacement of a force system by a single resultant force $\mathbf{R}$, at an arbitrary point D, and a single resultant couple $\mathbf{M}_R$ is considerably simplified if each of the forces is expressed in terms of its x and y components. The scalar components of the resultant force and the magnitude of the resultant force itself are, then, determined from the relationship $R_x = \sum F_x$, $R_y = \sum F_y$, and $R = \sqrt{R_x^2 + R_y^2}$. Furthermore, the moment of each force component is determined about an axis perpendicular to the page at point D, assigning a plus or minus sign depending on whether the moment is ccw or cw. The magnitude of the resultant couple is, then, found by an algebraic addition of all moments, including original couples in the given force system, by the relationship $\mathbf{M}_R = \sum \mathbf{M}$.

The concepts above are illustrated in the following example.

■ **Example 4.6**
A cantilevered member is subjected to a force system consisting of three forces and one couple, as shown in Figure E4.6(a). Replace this force system by a force-couple system at point A.

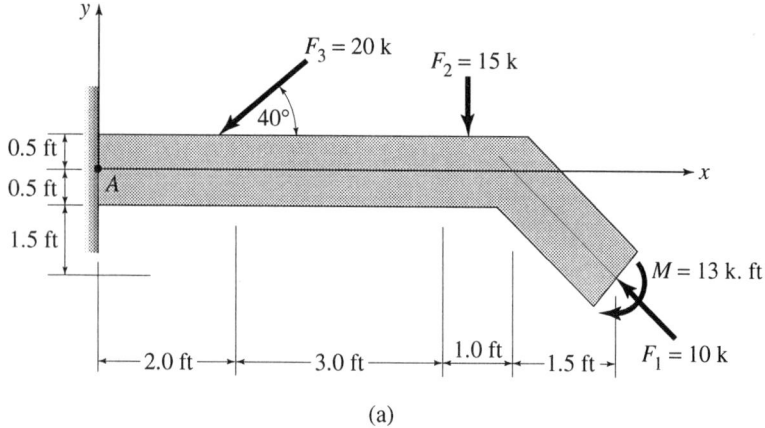

(a)

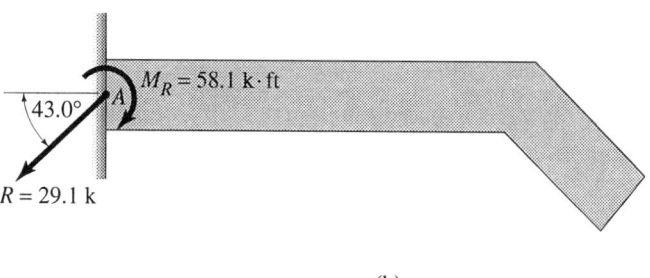

FIGURE E4.6.   (b)

Solution

Each force and couple in the force system is treated separately. It is also convenient to deal with each force in terms of its x and y components. This makes it easier to obtain the resultant force $\mathbf{R}$ and the resultant couple $\mathbf{M_R}$.

The force F_1 is replaced at A by a force F_1 whose components are $F_{1x} = -6$ k, $F_{1y} = 8$ k and a couple $M_1 = 8(7.5) - 6(2.0) = 48.0$ k·ft.

The force F_2 is replaced at A by a force F_2 whose components are $F_{2x} = 0$, $F_{2y} = -15$ k and a couple $M_2 = -15\,(5.0) = -75.0$ k·ft.

The force F_3 is replaced at A by a force F_3 whose components are $F_{3x} = -15.320$ k, $F_{3y} = -12.856$ k and a couple $M_3 = 15.320(0.5) - 12.856(2.0) = -18.1$ k·ft.

The applied couple $M = -13$ k·ft is simply moved to point A because it is a free vector. Therefore, the x and y components of the resultant force at A are

$$R_x = \sum F_x = -6 + 0 - 15.320 = -21.320 \text{ k}$$

and

$$R_y = \sum F_y = 8 - 15 - 12.856 = -19.856 \text{ k}.$$

The magnitude of the resultant **R** at A becomes

$$R = \sqrt{(21.320)^2 + (19.856)^2} = 29.1 \text{ k } \underset{\theta}{\diagup} \qquad \text{ANS.}$$

where the angle θ is found from the relationship

$$\theta = \tan^{-1}\frac{R_y}{R_x} = 43.0°. \qquad \text{ANS.}$$

The magnitude of the resultant couple M_R at A is found from the relationship

$$M_R = \sum M = 48.0 - 75.0 - 18.1 - 13.0 = -58.1 \text{ k·ft}$$
$$= 58.1 \text{ k·ft } \circlearrowleft. \qquad \text{ANS.}$$

The equivalent force-couple system consisting of **R** and M_R are shown in Figure E4.6(b).

■

4.5 Replacement of a Force System by a Single Force

It was shown in Section 4.3 that a single force acting on a two-dimensional rigid body may be replaced by an equivalent force-couple system at any arbitrary point. By reversing the steps used in Section 4.3, we can also show that a force-couple system acting at a given point on a two-dimensional rigid body may be replaced by a single equivalent force acting at some other point. Also, as shown in Section 4.4, because a general two-dimensional force system can be reduced to an equivalent force-couple system, it follows that this general force system may be reduced to a single equivalent force acting at a specified location.

Consider, for example, the general two-dimensional force system shown in Figure 4.13(a). As discussed in Section 4.4, this force system may be replaced by the force-couple system at point D consisting of the resultant force **R** and the resultant couple M_R, as shown in Figure 4.13(b). Using the condition that the moment of **R** about an axis perpendicular to the page at point D is equal to M_R, this force-couple system may then be reduced to the single force **R** shown in Figure 4.13(c). The force **R** may be positioned along the x axis by locating point E. Thus, because $\mathbf{M_D} = (Rd)\mathbf{k} = (M_R)\mathbf{k}$, it follows that

$$d = \frac{M_R}{R} \qquad (4.5)$$

where d is the perpendicular distance from point D to the line of action of the force **R**. Because $d = \bar{x}\sin\alpha$, we conclude that $\bar{x} = d/\sin\alpha$ which yields

$$\bar{x} = \frac{M_R}{R\sin\alpha} = \frac{M_R}{R_y}. \qquad (4.6)$$

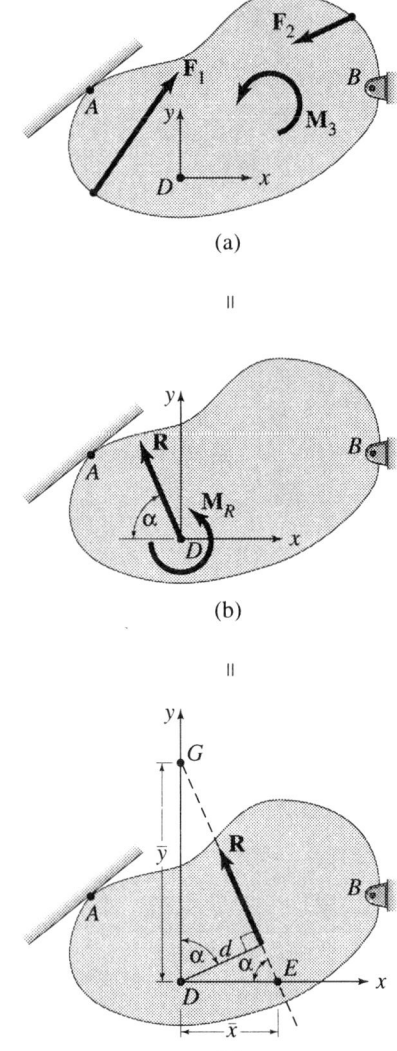

FIGURE 4.13.

In other words, if the resultant force **R** is placed at E along the x axis, the only component producing a moment about point D is $\mathbf{R}_y$ with a moment arm equal to $\bar{x}$. Alternatively, the force **R** may be positioned along the y axis by locating point G. Using Eq. (4.5) and the fact that $\bar{y} = d/\cos \alpha$, we conclude that

$$\bar{y} = \frac{M_R}{R \cos \alpha} = \frac{M_R}{R_x}. \qquad (4.7)$$

Thus, when the resultant force **R** is placed at G along the y axis, the only component producing a moment about point D is R_x with a moment arm equal to $\bar{y}$. Note that points E and G lie on the line of action of the resultant force **R**. Therefore, positioning this resultant force at either one of these two points or at any other point along line EG is consistent with the principle of transmissibility.

Unless required for a reason, it is not necessary to show the intermediate step represented by the force-couple system of Figure 4.13(b). Rather, we may go directly from the given force system of Figure 4.13(a) to the single resultant force **R** of Figure 4.13(c). The x and y scalar components of **R** are found from the relationships $R_x = \sum F_x$ and $R_y = \sum F_y$. The magnitude of the resultant force is determined from the equation $R = \sqrt{R_x^2 + R_y^2}$ and the angle α defining its direction from the relationship $\alpha = \tan^{-1}(R_y/R_x)$. Also, the magnitude of the resultant couple is found from the relation $M_R = \sum M$ at the origin of the x-y coordinate system. Finally, the position of the resultant is established by using either Eq. (4.6) or Eq. (4.7). Thus, we conclude that the resultant of a general two-dimensional force system is a single force **R**.

These concepts are illustrated in the solution of the following example.

■ **Example 4.7**

A frame is subjected to three forces and a couple as shown in Figure E4.7(a). Determine the magnitude, direction, and sense of a single equivalent resultant force **R**, and give its location (a) along the x axis

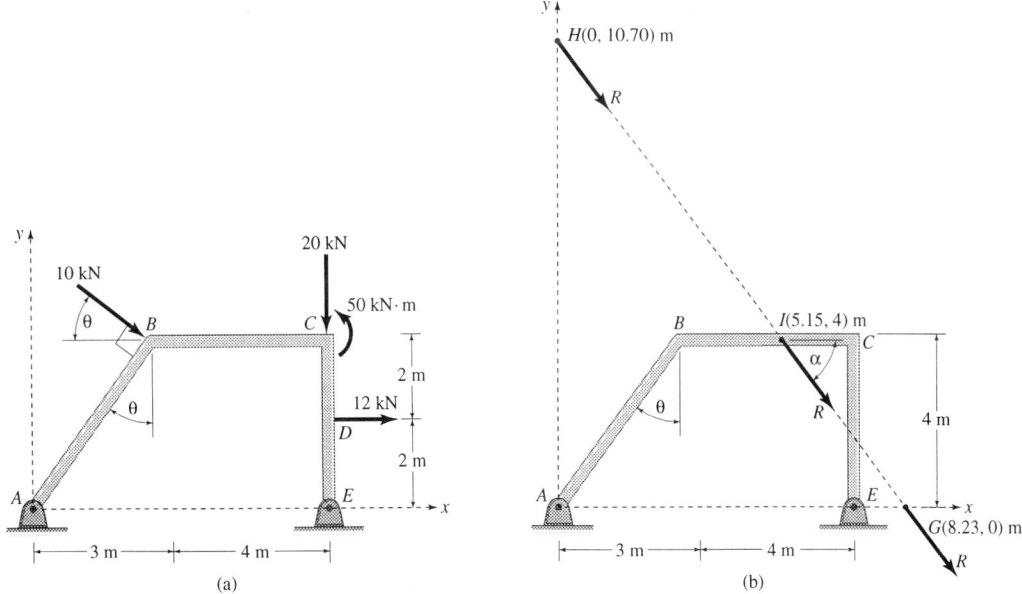

(a)

(b)

FIGURE E4.7.

measured from point A, (b) along the y axis measured from point A and (c) along member BC.

Solution

The x and y components of the resultant force as well as the resultant force itself are found as follows:

$$R_x = \sum F_x = 10 \cos \theta + 12 = 20 \text{ kN} \rightarrow,$$

and

$$R_y = \sum F_y = -10 \sin \theta - 20 = -26 \text{ kN} = 26 \text{ kN} \downarrow,$$

where, from the geometry given,

$$\theta = \tan^{-1} \frac{3}{4} = 36.87°.$$

Also,

$$R = \sqrt{R_x^2 + R_y^2} = 32.8 \text{ kN},$$

and

$$\alpha = \tan^{-1} \frac{R_y}{R_x}.$$

The resultant couple, created as the given force system is moved to point A, becomes

$$M_R = -10(5) - 20(7) - 12(2) = -214 \text{ k·ft} = 214 \text{ k·ft} \circlearrowright.$$

(a) By Eq. (4.6), we can locate point G on the x axis. Thus,

$$\bar{x} = \frac{M_R}{R_y} = \frac{214}{26} = 8.23 \text{ m} \quad \text{right of point A.} \quad\quad\quad \text{ANS.}$$

The resultant force **R** and its location on the x axis are shown in Figure E4.7(b).

(b) Point H on the y axis may be located by Eq. (4.7). Thus,

$$\bar{y} = \frac{M_R}{R_x} = \frac{214}{20} = 10.70 \text{ m} \quad \text{above point A.} \quad\quad\quad \text{ANS.}$$

The location of point H on the y axis is shown in Figure E4.7(b).

(c) The position of the resultant force **R** on member BC is defined by point I $(x, 4)$ m. When the resultant force **R** is placed at I, both of its components produce moments about point A. Thus,

$$M_A = -R_x(4) - R_y(x) = -214.$$

Therefore,

$$\bar{x} = \frac{214 - R_x(4)}{R_y} = \frac{214 - 20(4)}{26} = 5.15 \text{ m} \qquad \text{right of point A. \quad ANS.}$$

The location of point I on member BC of the frame is shown in Figure E4.7(b).

■ Example 4.8

A machine member is subjected to the force system shown in Figure E4.8(a) in which the force Q is unknown in magnitude. Determine the magnitude of the force Q so that the resultant of the applied force system is a force that passes through point O, the origin of the x-y coordinate system. Also, find the magnitude, direction, and sense of this resultant force.

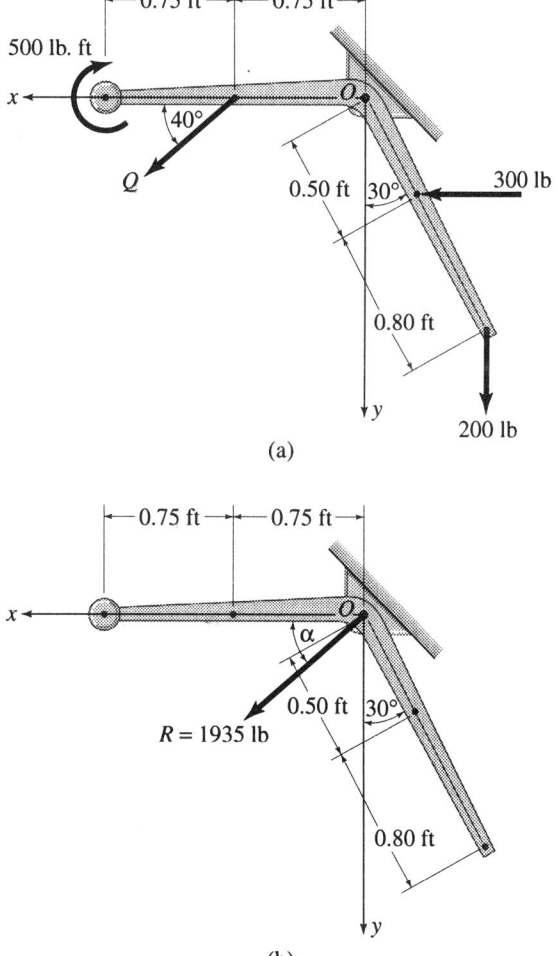

(a)

(b)   FIGURE E4.8.

Solution   The x and y components of the resultant force as well as the resultant force itself are found from the following relationships

$$R_x = \sum F_x = 300 + Q \cos 40°$$
$$R_y = \sum F_x = 200 + Q \sin 40°$$

Because the resultant force R passes through point O, it follows that the resultant couple M_R with respect to this point must vanish. Therefore,

$$M_R = 0 = (Q \sin 40°)(0.75) - 300(0.50 \cos 30°)$$
$$- 200(1.30 \sin 30°) - 500$$

from which

$$Q = 1576 \text{ lb.}$$ ANS.

Thus,

$$R_x = 300 + 1576 \cos 40° = 1507 \text{ lb} \leftarrow,$$
$$R_y = 200 + 1576 \sin 40° = 1213 \text{ lb} \downarrow,$$

and

$$R = \sqrt{R_y^2 + R_y^2} = 1935 \text{ lb} \; \overset{\alpha}{\diagup}$$ ANS.

where the angle α is found from the relationship

$$\alpha = \tan^{-1} \frac{R_y}{R_x} = 38.8°$$ ANS.

The resultant force R is shown in Figure E4.8(b).

■

Problems

4.26 A cantilever beam is subjected to a force of 7.5k as shown in Figure P4.26. Replace this force by a force-couple system acting at A.

4.27 A wrench is used to turn the bolt at B by applying a force of 100 N as shown in Figure P4.27. Replace this force by a force-couple system acting at B.

4.28 A column is subjected to an eccentric force of 25 k as shown in Figure P4.28. Replace this force by a force-couple system at C.

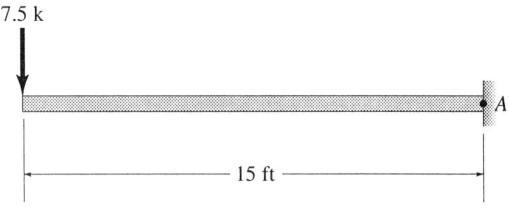

FIGURE P4.26.

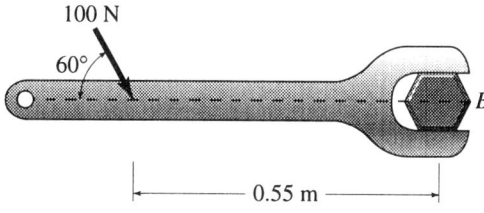

FIGURE P4.27.

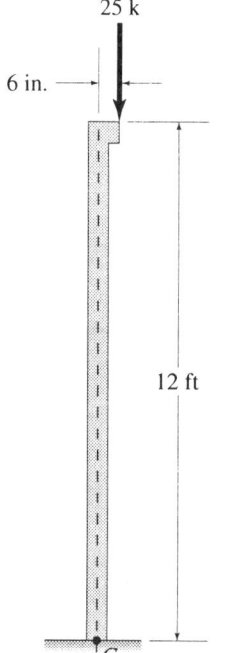

FIGURE P4.28.

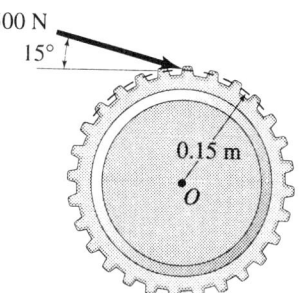

FIGURE P4.29.

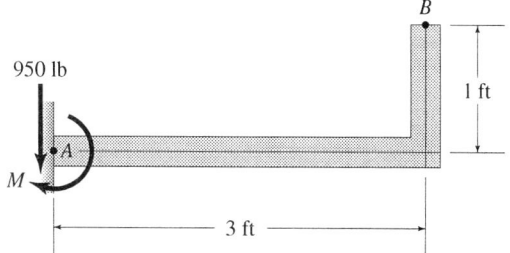

FIGURE P4.30.

4.29 A gear is subjected to a tooth force of 500 N as shown in Figure P4.29. Replace this force by a force-couple system acting at point O which represents the axis of rotation of the gear.

4.30 A force-couple system consisting of a 950-lb force and a couple M is shown acting at point A on the frame of Figure P4.30. Find the value of M if the given system can be replaced by a single vertical force at B.

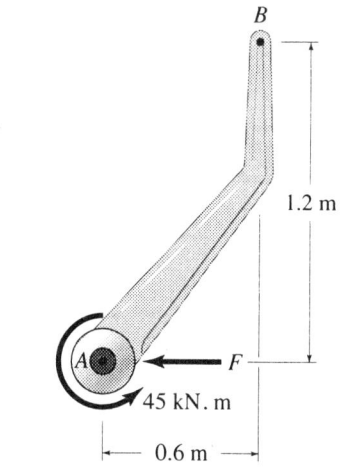

FIGURE P4.31.

4.31 A force-couple system consisting of a 45-kN·m couple and a force F is shown acting at point A on the control lever shown in Figure P4.31. Determine the value of F

if the given system can be replaced by a single horizontal force at B.

4.32 A rectangular plate is subjected to four coplanar forces as shown in Figure P4.32. Replace these four forces by a force-couple system acting at point A.

4.33 Replace the four forces acting on the plate of Figure P4.32 by a force-couple system acting at point B which is the geometric center of the rectangular plate.

4.34 A cantilever beam is subjected to a force system consisting of three parallel forces and a couple as shown in Figure P4.34. Replace this force system by a force-couple system acting at point A.

4.35 Replace the force system acting on the beam of Figure P4.34 by a force-couple system acting at point B.

4.36 A frame is subjected to four forces and two couples as shown in Figure P4.36. Replace these loads by a force-couple system acting at point A.

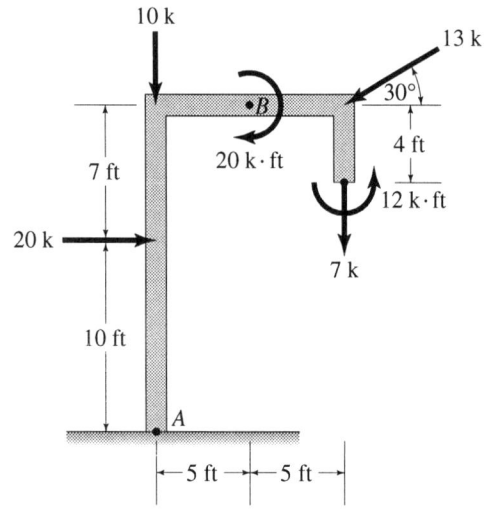

FIGURE P4.36.

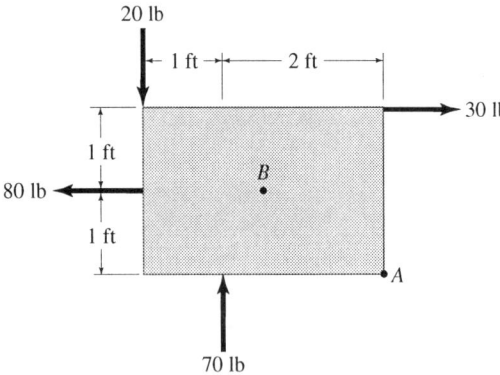

FIGURE P4.32.

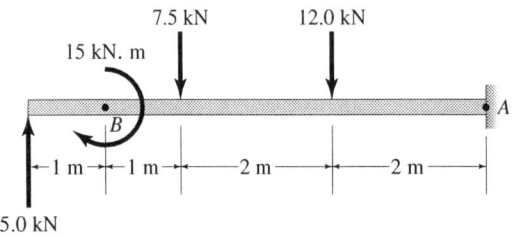

FIGURE P4.34.

4.37 Replace the four forces and the two couples acting on the frame of Figure P4.36 by a force-couple system acting at point B.

4.38 A triangular plate is subjected to the coplanar force system shown in Figure P4.38. Let $P_1 = 7$ kN, $P_2 = 10$ kN, $P_3 = 15$ kN and $M = 7$ kN·m. Replace this force system by a single resultant force providing its magnitude, direction, and sense and specifying its location (a) along the x axis and (b) along the y axis.

4.39 Refer to the force system shown in Figure P4.38. Let $P_2 = 650$ N, $P_3 = 575$ N and $M = 800$ N·m. Determine the magnitude and sense of P_1 so that the resultant of

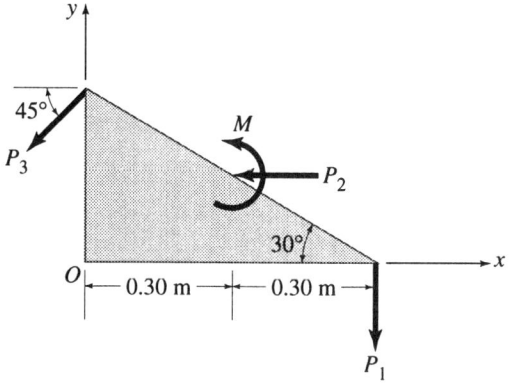

FIGURE P4.38.

the applied force system passes through point O. Also, find the magnitude, direction, and sense of this resultant force.

4.40 The frame ABC is subjected to a coplanar force system as shown in Figure P4.40. Let $Q_1 = 5$ k, $Q_2 = 15$ k, $Q_3 = 25$ k, and $M = 27$ k·ft. Replace this force system by a single resultant force specifying its magnitude, direction, and sense, and give its location on member AB or AB extended, measuring its position from point A.

4.41 Refer to frame ABC of Figure P4.40. Let $Q_1 = 7.5$ k, $Q_2 = 13.0$ k, and $Q_3 = 17.3$ k. Determine the magnitude and sense of M so that the resultant of the applied force system passes through point B. Also find, the magnitude, direction, and sense of this resultant force.

4.42 The applied force system shown acting on the member of Figure P4.42 may be replaced by a horizontal force of 5 kN acting at A and a cw couple of 30 kN·m. Determine Q, α, and M_1.

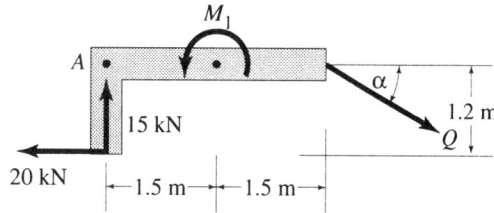

FIGURE P4.42.

4.43 The bridge truss shown in Figure P4.43 is subjected to a wind load of 8 kN and traffic loads of 12 kN and 6 kN at a given instant. Replace this system of forces by a single resultant force, specifying its magnitude, direction, and sense. Also, determine its location along line AB measuring its position from point A.

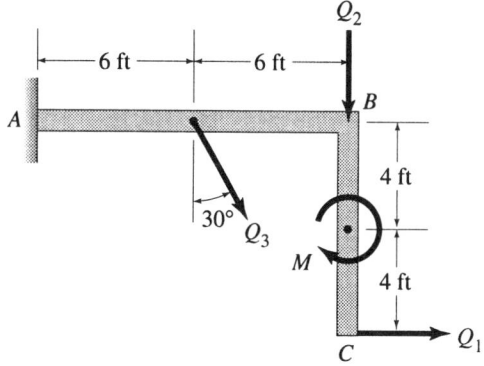

FIGURE P4.40.

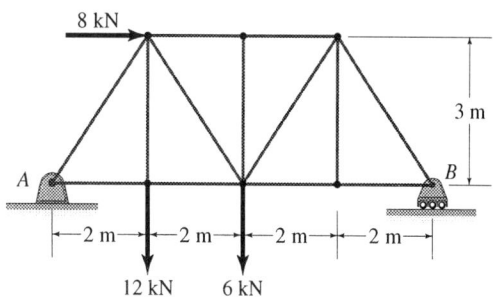

FIGURE P4.43.

4.44 A disabled pickup truck is towed at constant velocity to the garage by applying a towing force P as shown in Figure P4.44. The weight of the truck is 4500 lb and the weight of its cargo is 1200 lb. Of necessity, the resultant of this applied force system is a force passing through point A, the point of contact between the rear wheels and the load. What must be the magnitude of force P? Also, find the magnitude, direction, and sense of the resultant of the applied force system.

4.45 Consider the machine member in Figure P4.45 subjected to four forces as shown. Determine the proper magnitude of the dimension a so that the resultant of the four applied forces is a force that passes through the hinge at point A. Also, find the magnitude, direction, and sense of this resultant.

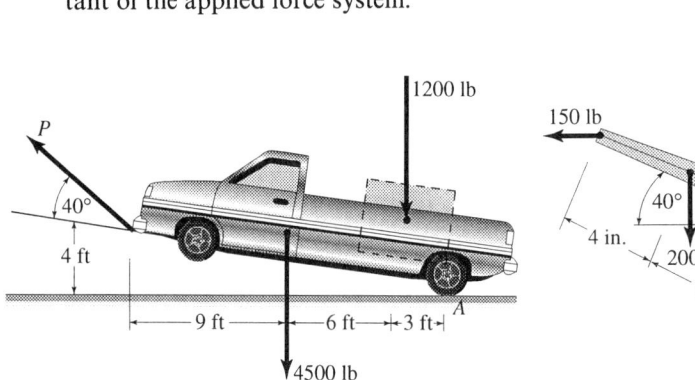

FIGURE P4.44.

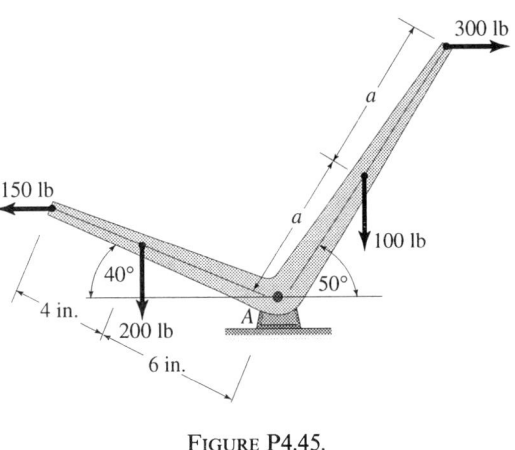

FIGURE P4.45.

4.6 Replacement of a Distributed Force System by a Single Force

A force is said to be *concentrated* if it acts over a relatively small area or volume. In practice, a concentrated force is assumed to act at a specific point in a body. In reality, however, concentrated forces do not exist for the simple reason that a force, of necessity, must be applied over a finite area or volume. Nevertheless, if a force is applied over an area or volume which is very small in comparison to the other dimensions of the body on which the force acts, it may justifiably be considered a concentrated force.

Although we dealt exclusively with concentrated forces in previous analyses, we often encounter forces that are *distributed* over large areas or volumes. Examples of *distributed forces* include those produced by the wind on the sail of a boat, by water on the inside surfaces of its container, by sand on the surface on which it is piled, and by the pull of gravity on any object on or near the surface of the earth.

Often, the distributed force varies in one direction only. In such cases, the intensity of the distributed force may be conveniently plotted

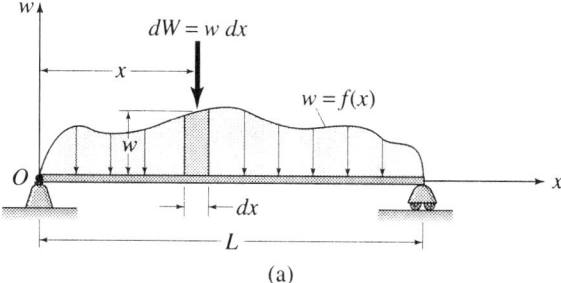

(a)

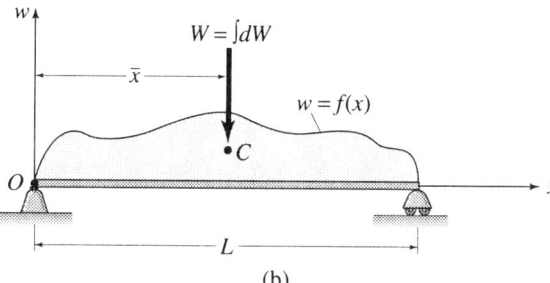

(b)

FIGURE 4.14.

with respect to the direction or axis of variation. For example, Figure 4.14(a) indicates *a force distribution diagram* showing how the intensity w of a distributed force varies along the x axis which coincides with the longitudinal axis of a floor beam. The intensity w at any point along the beam is expressed in terms of force per unit of length of the beam. Thus, such units as lb/ft (or k/ft) in the U.S. Customary system and N/m (or kN/m) in the SI system are common for the load intensity w. Therefore, the differential force dW acting over a differential length dx along the beam may be determined as the product of the intensity w and the length dx. Thus, $dW = w\,dx$ where $w\,dx$ represents the area dA of the differential strip shown shaded in Figure 4.14(a). Therefore, the distributed force acting on the beam may be thought of as an infinitely large number of parallel forces each of magnitude $dW = w\,dx = dA$. Consequently, the distributed force may be replaced by a single equivalent resultant force whose magnitude W is obtained from the relationship

$$W = \int_0^L w\,dx = \int dW = \int dA = A \tag{4.8}$$

where A represents the area under the force distribution diagram.

The precise position of the resultant W on the beam measured from point O may be determined by Varignon's theorem. We may recall that this theorem states that the moment of a resultant force about some axis, is equal to the sum of the moments of all of its components about the same axis. Thus, if $\bar{x}$ in Figure 4.14(b) is the perpendicular distance from point O to the line of action of the resultant W, it follows that magnitude of the moment of W about an axis through point O perpendicular to the page is $\bar{x}W = \bar{x}\int dW$. Also, the magnitude of the moment of each individual component dW about the same axis is $x\,dW$ and the magnitude of the moment of the entire force distribution becomes $\int x\,dW$. It follows, therefore, by Varignon's theorem that $\bar{x}\int dW = \int x\,dW$, from which

$$\bar{x} = \frac{\int x\,dW}{\int dW}.$$

(4.9)

Also, because by Eq. (4.8) $dW = dA$, the element of area under the force distribution, it follows that

$$\bar{x} = \frac{\int x\,dA}{\int dA}.$$

(4.10)

As will be shown in Chapter 10, Eq. (4.10) yields the x coordinate for the location of the *centroid* of an area. The concept of the centroid of an area will be discussed in detail in Chapter 10. For our present purposes, we may think of the centroid of an area as a point where the entire area may be imagined to be concentrated. Appendix A tabulates some of the most commonly encountered areas and gives their values and the location of their centroids. This information is very useful in solving some of the problems at the end of this section. Occasionally, however, we encounter areas which are not included in Appendix A. In such cases, the use of integration together with Varignon's theorem becomes necessary. Appendix E provides a list of commonly encountered integrals and their equivalents that may be used in solving problems. The conclusions reached on the basis of Eqs. (4.8) and (4.10) may be summarized as follows:

The resultant **W** *of a distributed force has a magnitude W which is equal to the area under the force distribution diagram. This resultant force has a direction and sense which are the same as those of the distributed force and acts through the centroid of the area under the force distribution.*

Occasionally, the force distribution diagram is a composite of two or more simple geometric diagrams. In such cases, it is convenient to break down the composite diagram into its simple geometric component parts. Each of these parts represents a force with a magnitude equal to its area and a line of action that passes through the centroid of this area. Thus, the composite distribution is reduced to a system of two or more parallel concentrated forces whose resultant is, then, determined by the methods developed in Section 4.5. Some of the concepts above are illustrated in the following examples.

■ **Example 4.9**

Replace the distributed force system shown acting on the beam in Figure E4.9(a) by a single resultant force. Locate this resultant force from the support at A.

Solution

The force distribution diagram shown in Figure E4.9(a) may be viewed as a composite distribution diagram consisting of two distribution

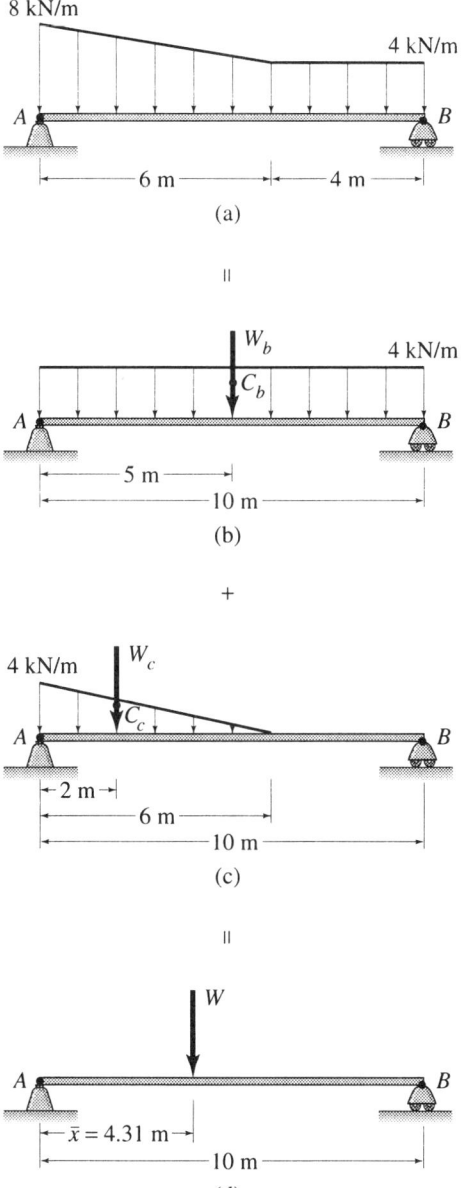

FIGURE E4.9.

diagrams which are simpler geometrically and whose areas and centroid locations may be obtained directly by reference to Appendix A. The first of these two force distribution diagrams is a rectangle, as shown in Figure E4.9(b), and the second is a triangle, as shown in Figure E4.9(c). The magnitude of the resultant of each of these two distributions is equal to the area of, and passes through the centroid of, the corresponding distribution. Thus,

$$W_b = A_b = 4(10) = 40 \text{ kN} \quad \text{at 5 m from point A, as shown}$$
in Figure E4.9(b).

Also

$$W_c = A_c = \tfrac{1}{2}4(6) = 12 \text{ kN} \quad \text{at 2 m from point A, as shown}$$
in Figure E4.9(b).

The resultant W of these two parallel forces has a magnitude W given by

$$W = W_b + W_c = 40 + 12 = 52.0 \text{ kN} \qquad \text{ANS.}$$

The location $\bar{x}$ of this resultant force is determined by applying Varignon's theorem. Thus, equating moments about an axis through point A perpendicular to the page,

$$-\bar{x}W = -5W_b - 2W_c.$$

Substituting the known values of W, W_b, and W_c and solving for $\bar{x}$,

$$\bar{x} = 4.31 \text{ m.} \qquad \text{ANS.}$$

The resultant W is shown properly positioned in Figure E4.9(b).

■ **Example 4.10**

A distributed force system acting on a beam is described by the equation $w = 5\sin(\pi x/40)$, as shown in Figure E4.10(a), where w represents the force intensity in k/ft. Replace this force distribution by a single resultant force specifying its magnitude and location measured from the support at A.

Solution

The area representing our force distribution is not listed in Appendix A. Therefore, integration along with Varignon's theorem will be used to solve the problem. Thus, by Eq. (4.8), the resultant W has a magnitude given by

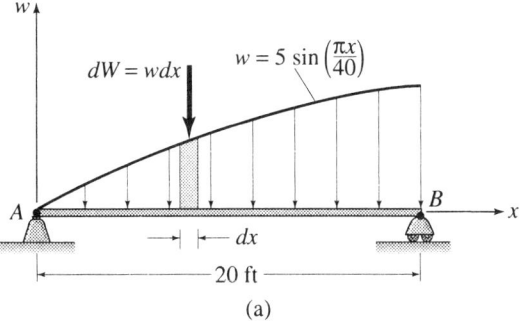

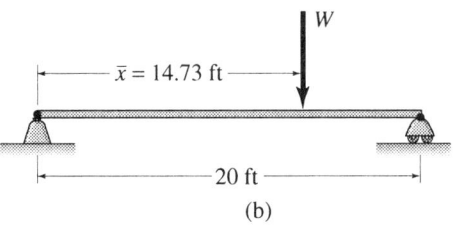

FIGURE E4.10.

$$W = \int dA = \int_0^{20} w\,dx = \int_0^{20} 5\sin\left(\frac{\pi x}{40}\right)dx$$

$$= -5\left(\frac{40}{\pi}\right)\cos\left(\frac{\pi x}{40}\right)\Big|_0^{20}$$

$$= 63.66 \approx 63.7 \text{ lb.} \qquad\qquad \text{ANS.}$$

The location $\bar{x}$ of the resultant, measured from the support at A, is found by Varignon's theorem which leads to Eq. (4.10), namely $\bar{x} = \int x\,dA / \int dA$. In this equation, $\int dA = 63.66$ lb as found above and

$$\int x\,dA = \int_0^{20} x(5)\sin\left(\frac{\pi x}{40}\right)dx$$

$$= 5\left[\left(\frac{40}{\pi}\right)^2 \sin\left(\frac{\pi x}{40}\right) - \left(\frac{4x}{\pi}\right)\cos\left(\frac{\pi x}{40}\right)\right]_0^{20}$$

$$= 937.89 \text{ ft·lb.}$$

Therefore, by Eq. (4.10),

$$\bar{x} = \frac{\int x\,dA}{\int dA} = \frac{937.89}{63.66} = 14.73 \text{ ft.} \qquad\qquad \text{ANS.}$$

The resultant **W** is shown properly positioned in Figure E4.10(b). Note that the integrations above were performed with the aid of the table of integrals in Appendix E.

■ Example 4.11

The wind load on a 20-m building was obtained by actual measurements at 2-m intervals along the building and the results are given in Figure E4.11(a) where the measured values are connected by straight-line segments. Replace this force distribution by an approximate resultant force specifying its magnitude, direction, and sense and its location from point O, the base of the building.

Solution

This type of problem where the input data is obtained by measurements occurs frequently in engineering practice. Because of the approximate nature of the data, the problem does not lend itself to a closed-form solution, and resort is made to numerical techniques that, although approximate, yield answers acceptable in engineering practice.

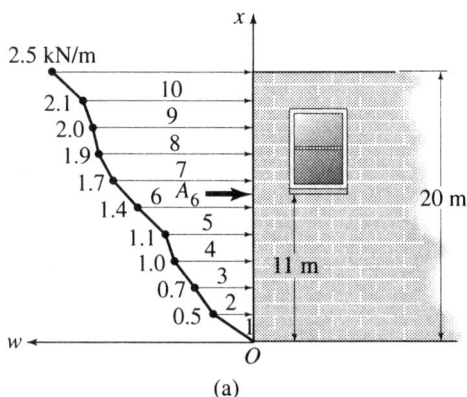

(a)

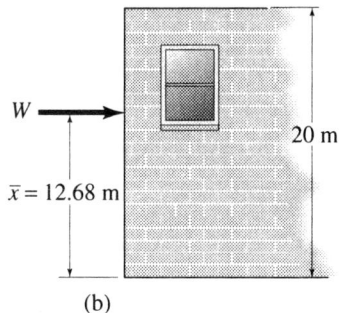

(b)

FIGURE E4.11.

To determine the resultant force W, we need to find the area A under the w-x curve. Because this curve is not described by a mathematical equation, we can obtain a good approximation for the area by either the trapezoidal rule or Simpson's rule. In this solution, the trapezoidal rule is used and, from Appendix C,

$$W = A = \Delta x(\tfrac{1}{2}x_0 + x_1 + x_2 + \cdots + x_{n-1} + \tfrac{1}{2}x_n)$$
$$= 2[\tfrac{1}{2}(0) + 0.5 + 0.7 + 1.0 + 1.1 + 1.4$$
$$+ 1.7 + 1.9 + 2.0 + 2.1 + \tfrac{1}{2}(2.5)]$$
$$= 27.3 \text{ kN.} \qquad \text{ANS.}$$

The location of this resultant is obtained by Varignon's theorem or the principle of moments which states that the moment of the resultant W about an axis through point O is equal to the sum of the moments of the components of the resultant about the same axis. In this solution, the components are defined as the resultants of the ten trapezoidal areas in Figure E4.11(a). Thus, for example, the resultant corresponding to trapezoidal area #6 is $A_6 = 1/2(1.4 + 1.7)(2) = 3.1$ kN and its moment arm is 11 m as shown in Figure E4.11(a). Thus, applying the principle of moments about point O,

$$-27.3\bar{x} = -\tfrac{1}{2}(0 + 0.5)(2)(1) - \tfrac{1}{2}(0.5 + 0.7)(2)(3) - \tfrac{1}{2}(0.7 + 1.0)(2)(5)$$
$$- \tfrac{1}{2}(1.0 + 1.1)(2)(7) - \tfrac{1}{2}(1.1 + 1.4)(2)(9) - \tfrac{1}{2}(1.4 + 1.7)(2)(11)$$
$$- \tfrac{1}{2}(1.7 + 1.9)(2)(13) - \tfrac{1}{2}(1.9 + 2.0)(2)(15)$$
$$- \tfrac{1}{2}(2.0 + 2.1)(2)(17) - \tfrac{1}{2}(2.1 + 2.5)(2)(19).$$

The solution of this equation leads to

$$\bar{x} = 12.68 \text{ m.} \qquad \text{ANS.}$$

The resultant W is shown properly positioned in Figure E4.11(b).

Problems

4.46 Replace the distributed force shown acting on the simply supported beam of Figure P4.46 by a single resultant force. Specify its magnitude, direction, and sense, and give its location with respect to the support at A.

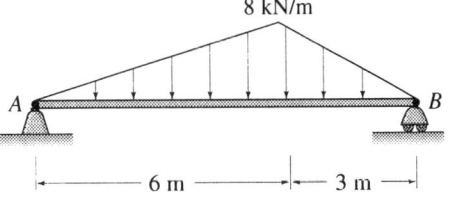

FIGURE P4.46.

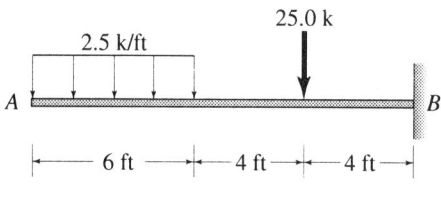

FIGURE P4.47.

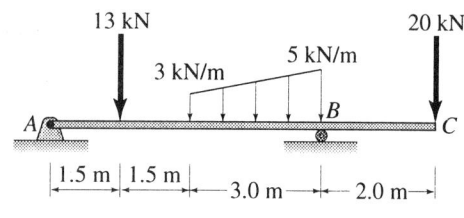

FIGURE P4.50.

4.47 Replace the force system acting on the cantilever beam of Figure P4.47 by a force and couple at support B.

4.48 Determine the resultant of the distributed force acting on the simply supported beam of Figure P4.48. Specify its magnitude, direction and sense, and give its location measured from the support at A.

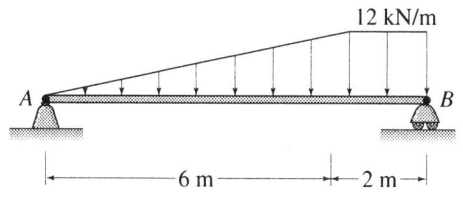

FIGURE P4.48.

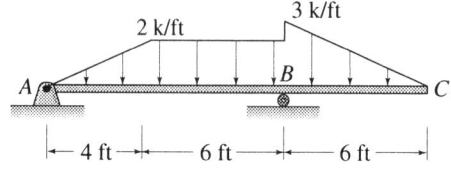

FIGURE P4.49.

4.49 Replace the distributed force system acting on the beam of Figure P4.49 by a force-couple system at support A.

4.50 Replace the force system shown acting on the beam of Figure P4.50 by a single resultant force. Give its magnitude, direction, and sense, and specify its location from point A.

4.51 Replace the force system acting on the cantilever beam of Figure P4.51 by a force-couple system acting at support A.

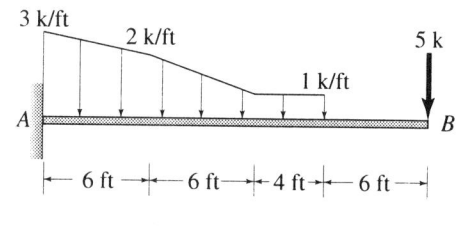

FIGURE P4.51.

4.52 A vertical concrete wall AB separates two bodies of water as shown in Figure P4.52. The hydrostatic forces due to water pressure on both sides of the wall are indicated by the triangular force distributions. Determine the resultant of this force system giving its magnitude, direction, and sense, and specify its location from point A.

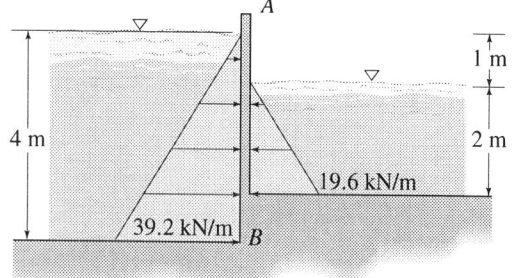

FIGURE P4.52.

4.53 The wings of an experimental fighter plane in level flight are subjected to distributed lift forces as shown in Figure P4.53. The magnitude of the loading is approximated by the relationship $w = x^2/25$ where x is in meters and w in kN/m. The weight of each wing including the contained fuel is $W = 8$ kN as shown. Consider the left wing AB and replace the force system acting on it by a force-couple system at B where the wing is joined to the fuselage. Make use of the information contained in Appendix A relative to a parabolic area.

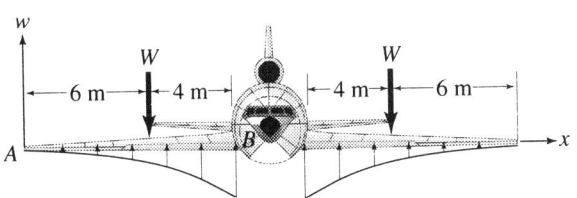

FIGURE P4.53.

4.54 The wind loading on a building is approximated by the force distributions shown in Figure P4.54. Replace this force system by a single resultant force specifying its magnitude, direction, and sense. Position this resultant force along wall AB (or AB extended).

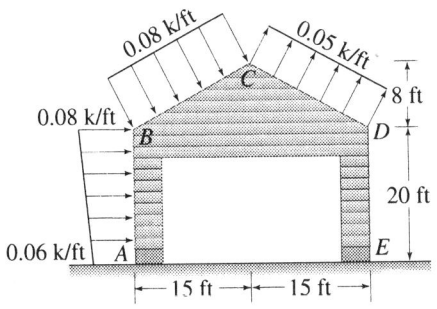

FIGURE P4.54.

4.55 A retaining wall is subjected to soil pressures that lead to force distributions approximated, as shown in Figure P4.55. Replace these force distributions by a force-couple system acting at point A which is at the mid-thickness of the wall.

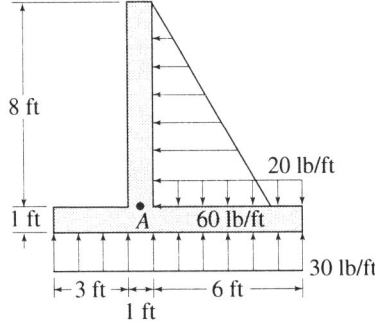

FIGURE P4.55.

4.56 A beam is subjected to force distributions, as shown in Figure P4.56. Determine the dimensions a and b if the resultant of the distributed force system is a downward vertical force passing through point B with a magnitude of 10 kN.

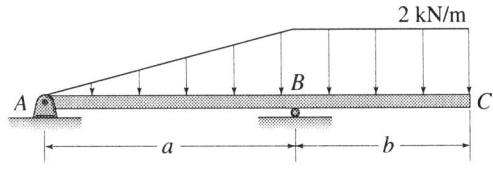

FIGURE P4.56.

4.57 Find the quantities w and a so that the beam of Figure P4.57 experiences no resultant force and no resultant couple.

4.58 Consider the distributed force acting on the beam of Figure P4.58. Determine the magnitude, direction, sense and location on the beam of the resultant force using

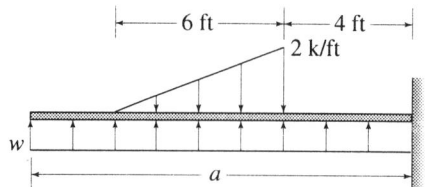

FIGURE P4.57.

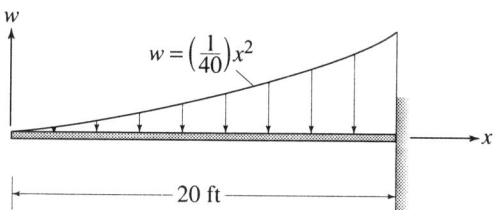

FIGURE P4.59.

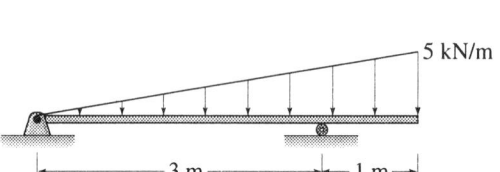

FIGURE P4.58.

(a) the information contained in Appendix A and (b) integration and Varignon's theorem.

4.59 The distributed force acting on the beam of Figure P4.59 is described by the relationship $w = x^2/40$ where w is in k/ft and x is in ft. Determine the magnitude, direction, sense, and location of the resultant force on the beam using (a) the information contained in Appendix A and (b) integration and Varignon's theorem.

4.60 The distributed force acting on the beam of Figure P4.60 is described by the relationship $w = 8\cos\left(\dfrac{\pi x}{10}\right)$ where w is in kN/m and x is in m. Use integration and Varignon's theorem to find the magnitude, direction, sense, and location of the resultant force on the beam.

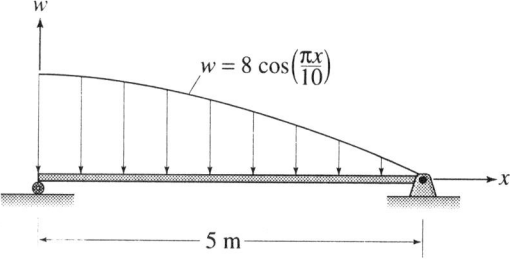

FIGURE P4.60.

4.7
Supports and
Connections

The concept of the free-body diagram was introduced in Chapter 2 in connection with the equilibrium of a particle. This concept will be extended in Section 4.8 to include the case of rigid bodies in equilibrium. As discussed in Chapter 2, before we can properly construct a free-body diagram, we need to be able to identify the reactive forces at supports and connections. Several of these supports and connections were discussed in Chapters 2 and 3 and are summarized in Appendix F. Three other types of supports that are frequently encountered in the treatment of two-dimensional equilibrium problems are discussed here and summarized in Appendix F as cases 7, 8, and 9.

Rollers or Rockers

Rollers or rockers are used often in the support of beams to allow movement in a given direction due to such factors as temperature changes. Examples of such supports are shown in Figure 4.15(a). In all instances, the supports allow movement in a direction parallel to the plane supporting the roller or rocker. Therefore, this type of support cannot sustain (resist) forces parallel to the plane. This conclusion is based upon the assumption that pure rolling without sliding occurs during any movement. However, the roller or rocker support system does not permit movement in a direction perpendicular to the plane, thus developing a reactive force N_B, normal to the plane, as shown in Figure 4.15(b) which represents all three of the cases shown in Figure 4.15(a). Thus, this type of support is effectively the same as the frictionless plane discussed in Chapter 2 and shown in case 5 of Appendix F.

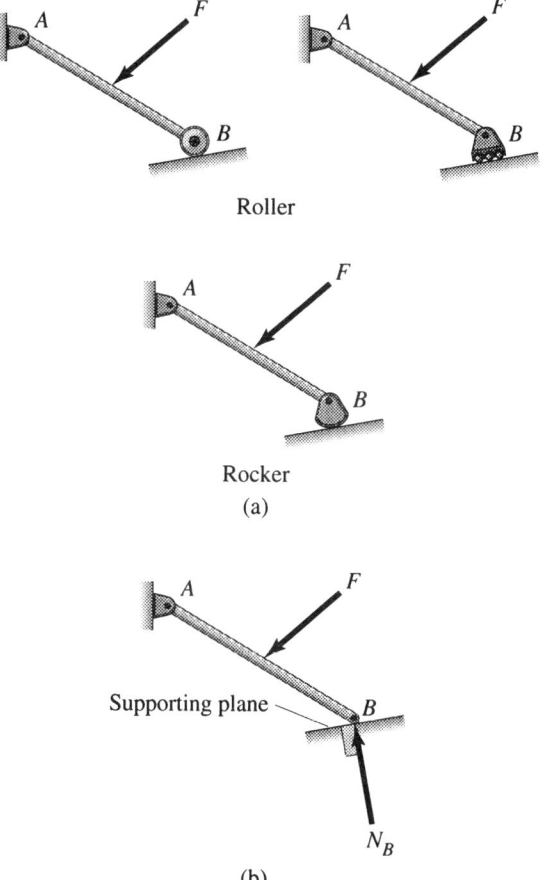

Roller

Rocker

(a)

(b)

FIGURE 4.15.

Collars or Sliders

Collars on smooth rods or sliders in frictionless slots are frequently used with machine or structural members to allow freedom of movement along the axis of the rod or the slot. Examples of these two types of supports are shown in Figure 4.16(a). As in the case of rollers and rockers, collars and sliders, cannot resist forces along the axis of the rod on which the collar is mounted or along the axis of the slot in which the slider fits. Of course, the assumption is made that frictionless conditions exist between the collar or the slider and their contacting surfaces. Also, as in the case of rollers and rockers, only a force N_A normal to the axis of movement can be transmitted, as indicated in Figure 4.16(b), which represents both of the cases shown in Figure 4.16(a).

Fixed Supports

Fixed supports are generally used with beams where it is desired to prevent movement in any direction or rotation about an axis perpendicular to the plane in which the beam is loaded. An example of a fixed support is shown at A in Figure 4.17(a). The symbolic representation used for the fixed support implies that the beam is imbedded deeply into the wall (or any other supporting structure). Thus, under ideal conditions, the beam is not able to translate in either the x or the y direction or to rotate about a z axis at the support. Preventing these translations and rotation requires that the fixed support develop reac-

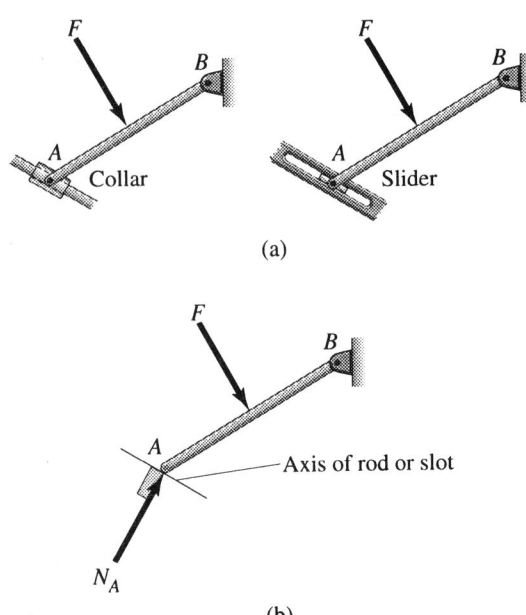

(a)

(b)

FIGURE 4.16.

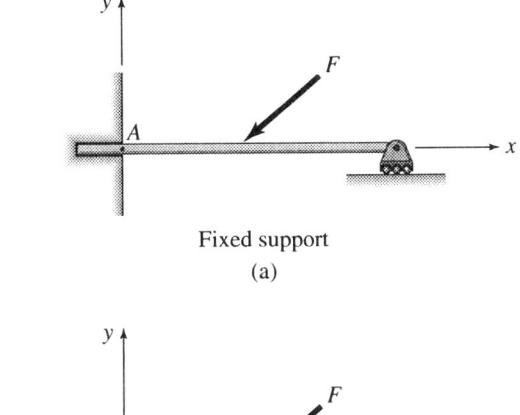

Fixed support

(a)

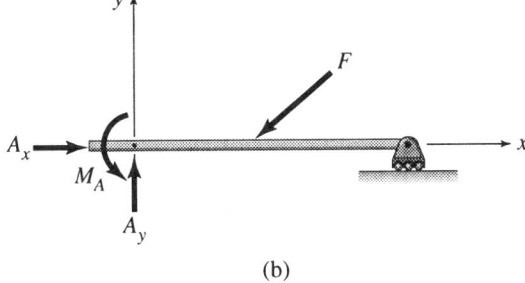

(b)   FIGURE 4.17.

tive force components A_x and A_y and reactive moment M_A, as shown in Figure 4.17(b). It should be pointed out that the symbols A_x and A_y represent the resultants of reactive force distributions acting on that part of the beam that is imbedded in the support.

The use of the above supports as well as others discussed previously will be illustrated in the examples at the end of Section 4.8.

4.8
The
Free-Body
Diagram

By definition, a free-body diagram is a sketch showing one body or a group of connected bodies, which has been isolated as a single unit away from its supports and other connections attaching it to the entire system of bodies representing a structure or a machine. The diagram should indicate clearly *all* of the forces acting on the unit. These forces should include the weight of the isolated unit which represents the gravitational pull of Earth, acts through the unit's *center of gravity*, and is directed toward Earth's center of gravity.

The concept of the center of gravity and its determination is discussed in Chapter 10. For our purposes here, however, the center of gravity of a body is a point in the body through which its weight is assumed to act. Such a force is known as a *gravitational force* usually given the symbol W. Other forces that should be included are those *externally applied* to the unit and *contact forces* exerted on the unit by its supports and other neighboring bodies attached to it. It is essential to ensure that the free-body diagram does *not* include forces with no direct bearing on the isolated unit even though they act at other locations in the system of which the unit is a part. Known forces should be shown fully in magnitude, direction, and sense. Unknown forces are indicated by directed line segments with the unknown magnitude and/ or sense represented by a suitable symbol. The sense of an unknown force may be assumed arbitrarily; a positive solution would indicate a correct assumption and a negative solution an incorrect one. Additionally, if the isolated unit consists of two or more connected bodies, it is essential *not* to include the internal forces of action and reaction exerted by the connected bodies on each other at the connection. These internal forces of action and reaction occur in pairs and cancel each other.

Because of the importance of the free-body diagram, the systematic procedure for a particle, introduced in Chapter 2, is repeated below, with modification, where appropriate.

1. A decision is made at the very outset as to which unit (a body or system of connected bodies) is to be isolated. This unit is then separated from the rest of the entire system and its external outline is shown neatly in a separate diagram.
2. If a coordinate system has not been provided, one should be clearly indicated in the free-body diagram along with all geometric features (known dimensions and angles) necessary for a correct application of the equations of equilibrium.
3. All forces acting *on* the isolated unit are then shown at their proper locations. These should include the weight of the unit and the externally applied forces, as well as those representing the action of removed connections and supports on the isolated unit. Known forces should be clearly shown in magnitude, direction, and sense whereas unknown forces are shown, with an assumed sense, by a directed line segment using a convenient symbol (including subscripting if needed) to represent its unknown magnitude and/or sense. It is essential that known and unknown forces be shown on the free-body diagram. It is also important to note that *all* of these forces are external to the isolated unit.

The concepts above dealing with the construction of free-body diagrams are illustrated in the following examples.

■ Example 4.12

The rigid angular plate shown in Figure E4.12(a) is hinged at A and supported by a roller at B. The plate may be assumed weightless and is

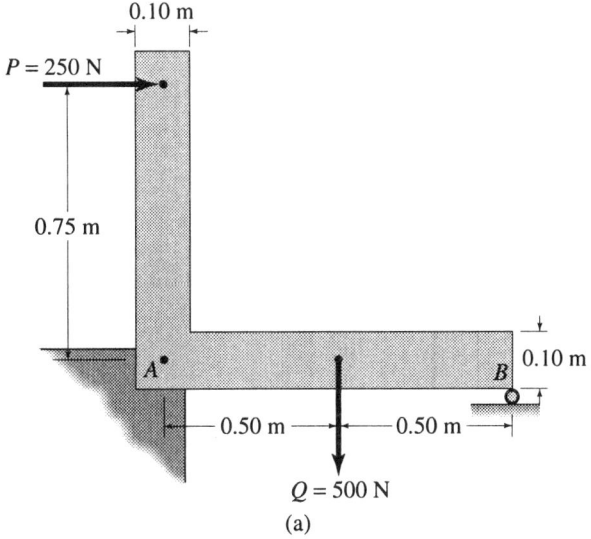

0.10 m

$P = 250$ N

0.75 m

0.10 m

A

B

0.50 m   0.50 m

$Q = 500$ N

(a)

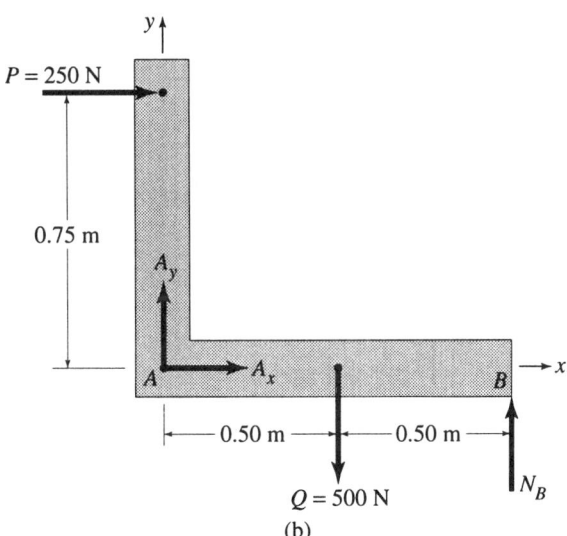

y

$P = 250$ N

0.75 m

A_y

A A_x B x

0.50 m   0.50 m

$Q = 500$ N

N_B

(b)   Figure E4.12.

subjected to the two forces P and Q, as shown. Construct the free-body diagram of the plate.

Solution

The free-body diagram of the angular plate, shown in Figure E4.12(b), was constructed by the three-step procedure stated above. The following comments are made:

1. After detachment from its hinge at A and its roller at B, the plate is acted upon by the two applied forces P and Q and by the reactive

forces at the hinge and the roller supports. The hinge is capable of resisting a force in some direction in the plane of the plate. Such a force is usually expressed in terms of two rectangular components. The roller can resist only a force perpendicular to the plane supporting the roller.

2. The known forces P and Q are shown in their proper locations and with proper magnitude, direction, and sense. With the x-y coordinate system, as shown, the unknown reactive force components at hinge A may be denoted as A_x and A_y. It is good practice to assume that the sense of such unknown forces is in the positive directions of the corresponding coordinate axes. The reactive force N_B at the roller support (point B) is perpendicular to the plane supporting the roller. Because the plane is horizontal in this case, it follows that N_B must be in the vertical direction or along the y axis. For this reason, another acceptable notation for the support reaction at B is B_y. For the given conditions, it is obvious that the reactive force N_B must have a sense upward. In less obvious cases, the sense of the reactive force may be assumed either along the inward or outward normal to the plane. A positive solution for any of the three unknown forces A_x, A_y, and N_B indicates a correct assumption and a negative solution an incorrect one.

3. To complete the construction of the free-body diagram of the plate, its geometry must be specified. In our particular problem, the required geometry consists of the dimensions of the plate, particularly, the distances between points of force application as shown in Figure E4.12(b).

■ Example 4.13

Refer to the frame shown in Figure E4.13(a) and construct the free-body diagram of (a) the entire frame and (b) the horizontal member GEB including the pulley. Assume that all members of the frame are weightless.

Solution

(a) The free-body diagram of the entire frame is shown in Figure E4.13(b). It was constructed by the three-step procedure outlined earlier. The following observations are made:

1. After detachment from its hinges at A and F, an outline of the frame is shown ignoring any unnecessary internal details. The frame is acted upon by the three applied forces P_1, P_2, and P_3 and by the reactive forces at the two hinges. The support reaction at the hinge may be expressed in terms of two rectangular components.

2. The known applied forces P_1, P_2, and P_3 are shown in proper location and with proper magnitude, direction, and sense. Using the x-y coordinate system as shown, the reaction components at hinges A

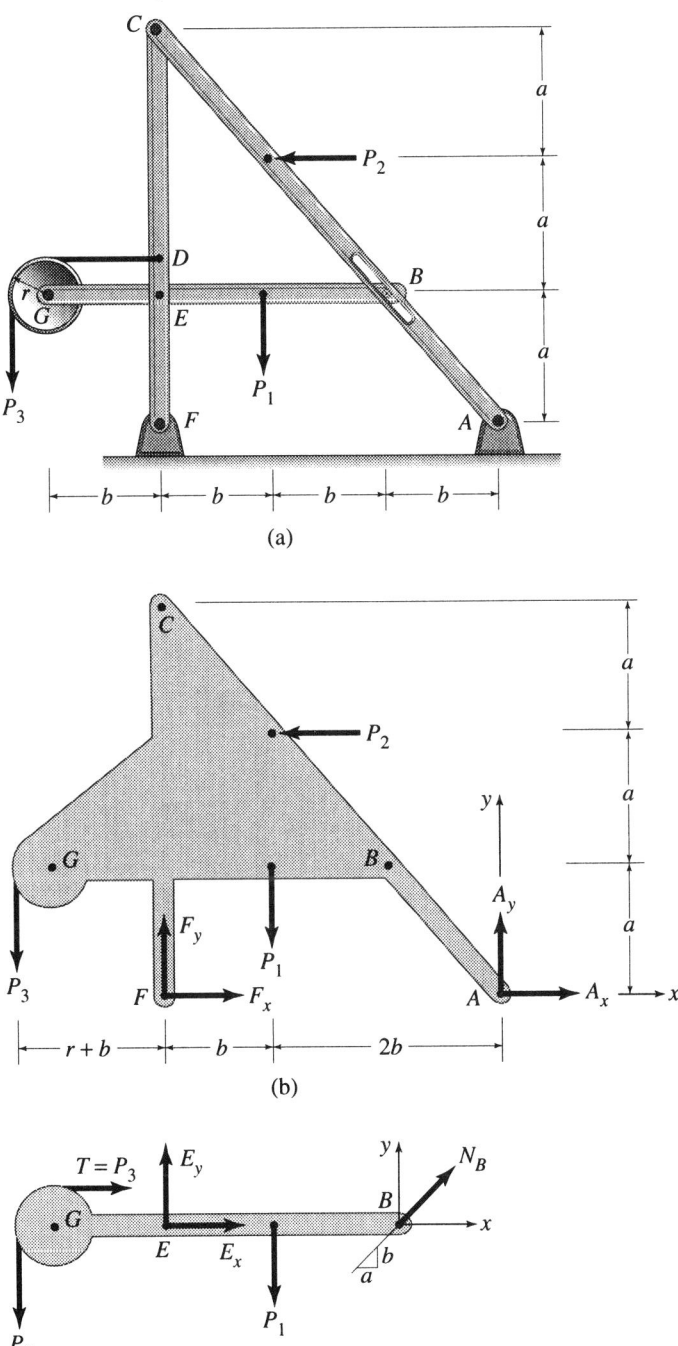

FIGURE 4.13.

and F are represented, respectively, by the symbols A_x, A_y, F_x, and F_y assumed to have senses along the positive directions of their respective axes. This assumption is confirmed or denied once a solution is obtained.

3. All significant dimensions are also indicated on the free-body diagram to facilitate application of the conditions of equilibrium as will be demonstrated in the next section.

(b) Again, by the three-step procedure, the free-body diagram of member GEB is constructed as shown in Figure E4.13(c). The following comments are made:

1. After detachment from the hinge at E, the cable at D, and the slider at B, an outline of the member GEB is drawn including the pulley. This member-pulley system is acted upon by the externally applied forces P_1 and P_3 and by the reactive forces at hinge E, slider B, and the cable tension at D. The support reaction at the hinge may be expressed in terms of two rectangular components. The reaction at the slider is a force normal to the axis of the slot and the cable tension is, of course, along the axis of the cable.

2. The known applied forces P_1 and P_3 are shown in proper location and with proper magnitude, direction, and sense. Using the x-y coordinate system as shown, the reaction components at hinge E are denoted by the symbols E_x and E_y. The reaction at slider B, N_B, is normal to the axis of the slot. Because the slot is along member AC, whose slope is known, the direction of N_B can be specified as shown in Figure E4.13(c). Note that the sense of N_B is assumed upward and to the right. A positive solution would indicate a correct assumption and a negative solution an incorrect one. Because a pulley changes only the direction but not the magnitude of the force in the cable, it follows that the cable tension T acting on the pulley from the anchor point at D must be equal in magnitude to P_3.

3. Finally, all significant dimensions needed for correct application of the equations of equilibrium are included on the free-body diagram, as shown in Figure E4.13(c).

■

Problems

4.61 Construct the free-body diagram of the uniform rod AB shown in Figure P4.61. The weight of the rod $W = 60$ lb is assumed to act through the center of gravity G located at the geometric center of the rod. Note that the frictionless roller at the top enables the spring to remain vertical.

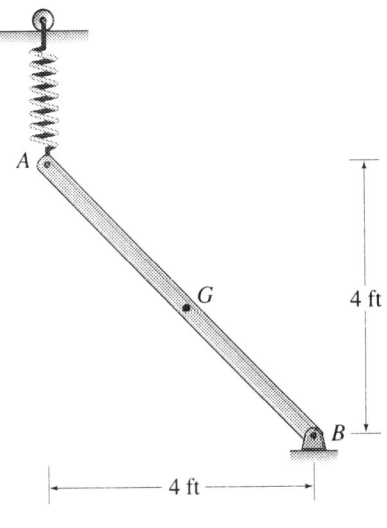

FIGURE P4.61.

4.62 The uniform member AB leans against a frictionless vertical wall at A and is hinged at B, as shown in Figure P4.62. The weight of the member is 250 N and may be assumed to act through its geometric center. Construct the free-body diagram of member AB.

4.63 The uniform rectangular plate ABCD is hinged at A and supported at B by cable

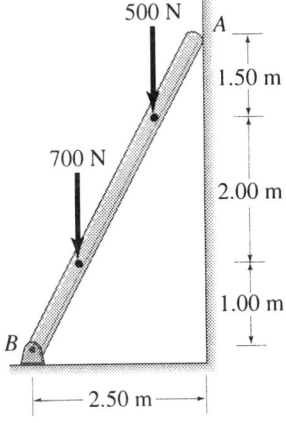

FIGURE P4.62.

BE. A force F is applied at C by a pulley-cable system, as shown in Figure P4.63. The weight of the plate is W and may be assumed to act at the geometric center of the plate. Construct the free-body diagram of plate ABCD.

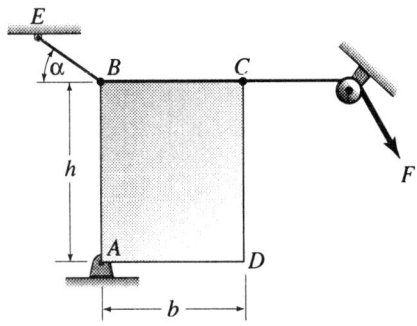

FIGURE P4.63.

4.64 Beam ABC is fixed at A and is supported by a roller on a horizontal surface at C, as shown in Figure P4.64. Assume that the beam is weightless and construct its free-body diagram.

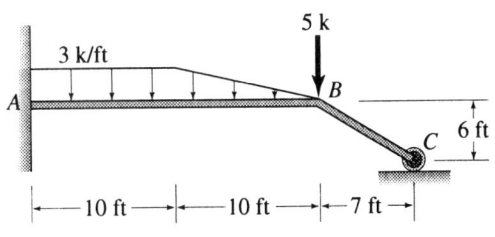

FIGURE P4.64.

4.65 Beam ABC is hinged at A and is supported by a roller at B as shown in Figure P4.65. Assume that the beam is weightless, and construct its free-body diagram.

4.66 Beam AB is hinged at A and is supported by a roller at B, as shown in Figure P4.66.

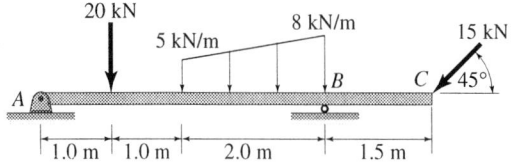

FIGURE P4.65.

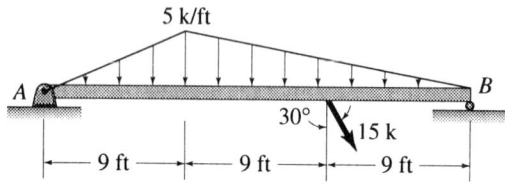

FIGURE P4.66.

Assume that the beam is weightless, and construct its free-body diagram.

4.67 A truss is hinged at A and supported by a roller system at B, as shown in Figure P4.67. Assume that the weight of the truss is 5 kN and acts at C. Construct the free-body diagram of the truss.

4.68 A schematic representation of a garden wheel barrow is shown in Figure P4.68. Assume that the weight of the wheelbarrow, including the load in the bucket,

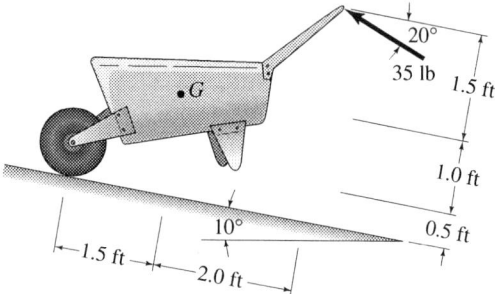

FIGURE P4.68.

is 81 lb and acts through the center of gravity at G. The force applied by the gardener to move the wheel barrow at constant speed up the 10° inclined plane is 35 lb, as shown. Construct the free-body diagram of the wheel-barrow. Assume that the 10° inclined plane is rough.

4.69 The uniform member AB shown in Figure P4.69 has a weight of 500 N and is attached to a collar at B that can slide freely on the vertical weightless post CD. It is supported at A by a frictionless plane inclined to the horizontal at the angle α. Assume that the weight of member AB acts through its geometric center. Construct the free-body diagram (a) of member AB and (b) weightless post CD.

4.70 Refer to the mechanism shown in Figure P4.70. By applying the proper value of

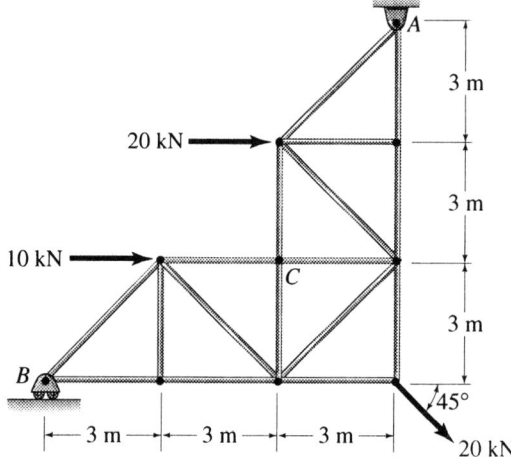

FIGURE P4.67.

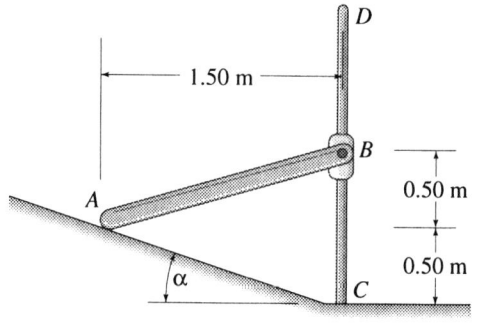

FIGURE P4.69.

the load P, the cable AD may be stretched by any desired amount. Member BD is hinged at both ends and may be assumed to be a weightless link. The slider at D moves in a frictionless slot. For the case when $P = 1.2$ k, construct the free-body diagram of (a) member ABC, (b) link BD, and (c) the slider at D.

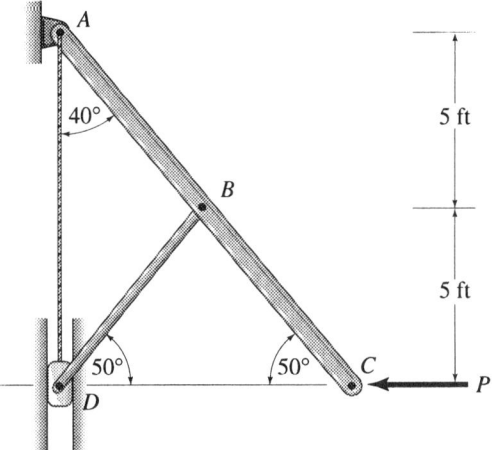

FIGURE P4.70.

4.71 A schematic representation of a pair of pliers used to grip rod C is shown in Figure P4.71. Assume that the hinge at B is frictionless, and construct the free-body diagram of members AB and DB, and rod C. Place all three free-body diagrams in proper position with respect to each

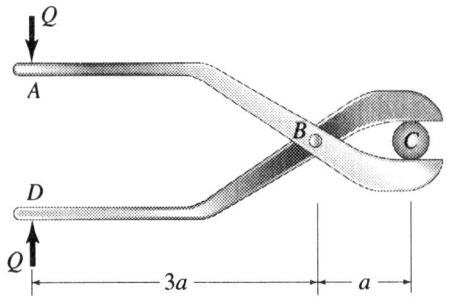

FIGURE P4.71.

other to show the force interactions among them.

4.72 The frame shown in Figure P4.72 is hinged at A and supported by a roller system at E. Construct the free-body diagram of (a) the entire frame and (b) member EDC. Assume weightless members and consider that member BD acts as a short link.

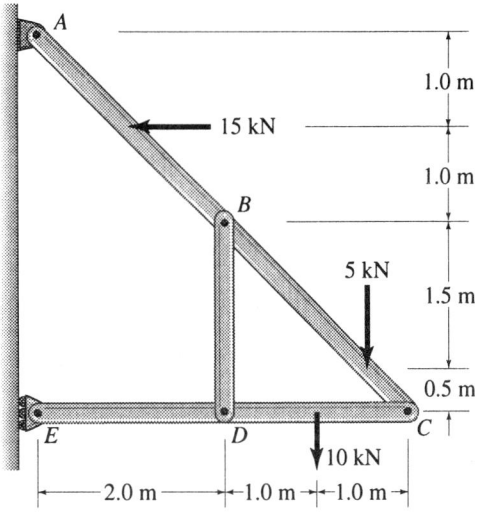

FIGURE P4.72.

4.73 Refer to Problem 4.72 and construct the free-body diagram of (a) member ABC and (b) member DB which may be assumed to act as a short link.

4.74 The frame of Example 4.13 is repeated in Figure P4.74 for convenience. Review the solution in Example 4.13, and construct the free-body diagram of (a) member ABC and (b) member CDEF.

4.75 A toggle press is shown in Figure P4.75. Assume that the hinges at A, B and C are frictionless, and construct the free-body diagram of (a) member BCD and (b) member AB which may be assumed to act as a short link. Assume that all members are weightless.

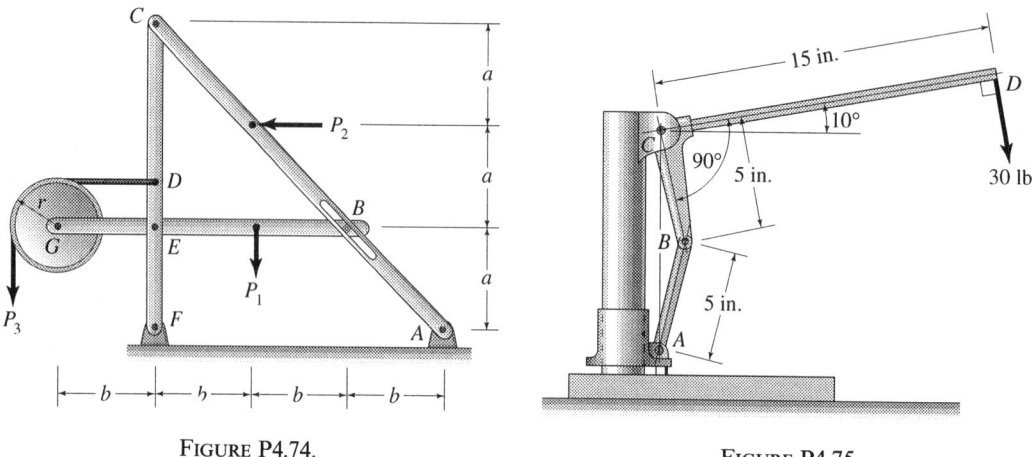

FIGURE P4.74.   FIGURE P4.75.

4.9
Equilibrium
Conditions
and
Applications

Conditions for Equilibrium

The concept of equilibrium was introduced in Chapters 2 and 3 where it was discussed in relation to a single particle. In this chapter, this concept is developed further in connection with a rigid body.

As discussed in relation to a particle, *equilibrium* requires that the resultant of the force system acting on the particle must be zero. In the case of a rigid body, we learned in Section 4.4, that the most general two-dimensional force system may be replaced by a resultant force $\mathbf{R}$ and a resultant couple $\mathbf{M_R}$. Thus, equilibrium of a rigid body in two-dimensions requires that both $\mathbf{R}$ and $\mathbf{M_R}$ be equal to zero independently. In the two-dimensional case defined by an x-y coordinate system, $\mathbf{R} = \sum \mathbf{F} = (\sum F_x)\mathbf{i} + (\sum F_y)\mathbf{j}$, and $\mathbf{M_R} = \sum \mathbf{M} = (\sum M_{zA})\mathbf{k}$. The notation $\sum M_{zA}$ represents the magnitude of the sum of all the moments about a z axis through an arbitrary point A. In subsequent work, this notation is simplified by deleting the subscript z and writing only $\sum M_A$. This is justified on the basis that no confusion is likely in the two-dimensional case defined by an x-y coordinate system because forces in this system produce moments *only* about z axes. Hence, the two-dimensional equilibrium of a rigid body may be stated mathematically by the relationships

$$\mathbf{R} = \sum \mathbf{F} = (\sum F_x)\mathbf{i} + (\sum F_y)\mathbf{j} = 0 \qquad (4.11a)$$

and

$$\mathbf{M_R} = \sum \mathbf{M} = (\sum M_A)\mathbf{k} = 0. \qquad (4.11b)$$

The vector relationships expressed in Eqs. (4.11) lead to the following three scalar equations that are both necessary and sufficient for the equilibrium of a rigid body in two dimensions:

$$\left.\begin{array}{l} \sum F_x = 0, \\[4pt] \sum F_y = 0, \\[4pt] \sum M_A = 0. \end{array}\right\} \qquad (4.12)$$

and

The first two of the equations above show that, if a force system is to be in equilibrium insofar as translation in the x and y directions, then, the components of all forces in the x and y directions, respectively, must be in perfect balance. The last of the three equations above states that, if a force system is to be in equilibrium insofar as rotation about a z axis at point A, then, the moments of all forces about the z axis must be in perfect balance. Therefore, if a rigid body is to be in complete equilibrium in two dimensions, a perfect and complete balance of forces must exist in each of the x and y directions and of moments about a z axis passing through any point A in the x-y plane. Note further that the three conditions of equilibrium expressed in Eq. (4.12) are independent of each other and that one may be satisfied and not the others. Thus, for example, it is possible in a given case for $\sum F_x = 0$ and $\sum F_y = 0$ whereas $\sum M_A \neq 0$. In such a case, the rigid body does not translate but will have rotational acceleration about a z axis through point A in the x-y plane.

Now, consider the two-dimensional force system shown in Figure 4.18(a) and refer back to the three conditions of equilibrium expressed in Eq. (4.12). It is important to note that there are two alternate ways of expressing these three conditions in solving specific two-dimensional problems. The first way consists of one force equation and two moment equations. Thus

$$\left.\begin{array}{l} \sum F_y = 0, \\[4pt] \sum M_A = 0, \\[4pt] \sum M_B = 0. \end{array}\right\} \qquad (4.13)$$

and

As discussed in Section 4.5, a general two-dimensional force system of the type shown in Figure 4.18(a) may be replaced by a single resultant force $\mathbf{R}$ at point A as shown in Figure 4.18(b). The condition $\sum F_y = 0$ in Eq. (4.13) requires that this resultant, if it exists, must be perpendicular to the y axis, i.e., parallel to the x axis, as shown in Figure 4.18(b). Because the resultant $\mathbf{R}$ passes through point A, the condition $\sum M_A = 0$ is obviously satisfied. Finally, the condition $\sum M_B = 0$ is satisfied if point B is along the x axis or if the resultant $\mathbf{R}$ does not exist, in which case, the force system is in equilibrium. Therefore, the three conditions shown in Eq. (4.13) are not only necessary but also sufficient for the equilibrium of a two-dimensional force system, *provided that the line joining points A and B is not parallel to the x axis.*

The second alternate way of expressing the three conditions of equilibrium consists of three moment equations. Thus,

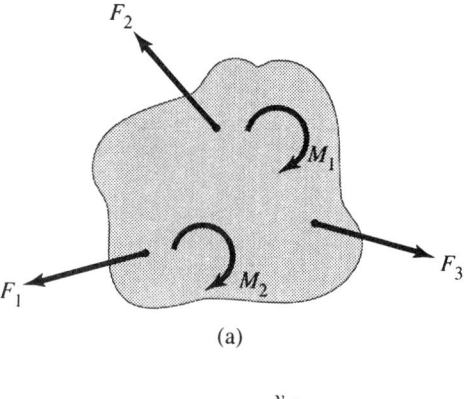

(a)

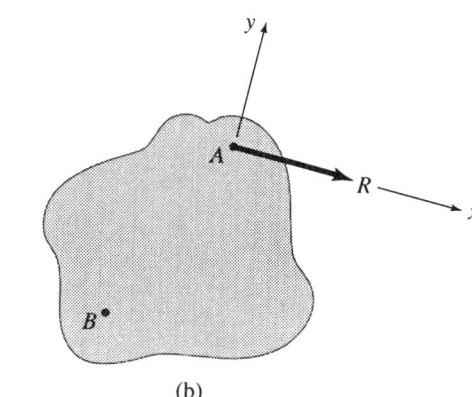

FIGURE 4.18.   (b)

$$\begin{array}{r}
\sum M_A = 0, \\
\sum M_B = 0, \\
\sum M_C = 0.
\end{array}\Bigg\} \qquad (4.14)$$

and

These three conditions require that, if the force system has a resultant
R, it must pass through the three points A, B, and C. Obviously, this is
only possible if the three points lie on the same straight line. Thus, the
three moment conditions expressed in Eq. (4.14) are necessary and
sufficient for the equilibrium of a two-dimensional force system *provided that points A, B, and C do not lie on the same straight line.*

   It is important to know that, regardless of which set of equilibrium
conditions is used (i.e., Eq. (4.12), (4.13) or (4.14)), only three unknown
quantities may be solved for in any specific free-body diagram. If only
three unknown quantities are present in a two-dimensional free-body
diagram which is prevented from any type of motion, the force system is
known as *statically determinate.* If more than three unknown quantities
exist, the force system is called *statically indeterminate.* Only statically
determinate force systems are dealt with in this chapter. However, a

brief discussion of determinacy and constraints is provided in Section 5.10.

Four special cases of two-dimensional force systems are identified and discussed as follows:

Two-Force Member

By definition, a two-force member is one that carries no moments and is subjected to forces only at two locations. This condition requires that the weight of the member be neglected, an assumption which is generally justified. Such members are encountered frequently in structures and machines. Consider, for example, the member shown in Figure 4.19(a) which is subjected to forces at points A and B only. The forces F_A and F_B may each represent the resultant of several forces acting at points A and B, respectively. Rotational equilibrium means that the sum of the moments about any axis is zero. Thus, $\sum M_A = 0$ requires that F_B passes through point A as in Figure 4.19(b), and $\sum M_B = 0$ necessitates that F_A passes through point B as in Figure 4.19(c). Also, if we arbitrarily chose the x axis to be in a direction from A to B as in Figure 4.19(c), the condition $\sum F_x = 0$ leads to the conclusion that F_A and F_B must be equal in magnitude and opposite in sense.

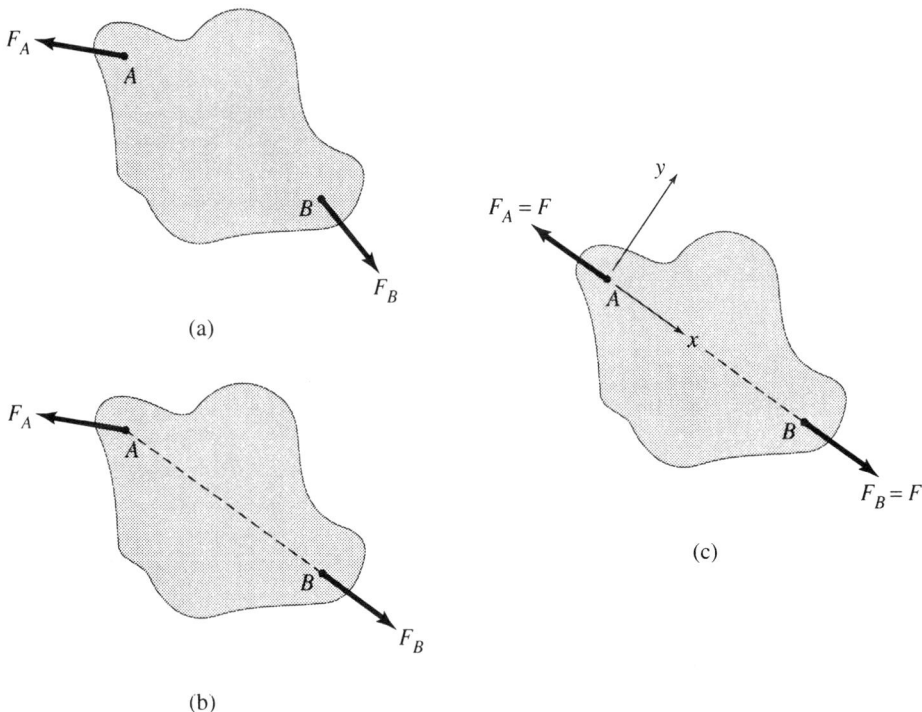

FIGURE 4.19.

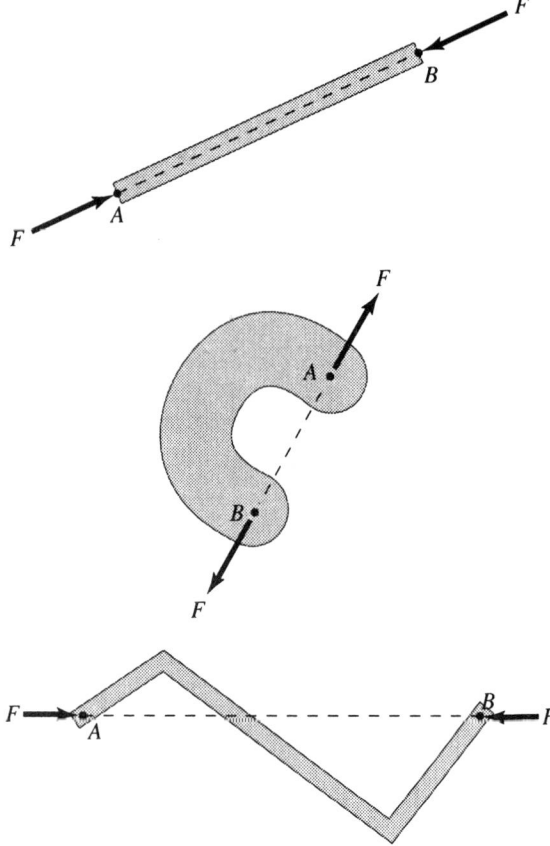

FIGURE 4.20.

The remaining equilibrium condition $\sum F_y = 0$ is neither necessary nor applicable in this case. Therefore, we conclude that in the case of a member subjected to forces at two locations only (a two-force member), *the two forces must be equal in magnitude, opposite in sense, and must be directed along the line connecting the two points of force application for the member to be in equilibrium.* This conclusion is valid regardless of the shape of the two-force member. Other examples of two-force members are shown in Figure 4.20.

Parallel Force System

An example of such a force system is shown in Figure 4.21. Note that the coordinate system was arbitrarily chosen so that the forces are parallel to the x axis. Because such a force system does not have components in the y direction, the second force condition in Eq. (4.12) $\sum F_y = 0$ is neither applicable nor necessary, and the set of three equilibrium conditions in Eq. (4.12) reduces to $\sum F_x = 0$ and $\sum M_A = 0$.

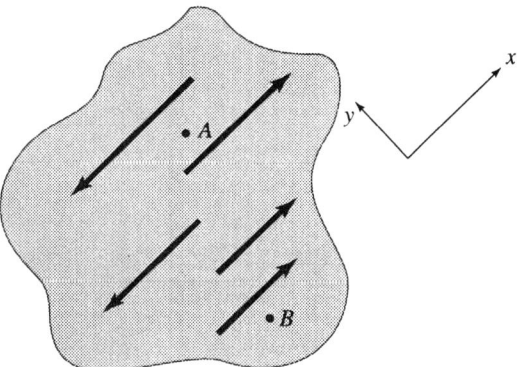

FIGURE 4.21.

Alternatively, two of the three moment conditions expressed in Eq. (4.14) as, for example, $\sum M_A = 0$ and $\sum M_B = 0$ may be used. However, points A and B must be any two points that *do not* lie on a line parallel to the force system, as shown in Figure 4.21. In either case (Eq. (4.12) or Eq. (4.14)) only a maximum of two unknown quantities may be determined for a given free-body diagram of a parallel force system.

Concurrent Force System

A concurrent force system is shown in Figure 4.22 where an arbitrary x-y coordinate system has been selected. Only the first two of the equilibrium conditions of Eq. (4.12), $\sum F_x = 0$ and $\sum F_y = 0$, are needed

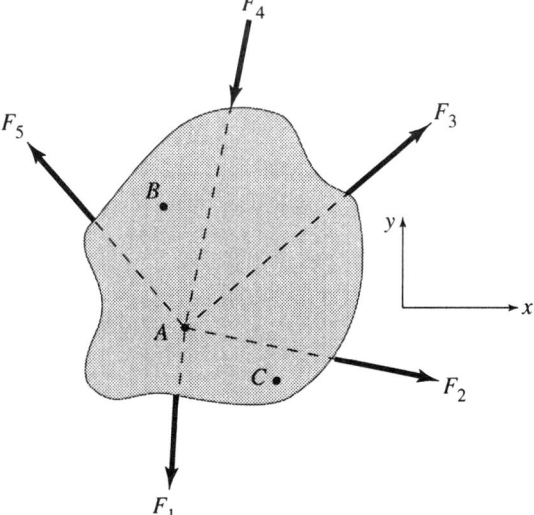

FIGURE 4.22.

because all of the forces pass through the same point A and, therefore, the third equilibrium condition $\sum M_A = 0$ is automatically satisfied. Alternatively, two of the three moment conditions of Eq. (4.14), for example, $\sum M_B = 0$ and $\sum M_C = 0$, may be used, where B and C are any two points *not* collinear with the point of concurrency, as shown in Figure 4.22. Note that, whether Eq. (4.12) or Eq. (4.14) is used, only a maximum of two unknown quantities may be solved for in the case of a given free-body diagram of a concurrent force system.

Three-Force Member

A special case of a concurrent two-dimensional force system is a member that carries no moments and is subjected to forces at only three locations. This member is known as a *three-force member* which can be in equilibrium only if the three forces acting on it are concurrent. Consider, for example, the member shown in Figure 4.23(a) which is subjected to three forces F_1, F_2, and F_3 at three points D, E, and G, respectively. If the point of concurrency of any two of the three forces, such as F_1 and F_2, is A, then, F_3 must pass through this point if the three forces are to be in complete equilibrium. Otherwise, F_3 will produce a moment about a z axis through point A. This conclusion is very useful in graphical solutions of equilibrium problems *but not in the analytical solutions* pursued in this chapter. Such analytical solutions are based on either Eq. (4.12), $\sum F_x = 0$ and $\sum F_y = 0$, or Eq. (4.14), $\sum M_B = 0$ and $\sum M_C = 0$ where, as stated earlier, points B and C are

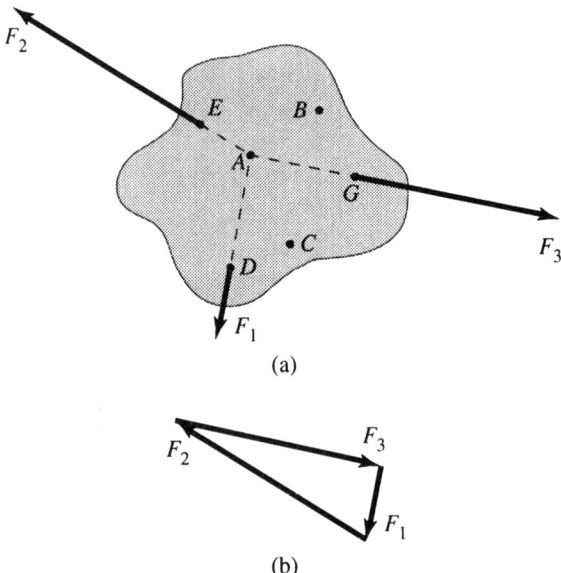

(a)

(b)

FIGURE 4.23.

any two points *not* collinear with the point of concurrency, as shown in Figure 4.23(a). Note that any of the three forces F_1, F_2, and F_3 represents a single force or the resultant of several forces acting at points D, E, or F, respectively.

Occasionally, the force-polygon method is extremely helpful in solving problems dealing with three-force members because the force polygon for a three-force member in equilibrium is a closed triangle whose geometry may be analyzed to determine the relationships among the three forces. Thus, for example, the forces F_1, F_2, and F_3 acting on the three-force member of Figure 4.23(a) would have to form the closed force triangle shown in Figure 4.23(b) if the three-force member is in equilibrium.

The following examples illustrate some of the concepts discussed above.

■ Example 4.14

Beam AB is hinged at A and supported by a roller at B as shown in Figure E4.14(a). Determine the reaction components at A and B.

Solution

The free-body diagram of the beam, shown in Figure E4.14(b), indicates three unknown reactive forces, A_x, A_y, and B_y. Note that the uniform distribution has been replaced by its resultant of 20.0 k acting through the centroid of the distribution, located 5 ft to the right of support A. An appropriate coordinate system indicating the positive directions for x and y and for moments is shown in Figure E4.14(c).

Because three independent conditions of equilibrium are available, the force system is statically determinate, and a complete solution for the unknown reactive forces is possible. Using the three conditions expressed in Eq. (4.12),

$$\sum F_x = 0, \quad A_x - 15.0(\cos 60°) = 0,$$

$$A_x = 7.50 \text{ k} \ \rightarrow . \qquad \text{ANS.}$$

$$\sum M_A = 0, \quad 20B_y - 15.0(\sin 60°)(15) - 20.0(5) = 0,$$

$$B_y = 14.74 \text{ k} \ \uparrow . \qquad \text{ANS.}$$

$$\sum F_y = 0, \quad A_y + 14.74 - 20 - 15.0(\sin 60°) = 0,$$

$$A_y = 18.25 \text{ k} \ \uparrow . \qquad \text{ANS.}$$

The positive answers indicate that A_x, A_y, and B_y were assumed correctly on the free-body diagram. The same answers are obtained by the three equilibrium conditions expressed in Eq. (4.13). Thus,

$$\sum M_A = 0, \quad 20B_y - 15.0(\sin 60°)(15) - 20.0(5) = 0,$$

$$B_y = 14.74 \text{ k} \ \uparrow . \qquad \text{ANS.}$$

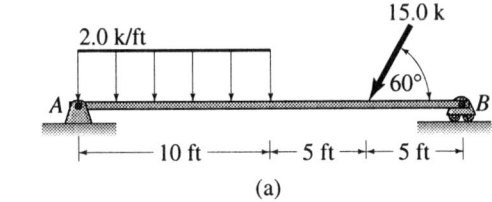

(a)

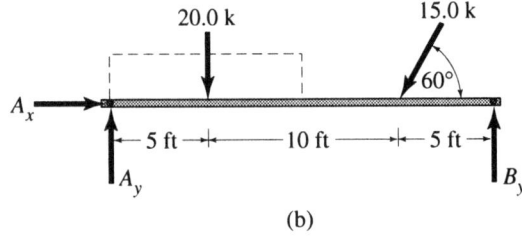

(b)

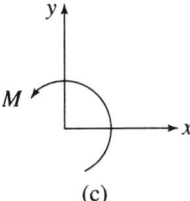

(c)

FIGURE E4.14.

$$\sum M_B = 0, \quad 15.0(\sin 60°)(5) + 20.0(15) - 20A_y = 0,$$

$$A_y = 18.25 \text{ k} \uparrow. \qquad \text{ANS.}$$

$$\sum F_x = 0, \quad A_x - 15.0(\cos 60°) = 0,$$

$$A_x = 7.50 \text{ k} \rightarrow. \qquad \text{ANS.}$$

Obviously the answers obtained by the two sets of equilibrium conditions are identical. Under certain conditions, one set may be more convenient to use than the other. It is desirable to perform a check on a given solution. For example, after solving the problem above using the conditions in Eq. (4.12) which include $\sum M_A = 0$, we may perform a partial check by the use of $\sum M_B = 0$ from Eq. (4.13). Also note that the three equilibrium conditions given in Eq. (4.14) could be used but this would require establishing a third moment center that does not lie along the line joining A and B.

■ **Example 4.15**   A loaded cantilevered structure is shown in Figure E4.15(a). Determine the reaction components at the fixed support at A.

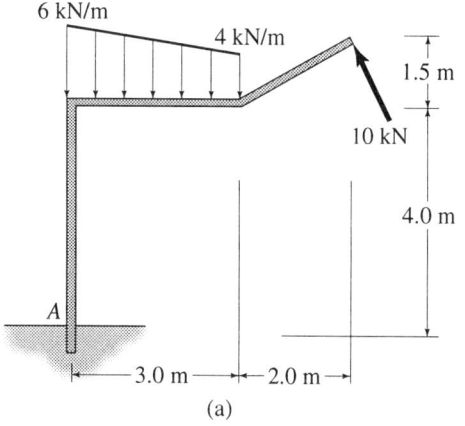

(a)

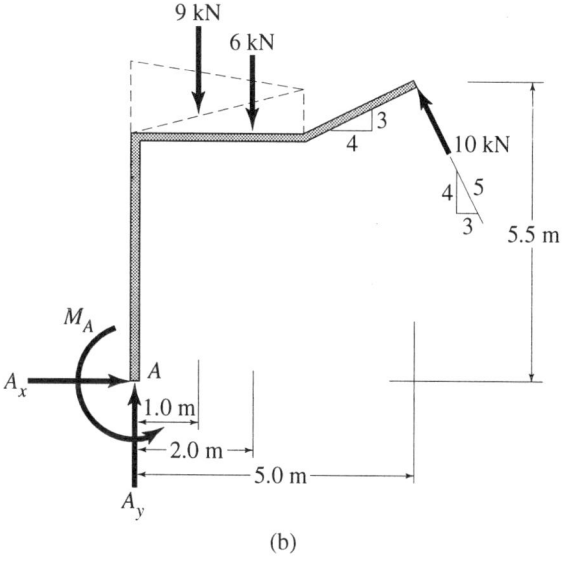

(b)

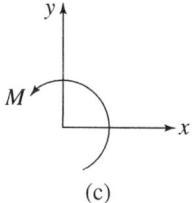

(c)

FIGURE E4.15.

Solution

The free-body diagram of the cantilevered structure, shown in Figure E4.15(b), indicates three reaction components A_x, A_y, and M_A at the fixed support. Note that the trapezoidal force distribution was considered to consist of two triangular distributions which were, then, replaced by their respective resultants acting through their corresponding centroids. A coordinate system indicating the positive senses of the x and y axes as well as the positive sense of the moment is shown in Figure E4.15(c).

The three conditions of equilibrium expressed in Eq. (4.12) are used to determine the three unknown reactive components. Thus,

$$\sum F_x = 0, \quad A_x - \tfrac{3}{5}(10) = 0,$$

$$A_x = 6.00 \text{ kN} \rightarrow. \qquad \text{ANS.}$$

$$\sum F_y = 0, \quad A_y - 9 - 6 + \tfrac{4}{5}(10) = 0,$$

$$A_y = 7.00 \text{ kN} \uparrow. \qquad \text{ANS.}$$

$$\sum M_A = 0, \quad \tfrac{3}{5}(10)(5.5) + \tfrac{4}{5}(10)(5.0) + M_A - 6(2.0) - 9(1.0) = 0,$$

$$M_A = -52.0 = 52.0 \text{ kN·m} \; \circlearrowright. \qquad \text{ANS.}$$

The positive answers confirm the assumed senses for A_x and A_y whereas the negative answer indicates that the assumed sense for M_A is incorrect and must be reversed.

■ Example 4.16

Consider the structure shown in Figure E4.16(a). Assume weightless members, and determine the reaction components at hinge supports A and C.

Solution

The free-body diagram of member BCD is shown in Figure E4.16(b). An appropriate coordinate system indicating the positive senses for x and y and for moments is shown in Figure E4.16(c). The hinge support at C develops the two reaction components C_x and C_y. Recognizing that member AB is a two-force member, we conclude that the force F at B on member BCD must be directed along the line joining hinges A and B. Thus, whereas the magnitude and sense of the force F at joint B are unknown, its direction is known and its two components F_x and F_y may be expressed in terms of the single unknown F. Thus, the free-body diagram shown in Figure E4.16(b) contains three unknown quantities, F, C_x, and C_y, and three conditions of equilibrium are available for their determination. Using the equilibrium conditions expressed in Eq. (4.12),

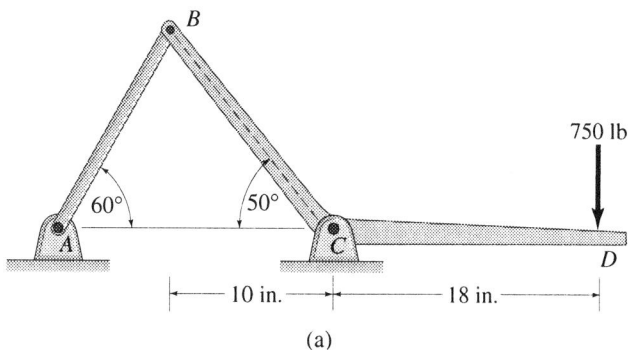

(a)

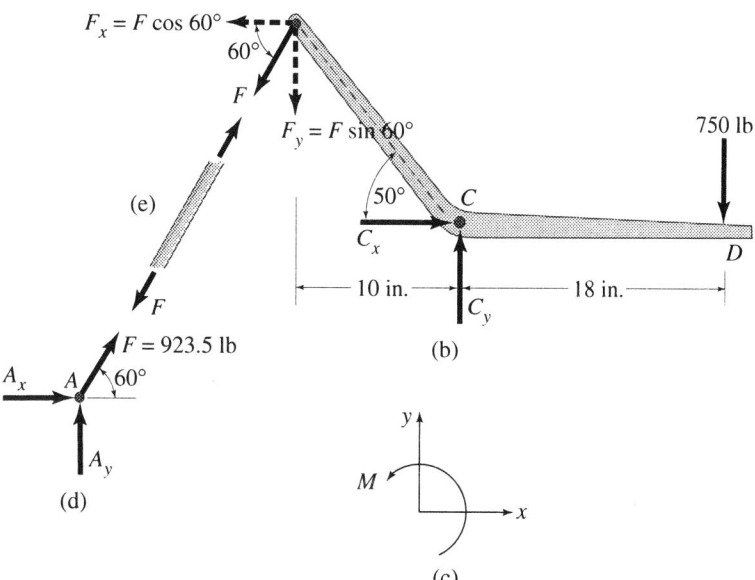

FIGURE E4.16.

$$\sum M_C = 0, \quad (F \cos 60°)(10 \tan 50°) + (F \sin 60°)(10) - 750(18) = 0,$$

$$F = 923.5 \text{ lb}.$$

$$\sum F_x = 0, \quad C_x - 923.5(\cos 60°) = 0,$$

$$C_x = 462 \text{ lb} \rightarrow . \qquad\qquad \text{ANS.}$$

$$\sum F_y = 0, \quad C_y - 750 - 923.5 (\sin 60°) = 0,$$

$$C_y = 1550 \text{ lb} \uparrow . \qquad\qquad \text{ANS.}$$

The components of the support reaction at the hinge at A may be obtained by considering the free-body diagram of the pin at this support as shown in Figure E4.16(d). This free-body diagram shows that there are two unknown reaction components, A_x and A_y. Two equations of equilibrium are available for their determination. Note the free-body diagram of a segment of member AB shown in Figure E4.16(e). Thus,

$$\sum F_x = 0, \quad A_x + 923.5(\cos 60°) = 0,$$

$$A_x = -462 = 462 \text{ lb} \leftarrow. \qquad\qquad \text{ANS.}$$

$$\sum F_y = 0, \quad A_y + 923.5(\sin 60°) = 0,$$

$$A_y = -800 = 800 \text{ lb} \downarrow. \qquad\qquad \text{ANS.}$$

■ **Example 4.17**

The motor drive shown in Figure E4.17(a) is pivoted at A and the weight W of the motor maintains tension in the belt around the pulley. Let the tension $T_1 = 750$ N, and assume that the motor is at rest. Determine the tension T_2, the weight W, and the reactive components at the pivot at A. Assume that the weight W acts through the motor shaft at C and that the supporting platform AB has negligible weight. The diameter of the drive pulley is 0.4 m.

Solution

To determine the tension T_2, a free-body diagram of the pulley is constructed as shown in Figure E4.17(b). An appropriate coordinate system is shown in Figure E4.16(c). This free-body diagram contains four unknown quantities, W, T_2, C_x, and C_y. Because only three conditions of equilibrium are available, not all of these unknown quantities may be determined. However, only T_2 is needed and, therefore, using the third of Eqs. (4.12),

$$\sum M_C = 0, \quad 750(0.2) - T_2(0.20) = 0,$$

$$T_2 = 750 \text{ N}. \qquad\qquad \text{ANS.}$$

Thus, under static conditions, the tensions T_1 and T_2 are identical.

To determine the weight W and the reaction components at A, we construct the free-body diagram of the entire system consisting of the motor and its platform as shown in Figure E4.17(d). The coordinate system shown in Figure E4.17(c) will be used. The free-body diagram of Figure E4.17(d) indicates three unknown quantities, W, A_x, and A_y. Because three conditions of equilibrium are available, these three unknown quantities may be determined. Thus, by Eq. (4.12),

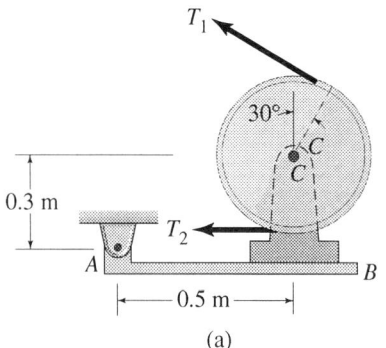

(a)

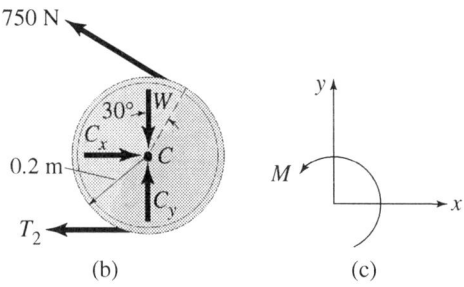

(b)   (c)

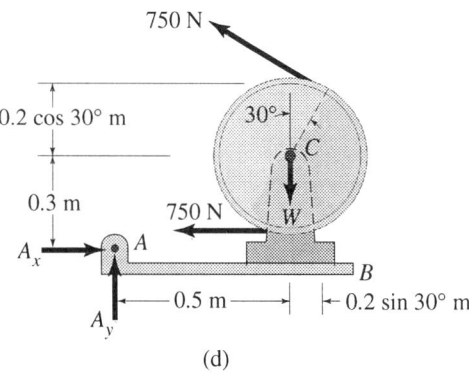

(d)   FIGURE E4.17.

$$\sum M_A = 0, \quad 750(0.1) + (750\cos 30°)(0.30 + 0.2\cos 30°)$$
$$+ (750\sin 30°)(0.5 + 0.2\sin 30°) - 0.5W = 0,$$
$$W = 1215 \text{ N.} \qquad \text{ANS.}$$

$$\sum F_x = 0, \quad A_x - 750 - 750\cos 30° = 0,$$
$$A_x = 1400 \text{ N} \;\rightarrow. \qquad \text{ANS.}$$

$$\sum F_y = 0, \quad A_y + 750\sin 60° - W = 0,$$
$$A_y = 840 \text{ N} \uparrow. \qquad \text{ANS.}$$

■ **Example 4.18**   Two smooth cylindrical containers weighing $W_E = 500$ lb and $W_G = 300$ lb are placed inside two frictionless planes, as shown in Figure E4.18(a). Determine the reactive forces developed at support points A, B, and C.

Solution   Consider the geometry of the entire system shown in Figure E4.18(b). The orientation of straight line EDG from the horizontal is obtained as follows:

$$\alpha = \sin^{-1} \frac{0.5}{2.5} = 11.54°.$$

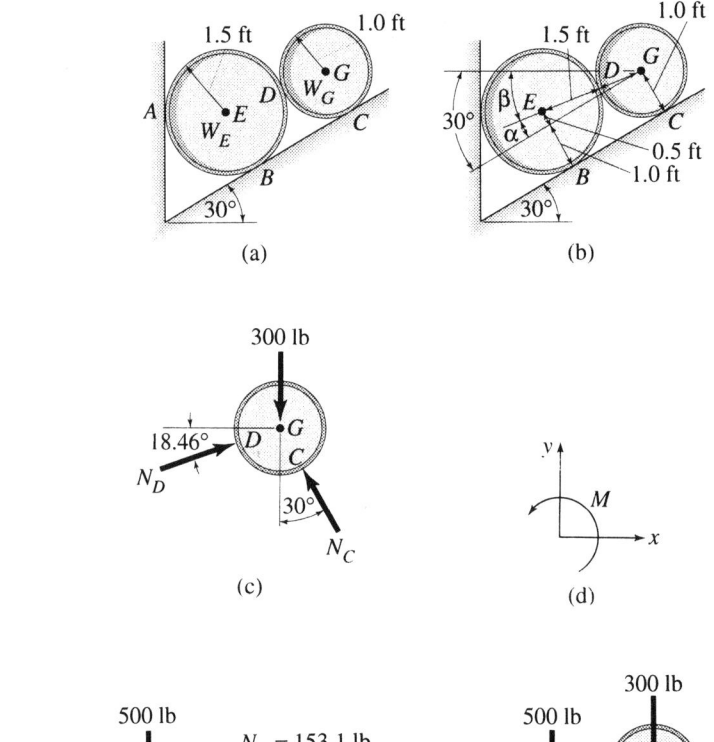

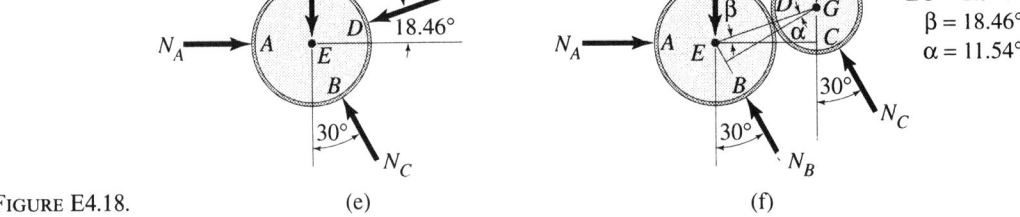

Therefore,

$$\beta = 30 - \alpha = 18.46°.$$

Two solutions are presented.

Solution I

This solution consists of analyzing the separate free-body diagrams of the two containers. The free-body diagram of the top container is shown in Figure E4.18(c) where the quantities N_C and N_D represent, respectively, the unknown normal reactive forces at smooth contact points C and D. Because the three forces in this free-body diagram are concurrent, two conditions of equilibrium are available to determine the two unknowns N_C and N_D. An appropriate coordinate system is shown in Figure E4.18(d). Using the first two equilibrium conditions expressed in Eq. (4.12),

$$\sum F_x = 0, \quad N_D \cos 18.46° - N_C \sin 30° = 0,$$

$$N_D = 0.527 N_C. \tag{a}$$

$$\sum F_y = 0, \quad N_D \sin 18.46° - N_C \cos 30° - 300 = 0. \tag{b}$$

Substituting Eq. (a) in Eq. (b) and solving for the unknowns,

$$N_C = 290 \text{ lb } {}_{60°} \!\!\diagup\!\!\triangle \; . \qquad\qquad \text{ANS.}$$

and

$$N_D = 153.1 \text{ lb}.$$

Now consider the free-body diagram of the bottom cylinder shown in Figure E4.18(e) where the quantities N_A and N_B represent the unknown normal reactive forces at smooth contact points A and B, respectively. Again, the force system in this free-body diagram is concurrent, and the first two equilibrium conditions in Eq. (4.12) are used, along with the coordinate system of Figure E4.18(d). Thus,

$$\sum F_y = 0, \quad N_B \cos 30° - 500 - 153.1 \sin 18.46° = 0,$$

$$N_B = 633 \text{ lb } {}_{60°}\!\!\diagup\!\!\triangle \; . \qquad\qquad \text{ANS.}$$

$$\sum F_x = 0, \quad N_A - 153.1 \cos 18.54° - N_B \sin 30° = 0,$$

$$N_A = 462 \text{ lb } \rightarrow . \qquad\qquad \text{ANS.}$$

Solution II

The second solution uses the free-body diagram of the entire system consisting of the two cylindrical containers as shown in Figure E4.18(f). The quantities N_A, N_B, and N_C, once again, represent the unknown normal reactive forces at smooth contact points A, B, and C, respectively. The three conditions of equilibrium expressed in Eq. (4.12) are available and may be used, along with the coordinate system of Figure

E4.18(d), to solve for the three unknown quantities. Thus,

$$\sum M_E = 0, \quad N_C(2.5 \cos \alpha) - 300(2.5 \cos \beta) = 0,$$

$$N_C = 290 \text{ lb } \overset{60°}{\diagdown} .$$ ANS.

$$\sum F_y = 0, \quad N_B \cos 30° + N_C \cos 30° - 500 - 300 = 0,$$

$$N_B = 633 \text{ lb } \overset{60°}{\diagdown} .$$ ANS.

$$\sum F_x = 0, \quad N_A - N_B \sin 30° - N_C \sin 30° = 0,$$

$$N_A = 462 \text{ lb } \rightarrow .$$ ANS.

Obviously, the answers obtained by the second solution are identical to those obtained by the first solution. The second solution, however, is a more direct method of obtaining the required reactions. If, on the other hand, the reaction at D between the two containers is needed, the first solution becomes necessary.

■ **Example 4.19**

Beam AB is hinged at A and supported by a roller at B as shown in Figure E4.19(a). It is subjected to the uniformly distributed load of intensity w and to the concentrated force P equal in magnitude to wL and located at a variable distance x from point A. Determine the support reactions at A and B in terms of w, L, and x. Investigate the variation of these reactions with x.

Solution

The free-body diagram of beam AB is shown in Figure E4.19(b). Note that, although the frictionless hinge at A is capable of supporting a horizontal as well as a vertical force component, only the vertical component A_y is needed because there are no applied forces to mobilize a horizontal reaction. Thus, as seen from the free-body diagram, we are dealing with a parallel, two-dimensional force system with two unknown quantities, A_y and B_y. Thus, using Eqs. (4.12),

$$\sum F_y = 0, \quad A_y + B_y - \frac{wL}{2} - wL = 0,$$

$$A_y + B_y - \frac{3wL}{2} = 0,$$ (a)

and

$$\sum M_A = 0, \quad B_y(L) - wL(x) - \frac{wL}{2}\left(\frac{L}{4}\right) = 0,$$

$$B_y(L) - wL(x) - \frac{wL^2}{8} = 0.$$ (b)

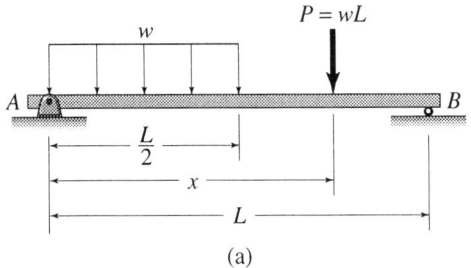

(a)

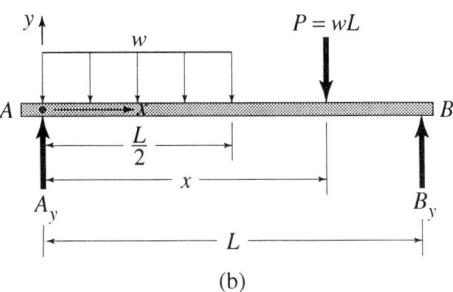

(b)

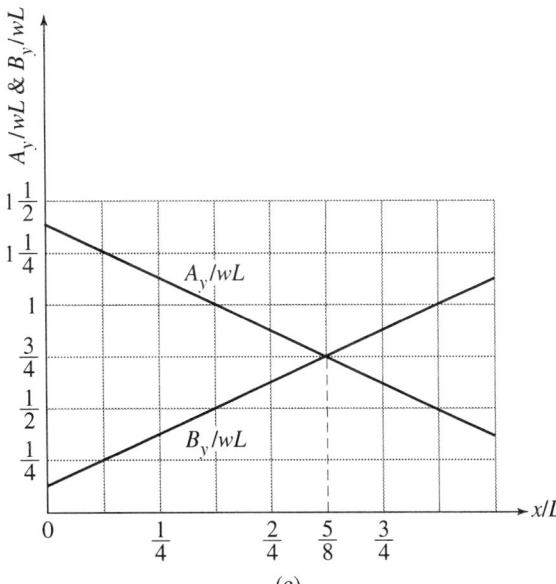

(c)

FIGURE E4.19.

The simultaneous solution of Eqs, (a) and (b) yields

$$A_y = \frac{11wL}{8} - wx \qquad \text{ANS.}$$

and

$$B_y = \frac{wL}{8} + wx. \qquad \text{ANS.}$$

Note that the same results would be obtained by using Eqs. (4.13) or Eqs. (4.14) instead of Eqs. (4.12). In general, however, the use of moment equations rather than force equations may eliminate the need for solving simultaneous equation, particularly if the moment centers are judiciously chosen so as to eliminate as many of the unknowns as possible.

A very convenient way to explore the variations in the values of A_y and B_y, as the variable x changes, is to form the dimensionless ratios A_y/wL and B_y/wL. Thus,

$$\frac{A_y}{wL} = \frac{11}{8} - \frac{x}{L},$$

and

$$\frac{B_y}{wL} = \frac{1}{8} + \frac{x}{L}.$$

Plots of these two relationships are shown in Figure E4.19(c). Such plots give the analyst a broader view of the solution than would have been possible if x was initially given a fixed value. With the plots in Figure E4.19(c), we can see at a glance how A_y and B_y vary with x and can also determine values for these two reactive components for any specific value of x. If we were interested, for example, in the position of the concentrated force (i.e., the value of x) that would make the magnitudes of A_y and B_y equal, all we would need to do is determine the value of x at the intersection of the two curves, $x = \frac{5L}{8}$.

■

Problems

4.76 The overhanging beam shown in Figure 4.76 is hinged at A and supported by a roller at B. If $w = 10$ kN/m, $L = 2$ m and $x = 2$ m, determine the reaction components at the two supports A and B.

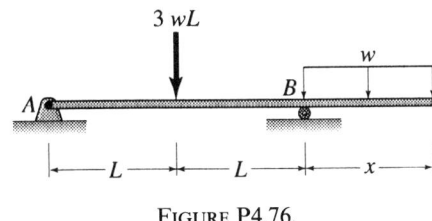

FIGURE P4.76.

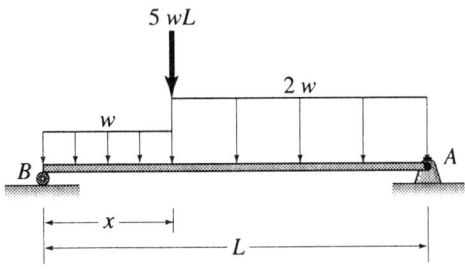

FIGURE P4.79.

4.77   Refer to the beam of Figure P4.76, and, without assigning numerical values to w, L, and x, determine, in terms of these quantities, the reaction components A_y and B_y at supports A and B, respectively. Examine the variation of the ratios A_y/wL and B_y/wL with the variable x for $0 \le x \le 3L$.

4.78   The beam shown in Figure P4.78 is hinged at A and supported by a roller at B. Determine (a) the value of x so that the vertical component of the reaction at A is zero and (b) the components of the reactions at A and B for the value of x found in part (a).

terms of these quantities, the reaction components A_y and B_y at supports A and B, respectively. Examine the variation of the ratios A_y/wL and B_y/wL with the variable x for $0 \le x \le L$.

4.81   A simply supported beam is hinged at A and supported by a roller at B as shown in Figure P4.81. Determine the dimension x which positions the distributed loading on the beam so that the vertical components of the reactions at A and B are identical. What are the reaction components at A and B for this value of x?

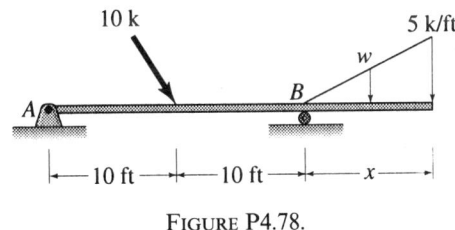

FIGURE P4.78.

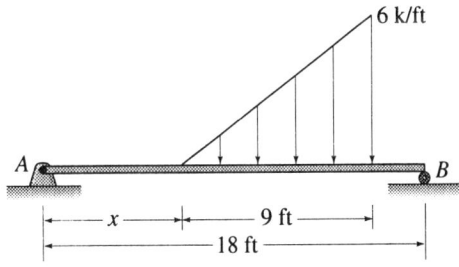

FIGURE P4.81.

4.79   A simply supported beam is hinged at A and supported by a roller at B as shown in Figure P4.79. Let $w = 6$ kN/m, $L = 5$ m, and $x = 3$ m, and determine the reaction components at supports A and B.

4.80   Refer to the beam shown in Figure P4.79, and, without assigning numerical values to w, L, and x, determine, in

4.82   A cantilever beam is loaded as shown in Figure P4.82 where $w = 4$ kN/m, $L = 4$ m, and $\theta = 30°$. Determine the reaction components A_x, A_y, and M_A at the fixed support at A.

4.83   Refer to the cantilever beam shown in Figure P4.82 and, without assigning

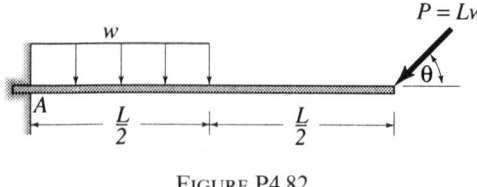

FIGURE P4.82.

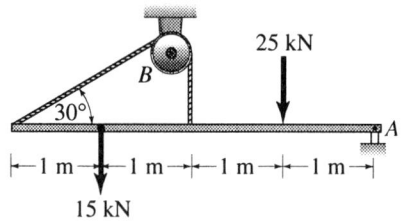

FIGURE P4.85.

numerical values to w, L, and θ, determine, in terms of these quantities, the reaction components A_x, A_y, and M_A. Form the ratios A_x/wL, A_y/wL, and M_A/wL^2, and examine their variation with θ for $0 \le \theta \le \pi$.

4.84   A cantilevered frame is loaded as shown in Figure P4.84. Determine the reaction components A_x, A_y, and M_A at the fixed support at A.

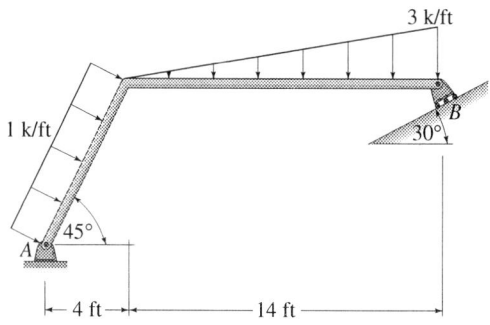

FIGURE P4.86.

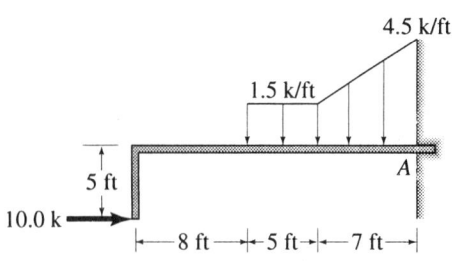

FIGURE P4.84.

4.85   A beam is hinged A and is supported by a continuous cable that passes over a frictionless pulley at B, as shown in Figure P4.85. Determine the tension in the cable and the reaction components at A.

4.86   The frame shown in Figure P4.86 is hinged at A and supported by a roller at B. Determine the reaction components at these two supports.

4.87   The homogeneous bar AB of Figure P4.87 weighs 20 lb and this weight may be assumed to act through G, the geometric center of the bar. In the position

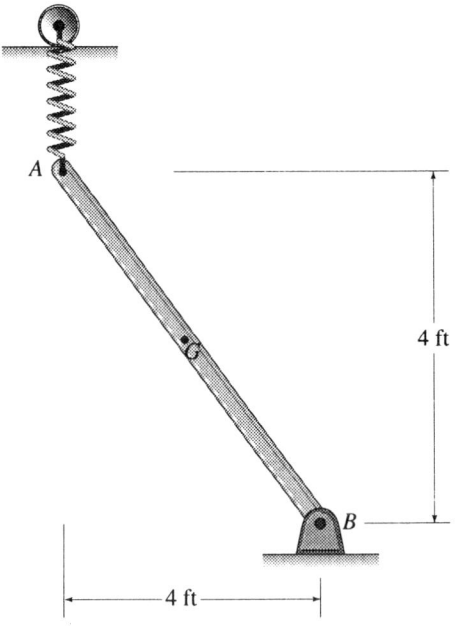

FIGURE P4.87.

shown, the deformation of the spring was measured as 0.5 in. Determine the spring constant k and the reaction components at the hinge support at B. Note that the frictionless roller at the top allows the spring to maintain a vertical position.

4.88   The uniform member AB leans against a frictionless vertical wall at A and is hinged at B as shown in Figure P4.88. The weight of the member is 250 N and may be assumed to act through its geometric center. Determine the reaction components at A and at B.

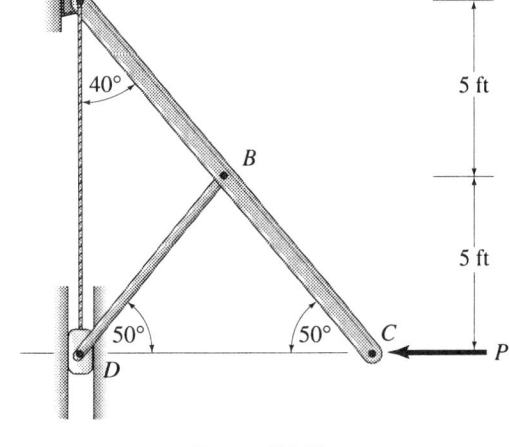

FIGURE P4.89.

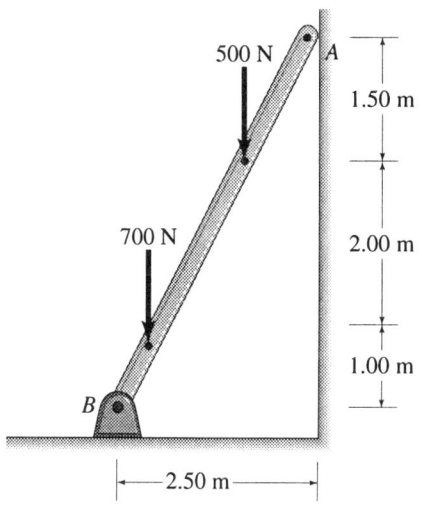

FIGURE P4.88.

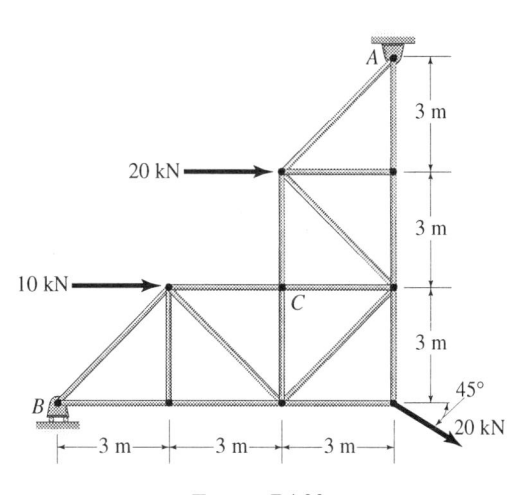

FIGURE P4.90.

4.89   The frame shown in Figure P4.89 is supported at A by a frictionless hinge and at D by a frictionless slider. One end of a cable is attached to the pin at A and the other to the slider at D. If $P = 1.2$ k, determine the support reactions at A and the forces acting on the slider at D.

4.90   The truss shown in Figure P4.90 is supported at A by a frictionless hinge and at B by a system of rollers. For the applied loading shown, determine the components of the support reactions at A and at B.

4.91   A schematic representation of a garden wheel barrow is shown in Figure P4.91. Assume that the weight of the wheel barrow, including the load in the bucket is W and acts through the center of gravity at G. The force applied by the

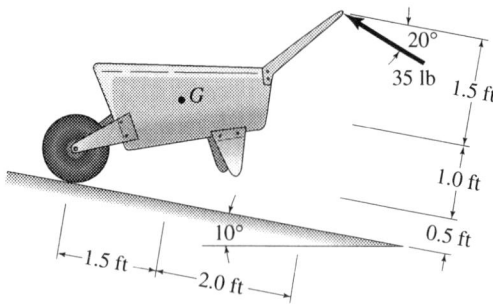

FIGURE P4.91.

4.93   Refer to Problem 4.92 and determine the reaction components at B and C on post CD.

4.94   A force of 30 lb is applied at the end of the handle of the press, as shown in Figure P4.94. Determine (a) the reaction components at the frictionless hinges at B and C and (b) the reaction components at the frictionless hinge at A.

gardener to move the wheel barrow up the 10° inclined plane at *constant speed* is 35 lb as shown. Determine the weight W and the reaction components between the wheel and the rough inclined plane.

4.92   The uniform member AB shown in Figure P4.92 weighs 500 N and is attached to a collar at B that can slide freely on the vertical, weightless post CD. It is supported at A by a frictionless plane inclined to the horizontal at an angle α. Assume that the weight of member AB acts through its geometric center. Determine the reaction components at A and B on member AB and the angle α.

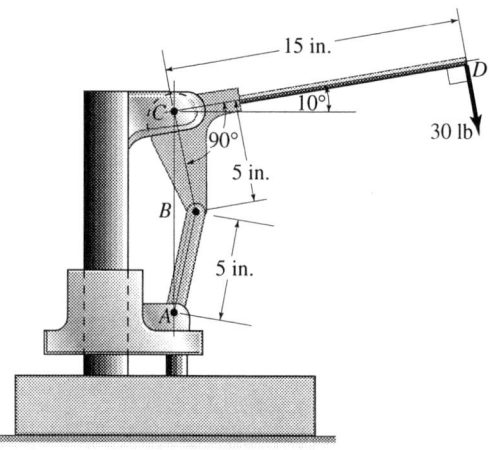

FIGURE P4.94.

4.95   The frame shown in Figure P4.95 is subjected to the forces $P_1 = 5$ kN, $P_2 = 7$ kN, and $P_3 = 3$ kN as shown. Construct the free-body diagram of member GEB, including the pulley. Then, determine the reaction components at the frictionless hinge E and at the slider at B. Let $a = b = 3r = 1.5$ m.

4.96   Refer to Problem 4.95, and determine the force components acting at frictionless hinges A and F. *Hint: Use the reaction components at hinge E determined in Problem 4.95.*

4.97   Refer to Problem 4.95 and construct the free-body diagram of member ABC, and determine the force components at the

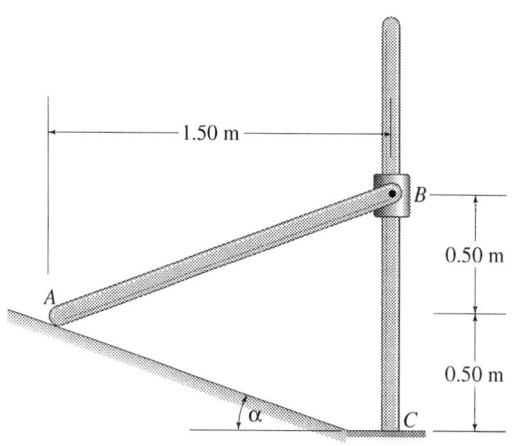

FIGURE P4.92.

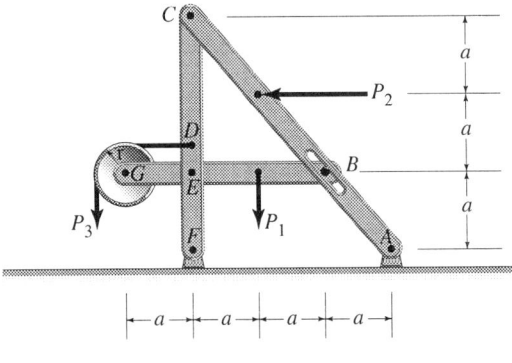

FIGURE P4.95.

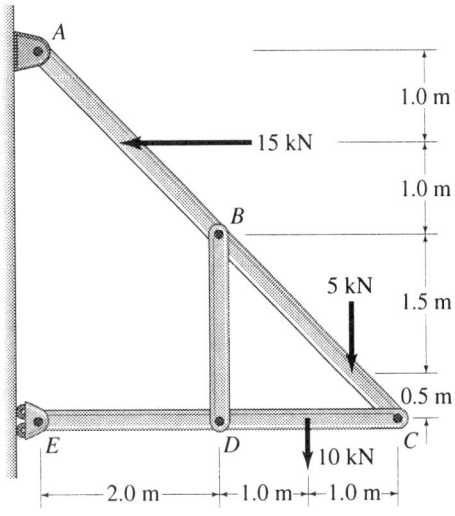

FIGURE P4.99.

frictionless hinge at C and at the frictionless slider at B. *Hint: Use the reaction components at hinge A determined in Problem 4.96.*

4.98   The forces $Q = 15$ lb are applied to the pliers as shown in Figure P4.98. Separate the two parts of the pliers from their connecting frictionless pin at B and construct their free-body diagrams. Assume that the bolt shank at C is frictionless, and determine (a) the force applied to the bolt shank and (b) the reaction components at B. Let $a = 2$ in.

4.100   Refer to Problem 4.99 and detach member EDC from its frictionless connecting pins at E, D, and C. Construct its free-body diagram, and determine the force components at D and C. *Hint: Use the value of the support reaction at E found in Problem 4.99.*

4.101   The frame shown in Figure P4.101 is supported by frictionless hinges at D and C. Assume that all members are weightless, and determine the reaction

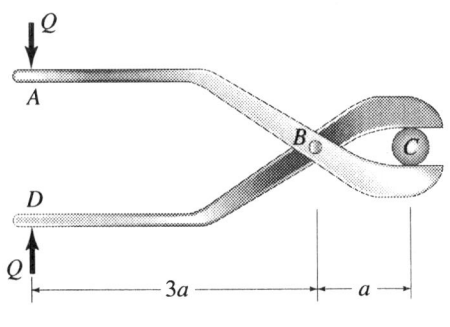

FIGURE P4.98.

4.99   The frame shown in Figure P4.99 is supported at A by a frictionless hinge and at E by a system of rollers. Determine the reaction components at A and at E.

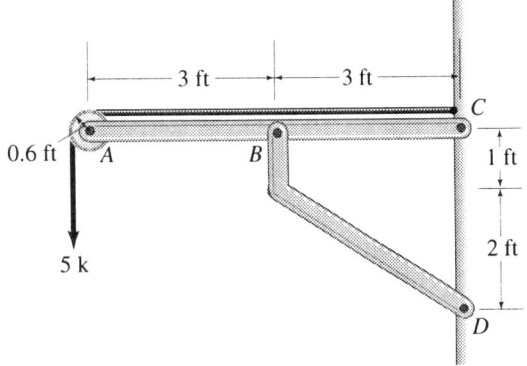

FIGURE P4.101.

components at support C and at the frictionless pin connection at B.

4.102 Refer to Problem 4.101, detach the pulley from its frictionless hinge support at A, and construct its free-body diagram. Determine the reaction components at support A.

4.103 A belt passes over two pulleys and a tension of 650 N is maintained in it, as shown in Figure P4.103. Determine the reaction components at the frictionless hinges A and B.

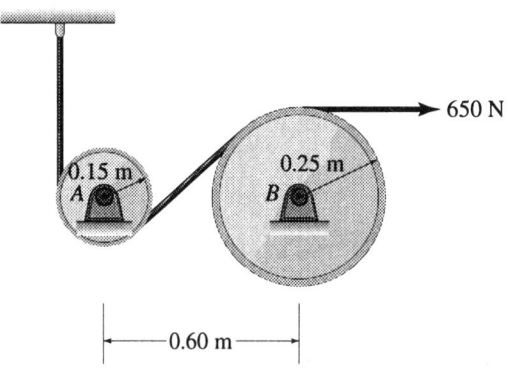

FIGURE P4.103.

4.104 A workman lowers himself at *constant speed* by applying the necessary force on the cable, as shown in Figure P4.104. If the spring scale at A indicates a reading of 35 lb, determine the weight of the workman. Assume that all bodies are very small.

4.105 A drum with a mass of 200 kg is placed between two smooth retaining walls, as shown in Figure P4.105. Determine the normal reactions at contact points A and B.

4.106 Two cylindrical containers are placed between two smooth retaining walls as shown in Figure P4.106. Determine the reaction components at contact points A and D.

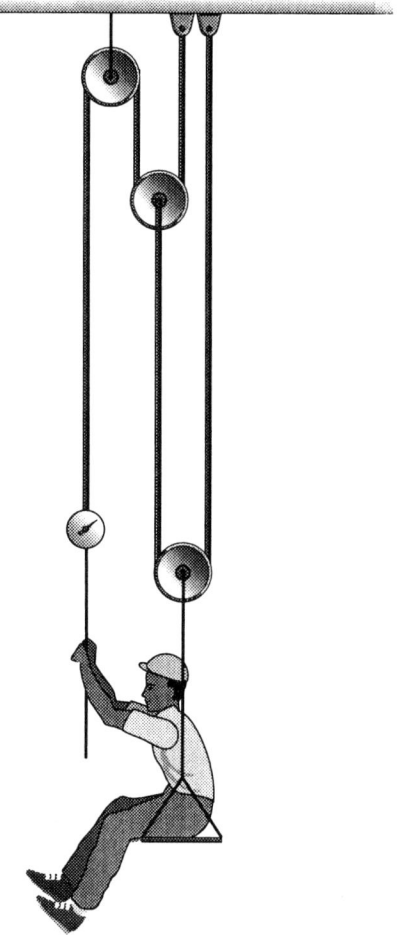

FIGURE P4.104.

4.107 Refer to Problem 4.106 and determine the normal reactions at contact points B and C. *Hint: Use the value of the normal reaction at contact point D found in Problem 4.106.*

4.108 Two identical cylinders of weight W each are placed inside the opening between two smooth vertical walls as shown in Figure P4.108. If the reaction at D is equal to $1/2$ W, determine the horizontal spacing b between the two walls. Express b in terms of the radius R.

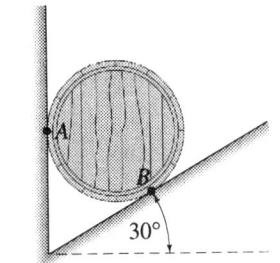

FIGURE P4.105.

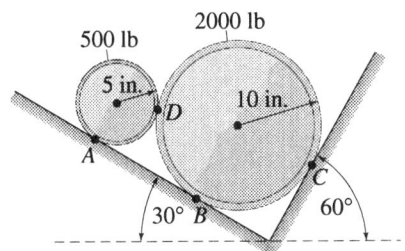

FIGURE P4.106.

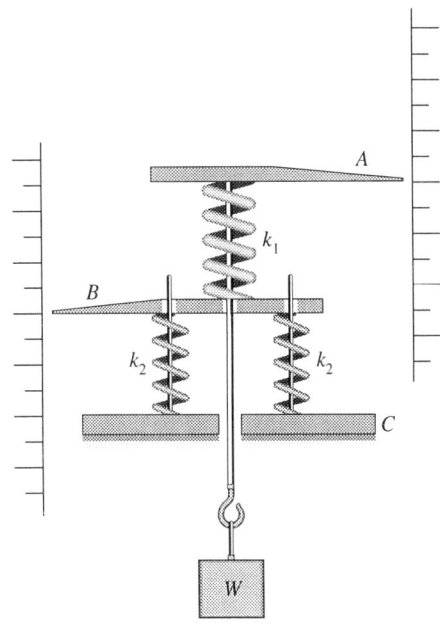

FIGURE P4.109.

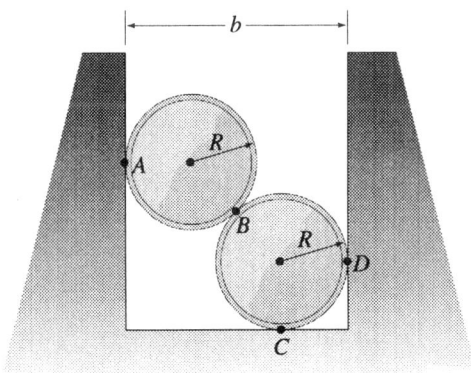

FIGURE P4.108.

4.109 An experimental scale is constructed as shown in Figure P4.109. It consists of a long rod fastened rigidly to a pointer A. A linear spring of spring constant k_1 is placed between pointers A and B and two smaller identical springs of spring constant k_2 are placed between pointer B and a fixed support at C. Assume that all members are weightless and, except for the springs, all members are rigid. The springs are unstretched when the weight W is removed. Develop expressions for the vertical movements of pointers A and B in terms of W, k_1, and k_2. When $k_1 = k_2$, which of the two scales A or B is more accurate? State why. Assume that the three rods pass freely through holes in plates B and C as shown.

4.110 A car of weight $W = 4000$ lb is driven onto a platform hinged at A and attached to a cable at B, as shown in Figure P4.110. This cable is wrapped around a motorized drum used to raise the platform as shown. Before raising, the rear wheels of the car are locked against rotation whereas the front

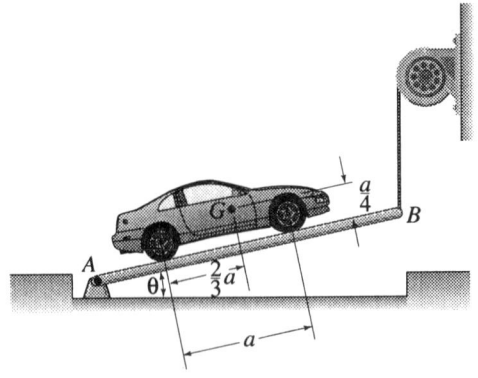

FIGURE P4.110.

wheels remain free to rotate. If the rough platform is capable of developing a maximum frictional force equal to 500 lb, parallel to the platform, at each of the rear wheels, determine the maximum value of θ that can be reached. Also, determine the normal forces at each of the four wheels for this value of θ.

4.111 A small jet aircraft in level flight at *constant speed* is shown in Figure P4.111. The forces acting on such a craft consist of the weight W that may be assumed to act through point G, the propelling force or thrust T, the lift L provided by the wings and the air resistance or drag D. Develop an expression for the dimension h in terms of the known quantities W, T, and d. What are the values of L and D in terms of W and T?

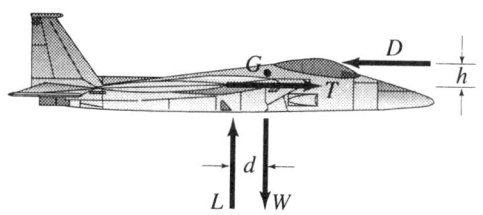

FIGURE P4.111.

4.112 An experimental missile launcher is shown in Figure P4.112. The launch position is adjusted by changing the angle α which defines the orientation of the two hydraulic cylinders AB (only one is shown). The weight of the missile is W and may be assumed to act at point G. The two hydraulic cylinders share equally in supporting W. Develop an expression for the total force F in both hydraulic cylinders in terms of the weight W and the angle α. Except for the missile, assume that all other members are weightless. Specialize this expression for the case when $\alpha = 60°$.

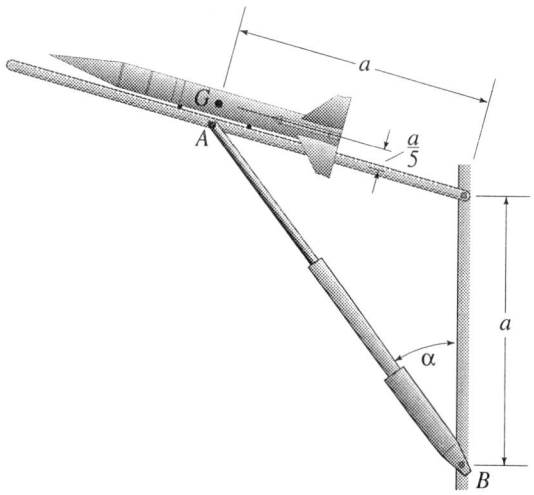

FIGURE P4.112.

4.113 A 2-m diameter cylindrical container with a mass of 280 kg must be rolled over the step at A as shown in Figure P4.113, using a cable for which the tension T cannot exceed 1000 N. If the step at A is rough, determine the minimum value of the angle α defining the direction of the cable with the horizontal. Assume that the weight of the container

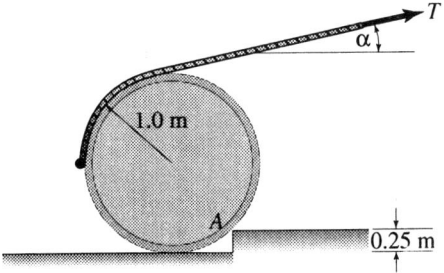

FIGURE P4.113.

acts through its geometric center. Also, determine the normal and tangential (frictional) components of the reaction at A.

4.114 Rod AB of length $L = 14$ in., weighing 15 lb., is supported at A by a frictionless

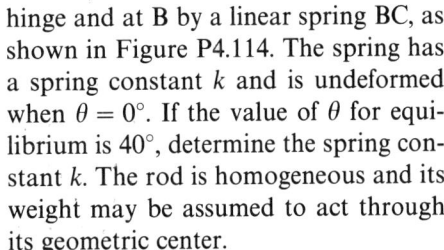

hinge and at B by a linear spring BC, as shown in Figure P4.114. The spring has a spring constant k and is undeformed when $\theta = 0°$. If the value of θ for equilibrium is $40°$, determine the spring constant k. The rod is homogeneous and its weight may be assumed to act through its geometric center.

4.115 The homogeneous member AB with a mass of 10 kg has its ends supported by frictionless slots as shown in Figure P4.115. A spring AC with a spring constant $k = 75$ N/m is attached at end A, as shown. Assume that the weight of AB acts through its geometric center, and determine the value(s) of the angle θ for equilibrium. The spring is undeformed when $\theta = 0°$.

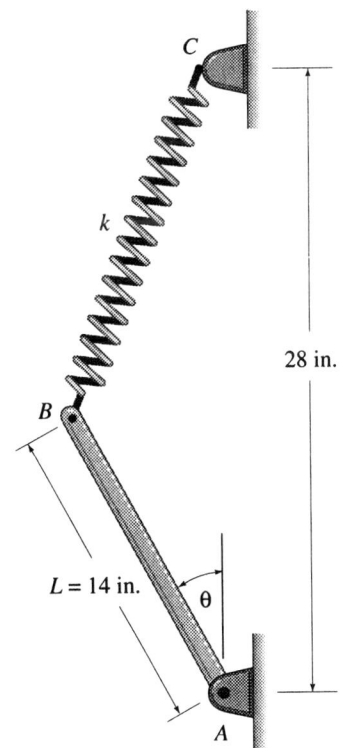

FIGURE P4.114.

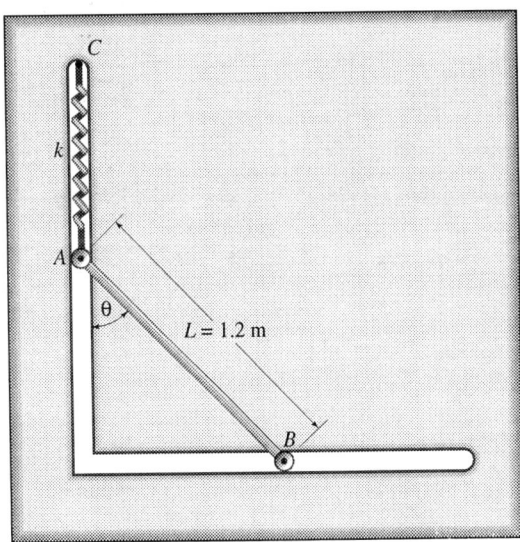

FIGURE P4.115.

4.116 An assembly consisting of a pulley and an attached arm AB weighs 2500 lb and is supported by a frictionless hinge at O, as shown in Figure P4.116. Assume that the weight of the assembly acts through

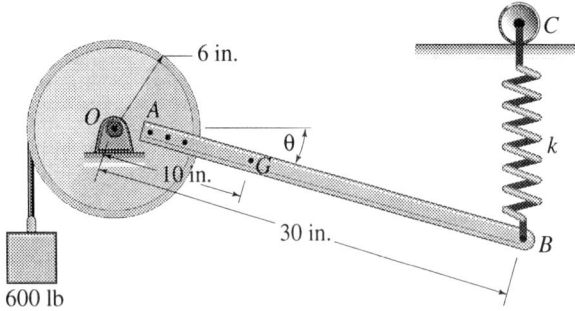

point G. A linear spring of constant $k = $ 200 lb/in. is attached at the end **B** of arm **AB** and the frictionless roller at **C** allows the spring to remain vertical. Assume that the spring is undeformed when $\theta = 0°$. Determine the value(s) of the angle θ for equilibrium (a) when the 600-lb weight is suspended as shown and (b) when the 600-lb weight is eliminated.

FIGURE P4.116.

Review Problems

4.117 The strap wrench shown in Figure P4.117 is used to loosen the automobile oil filter. If a moment of 12.5 lb. ft is needed for this purpose, determine the magnitude of the force *P* that a mechanic must apply at the end of the handle.

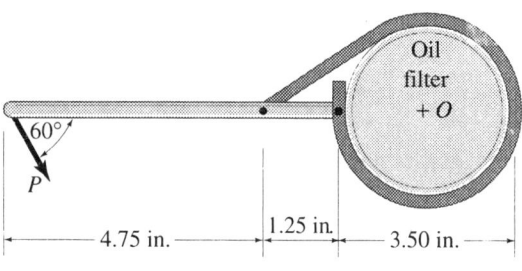

FIGURE P4.117.

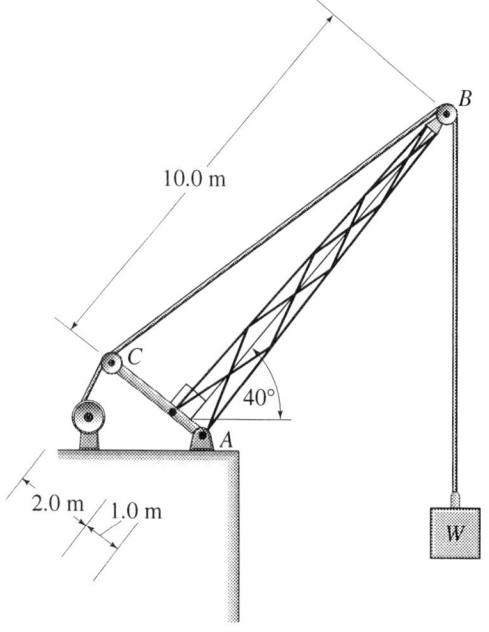

FIGURE P4.118.

4.118 The crane shown in Figure P4.118 is used to lift the crate of weight $W = $ 2.5 kN. Determine the magnitude, direction, and sense of the moment that this weight produces about (a) the hinge at A and (b) the hinge at C. Note that member AC is attached to the crane and is perpendicular to it.

4.119 Two 1-ft diameter pulleys are attached to a plate, as shown in Figure P4.119. A belt is wound around the two pulleys and a force $P = 5$ k is applied as shown.

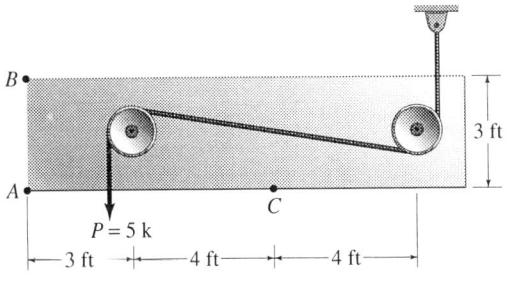

FIGURE P4.119.

(a) Find the magnitudes and senses of two equal and horizontal forces applied at A and B that would create a couple of

equal magnitude and direction but opposite in sense. (b) Repeat part (a) for two equal and vertical forces applied at A and C.

4.120 The cantilever beam of Figure P4.120 is subjected to two couples as shown. Determine the magnitude, direction, and sense of the resultant couple (a) by finding the moment of each couple and adding them, (b) by finding the moment of each of the four forces about A and adding them, and (c) by finding the moment of each of the four forces about B and adding them.

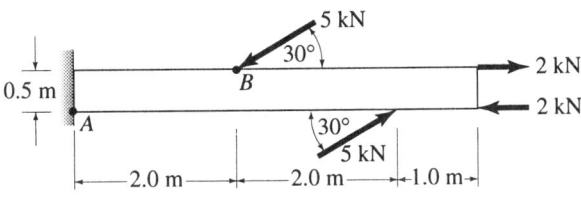

FIGURE P4.120.

4.121 A structural member is subjected to loads as shown in Figure P4.121. (a) Replace the given loads by a force-couple system acting at O. (b) Replace the given loads by a single force specifying its

magnitude, direction, sense, and its location along the x axis.

4.122 A machine member is to be subjected to loads as shown in Figure P4.122. Determine the dimension b so that the resultant of the given load system is a force passing through point O. Also, find the magnitude, direction, and sense of this resultant force.

4.123 A structure is subjected to forces, as shown in Figure P4.123. Replace these forces by a single resultant force specifying its magnitude, direction, and sense. Position this resultant force along member AB (or AB extended).

4.124 Find the quantities a and w so that the beam of Figure P4.124 experiences no resultant force and no resultant moment.

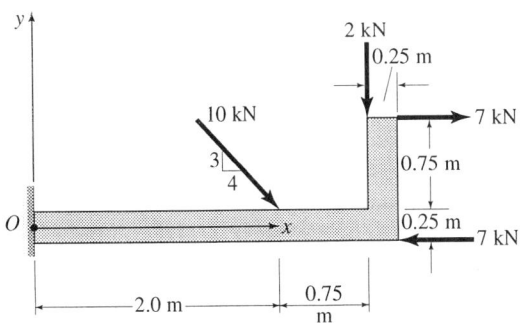

FIGURE P4.121.

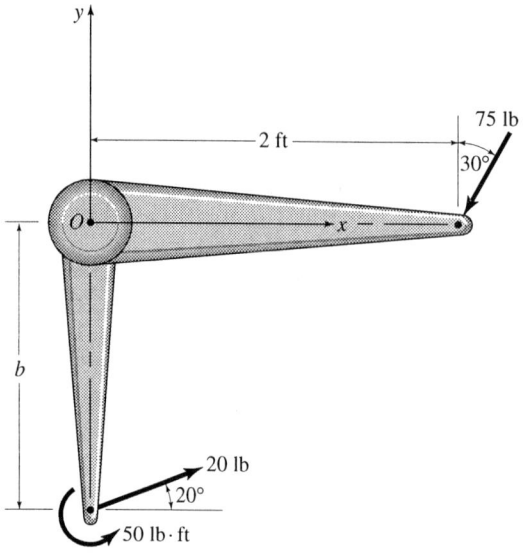

FIGURE P4.122.

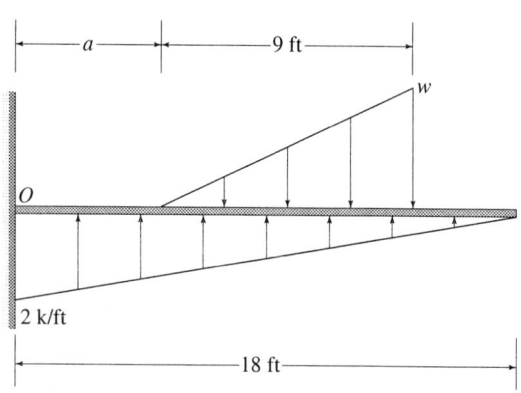

FIGURE P4.124.

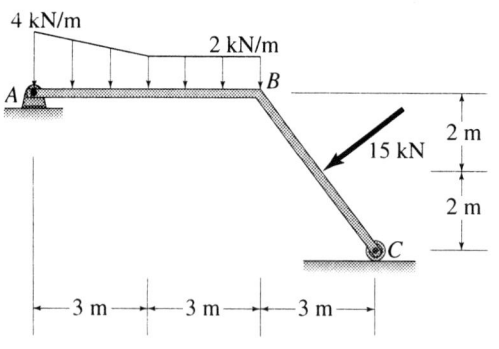

FIGURE P4.123.

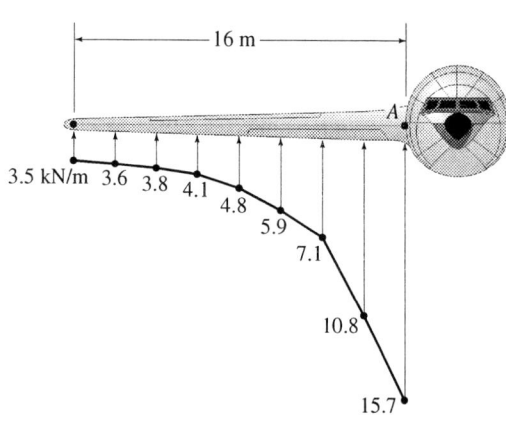

FIGURE P4.125.

4.125 The lift on the wing of an airplane was obtained by actual measurements at 2-m intervals along the length of the wing. The results are given in Figure P4.125 where the measured values are connected by straight line segments. Replace this force distribution by an approximate resultant force specifying its magnitude, direction, sense and its distance from point A, the point at which the wing is connected to the fuselage. Use the trapezoidal rule.

4.126 A beam is loaded as shown in Figure P4.126. If the magnitude of the resultant reaction at either A or B cannot exceed 20 k, determine the magnitude of the load intensity w.

4.127 A beam is hinged at A and supported by a continuous cable that passes over two frictionless pulleys, as shown in Figure P4.127. Find the tension in the cable and the reaction components at A.

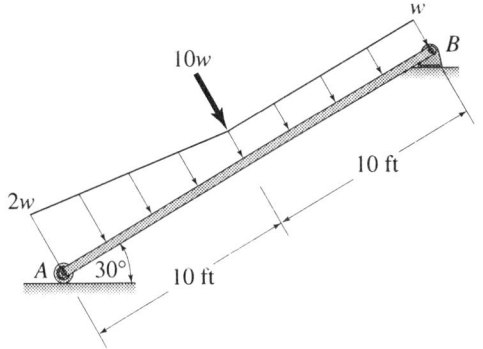

FIGURE P4.126.

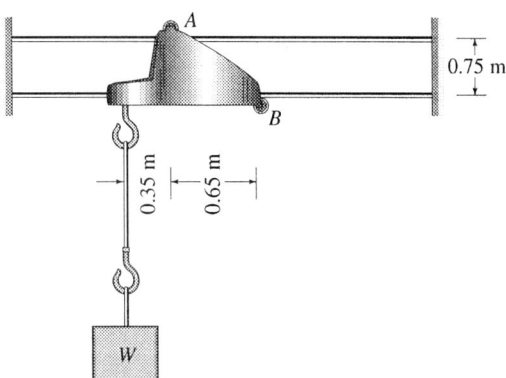

FIGURE P4.128.

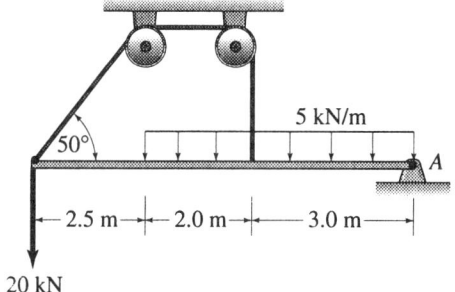

FIGURE P4.127.

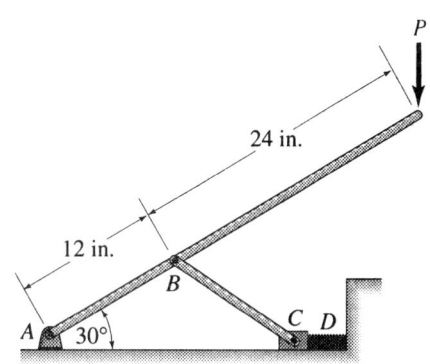

FIGURE P4.129.

4.128 The overhead trolley crane moves horizontally on a beam, as shown in Figure P4.128. It is used to transport equipment of weight W from one location to another in a factory. If the reaction between the beam and the roller at A is to be limited to 15 kN, determine the maximum allowable weight W that may be carried. What are the reactions between the beam and the two rollers for maximum W? Assume frictionless conditions.

4.129 Refer to the mechanism shown in Figure P4.129. By applying a vertical force P, a compressive force is exerted on block D. Assume frictionless conditions, and determine the compressive force on block D if $P = 40$ lb. The length of member BC is 10 in.

4.130 Refer to the overhead crane shown in Figure P4.130, and determine the tension in cable AC and the reaction components at B in terms of the variable x that locates the load $W = 15$ k. The beam has a height of 2 ft and weighs 100 lb/ft. Specialize your answers for the case when $x = 0$, $x = 8$ ft, and $x = 20$ ft.

4.131 A cylinder with a mass $m = 100$ kg and a radius $R = 0.15$ m is placed between a frictionless inclined plane and a frictionless vertical plate as shown in Figure P4.131. The vertical plate is supported

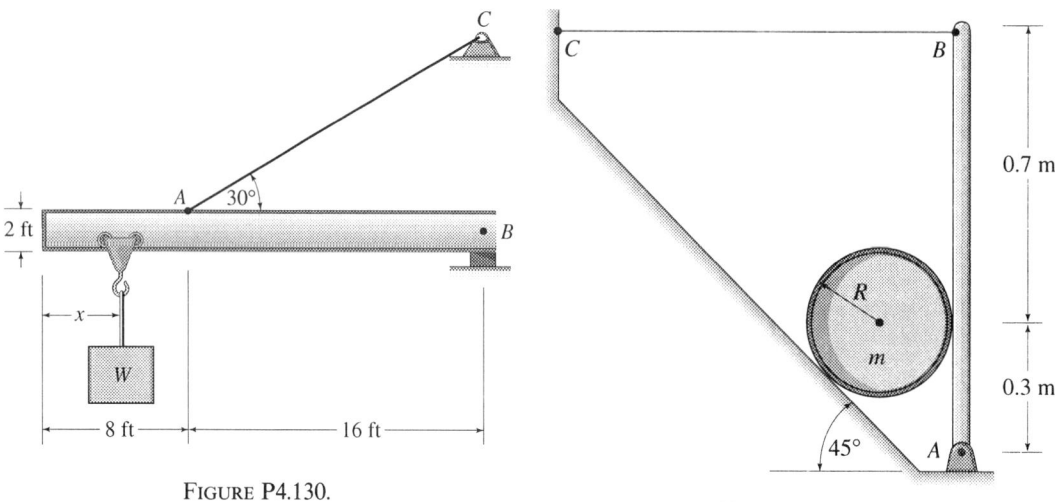

FIGURE P4.130.

FIGURE P4.131.

at A by two frictionless hinges (only one is shown) and two horizontal cables BC (only one is shown). Determine the tension in each of the two cables and the reaction components in each of the two hinges.

4.132 Two cylinders are placed in the space between two frictionless inclined planes as shown in Figure P4.132. The inclination of the left plane is fixed at 45°, but that of the right plane, defined by the angle α, is adjustable. (a) Derive expres-

sions for the reactions at C and D in terms of angle α. (b) What is the magnitude of α for a zero reaction at C. What happens physically for values of α less than this? What is the reaction at D corresponding to this critical value of α? (c) Find the reactions at C and D for α = 30°, 60°, and 90°. The upper cylinder has a mass of 20 kg and the lower cylinder a mass of 10 kg.

4.133 A dump truck is shown in Figure P4.133. The inclination angle θ of the truck is

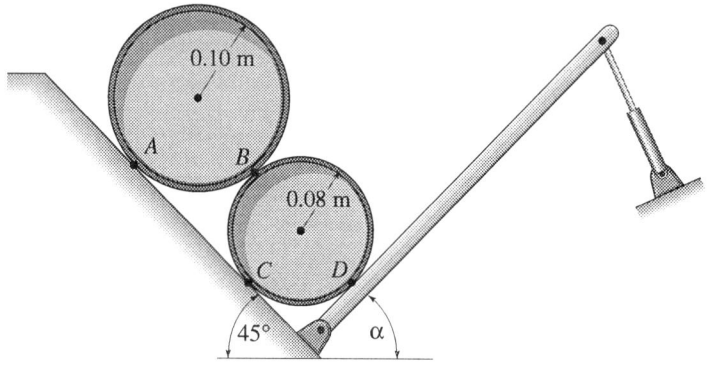

FIGURE P4.132.

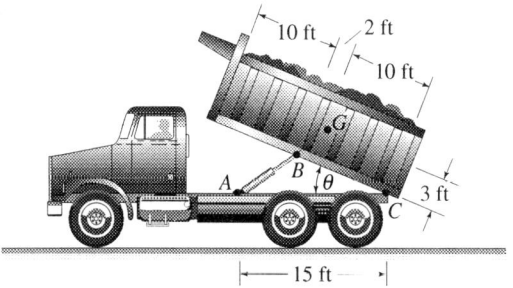

10 ft   2 ft
10 ft
.G
A
B θ
3 ft
C
15 ft

FIGURE P4.133.

changed by activating the two hydraulic cylinders **AB** (one on each side of the truck). The dump truck and its contents weigh 6000 lb, the line of action for which may be assumed to act through point G. (a) Develop an expression for the force F_{AB} in each hydraulic cylinder as a function of the angle θ. (b) Plot the force F_{AB} vs. the angle θ for the range $0° < \theta \leq 25°$.

5

Equilibrium of Rigid Bodies in Three Dimensions

The Roman Colosseum in Italy, still standing after almost 2000 years.

The design of an engineering system requires intimate knowledge of the forces acting on the system. The fundamental concepts of force equilibrium applied to particles are equally applicable to the analysis of rigid bodies. However, some new concepts are needed to simplify the analysis.

Chapter 4 was devoted to a discussion of the equilibrium of rigid bodies in two dimensions. This discussion is extended in Chapter 5 to include three-dimensional forces acting on a rigid body. As stated in earlier chapters, the analysis of three-dimensional force systems is much more conveniently performed with vector algebra than with scalar algebra. Thus, our knowledge of vector algebra is expanded in Sections 5.1 and 5.2 where the concept of the cross (vector) product is introduced. This concept is, then, utilized to develop several ideas and techniques that are useful in solving three-dimensional equilibrium problems. Thus, using the cross product, the moment of a force about a point is defined in Section 5.3, Varignon's theorem is developed in Section 5.4, and the moment of a force about any axis is discussed in Section 5.5 which leads to the introduction of the mixed triple product. Also, using the cross product, the concept of the couple is represented in vector form in Section 5.6, and a general three-dimensional force system is replaced by a force and a couple in Sections 5.7 and 5.8. Then, in Section 5.9, the conditions of three-dimensional equilibrium of a rigid body are developed and applied to solving three-dimensional equilibrium problems. Finally, a brief discussion is given in Section 5.10, dealing with the question of determinacy and constraints.

5.1 Definition of the Cross (Vector) Product

In dealing with three-dimensional force systems acting on rigid bodies, moments of forces about points in space must be determined. This determination is facilitated by the use of the *cross* or *vector* product of two vector quantities.

The cross-product of any two vectors **V** and **W** is written as **V** × **W** resulting in a vector quantity **U**. Thus, by definition,

$$\mathbf{U} = \mathbf{V} \times \mathbf{W}. \tag{5.1}$$

As a vector quantity, the cross-product **U** possesses a magnitude, direction, and sense. The magnitude U of the cross product **U** is defined by the expression

$$U = VW \sin \theta \tag{5.2}$$

where V and W are, respectively, the magnitudes of vectors **V** and **W** and θ is the angle between them. By definition, the direction of vector **U** is perpendicular to the plane formed by the two vectors **V** and **W** and its sense is established by the *right-hand* rule. According to this rule, if the four fingers of the right hand are curled in the direction from vector **V** to vector **W**, (i.e., from the first to the second vector in the

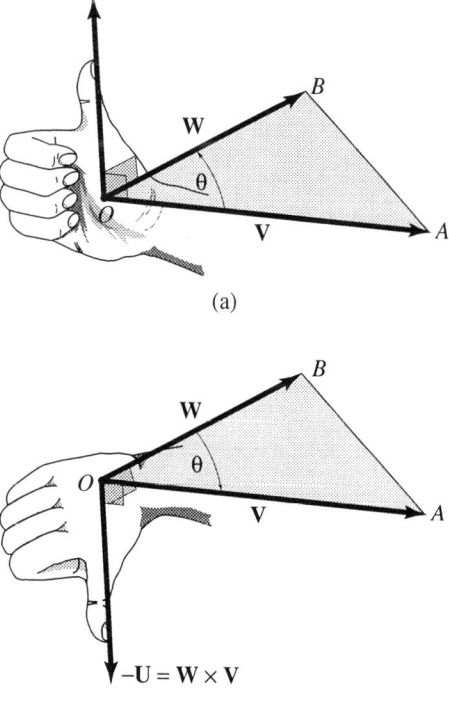

(a)

(b)

FIGURE 5.1.

product) the thumb would point in the correct sense of the cross-product $\mathbf{U}$. Thus, in Figure 5.1(a), the cross-product $\mathbf{U} = \mathbf{V} \times \mathbf{W}$ is perpendicular to plane OAB formed by vectors $\mathbf{V}$ and $\mathbf{W}$ and, according to the right-hand rule, its sense is upward. As a consequence of the right-hand rule, note that the cross-product of two vectors is not commutative. For example, as seen in Figure 5.1(b), the cross-product $\mathbf{W} \times \mathbf{V}$, although still perpendicular to plane OAB, has a sense which is downward or opposite to that of the product $\mathbf{V} \times \mathbf{W}$. In other words,

$$\mathbf{W} \times \mathbf{V} = -\mathbf{V} \times \mathbf{W}. \tag{5.3}$$

If a unit vector λ is defined in the direction and sense of $\mathbf{U} = \mathbf{V} \times \mathbf{W}$, then, using Eq. (5.2), the cross-product $\mathbf{U}$ may be expressed as a magnitude multiplying a unit vector. Thus,

$$\mathbf{U} = (VW\sin\theta)\lambda. \tag{5.4}$$

5.2
The Cross Product in Terms of Rectangular Components

The cross product $\mathbf{U} = \mathbf{V} \times \mathbf{W}$ may also be expressed in terms of its rectangular components. To this end, we will determine the cross products of any two of the three unit vectors $\mathbf{i}$, $\mathbf{j}$, and $\mathbf{k}$. You may recall that the unit vectors $\mathbf{i}$, $\mathbf{j}$, and $\mathbf{k}$ are defined along the rectangular x, y, and z coordinate axes, respectively, as shown in Figure 5.2. Thus, using the definition of the cross-product given in Eq. (5.4) we find, for example, that $\mathbf{i} \times \mathbf{j} = [(1)(1)\sin 90°]\mathbf{k} = \mathbf{k}$ as shown in Figure 5.2. Similarly, the following cross products are obtained:

$$\left.\begin{array}{l} \mathbf{i} \times \mathbf{i} = \mathbf{j} \times \mathbf{j} = \mathbf{k} \times \mathbf{k} = 0. \\ \mathbf{i} \times \mathbf{j} = \mathbf{k}; \quad \mathbf{i} \times \mathbf{k} = -\mathbf{j}; \quad \mathbf{j} \times \mathbf{k} = \mathbf{i}; \\ \mathbf{j} \times \mathbf{i} = -\mathbf{k}; \quad \mathbf{k} \times \mathbf{i} = \mathbf{j}; \quad \mathbf{k} \times \mathbf{j} = -\mathbf{i}. \end{array}\right\} \tag{5.5}$$

No effort should be made to memorize the relationships expressed in Eq. (5.5). They are easily developed from the ideas presented in Figure

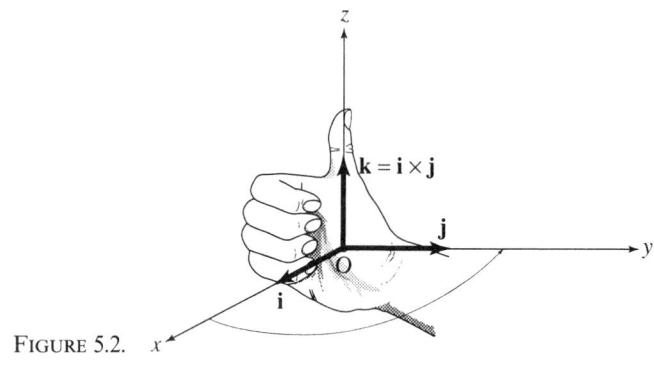

FIGURE 5.2.

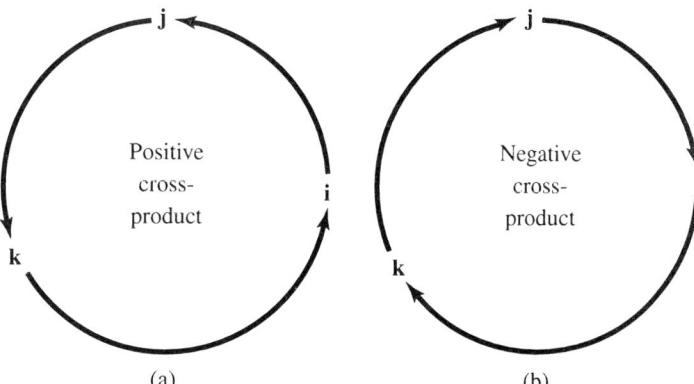

Positive cross-product

Negative cross-product

FIGURE 5.3.   (a)   (b)

5.2 for determining the cross product $\mathbf{i} \times \mathbf{j} = \mathbf{k}$. Also, Figure 5.3 serves as an aid in obtaining the sign of the cross product of any two of the three unit vectors. If the two unit vectors follow each other in a ccw manner, as in Figure 5.3(a), the sign of their cross product is positive. If, on the other hand, the two unit vectors follow each other in a cw manner, as in Figure 5.3(b), the sign of their cross product is negative.

Having determined the cross products of any two of the three rectangular unit vectors as expressed in Eq. (5.5), we may now proceed to express the cross product of any two vectors $\mathbf{V}$ and $\mathbf{W}$ in terms of their rectangular components. Thus, if $\mathbf{V} = V_x\mathbf{i} + V_y\mathbf{j} + V_z\mathbf{k}$ and $\mathbf{W} = W_x\mathbf{i} + W_y\mathbf{j} + W_z\mathbf{k}$ using Eq. (5.5), it follows that

$$\mathbf{U} = \mathbf{V} \times \mathbf{W}$$

$$= (V_x\mathbf{i} + V_y\mathbf{j} + V_z\mathbf{k}) \times (W_x\mathbf{i} + W_y\mathbf{j} + W_z\mathbf{k})$$

$$= (V_y W_z - V_z W_y)\mathbf{i} + (V_z W_x - V_x W_z)\mathbf{j} + (V_x W_y - V_y W_x)\mathbf{k}. \qquad (5.6)$$

Therefore, the scalar rectangular components of the cross product $\mathbf{U} = \mathbf{V} \times \mathbf{W}$ are

$$\left.\begin{aligned} U_x &= V_y W_z - V_z W_y \\ U_y &= V_z W_x - V_x W_z \\ U_z &= V_x W_y - V_y W_x \end{aligned}\right\} \qquad (5.7)$$

Again, no attempt should be made to memorize the relationships expressed in Eq. (5.6) or Eq. (5.7) as these relations can always be developed using the procedure outlined above. Furthermore, we note that Eq. (5.6) may be compactly represented by a determinant which may very easily be expanded when needed. Thus, Eq. (5.6) may be expressed by the following determinant:

$$\mathbf{U} = \mathbf{V} \times \mathbf{W} = \begin{vmatrix} \mathbf{i} & \mathbf{j} & \mathbf{k} \\ V_x & V_y & V_z \\ W_x & W_y & W_z \end{vmatrix}. \qquad (5.8)$$

You may recall that any determinant consisting of three rows and three columns, just as the one in Eq. (5.8), may be expanded by one of two methods. In the first method, known as *Laplace's* expansion, the determinant is obtained as the sum of the products of the elements in any row or column and their corresponding cofactors. Thus, using the elements of the first row of the determinants in Eq. (5.8) and the corresponding cofactors,

$$\mathbf{U} = \mathbf{i}(-1)^{1+1}\begin{vmatrix} V_y & V_z \\ W_y & W_z \end{vmatrix} + \mathbf{j}(-1)^{1+2}\begin{vmatrix} V_x & V_z \\ W_x & W_z \end{vmatrix} + \mathbf{k}(-1)^{1+3}\begin{vmatrix} V_x & V_y \\ W_x & W_y \end{vmatrix}$$

which reduces to what we have in Eq. (5.6). In the second method, the first two columns are repeated as shown below. The determinant is, then, obtained by subtracting the sum of the products along diagonal lines marked ② from that along diagonal lines marked ①:

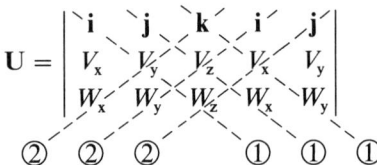

Thus, the determinant in Eq. (5.8) becomes

$$\mathbf{U} = \mathbf{i}(V_yW_z) + \mathbf{j}(V_zW_x) + \mathbf{k}(V_xW_y) - \mathbf{i}(W_yV_z) - \mathbf{j}(W_zV_x) - \mathbf{k}(W_xV_y).$$

After combining terms, the relationship above again reduces to Eq. (5.6). The question as to which of the two methods should be used to evaluate a three-by-three determinant remains a matter of personal preference because there are no distinct advantages of one over the other.

The following example illustrates some of the concepts discussed above.

■ **Example 5.1**

(a) Compute the cross product $\mathbf{C} = \mathbf{A} \times \mathbf{B}$ where $\mathbf{A} = 5\mathbf{i} - 3\mathbf{j} + 7\mathbf{k}$ and $\mathbf{B} = -10\mathbf{i} - 9\mathbf{j} + 11\mathbf{k}$.

(b) Derive an expression for a unit vector perpendicular to the plane formed by vectors $\mathbf{A}$ and $\mathbf{B}$ given in part (a).

Solution

(a) By Eq. (5.8),

$$C = A \times B$$

$$= \begin{vmatrix} i & j & k \\ 5 & -3 & 7 \\ -10 & -9 & 11 \end{vmatrix}$$

$$= i \begin{vmatrix} -3 & 7 \\ -9 & 11 \end{vmatrix} - j \begin{vmatrix} 5 & 7 \\ -10 & 11 \end{vmatrix} + k \begin{vmatrix} 5 & -3 \\ -10 & -9 \end{vmatrix}$$

$$= 30i - 125j - 75k. \qquad \text{ANS.}$$

(b) It follows from the properties of the cross product that vector **C** must be perpendicular to the plane defined by vectors **A** and **B**. Also, vector **C** may be expressed as a magnitude multiplying a unit vector λ in the direction of vector **C**. Thus, because

$$C = \sqrt{30^2 + 125^2 + 75^2} = 148.8$$

it follows that

$$C = 148.8 \left(\frac{30}{148.8} i - \frac{125}{148.8} j - \frac{75}{148.8} k \right)$$

$$= 148.8(0.202i - 0.840j - 0.504k)$$

$$= 148.8\lambda$$

where

$$\lambda = (0.202i - 0.840j - 0.504k) \qquad \text{ANS.}$$

is a unit vector perpendicular to the plane of vectors **A** and **B**.

■

5.3
Vector Representation of the Moment of a Force

The moment of a force about an axis was defined in Chapter 4 as the turning effect of the force about this axis. This turning effect has a magnitude given by Eq. (4.1). This equation states that the magnitude of the turning effect is equal to the product of the magnitude of the force multiplied by the perpendicular distance (i.e., moment arm) from the line of action of the force to the axis of rotation. In dealing with three-dimensional problems, determining the moment arm becomes involved and, in some instances, rather complex, and resort is made to vector algebra to facilitate the solution.

Consider, for example, the force **F** acting on a rigid body as shown in Figure 5.4(a). The moment of this force about axis OB, normal to the plane formed by point O and the force **F**, is referred to as the *moment of the force* **F** *about point* O and written as M_O. This moment

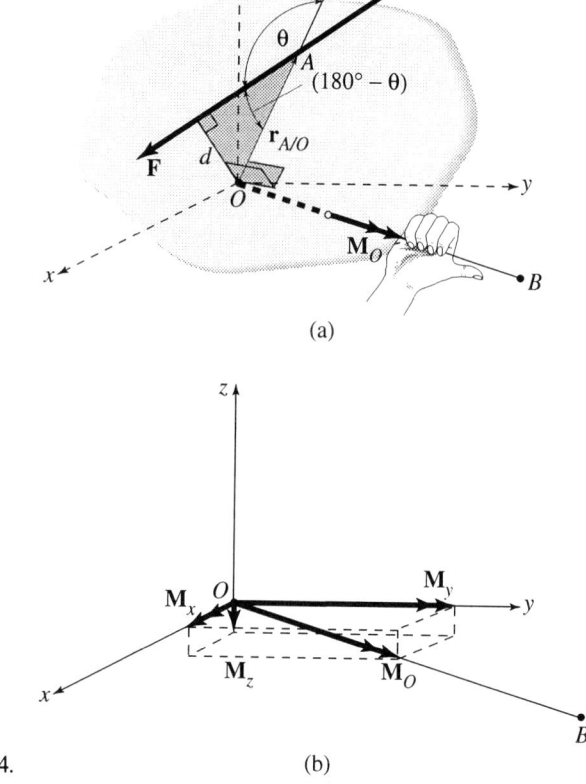

FIGURE 5.4.   (b)

is defined by the vector expression

$$\mathbf{M_O} = \mathbf{r}_{A/O} \times \mathbf{F} \qquad (5.9)$$

where $\mathbf{r}_{A/O}$ is the position vector from point O on the axis of rotation to point A which is *any* point on the line of action of the force $\mathbf{F}$. The vector representation given in Eq. (5.9) provides the correct magnitude, direction, and sense for the moment $\mathbf{M_O}$. Thus, using Eq. (5.2), the magnitude of $\mathbf{M_O}$ becomes

$$M = F[r_{A/O} \sin(180° - \theta)] = F(r_{A/O} \sin \theta)$$

$$= Fd \qquad (5.10)$$

where $d = r_{A/O} \sin \theta$, as seen from Figure 5.4(a), represents the moment arm from point O to the line of action of the force $\mathbf{F}$. Note that Eq. (5.10) is identical to Eq. (4.1) originally introduced in Chapter 4 to provide the magnitude of the moment of a force. Therefore, Eq. (5.9) provides the proper magnitude of the moment $\mathbf{M_O}$. According to the definition of the cross product, the direction of $\mathbf{M_O}$ is perpendicular to the plane formed by the vectors $r_{A/O}$ and F. Thus, vector M_O is parallel to the axis of rotation OB. Finally the sense of the vector M_O is given by the right-hand rule, as indicated in Figure 5.4(a). In other words, if the body were free to rotate about axis OB, an observer, positioned at B and looking toward O, would see the body rotating in a ccw sense.

Note that, because of the definition of the cross product, the exact location of point A on the line of action of the force F is inconsequential. However, it is important to note that the magnitude, direction, and sense of vector M_O depend on the location of point O.

As a vector, M_O may be expressed in one of two ways. The first expresses it as a magnitude multiplying a unit vector λ in its direction. Thus,

$$M_O = [F(r_{A/O} \sin \theta)]\lambda = (Fd)\lambda \qquad (5.11)$$

where λ is a unit vector perpendicular to the plane formed by vectors $r_{A/O}$ and F. In other words, λ is a unit vector along line OB. The second expresses the vector M_O in terms of its x, y, and z components. To this end, the vectors $r_{A/O}$ and F are expressed in terms of their respective x, y, and z components. Thus,

and
$$\left. \begin{array}{l} r_{A/O} = x_A i + y_A j + z_A k, \\ F = F_x i + F_y j + F_z k. \end{array} \right\} \qquad (5.12)$$

Substituting Eqs. (5.12) in Eq. (5.8) for the cross product of two vectors,

$$M_O = r_{A/O} \times F = \begin{vmatrix} i & j & k \\ x_A & y_A & z_A \\ F_x & F_y & F_z \end{vmatrix}. \qquad (5.13)$$

Expanding the determinant in Eq. (5.13) leads us to the following expression for M_O:

$$\left. \begin{array}{l} M_O = (y_A F_z - z_A F_y)i + (z_A F_x - x_A F_z)j + (x_A F_y - y_A F_x)k \\ = M_x i + M_y j + M_z k \end{array} \right\} \qquad (5.14)$$

where $M_x = y_A F_z - z_A F_y$, $M_y = z_A F_x - x_A F_z$ and $M_z = x_A F_y - y_A F_x$ are, respectively, the x, y, and z scalar components of M_O. For the sake of clarity, the vector M_O is redrawn in Figure 5.4(b) along with its three rectangular components, M_x, M_y, and M_z. Physically, M_O represents the tendency for rotation about axis OB produced by the force F on the rigid body. Also, the vectors M_x, M_y, and M_z represent the tendency for rotation about the rectangular axes x, y, and z, respectively, produced by the force F on the rigid body. In each case, the sense of this tendency for rotation is ascertained by applying the right-hand rule. Thus, for example, if we place the thumb of the right hand along vector M_x, the curled four fingers of this hand indicate the sense in which rotation would take place about the x axis. For the case of $\mathbf{M}_x$ in Figure 5.4(b), this sense would be ccw, as viewed by an observer on the positive x axis looking toward the origin of the coordinate system. Similarly, for $\mathbf{M}_y$, the sense would be ccw, as viewed by an observer on the positive y axis looking toward the origin. However, for $\mathbf{M}_z$, the sense would be cw when viewed from the positive z axis toward the origin.

In the general case, when point O does not coincide with the origin of the coordinate system, as shown in Figure 5.5, the moment $\mathbf{M}_O$ is still determined by Eq. (5.9). However, the position vector $\mathbf{r}_{A/O}$ is now given by the expression

$$\mathbf{r}_{A/O} = (x_A - x_O)\mathbf{i} + (y_A - y_O)\mathbf{j} + (z_A - z_O)\mathbf{k}.$$

Thus,

$$\mathbf{M}_O = \mathbf{r}_{A/O} \times \mathbf{F} = \begin{vmatrix} \mathbf{i} & \mathbf{j} & \mathbf{k} \\ (x_A - x_O) & (y_A - y_O) & (z_A - z_O) \\ F_x & F_y & F_z \end{vmatrix}. \quad (5.15)$$

Note that Eq. (5.15) reduces to Eq. (5.13) for the special case when point O coincides with point O', the origin of the coordinate system because, for such a case, $x_O = y_O = z_O = 0$.

5.4
Varignon's
Theorem

Varignon's theorem, or the principle of moments, was introduced in Chapter 4 for the case of two-dimensional force systems. This theorem still holds when dealing with three-dimensional force systems and states that *the moment of a force about any point in space is equal to the sum of the moments of its components about the same point.* Before, we can prove this theorem, we need to establish the fact that the *distributive law* is applicable to the cross product of two vectors. Thus, con-

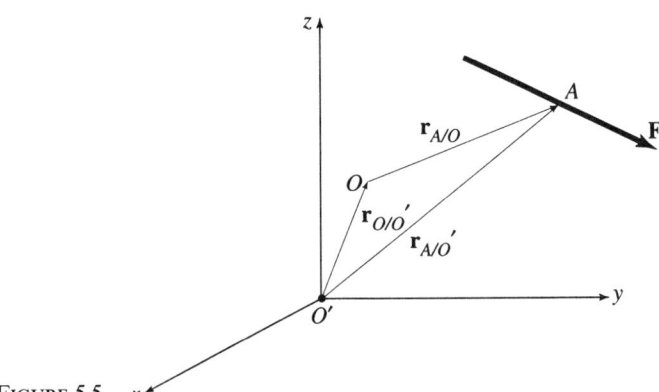

FIGURE 5.5.

sider the three vectors **A**, **B**, and **C** and show that

$$\mathbf{A} \times (\mathbf{B} + \mathbf{C}) = \mathbf{A} \times \mathbf{B} + \mathbf{A} \times \mathbf{C}. \tag{5.16}$$

The distributive law expressed in Eq. (5.16) may be proven as follows:

$$
\begin{aligned}
\mathbf{A} \times (\mathbf{B} + \mathbf{C}) &= (A_x\mathbf{i} + A_x\mathbf{j} + A_x\mathbf{k}) \\
&\quad \times [(B_x + C_x)\mathbf{i} + (B_y + C_y)\mathbf{j} + (B_z + C_z)\mathbf{k}] \\
&= [A_y(B_x + C_x) - A_z(B_y + C_y)]\mathbf{i} \\
&\quad + [A_z(B_x + C_x) - A_x(B_z + C_z)]\mathbf{j} \\
&\quad + [A_x(B_y + C_y) - A_y(B_x + C_x)]\mathbf{k} \\
&= (A_yB_z - A_zB_y)\mathbf{i} + (A_zB_x - A_xB_z)\mathbf{j} + (A_xB_y - A_yB_x)\mathbf{k} \\
&\quad + (A_yC_z - A_zC_y)\mathbf{i} + (A_zC_x - A_xC_z)\mathbf{j} \\
&\quad + (A_xC_y - A_yC_x)\mathbf{k}
\end{aligned}
$$

which expresses the equality given in Figure (5.16). The distributive property of the cross product of two vectors simplifies the proof of Varignon's theorem, particularly, for three-dimensional force systems. Thus, consider the concurrent forces $\mathbf{F}_1, \mathbf{F}_2, \ldots, \mathbf{F}_n$ shown in Figure 5.6 which are components of the force **F**. The sum of the moments of the components $\mathbf{F}_1, \mathbf{F}_2, \ldots, \mathbf{F}_n$ about point O is given by the relationship

$$
\begin{aligned}
\mathbf{M}_O &= \mathbf{r}_{A/O} \times \mathbf{F}_1 + \mathbf{r}_{A/O} \times \mathbf{F}_2 + \cdots + \mathbf{r}_{A/O} \times \mathbf{F}_n \\
&= \mathbf{r}_{A/O} \times (\mathbf{F}_1 + \mathbf{F}_2 + \cdots + \mathbf{F}_n) \\
&= \mathbf{r}_{A/O} \times \mathbf{F} \tag{5.17}
\end{aligned}
$$

where $\mathbf{F} = \mathbf{F}_1 + \mathbf{F}_2 + \cdots + \mathbf{F}_n$. Equation (5.17) is a symbolic statement of Varignon's theorem.

The following examples illustrate some of the concepts discussed in Sections 5.3 and 5.4.

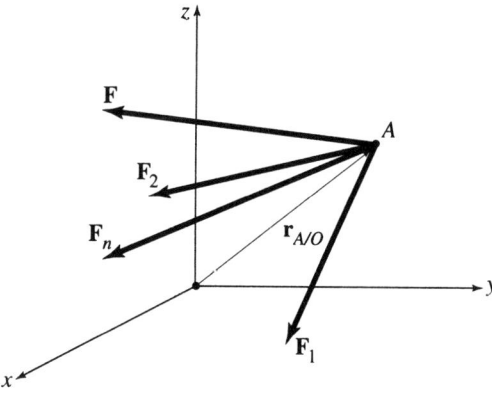

FIGURE 5.6.

■ **Example 5.2**

A 7 ft by 10 ft rectangular plate is partially supported in a horizontal position by cables AB and AC, as shown in Figure E5.2(a). If the tension in cable AB, $F_{AB} = 950$ lb and that in AC, $F_{AC} = 700$ lb, determine (a) the resultant **R** of $\mathbf{F}_{AB}$ and $\mathbf{F}_{AC}$, then compute the moment of **R** about point O; (b) separately, the moments of $\mathbf{F}_{AB}$ and $\mathbf{F}_{AC}$ about point O. Find their sum and compare to the answer obtained in part (a).

Solution

(a) The tensions $\mathbf{F}_{AB}$ and $\mathbf{F}_{AC}$ are expressed in vector form. Thus,

$$\mathbf{F}_{AB} = 950 \left[\frac{(8-7)\mathbf{i} + (-2-10)\mathbf{j} + (10-0)\mathbf{k}}{\sqrt{1^2 + 12^2 + 10^2}} \right]$$

$$= 60.7\mathbf{i} - 728.3\mathbf{j} + 606.9\mathbf{k},$$

and

$$\mathbf{F}_{AC} = 700 \left[\frac{(0-7)\mathbf{i} + (4-10)\mathbf{j} + (10-0)\mathbf{k}}{\sqrt{7^2 + 6^2 + 10^2}} \right]$$

$$= -360.3\mathbf{i} - 308.8\mathbf{j} + 514.7\mathbf{k}.$$

Therefore,

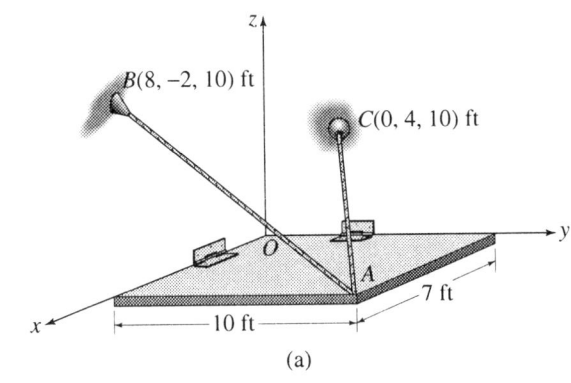

(a)

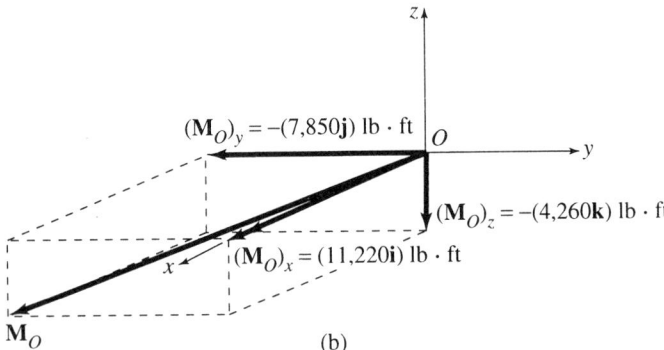

FIGURE E5.2. $\mathbf{M}_O$ (b)

$$\mathbf{R} = \mathbf{F}_{AB} + \mathbf{F}_{AC} = -299.6\mathbf{i} - 1037.1\mathbf{j} + 1121.6\mathbf{k}.$$

Now, by Eq. (5.9),

$$\mathbf{M}_O = \mathbf{r}_{A/O} \times \mathbf{R} = (7\mathbf{i} + 10\mathbf{j}) \times (-299.6\mathbf{i} - 1037.1\mathbf{j} + 1121.6\mathbf{k})$$

$$= \begin{vmatrix} \mathbf{i} & \mathbf{j} & \mathbf{k} \\ 7 & 10 & 0 \\ -299.6 & -1037.1 & 1121.6 \end{vmatrix}$$

$$= (11{,}220\mathbf{i} - 7{,}850\mathbf{j} - 4{,}260\mathbf{k}) \text{ lb·ft.} \qquad \text{ANS.}$$

(b) For $\mathbf{F}_{AB}$,

$$(\mathbf{M}_O)_{AB} = \mathbf{r}_{A/O} \times \mathbf{F}_{AB} = (7\mathbf{i} + 10\mathbf{j}) \times (60.7\mathbf{i} - 728.3\mathbf{j} + 606.9\mathbf{k})$$

$$= \begin{vmatrix} \mathbf{i} & \mathbf{j} & \mathbf{k} \\ 7 & 10 & 0 \\ 60.7 & -728.3 & 606.9 \end{vmatrix}$$

$$= 6{,}069.0\mathbf{i} + 4{,}248.3\mathbf{j} + 5{,}705.1\mathbf{k}.$$

For $\mathbf{F}_{AC}$,

$$(\mathbf{M}_O)_{AC} = \mathbf{r}_{A/O} \times \mathbf{F}_{AC} = (7\mathbf{i} + 10\mathbf{j}) \times (-360.3\mathbf{i} - 308.8\mathbf{j} + 514.7\mathbf{k})$$

$$= \begin{vmatrix} \mathbf{i} & \mathbf{j} & \mathbf{k} \\ 7 & 10 & 0 \\ -360.3 & -308.8 & 514.7 \end{vmatrix}$$

$$= 5{,}147.0\mathbf{i} - 3{,}602.9\mathbf{j} + 1{,}441.4\mathbf{k}.$$

Thus,

$$\mathbf{M}_O = (\mathbf{M}_O)_{AB} + (\mathbf{M}_O)_{AC}$$
$$= (11{,}220\mathbf{i} - 7{,}850\mathbf{j} - 4{,}260\mathbf{k}) \text{ lb·ft} \qquad \text{ANS.}$$

which, of course, is identical with the answer obtained in part (a). Why?

The moment $\mathbf{M}_O$ along with its x, y, and z components, is shown in Figure E5.2(b).

■ **Example 5.3**

A force $\mathbf{F} = (80\mathbf{i} - 120\mathbf{j} - 100\mathbf{k})$ N passes through point B as shown in Figure E5.3.

(a) Determine the moment of $\mathbf{F}$ about point O.
(b) Find the perpendicular distance d from point O to the line of action of $\mathbf{F}$.
(c) Identify the axis of rotation OC by specifying a unit vector λ_{OC} in its direction.

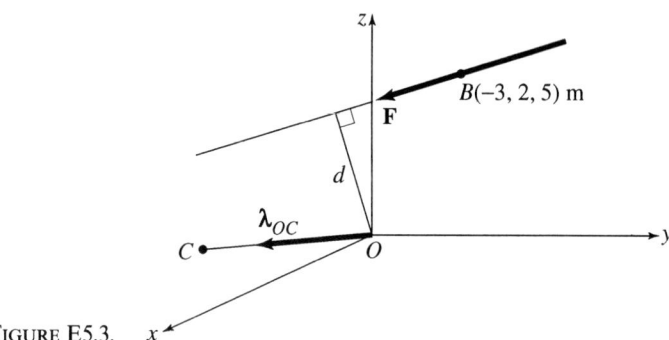

FIGURE E5.3. x

Solution

(a) By Eq. (5.9),

$$\mathbf{M}_O = \mathbf{r}_{B/O} \times \mathbf{F}$$

where

$$\mathbf{r}_{B/O} = -3\mathbf{i} + 2\mathbf{j} + 5\mathbf{k}.$$

Thus,

$$\mathbf{M}_O = (-3\mathbf{i} + 2\mathbf{j} + 5\mathbf{k}) \times (80\mathbf{i} - 120\mathbf{j} - 100\mathbf{k})$$

$$= \begin{vmatrix} \mathbf{i} & \mathbf{j} & \mathbf{k} \\ -3 & 2 & 5 \\ 80 & -120 & -100 \end{vmatrix}$$

$$= (400\mathbf{i} + 100\mathbf{j} + 200\mathbf{k}) \text{ N·m.} \qquad \text{ANS.}$$

(b) The magnitudes of $\mathbf{F}$ and $\mathbf{M}_O$ are found as follows:

$$F = \sqrt{80^2 + 120^2 + 100^2} = 175.499 \text{ N,}$$

and

$$M_O = \sqrt{400^2 + 100^2 + 200^2} = 458.258 \text{ N·m.}$$

By Eq. (5.10),

$$d = \frac{M_O}{F} = 2.61 \text{ m.} \qquad \text{ANS.}$$

(c)

$$\lambda_{OC} = \frac{\mathbf{M}_O}{M_O} = 0.873\mathbf{i} + 0.218\mathbf{j} + 0.436\mathbf{k}. \qquad \text{ANS.}$$

The unit vector λ_{OC} is indicated in Figure E5.3.

■

Problems

5.1 A force $\mathbf{F} = (25\mathbf{i} - 40\mathbf{j} + 571\mathbf{k})$ lb passes through point A whose position vector from the origin of the coordinate system is $\mathbf{r}_{A/O} = (-3\mathbf{i} + 4\mathbf{j} + 5\mathbf{k})$ ft. Determine (a) the moment of $\mathbf{F}$ about point O and (b) the moment of $\mathbf{F}$ about point B where $\mathbf{r}_{B/O} = (5\mathbf{i} + 7\mathbf{j} - 8\mathbf{k})$ ft.

5.2 Refer to Problem 5.1 and determine (a) the perpendicular distance from point O to the line of action of $\mathbf{F}$ and (b) the perpendicular distance from point B to the line of action of $\mathbf{F}$.

5.3 A force $\mathbf{F} = (-10\mathbf{i} + 22\mathbf{j} - 8\mathbf{k})$ N acts at point A as shown in Figure P5.3. Compute (a) the moment of $\mathbf{F}$ about point O and (b) the moment of $\mathbf{F}$ about point B.

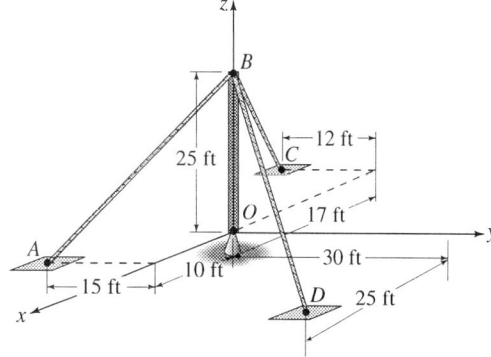

FIGURE P5.5.

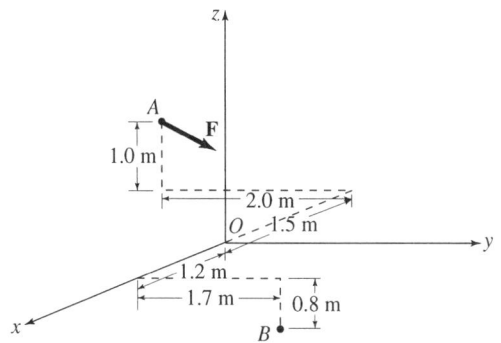

FIGURE P5.3.

5.4 Refer to Problem 5.3 and find (a) the perpendicular distance from point O to the line of action of $\mathbf{F}$ and (b) the perpendicular distance from point B to the line of action of $\mathbf{F}$.

5.5 Mast OB is supported by three guy cables, as shown in Figure P5.5. The tension in cable BD $T_{BD} = 750$ lb. Determine the moment $\mathbf{M_O}$ of this force about

point O. Find the directional angles α, β, and γ made by $\mathbf{M_O}$ with the x, y, and z axes, respectively.

5.6 Refer to Problem 5.5. Determine the moment $\mathbf{M_C}$ produced by T_{BD} about point C. What is the perpendicular distance from point C to cable BD?

5.7 A force $P = 20$ k is applied as shown in Figure P5.7. Determine the moment of this force about point A. Find the per-

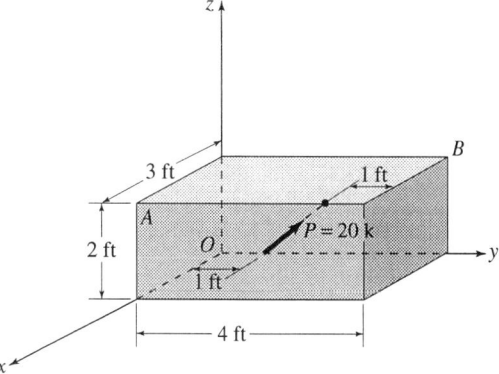

FIGURE P5.7.

pendicular distance from point A to the line of action of the force **P**.

5.8 Refer to Problem 5.7 and determine the moment M_B of the force **P** about point B. Express M_B as a magnitude multiplying a unit vector. What is the significance of this unit vector?

5.9 A force $F = 10$ kN is applied as shown in Figure P5.9. Find the moment M_A of this force about point A. Determine the directional angles α, β, and γ made by M_A with the x, y, and z axes, respectively.

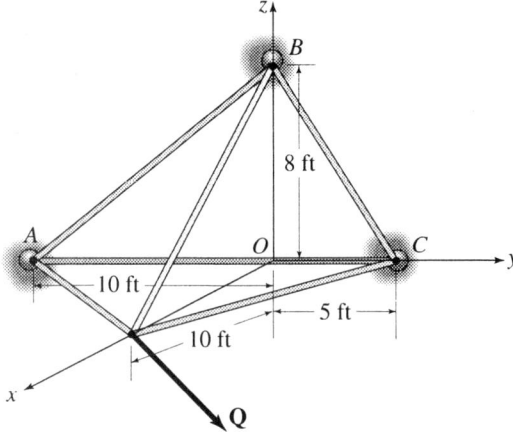

FIGURE P5.11.

tensions $T_1 = (-200i + 300j - 150k)$ N and $T_2 = (150i - 170j - 200k)$ N as shown in Figure P5.13(a). (a) Determine the resultant **R** of T_1 and T_2, and compute the moment of this resultant about point O. (b) Find, separately, the moments of T_1 and T_2 about point O. Find their sum and compare it to the answer obtained in part (a).

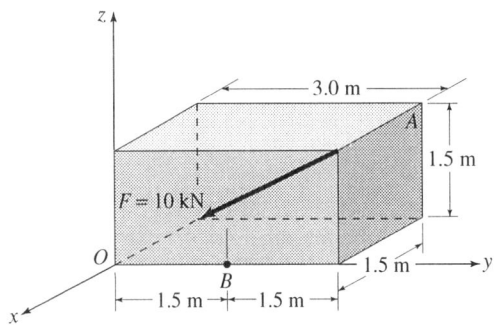

FIGURE P5.9.

5.10 Refer to Problem 5.9 and compute the moment M_B of the force **F** about point B. Determine the perpendicular distance from point B to the line of action of the force **F**.

5.11 The frame shown in Figure P5.11 is subjected to the force $Q = (5i + 7j - 6k)$ k. Find the moment of this force about point B. Also, find the perpendicular distance from point B to the line of action of the force **Q**.

5.12 Refer to Problem 5.11 and determine the moment M_C of the force **Q** about point C. Find the directional angles α, β, and γ made by M_C with the x, y, and z coordinate axes, respectively.

5.13 A telephone pole is subjected to the cable

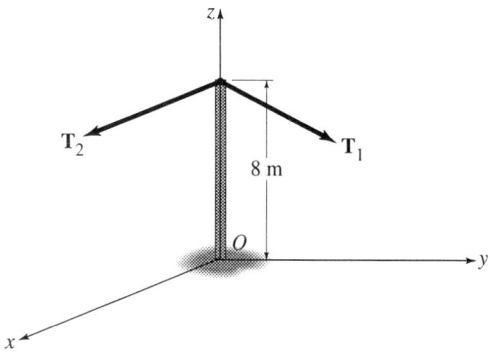

FIGURE P5.13.

5.14 The moment of a force **F** about the origin of the coordinate system M_O =

$(-65\mathbf{i} + 35\mathbf{j} + 5\mathbf{k})$ lb·ft. It is known that the magnitude of $\mathbf{F}$ is 12 lb and that it passes through point A where the position vector $\mathbf{r}_{A/O} = (2\mathbf{i} + 3\mathbf{j} + 5\mathbf{k})$ ft. Determine the x, y, and z components of the force $\mathbf{F}$.

5.15 The force $\mathbf{P} = (-40\mathbf{i} + 60\mathbf{j} - 30\mathbf{k})$ N produces a moment about point O, the origin of the coordinate system, such that $\mathbf{M}_O = (-60\mathbf{i} - 60\mathbf{j} - 40\mathbf{k})$ N·m. It is known that the force P passes through point A where $r_{A/O} = 5.4$ m. Determine the coordinates of point A.

5.5 Moment of a Force About a Specific Axis

In Section 5.3, we learned how to determine the moment of a force about a given point. In reality, this moment represents the turning effect of the force about a very specific axis passing through the point. In particular, as shown in Figure 5.7, the moment $\mathbf{M}_O$ of the force $\mathbf{F}$ about point O (given by Eq. (5.9)) represents the turning effect of the force about axis OB which is perpendicular to the plane formed by $\mathbf{F}$ and point O. Note that the axis OB also coincides with the direction of the moment vector $\mathbf{M}_O$.

Frequently, it becomes necessary to determine the turning effect of the force $\mathbf{F}$ about an arbitrary axis different from axis OB. This may be accomplished by finding the component of vector $\mathbf{M}_O$ along the desired axis. For example, the scalar component of $\mathbf{M}_O$ along axis OC in Figure 5.7, may be found by the dot product expressed in Eq. (3.18). Thus

$$M_{OC} = \boldsymbol{\lambda} \cdot \mathbf{M}_O \tag{5.18}$$

where M_{OC} represents the scalar component of $\mathbf{M}_O$ along axis OC and $\boldsymbol{\lambda}$ is a unit vector along this axis. Note that M_{OC} in Eq. (5.18) represents the magnitude of the turning effect of the force $\mathbf{F}$ about axis OC.

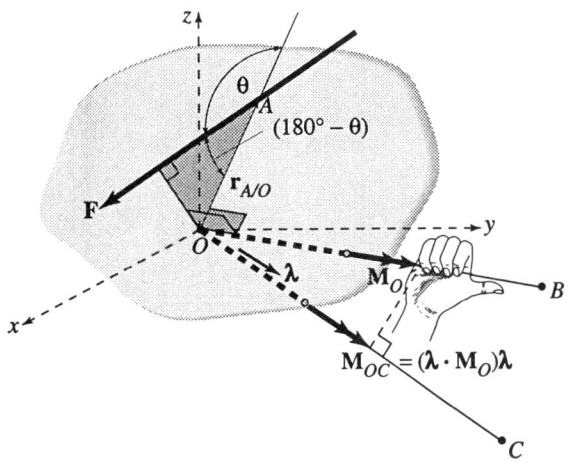

FIGURE 5.7.

Because $\mathbf{M}_O = \mathbf{r}_{A/O} \times \mathbf{F}$, it follows from Eq. (5.18) that

$$\mathbf{M}_{OC} = \boldsymbol{\lambda} \cdot (\mathbf{r}_{A/O} \times \mathbf{F}) \tag{5.19}$$

where point A represents *any* point on the line of action of the force $\mathbf{F}$ (see Figure 5.7). The product given in Eq. (5.19) is referred to as the mixed triple product because it represents a mixture between the dot and the cross-products. Using the results expressed in Eqs. (3.20) and (5.14), Eq. (5.19) becomes

$$M_{OC} = \lambda_x(y_A F_z - z_A F_y) + \lambda_y(z_A F_x - x_A F_z) + \lambda_z(x_A F_y - y_A F_x) \tag{5.20}$$

where λ_x, λ_y, and λ_z represent, respectively, the scalar components of the unit vector $\boldsymbol{\lambda}$ in the x, y, and z directions. Equation (5.20) may be written more compactly in the following determinant form:

$$M_{OC} = \begin{vmatrix} \lambda_x & \lambda_y & \lambda_z \\ x_A & y_A & z_A \\ F_x & F_y & F_z \end{vmatrix}. \tag{5.21}$$

In summary, the magnitude of the turning effect M_{OC} of a force $\mathbf{F}$ about an arbitrary axis OC may be obtained by, first, finding the moment $\mathbf{M}_O$ of the force $\mathbf{F}$ about *any* point O on this axis and, then, dot multiplying a unit vector $\boldsymbol{\lambda}$ along axis OC with the vector $\mathbf{M}_O$. Note that, after the scalar component M_{OC} is determined, it may be expressed in vector form by the relationship

$$\mathbf{M}_{OC} = M_{OC}\boldsymbol{\lambda}. \tag{5.22}$$

The following example illustrates the use of the concepts above.

■ Example 5.4

A three-dimensional truss is subjected to a force $\mathbf{F} = (-10\mathbf{i} + 15\mathbf{j} + 6\mathbf{k})$ k at joint A as shown in Figure E5.4(a). Compute the moment of this force (a) about the x, y, and z axes and (b) about axis BC.

Solution

(a) Because point O is common to the x, y, and z axes, the moment of $\mathbf{F}$ about these three axes may be obtained first, by, finding its moment about point O. Thus,

$$\mathbf{M}_O = \mathbf{r}_{A/O} \times \mathbf{F} = (12\mathbf{i} + 5\mathbf{j} + 25\mathbf{k}) \times (-10\mathbf{i} + 15\mathbf{j} + 6\mathbf{k})$$

$$= \begin{vmatrix} \mathbf{i} & \mathbf{j} & \mathbf{k} \\ 12 & 5 & 25 \\ -10 & 15 & 6 \end{vmatrix}$$

$$= -345\mathbf{i} - 322\mathbf{j} + 230\mathbf{k}.$$

The moments produced by the force $\mathbf{F}$ about the three coordinate axes

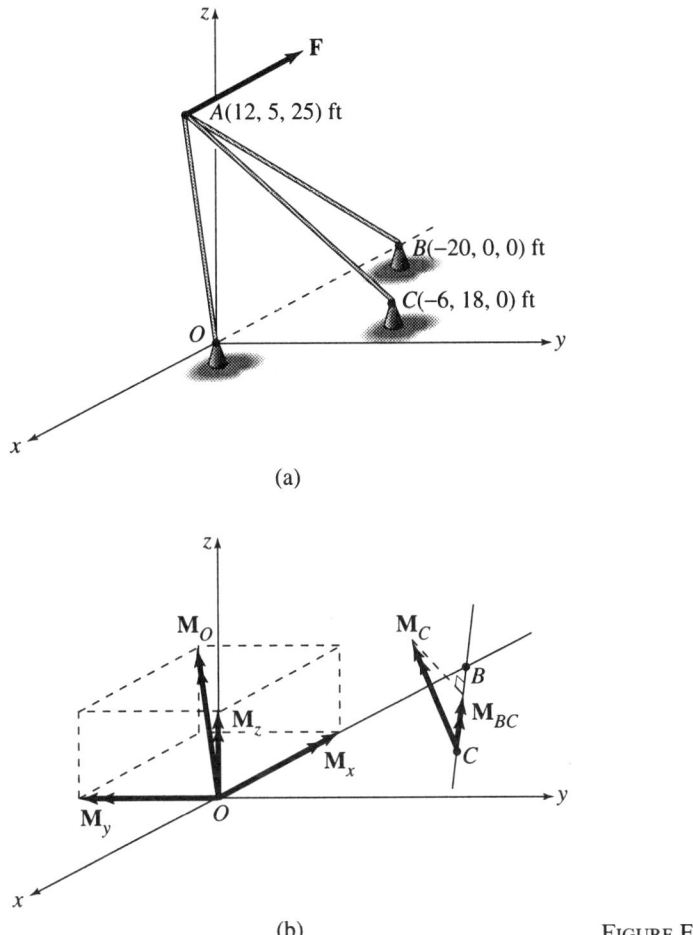

(a)

(b)

FIGURE E5.4.

are given by the components of $\mathbf{M}_O$ about these axes. Therefore,

$$\left.\begin{array}{l}\mathbf{M}_x = (-345\mathbf{i})\ k\cdot ft, \\[4pt] \mathbf{M}_y = (-322\mathbf{j})\ k\cdot ft, \\[4pt] \mathbf{M}_z = (230\mathbf{k})\ k\cdot ft.\end{array}\right\} \qquad \text{ANS.}$$

and

Clearly, the same answers are obtained by the use of Eqs. (5.18) and (5.22). Thus, for example, from Eq. (5.18),

$$M_x = \boldsymbol{\lambda}\cdot\mathbf{M}_O = \mathbf{i}\cdot\mathbf{M}_O = (-345\mathbf{i})\ k\cdot ft,$$

and, by Eq. (5.22),

$$\mathbf{M}_x = (-345\mathbf{i})\ k\cdot ft$$

which is identical with the answer found above. The moments $\mathbf{M}_x$, $\mathbf{M}_y$, and $\mathbf{M}_z$ are shown in Figure E5.4(b) along with their resultant $\mathbf{M}_O$.

(b) To determine the moment of the force $\mathbf{F}$ about axis BC, we select any point, such as C, on this axis and find the moment of $\mathbf{F}$ about this point. Thus,

$$\mathbf{M}_C = \mathbf{r}_{A/C} \times \mathbf{F} = (18\mathbf{i} - 13\mathbf{j} + 25\mathbf{k}) \times (-10\mathbf{i} + 15\mathbf{j} + 6\mathbf{k})$$

$$= \begin{vmatrix} \mathbf{i} & \mathbf{j} & \mathbf{k} \\ 18 & -13 & 25 \\ -10 & 15 & 6 \end{vmatrix}$$

$$= -453\mathbf{i} - 358\mathbf{j} + 140\mathbf{k}. \tag{a}$$

We now define a unit vector λ_{BC} along axis BC. The unit vector chosen is defined from B to C, although one defined from C to B would lead to the same magnitude, but with the opposite sign. Therefore, by Eq. (5.19),

$$M_{BC} = \lambda_{BC} \cdot \mathbf{M}_C$$

where

$$\lambda_{BC} = \frac{\mathbf{r}_{C/B}}{r_{C/B}} = \frac{14\mathbf{i} + 18\mathbf{j}}{\sqrt{14^2 + 18^2}} = 0.614\mathbf{i} + 0.789\mathbf{j}.$$

Therefore,

$$M_{BC} = (0.614\mathbf{i} + 0.789\mathbf{j}) \cdot (-453\mathbf{i} - 458\mathbf{j} + 140\mathbf{k}) = -560.6 \text{ k·ft.} \tag{b}$$

The two separate operations indicated in Eqs. (a) and (b) above could be combined into a single operation as suggested by Eq. (5.21). Thus,

$$M_{BC} = \begin{vmatrix} 0.614 & 0.784 & 0 \\ 18 & -13 & 25 \\ -10 & 15 & 6 \end{vmatrix} = -560.6 \text{ k·ft}$$

which, of course, is identical to that obtained in Eq. (b). Using Eq. (5.22), we can express the moment about axis BC in vector form. Thus,

$$\mathbf{M}_{BC} = (-560.6)(0.614\mathbf{i} + 0.789\mathbf{j}) = -(344\mathbf{i} + 442\mathbf{j}) \text{ k·ft.} \quad \text{ANS.}$$

The minus sign on $\mathbf{M}_{BC}$ indicates that this moment vector is directed from C to B and not from B to C as assumed by the chosen unit vector λ_{BC}. The moment $\mathbf{M}_C$ and its component $\mathbf{M}_{BC}$ are also indicated in Figure E5.4(b).

■ **Example 5.5**

Use the moment $|M_{BC}| = 560.6$ k·ft found in part (b) of Example 5.4 to determine the perpendicular distance between the line of action of the force $\mathbf{F}$ and the axis BC.

Solution

The angle θ between the line of action of **F** and the axis BC is determined by Eq. (3.21). Thus

$$\theta = \cos^{-1}\left[\frac{\mathbf{F}\cdot\mathbf{r}_{C/B}}{(F)(r_{C/B})}\right]$$

where

$$\mathbf{F} = -10\mathbf{i} + 15\mathbf{j} + 6\mathbf{k},$$

$$F = \sqrt{10^2 + 15^2 + 6^2} = 19.0,$$

$$\mathbf{r}_{C/B} = 14\mathbf{i} + 18\mathbf{j},$$

$$\mathbf{r}_{C/B} = \sqrt{14^2 + 18^2} = 22.8.$$

Therefore,

$$\theta = \cos^{-1}\left[\frac{(-10\mathbf{i} + 15\mathbf{j} + 6\mathbf{k})\cdot(14\mathbf{i} + 18\mathbf{j})}{(19.0)(22.8)}\right]$$

$$= \cos^{-1}(0.3001)$$

$$= 72.5°.$$

The component of **F** parallel to the axis BC is $F\cos\theta$ which, obviously, produces no moment about this axis. However, the component $F\sin\theta$, which lies in a plane perpendicular to axis BC, as shown in Figure E5.5, produces a moment about this axis of magnitude $M_{BC} = (F\sin\theta)d$ where $d = $ ED, is the required perpendicular distance. Thus,

$$d = \frac{M_{BC}}{F\sin\theta} = \frac{560.6}{(19.0)\sin 72.5°} = 30.9 \text{ ft.} \qquad \text{ANS.}$$

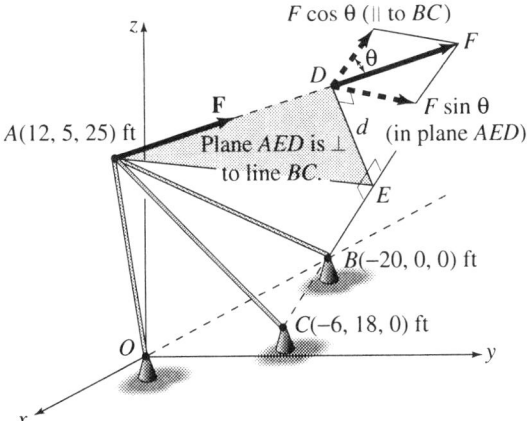

FIGURE E5.5.

Problems

5.16 Find the mixed triple products (a) $\mathbf{A}\cdot(\mathbf{B}\times\mathbf{C})$, (b) $(\mathbf{B}\times\mathbf{C})\cdot\mathbf{A}$, and (c) $\mathbf{B}\cdot(\mathbf{A}\times\mathbf{C})$ if $\mathbf{A} = 2\mathbf{i} - 4\mathbf{j} + 7\mathbf{k}$, $\mathbf{B} = -5\mathbf{i} - 3\mathbf{j} + 8\mathbf{k}$ and $\mathbf{C} = 4\mathbf{i} + 6\mathbf{j} - 3\mathbf{k}$.

5.17 Use the properties of the dot and of the cross-products to show that the mixed triple product of any three coplanar vectors (vectors that lie in the same plane) is zero.

5.18 Let $\mathbf{U} = 5\mathbf{i} + 3\mathbf{j} + 2\mathbf{k}$, $\mathbf{V} = 2\mathbf{i} + V_y\mathbf{j} - 5\mathbf{k}$ and $\mathbf{W} = 6\mathbf{i} + 4\mathbf{j} + 10\mathbf{k}$. Determine the scalar component V_y if the three given vectors are coplanar. Hint: *Use the result obtained in Problem 5.17.*

5.19 Let $\mathbf{U} = U_x\mathbf{i} + 25\mathbf{j} + U_z\mathbf{k}$, $\mathbf{V} = -7\mathbf{i} - 6\mathbf{j} + 12\mathbf{k}$, and $\mathbf{W} = 4\mathbf{i} + 3\mathbf{j} - 5\mathbf{k}$. Determine the scalar components U_x and U_z if the three given vectors are coplanar and if $U = 60$. Hint: *Use the result obtained in Problem 5.17.*

5.20 A force $\mathbf{F} = (-5\mathbf{i} + 10\mathbf{j} - 6\mathbf{k})$ k acts at point A, as shown in Figure P5.20. Compute the moment of this force about (a) the x, y, and z coordinate axes and (b) the axis OB.

5.21 A force $F = 10$ kN is applied as shown in Figure P5.21. Determine the moment of

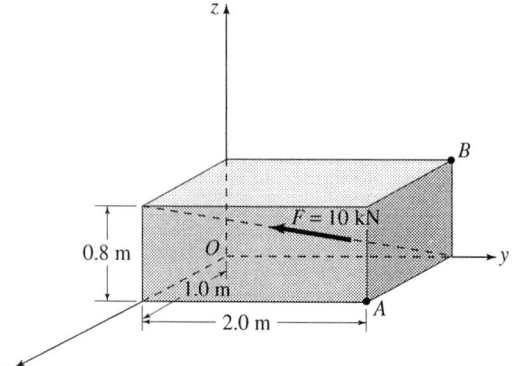

this force about (a) the x, y, and z coordinate axes and (b) the axis AB.

5.22 A force $P = 20$ k is applied, as shown in Figure P5.22. Determine the moment of this force about (a) the axis OA and (b) the axis AB.

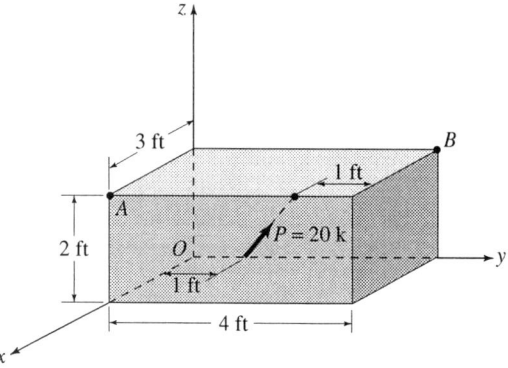

FIGURE P5.22.

5.23 A force $\mathbf{F} = (-12\mathbf{i} + 7\mathbf{j} - 5\mathbf{k})$ kN is applied at joint D of the three-dimensional truss shown in Figure P5.23. Find the

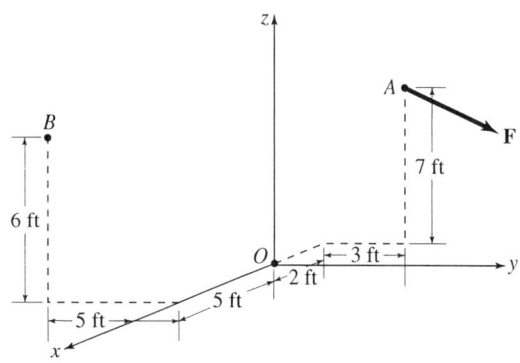

FIGURE P5.20.

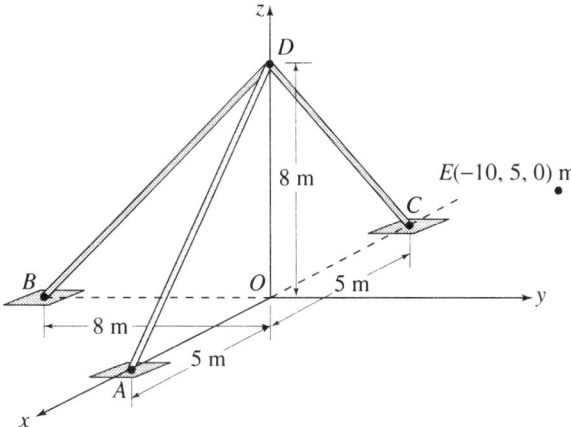

FIGURE P5.23.

moment of this force about (a) axis AB and (b) axis BC.

5.24 Refer to the three-dimensional truss shown in Figure P5.23. A force **P** is applied at joint D and is directed from D to point E $(-10, 5, 0)$ m. Find the largest magnitude P of the force **P** if the magnitude of the moments it produces about the x and y axes cannot exceed 100 kN·m and 150 kN·m, respectively.

5.25 Refer to the three-dimensional truss shown in Figure P5.23. A force **Q** is applied to joint D and is directed from D to joint E $(-10, 5, 0)$ m. Find the magnitude Q of the force **Q** if the moment produced by this force about axis AC has a magnitude of 200 kN·m.

5.26 A force $\mathbf{F} = F_x\mathbf{i} + F_y\mathbf{j} + F_z\mathbf{k}$ is applied at joint B of the pipe assembly OABC shown in Figure P5.26. Show that this force produces moments $\mathbf{M}_x$, $\mathbf{M}_y$, and $\mathbf{M}_z$ whose magnitudes are related by the expression $15\mathbf{M}_x + 8\mathbf{M}_x - 6\ \mathbf{M}_x = 0$. Note that segments AB and BC of the pipe assembly lie in a plane parallel to the y-z plane.

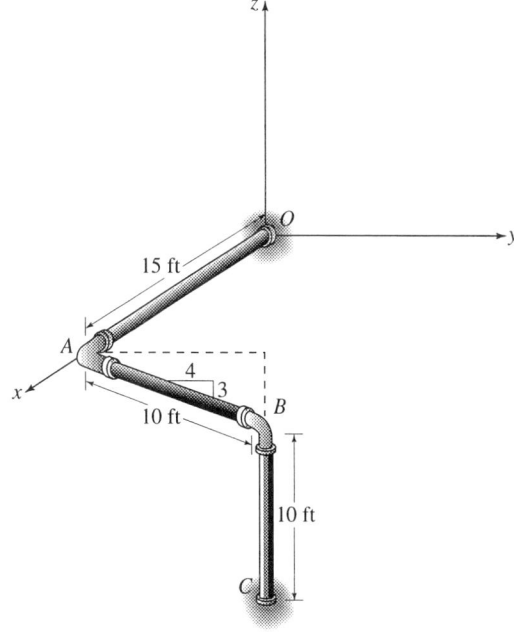

FIGURE P5.26.

5.27 Refer to the pipe assembly shown in Figure P5.26. A force $P = 75$ lb is applied at joint A and directed from A to C. Determine the moment of this force (a) about the x, y, and z coordinate axes and (b) about the axis OB.

5.28 A plate in the form of a right-angle isosceles triangle is hinged along its side OB and supported by cable AC, as shown in

Figure P5.28. For the position shown, the tension in cable AC is 750 N. Determine the moment produced by the cable tension (a) about the x, y, and z coordinate axes and (b) about the hinged axis OB.

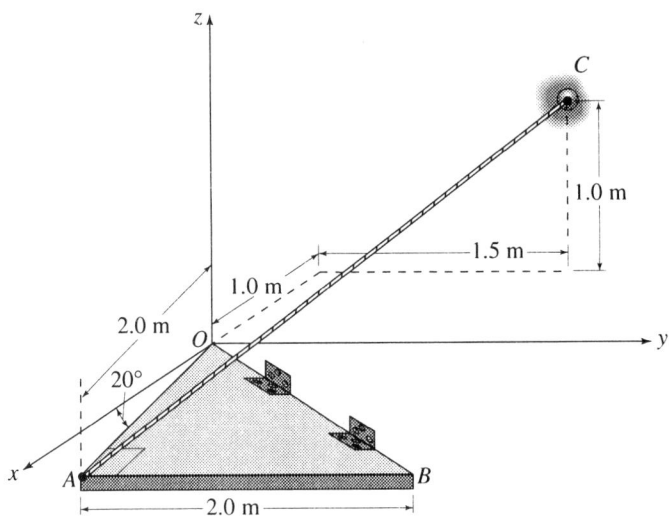

FIGURE P5.28.

5.29 Refer to the plate shown in Figure P5.28, and determine the perpendicular distance between the cable AC and the hinged axis OB. *Hint: Use the result obtained in part (b) of Problem 5.28.*

5.30 Refer to the plate shown in Figure P5.28, and determine the perpendicular distance between the cable AC and the y axis. *Hint: Use the result obtained in part (a) of Problem 5.28.*

5.31 A force $Q = (10i + 7j - 15k)$ k is applied at joint B of the rigid frame shown in Figure P5.31. Find the moment of this force (a) about the axis AD and (b) about the axis FE.

5.32 Refer to the frame shown in Figure P5.31, and determine the perpendicular distance

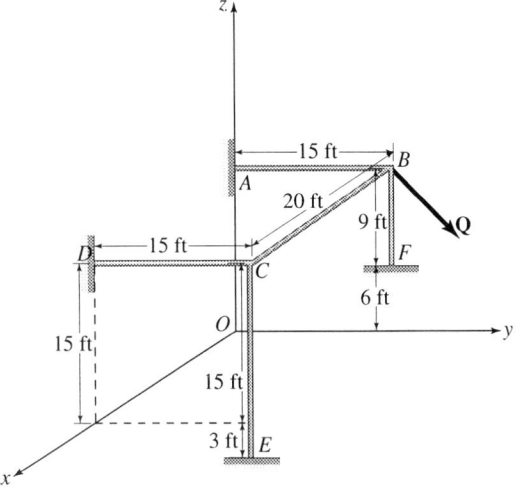

FIGURE P5.31.

between the line of action of the force **Q** and the axis AD. *Hint: Use the result obtained in part (a) of Problem 5.31.*

5.33 Refer to the frame shown in Figure P5.31, and determine the perpendicular distance between the line of action of the force **Q** and the axis FE. *Hint: Use the result obtained in part (b) of Problem 5.31.*

5.6 Vector Representation of a Couple

The concept of the *couple* was introduced in Chapter 4 in relation to two-dimensional force systems. This concept will be expanded here to cover the case of three-dimensional force systems.

As stated in Chapter 4, a *couple* consists of two parallel forces, equal in magnitude, opposite in sense, and noncollinear. Consider, for example, the two forces **F** and −**F** shown in Figure 5.8. Such a force system produces no translation because the sum of the components of the two forces in any direction is zero. However, as will be shown shortly, a couple leads to a rotating tendency due to the *moment of the couple* **M** whose magnitude, direction, and sense are invariant regardless of the location of the point about which moments are taken. Thus, in Figure 5.8, point O is selected arbitrarily to serve as the origin of a coordinate system. The moment of the couple consisting of forces **F** and −**F** may be determined about point O by adding the moments produced, separately, about this point by the two forces **F** and −**F**. Thus, arbitrarily selecting points A and B on the lines of action of forces **F** and −**F**, respectively,

$$\mathbf{M} = \mathbf{r}_{A/O} \times \mathbf{F} + \mathbf{r}_{B/O} \times (-\mathbf{F})$$

$$= (\mathbf{r}_{A/O} - \mathbf{r}_{B/O}) \times \mathbf{F}$$

$$= \mathbf{r}_{A/B} \times \mathbf{F} \tag{5.23}$$

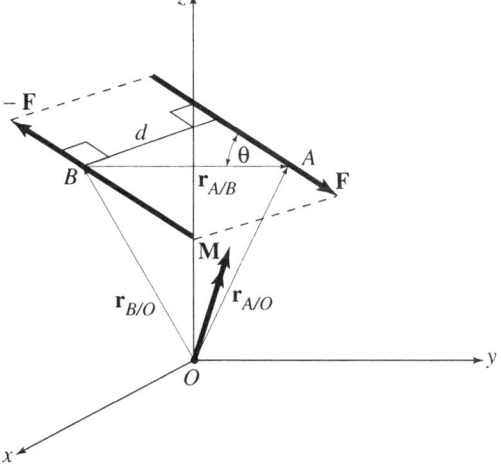

FIGURE 5.8.

where $r_{A/B} = r_{A/O} - r_{B/O}$ is a vector in the plane formed by F and $-F$. Equation (5.23) shows that the moment of the couple M is independent of the location of point O. For example, if point O is chosen to coincide with point B in Figure 5.8, then, $r_{B/O} = 0$, $r_{A/B} = r_{A/O}$, and the moment of the couple M reduces once again to $r_{A/B} \times F$ as given by Eq. (5.23). It follows, therefore, that the moment of a couple M is a *free vector* that may occupy any position in space provided that its magnitude, direction, and sense are preserved. For convenience, the moment of the couple M in Figure 5.8 is shown at point O, although it could equally be placed at any other point in space.

Equation (5.23) also shows that the moment of a couple is definable by the cross-product of two vectors. Therefore, by Eq. (5.2), the magnitude of the couple M is given by

$$M = F(r_{A/B}\sin\theta) = Fd \tag{5.24}$$

where $d = (r_{A/B})\sin\theta$ is the perpendicular distance between the lines of action of the two forces of the couple, as shown in Figure 5.8. Also, the direction of M is perpendicular to the plane formed by the forces of the couple F and $-F$, and its sense is given by the right-hand rule.

As a vector, the moment of a couple M obeys all of the rules of vector operations. Thus, for example, several couples $M_1, M_2, \ldots, M_n$ may be added by the methods of Chapter 3 to obtain the resultant couple M_R. Accordingly,

$$M_R = M_1 + M_2 + \cdots + M_n = \sum_{i=1}^{n} M_i \tag{5.25}$$

If each of the couples M_1 to M_n is expressed in terms of rectangular (i, j, and k) components by Eq. (5.14) and the respective i, j, and k terms are added,

$$M_R = \left(\sum_{i=1}^{n} M_{xi}\right)i + \left(\sum_{i=1}^{n} M_{yi}\right)j + \left(\sum_{i=1}^{n} M_{zi}\right)k$$

$$= M_x i + M_y j + M_y k. \tag{5.26}$$

The symbol $M_x = \sum M_{xi}$ is the x component of the resultant couple M_R and may be found by adding, algebraically, *all* of the scalar couple components about the x axis. Similarly, the symbols $M_y = \sum M_{yi}$ and $M_z = \sum M_{zi}$ represent, respectively, the y and z components of the resultant couple M_R obtained by algebraic addition of all of their respective scalar components. As another example, the component of a couple may be determined along any arbitrary axis by using the dot product expressed in Eq. (5.18).

Two or more couples are said to be *equivalent* if they possess the same magnitude, direction, and sense, regardless of their locations in

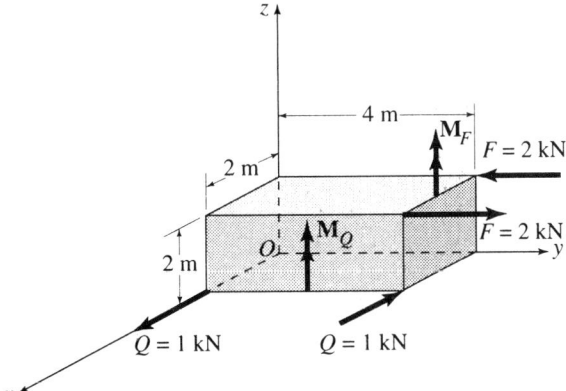

FIGURE 5.9.

space. Thus, the couple $M_F = (4\mathbf{k})$ kN·m, consisting of the two forces
$\mathbf{F}$ each of magnitude $F = 2$ kN is equivalent to the couple $M_Q =$
$(4\mathbf{k})$ kN·m, consisting of the two forces $\mathbf{Q}$ each of magnitude $Q = 1$ kN
as shown in Figure 5.9. These two couples produce identical effects on
the rigid body on which they act.

In Chapter 4, a *null couple* was defined in the two-dimensional case
as a couple in which the perpendicular distance between the two forces
of the couple is zero. In the three-dimensional case, a null couple is one
for which the position vector between the two forces vanishes. Thus, in
Eq. (5.23), $\mathbf{r}_{A/B} = \mathbf{0}$ and the two forces $\mathbf{F}$ and $-\mathbf{F}$ of the null couple are
collinear, as shown in Figure 5.10. Thus, a null couple produces neither
a translating nor a rotating tendency on the rigid body on which it
acts. As will be shown in the following sections, null couples are useful
in transforming a complex force system into another that is simpler but
equivalent to the first.

The following examples illustrate some of the concepts discussed
above.

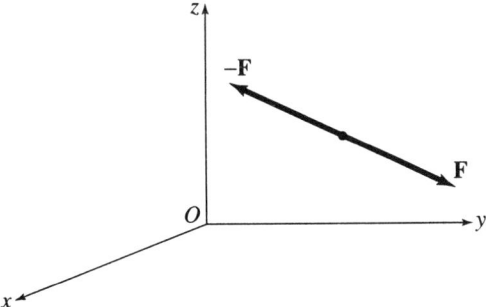

FIGURE 5.10.

■ **Example 5.6**

The magnitude of the forces $\mathbf{Q}$ and $-\mathbf{Q}$ shown in Figure E5.6 is $Q = 200$ N. Determine the moment $\mathbf{M}$ of the couple produced by $\mathbf{Q}$ and $-\mathbf{Q}$. Use vector algebra. Also, determine the magnitude of $\mathbf{M}$ and the directional angles that it makes with the x, y, and z axes, respectively.

Solution

Express $\mathbf{Q}$ in terms of its rectangular components. Thus,

$$\mathbf{Q} = Q\left(\frac{\mathbf{r}_{B/A}}{r_{B/A}}\right) = 200\left(\frac{1.5\mathbf{i} - 0.7\mathbf{k}}{\sqrt{1.5^2 + 0.7^2}}\right) = 181.2\mathbf{i} - 84.6\mathbf{k}.$$

A position vector between the lines of action of the two forces of the couple is now defined. This position vector could originate at *any* point on the line of $-\mathbf{Q}$ and terminate at any point on the line of action of $\mathbf{Q}$. This freedom allows us to choose the most convenient position vector. In our case, either $\mathbf{r}_{A/C}$ or $\mathbf{r}_{B/D}$ is convenient. Using $\mathbf{r}_{A/C} = 2\mathbf{j}$, by Eq. (5.23),

$$\mathbf{M} = \mathbf{r}_{A/C} \times \mathbf{Q} = 2\mathbf{j} \times (181.2\mathbf{i} - 84.6\mathbf{k})$$

$$= -(169.2\mathbf{i} + 362.4\mathbf{k})\ \text{N·m}. \qquad \text{ANS.}$$

Note that $\mathbf{M}$ has no component in the y direction and, therefore, lies entirely in the x-z plane, as shown in Figure E5.6. Although the vector $\mathbf{M}$ is placed at point O in Figure E5.6, it could have been placed at any other point in space.

The magnitude M is found from a relationship comparable to that of Eq. (3.5). Thus,

$$M = \sqrt{M_x^2 + M_y^2 + M_z^2} = 400\ \text{N·m}. \qquad \text{ANS.}$$

It should be observed that, because of the simple geometry in Figure E5.6, the magnitude M could also be obtained by Eq. (5.24), $M = Fd$, where $F = 200$ N and $d = 2.0$ m.

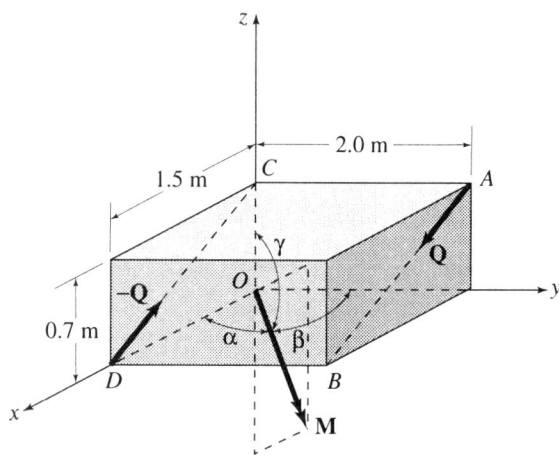

FIGURE E5.6.

The directional cosines are determined from relationships similar to those in Eq. (3.4). Thus,

$$\alpha = \cos^{-1}\frac{M_x}{M} = \cos^{-1}\left(\frac{-169.2}{400}\right) = 115.0°, \qquad \text{ANS.}$$

$$\beta = \cos^{-1}\frac{M_y}{M} = \cos^{-1}\left(\frac{0}{400}\right) = 90.0°, \qquad \text{ANS.}$$

and

$$\gamma = \cos^{-1}\frac{M_z}{M} = \cos^{-1}\left(\frac{-362.4}{400}\right) = 155.0°. \qquad \text{ANS.}$$

■ **Example 5.7**

Two couples ($\mathbf{F}$ and $-\mathbf{F}$) and ($\mathbf{P}$ and $-\mathbf{P}$) act as shown in Figure E5.7(a). The force $\mathbf{F}$ has a magnitude $F = 150$ lb and the force $\mathbf{P}$ a magnitude $P = 100$ lb. (a) Find the resultant couple $\mathbf{M_R}$ of these two couples and express it as a magnitude multiplying a unit vector. (b) Determine the component of $\mathbf{M_R}$ along axis OE.

Solution

(a) Express the forces $\mathbf{F}$ and $\mathbf{P}$ in terms of their rectangular components. Thus,

$$\mathbf{F} = {}^F\!\left(\frac{\mathbf{r}_{B/E}}{r_{B/E}}\right) = 150\left(\frac{5\mathbf{j} - 4\mathbf{k}}{\sqrt{5^2 + 4^2}}\right) = 117.1\mathbf{j} - 93.7\mathbf{k},$$

and

$$\mathbf{P} = P\left(\frac{\mathbf{r}_{G/D}}{r_{G/D}}\right) = 100\left(\frac{3\mathbf{j} - 1\mathbf{k}}{\sqrt{3^2 + 1^2}}\right) = 94.9\mathbf{j} - 31.6\mathbf{k}.$$

Using Eq. (5.23), we find the moments of the two couples as follows:

$$\mathbf{M}_F = \mathbf{r}_{B/A} \times \mathbf{F} = 6\mathbf{i} \times (117.1\mathbf{j} - 93.7\mathbf{k})$$

$$= 562.2\mathbf{j} + 702.6\mathbf{k},$$

and

$$\mathbf{M}_P = \mathbf{r}_{D/C} \times \mathbf{P} = -5\mathbf{j} \times (94.9\mathbf{j} - 31.6\mathbf{k})$$

$$= 158.0\mathbf{j} + 474.5\mathbf{k}.$$

The resultant $\mathbf{M_R}$ of these two couples is obtained by Eq. (5.25). Thus,

$$\mathbf{M_R} = \mathbf{M}_F + \mathbf{M}_P = (158.0\mathbf{i} + 562.2\mathbf{j} + 1177.1\mathbf{k}) \text{ lb·ft.}$$

The magnitude M_R of the resultant couple becomes

$$M_R = \sqrt{158.0^2 + 562.2^2 + 1177.1^2} = 1314 \text{ lb·ft.}$$

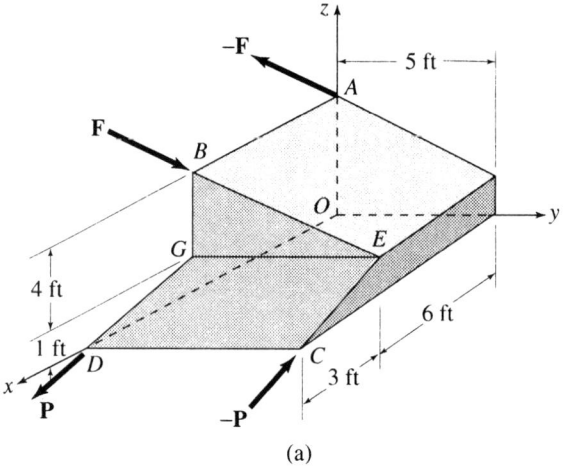

(a)

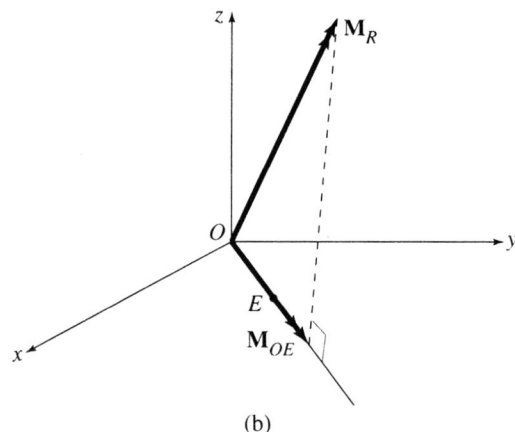

FIGURE E5.7.   (b)

Therefore,

$$\mathbf{M_R} = M_R\left(\frac{\mathbf{M_R}}{M_R}\right) = 1314 \text{ lb·ft}(0.120\mathbf{i} + 0.428\mathbf{j} + 0.896\mathbf{k}). \quad \text{ANS.}$$

The resultant couple $\mathbf{M_R}$ is shown in Figure E5.7(b) located arbitrarily at the origin of the coordinate system.

(b) A unit vector λ_{OE} is determined as follows:

$$\lambda_{OE} = \frac{\mathbf{r}_{E/O}}{r_{E/O}} = \frac{6\mathbf{i} + 5\mathbf{j} + 1\mathbf{k}}{\sqrt{6^2 + 5^2 + 1^2}}$$

$$= 0.762\mathbf{i} + 0.635\mathbf{j} + 0.127\mathbf{k}.$$

Therefore

$$M_{OE} = \lambda_{OE} \cdot \mathbf{M_R} = 627 \text{ lb·ft,}$$

and

$$\mathbf{M}_{OE} = 627 \text{ lb·ft}(0.762\mathbf{i} + 0.635\mathbf{j} + 0.127\mathbf{k}). \qquad \text{ANS.}$$

The component $\mathbf{M}_{OE}$ is also shown in Figure E5.7(b)). This component represents the turning effect that the two given couples produce about axis OE.

■

5.7 Replacement of a Single Force by a Force and a Couple

The process of replacing a single force acting at a given point on a body by an equivalent force-couple system at a second point was developed in Chapter 4 for the two-dimensional case. This process will be generalized to cover the case of three-dimensional force systems.

Consider the rigid body subjected to the force $\mathbf{F}$ applied at point A as shown in Figure 5.11(a). It is desired to replace this force by an equivalent force-couple system at point B. This may be accomplished by applying a null couple at point B whose forces are $\mathbf{F}$ and $-\mathbf{F}$, as shown in Figure 5.11(b). The two forces marked with the small circles in Figure 5.11(b) constitute a couple $\mathbf{M}_B = \mathbf{r}_{A/B} \times \mathbf{F}$. This leaves the unmarked force $\mathbf{F}$ acting at point B. Therefore, the resulting force system consists of a couple $\mathbf{M}_B = \mathbf{r}_{A/B} \times \mathbf{F}$ plus the force $\mathbf{F}$, which are perpendicular to each other, as shown in Figure 5.11(c). It should be noted that the force systems shown in Figures 5.11(a), (b), and (c) are equivalent force systems because they produce the same effects on the rigid body on which they act. These effects refer to support reactions if the rigid body is in equilibrium or to its linear and angular accelerations if the rigid body is in motion.

5.8 Replacement of a General Force System by a Force and a Couple

Consider the general force system, shown in Figure 5.12(a), consisting of two forces $\mathbf{F}_1$ and $\mathbf{F}_2$ acting at points A_1 and A_2, respectively, and two couples $\mathbf{M}_3$ and $\mathbf{M}_4$. It is desired to replace the given force system by an equivalent force-couple system at point B. To this end, each of the two forces $\mathbf{F}_1$ and $\mathbf{F}_2$ (and any others that the force system may include) may be replaced by a force-couple system at point B using the method of Section 5.7. Thus, $\mathbf{F}_1$ at A_1 is replaced by the force-couple system $\mathbf{F}_1$ and $\mathbf{M}_1 = \mathbf{r}_{A1/B} \times \mathbf{F}_1$ at point B and, $\mathbf{F}_2$ at A_2 is replaced by the force-couple system $\mathbf{F}_2$ and $\mathbf{M}_2 = \mathbf{r}_{A2/B} \times \mathbf{F}_2$ at the same point. The couples $\mathbf{M}_3$ and $\mathbf{M}_4$ (and any others that the force system may contain) are simply shifted to point B because they are free vectors. All of these forces and couples are shown in Figure 5.12(b). The forces $\mathbf{F}_1$ and $\mathbf{F}_2$ plus any others in the given force system constitute a concurrent force system whose resultant is obtained by Eqs. (3.24) and (3.25), namely, $\mathbf{R} = \sum \mathbf{F}_i = (\sum F_x)\mathbf{i} + (\sum F_y)\mathbf{j} + (\sum F_z)\mathbf{k}$. Also, the couples $\mathbf{M}_1$ to $\mathbf{M}_4$ plus any others in the given system may be added by Eq. (5.25), namely

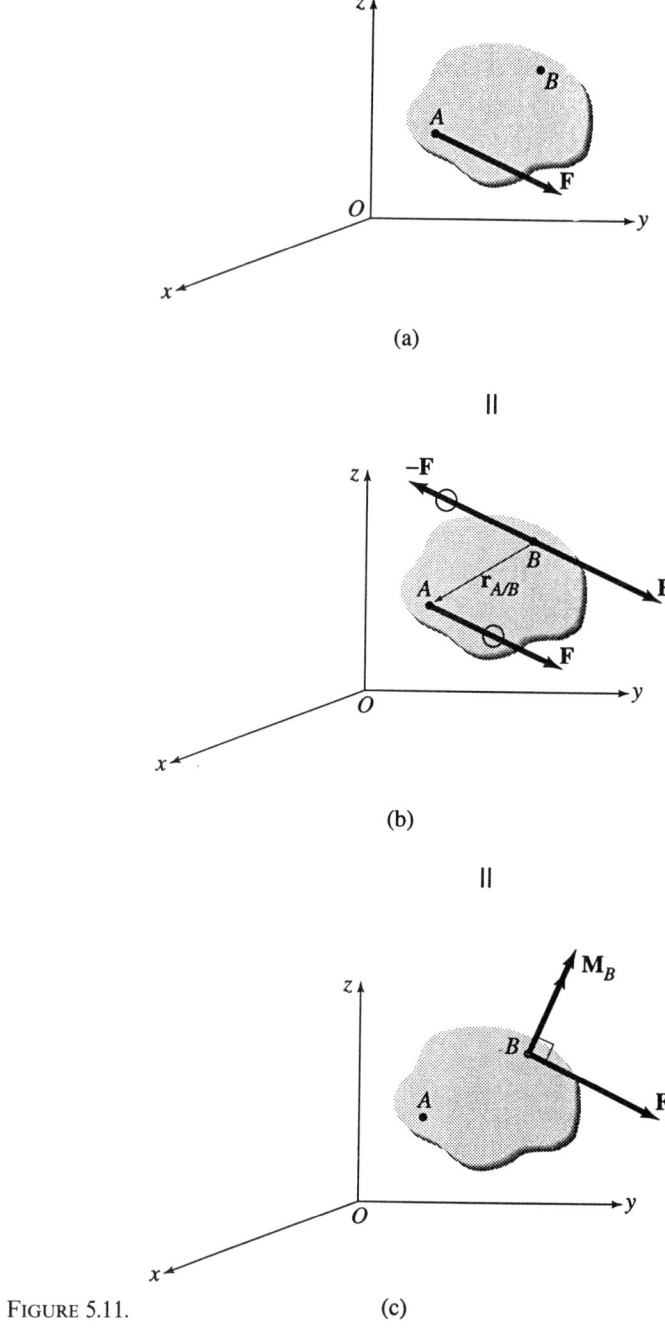

FIGURE 5.11.

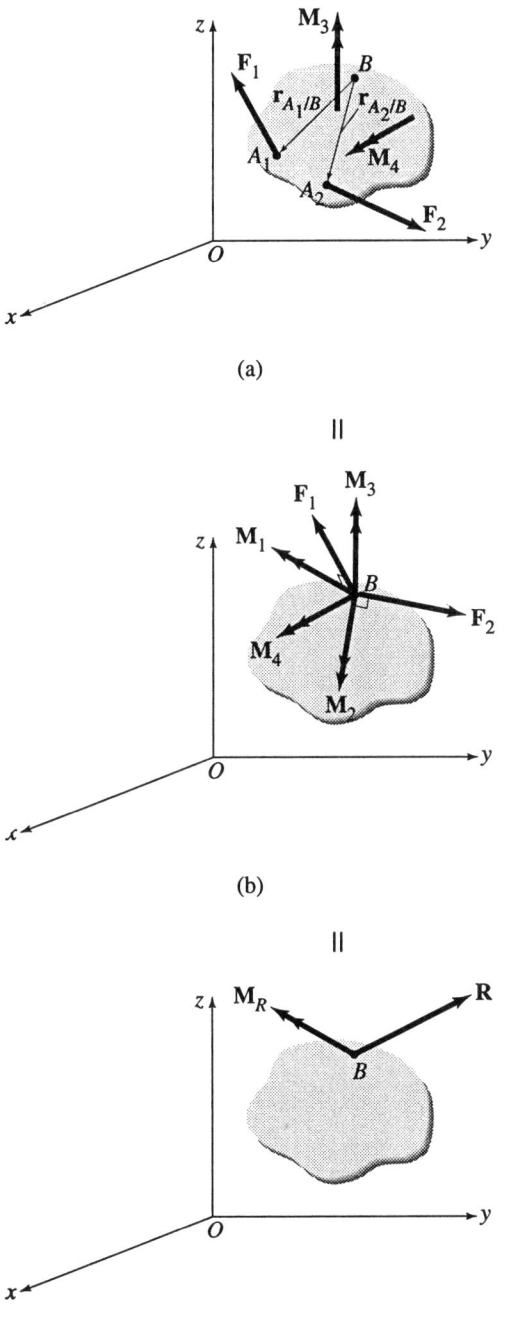

(a)

||

(b)

||

(c)   FIGURE 5.12.

$M_R = \sum M_i$. Thus, any general force system may be replaced by an equivalent force-couple system at any arbitrary point B. This force-couple system consists of a resultant force **R** given by Eq. (3.24) or Eq. (3.25) and a resultant couple M_R given by Eq. (5.25) or Eq. (5.26). Four cases of special interest are identified below:

Equilibrated Force System

This is the case in which the force-couple system consists of a resultant force **R** and a resultant couple M_R both of which vanish. Such a force system is said to be in *perfect balance* and the equations **R** = **0** and M_R = **0** describe, symbolically, the conditions of *equilibrium* which will be developed and discussed in detail in Section 5.9. It should be noted, however, that, even though a force system is in perfect balance, it may not be stable and any slight disturbance will cause the system to change its state.

Concurrent Force System

By definition, a concurrent force system is one in which all of the forces in the system intersect at the same point in space. In the light of the process developed above, it is clear that a three-dimensional concurrent force system may be replaced by an equivalent force-couple system at the point of concurrency consisting of a resultant force $R = \sum F$ and a resultant couple M_R = **0**. Obviously, this conclusion is in agreement with the result obtained in Chapter 3 where three-dimensional concurrent force systems were discussed.

Two-Dimensional Force System

Two-dimensional or coplanar force systems were discussed in detail in Chapter 4 and the student is urged to review that material, particularly, Section 4.4. In that section, we concluded that a general two-dimensional force system may be replaced by an equivalent force-couple system at a given point, consisting of a resultant force **R** and a resultant couple M_R perpendicular to one another. Furthermore, we discovered that a unique point exists in the plane of the force-couple system at which the resultant couple vanishes and the given general two-dimensional force system may be replaced by an equivalent single resultant force **R**.

Parallel Force System

Consider the parallel force system shown in Figure 5.13(a) in which all of the forces are parallel to the z axis. Only three forces are shown but the method outlined here applies to any number of parallel forces. The three forces F_1, F_2, and F_3 pass through the points A_1, A_2, and A_3, respectively, which lie in the x-y plane. This parallel force system may be replaced by an equivalent force-couple system at any point B in the x-y plane, consisting of a resultant force **R** and a resultant couple M_R as shown in Figure 5.13(b). Note that **R** and M_R are mutually perpendicular because, as each of the given forces is moved to point B, a couple is created that lies in the x-y plane. Thus, the resultant couple M_R must lie in the x-y plane or be perpendicular to the resultant force **R** which is parallel to the z axis. Because **R** and M_R are perpendicular

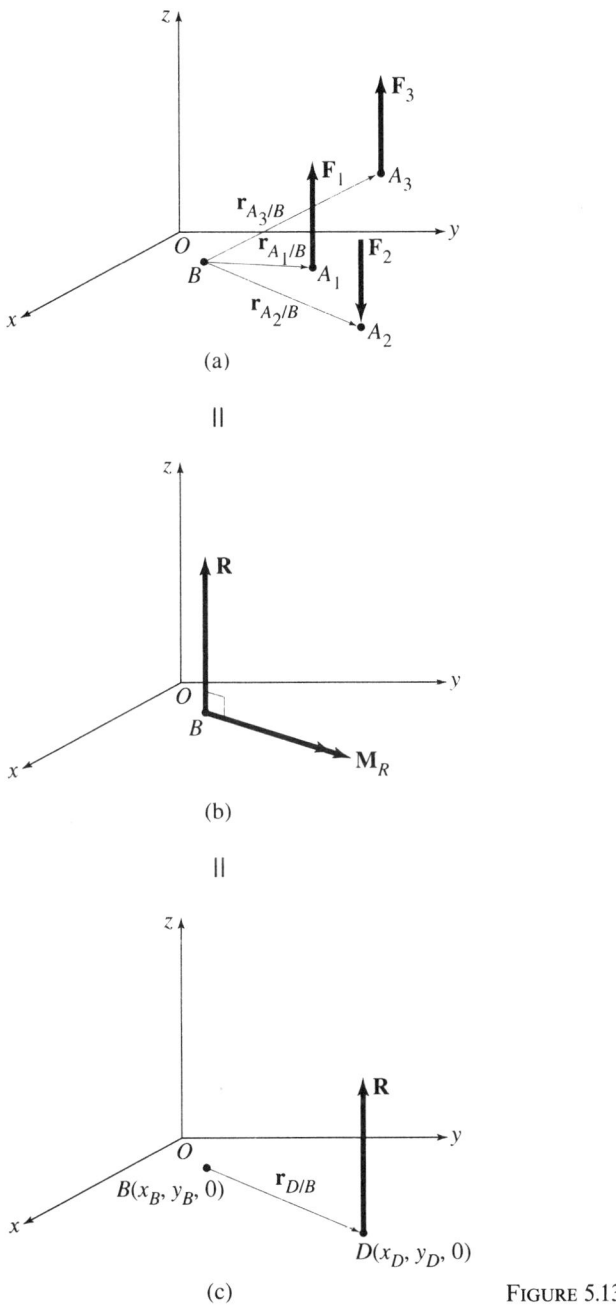

(a)

||

(b)

||

(c)

FIGURE 5.13.

to one another, a unique point D may be found in the x-y plane at which the resultant couple vanishes and the given parallel force system may be replaced by an equivalent single resultant force $\mathbf{R}$, as shown in Figure 5.13(c). The exact location of point D is determined from the condition that the moment of $\mathbf{R}$ about point B is equal to $\mathbf{M_R}$. Thus, referring to Figure 5.13(c),

$$\mathbf{r_{D/B}} \times \mathbf{R} = \mathbf{M_R} \qquad (5.27a)$$

where $\mathbf{r_{D/B}} = (x_D - x_B)\mathbf{i} + (y_D - y_B)\mathbf{j}$, $\mathbf{R} = \sum \mathbf{F} = R_z\mathbf{k}$ and $\mathbf{M_R} = M_x\mathbf{i} + M_y\mathbf{j}$. Substituting these quantities in Eq. (5.27a), performing the indicated cross product, equating the coefficients of the $\mathbf{i}$ and $\mathbf{j}$ unit vectors, and simplifying,

and

$$\left.\begin{aligned} x_D &= x_B - \frac{M_y}{R_z} \\[2mm] y_D &= y_B + \frac{M_x}{R_z}. \end{aligned}\right\} \qquad (5.27b)$$

The development above assumes that $\mathbf{R} \neq \mathbf{0}$. If $\mathbf{R} = \mathbf{0}$, it is obvious that the given parallel force system is replaceable by an equivalent resultant couple $\mathbf{M_R}$. In other words, because the resultant force vanishes, the given parallel force system may be replaced by an equivalent single resultant couple $\mathbf{M_R}$.

In each of the above four special cases, the equivalent force-couple system consisting of a resultant force $\mathbf{R}$ and a resultant couple $\mathbf{M_R}$ is such that one of the following four conditions was present:

1. $\mathbf{R} = \mathbf{0}$ and $\mathbf{M_R} = \mathbf{0}$.
2. $\mathbf{R} \neq \mathbf{0}$ and $\mathbf{M_R} = \mathbf{0}$.
3. $\mathbf{R} = \mathbf{0}$ and $\mathbf{M_R} \neq \mathbf{0}$.
4. $\mathbf{R} \neq \mathbf{0}$ and $\mathbf{M_R} \neq \mathbf{0}$ with the two vectors perpendicular to one another.

Note that $\mathbf{R}$ and $\mathbf{M_R}$ are perpendicular to each other if their dot product vanishes.

The general case when $\mathbf{R} \neq \mathbf{0}$ and $\mathbf{M_R} \neq \mathbf{0}$, with the two vectors *not* perpendicular to one another may be replaced by a *wrench* which consists of a resultant force and a collinear resultant couple. Thus, consider the general force-couple system at B, as shown in Figure 5.14(a), consisting of a resultant force $\mathbf{R}$ and a resultant couple $\mathbf{M_R}$ which are not mutually perpendicular. The resultant couple $\mathbf{M_R}$ may be decomposed into the two components $\mathbf{M_P}$, perpendicular to $\mathbf{R}$, and $\mathbf{M_C}$, collinear with $\mathbf{R}$, as shown in Figure 5.14(b). The combination $\mathbf{M_P}$ and $\mathbf{R}$ may be replaced by a single force $\mathbf{R}$ with a new line of action, in a manner similar to that used for the special case of the parallel force system. This new line of action is, of course, parallel to $\mathbf{R}$ and its exact

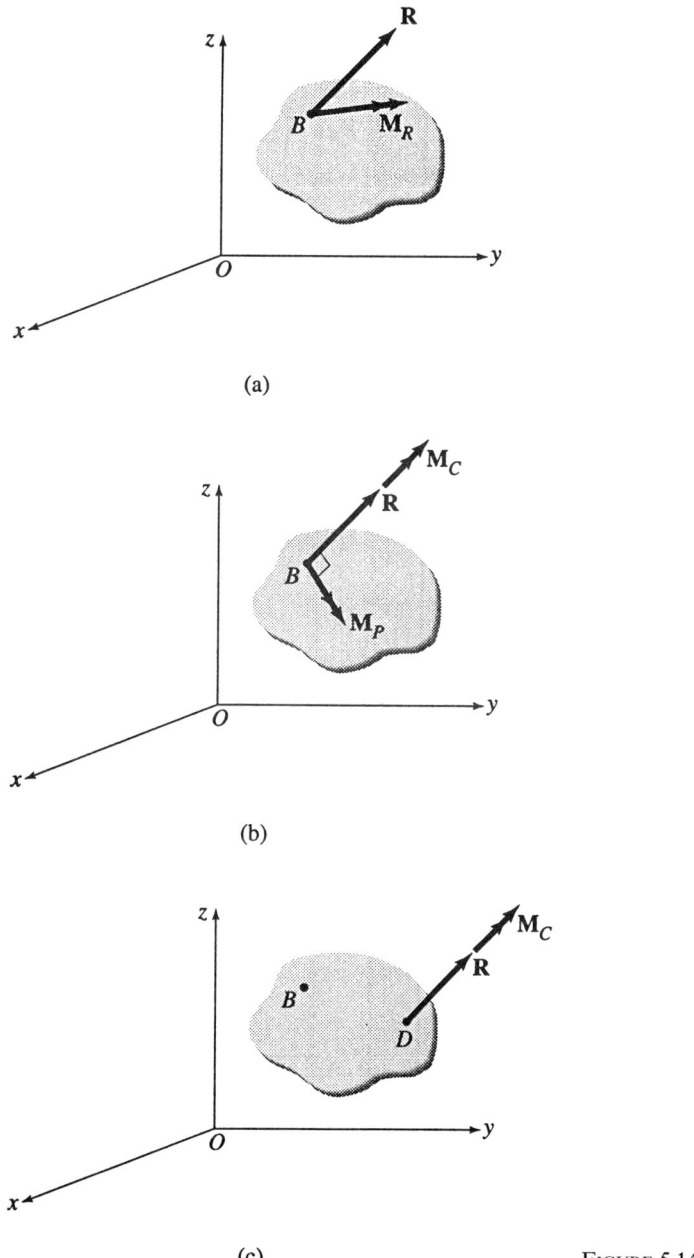

(a)

(b)

(c)

FIGURE 5.14.

position in space may be located by establishing its intersection, point D, with one of the three coordinate planes. Being a free vector, the couple $\mathbf{M_C}$ may be moved to point D yielding the *wrench* which consists of the force $\mathbf{R}$ and the collinear couple $\mathbf{M_C}$, as shown in Figure 5.14(c). Thus, the most general three-dimensional force system may be replaced by a wrench which is a special type of force-couple system in which the couple is collinear with the force. If point D represents, for example, the point of intersection of the line of action of the wrench with the *y-z* plane, then $x = 0$ and its *y* and *z* coordinates may be determined by equating the moment of the wrench about point B, to the resultant couple $\mathbf{M_R}$. Thus,

$$\mathbf{M_R} = \mathbf{r}_{D/B} \times \mathbf{R} + \mathbf{M_C}. \qquad (5.28)$$

Equation (5.28) yields two independent scalar relations that may be used to solve for the coordinates *y* and *z*.

The following examples illustrate some of the concepts above.

■ Example 5.8

The force $\mathbf{F}$ acting on the machine member shown in Figure E5.8(a) has a magnitude $F = 10.0$ kN. Replace this force by an equivalent force-couple system acting at (a) point A and (b) point O. Note that both the handle AB and the force $\mathbf{F}$ lie in a plane parallel to the *y-z* plane.

Solution

(a) The equivalent force-couple system at point A consists of a force $\mathbf{F}$ and a couple $\mathbf{M_A}$ as follows:

$$\mathbf{F} = (-10.00\mathbf{k}) \text{ kN}, \qquad \text{ANS.}$$

and

$$\mathbf{M_A} = \mathbf{r}_{B/A} \times \mathbf{F} = (0.866\mathbf{j} - 0.500\mathbf{k}) \times (-10.0\mathbf{k})$$

$$= -(8.66\mathbf{i}) \text{ kN·m.} \qquad \text{ANS.}$$

This equivalent force-couple system is shown in Figure E5.8(b).

(b) The equivalent force-couple system at point O consists of a force $\mathbf{F}$ and a couple $\mathbf{M_O}$ as follows:

$$\mathbf{F} = (-10.00\mathbf{k}) \text{ kN}, \qquad \text{ANS.}$$

and

$$\mathbf{M_O} = \mathbf{r}_{B/O} \times \mathbf{F} = (2.000\mathbf{j} + 0.866\mathbf{j} - 0.500\mathbf{k}) \times (-10.0\mathbf{k})$$

$$= (-8.66\mathbf{i} + 20.0\mathbf{j}) \text{ kN·m.} \qquad \text{ANS.}$$

This equivalent force-couple system is shown in Figure E5.8(c). Note that, for clarity, the components $\mathbf{M_x} = (-8.66\mathbf{i})$ kN·m and $\mathbf{M_y} = (20.0\mathbf{j})$ kN·m of the couple $\mathbf{M_O}$ are also shown in Figure E5.8(c).

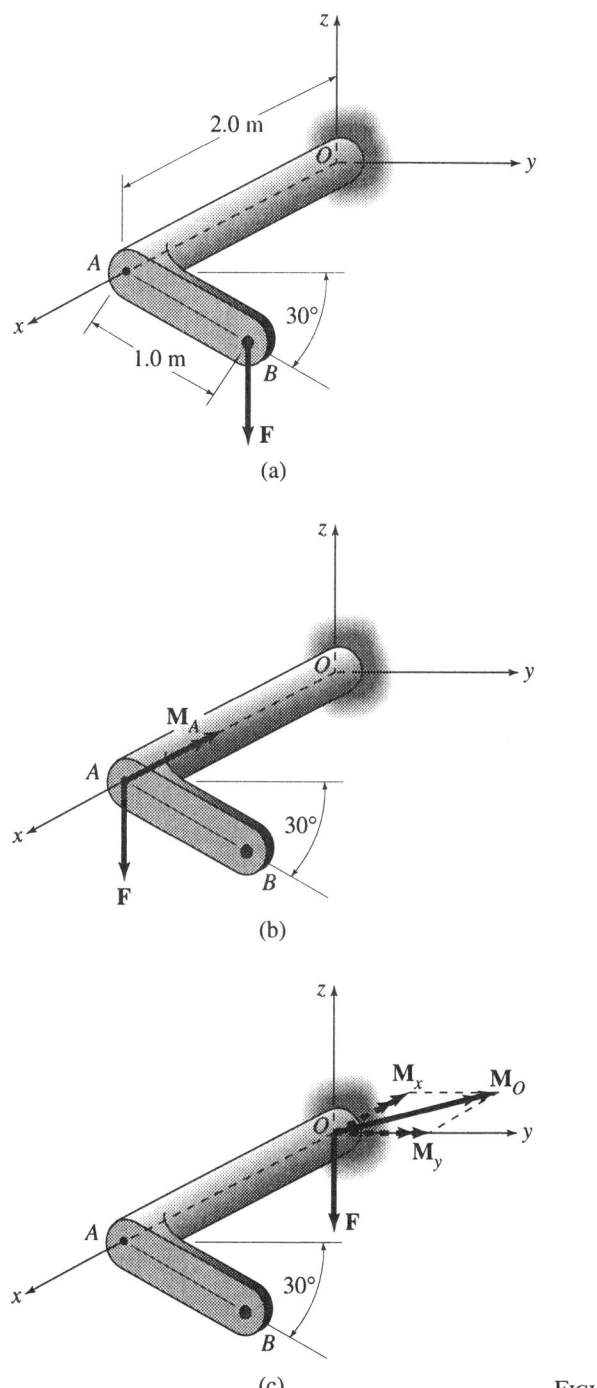

■ **Example 5.9**

A 1 ft × **8 ft** × 16 ft plate is subjected to forces $\mathbf{F}_1$ and $\mathbf{F}_2$ and to a couple $\mathbf{M}$ as shown in Figure E5.9(a). Let $F_1 = 5$ k, $F_2 = 7$ k and $M = 10$ k·ft. Replace the given force system by an equivalent force-couple system at point E.

Solution

The forces $\mathbf{F}_1$ and $\mathbf{F}_2$ are expressed in terms of their x, y, and z components as follows:

$$\mathbf{F}_1 = F_1 \left(\frac{\mathbf{r}_{B/A}}{r_{B/A}} \right) = 5 \left(\frac{4\mathbf{i} - 10\mathbf{j} - 16\mathbf{k}}{\sqrt{4^2 + 10^2 + 16^2}} \right)$$

$$= 1.037\mathbf{i} - 2.592\mathbf{j} - 4.148\mathbf{k},$$

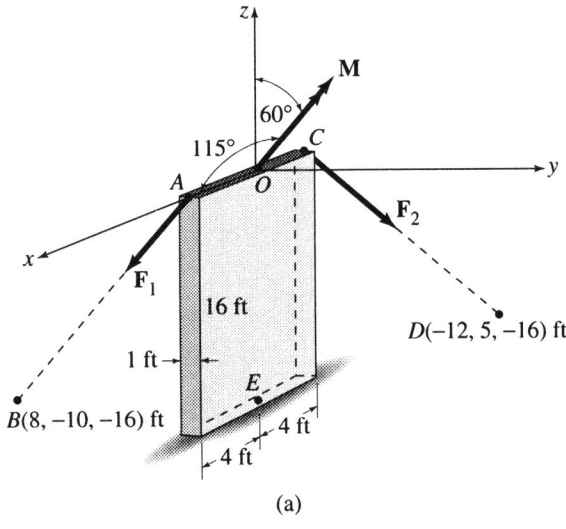

(a)

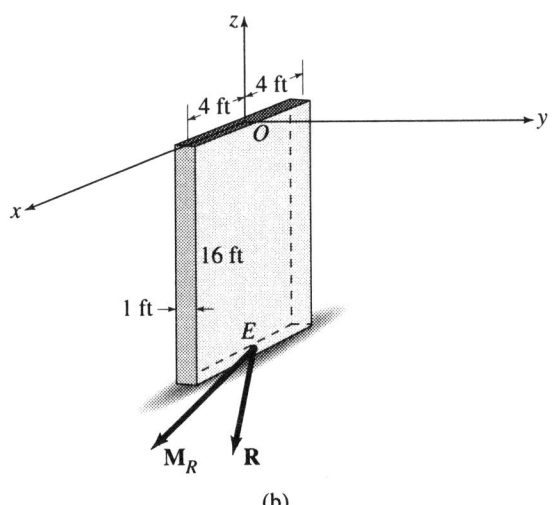

FIGURE E5.9.   (b)

and

$$\mathbf{F}_2 = F_2\left(\frac{\mathbf{r}_{D/C}}{r_{D/C}}\right) = 7\left(\frac{-8\mathbf{i} + 5\mathbf{j} - 16\mathbf{k}}{\sqrt{8^2 + 5^2 + 16^2}}\right)$$

$$= -3.015\mathbf{i} + 1.884\mathbf{j} - 6.030\mathbf{k}.$$

Because the angles $\alpha = 115°$ and $\gamma = 60°$ are given, the angle β between $\mathbf{M}$ and the y axis may be determined by Eq. (3.7). Thus,

$$\cos^2 115° + \cos^2 \beta + \cos^2 60° = 1$$

from which $\beta = 40.9°$. Therefore, the given couple $\mathbf{M}$ may be expressed in terms of its rectangular components as follows:

$$\mathbf{M} = (10\cos^2 115°)\mathbf{i} + (10\cos^2 40.9°)\mathbf{j} + (10\cos^2 60°)\mathbf{k}$$

$$= -4.226\mathbf{i} + 7.559\mathbf{j} + 5.000\mathbf{k}.$$

The equivalent force-couple system at point E consists of a resultant force $\mathbf{R}$ found by Eq. (3.24) and a resultant couple $\mathbf{M}_R$ found by Eq. (5.25). Thus, by Eq. (3.24),

$$\mathbf{R} = \mathbf{F}_1 + \mathbf{F}_2$$

$$= (1.037 - 3.015)\mathbf{i} + (-2.592 + 1.884)\mathbf{j} + (-4.148 - 6.030)\mathbf{k}$$

$$= (-1.978\mathbf{i} - 0.708\mathbf{j} - 10.18\mathbf{k})\text{ k.} \qquad \text{ANS.}$$

Before Eq. (5.25) can be used to find $\mathbf{M}_R$ we need to find the couples produced by $\mathbf{F}_1$ and $\mathbf{F}_2$ as they are moved to point E. Thus, due to $\mathbf{F}_1$,

$$\mathbf{M}_1 = \mathbf{r}_{A/E} \times \mathbf{F}_1$$

$$= (4\mathbf{i} + 16\mathbf{k}) \times (1.037\mathbf{i} - 2.592\mathbf{j} - 4.148\mathbf{k})$$

$$= 41.472\mathbf{i} + 33.184\mathbf{j} - 10.368\mathbf{k}.$$

Also, due to $\mathbf{F}_2$,

$$\mathbf{M}_2 = \mathbf{r}_{C/E} \times \mathbf{F}_2$$

$$= (-4\mathbf{i} + 16\mathbf{k}) \times (-3.015\mathbf{i} + 1.884\mathbf{j} - 6.030\mathbf{k})$$

$$= -30.144\mathbf{i} - 72.360\mathbf{j} - 7.536\mathbf{k}.$$

Therefore, using Eq. (5.25),

$$\mathbf{M}_R = \mathbf{M} + \mathbf{M}_1 + \mathbf{M}_2$$

$$= (-4.226 + 41.472 - 30.144)\mathbf{i} + (7.559 + 33.184 - 72.360)\mathbf{j}$$

$$+ (5.000 - 10.363 - 7.536)\mathbf{k}$$

$$= (7.10\mathbf{i} - 31.6\mathbf{j} - 12.90\mathbf{k})\text{ k}\cdot\text{ft.} \qquad \text{ANS.}$$

The equivalent force-couple system at point E consisting of the resultant force $\mathbf{R}$ and the resultant couple $\mathbf{M}_R$ is shown schematically in Figure E5.9(b).

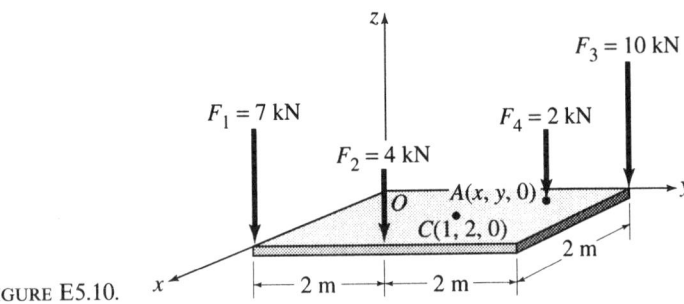

FIGURE E5.10.

■ **Example 5.10**

A 2 m × 4 m rectangular plate is subjected to a system of parallel forces as shown in Figure E5.10. Determine the x and y coordinates of point A which define the line of action of the force $\mathbf{F_4}$, if the parallel force system can be replaced by a single resultant force $\mathbf{R}$ passing through point C, the geometric enter of the plate. Also, find the resultant force $\mathbf{R}$.

Solution

Any parallel force system may be replaced by a force-couple system at any point, consisting of a resultant force $\mathbf{R}$ and a resultant couple $\mathbf{M_R}$. At point C, however, because this parallel force system is replaceable by a single resultant force $\mathbf{R}$, it follows that $\mathbf{M_R} = \mathbf{0}$. Thus,

$$\mathbf{R} = \sum \mathbf{F} = -7\mathbf{k} - 4\mathbf{k} - 10\mathbf{k} - 2\mathbf{k} = (-23.0\mathbf{k}) \text{ kN}, \qquad \text{ANS.}$$

and

$$\mathbf{M_R} = \sum \mathbf{r} \times \mathbf{F}$$
$$= (1\mathbf{i} - 2\mathbf{j}) \times (-7\mathbf{k}) + (\mathbf{i}) \times (-4\mathbf{k}) + (-\mathbf{i} + 2\mathbf{j}) \times (-10\mathbf{k})$$
$$+ [(x - 1)\mathbf{i} + (y - 2)\mathbf{j}] \times (-2\mathbf{k})$$
$$= -(2y + 2)\mathbf{i} + (2x - 1)\mathbf{j} = \mathbf{0}.$$

Therefore,

$$2y + 2 = 0,$$
$$y = -1.000 \text{ m}, \qquad \text{ANS.}$$

and

$$2x - 1 = 0,$$
$$x = 0.5000 \text{ m}. \qquad \text{ANS.}$$

■ **Example 5.11**

A gear box, shown schematically in Figure E5.11(a), is subjected to a force system consisting of two forces and two couples. Replace this

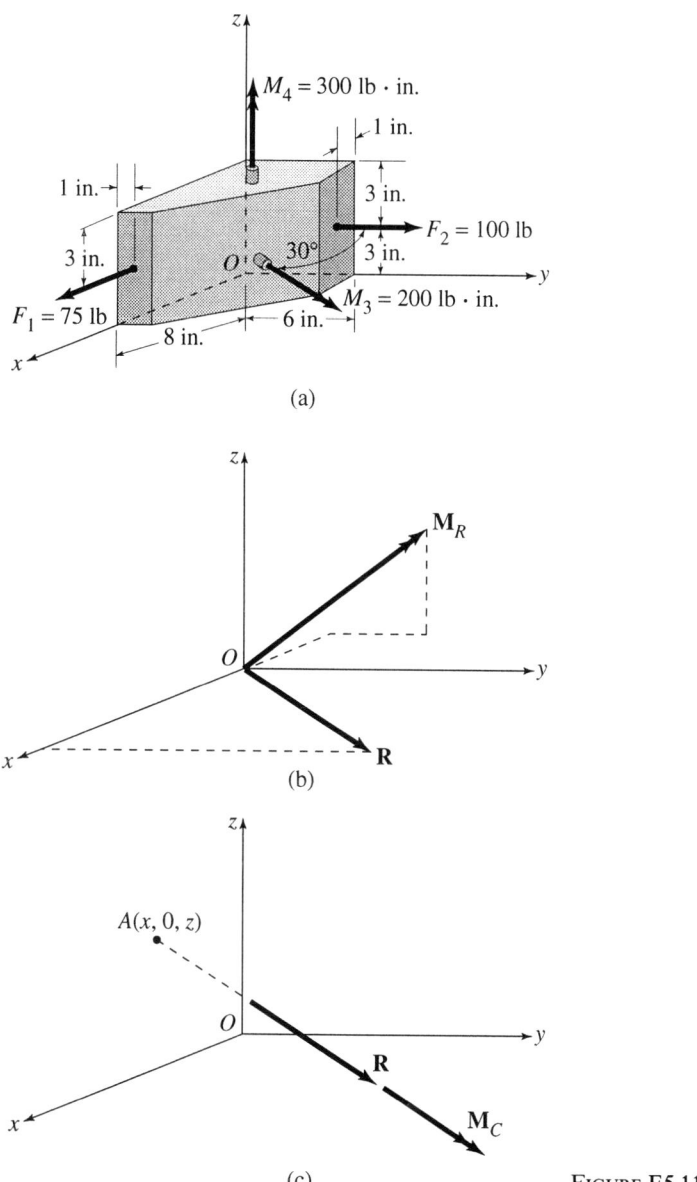

FIGURE E5.11.

force system by a wrench specifying the magnitudes of its force and collinear couple as well as its line of action. Note that the forces $\mathbf{F}_1$ and $\mathbf{F}_2$ and the couple $\mathbf{M}_3$ all lie in the same plane parallel to the x-y plane.

Solution

First, we replace the given general force system by an equivalent force-couple system at point O. The equivalent force-couple system consists of a resultant force $\mathbf{R}$ and a resultant couple $\mathbf{M_R}$ which are shown in Figure E5.11(b) and are determined as follows:

$$\mathbf{R} = \mathbf{F_1} + \mathbf{F_2}$$
$$= 75\mathbf{i} + 100\mathbf{j},$$

and

$$\mathbf{M_R} = \mathbf{M_1} + \mathbf{M_2} + \mathbf{M_3} + \mathbf{M_4}$$

where

$$\mathbf{M_1} = (8\mathbf{i} + 1\mathbf{j} + 3\mathbf{k}) \times 75\mathbf{i} = 225\mathbf{j} - 75\mathbf{k},$$
$$\mathbf{M_2} = (1\mathbf{i} + 6\mathbf{j} + 3\mathbf{k}) \times 100\mathbf{i} = -300\mathbf{j} + 100\mathbf{k},$$
$$\mathbf{M_3} = 100\mathbf{i} + 173\mathbf{j},$$

and

$$\mathbf{M_4} = 300\mathbf{k}.$$

Therefore

$$\mathbf{M_R} = -200\mathbf{i} + 398\mathbf{j} + 325\mathbf{k}.$$

We now determine the component $\mathbf{M_C}$ of the couple $\mathbf{M_R}$ collinear with the force $\mathbf{R}$. Thus

$$\lambda_\mathbf{R} = \frac{\mathbf{R}}{R} = 0.6\mathbf{i} + 0.8\mathbf{j},$$

and

$$\mathbf{M_C} = (\lambda_\mathbf{R} \cdot \mathbf{M_R})\lambda_\mathbf{R}$$
$$= [(0.6\mathbf{i} + 0.8\mathbf{j}) \cdot (-200\mathbf{i} + 398\mathbf{j} + 325\mathbf{k})](0.6\mathbf{i} + 0.8\mathbf{j})$$
$$= 119.0\mathbf{i} + 158.7\mathbf{j}.$$

Therefore, the required wrench consists of a force $\mathbf{R}$ and a couple $\mathbf{M_C}$ given by

$$\mathbf{R} = (75.0\mathbf{i} + 100.0\mathbf{j}) \text{ lb} \qquad \text{ANS.}$$

and

$$\mathbf{M_C} = (119.0\mathbf{i} + 158.7\mathbf{j}) \text{ lb·in.} \qquad \text{ANS.}$$

This wrench is shown in Figure E5.11(c) intersecting the x-z plane at point A $(x, 0, z)$ where the coordinates x and z may be found by the use of Eq. (5.28). Thus,

$$\mathbf{M_O} = \mathbf{M_R} = \mathbf{r_{A/O}} \times \mathbf{R} + \mathbf{M_C}.$$

Substituting gives

$$-200\mathbf{i} + 398\mathbf{j} + 325\mathbf{k} = (x\mathbf{i} + z\mathbf{k}) \times (75\mathbf{i} + 100\mathbf{j}) + 119.0\mathbf{i} + 158.7\mathbf{j}.$$

Equating the **i**, **j**, and **k** terms on both sides of the equation,

$$-200 = 119.0 - 100z, \tag{a}$$

$$398 = 158.7 + 75z, \tag{b}$$

and

$$325 = 100x. \tag{c}$$

Equation (a) or Eq. (b) yields the value of z and Eq. (c) the value of x. Thus,

$$z = 3.19 \text{ in.}, \tag{ANS.}$$

and

$$x = 3.25 \text{ in.} \tag{ANS.}$$

■

Problems

5.34 Find the couple **M** produced by the forces **F** and $-\mathbf{F}$ acting on the pipe assembly shown in Figure P5.34. Express **M** in terms of its rectangular components, and find the directional angles α, β, and γ that it makes with the x, y, and z axes, respectively, given $\mathbf{F} = (-200\mathbf{i} + 50\mathbf{j} - 350\mathbf{k})$ N.

5.35 Replace the two couples shown in Figure P5.35 by a single resultant couple $\mathbf{M_R}$. Express $\mathbf{M_R}$ as a magnitude multiplying a unit vector. What are the directional angles α, β, and γ that $\mathbf{M_R}$ makes with the x, y, and z axes, respectively.

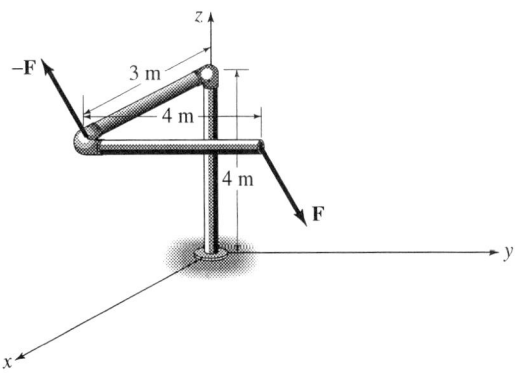

FIGURE P5.34.

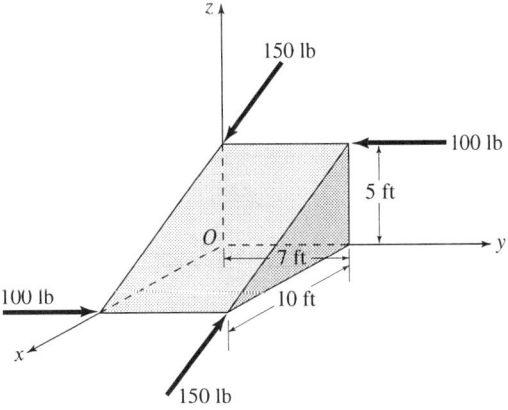

FIGURE P5.35.

5.36 A gear system, shown in Figure P5.36, is subjected to the three couples $\mathbf{M}_1$, $\mathbf{M}_2$, and $\mathbf{M}_3$ whose magnitudes are $M_1 = 50$ N·m, $M_2 = 75$ N·m, and $M_3 = 40$ N·m. Determine the resultant couple $\mathbf{M}_R$ expressing it in terms of its three rectangular components. Also, find the directional angles α, β, and γ that $\mathbf{M}_R$ makes with the x, y, and z axes, respectively. Note that both $\mathbf{M}_2$ and $\mathbf{M}_3$ lie in the y-z plane.

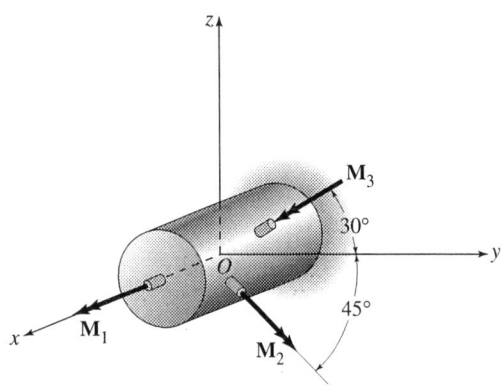

FIGURE P5.36.

5.37 Prove that the two couples ($\mathbf{F}$ and $-\mathbf{F}$) and ($\mathbf{P}$ and $-\mathbf{P}$) in Figure P5.37 are equivalent couples. Use vector algebra in

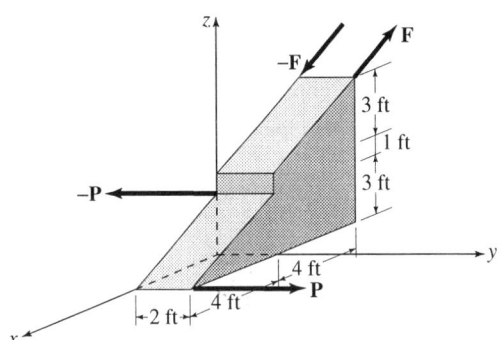

FIGURE P5.37.

your analysis. The magnitude of $\mathbf{F}$, $F = 100$ lb, and that of $\mathbf{P}$, $P = 40$ lb.

5.38 A frame is subjected to three couples, as shown in Figure P5.38. The resultant of the three couples $\mathbf{M}_R = (-25\mathbf{i} + 3\mathbf{j} - 29\mathbf{k})$ kN·m. Determine the magnitudes of the forces $\mathbf{F}$ and $\mathbf{Q}$.

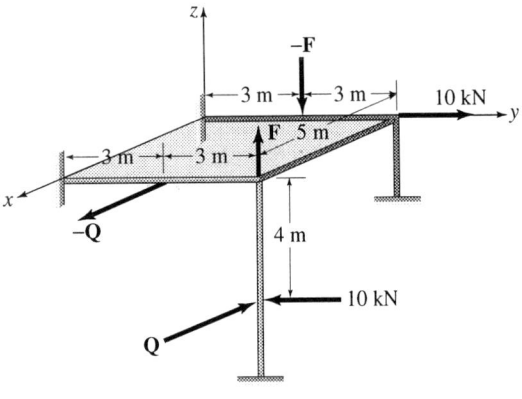

FIGURE P5.38.

5.39 A wall bracket is subjected to a cable force $\mathbf{F}$, as shown in Figure P5.39. The magnitude of the force $F = 500$ lb, and it is directed from A to D. Replace the

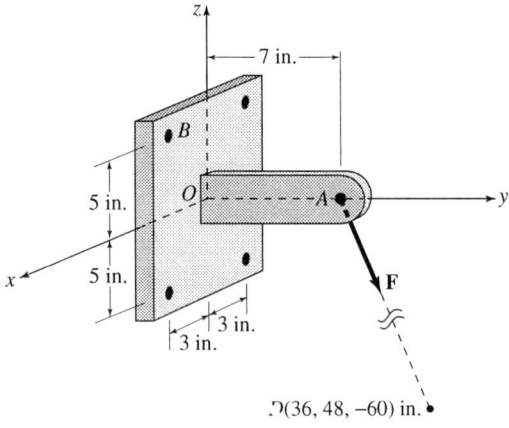

FIGURE P5.39.

given force by an equivalent force-couple system at point O. What are the turning tendencies of the force **F** about the *x*, *y*, and *z* axes?

5.40 Refer to the bracket in Problem 5.39 and replace the given cable force by an equivalent force-couple system at support point B which is located at the mid-thickness of the support plate.

5.41 A three-dimensional truss, shown in Figure P5.41, is subjected to the force **P** = $(-6\mathbf{i} + 8\mathbf{j} + 3\mathbf{k})$ kN. Replace this force by an equivalent force-couple system at support O. How much turning tendency does the force **P** produce about the *x*, *y*, and *z* axes.

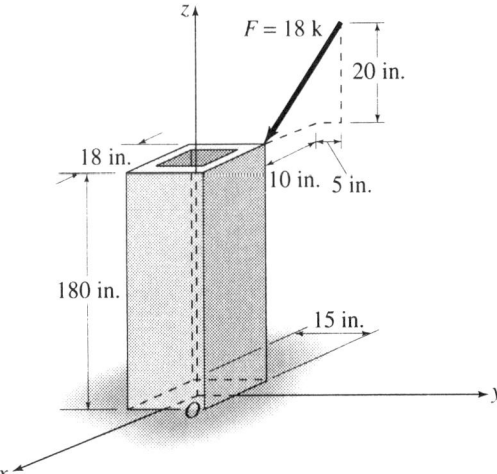

FIGURE P5.43.

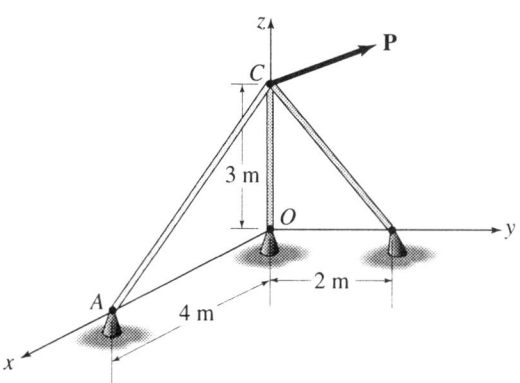

FIGURE P5.41.

5.42 Refer to the three-dimensional truss of Problem 5.41 and replace the given force **P** by an equivalent force-couple system at support A.

5.43 A hollow box section is used as a column and subjected to the force $F = 18$ k as shown in Figure P5.43. Replace this force by an equivalent force-couple system at point O. Also, find the turning tendency produced by the force about the *x*, *y*, and *z* axes.

5.44 A girder is being hoisted into position at constant speed, as shown in Figure P5.44. The tension in cable ABD is 10 k. Replace the force acting on the girder at A by an equivalent force-couple system at (a) point C and (b) point O.

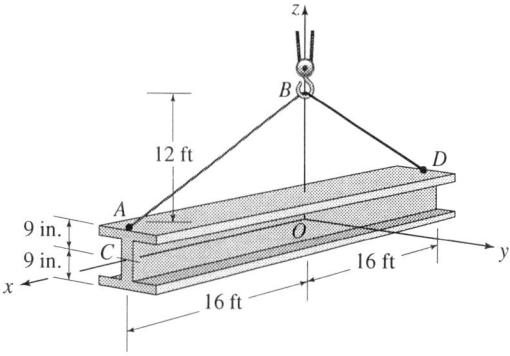

FIGURE P5.44.

5.45 Three cable tensions are applied to a vertical post, as shown in Figure P5.45. Replace the three given forces by an

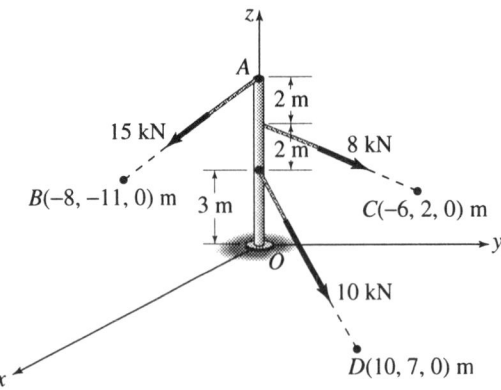

FIGURE P5.45.

equivalent force-couple system acting at point O.

5.46 Refer to Problem 5.45 and replace the three given forces by an equivalent force-couple system acting at point B.

5.47 A pipe assembly is subjected to the force system shown in Figure P5.47. Replace this force system by an equivalent force-couple system at point O. How much

turning tendency exists in pipe segment OA about its axis?

5.48 Refer to Problem 5.47 and replace the given force system by an equivalent force-couple system at point B. How much turning tendency exists in pipe segment BC about its axis?

5.49 A frame is subjected to the loading system shown in Figure P5.49 in which $F_1 = 25$ kN, $F_2 = 12$ kN, and $M_3 = 60$ kN·m. Replace this loading system by an equivalent force-couple system at point A. What is the turning tendency in member AC about its axis?

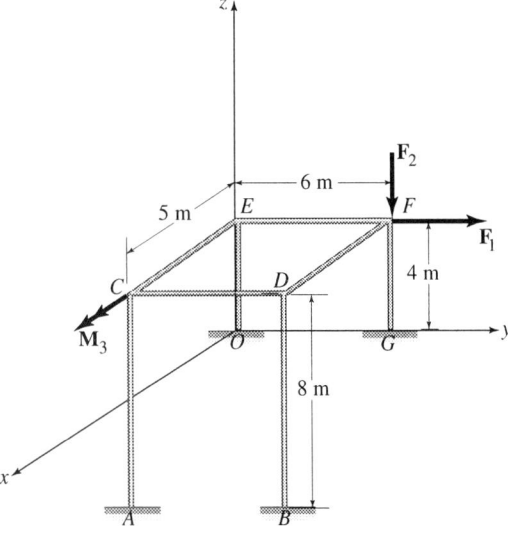

FIGURE P5.49.

5.50 Refer to the frame shown in Figure P5.49. The given loading system may be replaced by a force-couple system at point B consisting of a force **R** and a couple $M_R = (10\mathbf{i} - 20\mathbf{j} - 30\mathbf{k})$ kN·m. Find the magnitudes of $\mathbf{F}_1$, $\mathbf{F}_2$, and $\mathbf{M}_3$. Also, find the force **R**, and express it in terms of its rectangular components.

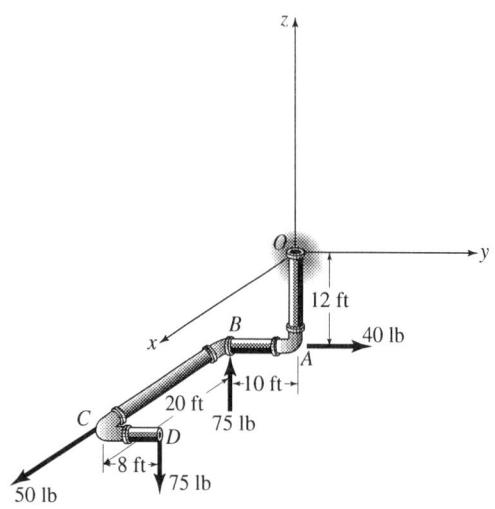

FIGURE P5.47.

5.51 A 30 ft × 40 ft floor of a building is subjected to the column loads shown in Figure P5.51. Replace the given parallel loads by a single resultant force **R** specifying its location.

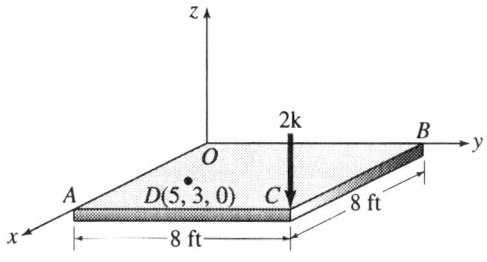

FIGURE P5.53.

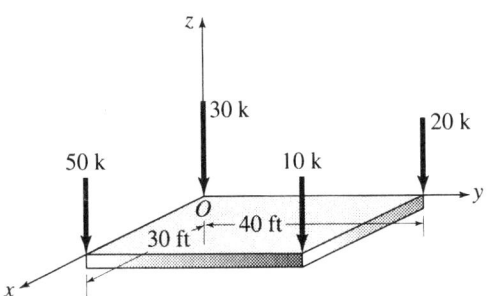

FIGURE P5.51.

5.52 A 10-m diameter concrete floor is subjected to three vertical loads as shown in Figure P5.52. Determine r and θ defining the position of the 25 kN load, so that the given parallel force system can be replaced by a single resultant force **R** at point O.

termine the magnitude and sense of the loads at O, A, and B so that the load system is replaceable by a single resultant force $\mathbf{R} = (-12\mathbf{k})$ k acting at point D.

5.54 A vertical wall is subjected to three horizontal forces, as shown in Figure P5.54. Two additional horizontal forces $\mathbf{F}_A$ and $\mathbf{F}_B$ applied at A and B are unknown in magnitude and sense. Determine $\mathbf{F}_A$ and $\mathbf{F}_B$ so that the five applied loads are replaceable by a single resultant horizontal force **R** acting at point C. Determine **R**.

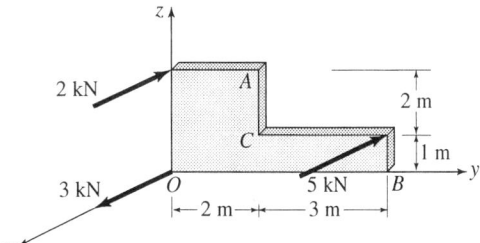

FIGURE P5.54.

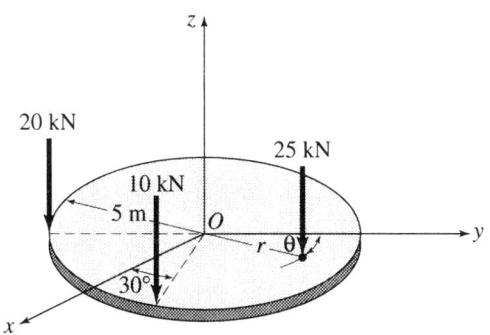

FIGURE P5.52.

5.53 The horizontal plate shown in Figure P5.53 is subjected to vertical loads at O, A, B, and C. The load at C is downward with a magnitude of 2 k, as shown. De-

5.55 Refer to the loading system shown in Figure P5.51. Determine the location of an additional downward load of 100 k that must be applied to the building floor so that the five vertical loads can be replaced by a single resultant force **R** at the geometric center of the floor.

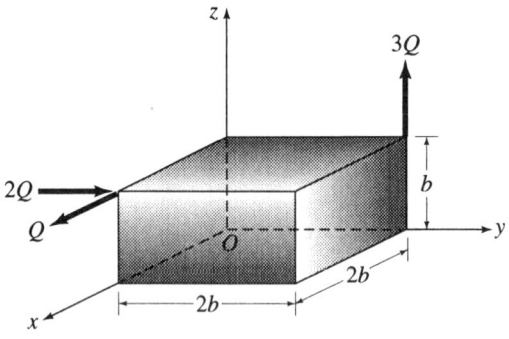

FIGURE P5.56.

5.56 Replace the force system shown in Figure P5.56 by a wrench. Specify the force and couple of the wrench. Also, state its direction, and locate the point of intersection of its line of action with the *y-z* coordinate plane.

5.57 Replace the load system shown in Figure P5.45 by a wrench. Specify the force and couple of the wrench. Also, state its direction and locate the point of intersection of its line of action with the *y-z* coordinate plane. *Hint: Begin your solution with the equivalent force-couple system obtained in Problem 5.45.*

5.58 Replace the force system shown in Figure P5.47 by a wrench. Specify the force and couple of the wrench. Also, state its direction, and locate the point of intersection of its line of action with the x-z coordinate plane. *Hint: Begin your solution with the equivalent force-couple system obtained in Problem 5.47.*

5.59 Replace the force system, shown in Figure P5.49, by a wrench. Specify the force and the couple of the wrench. Also, state its direction, and locate the point of intersection of its line of action with the x-y coordinate plane. *Hint: Begin your solution with the equivalent force-couple system obtained in Problem 5.49.*

5.9
Equilibrium
Conditions
and
Applications

As discussed in Chapter 4 for the two-dimensional equilibrium of a rigid body, the conditions for the three-dimensional equilibrium of a rigid body are achieved by ensuring that both the resultant force $\mathbf{R}$ and the resultant couple $\mathbf{M_R}$ acting on the rigid body vanish. It follows, therefore, that

and

$$\left.\begin{array}{c} \mathbf{R} = \sum \mathbf{F} = 0 \\ \mathbf{M_R} = \sum \mathbf{M} = 0 \end{array}\right\} \qquad (5.29)$$

or

and

$$\left.\begin{array}{c} \mathbf{R} = \left(\sum F_x\right)\mathbf{i} + \left(\sum F_y\right)\mathbf{j} + \left(\sum F_z\right)\mathbf{k} = 0 \\ \mathbf{M_R} = \left(\sum M_x\right)\mathbf{i} + \left(\sum M_y\right)\mathbf{j} + \left(\sum M_z\right)\mathbf{k} = 0. \end{array}\right\} \qquad (5.30)$$

The only way that Eqs. (5.30) can be satisfied is for each of the three terms in both equations to vanish separately. This, obviously, means that the coefficients of $\mathbf{i}$, $\mathbf{j}$, and $\mathbf{k}$ in each of the two equations must be equal to zero. Thus, equilibrium of a rigid body in three dimensions requires that the following six scalar equations be satisfied:

$$\left. \begin{aligned} \sum F_x = 0; && \sum M_x = 0; \\ \sum F_y = 0; && \sum M_y = 0; \\ \sum F_z = 0; && \sum M_z = 0. \end{aligned} \right\} \qquad (5.31)$$

Note that the six conditions expressed in Eq. (5.31) are both *necessary* and *sufficient* for the equilibrium of a rigid body in space. Note also that these six conditions are entirely independent of one another and that any one of them may be satisfied even though the other five may not be. Thus, for complete equilibrium of a rigid body in space, all six of the conditions expressed in Eq. (5.31) must be satisfied.

The fact that we have available six independent equations for the three-dimensional equilibrium of a rigid body means that we can solve for no more than six unknown quantities in a given problem. In solving for the six unknown quantities, we have the choice of proceeding directly to the six scalar conditions of Eq. (5.31) or beginning the solution with vector Eqs. (5.29). In general, in dealing with three-dimensional force systems, it is much more practical to use the vector approach and begin the solution with Eqs. (5.29), thus avoiding the difficult task of having to visualize geometric relationships in three dimensions.

As in the case of two dimensions, solving three-dimensional rigid body equilibrium problems requires the construction of an appropriate free-body diagram. This construction is accomplished by the same three-step procedure introduced for particles in Chapter 2 and developed further in conjunction with rigid bodies in Chapter 4. Therefore, the student is urged to review this material in Chapters 2 and 4 before proceeding with the study of three-dimensional equilibrium of rigid bodies.

The construction of free-body diagrams requires knowledge of the reactive forces produced at supports and connections. In addition to the supports and connections that have already been discussed in preceding chapters, we will discuss, here, a few others necessary for the construction of three-dimensional structures and machines. These additional supports will be discussed in some detail in the following pages and are also summarized in cases 10 to 13 in Appendix F.

Universal Joint

The universal joint is a device that is occasionally used to connect two members which require relative rotations with respect to each other about two axes. However, such a device prevents relative rotation about the third axis and relative translation about any axis. A schematic sketch of a universal joint, similar to the one used in the drive shaft of an automobile, is shown in Figure 5.15(a) connecting segments AB and BC of the shaft. The joint is designed to allow relative rotation between the two segments of the shaft about the x and z axes but not

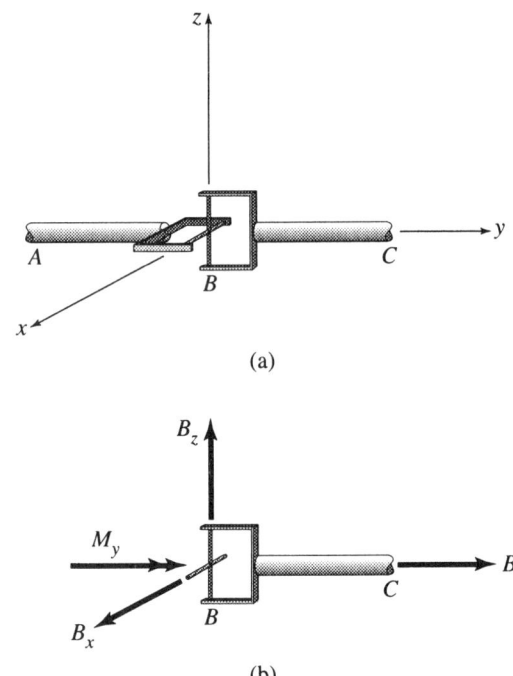

FIGURE 5.15.   (b)

about the y axis. Thus, a moment M_y may be transmitted from segment AB to segment BC. However, because of the freedom of rotation about the x and z axes, both M_x and M_z are zero. Also, because the joint allows no relative translation between the two segments of the shaft, it follows that reactive forces B_x, B_y, and B_z must develop to prevent translation in the x, y, and z axes, respectively. A sketch of segment BC of the shaft is depicted in Figure 5.15(b) showing the reactive forces B_x, B_y, B_z, and the reactive moment M_y that segment AB is capable of exerting on segment BC.

Bearing and Hinge Supports

A bearing support may be designed in one of two ways. In the first type, the bearing is designed to resist only forces along radial axes and moments about these same axes. An example of such a bearing is shown in Figure 5.16(a). The shaft, which is supported in the bearing support, is assumed frictionless and may rotate freely about its axis. Also, no provisions are made to prevent the shaft from translating along its axis in the bearing. Thus, this type of bearing cannot resist a moment about the axis of the bearing (i.e., $M_x = 0$) nor a force along this axis (i.e., $F_x = 0$). However, the bearing will resist any tendency the shaft may have to translate along, or to rotate about, any radial line, such as the y or the z axis of Figure 5.16(a). Therefore, resisting forces

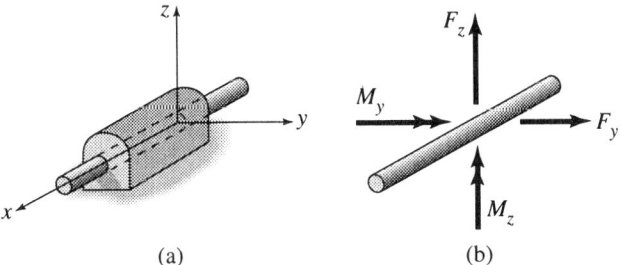

(a)   (b)   FIGURE 5.16.

may exist in the y and z directions as well as resisting moments about these two axes. Figure 5.16(b) shows a diagram of the shaft indicating the reactive forces and moments that the bearing is capable of exerting on it. Occasionally, this type of bearing is designed to provide little or no resistance against moments so that $M_y = M_z = 0$ in Figure 5.16(b). Such a bearing resists only radial forces and is similar in action to the two-dimensional hinge support shown in case 4 of Appendix F.

In the second type of bearing, a provision is made in the design to prevent translation of the shaft along its axis. Thus, a resisting force can develop along the axis of the shaft, called the *axial thrust*, and the only allowable movement in this type of bearing is rotation of the shaft about the axis of the bearing. Therefore, the only difference between the two types of bearing supports is the existence of an axial thrust in the latter type. An illustration of this type of bearing support is shown in Figure 5.17(a) and the corresponding diagram of the shaft in Figure 5.17(b) showing the reactive forces and moments that the bearing is capable of exerting on it.

The three-dimensional *hinge* shown in Figure 5.18(a) is identical in action to the bearing support shown in Figure 5.17(a). As in the case of the bearing support with axial thrust, the only permissible motion is rotation about the axis of the hinge. Therefore, resisting forces can exist in all three directions but resisting moments only about two axes. A sketch of the lower part of the hinge is shown in Figure 5.18(b), indicating three reactive forces and two reactive moments that the

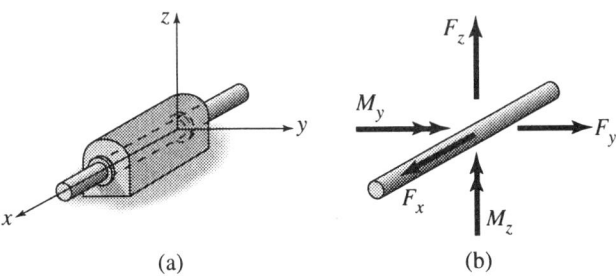

(a)   (b)   FIGURE 5.17.

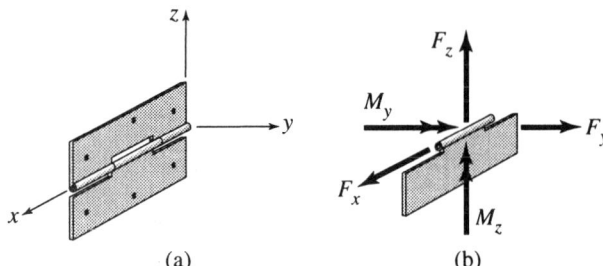

FIGURE 5.18.   (a)   (b)

shaft of the hinge is capable of exerting on it. The hinge support or
connection, discussed in Chapter 4 and summarized in case 4 of
Appendix F, is a special case of the three-dimensional hinge shown
in Figure 5.18.

Note that the reactive forces and couples given in Figures 5.16, 5.17,
and 5.18 apply *only* to the cases of individual bearings and hinges.
When used in combination with other bearings and hinges, these supports will not develop any reactive couples *if* the set of bearings or
hinges used to support a rigid body is carefully and properly aligned.
In such a case, the rigid body is prevented from unwarranted rotations
by the reactive forces, thus eliminating the need for any reactive
couples.

Fixed Support

A three-dimensional fixed support is shown in Figure 5.19(a). The
sketch is intended to convey the idea that the member is imbedded into
the supporting frame which in this case is a wall. In theory, such a
support does not allow any translation along the x, y, and z axis nor
will it permit rotation about any of these three axes. Thus, the three-
dimensional fixed support is capable of developing three resisting forces,
F_x, F_y, and F_z and three resisting moments M_x, M_y, and M_z. A diagram
of the supported member is shown in Figure 5.19(b) indicating the
three resisting forces as well as the three resisting moments. The fixed
support, discussed in Chapter 4 and summarized in case 9 of Appendix F, is a special case of the three-dimensional fixed support of
Figure 5.19.

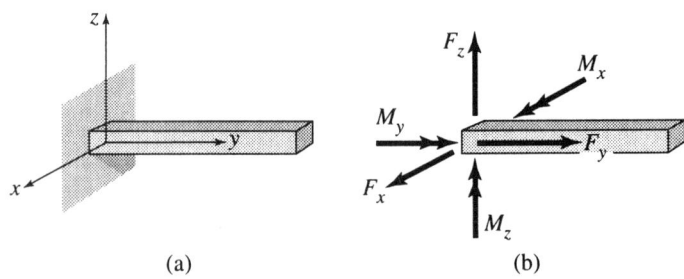

FIGURE 5.19.   (a)   (b)

The following examples illustrate the concepts discussed above.

■ **Example 5.12**

A homogeneous plate of negligible thickness weighing 2000 N is supported in a horizontal position by two vertical cables AD and BE and by a ball and socket at point C, as shown in Figure E5.12(a). Assume that the weight of the plate acts through its geometric center, and determine the tensions in the two cables and the reaction components in the ball and socket.

Solution

The free-body diagram of the plate is shown in Figure E5.12(b). Note that the two cable tensions T_{AD} and T_{BE} act along the respective axes of the two cables and the ball and socket is capable of developing three reactive components C_x, C_y, and C_z. Thus, the force system in Figure E5.12(b) has five unknown quantities.* Note also that the geometric information provided in Figure E5.12(a) has been expressed in terms of

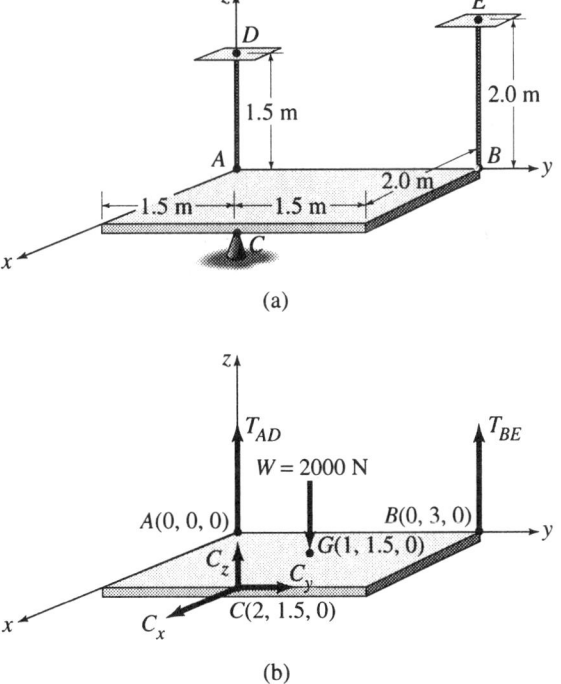

(a)

(b)   FIGURE E5.12.

---

point coordinates in the free-body diagram and, because of the negligible thickness of the plate, point C is shown on top of the plate.

Although the solution of this relatively simple force system may be executed by scalar algebra, we will perform the solution with the vector approach to gain facility with it. To this end, we will express all forces acting on the plate in vector form. Thus,

$$\mathbf{W} = -2000\mathbf{k},$$

$$\mathbf{T_{AD}} = T_{AD}\mathbf{k},$$

$$\mathbf{T_{BE}} = T_{BE}\mathbf{k},$$

and

$$\mathbf{C} = C_x\mathbf{i} + C_y\mathbf{j} + C_z\mathbf{k}$$

Now, we apply the first condition of equilibrium expressed in Eq. (5.29). Thus,

$$\sum \mathbf{F} = \mathbf{0}, \quad -2000\mathbf{k} + T_{AD}\mathbf{k} + T_{BE}\mathbf{k} + C_x\mathbf{i} + C_y\mathbf{j} + C_z\mathbf{k} = \mathbf{0}.$$

Combining terms yields

$$C_x\mathbf{i} + C_y\mathbf{j} + (C_z + T_{AD} + T_{BE} - 2000)\mathbf{k} = \mathbf{0}$$

which reduces to the following three scalar equations:

$$C_x = 0, \qquad\qquad\qquad\qquad \text{ANS.}$$

$$C_y = 0, \qquad\qquad\qquad\qquad \text{ANS.}$$

and

$$C_z + T_{AD} + T_{BE} - 2000 = 0. \qquad\qquad (a)$$

The second equilibrium condition expressed in Eq. (5.29) is now applied using point A as a moment center. Thus,

$$\sum \mathbf{M_A} = \mathbf{0},$$

$$\mathbf{r}_{G/A} \times \mathbf{W} + \mathbf{r}_{B/A} \times \mathbf{T_{BE}} + \mathbf{r}_{C/A} \times \mathbf{C} = \mathbf{0}.$$

Substituting values for $\mathbf{r}_{G/A}$, $\mathbf{r}_{B/A}$, and $\mathbf{r}_{C/A}$ and remembering that $C_x = C_y = 0$,

$$((1.0\mathbf{i} + 1.5\mathbf{j}) \times (-2000\mathbf{k}) + (3.0\mathbf{j}) \times (T_{BE}\mathbf{k}) + (2.0\mathbf{i} + 1.5\mathbf{j}) \times (C_z\mathbf{k}) = \mathbf{0}.$$

Performing the indicated vector operations, combining the $\mathbf{i}$ and $\mathbf{j}$ terms and setting their scalar coefficients separately equal to zero,

$$1.5C_z + 3.0T_{BE} - 3000 = 0, \qquad\qquad (b)$$

$$2000 - 2C_z = 0. \qquad\qquad (c)$$

The simultaneous solution of Eqs. (a), (b), and (c) yields the following values:

$$C_z = 1000 \text{ N } \uparrow, \qquad\qquad \text{ANS.}$$

$$T_{BE} = 500 \text{ N } \uparrow, \qquad\qquad \text{ANS.}$$

and

$$T_{AD} = 500 \text{ N } \uparrow. \qquad\qquad \text{ANS.}$$

■ **Example 5.13**   The frame shown in Figure E5.13(a) is supported at O by a smooth plane and at B and C by ball-and-socket supports. The two forces applied to the frame are $\mathbf{F} = (10\mathbf{i} - 15\mathbf{j} - 5\mathbf{k}) \text{ k}$ and $\mathbf{Q} = (-7\mathbf{i} + 12\mathbf{j} - 8\mathbf{k}) \text{ k}$. Determine the magnitude of the support reaction at O.

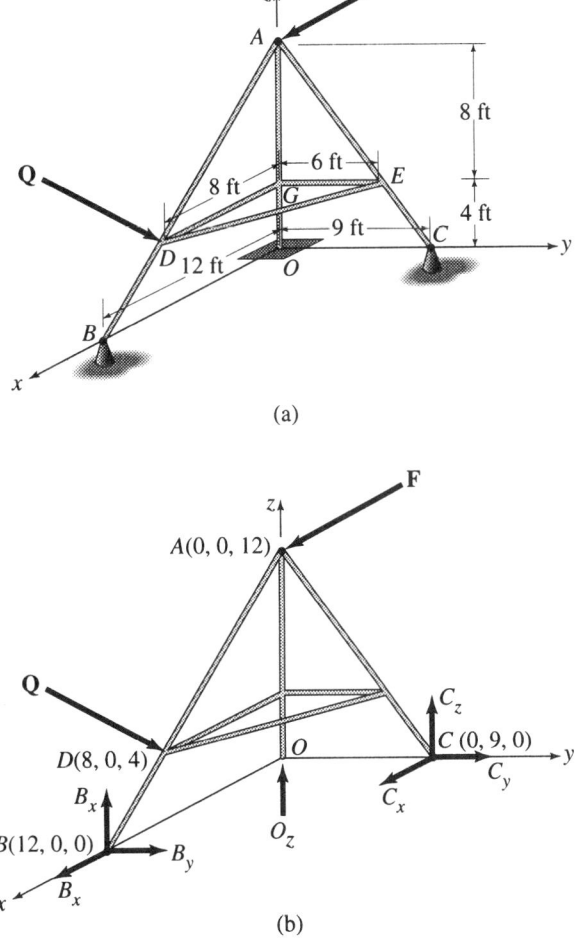

(a)

(b)

FIGURE E5.13.

Solution

The free-body diagram of the frame is shown in Figure E5.13(b). It is obvious that seven unknown quantities exist, namely, O_z, B_x, B_y, B_z, C_x, C_y, and C_z. Because only six conditions of equilibrium are available, it is not possible to obtain all of the seven unknown quantities and the force system is said to be *statically indeterminate*. However, in this case, only the support reaction O_z is needed. This may be obtained by writing one moment equation about axis BC. Such an equation would contain O_z as the only unknown because the reaction components at supports B and C produce no moments about axis BC. Thus,

$$M_{BC} = \lambda_{BC} \cdot \left(\sum M_B \right) = 0. \tag{a}$$

Application of Eq. (a) requires that we find the following vector quantities

$$\lambda_{BC} = \frac{-12i + 9j}{\sqrt{12^2 + 9^2}} = -0.8i + 0.6j,$$

$$r_{A/B} = -12i + 12j,$$

$$r_{D/B} = -4i + 4k,$$

$$r_{O/B} = -12i,$$

$$C = C_x i + C_y j + C_z k,$$

and

$$O = O_z k.$$

Therefore,

$$M_{BC} = \lambda_{BC} \cdot (r_{A/B} \times F + r_{D/B} \times Q + r_{C/B} \times C + r_{O/B} \times O)$$

$$= (-0.8i + 0.6j) \cdot \begin{bmatrix} (-12i + 12k) \times (10i - 15k - 5k) \\ +(-4i + 4k) \times (-7i + 12j - 8k) \\ +(-12i + 9k) \times (C_x i + C_y j + C_z k) \\ +(-12i) \times (O_z k) \end{bmatrix}$$

$$= 0.$$

Performing the indicated operations and combining terms,

$$(-0.8i + 0.6j)$$

$$\cdot [(132 + 9C_z)i + (12O_z + 12C_z)j + (132 - 9C_x - 12C_y)k] = 0$$

which reduces to

$$7.2O_z - 105.6 = 0$$

from which

$$O_z = 14.67 \text{ k } \uparrow. \qquad \text{ANS.}$$

■ Example 5.14

The homogeneous access gate shown in Figure E5.14(a) is supported in the horizontal position by a cable CD and by hinges at A and B which are properly aligned and, consequently, will not develop reactive couples. The hinge at A can resist axial and radial forces and the hinge at B only radial forces. If the weight of the gate $W = 2000$ N, determine the tension in cable CD and the support reactions at hinges A and B. Assume that the weight of the gate acts through its geometric center.

Solution

The free-body diagram of the access gate is shown in Figure E5.14(b). Based on the information provided in the statement of the problem,

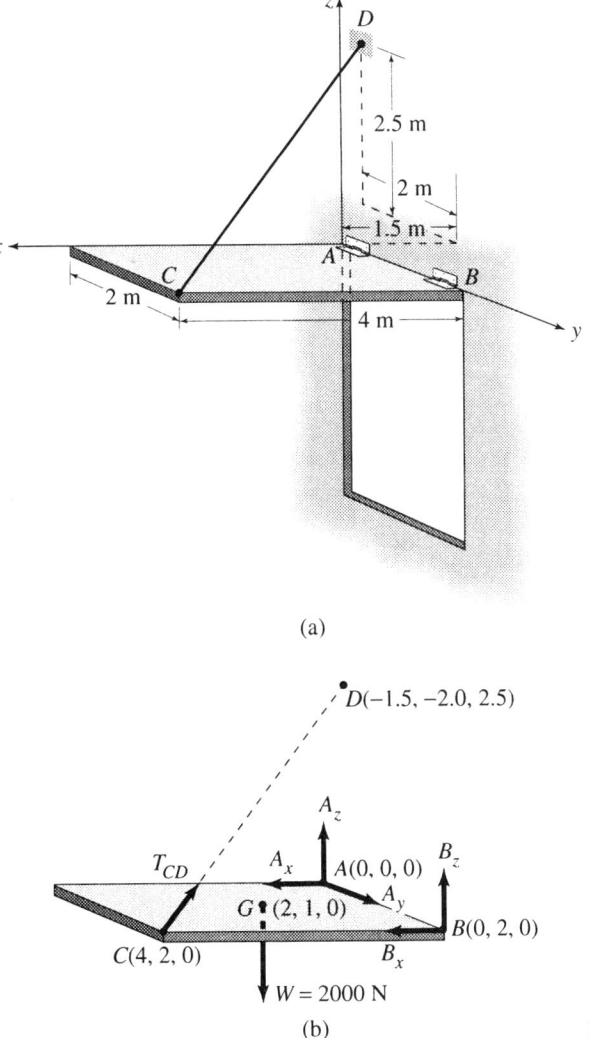

(a)

(b)   FIGURE E5.14.

the hinge at A can develop three reaction components A_x, A_y, and A_z whereas the hinge at B only two, B_x and B_z. In addition to the foregoing unknown reaction components, there is the tension in the cable, T_{CD}. Thus, there are a total of six unknown quantities which may be found by using six available conditions of equilibrium. Therefore, the force system shown in Figure E5.14(b) is *statically determinate*. Note that, in the free-body diagram, the given geometric information is expressed in terms of point coordinates.

To apply the conditions of equilibrium, we need to express all of the forces acting on the access gate in vector form. Thus,

$$\mathbf{T}_{CD} = T_{CD}\left(\frac{-5.5\mathbf{i} - 4.0\mathbf{j} + 2.5\mathbf{k}}{\sqrt{5.5^2 + 4.0^2 + 2.5^2}}\right) = T_{CD}(-0.759\mathbf{i} - 0.552\mathbf{j} + 0.345\mathbf{k}),$$

$$\mathbf{A} = A_x\mathbf{i} + A_y\mathbf{j} + A_z\mathbf{k},$$

$$\mathbf{B} = B_x\mathbf{i} + B_z\mathbf{k},$$

and

$$\mathbf{W} = -200\mathbf{k}.$$

According to the second equilibrium condition of Eq. (5.29) applied about point A,

$$\sum \mathbf{M}_A = \mathbf{0},$$

$$\mathbf{r}_{C/A} \times \mathbf{T}_{CD} + \mathbf{r}_{B/A} \times \mathbf{B} + \mathbf{r}_{G/A} \times \mathbf{W} = \mathbf{0},$$

$$(4.0\mathbf{i} + 2.0\mathbf{j}) \times [T_{CD}(-0.759\mathbf{i} - 0.552\mathbf{j} + 0.345\mathbf{k})$$

$$+ (2.0\mathbf{j}) \times (B_x\mathbf{i} + B_z\mathbf{k}) + (2.0\mathbf{i} + 1.0\mathbf{j}) \times (-2000\mathbf{k})] = \mathbf{0}.$$

Performing the indicated vector operations, combining the $\mathbf{i}$, $\mathbf{j}$, and $\mathbf{k}$ terms, and setting their coefficients separately equal to zero,

$$0.690T_{CD} + 2.000B_z = 2000, \tag{a}$$

$$1.380T_{CD} = 4000, \tag{b}$$

and

$$0.690T_{CD} + 2.000B_x = 0. \tag{c}$$

Simultaneous solution of Eqs. (a), (b), and (c) yields

$$T_{CD} = 2900 \text{ N}, \qquad \text{ANS.}$$

$$B_x = -1000 \text{ N}, \qquad \text{ANS.}$$

and

$$B_z = 0. \qquad \text{ANS.}$$

Using the first equilibrium condition of Eq. (5.29),

$$\sum \mathbf{F} = \mathbf{0},$$

$$-(0.759T_{CD})\mathbf{i} - (0.552T_{CD})\mathbf{j} + (0.345T_{CD})\mathbf{k}$$

$$+ A_x\mathbf{i} + A_y\mathbf{j} + A_z\mathbf{k} + B_x\mathbf{i} + B_z\mathbf{k} - 2000\mathbf{k} = \mathbf{0}.$$

Combining the $\mathbf{i}$, $\mathbf{j}$, and $\mathbf{k}$ terms and setting their scalar coefficients, separately, equal to zero,

$$-0.759T_{CD} + A_x + B_x = 0. \tag{d}$$

$$-0.552T_{CD} + A_y = 0. \tag{e}$$

$$0.345T_{CD} + A_z + B_z = 0. \tag{f}$$

Substituting the known values for T_{CD}, B_x, and B_z, we may solve Eqs. (d), (e) and (f) for the remaining unknown forces to obtain

$$A_z = 1000 \text{ N}, \qquad\qquad\qquad \text{ANS.}$$

$$A_y = 1600 \text{ N}, \qquad\qquad\qquad \text{ANS.}$$

and

$$A_x = 3200 \text{ N}. \qquad\qquad\qquad \text{ANS.}$$

■

Problems

5.60 A rectangular plate of negligible thickness with a mass of 150 kg is supported at O by a ball and socket and at A and B by two vertical cables as shown in Figure P5.60. Assume that the weight of the plate acts through its geometric center, and determine the reaction components in the ball and socket and the tensions in the two cables.

5.61 A reinforced concrete platform with a radius of 10 ft and a thickness of 1.5 ft is supported by three footings as shown in Figure P5.61. Reinforced concrete weighs 150 lb/ft^3 and the weight of the platform may be assumed to act through its geometric center. If the footings are assumed to be smooth, determine the forces carried by each of the three footings.

5.62 A triangular three-wheel dolly, shown in Figure P5.62, has a platform which may

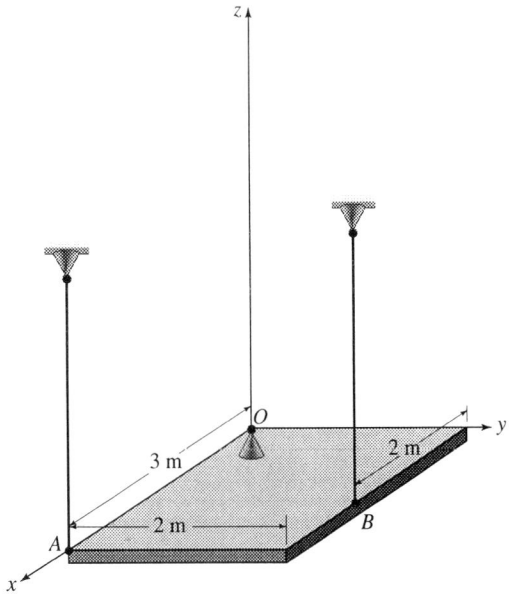

FIGURE P5.60.

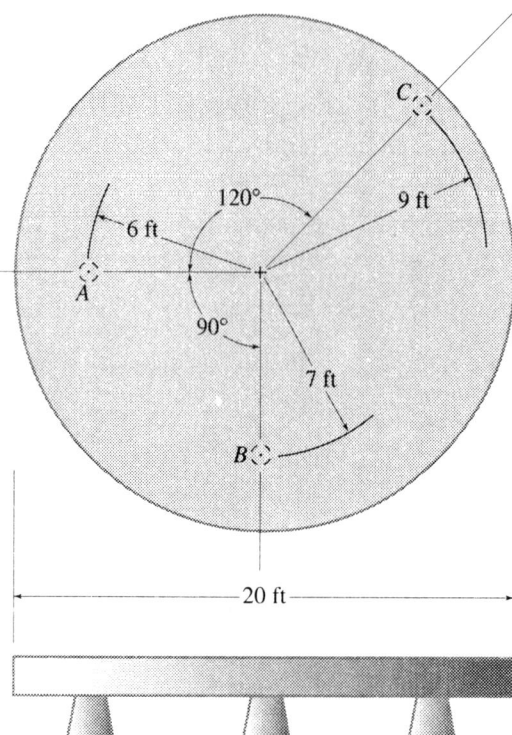

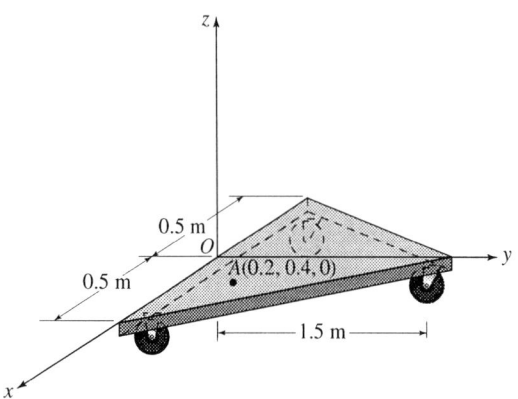

FIGURE P5.61.

FIGURE P5.62.

may be assumed to act at point A. Determine the reaction of the floor on each of the three wheels.

5.63 A circular plate with a radius of 5 ft is suspended by three identical cables AD, BE, and CF as shown in Figure P5.63. The plate is nonhomogeneous with a weight of 2000 lb which acts at point G, a distance d from point O, the geometric center of the plate. If the tension in cable BE is 500 lb, determine the distance d and the forces in the other two cables.

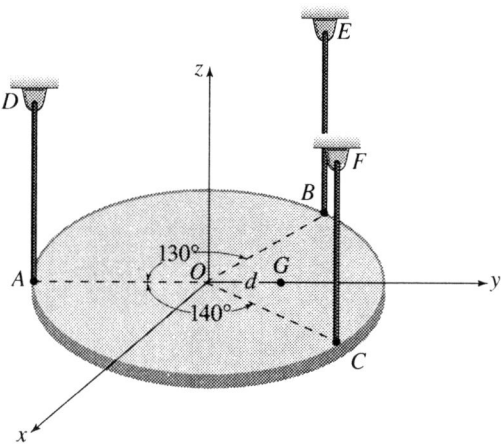

FIGURE P5.63.

5.64 A homogeneous plate ACE with a mass of 525 kg is suspended by three vertical springs AB, CD, and EF, as shown in Figure P5.64. Spring AB has a spring constant $k_{AB} = 2$ kN/m. Determine the necessary spring constants for springs CD and EF so that the plate hangs horizontally.

5.65 The vertical mast shown in Figure P5.65 is supported at A by a ball and socket and at D by two cables DB and DC. The applied forces are $\mathbf{P} = (7\mathbf{i} + 9\mathbf{j} + 4\mathbf{k})$ k and $\mathbf{Q} = (3\mathbf{i} + 6\mathbf{j} + 5\mathbf{k})$ k. Determine the tension in cable DC by writing a single

be assumed weightless. A 500-kg crate is placed on the platform so that its weight

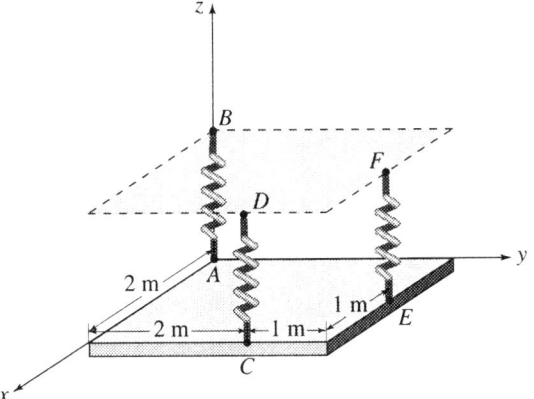

FIGURE P5.64.

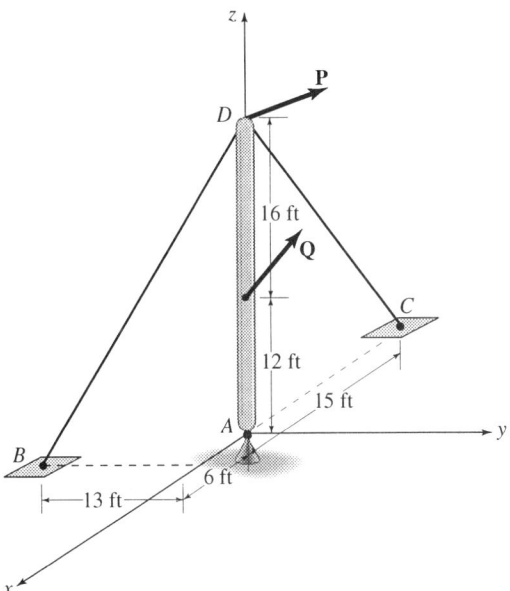

FIGURE P5.65.

moment equation about axis BA. The vertical mast AB may be assumed to be weightless.

5.66 Consider the mast shown in Figure P5.65. Let $\mathbf{P} = (7\mathbf{i} + 9\mathbf{j} + 4\mathbf{k})$ k. and $\mathbf{Q} = (3\mathbf{i} + 6\mathbf{j} + 5\mathbf{k})$ k. Determine the reaction com-

ponents at the ball and socket at A and the tension in cables DB and DC. Assume that the vertical mast AC is weightless.

5.67 An access door with a mass of 100 kg is supported in the position shown by hinges at A and B and by cable CD as in Figure P5.67. Determine the tension in cable CD by writing a single moment equation about axis AB. Neither of the two hinges at A and B can resist moments about any axis. The hinge at A can resist only radial forces whereas the hinge at B can resist radial forces as well as an axial thrust. Assume that the weight of the gate acts through its geometric center.

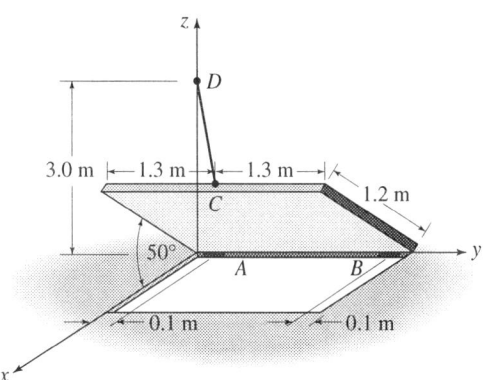

FIGURE P5.67.

5.68 Consider the access door shown in Figure P5.67 whose mass is 100 kg. It is supported in the position shown by hinges A and B and by cable CD. Determine the reaction components at the two hinges A and B and the tension in cable CD. Neither of the two hinges can resist moments about any axis. The hinge at A can resist only radial forces whereas that at B can resist radial as well as axial forces. Assume that the weight of the gate to act through its geometric center.

5.69 The frame shown in Figure P5.69 supports a 5000-lb sign that hangs from point A by two identical cables AE and AF. The supports at C and D are balls and sockets whereas the support at B is a roller that can resist only a force in the y direction. Determine the force at support B.

5.70 A frame is constructed from steel pipe and supported as shown in Figure P5.70. The support at C is a bearing that can resist only radial forces but no moments, and the support at E is a ball and socket. The given loads are $P = (5i + 10j - 4k)$ kN and $M = (20k)$ kN·m. Determine the magnitude and sense of a vertical force at A that is needed for equilibrium of the frame. Solve the problem by writing a single moment equation about axis EC. Assume that the pipe is weightless.

5.71 Refer to Problem 5.70 and determine the reaction components at supports C and E, using the result obtained in problem 5.70 for the vertical force at A.

5.72 A frame is constructed from steel pipe and supported as shown in Figure P5.72. The support at C is a bearing that can support only radial forces and a moment about the y axis. The support at G is a ball and socket. The loading on the frame is a force F at A given by the expression $F = (-7.5i + 10.0k)$ k. Determine the reactive moment at C by writing a single moment equation about axis CG. Assume that the pipe is weightless.

5.73 A frame is constructed from steel pipe and supported as shown in Figure P5.72. The support at C is a bearing that can support only radial forces and a moment about the y axis. The support at G is a ball and socket. The loading on the frame is a force F at A given by the expression

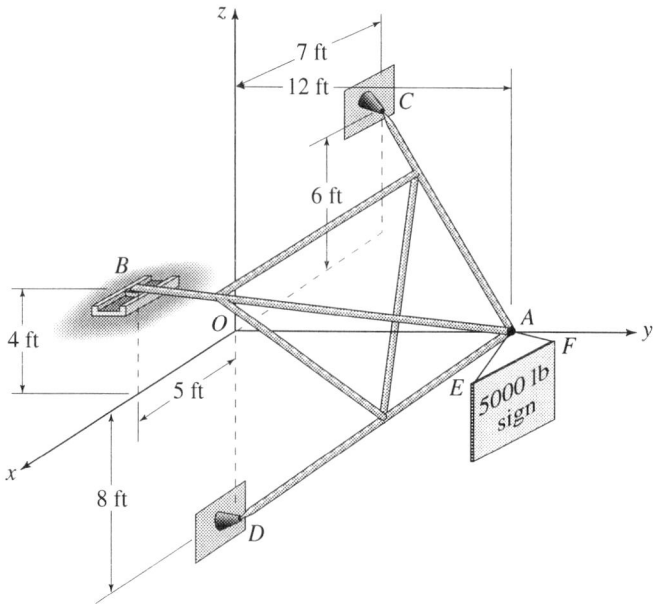

FIGURE P5.69.

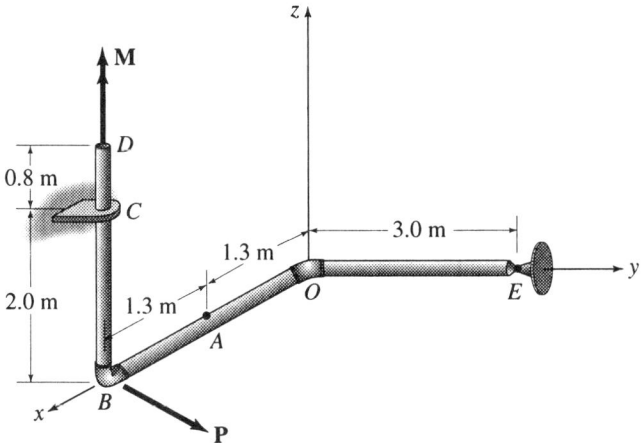

FIGURE P5.70.

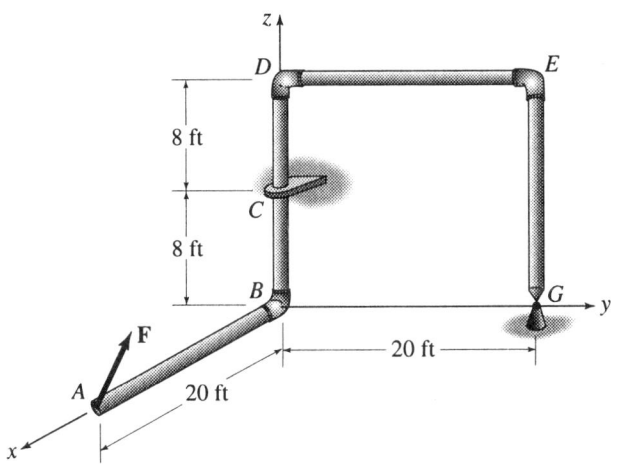

FIGURE P5.72.

$\mathbf{F} = (-7.5\mathbf{i} + 10.0\mathbf{k})$ k. Determine the reaction components at supports C and G.

5.74 The frame shown in Figure P5.74 is supported at A and E by balls and sockets and at C by cable CG. Determine the tension in cable CG when the frame is subjected to the loads shown. Solve by writing a single moment equation about axis AE. Assume that the frame is weightless.

5.75 Refer to the frame of Problem 5.74. Point G can be placed any where along line HI which is parallel to the x axis. Determine the exact position of point G along line HI so that the tension in cable CG has a minimum value. What is the minimum tension in the cable? *Hint: Write a moment equation about axis AE leading to an expression relating the tension T in the*

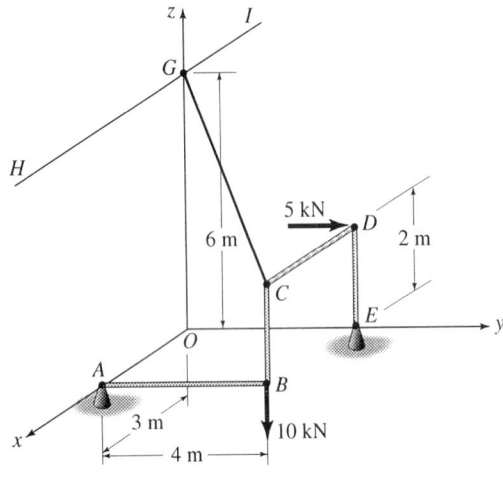

FIGURE P5.74.

cable to the variable x that locates point G. Minimize this expression.

5.76 The shaft shown in Figure P5.76 is in equilibrium under the action of the given belt tensions and the reactive forces in the two bearings at A and B. The bearing at A can resist radial forces as well as axial thrust but no moments. The bearing at B

can resist *only* radial forces. Determine the unknown belt tension T and the reaction components in the two bearings.

5.77 The system shown in Figure P5.77 is used to raise the weight W by applying the force F to the handle at C. For the position shown and for $F = 100$ lb, determine the weight W for the system to be in equilibrium. Determine also the reaction components at the two bearings A and B. The bearing at A can resist radial as well as axial forces but no moments. The bearing at B can resist *only* radial forces.

5.78 A member is bent into the shape shown in Figure P5.78. It is supported at A by a ball and socket and at B by a bearing which is capable of resisting only radial forces and a moment about an x axis through B. Let $\mathbf{F} = (-2\mathbf{i} + 3\mathbf{j} + 5\mathbf{k})$ kN, and determine the reaction components at supports A and B.

5.79 Member AD shown in Figure P5.79 is assumed weightless. It is supported at A by a ball and socket and at B and D by cables BE and DF, respectively. A weight W is suspended from C as shown. In

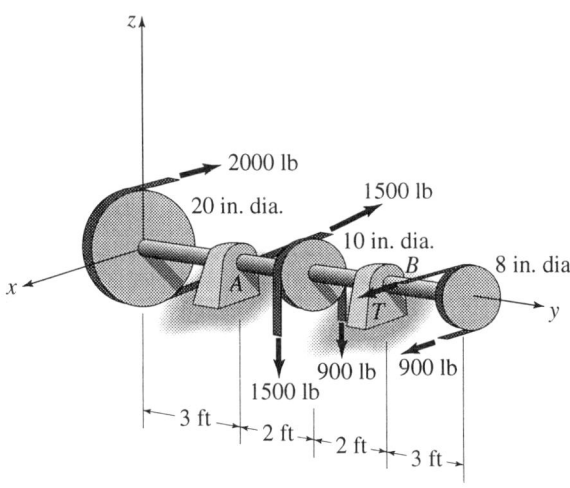

FIGURE P5.76.

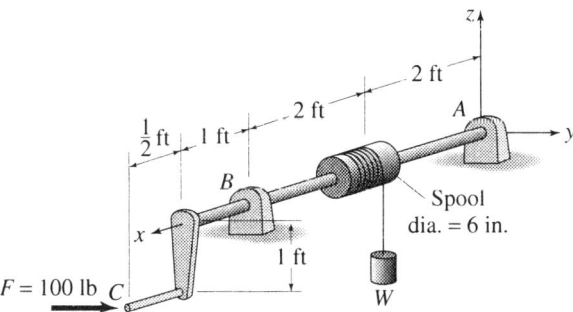

FIGURE P5.77.

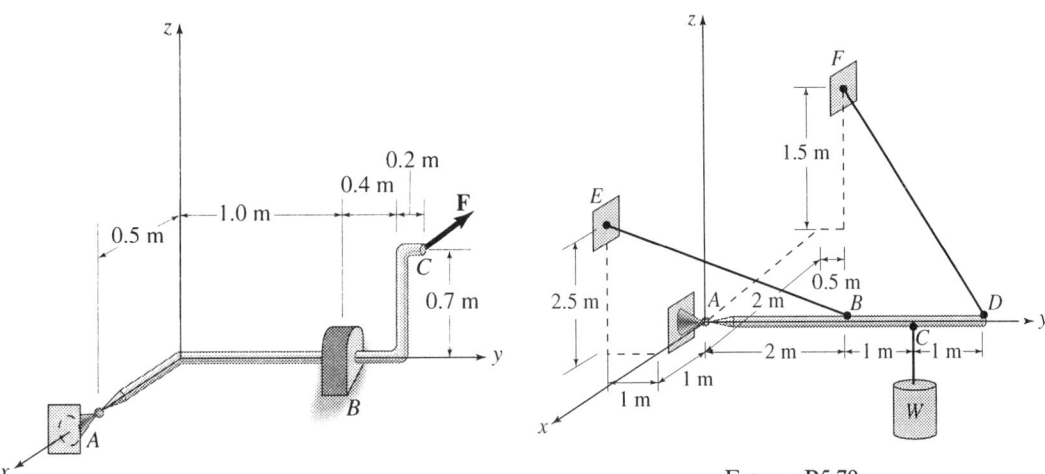

FIGURE P5.78.

FIGURE P5.79.

terms of *W*, determine the reaction components at A and the tensions in cables
BE and DF.

5.80 Refer to Problem 5.79. Assume that the
tension in cable DF is not to exceed 10
kN. Determine the largest weight *W* that
may be supported at C. What are the tension in cable BE and the reaction components at A?

5.81 The homogeneous plate, shown in Figure
P5.81, weighs 5 k. This weight may be
assumed to act through the geometric
center of the plate. The support at A is

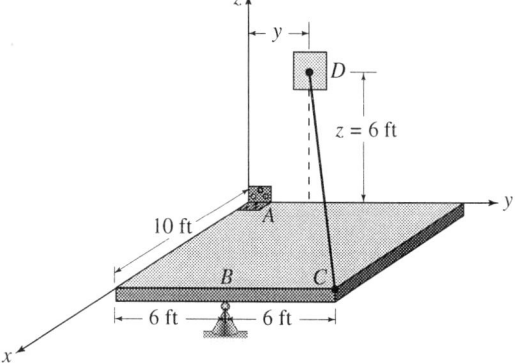

FIGURE P5.81.

a hinge capable of resisting only radial force whereas the support at B is a ball and socket. The plate is also partially supported by cable CD. Let $y = 4$ ft and determine the reaction components at A and B and the tension in cable CD. Note that D lies in the y-z plane.

5.82 Refer to Problem 5.81. Assume that point D, which lies in the y-z plane, may be positioned at any point along a line de-fined by the equation $z = 6$ ft. Determine the exact position of point D in the y-z plane so that the tension in cable CD is the least possible. What is the magnitude of this least possible tension? *Hint: Write a moment equation about axis AB leading to an expression relating the tension T in the cable to the variable y that locates point D. Minimize this expression.*

5.10 Determinacy and Constraints

All of the problems that have been encountered, thus far in this book, have dealt with systems in which the constraints (i.e., supports needed to prevent movement) developed the minimum number of unknown reactive force components needed to establish and sustain equilibrium. In such cases, the available equations of equilibrium are necessary and just sufficient for solving the unknown reactive force components. Thus, in the two-dimensional force systems for example, three scalar equations of equilibrium are available and, consequently, a maximum of three unknown reactive force components may be found. Such force systems, for which the available equilibrium equations are both necessary and sufficient for determining the unknown constraining reactive forces, are known as *statically determinate* force systems.

There are, however, many cases of force systems in which the supports develop a number of unknown reactive components which exceeds the number of available equations of equilibrium. Such force systems are said to be *statically indeterminate* because the unknown reactive force components cannot all be determined by using only the equations of statics. Consider, for example, the case of the two-dimensional member shown in Figure 5.20(a) which is supported by a roller at A

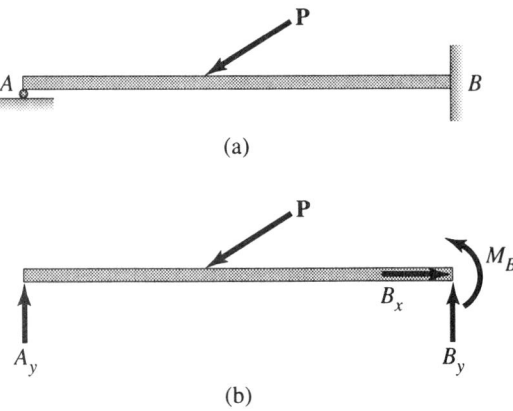

(a)

(b)

FIGURE 5.20.

and a fixed support at B and subjected to the load **P**. The free-body diagram of the member is shown in Figure 5.20(b). It indicates that there are four unknown reactive force components A_y, B_x, B_y, and M_B. Because only three equations of equilibrium are available in a two-dimensional force system, it follows that the system under consideration is statically indeterminate and that it has one more unknown reactive force component than the available equilibrium equations. Thus, this system is said to be statically indeterminate to the first degree or that it has one *redundant* reactive force component. The term *redundant* means 'extraneous' or 'unnecessary' for the equilibrium of the system. As a matter of fact, one may eliminate one redundant reactive force component by removing, for example, the roller support at A and still have a member perfectly capable of sustaining equilibrium under the action of the load **P** or any other loading system.

If one were to substitute a fixed support, instead of the roller support at A, so that the member is now fixed at both ends, as shown in Figure 5.21(a), the resulting free-body diagram would be as shown in Figure 5.21(b). Note that, in this case, there is a total of six unknown reactive force components, A_x, A_y, M_A, B_x, B_y, and M_B. Because the number of available equations of equilibrium is still only three, it follows that this member is statically indeterminate to the third degree. Another way of expressing the same thought is to say that the member has three redundant reactive force components. These three redundant reactive force components may be eliminated by removing, for example, either of the two fixed supports and, still, the member would be perfectly capable of sustaining equilibrium under any loading such as **P**. Thus, the degree of indeterminacy or the number of redundants of a given force system is found by taking the difference between the total number of unknown reactive force components and the number of equations of equilibrium available.

Therefore, the solution for all of the unknown reactive force compo-

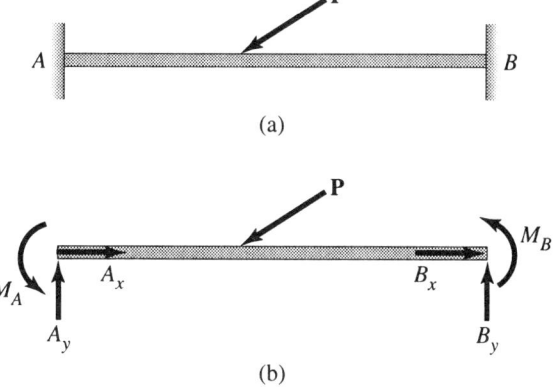

(a)

(b)   FIGURE 5.21.

nents in a statically indeterminate force system cannot be accomplished by using only the available equations of equilibrium because, although necessary, they are not sufficient. We need to supplement these equations by additional relationships derived from the deformational characteristics of the system. Such solutions are not treated in this textbook, but interested students may obtain pertinent information by consulting textbooks on strength or mechanics of materials.

Even if the number of unknown reactive force components in a given force system is the minimum required for equilibrium, cases may exist in which a member is *improperly constrained* because of the geometric arrangement of the supports. Consider, for example, the two-dimensional member shown in Figure 5.22 which is supported by the three short links AB, CD, and EF and subjected to a general loading **P**. The three short links are hinged at their ends and serve as two-force members capable of carrying loads (tension or compression) directed along their respective axes. With the support arrangement shown in Figure 5.22, the member is prevented from translating in any direction within the x-y plane and from rotating about any axis parallel to the z axis. Thus, the two-dimensional equilibrium of the member is assured. It is obvious that one can find an infinite number of geometric arrangements of the three links that will ensure the two-dimensional equilibrium of the member of Figure 5.22. However, there are two special geometric arrangements of the three link supports that would *not* ensure the two-dimensional equilibrium of the member. These two special geometric arrangements of the three links occur when the reactive forces in them either intersect at the same point or are parallel. In such cases, the member is said to be *improperly constrained* because the resulting force system does not ensure complete two-dimensional fixity of the member. Consider, for example, the geometric arrangement of the three links shown in Figure 5.23. In this case, all of the reactive forces intersect at the same point G. Although the number of reactive forces (three) is the minimum required for equilibrium, the geometric arrangement of the supports is such that a small initial rotation is possible about point G under the influence of a general force **P** unless,

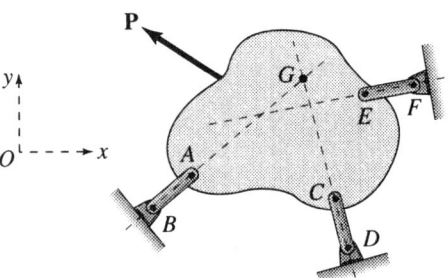

FIGURE 5.22.

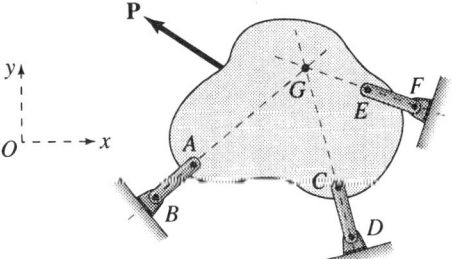

FIGURE 5.23.

of course, the action line of the applied force **P** also passes through point G. Thus, the member of Figure 5.23 is improperly constrained. Also, consider the geometric arrangement of the three links shown in Figure 5.24. In this case, all of the reactive forces are parallel. Here again, even though the three reactive forces represent the minimum number needed for equilibrium, the geometric arrangement of the supports is such that, under a general force **P**, the member can have a small initial translation in a direction perpendicular to the links unless, of course, the action line of the force **P** is also parallel to the reactive forces (i.e., the y axis). Thus, the member of Figure 5.24 is improperly constrained. Note that, for special orientations of the load **P**, both of the force systems shown in Figures 5.23 and 5.24 may be statically indeterminate. In Figure 5.23, with the load **P** passing through point G, the equilibrium equation $\sum M_G = 0$ does not lead to any useful information and only two equilibrium conditions, $\sum F_x = 0$ and $\sum F_y = 0$, are available to solve for three unknown quantities. Also, in Figure 5.24 with the load **P** parallel to the y axis, the equilibrium equation $\sum F_x = 0$ does not provide any useful information, and only two conditions of equilibrium, $\sum F_y = 0$ and $\sum M_z = 0$ are available to solve for three unknown forces.

Cases of improper constraints can also exist in three-dimensional situations if the geometric configuration of the supports is such that the induced reactive forces either intersect points on a common axis or

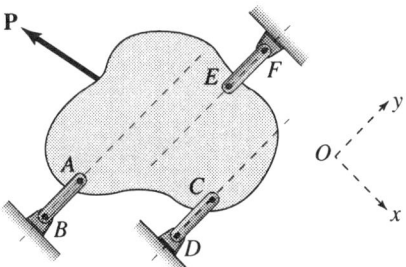

FIGURE 5.24.

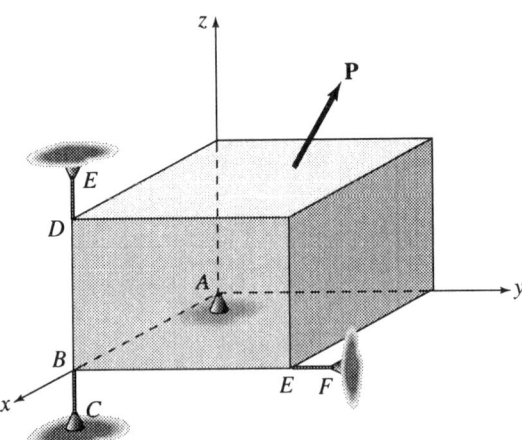

FIGURE 5.25.

are parallel to only one or two of the three coordinate axes. Consider, for example, the three dimensional member shown in Figure 5.25. This member is supported by a ball and socket at A and by short links BC, DE, and EF. Because the three short links are assumed hinged at their ends and serve as two-force members, the reactive forces in them will be along their respective axes and would intersect at point B. Furthermore, the ball-and-socket support is capable of developing reactive components in the x, y, and z directions which, obviously, are concurrent at point A. Therefore, all of the six reactive force components intersect the common axis AB. Even though the number of reactive force components is the minimum needed for equilibrium (six in this case), the member is improperly constrained because it can have a small initial rotation about the axis AB under the action of a general force **P**. The only way that this small rotation may be avoided is for the force **P** to intersect the axis AB, also. Now, consider the member shown in Figure 5.26. This member is supported by the six short links AB, CD, EF, GH, IJ, and KL which serve as two-force members. Note that the three links GH, IJ, and KL are parallel to the x axis and that AB, CD, and EF are parallel to the y axis. Thus, there are, no reactive forces parallel to the z axis, and the member can have a small initial translation in the z direction under the action of a general force **P**. The only way that this small translation can be prevented is for the force **P** to have no component in the z direction. Therefore, despite the fact that the number of reactive force components is six, the minimum number required for equilibrium, the member is improperly constrained. Also, note that, for special orientations of the load **P**, both of the force systems in Figures 5.25 and 5.26 may be statically indeterminate. In Figure 5.25, with the load **P** passing through line AB, the equilibrium condition $\sum M_x = 0$ does not yield any useful information,

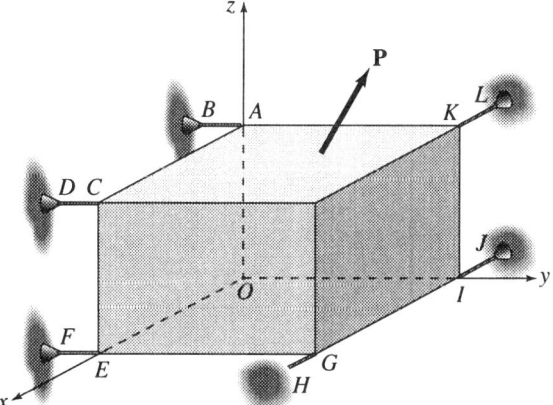

FIGURE 5.26.

and there are only five equilibrium conditions ($\sum F_x = 0$, $\sum F_y = 0$, $\sum F_z = 0$, $\sum M_y = 0$, and $\sum M_z = 0$) available to solve for six unknown quantities. In Figure 5.26 with the load **P** entirely in the x-y plane, the equilibrium equation $\sum F_z = 0$ does not provide any useful information, and there are only five equilibrium equations ($\sum F_x = 0$, $\sum F_y = 0$, $\sum M_x = 0$, $\sum M_y = 0$, and $\sum M_z = 0$) available to solve for six unknown quantities.

Under certain conditions, the number of available equations of equilibrium exceeds the number of reactive force components. This is the case when a member does not have a sufficient number of constraining

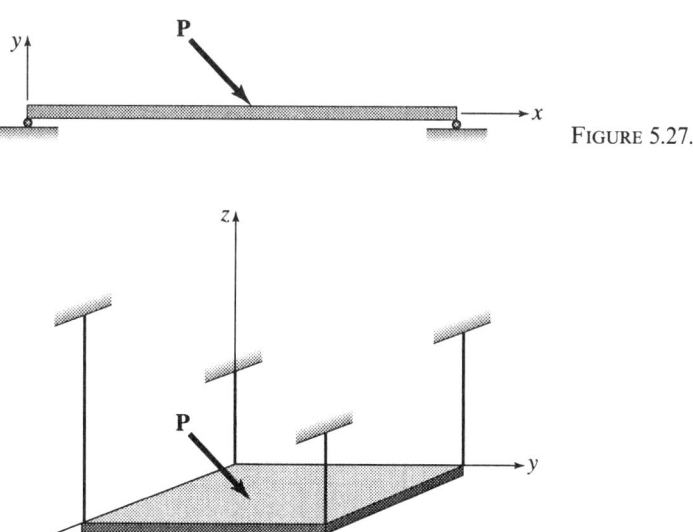

FIGURE 5.27.

FIGURE 5.28.

supports to ensure its complete equilibrium. Such a member is said to be *partially constrained*. An example of a two-dimensional partially constrained beam is shown in Figure 5.27. Under a general load **P**, the support system will be unable to keep the beam from translating in the *x* direction. An example of a three-dimensional partially constrained plate is shown in Figure 5.28. When subjected to a general load **P**, the support system (the four cables in this case) are not able to prevent the plate from translating in either the *x* or *y* direction or from rotating about the *z* axis. Such systems are *unstable* under general loading conditions.

Problems

5.83 to 5.95 Classify the force system shown in Figures P5.83 to P5.95 as to static determinacy and type of constraints. In each case, indicate the degree of redundancy if the system is statically indeterminate, and state whether the system is improperly or partially constrained. Assume weightless members in all cases.

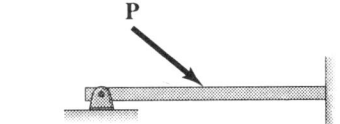

FIGURE P5.85.

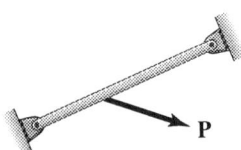

FIGURE P5.83.

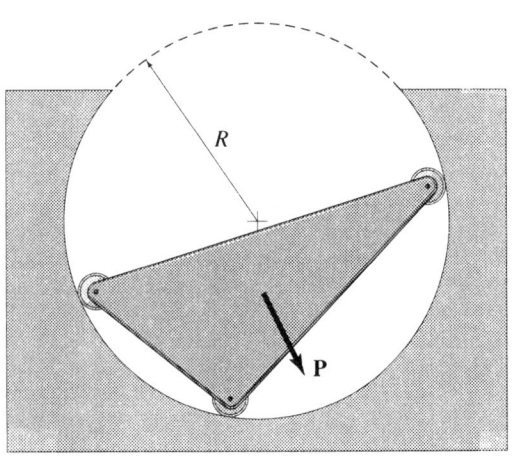

FIGURE P5.86.

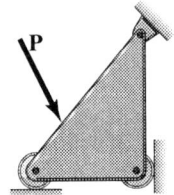

FIGURE P5.84.

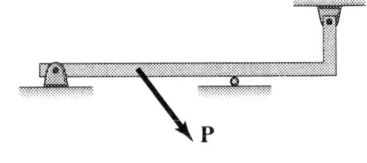

FIGURE P5.87.

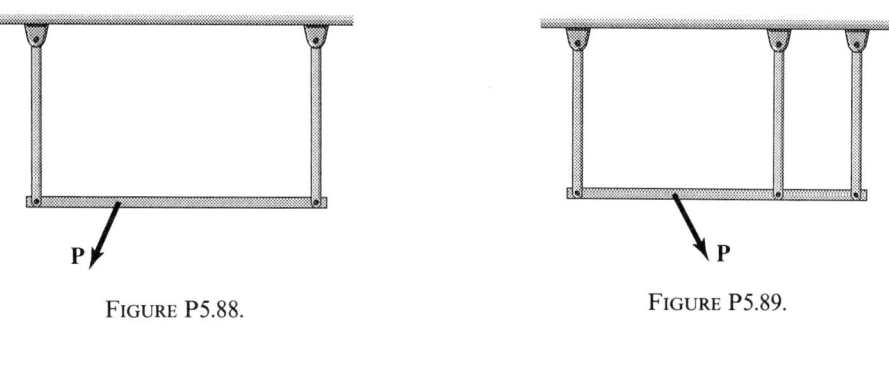

FIGURE P5.88.

FIGURE P5.89.

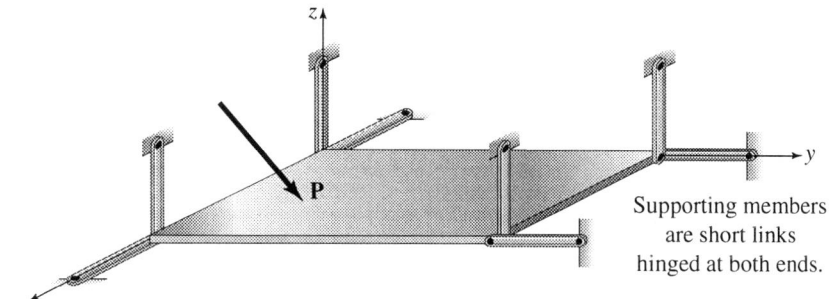

Supporting members
are short links
hinged at both ends.

FIGURE P5.90.

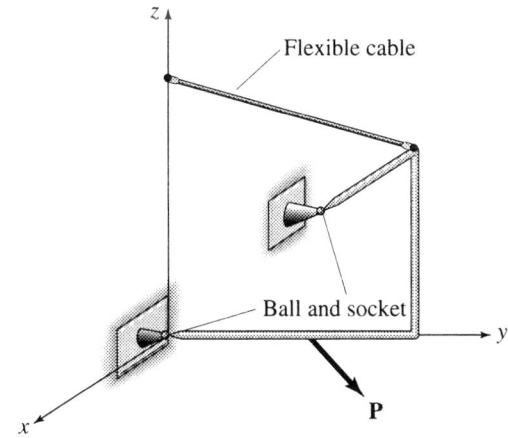

Flexible cable

Ball and socket

FIGURE P5.91.

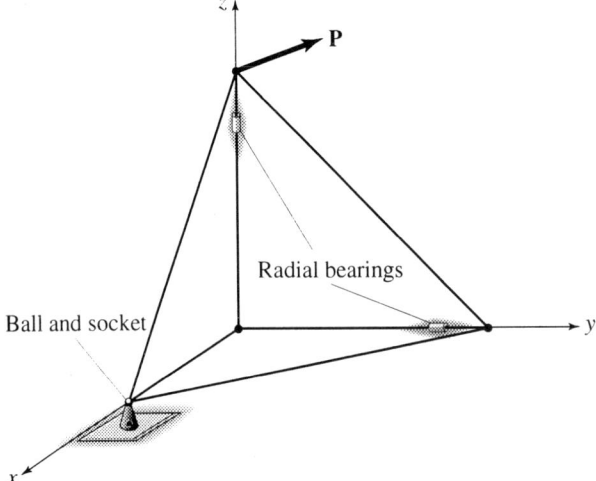

FIGURE P5.92.

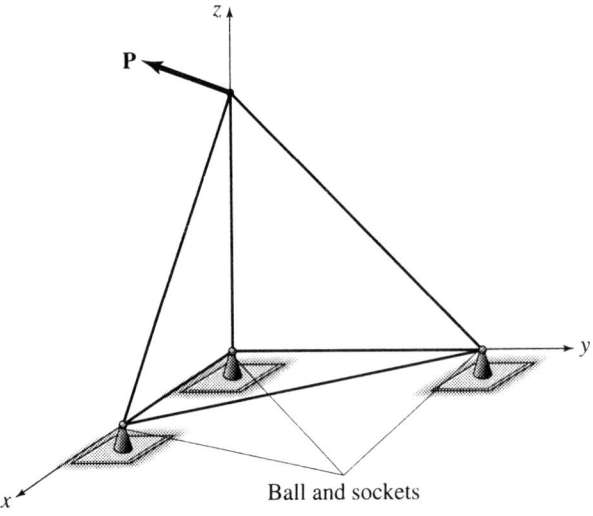

FIGURE P5.93.

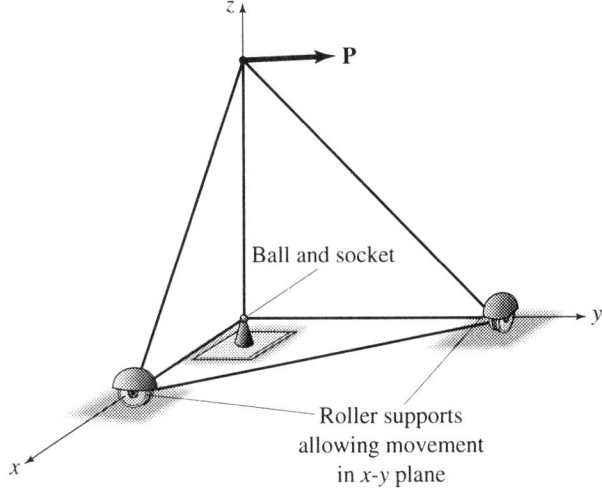

FIGURE P5.94.

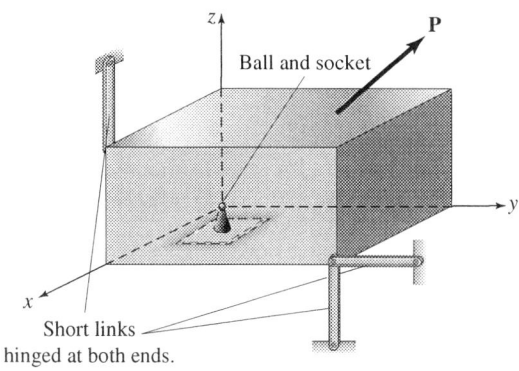

FIGURE P5.95.

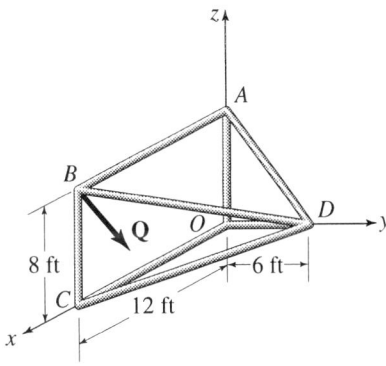

FIGURE P5.96.

Review Problems

5.96 The frame shown in Figure P5.96 is sub-
jected to a force $\mathbf{Q} = (-50\mathbf{i} - 70\mathbf{j} - 100\mathbf{k})$ lb. Find the perpendicular distance from point D to the line of action
of the force $\mathbf{Q}$.

5.97 The moment of a force $\mathbf{F}$ about the ori-
gin of the coordinate system is $\mathbf{M_O} = (-50\mathbf{i} + 30\mathbf{j} + 10\mathbf{k})$ N·m. The magnitude of $\mathbf{F}$ is 10 N and it passes through
point A where the position vector $\mathbf{r}_{A/O} =$

$(3\mathbf{i} + 4\mathbf{j} + 6\mathbf{k})$ m. Find the x, y, and z components of the force $\mathbf{F}$.

5.98 (a) Refer to Problem 5.96 and determine the moment of $\mathbf{Q}$ about member AD of the frame.
(b) Find the perpendicular distance from the line of action of $\mathbf{Q}$ to member AD.

5.99 Let $\mathbf{U} = 3\mathbf{i} + 2\mathbf{j} - 5\mathbf{k}$, $\mathbf{V} = -5\mathbf{i} - 2\mathbf{j} + 3\mathbf{k}$, and $\mathbf{W} = W_x\mathbf{i} + W_y\mathbf{j} + 7\mathbf{k}$. Determine the scalar components W_x and W_y

if the three given vectors are coplanar and if $W = 15$.

5.100 The pipe assembly shown in Figure P5.100 is subjected to the force $\mathbf{P}$ applied at C and directed as shown. Determine the magnitude of the force $\mathbf{P}$ so that the moment it produces about axis OA has a magnitude of 25 kN·m.

5.101 A wall bracket is subjected to a force $P = 60$ lb, as shown in Figure P5.101. Replace this force by an equivalent force-couple system acting at point O.

5.102 Refer to the wall bracket shown in Figure P5.101. Replace the given force by a force-couple system acting at point C.

5.103 The horizontal plate shown in Figure P5.103 is subjected to vertical loads at O, A, B, and C. The load at O is downward with a magnitude of 10 kN, as shown. Find the magnitude and sense of the loads at A, B, and C so that the load system is replaceable by a single resultant force $\mathbf{R} = (-30\mathbf{k})$ kN acting at point D.

5.104 Replace the force system shown in Figure P5.104 by a wrench. Specify the

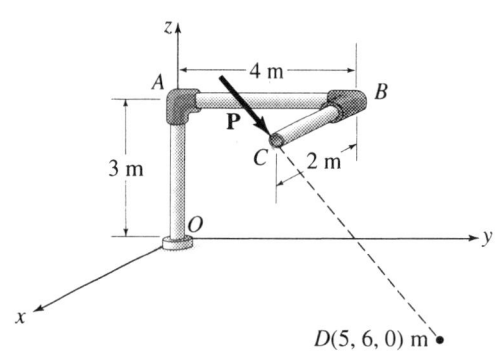

Figure P5.100.

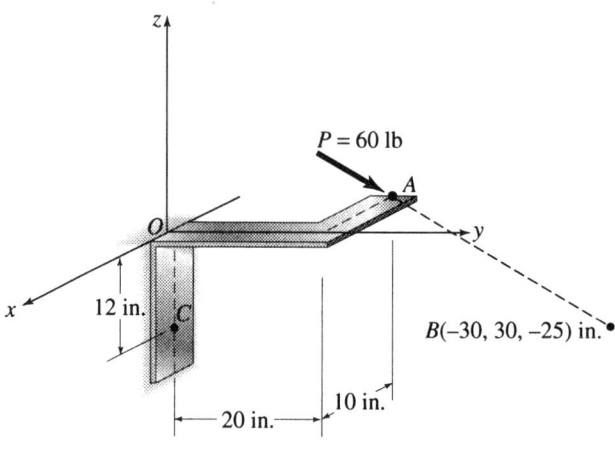

Figure P5.101.

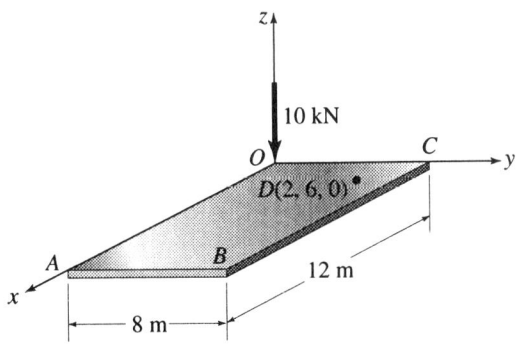

FIGURE P5.103.

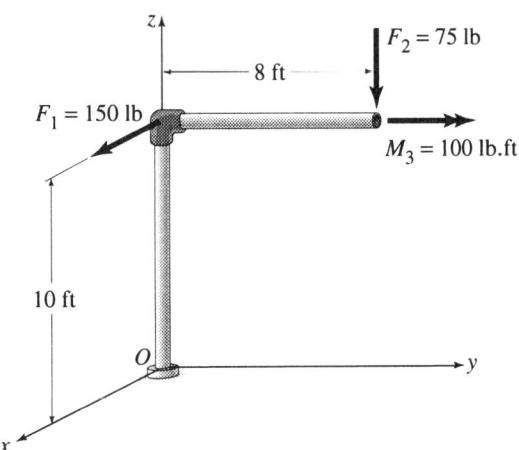

FIGURE P5.104.

force and couple of the wrench. Also, state its direction, and locate the point of intersection of its line of action with the y-z coordinate plane.

5.105 The homogeneous plate shown in Figure P5.105 has a mass $m = 80$ kg and is supported in the position shown by cable AB and by the two hinges at D and C. Assume that the weight of the plate acts through its geometric center, and determine the tension in cable AB.

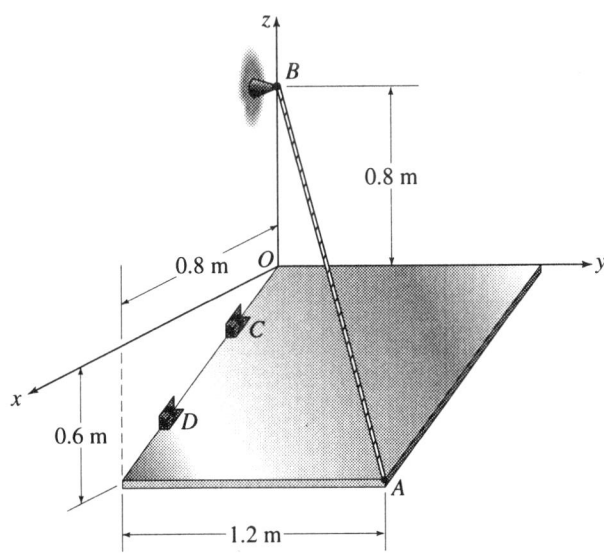

FIGURE P5.105.

5.106 Refer to Problem 5.105. Assume that point B may be placed at any position in the y-z plane along a line defined by the equation $z = 0.8$ m. Determine the exact location of point B in the y-z plane so that the tension in cable AB is the least possible. What is the magnitude of this minimum tension?

5.107 A frame is constructed from heavy pipe and supported as shown in Figure P5.107. The support at A is a ball and socket, and that at B is a bearing that can resist only radial forces but no moments. The frame is also supported by cable CD. Determine the tension in cable CD by writing a single moment equation about axis AB. Let $\mathbf{Q} = (5\mathbf{i} + 8\mathbf{j} + 10\mathbf{k})$ k. Assume that the frame is weightless.

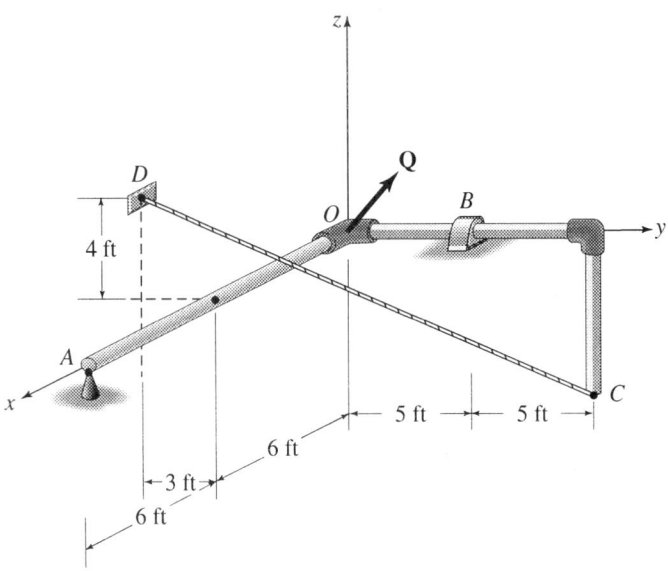

FIGURE P5.107.

5.108 A frame is constructed from heavy pipe and supported as shown in Figure P5.107. The support at A is a ball and socket and that at B is a bearing that can resist only radial forces but no moments. The frame is also supported by cable CD. Determine the reaction components at supports A and B and the tension in cable CD. Assume that the frame is weightless. Let $\mathbf{Q} = (5\mathbf{i} + 8\mathbf{j} + 10\mathbf{k})$ k.

5.109 The system, shown in Figure P5.109, is in equilibrium under the action of the given belt tensions and the reactive forces in the two bearings at A and B. The bearing at A can resist only radial forces, and that at B can resist radial forces as well as axial thrust. Neither bearing is designed to resist moments. Find the unknown belt tension T and the reaction components in the two bearings.

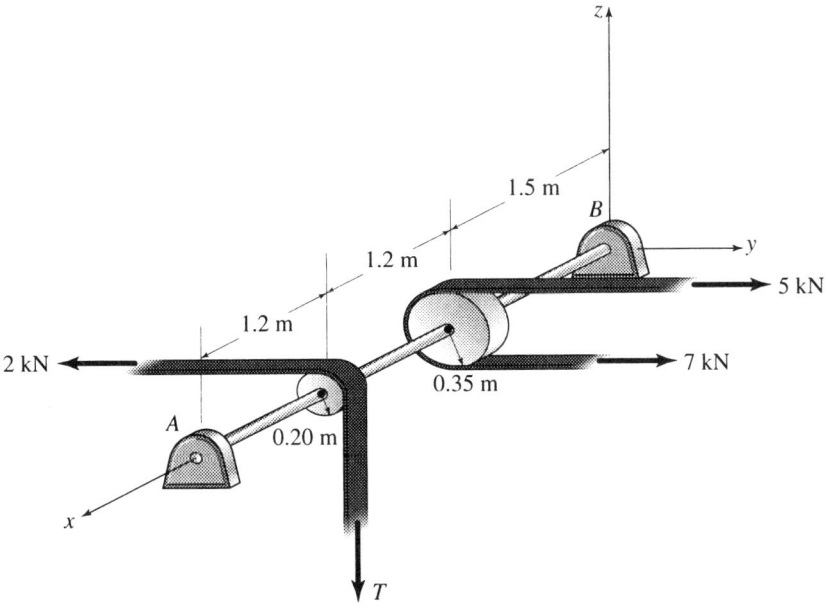

FIGURE P5.109.

6

Truss Analysis

This photograph shows various types of trusses needed for the launch of Apollo 11.

Visionary engineers have always led the way for the advancement and betterment of humankind. Using only existing technology and materials, humankind could establish permanent habitats in space before the end of the twentieth century. Such colonies would help relieve the problems of overcrowding, pollution, and energy shortages here on the Earth.

Space colonists of the mid-twenty-first century may find themselves surrounded by the immensity and convenience of a paired system of rotating pressure vessels and trusses 15–20 miles in length. Depending on the size of each colony, most colonies will be prefabricated from low-cost lunar materials using energy-saving, zero-gravity building techniques.

Currently, trusses of all shapes, sizes, and types play an enormously important role in buildings, bridges, machines, airplanes, and other types of structures. This chapter will provide the student with the basic knowledge of the different types of trusses and explores alternative methods of analysis.

6.1
Analysis of
Simple
Trusses

Forces applied to structures are termed *loads and* they are applied by either man or nature. In an apartment building, the loads from furniture and occupancy of humans are applied by man whereas loads from snow accumulation would be applied by nature through the building's roof. In general, structures provide pathways to the earth for loadings. In Figure 6.1, a mill building framework and a highway bridge provide typical examples of structures. Load transmission pathways are shown for wind applied to the roof and walls of the building and for vehicle loads applied to the highway bridge slab. Key structural parts of both the building and the bridge are their trusses. Some typical examples of

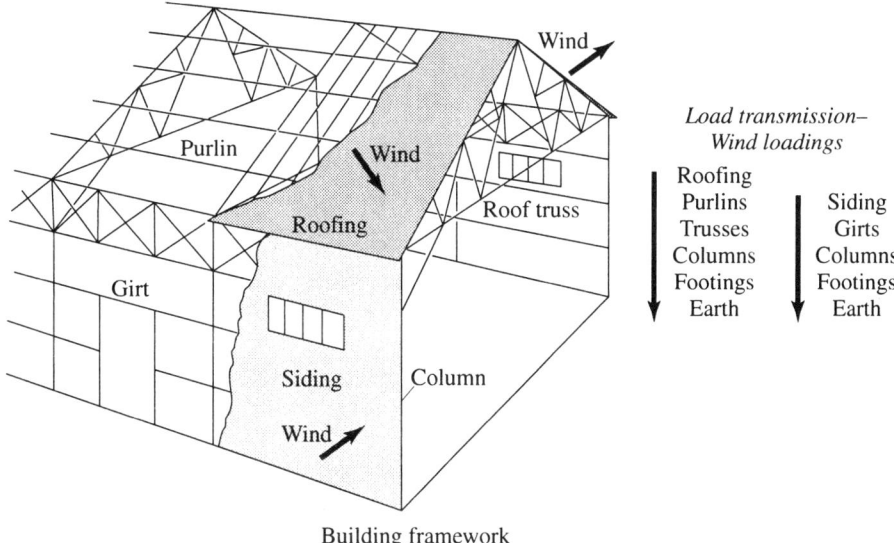

Building framework

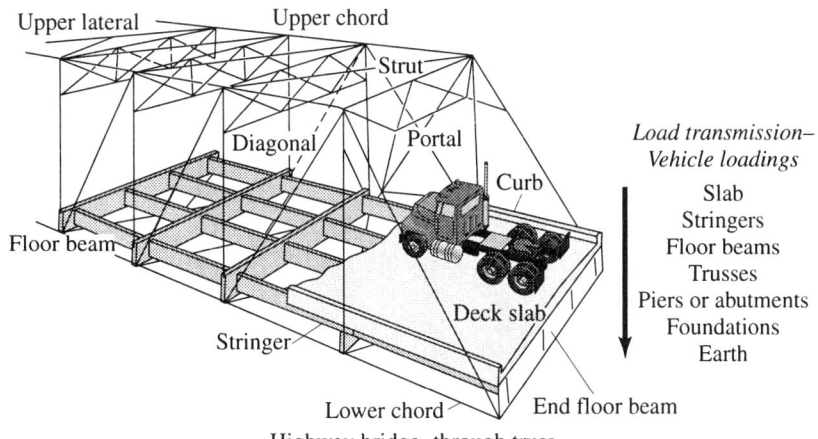

FIGURE 6.1.   Highway bridge–through truss

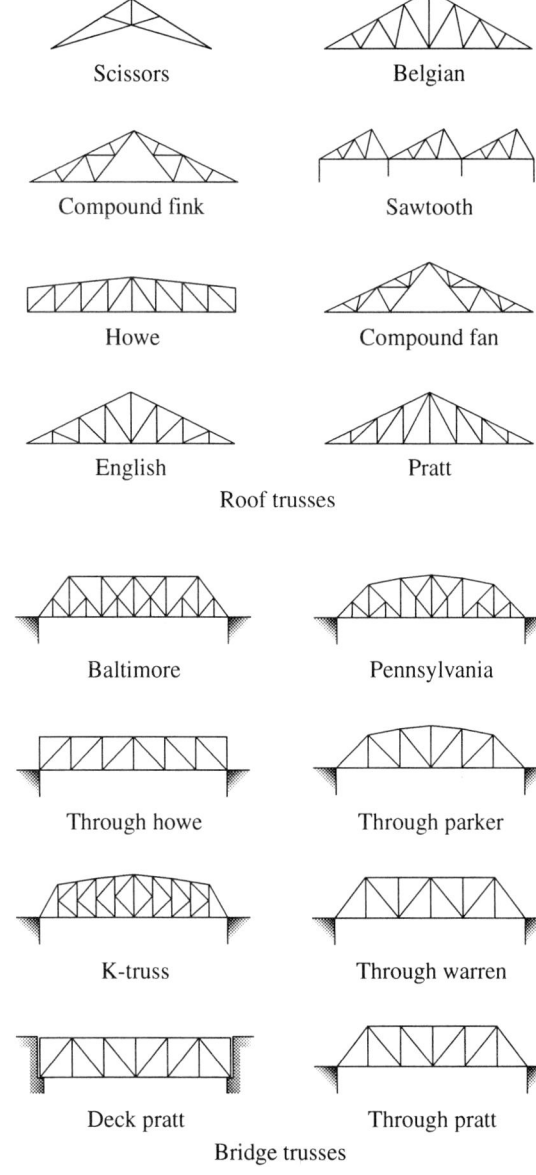

Scissors	Belgian
Compound fink	Sawtooth
Howe	Compound fan
English	Pratt

Roof trusses

Baltimore	Pennsylvania
Through howe	Through parker
K-truss	Through warren
Deck pratt	Through pratt

FIGURE 6.2.   Bridge trusses

roof and bridge trusses are shown in Figure 6.2. These represent only a small number of truss types employed by creative engineers. Only the human imagination limits the use of trusses of widely varying shapes and sizes. It is unimportant to memorize the names of these trusses. They have names because, historically, trusses were once patentable. All patents have expired, and trusses are no longer subject to patent.

Once the overall geometry and loadings are known, the equations of equilibrium enable us to calculate the external reactions and the internal forces. The foregoing chapters have dealt mainly with single rigid bodies in equilibrium. In this chapter we consider systems of connected rigid bodies and utilize the following basic concept for determining internal and external forces acting on the component rigid bodies of the systems. *If a system of connected rigid bodies is in equilibrium, then, any part of the system, which we care to consider, will also be in equilibrium.*

The engineering theory for the analysis of trusses is based upon simplifying assumptions which are used in practice for analyzing and designing almost all trusses.

Assumption 1

Members of a truss are connected together at the joints with frictionless pins. Reality: Pin-connected eyebar trusses, such as the one shown in Figure 6.3, are primarily of historical interest. Of course, it never was possible to fit pins into holes of eyebars and have no friction forces acting, when the members rotated with respect to the pins, as loads built up on the trusses. Modern truss members are usually joined together by welding, as shown in Figure 6.4, or bolted together, as shown in Figure 6.5. Connection material extends over a region around a joint rather than being localized in a single pin centered at a joint.

Assumption 2

Members of trusses are straight two-force members. Such a member is shown in Figure 6.6(a). The member is assumed in tension and is subjected to equal and opposite force resultants T at its two ends. End pins exert these forces on the member, and, in turn, the member exerts equal and opposite forces T on the end pins. The action lines of these tensile forces T are directed along the line joining the pin centers which

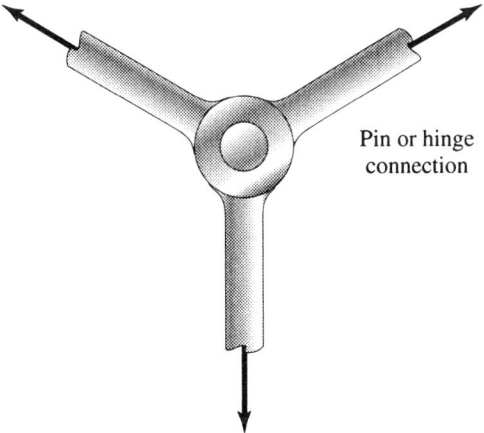

Pin or hinge connection

FIGURE 6.3.

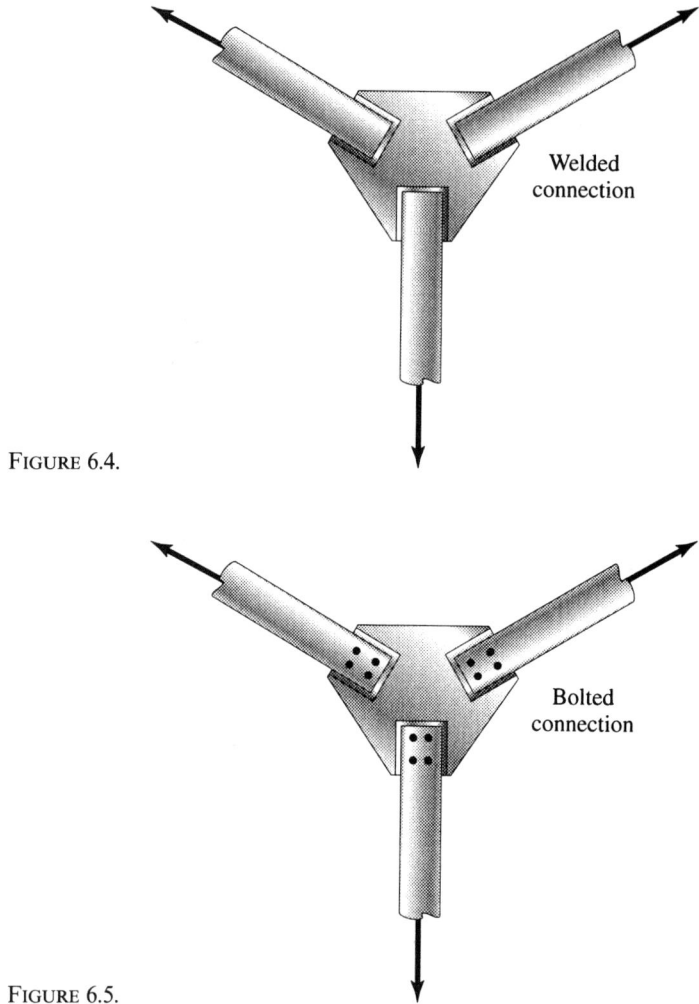

Welded
connection

FIGURE 6.4.

Bolted
connection

FIGURE 6.5.

is also the center line of the straight member. Member forces are actually distributed over member cross-sections as shown in Figure 6.6(b). In our analyses, we will be interested in determining the force resultants of the force distributions in truss members and whether they are tensile (T) or compressive (C) forces. A tensile force tends to lengthen a member and a compressive force tends to shorten it. Reality: Truss members are fabricated as straight as possible. However, the weights of truss members are distributed throughout the volume of the member, and because these forces act in addition to the end forces applied by the pins, the truss members are not really two-force members. Because member weights are usually relatively small compared to other truss

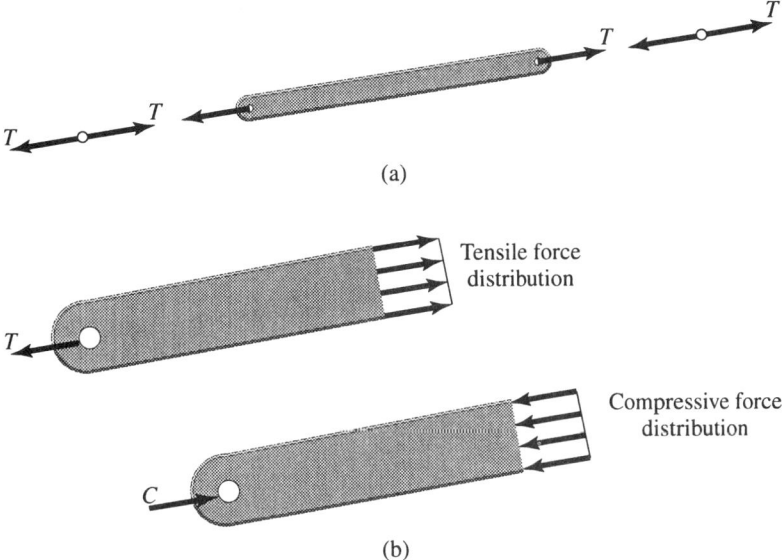

(a)

(b)

FIGURE 6.6.

loadings, we are able to allow for member weights by applying one half a member's weight at each of its two end joints. If we accept this replacement of distributed member weights by concentrated loads at truss joints, then, this straight two-force member assumption is reasonable.

Assumption 3

Loads and reactions are applied only at the joints. Reality: Members which frame into trusses deliver loads to the trusses but the necessary connection material cannot be concentrated at one joint. In Figure 6.1, the floor beams transmit loads to the lower chords of the highway bridge trusses. These loadings are distributed throughout the connection material and fasteners at the floor beam ends. End reactions of these highway bridge trusses are transmitted to piers and abutments, and, once again, the necessary hardware cannot be concentrated at one point. In effect, this assumption means that we are replacing a distributed system of forces at each joint by a single resultant force acting at that joint, idealized as a point.

Assumption 4

Planar trusses are assumed to be rigid bodies. Reality: The concept of a perfectly rigid body is an idealization. All actual trusses deform when subjected to loadings. However, these deformations are usually very small for trusses fabricated of materials used for structures. We formulate the equations of equilibrium based upon the undeformed truss, and this is almost always an acceptable procedure.

We have stated and critiqued the engineering theory of trusses, and the student should realize that more elaborate theories have and could be developed by relaxing one or more of the basic assumptions. All engineering analyses are based upon similar idealizations because the physical world can only be represented approximately by mathematical models. In an age in which computer solutions are widely employed, we should always be aware of or raise questions about the nature of the assumptions made and limitations inherent in any given computer program. No matter how sophisticated the program, it is based upon assumptions and, therefore, has limitations. An engineer who blindly employs such programs is very ill-advised because such usage of computer solutions may lead to the loss of many lives and much property.

Construction of what we term a *simple truss* is shown pictorially in Figure 6.7(a). We begin with three members and three pins in triangu-

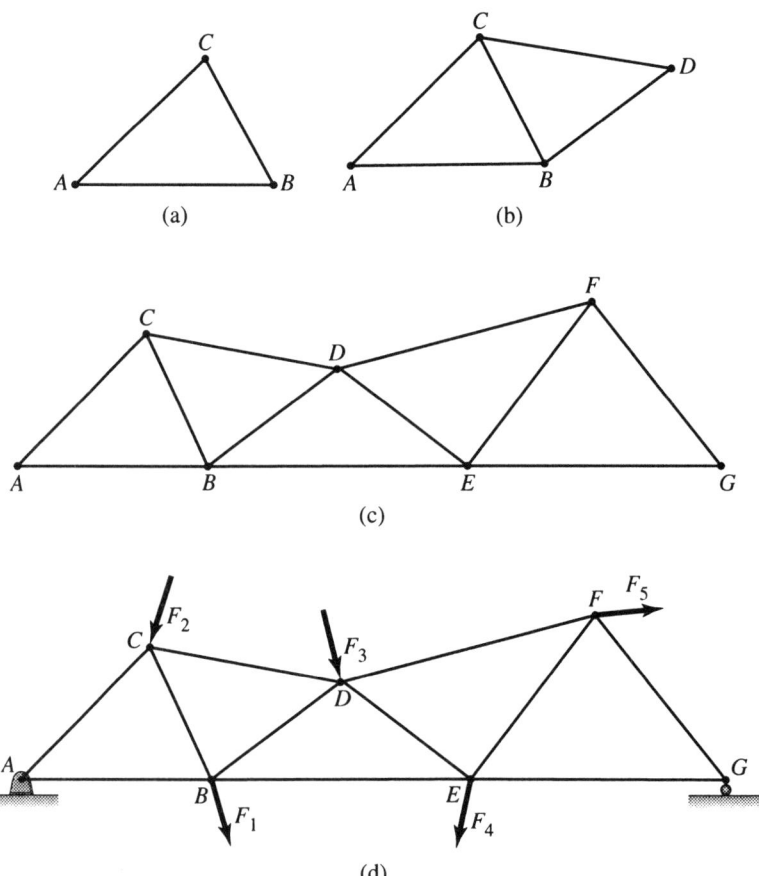

FIGURE 6.7.

lar form which is the rigid elemental body of which simple trusses are constructed. If we were to start with four bars and four pins to form a quadrilateral, it would collapse as a mechanism when subjected to forces at the pins but the triangular shape does not collapse under loading. To extend our initial triangular shape, we add two bars and one pin to form two connected triangular forms as shown in Figure 6.7(b). If we continue to add two bars and one pin to form additional triangles, we have constructed a simple truss as shown in Figure 6.7(c). If we apply loads only at the joints and provide reactions only at joints, as shown in Figure 6.7(d), we have what is termed an *ideal, simple truss*.

6.2 Member Forces Using the Method of Joints

Because an entire truss under the action of external forces is in equilibrium, *then, any part of the truss, which we care to isolate, is also in equilibrium*. The method of joints uses free-body diagrams of the individual pins of a truss to determine the forces in the truss members. Suppose we wish to find the member forces in all members of the truss shown in Figure 6.8. The first step is to determine the reactions R_{1x}, R_{1y}, and R_{6y} by writing the three equations of equilibrium for the free-body diagram of the entire truss shown in Figure 6.8(a). Regarding P_1, P_2, Q_1, Q_2, p, and h as known quantities,

$$\sum F_x = 0, \quad Q_1 + Q_2 + R_{1x} = 0$$

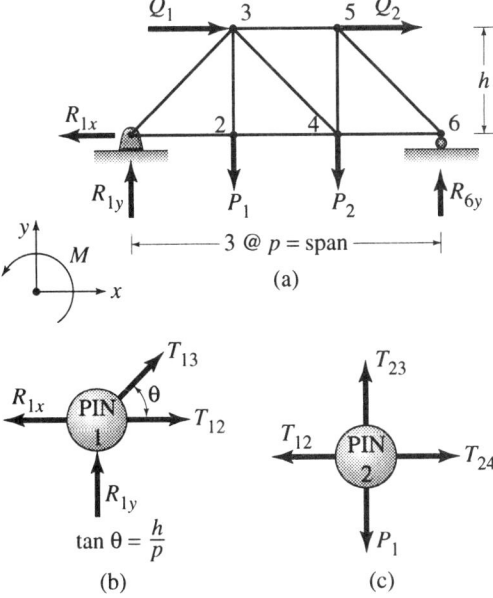

FIGURE 6.8.

$$R_{1x} = -Q_1 - Q_2 = (Q_1 + Q_2)$$

$$\sum M_6 = 0, \quad P_1(2p) + P_2(p) - R_{1y}(3p) - (Q_1 + Q_2)h = 0$$

$$R_{1y} = \frac{1}{3}(2P_1 + P_2) - \frac{h}{3p}(Q_1 + Q_2),$$

and

$$\sum F_y = 0, \quad R_{1y} + R_{6y} - P_1 - P_2 = 0$$

$$R_{6y} = P_1 + P_2 - R_{1y} = R_{1y} = \frac{1}{3}(P_1 + 2P_2) + \frac{h}{3p}(Q_1 + Q_2).$$

Next, we draw a free-body diagram of pin 1 at the left end of the truss, as shown in Figure 6.8(b). The reactions R_{1x} and R_{1y}, which are now known, act on this pin. Two members 1–2 and 1–3 are also attached to this pin. These members exert forces on pin 1 which are designated as T_{12} and T_{13}. Both of these member forces are unknown at this stage. The members are assumed to be in tension and the force vectors, shown on the same side of the pin as the members lie, are assumed to pull away from the center of the pin. Conventionally, we assume that all unknown member forces are tensile, and, if the answer is positive, then, the member force is tensile. If the answer is negative, then, the member force is compressive. The system of forces acting on pin 1 is a coplanar, concurrent system, and, thus, two equations are available to solve for the unknown member forces T_{12} and T_{13}. Thus,

$$\sum F_x = 0, \quad T_{12} + T_{13}\cos\theta - R_{1x} = 0,$$

and

$$\sum F_y = 0, \quad T_{13}\sin\theta + R_{1y} = 0.$$

This basic procedure of drawing free-body diagrams of truss pins and writing the equations $\sum F_x = 0$ and $\sum F_y = 0$ to solve for a maximum of two unknown forces at each joint is known as *the method of joints*. For the problem of Figure 6.8, we next construct a free-body diagram of pin 2 as shown in Figure 6.8(c). Again, write the equations of equilibrium for the pin, because it must be in equilibrium if the entire truss is in equilibrium.

$$\sum F_x = 0, \quad T_{24} - T_{12} = 0$$

$$\sum F_y = 0, \quad T_{23} - P_1 = 0.$$

Because P_1 is given and T_{12} was found above, we solve for T_{24} and T_{23}. In similar fashion, we proceed across the truss considering the joints in the following order: 1, 2, 3, 4, 5, and 6. The only criterion to be kept in mind is that we may always solve for a maximum of two unknown forces at a given joint. Considering the pins in the order 1 through 6 is not the only feasible order for solving the problem. In fact, all forces are known after pin 5 has been considered, and, hence, the equations for pin 6 need not be written. These equations would enable us to check the forces in members 5–6 and 4–6 because R_{6y} was determined earlier.

We need not always solve for the reactions acting on a truss as the first step in a method of joints analysis. Example 6.2 dealing with a cantilever truss illustrates a case where we are able to determine all member forces without first finding the reactions.

The following examples illustrate using the method of joints in analyzing trusses.

■ **Example 6.1**

Refer to the truss free-body diagram shown in Figure E6.1(a) and find the reactions at joints 1 and 6. Use the method of joints to determine the forces in all members of the truss.

Solution

Using the free-body diagram of the truss along with the coordinate system given in Figure E6.1(a),

$$\sum F_x = 0, \quad 10 + R_{1x} = 0,$$

$$R_{1x} = -10.00 \text{ k} = 10.00 \text{ k} \qquad \text{ANS.}$$

$$\sum M_1 = 0, \quad R_{6y}(80) - 50(40) - 10(30) = 0,$$

$$R_{6y} = 28.75 \approx 28.8 \text{ k}, \qquad \text{ANS.}$$

and

$$\sum F_y = 0, \quad R_{1y} + 28.75 - 50 = 0,$$

$$R_{1y} = 21.25 \approx 21.3 \text{ k}. \qquad \text{ANS.}$$

Consider free-body diagrams of the truss pins in the order 1, 2, 3, 4, 5, 6 to solve for all bar forces in the members of the truss. Refer to Figure E6.1(b) showing a free-body diagram of the pin at joint number 1.

$$\sum F_x = 0, \quad T_{14} - 10.00 = 0,$$

$$T_{14} = 10.00 \text{ k} \quad \text{(T)}. \qquad \text{ANS.}$$

The symbol (T) indicates that the force is tensile, because the member is assumed to be in tension and the answer is positive.

$$\sum F_y = 0, \quad 21.25 + T_{12} = 0,$$

$$T_{12} = -21.25 \approx 21.3 \text{ k} \quad \text{(C)}. \qquad \text{ANS.}$$

The symbol (C) indicates that the force is compressive, because the member is assumed to be in tension and the answer is negative.

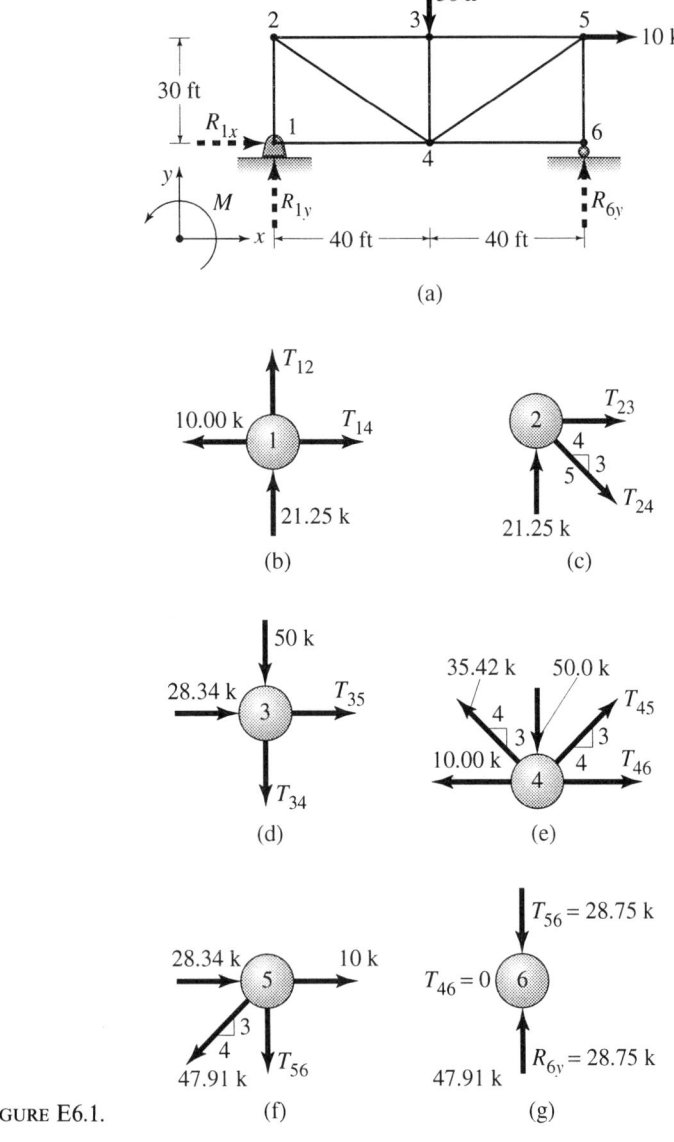

FIGURE E6.1.

Refer to the free-body diagram of pin 2 shown in Figure E6.1(c). Note that the force in member 1–2 is known from the foregoing calculations and is shown in compression with a value of 21.25 k. Thus,

$$\sum F_y = 0, \quad 21.25 - \left(\frac{3}{5}\right) T_{24} = 0,$$

$$T_{24} = 35.42 \approx 35.4 \text{ k} \quad (\text{T}), \qquad\qquad \text{ANS.}$$

and

$$\sum F_x = 0, \quad T_{23} + \left(\frac{4}{5}\right)(35.42) = 0,$$

$$T_{23} = -28.34 \approx 28.3 \text{ k} \quad \text{(C)}. \qquad \text{ANS.}$$

Refer to the free-body diagram of pin 3 shown in Figure E6.1(d). Note that the force in member 2–3 is shown as 28.34 k in compression and that the externally applied force of 50 k downward is shown acting on the pin. Thus,

$$\sum F_x = 0, \quad 28.34 + T_{35} = 0,$$

$$T_{35} = -28.34 \approx 28.3 \text{ k} \quad \text{(C)}, \qquad \text{ANS.}$$

and

$$\sum F_y = 0, \quad -50 - T_{34} = 0,$$

$$T_{34} = -50 = -50.0 \text{ k} \quad \text{(C)}. \qquad \text{ANS.}$$

Refer to the free-body diagram of pin 4 shown in Figure E6.1(e). Note that the force in member 1–4 is shown as 10.00 k in tension, the force in 2–4 is shown as 35.42 k in tension, and the force in 3–4 is shown as 50.0 k in compression. Hence,

$$\sum F_y = 0, \quad \left(\frac{3}{5}\right)(35.42) - 50 + \left(\frac{3}{5}\right)T_{45} = 0,$$

$$T_{45} = 47.91 \approx 47.9 \text{ k} \quad \text{(T)}, \qquad \text{ANS.}$$

and

$$\sum F_x = 0, \quad T_{46} + \left(\frac{4}{5}\right)(47.91) - \left(\frac{4}{5}\right)(35.42) - 10.00 = 0,$$

$$T_{46} = 0.008 \approx 0. \qquad \text{ANS.}$$

For this particular loading there is no force in member 4–6, but we should keep in mind that structures are designed for many different loadings. For another loading, the force in this member may well take on a nonzero value.

Refer to the free-body diagram of pin 5 shown in Figure E6.1(f). Note that the force in member 3–5 is shown as 28.34 k in compression and the force in 4–5 is shown as 47.91 k in tension. Thus,

$$\sum F_y = 0, \quad -T_{56} - \left(\frac{3}{5}\right)(47.91) = 0,$$

$$T_{56} = -28.75 \approx 28.8 \text{ k} \quad \text{(C)}. \qquad \text{ANS.}$$

At this stage, we have solved for the bar forces in all members of the truss and could terminate the solution. However, we will write the one remaining equation for pin 5 and use the two equations for pin 6 as a check on the foregoing analysis. Returning to pin 5,

$$\sum F_x = 0, \quad 10 + 28.34 - \frac{4}{5}(47.91) = 0.012 \approx 0. \qquad \text{check}$$

Refer to the free-body diagram of pin 6 shown in Figure E6.1(g). Without formally writing the equations of equilibrium, we observe that the pin is in equilibrium under the action of R_{6y} and the forces in members 4–6 and 5–6.

■ **Example 6.2**   A cantilever truss is shown in Figure E6.2(a).

(a) Use the method of joints to determine the forces in all members of this truss in terms of P. Regard P and b as known quantities.

(b) Let $P = 20$ kN and determine numerical values for the forces in all members of this truss.

Solution   (a) Refer to Figure E6.2(b) and write the equations of equilibrium for pin 1 using the given x-y coordinate system.

$$\sum F_x = 0, \quad P + \frac{1}{\sqrt{2}} T_{13} = 0,$$

$$T_{13} = \sqrt{2}P \quad \text{(C)}. \qquad \text{ANS.}$$

$$\sum F_y = 0, \quad -T_{12} - \frac{1}{\sqrt{2}}(-\sqrt{2}P) = 0,$$

$$T_{12} = P \quad \text{(T)}. \qquad \text{ANS.}$$

Next, refer to the free-body diagram of pin 2 shown in Figure E6.2(c) and note that the force in member 1–2 is shown as P in tension. Hence,

$$\sum F_x = 0, \quad T_{23} + 2P = 0,$$

$$T_{23} = -2P = 2P \quad \text{(C)}. \qquad \text{ANS.}$$

$$\sum F_y = 0, \quad P - T_{25} = 0,$$

$$T_{25} = P \quad \text{(T)}. \qquad \text{ANS.}$$

A free-body diagram of pin 3 is shown in Figure E6.2(d). The force in member 1–3 is shown as $\sqrt{2}P$ in compression and the force for 2–3 is shown as $2P$ in compression. Thus,

$$\sum F_x = 0, \quad \frac{1}{\sqrt{2}}(\sqrt{2}P) + 2P - \frac{1}{\sqrt{2}}T_{35} = 0,$$

$$T_{35} = 3\sqrt{2}P \quad \text{(T)}. \qquad \text{ANS.}$$

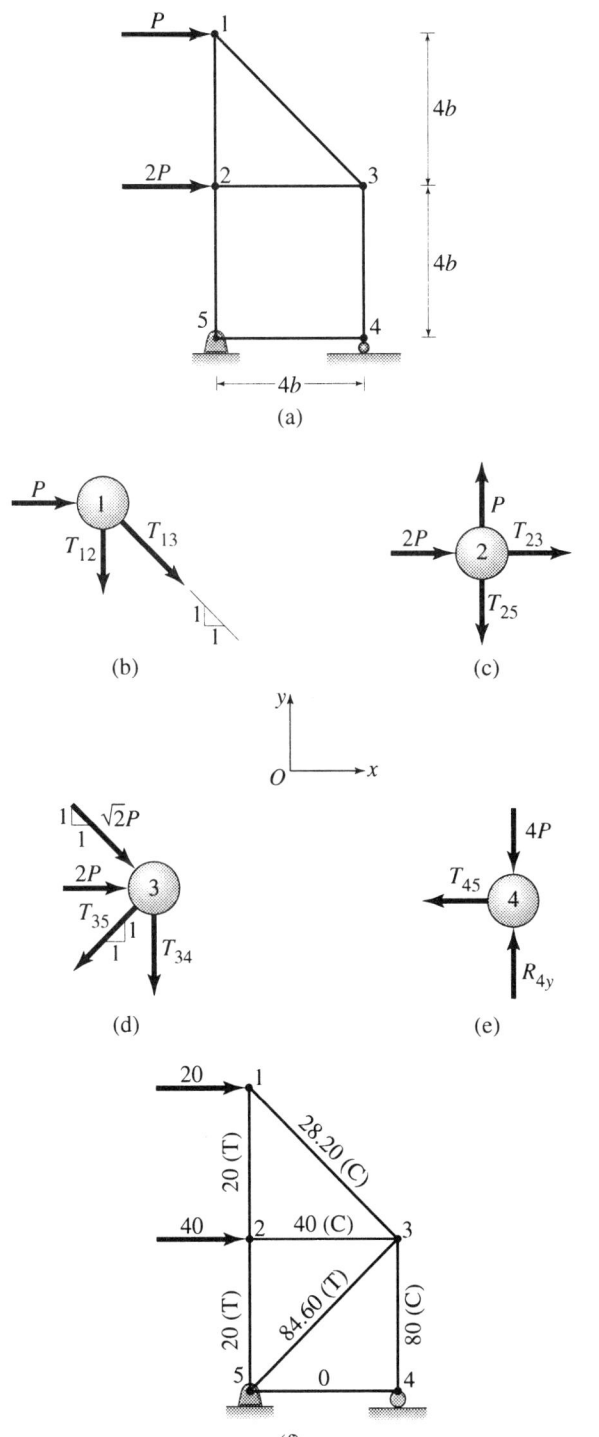

FIGURE E6.2.

$$\sum F_y = 0, \quad -\frac{1}{\sqrt{2}}(\sqrt{2}P) - \frac{1}{\sqrt{2}}T_{35} - T_{34} = 0,$$

$$T_{34} = -4P = 4P \quad \text{(C)}. \qquad \text{ANS.}$$

Refer to the free-body diagram of pin 4 shown in Figure E6.2(e). The force in member 3–4 is shown as $4P$ in compression. Hence,

$$\sum F_x = 0, \quad -T_{45} = 0,$$

$$T_{45} = 0. \qquad \text{ANS.}$$

For the given loading the force in member 4–5 vanishes. This concludes the solution for all bar forces in this cantilever truss. Note that we did not need to determine the reactions acting on the truss nor make use of the free-body diagram of pin 5. It is left as an exercise for the reader to determine the reactions and use the equation $\sum F_y = 0$ for pin 4 and two equations for pin 5 as checks on the foregoing analysis.

(b) For $P = 20$ kN the member forces become those shown in Figure E6.2(f). All member forces are expressed in kN. Students are encouraged to state results for truss problems in a similar fashion.

■ **Example 6.3**

A 24-ton truss is being lifted into position during erection of a structure as shown in Figure E6.3(a). Using the method of joints, determine the member forces in all truss members due to the weight of the truss. Assume that the weights of the members act at the top and bottom chord panel points as shown.

Solution

In general, impact and accelerated motion would be considered, but we confine our analysis to static or constant velocity motions.

By symmetry, the cable tensions T_A and T_B are each equal to one-half the total weight of the truss. Thus,

$$T_A = T_B = \frac{1}{2}(24 \text{ tons})\left(\frac{2 \text{ k}}{1 \text{ ton}}\right) = 24 \text{ k} \uparrow. \qquad \text{ANS.}$$

A free-body diagram of pin 1 is shown in Figure E6.3(b). To begin the solution it was necessary to choose a pin on which only two unknown forces act. The pin equilibrium equations enable us to solve for these unknown forces. Thus, using the coordinate system shown,

$$\sum F_x = 0, \quad T_{12} = 0. \qquad \text{ANS.}$$

A zero result for this member force does not mean that we could omit this member from the truss. It means that for this particular loading member 1–2 carries no force, but structures, in general, are designed to resist a number of loading conditions.

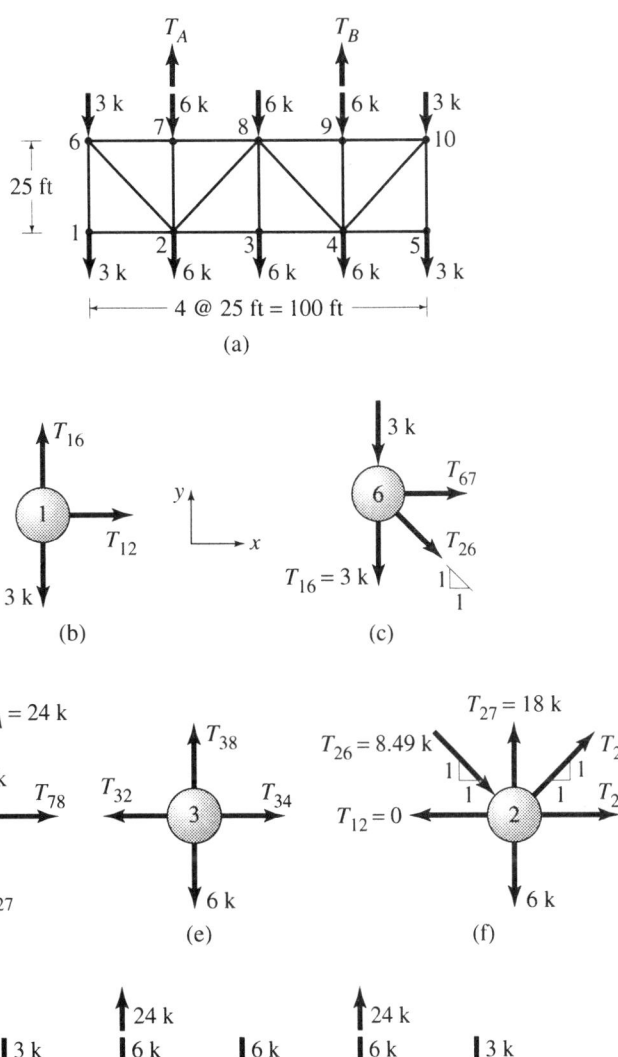

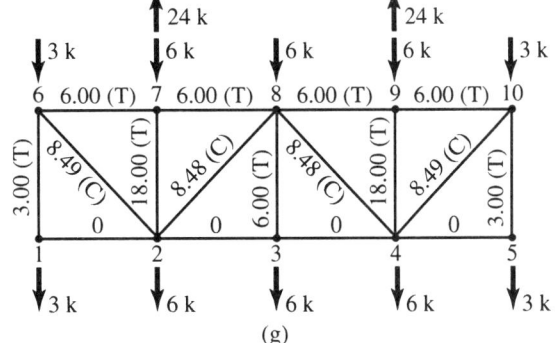

FIGURE E6.3.

$$\sum F_y = 0, \quad T_{16} - 3 = 0,$$

$$T_{16} = 3.00 \text{ k} \quad (\text{T}). \qquad\qquad \text{ANS.}$$

Next we consider free-body diagrams for either joint 2 or joint 6. If we choose joint 2, there will be 4 unknown forces, but, if we choose joint 6, there will be only two unknown forces. Because only two equations of equilibrium are available for each pin, we refer to the free-body diagram of pin 6 in Figure E6.3(c). Note that we show $T_{16} = 3$ k acting in tension. Thus,

$$\sum F_y = 0, \quad 3 + 3 + \frac{1}{\sqrt{2}} T_{26} = 0,$$

$$T_{26} = -8.49 \text{ k} = 8.49 \text{ k} \quad (\text{C}). \qquad\qquad \text{ANS.}$$

$$\sum F_x = 0, \quad T_{67} + \frac{1}{\sqrt{2}}(-8.49) = 0,$$

$$T_{67} = 6.00 \text{ k} \quad (\text{T}). \qquad\qquad \text{ANS.}$$

Next, we must choose between joints 2 and 7. At joint 7, two unknowns remain and at joint 2, three unknowns remain, which leads to the choice of joint 7 with a free-body diagram shown in Figure E6.3(d).

$$\sum F_y = 0, \quad 24 - 6 - T_{27} = 0,$$

$$T_{27} = 18.00 \text{ k} \quad (\text{T}).$$

$$\sum F_x = 0, \quad T_{78} - 6 = 0,$$

$$T_{78} = 6.00 \text{ k} \quad (\text{T}). \qquad\qquad \text{ANS.}$$

Now, consider pin 3 whose free-body diagram is shown in Figure E6.3(e). Thus,

$$\sum F_y = 0, \quad T_{38} - 6 = 0,$$

$$T_{38} = 6.00 \text{ k} \quad (\text{T}). \qquad\qquad \text{ANS.}$$

Because the truss geometry, the cable supports and the loading are all symmetric, the solution can be completed by considering the free-body diagram of pin 2 shown in Figure E6.3(f). Hence,

$$\sum F_y = 0, \quad 18 + \left(\frac{1}{\sqrt{2}}\right) T_{28} - 6 - \left(\frac{1}{\sqrt{2}}\right)(8.49) = 0,$$

$$T_{28} = -8.48 = 8.48 \text{ k} \quad (\text{C}). \qquad\qquad \text{ANS.}$$

$$\sum F_x = 0, \quad T_{23} + \left(\frac{1}{\sqrt{2}}\right)(-8.48) + \left(\frac{1}{\sqrt{2}}\right)(8.49) = 0,$$

$$T_{23} = 0.007 \approx 0. \qquad\qquad \text{ANS.}$$

For this loading all member forces in the truss are shown in Figure E6.3(g). Erection studies like this are always required because these conditions may govern the design of some components of the overall structure.

■

Problems

P6.1–6.12 In each of the following problems, refer to the associated figure and find the forces in all members of the pin-connected trusses by the method of joints. Also, determine the reactions acting on the trusses. Express all answers in terms of P, and show your answers on a sketch of the truss. Be careful to state whether a given member is in tension or compression.

6.1 Figure P6.1. In addition, let $P = 6$ kN, and show bar forces on a truss sketch.

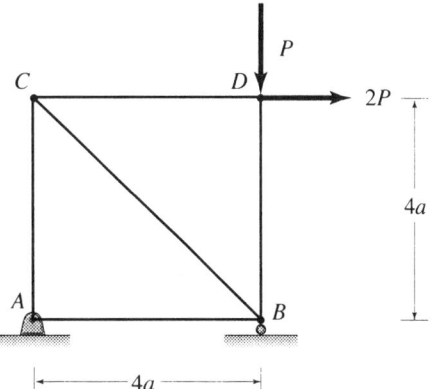

FIGURE P6.2.

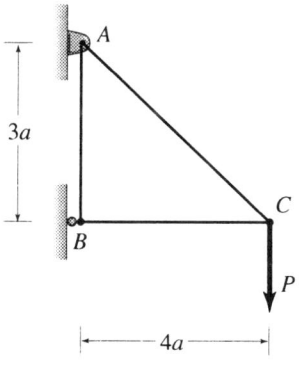

FIGURE P6.1.

6.2 Figure P6.2.
6.3 Figure P6.3. In addition, let $P = 8$ k, and show bar forces on a truss sketch.
6.4 Figure P6.4.
6.5 Figure P6.5.
6.6 Figure P6.6. In addition, let $P = 15$ kN, and show bar forces on a truss sketch.

6.7 Figure P6.7.
6.8 Figure P6.8. In addition, let $P = 10$ k, and show bar forces on a truss sketch.
6.9 Figure P6.9. In addition, let $P = 20$ k, and show bar forces on a truss sketch.
6.10 Figure P6.10. In addition, let $P = 30$ kN, and show bar forces on a truss sketch.
6.11 Figure P6.11.
6.12 Figure P6.12.
6.13 Without finding the reactions acting on the vertical cantilever truss shown in Figure P6.13, determine the forces in members 1–2, 1–3, 2–3, 2–4 and 3–4, in terms of Q. Use the method of joints. If $Q = 20$ kN and $c = 1$ m, state numerical values for these bar forces.
6.14 Refer to Figure P6.14, and determine the bar forces in members 1–2 and 1–3 of

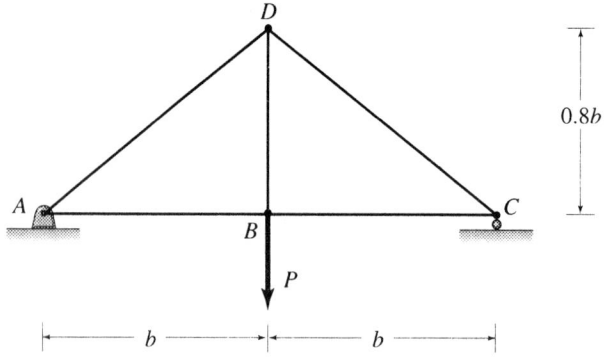

FIGURE P6.3.

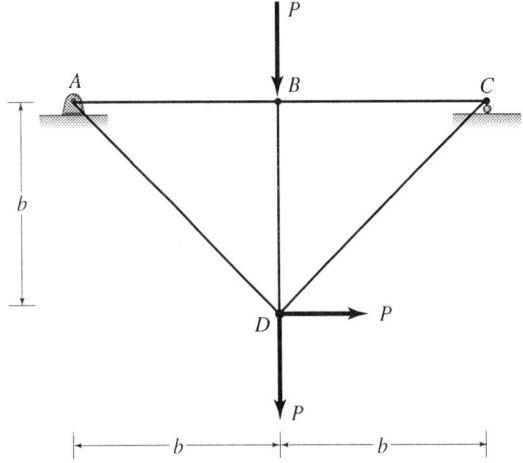

FIGURE P6.4.

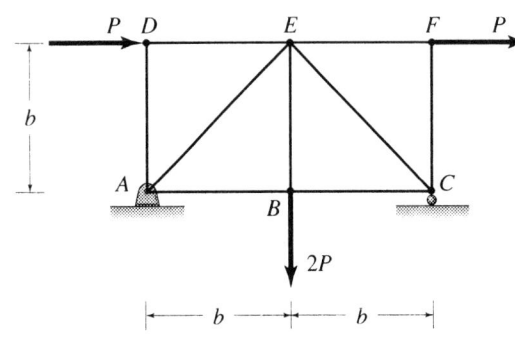

FIGURE P6.5.

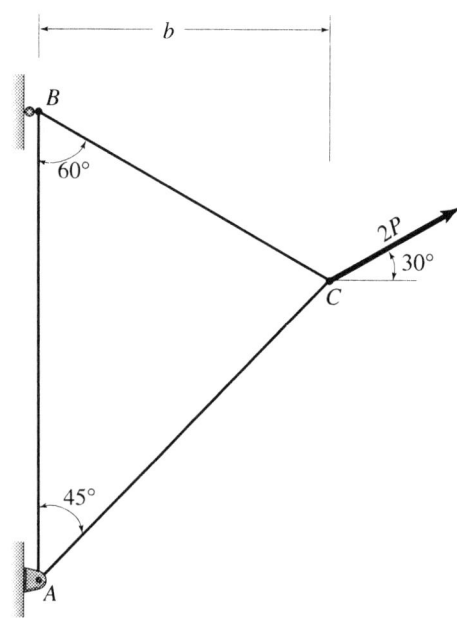

FIGURE P6.6.

this truss as a function of Q and β. Use the method of joints. If $\beta = 30°$ and $Q = 10$ k, state numerical values for these bar forces.

6.15 Determine the bar forces in members 1–2 and 1–3 of the truss shown in Figure

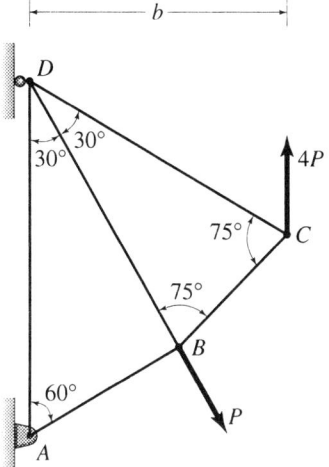

FIGURE P6.7.

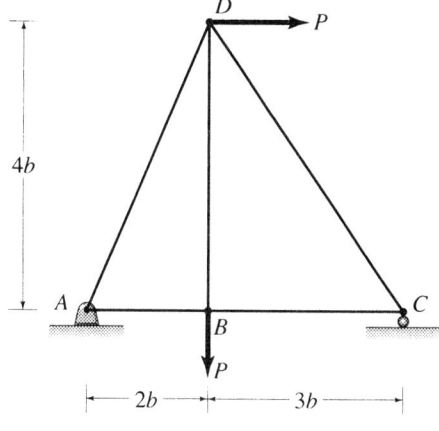

FIGURE P6.10.

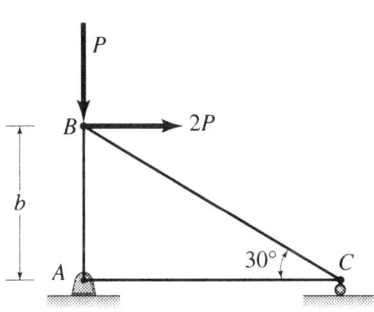

FIGURE P6.8.

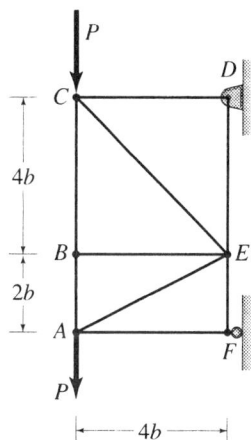

FIGURE P6.11.

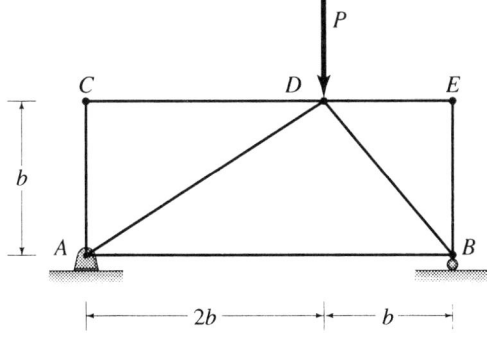

FIGURE P6.9.

P6.15 in terms of Q and β. Do your answers depend upon the length b? Specialize your answers for $\beta = 60°$.

6.16 Use the method of joints to determine the bar forces in members 3–5, 4–5, 3–4, and 2–4 of the truss shown in Figure P6.16. Express your answers in terms of P. Note that you need not determine the reactions at joints 1 and 2. The roller at joint 2 is capable of applying an upward or downward reacting force to this truss. If $P =$

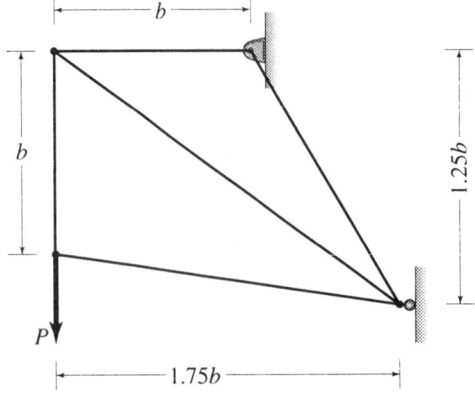

FIGURE P6.12.

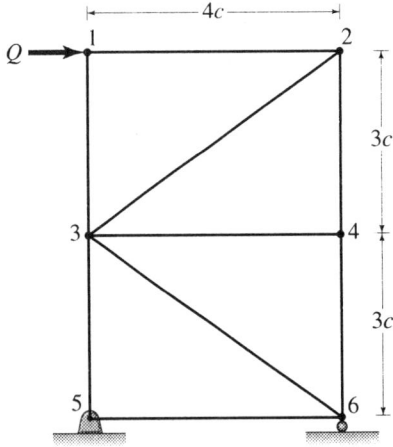

FIGURE P6.13.

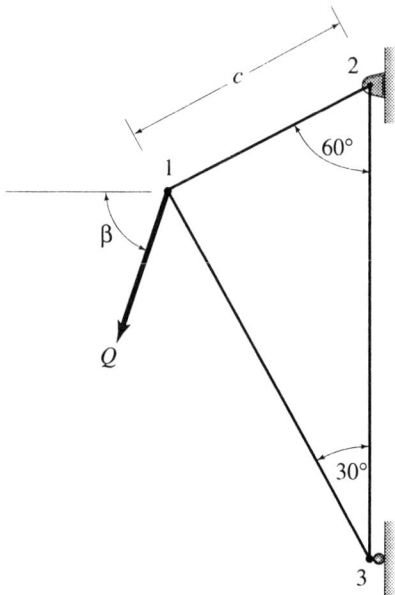

FIGURE P6.14.

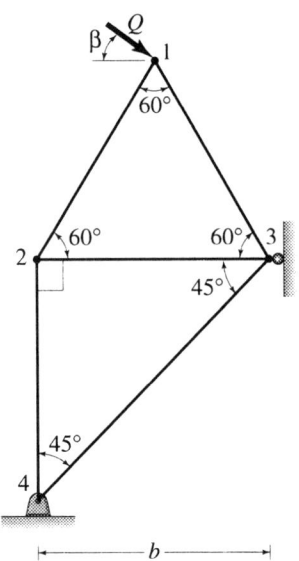

FIGURE P6.15.

10 kN and $b = 2$ m, determine numerical values for these bar forces. Do your answers depend upon the value of b?

6.17 Refer to the through-bridge truss depicted in Figure P6.17, and determine the reactions at joints 1 and 7. Then use the method of joints to find the forces in members 1–2, 1–8, 2–8, and 8–9.

6.18 The bar forces in members 3–4 and 4–5 are equal, by symmetry of the through-bridge truss and loading of Figure P6.17.

Each of these bar forces is 60.92 k in compression. Draw a free-body diagram of the pin at joint 4, and determine the force in bar 4–10 of this truss.

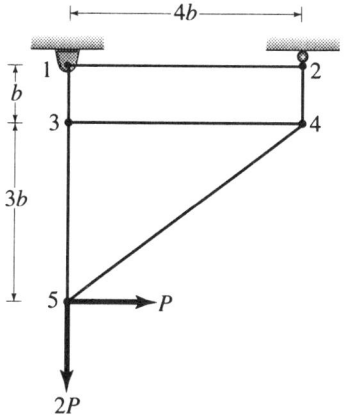

FIGURE P6.16.

6.19 Find the reactions acting at joints 6 and 10 on the deck-bridge truss shown in Figure P6.19. Then, use the method of joints to determine the forces in members 1–6, 2–6, and 6–7.

6.20 By considering free-body diagrams of joints 5 and 9 of the deck-bridge truss shown in Figure P6.19, it is easy to show that bars 4–5 and 4–9 each carry no force under this loading. Given that bar 3–4 carries a compressive force of 160 kN, draw a free-body diagram of pin 4, and determine the bar forces in members 4–8 and 4–10 by the method of joints.

6.21 Refer to the roof truss depicted in Figure P6.21, and find the reactions at joints 1 and 5 in terms of P. Draw a free-body diagram of pin 1, and determine the bar forces in members 1–2 and 1–6 in terms of P by the method of joints.

6.22 Find the reactions at joints 8 and 14

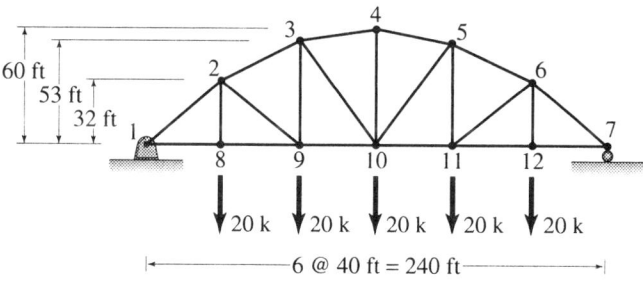

FIGURE P6.17.

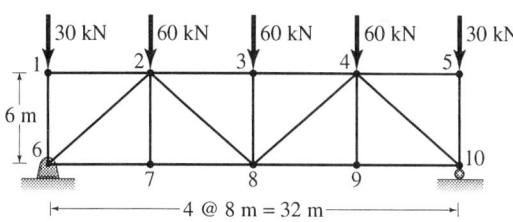

FIGURE P6.19.

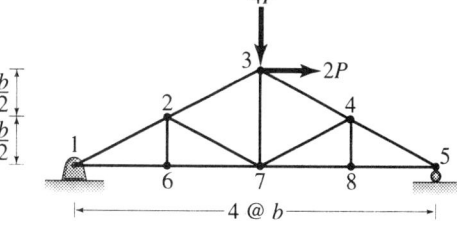

FIGURE P6.21.

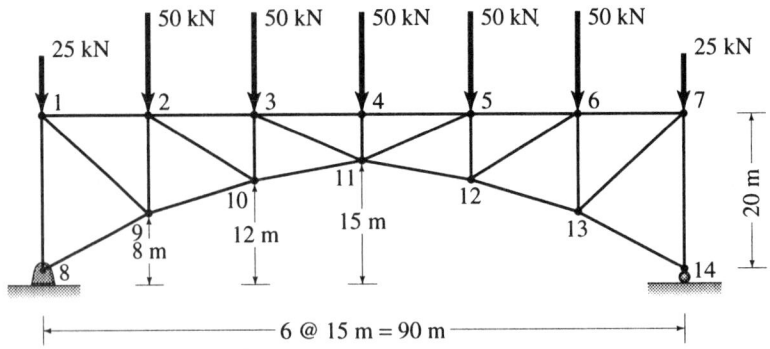

FIGURE P6.22.

acting on the structure shown in Figure P6.22. Then, construct a free-body diagram of pin 14 to find the bar forces in members 7–14 and 13–14. Finally, draw a free-body diagram of pin 7 to determine the bar forces in members 6–7 and 7–13.

6.23 A *Baltimore through-bridge truss* is shown in Figure P6.23. Such trusses are only economical for relatively long spans. Let the lower chord panel point loadings *P* each equal 20 k, and find the reaction at A. Then, by considering joints A, B, and

D in order, determine the bar forces in members AB, AD, BC, BD, CD, and DE.

6.24 The Pratt roof truss shown in Figure P6.24 is loaded with forces $Q = 8$ kN at each upper chord panel point. Determine the left reaction at A. Then, consider joints A and B to determine bar forces in members AB, AC, BC, and BD. If you were to find the forces in all members of this truss, which joint would be next, C or D? Why?

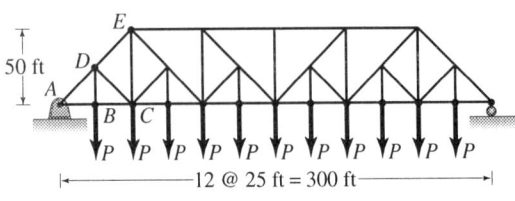

FIGURE P6.23.

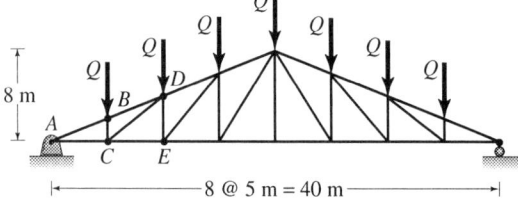

FIGURE P6.24.

6.3
Members
Carrying No
Forces

Whether a truss is hidden from view inside an airplane or is in full view as part of a highway bridge, it would have been designed for a number of loading conditions. Some members carry no forces under a given loading condition, but they will carry nonzero forces for some or many other loading conditions. Analysis of a truss to determine bar forces will be accomplished more rapidly if we are able to identify *zero-force members*. Such knowledge may be useful in preliminary design or in checking computer output for truss analysis. Two propositions form the basis for finding *zero-force members by inspection in* planar trusses: *(1) At an unloaded truss joint, if only two members meet and their axes do not lie along the same straight line, then, both members carry zero force. (2) At an unloaded truss joint, if only three members meet and if two, but not three, of their axes lie along a common straight line, then, the member whose axis does not lie along this common line will carry zero force.*

These statements will be proven by applying the equations of equilibrium by the method of joints. Students are urged to write the equations in each case after visually locating zero-force members.

In Figure 6.9(a), two members meet at a joint where no external load is applied. We arbitrarily orient the x and y axes as shown and write

$$\sum F_x = 0, \quad T_1 \cos \theta = 0.$$

Provided $\cos \theta \neq 0$, the solution is $T_1 = 0$. Also,

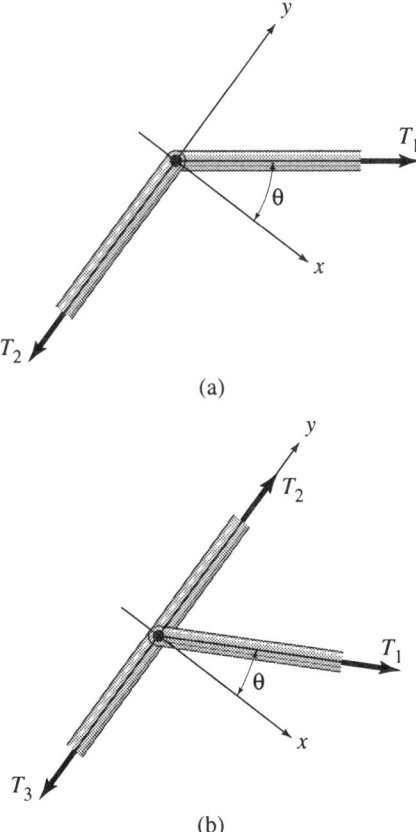

(a)

(b)   FIGURE 6.9.

$$\sum F_y = 0, \quad T_1 \sin \theta - T_2 = 0.$$

Because $T_1 = 0$, it follows that T_2 also equals zero.

Figure 6.9(b) shows three members meeting at a joint where no external load is applied. We arbitrarily orient the x and y axes as shown and write

$$\sum F_x = 0, \quad T_1 \cos \theta = 0.$$

Provided that $\cos \theta \neq 0$, it follows that $T_1 = 0$. Also,

$$\sum F_y = 0, \quad T_2 + T_1 \sin \theta - T_3 = 0.$$

Because $T_1 = 0$, it follows that $T_2 = T_3$.

■ **Example 6.4**

Consider the trusses shown in Figures E6.4(a) and (b). Determine the members which carry no forces under the given loadings.

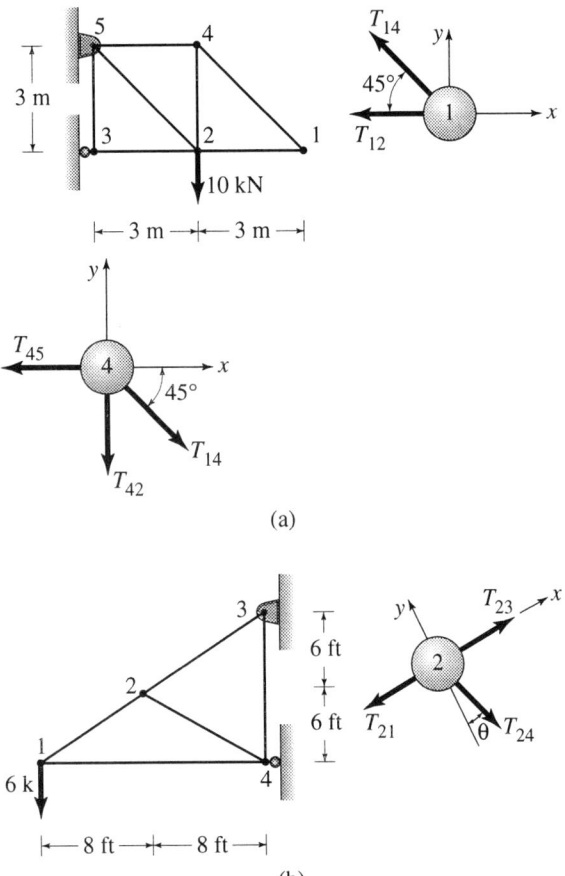

FIGURE E6.4.

Solution

Consider the truss shown in Figure E6.4(a) and note that joints 1 and 4 are not subjected to loading.

Joint 1

Because two bars meet at this joint and they do not have a common line of action, we conclude that $T_{12} = 0$ and $T_{14} = 0$. To confirm this conclusion, we refer to the free-body diagram of pin 1 and write

$$\sum F_y = 0, \quad T_{14} \sin 45° = 0,$$

$$T_{14} = 0, \qquad \text{ANS.}$$

and

$$\sum F_x = 0, \quad -T_{12} - T_{14} \cos 45° = 0,$$

$$T_{12} = 0. \qquad \text{ANS.}$$

Joint 4

Because $T_{14} = 0$ from above and the remaining two members at joint 4 are perpendicular to each other, we conclude that $T_{45} = 0$ and $T_{42} = 0$. Again, we verify these results by considering the free-body diagram of pin 4. Thus,

$$\sum F_x = 0, \quad -T_{45} + T_{14} \cos 45° = 0.$$

Because $T_{14} = 0$, it follows that

$$T_{45} = 0, \qquad \text{ANS.}$$

and

$$\sum F_y = 0, \quad -T_{42} - T_{14} \sin 45° = 0.$$

Again, because $T_{14} = 0$, we conclude that

$$T_{42} = 0. \qquad \text{ANS.}$$

Now, consider the truss shown in Figure E6.4(b), and note that joint 2 is not subjected to external loads.

Joint 2

Because the action lines of members 2–1 and 2–3 lie along a common line and member 2–4 is not directed along this same line, we conclude that $T_{24} = 0$. Again, we verify this conclusion by referring to the free-body diagram of pin 2. Thus,

$$\sum F_y = 0, \quad -T_{24} \cos \theta = 0.$$

Because $\cos \theta = 0$, we conclude that

$$T_{24} = 0. \qquad \text{ANS.}$$

■ **Example 6.5**

Refer to the Howe truss shown in Figure E6.5(a), and determine all members which carry zero force. Note that loads are applied only at joint 7. Describe the structure which carries nonzero forces.

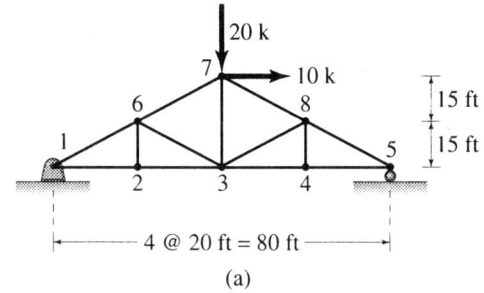

(a)

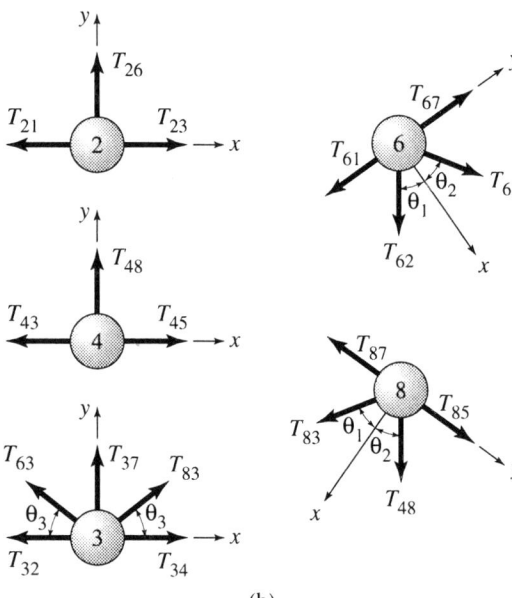

FIGURE E6.5.   (b)

Solution

Free-body diagrams of pins 2, 6, 4, 8, and 3 (considered in that sequence) are shown in Figure E6.5(b) together with coordinate axes. In each case, we have selected joints where no external loads or reactions are applied.

Joint 2

Note that T_{21} and T_{23} both act along the x axis.
$$\sum F_y = 0, \quad T_{26} = 0.$$ ANS.

Joint 6

Note that T_{61} and T_{67} both act along the y axis.
$$\sum F_x = 0, \quad T_{63} \cos\theta_2 + T_{26} \cos\theta_1 = 0.$$
Because $\cos\theta_2 \neq 0$ and $T_{26} = 0$, then,
$$T_{63} = 0$$ ANS.

Joint 4   Note that T_{43} and T_{45} both act along the x axis.

$$\sum F_y = 0, \quad T_{48} = 0. \qquad\qquad \text{ANS.}$$

Joint 8   Note that T_{87} and T_{85} both act along the y axis.

$$\sum F_x = 0, \quad T_{83}\cos\theta_1 + T_{48}\cos\theta_2 = 0.$$

Because $\cos\theta_1 \neq 0$ and $T_{48} = 0$, then,

$$T_{83} = 0. \qquad\qquad \text{ANS.}$$

Joint 3   Note that T_{32} and T_{34} both act along the x axis.

$$\sum F_y = 0, \quad T_{63}\sin\theta_3 + T_{83}\sin\theta_3 + T_{37} = 0.$$

Because $T_{63} = 0$ and $T_{83} = 0$, then,

$$T_{23} = 0. \qquad\qquad \text{ANS.}$$

Only the large outer triangle defined by joints 1, 7 and 5 carries the applied loads. All members interior to this triangle are zero-force members.

■

Problems

6.25 Correct reactions are shown acting at joints 3 and 5 of the truss shown in Figure P6.25. Determine which members carry zero force. Complete your analysis for the remaining bar forces, and describe the structure which resists the applied loading of 10k downward at joint 4.

6.26 Refer to the loaded truss shown in Figure P6.26. Determine which members

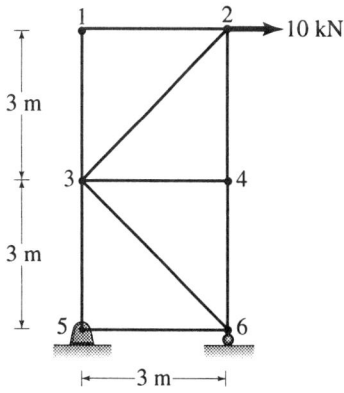

FIGURE P6.26.

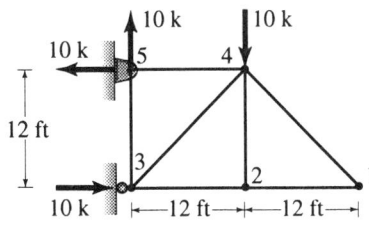

FIGURE P6.25.

carry zero force. Verify these zero bar forces by writing appropriate equations of equilibrium.

6.27 Determine the zero-force members of the truss depicted in Figure P6.27. Verify your answers by writing appropriate equations of equilibrium. Note that you need not determine the reactions at joints A and C.

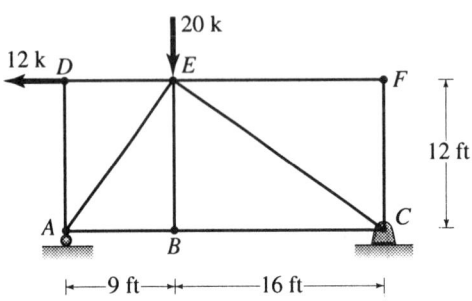

FIGURE P6.27.

6.28 Two members of the truss shown in Figure P6.28 carry no forces. Find them by inspection, and, then, verify your answers with equations of equilibrium.

6.29 Refer to the truss shown in Figure P6.28, and determine all nonzero bar forces. Correct reactions are given at joints 3 and 4.

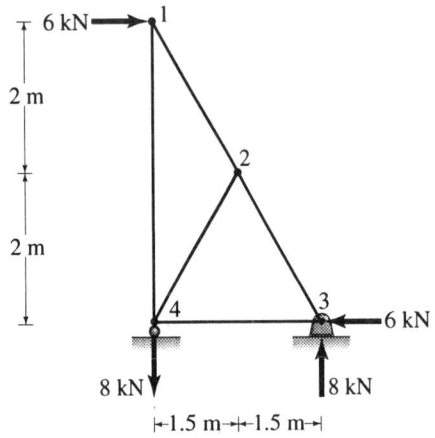

FIGURE P6.28.

6.30 Let $P_1 = 0$, $P_2 = 10$ k, and $P_3 = 0$ in Figure P6.30, and determine all zero-force members for this truss. Verify your answers by writing appropriate equations of equilibrium. Note that you need not determine the reactions at joints A and E.

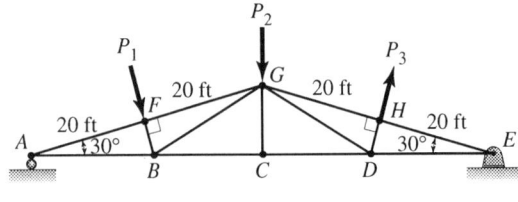

FIGURE P6.30.

6.31 Refer to the truss shown in Figure P6.30 and let $P_1 = 0$, $P_2 = 0$, and $P_3 = 10$ k. Determine all zero-force members of this truss. Verify your answers by writing appropriate equations of equilibrium. Note that you need not find the reactions at joints A and E.

6.32 For the truss shown in Figure P6.30, let $P_1 = 10$ k, $P_2 = 0$, and $P_3 = 12$ k. Determine all zero-force members of this truss, and verify your answers by writing appropriate equations of equilibrium. Note that you need not find the reactions at A and E.

6.33 A downward force of 80 k is applied at joint C of the truss depicted in Figure P6.33. Find which members carry zero force, and verify your answers by writing

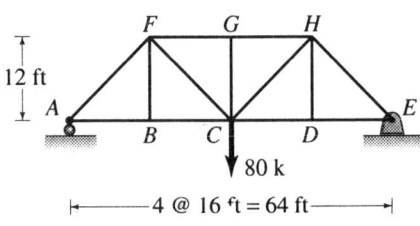

FIGURE P6.33.

appropriate equations of equilibrium. You need not find the reactions at joints A and E.

6.34 In Figure P6.34 let $P_1 = 0$, $P_2 = P_3 = P_4 = 10$ k, and $P_5 = 0$. Find all zero-force members for this truss, and verify your answers by writing appropriate equations of equilibrium. It will not be necessary to find the end reactions.

6.35 Let $P_1 = 5$ k and $P_2 = P_3 = P_4 = P_5 = 0$ for the truss of Figure P6.34, and determine all zero-force members. Write appropriate equations of equilibrium to verify your answers. End reactions will not be required.

6.36 A Howe roof truss is shown in Figure P6.36. Find which members carry zero force, and verify your answers by writing appropriate equations of equilibrium. Describe the structure composed of members with non-zero bar forces. You need not find the reactions at A and G.

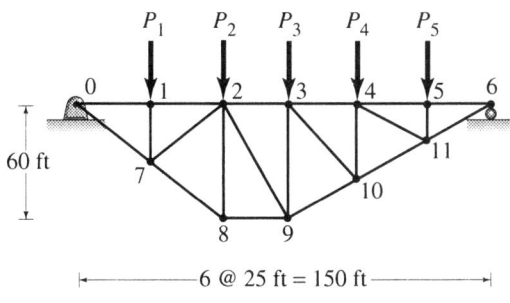

FIGURE P6.34.

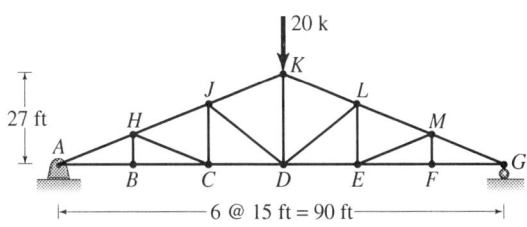

FIGURE P6.36.

6.4
Member Forces Using the Method of Sections

An alternate method for determining member forces in trusses is termed *the method of sections*. This method is particularly useful for finding the forces in a limited number of members which may be located anywhere in the truss. Both this method and the method of joints are valuable for checking computer outputs, but the method of sections may be conveniently used to check forces independently in members located anywhere in the truss.

Initially, we consider a free-body diagram of the entire truss and determine the reactions at both supports. Both reactions may not be required. For example, if we wish to deal with one part of the truss in solving for unknown forces, we need to find only the corresponding reaction.

Refer to the truss of Figure 6.10(a). Suppose we wish to determine the forces in members CD, JD, and JK. We may imagine the truss divided by a cutting plane into two parts so that we cut through those members whose forces we wish to determine. We may cut through any number of members provided there are only three unknown forces among the cut members.

Each part of the truss, left or right, is a rigid body in equilibrium

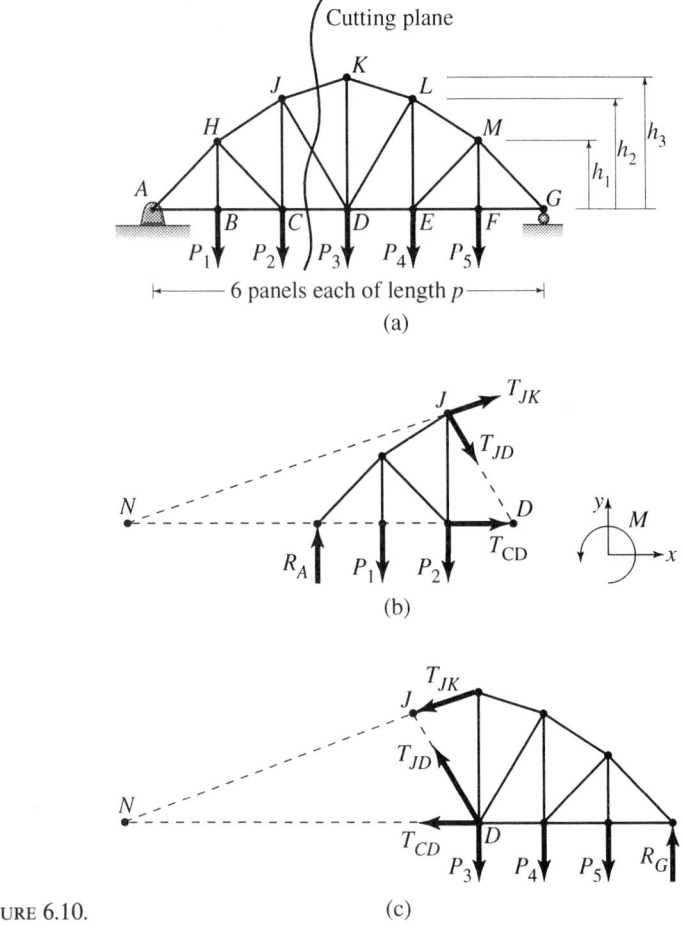

FIGURE 6.10.

under the action of external forces. These external forces now include the forces in the cut members, three of which may be unknowns. Note that internal forces are exposed and become external when the members are cut.

As stated in Section 6.2, if an entire truss is in equilibrium, then any part of the truss, which we care to isolate, is also in equilibrium. Refer to the free-body of the left part of the truss shown in Figure 6.10(b) and note that we have shown the reaction R_A and the applied forces P_1 and P_2 as well as the desired three unknown forces T_{CD}, T_{JD}, and T_{JK}. By showing the sense of the three unknown bar forces acting away from the free-body, we have assumed that they are in tension. The force system is coplanar and nonconcurrent which means that we have three independent equations of equilibrium available to solve for the three unknown forces. To select an appropriate moment equation which may be solved for a single unknown bar force, we ask the following question: Where do the action lines of the other two forces intersect? For example, suppose we wish to choose the appropriate moment equation to solve for the force T_{JK} of Figure 6.10(b). We ask the question: Where do the action lines of T_{JD} and T_{CD} intersect? The answer is that their action lines intersect at point D. The appropriate equation is $\sum M_D = 0$ because this equation will contain only the unknown force T_{JK}. Similarly, to determine T_{CD}, we ask where the action lines of T_{JK} and T_{JD} intersect. Because their action lines intersect at point J, we write $\sum M_J = 0$. The equation will contain the single unknown T_{CD}. Finally, if this key question is asked to find the force T_{JD}, the action lines of T_{JK} and T_{CD} must be extended until they intersect at point N. The appropriate equation $\sum M_N = 0$ contains the single unknown force T_{JD}. The fact that N lies off the free-body diagram is immaterial. Forces in members which have not been cut are internal forces. These internal forces occur in equal and opposite pairs and their effects are canceled out. Only the external forces R_A, P_1, P_2, and the three unknown forces act on the left free-body diagram. The right free-body diagram of Figure 6.10(c) may also be used to determine the same three unknowns. The external forces R_G, P_3, P_4, and P_5 and the three unknown forces act on the right free-body diagram. Note that the same three moment axes, perpendicular to the plane of the forces through the points D, J, and N, are appropriate choices enabling us to solve for T_{JK}, T_{CD}, and T_{JD} in that order. Should we choose the left or right free-body diagram? Time and effort may be saved by choosing the free-body which has fewer forces acting on it. In this case, there is a slight preference for the left free-body diagram.

Sometimes the action lines of the two unknowns, which we do not wish to find, do not intersect but are parallel. In such a case, we sum forces in a direction perpendicular to the parallel action lines of these two forces. For example, consider the parallel chord truss shown in Figure 6.11(a), and suppose we wish to determine the force in member DL. After imagining that the truss is cut by the cutting plane shown in Figure 6.11(a), we may choose either the left free-body diagram of Figure 6.11(b) or the right free-body diagram of Figure 6.11(c). There is a slight preference for the right free-body diagram because it contains fewer forces. We ask the question: Where do the action lines of T_{KL} and T_{DE} intersect? The answer, of course, is that they are parallel and intersect only at infinity. Therefore, we sum forces in a direction perpendicular to their action lines because the only unknown entering this equation will be T_{DL}. The appropriate equation is $\sum F_y = 0$ if we assume that the reaction R_G has been determined from a free-body diagram of the entire truss. To find T_{KL}, we write $\sum M_D = 0$ because the action lines of T_{KL} and T_{DE} intersect at D. Similarly, to find T_{DE},

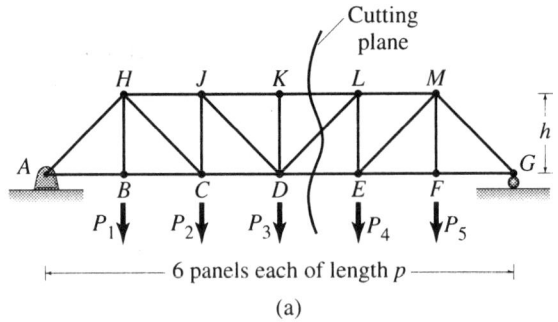

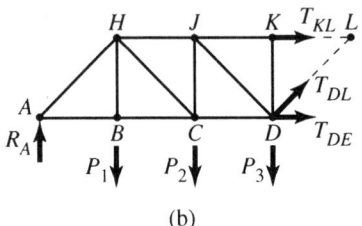

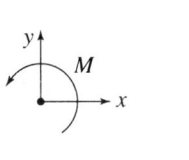
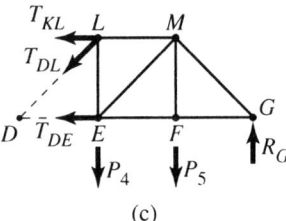

Figure 6.11.   (c)

we write $\sum M_L = 0$ because the action lines of T_{KL} and T_{DL} intersect at L.

The following examples illustrate some of the concepts above.

■ **Example 6.6**

A cantilever truss is shown in Figure E6.6(a). Determine the bar forces in members BC, BF, and GF using the method of sections.

Solution

We imagine a horizontal plane cutting through members CB, BF, and FG as shown. It is convenient to choose an upper free-body diagram as shown in Figure E6.6(b) because the reactions at A and H need not be determined. To determine T_{CB}, we sum moments about point F because the unknown forces T_{CB} and T_{FG} intersect at this point. Thus,

$$\sum M_F = 0, \quad T_{CB}(5) - 80(5) = 0,$$

$$T_{CB} = 80 \text{ k} \quad \text{(T)}. \qquad\qquad \text{ANS.}$$

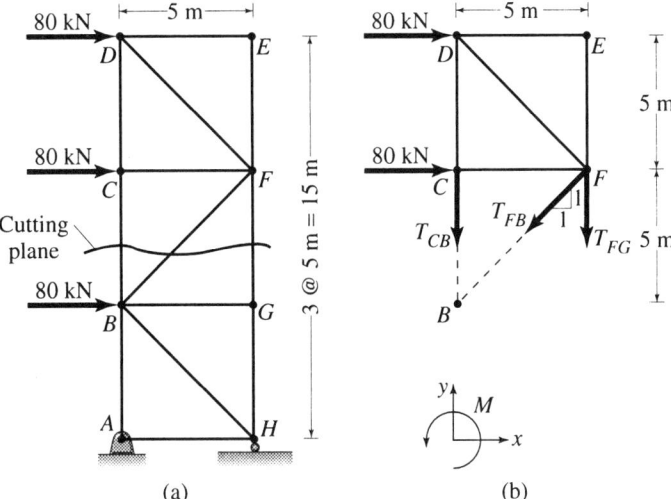

FIGURE E6.6.

(a)   (b)

Because the unknown forces T_{CB} and T_{FG} have action lines parallel to the y axis, we sum forces in the x direction. Thus,

$$\sum F_x = 0, \quad 80 + 80 - \frac{1}{\sqrt{2}} T_{FB} = 0,$$

$$T_{FB} = 226 \text{ k} \quad \text{(T)}. \qquad \text{ANS.}$$

Next, we sum moments about B in order to determine T_{FG}. Thus,

$$\sum M_B = 0, \quad -T_{GF}(5) - 80(5) - 80(10) = 0,$$

$$T_{FG} = -240 \text{ k} = 240 \text{ k} \quad \text{(C)}. \qquad \text{ANS.}$$

An alternative way to find T_{FG}, once we know T_{CB} and T_{FB}, is to sum forces in the y direction. Thus,

$$\sum F_y = 0, \quad -T_{FG} - 80 - \frac{1}{\sqrt{2}}(226) = 0,$$

$$T_{FG} = -240 \text{ k} = 240 \text{ k} \quad \text{(C)} \quad \text{as above.}$$

■ **Example 6.7**

Refer to the through-bridge truss of Figure E6.7(a), and determine the forces in members GH, CH, and CD using the method of sections.

Solution

First, determine the right reaction R_E by considering the free-body diagram of the entire truss (not shown), and write

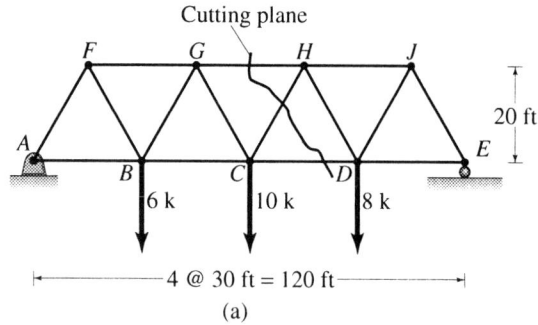

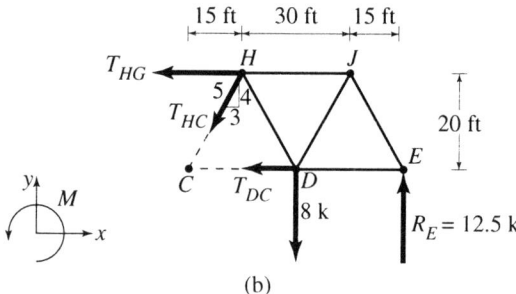

FIGURE E6.7.   (b)

$$\sum M_A = 0, \quad R_E(120) - 6(30) - 10(60) - 8(90) = 0,$$

$$R_E = 12.5 \text{ k} \quad (\uparrow).$$

As shown in Figure 6.7(a), we imagine a plane that cuts through members GH, CH, and CD. The right free-body diagram of Figure E6.7(b) is clearly preferable in this case. To find T_{GH}, we ask the question: Where do the action lines of T_{CH} and T_{CD} intersect? The answer is point C, and we write

$$\sum M_C = 0, \quad T_{HG}(20) - 8(30) + 12.5(60) = 0,$$

$$T_{HG} = -25.5 \text{ k} = 25.5 \quad (C). \qquad\qquad \text{ANS.}$$

The action lines of T_{HG} and T_{DC} are parallel, and, to find T_{HC}, we write

$$\sum F_y = 0, \quad -\frac{4}{5} T_{HC} - 8 + 12.5 = 0,$$

$$T_{GH} = 5.63 \text{ k} \quad (T). \qquad\qquad \text{ANS.}$$

Next, we sum moments about H to find T_{DC}. Thus,

$$\sum M_H = 0, \quad 12.5(45) - 8(15) - T_{CD}(20) = 0,$$

$$T_{CD} = 22.1 \text{ k} \quad (T). \qquad\qquad \text{ANS.}$$

As a check of these results,

$$\sum F_x = 0, \quad -T_{HG} - \frac{3}{5}T_{HC} - T_{DC} \stackrel{?}{=} 0,$$

$$-(-25.5) - \frac{3}{5}(5.63) - 22.1 = 0.022 \approx 0. \quad \text{Check.}$$

■ **Example 6.8**

Refer to the roof truss shown in Figure E6.8(a) and determine the forces in members LN, LM, and CD by using the method of sections.

Solution

The right free-body diagram shown in Figure E6.8(b) is preferred for determining these three unknowns. We note that the right reaction F_y is required. Using a free-body diagram of the entire truss (not shown),

$$\sum M_A = 0, \quad F_y(27.71) - 10(4) - 10(8) - 10(12) - 5(16) = 0,$$

$$F_y = 11.55 \text{ kN.} \qquad \text{ANS.}$$

To determine T_{NL}, we observe that the action lines of T_{ML} and T_{DC} intersect at point D and, therefore, we write a moment equation about

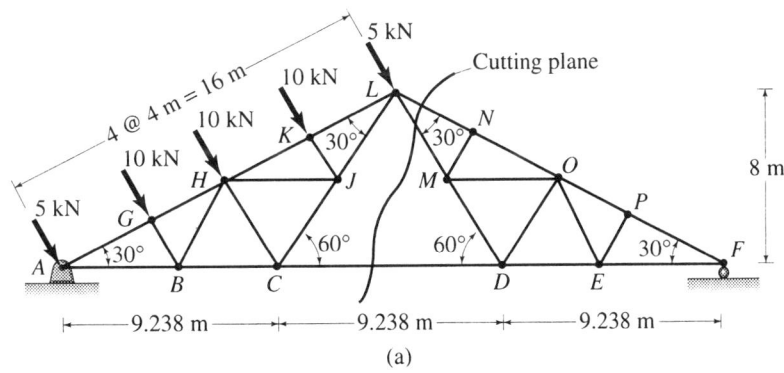

(a)

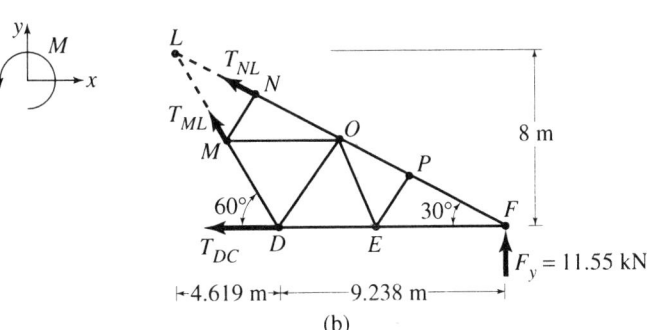

(b)

FIGURE E6.8.

this point. In writing this equation, we break T_{NL} into x and y components at point F. Only the y component of T_{NL} enters the equation because the action line of its x component passes through D. Thus,

$$\sum M_D = 0, \quad 11.55(9.238) + 9.238\,T_{NL}\sin 30° = 0,$$

$$T_{NL} = -23.1 \text{ kN} = 23.1 \text{ kN} \quad \text{(C)}. \qquad \text{ANS.}$$

To find T_{ML}, we write a moment equation about point F. We break T_{ML} into x and y components at point D. Thus,

$$\sum M_F = 0, \quad 9.238\,T_{ML}\sin 60° = 0,$$

$$T_{ML} = 0. \qquad \text{ANS.}$$

Of course, member LM may carry a nonzero force for other loadings applied to the truss, and we should bear in mind that structures are normally designed for various loading conditions.

To find T_{DC}, we sum moments about point L. Thus,

$$\sum M_L = 0, \quad 11.55(13.857) - T_{CD}(8) = 0,$$

$$T_{DC} = 20.0 \text{ kN} \quad \text{(T)}.$$

The force summation equations $\sum F_x = 0$ and $\sum F_y = 0$ could be used to check these results. The check is left as an exercise for the student.

■

Problems

6.37 Refer to Figure P6.37, let $P = 10$ k, and use the method of sections to find the forces in members BC, BF, and EF.

6.38 Refer to Figure P6.38, and apply the method of sections to express the forces

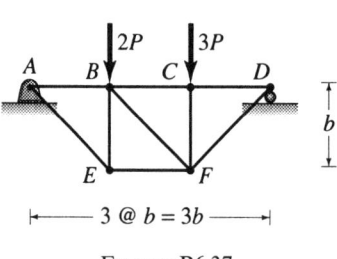

FIGURE P6.37.

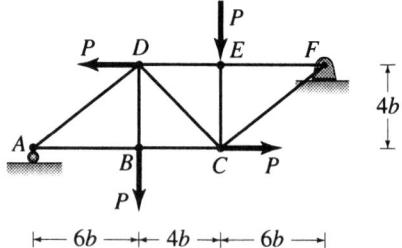

FIGURE P6.38.

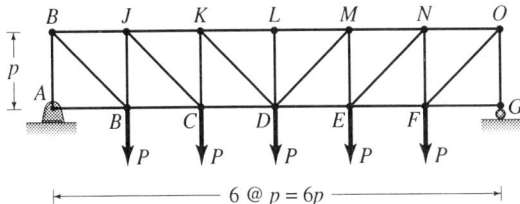

FIGURE P6.39.

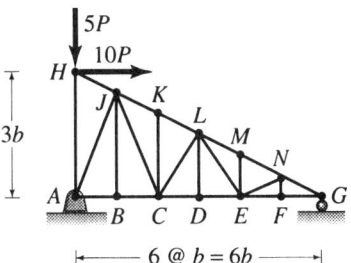

FIGURE P6.43.

in members DE, DC, and BC in terms of *P*.

6.39 Refer to the *Deck Pratt Truss* of Figure P6.39, and determine the bar forces in members KL, KD, and CD in terms of *P*. Use the method of sections.

6.40 Let *P* = 4 kN for the *Deck Pratt Truss* shown in Figure P6.39, and apply the method of sections to determine the bar forces in members MN, ME, and DE.

6.41 Use the method of sections to find the forces in members FG, FC, and BC of the roof truss shown in Figure P6.41.

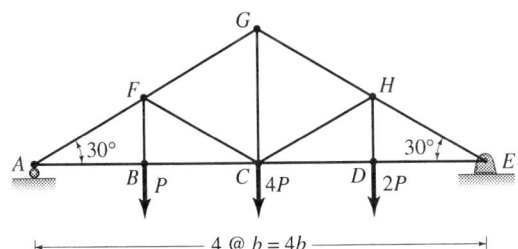

FIGURE P6.41.

6.42 In Figure P6.41, let *P* = 6 k, and use the method of sections to determine the forces in members GH, CH, and CD.

6.43 The forces in members KL, CL, and CD of the roof truss of Figure P6.43 are required in terms of *P*. Use the method of sections to determine them.

6.44 Refer to the roof truss of Figure P6.43. Let *P* = 1 kN, and use the method of sections to determine the bar forces in LM, LE, and DE.

6.45 Refer to the truss depicted in Figure P6.45, and express the bar forces in members HJ, HE, and DE in terms of *P*. Apply the method of sections.

6.46 Let *P* = 2 kN for the truss of Figure P6.45, and apply the method of sections to find the bar forces in members EJ, EM, and EF.

6.47 Determine the forces in all bars of the truss shown in Figure P6.47 for *P* = 2 k. Carefully show that joints A and B are in equilibrium under the action of reactions and bar forces. Note that the method of joints is appropriate for solving this problem.

6.48 Bar forces in members FG, GC, and BC are required for the truss of Figure P6.48 for *Q* = 6 kN. Apply the method of sections to this cantilever truss.

6.49 Express the forces in members EF, FA, and AB in terms of *Q* for the truss of Figure P6.48. Use the method of sections for this cantilever truss.

6.50 Let *P* = 5 k and solve for the forces in members LM, EM, and EF of the truss depicted in Figure P6.50. Use the method of sections for solving this problem.

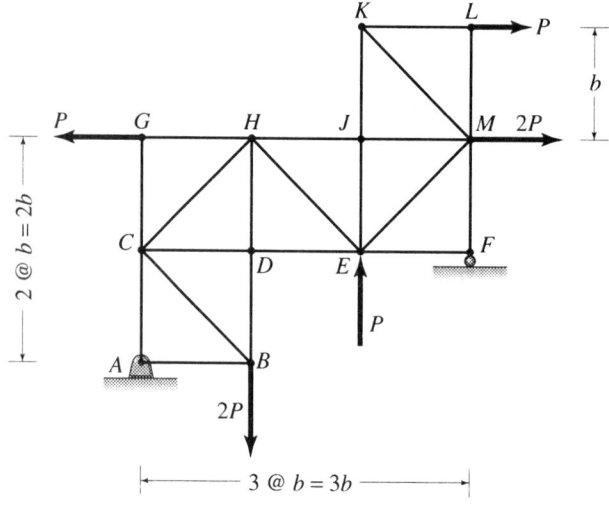

FIGURE P6.45.

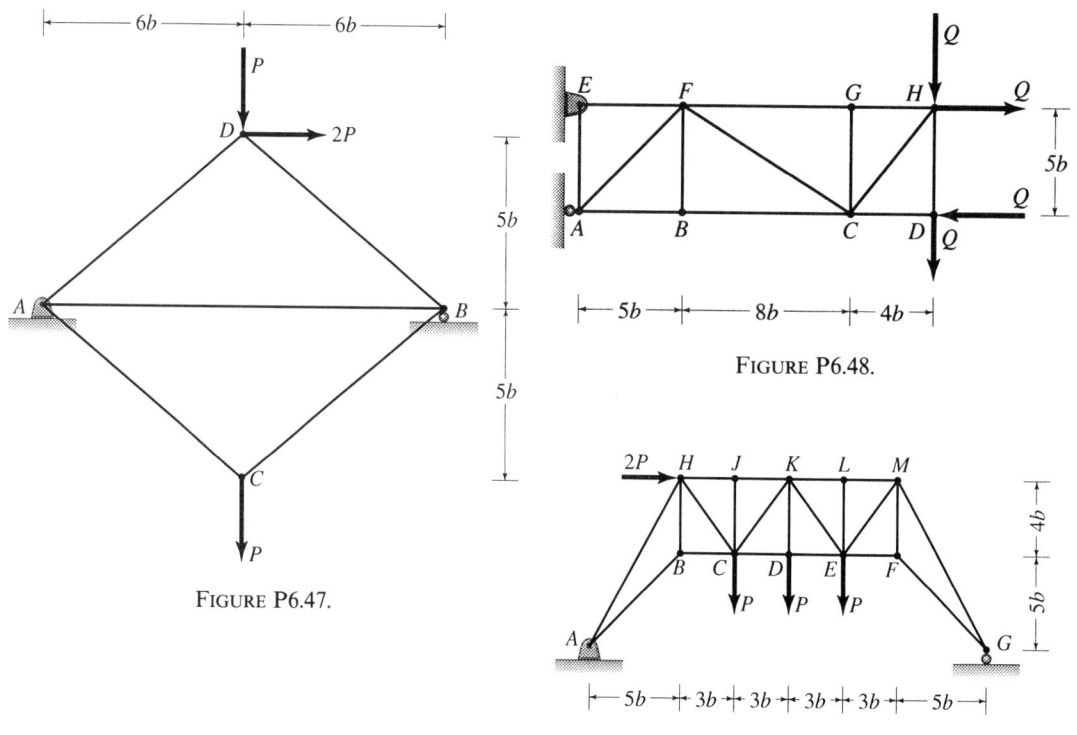

FIGURE P6.47.

FIGURE P6.48.

FIGURE P6.50.

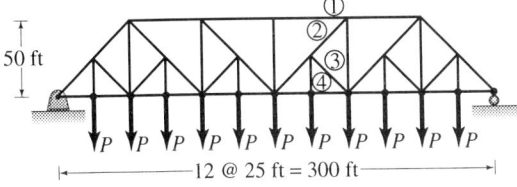

FIGURE P6.51.

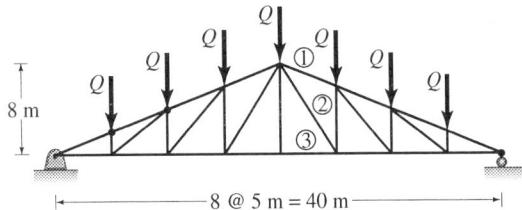

FIGURE P6.52.

6.51 Refer to the *Baltimore* truss shown in Figure P6.51 and let $P = 20$ k at each lower chord panel point. It is well to note that the *Baltimore* truss is economical only for relatively long spans. Use the method of sections to determine the bar forces in the members marked 1, 2, 3, and 4. (Hints: Because more than three unknown bar forces are requested, you will need to consider two different sections. It will be simpler to use the right free-body diagram in each case.)

6.52 Refer to the *Pratt* roof truss shown in Figure P6.52 and let $Q = 8$ kN at each upper chord panel point. Use the method of sections to find the bar forces in the members marked 1, 2, and 3. (Hint: De-

termine the right reaction and use a right free-body diagram.)

6.53 Determine bar forces in JK, CK, and CD of the truss depicted in Figure P6.50 in terms of P. Use the method of sections.

6.54 Refer to Figure P6.54, and use the method of sections to determine the bar forces in members 3–4, 3–8, and 7–8 in terms of P, a, and b. Specialize your results for b/a equal to 1/2 and 2.

6.55 The central diagonals 4–11 and 5–10 of the truss shown in Figure P6.55 are connected at their centers. Use the method of sections to determine the bar forces in members 5–6, 6–11, and 11–12 of this truss in terms of P and Q. Specialize your results for $P = 1$ k and $Q = 2$ k.

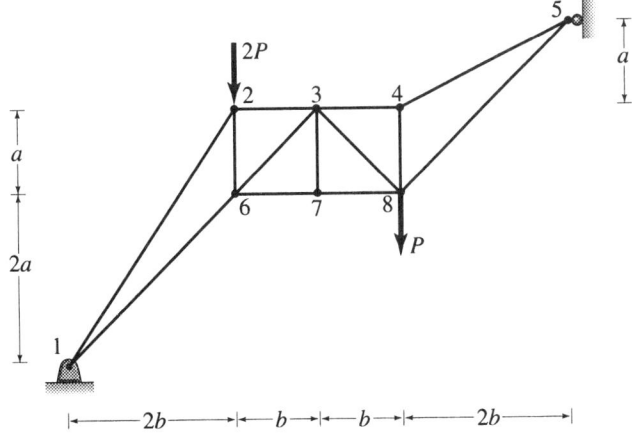

FIGURE P6.54.

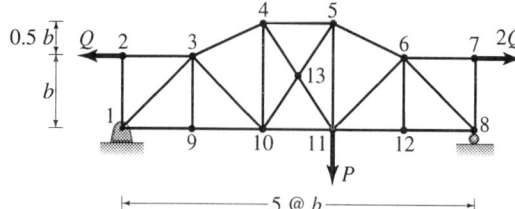

FIGURE P6.55.

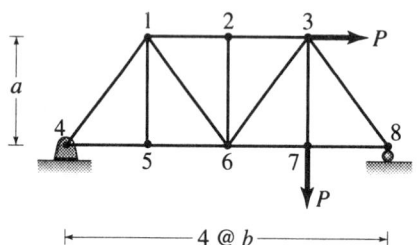

FIGURE P6.56.

6.56 Refer to Figure P6.56, and use the method of sections to express the bar forces in members 2–3, 3–6, and 6–7 as functions of P and the ratio of b to a. Sketch these functions for $(0 \leq b/a \leq 4)$. Let $P = 10$ kN and $b/a = 2$, then, state numerical values for these bar forces.

6.57 Find the forces in the members marked, 1, 2, and 3 in Figure P6.57. Express your answers in terms of P. If $P = 10$ k, state numerical values for these bar forces.

6.58 Determine the bar forces in the members marked 1, 2, and 3 in Figure P6.58. Express your answers as functions of P, Q, and the ratio of b to a. Find numerical values for these bar forces when $P = Q = 10$ k and $a = b = 10$ ft.

6.59 An aluminum cantilever truss of an aircraft structure is shown in Figure P6.59. Use the method of sections to find the forces in members 3–4, 3–9, and 8–9 of this truss.

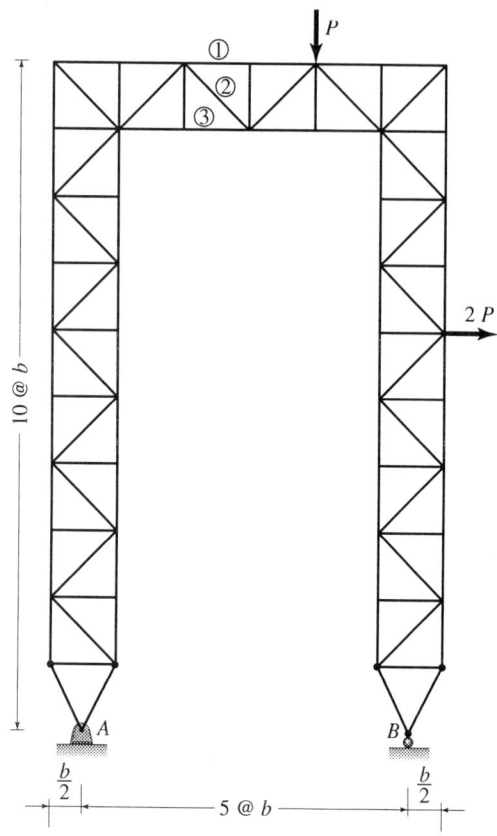

FIGURE P6.57.

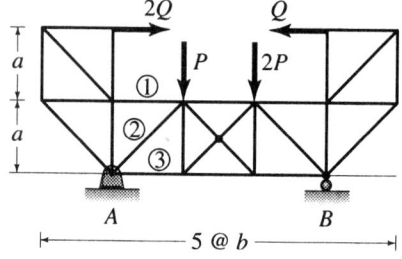

FIGURE P6.58.

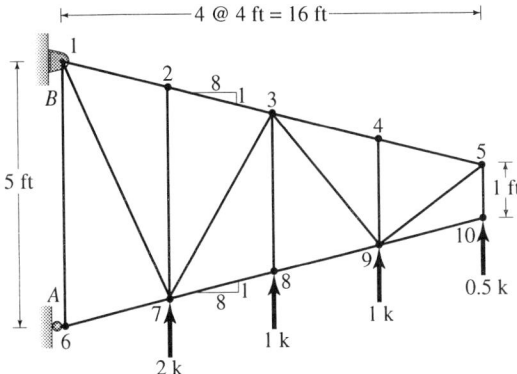

FIGURE P6.59.

6.60 A deck truss for a pedestrian walkway is shown in Figure P6.60. Determine the forces in the members marked 1, 2, and 3 using the method of sections.

6.61 Long span trusses such as the one shown in Figure P6.61 are often used to support roofs to provide column-free space for civic centers where sporting and theatrical events are staged. Find the bar forces in the truss members marked 1, 2, and 3 by applying the method of sections.

6.62 Determine the bar forces in the members marked 1, 2, and 3 of the structure shown in Figure P6.62. Use the method of sections, clearly show the free-body diagrams required, and state whether each member is in tension or compression.

6.63 Iron ore is stockpiled near steel mills. Overhead cranes, supported by ore bridges, are used to move this ore. A planar truss of such an ore bridge is shown in Figure P6.63. For a lift of 100 k at the right end of this structure, find the forces in the members marked 1, 2, and 3.

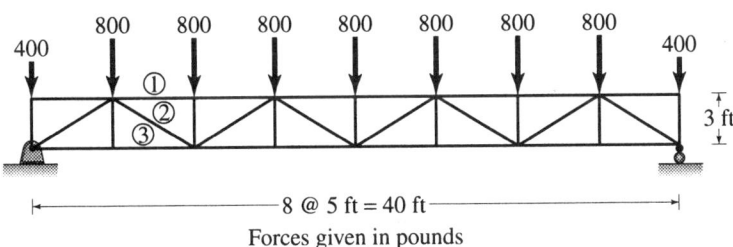

Forces given in pounds

FIGURE P6.60.

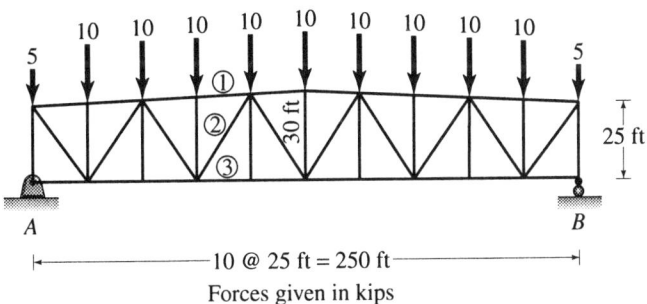

Forces given in kips

FIGURE P6.61.

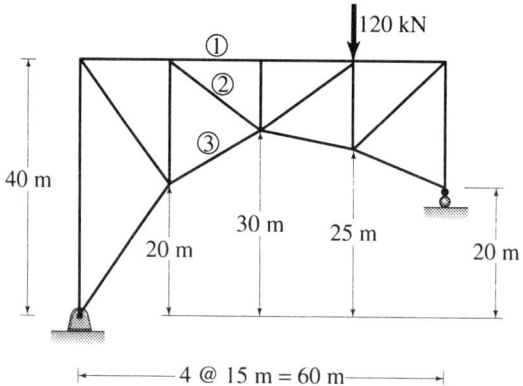

FIGURE P6.62.

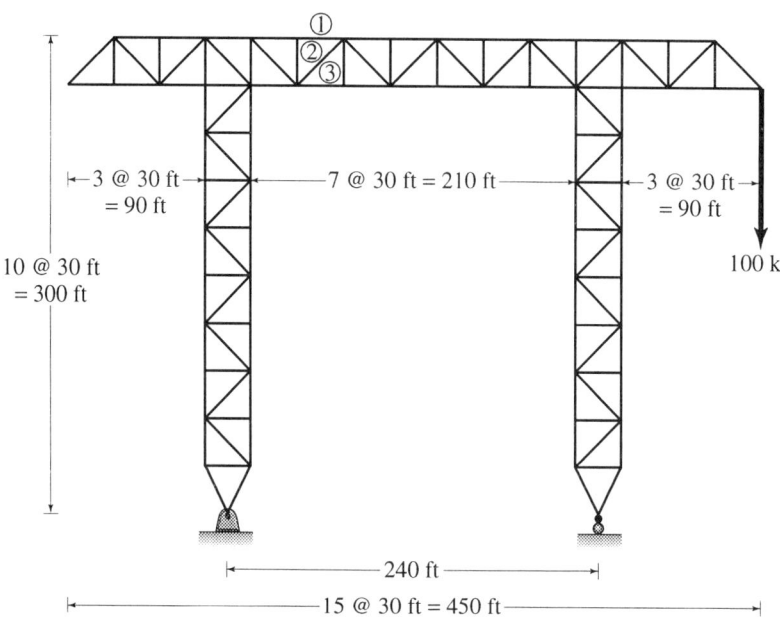

FIGURE P6.63.

6.5*
Determinacy
and
Constraints

Engineering applications often involve systems for which the equations of equilibrium are insufficient for a complete solution. These equations of statics are necessary but when they are insufficient, we refer to the system as *statically indeterminate*. In other words, the equations of equilibrium must be supplemented by additional equations to solve statically indeterminate problems. In this text, we will solve only stati-

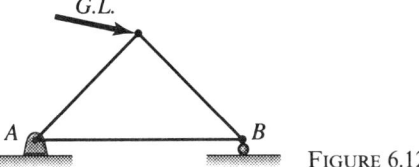

FIGURE 6.12.

cally determinate problems, but it is important that we learn to classify systems as statically determinate or statically indeterminate.

We begin our discussion with the truss shown in Figure 6.12. This truss consists of three members and we will use the symbol m to denote the number of members of any truss. For this truss, $m = 3$. The three members are connected by pins at joints A, B, and C. We will use the symbol j to denote the number of joints of any truss. For this truss, $j = 3$. The members are arranged in a rigid triangular pattern and constrained by a pin support attached to the ground at point A and by a frictionless roller at joint B. Thus, there are three reaction components available, two at A and one at B, which prevent the truss from moving. We use the symbol r to denote the number of scalar reaction components. For this truss, $r = 3$. We imagine the truss subjected to general loading denoted by symbol G.L. Truss loadings are applied at the joints and, for larger trusses, we will show loadings at a number of joints but will use the same symbol G.L. once to denote a general loading.

To decide whether a given truss is statically determinate or not, we need to compare the number of unknown forces with the number of equations of equilibrium available to solve for these unknowns. The total number of force unknowns equals the number of scalar reaction components added to the number of internal bar forces. Because each truss member is associated with a single unknown force, the number of internal bar forces equals the number of members. Thus, the number of force unknowns in a given truss is $r + m$. Because two equations of equilibrium are available at each joint for a two-dimensional truss, the total number of available equations is $2j$. For statical determinateness of a truss, the number of unknown forces must equal the number of available equations of equilibrium. Thus, a two-dimensional truss is statically determinate if

$$r + m = 2j. \tag{6.1}$$

Equation (6.1) is necessary but not sufficient for a truss to be statically determinate and completely constrained. In addition to satisfying this equation, the members of the truss and the support must be properly arranged.

Let us refer to Figure 6.13, and consider determinacy as well as the

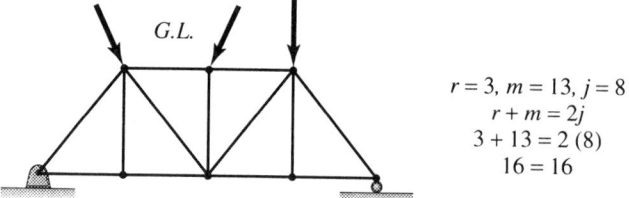

$r = 3, m = 13, j = 8$
$r + m = 2j$
$3 + 13 = 2 (8)$
$16 = 16$

(a) Statically determinate and properly constrained

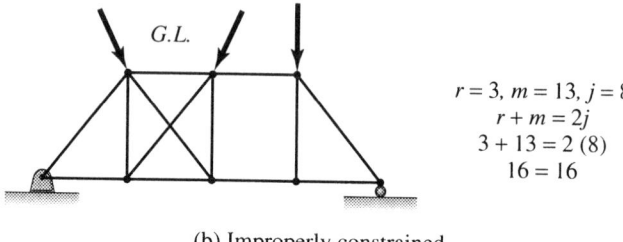

$r = 3, m = 13, j = 8$
$r + m = 2j$
$3 + 13 = 2 (8)$
$16 = 16$

(b) Improperly constrained

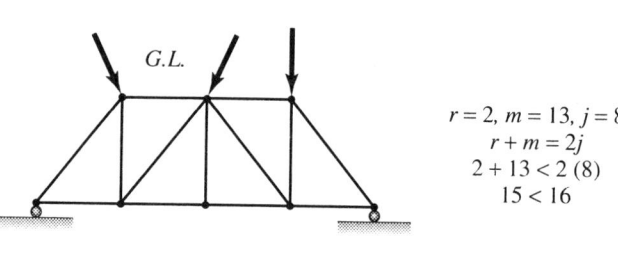

$r = 2, m = 13, j = 8$
$r + m = 2j$
$2 + 13 < 2 (8)$
$15 < 16$

FIGURE 6.13.   (c) Partially constrained

arrangement of the truss members and its support. The trusses shown have a constant number of joints $j = 8$, and each will be subjected to a general loading G.L. The simple truss of Figure 6.13(a) satisfies the equation $r + m = 2j$ and is composed of rigid planar triangles as well as having the reacting pin and roller arranged to prevent motion. This enables us to conclude that the truss is statically determinate and properly constrained. We have established in advance, without performing calculations, that this and similar problems are solvable by applying the equations of statics. Referring to Figure 6.13(b), we note that the equation $r + m = 2j$ is, again, satisfied, and the reactions are arranged as before, but the third panel from the left is a quadrilateral pinned at its four vertices. This truss will not remain in the position shown because the quadrilateral panel is nonrigid and will collapse under a general loading. Even though the equation $r + m = 2j$ is satisfied, the members are arranged so that the truss is unstable because it is *improperly constrained*. In Figure 6.13(c), the support systems is

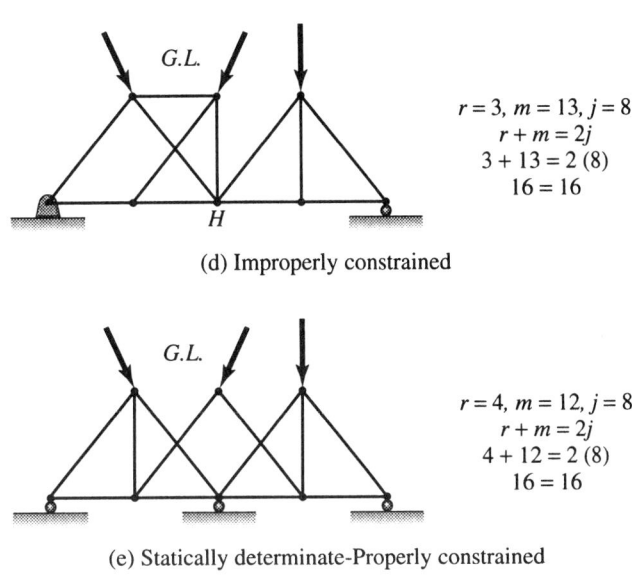

$$r = 3, m = 13, j = 8$$
$$r + m = 2j$$
$$3 + 13 = 2 (8)$$
$$16 = 16$$

(d) Improperly constrained

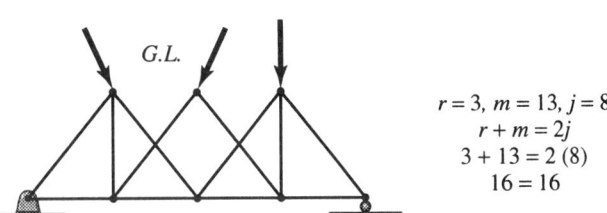

$$r = 4, m = 12, j = 8$$
$$r + m = 2j$$
$$4 + 12 = 2 (8)$$
$$16 = 16$$

(e) Statically determinate-Properly constrained

$$r = 3, m = 13, j = 8$$
$$r + m = 2j$$
$$3 + 13 = 2 (8)$$
$$16 = 16$$

FIGURE 6.13 *(Cont.)*.   (f) Statically determinate-Properly constrained

changed and the number of reactions has been reduced to $r = 2$. These frictionless rollers cannot prevent the truss from moving horizontally. Even if we were to add another diagonal member in either the second or third panel to satisfy the equation $r + m = 2j$, the truss would still be *partially constrained*. Structures are designed to provide pathways to the ground for forces, and we wish to rigidly attach them to the earth. Therefore, motion with respect to the earth is undesirable. Machines differ from structures in that we desire machines to move. For example, if we were considering a highway vehicle, instead of this truss, then, we would want the vehicle free to move horizontally, except when stopping.

In Figure 6.13(d), the equation $r + m = 2j$ is, again, satisfied, but the truss is improperly constrained because either part of the truss to the left or right of hinge H cannot resist a general loading without moving from the position shown. Therefore, even though the equation $r + m = 2j$ is satisfied, the truss is unstable because it is *improperly constrained*.

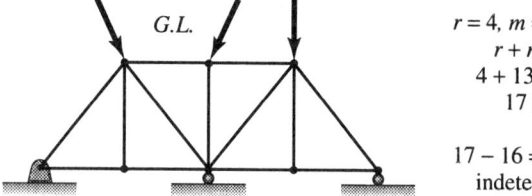

$r = 4, m = 13, j = 8$
$r + m > 2j$
$4 + 13 > 2 (8)$
$17 > 16$

$17 - 16 = 1$ degree
indeterminate

(g) Statically indeterminate externally-First degree

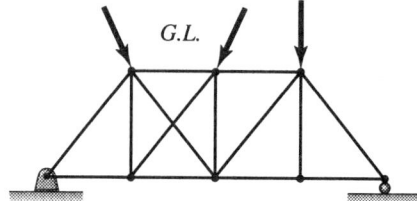

$r = 3, m = 14, j = 8$
$r + m > 2j$
$3 + 14 > 2 (8)$
$17 > 16$

$17 - 16 = 1$ degree
indeterminate

(h) Statically indeterminate internally-First degree

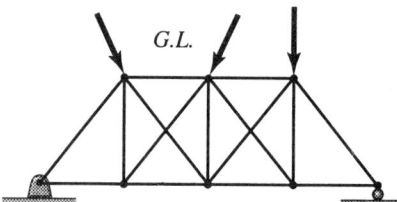

$r = 3, m = 15, j = 8$
$r + m > 2j$
$3 + 15 > 2 (8)$
$18 > 16$

$18 - 16 = 2$ degrees
indeterminate

FIGURE 6.13 (*Cont.*).   (i) Statically indeterminate internally-Second degree

The trusses of Figures 6.13(e) and (f) are both determinate and properly constrained. In each case, it is possible to apply the method of joints and determine member forces in all members of these trusses for any specified general loading. If a given truss is improperly constrained, it will not be possible to obtain a solution in which all member forces are in balance at all of the truss joints. Although we may not immediately see that a truss is improperly constrained, we can be sure that attempting to solve for the member forces will reveal this fact. This may be time consuming, but it is reassuring to know that we are likely to discover instability because of improper constraints, provided our chosen loading is not of a very special character.

Trusses shown in Figures 6.13(g), (h), (i), and (j) are statically indeterminate and properly constrained. In each case, the number of unknowns exceeds the number of equations of equilibrium available to solve for those unknowns. The phrase *statically indeterminate* has modifiers of *internal* and *external* which refer to member forces and

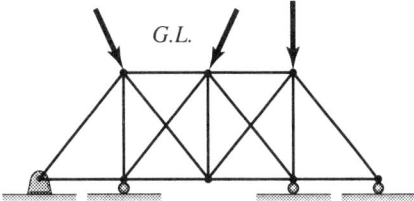

$$r = 5, m = 15, j = 8$$
$$r + m > 2j$$
$$5 + 15 > 2\,(8)$$
$$20 > 16$$

$$20 - 16 = 4 \text{ degrees}$$
indeterminate

(j) Statically indeterminate-Fourth degree
Twice external and twice internal

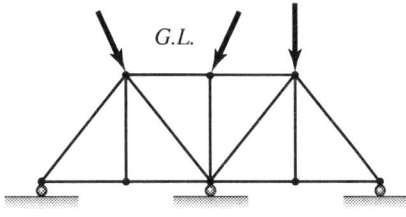

$$r = 3, m = 13, j = 8$$
$$r + m = 2j$$
$$3 + 13 = 2\,(8)$$
$$16 = 16$$

(k) Partially constrained

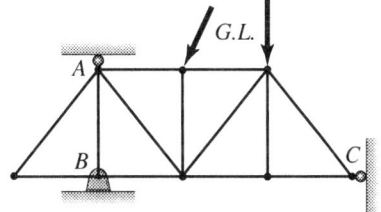

$$r = 4, m = 13, j = 8$$
$$r + m > 2j$$
$$4 + 13 > 2\,(8)$$
$$17 > 16$$

FIGURE 6.13 (*Cont.*).   (l) Improperly constrained

reactions, respectively. For a two-dimensional truss, the overall degree of indeterminacy is given by subtracting $2j$ from $r + m$. The external degree of indeterminacy is given by subtracting three form the total number of reaction components. The internal degree of indeterminacy is given by subtracting the external degree of indeterminacy from the overall degree of indeterminacy. For the indeterminate trusses of Figures 6.13(g), (h), (i), and (j), the degree of indeterminacy is shown in the following table.

Degree of Indeterminacy for the Trusses of Figure 6.13

Figure 6.13	$r + m$	$2j$	Overall	Internal	External
(g)	17	16	1	0	$4 - 3 = 1$
(h)	17	16	1	1	$3 - 3 = 0$
(i)	18	16	2	2	$3 - 3 = 0$
(j)	20	16	4	2	$5 - 3 = 2$

The truss shown in Figure 6.13(k) is *partially constrained* even though the equation $r + m = 2j$ is satisfied. The three frictionless rollers cannot prevent the truss from moving horizontally. Because this truss could resist vertical, but not horizontal, forces effectively, we say that it is partially constrained. Referring to the truss of Figure 6.13(*l*), we note that the action lines of the three reactions at A, B, and C meet at the pin centered at joint B. If a single force of the general loading is applied to this truss so that its action line does not intersect joint B, the truss will rotate about an axis through B. Because the truss will move from the position shown under a general loading, we say that it is improperly constrained.

Problems

Consider a general loading applied to the truss. Connected diagonals are indicated by **X** *whereas unconnected ones are indicated by* X.

6.64–6.68 Classify the trusses of each of the figures as properly, partially, or im-properly constrained. If a given truss is properly constrained, state whether it is statically determinate or indeterminate. Further classify the indeterminate trusses as to degree of indeterminacy. For each truss, compare $r + m$ to $2j$.

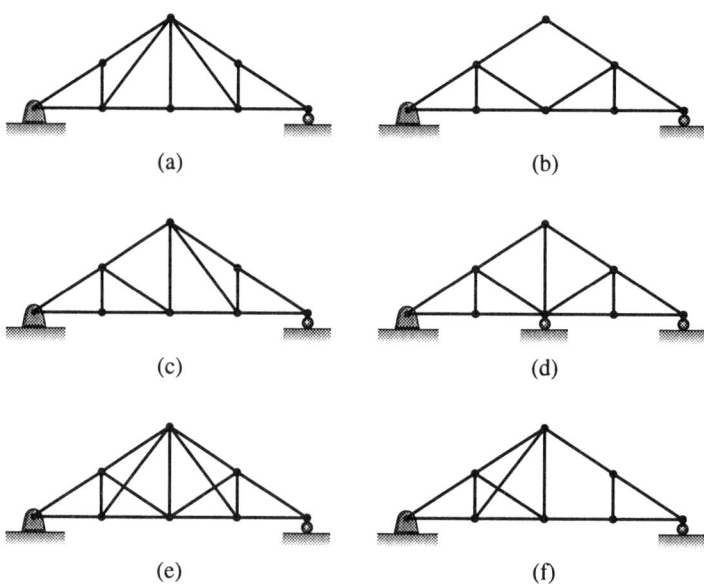

(a)

(b)

(c)

(d)

(e)

(f)

FIGURE P6.64.

6.64 Figure P6.64.
6.65 Figure P6.65.
6.66 Figure P6.66.
6.67 Figure P6.67.
6.68 Figure P6.68.
6.69 The truss shown in Figure P6.69 is improperly constrained. For the loading shown, apply the equations of equilibrium, and attempt to solve for all member forces and reactions for this truss. Comment on why you were unable to continue your solution.

6.70–6.72 Classify the trusses of each of the figures as properly, partially, or improperly constrained. If a given truss is properly constrained, state whether it is statically determinate or indeterminate. Further classify the indeterminate trusses as to degree of indeterminacy. For each truss, compare $r + m$ to $2j$.
6.70 Figure P6.70.
6.71 Figure P6.71.
6.72 Figure P6.72.

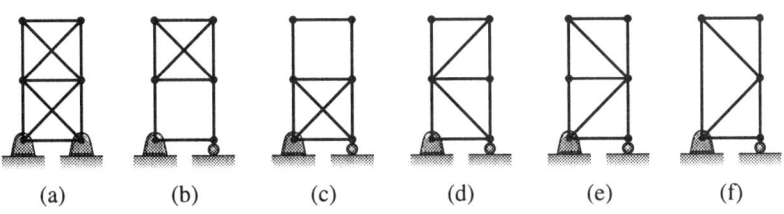

(a) (b) (c) (d) (e) (f)

FIGURE P6.65.

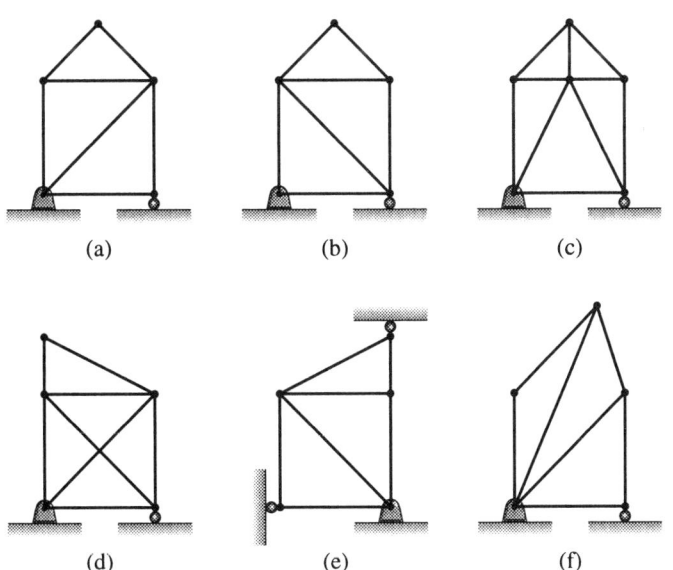

(a) (b) (c)

(d) (e) (f)

FIGURE P6.66.

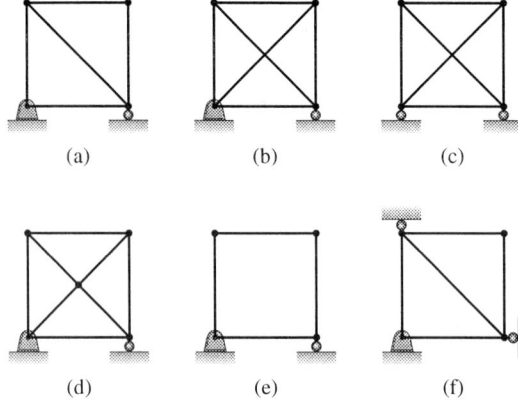

FIGURE P6.67.

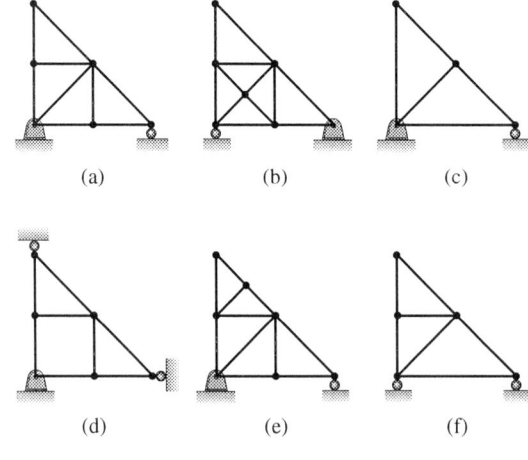

FIGURE P6.68.

6.73 Although the number of available equations equals the number of unknown forces for the truss shown in Figure P6.73, you will find that the truss is im- properly constrained. Attempt to solve for all bar forces and reactions for this truss. Comment on why you were unable to continue your solution.

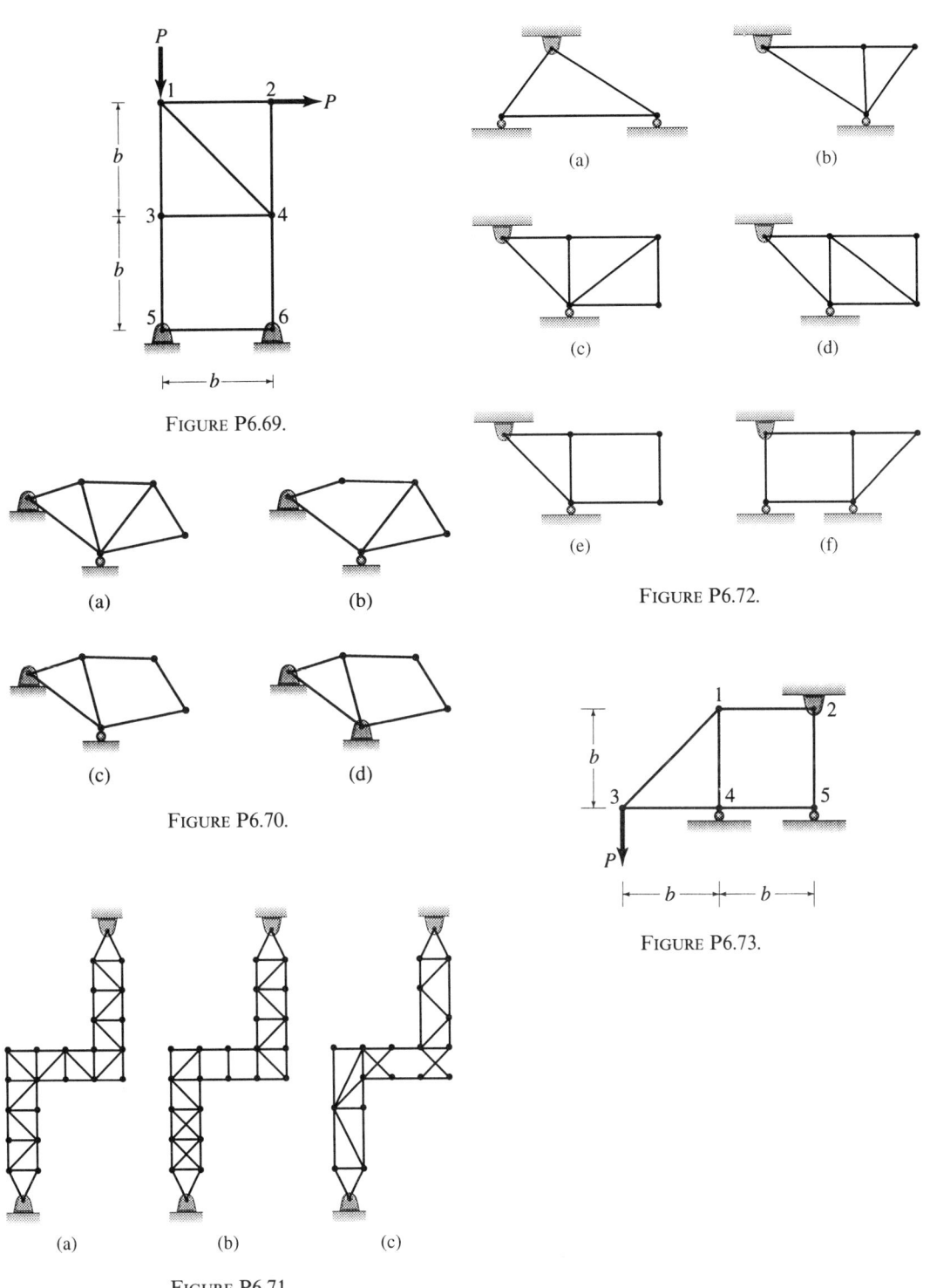

FIGURE P6.69.

FIGURE P6.70.

FIGURE P6.71.

FIGURE P6.72.

FIGURE P6.73.

6.6
Compound
Trusses

Compound trusses are formed by fastening two or more simple trusses together. Refer to Figure 6.14 which is carefully captioned to describe the formation of compound trusses from the two simple trusses of Figure 6.14(a). A common pin at joints 5 and 9 fastens these simple trusses together as shown in Figure 6.14(b). Pin supports to transmit forces to ground are provided at joints 1 and 13, and, although we have formed and supported a compound truss, it is incompletely constrained. Each simple truss is rigid, but a single pin in common at joints 5 and 9 is not sufficient to form a rigid compound truss. Rotation can take place at this common pin. Of course, all trusses in the physical world deform under the action of forces, but, in our theory, we are dealing with perfectly rigid trusses and cannot admit this relative rotation between the two truss halves at the center joint.

In Figure 6.14(c), we add a member between joints 8 and 16 to form a stable compound truss. The resulting truss is statically determinate and properly constrained. Because $r = 3$, $m = 27$, and $j = 15$, the equation $r + m = 2j$ is satisfied. This compound truss cannot be assembled by the procedure for constructing simple trusses discussed earlier.

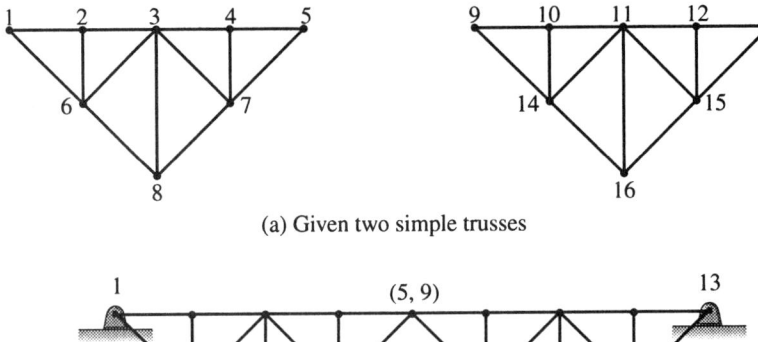

(a) Given two simple trusses

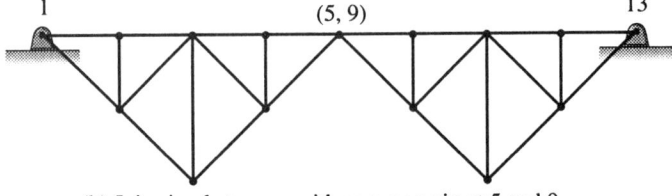

(b) Join simple trusses with common pin at 5 and 9.
Provide pin supports at 1 and 13.

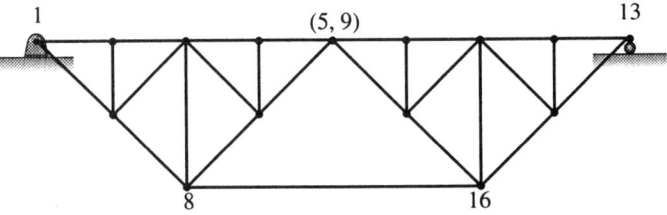

FIGURE 6.14.

(c) and a member connecting 8 to 16.

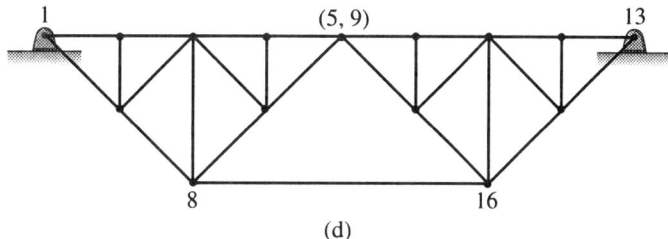

(d)

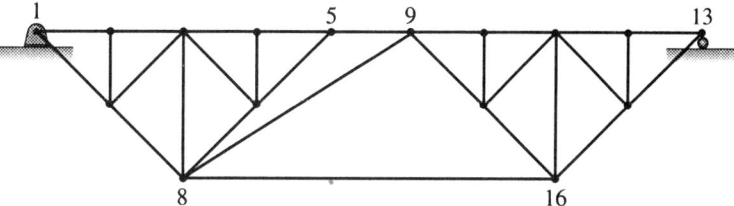

(e) Join simple trusses with three members, 5–9,
8–9 and 8–16. Provide a pin support at 1
and a roller at 13.

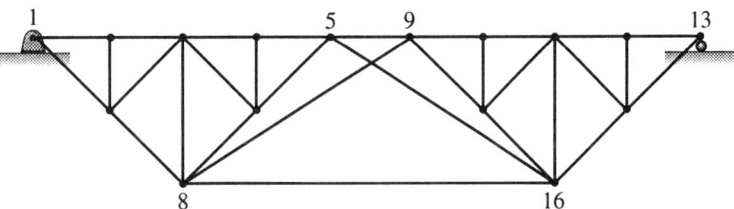

(f) Join simple trusses with four bars, 5–9, 5–16,
8–9 and 8–16. Provide a pin support at 1
and a roller at 13.

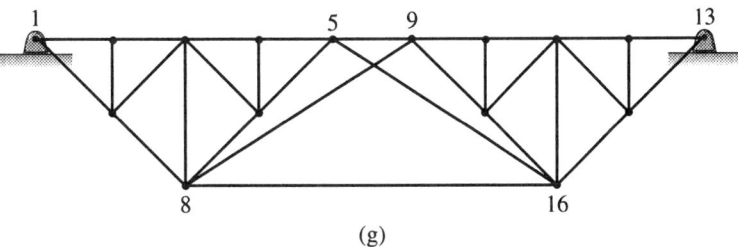

(g)

FIGURE 6.14 (*Cont.*).

In Figure 6.14(c), if we replace the roller at joint 13 by a pin support, the truss shown in Figure 6.14(d) is the result. Because $r = 4$, $m = 27$, and $j = 15$, we find that $r + m = 31$ and $2j = 30$. There is one more unknown than there are equations which means that this compound truss is statically indeterminate externally to the first degree. This compound truss is properly constrained, but the equations of equilibrium, although necessary for a solution, are not sufficient for a complete solution for all reactions and bar forces. In more advanced courses, the equations of equilibrium must be supplemented by equations involving truss deformations to determine a complete solution.

The two simple trusses of Figure 6.14(a) are joined together by three members 5–9, 8–9, and 8–16 to form the compound truss of Figure 6.14(e) which is supported by a pin at joint 1 and a roller at joint 13. This resulting truss is statically determinate and properly constrained. Note that $r = 3$, $m = 29$, and $j = 16$ which means that $r + m = 2j$ is satisfied.

Four members, 5–9, 5–16, 8–9, and 8–16 were used to join two simple trusses to create the compound truss shown in Figure 6.14(f). In this case, $r = 3$, $m = 30$ and $j = 16$ which means that $m + r = 33$ whereas $2j = 32$. This compound truss is properly constrained and statically indeterminate to the first degree, internally.

If we replace the roller at joint 13 with a pin in the truss shown in Figure 6.14(f), we obtain the truss of Figure 6.14(g) which is properly constrained and statically indeterminate to the second degree, once internally and once externally.

The following example illustrates the procedure used in solving compound trusses.

■ **Example 6.9**

A compound truss, with a downward load P, is shown in Figure E6.9(a). The support reactions are also shown. In terms of P, determine (a) the forces in members 4–5, 5–11, 11–16, and 15–16 and (b) numerical values for these forces for $P = 250$ kN and $b = 5$ m.

Solution

(a) We begin by solving for the force in member 3–4. This is accomplished by cutting the truss into two sections and constructing the free-body diagram shown in Figure E6.9(b). Thus,

$$\sum M_3 = 0, \quad T_{1516}(3b) - \frac{5}{8}P(4b) = 0,$$

$$T_{1516} = 0.833P \quad \text{(T)}. \qquad\qquad \text{ANS.}$$

Now that we know the force in member 15–16, we pass another cutting plane through members 4–5, 11–15, 11–16, and 15–16 and create

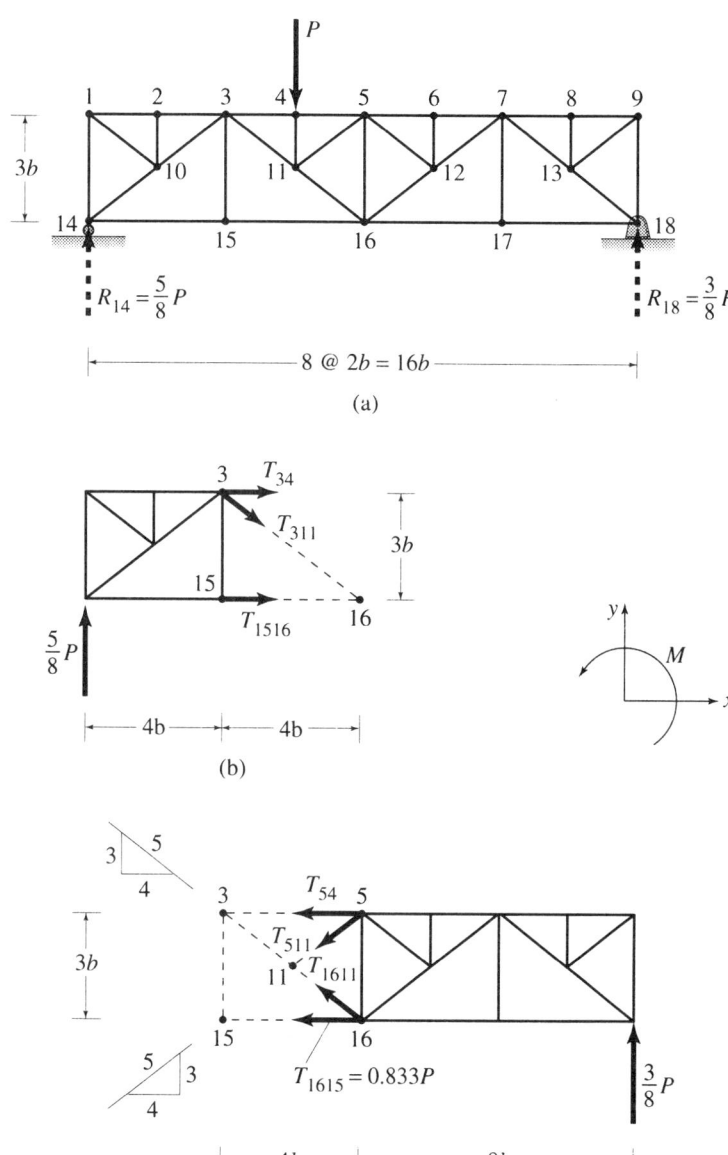

FIGURE E6.9.

the free-body diagram shown in Figure E6.9(c). Thus,

$$\sum M_5 = 0, \quad \frac{3}{8}P(8b) - 0.833P(3b) - \left(\frac{4}{5}\right)T_{1611}(3b) = 0,$$

$$T_{1611} = 0.209P(T).$$

$$\sum F_y = 0, \quad \frac{3}{8}P + \frac{3}{5}(0.209P) - \frac{3}{5}T_{511} = 0,$$

$$T_{511} = 0.834P \quad (T). \qquad\qquad \text{ANS.}$$

$$\sum M_{16} = 0, \quad \frac{3}{8}P(8b) + T_{54}(3b) + \frac{4}{5}(0.834P)(3b) = 0,$$

$$T_A = -1.667P = 1.667P \quad (C). \qquad\qquad \text{ANS.}$$

To check these results, we will sum the moments of the forces with respect to axes through points 3 and 15.

$$\sum M_3 = 0, \quad \frac{3}{8}P(12b) - \frac{3}{5}(0.834P)(4b) - 0.833(3b) \approx 0. \quad \text{Check.}$$

$$\sum M_{15} = 0, \quad \frac{3}{8}P(12b) + \frac{3}{5}(0.209P)(4b) - 1.667(3b) \approx 0. \quad \text{Check.}$$

Interested readers may write $\sum M_{18} = 0$ as another check equation.

(b) Note that member forces are independent of the value of b. For $P = 250 \text{ kN}$,

$$T_{45} = 1.667P \quad (C) = 417 \text{ kN} \quad (C), \qquad\qquad \text{ANS.}$$

$$T_{511} = 0.834P \quad (T) = 209 \text{ kN} \quad (T), \qquad\qquad \text{ANS.}$$

$$T_{1116} = 0.209P \quad (T) = 52.0 \text{ kN} \quad (T), \qquad\qquad \text{ANS.}$$

and

$$T_{1516} = 0.833P \quad (T) = 208 \text{ kN} \quad (T). \qquad\qquad \text{ANS.}$$

■

Problems

6.74 Two simple trusses were connected at pin B, and member DE was added to form the compound truss shown in Figure P6.74. It is supported by a pin at A and a roller at C. Determine the reaction at C and the force in member DE. Then, determine the forces in the members denoted 1, 2, and 3.

6.75 In Figure P6.75, the center truss was connected to the side ones by pins at 10 and

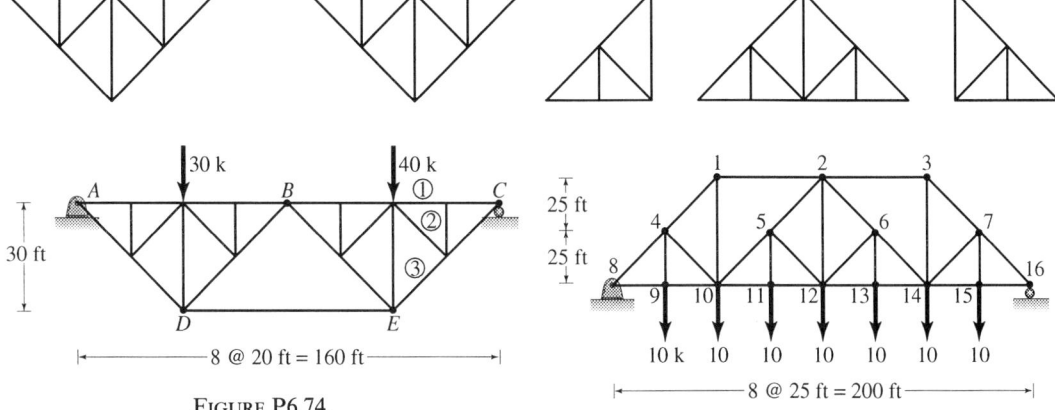

FIGURE P6.74.

FIGURE P6.75.

14 and by the addition of members 1–2 and 2–3 to form the compound truss shown. Find the bar forces in members 1–2, 2–5, 5–12, and 11–12 of this truss.

6.76 Refer to the compound truss shown in Figure P6.75, and determine the bar forces in members 2–3, 2–6, 6–12, and 12–13.

6.77 In Figure P6.77, the side trusses were connected to the center truss by pins at 11 and 15 and by the addition of members 3–4 and 4–5 to form the compound

truss shown. In terms of P, find the bar forces in members 3–4 and 4–5 of this truss.

6.78 Let $P = 30$ k and $b = 25$ ft. for the compound truss of Figure P6.77, and, then, determine the bar forces in members 5–6, 6–15, and 15–16.

6.79 The truss of Figure P6.79 was created by fastening the center truss to the outside ones with pins at 3 and 7 and by adding the members 15–16 and 16–17. In terms

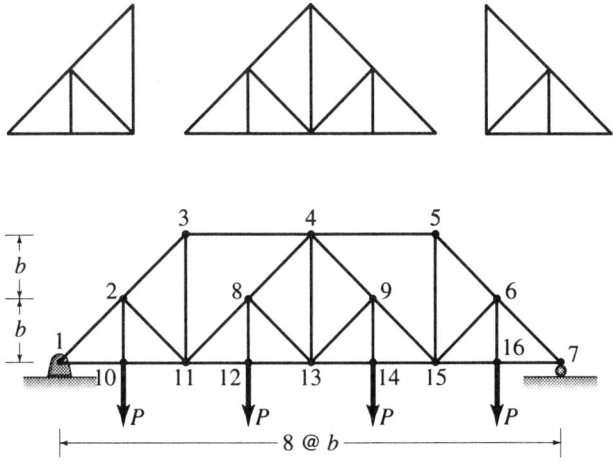

FIGURE P6.77.

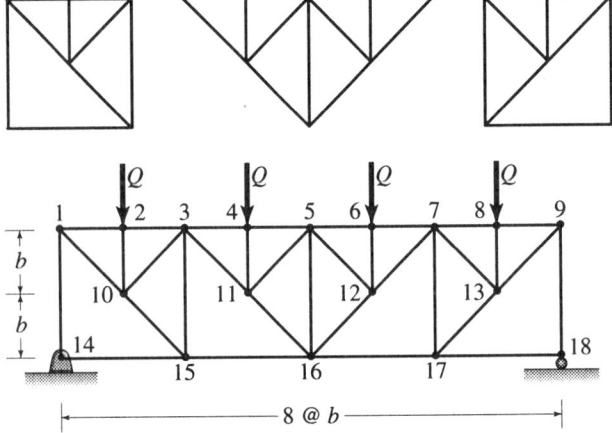

FIGURE P6.79.

of Q, determine the bar forces in members 5–6, 5–12, 12–16, and 16–17.

6.80 Let $Q = 150$ kN and $b = 10$ m for the compound truss of Figure P6.79, and, then, find member forces for 6–7, 7–12, and 16–17.

6.81 Two simple trusses were joined at pin 5, and member 14–15 was added to form the compound truss of Figure P6.81. It is supported by a pin at 1 and a roller at 9. In terms of P, determine the bar forces in members 4–5, 5–11, and 14–15.

6.82 Let $P = 100$ kN and $p = 8$ m for the compound truss of Figure P6.81, and, then, determine the bar forces in members 5–6, 5–12, and 14–15.

6.83 Three trusses are joined by a pin at joint 4 and bars 3–8 and 5–12 to form the

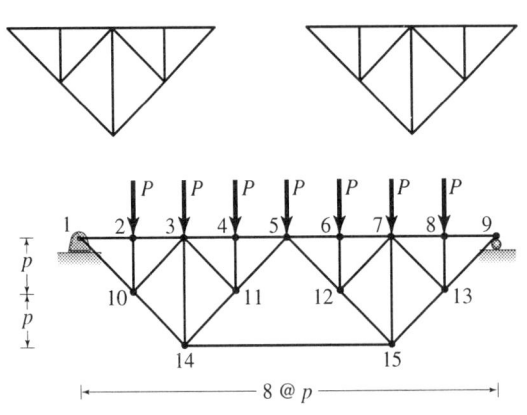

FIGURE P6.81.

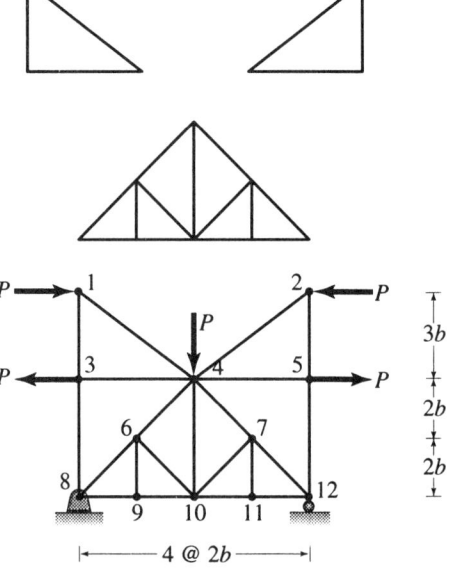

FIGURE P6.83.

compound truss shown in Figure P6.83. In terms of P, determine the bar forces in all members of this truss. Show your answers on a sketch of the truss.

6.84 Assign P a value of 5 k and b a value of 10 ft. for the compound truss of Figure P6.83, and, then, determine the bar forces in all members of this truss. Show your answers on a sketch of the truss.

6.85 In Figure P6.85, the inner triangular truss 2–3–4 is connected to the outer triangular truss 1–5–6 with the three bars 1–2, 3–6, and 4–5 to form the compound truss shown. Determine the reactions at joints 5 and 6. After the reactions are known in terms of Q, there are 3 unknown bar forces at all joints of this truss which means we cannot start at a joint, solve for two unknowns, and proceed to solve for the remaining unknowns. Write the equilibrium equations for pins 1, 2, 3, and 4. You need not solve these equations but you will note that they form a system of 8 equations in 8 unknowns. If we were to solve them, then, the horizontal equilibrium equation for joint 6 would enable us to determine the final bar force in member 5–6. Joint equilibrium equations written for pin 5 would provide

a check of the bar forces and reaction components.

6.86 Two simple trusses were connected at pin 11, and member 1–2 was added to form the compound truss shown in Figure P6.86. It is supported by a pin at joint 7 and a roller at joint 15. Determine the bar forces in members 1–2, 2–5, 5–13, and 12–13.

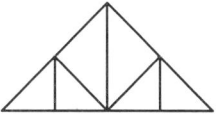 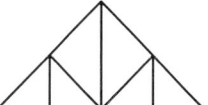

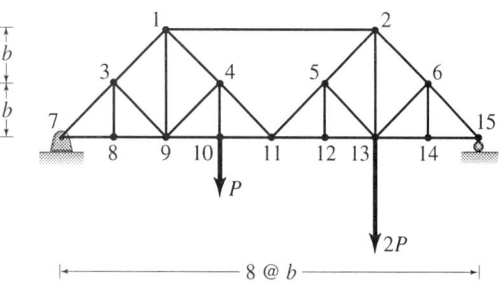

FIGURE P6.86.

6.87 In Figure P6.87, a compound truss is formed by connecting two simple trusses with three bars 5–6, 6–15, and 15–16.

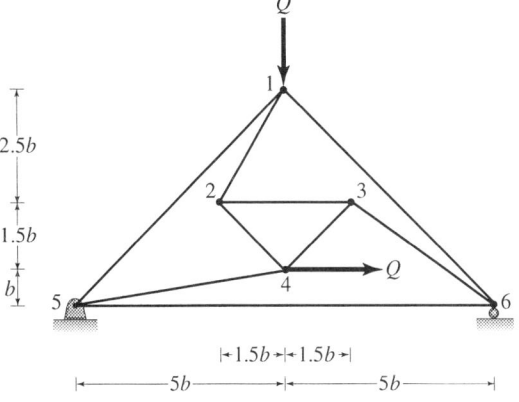

FIGURE P6.85.

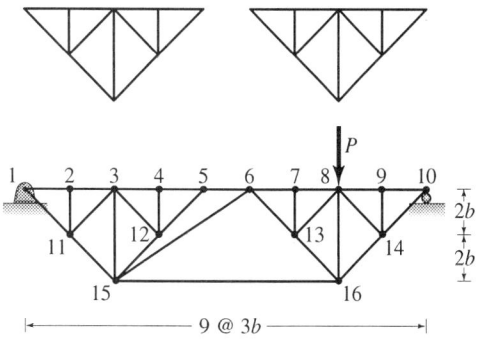

FIGURE P6.87.

This truss is supported by a pin at 1 and a roller at 10. Assign P a value of 50 k and b a value of 10 ft., and determine the bar forces in these three members which connect the simple trusses.

6.88 Two simple trusses are fastened together by three bars 5–6, 5–16, and 15–16 to form the compound truss shown in Figure P6.88. Assign Q a value of 120 kN and b a value of 5 m, and determine the forces in these three members which connect the simple trusses.

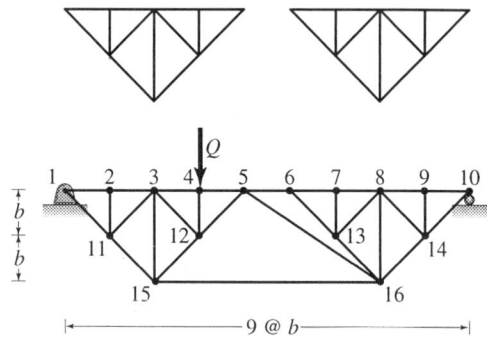

FIGURE P6.88.

6.7*
Three-Dimensional Trusses: Member Forces Using the Method of Joints

In this section we extend our study to three-dimensional trusses. Although the trusses considered here have few members, the basic concepts and the equations of equilibrium are the same for more practical trusses of many members. Computer methods are required for these more practical structures because they contain many more members, and numerous loading conditions must be examined. Personal or mainframe computers are required for efficient analysis and design of such structures. Three members forming a triangle is the elemental rigid body of which two-dimensional, simple trusses are constructed. In a similar fashion, six members forming a tetrahedron is the elemental rigid body of which three-dimensional, simple trusses are constructed. In Figure 6.15, a three-dimensional or space truss is represented. It consists of six members fastened together at four joints A,

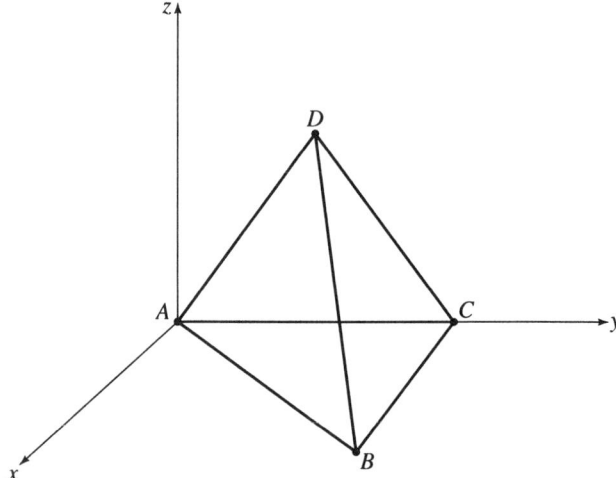

FIGURE 6.15.

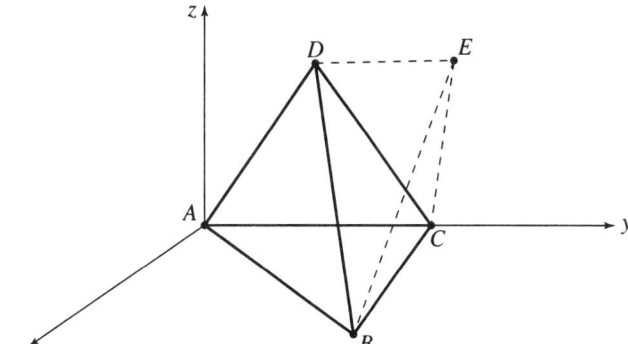

FIGURE 6.16.

B, C, and D to form a tetrahedron. A simple space truss is defined
as one constructed by starting with a single tetrahedron ABCD and
adding additional members and joints to form an assemblage of
tetrahedrons. This is accomplished by a procedure illustrated in Figure
6.16. Starting with tetrahedron ABCD, we add the members BE, CE,
and DE and fasten them together at an added joint E to form
tetrahedron BCDE. This procedure of adding three members connected at a single added joint can be repeated to form simple space
trusses consisting of any desired number of tetrahedrons. The simple
space truss depicted in Figure 6.16 consists of two tetrahedrons composed of nine members and five joints.

The following assumptions are made for the analysis of space trusses:

1. Members are connected together at the joints with frictionless ball-
   and-socket joints.
2. Members of the truss are straight two-force members.
3. Member weights may be considered by applying concentrated
   forces at the joints. Loads and reactions are applied only at the
   joints.
4. Space trusses are rigid bodies.

Of course, these assumptions are only satisfied approximately by
actual space trusses. A critique could be made of them, but it would
be quite similar to the one made in Sec, 6.1 for planar trusses.

As shown in Figure 6.17(a), determinacy of simple space trusses can
be summarized by a single equation: $r + m = 3j$. Six properly placed,
reaction components are required to fully constrain a rigid body in
space. The symbol r refers to the number of these reaction components.
Figure 6.17(b) shows the free-body diagram of this space truss. The ball
and socket at A prevents motion of joint A in the x, y, and z directions.
The roller at B prevents motion of joint B in the y and z directions and
allows motion in the x direction. The ball at C prevents motion of C in

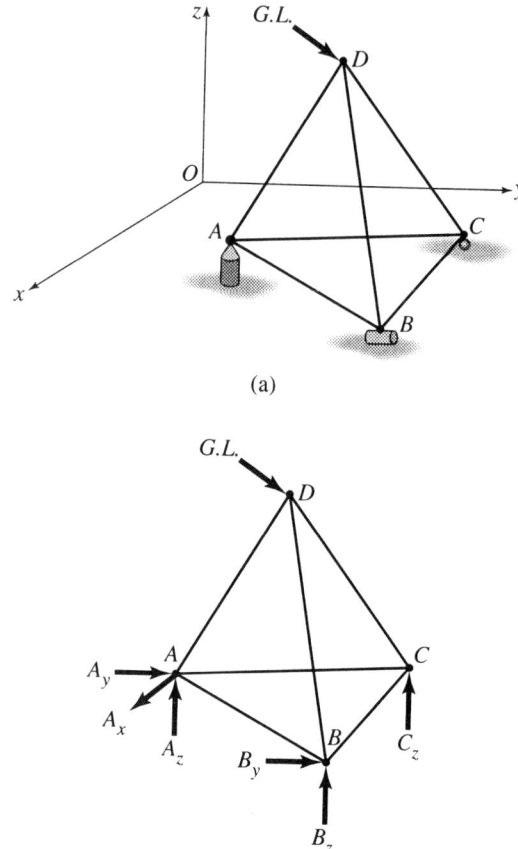

FIGURE 6.17.   (b)

the z direction and allows motion in the x and y directions. The symbol m refers to the number of members of the space truss. Bar forces are directed along the members, so that only the magnitude of each force is unknown. The total number of unknowns is, therefore, $r + m$. The symbol j refers to the number of joints of the space truss. Forces acting at a given joint form a three-dimensional, concurrent system of forces for which three equations of equilibrium are available. The total number of available equations of equilibrium is obviously $3j$. For a statically determinate simple space truss,

$$r + m = 3j. \tag{6.2}$$

The method of joints of Section 6.2 is extended here to space trusses. The resultant **R** of a three-dimensional, concurrent system of forces is given by

$$\mathbf{R} = (\textstyle\sum F_x)\mathbf{i} + (\sum F_y)\mathbf{j} + (\sum F_z)\mathbf{k}.$$

Because **R** equals a null force for the equilibrium of each joint, there are three scalar equations of equilibrium for each joint. Thus,

$$\sum F_x = 0,$$
$$\sum F_y = 0,$$
$$\sum F_z = 0.$$

In the following examples, the vector method is used, because the approach is efficient for three-dimensional problems.

■ **Example 6.10**

A space truss consisting of three members is subjected to the external force $\mathbf{P} = (30\mathbf{i} + 20\mathbf{j} + 30\mathbf{k})$ kN as shown in Figure E6.10(a). Deter-

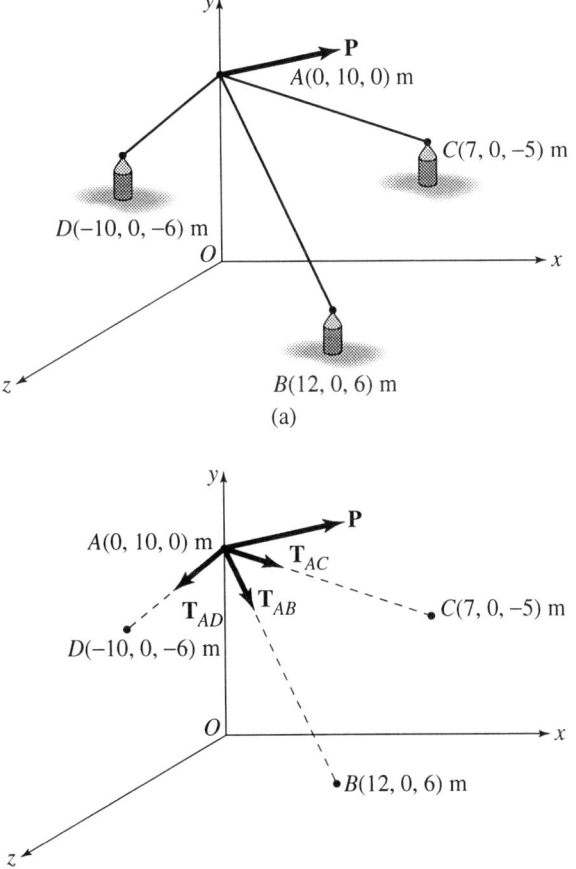

FIGURE E6.10.

mine the forces in the bars AB, AC, and AD. The joints at B, C and D
are ball-and-socket joints. Joint A is free to move in the x, y, and z
directions, but movement is prevented at B, C, and D.

Solution

Figure E6.10(b) shows the free-body diagram of joint A. We express all
forces acting at the joint in vector force. Thus,

$$\mathbf{T}_{AB} = T_{AB}\lambda_{AB} = 0.717T_{AB}\mathbf{i} - 0.598T_{AB}\mathbf{j} - 0.359T_{AB}\mathbf{k},$$

$$\mathbf{T}_{AC} = T_{AC}\lambda_{AC} = 0.531T_{AC}\mathbf{i} - 0.758T_{AC}\mathbf{j} - 0.379T_{AC}\mathbf{k},$$

$$\mathbf{T}_{AD} = T_{AD}\lambda_{AD} = -0.651T_{AD}\mathbf{i} - 0.651T_{AD}\mathbf{j} - 0.391T_{AD}\mathbf{k},$$

and

$$\mathbf{P} = 30\mathbf{i} + 20\mathbf{j} + 30\mathbf{k}.$$

The sum of these forces equals a null vector because the space truss is
in equilibrium. Thus,

$$\sum \mathbf{F} = \mathbf{0}, \quad \mathbf{T}_{AB} + \mathbf{T}_{AC} + \mathbf{T}_{AD} + \mathbf{P} = \mathbf{0}.$$

This single vector equation leads to three scalar, simultaneous, algebraic equations in the three unknowns T_{AB}, T_{AC}, and T_{AD}. Thus,

$$0.717T_{AB} + 0.531T_{AC} - 0.651T_{AD} + 30 = 0,$$

$$-0.598T_{AB} - 0.758T_{AC} - 0.651T_{AD} + 20 = 0,$$

and

$$-0.359T_{AB} - 0.379T_{AC} - 0.391T_{AD} + 30 = 0.$$

Solving these equations simultaneously yields

$$T_{AB} = -27.2 \text{ kN} = 27.2 \text{ kN} \quad (C). \qquad\qquad \text{ANS.}$$

$$T_{AC} = 19.96 \text{ kN} \quad (T), \qquad\qquad \text{ANS.}$$

and

$$T_{AD} = 32.4 \text{ kN} \quad (T). \qquad\qquad \text{ANS.}$$

■ **Example 6.11**

A space truss consisting of six members is subjected to the force $\mathbf{P} = (30\mathbf{i} + 20\mathbf{j} + 30\mathbf{k})$ k shown in Figure E6.11(a). These members form a
tetrahedron. Show that the truss is statically determinate, and use the
method of joints to determine all member forces and all reaction components. Note that the support at B is a ball and socket, the roller
support at C allows motion in the x direction, and the ball at D allows
motion in the x-z plane. Members are fastened with ball-and-socket
joints at A, which is free to move in the x, y, and z directions.

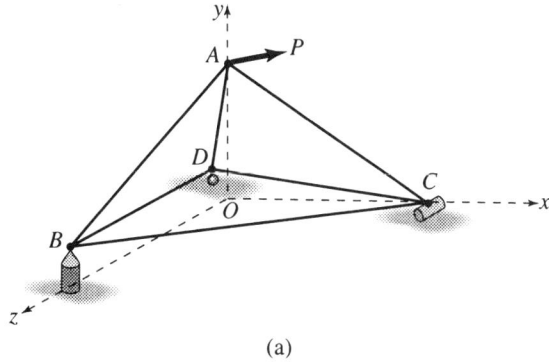

(a)

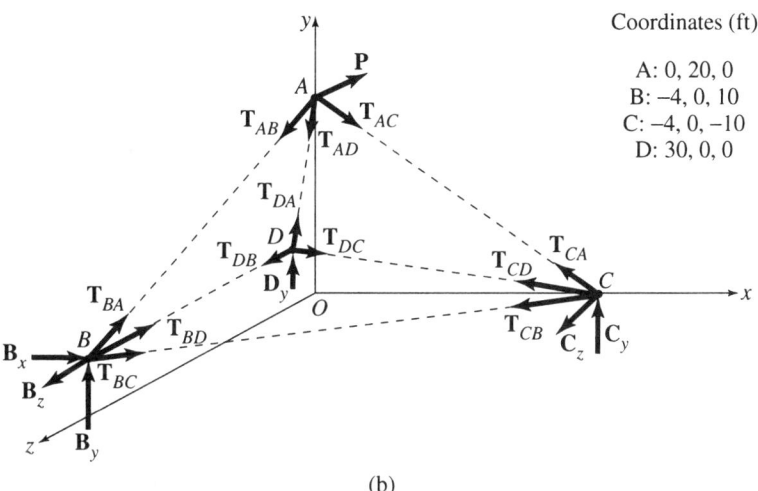

Coordinates (ft)

A: 0, 20, 0
B: −4, 0, 10
C: −4, 0, −10
D: 30, 0, 0

FIGURE E6.11.   (b)

Solution

Free-body diagrams of the four joints A, B, C, and D are shown in Figure E6.11(b). There are six unknown forces in the six members of the truss and there are six unknown reaction components (B_x, B_y, B_z, C_y, C_z, and D_y) which makes a total of 12 scalar unknowns. At each of four joints, we have one vector equation or 3 scalar equations of equilibrium which makes a total of 12 scalar equations available. Because there are 12 equations and 12 unknowns, the space truss is statically determinate.

The method of joints is applied to the four joints in the order A, D, C, and B as follows:

Joint A

$$\mathbf{P} = 30\mathbf{i} + 20\mathbf{j} + 30\mathbf{k},$$

$$\mathbf{T}_{AB} = T_{AB}\boldsymbol{\lambda}_{AB} = -0.176T_{AB}\mathbf{i} - 0.880T_{AB}\mathbf{j} + 0.440T_{AB}\mathbf{k},$$

$$\mathbf{T}_{AC} = T_{AC}\boldsymbol{\lambda}_{AC} = 0.832T_{AC}\mathbf{i} - 0.555T_{AB}\mathbf{j}$$

$$T_{AD} = T_{AD}\lambda_{AD} = -0.176T_{AD}\mathbf{i} - 0.880T_{AD}\mathbf{j} - 0.440T_{AD}\mathbf{k}.$$

$$\sum \mathbf{F} = \mathbf{0}, \quad \mathbf{P} + \mathbf{T}_{AB} + \mathbf{T}_{AC} + \mathbf{T}_{AD} = \mathbf{0}.$$

The three scalar equations become

$$30 - 0.176T_{AB} + 0.832T_{AC} - 0.176T_{AD} = 0,$$

$$20 - 0.880T_{AB} - 0.555T_{AC} - 0.880T_{AD} = 0,$$

and

$$30 + 0.440T_{AB} - 0.440T_{AD} = 0.$$

Solving simultaneously,

$$T_{AB} = -14.03 \text{ k} \approx 14.0 \text{ k} \quad (C),$$ ANS.

$$T_{AC} = -27.57 \text{ k} \approx 27.6 \text{ k} \quad (C),$$ ANS.

and

$$T_{AD} = 54.15 \text{ k} \approx 54.2 \text{ k} \quad (T).$$ ANS.

Joint D

$$T_{AD} = -T_{AD}\lambda_{AD} = 0.176T_{AD}\mathbf{i} + 0.880T_{AD}\mathbf{j} + 0.440T_{AD}\mathbf{k},$$

$$T_{DB} = T_{DB}\lambda_{DB} = T_{DB}\mathbf{k},$$

$$T_{DC} = T_{DC}\lambda_{DC} = 0.959T_{DC}\mathbf{i} + 0.282T_{DC}\mathbf{k},$$

and

$$\mathbf{D} = D_y\mathbf{j}.$$

Because $T_{AD} = 54.15$ k (T), the three scalar equations become

$$0.176(54.15) + 0.959T_{DC} = 0,$$

$$0.880(54.15) + D_y = 0,$$

and

$$T_{DB} + 0.440(54.15) + 0.282T_{DC} = 0.$$

Solving,

$$T_{DC} = -9.94 \text{ k} = 9.94 \text{ k} \quad (C),$$ ANS.

$$D_y = -47.65 \text{ k},$$ ANS.

and

$$T_{DB} = -21.02 \text{ k} \approx 21.0 \text{ k} \quad (C).$$ ANS.

Joint C

$$T_{CA} = -T_{AC} = -0.832(-27.57)\mathbf{i} + 0.555(-27.57)\mathbf{j},$$

$$T_{CD} = -T_{DC} = -0.959(-9.94)\mathbf{i} - 0.282(-9.94)\mathbf{k},$$

$$T_{CB} = T_{CB}\lambda_{CB} = -0.959T_{CB}\mathbf{i} + 0.282T_{CB}\mathbf{k},$$

and

$$\mathbf{R_C} = C_y\mathbf{j} + C_z\mathbf{k}.$$

The corresponding scalar equations are

$$-0.832(-27.57) - 0.959(-9.94) - 0.959T_{CB} = 0,$$

$$0.555(-27.57) + C_y = 0,$$

and

$$-0.282(-9.94) + 0.282T_{CB} + C_z = 0.$$

Solving,

$$T_{CB} = 33.86 \text{ k} \approx 33.9 \text{ k} \quad \text{(T)}, \qquad \text{ANS.}$$

$$C_y = 15.30 \text{ k}, \qquad \text{ANS.}$$

and

$$C_z = -12.35 \text{ k}. \qquad \text{ANS.}$$

Joint B

$$\mathbf{T_{BA}} = -\mathbf{T_{AB}} = 0.176(-14.03)\mathbf{i} + 0.880(-14.03)\mathbf{j} - 0.440(-14.03)\mathbf{k},$$

$$\mathbf{T_{BC}} = -\mathbf{T_{CB}} = 0.959(33.86)\mathbf{i} - 0.282(33.86)\mathbf{k},$$

$$\mathbf{T_{BD}} = -\mathbf{T_{DB}} = -(-21.02)\mathbf{k},$$

and

$$\mathbf{B} = B_x\mathbf{i} + B_y\mathbf{j} + B_z\mathbf{k}.$$

The corresponding scalar equations are

$$0.176(-14.03) + 0.959(33.86) + B_x = 0,$$

$$0.880(-14.03) + B_y = 0,$$

and

$$-0.440(-14.03) - 0.282(33.86) - (-21.02) + B_z = 0.$$

Solving,

$$B_x = -30.0 \text{ k}, \qquad \text{ANS.}$$

$$B_y = 12.35 \text{ k}, \qquad \text{ANS.}$$

and

$$B_z = -17.65 \text{ k}. \qquad \text{ANS.}$$

As a check, we could consider the free-body diagram of the entire space truss. The member forces would become internal, and we need consider only P and the reactions. The details of this check are left as an exercise for the student to carry out.

Problems

6.89 The free-body diagram of a truss is shown in Figure P6.89. It consists of six members which form a tetrahedron. Show that this space truss is, overall, statically determinate. Determine the unknown reaction components for given coordinates of D $(2, 6, 8)$ expressed in meters.

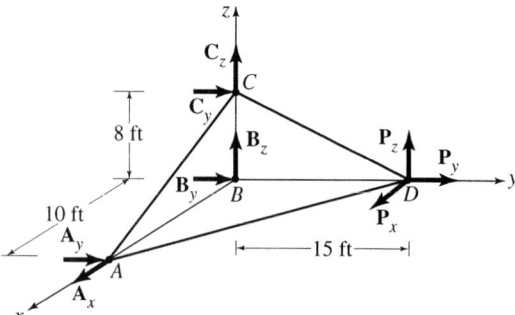

FIGURE P6.91.

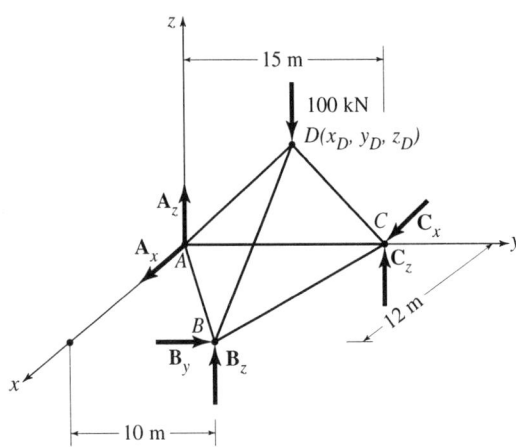

FIGURE P6.89.

6.90 Solve Problem 6.89 for the coordinates of D $(3, 5, 6)$ expressed in meters.

6.91 The free-body diagram of a six-member space truss is shown in Figure P6.91. Show that it is, overall, statically determinate. The three components of force P at D in x, y, and z order and expressed in pounds are 0, 2000, and 6000. Assume $B_z = C_z$ when solving for the member forces and reactions.

6.92 Solve Problem 6.91 for the three force components of P acting at D in x, y, and z order and expressed in pounds as 1500, 2500, and 0. Assume $B_z = 0$.

6.93 The free-body diagram of a six-member

space truss is shown in Figure P6.93. Determine all six reaction components, given the coordinates of D $(1.5, 2, -5)$ in meters.

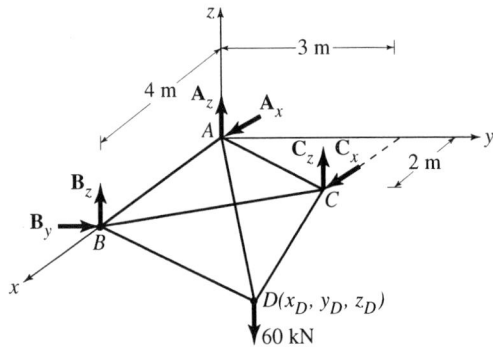

FIGURE P6.93.

6.94 Solve Problem 6.93 for the coordinates of D $(2.5, 1, -8)$ in meters.

6.95 The free-body diagram of an eight-member space truss is shown in Figure P6.95. The applied force of 60 k is directed in the positive x sense.
(a) Show that this space truss is statically determinate from an overall standpoint.

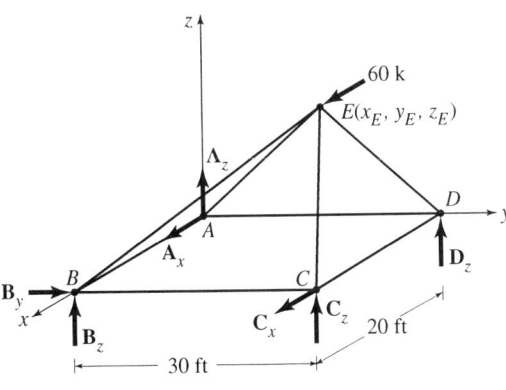

FIGURE P6.95.

(b) If we consider the entire truss as a rigid body and write moment and force

summation equations, then, there are seven reaction components but only six scalar equations available. We will not be able to find all seven reactions components from these equations. However, we will be able to solve for some of these unknown reaction components. As an illustration of this, write $\sum \mathbf{F_x} = \mathbf{0}$ and $\sum \mathbf{M_z} = \mathbf{0}$ to find A_x and C_x. Use the coordinates of E (8, 20, 15) expressed in ft.

6.96   Solve Problem 6.95 for the coordinates of E (6, 24, 10) expressed in ft.

6.97–6.99 A three member space truss is shown in Figure P6.97. Determine the axial force in each bar for the following given information:

Problem Number	A (ft)	B (ft)	C (ft)	P_x (lb)	P_y (lb)	P_z (lb)
6.97	6, 6, 6	8, 10, 12	−6, 8, 4	400	−600	200
6.98	4, 5, 6	7, 12, 15	−10, 6, 2	800	−400	800
6.99	8, 10, 16	8, 20, 14	−8, 12, −6	1000	−1000	600

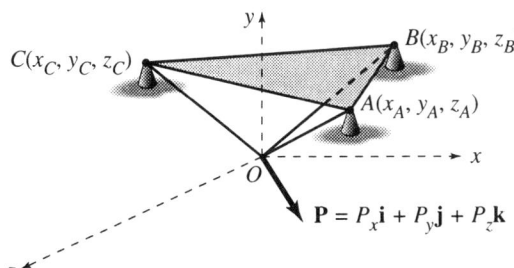

FIGURE P6.97.

6.100 A six member space truss is shown in Figure P6.100. The members are connected to a ball and socket at A, joint B is constrained to move along the x axis, and Joint D is free to move in the x-z plane. Let $\mathbf{P} = (10\mathbf{i} + 15\mathbf{j} - 20\mathbf{k})$ kN.

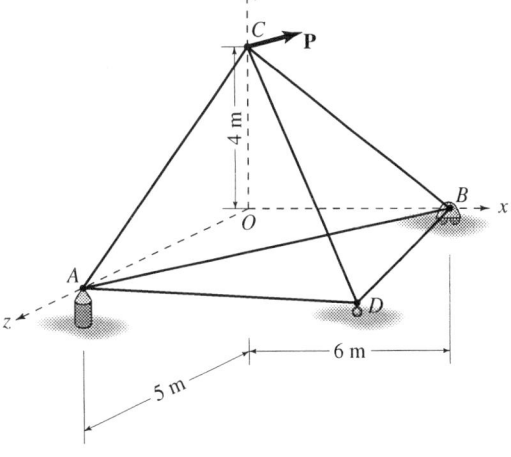

FIGURE P6.100.

(a) Show that the truss is statically determinate.

(b) Find the reaction components at A, B, and D.

(c) Find the bar forces in member AB, AC, and AD.

Review Problems

6.101–6.105 Use the method of joints to determine the forces in the members indicated, specifying whether the members are in tension or compression.

6.101 Members 1–3 and 2–3 in Figure P6.101.

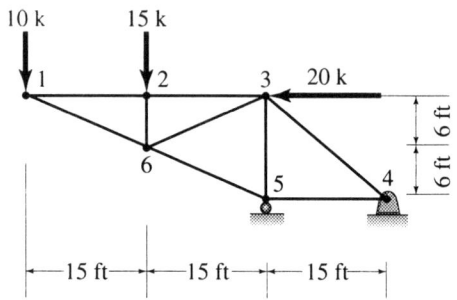

FIGURE P6.102.

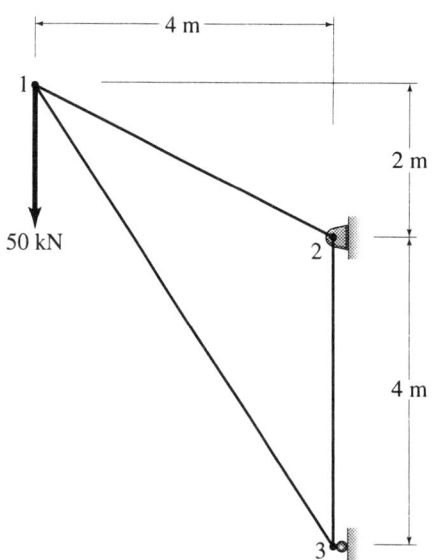

FIGURE P6.101.

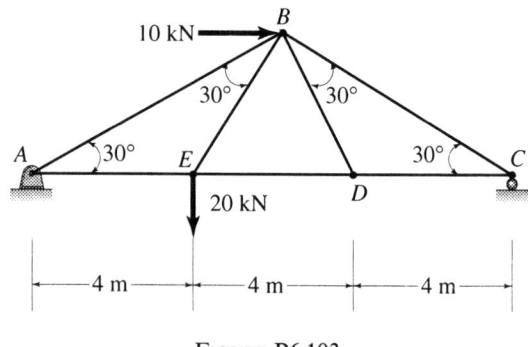

FIGURE P6.103.

6.102 Members 1–2, 1–6, 2–3, 2–6, 6–3, and 6–5 in Figure P6.102.

6.103 All members in Figure P6.103.

6.104 All members in Figure P6.104.

6.105 All members in Figure P6.105.

6.106–6.110 Use the method of sections to determine the forces in the members indicated, specifying whether the members are in tension or compression.

6.106 Members BC and FE in Figure P6.106.

6.107 Members FG and FB in Figure P6.106.

6.108 Members 3–4 and 8–7 in Figure P6.108.

6.109 Members 9–10 and 2–3 in Figure P6.109. Hint: Use a plane that cuts through 3–2, 3–6, 6–9, and 9–10.

6.110 Member 3–6 in Figure P6.109. Hint: Use a plane that cuts through 3–2, 3–6, 5–9, and 5–8 and the result obtained for member 3–2 in Problem 6.109.

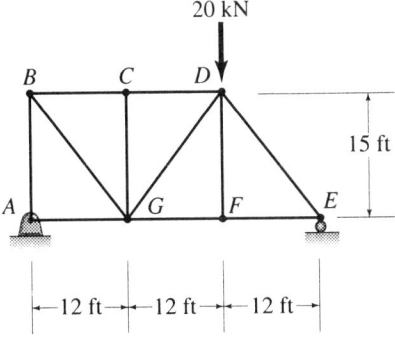

FIGURE P6.104.

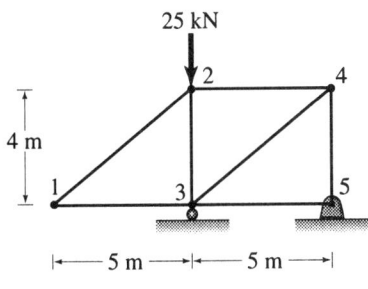

FIGURE P6.105.

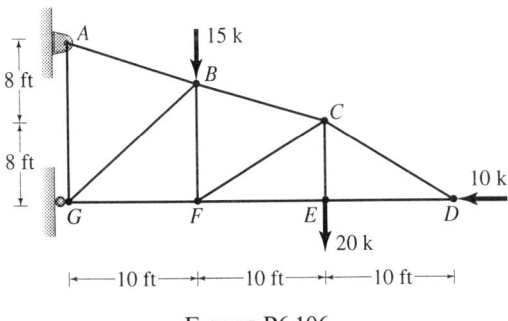

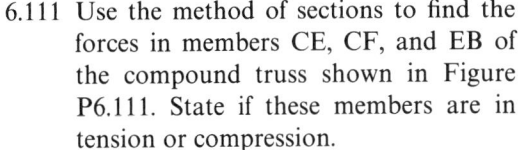

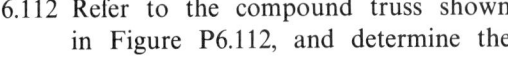

FIGURE P6.106.

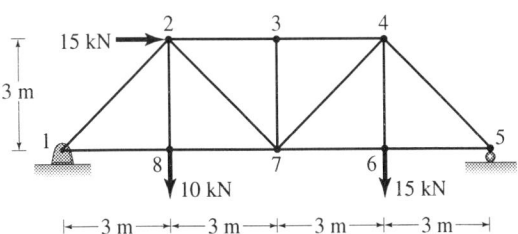

FIGURE P6.108.

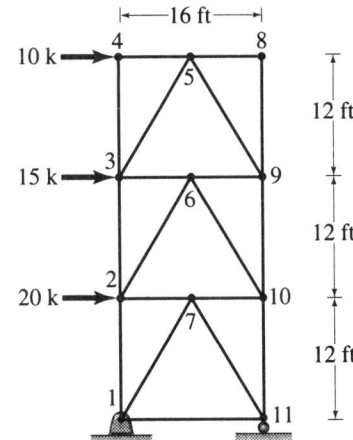

FIGURE P6.109.

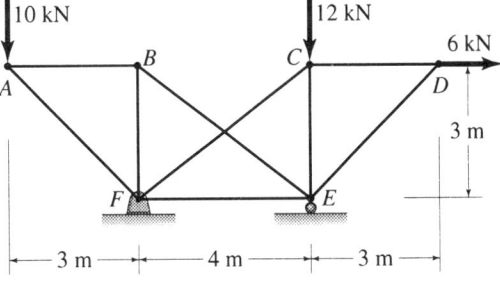

FIGURE P6.111.

6.111 Use the method of sections to find the forces in members CE, CF, and EB of the compound truss shown in Figure P6.111. State if these members are in tension or compression.

6.112 Refer to the compound truss shown in Figure P6.112, and determine the forces in members 2–3, 3–5 and 5–6, stating whether they are in tension or compression.

6.113 Determine the forces in members AD, BE, and CF in the compound truss

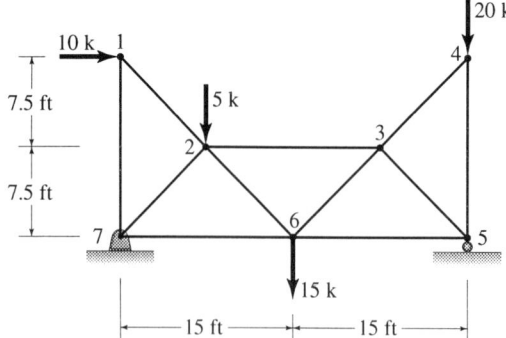

FIGURE P6.112.

shown in Figure P6.113, and state whether these members are in tension or compression. Hint: Construct the free-body diagram of truss DEF.

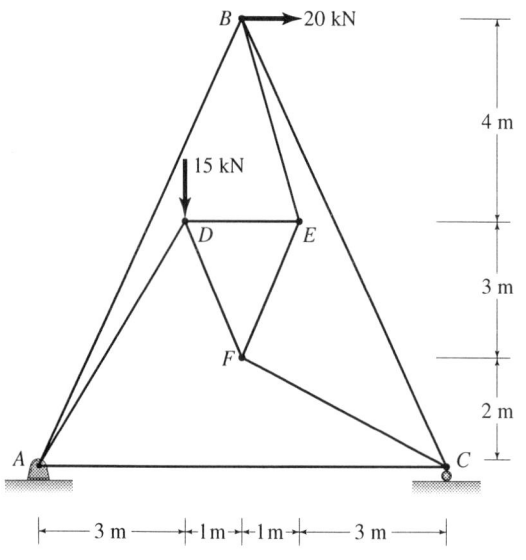

FIGURE P6.113.

6.114 Determine the six reaction components at supports A (ball), B (ball and socket) and C (roller) that support the six-member space truss shown in Figure P6.114. Let **P** = 20**i** − 10**j** + 5**k**) kN, and use vector algebra.

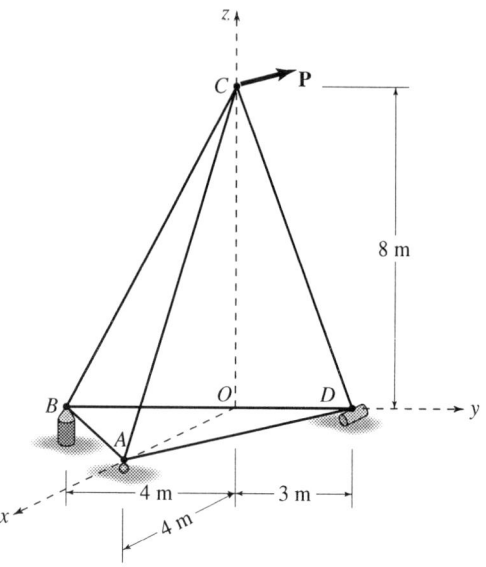

FIGURE P6.114.

6.115 Refer to the space truss of Figure P6.114, and, using vector algebra, determine the forces in members BC, BD, and BA. State whether these members are in tension or compression. Hint: Use the support reactions obtained in Problem 6.114.

7

Frames and Machines

A water wheel— the oldest machine still in use.

According to the Guinness Book of World Records, the oldest machine still in use today is the "dalu"—a water-raising instrument known to have been in use in the Sumerian civilization, which originated 5500 years ago in what is now lower Iraq. The complexity of modern machines is matched only by the ingenuity of engineers whose spirit to conquer has lived throughout the ages.

Complex machines are found everywhere on this planet but the ancient Greeks had a knowledge of the six fundamental machines discussed in Chapter 9. A very interesting fact is that the most complex machines known to man are combinations of many of these six fundamental machines. The marvelous machines we use today, ranging from supersonic jets to magnificent ships, reflect the degree to which engineering science has evolved.

The student is advised to carefully study the examples and to solve a large variety of problems to master the analysis of frames and machines. Qualitative analysis can be used to supplement more time-consuming quantitative analysis. Draw the free-body diagram and compare the number of unknowns to the number of equations available. Outline the solution by stating which unknown(s) can be solved from which equations. In other words, outline a method of attacking the problem before actually writing and solving the equations in detail.

7.1
Multiforce
Members

The concept of two- and three-force members was introduced in Chapter 4, and the case of two-force members was discussed further in connection with truss analysis in Chapter 6. In this section, we discuss *multi-force members* which are essential components of frames and machines. By definition, multiforce members are those that may carry moments and are subjected to forces at three or more locations. Thus, unlike a two-force member in which the applied forces must act along the axis of the member, the forces applied to a multiforce member do *not*, in general, act along the axis of the member. Therefore, whereas trusses are composed exclusively of two-force members, frames and machines contain multiforce members.

A three-force member is a special case of multiforce members and often encountered in the analysis of frames and machines. Such a member was discussed briefly in Chapter 4 where we concluded that the three forces acting on the member must be concurrent if the member is to be in equilibrium. Two cases are identified. In the first case, the point of concurrence is determinate, as shown in Figure 7.1(a), in which case the three forces are nonparallel. In the second case, the point of concurrence is indeterminate as shown in Figure 7.2, in which case the three forces are parallel and their point of concurrence is at infinity. In either case, we are dealing with a two-dimensional, concurrent, force system in equilibrium and, therefore, two independent equations of equilibrium are available for determining two unknown quantities. In certain cases where the point of concurrence of the three forces is determinate, it is convenient to use the polygon method instead of the equations of equilibrium. In such a case, because the three forces are in equilibrium, the force polygon reduces to a closed triangle which may be conveniently analyzed by the trigonometric properties of triangles to determine the unknown quantities. As an illustration, the force triangle for the three-force member of Figure 7.1(a) is shown in Figure 7.1(b).

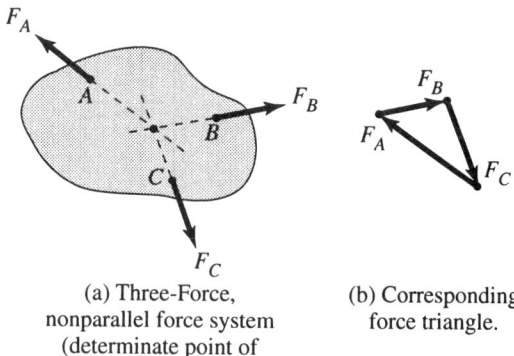

(a) Three-Force,
nonparallel force system
(determinate point of
concurrence).

(b) Corresponding
force triangle.

FIGURE 7.1.

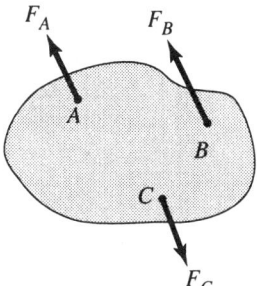

Three-Force,
parallel force system
(indeterminate point
of concurrence).   FIGURE 7.2.

If a multiforce member is subjected to forces at more than three locations, the force system is a general two-dimensional force system in which three independent equations of equilibrium are available for determining three unknown quantities.

The following examples illustrate some of the concepts discussed above.

■ **Example 7.1**

A load $P = 10$ kN is applied at A to member ABD as shown in Figure E7.1(a). The member is supported by a cable BC and a frictionless pin at D. Determine the cable tension and the pin reaction at D using (a) the equations of equilibrium and (b) the force polygon method.

Solution

(a) The free-body diagram of member ABD is shown in Figure E7.1(b) along with an appropriate coordinate system. Note that $\theta = \tan^{-1}(2/3) = 33.7°$. Thus,

$$\sum M_{\mathrm{D}} = 0, \quad 10(5) - T \sin 33.7°(3) = 0,$$

$$T = 30.0 \text{ kN}. \qquad \text{ANS.}$$

$$\sum F_x = 0, \quad T \cos 33.7° + D_x = 0,$$

$$D_x = -25.0 = 25.0 \text{ kN} \leftarrow. \qquad \text{ANS.}$$

$$\sum F_y = 0, \quad -10 + T \sin 33.7° + D_y = 0,$$

$$D_y = -6.67 = 6.67 \text{ kN} \downarrow. \qquad \text{ANS.}$$

Combining these components to obtain the resultant pin reaction at D,

$$R_{\mathrm{D}} = \sqrt{D_x^2 + D_y^2} = \sqrt{25.0^2 + 6.67^2}$$

$$= 25.9 \text{ kN}. \qquad \text{ANS.}$$

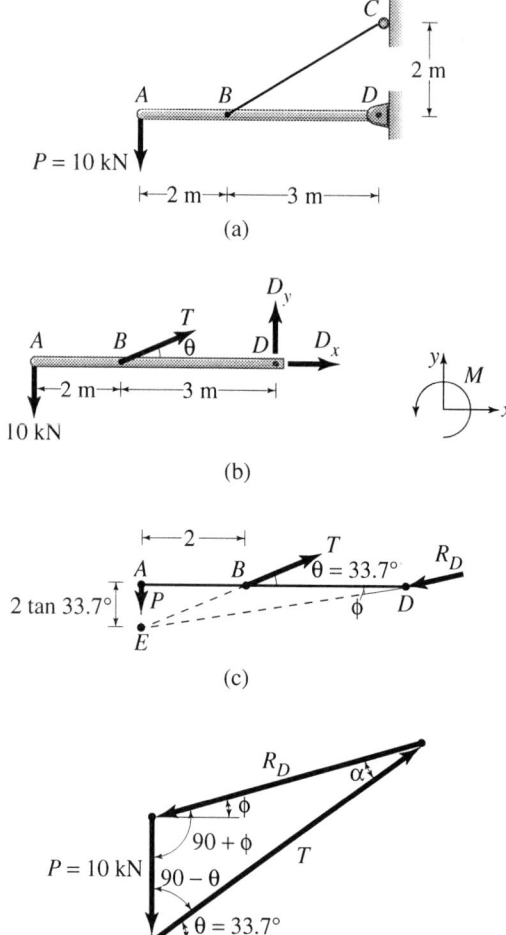

FIGURE E7.1.

where

$$\phi = \tan^{-1}\left(\frac{6.67}{25.0}\right) = 14.94°.$$

(b) Forces act on ABD at A, B, and D. Because it is a three-force member in equilibrium, the action lines of these forces must be concurrent. The point of concurrence is denoted by E in the geometric diagram of Figure E7.1(c). It is located by extending the lines of action of the two forces of known direction, the applied load P and the cable tension T, until they intersect at E. Of necessity, the action line of R_D must also pass through point E. This enables us to determine the angle ϕ which defines the direction of R_D. Thus,

$$\tan \phi = \frac{2 \tan 33.7^\circ}{5} = 0.2668 \Rightarrow \phi = 14.94^\circ$$

Because the directions of all three forces are known, the force triangle, shown in Figure E7.1(d) together with the law of sines will enable us to determine the cable tension T and the reaction R_D. The angle $\alpha = 180^\circ - (90^\circ + \phi) - (90^\circ - \theta) = 18.76^\circ$. Hence,

$$\frac{T}{\sin(90^\circ + \phi)} = \frac{10}{\sin \alpha} = \frac{R_D}{\sin(90^\circ - \theta)}$$

Substituting $\phi = 14.94^\circ$, $\theta = 33.7^\circ$, and $\alpha = 18.76^\circ$ gives

$$T = 30.0 \text{ kN}$$

and

$$R_D = 25.9 \text{ kN}. \hspace{3cm} \text{ANS.}$$

These values check those from part (a). The choice of this second method depends on whether or not the trigonometric problems can be easily solved.

■ **Example 7.2**

A bracket is constructed by attaching member ABC to wall CD with a frictionless hinge at C and a horizontal cable at A, as shown in Figure E7.2(a). A cylinder of weight $W = 1.20$ kN is placed in the bracket, as shown. Assume frictionless conditions and determine (a) the force acting on the cylinder at contact points B and D and (b) the tension in the cable and the reaction components at C.

Solution

(a) The free-body diagram of the cylinder is shown in Figure E7.2(b) which indicates that the cylinder is a three-force member in equilibrium. The force system is, therefore, concurrent at the geometric center of the cylinder. Although two equations of equilibrium are available to find N_B and N_D, it is more convenient to use the force-triangle method in this case. The force triangle corresponding to the force system of Figure E7.2(b) is shown in Figure E7.2(c). Applying the law of the sines,

$$\frac{N_D}{\sin 40^\circ} = \frac{N_B}{\sin 65^\circ} = \frac{1.20}{\sin 75^\circ}.$$

Solving for N_B and N_D,

$$N_B = 1.126 \text{ kN}, \hspace{3cm} \text{ANS.}$$

and

$$N_D = 0.799 \text{ kN}. \hspace{3cm} \text{ANS.}$$

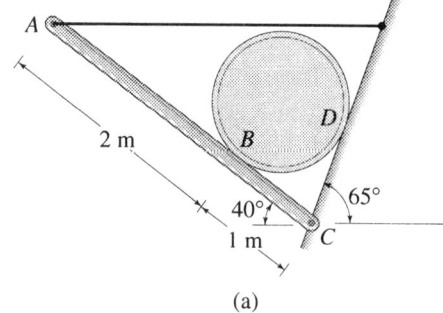

(a)

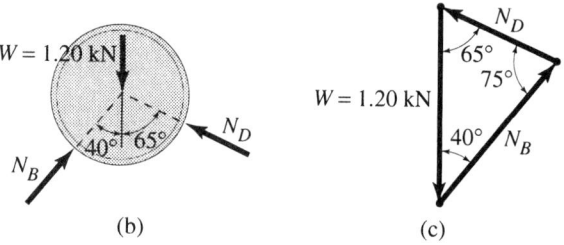

(b)   (c)

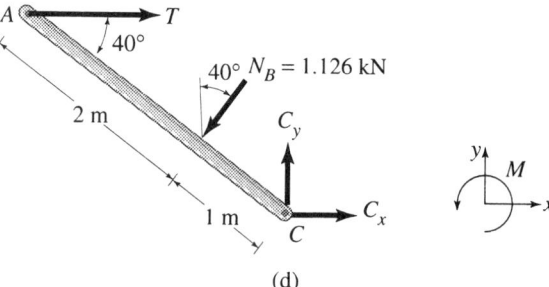

FIGURE E7.2.   (d)

(b) The free-body diagram of member ABC is shown in Figure E7.2(d) along with a convenient coordinate system. It is obvious that ABC is a three-force member and that the force-triangle method may also be used in this case. However, it is more convenient to use the equations of equilibrium because the point of concurrence of the three forces is not immediately obvious. Thus,

$$\sum M_C = 0, \quad 1.126(1) - (T\sin 40°)(3) = 0,$$

$$T = 0.584 \text{ kN.} \qquad\qquad\qquad \text{ANS.}$$

$$\sum F_x = 0, \quad 0.584 - 1.126 \sin 40° + C_x = 0,$$

$$C_x = 0.1398 \text{ kN.} \qquad\qquad \text{ANS.}$$

$$\sum F_y = 0, \quad C_y - 1.126 \cos 40° = 0,$$

$$C_y = 0.1398 \text{ kN.} \qquad\qquad \text{ANS.}$$

■ **Example 7.3**   A beam is supported and loaded as shown in Figure E7.3(a). Determine the reactions at supports A and B.

Solution   The free-body diagram of beam ABC is shown in Figure E7.3(b) along with a convenient coordinate system. Note that the beam is a multiforce member subjected to forces at more than three locations. Therefore, the force system acting on the beam is a general two-dimensional force system in equilibrium and three equations are available for the solution. Thus,

$$\sum M_A = 0, \quad N_B(10) - 20(5) - 10(16) = 0,$$

$$N_B = 26.0 \text{ k} \overset{4}{\underset{}{\diagdown}}{}_{3}. \qquad\qquad \text{ANS.}$$

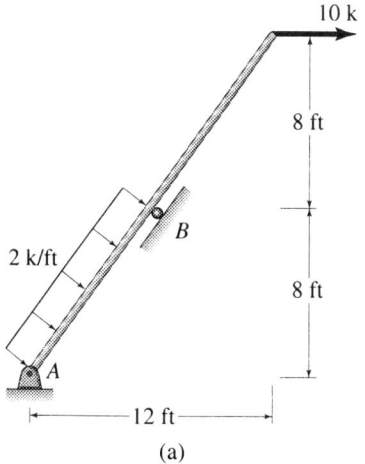

(a)

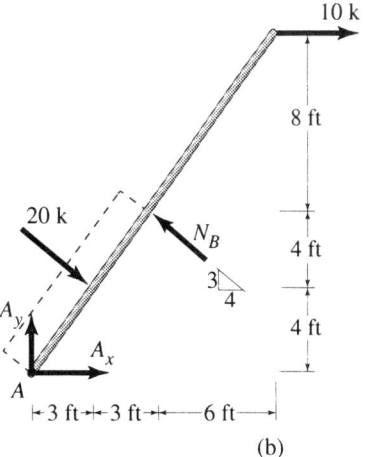

(b)

FIGURE E7.3.

$$\sum F_x = 0, \quad A_x + \left(\frac{4}{5}\right)(20) - \left(\frac{4}{5}\right)(26.0) + 10 = 0,$$

$$A_x = -5.2 = 5.2 \text{ k} \leftarrow.$$

$$\sum F_y = 0, \quad A_y + \left(\frac{3}{5}\right)(26.0) - \left(\frac{3}{5}\right)(20) = 0,$$

$$A_y = -3.6 = 3.6 \text{ k} \downarrow. \qquad\qquad\qquad \text{ANS.}$$

$$A = \sqrt{A_x^2 + A_y^2} = 6.32 \text{ k} \; \overline{\theta \nearrow} \qquad\qquad \text{ANS.}$$

where

$$\theta = \tan^{-1}\left(\frac{3.6}{5.2}\right) = 34.7°.$$

■

Problems

7.1   A uniform steel beam of weight W and length L is to be suspended horizontally by attaching cables at its ends, as shown in Figure P7.1. Express the cable tensions T as a function of W and θ. Supposing this beam weighs 50 lb/ft and is 20 ft long, express the end cable tensions T as a function of θ, and prepare a neat plot of T vs. θ. Let θ vary from $0°$ to $90°$ in $30°$ increments. Is a zero value of θ permissible. Explain? The weight W acts through the beam's geometric center.

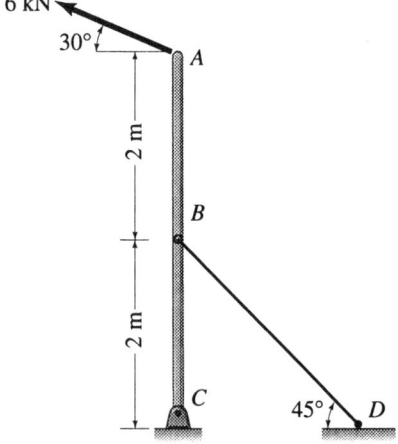

FIGURE P7.2.

7.2   Vertical member ABC of Figure P7.2 is held in place by cable BD and a pin at C. Determine the cable tension T and the pin reaction resultant at C by
(a) finding the point of concurrence and using the force triangle and
(b) writing the equations of equilibrium.

7.3   A cylinder of weight $W = 99.5$ lb and length L is held in a bracket, as shown in

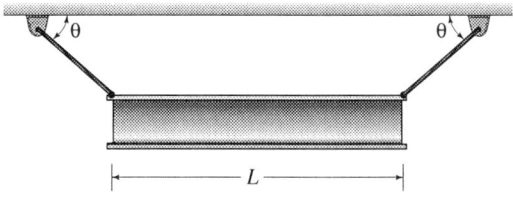

FIGURE P7.1.

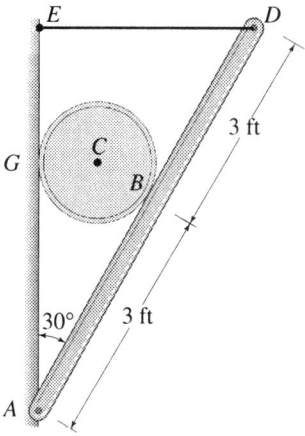

FIGURE P7.3.

Figure P7.3. Determine the radius of the cylinder and its length L if its density is 490 lb/ft³. Assume that the weight of the cylinder acts through its geometric center at C, and determine the reactions at points B and G. Also, determine the force in cable DE and the reaction components at hinge A. Assume frictionless conditions.

7.4   A sphere of weight W rests between two smooth surfaces as shown in Figure P7.4. Draw the force triangle for the three forces acting on the sphere, and express the normals at A and B as functions of θ, ϕ, and W.

7.5   Solve problem 7.4 for the special case of $W = 600$ N, $\theta = 60°$, and $\phi = 30°$.

7.6   Refer to the parallel force system shown in Figure P7.6. The pulley is frictionless and has a radius of 0.10 m. Determine all unknown forces of this system if $Q = 2$ kN and $P = 4$ kN. Determine the tension in the cable and the reactions at support A, D, F, and G.

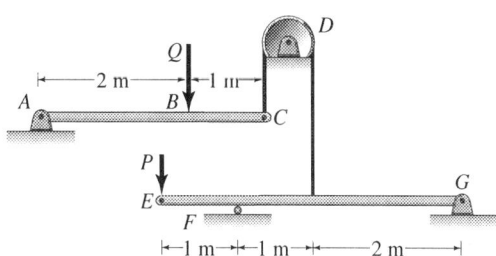

FIGURE P7.6.

7.7   The lawn roller depicted in Figure P7.7 has a diameter of 1.00 ft and weighs 225 lb. Determine the minimum force P and the angle ϕ so that the roller will begin to move over the 0.8 in. obstruction.

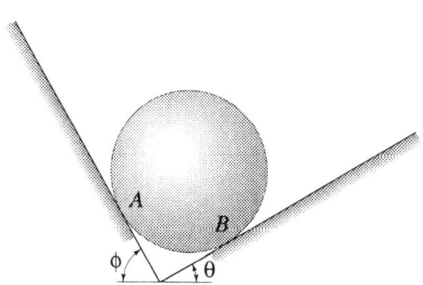

FIGURE P7.4.

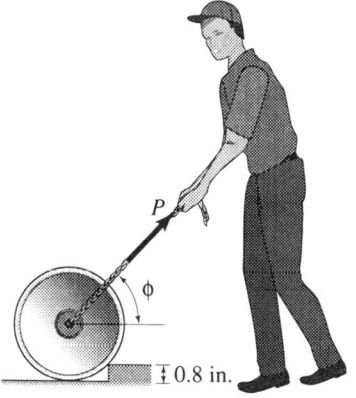

FIGURE P7.7.

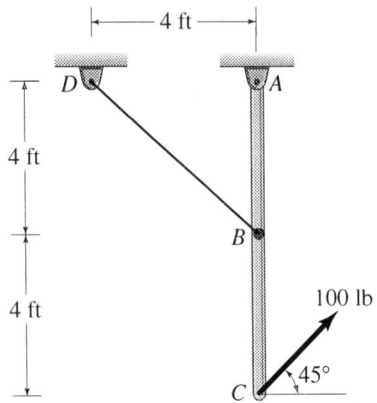

FIGURE P7.8.

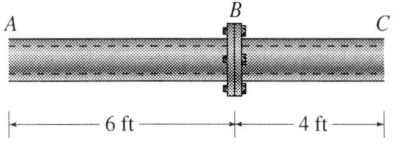

FIGURE P7.9.

7.8 Vertical member ABC of Figure P7.8 is held in place by cable BD and a pin at A. Determine the cable tension T and the pin reaction resultant at A by
(a) finding the point of concurrence and using the force triangle and
(b) writing the equations of equilibrium.

7.9 Two pieces of pipe are bolted together at B, as shown in Figure P7.9. Segment AB weighs 25 lb/ft, and segment BC weighs 35 lb/ft. The center of gravity, through which the weight of this body acts, lies 5.414 ft to the right of A. If it is to be suspended horizontally from vertical cables at A and C, what will the tension be in each cable? Ignore the connecting material's weight.

7.10 Solve problem 7.9 if the cable attached at C makes an angle of 45° with the horizontal. What angle does the cable attached at A make with the horizontal? What is the tension in each cable if the pipe is in equilibrium in a horizontal position?

7.11 The uniform slender rod BC of Figure P7.11 weighs 50 lb and is 20 ft long. Determine the tension in cable AB and in cable CD and the angle ϕ so that the rod will be in equilibrium as shown. Assume that the weight of rod BC acts through its geometric center.

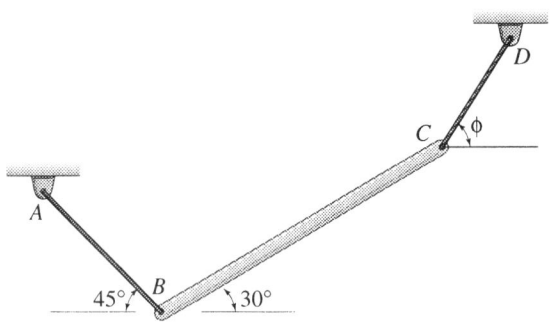

FIGURE P7.11.

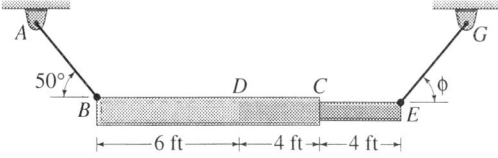

FIGURE P7.12.

7.12 In Figure P7.12, a rod DE weighing 6.0 lb/ft is tightly fitted inside a pipe BC weighing 3.6 lb/ft. This body is held in equilibrium by cables AB and EG. Determine the tensions in cables AB and EG and the angle ϕ by writing the equations of equilibrium. Consider the weights of pipe BC and rod DE as separate forces.

7.13 Solve problem 7.12 knowing that the center of gravity of this composite body lies 7.857 ft to the right of B. Locate the point of concurrence and use the force triangle to find the cable tensions and ϕ.

7.14 In Figure P7.14, three identical disks are shown in plan view on a frictionless horizontal surface. In each case, determine a third force, which must be added to the two shown, to hold each disk in equilibrium. Show the force triangle for each case, and clearly indicate the required force in magnitude and direction on a sketch of each disk. The action lines of all forces pass through the disk centers C.

7.15 A uniform platform weighing 4000 lb is shown in Figure P7.15. It is temporarily supported by two links CD and EG and a cable AD. If the breaking strength of the cable is 5000 lb, what force P applied to this platform would cause the cable to break? What forces must be carried by the links at this stage? Assume that these links are weightless and that the weight of the platform acts through its geometric center.

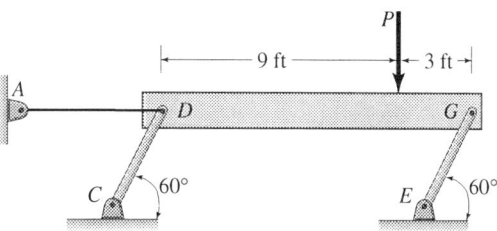

FIGURE P7.15.

7.16 A thin rectangular plate rests on a smooth surface and is subjected to the two forces shown in Figure P7.16, which is a plan view. Determine a third force required for equilibrium of this plate. Find the point of concurrence and use the force triangle to solve for this force.

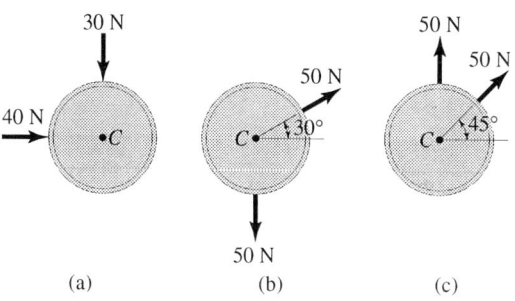

FIGURE P7.14.

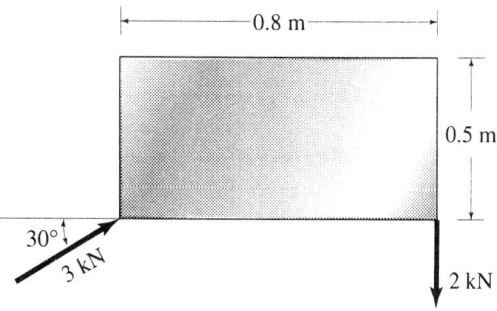

FIGURE P7.16.

7.17 In Figure P7.17, a rod DE weighing 6.0 lb/ft is tightly fitted inside a pipe BC weighing 3.6 lb/ft. This body is held in equilibrium by cables AB and EG. Determine the tensions in cables AB and EG and the angle ϕ by writing the equations of equilibrium. Consider the weights of pipe BC and rod DE as separate forces.

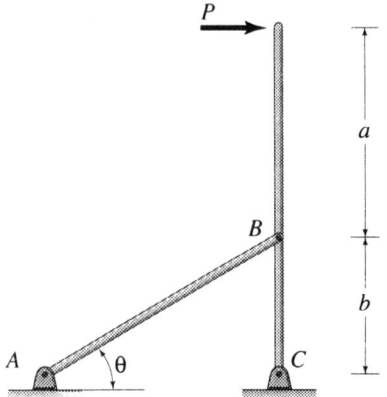

FIGURE P7.19.

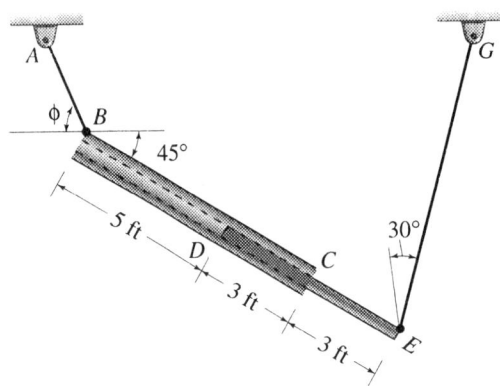

FIGURE P7.17.

7.18 Solve problem 7.17 knowing that the center of gravity of this composite body lies 6.222 ft from B measured along BC. Locate the point of concurrence, and use the force triangle to find the cable tensions and ϕ.

7.19 The frictionless pin at A, shown in Figure P7.19, may be moved completely around a circle centered at B which changes the angle θ. Select the angle θ to minimize the force in member AB. Find the corresponding force in AB and the pin reaction at C. Regard P, a, and b as known quantities.

7.20 Three uniform cylinders of equal diameter but different weights are shown in

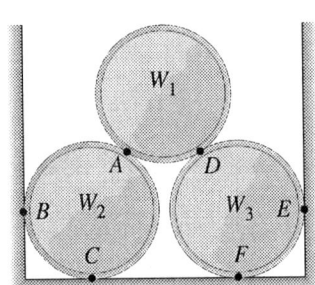

FIGURE P7.20.

Figure P7.20. The system is frictionless, and the bottom cylinders are set near each other without touching. Determine all forces acting on each of the cylinders in terms of W_1, W_2, and W_3. Show your answers on sketches of each of the cylinders.

7.21 Solve problem 7.20 for $W_1 = 100$ N, $W_2 = 120$ N, and $W_3 = 160$ N.

7.22 Solve problem 7.20 for $W_1 = 60$ lb, $W_2 = 80$ lb, and $W_3 = 100$ lb.

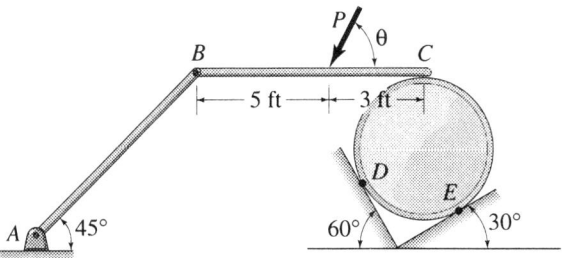

FIGURE P7.23.

7.23 The system shown in Figure P7.23 is frictionless and weightless. Determine the angle θ corresponding to equilibrium and the force in member AB for $P = 200$ lb. What forces are exerted on the cylinder at points C, D, and E? Show your answers on neat sketches of BC and the cylinder.

7.24 A right triangular plate weighing 60 lb is held in equilibrium by a pin at B and a pin-connected rod CD, as shown in Figure P7.24. Determine the pin reaction components acting on the plate at B and the force exerted on the plate by the rod. The weight of the plate acts through point G.

7.25 Three cylinders are in equilibrium as shown in Figure P7.25. If the cylinders are of constant density and have equal lengths, determine the ratio of the diameters of W_1 and W_2 if $W_1 = 2W_2$. Find the forces acting on the cylinders at A, B, C, D, E and F in terms of W_1.

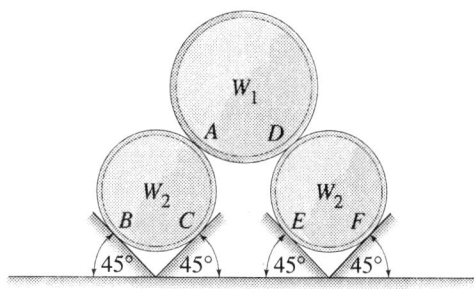

FIGURE P7.25.

7.26 Solve problem 7.25 for $W_1 = 200$ lb and $W_2 = 100$ lb.

7.27 Solve problem 7.25 for $W_1 = 80$ N and $W_2 = 40$ N.

7.28 Two vertical rods support a homogeneous rectangular plate, as shown in Figure P7.28. If the tension in rod CD is $0.4W$ where W is the weight of the plate, determine the distance x which positions

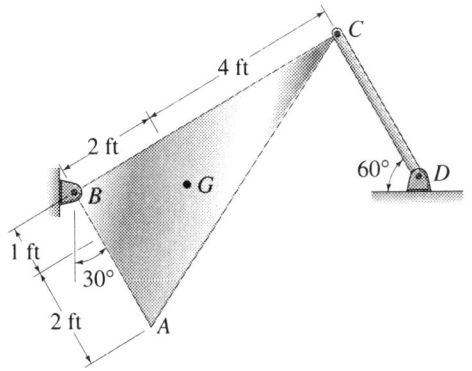

FIGURE P7.24.

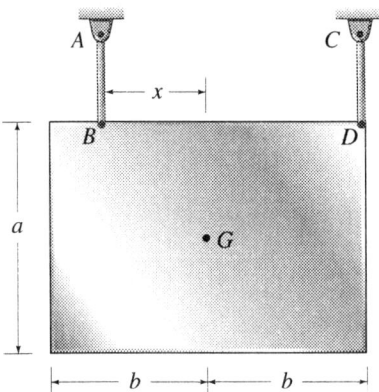

FIGURE P7.28.

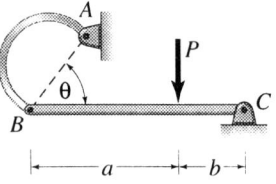

FIGURE P7.30.

rod AB and the tension in rod AB in terms of *W*. Assume that the weight of the plate acts through point G.

7.29 Slender rod BC in Figure 7.29 has a weight *W* and is supported at B by cable AB and at C by the weightless link CD. Determine the angle θ consistent with equilibrium of this rod, and express the cable and link forces in terms of the weight *W*. The weight of rod BC acts through its geometric center.

(a) Determine the force exerted on BC by the pin at B and the pin reaction components at C as functions of *P*, *a*, *b*, and θ.
(b) Find numerical values of these forces for *P* = 1000 lb, *a* = 6 ft, *b* = 4 ft, and $\theta = 60°$. Neglect the weights of both members.

7.31 In Figure P7.31, a uniform square plate of weight *W* is suspended by cables AB and CD. It is subjected to a horizontal load *P* as shown. If the angles θ are each equal to 60°, determine the force *P* in terms of *W* consistent with equilibrium and the cable tensions. Assume that the weight *W* acts through the plate's geometric center.

7.32 Solve problem 7.31 for *W* = 500 N.

7.33 Solve problem 7.31 for *W* = 200 lb.

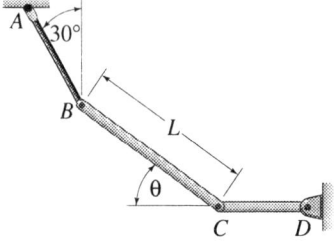

FIGURE P7.29.

7.30 Horizontal member BC in Figure P7.30 is supported by the curved member AB on the left and by the pin C on the right.

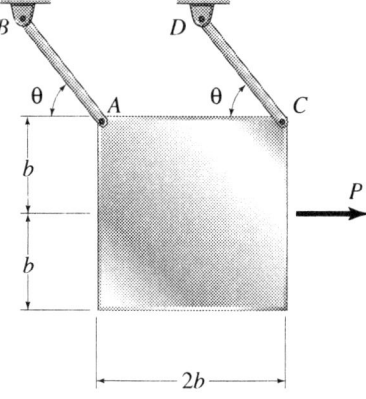

FIGURE P7.31.

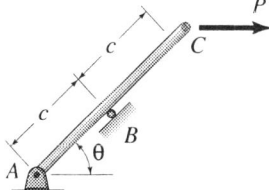

FIGURE P7.34.

7.34 In Figure P7.34, the inclined member AC of weight W is supported by pin A and by a fixed, smooth peg B. Express the force exerted on this member by the peg B and the pin reaction components at A as functions of the member weight W, the applied force P, and the angle θ. Consider P of magnitude W but directed to the left instead of to the right and find the angle θ corresponding to a zero force acting on the member at B. The weight W acts through the geometric center of member AC.

7.35 Refer to problem 7.34 and let $W = 200$ N, $P = 500$ N, and $\theta = 60°$. Determine the force exerted on member AC by the peg at B, and find the pin reaction component at A.

7.36 The plate shown in Figure P7.36 is 1 in. thick and is fabricated of a material which weighs 160 lb/ft^3. Determine the angle θ consistent with equilibrium and the corresponding tensions in the cables AB and CD. Hint: Divide the plate into a rectangle and a semicircle. Assume that the weight of the rectangle acts through its geometric center and that of the semicircle through point G.

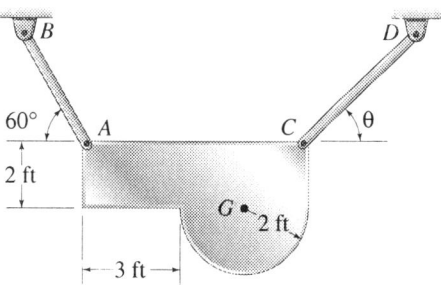

FIGURE P7.36.

7.2
Frame
Analysis

Frames may be distinguished from trusses by noting that trusses consist entirely of two-force members whereas frames consist of interconnected, multiforce members (i.e., three or more forces) as well as two-force members and pulleys. The analysis of a frame requires that we dismember it (i.e., separate its various members) and construct the free-body diagram of each and every member. This process exposes all of the unknown forces in the frame, consisting of those that hold the various members together and those that act on the frame at the points connecting it to the ground.

Four qualitative and two quantitative examples are presented to illustrate the process of frame analysis. A major objective of the two qualitative examples is to identify all of the unknown forces in a frame and to compare their total number to the number of equilibrium equations available for their solution. Developing a facility for outlining the solution of such problems is of interest not solely in statics but in all engineering analysis.

■ **Example 7.4**
Refer to the frame shown in Figure E7.4(a), and outline a solution that would provide the force in member AC and the reaction components at hinge B. Regard P, a, and b as known quantities.

Solution
Member AC is a two-force member whose free-body diagram is shown in Figure E7.4(b) where T_{AC} represents the magnitude of the force in the member. The free-body diagram of member AB is shown in Figure E7.4(c) where T_{AC} is known in direction because the angle θ may be

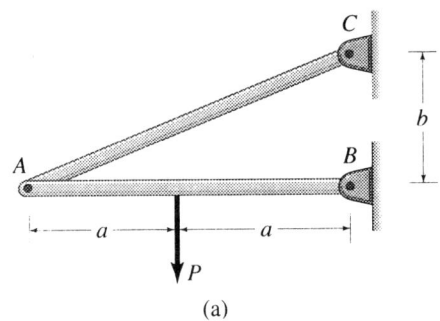

(a)

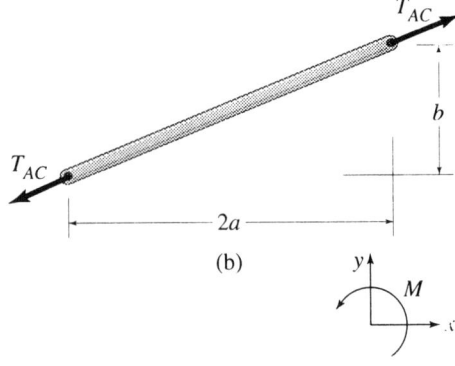

(b)

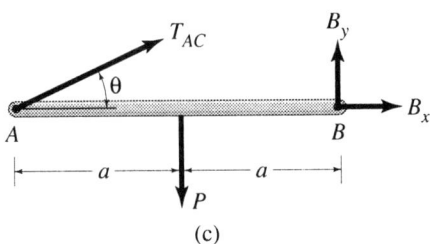

FIGURE E7.4.

(c)

determined from the equation $\tan \theta = b/2a$. Thus, Figure E7.4(c) contains the three unknown quantities that we need to determine, namely, the magnitudes of T_{AC}, Bx, and By. Also, three independent equations are available to solve for these unknowns. Thus,

$$\sum M_B = 0 \quad \text{will enable us to solve for } T_{AC},$$

$$\sum F_x = 0 \quad \text{will enable us to solve for } B_x,$$
$$\text{because we now know } T_{AC}, \text{ and}$$

$$\sum F_y = 0 \quad \text{will enable us to solve for } B_y,$$
$$\text{because we now know } T_{AC}.$$

We observe that the equations were written for the free-body diagram of Figure E7.4(c) and that the free-body diagram of Figure E7.4(b) served only to remind us that AC is a two-force member. If we had not recognized that AC is a two-force member, we would have had to deal with a total of four unknown quantities, two unknown pin reaction components at each of pins A and C. This would have lengthened the solution or perhaps led to an impasse. Identifying two-force members of frames greatly reduces the solution time, in general.

■ **Example 7.5**

Refer to the frame shown in Figure E7.5(a) and regard W, P, b, c, and θ as known quantities. Outline a solution that would yield the tensions on both sides of the cable, the reaction components at A, and the reaction components at C.

Solution

Refer to Figure E7.5(b), (c), and (d), and enumerate all unknown forces T_1, T_2, A_x, A_y, C_x, C_y, and M_C. There are a total of seven unknown forces. Enumerating the available equations reveals that we have one equation in Figure E7.5(b), three equations in Figure E7.5(c), and three equations in Figure E7.5(d).

We conclude that there are seven equations available to solve for seven unknowns and the frame is, therefore, statically determinate.

The following tabulation outlines the complete solution:

Figure	Equation	Enables us to solve for
E7.5(b)	$\sum F_y = 0$	T_1
E7.5(c)	$\sum M_A = 0$	T_2
E7.5(c)	$\sum F_x = 0$	A_x
E7.5(c)	$\sum F_y = 0$	A_y
E7.5(d)	$\sum F_x = 0$	C_x
E7.5(d)	$\sum F_y = 0$	C_y
E7.5(d)	$\sum M_C = 0$	M_C

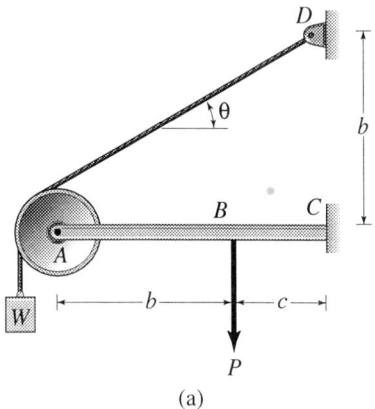

(a)

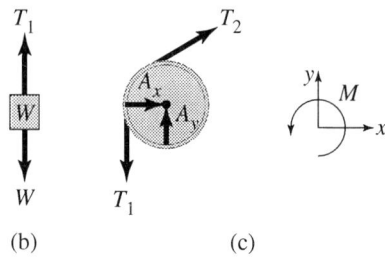

(b)   (c)

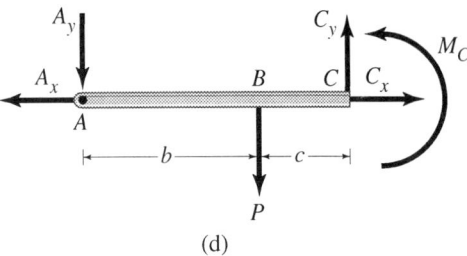

(d)

Figure E7.5.

■ **Example 7.6**

Refer to the frame shown in Figure E7.6(a), and outline the solution for all unknown forces acting on members ACE, BDE, and CD. Regard P, b, c, d, and e as known inputs.

Solution

Referring to Figure E7.6(a), we calculate

$$\theta = \tan^{-1}\left(\frac{b}{d}\right)$$

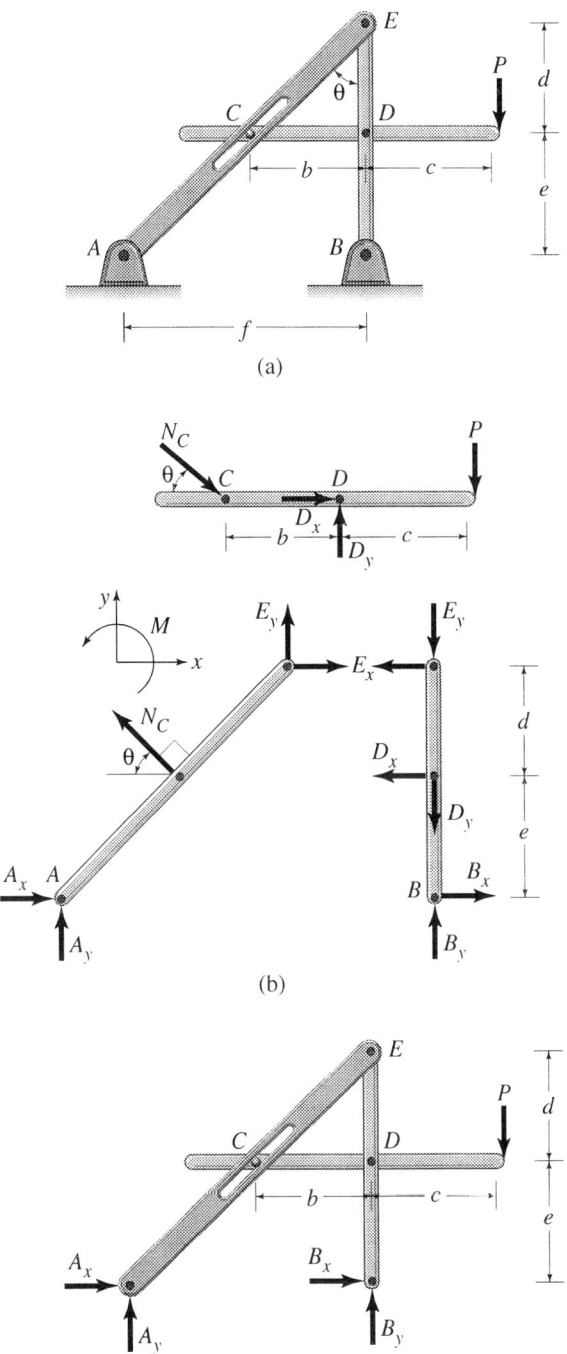

(a)

(b)

(c)

Figure E7.6.

and

$$f = (d + e)\tan\theta.$$

Note that the force components shown on the various free-body diagrams of Figure E7.6(b) are consistent with Newton's third law. For example, E_x and E_y are shown acting to the right and upward on member ACE and to the left and downward on member EDB.

Referring to the free-body diagrams of Figures E7.6(b), we enumerate the unknowns D_x, D_y, N_C, B_x, B_y, E_x, E_y, A_x, and A_y. There are nine scalar unknowns and, for each of the three free-body diagrams of Figure E7.6(b), we may write three equations of equilibrium. Thus, we have a total of nine equations to solve for the nine unknowns and the system is, therefore, statically determinate. It is possible to deal with only these free-bodies, but it will simplify the solution if we use the free-body diagram of the entire frame shown in Figure E7.6(c). We note that four unknown forces act on the frame, namely, A_x, A_y, B_x, and B_y. Because only three equations are available, we will be unable to solve for all four of these unknowns, but two moment equations will enable us to find A_y and B_y. Thus, using Figure E7.6(c),

$$\sum M_B = 0 \quad \text{yields } A_y \quad \text{and}$$
$$\sum M_A = 0 \quad \text{yields } B_y.$$

Consider the free-body diagram of member CD shown in Figure E7.6(b). Assuming a smooth slot, the force N_C acts in a direction perpendicular to it, as shown. Thus,

$$\sum M_D = 0 \quad \text{yields } N_C,$$
$$\sum F_x = 0 \quad \text{yields } D_x, \quad \text{and}$$
$$\sum F_y = 0 \quad \text{yields } D_y.$$

Next, consider the free-body diagram of member BDE shown in Figure E7.6(b). Because D_x, D_y, and B_y are known, the three remaining unknowns are B_x, E_x, and E_y. Thus,

$$\sum M_E = 0 \quad \text{yields } B_x,$$
$$\sum F_x = 0 \quad \text{yields } E_x, \quad \text{and}$$
$$\sum F_y = 0 \quad \text{yields } E_y.$$

Referring to the free-body diagram of member ACE, shown in Figure E7.6(b), only A_x remains unknown. Thus,

$$\sum F_x = 0, \quad \text{yields } A_x.$$

At this stage we have outlined the solution for all unknown forces. Various alternate approaches could have been adopted for solving the problem but the one outlined above is rather compact and direct.

Other equations may be used to check one or more of these results. For example, after finding A_x above, we could refer to Figure E7.6(c), and check its value using the equation $\sum F_x = 0$. Thus,

$$A_x + B_x = 0,$$

or

$$A_x = -B_x.$$

Of course, if A_x and B_x are not of equal magnitude and opposite sense, we have made an error in writing one or more of the original equations. At this stage, if we are dealing with numerical values it is advisable to construct new diagrams showing all forces and the proper sense for each. It is, then, relatively easy to detect errors visually by noting whether or not the bodies are in equilibrium under the action of the calculated forces.

■ **Example 7.7**

Refer to Figure E7.7(a), and outline a complete solution for all unknown forces acting on the frame. Regard P, b, and c as known quantities.

Solution

The given frame is dismembered, and the free-body diagram of its members is shown in Figure E7.7(b). Members BC and BE are two-force members, and, therefore, the directions of T_{BC} and T_{BE} are known. Let us enumerate the unknown forces: A_x, A_y, T_{BE}, T_{BC}, D_x, D_y, D_x', D_y', D_x'', D_y'', and F_y. There are a total of 11 scalar force unknowns. The number of available equations of equilibrium is summarized in the following tabulation:

Member	Number of Equations Available
CDG	3
ABD	3
DEF	3
Pin D	2
Total	11

We conclude that the system is statically determinate.

Although we did not utilize the free-body diagram of the entire frame in our overall assessment of this problem, we begin our solution most conveniently by referring to this free-body diagram which is shown in Figure E7.7(c). Thus,

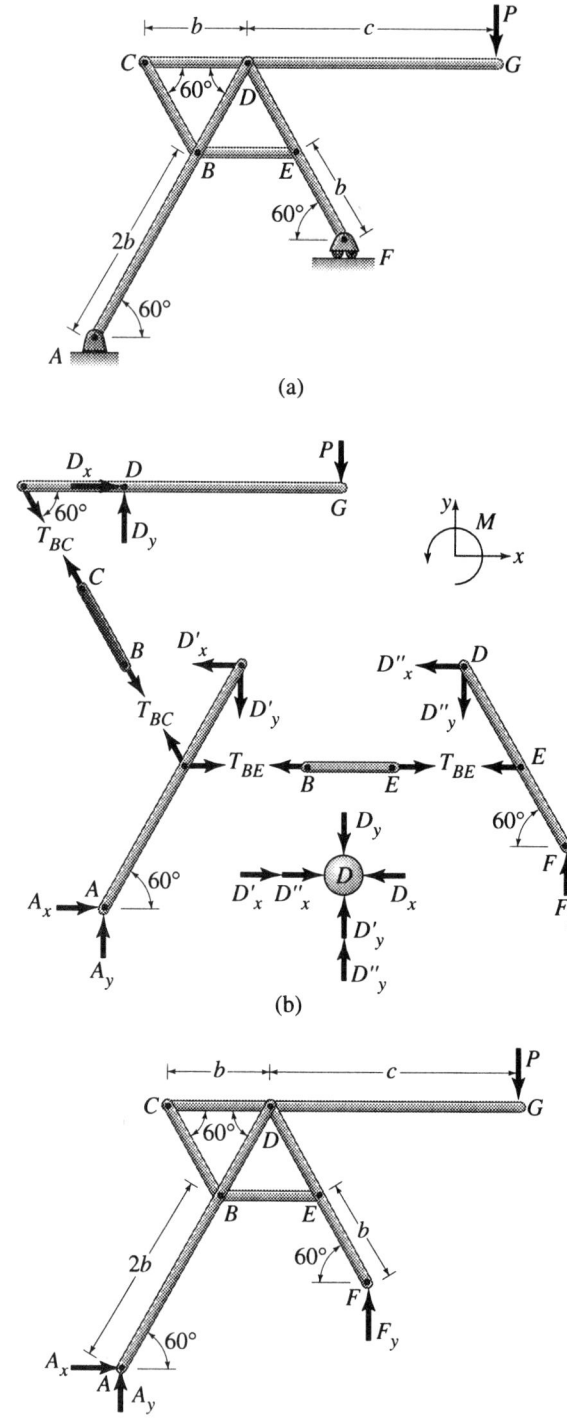

FIGURE E7.7.

$$\sum M_{\mathrm{A}} = 0 \quad \text{enables us to solve for } F_{\mathrm{y}},$$

$$\sum F_{\mathrm{y}} = 0 \quad \text{yields } A_{\mathrm{y}}, \quad \text{and}$$

$$\sum F_{\mathrm{x}} = 0 \quad \text{yields } A_{\mathrm{x}}.$$

Referring to the free-body diagram of member CDG, shown in Figure E7.7(b), we note that only three unknowns are acting. Thus,

$$\sum M_{\mathrm{D}} = 0 \quad \text{yields } T_{\mathrm{BC}},$$

$$\sum F_{\mathrm{x}} = 0 \quad \text{yields } D_{\mathrm{x}}, \quad \text{and}$$

$$\sum F_{\mathrm{y}} = 0 \quad \text{yields } D_{\mathrm{y}}.$$

Now we observe that the free-body diagram of member ABD has only three unknowns acting on it because A_{x}, A_{y} and T_{BC} have been determined. Therefore,

$$\sum M_{\mathrm{D}} = 0 \quad \text{yields } T_{\mathrm{BE}},$$

$$\sum F_{\mathrm{x}} = 0 \quad \text{yields } D'_{\mathrm{x}},$$

and

$$\sum F_{\mathrm{y}} = 0 \quad \text{yields } D'_{\mathrm{y}}.$$

Next, consider the free-body diagram of member DEF. Only D''_{x} and D''_{y} are unknown because F_{y} and T_{BE} have been determined. Thus,

$$\sum F_{\mathrm{x}} = 0 \quad \text{determines } D''_{\mathrm{x}}, \quad \text{and}$$

$$\sum F_{\mathrm{y}} = 0 \quad \text{determines } D''_{\mathrm{y}}.$$

We have now determined all 11 unknown force magnitudes and, thus, may use the free-body diagram of Pin D to check our results. If pin D is not in equilibrium, then, we have erred in our solution. The solution outlined above is not the only possible one for this frame problem, but it is a reasonably efficient one revealing the action of all forces in the system. When numerical results are obtained, it is desirable to sketch each body of the system and show the forces as they actually act on these bodies. This enables us to detect errors at a glance.

■ **Example 7.8**

Refer to the frame shown in Figure E7.8(a), Let $T_1 = 500$ lb, $a = 4$ ft, $b = 3$ ft, and $r = 0.8$ ft. Determine all forces acting on member CBD.

Solution

The free-body diagram of the pulley is shown in Figure E7.8(b) and that of member CBD in Figure E7.8(c). Because member AB is a two-force member, the direction of force T_{AB}, defined by the angle θ, is

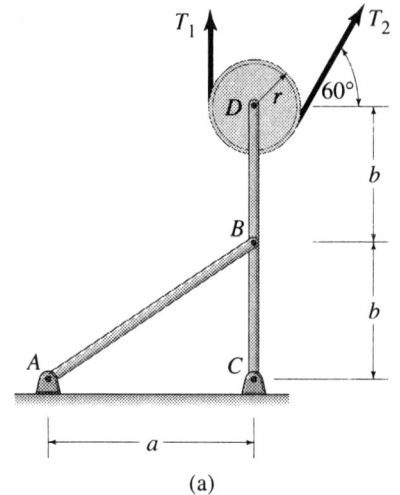

(a)

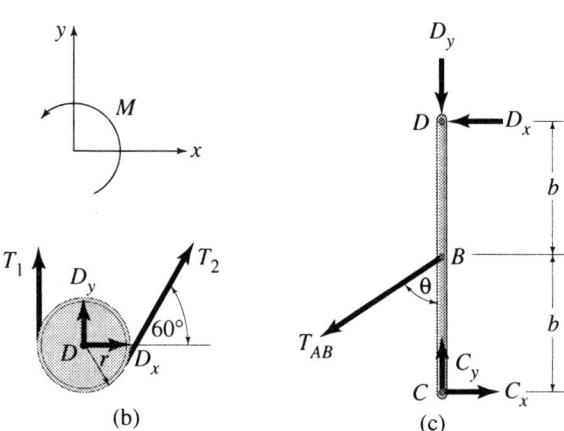

FIGURE E7.8.   (b)   (c)

known because $\theta = \tan^{-1}(a/b) = 53.1°$. Thus, there are six unknown quantities, six equations of equilibrium, and the system is statically determinate.

Considering the free-body diagram of Figure E7.8(b),

$$\sum M_D = 0, \quad T_1 r - T_2 r = 0,$$

$$T_2 = T_1 = 500 \text{ lb.} \qquad\qquad \text{ANS.}$$

$$\sum F_x = 0, \quad T_2 \cos 60° + D_x = 0,$$

$$D_x = -250 = 250 \text{ lb} \leftarrow \text{ on pulley,} \qquad \text{ANS.}$$

and

$$\sum F_y = 0, \quad T_1 - D_y + T_2 \sin 60° = 0,$$

$$D_y = -933 = 933 \downarrow \text{ on pulley.} \qquad \text{ANS.}$$

Considering the free-body diagram of Figure E7.8(c),

$$\sum M_C = 0, \quad (T_{AB} \sin \theta)b + D_x(2b) = 0,$$

$$T_{AB} = -\frac{2D_x}{\sin \theta} = 625 \text{ lb} \quad (T), \qquad \text{ANS.}$$

$$\sum F_x = 0, \quad -D_x - T_{AB} \sin \theta + C_x = 0,$$

$$C_x = D_x + T_{AB} \sin \theta = 250 \text{ lb} \rightarrow, \qquad \text{ANS.}$$

and

$$\sum F_y = 0, \quad -D_y - T_{AB} \cos \theta + C_y = 0,$$

$$C_y = D_y + T_{AB} \sin \theta = -558 \text{ lb} = 558 \text{ lb} \downarrow. \qquad \text{ANS.}$$

The reader may wish to sketch the pulley and the vertical member CBD, show all forces acting on them, and conclude that they are, in fact, in equilibrium.

■ **Example 7.9**

Refer to the frame shown in Figure E7.9(a), and determine the forces acting on members AD, EC and BD in terms of the known quantities P, a, b, and c. Find numerical values for these forces if $P = 5$ kN, $a = 0.5$ m, $b = 1.0$ m, and $c = 1.0$ m.

Solution

The free-body diagrams of members AD, EC, and BD are shown in Figures E7.9(b). The unknown quantities are R_E, C_x, C_y, A_x, A_y, D_x, D_y, B_x, and B_y. There are nine scalar unknown force components and nine independent equations of equilibrium available for solving this problem. Note that

$$\theta = \tan^{-1}\left(\frac{b}{2c}\right).$$

Referring to the free-body diagram of member EC,

$$\sum M_C = 0, \quad 2Pa - Pa - R_E b = 0,$$

$$R_E = \frac{a}{b}P \quad (\leftarrow \text{ on EC}), \qquad \text{ANS.}$$

$$\sum F_x = 0, \quad P + C_x - R_E = 0,$$

$$C_x = \left(\frac{a-b}{b}\right)P \quad (\rightarrow \text{ on EC provided } a > b), \qquad \text{ANS.}$$

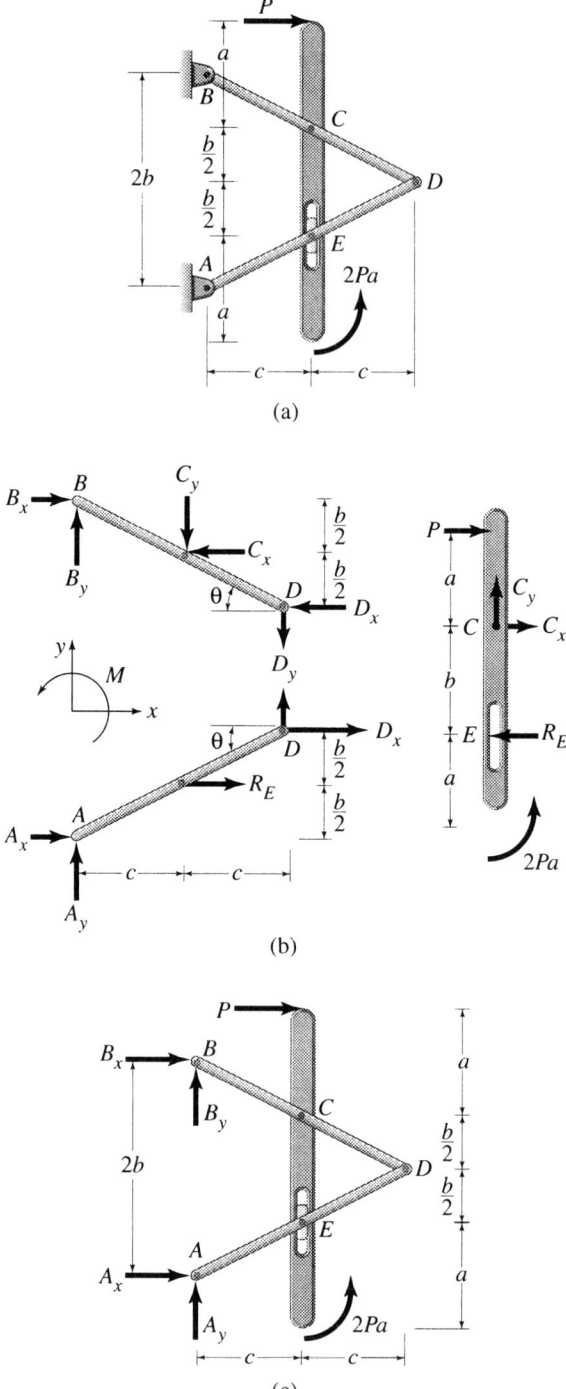

and

$$\sum F_y = 0, \quad C_y = 0. \qquad \text{ANS.}$$

Referring to the free-body diagram of the entire frame shown in Figure E7.9(c),

$$\sum M_B = 0, \quad 2Pa + A_x(2b) - P(2a + b - 2b)/2 = 0,$$

$$A_x = -P\left(\frac{2a + b}{4b}\right) = P\left(\frac{2a + b}{4b}\right) \quad (\leftarrow \text{ on AD}), \qquad \text{ANS.}$$

$$\sum F_x = 0, \quad P + B_x + A_x = 0,$$

$$B_x = -A_x - P,$$

$$= \left(\frac{2a - 3b}{4b}\right)P \quad (\rightarrow \text{ on BD provided } 2a > 3b). \qquad \text{ANS.}$$

Refer to the free-body diagram of member AD shown in Figure E7.9(b) and write

$$\sum M_D = 0, \quad A_x(b) + R_E(b/2) - A_y(2c) = 0.$$

Substituting for A_x and R_E and solving for A_y,

$$A_y = -\frac{b}{8c}P = \frac{b}{8c}P \quad (\downarrow \text{ on AD}), \qquad \text{ANS.}$$

$$\sum F_y = 0, \quad A_y + D_y = 0,$$

$$D_y = -A_y = \frac{b}{8c}P \quad (\uparrow \text{ on AD}), \qquad \text{ANS.}$$

and

$$\sum F_x = 0, \quad A_x + R_E + D_x = 0.$$

Substituting for R_E and A_x and solving for D_x,

$$D_x = \left(\frac{b - 2a}{4b}\right)P \quad (\rightarrow \text{ on AD provided } b > 2a). \qquad \text{ANS.}$$

Using the free-body diagram of the entire frame (Fig. E7.9(c)),

$$\sum F_y = 0, \quad A_y + B_y = 0,$$

$$B_y = -A_y = \frac{b}{8c}P \quad (\uparrow \text{ on BD}). \qquad \text{ANS.}$$

Substitution of the given numerical values yields the following results:

$$R_E = \frac{a}{b}P = 2.50 \text{ kN} \quad (\leftarrow \text{ on EC}). \qquad \text{ANS.}$$

$$C_x = \left(\frac{a-b}{b}\right)P = -2.50 \text{ kN} = 2.5 \text{ kN} \quad (\leftarrow \text{ on EC}). \qquad \text{ANS.}$$

$$C_y = 0. \qquad \text{ANS.}$$

$$A_x = P\frac{a}{b}\left(\frac{2a+b}{4b}\right) = 2.50 \text{ kN} \quad (\leftarrow \text{ on AD}). \qquad \text{ANS.}$$

$$B_x = \left(\frac{2a-3b}{4b}\right)P = -2.50 = 2.50 \text{ kN} \quad (\leftarrow \text{ on BD}). \qquad \text{ANS.}$$

$$A_y = -\frac{b}{8c}P = -0.625 = 0.625 \text{ kN} \quad (\downarrow \text{ on BD}). \qquad \text{ANS.}$$

$$D_y = \frac{b}{8c}P = 0.625 \text{ kN} \quad (\uparrow \text{ on AD}). \qquad \text{ANS.}$$

$$D_x = \left(\frac{b-2a}{4b}\right)P = 0. \qquad \text{ANS.}$$

$$B_y = -A_y = 0.625 \text{ kN} \quad (\uparrow \text{ on BD}). \qquad \text{ANS.}$$

■

Problems

7.37 Refer to Example 7.4, let $a = 0.4$ m, $b = 0.6$ m, and $P = 2$ kN. Solve for all unknown forces, and show them acting on a free-body diagram of member AB.

7.38 Refer to Example 7.5, let $b = 4$ ft, $c = 3$ ft, $r = 1$ ft, $\theta = 30°$, $W = 200$ lb, and $P = 800$ lb. Solve for all unknown forces and show them acting on free-body diagrams of the pulley and member ABC.

7.39 Refer to Example 7.6, let $b = 0.6$ m, $c = 1.2$ m, $d = 1.5$ m, $e = 2.0$ m, and $P = 25$ kN. Solve for all unknown forces, and show them acting on complete free-body diagrams of the three members.

7.40 Refer to Example 7.7, let $b = 1$ ft, $c = 3$ ft, and $P = 200$ lb. Solve for all unknown forces, and show them acting on complete free-body diagrams of the three members, CDG, ABD, and DEF.

7.41 Refer to Example 7.8, let $T_1 = 800$ N, $a = 1$ m, $b = 0.6$ m, and $r = 0.1$ m. Determine all unknown forces, and show them acting on free-body diagrams of the pulley and member CBD.

7.42 Refer to Example 7.9, let $P = 800$ lb, $a = 0.8$ ft, $b = 1.0$ ft, and $c = 1.2$ ft. Determine all unknown forces, and show them acting on free-body diagrams of members AD, EC, and BD.

7.43 Refer to Figure P7.43, and let $Q = 0$, $P = 3$ k, $a = 3$ ft, $b = 2$ ft, $c = 2.5$ ft, and $\theta = 30°$. The slot in the horizontal member CD is frictionless. Determine all forces acting on the three members of this two-dimensional frame.

7.44 Refer to Figure P7.43, and let $Q = 4$ kN, $P = 6$ kN, $a = 1$ m, $b = 0.6$ m, $c = 0.8$ m, and $\theta = 30°$. The slot in the horizontal

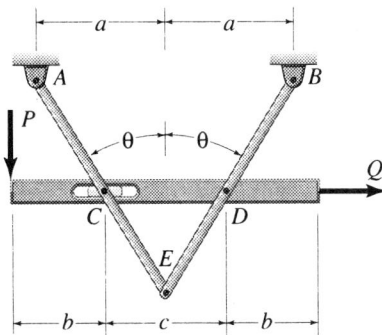

FIGURE P7.43.

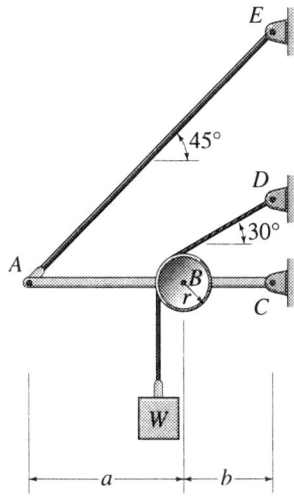

FIGURE P7.45.

member CD is frictionless. Determine all forces acting on the three members of this two-dimensional frame.

7.45   Refer to Figure P7.45, and let $W =$ 800 lb, $r = 0.2$ ft, $a = 2$ ft, and $b = 3$ ft. Determine the forces acting on the pulley at B. Then, find the tension in cable AE and the pin reaction components at C acting on ABC. The pulley is frictionless.

7.46   Refer to Figure P7.45, and let $W =$ 1200 N, $r = 0.1$ m, $a = 1$ m, and $b =$ 0.8 m. Determine the forces acting on

the pulley at B. Then, find the tension in cable AE and the pin reaction components at C acting on ABC. The pulley is frictionless.

7.47   Refer to Figure P7.47 and regard P, Q, a, b, and c as known inputs. Determine the force in member BC and the pin reaction components at A and C in terms of these inputs.

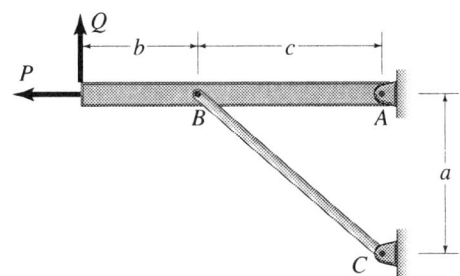

FIGURE P7.47.

7.48   Refer to Figure P7.48, and regard W, r, a, b, and c as known inputs. Draw a free-body diagram of the pulley, and find the force components it exerts on the vertical member ABC in terms of W. Determine the pin reaction components at C and the force in member BD as a function of the input quantities.

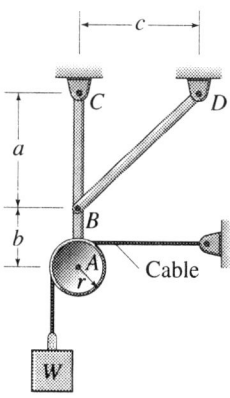

FIGURE P7.48.

7.49   Determine the pin reaction components at A, B, and E for the frame shown in Figure P7.49. Express your answers in terms of the quantities P, Q, a, and b. Show these components on a free-body diagram of each of the members ABC and DBE.

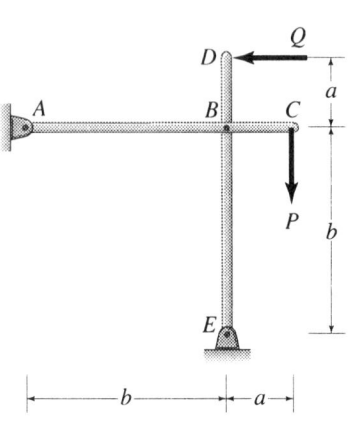

FIGURE P7.49.

7.50   Refer to Figure P7.50 and find all forces acting on members ABC and DBE in terms of W. Show horizontal and vertical pin reaction components on free-body diagrams of each of these members.

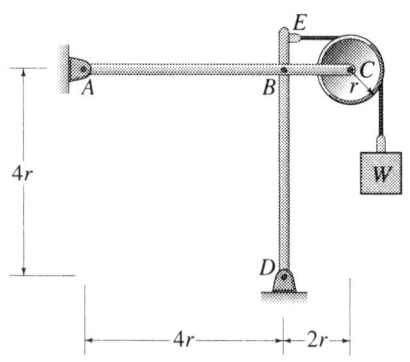

FIGURE P7.50.

7.51   Determine all forces acting on members AC and ADE of the frame in Figure P7.51. Express these forces in terms of P, and show them acting on free-body diagrams of each of these members.

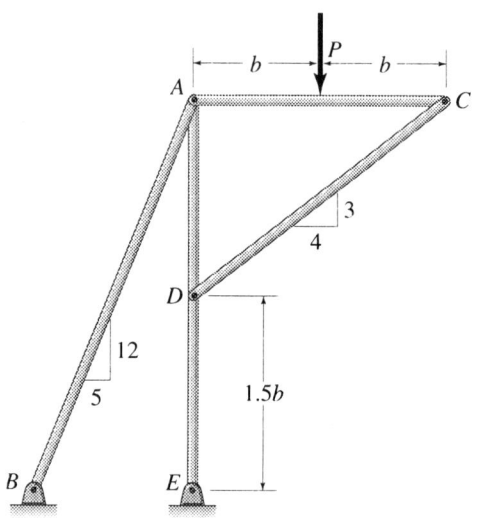

FIGURE P7.51.

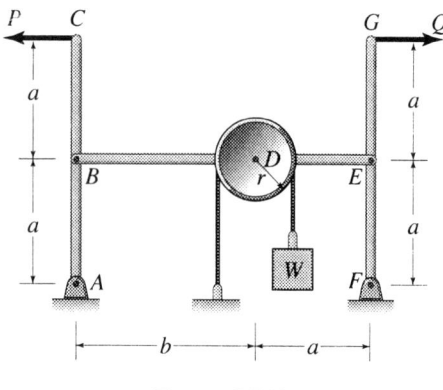

FIGURE P7.52.

7.52   Draw free-body diagrams of the pulley and members ABC, BDE, and FEG of the frame shown in Figure P7.52. Before solving for the unknown forces, compare the total number of unknowns to the total number of equations available to establish the statical determinacy of the system. Express unknown forces in terms of given quantities P, Q, a, b, and W. Show your answers on free-body diagrams of the pulley and members ABC, BDE, and FEG.

7.53 Draw free-body diagrams of members AB and BC of the frame shown in Figure P7.53, and determine all pin force components for $w = 200$ lb/ft and $b = 10$ ft. Sketch each member, and show all forces as they act on these members.

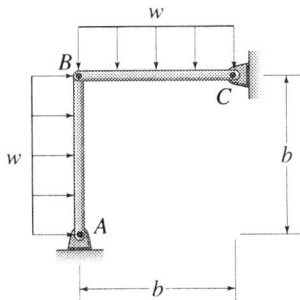

FIGURE P7.53.

7.54 Refer to the frame shown in Figure P7.54, and let $P = 20$ kN and $b = 2$ m. Determine the vertical and horizontal force components at pins A, B, and C. Sketch each member, and show all forces as they act on these members.

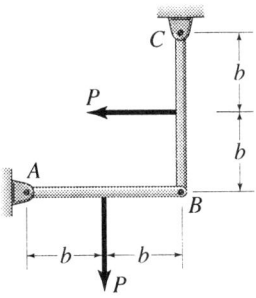

FIGURE P7.54.

7.55 Refer to Figure P7.55, and construct free-body diagrams for members AB and BC. Determine all forces acting on them.

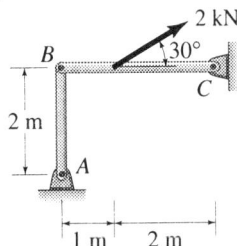

FIGURE P7.55.

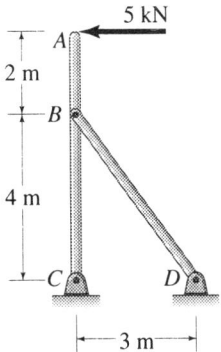

FIGURE P7.56.

7.56 Refer to Figure P7.56, and determine all forces acting on member ABC.

7.57 Let $w = 200$ lb/ft and $b = 4$ ft for the frame shown in Figure P7.57. Find the

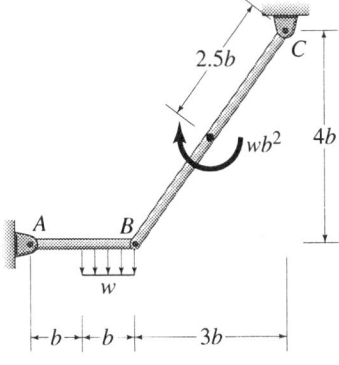

FIGURE P7.57.

pin force components at A, B, and C. Sketch each member, and show all forces as they act on these members.

7.58 Find all forces acting on member AB shown in Figure P7.58. Express your answers in terms of w and b.

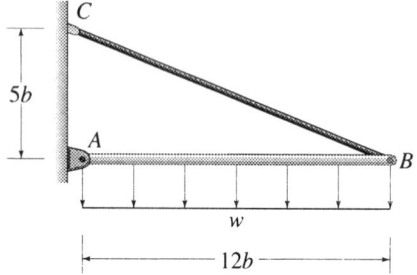

FIGURE P7.58.

7.59 Refer to Figure P7.59, and find the pin force components at A, B, and E. Sketch each member, and show all forces as they act on these members.

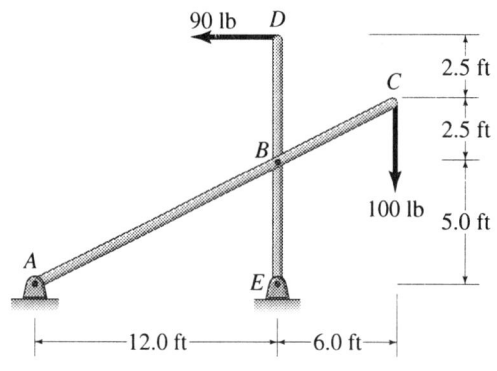

FIGURE P7.59.

7.60 The pulley shown in Figure P7.60 is frictionless. In terms of W, determine all forces acting on member ABC. Why are your answers independent of b? Would your answers be different if the pulley radius were changed to $0.1b$?

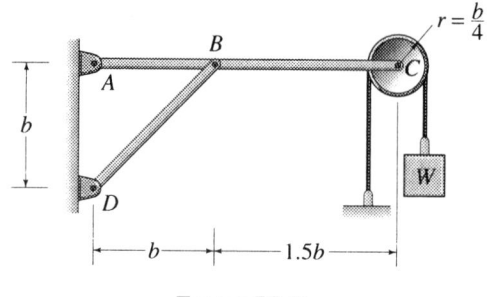

FIGURE P7.60.

7.61 Determine all force components acting on members ABC and CED of the frame depicted in Figure P7.61. Sketch each member, and show all forces as they act on these members.

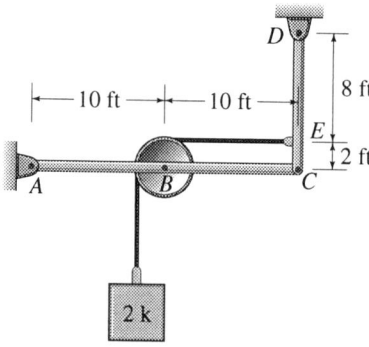

FIGURE P7.61.

7.62 The member ABC shown in Figure P7.62 is fixed at its base C. Determine the reaction components at C.

7.63 Three members are pin connected to form the frame of Figure P7.63. Determine the pin reaction components acting at A, B, C, D, and E. Show a sketch of each member with all forces acting on it. Check to see that they are each in equilibrium. The slot in ACE is frictionless.

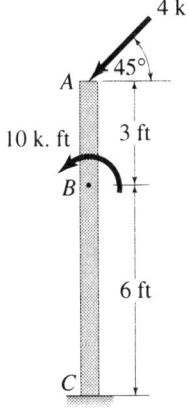

FIGURE P7.62.

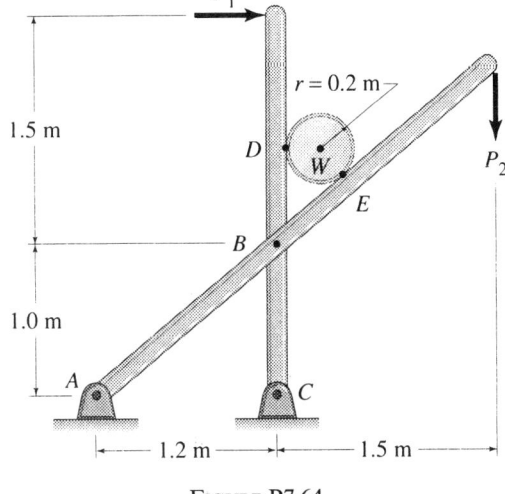

FIGURE P7.64.

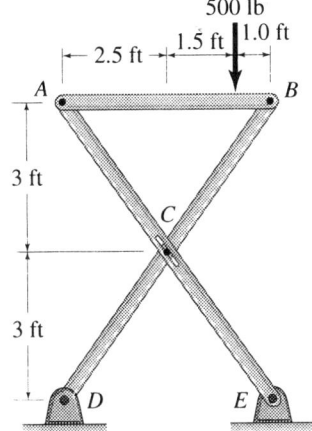

FIGURE P7.63.

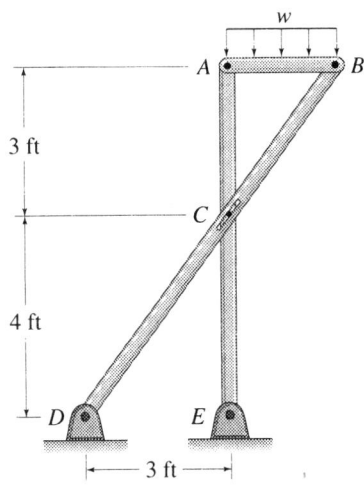

FIGURE P7.65.

7.64   Let $P_1 = 800$ N, $P_2 = 600$ N, and $W = 400$ N for the system shown in Figure P7.64. The cylinder has a diameter of 0.4 m, and the system is frictionless. Determine all forces acting on each component of the system, and draw a sketch of each one showing the forces as they actually act.

7.65   Refer to Figure P7.65, and let $w = 500$ lb/ft. Determine all forces acting on each member of the system, and draw a sketch of each showing the forces as they actually act. The slot in DCB is frictionless.

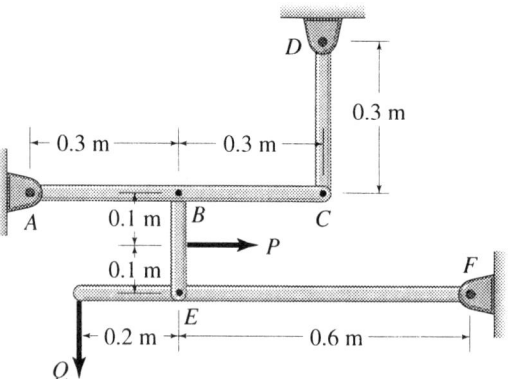

FIGURE P7.66.

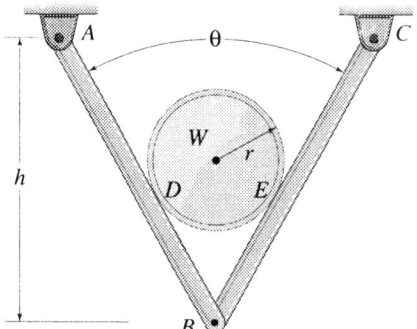

FIGURE P7.67.

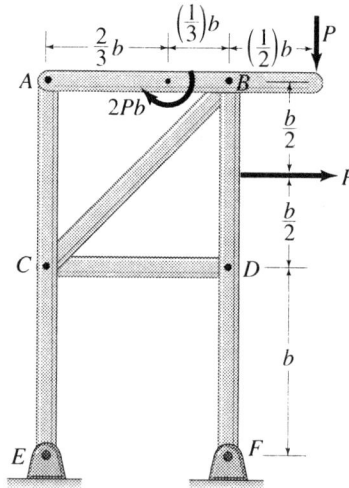

FIGURE P7.68.

7.66 If $P = 800$ N and $Q = 600$ N, determine all pin force components for the system shown in Figure P7.66. Sketch each member and show the forces as they actually act on it.

7.67 The frictionless system shown in Figure P7.67 is symmetric about a vertical line through point B. Regard W, θ, r, and h as known quantities, and express the pin force components at A and B in terms of these quantities.

7.68 In terms of P, determine all forces which act on members ACE and BDF of the frame shown in Figure P7.68. Would your answers change if the couple were applied elsewhere on the member AB?

7.69 Refer to the frame depicted in Figure P7.69, and find the forces which act on members BCE and CDF in terms of w and a.

7.70 Solve problem 7.69 for $w = 200$ N/m and $a = 4$ m.

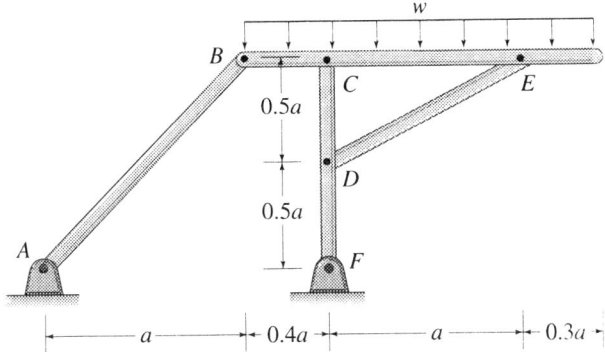

FIGURE P7.69.

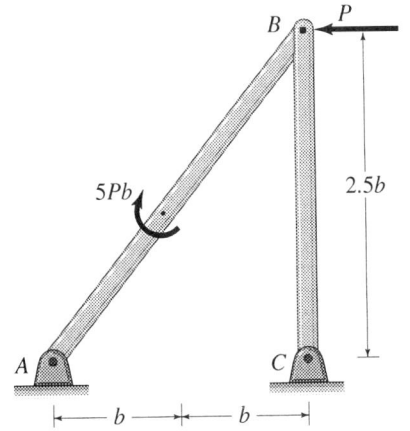

FIGURE P7.72.

7.71   Solve problem 7.69 for $w = 40$ lb/ft and $a = 10$ ft.

7.72   Given $P = 500$ lb and $b = 4$ ft, determine all forces acting on member AB of the frame shown in Figure P7.72.

7.73   Express all forces acting on member AB of the frame shown in Figure P7.72 in terms of P. Would your answers change if the couple were applied elsewhere on AB?

7.74   Solve problem 7.73 for $P = 400$ N.

7.75   Determine all forces acting on the horizontal member of the frame shown in Figure P7.75. Express answers in terms of P and θ.

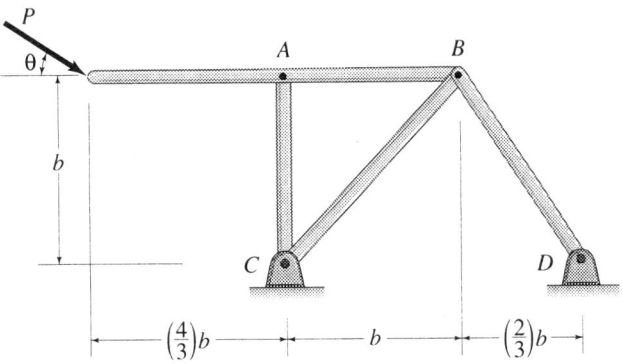

FIGURE P7.75.

7.76 Solve problem 7.75 for $P = 200$ lb and $\theta = 60°$.

7.77 Find all forces acting on member BCD in terms of w and b for the frame shown in Figure P7.77.

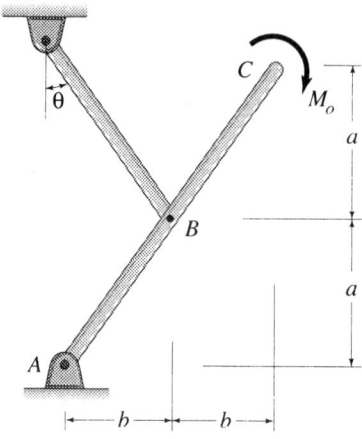

FIGURE P7.80.

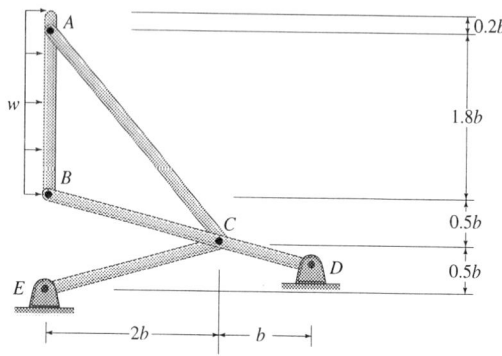

FIGURE P7.77.

7.78 Let $w = 800$ N/m and $b = 1$ m, and determine all forces acting on member BCD of the frame shown in Figure P7.77.

7.79 Solve problem 7.77 for $w = 60$ lb/ft and $b = 2$ ft.

7.80 Regard M_O, θ, a, and b as known quantities for the frame of Figure P7.80. Determine all forces acting on member ABC in terms of these quantities.

7.81 Solve problem 7.80 for $M_O = 2.00$ k·ft, $\theta = 45°$, $a = 2$ ft, and $b = 3$ ft.

7.82 Let $M_O = 20.0$ N·m, $\theta = 30°$, $a = 0.5$ m, and $b = 1.0$ m, and find all forces acting on member ABC of Figure P7.80.

7.83 In terms of Q, determine all forces acting on members ABC and DBE of the frame shown in Figure P7.83.

7.84 Solve problem 7.83 for $Q = 400$ N.

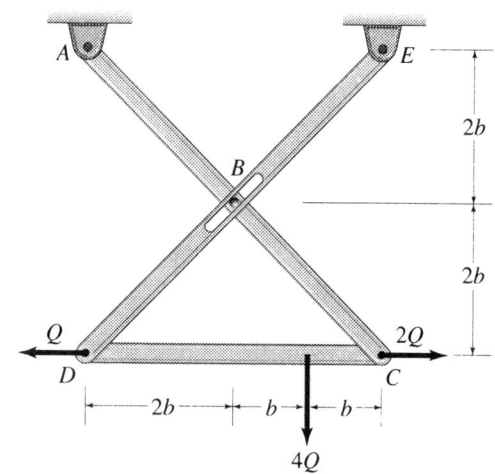

FIGURE P7.83.

Explain why a numerical value of b is not required.

7.85 Solve problem 7.83 for $Q = 100$ lb.

7.86 Refer to Figure P7.86, and find all forces acting on the horizontal and vertical members in terms of P.

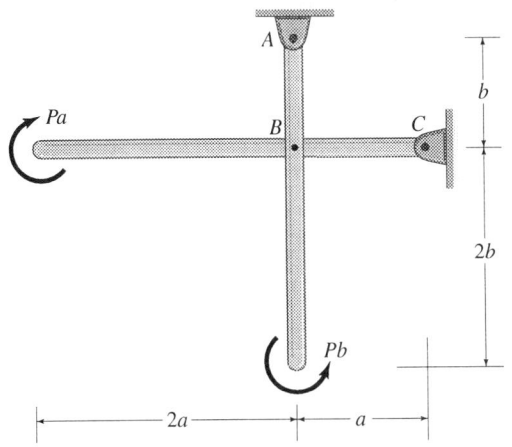

FIGURE P7.86.

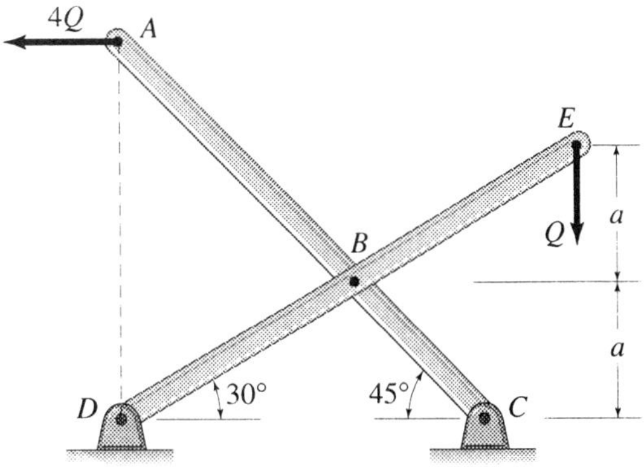

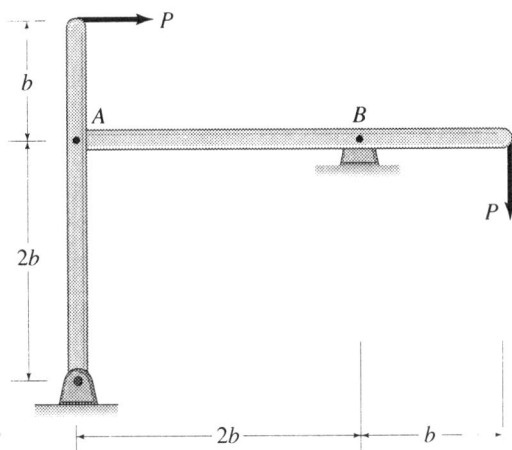

FIGURE P7.87.

FIGURE P7.88.

7.87 Determine all forces acting on the horizontal and vertical members of the frame shown in Figure P7.87. Express answers in terms of P.

7.88 In terms of Q, find all forces acting on members ABC and DBE shown in Figure P7.88.

7.89 A uniform load per unit length w is applied perpendicular to the inclined member shown in Figure P7.89. Deter-

mine the force in member BC and the pin reaction components at A in terms of the known quantities w, b, and θ.

7.90 Solve problem 7.89 for w = 100 N/m, b = 1.0 m, and θ = 30°.

7.91 Solve problem 7.89 for w = 50 lb/ft, b = 2.0 ft, and θ = 20°.

7.92 Find all forces acting on member CDE of Figure P7.92. Express your answers in terms of P and θ.

7.93 Let P = 500 N and θ = 45°, and solve problem 7.92.

7.94 Solve problem 7.92 for P = 200 lb and θ = 30°.

7.95 Make an overall qualitative analysis of the frame shown in Figure P7.95. Regard W and b as known quantities. Draw free-body diagrams of each member, count unknown forces, and compare them to the number of available independent equations of equilibrium. Write and solve these equations for all unknowns expressing them in terms of W.

7.96 Determine all forces acting on member ABCD shown in Figure P7.96. Express these forces in terms of W. Sketch this member, show each force component as it acts, and check to see that ABCD is in equilibrium.

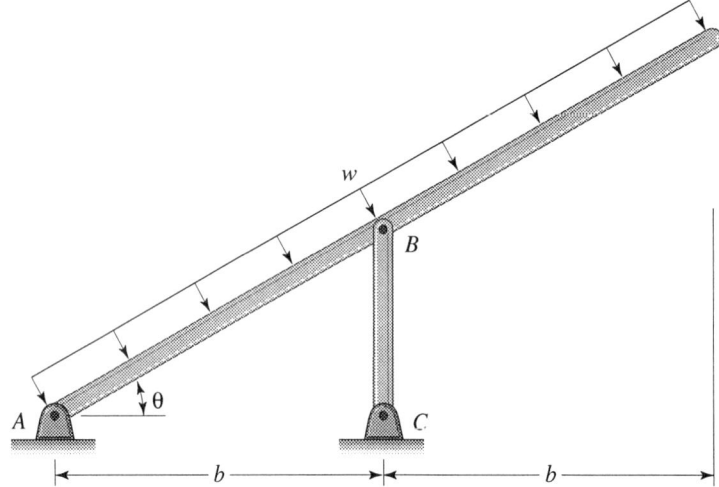

FIGURE P7.89.

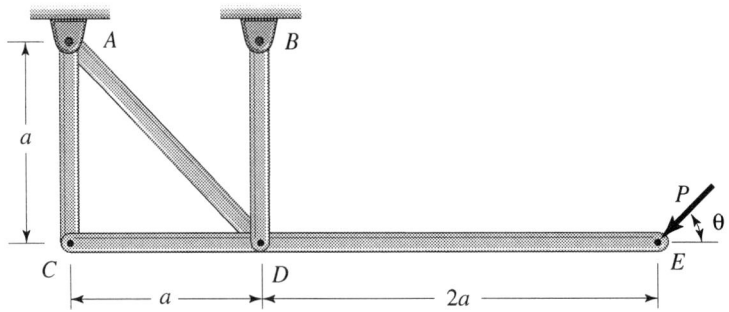

FIGURE P7.92.

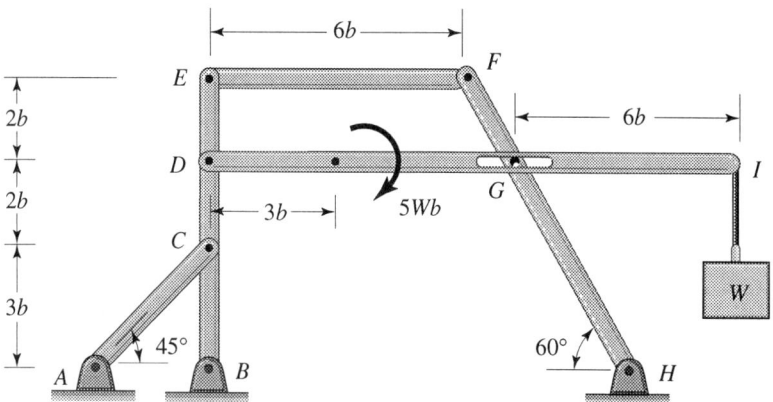

FIGURE P7.95.

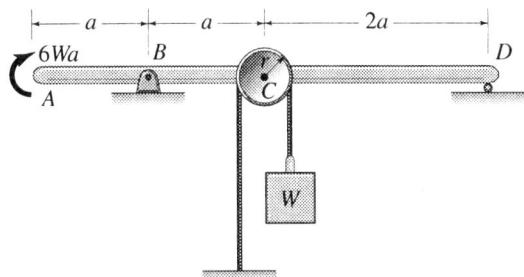

FIGURE P7.96.

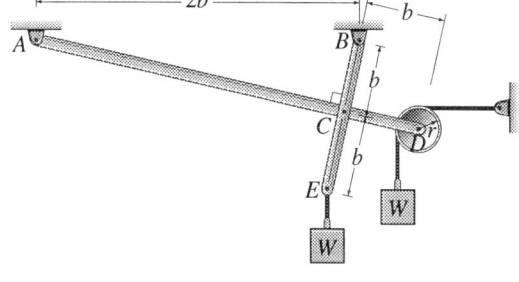

FIGURE P7.98.

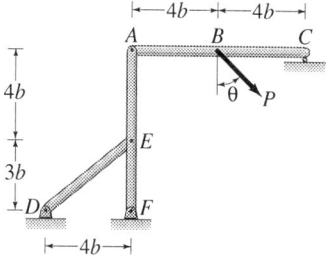

FIGURE P7.97.

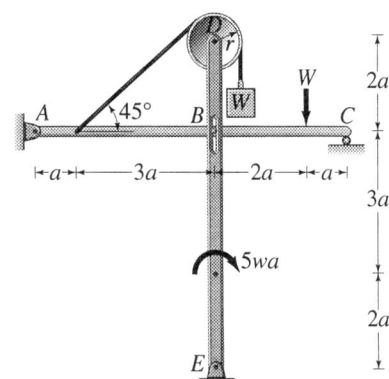

FIGURE P7.99.

7.97 Refer to Figure P7.97, and, in terms of P and θ, determine all force components acting on members ABC and AEF. Specialize your results for $\theta = 0°$ and $\theta = 90°$. For each of these special cases, sketch the two members, and show each force component as it acts on that member. Check to see that each member is in equilibrium.

7.98 Refer to Figure P7.98, and, in terms of W, determine all unknown components of the forces acting on members ACD and BCE. Sketch each member, and show each force component as it acts on that member. Check to see that each member is in equilibrium.

7.99 In terms of W, find all unknown force components acting on members ABC and DBE of the frame shown in Figure P7.99. Sketch each member, and show the force components acting on it. Check to see that each member is in equilibrium.

7.100 In terms of W, determine all forces acting on member BCD of Figure P7.100. Sketch this member, and show the force components acting on it.

7.101 Refer to Figure P7.101, and, in term of P, a, and b, find all forces acting on vertical member AB.

7.102 Analyze the frame of Figure P7.102

qualitatively. Draw free-body diagrams of members BCD and CEF and the pulley attached at D. Count the unknown forces, and compare them to the number of available, independent equations of equilibrium. Write and solve these equa-

tions which will enable you to express all unknowns in terms of W.

7.103 Find all unknown force components acting on members ABC and DBE of the frame shown in Figure P7.103. Express all answers in terms of P. Then, let $P = 10$ kN and $b = 1$ m, and find

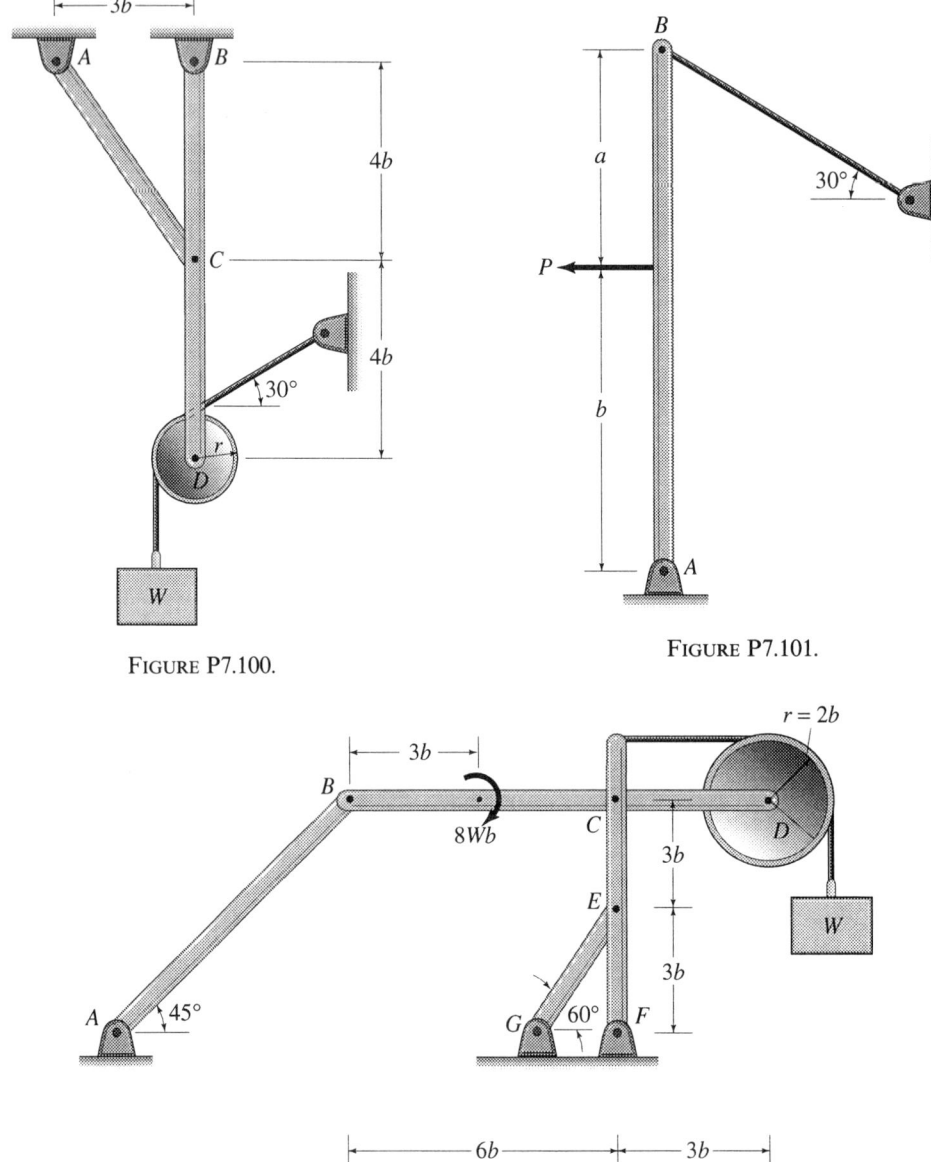

FIGURE P7.100.

FIGURE P7.101.

FIGURE P7.102.

numerical values for all forces acting on these members. Check to see that each member is in equilibrium.

7.104 Two members AB and BC are attached to each other at hinge B to form the structure shown in Figure P7.104. Hinges at A and C connect this structure to ground. (a) Apply a cw known couple M_O anywhere on member AB, and determine the horizontal and verti-cal force components at the hinges, A, B, and C. (b) Repeat part (a) for a cw cou-ple M_O applied anywhere on member BC. (c) Do the answers to parts (a) and (b) differ? Explain.

7.105 Solve Problem 7.104 for the special case $b = a$.

7.106 Solve Problem 7.104 for the following numerical inputs: $M_O = 30.0$ k·ft., $a = 10.0$ ft, and $b = 20.0$ ft.

7.107 The structure shown in Figure P7.107 has pinned connections at A, B, and C.

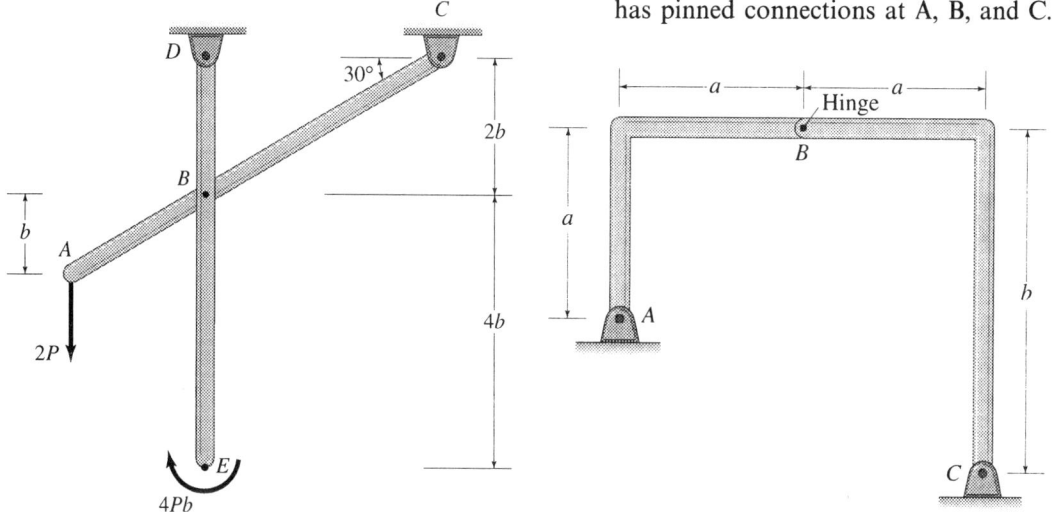

FIGURE P7.103.

FIGURE P7.104.

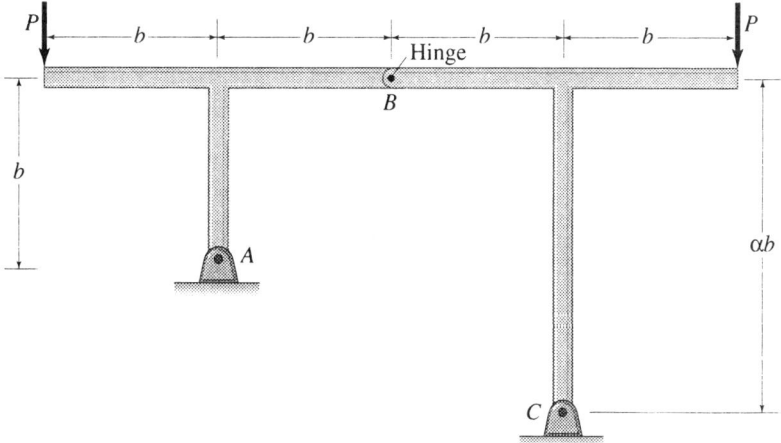

FIGURE P7.107.

Regard P, b, and α as known quantities. Determine horizontal and vertical force components at A, B, and C. Show sketches of each body showing these force components acting with their proper senses.

7.108 Solve Problem 7.107 for $\alpha = 3.0°$.

7.109 Member BC in the frame shown in Figure P7.109 is in the form of a quarter circle. Determine the components of the reactions at A, B, and C.

7.110 (a) Analyze the frame shown in Figure P7.110 to determine the reaction components at A, B, and C in terms of w, a, and θ.

(b) Specialize the answers obtained in part (a) for the case where $w = 3$ k/ft, $a = 4$ ft, and $\theta = 30°$.

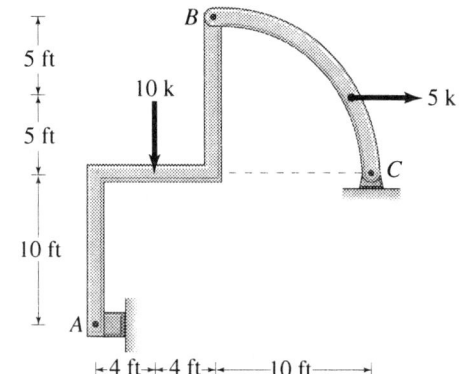

FIGURE P7.109.

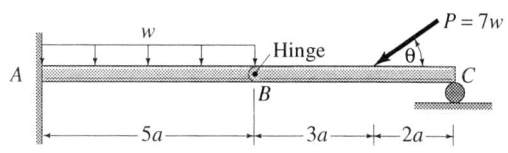

FIGURE P7.110.

7.3
Machine
Analysis

Machines, whether simple or complex, may be defined as structures (frames) with moving components designed to transmit and modify forces. Thus, the analysis of machines in equilibrium is identical to the analysis of frames discussed in Section 7.2. As in the case of frames, a machine in a given equilibrium position is dismantled to create a number of free-body diagrams sufficient to solve for unknown forces. As was done in section 7.2, it is advisable to compare the number of all unknown quantities to the number of available equations of equilibrium to insure that we are dealing with a statically determinate system.

As stated earlier, a machine is designed to alter an *input force* into an *output force*. By definition, the *mechanical advantage* of a given machine is the ratio of the output force to the input force.

The following examples illustrates the type of analysis required in solving a given machine problem.

■ Example 7.10

Parallel action pliers are shown in Figure E7.10(a). The slots at E and C are assumed to be frictionless.

(a) Enumerate unknowns and available equations given that P is known.

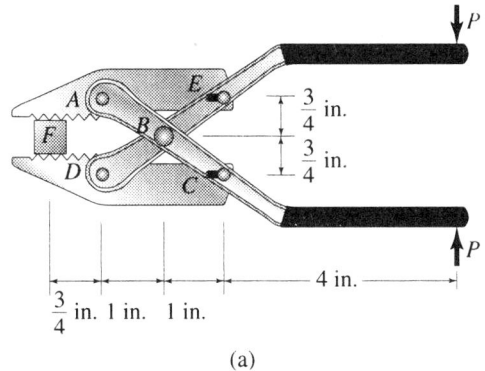

(a)

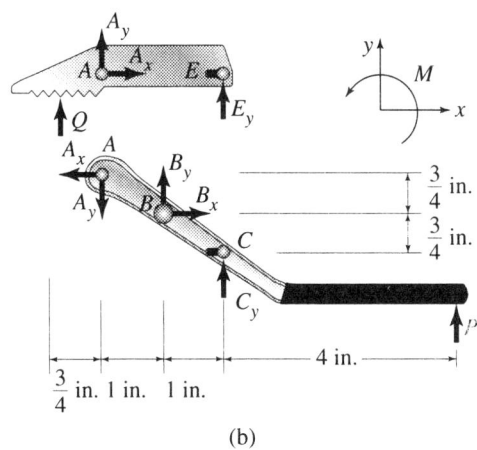

(b)

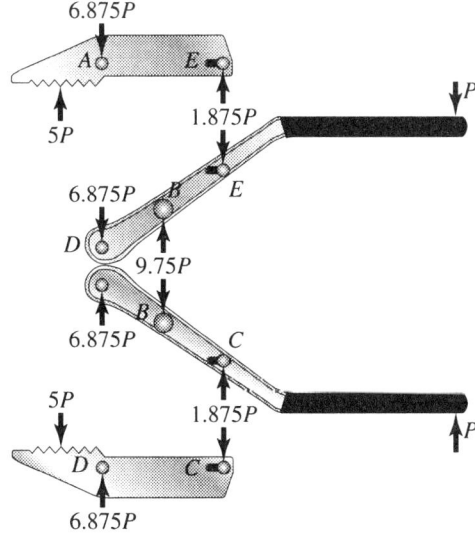

(c)

FIGURE E7.10.

(b) In terms of P, find the force exerted by the pliers on block F and the reaction components at A, B, C, D, and E.

(c) What is the mechanical advantage of these pliers?

Solution

(a) The free-body diagrams of the upper jaw and the lower handle are shown in Figure E7.10(b). The unknown quantities are Q, A_x, A_y, E_y, B_x, B_y, and C_y, where Q is the force exerted by block F on the upper jaw. By symmetry, however, we conclude that E_y and C_y are identical in magnitude and, therefore, we have only a total of six unknown quantities. Because there are three equations of equilibrium available for each of the two free-body diagrams, we have a total of six available equations, and, therefore, we conclude that the system shown in Figure E7.10(b) is statically determinate. Also, by symmetry or by considering the free-body diagrams of the lower jaw and the upper handle, we conclude not only that the magnitudes of E_y and C_y are equal as stated earlier but that the magnitudes of D_x and A_x and those of D_y and A_y are identical.

(b) Referring to the free-body diagram of the upper jaw,

$$\sum F_x = 0, \quad A_x = 0, \qquad \text{(a) ANS.}$$

$$\sum M_A = 0, \quad E_y(2) - Q\left(\frac{3}{4}\right) = 0,$$

$$E_y = 0.375Q, \qquad \text{(b)}$$

$$\sum F_y = 0, \quad Q + A_y + E_y = 0. \qquad \text{(c)}$$

Substituting Eq. (b) in Eq. (c) and solving for A_y,

$$A_y = -1.375Q. \qquad \text{(d)}$$

Refer now to the free-body diagram of the lower handle and write

$$\sum F_x = 0, \quad B_x - A_x = 0,$$

$$B_x = A_x = 0, \qquad \text{(e) ANS.}$$

$$\sum M_B = 0, \quad A_y(1) + A_x\left(\frac{3}{4}\right) + C_y(1) + P(5) = 0.$$

Substituting Eqs. (a), (b) and (d) in Eq. (f), we obtain a relationship between Q and P from which

$$Q = 5P. \qquad \text{(g) ANS.}$$

Thus, using Eqs. (b) and (d), we conclude that

$$E_y = 1.875P \quad (\uparrow \text{ on upper jaw}), \qquad \text{ANS.}$$

$$A_y = -6.875P = 6.875P \quad (\downarrow \text{ on upper jaw}), \qquad \text{(h) ANS.}$$

and

$$\sum F_y = 0, \quad B_y + C_y + P - A_y = 0. \tag{i}$$

Substituting Eqs. (b) and (d) in Eq. (i) and solving for B_y,

$$B_y = -9.75P = 9.75P \quad (\downarrow \text{ on lower handle}). \qquad \text{ANS.}$$

These force components, as well as those acting on the lower jaw and upper handle, are shown on the free-body diagrams of Figure E7.10(c).

(c) Mechanical Advantage $= \dfrac{Q}{P} = \dfrac{5P}{P} = 5.$ \qquad ANS.

■ **Example 7.11** A slider-crank mechanism is shown in Figure E7.11(a). It transforms straight line motion into rotary motion or vice versa. Block B is the slider which moves along a straight line and the crank OA is a line on the disk of weight W which rotates about a smooth axis at O. This interesting mechanism will be studied further in dynamics but we are concerned here with its equilibrium positions. For the position shown,

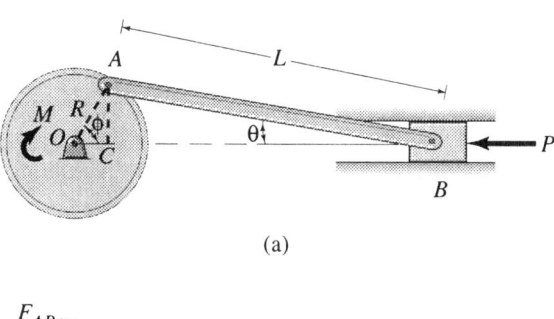

(a)

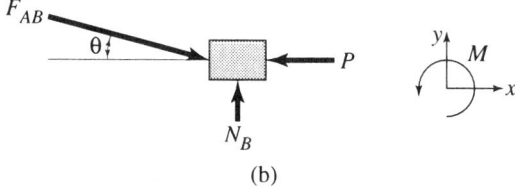

(b)

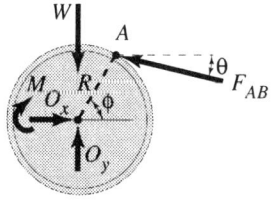

(c)   FIGURE E7.11.

(a) Find the force in the connecting rod AB and the force exerted on block B by its smooth guide. Also, find the reaction components at O.

(b) Determine the couple M required for equilibrium as a function of the given quantities P, L, R, and θ.

(c) Determine numerical values for all unknowns, given $P = 8$ kN, $W = 2$ kN, $L = 0.60$ m, $R = 0.2$ m, and $\theta = 15°$.

Solution

(a) The free-body diagram of the slider block B is shown in Figure E7.11(b) where F_{AB} is the force in connecting rod AB and N_B is the force exerted on the block by its smooth guide. Thus,

$$\sum F_x = 0, \quad F_{AB} \cos\theta - P = 0,$$

$$F_{AB} = \frac{P}{\cos\theta}, \qquad \text{ANS.}$$

and

$$\sum F_y = 0, \quad N_B - F_{AB} \sin\theta = 0,$$

$$N_B = F_{AB} \sin\theta = P \tan\theta. \qquad \text{ANS.}$$

(b) The free-body diagram of the disk is shown in Figure E7.11(c) where the angle ϕ may be expressed in terms of the given angle θ by using the geometry provided in Figure E7.11(a). Thus,

$$AC = R \sin\phi = L \sin\theta,$$

$$\phi = \sin^{-1}\left(\frac{L \sin\theta}{R}\right), \qquad \text{(a)}$$

Applying the conditions of equilibrium to the disk,

$$\sum M_O = 0, \quad F_{AB} \sin\theta (R \cos\phi) + F_{AB} \cos\theta (R \sin\phi) - M = 0,$$

$$M = R F_{AB}(\sin\theta \cos\phi + \cos\theta \sin\phi)$$

$$= R\left(\frac{P}{\cos\theta}\right)\sin(\theta + \phi) \qquad \text{ANS.}$$

where ϕ is given by Eq. (a).

$$\sum F_x = 0, \quad O_x - F_{AB} \cos\theta = 0,$$

$$O_x = F_{AB} \cos\theta = P, \qquad \text{ANS.}$$

and

$$\sum F_y = 0, \quad O_y + F_{AB} \sin\theta - W = 0,$$

$$O_y = W - F_{AB} \sin\theta = W - P \tan\theta. \qquad \text{ANS.}$$

(c) Substituting given numerical quantities, after using Eq. (a) to find $\phi = 50.94°$,

$$\left.\begin{array}{l} F_{AB} = 8.28 \text{ kN,} \\ N_B = 2.14 \text{ kN,} \\ M = 1.512 \text{ kN·m,} \\ O_x = 8.00 \text{ kN,} \\ O_y = -0.1436 \text{ kN.} \end{array}\right\}$$ ANS.

and

Problems

7.111 A compaction device is shown in Figure P7.111. As a function of P, θ, d, and c, determine the vertical compaction force exerted by piston A which moves in frictionless guides.

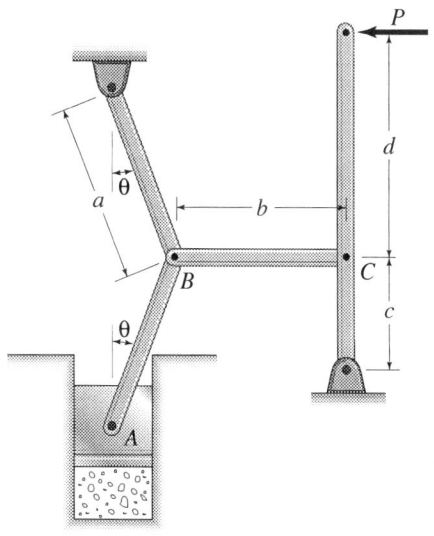

FIGURE P7.111.

7.112 Solve problem 7.111 for $P = 100$ lb, $\theta = 8°$, $d = 3$ ft, and $c = 1$ ft. Briefly explain why numerical values are not required

for a and b. What is the ratio of the compaction force to the applied force P?

7.113 Solve problem 7.111 for $P = 50$ lb, $\theta = 5°$, $d = 3$ ft, and $c = 1$ ft. Determine the ratio of the compaction force to the applied force P.

7.114 A hoisting mechanism for truck engines is shown in Figure P7.114. Given an engine weight $W = 5$ kN, $\theta = 60°$, $a = 2$ m, $b = 1.5$ m, $c = 2$ m, and $d = 4$ m, determine (a) the force in the hydraulic cylinder BD and (b) the pin reaction components at C acting on ABC.

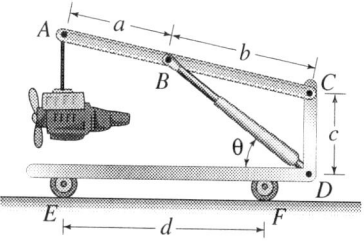

FIGURE P7.114.

7.115 Solve problem 7.114 for $W = 6$ k, $\theta = 60°$, $a = 5$ ft, $b = 4.2$ ft, $c = 5.4$ ft, and $d = 11.4$ ft. Also, find the vertical reactions at F and E.

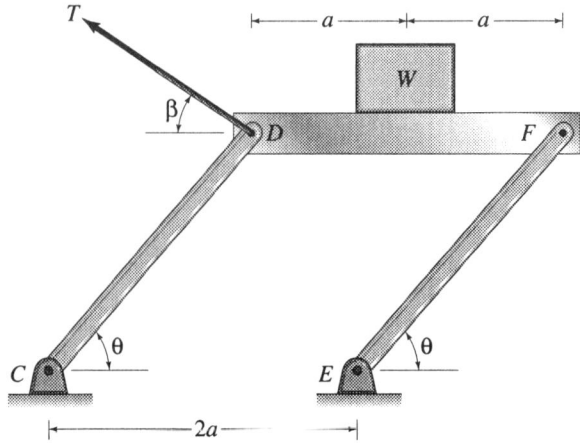

FIGURE P7.116.

7.116 A four bar linkage CDEF used to raise heavy weights from one level to another is shown in Figure P7.116. Given that $W = 5000$ lb, $a = 6$ ft, $\theta = 45°$, and $\beta = 30°$, determine the cable tension T and the axial forces in links CD and EF.

7.117 Solve problem 7.116 for $W = 8000$ lb, $a = 8$ ft, $\theta = 30°$, and $\beta = 45°$.

7.118 A revolving head punch is shown in Figure P7.118. It is used to punch various sized holes in leather and other soft materials. If a force of 90 lb is required to punch a certain hole, what force P must be exerted on each handle? Also, find the components of the reaction at the hinge at B.

7.119 The pipe pliers shown in Figure P7.119 are used to grip pipe work. If a force of 30 lb is exerted on each handle, what force is exerted on each side of the pipe? What are the components of the reactions at hinge A.

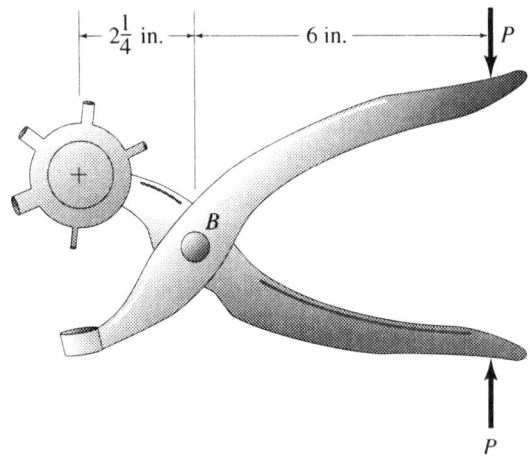

FIGURE P7.118.

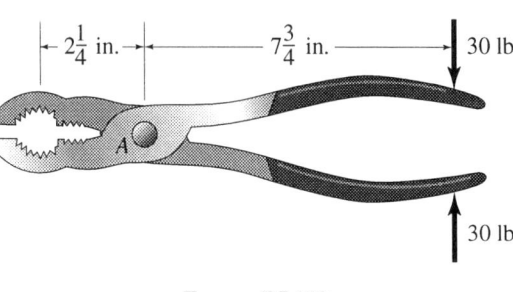

FIGURE P7.119.

7.120 The lopping shears shown in Figure P7.120 are used to prune shrubs. If a

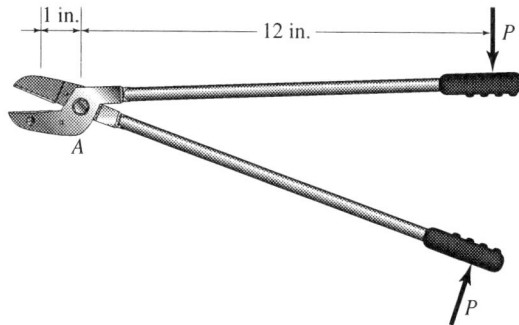

FIGURE P7.120.

force of 240 lb is required to cut a shrub main stem, what force P must be exerted on each side of the grips by the operator? Find the reaction components at hinge A.

7.121 The slider crank mechanism shown in Figure P7.121 is frictionless. Given that $P = 500$ lb, $b = 6$ in, $\theta = 30°$, and $a = 12$ in., determine the couple C required for equilibrium.

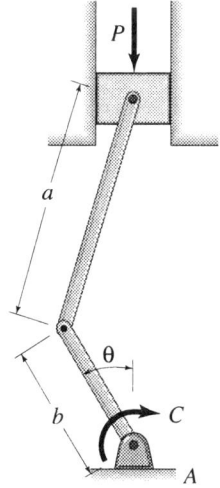

FIGURE P7.121.

7.122 Refer to Figure P7.121, and let $C = 2500$ lb·in. Determine the force P exerted on the piston if $a = 10$ in., $b = 4$ in., and $\theta = 45°$. Also, find the normal reaction exerted on the piston by the cylinder. Assume frictionless conditions.

7.123 In Figure P7.121, $P = 4.00$ kN, $b = 0.20$ m, $\theta = 60°$, and $a = 0.35$ m. Find the couple C consistent with equilibrium of this slider-crank mechanism. Also, find the reaction components at hinge A.

7.124 For the mechanism depicted in Figure P7.124, determine the horizontal reaction R as a function of the applied forces P and Q and the angle θ.

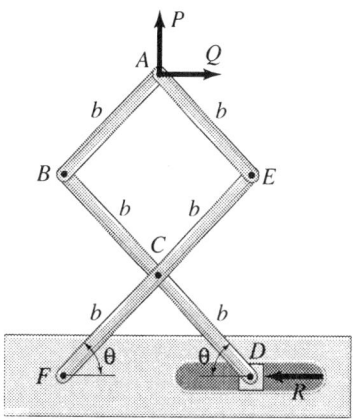

FIGURE P7.124.

7.125 Solve problem 7.124 for $P = 800$ N, $Q = 0$, and $\theta = 60°$.

7.126 Solve problem 7.124 for $P = 0$, $Q = 600$ lb, and $\theta = 45°$. Also, find the reaction components at hinge C.

7.127 Forces in the left front of a vehicle frame arc transmitted to ground by the assembly shown in Figure P7.127. Find the force in link CD and the joint reaction components at A consistent with equilibrium of this system for $R = 10.0$ kN.

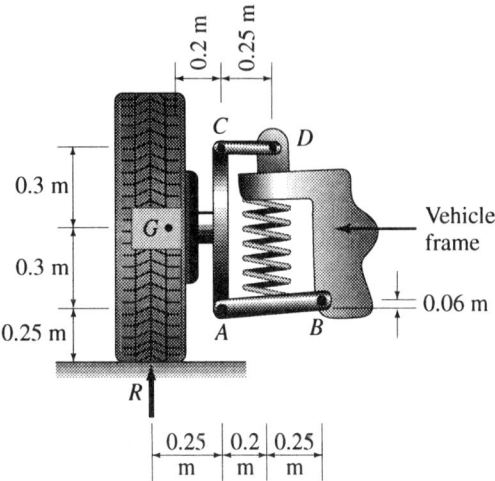

FIGURE P7.127.

This assembly weighs 0.6 kN which acts at G.

7.128 Refer to problem 7.127, and determine the force in the spring and the joint reaction component at B.

7.129 A four bar linkage is depicted in Figure P7.129. Determine the couple M as a function of P, a, b, c, and θ consistent with equilibrium of this mechanism.

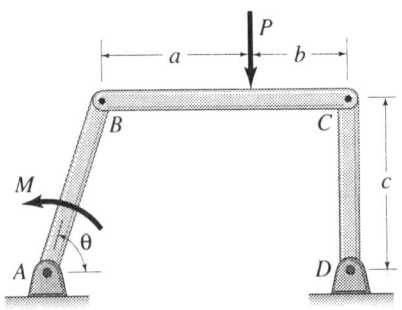

FIGURE P7.129.

7.130 Solve problem 7.129 for $P = 600$ lb, $a = 20$ in., $b = 10$ in., $c = 15$ in., and $\theta = 60°$. Also, find the reaction components at hinges A and D.

7.131 Solve problem 7.129 for $P = 4000$ N, $a = 0.5$ m, $b = 0.4$ m, $c = 0.6$ m, and $\theta = 45°$.

7.132 Parallel action pliers are shown in Figure P7.132. The jaws remain parallel as they open and close. If a user exerts a force $P = 100$ N on each handle, what force is exerted by the jaws on block Q? Given dimensions are $a = 0.020$ m, $b = 0.025$ m, $c = 0.030$ m, and $d = 0.120$ m. Pins C and E are free to move in frictionless slots in the handles.

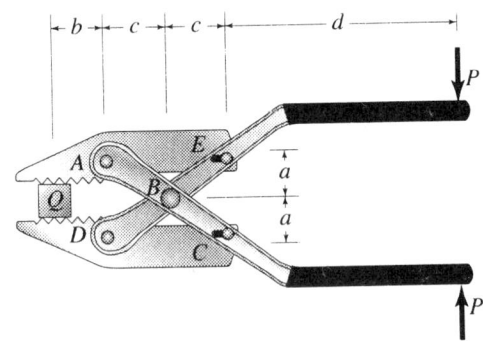

FIGURE P7.132.

7.133 Solve problem 7.132, given $P = 25$ lb, $a = 1.00$ in., $b = 1.50$ in., $c = 1.20$ in., and $d = 5.00$ in. Also, find the reaction components at hinges A, B, and C.

7.134 The shovel depicted in Figure P7.134 weighs 400 kN which is assumed to act through point G. The pressure exerted upward by the pavement on the shovel is given by

$$p = p_0\left[1 - \left(\frac{x}{4}\right)^2\right]$$

where x is measured from point A. Determine p_0 which is the pressure at point A. These pressures are exerted on two treads, each of which measures 4 m × 0.4 m. Also, find the maximum permissible length L.

7.135 Solve problem 7.134 for a linear variation of pressure given by

$$p = p_0\left[1 - \left(\frac{x}{4}\right)\right].$$

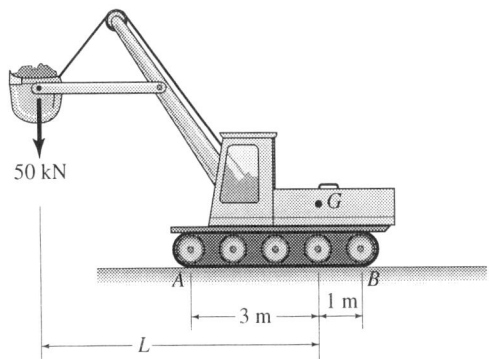

FIGURE P7.134.

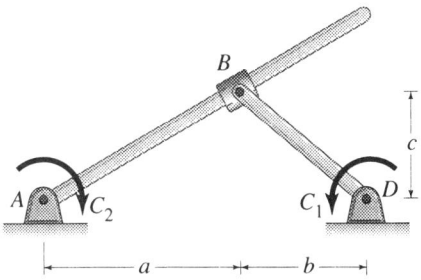

FIGURE P7.138.

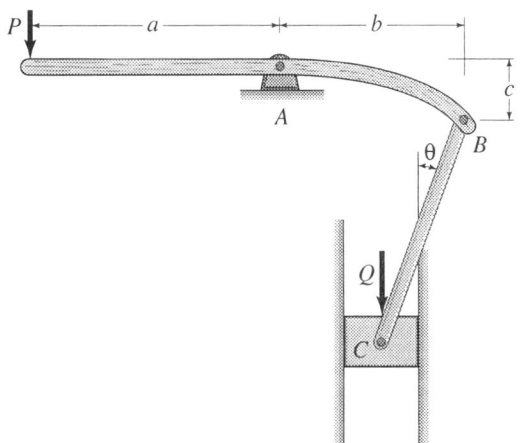

FIGURE P7.136.

7.136 A hand pump is depicted in Figure P7.136. Determine the force P exerted by the user, given that $Q = 150$ N, $\theta = 10°$, $a = 0.25$ m, $b = 0.15$ m, and $c = 0.06$ m. Assume a frictionless system.

7.137 Solve problem 7.136 given that $Q = 250$ N, $\theta = 15°$, $a = 0.30$ m, $b = 0.20$ m, and $c = 0.10$ m. Also, find the reaction components at hinge A.

7.138 Refer to Figure P7.138, and let couple $C_1 = 650$ lb·in. Find couple C_2 consistent with equilibrium given that $a = 12$ in., $b = 8$ in., and $c = 5$ in. The system is frictionless.

7.139 Solve problem 7.138 for $C_1 = 800$ lb·in., $a = 10$ in., $b = 5$ in., and $c = 5$ in.

7.140 An automobile lift mechanism is depicted in Figure P7.140. There are two parallel mechanisms each of which supports one half of the automobile weight $W = 15.8$ kN which is assumed to act through point G. Let $\theta = 45°$, $a = 1.60$ m, $b = 1.00$ m, $c = 2.40$ m, $d = 3.00$ m, and $e = 1.30$ m. Determine the force in each hydraulic cylinder CD and the force in each member BF.

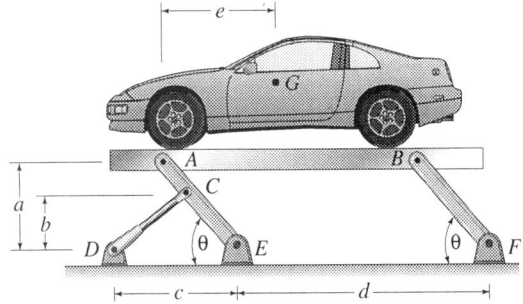

FIGURE P7.140.

7.141 Solve problem 7.140 for $W = 3200$ lb, $\theta = 45°$, $a = 60$ in., $b = 40$ in., $c = 90$ in., $d = 120$ in., and $e = 40$ in.

7.142 Flat-nosed pliers used for light work, such as bending wires, are depicted in Figure P7.142. If the force exerted on

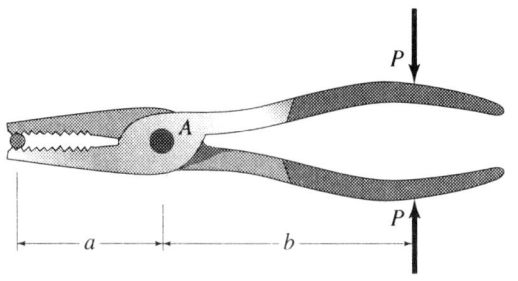

FIGURE P7.142.

the wire $Q = 40$ lb, $a = 3$ in., and $b = 4$ in., find P. What are the reaction components at hinge A?

7.143 The vehicle shown in Figure P7.143 is used to transport heavy loads of raw materials. The forward unit weighs 6 tons concentrated at G_1 and the loaded

aft unit weighs 40 tons concentrated at G_2. Ignore friction at the wheels, and determine the forces exerted on the tires at D and E on each side of the vehicle. Also, determine the force in each of the two hydraulic cylinders AB (one on each side).

7.144 Refer to Problem 7.143 and determine the reaction components at C on each side of the vehicle.

7.145 A Scotch yoke mechanism is shown in Figure P7.145. Determine the following quantities consistent with the equilibrium of this frictionless system: (a) the force exerted on pin B which is attached to the disk, (b) the vertical forces at D and E, (c) the couple C, and (d) the pin reaction components at A. Neglect the weights of the disk and slider.

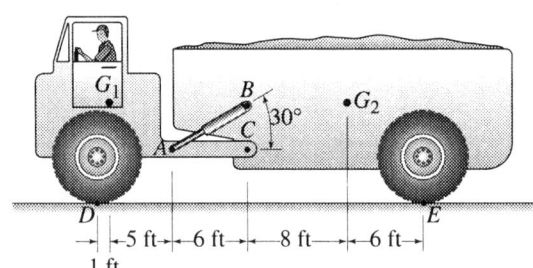

FIGURE P7.143.

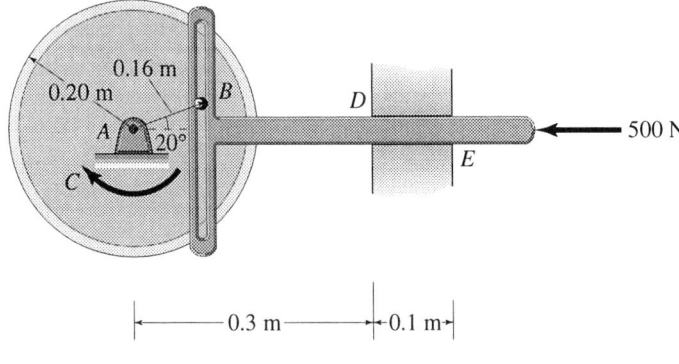

FIGURE P7.145.

Review Problems

7.146 Consider the Frame shown in Figure P7.146, and determine the resultant forces acting on member AB at A and B.

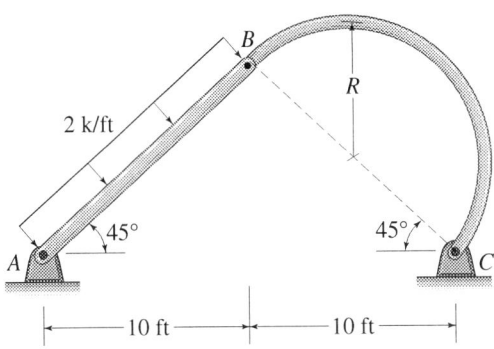

FIGURE P7.146.

7.147 Refer to the frame shown in Figure P7.147. Let $W = 15$ kN, $a = 3$ m, and $r = 0.5$ m, and determine the components of the reactions at C, D, and E acting on member CDE.

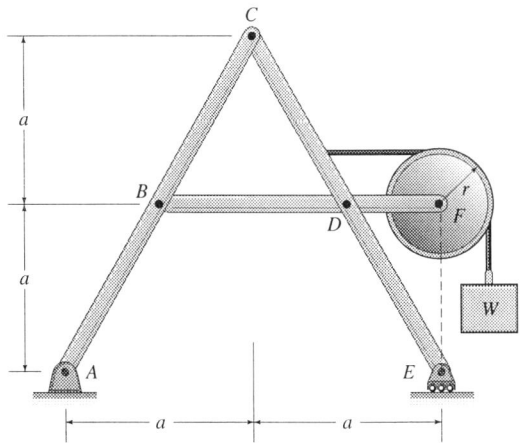

FIGURE P7.147.

7.148 Refer to the frame shown in Figure P7.147. Let $W = 10$ k, $a = 10$ ft., and $r = 1$ ft., and determine the components of the reactions at B, D, and F acting on member BDF.

7.149 Refer to the frame shown in Figure P7.147. Let $W = 10$ kN, $a = 4$ m, and $r = 0.75$ m, and determine the components of the reactions at A, B, and C acting on member ABC.

7.150 Consider the frame shown in Figure P7.150. Let $w = 3$ k/ft. and $a = 10$ ft., and determine the components of the reactions at A, B, and C acting on member ABC.

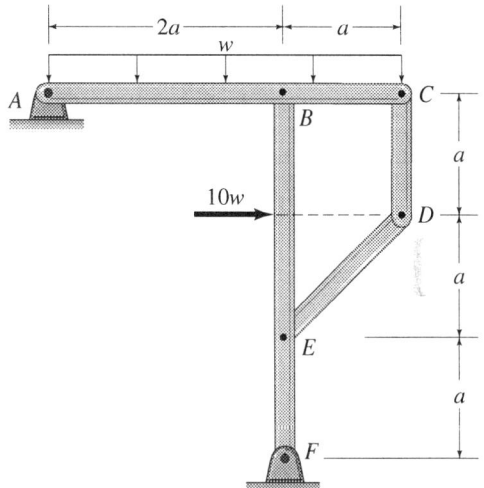

FIGURE P7.150.

7.151 Consider the frame shown in Figure P7.150. Let $w = 1.5$ kN/m and $a = 3$ m, and determine the components of the reactions at B, E and F acting on member BEF.

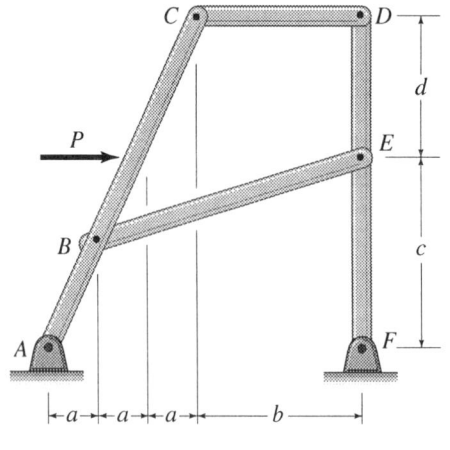

FIGURE P7.152.

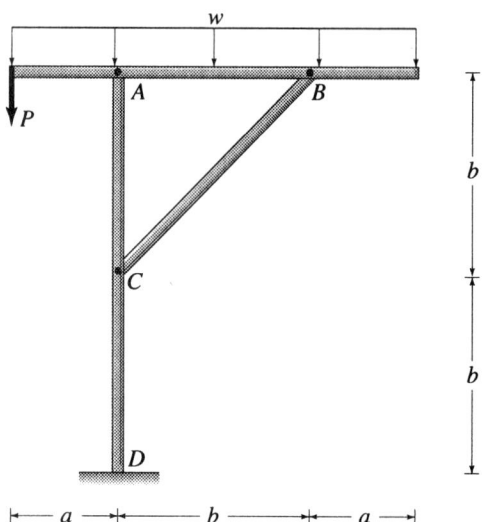

FIGURE P7.155.

7.152 Refer to the frame shown in Figure P7.152. Let $P = 15$ k, $a = 5$ ft., $b = 10$ ft., $c = 12$ ft., and $d = 8$ ft., and find all forces acting on member ABC.

7.153 Refer to the frame shown in Figure P7.152. Let $P = 20$ kN, $a = 2$ m, $b = 3$ m, $c = 4$ m, and $d = 3$ m, and determine all forces acting on member DEF.

7.154 Consider the frame shown in Figure P7.154 and determine the force components acting on member ABCD at A, B, C and D.

7.156 Refer to the frame shown in Figure P7.155. Let $P = 5$ k, $w = 1$ k/ft., $a = 4$ ft, and $b = 6$ ft, and determine the forces acting on member ACD at A, C, and D.

7.157 Consider the frame shown in Figure P7.157, and determine the force compo-

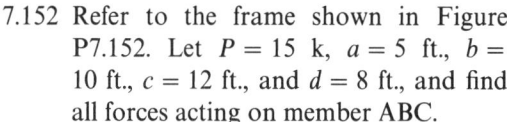

FIGURE P7.154.

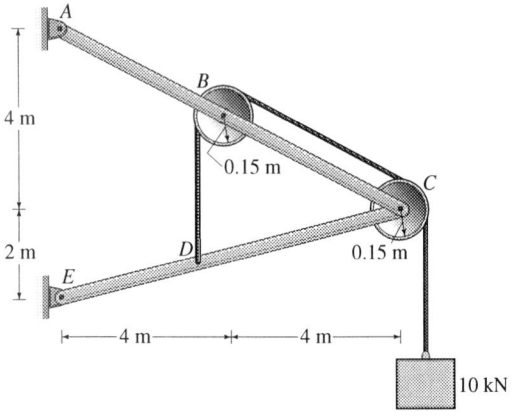

7.155 Refer to the frame shown in Figure P7.155. Let $P = 15$ kN, $w = 4$ kN/m, $a = 2$ m and, $b = 3$ m, and find the forces acting on member AB at A and B.

FIGURE P7.157.

nents acting on member ABC at A, B, and C.

7.158 Refer to Problem 7.157, and determine the force components acting on member EDC at E, D, and C.

7.159 The mechanism shown in Figure P7.159 represents a shearing machine used to cut 12 in. × 3 in. pieces of thin, sheet material. If the shearing resistance of the material is 100 lb./in., determine the force P needed to do the cutting.

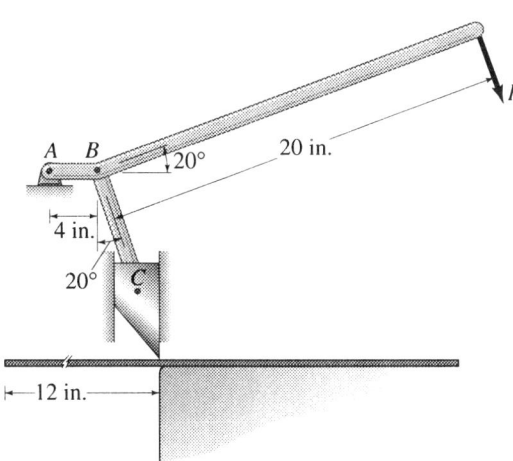

FIGURE P7.159.

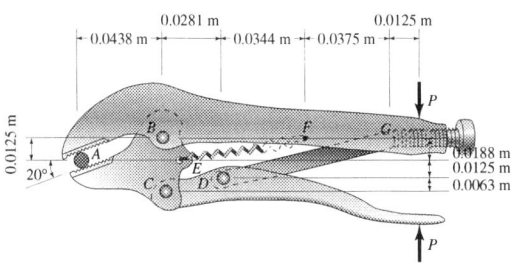

FIGURE P7.160.

7.160 If the force required to grip the 0.008-m diameter stock at A is 3600 N, determine the magnitude of the force P that

must be applied to the handles of the vise grip shown in Figure P7.160. The force in spring EF is sufficiently small compared to other forces and may be ignored.

7.161 Figure P7.161 represents one of two mechanisms that control the bucket of a front-end loader in a specific position during its operation. The total weight of the load and the bucket is 20 k and may be assumed to act through point G. Determine the forces acting on each of the two members labeled EDB. Note that hydraulic cylinder CD is horizontal.

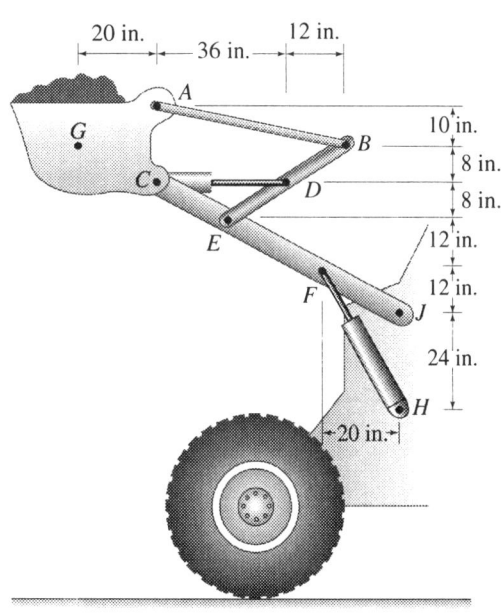

FIGURE P7.161.

7.162 Refer to Problem 7.161, and determine the forces acting on each of the two members labeled CEFJ. Hint: Use the results obtained in the solution of Problem 7.161.

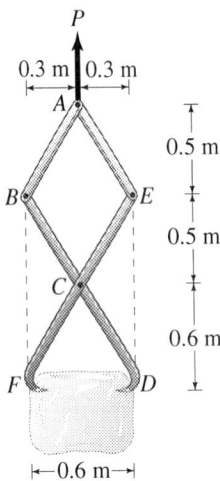

FIGURE P7.163.

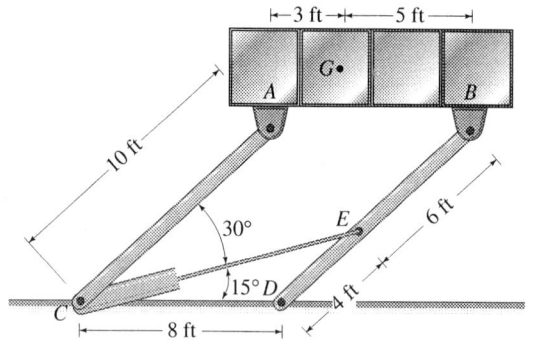

FIGURE P7.164.

7.163 A block of ice weighing 200 N is held by tongs, as shown in Figure P7.163. Find all the forces acting on member ECF.

7.164 The position of the platform is controlled by the hydraulic cylinder CE, as shown in Figure P7.164. If the weight of the platform and its contents is 2000 lb, which may be assumed to act through point G, determine the force in the hydraulic cylinder.

7.165 Determine the compressive force exerted on block E when the 80-N force is applied to the toggle press, as shown in Figure P7.165.

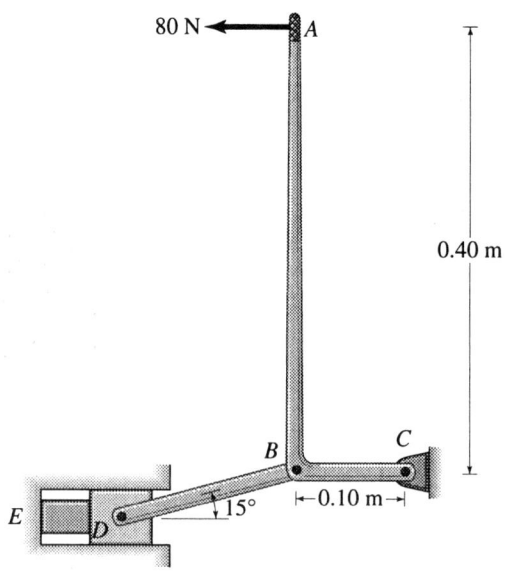

FIGURE P7.165.

8

Internal Forces in Members

The Golden Gate Bridge—a lasting tribute to American engineers from years past.

The design of bridges, ships, vehicles, airplanes, and other engineering systems requires knowledge of the forces, torques, shears, and moments associated with the system being designed. The evolution of technology throughout the ages has always been accompanied by new challenges. Newton's laws provided engineers with new mathematical tools that dramatically revolutionized engineering. However, our ability to tackle complex systems, such as the space shuttle, modern bridges, ..., etc., required new technological advances in computer software and hardware. This is because of our inability to perform a significant number of calculations at a sufficient speed. Consequently, the analysis of individual members composing a system is a prerequisite for analyzing and designing a system. Furthermore, a thorough understanding of basic principles governing equilibrium of forces is needed.

Engineers, who creatively design machines and structures, need to determine external forces acting on bodies in equilibrium. For engineers to safely design a system to resist external forces or loadings, they must have a thorough knowledge of the forces in each of its components. Internal forces are classified as axial forces, torques, shears, and moments, and this chapter is concerned with determining relationships between external and internal forces in a given system and its components.

8.1 Internal Forces

In previous chapters, we learned how to compute the forces acting on members at points of support and connection. Such forces are classified as *external*. We computed such forces for straight, two-force members in Chapter 6 and for multiforce and curved two-force members in Chapter 7. In this chapter, we will develop methods to find the forces that must exist within a given member (i.e., *internal forces*), when the member is subjected to external forces.

Two concepts are basic for determining internal forces in members. First, if an entire member is in equilibrium, then, any part of this member, which we care to consider, will also be in equilibrium. Secondly, when the adjectives *internal* and *external* modify the noun *forces* they are to be regarded as relative. An internal force for an entire member becomes an external force for a part of this same member.

The rigid body idealization will continue to be used for overall machines or structures and members or partial members. Another way to state this idealization is that the equations of equilibrium will be formulated for the undeformed body. Obviously, real machines and structures deform when subjected to forces, but, in most engineering applications, we may safely ignore these deformations when writing the equations of equilibrium. However, the rigid body idealization is not always acceptable. In Sections 8.8, 8.9, and 8.10, we will study cables and will assume them to be inextensible, but problems do arise, for example, in suspension bridge design, where this assumption is not satisfactory. Acceptance or rejection of analytical assumptions is based upon comparisons with experimental results, experience and, engineering judgment.

A straight two-force member AB is shown in Figure 8.1 subjected to the external forces Q which place the member in tension. Note that the same conclusion would be reached if the two-force member were subjected to compression. If we imagine that the member is cut at any position C into two parts by a cutting plane, each of the two parts AC and CB must be in equilibrium. Thus, if we consider part AC, equilib-

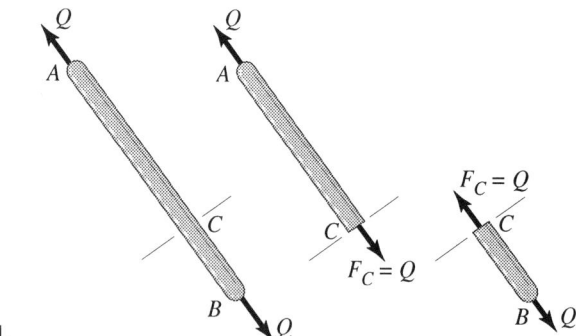

FIGURE 8.1.

rium dictates that the force F_C acting at C must be equal in magnitude to the force Q applied at A. Similarly, the force F_C at C acting on part CB must be equal in magnitude to the force Q applied at B. Thus, we conclude that the force $F_C = Q$, acting on part AC, represents the force exerted on it by part CB. Similarly, the force $F_C = Q$, acting on part CB, represents the force exerted on it by part AB. Therefore, because member AB is not *really* cut, the force $F_C = Q$ at C acts *internally* at location C in the loaded member.

Figure 8.2 shows a member subjected to the external couples Q represented by double-headed vectors which are interpreted by the right-hand rule. Note that, for convenience, a circular cross-section is shown for the member. However, other cross-sections would serve as well. Note also that a couple applied about the longitudinal axis of a member is referred to as a *torque*. If the member is cut at any location C into two parts by an imaginary cutting plane as shown, each of the two parts AC and CB must be in equilibrium. Following the logic used in the case of the two-force member, we conclude that the torque T_C at C must be equal in magnitude to the applied torque Q. Also, because member AB is not *really* cut, the torque $T_C = Q$ at C acts *internally* at location C in the loaded member.

Next, consider the case of the straight, multiforce member AB shown

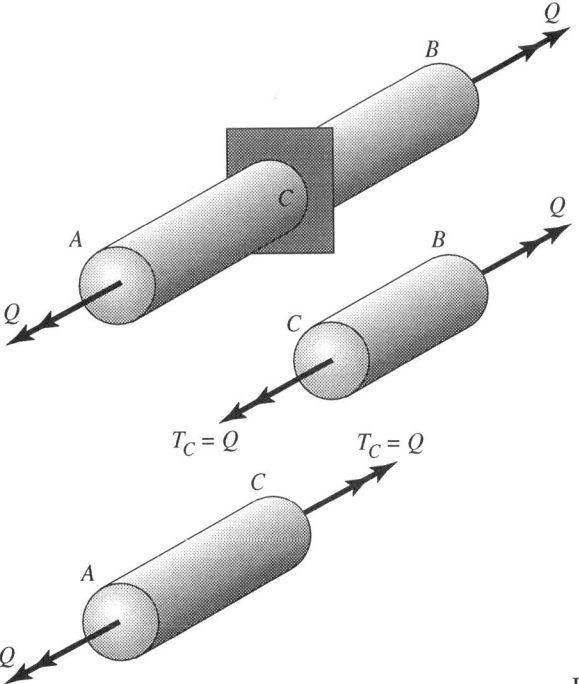

FIGURE 8.2.

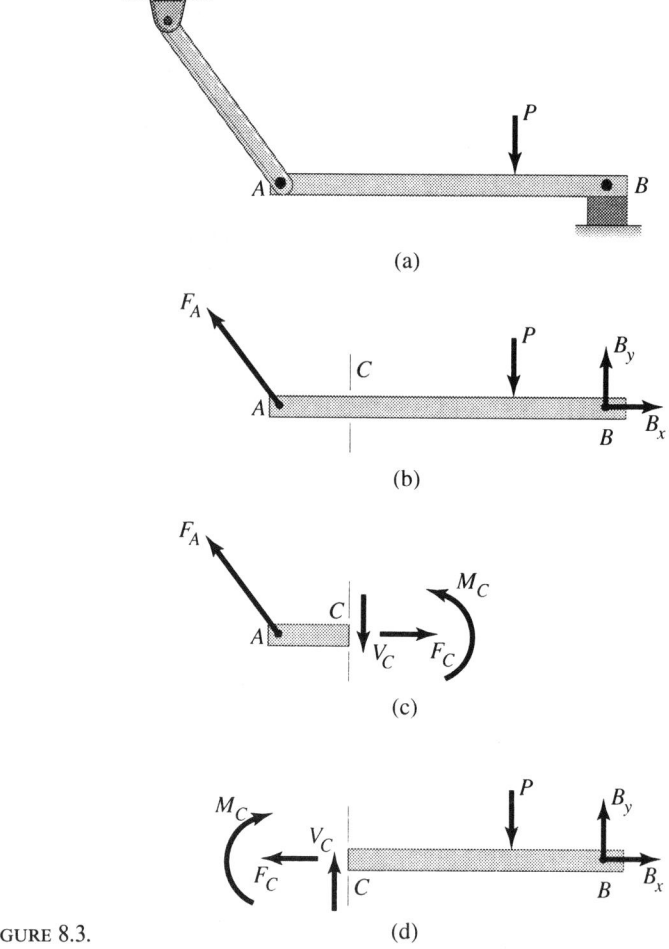

FIGURE 8.3.

in Figure 8.3(a). Such a member, known as a *beam*, serves a very useful purpose in structural applications. Beam AB is held in position by a two-force member at A and by the hinge at B and is subjected to the load P. If the beam is isolated from its supports and its free-body diagram is constructed, we obtain the system shown in Figure 8.3(b), where F_A, B_x, and B_y may be determined in terms of the load P. Now, if we imagine that the beam is cut at C into two parts by a cutting plane, each of the two parts AC and CB must be in equilibrium.

Let us consider the equilibrium of part AC whose free-body diagram is shown in Figure 8.3(c). Equilibrium of member AC requires that we have a force F_C at C to balance the horizontal component of F_A, a force V_C to balance the vertical component of F_A, and a moment M_C to balance the moment produced by the vertical component of F_A. In other words, the free-body diagram of part AC represents a two-dimensional force system of a rigid body and, therefore, we have three equations of equilibrium ($\sum F_x = 0$, $\sum F_y = 0$ and $\sum M = 0$) available to solve for the three unknown quantities F_C, V_C, and M_C. The three unknown quantities F_C, V_C, and M_C acting on part CB, shown in Figure 8.3(d), are determined in a similar fashion and are found to be equal in magnitude to the corresponding quantities acting on part AC. Because the forces F_A, B_x, and B_y are known in terms of P, it follows that F_C, V_C, and M_C can be determined in terms of the applied load P. Also, because beam AB is not *really* cut at C, the forces F_C, V_C, and M_C act *internally* at this location in the loaded beam.

The force F_C, known as the axial force at a specific position C in the beam, acts perpendicularly to the beam's cross-section producing

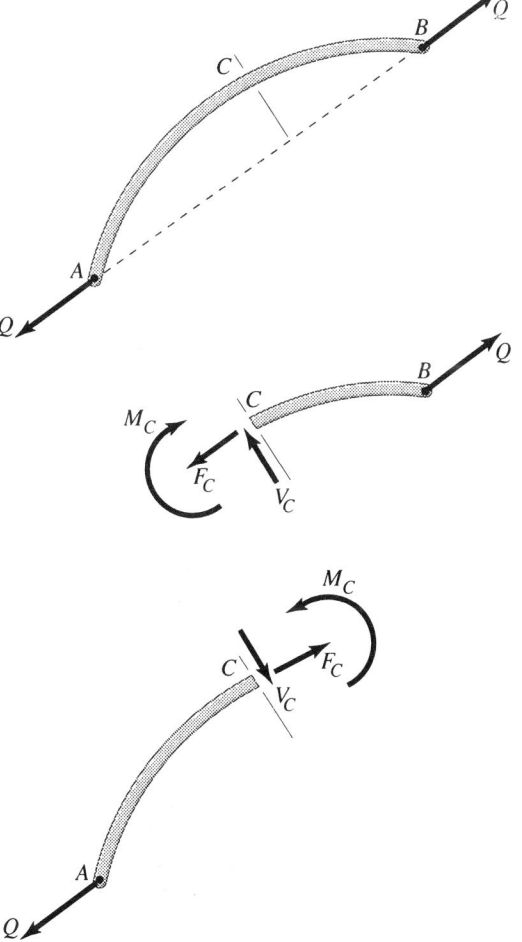

FIGURE 8.4.

either tension or compression. The force V_C referred to as the *shear force* at a given location C in the beam, acts in a direction parallel to the beam's cross-section, tending to produce shearing (sliding) action of one part of the beam relative to the other. The moment M_C, known as the *bending moment* at position C in the beam, *rotates* the beam about an axis at C perpendicular to the page causing a change in the curvature of the beam at C. These aspects of beam behavior are beyond the scope of this book but are dealt with in *mechanics of materials* books.

A curved two-force member AB is shown in Figure 8.4 subjected to the external forces Q. If the member is cut at any position C into two parts by an imaginary plane, each of the two parts AC and CB must be in equilibrium. Considering part AC and proceeding as we did in the case of the multiforce member shown in Figure 8.3, we concluded that, in general, the internal forces at any position C consist of an axial force F_C, a shear force V_C, and a bending moment M_C as shown in Figure 8.4. A similar conclusion is reached if we consider part CB, and, therefore, the forces, F_C, V_C, and M_C at C act *internally* at this location C in the loaded member.

8.2 Sign Conventions

In writing the equations of equilibrium in the preceding chapters, we have adopted and made use of what is termed an *equations sign convention*, which is arbitrarily chosen. In each case, our choice has been governed by convenience and adherence to the right-hand rule. Thus, in two-dimensional problems, for example, a free-body diagram was accompanied by an arbitrarily chosen coordinate system in which, in general, the x axis was chosen positive to the right, the y axis positive upward, and the z axis positive out of the page. Thus, according to the right-hand rule, a positive moment about the z-axis would have a ccw sense.

When we draw free-body diagrams to determine internal forces, we make use of what may be termed a *physical sign convention*. We assume that the axial force, torque, shear, and moment on a given free-body diagram are positive according to the physical sign conventions defined in Figure 8.5. In Figure 8.5(a) we show the convention generally used for a positive internal axial force F which acts normal to its plane and points away from it. This is the same as saying that a positive internal axial force F, by convention, produces tension. In Figure 8.5(b), we indicate the convention used in this textbook for a positive internal torque T, shown by a double-headed vector acting normal to its plane and pointing away from it. Use of the right-hand rule will enable us to determine the sense of rotation produced by the torque T. Finally, the standard sign conventions used for the shear V and the bending moment M are given in Figure 8.5(c). Because a beam, in general, is used

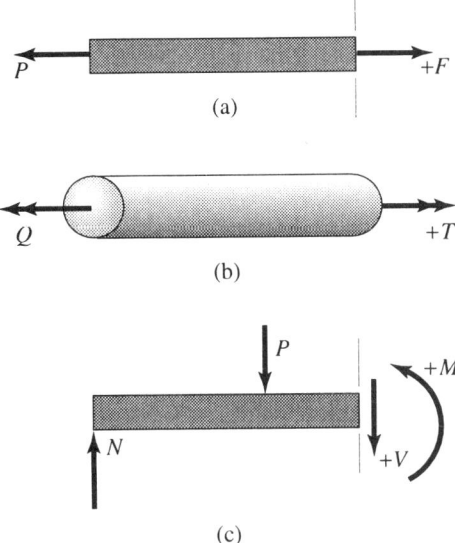

(a)

(b)

(c)

FIGURE 8.5.

in a horizontal position, an imaginary plane would cut it into a *left* and a *right* free-body diagram. With this in mind, we may say that a positive shear, by convention, is one that points downward on a left free-body diagram and upward on a right free-body diagram. A positive moment, by convention, would be ccw on a left free-body diagram and cw on a right free-body diagram.

Once these free-body diagrams are constructed assuming positive internal forces by the *physical sign convention*, we apply the equations of equilibrium using the arbitrarily chosen *equations sign convention*. If the solution indicates a positive answer, we conclude that the assumed sense of the internal force is correct. On the other hand, if the solution indicates a negative answer, we conclude that the assumed sense of the internal force is incorrect. For example, a positive answer for an axial force would confirm that it is tensile whereas a negative answer would indicate that it is compressive.

8.3
Axial Force
and Torque
Diagrams

Internal force diagrams are very useful devices that show, at a glance, the variation of the internal forces (axial, torque, shear, and moment) along the given member. In this section we will focus on the development of axial force and torque diagrams, and we will discuss the construction of shear and moment diagrams in Section 8.7.

In Section 8.1 we discussed the existence of internal axial forces F

and torques T in members subjected to external loads. To determine F and T at any specific cross-section in a member, we construct a free-body diagram containing this cross-section, showing F or T as positive quantities in accordance with the physical sign convention discussed in Section 8.2. After selecting a coordinate (usually x) measuring distances along the member from one of its two ends, the equations of equilibrium are applied ($\sum F_x = 0$ or $\sum T_x = 0$) using the arbitrarily chosen equations sign convention. The sign of the answer tells us whether or not the positive physical sense assumed in the free-body diagram is correct. This process of constructing a free-body diagram and applying the equations of equilibrium is repeated for as many cross-sections as needed. If F and T are plotted vs. the coordinate x, the resulting figures are referred to, respectively, as the *axial force* and *torque diagrams*.

The following examples illustrate the construction of axial force and torque diagrams.

■ **Example 8.1**

A straight member is subjected to the loads shown in Figure E8.1(a). The loads are applied through thin rigid plates at B, C, D, and E. Construct the axial force diagram.

Solution

The axial-force diagram is shown in Figure E8.1(b). It is good practice to show such a diagram adjacent to the given loaded member. The development of this diagram requires the construction of several free-body diagrams. Arbitrarily, we choose a longitudinal x coordinate measuring distances from the left end of the member and positive to the right.

The free-body diagram of the entire member is shown in Figure E8.1(c). Thus,

$$\sum F_x = 0, \quad 50 + 2(5) - 2(40) + 2(15) - 2(25) + F_G = 0,$$

$$F_G = 40 \text{ kN}$$

where F_G represents the reaction at the fixed support at G.

The free-body diagram for any cross section between A and B is shown in Figure E8.1(d). Hence,

$$\sum F_x = 0, \quad 50 + F_{AB} = 0,$$

$$F_{AB} = -50 \text{ kN} \quad (0 < x < 1 \text{ m}). \qquad \text{ANS.}$$

The free-body diagram containing any cross-section between B and C is shown in Figure E8.1(e). Hence,

$$\sum F_x = 0, \quad 50 + 2(5) + F_{BC} = 0,$$

$$F_{BC} = -60 \text{ kN} \quad (1 < x < 1.6 \text{ m}). \qquad \text{ANS.}$$

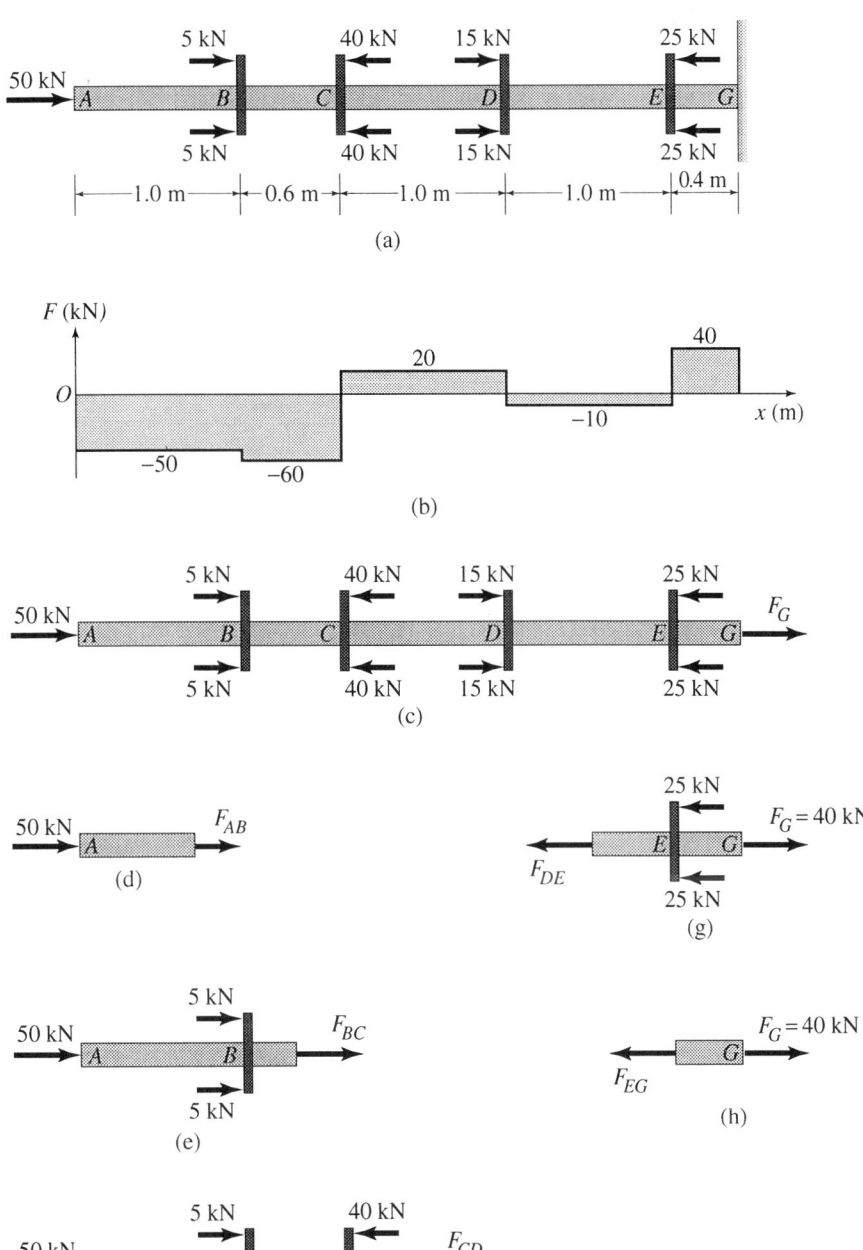

FIGURE E8.1.

The free-body diagram containing any cross section between C and D is shown in Figure E8.1(f). Thus,

$$\sum F_x = 0, \quad 50 + 2(5) - 2(40) + F_{CD} = 0,$$

$$F_{CD} = 20 \text{ kN} \quad (1.6 < x < 2.6). \qquad \text{ANS.}$$

Next, we construct the free-body diagram containing any cross-section between D and E which is shown in Figure E8.1(g). Hence,

$$\sum F_x = 0, \quad -F_{DE} - 2(25) + 40 = 0,$$

$$F_{DE} = -10 \text{ kN} \quad (2.6 < x < 3.6). \qquad \text{ANS.}$$

Finally, the free-body diagram containing any cross section between E and G is shown in Figure E8.1(h). Therefore,

$$\sum F_x = 0, \quad -F_{EG} + 40 = 0,$$

$$F_{EG} = 40 \text{ kN} \quad (3.6 < x < 4.0). \qquad \text{ANS.}$$

For convenience, the axial forces F_{AB}, F_{BC}, and F_{CD} were determined from left free-body diagrams whereas F_{DE} and F_{EG} were determined from right free-body diagrams. In each case, these axial forces were assumed in tension or acting away from their associated planes. A positive numerical result indicates tension and a negative result indicates compression. In the axial force diagram tensile values are plotted positive and compressive values are plotted negative.

■ Example 8.2

In Figure E8.2(a), a torsionally loaded member is shown which is fixed at its right end D. Construct the torque diagram for this member.

Solution

The torque diagram is shown adjacent to the given member in Figure E8.2(b). To develop this diagram, we need to deal with several free-body diagrams. Arbitrarily, we choose a longitudinal x coordinate measuring distances from the left end of the member and positive to the right.

The free-body diagram of the entire member is shown in Figure E8.2(c). Thus,

$$\sum T_x = 0, \quad T_D - 30 + 60 - 50 = 0,$$

$$T_D = 20 \text{ k·in.} \qquad \text{ANS.}$$

The free-body diagram containing any cross-section between A and B is shown in Figure E8.2(d). Therefore,

$$\sum T_x = 0, \quad T_{AB} - 50 = 0,$$

$$T_{AB} = 50 \text{ k·in.} \quad (0 < x < 20 \text{ in.}). \qquad \text{ANS.}$$

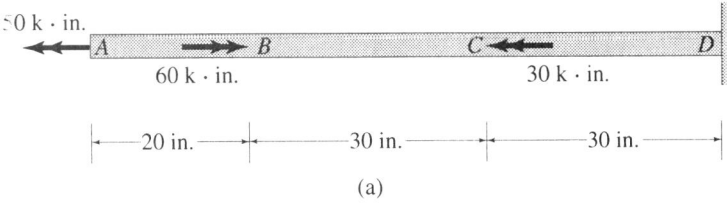

(a)

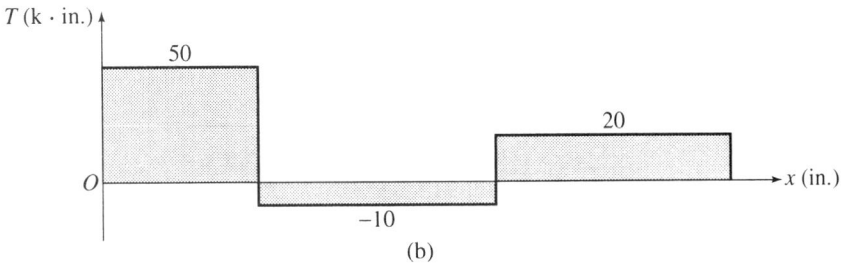

(b)

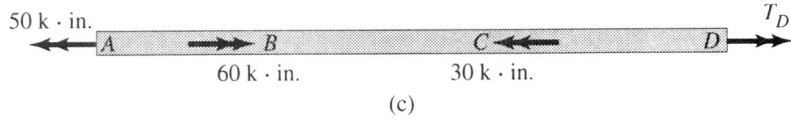

(c)

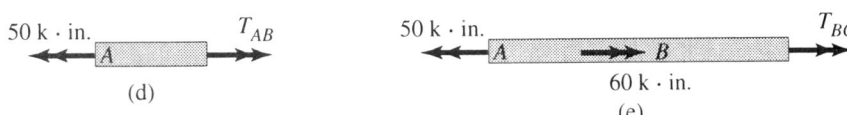

(d)   (e)

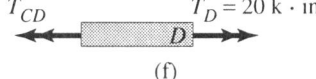

FIGURE E8.2.   (f)

The free-body diagram containing any cross-section between **B** and **C** is shown in Figure E8.2(e). Hence,

$$\sum T_x = 0, \quad T_{BC} + 60 - 50 = 0,$$

$$T_{BC} = -10 \text{ k·in.} \quad (20 < x < 50 \text{ in.}).\qquad \text{ANS.}$$

Finally, the free-body diagram containing any cross-section between **C** and **D** is shown in Figure E8.2(f). Thus,

$$\sum T_x = 0, \quad 20 - T_{CD} = 0,$$

$$T_{CD} = 20 \text{ k·in.} \quad (50 < x < 80 \text{ in.}).\qquad \text{ANS.}$$

For convenience, the torques T_{AB} and T_{BC} were determined from left free-body diagrams whereas T_{CD} was determined from a right free-body diagram. In each case, these torques were assumed positive because their double-headed vectors were shown acting away from their associated planes. A positive torque is plotted positive in the torque diagram and a negative one is plotted negative, as shown in Figure 8.2(b).

Problems

8.1 Construct an axial force diagram for the member shown in Figure P8.1. Determine the force R required to maintain the equilibrium of this member, and show all of the free-body diagrams required to construct the axial force diagram.

8.2 A building column is shown in Figure P8.2. Loads, which include an allowance for the column weight, are applied at the roof and floor levels. Construct an axial force diagram for this column.

8.3 Brackets, attached to a steel member, transfer axial loads to the member shown in Figure P8.3. Determine the overhead support reaction required to maintain the beam's equilibrium. Measure x positive downward from the top of this member. Construct the axial force diagram.

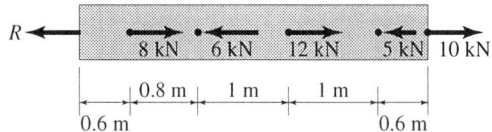

FIGURE P8.1.

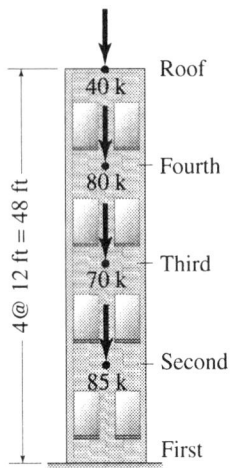

FIGURE P8.2.

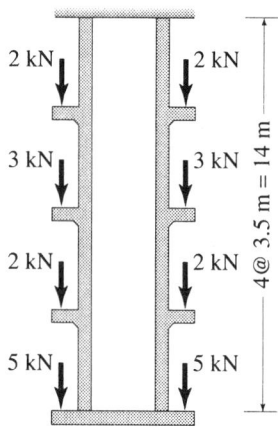

FIGURE P8.3.

8.4 The hollow circular shaft, shown in Figure P8.4, is in equilibrium. Construct a torque diagram for this shaft.

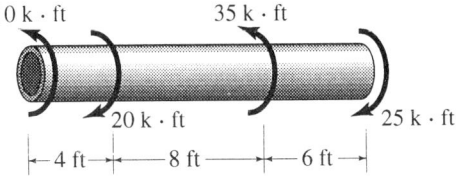

FIGURE P8.4.

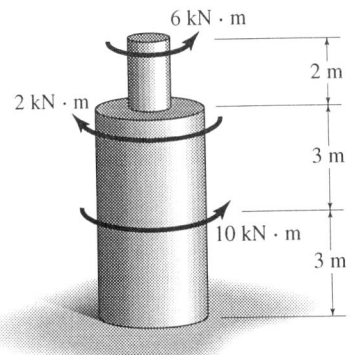

FIGURE P8.5.

8.5 The stepped shaft, shown in Figure P8.5, is fixed at its base. Determine the torque at the fixed support. Construct the torque diagram for this shaft.

8.6 The shaft of Figure P8.6 is in equilibrium. Construct the torque diagram for this shaft.

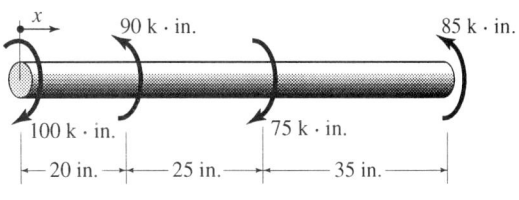

FIGURE P8.6.

8.4
Shear and Moment at Specified Cross-Sections

The free-body diagram of a simply supported beam is shown in Figure 8.6(a). The support reactions A_y and B_y may be determined by applying the equations of equilibrium to this free-body diagram.

Suppose that we want to find the internal shear V_C and the internal bending moment M_C at a section of the beam designated by C. To accomplish this task, we construct a free-body diagram of a part of the beam containing cross-section C. We may consider either a left free-body diagram, as shown in Figure 8.6(b), or a right free-body diagram, as shown in Figure 8.6(c). The choice of either a left or right free-body diagram is one of convenience. In this case, it would be simpler to choose the left free-body diagram because fewer forces would be involved in our calculations. The two equilibrium equations $\sum F_y = 0$ and $\sum M_C = 0$ would be applied to determine V_C and M_C. To check the

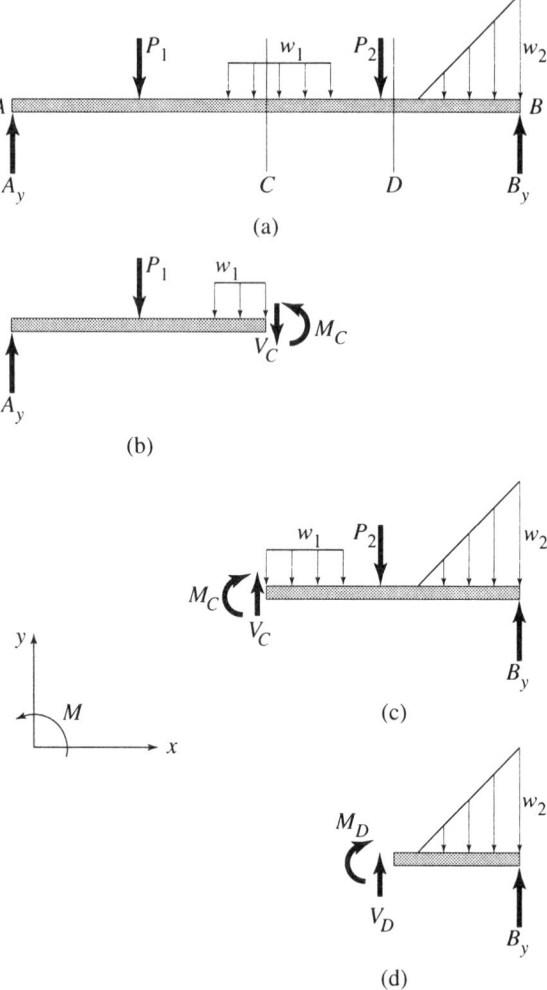

FIGURE 8.6.

accuracy of our calculations, we could use the right free-body diagram and, again, apply the equations of equilibrium to determine V_C and M_C. Of course, these values must agree with the values obtained by using the left free-body diagram, or we have incorrectly written or solved the equations. In many cases, this method of checking our results would be impractical. It would be simpler to use the same free-body diagram and write a moment equation about some other point to check the results. Note that the shear V_C and the moment M_C are assumed positive according to the physical sign convention discussed earlier. A positive sign for the numerical values of either of these quantities will indicate that the assumed sense is correct, and a negative sign will indicate that the assumed sense is incorrect.

Similarly, the internal shear V_D and internal bending moment M_D at any other cross-section D, may be determined by constructing a free-body diagram containing cross-section D. For cross-section D identified in Figure 8.6(a), it would be advantageous to work with the right free-body diagram, as shown in Figure 8.6(d).

The following examples illustrate the determination of the internal shear V and internal bending moment M at specified cross-sections in a given member.

■ **Example 8.3**

A loaded, simply supported beam is shown in Figure E8.3(a). Determine the shear and moment at cross-section 1 by using (a) a left free-body diagram and (b) a right free-body diagram.

Solution

The free-body diagram of the entire beam is shown in Figure E8.3(b) along with a convenient coordinate system. Thus,

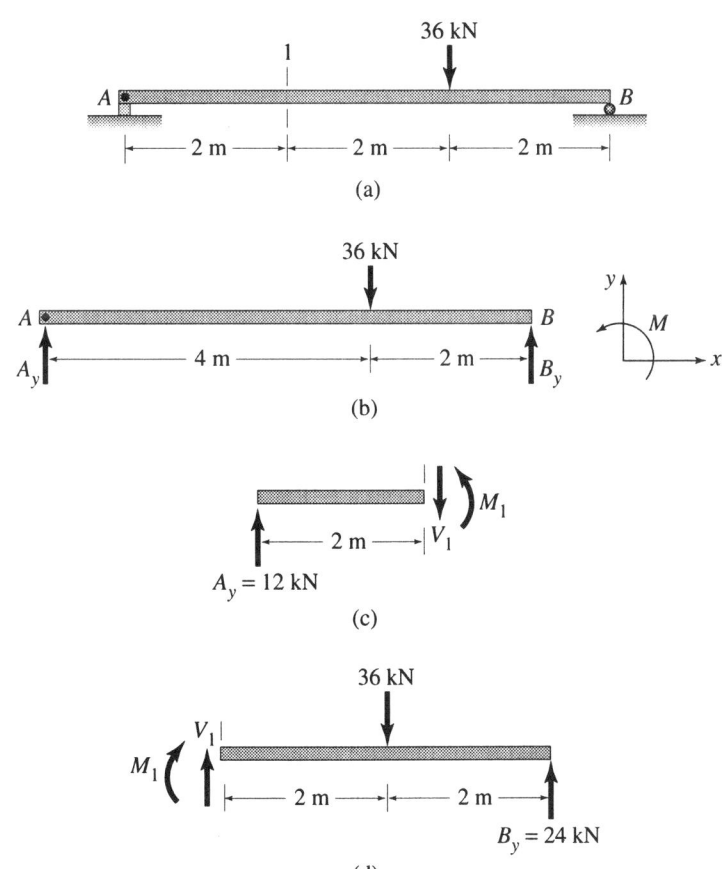

FIGURE E8.3.

$$\sum M_B = 0, \quad 36(2) - A_y(6) = 0,$$
$$A_y = 12 \text{ kN},$$

and

$$\sum F_y = 0, \quad A_y + B_y - 36 = 0,$$
$$B_y = 24 \text{ kN}.$$

(a) A left free-body diagram containing cross-section 1 is shown in Figure E8.3(c). Note that V_1 and M_1 are assumed positive by the physical sign convention. Thus, using the same coordinate system as above,

$$\sum F_y = 0, \quad 12 - V_1 = 0,$$
$$V_1 = 12.00 \text{ kN}, \qquad \text{ANS.}$$

and

$$\sum M_1 = 0, \quad M_1 - 12(2) = 0,$$
$$M_1 = 24.0 \text{ kN·m}. \qquad \text{ANS.}$$

(b) A right free-body diagram containing cross-section 1 is shown in Figure E8.3(d). Again, V_1 and M_1 are assumed positive. Thus,

$$\sum F_y = 0, \quad V_1 + 24 - 36 = 0,$$
$$V_1 = 12.00 \text{ kN}, \qquad \text{ANS.}$$

and

$$\sum M_1 = 0, \quad 24(4) - 36(2) - M_1 = 0,$$
$$M_1 = 24.0 \text{ kN·m}. \qquad \text{ANS.}$$

These answers are, of course, identical to those obtained in part (a).

■ Example 8.4

A loaded cantilever beam is shown in Figure E8.4(a). Determine the shear and moment (a) at the fixed support (cross section B) and (b) at cross-section C.

Solution

(a) A free-body diagram containing cross-section B would be that of the entire beam as shown in Figure E8.4(b). Note that both the shear and the moment were assumed positive by our physical sign convention. A convenvient coordinate system is also shown. Thus,

$$\sum F_y = 0, \quad -0.8(12) - V_B = 0,$$
$$V_B = -9.6 \text{ k}, \qquad \text{ANS.}$$

and

$$\sum M_B = 0, \quad M_B + 0.8(12)(6) = 0,$$
$$M_B = -57.6 \text{ k·ft}. \qquad \text{ANS.}$$

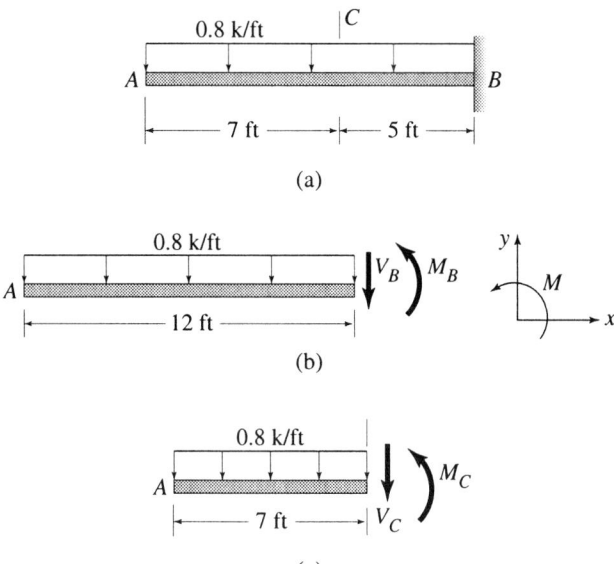

FIGURE E8.4.   (c)

(b) A free-body diagram containing cross-section C is shown in Figure E8.4(c). Note that, although a right free-body diagram is possible, we have chosen a left free-body diagram. Both the shear and the moment were assumed positive by our physical sign convention. Thus, using the same coordinate system as above,

$$\sum F_y = 0, \quad -0.8(7) - V_C = 0,$$

$$V_C = -5.60 \text{ k}, \qquad \text{ANS.}$$

$$\sum M_C = 0, \quad M_C + 0.8(7)(3.5) = 0,$$

$$M_C = -19.60 \text{ k·ft.} \qquad \text{ANS.}$$

■ **Example 8.5**

Consider the loaded overhanging beam shown in Figure E8.5(a), and determine the shear and moment at (a) cross-section 1 which is just to the left of point B and (b) cross-section 2 which is just to the right of point B.

Solution

The reaction at support C is determined from the free-body diagram of the entire beam shown in Figure E8.5(b). A convenient coordinate system is also shown. Thus,

$$\sum M_A = 0, \quad (\tfrac{1}{2})(10)(3) - 47 + R_C(8) = 0,$$

$$R_C = 4 \text{ kN.}$$

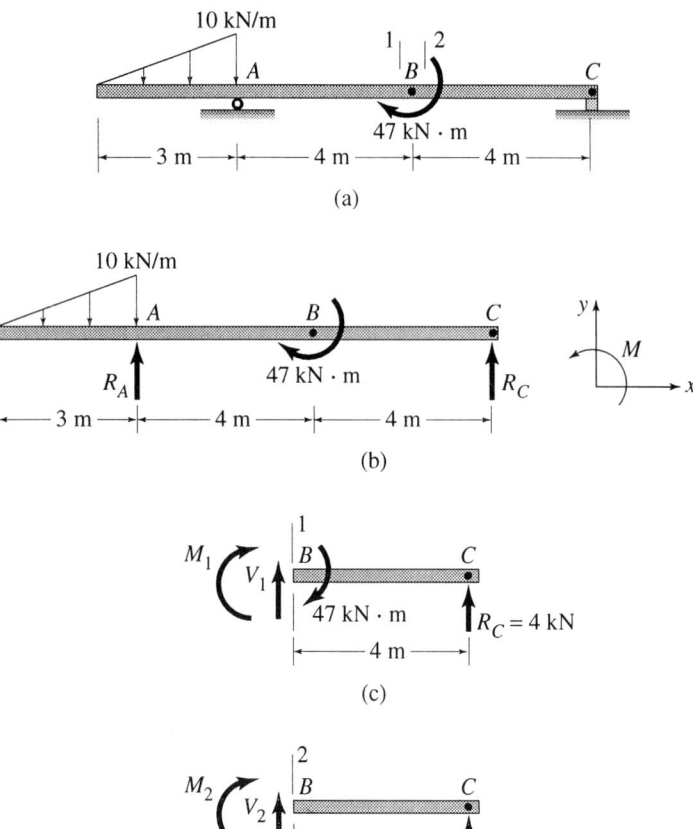

FIGURE E8.5.

(a) A right free-body diagram containing cross-section 1 is shown in Figure E8.5(c). Note that both the shear V_1 and the moment M_1 are assumed positive by our physical sign convention. Thus,

$$\Sigma F_y = 0, \quad V_1 + 4 = 0,$$

$$V_1 = -4.00 \text{ kN}, \qquad\qquad \text{ANS.}$$

and

$$\Sigma M_1 = 0, \quad 4(4) - 47 - M_1 = 0,$$

$$M_1 = -31.0 \text{ kN·m}. \qquad\qquad \text{ANS.}$$

(b) A right free-body diagram containing cross-section 2 is shown in Figure E8.5(d). Note that this free-body diagram does not show the

47-kN·m concentrated couple applied at B because the segment considered stops short of point B. Also, note that both V_2 and M_2 are assumed positive according to our physical sign convention. Thus,

$$\sum F_y = 0, \quad V_2 + 4 = 0,$$

$$V_2 = -4.00 \text{ kN}, \qquad\qquad \text{ANS.}$$

and

$$\sum M_2 = 0, \quad 4(4) - M_2 = 0,$$

$$M_2 = 16.00 \text{ kN·m}. \qquad\qquad \text{ANS.}$$

We observe that although the shear forces are equal at cross-sections 1 and 2, the moments are not. However, the change in moment $M_2 - M_1 = 16 - (-31) = 47$ kN·m is identical with the applied concentrated couple applied at B.

■ **Example 8.6**

The beam ABCD of Figure E8.6(a) is a combination of the overhanging beam ABC and the simply supported beam CD. The built-in hinge at C will transmit shear but will not transmit bending moment. Determine the shear and moment at cross-sections 1 and 2 in terms of the known quantities w and b. Note that cross-section 1 is just to the left of the concentrated load P and cross-section 2 is just to the right of it.

Solution

This beam is statically determinate, and we find the reactions by writing the equations of equilibrium for the free-body diagrams of ABC and CD shown in Figure E8.6(b). Also shown is a convenient coordinate system. Thus,

Free-body diagram of CD

$$\sum M_D = 0, \quad w(2b)(b) - V_C(2b) = 0,$$

$$V_C = wb \ \uparrow \text{ on CD}.$$

Free-body diagram of ABC

Because V_C acts upward on CD, it will act downward on ABC. Thus,

$$\sum M_B = 0, \quad 4Wb(b) - R_A(2b) - V_C(2b) = 0,$$

$$R_A = wb \ \uparrow,$$

and

$$\sum M_A = 0, \quad R_B(2b) - 4wb(b) - V_C(4b) = 0,$$

$$R_B = 4wb \ \uparrow.$$

For section 1, choose a left free-body diagram as shown in Figure E8.6(c), where V_1 and M_1 are both assumed positive according to the physical sign convention. Thus,

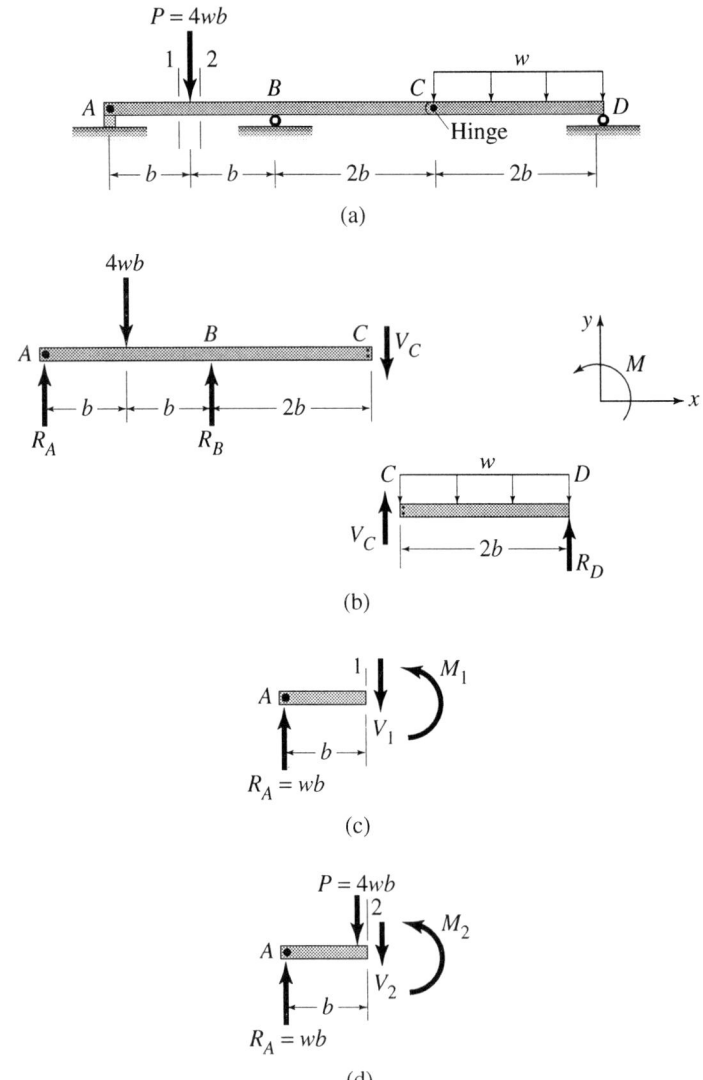

FIGURE E8.6.

$$\sum F_y = 0, \quad wb - V_1 = 0,$$

$$V_1 = wb, \qquad\qquad \text{ANS.}$$

and

$$\sum M_1 = 0, \quad M_1 - wb(b) = 0,$$

$$M_1 = wb^2 \qquad\qquad \text{ANS.}$$

For section 2, choose a left free-body diagram as shown in Figure E8.6(d). Again, the shear V_2 and the bending moment M_2 are both assumed positive by our physical sign convention. Thus,

$$\sum F_y = 0, \quad wb - 4wb - V_2 = 0,$$

$$V_2 = -3wb, \qquad\qquad \text{ANS.}$$

and

$$\sum M_2 = 0, \quad M_2 - wb(b) = 0,$$

$$M_2 = wb^2. \qquad\qquad \text{ANS.}$$

We observe that, although the moments are identical at cross sections 1 and 2, the shear forces are not. However, the change in shear $V_2 - V_1 = -3wb - wb = -4wb$ is equal in magnitude to the applied concentrated force P.

■

Problems

8.7   Both simply supported beams shown in Figure P8.7 are subjected to the same downward total force $P = wL$. Determine the shear and moment at the center of the uniformly loaded beam. Also, determine the shear and moment just to the left and just to the right of the concentrated load P. Briefly comment on your answers.

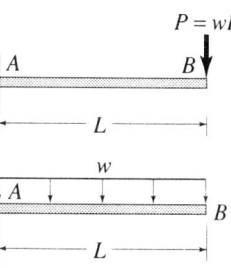

FIGURE P8.8.

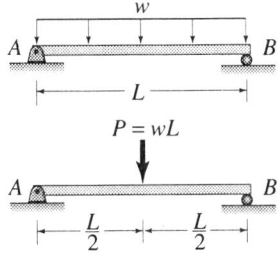

FIGURE P8.7.

8.8   Both cantilever beams in Figure P8.8 are subjected to the same downward total force $P = wL$. Find the shear and mo-

ment at the fixed end A for the cantilever loaded with an end concentrated load $P = wL$. Also, find the shear and moment at the fixed end A for the uniformly loaded cantilever. Briefly comment on your answers.

8.9   Refer to Figure P8.9, and determine the reactions at A and B for this simply supported beam. Determine the shear and moment at cross-section 1 in terms of known quantities P and b. First, use a left free-body diagram, then, use a right free-body diagram. Would you need to find

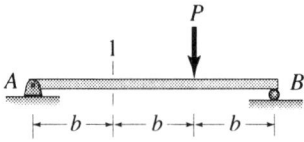

FIGURE P8.9.

both reactions if you were to choose a single free-body?

8.10 Refer to Figure P8.10, and determine the reactions at A and B for this simply supported beam. Determine the shear and moment at cross-section 1 in terms of known quantities P and b. First, use a left free-body diagram, then, use a right free-body diagram. Would you need to find both reactions if you were to choose a single free-body?

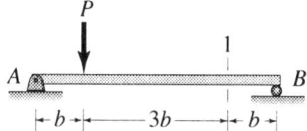

FIGURE P8.10.

8.11 Determine the shear and moment at cross-section 1 of the cantilever shown in Figure P8.11 in terms of the known quantities w and L without finding the reactions at the fixed end at A.

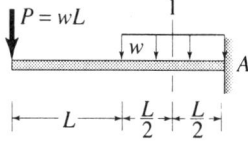

FIGURE P8.11.

8.12 Refer to the cantilever in Figure P8.11, and determine the shear and moment at

the fixed end at A in terms of the known quantities w and L.

8.13 The cantilever beam shown in Figure P8.13 is loaded with a linearly varying loading with an intensity w_O at the fixed end at A and an intensity of zero at the free-end at B. Choose a right free-body diagram, and find the shear and moment at cross-section 1 in terms of w_O and L.

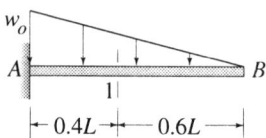

FIGURE P8.13.

8.14 Refer to the cantilever in Figure P8.13, and determine the shear and moment at the fixed end at A in terms of w_O and L. What is the shear and moment at B? Why?

8.15 Determine the shear and moment at the center of the beam shown in Figure P8.15. Express your answers in terms of known quantities P and L.

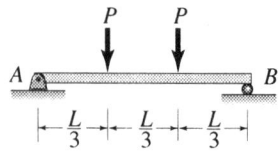

FIGURE P8.15.

8.16 A concentrated couple M_B is applied at B to the beam of Figure P8.16. Determine the shear and moment just to the left of B and just to the right of B in terms of M_B and L.

8.17 Refer to the overhanging beam of Figure P8.17, and determine the shear and mo-

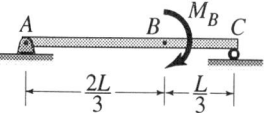

FIGURE P8.16.

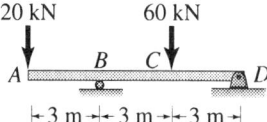

FIGURE P8.17.

ment at a cross-section (a) just to the left of B, (b) just to the right of B, and (c) just to the right of C.

8.18 Refer to Figure P8.18, and determine the shear and moment just to the left and just to the right of the support at B as functions of the known quantities w and L.

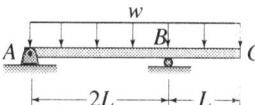

FIGURE P8.18.

8.19 Find the shear and moment at cross-section 1 of the overhanging beam of Figure P8.19. Use a right free-body, and, then, check your answers using a left free-body.

8.20 Refer to Figure P8.20, and determine the shear and moment at cross-section 1 of this simply supported beam.

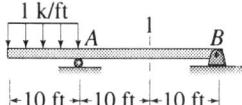

FIGURE P8.19.

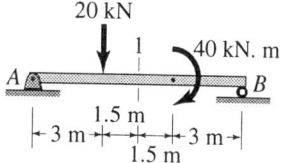

FIGURE P8.20.

8.21 Refer to Figure P8.21, and find the shear and moment at sections 1 and 2. What conclusions do you reach concerning the cross-sections between A and B?

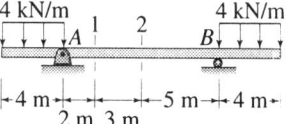

FIGURE P8.21.

8.22 Determine the shear and moment just to the left and just to the right of the roller support at A in the overhanging beam of Figure P8.22.

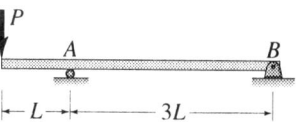

FIGURE P8.22.

8.23 The beam of Figure P8.23 is a combination of the overhanging beam ABC and

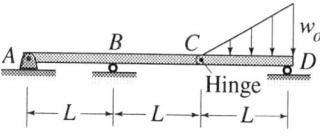

FIGURE P8.23.

the simply-supported beam CD joined together by a built-in hinge at C. Find the shear and moment just to the left and just to the right of the roller support at B in terms of w_O and L. (Hint: Refer to the solution of Example 8.6.)

8.24 A cantilever beam AB and a simply supported beam BC are fastened together with a built-in hinge at B to form the combination beam of Figure P8.24. Determine the shear and moment at cross-section 1 as a function of P and L. If $P = 10$ k and $L = 10$ ft, determine numerical values for the shear and moment. (Hint: Refer to the solution of Example 8.6.)

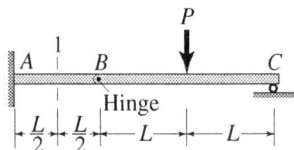

FIGURE P8.24.

8.25 The combination beam of Figure P8.25 consists of a simply supported beam AB attached to a cantilever beam BC with a built-in hinge at B. In terms of w and L, determine the shear and moment at cross-sections 1 and 2, where 2 is just to the left of the fixed support at C. (Hint: Refer to the solution of Example 8.6.)

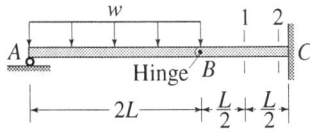

FIGURE P8.25.

8.26 Refer to the beam of Figure P8.26 and determine the shear and moment at cross section 1.

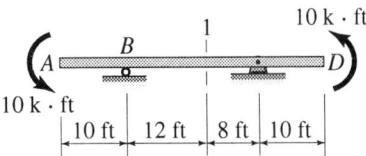

FIGURE P8.26.

8.27 Refer to the two beams shown in Figure P8.27. Determine the shear and moment at the center of each of these beams in terms of w_O and L. Add your results algebraically. Do these answers differ from the shear and moment at the center of a uniformly loaded simply supported beam? Briefly explain why.

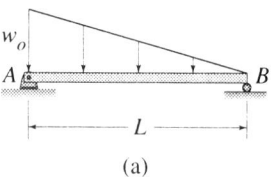

(a)

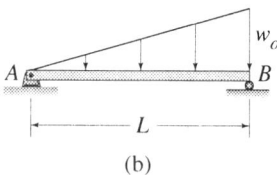

(b)

FIGURE P8.27.

8.28 Refer to the beam Figure P8.28, and determine the shear and moment at cross-section 1 in terms of w and L. This com-

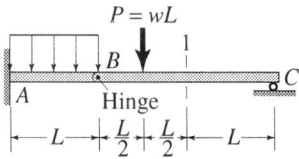

FIGURE P8.28.

bination beam consists of a cantilever AB attached to a simply supported beam BC with a built-in hinge at B. (Hint: Refer to the solution of Example 8.6.)

8.29 Refer to the beam of Figure P8.28, and determine the shear and moment at the fixed end A in terms of w and L. If $w = 4$ kN/m and $L = 5$ m, find numerical values for the shear and moment.

8.30 Refer to the cantilever of Figure P8.30, and determine the shear and moment at the fixed end A. Then find the shear and moment at cross-section 1 by considering both left and right free-body diagrams.

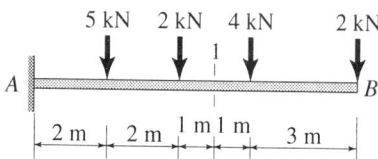

FIGURE P8.30.

8.31 Determine the shear and moment at the center of the simply supported beam shown in Figure P8.31.

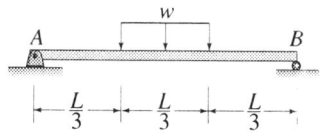

FIGURE P8.31.

8.32 Refer to the beam of Figure P8.32, and find the shear and moment at the cross-section where the applied load intensity is $1.6\ w_0$.

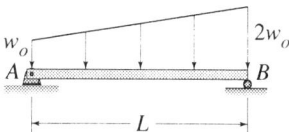

FIGURE P8.32.

8.33 Refer to Figure P8.33, and determine the shear and moment just to the left of the support at E. Note that C and D are built-in hinges. (Hint: Begin with a free-body diagram of CD, and find the shear transmitted by the built-in hinge at C and D. Then, solve for the reaction components at the remaining supports.)

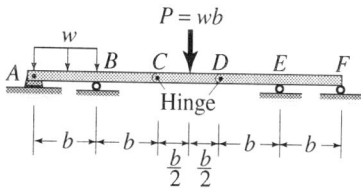

FIGURE P8.33.

8.5
Shear and Moment Equations

In Section 8.4, we learned how to find the shear V and moment M at a specific cross-section in a member and we discovered that, in general, these quantities change from cross-section to cross-section. In this section, we will learn how to write shear and moment functions that express these quantities in terms of a longitudinal coordinate x, measuring distances along the member from some convenient position. Such a position may be selected at a support, a member end, or a point of load application. These shear and moment functions may be used to plot *shear* and *moment* diagrams which show the variation of these quantities along the length of the member. An example of such plots is given in Example 8.10.

The determination of the shear and moment functions is similar to finding these quantities for a specific cross-section in a member. A free-body diagram is constructed containing the cross-section of interest whose location is defined by the variable x. The unknown shear V and moment M are assumed positive by our physical sign convention, and the equations of equilibrium are applied with a convenient equation sign convention. Solving these equations for V and M yields functions in terms of the variable x. Each shear and moment function is valid only for a restricted range of values for x which must be stated because of the discontinuities that occur in the values of the shear and moment at locations, such as points of load application. Such a discontinuity is encountered in Example 8.9 and illustrated in Example 8.10. Thus, whenever the character of the loading changes or a support reaction is encountered, we must consider a new free-body diagram to develop the shear and moment functions that apply in the new segment.

The following examples illustrate how shear and moment functions may be determined.

■ Example 8.7

Refer to the loaded cantilever beam AB shown in Figure E8.7(a), and develop the shear and moment equations for the entire beam, expressing these quantities in terms of a longitudinal coordinate x, measuring distances from the free end of the beam.

Solution

A cross-section in the beam is selected along its length at a distance x from the free end, and a left free-body containing this cross section is constructed as shown in Figure E8.7(b). Note that a right free-body

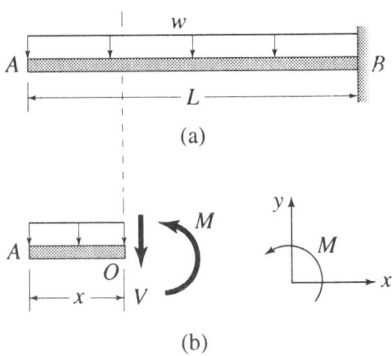

FIGURE E8.7.

diagram would require us to find the reactions at the fixed support. Note also that the quantities V and M are assumed positive according to our physical sign convention. An equations coordinate system is selected as shown. Although the coordinate system is shown separately from the free-body diagram, the variable x is measured from the left free end of the beam, positive to the right. Thus,

$$\sum F_y = 0, \quad -wx - V = 0,$$

$$V = -wx \quad (0 < x < L). \qquad \text{ANS.}$$

$$\sum M_O = 0, \quad M + wx(x/2) = 0,$$

$$M = -\frac{wx^2}{2} \quad (0 < x < L) \qquad \text{ANS.}$$

where O is a point in the cross-section of interest.

We observe that the shear V is a linear function of x and that the bending moment is a quadratic function of x. Note that, for both V and M, the permissible range of values for x is $(0 < x < L)$.

■ Example 8.8

A simply supported beam AB is loaded by a concentrated load P at C as shown in Figure E8.8(a). Develop the shear and moment functions for segments (a) AC and (b) CB.

Solution

The free body diagram of the entire beam is shown in Figure E8.8(b) along with a convenient equations coordinate system. Thus,

$$\sum M_A = 0, \quad B_y(L) - P\left(\frac{2}{3}L\right) = 0,$$

$$B_y = \frac{2}{3}P,$$

and

$$\sum F_y = 0, \quad A_y + B_y - P = 0,$$

$$A_y = \frac{1}{3}P.$$

(a) To develop the shear and moment functions for segment AC, we select any cross-section between A and C, measuring its location from the support at A with the longitudinal coordinate x. A left free-body diagram containing this cross-section is constructed, as shown in Figure E8.8(c). Using the coordinate system established earlier and remembering that x is measured from A, positive to the right,

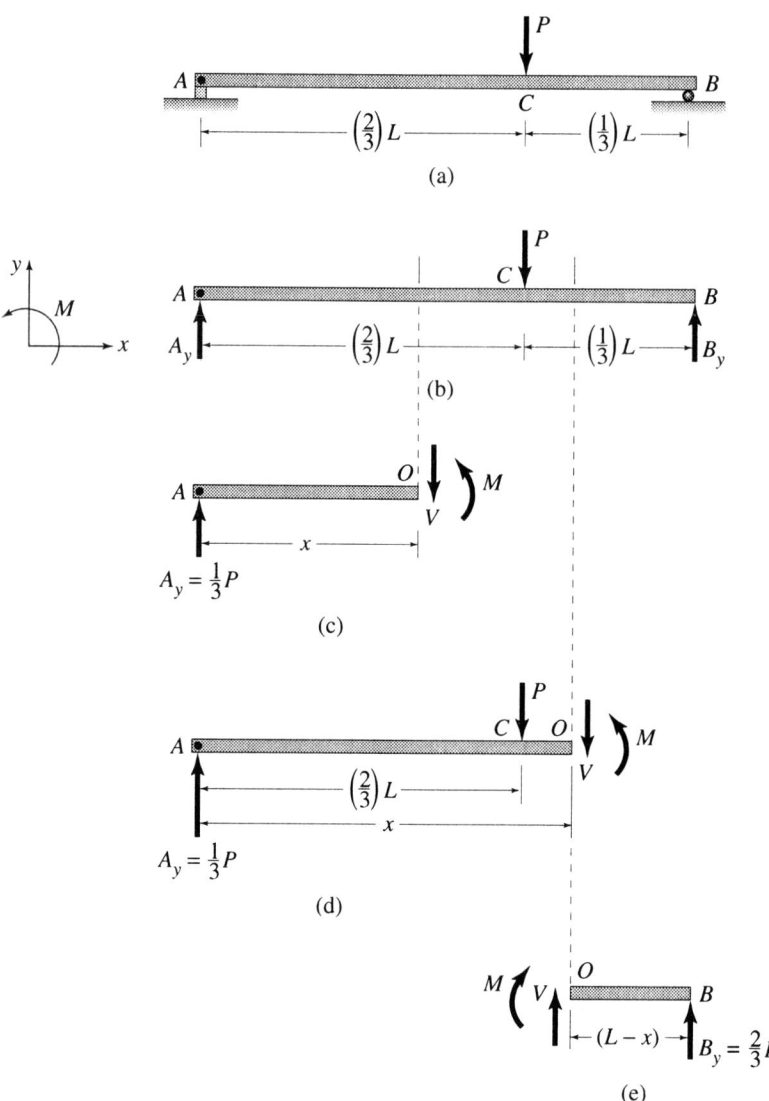

FIGURE E8.8.

$$\sum F_y = 0, \quad \frac{1}{3}P - V = 0,$$

$$V = \frac{1}{3}P \quad \left(0 < x < \frac{2}{3}L\right),$$ ANS.

and

$$\sum M_O = 0, \quad M - \frac{1}{3}P(x) = 0,$$

$$M = \frac{1}{3}P(x) \quad \left(0 < x < \frac{2}{3}L\right) \qquad \text{ANS.}$$

where O is a point in the selected cross-section.

(b) Two solutions will be presented based on the choice of a left or a right free-body diagram.

In the first solution, x is measured from A, positive to the right. A left free-body diagram is constructed to contain any cross-section between C and B as shown in Figure E8.8(d). Thus, using the same coordinate system as above,

$$\sum F_y = 0, \quad \frac{1}{3}P - P - V = 0,$$

$$V = -\frac{2}{3}P \quad \left(\frac{2}{3}L < x < L\right), \qquad \text{ANS.}$$

and

$$\sum M_O = 0, \quad M + P\left(x - \frac{2}{3}L\right) - \frac{1}{3}P(x) = 0,$$

$$M = \frac{2}{3}P(L - x) \quad \left(\frac{2}{3}L < x < L\right). \qquad \text{ANS.}$$

In the second solution, we still measure x from A, positive to the right. However, now, we choose a right free-body diagram containing the cross-section of interest as shown in Figure E8.8(e). Note that the length of the segment is $L - x$. Thus,

$$\sum F_y = 0, \quad V + \frac{2}{3}P = 0,$$

$$V = -\frac{2}{3}P \quad \left(\frac{2}{3}L < x < L\right), \qquad \text{ANS.}$$

and

$$\sum M_O = 0, \quad \frac{2}{3}P(x - L) - M = 0,$$

$$M = \frac{2}{3}P(L - x) \quad \left(\frac{2}{3}L < x < L\right). \qquad \text{ANS.}$$

Obviously, these answers are identical to those obtained from the first solution. However, it should be observed that the second free-body diagram and the corresponding equations are a little easier to handle.

■ **Example 8.9**

A simply supported beam AB is subjected to a uniform load of intensity 3k/ft for the entire length and to a concentrated couple of 40 k·ft applied at C, as shown in Figure E8.9(a). Derive the needed shear and moment equations for the entire beam.

Solution

The support reactions A_y and B_y are found from the free-body diagram of the entire beam, as shown in Figure E8.9(b). Thus, using the coordinate system shown,

$$\sum M_A = 0, \quad B_y(20) - 3(20)(10) - 40 = 0,$$

$$B_y = 32 \text{ k},$$

and

$$\sum F_y = 0, \quad A_y + B_y - 3(20) = 0,$$

$$A_y = 28 \text{ k}.$$

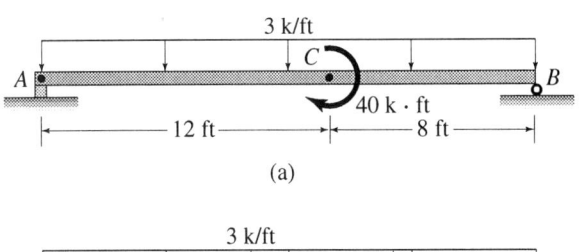

(a)

(b)

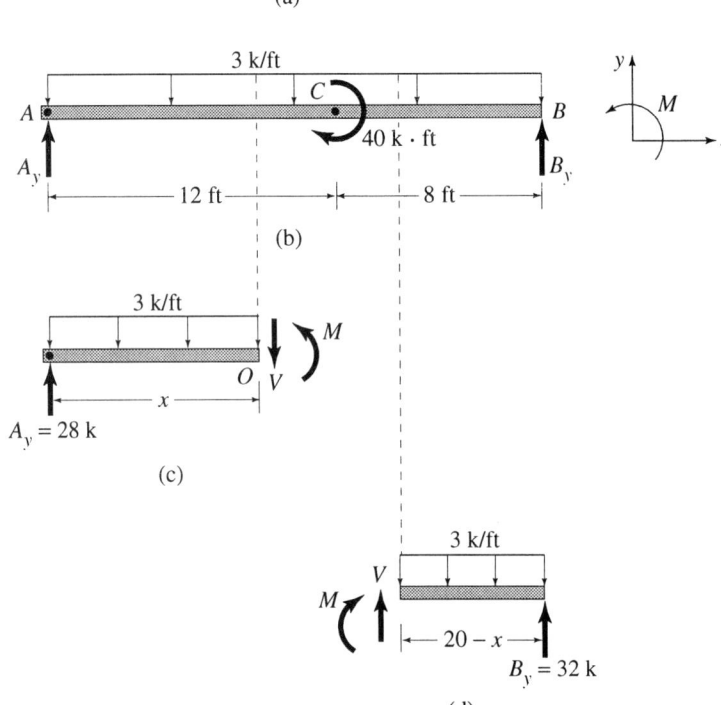

(c)

(d)

FIGURE E8.9.

A left free-body diagram is shown in Figure E8.9(c) containing a cross-section somewhere between A and C, whose position is measured from support A by the longitudinal coordinate x. Thus, using the same coordinate system as above,

$$\sum F_y = 0, \quad 28 - 3x - V = 0,$$
$$V = (28 - 3x) \, \text{k} \quad (0 < x < 12 \text{ ft}), \qquad \text{ANS.}$$

and

$$\sum M_O = 0, \quad M + 3x\left(\frac{x}{2}\right) - 28x = 0$$
$$M = (28x - 1.5x^2) \, \text{k·ft} \quad (0 < x < 12 \text{ ft}). \qquad \text{ANS.}$$

A right free-body diagram is shown in Figure E8.9(d) containing a cross-section between C and B. The coordinate x is still measured from A, and, therefore, the length of the segment is $20 - x$. Using the same coordinate system as above,

$$\sum F_y = 0, \quad V + 32 - 3(20 - x) = 0,$$
$$V = (28 - 3x) \, \text{k} \quad (12 < x < 20 \text{ ft}) \qquad \text{ANS.}$$

and

$$\sum M_O = 0, \quad 32(20 - x) - 3(20 - x)\left(\frac{1}{2}\right)(20 - x) - M = 0$$
$$M = 640 - 32x - 1.5(20 - x)^2$$
$$= (40 + 28x - 1.5x^2) \, \text{k·ft} \quad (12 < x < 20 \text{ ft}). \qquad \text{ANS.}$$

■ **Example 8.10**

Construct the shear and moment diagrams for the beam shown in Figure E8.10(a).

Solution

Note that this is the same beam, analyzed in Example 8.9, for which the shear and moment functions have already been developed. These functions will be plotted to obtain the required diagrams.

The shear functions obtained in Example 8.9 are repeated here for convenience. Thus,

$$V = (28 - 3x) \, \text{k} \quad (0 < x < 12 \text{ ft})$$

and

$$V = (28 - 3x) \, \text{k} \quad (12 < x < 20 \text{ ft}).$$

We note that the same shear function applies to both segments of the beam and that the shear is a linear function of x. Therefore, only the

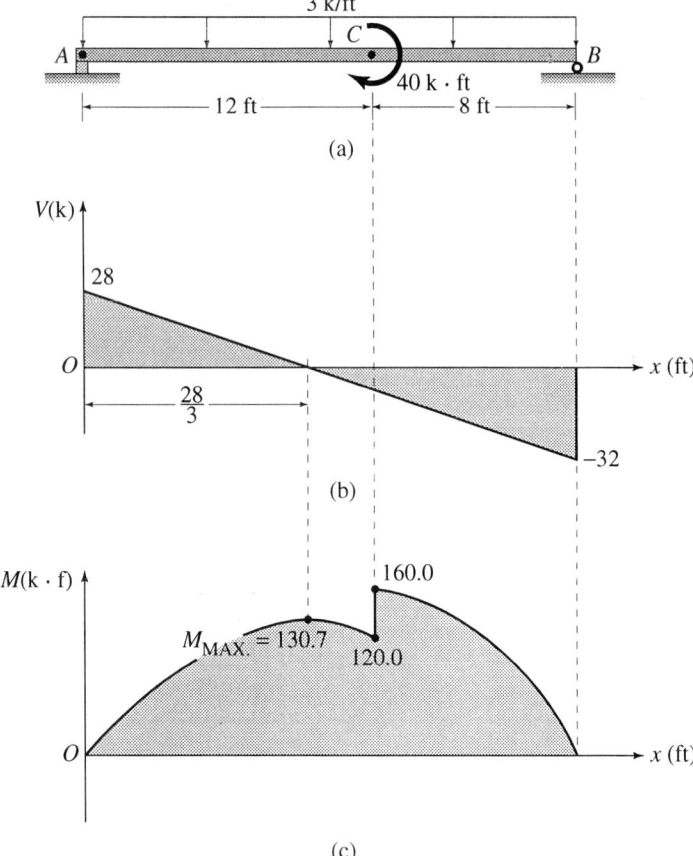

3 k/ft

A

C

40 k · ft

B

|— 12 ft —|— 8 ft —|

(a)

$V(k)$

28

O → x (ft)

$\dfrac{28}{3}$

−32

(b)

$M(k \cdot f)$

160.0

$M_{MAX.} = 130.7$

120.0

O → x (ft)

(c)

FIGURE E8.10.

end points are needed. Thus for $x \doteq 0$, $V = 28$ k and for $x \doteq 20$, $V = -32$ k. These quantities, plotted in Figure E8.10(b), are connected by a straight line, as shown. Note that the position on the beam where the shear vanishes may be obtained by setting $V = 0$ in the equation which leads to $x = 28/3$ ft. As will be shown in Section 8.6, the position where $V = 0$ corresponds to a relative maximum or minimum for the moment.

Again, the moment functions developed in Example 8.9 are repeated here for convenience. Thus,

$$M = (28x - 1.5x^2)\,\text{k·ft} \quad (0 < x < 12\ \text{ft}),$$

and

$$M = (40 + 28x - 1.5x^2)\,\text{k·ft} \quad (12 < x < 20\ \text{ft}).$$

We note that both equations represent second degree parabolic functions. We can sketch a reasonably accurate moment diagram by first finding the moments at the two end points for each segment, plus any other significant values within each segment, and connecting these points with smooth curves. As mentioned above, we can expect a relative maximum value for M within segment AC at $x = 28/3$ ft. Thus, for segment AC,

$$\text{at } x \doteq 0, M = 0$$

$$\text{at } x = 28/3 \text{ ft, } M_{\text{MAX}} = 130.7 \text{ k·ft,}$$

and

$$\text{at } x \doteq 12 \text{ ft, } M = 120.0 \text{ k·ft.}$$

Also, for segment CB,

$$\text{at } x \doteq 12 \text{ ft, } M = 160.0 \text{ k·ft}$$

and

$$\text{at } x \doteq 20 \text{ ft, } M = 0.$$

These values are plotted in Figure E8.10(c) and connected by smooth curves, as shown. Note that, at C, where the concentrated couple of 40 k·ft is applied, we have a sudden change of 40 k·ft in the magnitude of the moment, known as a discontinuity, due to the existence of the concentrated couple at that point.

■

Problems

(State the limits for x in each of the following problems.)

8.34 Refer to Figure P8.34, and write equations for shear and moment for segments AB and BC of this simply supported beam. Choose an origin for x at the left end.

8.35 A simply supported beam is loaded, as shown in Figure P8.35. Choose an origin at A, and write shear and moment equations for the three segments AB, BC, and CD.

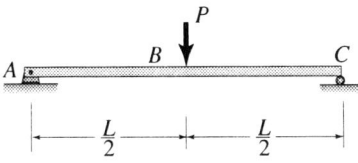

FIGURE P8.34.

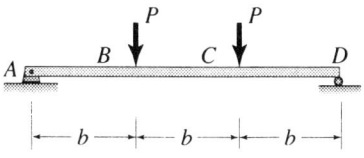

FIGURE P8.35.

8.36 Choose an origin at C for the variable x for the cantilever shown in Figure P8.36. Write the shear and moment as functions of x for the segments BC and AB.

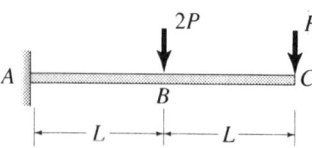

FIGURE P8.36.

8.37 For the cantilever shown in Figure P8.37, measure x from C, positive to the left. Determine the shear and moment as functions of x for the segments BC and AB.

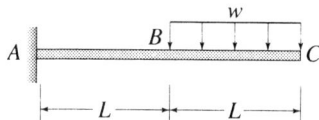

FIGURE P8.37.

8.38 The cantilever of Figure P8.38 is loaded by a linearly varying load. Express the shear and moment as a function of x measured from the left end A. Check the reactions at the fixed end B by letting $x = L$.

8.39 A uniform load w is applied to the right half of the beam shown in Figure P8.39. Determine the reactions at A and C, and

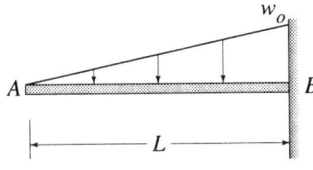

FIGURE P8.38.

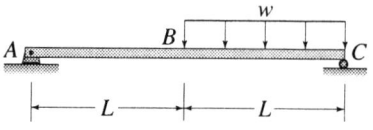

FIGURE P8.39.

write shear and moment functions for segments AB and BC. Choose an origin at A and measure x positive to the right. Use the shear equation for segment BC to find the reaction at C.

8.40 The cantilever shown in Figure P8.40 is loaded over its right half by a uniform load. Measure x positive to the right from the free end A, and write shear and moment equations for segments AB and BC. Recall that we are assuming a weightless beam.

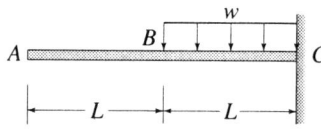

FIGURE P8.40.

8.41 The simply supported beam AB of Figure P8.41 is loaded by a linearly varying load. Determine the reactions at A and B. Then, express V and M as functions of x measured from A.

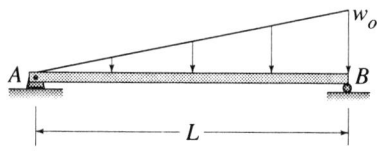

FIGURE P8.41.

8.42 A couple C_o is applied at the left end A of the cantilever of Figure P8.42. Choose an

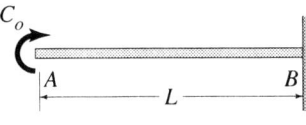

FIGURE P8.42.

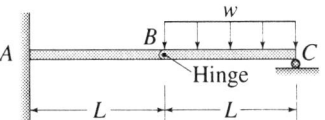

FIGURE P8.45.

origin at A, and express V and M as functions of x.

8.43 The overhanging beam of Figure P8.43 is loaded with a couple C_o at A. Determine the reactions at B and C. Choose an origin for x at the free end A, and write the shear and moment equations for segments AB and BC.

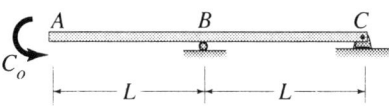

FIGURE P8.43.

8.44 A uniform loading is applied to segment BC of the overhanging beam of Figure P8.44. Find the reactions at A and B. Express the shear and moment in both segments AB and BC as functions of x measured from an origin at A.

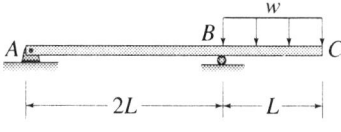

FIGURE P8.44.

8.45 A uniform loading is applied to span BC of the beam shown in Figure P8.45. Determine the reactions at A, B, and C. Choose an origin at A, and write the equations for shear and moment as functions of x. Note that B is a built-in hinge.

8.46 Refer to Figure P8.46, and find the reaction at the built-in hinge at B from a free-body diagram of span BC. Choose an origin at B. Measure x_1 positive to the left for span AB, and measure x_2 positive to the right for span BC. Determine the shear and moment equations for each of these spans.

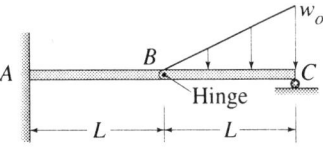

FIGURE P8.46.

8.47 Refer to Figure P8.47, and find the reaction at the built-in hinge at B from a free-body diagram of span BC. Choose an origin at B. Measure x_1 positive to the left on span AB, and measure x_2 positive to the right on span BC. Write equations for the shear and moment for each of these spans.

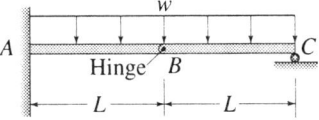

FIGURE P8.47.

8.48 Choose an origin at point A, and measure x positive to the right for the cantilever shown in Figure P8.48. Express the

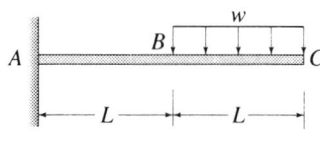

FIGURE P8.48.

shear and moment for segments AB and BC in terms of x.

8.49 Refer to Figure P8.49, and find the reactions at A and B in terms of P. Choose an origin at A and write the equations for the shear and moment for segments AB and BC in terms of x.

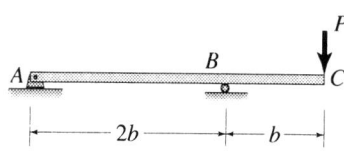

FIGURE P8.49.

8.50 Find the reaction at B in terms of P for the overhanging beam shown in Figure P8.50. For an origin at A, express V and M as functions of x for segments AB and BC.

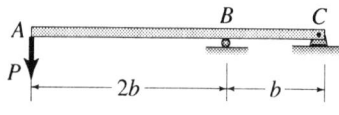

FIGURE P8.50.

8.51 Determine the left reaction for the simply supported beam shown in Figure P8.51.

Choose an origin at A and write shear and moment equations for segments AB and BC.

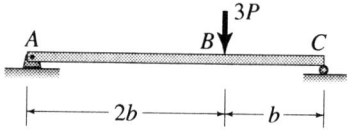

FIGURE P8.51.

8.52 A simply supported beam is symmetrically loaded, as shown in Figure P8.52. Choose an origin at A, and write the equations for shear and moment as functions of x for the three segments AB, BC, and CD.

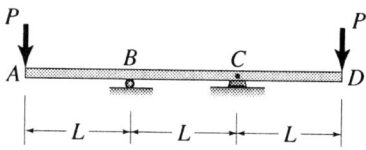

FIGURE P8.52.

8.53 Refer to the cantilever beam shown in Figure P8.53. Chose an origin at the free end A, and write shear and moment functions for segments AB and BC.

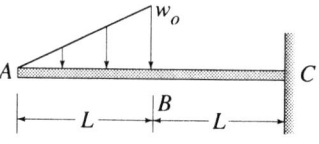

FIGURE P8.53.

8.6
Load, Shear, and Moment Relationships

The construction of shear and moment diagrams for a given beam may be accomplished, as stated in Section 8.5 and illustrated in Example 8.10, by, first, developing the shear and moment functions and, then, plotting these functions in a convenient location adjacent to the beam. However, if the beam is subjected to many concentrated loads or to segmented distributed loads, this method becomes lengthy and tedious. A more direct and less cumbersome approach uses the relationships that exist between the load acting on the beam and the shear V and between the shear V and the moment M.

Consider the beam shown in Figure 8.7(a) subjected to a general load $w = f(x)$. Let us isolate a small segment of differential length dx, as shown in Figure 8.7(b). On the left cross-section of this segment, the shear and moment are shown as V and M, respectively, whereas, on the right cross-section they are shown as $V + dV$ and $M + dM$. The quantities dV and dM represent the differential changes that occur in the shear and moment in the differential length dx. Note that, on both cross-sections, the shear and moment are shown as positive quantities by our physical sign convention. Thus, using the equations coordinate system shown in Figure 8.7(c)

$$\sum F_y = 0, \quad V - (V + dV) - w\,dx = 0,$$

$$dV = -w\,dx \tag{8.1}$$

which may be rewritten in the form

$$\frac{dV}{dx} = -w. \tag{8.2}$$

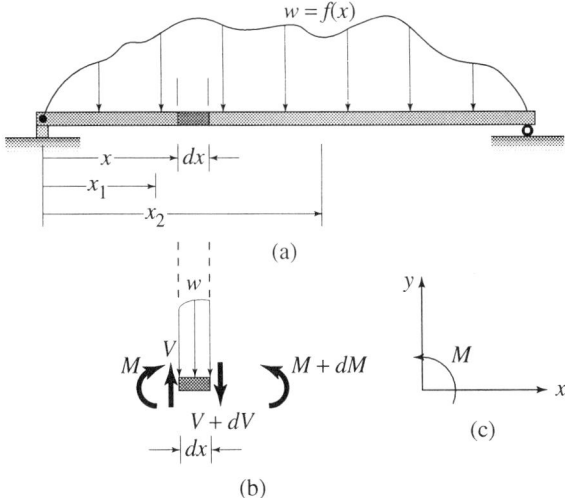

(a)

(b)

(c)

FIGURE 8.7.

Equation (8.2) states that the derivative of the shear function is equal to the negative of the load intensity w.

Let us now return to Eq. (8.1) and integrate it between any two positions defined by x_1 and x_2 as shown in Figure 8.7(a). The corresponding shear values are V_1 and V_2, respectively. Thus,

$$\int_{V_1}^{V_2} dV = -\int_{x_1}^{x_2} w\,dx$$

from which

$$V_2 - V_1 = -\int_{x_1}^{x_2} w\,dx. \qquad (8.3)$$

The left-hand side of Eq. (8.3) represents the change in shear $\Delta V = V_2 - V_1$, and the right-hand side is the area under the load function between x_1 and x_2. Thus,

$$\Delta V = \text{area under the load function } w \text{ between } x_1 \text{ and } x_2. \quad (8.4)$$

Let us now return to the free-body diagram shown in Figure 8.7(b) and sum moments about point O. Thus,

$$\sum M_O = 0, \quad M + dM + w\,dx\left(\frac{dx}{2}\right) - M - V\,dx = 0.$$

The third term in this equation is a second-order differential and may be ignored. The remaining terms reduce to

$$dM = V\,dx \qquad (8.5)$$

which may be rewritten in the form

$$\frac{dM}{dx} = V. \qquad (8.6)$$

Equation (8.6) shows that the derivative of the moment function with respect to x is equal to the shear V. It also shows that, for locations along the beam where the shear is zero, we have a relative maximum or minimum for the moment. This conclusion is useful in constructing moment diagrams, as was indicated in Example 8.10.

We now return to Eq. (8.5) and integrate it between x_1 and x_2 where the corresponding moments are M_1 and M_2, respectively. Thus,

$$\int_{M_1}^{M_2} dM = \int_{x_1}^{x_2} V\,dx$$

from which

$$M_2 - M_1 = \int_{x_1}^{x_2} V\,dx. \qquad (8.7)$$

The left-hand side of Eq. (8.7) represents the change in moment $\Delta M = M_2 - M_1$, and the right-hand side is the area under the shear function between x_1 and x_2. Thus,

$$\Delta M = \text{area under the shear diagram between } x_1 \text{ and } x_2. \quad (8.8)$$

We observe, here, that Eqs. (8.4) to (8.8) are valid only between points of discontinuities which exist at concentrated loads or points, where load changes occur, and at points of support reactions. We also note that Eqs. (8.4) and (8.8) are useful in constructing shear and moment diagrams, as will be illustrated in Section 8.7.

8.7 Shear and Moment Diagrams

As stated earlier, shear and moment diagrams graphically represent how the shear V and the moment M vary along a given beam. One method of developing these diagrams requires determining the shear and moment functions and plotting these functions, as was done in Example 8.10. A more direct and less tedious method uses Eqs. (8.4) and (8.8) developed in the previous section. The significance of these two equations is summarized in Figure 8.8 where a beam AB and its load $w = f(x)$ is shown in Figure 8.8(a), the corresponding shear diagram in Figure 8.8(b), and the corresponding moment diagram in Figure 8.8(c).

Equation (8.4) makes it possible to obtain the shear diagram (Fig. 8.8(b)) from the load diagram (Fig. 8.8(a)). According to Eq. (8.4), the negative of the area A_w under the load diagram between x_1 and x_2 is equal to the change in shear $\Delta V = V_2 - V_1$. Thus, if we know V_1, we can obtain V_2 from the relationship

$$V_2 = V_1 + \Delta V = V_1 - (A_w)_{1-2}. \quad (8.9)$$

Also, Eq. (8.8) helps us to obtain the moment diagram (Fig. 8.8(c)) from the previously obtained shear diagram (Fig. 8.8(b)). According to Eq. (8.8), the area A_v under the shear diagram between x_1 and x_2 is equal to the change in moment $\Delta M = M_2 - M_1$. Thus, if we know M_1 we can obtain M_2 from the relationship

$$M_2 = M_1 + \Delta M = M_1 + (A_v)_{1-2} \quad (8.10)$$

To use Eqs. (8.9) and (8.10) we need starting values for the shear and the moment to which we can add or subtract whatever changes occur in values of the shear and moment as we proceed from one position to the next along the beam. These starting values are generally the values of the shear and moment at the left end of the beam. For example, if the left end of the beam is hinged, the starting value for the shear is given by the support reaction at the hinge. This may be confirmed by constructing a left free-body diagram for a segment of length x from

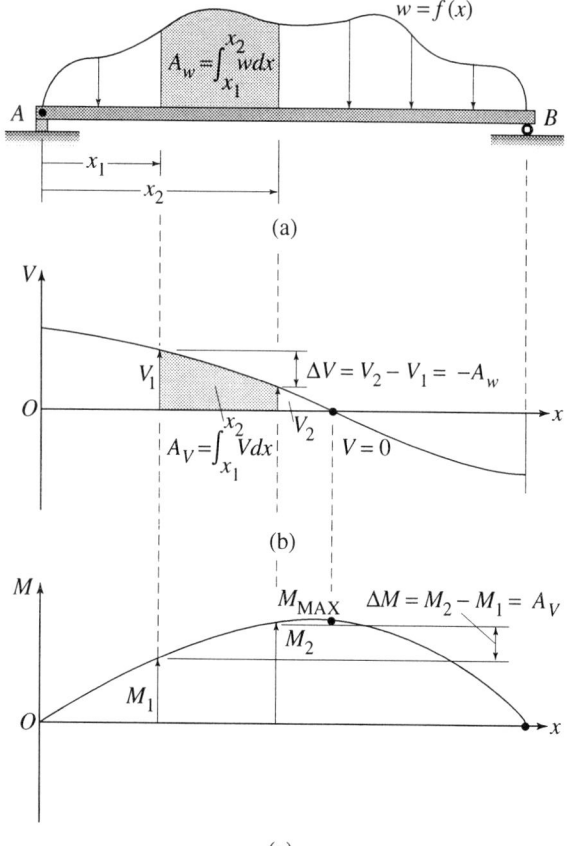

$w = f(x)$

$A_w = \int_{x_1}^{x_2} w\,dx$

(a)

V_1

$\Delta V = V_2 - V_1 = -A_w$

$A_V = \int_{x_1}^{x_2} V\,dx$ V_2 $V = 0$

(b)

M_{MAX} $\Delta M = M_2 - M_1 = A_V$

M_2

M_1

FIGURE 8.8.   (c)

the beam, shown in Figure 8.8(a), after determining the reaction at hinge A. If we let x approach zero in the resulting equation for the shear, we conclude that the shear just to the right of the hinge support at A is equal to the reaction there. Also, if we assume that there is no externally applied couple at A, the starting value for the moment is zero because an ideal hinge cannot support a moment.

The use of the above concepts in developing shear and moment diagrams is illustrated by the following examples.

■ **Example 8.11**   Use the relationships expressed in Eqs. (8.9) and (8.10) to develop the shear and moment diagrams for the uniformly loaded cantilever beam AB shown in Figure E8.11(a).

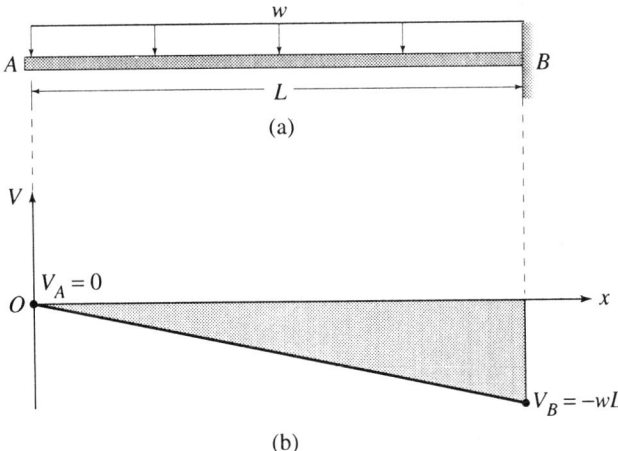

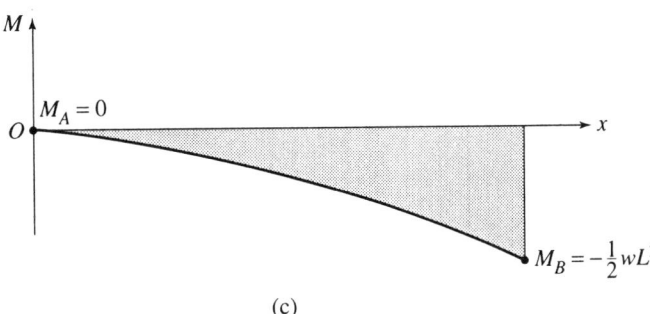

FIGURE E8.11.   (c)

Solution

The starting values for the shear and moment, which are both zero would be those at the free end A of the beam. This fact may be confirmed by referring to the shear and moment equations developed for this same beam in Example 8.7, namely, $V = -wx$ and $M = -\dfrac{wx^2}{2}$.

Because $x = 0$ at A, it follows that $V_A = M_A = 0$.

The value of the shear at B is obtained by Eq. (8.9) from its value at A. Thus,

$$V_B = V_A - (A_w)_{AB} = 0 - wL = -wL.$$

Thus, we have two points, $V_A = 0$ and $V_B = -wL$ that are plotted, as shown in Figure E8.11(b). A clue to the nature of the curve connecting these two points is obtained from Eq. (8.2). Ignoring the negative sign, this equation states that the slope of the shear diagram is equal to the intensity of the load at any point along the beam. Because this intensity is constant, it follows that the slope of the shear diagram is constant and that the curve connecting the shear values at A and B must be a straight line, as shown in Figure E8.11(b). This fact is confirmed by the shear equation of Example 8.7 (i.e., $V = -wx$) which states that V is a linear function of x.

The value of the moment at B is obtained from its value at A by Eq. (8.10). Thus,

$$M_B = M_A + (A_V)_{AB} = 0 + \left(-\frac{1}{2}(L)(wL) \right) = -\frac{1}{2}wL^2.$$

The two points we now have on the moment diagram (i.e., $M_A = 0$ and $M_B = -\frac{1}{2}wL^2$) are plotted, as shown in Figure E8.11(c). Equation (8.6) is consulted for a clue regarding the nature of the curve connecting these two points. This equation states that the slope of the moment diagram is equal to the shear at any point in the beam. Because the value of the shear decreases as we go from A to B, the slope of the moment diagram must decrease from A to B, as shown in Figure E8.11(c). As a matter of fact, if we refer to the moment equation of Example 8.7 $\left(\text{i.e., } M = \frac{wx^2}{2} \right)$, we conclude that the curve connecting A and B is a second-order parabolic function.

■ Example 8.12

Use the relationships expressed in Eqs. (8.9) and (8.10) to construct the shear and moment diagrams for the simply supported beam ABCD shown in Figure E8.12(a).

Solution

The free-body diagram for the entire beam is constructed and the support reactions at A and D are determined, as given in Figure E8.12(b). The starting value for the shear would be that just to the right of the support at A. This shear would be equal to the support reaction at A as discussed in Section 8.7. Thus, $V_A^R = 2wb$, where the notation V_A^R represents the shear just to the right of A. The shear values at B and just left of C are found by Eq. (8.9). Thus,

$$V_B = V_A - (A_w)_{AB} = 2wb - bw = wb$$

and

$$V_C^L = V_B - (A_w)_{BC} = bw - 0 = wb$$

where the notation V_C^L represents the value of the shear just to the left of point C. We have a concentrated load $P = 3.5wb$ at C that creates a discontinuity at this point, changing the shear abruptly. Thus the value of the shear just to the right of C becomes

$$V_C^R = bw - 3.5wb = -2.5wb$$

Finally, the shear just to the left of the support at D is given by

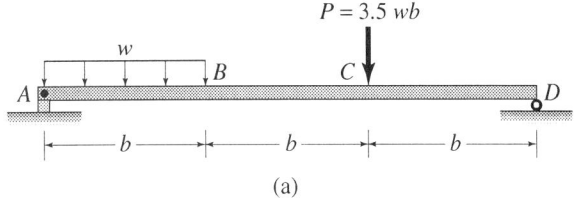

(a)

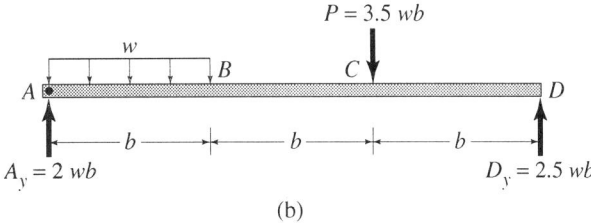

(b)

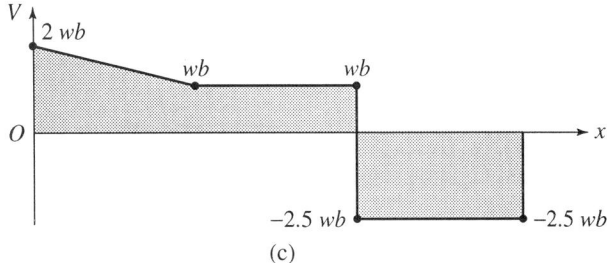

(c)

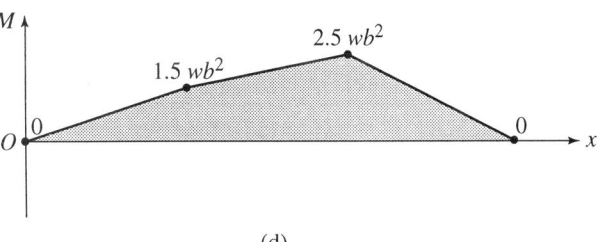

Figure E8.12.   (d)

Eq. (8.9). Thus,

$$V_D^L = V_C^R - (A_w)_{CD} = -2.5wb - 0 = -2.5wb.$$

These five shear values are plotted, as shown in Figure E8.12(c). Ignoring the negative sign, Eq. (8.2) states that the slope of the shear diagram is equal to the intensity of the load at any point along the beam. Thus, because this load intensity is constant from A to B, the curve connecting the shear values at A and B must be a straight line. Similarly, because there is no load intensity between B and C and between C and

D, the slopes of the shear diagram in these two regions must be zero, and the lines connecting the end points must be horizontal straight lines.

The starting value for the moment $M_A = 0$ because we assume an ideal hinge at A, and we have no externally applied couples. The moment values at B, C, and D are found from Eq. (8.10). Thus,

$$M_B = M_A + (A_V)_{AB} = 0 + \frac{1}{2}(2wb + wb)b = 1.5wb^2,$$

$$M_C = M_B + (A_V)_{BC} = 1.5wb^2 + (wb)b = 2.5wb^2,$$

and

$$M_D = M_C + (A_V)_{CD} = 2.5wb^2 + (-2.5wb)b = 0.$$

These three moment values are plotted in Figure E8.12(d). Equation (8.6) states that the slope of the moment diagram is equal to the shear at any point along the beam. Therefore, because the shear decreases from A to B in segment AB, the moment diagram must be a curve whose slope decreases as we go from A to B. Similarly, because the shear is constant in segments BC and CD, the moment diagram must be a straight line as shown.

The fact that M_D equals zero provides a good check on our computations because the moment does, in fact, vanish at an end roller support.

■

Problems

Using the method indicated, plot the shear and moment diagrams, and state the significant ordinates for the beams indicated in Problems 8.54 to 8.73.

8.54 Figure P8.54. Use Eqs. (8.9) and (8.10).
8.55 Figure P8.55. Use Eqs. (8.9) and (8.10).
8.56 Figure P8.56. Use Eqs. (8.9) and (8.10).
8.57 Figure P8.57. Write and plot the shear and moment equations.
8.58 Figure P8.58. Write and plot the shear and moment equations.
8.59 Figure P8.59. Use Eqs. (8.9) and (8.10).
8.60 Figure P8.60. Use Eqs. (8.9) and (8.10).

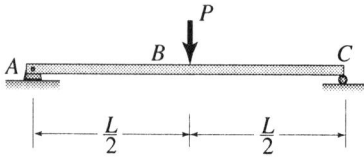

FIGURE P8.54.

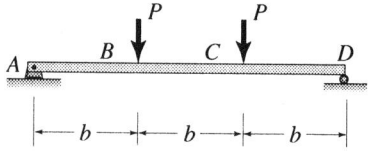

FIGURE P8.55.

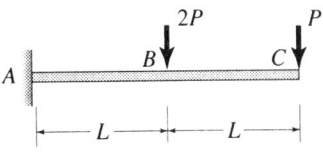

FIGURE P8.56.

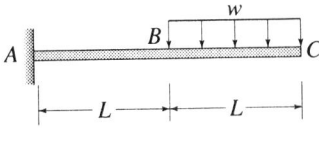

FIGURE P8.57.

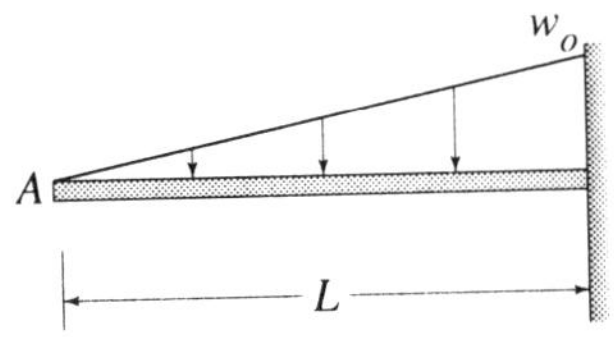

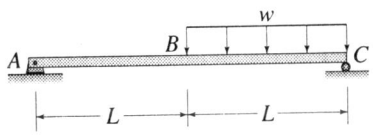

FIGURE P8.58.

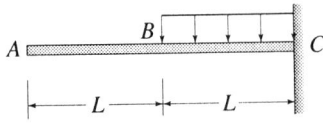

FIGURE P8.59.

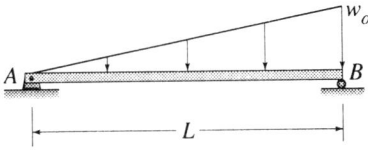

FIGURE P8.61.

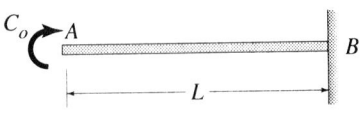

FIGURE P8.62.

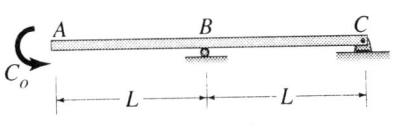

FIGURE P8.63.

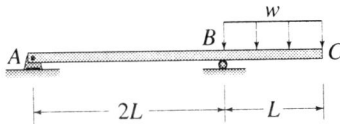

FIGURE P8.64.

FIGURE P8.60.

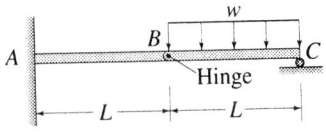

FIGURE P8.65.

8.61 Figure P8.61. Write and plot the shear and moment equations.
8.62 Figure P8.62. Use Eqs. (8.9) and (8.10).
8.63 Figure P8.63. Use Eqs. (8.9) and (8.10).
8.64 Figure P8.64. Use Eqs. (8.9) and (8.10).
8.65 Figure P8.65. Use Eqs. (8.9) and (8.10).
8.66 Figure P8.66. Write and plot the shear and moment equations.

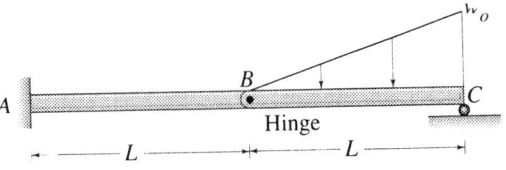

FIGURE P8.66.

8.67 Figure P8.67. Write and plot the shear and moment equations.

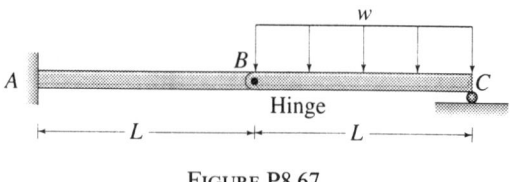

FIGURE P8.67.

8.68 Figure P8.68. Use Eqs. (8.9) and (8.10).

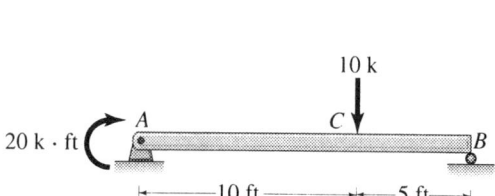

FIGURE P8.68.

8.69 Figure P8.69. Use Eqs. (8.9) and (8.10).

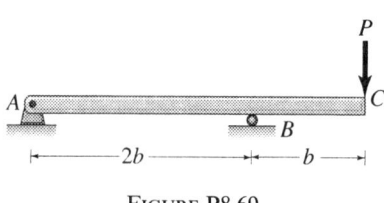

FIGURE P8.69.

8.70 Figure P8.70. Use Eqs. (8.9) and (8.10).

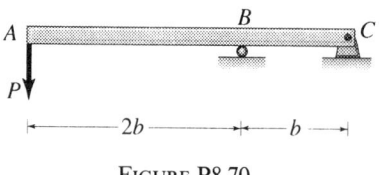

FIGURE P8.70.

8.71 Figure P8.71. Use Eqs. (8.9) and (8.10).

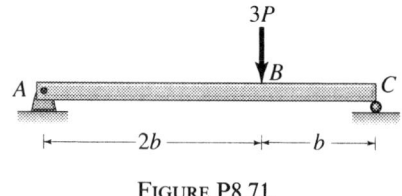

FIGURE P8.71.

8.72 Figure P8.72. Use Eqs. (8.9) and (8.10).

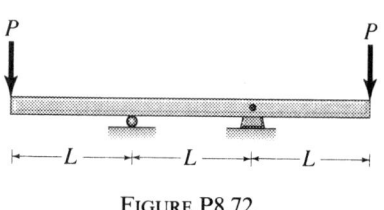

FIGURE P8.72.

8.73 Figure P8.73. Write and plot the shear and moment equations.

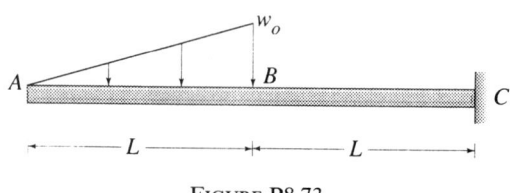

FIGURE P8.73.

8.8*
Cables Under Concentrated Loads

Cables are widely used in engineering structures, such as electrical transmission lines, guy wires for radio and television towers, suspension bridges, and roofs for sport arenas. We assume that cables are weightless and perfectly flexible and, therefore, provide no resistance to bending. Such cables resist applied loads by developing tensile axial forces.

Consider the cable, shown in Figure 8.9(a), held in place by the two fixed supports at A and B and subjected to the vertical concentrated loads P_1, P_2, and P_3. Because of the above assumptions, each segment of the cable may be handled as a straight two-force member. The free-body diagram of the entire cable, shown in Figure 8.9(b), indicates that we have four unknown quantities, A_x, A_y, E_x, and E_y. Because only three equilibrium equations are available, we need to supplement them by additional information to solve for the unknown reaction components. Such additional information comes from knowledge of the exact location of one of the loaded points, such as C, where the vertical distance h_C is known. Such a vertical distance measured from the chord AE to any point on the cable is known as the *sag*. Thus, by applying the equilibrium equation $\sum M_E = 0$, we obtain one relationship between the two unknowns A_x and A_y. A second relationship between A_x and A_y may be obtained by considering the free-body diagram of a segment of the cable from point A to a point just to the right of point C, as shown in Figure 8.9(c). Thus, writing the equilibrium equation $\sum M_C = 0$ yields a second relationship between A_x and A_y. Note that the distance $(h_C - y_C)$ is known because h_C is given and y_C may be determined from the given geometry. Once A_x and A_y are determined, we apply the equations $\sum F_x = 0$ and $\sum F_y = 0$ to the free-body diagram of the entire cable to obtain E_x and E_y. The next step in the solution consists of finding the tension in each segment of the cable. Thus, for example, using the free-body diagram of Figure 8.9(c), the components $(F_x)_{CD}$ and $(F_y)_{CD}$ of the cable tension in segment CD are found from the equilibrium equations $\sum F_x = 0$ and $\sum F_y = 0$. These components may, then, be combined to obtain F_{CD} by the relationship $F_{CD} = \sqrt{(F_x)_{CD}^2 + (F_y)_{CD}^2}$. Note that the equation $\sum F_x = 0$ leads to the conclusion that $(F_x)_{CD} = F_{CD} \cos \theta_{CD} = -A_x$. Using other similar free-body diagrams, we arrive at the conclusion that $(F_x)_{AB} = (F_x)_{BC} = (F_x)_{CD} = (F_x)_{DE} = -A_x =$ constant. This conclusion indicates that the horizontal component of the cable tension at any point in the cable is constant and will be assigned the symbol H. Thus, we may generalize the above conclusion by writing

$$F \cos \theta = H \qquad (8.11)$$

where F and θ represent the cable tension and the angle, respectively, made by the cable with the horizontal at any point along the cable.

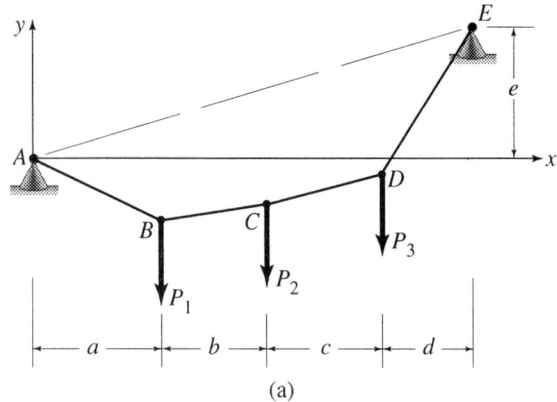

(a)

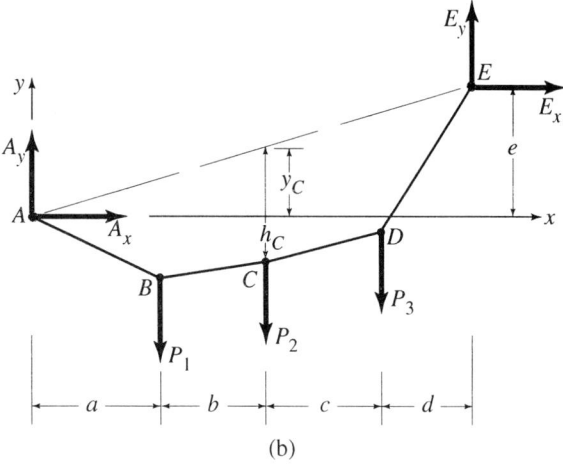

(b)

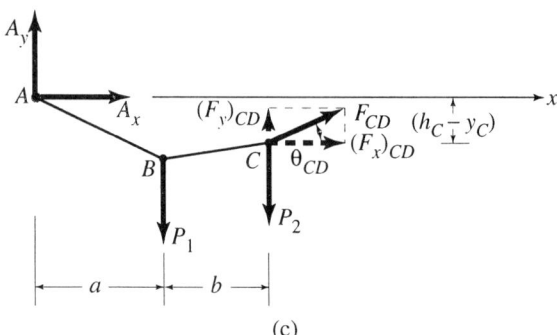

FIGURE 8.9.   (c)

From Eq. (8.11), we conclude that

$$F = H/\cos\theta \tag{8.12}$$

which shows, because H is a constant, that the tension F takes on its maximum value when $\cos\theta$ is the least. This means, of course, that the most highly stressed segment of the cable is the one that makes the largest angle θ with the horizontal. Obviously, this must be one of the two end segments of the cable. Determining the angle θ at any point along the cable requires analyzing the cable geometry, as will be illustrated in Example 8.13. Alternatively, once the angles θ are found, we may determine the cable tensions by considering each point of load application as a joint and writing the equations $\sum F_x = 0$ and $\sum F_y = 0$ at each joint. However, it should be noted that this method of analyzing cables can become tedious when dealing with several vertical loads and resort is made to the *general cable theorem* that will be developed in Section 8.9.

■ Example 8.13

Determine the maximum tension developed in the cable shown in Figure E8.13(a). Note that the sag at B is given as $h_B = 4.667$ m.

Solution

The free-body diagram of the entire cable is shown in Figure E8.13(b) along with a convenient coordinate system. Thus,

$$\sum M_D = 0, \quad -A_x(4) + 10(10) + 15(3) - A_y(15) = 0,$$

$$4A_x + 15A_y = 145. \tag{a}$$

Now let us consider the free-body diagram of a segment from A to a point just to the right of B as shown in Figure E8.13(c). Thus,

$$\sum M_B = 0, \quad -A_x(6) - A_y(5) = 0. \tag{b}$$

Solving Eqs. (a) and (b) simultaneously yields

$$A_x = -10.36 \text{ kN},$$

and

$$A_y = 12.43 \text{ kN}.$$

Now, returning to the free-body diagram of Figure E8.13(b),

$$\sum F_x = 0, \quad A_x + D_x = 0,$$

$$D_x = -A_x = 10.36 \text{ kN},$$

and

$$\sum F_y = 0, \quad A_y + D_y - 25 = 0,$$

$$D_y = 25 - A_y = 12.57 \text{ kN}.$$

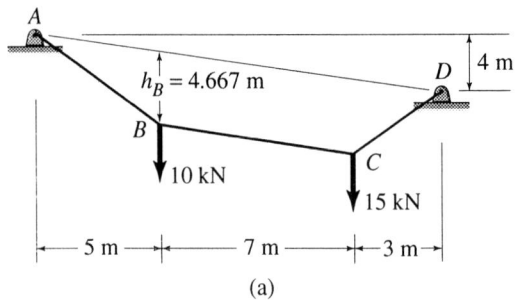

(a)

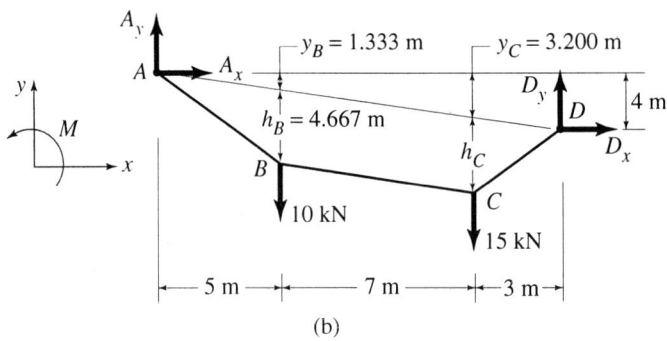

(b)

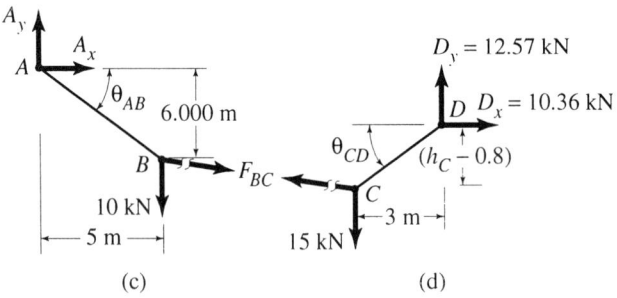

(c)   (d)

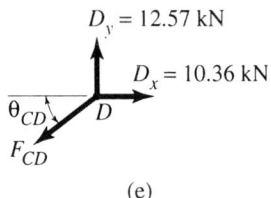

(e)

FIGURE E8.13.

The maximum tension occurs in the segment making the largest angle θ with the horizontal. This segment will be either AB or CD. The geometry, shown in Figure E8.13(b), enables us to find θ_{AB}. Thus,

$$\theta_{AB} = \tan^{-1}\left(\frac{y_B + h_B}{5}\right) = \tan^{-1}\left(\frac{6}{5}\right) = 50.19°.$$

Determination of angle θ_{CD} requires that we, first, find the sag h_C from the free-body diagram of a segment from D to just left of C, shown in Figure E8.13(d). Thus,

$$\sum M_C = 0, \quad 12.57(3) - 10.36(h_C - 0.8) = 0,$$

$$h_C = 4.44 \text{ m}.$$

Therefore,

$$\theta_{CD} = \tan^{-1}\left(\frac{h_C - 0.8}{3}\right) = \tan^{-1}\left(\frac{3.64}{3}\right) = 50.51°.$$

It follows, therefore, that the maximum tension occurs in segment CD and may be found by considering the free-body diagram of joint D, shown in Figure E8.13(e). Thus,

$$\sum F_x = 0, \quad 10.36 - F_{CD}\cos 50.51° = 0,$$

$$F_{max} = F_{CD} = 16.29 \text{ kN}. \qquad \text{ANS.}$$

Alternatively,

$$F_{max} = F_{CD} = \sqrt{(10.36)^2 + (12.57)^2} = 16.29 \text{ kN}. \qquad \text{ANS.}$$

8.9* General Cable Theorem

The general cable theorem is based upon an analogy existing between a cable and a horizontal beam of the same length and carrying the same vertical loads applied to the cable.

Consider the free-body diagram of a cable, as shown in Figure 8.10(a). Based on the conclusions reached in Section 8.8, the horizontal components of the reactions at A and E are indicated as H. Note that, although only three vertical loads are considered, the results obtained are general enough to apply to any number of vertical loads. For convenience, a coordinate system is given adjacent to the free-body diagram. Thus,

$$\sum M_E = 0, \quad P_1(L - a) + P_2(L - b) + P_3(L - c)$$
$$- A_y(L) - HL\tan\alpha = 0$$

from which

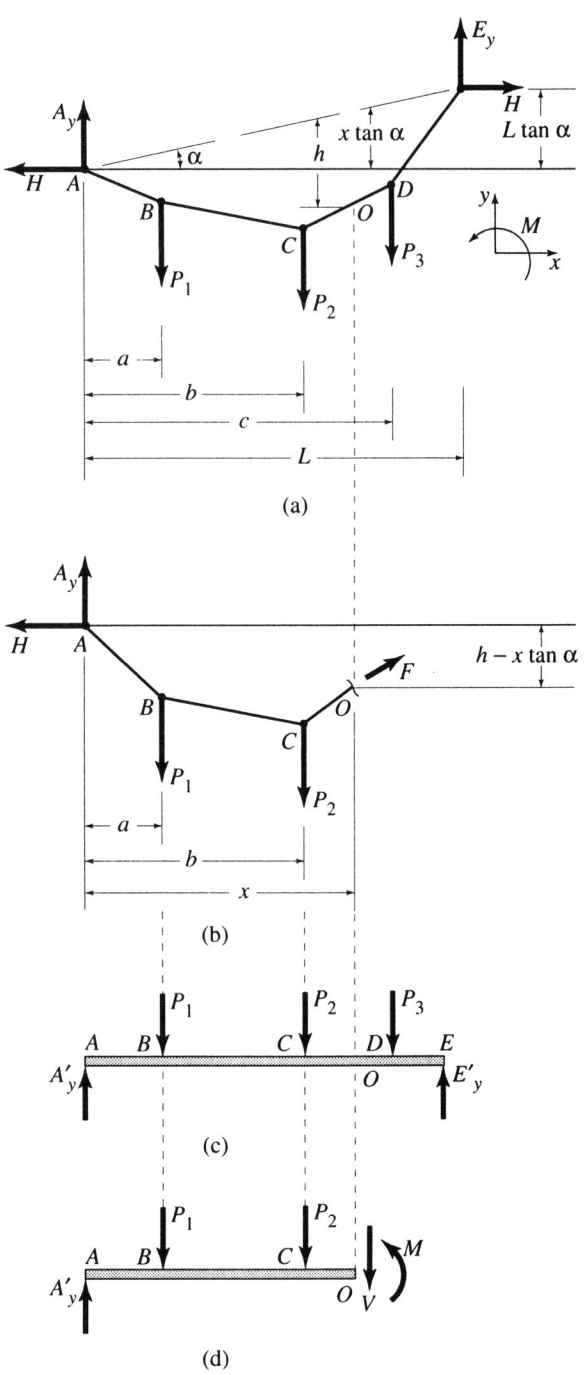

FIGURE 8.10.

$$A_y = \frac{\sum m_E - HL \tan \alpha}{L}. \tag{a}$$

In Eq. (a), the $\sum m_E = P_1(L-a) + P_2(L-b) + P_3(L-c)$ represents the sum of the moments about point E of all vertical loads left of this point.

Now consider the left free-body diagram of a segment of the cable of length x where the sag is h, as shown in Figure 8.10(b). Thus,

$$\sum M_O = 0,$$

$$H(h - x \tan \alpha) + P_1(x-a) + P_2(x-b) - A_y(x) = 0. \tag{b}$$

Substituting Eq. (a) in Eq. (b) and simplifying,

$$Hh + \sum m_O - \left(\frac{x}{L}\right) \sum m_E = 0 \tag{c}$$

where $\sum m_O = P_1(x-a) + P_2(x-b)$ represents the sum of the moments about point O of all vertical loads left of this point. Equation (c) may now be solved for the product Hh to obtain

$$Hh = \left(\frac{x}{L}\right) \sum m_E - \sum m_O \tag{d}$$

Now consider a horizontal simply supported beam AE, whose free-body diagram is shown in Figure 8.10(c), which has the same horizontal span as the cable and carries the same vertical loads. Thus,

$$\sum M_E = 0, \quad P_1(L-a) + P_2(L-b) + P_3(L-c) - A_y'(L) = 0,$$

$$A_y' = \frac{\sum m_E}{L} \tag{e}$$

where $\sum m_E$ has the same definition as given earlier. Next we consider a left free-body diagram of a segment of the beam of the length x, as shown in Figure 8.10(d). Thus,

$$\sum M_O = 0, \quad M + P_1(x-a) + P_2(x-b) - A_y'(x) = 0 \tag{f}$$

Substituting Eq. (e) in Eq. (f), simplifying, and solving for M,

$$M = \left(\frac{x}{L}\right) \sum m_E - \sum m_O \tag{g}$$

where $\sum m_O$ has the same definition given above. Comparing Eqs. (d) and (g) we conclude that

$$Hh = M. \tag{8.13}$$

Equation (8.13) is an expression of the general cable theorem, stating that the product of the horizontal component of the cable tension at any point multiplied by the cable sag at this point is equal in magnitude to the moment at the corresponding position in a simply supported beam of the same length and carrying the same vertical loads. We should observe, here, that, in general, the vertical components of the reaction A'_y and E'_y acting on the beam are not equal to the vertical components of the reaction A_y and E_y acting on the cable, except for the case where the two cable supports lie at the same elevation.

The use of the general cable theorem in analyzing cables subjected to concentrated loads is illustrated in the following examples.

■ **Example 8.14**

The cable shown in Figure E8.14(a) has supports A and E that lie at the same elevation. Point C on the cable is 4 m below the chord AE. Use the general cable theorem to find (a) the sags at B and D and (b) the maximum tension in the cable.

Solution

(a) The free-body diagram of a simply supported beam of the same span as the cable and carrying the same vertical loads is shown in

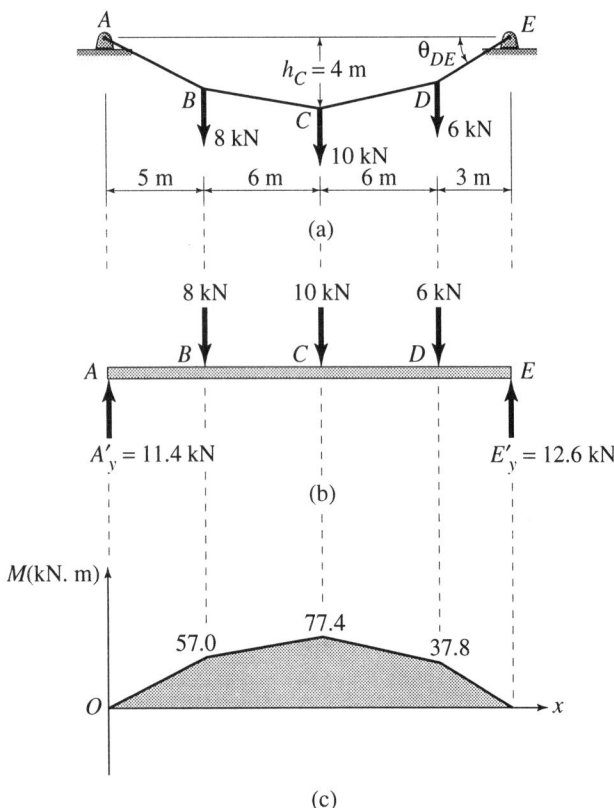

FIGURE E8.14.

Figure E8.14(b) where the vertical beam support reactions A'_y and E'_y have been determined. Note that, in this case, these will be identical to the vertical support reactions A_y and E_y of the cable because supports A and E are at the same elevation. The moment diagram for the companion beam is shown in Figure E8.14(c).

The general cable theorem, expressed in Eq. (8.13), is now applied to cable points C, B, and D. Thus,

Point C

$$Hh_C = M_C,$$

$$H = \frac{M_C}{h_C} = \frac{77.4}{4} = 19.35 \text{ kN}.$$

Point B

$$Hh_B = M_B,$$

$$h_B = \frac{M_B}{H} = \frac{57.0}{19.35} = 2.95 \text{ m}. \qquad \text{ANS.}$$

Point D

$$Hh_D = M_D,$$

$$h_D = \frac{M_D}{H} = \frac{37.8}{19.35} = 1.953 \text{ m}. \qquad \text{ANS.}$$

(b) Equation (8.12) shows that the maximum tension occurs in that portion of the cable making the largest angle θ with the horizontal. This portion is segment DE where

$$\theta_{DE} = \tan^{-1}\left(\frac{h_D}{3}\right) = \tan^{-1}\left(\frac{1.953}{3}\right) = 33.06°$$

By Eq. (8.12),

$$F_{DE} = \frac{H}{\cos \theta_{DE}} = \frac{19.35}{\cos 33.06°} = 23.1 \text{ kN}. \qquad \text{ANS.}$$

■ **Example 8.15**

Support A of the cable shown in Figure E8.15(a) is at an elevation 25 ft above support E, and point D is 10 ft below the chord AE. Use the general cable theorem to determine (a) the sags at B and C and (b) the maximum cable tension.

Solution

(a) The free-body diagram of a simply supported beam of the same span as the cable and carrying the same vertical loads is shown in Figure E8.15(b) where the vertical beam support reactions A'_y and E'_y have been determined. Note that, because the cable supports are not at the same elevation, these reactions will differ from the vertical cable support reactions. The moment diagram for the companion beam is shown in Figure E8.15(c).

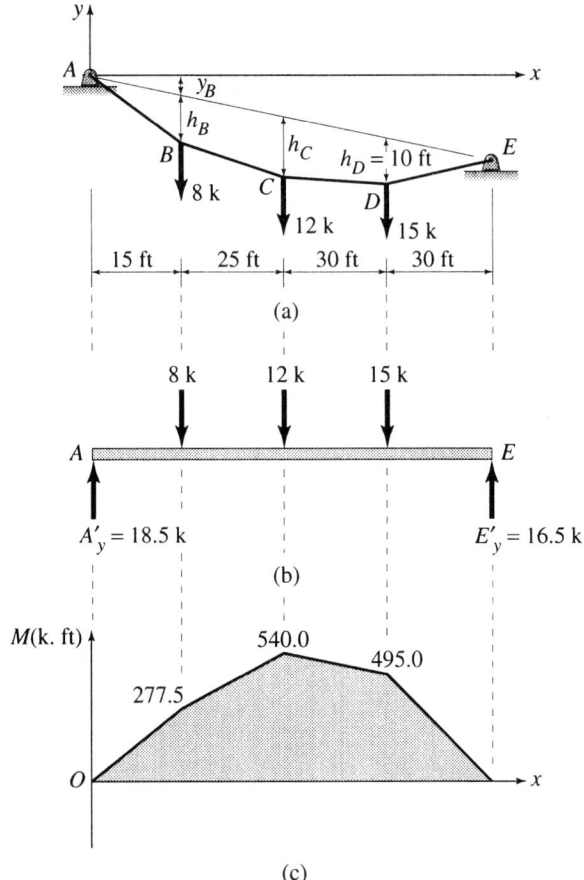

FIGURE E8.15.   (c)

The general cable theorem, expressed in Eq. (8.13), is now used for cable points D, B, and C. Thus,

Point D
$$Hh_D = M_D,$$

$$H = \frac{M_D}{h_D} = \frac{495}{10} = 49.5 \text{ k}.$$

Point B
$$Hh_B = M_B,$$

$$h_B = \frac{M_B}{H} = \frac{277.5}{49.5} = 5.61 \text{ ft}. \qquad \text{ANS.}$$

Point C
$$Hh_C = M_C,$$

$$h_C = \frac{M_B}{H} = \frac{540}{49.5} = 10.91 \text{ ft}. \qquad \text{ANS.}$$

(b) By Eq. (8.12), the maximum tension occurs in that segment of the cable making the largest angle with the horizontal. This segment is AB where

$$\theta_{AB} = \tan^{-1}\left(\frac{y_B + h_B}{15}\right) = \tan^{-1}\left(\frac{3.75 + 5.61}{15}\right) = 31.96°$$

Thus, by Eq. (8.12),

$$F_{AB} = \frac{H}{\cos \theta_{AB}} = \frac{49.5}{\cos 31.96°} = 58.3 \text{ k.} \qquad \text{ANS.}$$

Problems

8.74 A cable is loaded, as shown in Figure P8.74. (a) Determine the end reaction components at A and B. (b) Find the sag at point D on the cable.

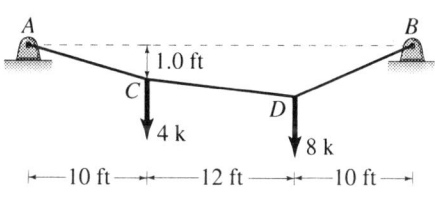

FIGURE P8.74.

8.75 The supports A and B are at the same elevation for the cable shown in Figure P8.75. (a) Find the horizontal and vertical reactions at A and B. (b) Determine the sag at point C. (c) Draw free-body

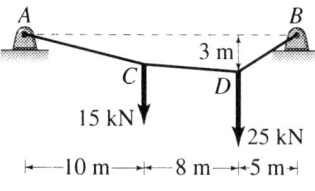

FIGURE P8.75.

diagrams of cable points D and B. Write the equations of equilibrium and use this information to check your results.

8.76 Refer to Figure P8.76, and find the reaction components at A and B and the sags at D and E. Also, find the maximum slope and the maximum tension in this cable.

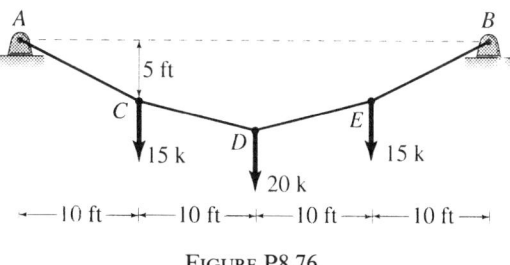

FIGURE P8.76.

8.77 Determine the load P and all reaction components for the cable shown in Figure P8.77. What is the maximum tension?

8.78 The end support B lies 50 ft below the end support A for the cable shown in Figure P8.78. Determine the end reaction components at A and B and the sags at C and D. Find the maximum slope of this cable and its maximum tensile force.

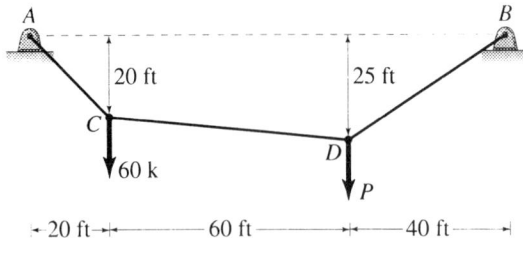

FIGURE P8.77.

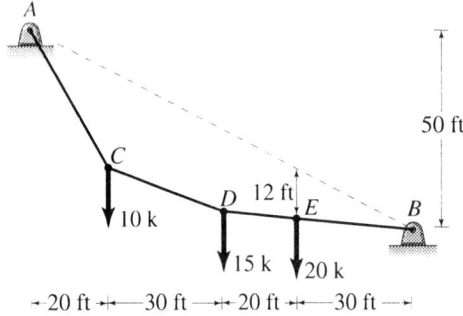

FIGURE P8.78.

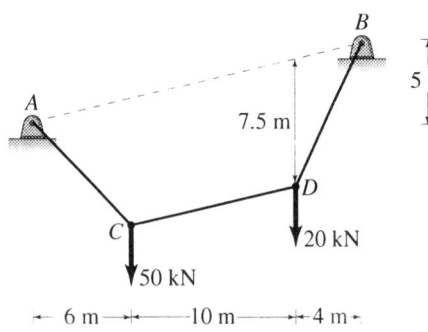

FIGURE P8.79.

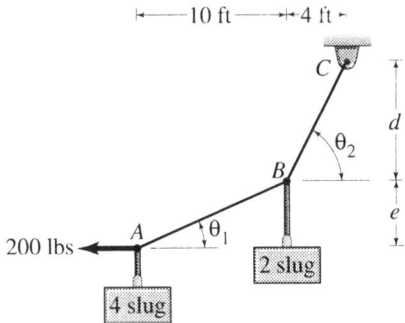

FIGURE P8.80.

and e. Hint: Draw free-body diagrams of points A and B, and write the equilibrium equations using the angles θ_1 and θ_2.

8.81 Refer to the loaded cable shown in Figure P8.81, and note that the supports A and B are at the same level. Points C and E are 3 m above the line AB and point D is 2 m below this line. Determine the horizontal and vertical reactions at A and B and the applied load P. Also, find the maximum tensions in the cable.

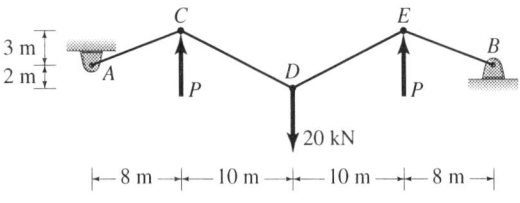

FIGURE P8.81.

8.79 Support B of the cable shown in Figure P8.79 is 5 m above support A. Determine the horizontal and vertical reaction components at both A and B. Find the sag at C, the maximum slope, and maximum tension in the cable.

8.80 The cable shown in Figure P8.80 is in equilibrium. Determine the forces in segments AB and BC and the distances d and e. Hint: Draw free-body diagrams of

8.82 Refer to the cable loaded as shown in Figure P8.82. Determine the force P applied upward at point D and the reaction components at A and B. Also, find the maximum tensions in the cable.

8.83 Determine the reaction components at A and B and the sag at D for the cable shown in Figure P8.83. Also, find the maximum slope and maximum cable tension.

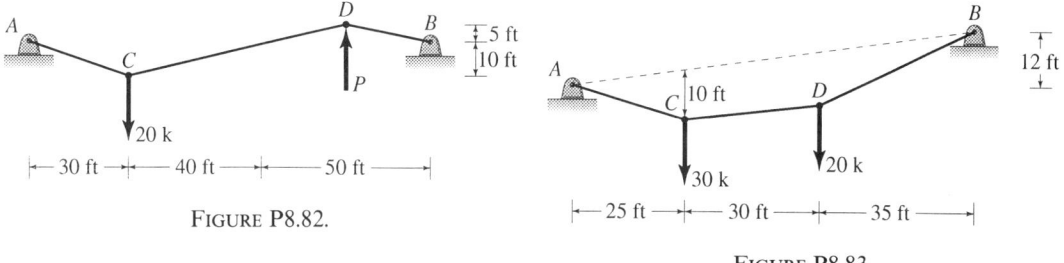

FIGURE P8.82.

FIGURE P8.83.

8.10*
Cables Under Uniform Loads

Uniform load over the horizontal— Parabola

A cable loaded uniformly over the horizontal is shown in Figure 8.11(a). It is supported at points A and B which are located a distance L apart measured horizontally and a distance d apart measured vertically. The distance h_C is measured vertically from chord AB to point C on the cable at midspan. We regard as known quantities the loading intensity w and the geometry specified by L, d, and h_C. A knowledge of these quantities will enable us to determine the end reactions (A_x, A_y, B_x, B_y), an equation expressing the sag h at any point as a function of x, the tension F at any point, and the position of the low point on the cable.

The free-body diagram of the entire cable is shown in Figure 8.11(b). Thus,

$$\sum F_x = 0, \quad B_x - A_x = 0,$$

$$B_x = A_x = H.$$

It can be shown that the general theorem for cables applies also to uniform loadings. Refer to Figures 8.11(c) and (d), and apply this theorem to point C to obtain

$$Hhc = \frac{wL^2}{8},$$

$$H = \frac{wL^2}{8h_C}. \tag{8.14}$$

At any distance x from the left end of the cable, the general theorem becomes

$$Hh = M = \frac{wL}{2}x - \frac{wx^2}{2}.$$

Substituting for H from Eq. (8.14) and solving for h,

$$h = \left(\frac{4h_C}{L^2}\right)x(L - x) \tag{8.15}$$

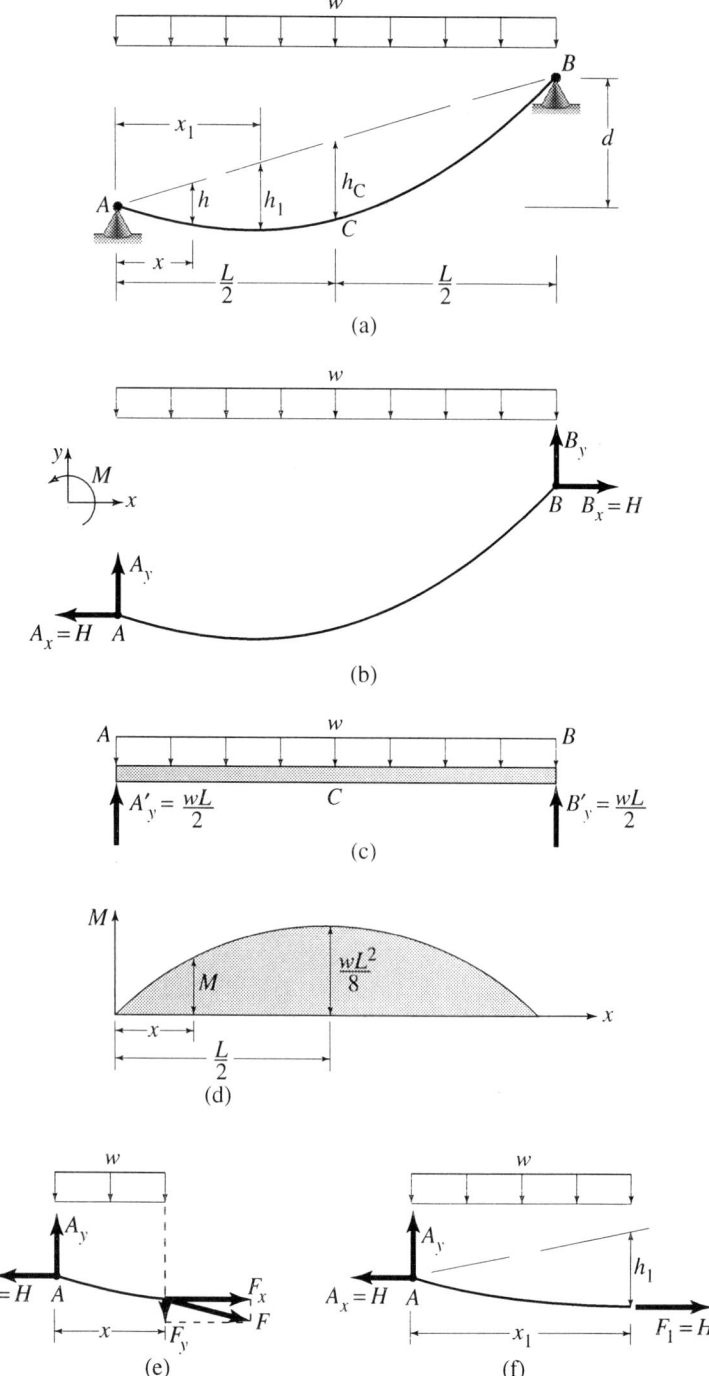

FIGURE 8.11.

which shows that the shape of a cable under uniform load over the horizontal is parabolic. Note that h measures the vertical distance from chord AB to the cable.

Referring to the cable free-body diagram of Figure 8.11(b),

$$\sum M_B = 0, \quad wL(L/2) - Hd - A_y L = 0.$$

Substituting for H and solving for A_y,

$$A_y = \frac{wL}{2}\left(1 - \frac{d}{4h_C}\right) \tag{8.16}$$

and

$$\sum F_y = 0, \quad A_y + B_y - wL = 0.$$

Substituting for A_y and solving for B_y,

$$B_y = \frac{wL}{2}\left(1 + \frac{d}{4h_C}\right). \tag{8.17}$$

To find the cable tension F at any distance x from the left support A, refer to Figure 8.11(e), and write the equations of equilibrium. Thus,

$$\sum F_x = 0, \quad F_x - H = 0,$$

$$F_x = H,$$

and

$$\sum F_y = 0, \quad A_y - wx - F_y = 0.$$

Substituting for A_y from Eq. (8.16) and solving for F_y,

$$F_y = \frac{wL}{2}\left(1 - \frac{2x}{L} - \frac{d}{4h_C}\right).$$

Therefore, the force F at any position in the cable becomes

$$F = \sqrt{F_x^2 + F_y^2},$$

$$F = \sqrt{H^2 + \left[\frac{wL}{2}\left(1 - \frac{2x}{L} - \frac{d}{4h_C}\right)\right]^2}. \tag{8.18}$$

At the low point of the cable, defined by the coordinates x_1 and h_1, the total tension will act horizontally and is equal to H, as shown in Figure 8.11(f). Thus,

$$\sum F_y = 0, \quad A_y - wx_1 = 0.$$

Substituting for A_y from Eq. (8.16) and solving for x_1,

$$x_1 = \frac{L}{2}\left(1 - \frac{d}{4h_C}\right). \tag{8.19}$$

Once x_1 is found, the corresponding value of h_1 is determined from Eq. (8.15).

In applying the general theorem for cables, it was convenient to measure x from the left support and h from chord AB, but to develop an equation for cable length, it is desirable to choose an origin O at the low point on the cable, as shown in Figure 8.12(a). In this coordinate system, the support points have coordinates $A(x_A, y_A)$ and $B(x_B, y_B)$. The x axis is tangent to the cable at O, and y is measured vertically upward from this point. Because x_1, h_1, d, and L are known, we express the coordinates of the support points in terms of these quantities. Thus,

$$\left. \begin{aligned} x_A &= -x_1, \\[2mm] y_A &= h_1 - \frac{x_1}{L}d, \\[2mm] x_B &= L - x_1, \end{aligned} \right\} \tag{8.20}$$

and

$$y_B = h_1 - \frac{x_1}{L}d + d = h_1 - d\left(1 - \frac{x_1}{L}\right).$$

The free-body diagram of a segment of the cable of length S is shown

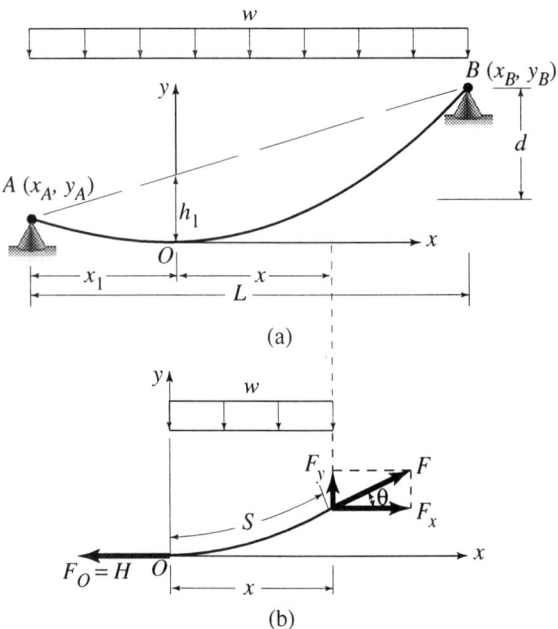

FIGURE 8.12.

in Figure 8.12(b). At the origin O, the low point, the cable tension is equal to H and acts horizontally. Thus,

$$\sum F_x = 0, \quad F_x - H = 0,$$

$$F_x = H,$$

and

$$\sum F_y = 0, \quad F_y - wx = 0,$$

$$F_y = wx.$$

Because the cable tension F acts tangent to the cable,

$$\tan \theta = \frac{F_y}{F_x} = \frac{wx}{H} = \frac{dy}{dx}.$$

This is the differential equation for the cable. It can be integrated by separating variables. Thus,

$$\int dy = \int \frac{wx}{H} dx,$$

$$y = \frac{wx^2}{2H} + C.$$

The constant of integration C equals zero because $y = 0$ when $x = 0$. Clearly, the cable has a parabolic shape given by

$$y = \frac{wx^2}{2H} \tag{8.21}$$

provided the loading per unit length w is uniform over the horizontal. To develop an expression for cable length, we recall from calculus that the differential arc length dS is given by

$$dS = \sqrt{1 + \left(\frac{dy}{dx}\right)^2} \, dx.$$

Substituting $\dfrac{dy}{dx} = \dfrac{w}{H} x$ and integrating with limits,

$$\int_0^S dS = \int_{x_A}^{x_B} \sqrt{1 + \left(\frac{wx}{H}\right)^2} \, dx$$

where S is the total length of the cable measured along its arc. To integrate the right hand side, we expand the integrand in a binomial series to give

$$S = \int_{x_A}^{x_B} \left(1 + \frac{w^2 x^2}{2H^2} - \frac{w^4 x^4}{8H^2} + \cdots\right) dx.$$

Term-by-term integration and substitution of upper and lower limits gives

$$S = x_B\left(1 + \frac{w^2 x_B^2}{6H^2} - \frac{w^4 x_B^4}{40H^2} + \cdots\right) - x_A\left(1 + \frac{w^2 x_A^2}{6H^2} - \frac{w^4 x_A^4}{40H^2} + \cdots\right)$$

From the equation of the parabola $\dfrac{wx^2}{2H} = y$,

$$\frac{wx_B^2}{2H} = y_B,$$

and

$$\frac{wx_A^2}{2H} = y_A.$$

Substituting in the above equation for S and simplifying,

$$S = x_B\left[1 + \frac{2}{3}\left(\frac{y_B}{x_B}\right)^2 - \frac{2}{5}\left(\frac{y_B}{x_B}\right)^2 + \cdots\right]$$

$$- x_A\left[1 + \frac{2}{3}\left(\frac{y_A}{x_A}\right)^2 - \frac{2}{5}\left(\frac{y_A}{x_A}\right)^4 + \cdots\right]. \tag{8.22}$$

Two terms of each series will usually provide sufficient accuracy because the series are rapidly convergent provided that $\left|\dfrac{y_B}{x_B}\right| < 0.5$ and $\left|\dfrac{y_A}{x_A}\right| < 0.5$, which is true for most practical applications.

Uniform load over the arc length— Catenary

A cable is supported at points A and B as shown in Figure 8.13(a) and is subjected to a uniform loading w over its arc length. As an example of practical interest, this loading could represent the weight per unit length of the cable itself. To simplify the equations, an origin O is chosen at a distance c below the low point of the cable.

A free-body diagram for a length S of cable is shown in Figure 8.13(b). The applied downward resultant loading is wS where S is measured along the arc length of the cable. At the low point, the cable tension H acts horizontally to the left, and, at point P the cable tension F acts along the tangent to the cable. Therefore,

$$\sum F_x = 0, \quad F\cos\theta - H = 0, \tag{a}$$

and

$$\sum F_y = 0, \quad F\sin\theta - wS = 0. \tag{b}$$

Divide Eq. (b) by Eq. (a) to give

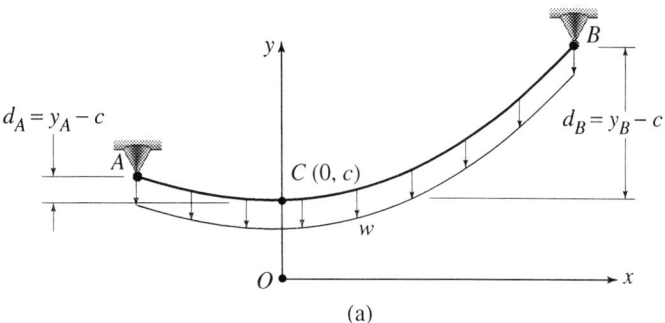

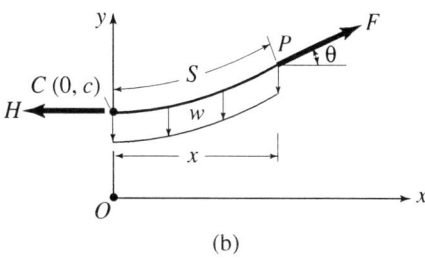

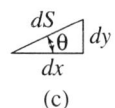

FIGURE 8.13.

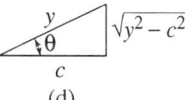

(d)

$$\tan \theta = \frac{wS}{H}$$

from which

$$S = \frac{H}{w}\tan \theta. \tag{c}$$

By the chain rule,

$$\frac{dy}{d\theta} = \frac{dy}{dS}\frac{dS}{d\theta} = \sin \theta \frac{dS}{d\theta} \tag{d}$$

because $\frac{dy}{dS} = \sin \theta$, as shown in Figure 8.13(c). Differentiate Eq. (c) with respect to θ to yield

$$\frac{dS}{d\theta} = \frac{H}{w} \sec^2\theta.$$

Substitute for $\frac{dS}{d\theta}$ in Eq. (d) to give

$$\frac{dy}{d\theta} = \sin\theta \left(\frac{H}{w} \sec^2\theta \right) = \frac{H}{w} \frac{\sin\theta}{\cos^2\theta}.$$

Multiply by $d\theta$ and integrate both sides of this equation to obtain

$$\int dy = \frac{H}{w} \int \cos^{-2}\theta \sin\theta \, d\theta.$$

Integration of this equation yields

$$y = \frac{H}{w} \cos^{-1}\theta + C_1 = \frac{H}{w} \sec\theta + C_1 \qquad \text{(e)}$$

where C_1 is a constant of integration and H and w are constant for a given cable. The constant C_1 is evaluated from the condition that at $x = 0$, $y = c$, and $\theta = 0$. Thus,

$$c = \frac{H}{w}(1) + C_1.$$

If we let $c = \frac{H}{w}$, then, $C_1 = 0$. Substituting in Eq. (e) gives

$$y = c \sec\theta = \frac{c}{\cos\theta}$$

from which

$$\cos\theta = \frac{c}{y}. \qquad \text{(f)}$$

Equation (f) may be interpreted as shown in Figure 8.13(d) from which

$$\frac{dy}{dx} = \tan\theta = \frac{\sqrt{y^2 - c^2}}{c}. \qquad \text{(g)}$$

Separating variables and integrating,

$$\int dx = c \int \frac{dy}{\sqrt{y^2 - c^2}},$$

$$x = c \cosh^{-1}\frac{y}{c} + C_2. \qquad \text{(h)}$$

Use the condition $x = 0$ and $y = c$ to find C_2. Thus,

$$0 = c \cosh^{-1}(1) + C_2.$$

Because $\cosh^{-1} 1 = 0$, then, $C_2 = 0$. Thus, Eq. (h) becomes

$$x = c \cosh^{-1} \frac{y}{c}$$

or

$$\frac{x}{c} = \cosh^{-1} \frac{y}{c}.$$

Operate with cosh on both sides of this equation to express y as a function of x. Thus,

$$y = c \cosh \frac{x}{c} \quad \text{where} \quad c = \frac{H}{w} \tag{8.23}$$

which is the equation of a cable loaded uniformly along its arc length, termed a *catenary*.

Because $c = \dfrac{H}{w}$, Eq. (c) may be written as $S = c \tan \theta$. From Eq. (g), we conclude that $\tan \theta = \dfrac{\sqrt{y^2 - c^2}}{c}$. Substitution for $\tan \theta$ in the equation for S gives

$$S = \sqrt{y^2 - c^2}$$

Square both sides and solve for y to obtain

$$y = \sqrt{S^2 + c^2}. \tag{8.24}$$

This equation expresses the ordinate y as a function of arc length S. Substitute Eq. (8.23) in Eq. (8.24) to obtain

$$c \cosh \frac{x}{c} = \sqrt{S^2 + c^2}.$$

Square both sides to obtain

$$c^2 \cosh^2 \frac{x}{c} = S^2 + c^2.$$

Solving for S^2 yields

$$S^2 = c^2 \left(\cosh^2 \frac{x}{c} - 1 \right).$$

Using the trigonometric identity $\cosh^2 \dfrac{x}{c} - \sinh^2 \dfrac{x}{c} = 1$,

$$S = c \sinh \frac{x}{c}. \tag{8.25}$$

This equation expresses the arc length S, measured from the point $(0, c)$, as a function of x.

Return to equilibrium Eqs. (a) and (b), square, and add to eliminate θ. Thus,

$$F^2 = H^2 + w^2 S^2.$$

Substitute for $H = cw$ and $S = c\sinh\dfrac{x}{c}$, and solve for F to give

$$F = cw\sqrt{1 + \sinh^2\left(\frac{x}{c}\right)},$$

$$F = cw\cosh\frac{x}{c}. \qquad (8.26)$$

The cable tension F acts tangent to the catenary and is a function of x because c and w are known for a given cable. Substituting Eq. (8.23) in Eq. (8.26) and simplifying yields

$$F = wy. \qquad (8.27)$$

Equation (8.27) states that the cable tension F at any point is proportional to the ordinate y to the catenary at this point.

The following examples illustrate some of the concepts discussed above.

■ **Example 8.16**

The uniform loading over the horizontal and the geometry for a cable are given in Figure E8.16. Determine (a) the horizontal component of the cable tension H, (b) the equation of the cable, $h = f(x)$, (c) the vertical reactions A_y and B_y, (d) the cable tension as a function of x and the tensions at the supports A and B, and (e) the coordinates of the low point, x_1 and h_1.

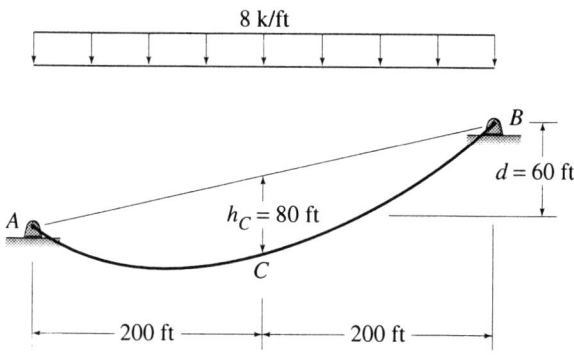

FIGURE E8.16.

Solution

Equations (8.14) through (8.19) are applicable.

(a) By Eq. (8.14),

$$H = \frac{wL^2}{8h_C} = \frac{8(400)^2}{8(80)} = 2000 \text{ k.} \qquad \text{ANS.}$$

(b) By Eq. (8.15),

$$h = \frac{4h_C x}{L^2}[L - x] = \frac{4(80)x}{(400)^2}[400 - x],$$

$$h = 0.8x - 0.002x^2. \qquad \text{ANS.}$$

A check on this equation is obtained by setting $x = 400$ ft at B and showing that at this point $h = 0$.

(c) By Eqs. (8.16) and (8.17),

$$A_y = \frac{wL}{2}\left(1 - \frac{d}{4h_C}\right),$$

$$A_y = \frac{8(400)}{2}\left[1 - \frac{60}{4(80)}\right] = 1300 \text{ k,}$$

and

$$B_y = \frac{wL}{2}\left[1 + \frac{d}{4h_C}\right] = \frac{8(400)}{2}\left[1 + \frac{60}{4(80)}\right] = 1900 \text{ k.} \qquad \text{ANS.}$$

(d) The cable tension as a function of x is given by Eq. (8.18). Thus,

$$F = \sqrt{H^2 + \left[\frac{wL}{2}\left(1 - \frac{2x}{L} - \frac{d}{4h_C}\right)\right]^2},$$

$$= \sqrt{(2000)^2 + \left[\frac{8(400)}{2}\left(1 - \frac{2x}{400} - \frac{60}{4(80)}\right)\right]^2},$$

$$= \sqrt{4.00 \times 10^6 + (1300 - 8x)^2}, \qquad \text{ANS.}$$

$$F_A = F_{x=0} = 2390 \text{ k,} \qquad \text{ANS.}$$

and

$$F_B = F_{x=400} = 2760 \text{ k.} \qquad \text{ANS.}$$

The end tension at B is the maximum cable tension because the cable slope has the greatest magnitude at this point.

(e) The low point coordinate x_1 is given by Eq. (8.19). Thus,

$$x_1 = \frac{L}{2}\left(1 - \frac{d}{4h_C}\right) = \frac{400}{2}\left[1 - \frac{60}{4(80)}\right] = 162.5 \text{ ft.} \qquad \text{ANS.}$$

From part (b) above, $h = 0.8x - 0.002x^2$. Substituting $x_1 = 162.5$ gives

$$h_1 = 0.8(162.5) - 0.002(162.5)^2 = 77.2 \text{ ft.} \qquad \text{ANS.}$$

Because $x_1 < 200$, the low point is located to the left of midspan.

■ **Example 8.17**   Refer to Example 8.16 and determine (a) the coordinates of the end points A and B for an origin at the low point, (b) the equation of the cable with an origin at the low point, and (c) the total length of the cable.

Solution   (a) By Eq. (8.20),

$$x_A = -x_1 = -162.5 \text{ ft,} \qquad \text{ANS.}$$

$$y_A = h_1 - \frac{x_1}{L} d = 77.2 - \frac{162.5}{400}(60) = 52.8 \text{ ft,} \qquad \text{ANS.}$$

$$x_B = L - x_1 = 237.5 \text{ ft,} \qquad \text{ANS.}$$

and

$$y_B = h_1 - d\left(1 - \frac{x_1}{L}\right) = y_A + d = 52.8 + 60 = 112.8 \text{ ft.} \quad \text{ANS.}$$

(b) By Eq. (8.21),

$$y = \frac{w}{2H}x^2 = \frac{8}{2(2000)}x^2$$

$$= 0.002x^2. \qquad \text{ANS.}$$

(c) The total length of the cable is given by Eq. (8.22). Thus,

$$S_T = x_B\left[1 + \frac{2}{3}\left(\frac{y_B}{x_B}\right)^2 - \frac{2}{5}\left(\frac{y_B}{x_B}\right)^4\right] - x_A\left[1 + \frac{2}{3}\left(\frac{y_A}{x_A}\right)^2 - \frac{2}{5}\left(\frac{y_A}{x_A}\right)^4\right]$$

$$= 237.5\left[1 + \frac{2}{3}\left(\frac{112.82}{237.5}\right)^2 - \frac{2}{5}\left(\frac{112.82}{237.5}\right)^4\right]$$

$$- (-162.5)\left[1 + \frac{2}{3}\left(\frac{52.82}{-162.5}\right)^2 - \frac{2}{5}\left(\frac{52.82}{-162.5}\right)^4\right]$$

$$= 268.38 + 173.21 = 442 \text{ ft.} \qquad \text{ANS.}$$

If we use two instead of three terms of the above series, the approximate total length is 447 ft, which is about 1% higher than the answer using three terms. A fourth term of this series would involve the eighth power of ratios less than 0.5 and would be very small. To check the convergence criteria, we note that

$$\left|\frac{y_B}{x_B}\right| = \frac{112.82}{237.5} = 0.475 < 0.5$$

and

$$\left|\frac{y_A}{x_A}\right| = \frac{52.82}{162.5} = 0.325 < 0.5$$

which confirms convergence of this series.

■ **Example 8.18**

During erection of the main span of a suspension bridge, a cable weighing 2.8 k/ft is supported on towers at A and B as shown in Figure E8.18(a). The span is 5000 ft and the center sag $h_C = 550$ ft at this stage. Determine (a) the equation of the cable profile, (b) the value of y for $x = 1250$ ft, (c) the length of the cable, and (d) the maximum and minimum cable tensions.

Solution

(a) Let us apply Eq. (8.23) to point B whose coordinates are $x_B = 2500$ and $y_B = 550 + c$. Thus,

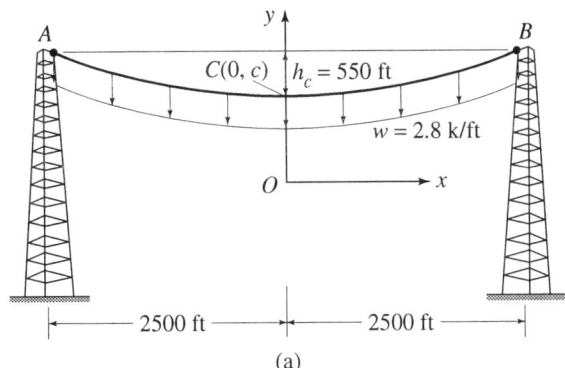

(a)

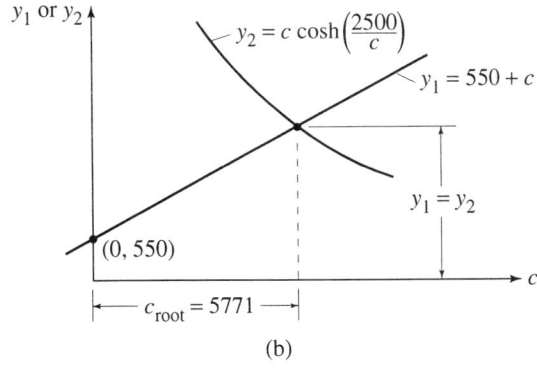

(b)

FIGURE E8.18.

$$550 + c = c \cosh \frac{2500}{c}.$$

This is a transcendental equation with c as the unknown. Iteration or 'trial and error' was used to solve it. Figure E8.8(b) is a sketch which enables us to understand the method. Let $y_1 = 550 + c$ and $y_2 = c \cosh \frac{2500}{c}$. The transcendental equation can be written as $y_1 = y_2$, which means that the ordinates to the straight line and the hyperbolic cosine functions are equal when we have chosen the correct c, the root of the equation. A number of trials, not shown, were required to determine $c = 5771$.

The equation of the cable profile thus becomes

$$y = 5771 \cosh \frac{x}{5771} \qquad \text{ANS.}$$

(b) For $x = 1250$ ft,

$$y = 5771 \cosh \frac{1250}{5771} \approx 5910 \text{ ft.} \qquad \text{ANS.}$$

(c) Eq. (8.25) for arc length is

$$S = c \sinh \frac{x}{c}.$$

If we substitute $x = 2500$ ft, this will give approximately one-half the total length of cable. Thus,

$$\text{the total cable length} = 2(5771) \sinh \frac{2500}{5771} \approx 5160 \text{ ft.} \qquad \text{ANS.}$$

(d) Eq. (8.27) for cable tension is

$$F = wy.$$

Because y is a minimum at the origin where $y = c = 5771$,

$$F_{\min} = 2.8(5771) = 16,160 \text{ k.} \qquad \text{ANS.}$$

Because y is a maximum at points A and B where $y_A = y_B = c + 550 = 5771 + 550 = 6321$,

$$F_{\max} = wy_A = 2.8(6221) = 17,700 \text{ k.} \qquad \text{ANS.}$$

■

Problems

8.84 A cable shown in Figure P8.84 is supported at A and B and subjected to a uniform load of intensity $w = 5$ k/ft over the horizontal. Determine (a) the equation of the cable $h = f(x)$, (b) the tension in the cable at support points A and B, and (c) the slope of the cable at support point B.

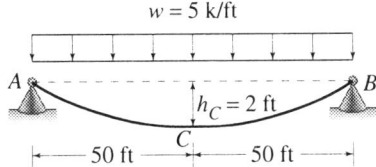

FIGURE P8.84.

8.85 A pedestrian suspension bridge has a span of 100 ft and a center sag $h_C = 20$ ft. One of its cables, shown in Figure P8.85, is subjected to a design loading intensity $w = 600$ lb/ft over the horizontal. Use the general theorem for cables to determine the horizontal component of the cable tension and the vertical distance measured from the chord AB to a point on the cable, 20 ft to the right or left of the center of the span.

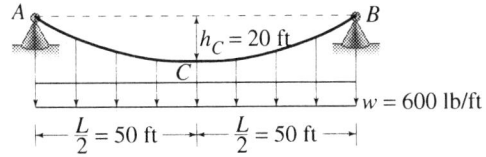

FIGURE P8.85.

8.86 A proposed suspension bridge is to have a main span of 4500 ft. At the towers of equal height, the design total tension is 40,000 k, and the center sag is 540 ft. Determine w, the design loading intensity over the horizontal span of the cable. What is the length of this cable?

8.87 Refer to Figure P8.87 which represents a main cable in a suspension bridge. Let $w = 6$ k/ft, $L = 500$ ft, $h_C = 50$ ft, and $d = 20$ ft. Determine (a) the horizontal component of cable tension H, (b) the equation for the cable $h = f(x)$, where h is measured vertically from chord AB to points on the cable and the origin for x is at the left side, (c) the vertical reaction components at A and B, and (d) the cable tension as a function of x and the tensions at the supports A and B.

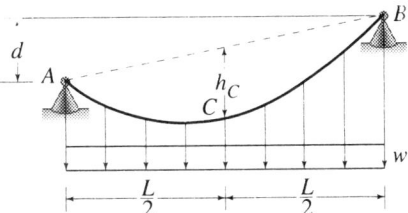

FIGURE P8.87.

8.88 Refer to Problem 8.87 and determine (a) the coordinate of the low point x_1 and h_1, (b) the coordinates of the support at A and B with respect to an origin at the low point, and (c) the total length S of the cable.

8.89 The low point O of the suspended cable, shown in Figure P8.89, is located 200 ft to the left of support B. Use an origin at O to locate the end supports A and B of this cable. Find the maximum and minimum cable tensions.

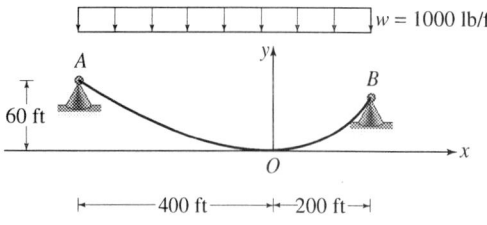

FIGURE P8.89.

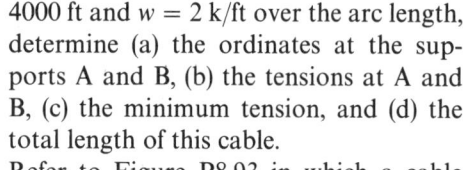

8.90 A cable weights 4 lb/ft and is suspended between two supports A and B, spanning 120 ft. The sag is 4 ft and the supports are at the same level. Assume the parabolic solution is valid and find (a) the length of the cable, (b) the end slopes, and (c) the maximum tension.

8.91 Solve Problem 8.90 using the catenary solution. Compare the parabolic and catenary solutions.

8.92 A side span of a suspension bridge cable is shown in Figure P8.92. Given $c =$ 4000 ft and $w = 2$ k/ft over the arc length, determine (a) the ordinates at the supports A and B, (b) the tensions at A and B, (c) the minimum tension, and (d) the total length of this cable.

8.93 Refer to Figure P8.93 in which a cable is loaded over its arc length with $w = 200$ lb/ft. The span $L = 260$ ft and $c = 300$ ft. Write the equation of this cable, and determine the maximum tension in it.

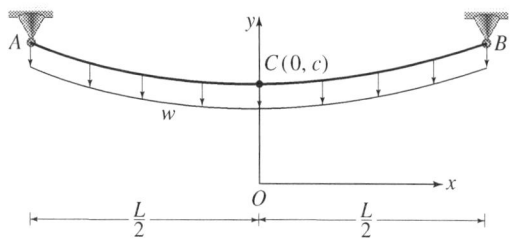

FIGURE P8.93.

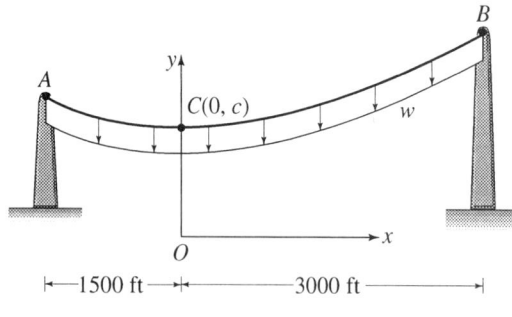

FIGURE P8.92.

8.94 A cable is loaded over its arc length with an intensity of $w = 150$ lb/ft as shown in Figure P8.93. The constant $c = 400$ ft and the span $L = 280$ ft. Determine the length of this cable and the minimum tension in it. Where does this minimum tension occur?

8.95 Refer to Problem 8.94 and write the equation of this cable. Determine the tension at the supports A and B.

8.11
Frames—
Internal
Forces at
Specified
Sections

In this section, two statically determinate structures will be analyzed for external reactions and internal forces. These structures together with appropriate free-body diagrams are shown in Figure 8.14.

A frame supported by a pin at A and a roller at B is shown in Figure 8.14(a). Its members are connected together so that we may idealize it as a single rigid-body. Three external reactions A_x, A_y, and B_y may be determined by writing the three equations of equilibrium available for each rigid body in two dimensions. The frame's overall geometry and applied loadings must be known if we wish to solve for these reactions.

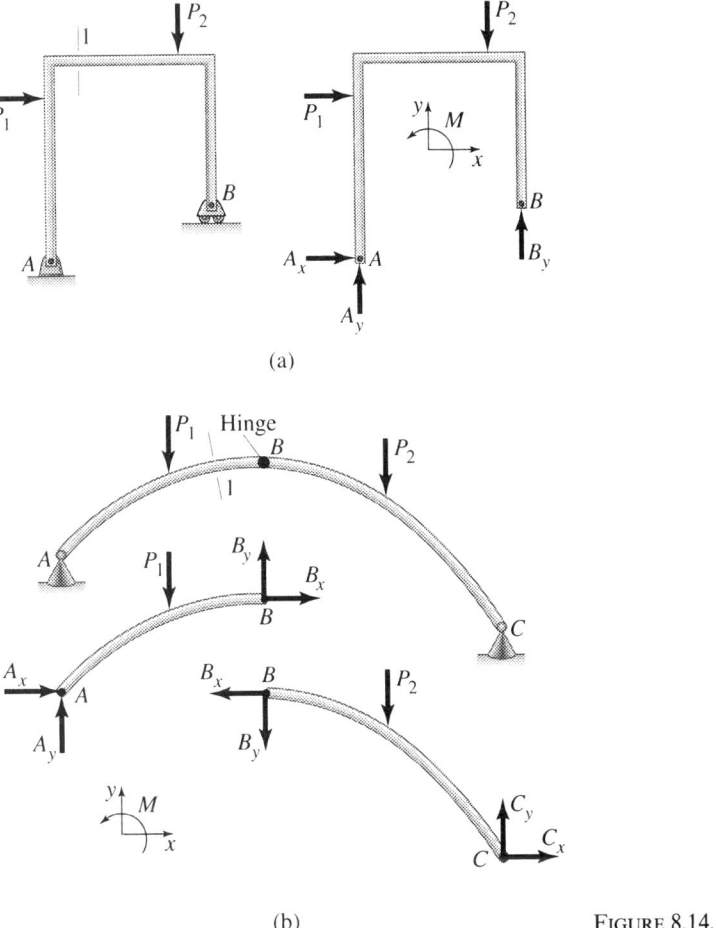

(a)

(b)   FIGURE 8.14.

A three-hinged arch is shown in Figure 8.14(b). The three hinges are located at A, B, and C. We can think of this arch as composed of two rigid bodies AB and BC connected by the hinge at B. The arch axis is curved to reduce shear and bending effects compared to axial forces.*

---

* It is shown in Example 8.21 (p. 516) that, if a parabolic arch is loaded uniformly over the horizontal, then, it will transfer loadings to the ground by axial compression, and both the shear and bending moment will vanish. Arches have been built since ancient times and, usually, these arches were fabricated of a material, such as stone, which could resist compression forces well but would crack when subjected to small tensile forces. The aqueduct at Segovia, Spain was built circa 100 A.D. It consists of one hundred double-tiered arches composed of granite blocks laid without lime or cement. It has survived for about 19 centuries because the internal forces are largely compressive rather than tensile.

Separate free bodies are drawn for the left part AB and right part BC of the arch, as shown in Figure 8.14(b). There are six unknown force components on these two free bodies: A_x, A_y, B_x, B_y, C_x, and C_y. A general procedure for the solution may be outlined as follows:

Free-body AB: $\sum M_A = 0$ $\Rightarrow$ one equation with B_x and B_y;
Free-body BC: $\sum M_C = 0$ $\Rightarrow$ a second equation with B_x and B_y.

These two equation may be solved for B_x and B_y.

Free-body AB: $\sum F_x = 0$ may be solved for A_x,
and $\sum F_y = 0$ may be solved for A_y.
Free-body BC: $\sum F_x = 0$ may be solved for C_x,
and $\sum F_y = 0$ may be solved for C_y.

In the special case when the hinges A and C are at the same elevation, it is simpler to consider the free-body diagram for the entire arch for which B_x and B_y will be internal forces. This procedure is outline below.

Free-body entire arch: $\sum M_C = 0$ may be solved for A_y,
and $\sum M_A = 0$ may be solved for C_y.
Free-body AB: $\sum M_B = 0$ may be solved for A_x.
Free-body BC: $\sum M_B = 0$ may be solved for C_x.

At this stage, we have written four equations to solve for the four reaction components at A and C. If the hinge forces at B are required, then,

Free-body AB: $\sum F_x = 0$ may be solved for B_x,
and $\sum F_y = 0$ may be solved for B_y.

Once the support reactions are known, we may wish to determine internal forces at any cross-section in a given frame. For example, we may wish to determine the internal axial force F_1, shear V_1, and moment M_1 at cross-section 1 in the frame of Figure 8.14(a) and in the arch of Figure 8.14(b). As was discussed in Sections 8.3 and 8.4, imaginary cutting planes normal to the members are used at the points of interest and free-body diagrams of one part of the frame are constructed. Such free-body diagrams are shown in Figure 8.15(a) for the frame and in Figure 8.15(b) for the arch. In both instances, left free-body diagrams are used, although right free-body diagrams would serve equally well. Using the coordinate system shown for each of the two free-body diagrams, we may write three equations of equilibrium for each, enabling us to find the three unknown quantities F_1, V_1, and M_1 in each case.

The following examples illustrate some of the concepts above.

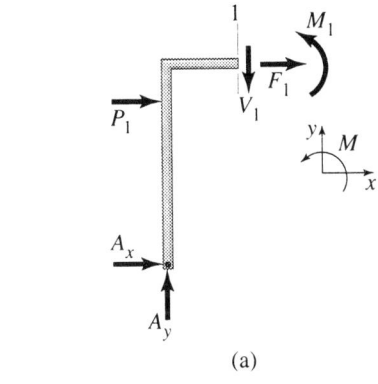

(a)

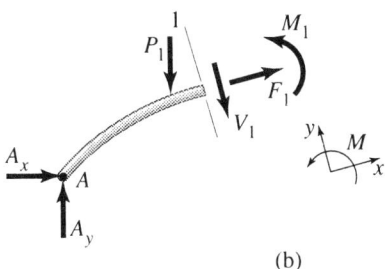

(b)   FIGURE 8.15.

■ **Example 8.19**

A rigid frame is shown in Figure E8.19(a). It is pinned at A and supported by a roller at D. Determine (a) the reactions at A and D, and (b) the axial force, shear and bending moment at cross-sections 1 and 2.

Solution

(a) The free-body diagram of the entire frame is shown in Figure E8.19(b) along with a convenient coordinate system.

$$\sum M_A = 0, \quad -4PL - 2PL - D_x(2L) = 0,$$

$$D_x = -3P, \qquad\qquad \text{ANS.}$$

$$\sum F_x = 0, \quad A_x + 4P + D_x = 0,$$

$$A_x = -P, \qquad\qquad \text{ANS.}$$

and

$$\sum F_y = 0, \quad A_y - 2P = 0,$$

$$A_y = 2P. \qquad\qquad \text{ANS.}$$

(b) The free-body diagram of a segment of the frame containing cross-section 1 is shown in Figure E8.19(c). Thus,

$$\sum F_y = 0, \quad 2P + F_1 = 0,$$

$$F_1 = -2P, \qquad\qquad \text{ANS.}$$

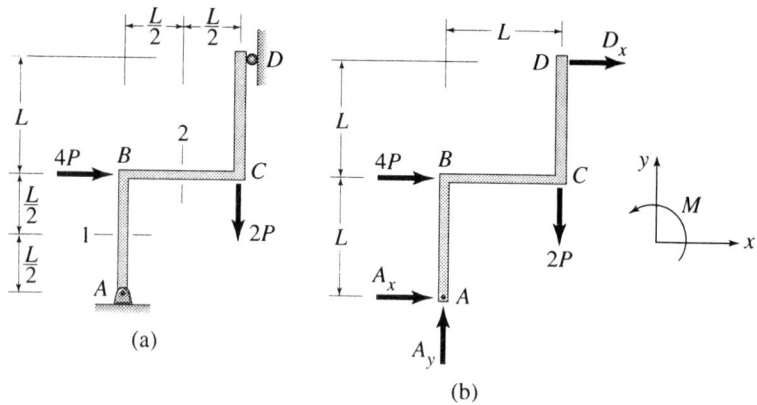

(a)

(b)

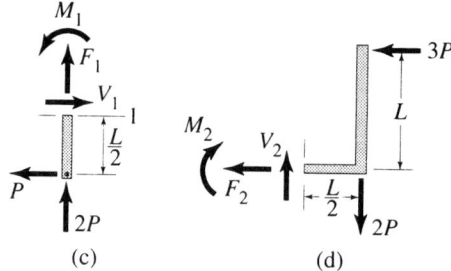

FIGURE E8.19.

(c)   (d)

$$\sum F_x = 0, \quad -P + V_1 = 0,$$

$$V_1 = P. \qquad \text{ANS.}$$

and

$$\sum M_1 = 0, \quad M_1 - P(L/2) = 0,$$

$$M_1 = PL/2. \qquad \text{ANS.}$$

The free-body diagram of a segment of the frame containing cross-section 2 is shown in Figure E8.19(d). Thus,

$$\sum F_x = 0, \quad -F_2 - 3P = 0,$$

$$F_2 = -3P, \qquad \text{ANS.}$$

$$\sum F_y = 0, \quad V_2 - 2P = 0,$$

$$V_2 = 2P, \qquad \text{ANS.}$$

and

$$\sum M_2 = 0, \quad -M_2 - 2P\left(\frac{L}{2}\right) + 3P(L) = 0,$$

$$M_2 = 2PL. \qquad \text{ANS.}$$

■ **Example 8.20**

A loaded frame is shown in Figure E8.20(a). It is pinned at A and supported by a frictionless roller at E. Determine (a) the reactions at A and E, and (b) the axial force, shear, and moment at cross-section 1 just to the right of B and at cross-section 2 just to the right of C.

Solution

(a) The free-body diagram for the entire frame is shown in Figure E8.20(b) along with a convenient coordinate system. Thus,

$$\sum F_x = 0, \quad wL + A_x = 0,$$

$$A_x = -wL, \qquad \text{ANS.}$$

$$\sum M_E = 0, \quad 1.5wL(L) - wL\left(\frac{L}{2}\right) - A_y(2L) = 0,$$

$$A_y = 0.5wL, \qquad \text{ANS.}$$

and

$$\sum F_y = 0, \quad A_y - 1.5wL + E_y = 0,$$

$$E_y = wL. \qquad \text{ANS.}$$

(b) The free-body diagram for a segment of the frame containing cross-section 1 is shown in Figure E8.20(c). Thus,

$$\sum F_x = 0, \quad F_1 + wL - wL = 0,$$

$$F_1 = 0, \qquad \text{ANS.}$$

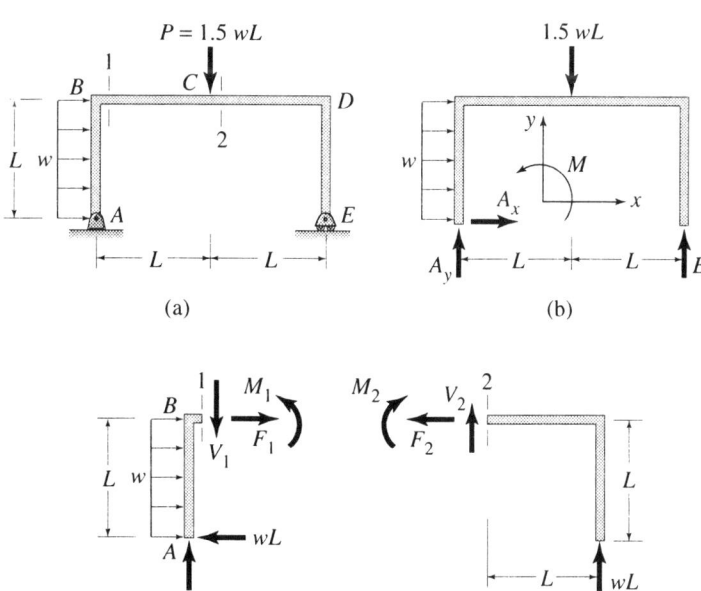

FIGURE E8.20.

$$\sum F_y = 0, \quad 0.5wL - V_1 = 0,$$

$$V_1 = 0.5wL, \qquad \text{ANS.}$$

and

$$\sum M_1 = 0, \quad wL(L/2) + M_1 - wL(L) = 0,$$

$$M_1 = wL^2/2 \qquad \text{ANS.}$$

The free-body diagram for a segment of the frame containing cross-section 2 is shown in Figure E8.20(d). Thus,

$$\sum F_x = 0, \quad F_2 = 0, \qquad \text{ANS.}$$

$$\sum F_y = 0, \quad V_2 + wL = 0,$$

$$V_2 = -wL, \qquad \text{ANS.}$$

and

$$\sum M_2 = 0, \quad wL(L) - M_2 = 0,$$

$$M_2 = wL^2. \qquad \text{ANS.}$$

■ Example 8.21

A three-hinged arch ABC is shown in Figure E8.21(a). It is uniformly load over the horizontal and has a parabolic shape given by the relationship $y = -\left(\dfrac{4h}{L}\right)x^2$. (a) In terms of w, h, and L, determine the reaction components at A, B, and C. (b) Show that both the bending moment and the shear force vanish at any cross-section in the arch.

Solution

(a) The free-body diagram of the entire arch is shown in Figure E8.21(b). Thus,

$$\sum M_C = 0, \quad wL\left(\frac{L}{2}\right) - A_y(L) = 0,$$

$$A_y = w\left(\frac{L}{2}\right), \qquad \text{ANS.}$$

$$\sum F_y = 0, \quad A_y + C_y - wL = 0,$$

$$C_y = \frac{wL}{2}, \qquad \text{ANS.}$$

and

$$\sum F_x = 0, \quad A_x + C_x = 0,$$

$$A_x = -C_x. \qquad \text{(a)}$$

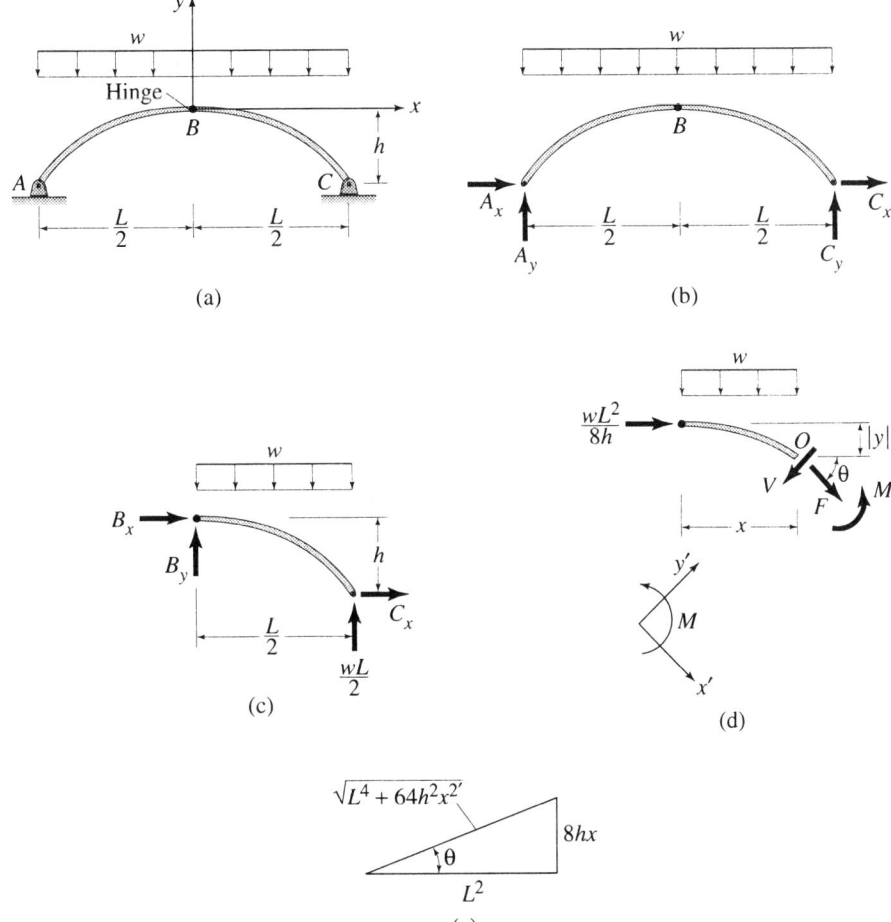

FIGURE E8.21.

The free-body diagram of segment BC of the arch is shown in Figure E8.21(c). Hence

$$\sum M_B = 0, \quad C_x h - w\frac{L}{2}\left(\frac{L}{4}\right) + \frac{wL}{2}\left(\frac{L}{2}\right) = 0,$$

$$C_x = -\frac{wL^2}{8h}, \qquad \text{ANS.}$$

$$\sum F_x = 0, \quad B_x + C_x = 0,$$

$$B_x = \frac{wL^2}{8h}, \qquad \text{ANS.}$$

and

$$\sum F_y = 0, \quad \frac{wL}{2} + B_y - w\left(\frac{L}{2}\right) = 0,$$

$$B_y = 0. \qquad\qquad \text{ANS.}$$

From Eq. (a), we conclude that

$$A_x = \frac{wL^2}{8h}. \qquad\qquad \text{ANS.}$$

(b) The free-body diagram of a segment BO of length x measured horizontally is shown in Figure E8.21(d). A new x'-y' coordinate system is also shown. Thus,

$$\sum M_O = 0, \quad M + wx\left(\frac{x}{2}\right) - \frac{wL^2}{8h}(|y|) = 0$$

Substituting for $|y|$ from the given parabolic function and solving for M,

$$M = \frac{wL^2}{8h}\left(\frac{4h}{L^2}\right)x^2 - \frac{wx^2}{2}$$

$$= \frac{wx^2}{2} - \frac{wx^2}{2} = 0, \qquad\qquad \text{ANS.}$$

and

$$\sum F_{y'} = 0, \quad \frac{wL^2}{8h}\sin\theta - wx\cos\theta - V = 0,$$

$$V = \frac{wL^2}{8h}\sin\theta - wx\cos\theta. \qquad\qquad \text{(b)}$$

The quantities $\sin\theta$ and $\cos\theta$ are determined from the given function $y = -\left(\frac{4h}{L^2}\right)x^2$. Thus,

$$\tan\theta = \frac{dy}{dx} = \left|\frac{8hx}{L^2}\right|$$

This function is interpreted as shown in Figure E8.21(e) from which $\sin\theta = \dfrac{8hx}{\sqrt{L^4 + 64h^2x^2}}$ and $\cos\theta = \dfrac{L^2}{\sqrt{L^4 + 64h^2x^2}}$. Substituting these functions in Eq. (b) and simplifying,

$$V = \frac{wL^2}{8h}\left(\frac{8hx}{\sqrt{L^4 + 64h^2x^2}}\right) - wx\left(\frac{L^2}{\sqrt{L^4 + 64h^2x^2}}\right)$$

$$= \frac{wL^2x}{\sqrt{L^4 + 64h^2x^2}} - \frac{wL^2x}{\sqrt{L^4 + 64h^2x^2}} = 0. \qquad\qquad \text{ANS.}$$

Thus, a parabolic arch under a uniform load over the horizontal is subjected only to an axial force F because both the shear V and the moment M vanish at any cross-section. It can be shown that the axial force F is a compressive force which may be expressed in the form

$$F = -\left(\frac{w}{8h}\right)\sqrt{L^4 + 64h^2x^2}$$

As an exercise, the student can confirm this equation by applying the condition $\sum F_{x'} = 0$.

■ **Example 8.22**

Two rigid-bodies AB and BC are fastened together with the hinge at B to form the frame shown in Figure E8.22(a). Regarding w and L as known quantities, determine (a) the reactions at A and C and the

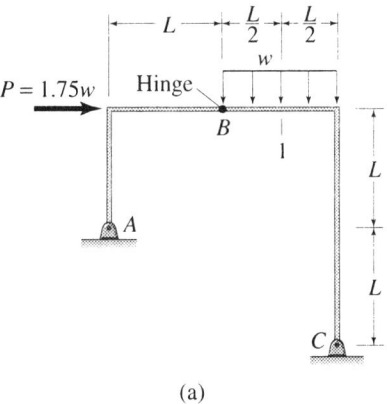

(a)

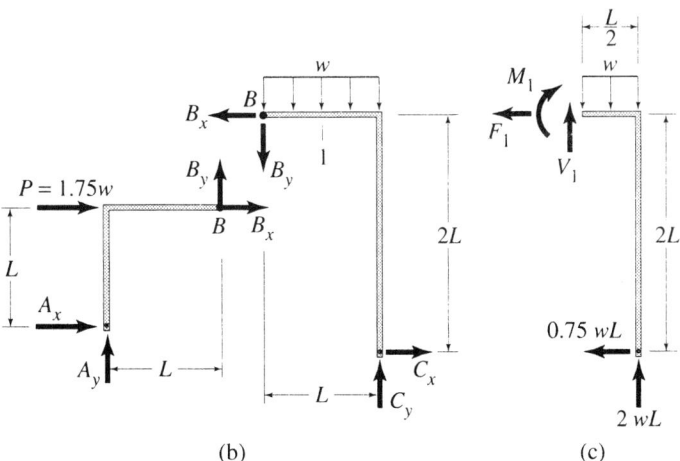

FIGURE E8.22.

(b)

(c)

hinge forces at B, and (b) the axial force, shear, and moment at cross-section 1.

Solution   (a) Refer to free-body diagrams for AB and BC, and write the equations of equilibrium.

AB: $\sum M_A = 0$, $B_y L - B_x L - 1.75wL(L) = 0$,

$$B_y - B_x = 1.75wL. \qquad \text{(a)}$$

BC: $\sum M_C = 0$, $B_y L + B_x(2L) + wL(L/2) = 0$,

$$B_y + B_x = -0.5wL. \qquad \text{(b)}$$

Solving Eqs. (a) and (b) simultaneously,

$$B_x = -0.75wL \qquad \text{ANS.}$$

and

$$B_y = wL. \qquad \text{ANS.}$$

AB: $\sum F_x = 0$, $1.75wL + A_x + B_x = 0$,

$$A_x = -wL, \qquad \text{ANS.}$$

and

$$\sum F_y = 0, \quad A_y + B_y = 0,$$
$$A_y = -wL. \qquad \text{ANS.}$$

BC: $\sum F_x = 0$, $C_x - B_x = 0$,

$$C_x = -0.75wL, \qquad \text{ANS.}$$

and

$$\sum F_y = 0, \quad C_y - B_y - wL = 0,$$
$$C_y = 2wL \qquad \text{ANS.}$$

(b) Refer to the free-body diagram for cross-section 1 shown in Figure E8.22(c). Thus,

$$\sum F_x = 0, \quad -0.75wL - F_1 = 0,$$
$$F_1 = -0.75wL, \qquad \text{ANS.}$$
$$\sum F_y = 0, \quad 2wL + V_1 - 0.5wL = 0,$$
$$V_1 = 1.500wL, \qquad \text{ANS.}$$

and

$$\sum M_1 = 0, \quad 2wL\left(\frac{L}{2}\right) - 0.75wL(2L) - w\left(\frac{L}{2}\right)\left(\frac{L}{4}\right) - M_1 = 0,$$

$$M_1 = -0.625wL^2. \qquad\qquad \text{ANS.}$$

Problems

8.96   A frame is loaded and supported as in Figure P8.96. Find the reactions at A and B. In terms of w and L, determine the axial force, shear, and moment at cross-sections 1 and 2.

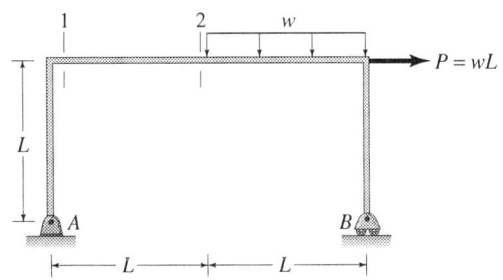

FIGURE P8.96.

8.97   Find the reactions at A and B acting on the frame of Figure P8.97. Determine the axial force, shear, and moment at cross-section 1 in terms of w and L.

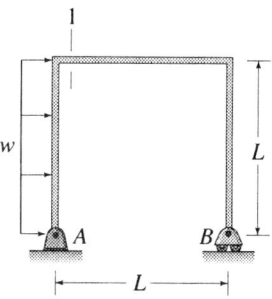

FIGURE P8.97.

8.98   Refer to the frame shown in Figure P8.98, and determine the reactions at A and B. Find the axial force, shear, and moment at cross-sections 1 and 2 in terms of P and L.

8.99   A frame is shown in Figure P8.99. Find the reactions at supports A and B. Determine the axial force, shear, and moment at cross-sections 1 and 2 in terms of P and L.

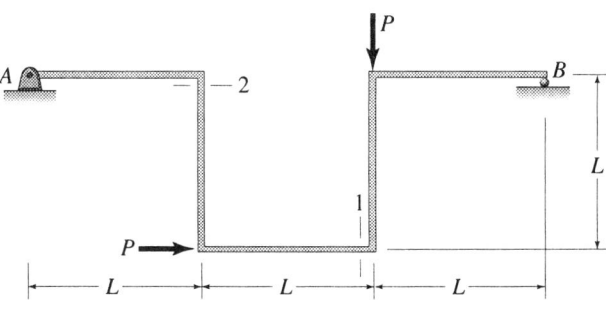

FIGURE P8.98.

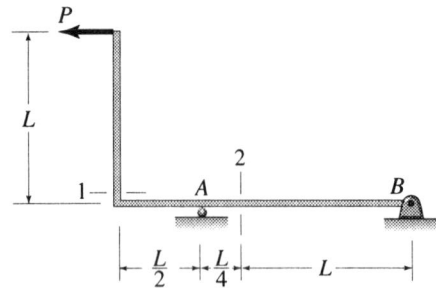

FIGURE P8.99.

8.100 Refer to the frame shown in Figure P8.100, and find the reactions at A and B. Determine the axial force, shear, and moment at cross-sections 1 and 2 in terms of P and L.

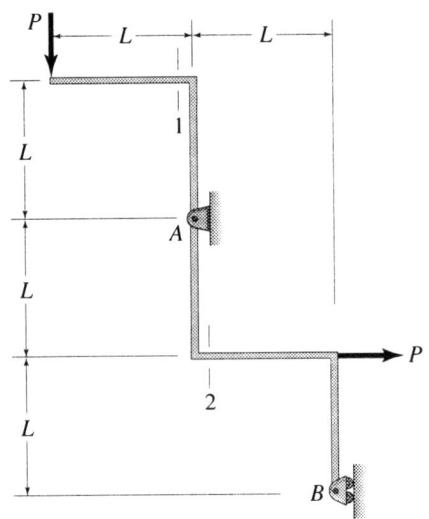

FIGURE P8.101.

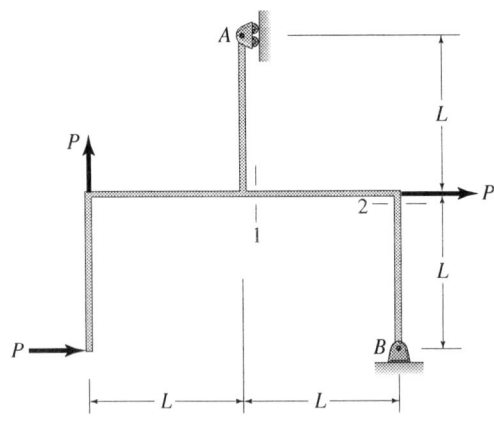

FIGURE P8.100.

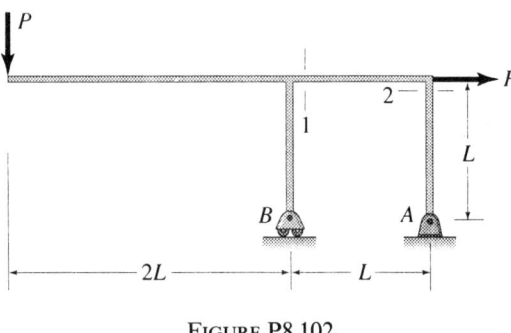

FIGURE P8.102.

8.101 Determine the reactions at pin A and roller B which act on the frame of Figure P8.101. Find the axial force, shear and moment at cross-sections 1 and 2 in terms of P and L.

8.102 Refer to the frame of Figure P8.102, and find the reactions at A and B. Find the axial force, shear, and moment at cross-sections 1 and 2 in terms of P and L.

8.103 Determine the axial force, shear, and moment at cross-sections 1 and 2 of the frame shown in Figure P8.103 in terms of P and L.

8.104 Refer to the frame of Figure P8.104, and find the axial force, shear, and moment at cross-sections 1 and 2 in terms of w and L.

8.105 Refer to the frame shown in Figure P8.105, and find the axial force, shear, and moment at cross-sections 1 and 2 in terms of w and L.

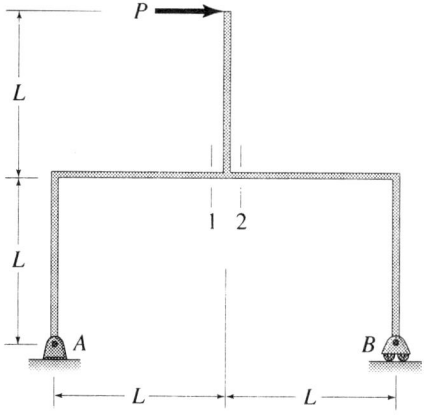

FIGURE P8.103.

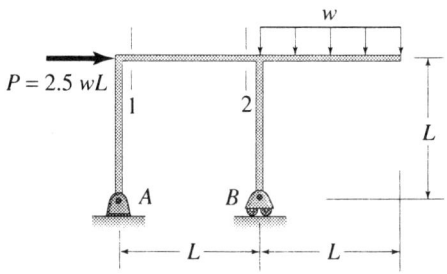

FIGURE P8.104.

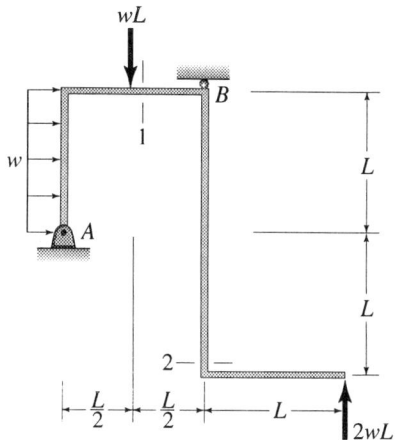

FIGURE P8.105.

8.106 Determine the reactions at A and B for the frame shown in Figure P8.106. Then, find the axial force, shear, and moment at cross-sections 1 and 2 in terms of P and b.

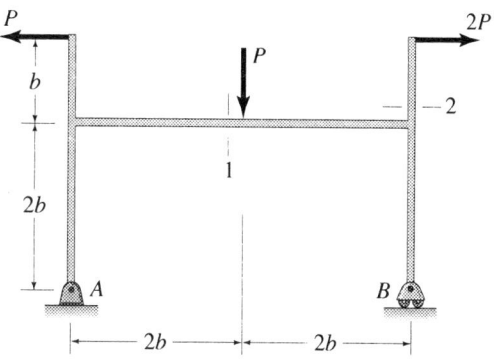

FIGURE P8.106.

8.107 Express the axial force, shear, and moment at cross-sections 1 and 2 in terms of P and L for the frame shown in Figure P8.107.

8.108 A uniformly loaded circular arch of radius R has hinges at A, B, and C as shown in Figure P8.108. Prove that the

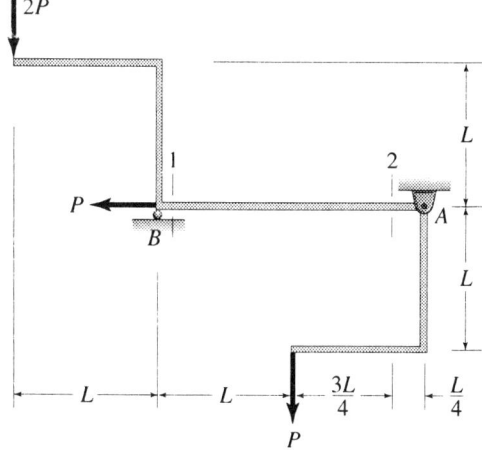

FIGURE P8.107.

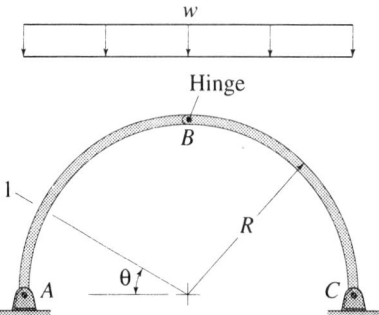

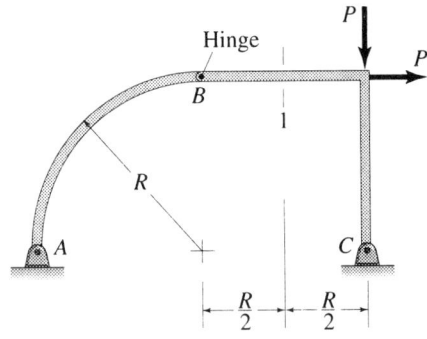

FIGURE P8.108.

FIGURE P8.111.

reactions are $wR/2$ right at A, wR up at A, $wR/2$ left at C, and wR up at C. Find the forces exerted by pin B on the left half of this arch.

8.109 Refer to Figure P8.108, and express the shear V_1, the axial force F_1, and the moment M_1 as functions of w, R, and θ. Use a left free-body diagram.

8.110 Refer to the circular arch of radius R in Figure P8.110 which is loaded by two radially directed forces of magnitude P. Determine the component reactions at A and C in terms of P and θ. Check the limiting cases for $\theta = 0°$ and $\theta = 90°$.

and moment at cross-section 1 in terms of P and R.

8.112 A uniformly loaded parabolic arch is shown in Figure P8.112. (a) Write the equation of this three-hinged arch using an origin at A as indicated. (b) Draw appropriate free-body diagrams, and determine the reactions at A and C. (c) Does the hinge at B transmit a shear force? Prove your answer.

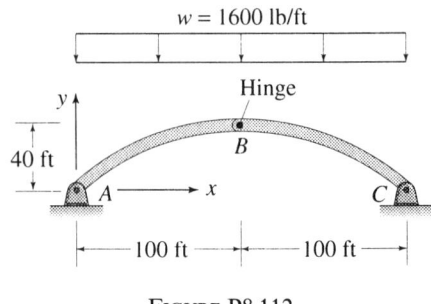

FIGURE P8.112.

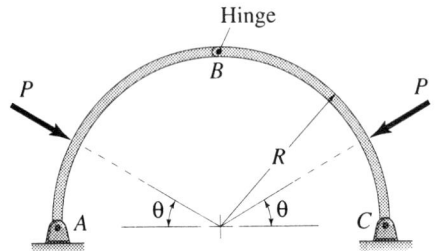

FIGURE P8.110.

8.111 The structure shown in Figure P8.111 has hinges at A, B, and C. Determine the reaction components at A and C in terms of P. Find the axial force, shear,

8.113 Refer to Figure P8.112, and determine the reactions at A and C. Draw a free-body diagram of the left half of segment AB which has a horizontal projection of 50 ft, and show that the bending moment at the cut cross-section will vanish.

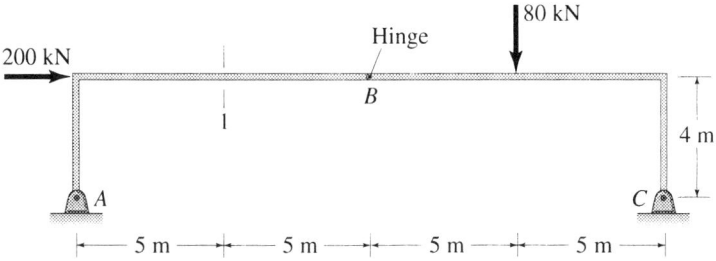

FIGURE P8.114.

8.114 The frame ABC of Figure P8.114 consists of two rigid bodies AB and BC which are connected with hinge B. Determine the axial force, shear, and moment at cross-section 1.

8.115 The frame of Figure P8.115 consists of two rigid bodies AB and AC which are connected with hinge A. Regard w and L as known quantities, and determine the axial force, shear, and moment at cross-section 1.

8.116 Determine the pin reaction components at A and C and the axial force, shear, and bending moment at cross-section 1 of the frame shown in Figure P8.116 which consists of rigid bodies AB and BC joined at hinge B.

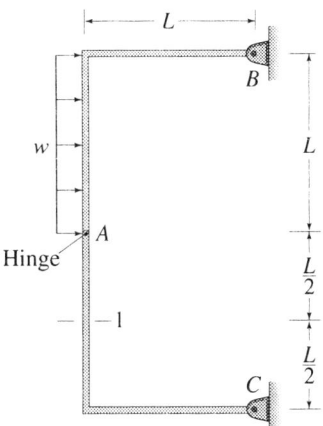

FIGURE P8.115.

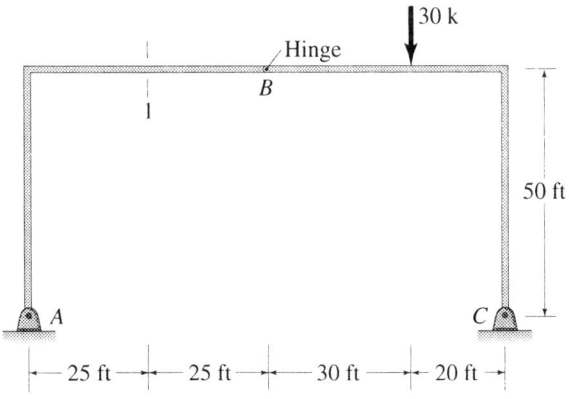

FIGURE P8.116.

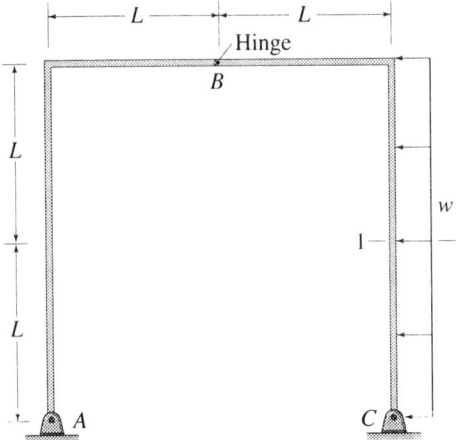

FIGURE P8.117.

8.117 Two rigid bodies AB and BC are joined by a pin at B to form the frame shown in Figure P8.117. Regard w and L as known quantities, and determine the pin reaction components at A and C. Find the axial force, shear, and bending moment at cross section 1.

8.118 A symmetric frame, which is loaded symmetrically, is shown in Figure P8.118. Determine the pin reaction components at A and C. Express the axial force, shear, and moment at cross-section 1 as functions of P and L.

8.119 Determine the axial force, shear, and bending moment at cross-section 1 of the frame shown in Figure P8.119 which

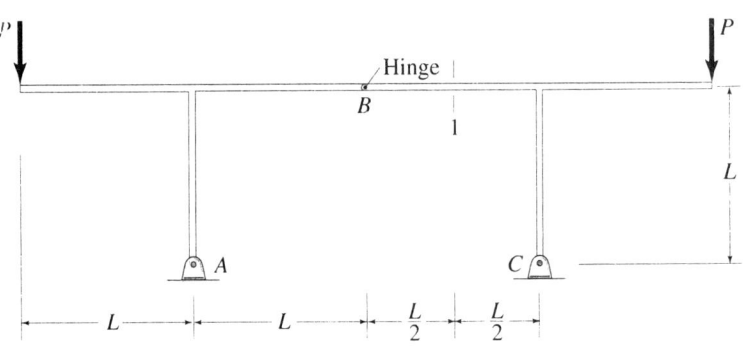

FIGURE P8.118.

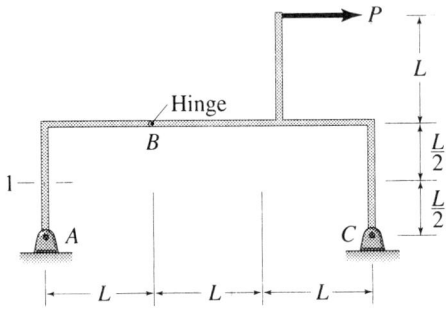

FIGURE P8.119.

consists of two parts AB and BC joined at hinge B. Express your answers in terms of P and L.

8.120 The frame shown in Figure P8.120 consists of two rigid bodies AB and BC joined by the pin at B. Regard P and L as given quantities. Determine the reactions at pins A and C in terms of P. Then find the axial force, shear, and bending moment at cross-section 1.

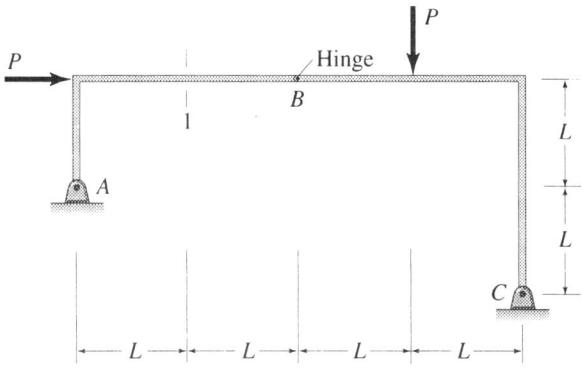

FIGURE P8.120.

8.121 Determine the pin reaction components at A and C acting on the frame shown in Figure P8.121. Find the axial force, shear, and bending moment at cross-section 1 in terms of the known quantities P and L. Consider that the horizontal member of length $2L$ is reduced to a length L and pin support C moved a distance L to the left. Attempt to determine the reactions acting on this revised frame in which all members are of length L. What's wrong?

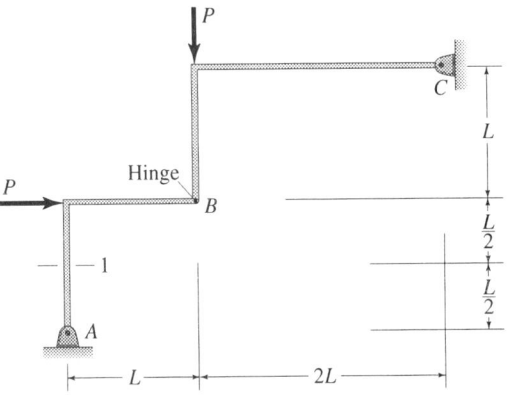

FIGURE P8.121.

8.12
Internal Force Diagrams for Two-Dimensional Frames

In Sections 8.3 and 8.7, we learned how to construct internal force diagrams for individual members. In this section we extend this knowledge to the construction of internal force diagrams for members of frames. A comprehensive analysis for all external and internal forces is typical of engineering practice because safe and economical designs are based on such information.

A given frame, such as that shown in Figure 8.16(a), is analyzed to determine the support reactions at C and D and the components of the forces at points of connection between members at A and B. This analysis, of course, requires dismembering the frame into components, as shown in Figure 8.16(b), which represents the free-body diagrams of the three components of the frame. Once the unknown forces on these free-body diagrams are known, the axial force F, shear V, and moment M diagrams can be constructed. These diagrams are shown for the

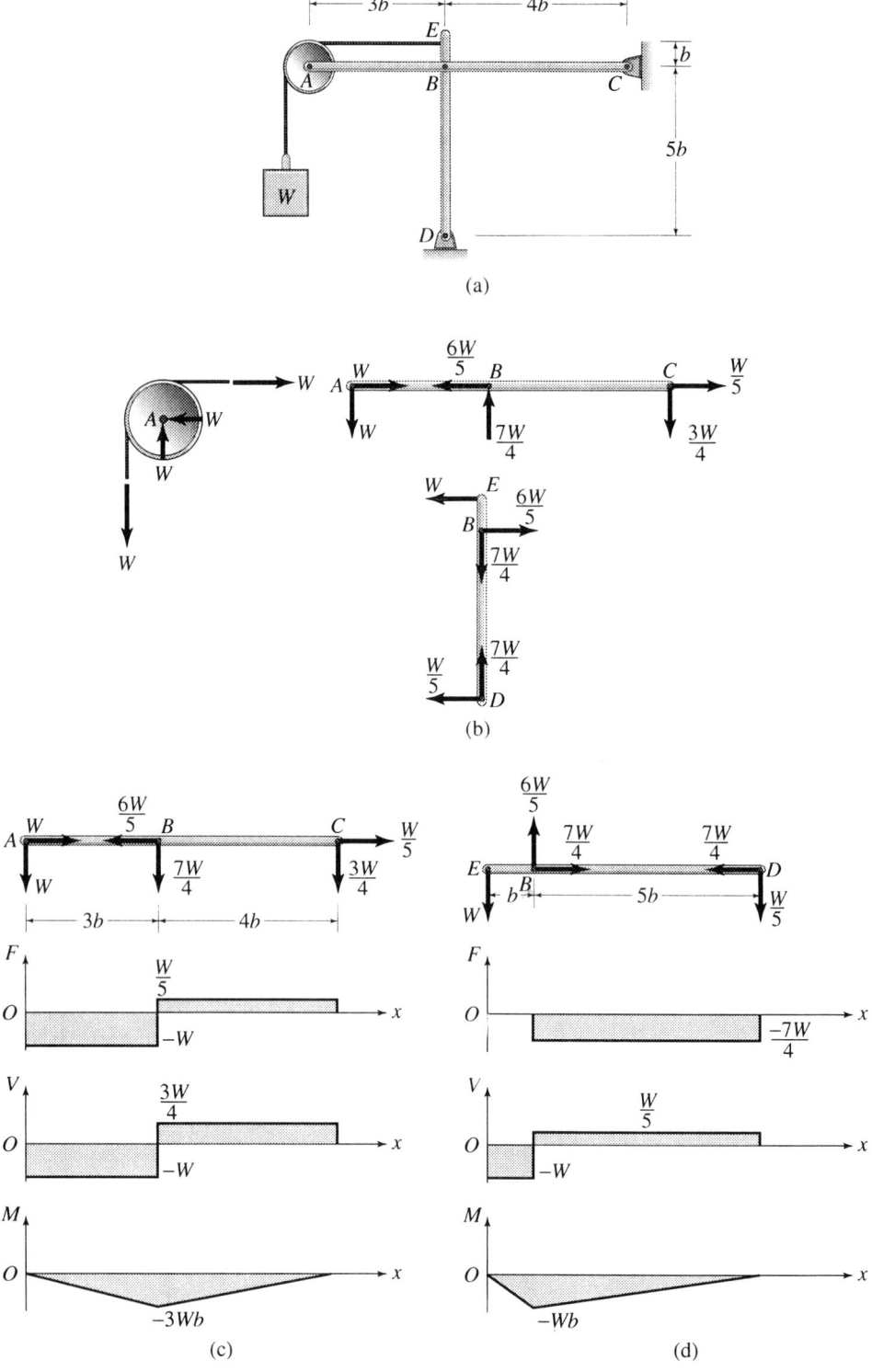

FIGURE 8.16.

horizontal member ABC in Figure 8.16(c). For inclined or vertical members, we adopt the convention of rotating them into a horizontal position before constructing their internal force diagrams. The sense of rotation, cw or ccw, is immaterial as long as the action lines of all forces are rotated through the same angle and in the same sense. Care should be taken to label points correctly and to show the correct geometry. Vertical member EBD and the forces acting on it were rotated 90° ccw before the internal force diagrams were constructed, as shown in Figure 8.16(d).

The following example illustrates some of the concepts discussed above.

■ **Example 8.23**

A two-dimensional frame is shown in Figure E8.23(a). Determine all unknown force components of this system, and construct internal force

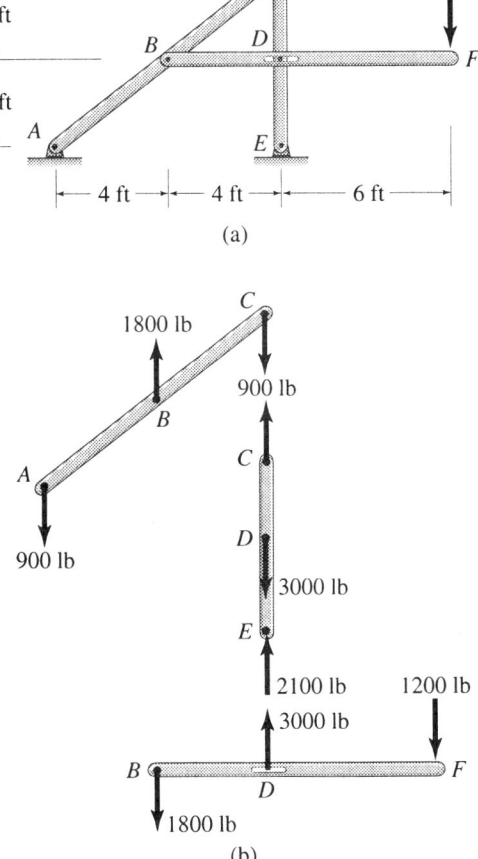

Figure E8.23.

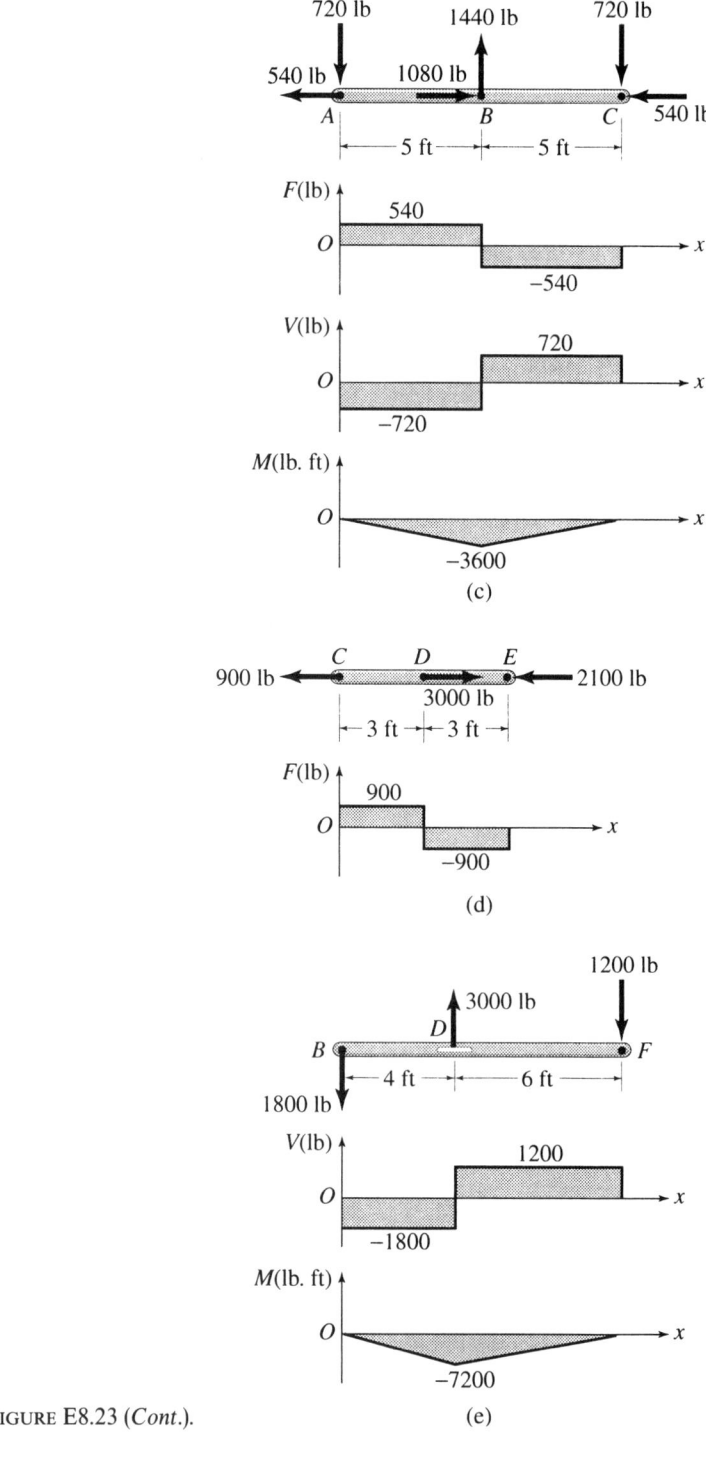

diagrams for the three component members, ABC, CDE, and BDF. A horizontal slot is cut in member BDF and a pin at D is fastened to member CDE. Assume frictionless conditions.

Solution

The frame is dismembered, and the conditions of equilibrium are applied to determine the unknown force components at A, B, C, D, and E. The results of such analysis are summarized in the free-body diagrams shown in Figure E8.23(b).

Internal force diagrams for member ABC are shown in Figure E8.23(c). Before constructing these diagrams, the member was rotated cw into a horizontal position, and the forces acting at A, B, and C were resolved into components parallel and perpendicular to the member. At point A, a downward force of 900 lb has a component of $\frac{3}{5}(900) = 540$ lb along this member and a component of $\frac{4}{5}(900) = 720$ lb perpendicular to this member, as shown. Forces at B and C were, similarly, resolved into components.

Vertical member CDE was rotated ccw through 90° into a horizontal position before constructing the axial force diagram of Figure E8.23(d). This member is free of shear and bending.

The internal force diagrams for horizontal member BDF are shown in Figure E8.23(e). This component is free of axial force.

■

Problems

8.122 Construct internal force diagrams for members AB and BC of the frame shown in Figure P8.122.

8.123 Refer to the frame of Figure P8.123, and draw internal force diagrams for member ABC.

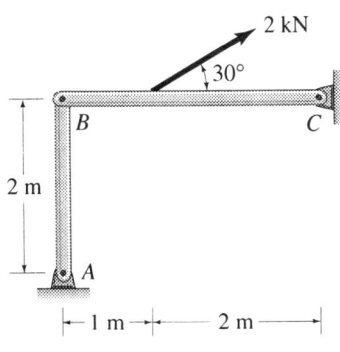

FIGURE P8.122.

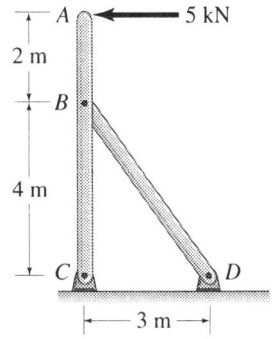

FIGURE P8.123.

8.124 Draw internal force diagrams for member AB of the frame of Figure P8.124.

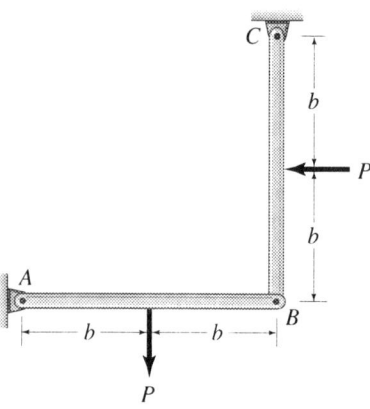

FIGURE P8.124.

8.125 Draw internal force diagrams for member ABC of Figure P8.125.

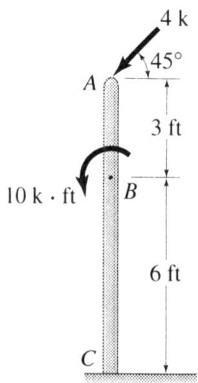

FIGURE P8.125.

8.126 Refer to the frame shown in Figure P8.126, and construct internal force diagrams for member ABC.

8.127 Draw internal force diagrams for member AB of the frame shown in Figure P8.127.

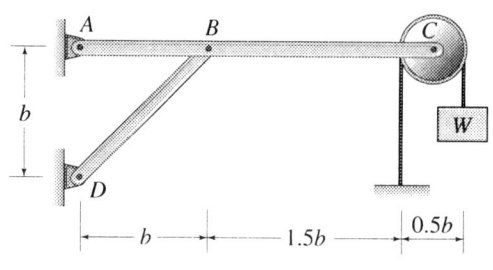

FIGURE P8.126.

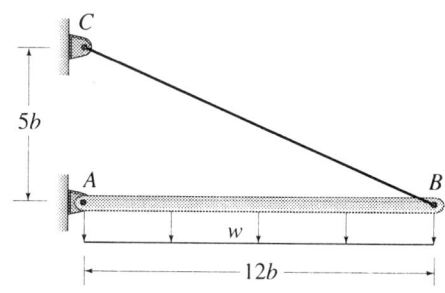

FIGURE P8.127.

8.128 Solve problem 8.127 for $w = 200$ lb/ft and $b = 1$ ft.

8.129 Draw internal force diagrams for member ABC of the frame shown in Figure P8.129.

8.130 Refer to the frame of problem 8.129 and construct the internal force diagrams for member EBD.

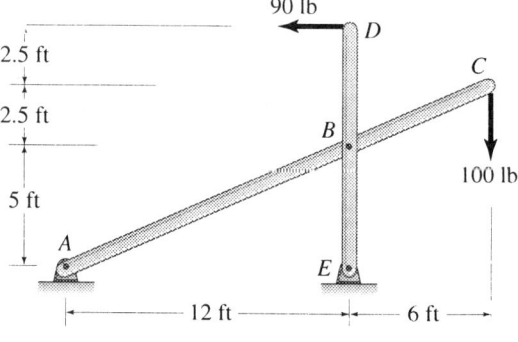

FIGURE P8.129.

8.131 A frame consisting of two members AB and BC is shown in Figure P8.131. Draw internal force diagrams for member BC. Let $w = 600$ lb/ft and $b = 4$ ft.

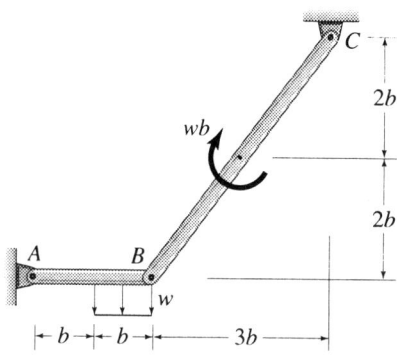

FIGURE P8.131.

8.132 Draw internal force diagrams for members ABC and CED of the frame shown in Figure P8.132.

8.133 Draw internal force diagrams for members AB and BC of the frame shown in Figure P8.133. Let $w = 1.5$ kN/m and $b = 10$ m.

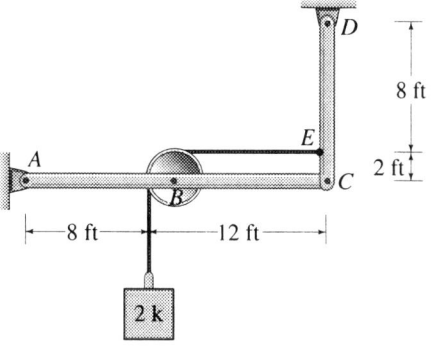

FIGURE P8.132.

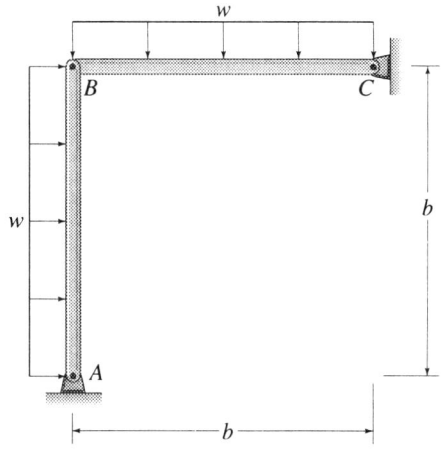

FIGURE P8.133.

Review Problems

8.134 Construct the axial force and torque diagrams for the member shown in Figure P8.134.

8.135 Find the shear and moment at cross-section 1 in the beam ABCD shown in Figure P8.135.

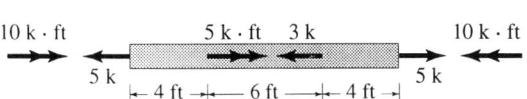

FIGURE P8.134.

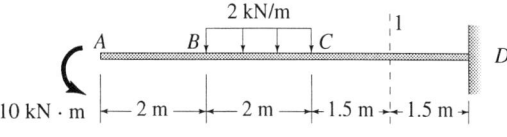

FIGURE P8.135.

8.136 Refer to the beam ABCD in Figure P8.135, establish an origin at A measuring x positive to the right, and write the equations for the shear and moment in segment BC.

8.137 Consider the beam ABCD shown in Figure P8.137, and determine the shear and moment at cross sections 1 and 2.

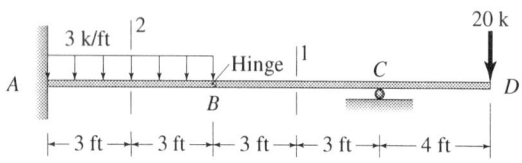

FIGURE P8.137.

8.138 Refer to beam ABCD in Figure P8.137, locate an origin at A measuring x positive to the right, and write the shear and moment equations for segment AB. Specialize these equations for cross-section 2, and compare them with the values obtained in Problem 8.137.

8.139 An empty tank weighing 0.6 kN/m is supported on the bed of a trailer truck, as shown in Figure P8.139. Determine the shear and moment at cross-section 1 of the tank.

8.140 Refer to the tank of Problem 8.139, locate an origin at its left end A measuring x positive to the right, and develop the

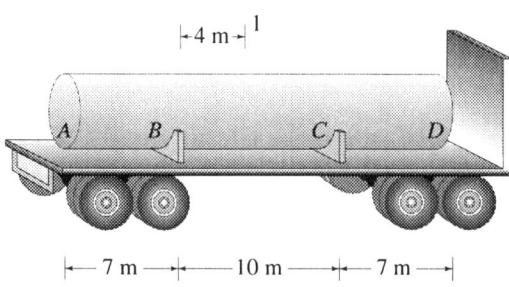

FIGURE P8.139.

shear and moment equations for segments AB, BC, and CD.

8.141 Member ABC is subjected to the distributed axial loads and torques as well as the two 15-k concentrated loads, as shown in Figure P8.141. Write the axial force and torque equations for segments AB and BC.

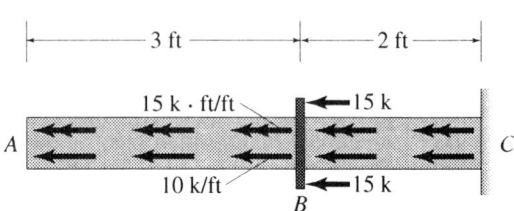

FIGURE P8.141.

8.142 Construct the shear and moment diagrams for the beam shown in Figure P8.142 by writing and plotting the shear and moment functions.

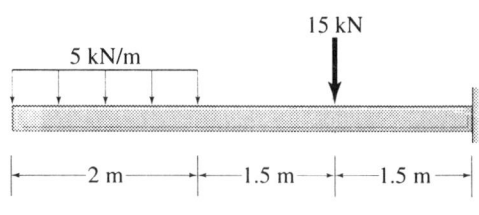

FIGURE P8.142.

8.143 By using Eqs. (8.9) and (8.10), construct the shear and moment diagrams for beam ABCD shown in Figure P8.143.

8.144 By using Eqs. (8.9) and (8.10), construct the shear and moment diagrams for beam AB, shown in Figure P8.144.

8.145 The main cable and its hangers, shown in Figure P8.145, have been dimensioned so that, when the main cable is tensioned, the force that develops in each hanger is 5 k. Determine the jacking force F that

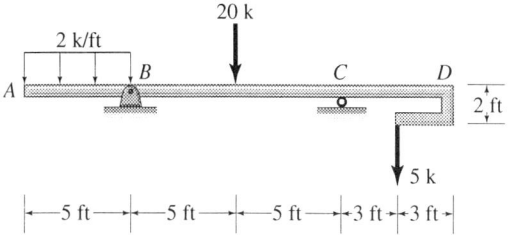

FIGURE P8.143.

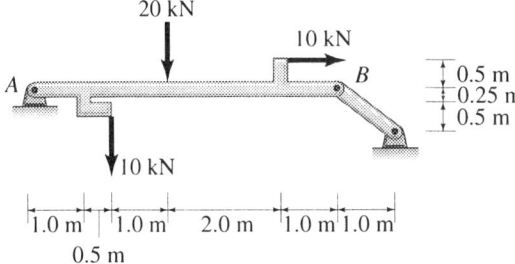

FIGURE P8.144.

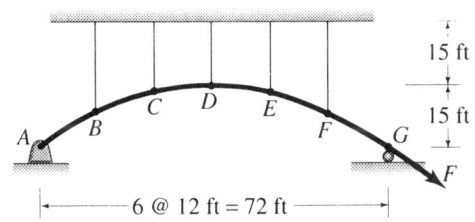

FIGURE P8.145.

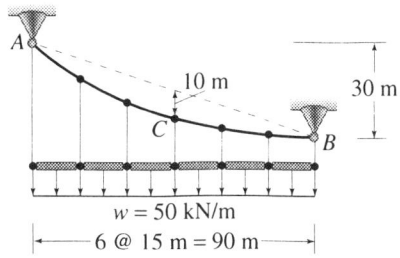

FIGURE P8.146.

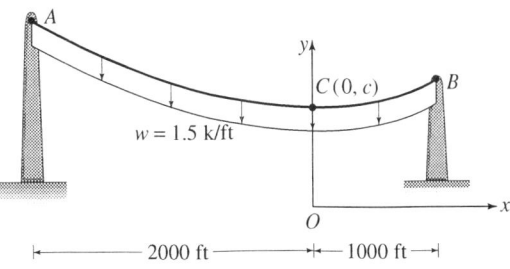

FIGURE P8.147.

must be applied at G to produce the needed hanger forces.

8.146 The main cable of a suspension bridge is shown in Figure P8.146. Assume that the lower end of the hangers serve as simple supports for the suspended beams. Determine (a) the maximum tension in the main cable and (b) the support reactions at A and B.

8.147 During erection of the main span of a suspension bridge, a cable weighing 1.5k/ft is supported on towers at A and

B, as shown in Figure P8.147. If $c = 3000$ ft, determine the maximum tension in the cable.

8.148 Refer to the frame shown in Figure P8.148, and determine the internal forces at cross-section 1.

8.149 Refer to the frame shown in Figure P8.148, and find the internal forces at cross-section 2.

8.150 A frame is loaded as shown in Figure P8.150. Determine the internal forces at cross-section 1.

8.151 Refer to the frame shown in Figure P8.150, and find the internal forces at cross-section 2.

8.152 Construct the internal force diagrams for member BC of the frame shown in Figure P8.152.

8.153 Construct the internal force diagrams for member AB of the frame shown in Figure P8.152.

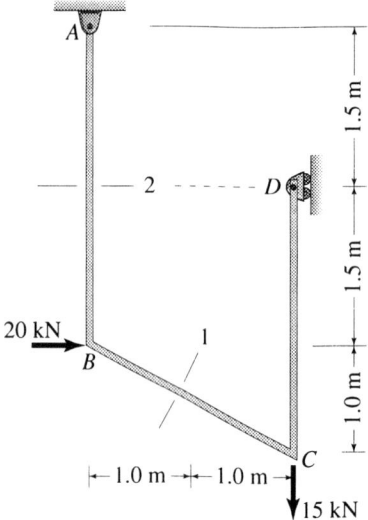

FIGURE P8.148.

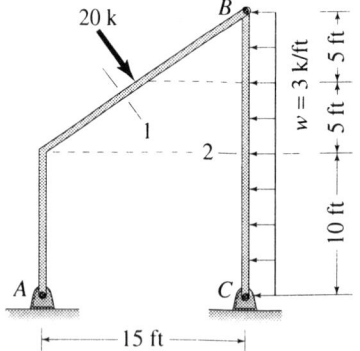

FIGURE P8.150.

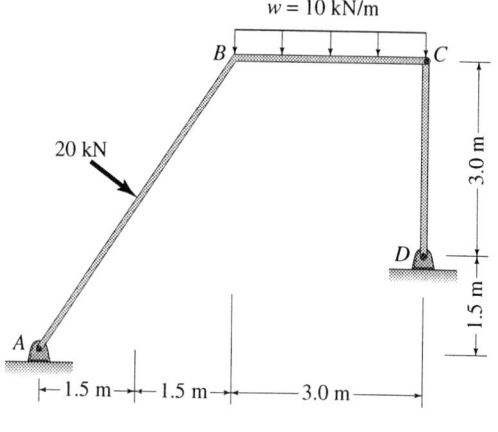

FIGURE P8.152.

9

Friction

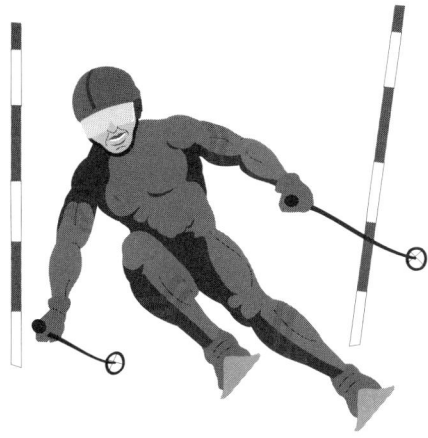

If it weren't for friction, it would be impossible to walk ... let alone ski!

The role of the engineer, as an innovator and practitioner, has not changed throughout the ages. The fundamental role has always been to make practical use of scientific discovery. One of the most dramatic developments ever made by humankind was the introduction of the wheel by the Sumerians to reduce friction, more than 5000 years ago.

The six fundamental types of machines were first recognized by the ancient Greeks more than 2500 years ago. A very interesting fact is that the most complex machines known to man today can be built from combinations of many of these six fundamental machines. The intricate machines in use today, from the space shuttle to bullet trains, reflect the immense techological progress made by the engineering profession.

Derivation and use of the equations for Vee and flat belts in Section 9.5 is a fascinating part of statics. By considering a differential element of a belt, we can write and solve a differential equation. The equations of equilibrium and the friction equation, again, enable us to derive the differential equation. Careful study of this chapter will pay dividends because the method of writing and solving a differential equation for a basic element will be repeated many times in other engineering and scientific courses of your curriculum. Pivot and collar bearings are analyzed in Section 9.6, and journal bearings are analyzed in Section 9.7.

The student is encouraged to understand the new concepts and to avoid superficial solutions based upon memorization. Draw free-body diagrams similar to those shown in the 22 examples. Then, carefully write and solve the equations. Check those solutions by back substitution of your answers into the equations.

9.1 Nature and Characteristics of Dry Friction

In solving problems previously, we ignored the existence of friction between contacting bodies and, in general, assumed that the surfaces in contact were perfectly *smooth* or *frictionless*. This idealization resulted in simplified solutions because, assuming smooth conditions, the resultant force that develops between contacting surfaces is perpendicular to these surfaces with no component parallel to them. In reality, however, frictionless conditions do *not* exist, and, whenever two bodies are in contact with each other, the resultant force that develops between the contacting surfaces has two components. One component, perpendicular to the contacting surfaces, is referred as the *normal* force and is given the symbol N. It is the only force that existed in previous idealizations. The second component is parallel to the contacting surfaces, always acts to oppose the motion of one body with respect to the other, and is known as the *frictional* force. The frictional force, which is given the symbol F, can be reduced by proper lubrication but can never be entirely eliminated.

The term *dry* or *Coulomb* friction refers to the condition in which contacting surfaces are *not* lubricated and is the only type of friction dealt with in this text. Another type of friction encountered in engineering practice is known as *fluid friction*. Fluid friction develops between contacting surfaces that have been lubricated and between particles of fluids (liquids or gases) flowing through pipes or channels. The topic of fluid friction is beyond the scope of this text. Interested readers are referred to textbooks on fluid mechanics where fluid friction is discussed.

The Coulomb theory of dry friction may be considered a macroscopic, rather than a microscopic theory. It was devised by the French engineer C.A. Coulomb (1736–1806) who, in 1781 presented his memoir titled: *Theory of Simple Machines*. His theory is still widely employed for approximate engineering calculations. Scientific work employing the electron microscope to study the condition of the contacting surfaces and the development of theories dealing with the fundamental nature of friction continues. Several factors limit the practicality of applying such knowledge in engineering practice. First, a number of variables are involved, and the resulting equations are too complex to apply. Secondly, important variables, such as the amount of lubrication of the surfaces in contact, will not usually be completely controllable. Although precise predictions are possible under controlled conditions, usually, engineers are unable to determine the necessarily precise inputs required to obtain the corresponding predictions. Hence, the simplified theory of Coulomb continues to be widely employed in engineering analyses.

The basic assumptions of the Coulomb theory of dry friction may be summarized as follows:

1. When motion impends, the frictional force is directly proportional to the normal force.
2. The constant of proportionality, which relates the frictional force to the normal force when motion impends, may be determined experimentally. This constant of proportionality is termed the *coefficient of friction* for which the symbol used is the Greek letter mu (μ). The subscript s for μ refers to the static coefficient and the subscript k for μ refers to the kinetic coefficient.
3. The frictional force is independent of the surface area in contact.
4. The coefficient of friction depends upon the materials in contact, the surface finish of these contacting surfaces, and the presence or absence of lubricants or other liquids on the contacting surfaces.

Coefficients of Friction

Let us now conduct a thought experiment by imagining that we have a block of weight W in contact with a horizontal surface and subjected to a horizontal force P as shown in Figure 9.1. When $P = 0$, the block does not tend to move to the right. The frictional force $F = 0$ and the normal force $N = W$, as shown. This condition is represented by point O in the graph of Figure 9.1. If P is increased somewhat, but not enough to cause the block to move, equilibrium dictates that $F = P$ and $N = W$. This condition is represented by any point on straight line OI in Figure 9.1, the exact location of which depends on the magnitude of P. If P is increased further, the block is brought to the *verge* of motion at which point the frictional force F reaches its maximum possible magnitude for the given contacting surfaces. This conditon is represented by point I in Figure 9.1. According to the Coulomb theory of dry friction, this maximum frictional force is given by $\mu_s N$, the

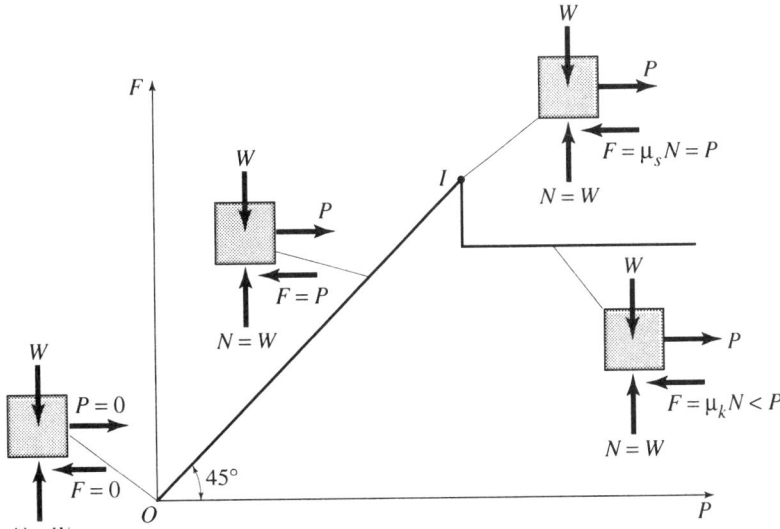

FIGURE 9.1. $N = W$

product of the static coefficient of friction μ_s and the normal force N. It is important to realize that *only* when the block is on the verge of motion (i.e., point I in Figure 9.1) is the frictional force given by

$$F = \mu_s N. \tag{9.1}$$

It is, again, emphasized that Eq. (9.1) is valid only for the condition when the force P is large enough to bring the block to the verge of motion, corresponding to point I in Figure 9.1. For any other value of the force P less than that corresponding to point I, the frictional force is less than its maximum possible value $\mu_s N$. Note that, as long as the block is in equilibrium, i.e., the condition of the block is describable by any point on straight line OI including the end points, the frictional force F is exactly equal to the applied force P.

When the force P is increased slightly beyond that corresponding to point I, the block begins to move to the right and the frictional force suddenly decreases to a value given by

$$F = \mu_k N \tag{9.2}$$

where μ_k is the dynamic coefficient of friction. As shown in Figure 9.1, the value of F given by Eq. (9.2) remains constant regardless of the magnitude of the applied force P.

Rules of Dry Friction

The significant conclusions resulting from the above thought experiment may be summarized as follows:

1. As long as the applied force P is not sufficiently large to cause the block to move, the frictional force F is *less than* its maximum possible value given by the product $\mu_s N$.
2. Only when the force P is large enough to bring the block to the verge of motion does the frictional force F reach its maximum value $\mu_s N$, as given by Eq. (9.1).
3. When the block begins to move, the frictional force F is less than its maximum value, and, according to Eq. (9.2), it is given by the product $\mu_k N$.

It should be emphasized that both the normal force N and the frictional force F represent the resultants of complex force distributions the exact nature of which are not easily determined and will not be needed in the solution of frictional problems. Consider, for example, the free-body diagram of a block of weight W placed on a horizontal plane and subjected to the force P as shown in Figure 9.2(a). The horizontal plane exerts the distribution f_N of forces normal to the contacting surfaces and the distribution f_F of forces parallel to these surfaces. As shown in Figure 9.2(b), the resultant of the distribution f_N is the normal force N, and the resultant of the distribution f_F is the frictional force F. Note that, when $P = 0$, $x = 0$. As P increases so that $F \leq \mu_s N$, x increases so that the ccw couple Wx is balanced by the cw

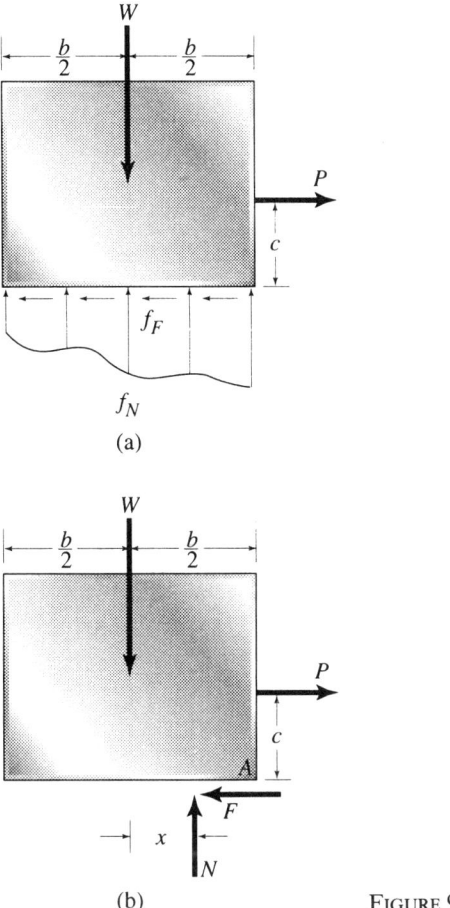

W

$\dfrac{b}{2}$ $\dfrac{b}{2}$

P

c

f_F

f_N

(a)

W

$\dfrac{b}{2}$ $\dfrac{b}{2}$

P

c

A

F

x

N

(b)   FIGURE 9.2.

couple Pc. The block will tip about the corner at A instead of sliding if x becomes equal to $b/2$ before the frictional force reaches its maximum value $\mu_s N$.

A tabulation of static and kinetic coefficients of friction is shown in Table 9.1. We note the great variation in these values caused by surface conditions varying widely, even when the materials in contact are nominally the same. Values should be selected with great discretion from this tabulation or similar ones. At best, these coefficients are very approximate and the results obtained from their use should be regarded as engineering estimates. If improved accuracy is required, engineers should consider a testing program to determine suitable coefficients. However, when surface and lubrication conditions change from those present during controlled experiments, then, the coefficients of friction will change.

TABLE 9.1. Selected Values of Coefficients of Friction

Materials in Contact	Static Coefficient of Friction μ_s	Kinetic Coefficient of Friction μ_k
Aluminum on steel	0.60	0.50
Brass on steel	0.50	0.45
Copper on cast iron	1.00	0.30
Copper on glass	0.70	0.50
Copper on steel	0.55	0.35
Glass on glass	0.90	0.40
Hemp rope on wood	0.50	0.40
Leather on wood	0.55	0.45
Oak on oak (parallel to grain)	0.65	0.50
Rubber tire on dry concrete	0.90	0.70
Rubber tire on wet concrete	0.70	0.50
Steel on steel	0.45	0.35
Steel on ice	0.10	0.05
Teflon on Teflon	0.04	0.04
Teflon on steel	0.05	0.05
Wood on wood (dry)	0.35	0.15

9.2 Angles of Static and Kinetic Friction

An alternate and, sometimes, convenient way of solving friction problems, is to deal with the resultant R of the normal N and the maximum frictional force $F = \mu_s N$. Consider, for example, the case of a block that is on the verge of motion along a horizontal plane, as shown in Figure 9.3(a). Equilibrium requires that $F = \mu_s N = P$ and $N = W$. If the two forces F and N are now combined into the resultant R, as shown in Figure 9.3(b), we conclude that

$$R \sin \phi_s = \mu_s N \tag{a}$$

and

$$R \cos \phi_s = N. \tag{b}$$

The angle ϕ_s is known as the *angle of static* friction, defined as the angle that R makes with the normal to the contacting surfaces. If we divide Eq. (a) by Eq. (b)

$$\tan \phi_s = \mu_s. \tag{9.3}$$

Equation (9.3) states that if the block is on the verge of motion, the angle of static friction ϕ_s has a tangent equal to the static coefficient of friction μ_s. If, however, the frictional force $F < \mu_s N$ and the block is not on the verge of motion, the angle ϕ between the resultant R and the normal is less that ϕ_s. On the other hand, if the block moves, as depicted in Figure 9.3(c), the frictional force $F = \mu_k N$,

$$R \sin \phi_k = \mu_k N, \tag{c}$$

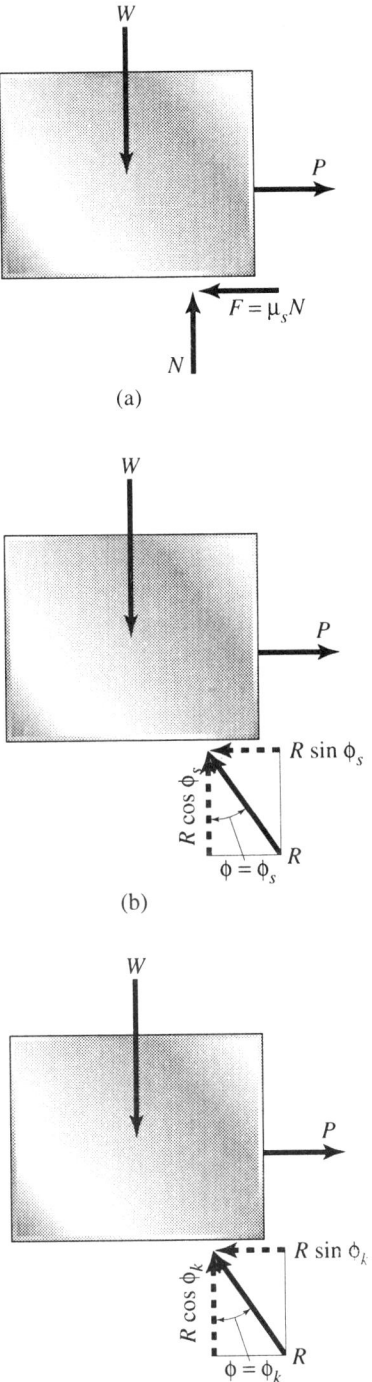

FIGURE 9.3.

and

$$R \cos \phi_k = N \qquad \text{(d)}$$

where the angle ϕ_k is referred to as the *angle of kinetic* friction. Dividing Eq. (c) by Eq. (d) yields

$$\tan \phi_k = \mu_k \qquad (9.4)$$

Note that, because $\mu_s > \mu_k$, it follows that $\phi_s > \phi_k$.

Angle of Repose

Let us now consider a block of weight W on an inclined plane whose angle θ with the horizontal may be varied at will. If the block is on the verge of motion, as shown in Figure 9.4(a), the angle ϕ between the resultant R and the normal to the inclined plane is ϕ_s. Because only two forces R and W act on the block, they must be equal and opposite for the block to be in equilibrium. This means, of course, that $R = W$ and acts vertically upward, as shown in Figure 9.4(a). Equilibrium dictates that $R \sin \phi_s = W \sin \theta$. Because $R = W$, it follows that, if the block is on the verge of motion down the plane, $\phi_s = \theta$. This conclusion is important because, along with Eq. (9.3), it provides a basis for an experimental technique to measure the coefficient of static friction. The angle θ for which motion of the block impends is known as the *angle of repose*.

If angle θ is changed to a value less than the angle of repose, the angle ϕ between the resultant R and the normal to the inclined plane is *less than* ϕ_s. The block is still in equilibrium and, therefore, $R = W$ and is directed vertically upward. Under these conditions, $F = R \sin \phi = W \sin \theta$ from which $\phi = \theta$. Also, $N = R \cos \phi = W \cos \theta$.

Finally, if angle θ is increased to a value larger than the angle of repose, as shown in Figure 9.4(b), the block will slide down the plane and the angle ϕ becomes equal to ϕ_k which is less than ϕ_s. Under these conditions, the block is *not* in equilibrium, and the resultant R is no longer directed vertically upward, as shown in Figure 9.4(b).

9.3 Applications of the Fundamental Equations

The five examples following illustrate how the equations of equilibrium, together with the friction equation $F \leq \mu_s N$, can be used to solve problems involving dry friction . In each case illustrated, the direction of impending translational or rotational motion is known. This enables us to construct free-body diagrams with frictional forces acting in the proper sense to oppose motion. Once we learn to show the friction forces properly on free-body diagrams, the solutions follow by applying the equations of equilibrium learned in earlier chapters.

We begin our solutions by constructing the required free-body diagrams showing all known and unknown forces. Frictional forces act to oppose motion and, if motion impends, we may use the condition

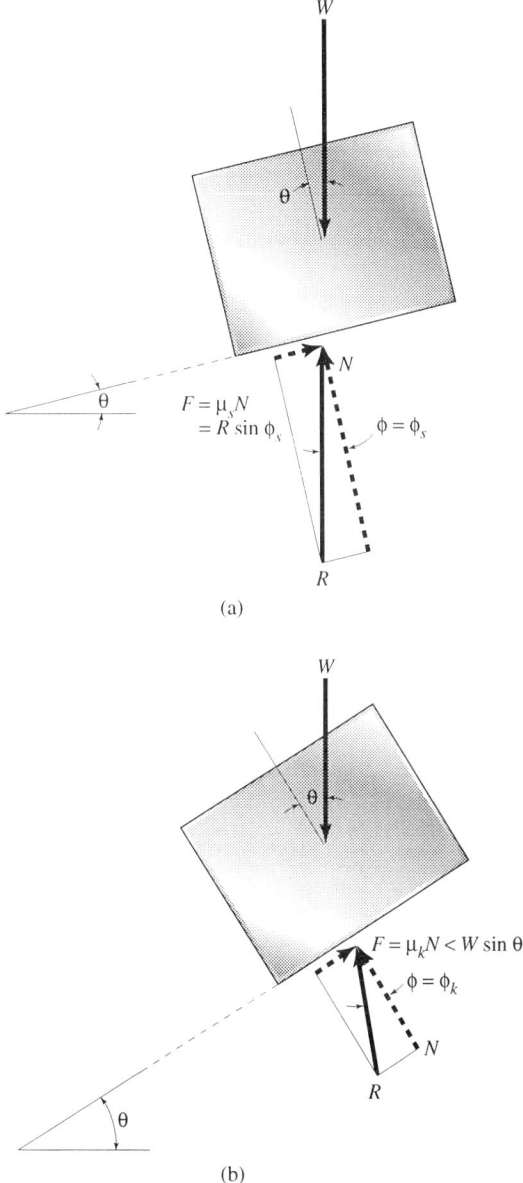

(a)

(b)   FIGURE 9.4.

$F = \mu_s N$. However, if motion does not impend, the frictional force F is less than $\mu_s N$ and must be determined from the conditions of equilibrium. It is always desirable to count unknowns and to count available equations. If these counts are equal, then, we can be confident that we will be able to solve for all unknowns by statics. Proper choice of moment axes will often enable us to eliminate unknowns which may not be required in a given situation.

■ **Example 9.1**

Block A shown on the inclined plane of Figure E9.1(a) weighs 400 N, and friction between the rope and the fixed surface at B is negligible. If the coefficient of static friction for the block and inclined plane is 0.4, determine the weight W and the resultant force R exerted by the plane on block A, for (a) impending motion of the block down the plane, and (b) impending motion of the block up the plane.

Solution

(a) Refer to the free-body diagram of Figure E9.1(b), and note that $T = W$ because surface B is frictionless. Use inclined axes for block A as shown, and observe that the frictional force F acts up the plane to oppose impending motion down the plane. Thus,

$$\sum F_x = 0, \quad T - 400\sin 60° + F = 0, \tag{a}$$

and

$$\sum F_y = 0, \quad N - 400\cos 60° = 0,$$

$$N = 200 \text{ N}. \tag{b}$$

Because motion impends,

$$F = \mu_s N = 0.4(200) = 80 \text{ N}. \tag{c}$$

Substituting this value for the frictional force in Eq. (a) and solving for T,

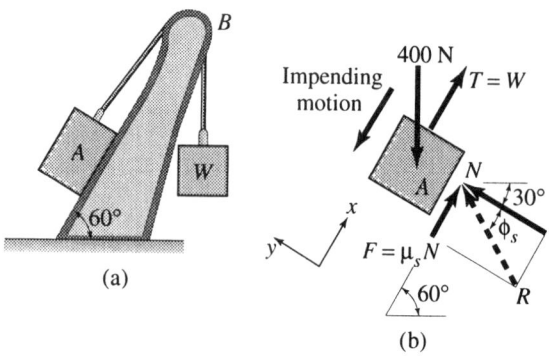

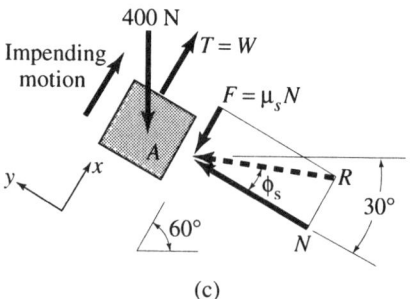

(c)

$$T = 266.4 \text{ N}.$$

Because $W = T$,

$$W = 266 \text{ N}. \qquad \text{ANS.}$$

Using Eqs. (b) and (c), we may solve for the resultant R. Thus,

$$R = \sqrt{N^2 + F^2} = 215.4 \text{ N}.$$

The angle between the normal force N and this resultant is given by the angle of static friction, which may be found from Eq. (9.3). Thus, $\phi_s = \tan^{-1} 0.4 = 21.8°$ measured cw from the normal to the plane. The angle between the vertical and the normal to the plane is equal to the slope angle $\theta = 60°$. Refer to Figure E9.1(b), and note that R acts up to the left at an angle of $\phi_s + 30° = 51.8°$ with the horizontal as indicated. Thus,

$$R = 215 \text{ N} \quad \overset{51.8°}{\diagup} \qquad \text{ANS.}$$

(b) Refer to the free-body diagram of Figure E9.1(c) and, again, note that $T = W$ because the surface at **B** is frictionless. Use inclined axes, as shown, and observe that the frictional force acts down the plane to oppose impending motion up the plane.

$$\sum F_x = 0, \quad T - 400 \sin 60° - F = 0, \qquad \text{(d)}$$

and

$$\sum F_y = 0, \quad N - 400 \cos 60° = 0,$$

$$N = 200 \text{ N}. \qquad \text{(e)}$$

Because motion impends,

$$F = \mu_s N = 0.4(200) = 80 \text{ N}. \qquad \text{(c)}$$

Substituting this value for the frictional force in Eq. (d) and solving for T,

$$T = 426.4 \text{ N}.$$

Because $W = T$,

$$W = 426 \text{ N}. \qquad \text{ANS.}$$

Using Eqs. (e) and (f), we solve for the resultant R. Hence,

$$R = \sqrt{N^2 + F^2} = 215.4 \text{ N}.$$

The angle between the normal force N and this resultant is given by the angle of friction $\phi_s = 21.8°$ measured ccw from the normal to the plane. The angle between the vertical and the normal to the plane is equal to the slope angle $\theta = 60°$. Refer to Figure E9.1(c), and note that R acts up to the left at an angle of $30° - \phi_s = 8.2°$ with the horizontal, as indicated. Thus,

$$R = 215 \text{ N} \quad \overset{8.2°}{\diagup} \qquad \text{ANS.}$$

Comments: Observe that, when motion of block A impends down the plane, the friction F and the tension T act together opposing the component of the weight of block A parallel to the plane. In other words, friction aids the rope tension in preventing block A from moving down the plane compared to motion impending up the plane.

When motion of block A impends up the plane, the friction F and the tension T act in opposite senses. The tension T must support the friction F and the component of the weight of block A parallel to the plane. This means that the tension T will be larger when motion impends up the plane compared to motion impending down the plane.

■ Example 9.2

A cast bracket fits on a fixed rod of square cross-section 0.10 m × 0.10 m, as shown in Figure E9.2(a). The bracket weighs 280 N and is subjected to a force P acting horizontally to the left. Motion of the bracket impends to the left, and the static coefficient of friction $\mu_s = 0.20$. Determine the magnitude of force P and the reaction forces at A and B. Assume that the weight of the bracket acts through point G.

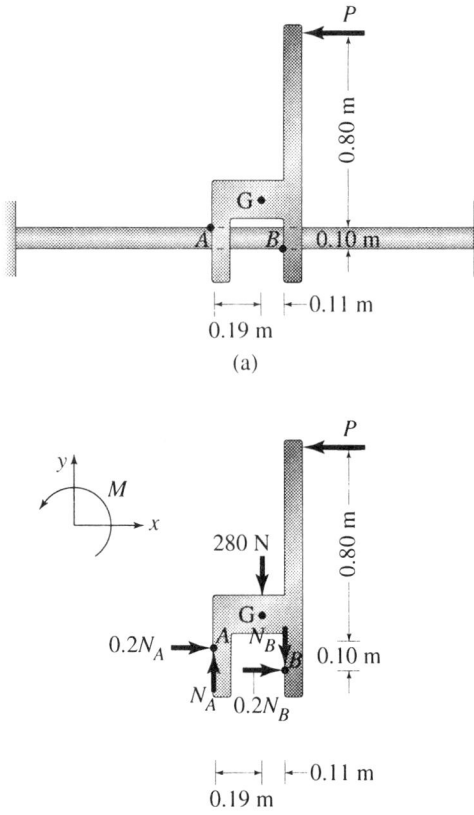

FIGURE E9.2.

Solution

Application of P rotates the cast bracket into contact with the fixed rod at points A and B. The free-body diagram of the bracket is shown in Figure E9.2(b). Thus, using the coordinate system shown,

$$\sum F_x = 0, \quad 0.2N_A + 0.2N_B - P = 0, \tag{a}$$

and

$$\sum F_y = 0, \quad N_A - N_B - 280 = 0. \tag{b}$$

Solving Eqs. (a) and (b) simultaneously for N_A and N_B in terms of P yields

$$N_A = 2.5P + 140, \tag{c}$$

and

$$N_B = 2.5P - 140. \tag{d}$$

$$\sum M_B = 0, \quad P(0.90) + 280(0.11) - N_A(0.30) - 0.2N_A(0.10) = 0. \tag{e}$$

Substituting for N_A from Eq. (c) in Eq. (e) gives

$$0.9P + 30.8 - 0.3(2.5P + 140) - 0.2(2.5P + 140)(0.10) = 0,$$

$$P = 140 \text{ N.} \qquad \text{ANS.}$$

Substituting this value of P in Eqs. (c) and (d) gives

$$N_A = 490 \text{ N},$$

and

$$N_B = 210 \text{ N}.$$

$$\phi_s = \tan^{-1} \mu_s = \tan^{-1} 0.2 = 11.31°.$$

Combining the normal and frictional forces at A and B gives

$$R_A = 500 \text{ N} \quad \overset{11.31°}{\diagup\!\!\!|} \quad, \qquad \text{ANS.}$$

and

$$R_B = 214 \text{ N} \quad \overset{11.31°}{\diagdown\!\!\!|} \quad. \qquad \text{ANS.}$$

■ **Example 9.3**

(a) Refer to Figure E9.3(a), and determine Q_1 so that the uniform crate of weight W is on the verge of tipping about its lower right edge. What is the maximum coefficient of friction so that sliding and tipping of the crate occur simultaneously? Assume that the weight W acts through point G which represents the geometric center of the crate.

(b) Refer to Figure E9.3(b), and determine Q_2 so that the uniform crate of weight W is on the verge of tipping about its lower right edge. A small block attached to the horizontal plane prevents the crate from sliding to the right. What forces are exerted on the crate at its lower

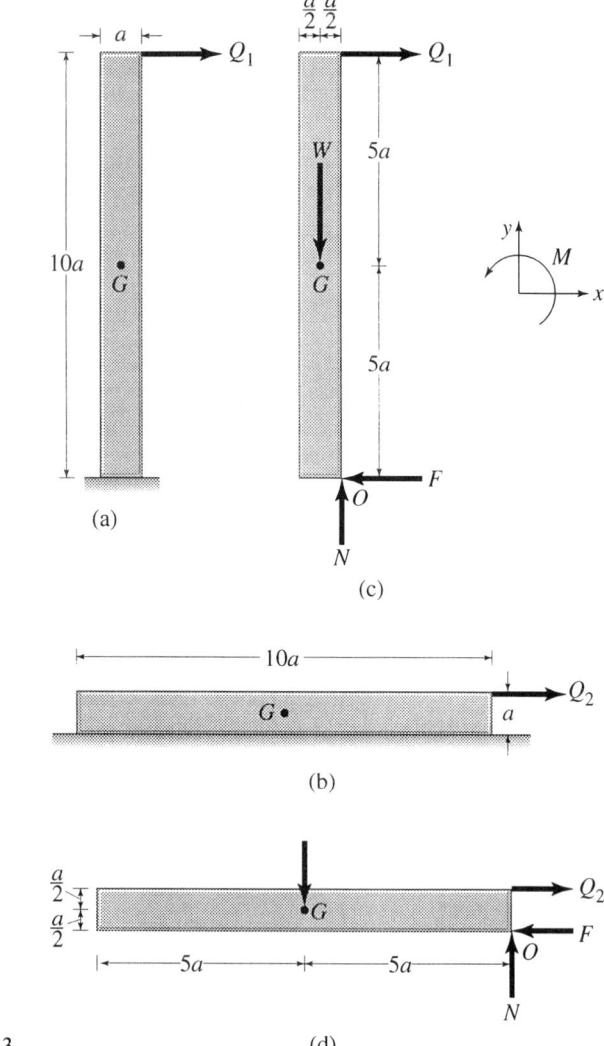

FIGURE E9.3.   (d)

right edge? Assume that the weight W acts through point G which represents the geometric center of the crate.

(c) Compare the answers to parts (a) and (b) above, and briefly discuss them.

Solution

(a) Refer to the free-body diagram shown in Figure E9.3(c) and note that the frictional force F and the normal force N act at the lower right edge O because tipping impends about this edge. In constructing this free-body diagram, we assume that the crate has rotated cw through an infinitesimal angle about O. This means that the crate contacts the plane along only the edge about which it is going to tip. Because tipping is not occurring about this edge, but only impends, the crate is in equilibrium, and we may validly apply the equations of statics. Two unknowns F and N will be eliminated by summing moments with respect to the axis through O. Thus, using the coordinate system shown in Figure E9.3(c)

$$\sum M_O = 0, \quad Q_1(10a) - W(a/2) = 0,$$

$$Q_1 = 0.05W. \qquad \qquad \text{ANS.}$$

The force summation equations will enable us to find the normal N and frictional force F. Thus,

$$\sum F_x = 0, \quad Q_1 - F = 0,$$

$$F = Q_1 = 0.05 \text{ W}, \qquad \qquad \text{ANS.}$$

and

$$\sum F_y = 0, \quad N - W = 0,$$

$$N = W. \qquad \qquad \text{ANS.}$$

From the definition of the coefficient of friction,

$$\mu_{\text{MAX}} = \mu_s = F/N = 0.05. \qquad \qquad \text{ANS.}$$

This is the maximum value of the coefficient of friction associated with impending simultaneous sliding and tipping of the crate. If μ_s were larger than 0.05, the crate would tip about its lower right edge, but sliding of the crate would not impend.

(b) Refer to the free-body diagram of the crate shown in Figure E9.3(d), and note that the force F exerted by the small fixed block and the normal force N both act at the lower right edge because tipping impends about this edge. Again, using the same coordinate system of Figure E9.3(c) and applying the equations of equilibrium,

$$\sum M_O = 0, \quad Q_2 a - W(5a) = 0,$$

$$Q_2 = 5W, \qquad \qquad \text{ANS.}$$

$$\sum F_x = 0, \quad Q_2 - F = 0,$$

$$F = Q_2 = 5W, \qquad \qquad \text{ANS.}$$

and

$$\sum F_y = 0, \quad N - W = 0,$$

$$N = W. \qquad \qquad \text{ANS.}$$

(c) Form the ratio of Q_2 to Q_1,

$$\frac{Q_2}{Q_1} = \frac{5W}{0.05W} = 100.$$

When the long side $10a$ of the crate is placed vertically, as shown in Figure E9.3(a), it is much more vulnerable to overturning about its lower right edge than when its long side $10a$ is placed horizontally as shown in Figure E9.3(b). A force 100 times as large is required to begin overturning the crate when the long side is placed horizontally compared to when the long side is placed vertically. Of course, we understand this intuitively but the statics equations have enabled us to quantify our *hunches*. This is an illustration of one main advantage of the technological revolution which has given us the power to provide quantitative answers to questions which, formerly, could be answered only qualitatively.

■ **Example 9.4**

A rigid block of weight W rests on a horizontal surface and is subjected to a horizontal force P as shown in Figure E9.4(a). When the block is on the verge of motion, its free-body diagram may be represented as shown in Figure 9.4(b). Derive the relationship that must exist between x and y which are defined in Figures E9.4(b). Plot and discuss this relationship.

Solution

Refer to the free-body diagram of Figure E9.4(b) which also shows a convenient coordinate system. Thus,

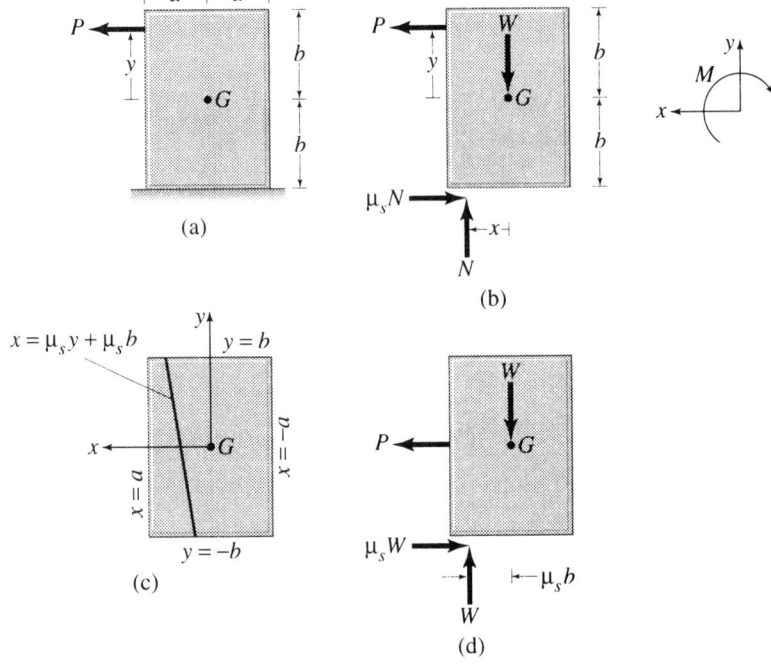

FIGURE E9.4.

$$\sum F_y = 0, \quad N - W = 0,$$

$$N = W, \tag{a}$$

$$\sum F_x = 0, \quad P - \mu_s N = 0,$$

$$P = \mu_s N = \mu_s W, \tag{b}$$

and

$$\sum M_G = 0, \quad Nx - \mu_s Nb - Py = 0. \tag{c}$$

Substituting Eqs. (a) and (b) in Eq. (c) and solving for x

$$x = \mu_s y + \mu_s b. \tag{d} \text{ ANS.}$$

Equation (d) represents a straight line plotted in Figure E9.4(c) with slope μ_s and x intercept $\mu_s b$. Because of the fact that the block has limited dimensions within which the forces must be applied, both x and y must be restricted so that $-a \le x \le a$ and $-b \le y \le b$. These restricted values of x and y are indicated in Figure E9.4(c) by a rectangle bounded by $x = \pm a$ and $y = \pm b$. Note that each point on the straight line defined by Eq. (d) represents a unique physical system. Thus, for example, when $y = 0$, $x = \mu_s b$. This unique physical system is shown in Figure E9.4(d).

It should be noted that, although the straight line defined by Eq. (d) represents an infinite number of unique physical systems, it is restricted to positive values of x only as Figure E9.4(c) indicates. Confirmation of this statement may be obtained by assuming some negative value for x, finding the corresponding value of y from Eq. (d), and constructing the related physical system. As an exercise, the student can construct such a physical system to show that the system cannot be in equilibrium.

Problems

9.1   The block W of Figure P9.1 weighs 80 lb, and motion impends down the inclined plane. Determine the normal force and the friction force acting on the block. What is the static coefficient of friction?

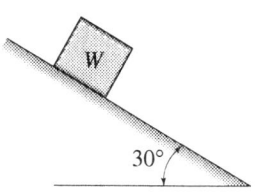

FIGURE P9.1.

9.2   The block W of Figure P9.1 weighs 500 N, and motion impends down the inclined plane. Determine the normal force and the friction force acting on the block. What is the static coefficient of friction?

9.3   In Figure P9.3, the block W weighs 500 lb, and, when $P = 100$ lb, motion of the block impends. Determine the normal force and the friction force acting on the block. What is the static coefficient of friction?

9.4   In Figure P9.3, the block W weighs 1200 N, and, when $P = 300$ N, motion of

FIGURE P9.3.

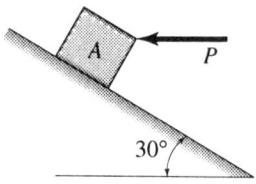

FIGURE P9.7.

the block impends. Determine the normal force and the friction force acting on the block. What is the static coefficient of friction?

9.5 Motion of block A of Figure P9.5 impends. If the block weighs 200 lb and the static coefficient of friction is 0.2, determine the applied force Q and the normal force exerted on the block by the plane.

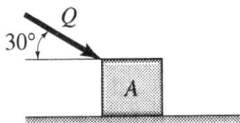

FIGURE P9.5.

9.6 Motion of block A of Figure P9.6 impends. If the block weighs 500 N and the static coefficient of friction is 0.2, determine the applied force Q and the normal force exerted on the block by the plane.

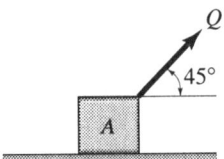

FIGURE P9.6.

9.7 Motion of block A of Figure P9.7 impends up the plane. If the block weighs 800 lb and the static coefficient of friction is 0.4, determine the horizontally applied force P and the normal force exerted on the block by the plane.

9.8 Refer to Figure P9.7. Let the weight of the block be 100 lb, and the horizontally applied force $P = 100$ lb. Determine the normal force and the static coefficient of friction consistent with impending motion of the block up the inclined plane.

9.9 A weight $W = 4000$ lb rests on the inclined plane shown in Figure P9.9. If the coefficient of static friction for the block and the plane is 0.3, determine the force Q which acts parallel to the plane to (a) begin to move the block up the plane and (b) just prevent the block from moving down the plane.

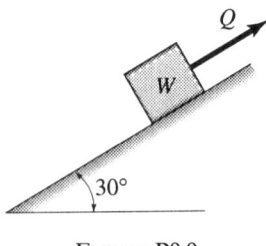

FIGURE P9.9.

9.10 Block A of Figure P9.10 weighs 100 N, and the coefficient of static friction is 0.5 for both blocks on their respective planes. Determine the weight of block B so that block A is on the verge of motion up the 45° inclined plane. Assume that the pulley at C is frictionless.

9.11 Block B of Figure P9.10 weighs 200 N, and the coefficient of static friction is 0.4

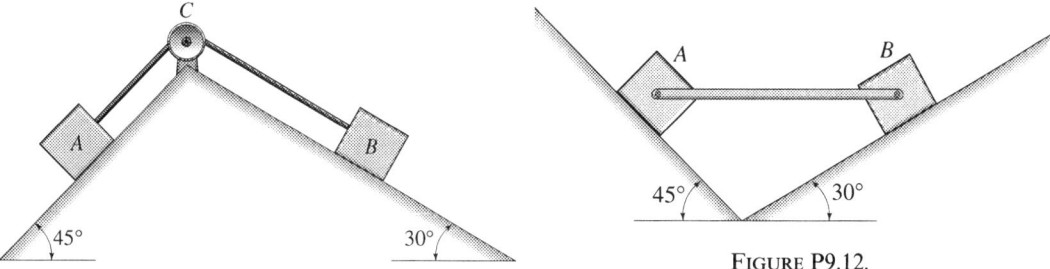

FIGURE P9.10.

FIGURE P9.12.

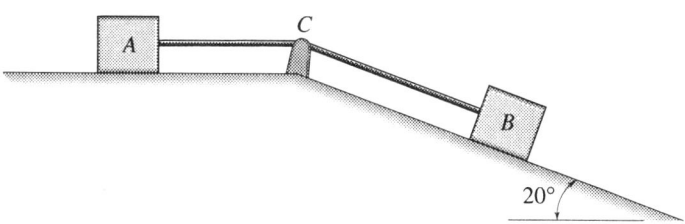

FIGURE P9.13.

for both blocks on their respective planes. Determine the weight of block A so that block B is on the verge of motion up the 30° inclined plane. Assume that the pulley at C is frictionless.

9.12 In Figure P9.12, the horizontal strut AB is pin connected to the blocks A and B. Block B weighs 100 lb and the coefficient of static friction is 0.2 for the planes and blocks. Determine the weight of block A corresponding to impending motion of block B up the plane on which it rests.

9.13 In Figure P9.13, block A rests on a horizontal plane and weighs 1600 N. If the coefficient of static friction for the blocks and planes is 0.2, determine the weight of block B so that it will begin to move down its plane. Ignore friction at C.

9.14 Refer to Figure P9.13, and let the weight of block B equal 600 lb. If the coefficient of static friction for the blocks and planes is 0.1, determine the weight of block A

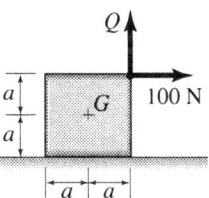

FIGURE P9.15.

just to prevent B from moving down its plane. Ignore friction at C.

9.15 Refer to Figure P9.15, and determine the force Q so that the uniform block is on the verge of tipping about its lower left edge. The block weighs 1000 N. For this force Q and for a coefficient of static friction of 0.4, answer the following questions: (a) What normal and frictional forces act at the lower left edge of the block? (b) Does sliding motion of the block impend? Assume that the weight of the block acts through point G.

9.16 Refer to Figure P9.16, and determine the force Q so that the uniform block is on the verge of tipping about its lower right edge. The block weighs 500 N. Determine the coefficient of friction μ_s so that sliding and tipping impend simultaneously. Assume that the weight of the block acts through point G.

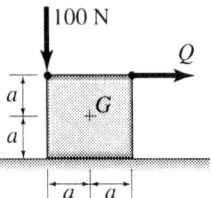

FIGURE P9.16.

9.17 Refer to Example 9.4 and let P act to the right rather than the left. With $a = 0.5$ ft, $b = 1$ ft, and $\mu_s = 0.25$, explore possible tipping of the body about its lower right edge. Is it possible for this body to tip about its lower left edge for this given set of values?

9.18 Refer to Figure P9.18, and let $a = 1$ ft, $b = 2$ ft, $c = 1.5$ ft, $W_1 = 150$ lb, $W_2 = 40$ lb, and $\mu_s = 0.4$. Ignore friction at the fixed peg B and answer the following questions. Be sure to support your answers with free-body diagrams, equations, and comments. (a) Does sliding motion of body A impend? (b) What frictional force

acts on body A? (c) Will body A tip about its lower right corner under these given conditions?

9.19 The cylinder shown in Figure P9.19 has a weight mg, a radius r, and is on the verge of rotating. Express the couple M as a function of mg, r, θ, and μ_s where μ_s is the static coefficient of friction at A and B.

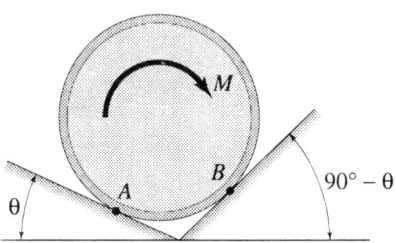

FIGURE P9.19.

9.20 Refer to Figure P9.19, and assign values as follows: $mg = 150$ lb, $r = 15$ in, $\theta = 30°$, and $\mu_s = 0.1$. Determine the normal and frictional forces at A and B and the couple M which brings the cylinder to the verge of rotation.

9.21 Refer to Figure P9.21. Let the weight of A equal 2400 N and the weight of B equal 750 N. If the coefficient of static friction is 0.6 under block A, and the system is on the verge of motion determine (a) the tension in the cable connecting these blocks and (b) the minimum value of the coefficient of static friction between block B and the plane.

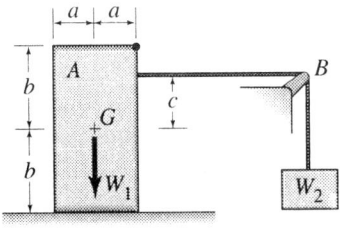

FIGURE P9.18.

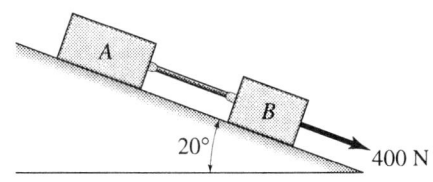

FIGURE P9.21.

9.22 In Figure P9.22, the semicircular cylinder C, hinged at O, weighs 1000 N, and motion impends between C and the plate A. If the coefficient of static friction is 0.2 between A and C and all other surfaces are frictionless, determine the force P applied vertically to the plate. Ignore the weight of the plate. Assume that the weight of the semicircular cylinder acts through point G.

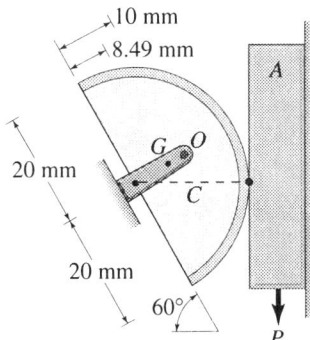

FIGURE P9.22.

9.23 In Figure P9.23, a homogeneous slender rod of weight W and length L rests against a vertical surface at A and on a horizontal surface at B. The static coefficient of friction is μ for both A and B. Assume that the weight of the rod acts through its geometric center. (a) Express

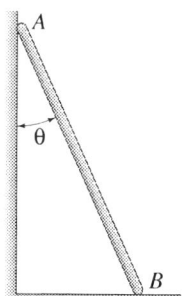

FIGURE P9.23.

the normal forces at A and B in terms of W and μ for impending motion of this rod. (b) Express the angle θ for impending motion as a function of μ. (c) Discuss the case of $\mu = 0$, and sketch the rod in equilibrium for this case.

9.24 Refer to Figure P9.23 and let $W = 120$ lb, $L = 12$ ft, $\theta = 30°$, $\mu_A = 0.1$, and $\mu_B = 0.3$. Determine the normal and frictional forces at A and B. Does slip impend first at A or at B?

9.25 Refer to Figure P9.25, and determine θ for impending motion of block C down the plane. Given: $W_B = 180$ N, $W_C = 240$ N, and $\mu = 0.2$ between blocks B and C. Ignore friction at pulley A and between block C and the inclined plane.

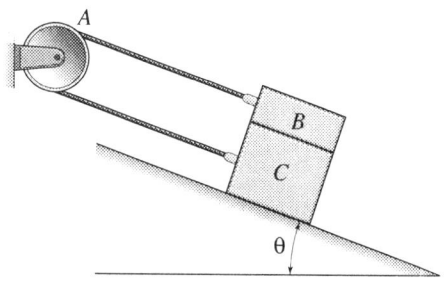

FIGURE P9.25.

9.26 Determine θ for impending motion of block C down the inclined plane for the system shown in Figure P9.25. Ignore friction at pulley A, and let the coefficient of static friction be 0.05 for all other surfaces in contact. Block B weighs 400 N, and block C weighs 650 N.

9.27 In Figure P9.27, block A weighs 100 lb, and block B weighs 150 lb. Determine P for motion of the system to impend to the left if the coefficient of static friction is 0.08 for all surfaces in contact. What is the corresponding force in the pin-connected strut joining the two blocks?

9.28 In Figure P9.28, determine the force P so that the drum, subjected to a couple C of 2000 N·m, is impending motion in a cw sense. The coefficient of static friction is 0.25 between the drum and the brake shoe B. Ignore friction elsewhere in this system.

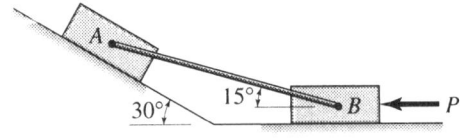

FIGURE P9.27.

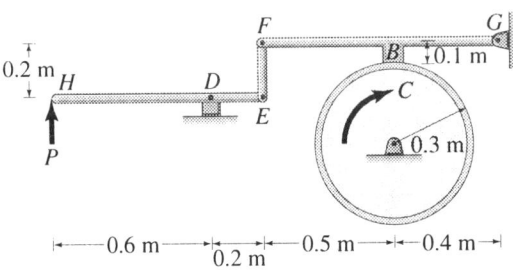

FIGURE P9.28.

9.29 Refer to the system depicted in Figure P9.28, and let $P = 3000$ N. Determine the couple C so that motion of the drum impends cw. The coefficient of static friction is 0.15 between the drum and the brake shoe B, and you may ignore friction elsewhere.

9.30 In Figure P9.30, the static coefficient of friction is 0.2 between the drum and the brake arm. If rotation of the drum impends, determine the weight W, the normal, and the frictional forces at D. Assume that the brake is weightless and the drum weighs 200 N. Assume that the weight of the drum acts through point O.

9.31 A slender rod of negligible weight rests against a fixed peg at B and on a horizontal surface at A, as shown in Figure P9.31. If $\theta = 30°$, determine the normal forces at A and B and the friction force at A consistent with equilibrium, given that $P = 100$ N. Let $\mu_B = 0$. If $\mu_A = 0.8$, will the rod be in equilibrium?

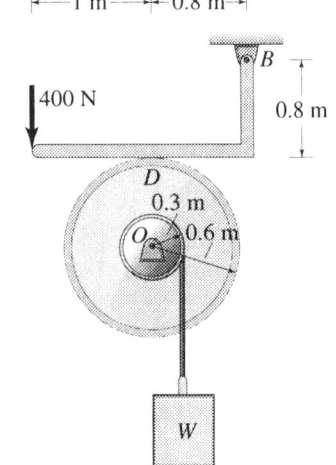

FIGURE P9.30.

9.32 Refer to Problem 9.31, and solve for the case where the force P is applied horizontally to the right instead of vertically upward as shown in Figure P9.31.

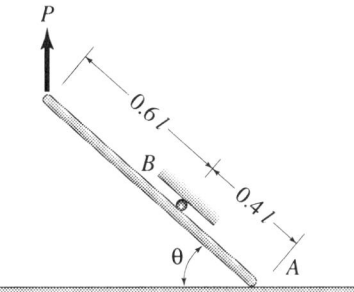

FIGURE P9.31.

FIGURE P9.35.

9.33 The body shown in Figure P9.33 weighs 500 N. For $\theta = 20°$, find the tension T and the coefficient of static friction required for equilibrium if the body is on the verge of motion. Assume that the weight of the body acts through its geometric center.

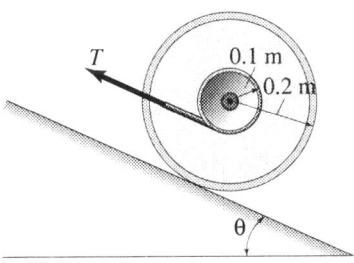

FIGURE P9.33.

9.34 Refer to Figure P9.33. Let $T = 200$ N and $\theta = 25°$. Determine the weight W of the body and the coefficient of static friction consistent with equilibrium if the body is on the verge of motion. The weight W acts through the geometric center of the body.

9.35 In Figure P9.35, a man weighing 680 N climbs a 10-m long ladder. It begins to slide when he reaches the position shown. Determine the static coefficient of friction for the ladder on the wall at **A** and for the ladder on the driveway at **B** assuming that they are equal.

9.36 A homogeneous block of weight W measures $2a \times 2b$, as shown in Figure P9.36. If this block is on the verge of slipping at **A** and **B** and the coefficient of static friction is μ, express the angle θ as a function of W, a, b, and μ. If $W = 800$ lb, $a = 4$ ft, $b = 1$ ft, and $\mu = 0.6$, determine the cor-

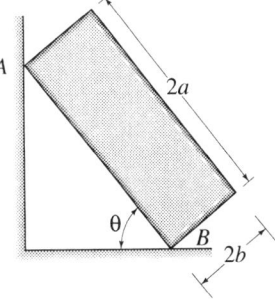

FIGURE P9.36.

responding θ. Assume that the weight W acts through the geometric center of the block.

9.37 A bracket supports a load P as shown in Figure P9.37. Determine the minimum coefficient of static friction so that the bracket will not slide down the vertical member which has a square cross-section. Ignore the weight of the bracket, and let $P = 150$ N.

9.39 Refer to Figure P9.38, and determine the force P consistent with impending motion of block B. Block A weighs 300 lb, and block B weighs 500 lb. The static coefficient of friction is 0.5.

9.40 In Figure P9.40, block A is constrained to move vertically, and links AB and BC are two-force members of negligible weight. Given $\theta = 60°$, $\mu = 0.2$ at A, and block A weighs 200 lb, determine the force in member AB and the applied force Q for impending motion of block A upward.

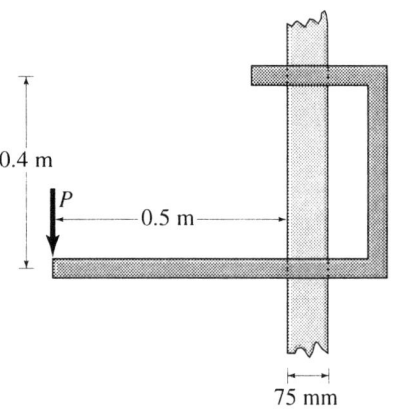

FIGURE P9.37.

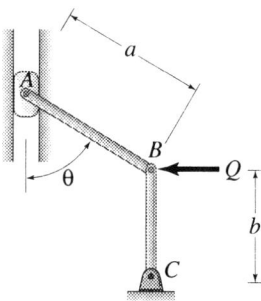

FIGURE P9.40.

9.38 In Figure P9.38, block A weighs 250 lb, and block B weighs 400 lb. Given that $P = 500$ lb and that the coefficient of friction is 0.40, determine whether block B moves.

9.41 Solve problem 9.40 for impending motion of block A downward.

9.42 Refer to Figure P9.42, and determine the ratio of Q/P in terms of the coefficient of static friction μ for impending motion of block C to the left. The pins at A, B and C are frictionless.

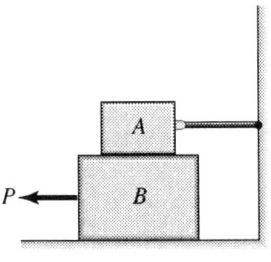

FIGURE P9.38.

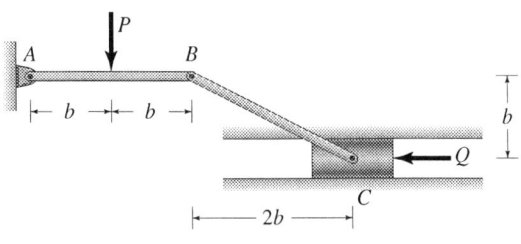

FIGURE P9.42.

9.43 Solve Problem 9.42 for impending motion of block C to the right.

9.44 Show that, for $\mu_s \geq \dfrac{2}{3\pi}$ the half-cylindrical body shown in Figure P9.44 will be in equilibrium, and determine the corresponding value of P for impending motion in terms of the body weight W.

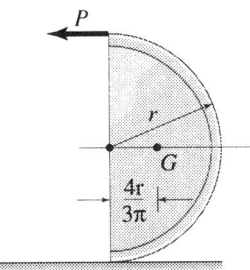

FIGURE P9.44.

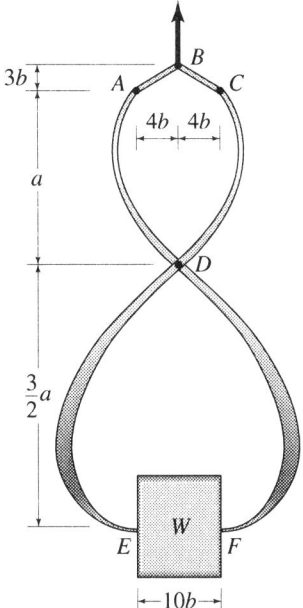

FIGURE P9.46.

9.45 Refer to Problem 9.44 and consider that the body is a hemisphere. Show that $\mu_s \geq 3/16$ for this case, and determine the corresponding value of P for impending motion in terms of the hemisphere weight W.

9.46 Friction tongs are used to lift a block of weight W as shown in Figure P9.46. Express the minimum coefficient of static friction μ as a function of the a/b ratio so that slipping of the body impends. Determine a numerical value of μ for $a/b = 10$. The pins at A, B, C, and D are frictionless. Neglect the weight of the tongs.

9.47 In Figure P9.47 the block weighs 25 lb. Q_y has a magnitude of 10 lb and $\mu_s = 0.6$. Find Q_x for impending motion of this block if $\theta = 30°$.

9.48 The cylinder centered at C in Figure P9.48 has a weight W, and $P = 5W$. Determine the static coefficient of friction μ_s for impending rotation of the cylinder. Consider only friction at D. Pins at A and B

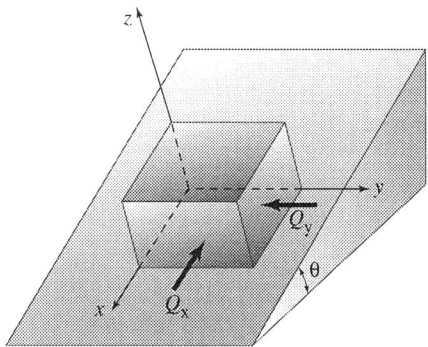

FIGURE P9.47.

are frictionless. Assume that the weight W acts through point C.

9.49 Solve Problem 9.48 for P in terms of W if μ_s is given as 0.35.

9.50 In Figure P9.50, block A weighs 100 lb, and the bars AB and BC are of negligible weight. Determine the force P so that the

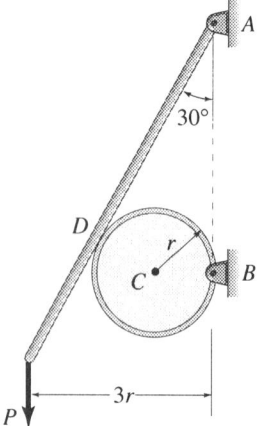

FIGURE P9.48.

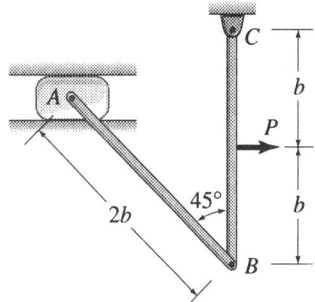

FIGURE P9.50.

motion of block A impends, if the coefficient of static friction is 0.20 between block A and the guides.

9.51 Bar AB rests on a rough inclined plane at A and on a frictionless roller at B, as shown in Figure P9.51. Motion of end A of this bar impends down the plane. Express the normal and frictional forces at A and the reaction at B in terms of the known quantities P, a, b, and θ. Note that, for impending motion, $\tan \theta = \mu_s$.

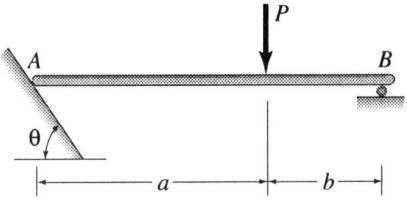

FIGURE P9.51.

9.52 Solve Problem 9.51 for $P = 200$ lb, $\theta = 45°$, $a = 6$ ft, and $b = 2$ ft.

9.53 A slender uniform rod of weight W rests against a wall at A and on a floor at B as shown in Figure P9.53. μ_s is the coefficient of friction between the wall and the rod, and the floor is frictionless. When the rod is on the verge of motion, determine the normal and frictional forces at A and the normal force at B in terms of the known quantities W, b, and a. If $a/b = 2$, determine the coefficient of friction μ_s.

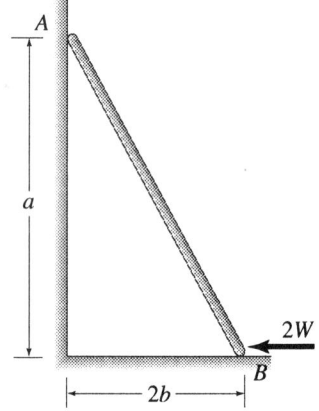

FIGURE P9.53.

9.4
The Six
Fundamental
Machines

The ancient Greeks knew of the six fundamental machines which are also called *simple machines*. What is remarkable is that all mechanical devices, known at present, can be reduced to combinations of these basic machines which are shown in Figure 9.5. Modern micro-electronic technology coupled with knowledge of combinations of these machines has led to the development of robots. In Japan, robots are already manufacturing robots, and, as a result humankind will soon face social problems rivaling those presented by the use of atomic energy and genetic engineering.

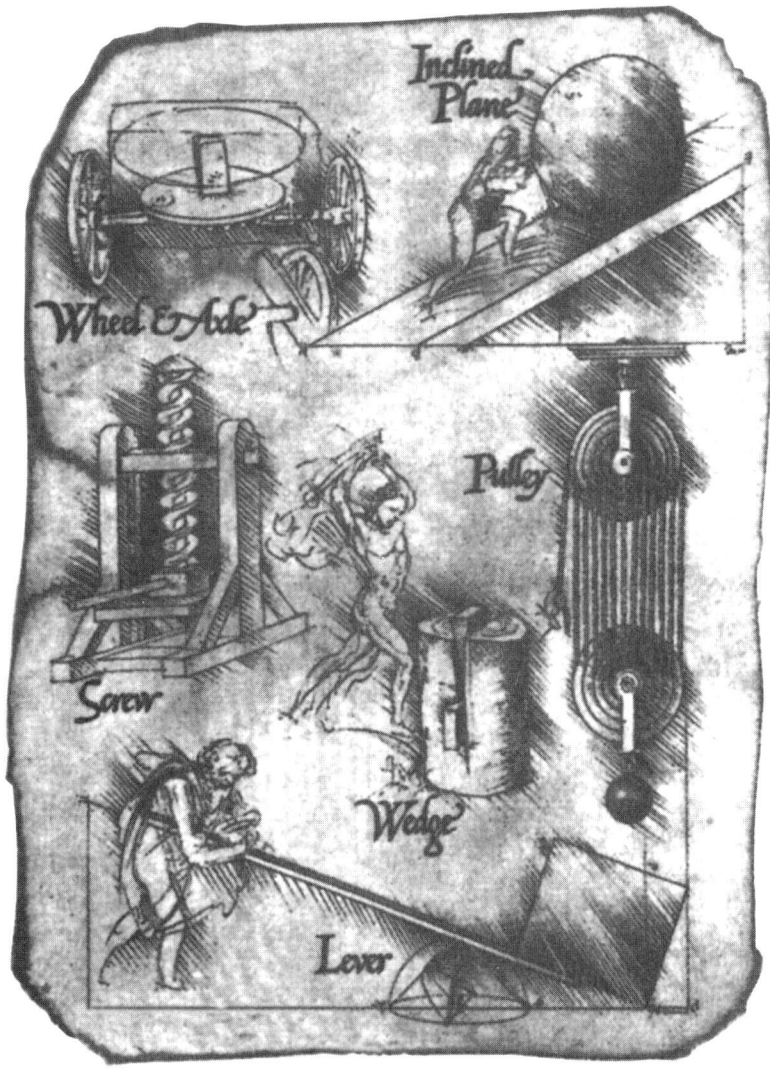

FIGURE 9.5. (Courtesy of the IBM Corporation.)

Examples following will discuss the statics of the six fundamental machines in the following order: levers, inclined planes, wedges, screws, pulleys, and wheels and axles. Some authors prefer to group inclined planes and wedges as a single basic machine and, then, refer to five fundamental machines.

The term *mechanical abvantage* (MA) is a useful concept in analyzing machines and may be defined as

$$MA = \frac{\text{Resisting force}}{\text{Applied force}}. \tag{9.5}$$

Although equations are given in symbolic form for these machines, students are urged to start with appropriate free-body diagrams and write the available equations of equilibrium for solving problems.

In this section, we study the statics of fundamental machines but this must be supplemented by a study of the dynamics of machines before a complete understanding is reached. Technical definitions of acceleration, velocity, work, energy, power, and efficiency are required to understand the dynamics of these machines.

■ **Example 9.5 (Levers)**

Three kinds of levers are shown in Figures E9.5(a), (b), and (c). Develop a symbolic expression for the mechanical advantage of the three types of levers and discuss it.

Solution

The free-body diagrams for the three kinds of lever are shown to the right of the given levers along with a convenient coordinate system. Note that, in all three cases, the resisting force is signified by the symbol W. Thus,

$$\sum M_O = 0, \quad Wc - Pd = 0,$$

$$W = \frac{d}{c}P.$$

Thus, by Eq. (9.5),

$$MA = \frac{\frac{d}{c}P}{P} = \frac{d}{c} \qquad \text{ANS.}$$

Note that, for levers of the first and second kind, $c < d$, whereas for levers of the third type, $c > d$. Therefore, for levers of the first and second kind, $MA > 1$ whereas, for a lever of the third kind, $MA < 1$. Although the mechanical advantage of levers of the third kind is less than unity, it should be noted that the ratio of the distance of travel of the resisting force to the distance of travel of the applied force is given

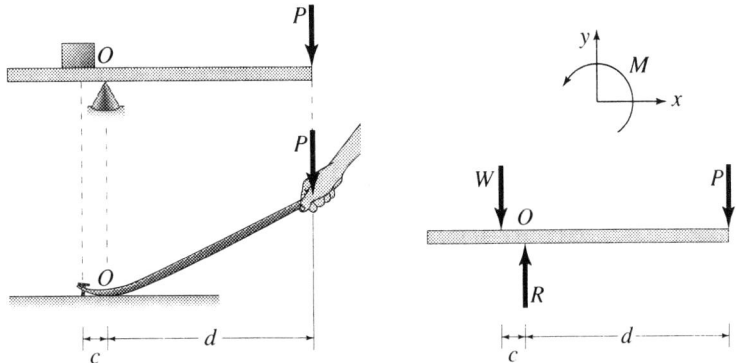

(a) Levers of the first kind. The fulcrum is between
the applied and the resisting forces.

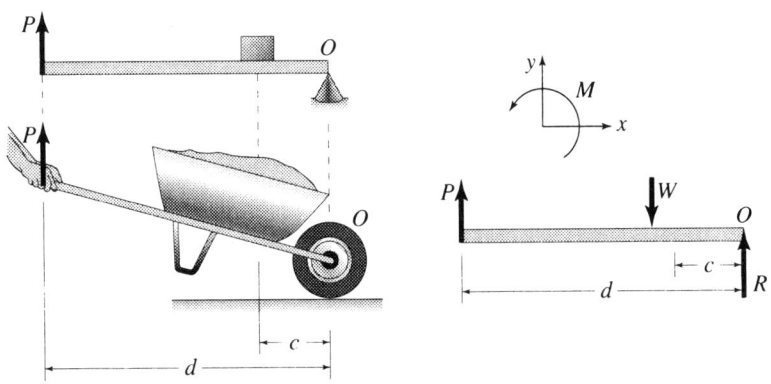

(b) Levers of the second kind. The resisting force
is between the applied force and the fulcrum.

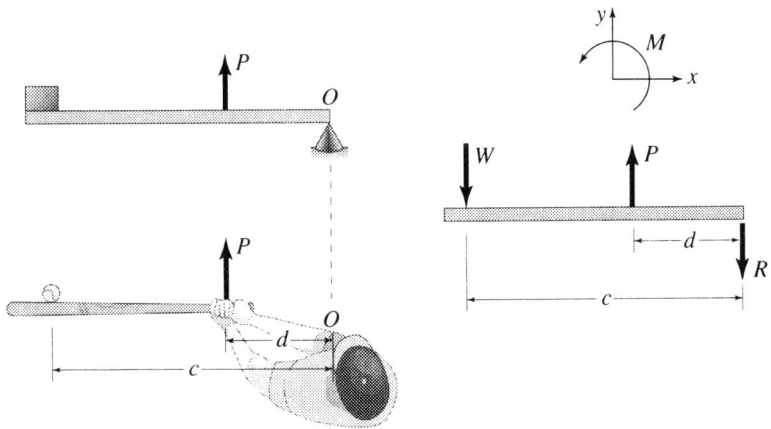

FIGURE E9.5.

(c) Levers of the third kind. The apllied force
is between the resisting force and the fulcrum.

by c/d which is greater than unity. It is this ratio which often makes levers of the third kind advantageous.

■ Example 9.6 (Inclined Planes)

Refer to the inclined planes shown in Figure E9.6(a) and (b), and determine, symbolically, the mechanical advantage for each of the two cases shown. The weight of each block is W and the coefficient of static friction for both cases is μ_s.

Solution

(a) The free-body diagram for the block of Figure E9.6(a) is shown to the right of the diagram along with a convenient coordinate system. Thus,

$$\sum F_y = 0, \quad N - W\cos\theta = 0,$$

$$N = W\cos\theta,$$

and

$$\sum F_x = 0, \quad P - \mu_s N - W\sin\theta = 0,$$

$$P = \mu_s N + W\sin\theta = W(\sin\theta + \mu_s\cos\theta).$$

Using Eq. (9.5), the mechanical advantage

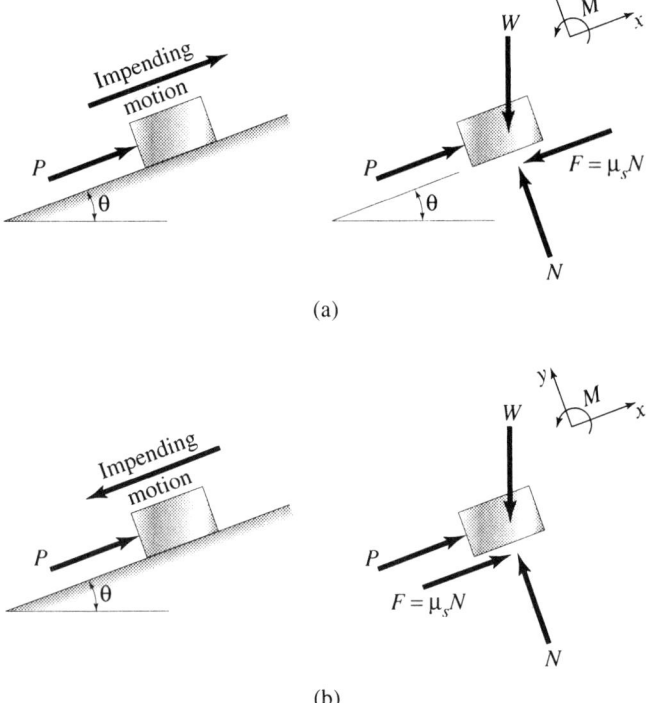

(a)

(b)

FIGURE E9.6.

$$\text{MA} = \frac{W}{P} = \frac{1}{\sin\theta + \mu_s\cos\theta}. \qquad \text{ANS.}$$

(b) The free-body diagram for the block of Figure E9.6(b) is shown to the right of the diagram along with a convenient coordinate system. Thus,

$$\sum F_y = 0, \quad N - W\cos\theta = 0,$$
$$N = W\cos\theta,$$

and

$$\sum F_x = 0, \quad P + \mu_s N - W\sin\theta = 0,$$
$$P = W\sin\theta - \mu_s N = W(\sin\theta - \mu_s\cos\theta).$$

By Eq. (9.5) the mechanical advantage becomes

$$\text{MA} = \frac{W}{P} = \frac{1}{\sin\theta - \mu_s\cos\theta}. \qquad \text{ANS.}$$

■ Example 9.7 (Wedge)

In Figure E9.7(a), a block of weight W is to be lifted by applying the force P to the wedge whose weight is negligible compared to W. The coefficient of static friction is μ_s for all surfaces in contact. Develop an expression for the mechanical advantage assuming that motion of the wedge impends to the right.

Solution

This example provides us with the opportunity to illustrate how the force-polygon technique may readily be utilized for solving frictional problems. In general, once the polygons are drawn, we utilize the trigonometric laws of sines and cosines to formulate appropriate equations.

A free-body diagram of the wedge and its associated force polygon is shown in Figure E9.7(b). Instead of showing both the normal and frictional forces, only their resultant R is given for each contacting surface. Motion of the wedge impends to the right and the frictional components of R_1 and R_2 act to resist this impending motion. The three forces acting on the wedge comprise a coplanar, concurrent system in equilibrium. The force polygon, a triangle in this case, must close because the wedge is in equilibrium. Construction of the force triangle is the key to solving this and similar problems although we could alternatively write and solve the equations of equilibrium. The force P acts to the right, and R_1 makes an angle of ϕ_s with the vertical, as shown in Figure E9.7(b). The force R_2 acts at an angle $(\phi_s + \theta)$ with the vertical, as shown.

By the law of the sines,

$$\frac{R_2}{\sin(90° - \phi_s)} = \frac{P}{\sin(2\phi_s + \theta)}$$

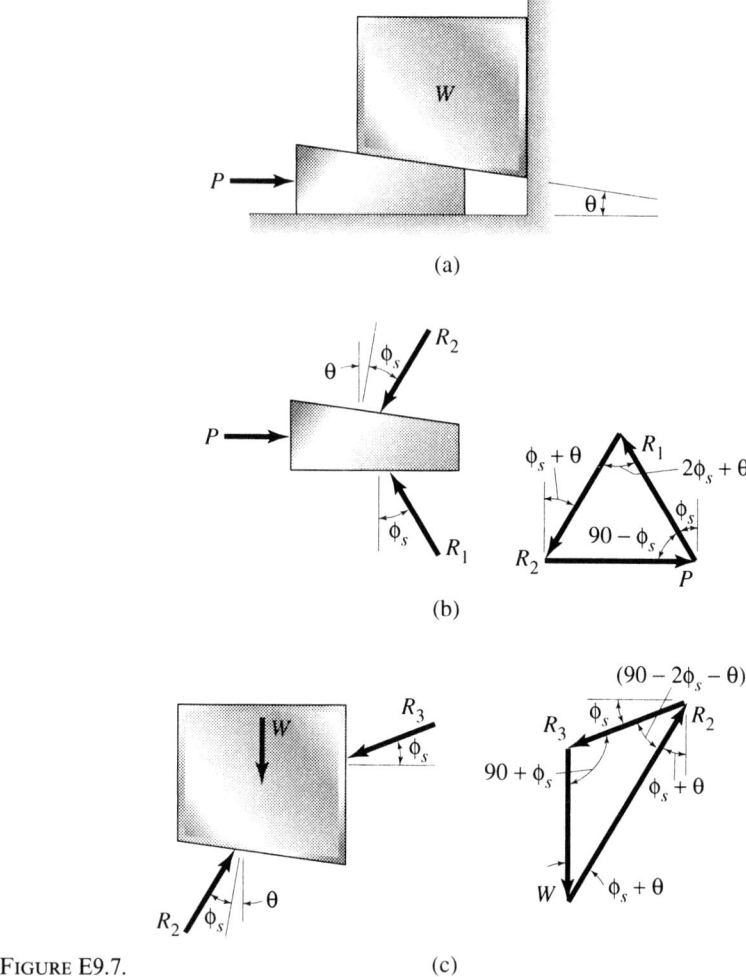

FIGURE E9.7.   (c)

where the angle ϕ_s is the angle of static friction which is given by Eq. (9.3) as

$$\phi_s = \tan^{-1} \mu_s.$$

Solving for R_2,

$$R_2 = \frac{\sin(90° - \phi_s)}{\sin(2\phi_s + \theta)} P. \tag{a}$$

Why did we choose R_2 instead of R_1? Because R_2 is common to the wedge and the weight to be raised. In other words, the force to raise the weight is transmitted from the wedge to the weight W through R_2.

Refer to the free-body diagram and force polygon for the block of weight W as shown in Figure E9.7(c). Because only three forces act on the block, which has impending motion, the block is in equilbrium and the force polygon is a closed triangle, as shown. Where the vertical fixed surface contacts the block, the resultant R_3 acts downward at an angle ϕ_s with the normal to these contacting surfaces. The resultant R_2 acting on the block is equal and opposite to the force exerted on the wedge by the block. By the law of the sines,

$$\frac{R_2}{\sin(90° + \phi_s)} = \frac{W}{\sin(90° - 2\phi_s - \theta)}$$

Solving for R_2,

$$R_2 = \frac{\sin(90° + \phi_s)}{\sin(90° - 2\phi_s - \theta)} W. \qquad \text{(b)}$$

Equating the above values of R_2 of Eqs. (a) and (b) and solving for P,

$$P = \frac{\sin(2\phi_s + \theta)\sin(90° + \phi_s)}{\sin(90° - 2\phi_s - \theta)\sin(90° - \phi_s)} W. \qquad \text{(c)}$$

The mechanical advantage may be found from Eq. (9.5). Thus,

$$\text{MA} = \frac{W}{P} = \frac{\sin(90° - 2\phi_s - \theta)\sin(90° - \phi_s)}{\sin(2\phi_s + \theta)\sin(90° + \phi_s)}. \qquad \text{ANS.}$$

■ Example 9.8 (Wedge)

Refer to Figure E9.8(a), and determine the force P, acting on wedge A, required for impending motion of block B to the right. Block B weighs 5 kN and the static coefficient of friction $\mu_s = 0.1$ for all surfaces. Neglect the weight of the wedge. Also, find the mechanical advantage of the system.

Solution

Construction of force triangles for wedge A and block B followed by application of the law of the sines will enable us to solve this problem. Because bodies A and B are in equilibrium, the force polygon for each body is closed.

By Eq. (9.3),

$$\phi_s = \tan^{-1}\mu_s = \tan^{-1}0.1 = 5.71°.$$

Refer to Figure E9.8(b) which shows the free-body diagram and force polygon for wedge A. Because motion of block B impends to the right, the motion of wedge A impends downward, and the frictional forces will act upward to oppose this impending motion. Note that, instead of showing both the normal and frictional forces, only their resultant R is given for each contacting surface. Thus, wedge A is

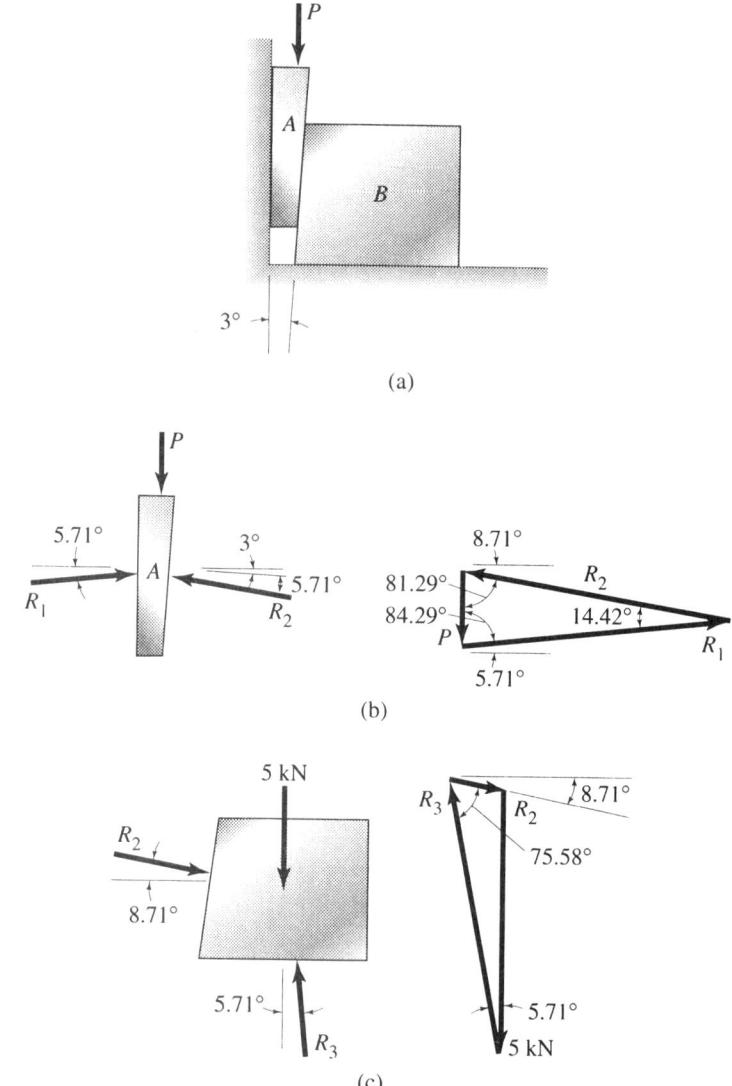

FIGURE E9.8.   (c)

subjected to only three forces, and the force polygon is a closed triangle. Applying the law of the sines,

$$\frac{R_2}{\sin 84.29°} = \frac{P}{\sin 14.42°},$$ (a)

$$R_2 = 3.996P.$$ (b)

Refer to Figure E9.8(c) which shows the free-body diagram and force
polygon for block B. The force R_2 acting on block B is equal and

opposite to R_2 acting on wedge A. The frictional component of R_3 acts to the left on B because the block has impending motion to the right. As in the case of the wedge, we conclude that the force polygon for the block is a closed triangle.

Applying the law of the sines,

$$\frac{R_2}{\sin 5.71°} = \frac{5}{\sin 75.58°}, \tag{c}$$

$$R_2 = 0.514 \text{ kN}. \tag{d}$$

Substituting for R_2 from Eq. (d) in Eq. (b) and solving for P,

$$P = 0.129 \text{ kN}, \qquad\qquad \text{ANS.}$$

$$\text{MA} = \frac{W}{P} = \frac{5}{0.129} = 38.8. \qquad\qquad \text{ANS.}$$

■ Example 9.9 (Screw)

The basic elements of a square-threaded jack screw are shown in Figure E9.9(a). It is used to raise or lower a weight W by applying the force $Q/2$ to the two ends of the handle of length $2L$. Let μ_s be the coefficient of static friction and r_m the mean radius of the screw. Also, let H represent the lead of the screw which is defined as the advance for one complete turn of the screw. In terms of W, r_m, L, μ_s, and H, determine the force Q and the mechanical advantage of the screw for raising the weight W. Note that the lead H may be expressed as the product np where n defines the type of screw (i.e., $n = 1$ for a single-threaded screw, $n = 2$ for a double-threaded screw, etc.) and p is the pitch which represents the advance for one complete turn of a single-threaded screw. Thus, for a single-threaded screw $H = p$.

Solution

A screw may be viewed as an inclined plane wound around a cylinder of radius r_m. If this inclined plane is unwound from the cylinder, we obtain the inclined plane shown in Figure E9.9(b) whose height is H and whose base is $2\pi r_m$. On this inclined plane, we have shown the free-body diagram of a block of weight W in contact with the inclined plane over a small area to represent the weight to be raised. This is permissible because, according to the Coulomb theory, the force of friction is independent of the contact area. The horizontal force P is the effective force produced by Q on this block and may be determined in terms of Q by summing moments about the vertical axis of the screw. Thus,

$$(Q/2)(2L) = Pr_m$$

from which

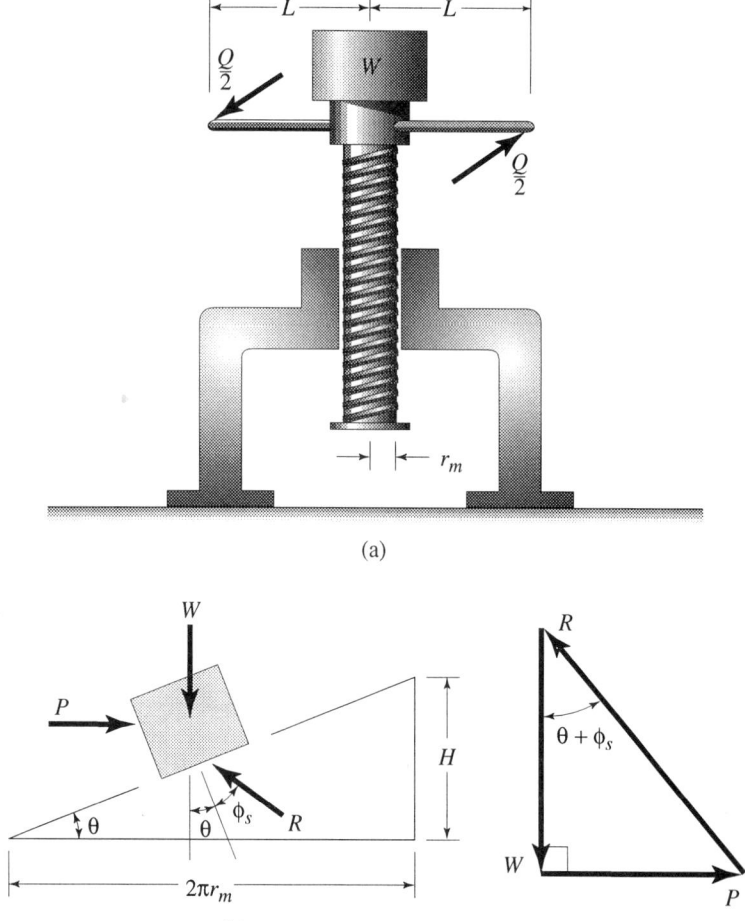

FIGURE E9.9.   (b)   (c)

$$P = \frac{QL}{r_m}.$$ (a)

Finally, the force R represents the resultant of the normal to the plane and the frictional force which acts down the plane on the block to oppose its impending motion. Note that $\theta = \tan^{-1}\left(\dfrac{H}{2\pi r_m}\right)$ and $\phi_s = \tan^{-1}\mu_s$.

Because the block is in equilibrium under the action of three forces, the force polygon in this case is a closed right triangle as shown in Figure E9.9(c). Thus, analysis of this right triangle yields

$$P = W\tan(\theta + \phi_s).$$ (b)

Substituting for P from Eq. (a) and solving for Q,

$$Q = \left(\frac{Wr_{\mathrm{m}}}{L}\right)\tan(\theta + \phi_{\mathrm{s}}).$$ ANS.

By Eq. (9.5), the mechanical advantage becomes

$$\mathrm{MA} = \frac{W}{Q} = \frac{L}{Wr_{\mathrm{m}}\tan(\theta + \phi_{\mathrm{s}})}$$ ANS.

where θ and ϕ_{s} were defined above in terms of known quantities.

■ **Example 9.10 (Screw)**

Refer to Example 9.9, and determine the force Q and the mechanical advantage of the screw for the case when the weight W is lowered.

Solution

The case of lowering the weight W is represented by the free-body diagram of the block shown in Figure 9.10(a) which is in contact with the inclined plane over a smaller area than the actual contact area between the male and female threads. As in Example 9.9, this is jus-

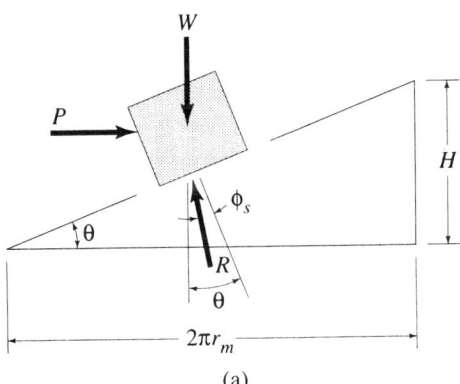

(a)

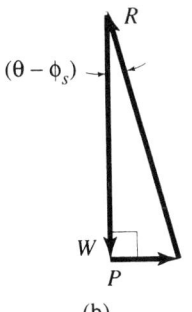

(b)   FIGURE E9.10.

tified by the Coulomb theory which states that the force of friction does not depend on the contact area. The force R is the resultant of the normal to the plane and the frictional force which acts up the plane to oppose the impending motion of the block. As in the case of Example 9.9, the force polygon, corresponding to this free-body diagram, is a closed right triangle, as shown in Figure E9.10(b). Therefore, analysis of the force triangle yields

$$P = W \tan(\theta - \phi_s). \tag{a}$$

Substituting for P from Eq. (a) of Example 9.9 and solving for Q,

$$Q = \left(\frac{Wr_m}{L}\right) \tan(\theta - \phi_s). \tag{b} \; \text{ANS.}$$

The mechanical advantage for this case becomes

$$\mathrm{MA} = \frac{W}{Q} = \frac{L}{Wr_m \tan(\theta - \phi_s)} \tag{ANS.}$$

where θ and ϕ_s were defined in Example 9.9.

Examination of Eq. (b) reveals that, if $\theta = \phi_s$, the force Q vanishes. If, however, $\theta < \phi_s$, the force Q becomes negative which implies, physically, that, to lower the weight W, we need to reverse the sense of Q from that shown in Figure 9.9(a). For such a case (i.e., $\theta < \phi_s$), we say that the jack screw is self-locking, which means that friction is sufficient to prevent the weight W from lowering. Thus, only when $\theta > \phi_s$, do we need to apply a force Q to prevent the weight W from lowering.

■ **Example 9.11**
(Pulley)

Three types of pulley arrangements used to lift a weight W are shown in Figures E9.11(a), (b), and (c). Determine the mechanical advantage for each arrangement. Assume frictionless conditions and weightless pulleys.

Solution

The needed free-body diagrams are shown to the right of the given pulley arrangements along with a convenient coordinate system.

(a)
$$\sum M_O = 0, \quad Pr - Wr = 0,$$
$$P = W,$$
$$\mathrm{MA} = \frac{W}{P} = 1. \tag{ANS.}$$

(b)
$$\sum M_O = 0, \quad Tr - Pr = 0,$$
$$T = P, \tag{a}$$

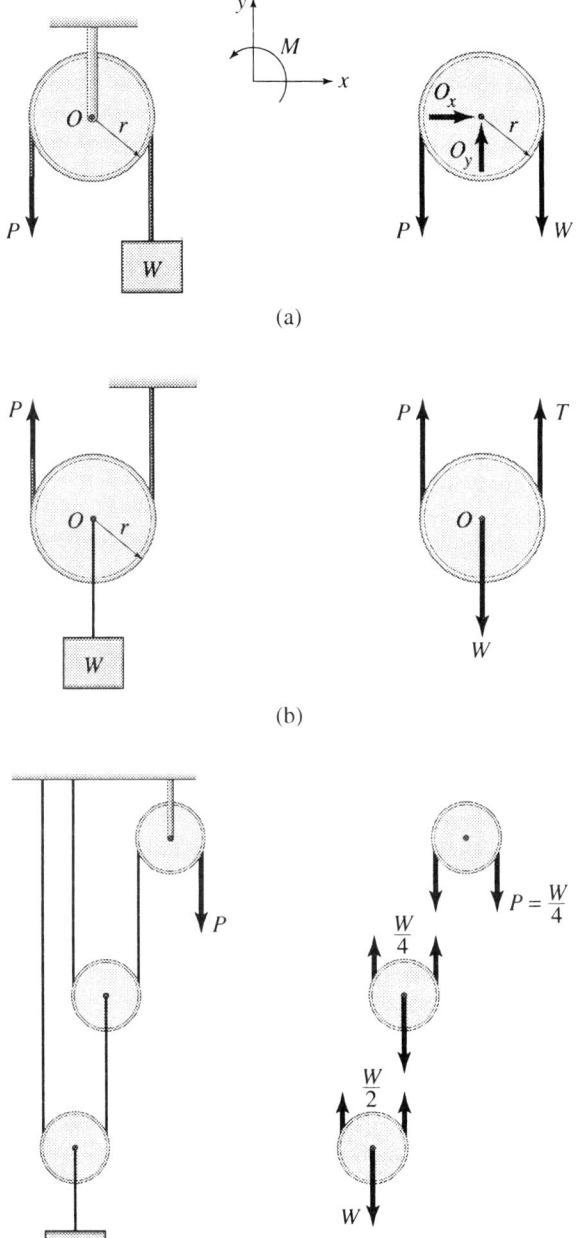

(a)

(b)

(c)

FIGURE E9.11.

and

$$\sum F_y = 0, \quad P + T - W = 0.$$

Substituting for T from Eq. (a) and solving for P yields

$$P = \frac{W}{2}$$

$$\text{MA} = \frac{W}{P} = 2. \qquad\qquad \text{ANS.}$$

(c) The free-body diagrams of the three pulleys in arrangement (c) are shown in their proper positions with the forces acting on them, as determined from the conditions of equilibrium. As an exercise, the student can verify these forces. Thus,

$$\text{MA} = \frac{W}{P} = 4. \qquad\qquad \text{ANS.}$$

We conclude, therefore, that, in theory, we could achieve any mechanical advantage we choose with pulley arrangements. This choice, however, is limited by practical considerations.

■ Example 9.12
(Wheels and
Axles)

Discuss the forces and couples acting on the wheels and axles shown in Figures E9.12(a), (b), and (c). Each wheel has a weight W and moves so that its mass center translates with constant velocity.

Solution

An idealized frictionless rigid wheel on a frictionless rigid plane is shown in Figure E9.12(a) along with the corresponding free-body diagram. Only the force W delivered by the axle to the wheel, which includes an allowance for the weight of the wheel, and the normal force N act on the wheel, and these forces are in balance. If these ideal conditions could be met, then, the wheel, after being placed in rolling motion, would continue to roll along the horizontal plane at constant angular velocity for an indefinite distance without loss of energy. The invention of the wheel probably occurred in prehistoric times and may have come from an astute observation of a rolling log.

In reality, both the wheel and the plane on which it rolls will deform, and rolling resistance develops, as indicated in Figure E9.12(b). A horizontal force Q must be applied to the wheel by the axle, as shown, to keep the wheel moving at constant angular velocity. Extensive research has been done on the nature of the deformed wheel and plane surfaces and the forces developed on the microscopic level, but a summary of these results is not possible due to the many variables involved. With the simplified theory presented here, we replace all forces between the wheel and plane by a resultant R. Because only three forces act on the

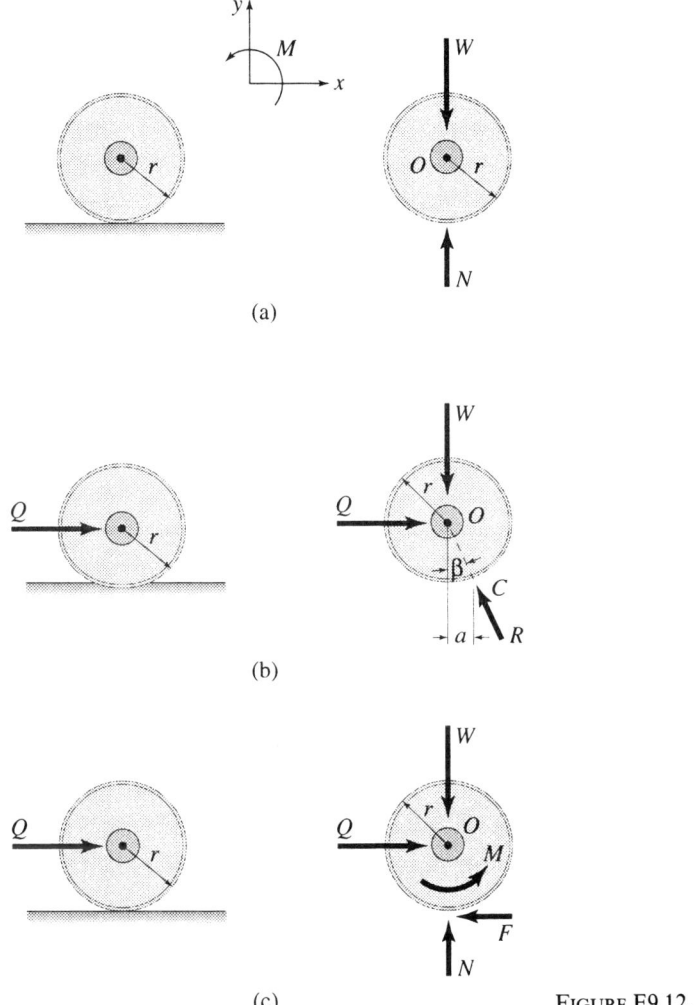

(a)

(b)

(c)   FIGURE E9.12.

wheel, the system must be concurrent at O the center of the wheel and axle, as shown in the free-body diagram of the wheel. The distance a, measured from a vertical through point O to the point C where R acts on the wheel, is termed the *coefficient of rolling resistance*. Values of a, determined experimentally or estimated, vary widely in magnitude.

Once a is known, the angle β follows from $\beta = \sin^{-1}(a/r)$. Thus, using the coordinate system shown,

$$\sum M_{\mathrm{C}} = 0, \quad Qr\cos\beta - Wa = 0,$$

$$Q = \frac{a}{r\cos\beta}\,W. \qquad\qquad \text{ANS.}$$

If the resultant R is required, we may determine it from

$$\Sigma F_y = 0, \quad R \cos \beta - W = 0,$$

$$R = \frac{W}{\cos \beta}.$$ ANS.

The case shown in Figure E9.12(c) corresponds to a rigid wheel on a rough rigid surface. Also shown is its free-body diagram. The moment M represents the resultant of the moments of the frictional forces distributed over the surface contact area between the wheel and the axle. The force Q must be applied to keep the wheel in motion at constant velocity. The forces N and F represent the normal and frictional forces, respectively, exerted by the rough rigid surface.

Using the coordinate system shown,

$$\Sigma M_O = 0, \quad M - Fr = 0,$$

$$F = M/r,$$ ANS.

$$\Sigma F_x = 0, \quad Q - F = 0,$$

$$Q = F = M/r,$$ ANS.

and

$$\Sigma F_y = 0, \quad N - W = 0,$$

$$N = W.$$ ANS.

Even when allowances are made for rolling resistance and axle friction, the wheel and axle enable us to transport weights much more efficiently than simply subjecting the weights to direct contact with friction forces on planar or curved surfaces.

■

Problems

9.54 The basic element ABC of a proposed tree pulling machine is shown in Figure P9.54. If a force of 250,000 pounds is required to pull trees from the ground, and the design force exerted by the hydraulic cylinder at C is 60,000 pounds, determine dimensions c and d subject to the limitation that their sum is not to exceed 8 ft.

9.55 A home owner is constructing a patio and desires to move two bags of cement

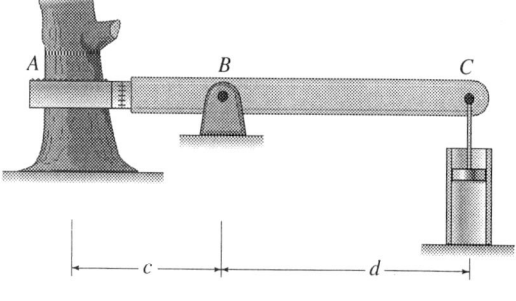

FIGURE P9.54.

with his wheelbarrow, as shown in Figure P9.55. Determine the vertical force applied through each of his arms at point A to just lift the wheelbarrow off the ground if each bag of cement weighs 425 N. Neglect the weight of the wheel barrow.

9.56 A large electric motor weighing 40 tons is to be pulled up the gradual inclined plane shown in Figure P9.56. Surface preparation has reduced the static coefficient of friction to 0.05. Determine the winch cable tension exerted parallel to the inclined plane required just to begin moving the motor to its permanent location.

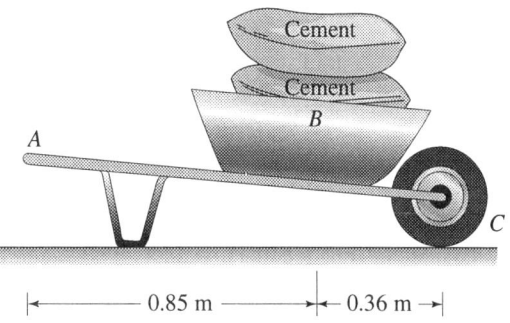

FIGURE P9.55.

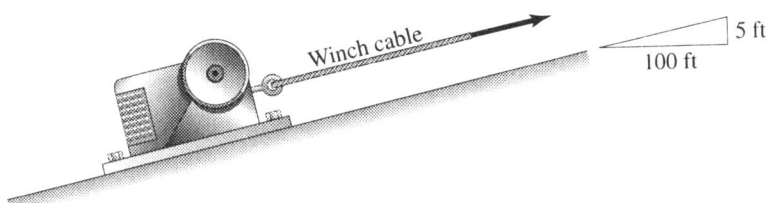

FIGURE P9.56.

9.57 Refer to Figure E9.6(a), and let $W = 5000$ N, $h = 1$ m, $d = 10$ m, and $\mu_s = 0.04$. Determine the force P required for impending motion of the body up the inclined plane. For the same given values, refer to Figure E9.6(b), and find the force P required for impending motion of the body down the inclined plane.

9.58 In Figure P9.58, block A weighs 2000 lb, and you are to neglect the weight of wedge B. Determine the force P applied to the wedge so that the motion of block A impends to the left. The coefficient of static friction is 0.2 for all surfaces.

9.59 Determine the force P applied to the wedge B required to move the two blocks labeled A in Figure P9.59. Neglect the weight of wedge B. Each block A weighs 12 kN and the coefficient of static friction is 0.15 for all surfaces.

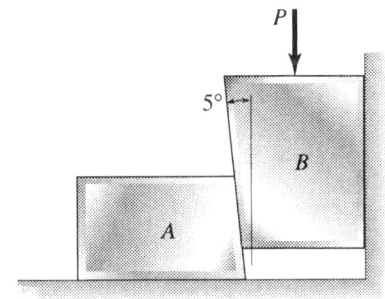

FIGURE P9.58.

9.60 Refer to the jack screw shown in Figure E9.9. If the screws are double-threaded (i.e., $n = 2$), the pitch p is 0.5 in., and the mean radius is 1.8 in., determine the maximum weight W which can be raised for $Q = 100$ lb, $L = 20$ in., and $\mu_s = 0.1$. Is

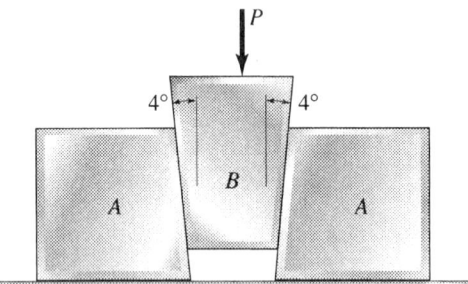

FIGURE P9.59.

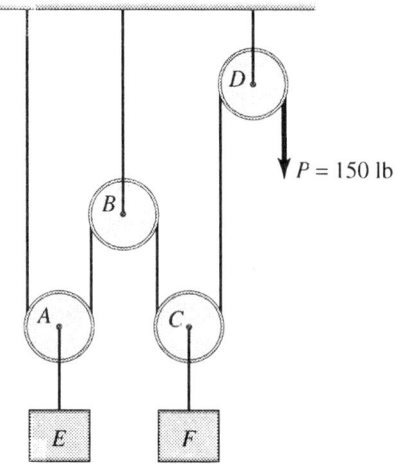

FIGURE P9.63.

the jack screw self-locking? Note that the lead $H = np$.

9.61 Refer to the jack screw shown in Figure E9.9. If the screws are triple-threaded (i.e., $n = 3$), the pitch p is 12 mm, and the mean radius is 50 mm, determine the maximum weight W which can be raised for $Q = 500$ N, $L = 0.5$ m, and $\mu_s = 0.12$. Is the jack screw self-locking? Note that the lead $H = np$.

9.62 Refer to Figure P9.62, and determine the weight of block D so that the motion of block A impends to the right. Block A weighs 2000 N, and the static coefficient of friction is 0.2 between Block A and the horizontal plane. Pulleys B and C are to be regarded as frictionless. Neglect the pulley weights.

9.63 As shown in Figure P9.63, $P = 150$ lb, and the pulleys A, B, C and D are assumed

frictionless and of negligible weight. Determine the weights of E and F that may be lifted at constant speed by the applied force P.

9.64 Refer to Figure P9.64, and find force Q required to move the lower wedge. The body weighs 3000 N and the coefficient of static friction is 0.25 between the two wedges and $\mu_s = 0.488$ for other surfaces of contact. The hinge support on the left is frictionless. Assume that the weight of the body acts through point G.

9.65 In Figure P9.65, determine the weight W which can be raised by the wedge and

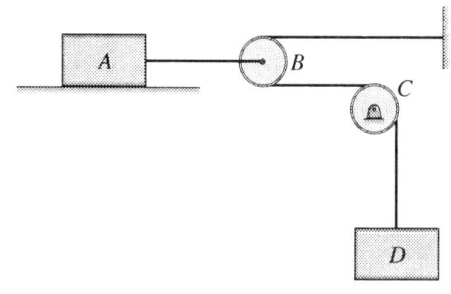

FIGURE P9.62.

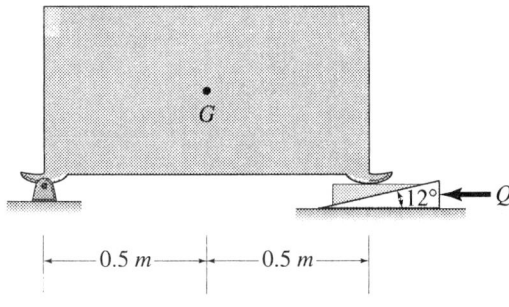

FIGURE P9.64.

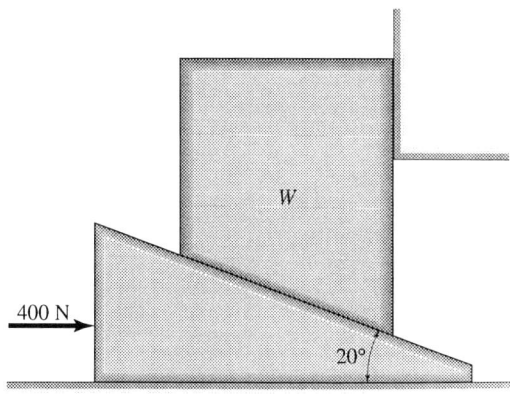

FIGURE P9.65.

the applied force shown. The coefficient of friction is 0.25 for all contacting surfaces.

9.66 Refer to Figure P9.66, and find the weight W which can be raised by applying the 100 N forces shown. The static coefficient of friction is 0.15 for all surfaces in contact.

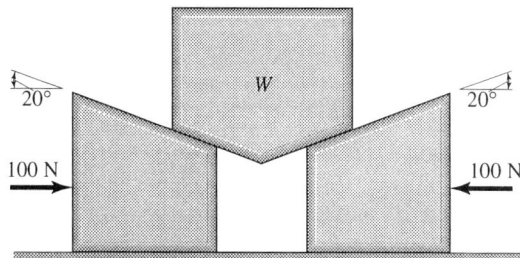

FIGURE P9.66.

9.67 Refer to Figure P9.67, and, if $\mu_s = 0.35$, determine the force P required to begin to move the block of weight $W = 1000$ N up the inclined plane. Also, determine P to prevent the block from moving down the inclined plane. Determine the mechanical advantage for each case.

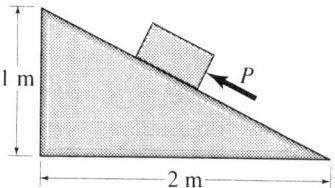

FIGURE P9.67.

9.68 Solve Problem 9.67 for $\mu_s = 0.60$. Explain why a force need not be applied to prevent motion impending down the inclined plane.

9.69 Refer to Figure P9.69, and determine the horizontal force P required for motion of the block to impend up the inclined plane. The block weighs 100 lb, and $\mu_s = 0.4$. Determine the mechanical advantage. Use the force-polygon method.

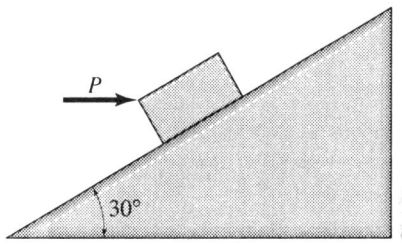

FIGURE P9.69.

9.70 (a) Solve Problem 9.69 for impending motion of the block down the inclined plane. (b) If $\mu > \tan 30°$, explain why the sense of P must be reversed, and provide a solution for $\mu_s = 0.65$.

9.71 Refer to Figure P9.71, and consider that the block weight is 200 lb and the coefficient of static friction is 0.15. Determine the smallest force P in magnitude and direction required (a) to keep the block from moving down the inclined plane and

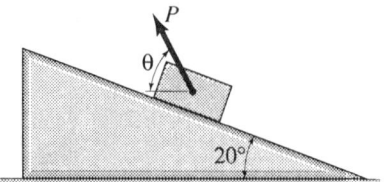

FIGURE P9.71.

(b) to have the motion of the block impend up the plane. Use the force-polygon method.

9.72 Refer to Figure P9.71, and consider that the block weight is 400 lb and the coefficient of static friction is $\mu_s = 0.10$. For $\theta = 45°$, find the smallest magnitude of P required (a) to prevent the block from moving down the inclined plane, (b) to have motion of the block impend up the inclined plane, and (c) to move the block up the inclined plane at constant speed. Use the force-polygon method. Let $\mu_k = 0.08$.

9.73 For the system shown in Figure P9.73, express the ratio of W_1 to W_2 as a function of θ and μ_s. Let $\mu_s = 0.1$ and plot W_1/W_2 vs. θ for the interval $0 < \theta < \pi/2$ using an increment of $\pi/6$ for calculating curve ordinates. Consider friction between W_2 and its plane, but ignore it elsewhere. Assume that the motion of W_2 impends up the plane.

9.74 Solve Problem 9.73 for impending motion of W_2 down the plane.

9.75 Refer to the system depicted in Figure P9.75, and express the ratio of W_2 to W_1 as a function of θ and μ_s. Consider friction between W_1 and the inclined plane, but ignore friction elsewhere. Assume that the motion of W_1 impends up the plane and that all pulleys are weightless.

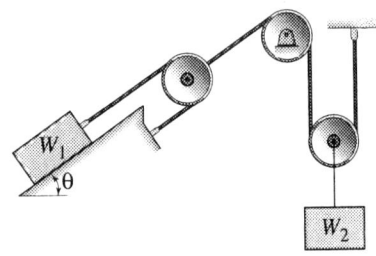

FIGURE P9.75.

9.76 Solve Problem 9.75 for impending motion of W_1 down the plane.

9.77 A block and tackle is shown in Figure P9.77. Determine the mechanical advantage of this system. Ignore friction.

9.78 The press shown in Figure P9.78 exerts a 2000 lb force on block A when a force $Q = 150$ lb is applied. Determine the distance L, given that $\mu_s = 0.40$, $r_m = 1.00$ in., and $H = np = 0.25$ in.

9.79 Solve Problem 9.78 if $\mu_s = 0.20$.

9.80 A vise is depicted in plan view in Figure P9.80. The screw is double-threaded (i.e., $n = 2$) with a pitch $p = 10$ mm and a mean diameter $2r_m = 0.1$ m. The coefficient of friction is 0.20. Find the force exerted on block B if a couple of 100 N·m is applied to the handle at A. Note that $H = np$.

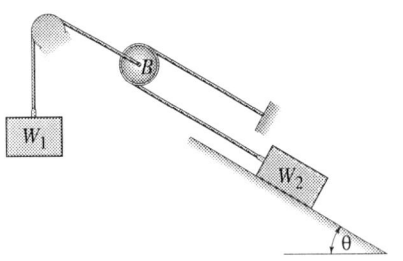

FIGURE P9.73.

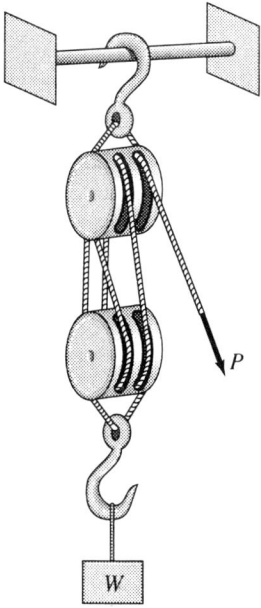

FIGURE P9.77.

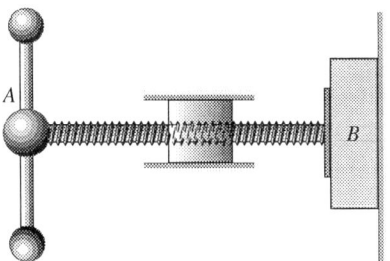

FIGURE P9.80.

9.81 The C-clamp shown in Figure P9.81 exerts a clamping force on the two blocks at A. Determine this clamping force if a couple of 50 N·m is exerted on the clamp handle. The clamp is single-threaded (i.e., $n = 1$) with a pitch $p = 4$ mm and a mean diameter $2r_m = 80$ mm. Use a coefficient of friction of 0.20. Note that $H = np$.

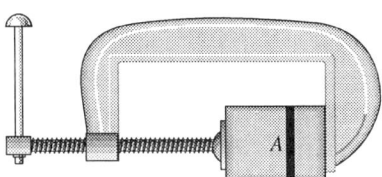

FIGURE P9.81.

9.82 Refer to Problem 9.81, and calculate the torque required to loosen the clamp if the clamping force is 3.5 kN.

9.83 A wedge is used, as shown in Figure P9.83, to split a block of timber. Determine the half wedge angle θ as a function of the coefficient of static friction μ so that the wedge will not slip out of the timber. Calculate θ for μ values of 0.4, 0.6, and 0.8. Neglect the weight of the wedge.

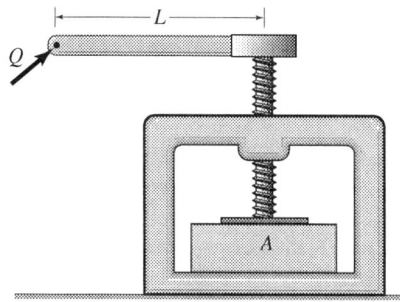

FIGURE P9.78.

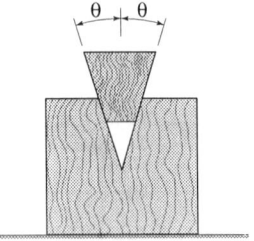

FIGURE P9.83.

9.84 A scissor jack supports an automobile of $W = 4000$ lb, as shown in Figure P9.84. The screw is double-threaded (i.e., $n = 2$) with a pitch $p = 0.20$ in. and a diameter of $2r_m = 4$ in. For a coefficient of friction $\mu_s = 0.4$, find the couple C to (a) raise the automobile and (b) lower the automobile. Neglect friction at pins D, E, F, and G and between the horizontal rod and the collar at F. Note that $H = np$.

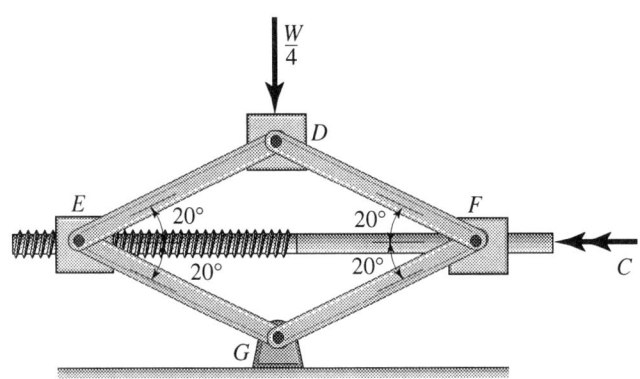

FIGURE P9.84.

9.5*
Friction on V-Belts and Flat Belts

The frictional forces acting on belts that pass over drums and pulleys, such as those found in band brakes and belt drives, must be considered for the proper design of these systems. Consider, for example, a V-belt whose cross-section is depicted in Figure 9.6(a). The belt passes over a stationary pulley, as shown in Figure 9.6(b), contacting the pulley over an arc length defined by the angle β, known as the *angle of wrap*. If we assume that the belt is on the verge of slipping in the groove of the pulley in a ccw sense, we conclude that the frictional force distributions f_F on the two inclined sides of the V-belt act to oppose the tendency of motion and are, therefore, pointed in a cw sense as shown in Figure 9.6(c). This condition leads us to conclude that $T_2 > T_1$. Also shown in Figure 9.6(c) are the normal force distributions f_N which act normally

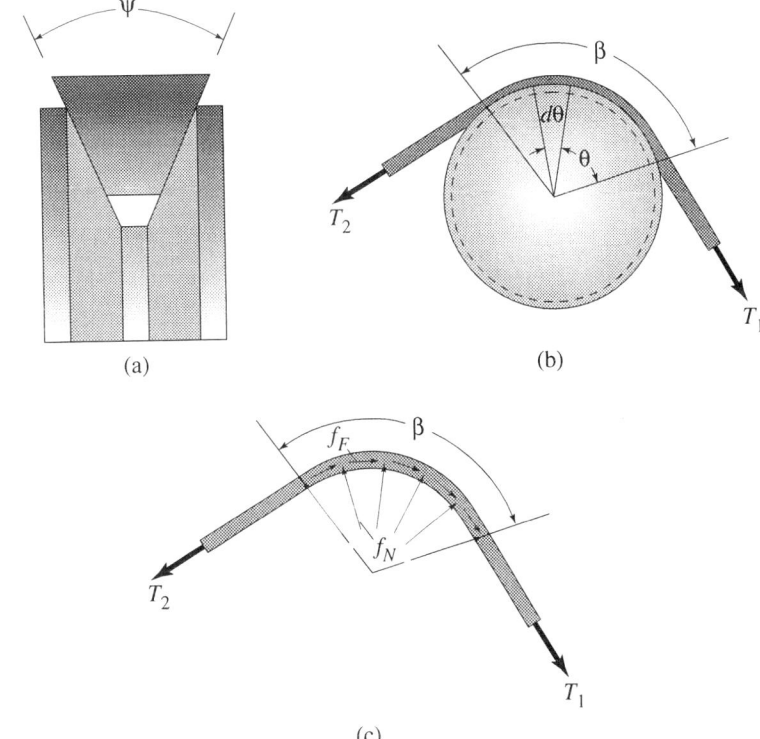

FIGURE 9.6.   (c)

to the two inclined sides of the V-belt. It should be pointed out that the
magnitudes of f_F and f_N vary from point to point within the arc of
contact defined by the angle β.

The free-body diagram of a small segment of the V-belt, defined in
Figure 9.6(b) by the angle $d\theta$, is shown in the three-dimensional view of
Figure 9.7. Because the tension in the V-belt increases from right to
left, we have identified the tension on the right face of the segment by T
and that on the left face by $T + dT$ where dT represents the increase in
T over the length of the segment defined by $d\theta$. A differential normal
force dN acts on each of the two inclined sides of the V-belt. Also,
because the belt is on the verge of motion, a frictional force $\mu_s\, dN$ acts
on each of the two inclined sides of the V-belt, as shown. However, for
clarity, only one force dN and one force $\mu_s\, dN$ are shown in Figure 9.7.
Using the coordinate system shown,

$$\sum F_z = 0, \quad 2\, dN \sin\left(\frac{\psi}{2}\right) - T\sin\left(\frac{d\theta}{2}\right) - (T + dT)\sin\left(\frac{d\theta}{2}\right) = 0. \quad \text{(a)}$$

Because $\dfrac{d\theta}{2}$ is extremely small, we may replace $\sin\left(\dfrac{d\theta}{2}\right)$ by $\dfrac{d\theta}{2}$ and

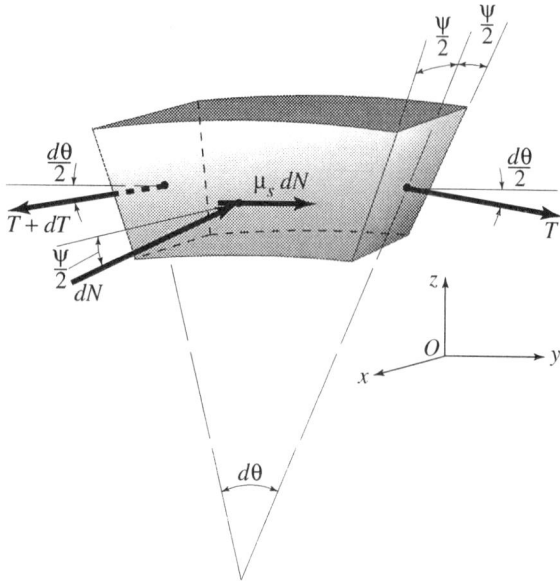

FIGURE 9.7.

simplify Eq. (a) to obtain

$$2dN \sin\left(\frac{\psi}{2}\right) - 2T\left(\frac{d\theta}{2}\right) - (dT)\left(\frac{d\theta}{2}\right) = 0. \qquad \text{(b)}$$

The third term in Eq. (b) is the product of two differential quantites and may be neglected to obtain

$$2dN \sin\left(\frac{\psi}{2}\right) = T\,d\theta, \qquad \text{(c)}$$

$$\sum F_y = 0, \quad T\cos\left(\frac{d\theta}{2}\right) + 2\mu_s\,dN - (T + dT)\cos\left(\frac{d\theta}{2}\right) = 0. \qquad \text{(d)}$$

Because $\dfrac{d\theta}{2}$ is extremely small, we may replace $\cos\left(\dfrac{d\theta}{2}\right)$ by unity and simplify Eq. (d) to obtain

$$dT = 2\mu_s\,dN \qquad \text{(e)}$$

Equations (c) and (e), the basic simultaneous differential relationships for the V-belt, represent the equilibrium of any differential segment of the belt when motion of the belt impends relative to the pulley.

Solving Eq. (c) for dN and substituting in Eq. (e),

$$dT = \left[\frac{\mu_s}{\sin\left(\dfrac{\psi}{2}\right)}\right] T\,d\theta. \qquad \text{(f)}$$

Separating variables and integrating,

$$\int_{T_1}^{T_2} \frac{dT}{T} = \frac{\mu_s}{\sin\left(\dfrac{\psi}{2}\right)} \int_0^\beta d\theta$$

from which

$$\ln T \Big|_{T_1}^{T_2} = \left[\frac{\mu_s}{\sin\left(\dfrac{\psi}{2}\right)}\right] \theta \Big|_0^\beta.$$

After substituting the limits and simplifying, the expression above may be written as

$$\ln \frac{T_2}{T_1} = \frac{\mu_s \beta}{\sin\left(\dfrac{\psi}{2}\right)} \tag{9.6a}$$

or

$$\frac{T_2}{T_1} = e^{\mu_s \beta / \sin(\psi/2)}. \tag{9.6b}$$

In summary, Eqs. 9.6 relate the tension T_2 to the tension T_1, where $T_2 > T_1$, in a V-belt that is on the verge of slippage relative to the pulley. The angle β is the angle of wrap which defines the arc of contact between the belt and the pulley, and the angle ψ is the angle of the V-belt.

Equations 9.6 may be specialized to the case of a flat belt for which $\psi = 180°$. Therefore, $\sin\left(\dfrac{\psi}{2}\right) = 1$ and Eqs. 9.6(a) and 9.6(b) become

$$\ln \frac{T_2}{T_1} = \mu_s \beta \tag{9.7a}$$

or

$$\frac{T_2}{T_1} = e^{\mu_s \beta}. \tag{9.7b}$$

The use of the above relationships in solving belt problems will be illustrated in the following examples.

■ **Example 9.13**

Refer to the system shown in Figure E9.13(a). If W_2 has impending downward motion, determine W_2 in terms of W_1 and the coefficient of static friction μ_s. Assume that the belt is flat (i.e., $\psi = \pi$). Specialize the resulting equation for the case where $W_1 = 100$ lb and $\mu_s = 0.5$.

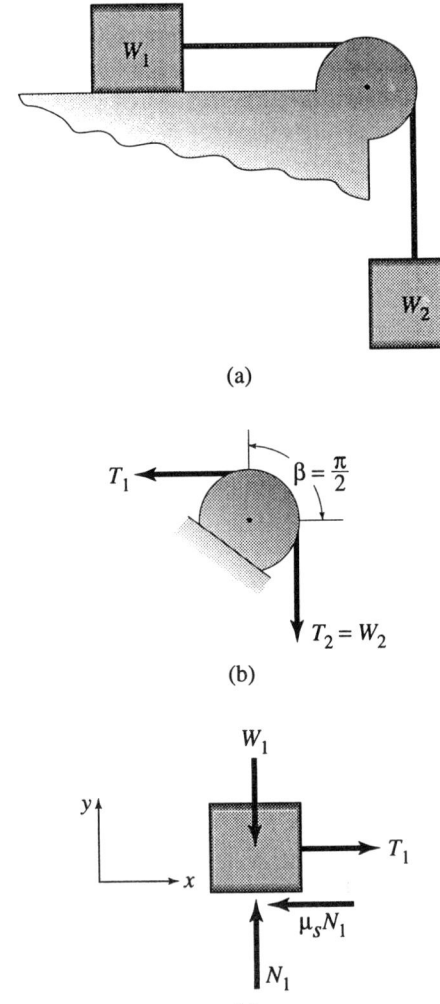

FIGURE E9.13.

Solution

Because W_2 has impending downward motion, frictional forces act to oppose this motion, and the belt tensions will be as shown in Figure E9.13(b) where the angle of wrap is $\beta = \pi/2$ and $T_2 = W_2$. Thus, by Eq. (9.7b),

$$\frac{W_2}{T_1} = e^{\mu_s(\pi/2)}. \tag{a}$$

The free-body diagram of block W_1 is shown in Figure E9.13(c) along with a convenient coordinate system. Thus,

$$\sum F_y = 0, \quad N_1 - W_1 = 0,$$

$$N_1 = W_1, \tag{b}$$

and

$$\sum F_x = 0, \quad N_1 - \mu_s N_1 = 0. \tag{c}$$

Substituting Eq. (b) in Eq. (c) and solving for T_1 yields

$$T_1 = \mu_s W_1 \tag{d}$$

Now, returning to Eq. (a), substituting for T_1 from Eq. (d), and solving for W_2

$$W_2 = \mu_s W_1 e^{\mu_s(\pi/2)}. \qquad \text{ANS.}$$

Substituting $\mu_s = 0.5$ and $W_1 = 100$ lb and solving for W_2 yields

$$W_2 = 109.7 \text{ lb.} \qquad \text{ANS.}$$

■ Example 9.14

Two weights are suspended from a rope which passes over two fixed cylindical pegs, as shown in Figure E9.14(a). The coefficient of static friction is μ_s and the weight W_2 is on the verge of moving downward. Express the ratio of W_2 to W_1 as a function of μ_s, and, thus, show that the result is independent of θ. Consider that the rope behaves like a flat belt.

Solution

Because W_2 is on the verge of moving downward, frictional forces over peg A act to oppose this motion, and the tensions in the rope on this cylinder will be as shown in Figure E9.14(b). Note that $T_2 = W_2$ and the angle of wrap $\beta = \pi/2 + \theta$. By Eq. (9.7b),

$$\frac{W_2}{T_1} = e^{\mu_s(\pi/2+\theta)}$$

from which

$$T_1 = \frac{W_2}{e^{\mu_s(\pi/2+\theta)}}. \tag{a}$$

The tensions in the rope on peg B will be as shown in Figure E9.14(c) because the frictonal forces tend to oppose the upward impending motion of block W_1. Note that $T_2' = T_1$, $T_1' = W_1$ and the angle of wrap $\beta = \pi/2 - \theta$. Thus, by Eq. (9.7b),

$$\frac{T_2'}{T_1'} = \frac{T_1}{W_1} = e^{\mu_s(\pi/2-\theta)}$$

from which

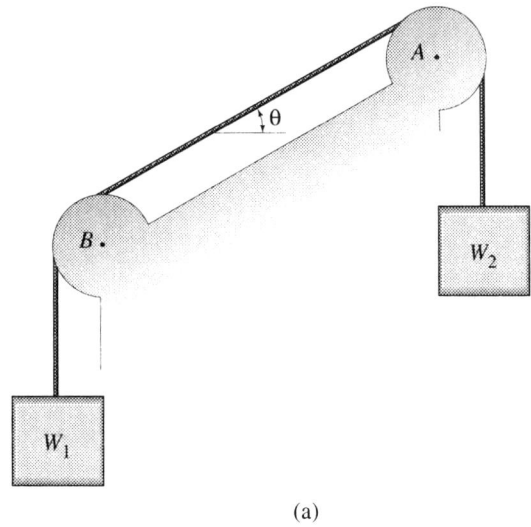

(a)

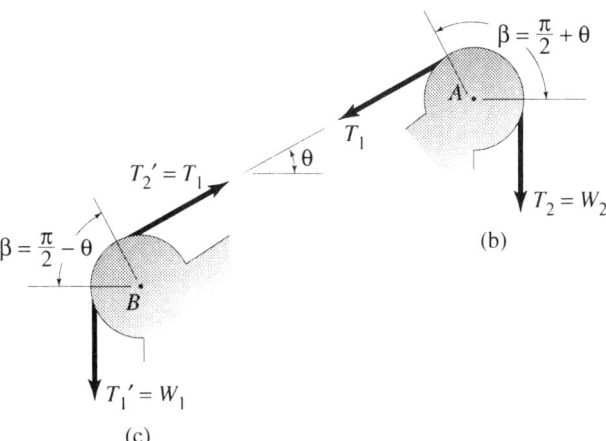

(b)

(c)

FIGURE E9.14.

$$T_1 = W_1 e^{\mu_s(\pi/2 - \theta)}. \tag{b}$$

Equating the values of T_1 from Eqs. (a) and (b) and solving for the ratio $\dfrac{W_2}{W_1}$ yields

$$\frac{W_2}{W_1} = e^{\mu_s(\pi/2 - \theta)} \cdot e^{\mu_s(\pi/2 + \theta)} = e^{\mu_s \pi}. \qquad \text{ANS.}$$

■ **Example 9.15**   Consider the system shown in Figure E9.15(a) in which $W_2 = 500$ lb and the coefficient of static friction $\mu_s = 0.2$ for all contacting surfaces.

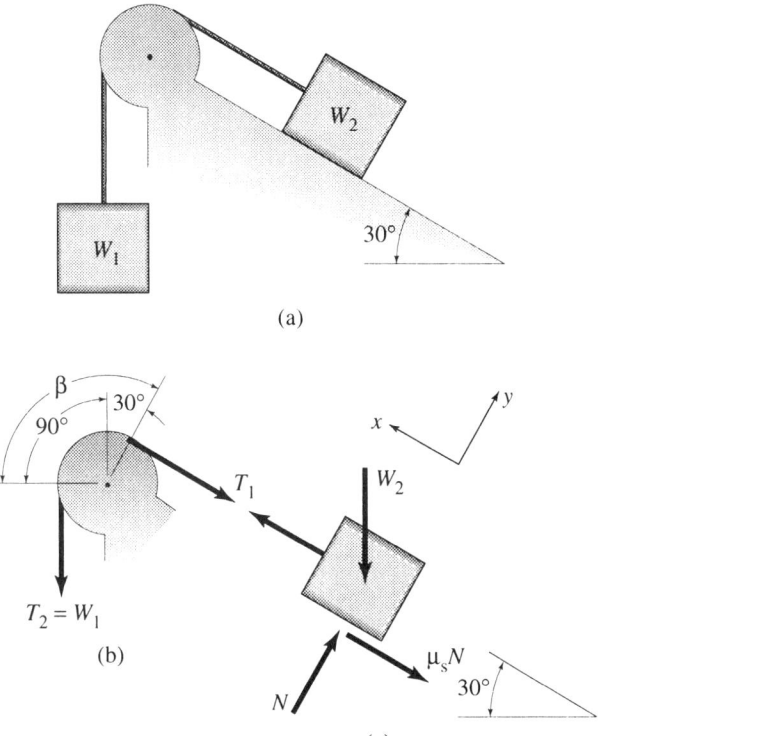

(a)

(b)

(c)

$\text{F\scriptsize IGURE}$ E9.15.

Find W_1 so that W_2 has impending motion up the inclined plane. Assume a V-belt with $\psi = \pi/2$.

Solution

Because motion of W_2 impends up the plane, frictional forces on the fixed peg act to impede this motion. The tensions in the V-belt on the peg will be as shown in Figure E9.15(b). Note that $T_2 = W_1$ and the angle of wrap $\beta = 90° + 30° = 120° = 2\pi/3$ rad. Thus, by Eq. (9.6b),

$$\frac{W_1}{T_1} = e^{0.2(2\pi/3)/\sin \pi/4} = 1.808.$$

Thus,

$$W_1 = 1.808\,T_1. \tag{a}$$

The free-body diagram of W_2 is shown in Figure E9.15(c) along with a convenient coordinate system. Thus

$$\sum F_y = 0, \quad N - W_2 \cos 30° = 0,$$

$$N = 0.866 W_2, \tag{b}$$

and

$$\sum F_x = 0, \quad T_1 - \mu N - W_2 \sin 30^\circ = 0,$$

$$T_1 = \mu N + 0.5 W_2. \tag{c}$$

Substituting from Eq. (b) in Eq. (c) yields

$$T_1 = 1.366 W_2. \tag{d}$$

Substituting Eq. (d) in Eq. (a) gives

$$W_1 = 2.470 W_2 = 2.470(500) = 1235 \text{ lb.} \qquad \text{ANS.}$$

■ **Example 9.16**   A band brake system is shown in Figure E9.16(a). The drum weights 0.4 kN and rotates at a constant cw angular velocity under the effect of

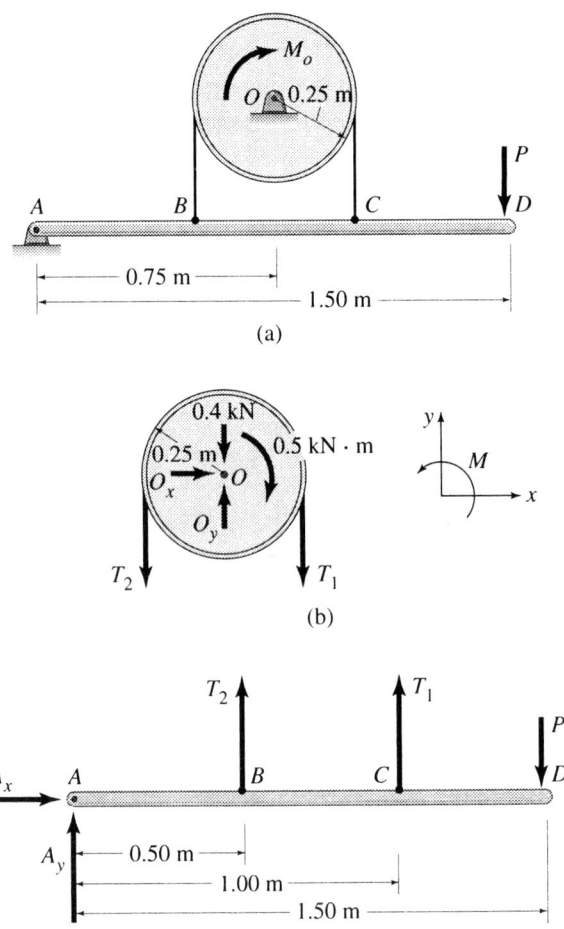

FIGURE E9.16.

the constant couple $M_O = 0.5$ kN·m. Determine the magnitude of the applied force P required to bring the drum to a complete stop if the coefficient of kinetic friction is 0.3 between the band and the drum. Assume that the band is a flat belt, and ignore friction at the hinge at A and at the shaft at O.

Solution

Because the drum rotates cw relative to the belt, frictional forces on it will act in a ccw sense to impede its rotation. By action and reaction, the frictional forces on the belt will act in a cw sense, and the tensions on the belt around the drum will be as shown in the free-body diagram of Figure E9.16(b). Note that the angle of wrap $\beta = \pi$. Also, because the drum rotates at a constant angular velocity, its angular acceleration is zero and the equations of statics are applicable. Using the coordinate system shown,

$$\sum M_O = 0, \quad T_2(0.25) - T_1(0.25) - 0.5 = 0,$$

$$T_2 - T_1 = 2 \tag{a}$$

Also, by Eq. (9.6b),

$$\frac{T_2}{T_1} = e^{0.3\pi} = 2.566$$

from which

$$T_2 = 2.566T_1. \tag{b}$$

Solving Eqs. (a) and (b) simultaneously yields

$$T_1 = 1.277 \text{ kN}$$

and

$$T_2 = 3.277 \text{ kN}.$$

Now consider the free-body diagram of the lever ABCD shown in Figure E9.16(c). Using the same coordinate system as above,

$$\sum M_A = 0, \quad 1.277(1.00) + 3.277(0.5) - P(1.50) = 0,$$

$$P = 1.944 \text{ kN}. \tag{ANS.}$$

∎

Problems

9.85 Motion of the V-belt ($\psi = \pi/2$ radians) impends cw with respect to the pulley shown in Figure P9.85. If the pulley weighs 20 lb and its radius is 0.5 ft, $T_1 =$ 100 lb, and $\mu_s = 0.3$ determine the belt tension T_2. Also, find the frictional couple M_O and the horizontal and vertical forces acting on the pulley at O.

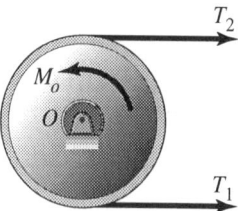

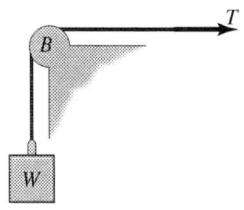

FIGURE P9.85.   FIGURE P9.87.

9.86 Refer to Figure P9.86, and determine T needed to keep the weight $W = 1200$ N from moving downward. A couple M_O keeps the cylinder from turning about its axle at A. Let $\mu_s = 0.2$. The rope is wrapped 2.25 times around the cylinder. Find the couple M_O and the resultant reaction at A. Assume that the rope is a flat belt.

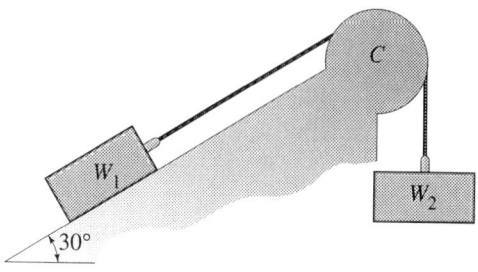

FIGURE P9.88.

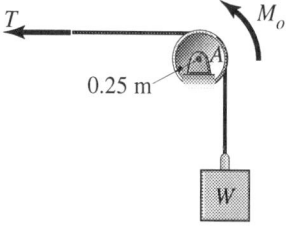

FIGURE P9.86.

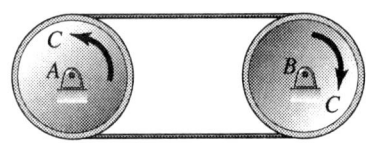

FIGURE P9.89.

9.87 Refer to Figure P9.87, and find T for $W = 2000$ N if $\mu_s = 0.1$ and if (a) motion of W impends downward and (b) motion of W impends upward. The cylinder at B is fixed, and the belt is flat.

9.88 $W_2 = 4000$ lb and $\mu_s = 0.2$ for the cable and fixed peg at C, as shown in Figure P9.88. If the inclined plane is assumed to be frictionless, determine W_1 for (a) impending downward motion of W_2 and (b) impending upward motion of W_2. Assumes that the belt is flat.

9.89 A V-belt for which $\psi = 45°$ connects pulley A to pulley B, as shown in Figure P9.89. The couples C each equals 20 kN·m, and the pulleys are both rotating cw at a constant angular velocity. Determine the belt tensions as slip of the belt impends on the pulleys for $\mu_s = 0.5$. Each pulley has a radius of 0.2 m. Because the pulleys rotate at constant angular velocity (i.e., the angular acceleration is zero), the equations of statics apply.

9.90 Refer to Figure P9.90, and let $\mu_s = 0.3$ for all surfaces. If $W = 100$ lb, find T so that (a) motion of W impends up the inclined plane and (b) motion of W impends down the inclined plane. Assume that the belt is flat.

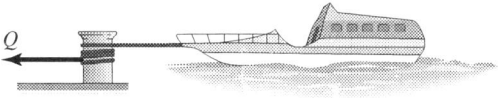

FIGURE P9.92.

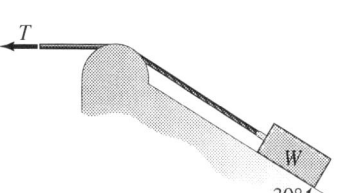

FIGURE P9.90.

9.91 Refer to Figure P9.91, and note that A and B are fixed cylinders. The rope is wrapped twice around A and 1.25 times around B, and the coefficient of static friction is 0.4. If $W = 500$ lb, determine T consistent with downward impending motion of W. Assume that the rope is a flat belt.

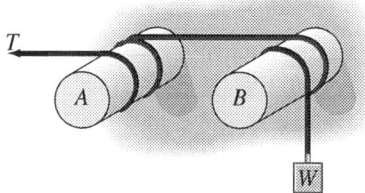

FIGURE P9.91.

9.92 A small yacht is temporarily moored, as shown in Figure P9.92. If it exerts a force of 20,000 lb on the mooring line and a person on the dock exerts a maximum force $Q = 100$ lb, determine the minimum number of turns of the moor-

ing line around the capstan for $\mu_s = 0.3$. Assume that the mooring line is a flat belt.

9.93 A heavy crate containing hardware weighing 40 kN is about to be lowered by the arrangement shown in Figure P9.93. The cylinders at A and B are fixed, and the coefficient of friction is 0.1 for the cable and cylinders. Regarding the cable as a flat belt, determine the forces P if the cable is wrapped 5.25 turns around each cylinder.

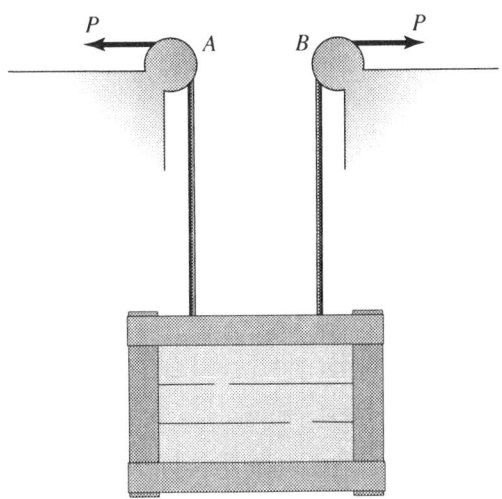

FIGURE P9.93.

9.94 A strong wind creates a force of 2000 lb in the mooring rope of the sail boat shown in Figure P9.94. The rope is wrapped twice around a tree, and we may assume that the coefficient of friction is 0.3 between the tree and rope.

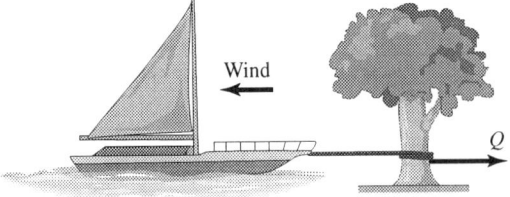

FIGURE P9.94.

Determine the force Q required just to prevent the sail boat from moving in the windward direction. Assume that the rope is a flat belt.

9.95   In Figure P9.95, block A weighs 400 N and block B weighs 1600 N. The coefficient of static friction is $\mu_s = 0.2$ for all surfaces. Determine P so that (a) the motion of B impends downward and (b) the motion of B impends upward. The connecting cable may be assumed to be a flat belt.

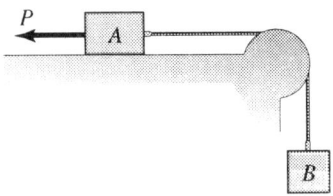

FIGURE P9.95.

9.96   Refer to Figure P9.93 showing a crate containing a shipment weighing 500 lb. The cylinders at A and B are fixed and are coated with a material which reduces the static coefficient of friction to 0.04. Regarding the cable as a flat belt, determine the forces P required to begin to move the crate upward. The cable is wrapped one-quarter turn around each cylinder.

9.97   In Figure P9.97, a cable is wrapped around a cylindrical peg at A and twice around each of the capstans B and C. If $\mu_s = 0.2$ and $Q = 100$ lb, determine the weight W which has an impending motion downward. Assume that the cable is a flat belt.

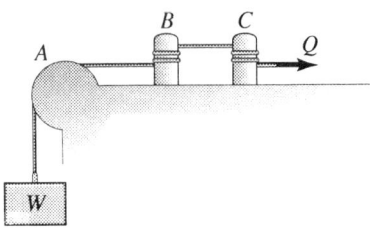

FIGURE P9.97.

9.98   Solve Problem 9.97 if the weight W is to be raised.

9.99   Refer to Figure P9.99, and determine the tensions in the brake band and the reaction components at pins A and B for the following inputs: $P = 62.5$ lb, $M = 2000$ lb·in and $\mu_s = 0.1$ between the brake band and the flywheel. The flywheel weight is 3000 lb, and you may neglect the weight and thickness of the vertical brake arm. Motion of the flywheel impends cw for the given values of P and M.

9.100 In Figure P9.100, two different schemes for positioning two fixed cylinders A and B are shown. Intuitively, which of these two schemes is preferred for raising the weight W? Perform analyses to find P and Q in terms of W. Was your intuitive answer correct? Briefly explain why. The ratio of r to d is equal to 0.25 and the static coefficient of friction is 0.3 between the flat belt and the fixed cylinders. (Hint: Prove that $\sin \theta = 2r/d$).

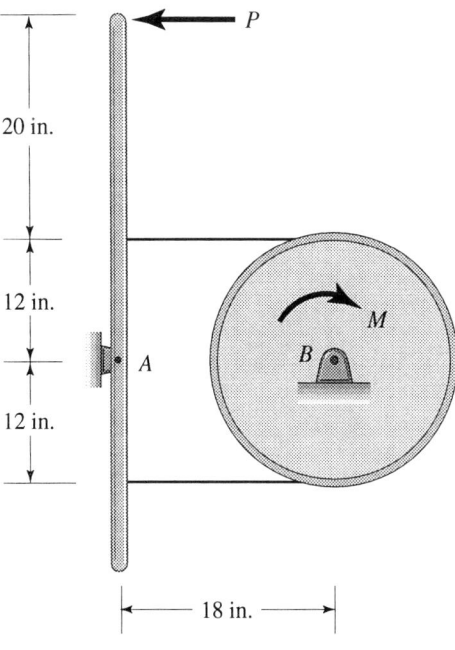

FIGURE P9.99.

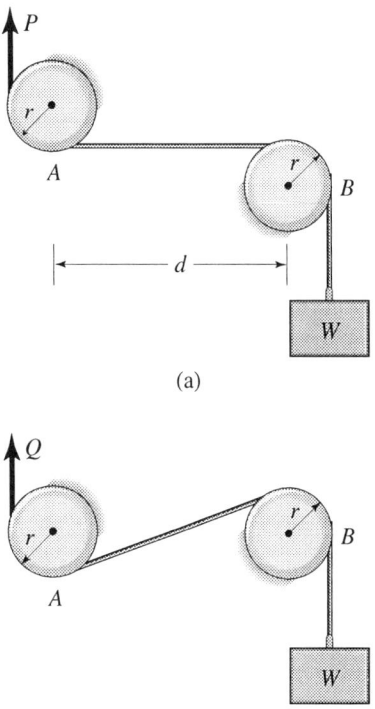

(a)

(b)

FIGURE P9.100.

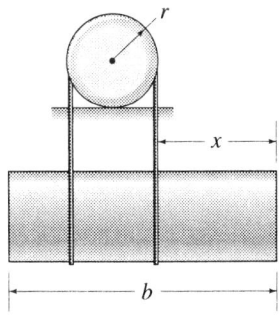

FIGURE P9.102.

9.101 Solve Problem 9.100 assuming that the weight W is to be lowered.

9.102 A steel ingot of weight W has cooled sufficiently to be suspended by a cable wrapped around a fixed overhead cylinder as shown in Figure P9.102. Determine the largest value of x as a function b, r and μ_s so that the ingot is on the verge of tipping cw. Assume that the weight of the ingot acts through its geometric center and that the cable is a flat belt. Use the frictionless case (i.e., let $\mu_s = 0$) as a check on your answer. For this special case, $x = b/2 - r$ which may be readily verified by drawing a free-body diagram of the ingot and letting each cable tension equal $W/2$.

9.103 Refer to the brake system shown in Figure P9.103, and note that all dimensions are given in terms of the radius r. Express the ratio of P to W as a function of

the static coeficient of friction μ_s for impending motion of W downward. Prepare a neat plot of P/W vs. μ_s for $0 < \mu_s \leq 1$. Ignore the weights of the drum and the brake arm.

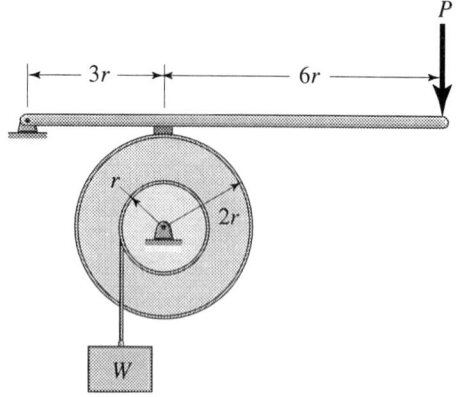

FIGURE P9.103.

9.104 Refer to Problem 9.103, and let $\mu_s = 0.5$ and $W = 400$ N. Solve for P and all other unknown forces of this system. Show your answers on free-body diagrams of the brake arm and the drum. Ignore the weights of the brake arm and the drum.

9.105 In Figure P9.105, the three cylinders A, B, and C are fixed, and the cable may be idealized as a flat belt. Express the ratio of Q to W as a function of the static coefficient of friction μ_s for (a) impending upward motion of W and (b) impending downward motion of W.

9.106 A disabled bus, shown in Figure P9.106, weighs 5000 lb, rests on a $10°$ inclined plane, and is prevented from moving by a rope wound around a guardrail post. If the person holding the rope exerts an 80-lb force, determine the number of wraps required around the post to prevent impending motion down the inclined plane. Assume $\mu_s = 0.2$ for the rope and post, and neglect frictional resistance at the bus wheels. Assume that the rope is a flat belt.

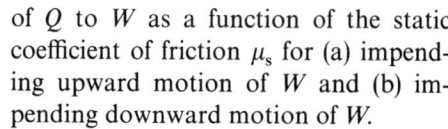

FIGURE P9.106.

9.107 A crate and its contents weigh 10,000 lb. It is to be lowered as shown in Figure P9.107. If each cable is wrapped 4.25

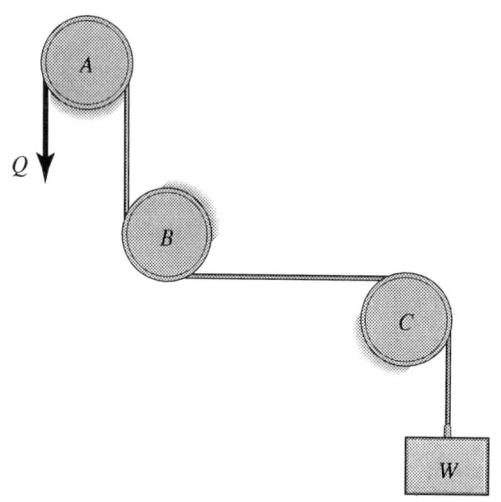

FIGURE P9.105.

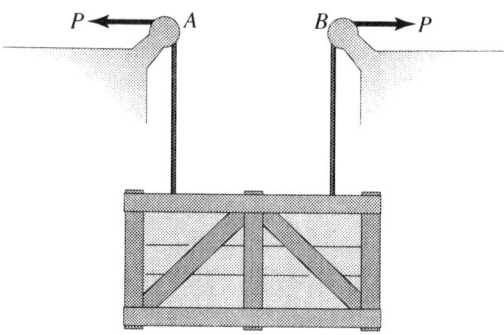

FIGURE P9.107.

times around the fixed cylinders A and
B and the coefficient of static friction
is 0.2, determine each force P applied
as the crate is on the verge of moving
downward.

9.108 Refer to Figure P9.108, and express the
couple C in terms of P and d. Motion
of the drum impends ccw, and the coefficient of static friction is 0.5 between the
flat belt and the drum. The pins at A
and B are frictionless.

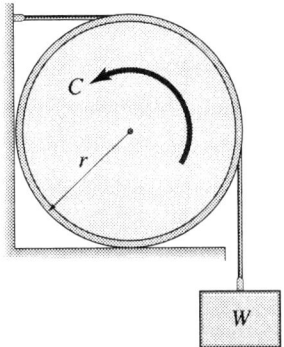

FIGURE P9.111.

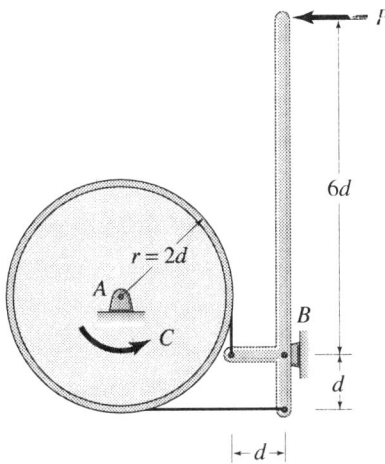

FIGURE P9.108.

9.109 Refer to Problem 9.108, and reverse the
sense of the couple C. Motion of the
drum now impends cw. Solve for the
couple C in terms of P and d.

9.110 Refer to Figure P9.108, and let $C =$
2000 lb·in., $P = 100$ lb, and $d = 4$ in.
Determine the static coefficient of friction μ_s for impending ccw rotation of
the drum. The pins at A and B are
frictionless.

9.111 Refer to Figure P9.111, and let the cylinder weight be $2W$. If the vertical surface
is frictionless and $\mu_s = 0.1$ for all other
surfaces, find the couple C as a function

of W and r so that rotation of the cylinder impends ccw. Assume that the
weight of the cylinder acts through its
geometric center and that the belt is flat.

9.112 Solve Problem 9.111 for $W = 500$ lb,
$r = 8$ in., and $\mu_s = 0.15$.

9.113 Refer to Figure P9.113, and assume that
the motion of W_1 impends down the
plane. Express W_2 as a function of W_1, θ,
and μ_s where μ_s is the coefficient of friction between the block and the plane
and between the cable and the fixed
peg A. Assume that the cable is a flat
belt.

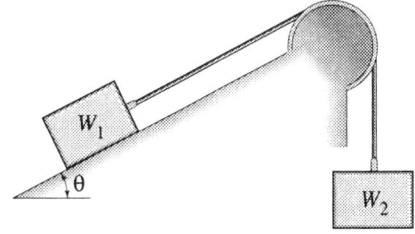

FIGURE P9.113.

9.114 Solve Problem 9.113 for impending motion of W_1 up the plane. Specialize the
solution for $W_1 = 800$ N, $\theta = 45°$, and
$\mu_s = 0.1$.

9.115 Solve Problem 9.113 for $W_1 = 1200$ N, $\theta = 10°$, and $\mu_s = 0.08$.

9.116 In Figure P9.116, the cylinder has a weight W_3, and there is no friction at A and B. Determine the applied couple C as a function of W_1, μ_s, and r when slip impends between the cable (flat belt) and the cylinder. Note that the couple C is independent of the angle θ. Why?

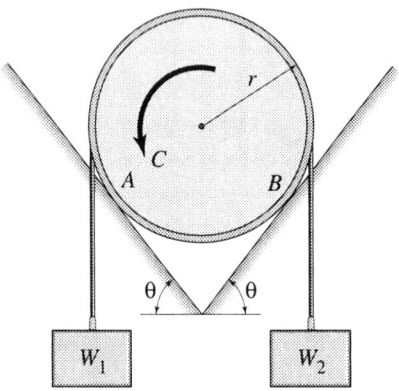

FIGURE P9.116.

9.117 (a) Solve problem 9.116 for the case when $W_2 = 500$ lb, $\mu_s = 0.15$, $r = 1.5$ ft, and $\theta = 45°$. (b) Find the normal reactions at A and B corresponding to the numerical values in part (a) and for $W_3 = 200$ lb.

9.118 (a) Solve Problem 9.116 for $W_1 = 200$ lb, $\mu_s = 0.35$, $\theta = 25°$, $r = 6$ in., and $W_3 = 800$ lb. (b) Find the normal reactions at A and B corresponding to the numerical values of part (a).

9.119 In Figure P9.119, the drum centered at A has a weight W_1, and the pins at A and B are frictionless. Determine the force P and the cable (flat belt) tensions in terms of W and the coefficient of static friction μ_s. Rotation of the drum impends ccw. The vertical cable is firmly attached to the drum.

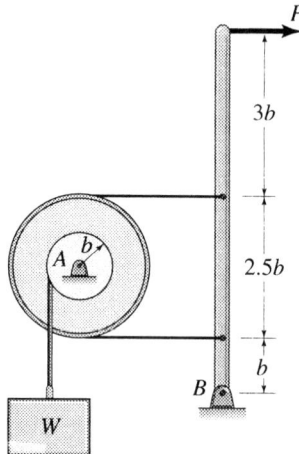

FIGURE P9.119.

9.120 Solve Problem 9.119 for $W = 200$ N and $\mu_s = 0.25$.

9.121 Refer to Figure P9.121 where the composite cylinder B weighs 1200 N. The horizontal plane and pulley A are frictionless. Determine the minimum coefficient of friction required to prevent slip between cylinder B and the inclined plane. Also, find the normal force between the cylinder and the horizontal plane. The cable is firmly attached to cylinder B.

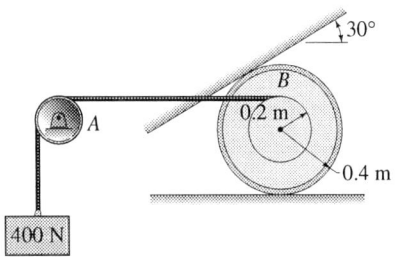

FIGURE P9.121.

9.122 In Figure P9.122, a tape passes over three fixed cylinders. If the tension at A is 10 lb, the coefficient of friction is 0.02 between the tape and the cylinders, and the motion of point B impends downward, find the tension T at B.

quantities M_O, r, a, b, c, and μ, the static coefficient of friction at the brake shoe. Express the reaction components at hinge C in terms of the same quantities.

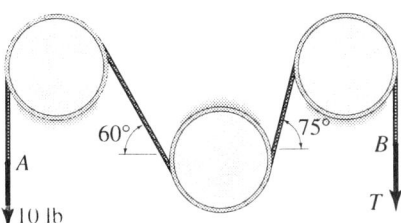

FIGURE P9.122.

9.123 Solve Problem 9.122 if the motion of point B impends upward.

9.124 A couple M_O is applied to the cylinder of radius r to keep it rotating at a constant ccw angular velocity as shown in Figure P9.124. Determine the couple M_O in terms of P, r, a, b, and μ, the kinetic coefficient of friction between the drum and the belt. Note that, because the cylinder rotates at a constant angular velocity, its angular acceleration is zero and the equations of statics apply.

9.125 The drum of radius r is kept from rotating under the effect of M_O by the brake system shown in Figure P9.125. Determine the force P in terms of the known

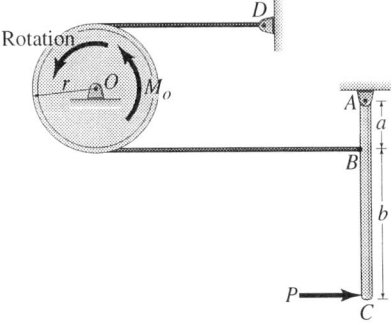

FIGURE P9.124.

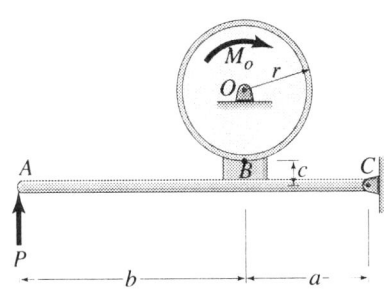

FIGURE P9.125.

9.6*
Friction on Pivot and Collar Bearings and Disks

Pivot and *collar* bearings are designed to support rotating shafts which must transmit axial forces and couples. Such bearings, referred to as *thrust bearings*, are illustrated in Figure 9.8(a) and (b). The pivot bearing shown in Figure 9.8(a) transmits its axial force P to a circular area of diameter D_O whereas the collar bearing, shown in Figure 9.8(b) transmits its axial force to an annular area of outside diameter D_O and inside diameter D_i. In each case, a *normal pressure p*, defined as force per unit area, develops and is generally assumed to be constant over the entire contact area A. Thus, $p = P/A$. The couple transmitted by a

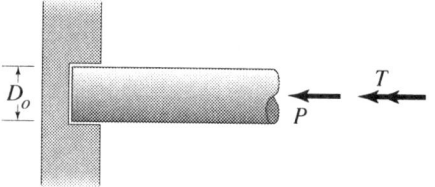

(a) Pivot bearing

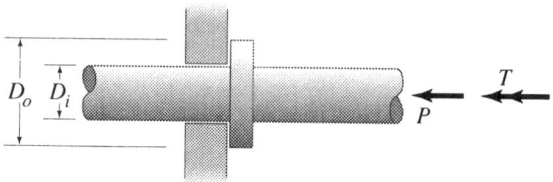

FIGURE 9.8.   (b) Collar bearing

shaft is known as a *torque* and is given the symbol T, as shown. During rotation, frictional forces develop which are distributed over the contact area A and act to resist the rotation of the shaft. This type of friction, referred to as *disk friction*, is found in a number of mechanical devices, including pivot and collar bearings as well as disk clutches and disk brakes.

It is desirable, therefore, to develop a general relationship that would provide the magnitude of the torque T in terms of the axial force P, the coefficient of kinetic friction μ_k, and the geometry involved. For this purpose, consider the hollow cylinder of diameters D_i and D_o, subjected to the axial force P and the torque T, as shown in Figure 9.9. Such a cylinder represents the conditions that exist in a collar bearing, but it could also represent a pivot bearing by letting $D_i = 0$.

Now, consider the differential element of area $dA = r\,dr\,d\theta$ shown in Figure 9.9. This area is subjected to a diferential normal force $dN = p\,dA = \left(\dfrac{P}{A}\right)dA$ and a differential friction force $dF = \mu_k\,dN = \mu_k\left(\dfrac{P}{A}\right)dA$. The differential couple dC produced by this differential friction force about the shaft axis, thus, becomes

$$dC = r\,dF = r\mu_k\left(\frac{P}{A}\right)dA. \tag{a}$$

Because $A = \dfrac{\pi}{4}(D_o^2 - D_i^2)$ and $dA = r\,dr\,d\theta$ for a hollow shaft, it follows from Eq. (a) that

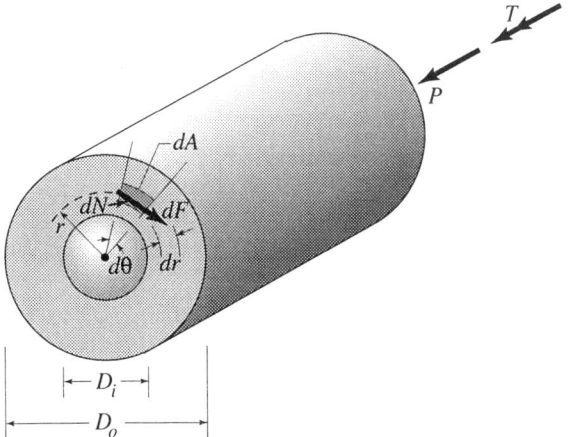

FIGURE 9.9.

$$dC = \frac{4\mu_k P}{\pi(D_O^2 - D_i^2)} r^2 \, dr \, d\theta. \tag{b}$$

Integration of Eq. (b) leads to the couple C produced by the frictional forces about the axis of the shaft. If the shaft rotates at a constant angular velocity, which is the case after the shaft has reached its operating speed, it follows, from the equilibrium of the system, that the couple C is equal to the applied torque T. Thus,

$$T = \int dC = \int_0^{2\pi} \int_{D_i/2}^{D_O/2} \frac{4\mu_k P}{\pi(D_O^2 - D_i^2)} r^2 \, dr \, d\theta,$$

$$= \frac{4\mu_k P}{\pi(D_O^2 - D_i^2)} \left[\frac{r^3}{3}\right]_{D_i/2}^{D_O/2} \int_0^{2\pi} d\theta,$$

$$= \frac{\mu_k P(D_O^3 - D_i^3)}{6\pi(D_O^2 - D_i^2)} [\theta]_0^{2\pi},$$

$$= \frac{\mu_k P}{3} \left[\frac{D_O^3 - D_i^3}{D_O^2 - D_i^2}\right]. \tag{9.8}$$

In the case of a pivot bearing or a disk for which $D_i = 0$, Eq. (9.8) reduces to

$$T = \frac{\mu_k P}{3} D_O. \tag{9.9}$$

Equations (9.8) and (9.9) provide the magnitude of the torque T needed to maintain rotation of the shaft at constant speed. By using μ_s instead of μ_k, these equations may also be used to give the torque T required for impending motion of the shaft.

The following examples illustrate some of the concepts discussed in this section.

■ **Example 9.17**

A pivot bearing is shown in Figure E9.17(a) and a collar bearing is shown in Figure E9.17(b). In each case, determine the torques T required for impending rotation and the torques to maintain rotation if the static and kinetic coefficients of friction are 0.25 and 0.20, respectively. Assume that the axial force P produces a pressure which is uniformly distributed for both of these bearings.

Solution

Equation (9.9) is applicable for the pivot bearing. Thus, for impending rotation,

$$T = \frac{\mu_s P D_O}{3} = \frac{0.25(800)(6)}{3} = 400 \text{ lb·in.} \qquad \text{ANS.}$$

To maintain rotation,

$$T = \frac{\mu_k P D_O}{3} = \frac{0.20(800)(6)}{3} = 360 \text{ lb·in.} \qquad \text{ANS.}$$

Equation (9.8) is applicable for the collar bearing. Thus, for impending rotation,

$$T = \frac{\mu_s P}{3}\left(\frac{D_O^3 - D_i^3}{D_O^2 - D_i^2}\right) = \frac{0.25(1200)}{3}\left(\frac{10^3 - 6^3}{10^2 - 6^2}\right)$$

$$= 1225 \text{ lb·in.} \qquad \text{ANS.}$$

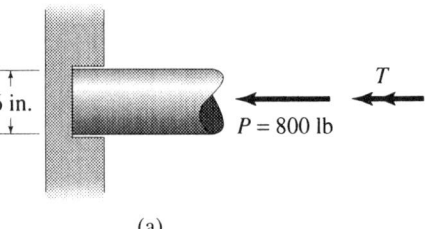

(a)

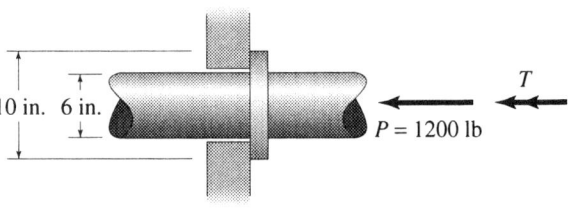

(b)

Figure E9.17.

To maintain rotation,

$$T = \frac{\mu_k P}{3}\left(\frac{D_O^3 - D_i^3}{D_O^2 - D_i^2}\right) = \frac{0.20(1200)}{3}\left(\frac{10^3 - 6^3}{10^2 - 6^2}\right)$$

$$= 980 \text{ lb·in.} \qquad\qquad \text{ANS.}$$

■ **Example 9.18**

Refer to the pivot bearing shown in Figure E9.18(a). The pressure p between the shaft and the bearing is not a constant as was the case in the development of Eqs. (9.8) and (9.9). It may, however, be assumed to vary linearly from p_O at the center to $0.75\,p_O$ at the outer circumference of the shaft, where p_O is a constant that may be determined. In terms of p, D_O, and the coefficient of static friction μ_s, develop an expression for the torque T needed for impending rotation of the shaft. Specialize this expression for the case where $P = 800$ lb, $\mu_s = 0.25$, and $D_O = 6$ in.

Solution

The linear variation of the pressure p along a radial line is given by the relationship

$$p = p_O\left(1 - \frac{r}{2D_O}\right) \qquad\qquad \text{(a)}$$

which agrees with the given information that $p = p_O$ at $r = 0$ and $p = 0.75 p_O$ at $r = D_O/2$. The pressure given by Eq. (a) is shown in Figure E9.18(b).

Using a differential element of area $dA = r\,dr\,d\theta$ as shown in Figure E9.18(c),

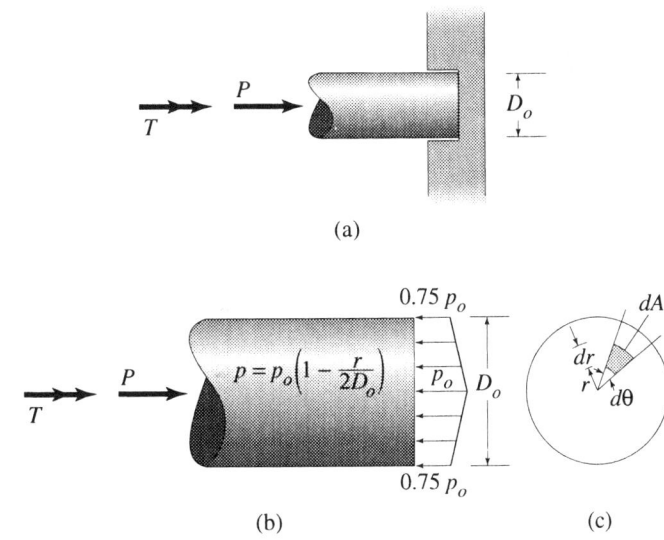

(a)

(b)   (c)

FIGURE E9.18.

$$dP = dN = p\,dA$$

$$= p_O\left(1 - \frac{r}{2D_O}\right)r\,dr\,d\theta. \qquad (b)$$

Thus,

$$P = \int dP = p_O \int_0^{\pi/2} \int_0^{D_O/2} \left(1 - \frac{r}{2D_O}\right)r\,dr\,d\theta$$

$$= \frac{5\pi D_O^2 p_O}{24}$$

from which

$$p_O = \frac{24P}{5\pi D_O^2}. \qquad (c)$$

To find an expression for the torque T, we proceed as in the development of Eqs. (9.8) and (9.9). Thus, from Eq. (b),

$$dN = p_O\left(1 - \frac{r}{2D_O}\right)r\,dr\,d\theta,$$

$$dF = \mu_s\,dN = \mu_s p_O\left(1 - \frac{r}{2D_O}\right)r\,dr\,d\theta,$$

$$dT = r\,dF = \mu_s p_O\left(1 - \frac{r}{2D_O}\right)r^2\,dr\,d\theta,$$

$$T = \int dT = \mu_s p_O \int_0^{2\pi} \int_0^{D_O/2} \left(1 - \frac{r}{2D_O}\right)r^2\,dr\,d\theta$$

$$= \frac{13\pi\mu_s D_O^3 p_O}{192}. \qquad (d)$$

Substituting for p_O from Eq. (c),

$$T = \frac{13\mu_s D_O P}{40}.$$

Substituting the given numerical values for P, μ_s, and D_O,

$$T = \frac{13(0.25)(800)(6)}{40} = 390\ \text{lb}\cdot\text{in}.$$

This compares to $T = 400$ lb·in. found in Example 9.17 for an identical shaft except that the pressure distribution was constant. Also, note that the expression for T in Eq. (d) may be used to determine the magnitude of the torque T needed to maintain rotation of the shaft. All we need to do is to replace μ_s by μ_k.

9.7*
Friction on Journal Bearings

Unlike pivot and collar bearings which are designed to transmit axial forces, journal bearings are designed to carry lateral forces. Two journal bearings A and B are used to support the mechanical system shown in Figure 9.10. Each of these journal bearings carries a downward vertical force W which accounts for weights of the shafts, gears, and vertical forces developed at the points of contact between the meshing gears. If the system is to rotate at constant speed, a torque must be supplied, possibly by an electric motor, as shown, to overcome the frictional couple applied to the shaft by the journal bearing. It is assumed that the bearings are not lubricated or lubricated lightly, so that the laws of Coulomb dry friction apply.

A cross-sectional view of the shaft in journal bearing A is shown in Figure 9.11(a). As the shaft begins to rotate in a ccw sense, it tends to roll up the inner surface of the bearing, and, when it reaches its operating speed, will make contact with this surface along a line represented by point C. The free-body diagram of the shaft is shown, magnified, in Figure 9.11(b) when it is rotating at a constant angular speed. Also shown is a convenient coordinate system. Note that T represents the torque that must be applied to the shaft in bearing A to keep it rotating at constant speed. Also, the force R represents the resultant of the normal force N and the friction force $\mu_k N$ at point C and makes the angle ϕ_k with the normal at this point, where $\phi_k = \tan^{-1} \mu_k$. We note that, because W acts vertically downward, the force R must be equal and opposite to W. Thus, the resultant R must point vertically upward. In other words,

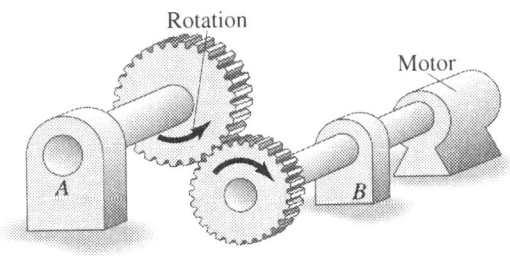

FIGURE 9.10.

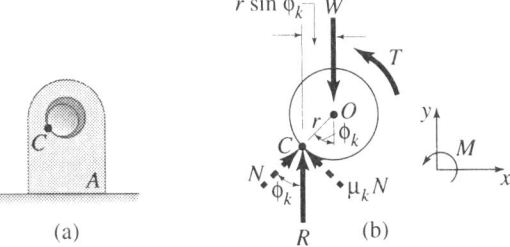

(a)   (b)   FIGURE 9.11.

$$\sum F_y = 0, \quad \Rightarrow \quad R = W \tag{9.10}$$

Also, the torque T is obtained from a summation of moments about point O. Thus,

$$\sum M_O = 0, \quad T - Rr \sin \phi_k = 0,$$

$$T = Rr \sin \phi_k.$$

Using Eq. (9.10), we conclude that

$$T = Wr \sin \phi_k. \tag{9.11}$$

Equation (9.11) provides the magnitude of the torque T required to maintain a constant rotation of the shaft. If we replace ϕ_k by ϕ_s, we can use Eq. (9.11) to determine the torque T needed for impending motion of the shaft.

■ **Example 9.19**

A shaft of radius $r = 3$ in. and weighing 30 lb/ft is attached rigidly to two pulleys each weighing 50 lb as shown in Figure E9.19. The two ends of the shaft are supported by journal bearings at A and B. If a torque of 100 lb·in. is needed, at one end of the shaft, to keep it rotating at constant speed, determine the coefficient of kinetic friction between the shaft and it bearings.

Solution

Because the system is symmetric, it may be assumed that each bearing has a weight W acting on it given by

$$W = \tfrac{1}{2}[30(8) + 2(50)] = 170 \text{ lb.}$$

Also, the torque T acting on each bearing is given by

$$T = \tfrac{1}{2}(100) = 50 \text{ lb·in.}$$

From Eq. (9.11),

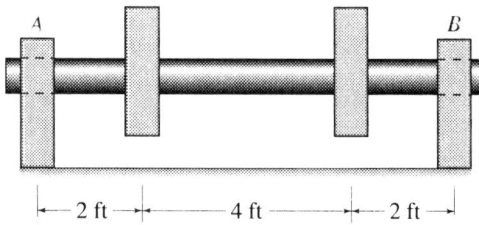

FIGURE E9.19. |— 2 ft —|— 4 ft —|— 2 ft —|

$$\sin \phi_k = \frac{T}{Wr} = \frac{50}{170(3)} = 0.098,$$

$$\phi_k = 5.63°.$$

Therefore,

$$\mu_k = \tan \phi_k = 0.0985. \qquad \text{ANS.}$$

■ Example 9.20

A segment of a beam weighing 1.5 kN/m fits loosely over a fixed cylinder of radius $r = 0.05$ m as shown in Figure E9.20(a). If $W_1 = 2$ kN, determine the magnitude of W_2 so that the beam is on the verge of rotating about the fixed cylinder (a) in a ccw sense and (b) in a cw sense. Assume that the weight of the beam segment acts through point G, and let $\mu_s = 0.15$.

Solution

(a) When the beam segment is on the verge of rotating in a ccw sense, its free-body diagram will be as shown in Figure E9.20(b). Thus, using the coordinate system shown,

$$\sum F_y = 0, \quad R - 2 - W_2 - 1.2 = 0,$$

$$R - W_2 = 3.2, \qquad (a)$$

and

$$\sum M_O = 0, \quad 1.2(0.1) + 2(0.5) - R(0.05) \sin \phi_s - W_2(0.3) = 0. \quad (b)$$

Because $\phi_s = \tan^{-1} 0.15 = 8.53°$, after simplification, Eq. (b) reduces to

$$R + 40.45 W_2 = 1.50. \qquad (c)$$

Solving Eqs. (a) and (c) simultaneously,

$$R = 6.77 \text{ kN},$$

and

$$W_2 = 3.57 \text{ kN}. \qquad \text{ANS.}$$

(b) When the beam segment is on the verge of rotating in a cw sense, its free-body diagram will be as shown in Figure E9.20(c). Using the coordinate system shown,

$$\sum F_y = 0, \quad R - 2 - W_2 - 1.2 = 0,$$

$$R - W_2 = 3.2, \qquad (d)$$

and

$$\sum M_O = 0, \quad 1.2(0.1) + 2(0.5) + R(0.05) \sin \phi_s - W_2(0.3) = 0. \quad (e)$$

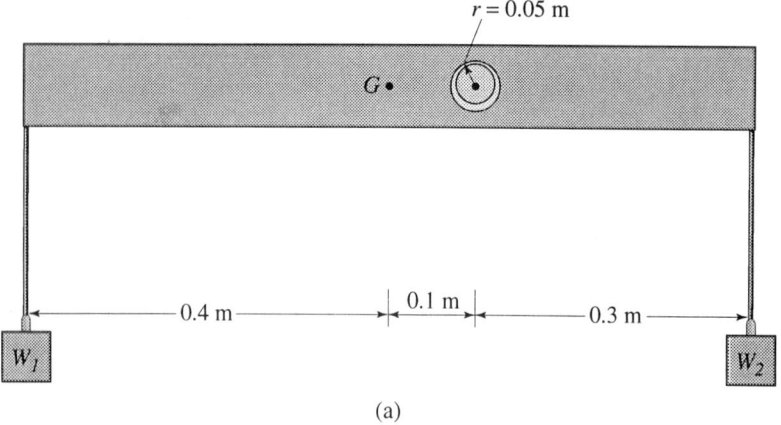

(a)

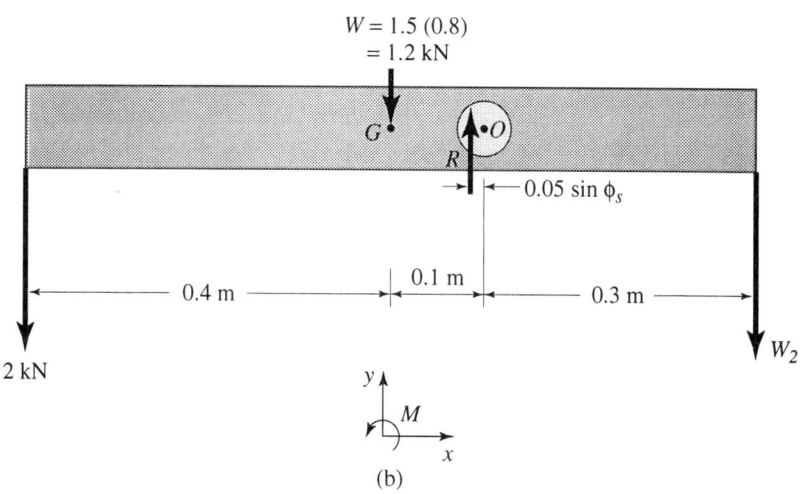

(b)

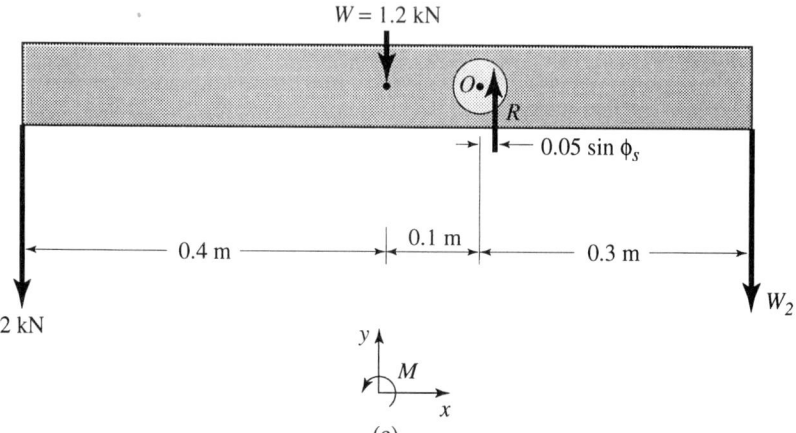

(c)

FIGURE E9.20.

Because $\phi_s = \tan^{-1} 0.15 = 8.53°$, after simplification, Eq. (e) reduces to

$$40.45W_2 - R = 151.0. \tag{f}$$

Solving Eqs. (d) and (f) simultaneously,

$$R = 7.11 \text{ kN},$$

and

$$W_2 = 3.91 \text{ kN}. \tag{ANS.}$$

Problems

9.126 A 4-in. diameter shaft fits in a pivot bearing. The compressive axial force in the shaft is 2000 lb, and the torque associated with impending rotation is 800 lb·in. If the pressure distribution is assumed uniform, determine the static coefficient of friction for the bearing surface.

9.127 A collar bearing has an outside diameter 1.6 times its inside diameter and is subjected to a uniform pressure of 204 psi. If the shaft carries an axial force of 4000 lb and the static coefficient of friction is 0.5, determine (a) the outside and inside diameters of the bearing and (b) the shaft torque associated with impending rotation.

9.128 A collar bearing carries a compressive axial force of 25,000 N. Its outside diameter is 0.18 m, and its inside diameter is 0.07 m. If the coefficient of static friction is 0.2, find the torque needed for impending rotation.

9.129 The bearing shown in Figure P9.129 has the shape of a frustum of a cone. Assume that the normal pressure p is uniformly distributed over the contact area. Express (a) the pressure p as a function

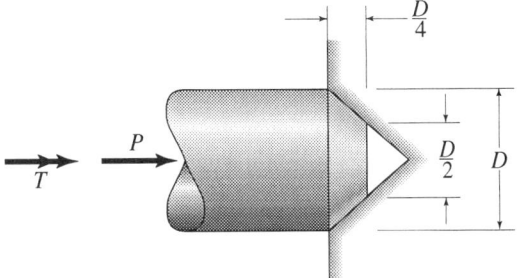

FIGURE P9.129.

of the shaft axial force P and the diameter D and (b) the torque T, associated with impending rotation, as a function of P, D, and μ_s, the static coefficient of friction for the bearing surface.

9.130 In Figure P9.130, the vertical shaft is supported by a movable collar bearing at A and by an end or pivot bearing at B. Compressed springs exert an upward total force of 250 lb on A when the supported weight W is 800 lb. Plate A is not free to rotate. At the collar bearing, the static coefficient of friction is 0.2, and, at the end bearing, this coefficient is 0.12.

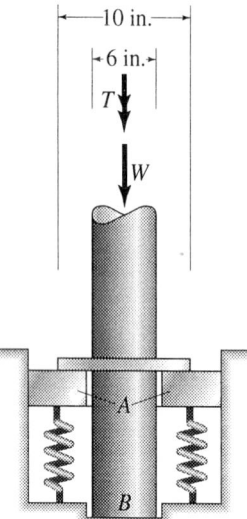

FIGURE P9.130.

Determine the applied torque T associated with impending rotation of the shaft about its vertical axis. What percentage of this total torque is resisted by the collar bearing?

9.131 Assume that the disk sander is pushed against the wooden floor, as shown in Figure P9.131, so that the normal pressure is uniformly distributed over the area of diameter D. The total normal force exerted on the floor is 16 lb, and

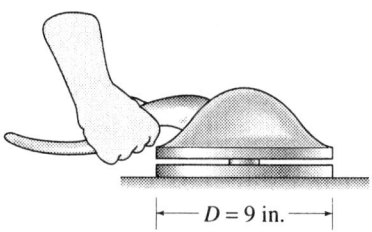

FIGURE P9.131.

the coefficient of friction is 0.6 between the sandpaper and the floor. Determine the output torque which the electric motor must exert to balance this frictional resistance.

9.132 In Figure P9.132, the pressure exerted by the force P may be assumed to be uniformly distributed over a circular area of diameter D. Determine the applied torque T associated with impending rotation for the following inputs: $P = 500$ lb, $D = 6$ in., and $\mu_s = 0.25$.

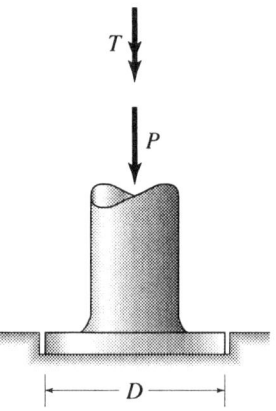

FIGURE P9.132.

9.133 Solve Problem 9.132 assuming that the pressure exerted by the force P is distributed over the circular area, so that the pressure intensity at the center is p_O and at the outside circumference is 0.8 p_O with a linear variation . Hint: Determine the value of p_O by summing vertical forces, then, express the pressure intensity as a function of a radial coordinate measured from the center, as in Example 9.18.

9.134 In Figure P9.134, the pressure produced by the force P is to be assumed uniformly distributed over an annular area of outside diameter D_O and inside diameter D_i. Determine the applied torque T associated with impending rotation for the following data: $P = 1000$ lb, $D_O = 12$ in., $D_i = 8$ in., and $\mu_s = 0.35$.

9.135 A bell crank mechanism is shown in Figure P9.135. It is supported by a fixed shaft centered at C. The coefficient of friction is 0.15 between the shaft and this mechanism. Determine the force F_2 consistent with (a) cw impending rotation of the mechanism and (b) ccw impending rotation of the mechanism.

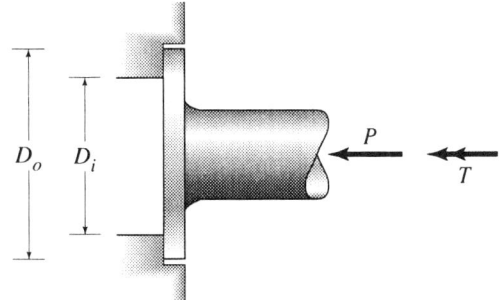

FIGURE P9.134.

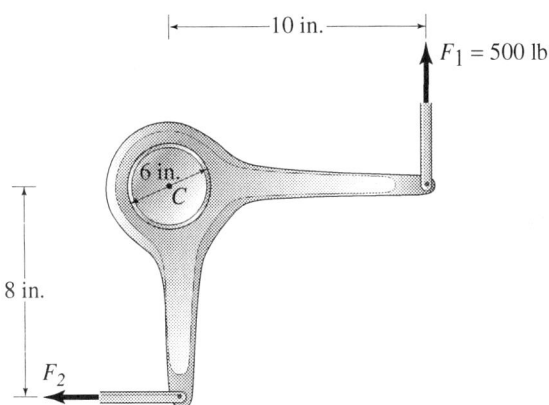

FIGURE P9.135.

9.136 Refer to Figure P9.136, and determine the couple C required for ccw rotation of the pulley at constant angular velocity. The weight $W_1 = 2000$ lb and the kinetic coefficient of friction is 0.30 for the bearing. An allowance of 200 lb is to be made for the pulley and shaft weights. The shaft has a diameter of 1.5 in.

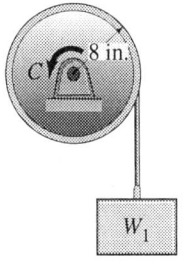

FIGURE P9.136.

9.137 Solve Problem 9.135 given $F_1 = 1000$ lb. All other information remains unchanged.

9.138 Solve Problem 9.136 given a 2.0-in. diameter shaft. All other information remains unchanged.

9.139 Solve Problem 9.136 given a kinetic coefficient of friction of 0.10 for the bearing. All other information remains unchanged.

9.8 Problems in Which Motion Is Not Pre-determined

All of the problems encountered previously in this chapter dealt with systems in which the type and nature of the motion was either given or was obvious from the information given in the problem. There are cases, however, in which the motion of the systems is not prescribed. In such cases, two or more different types of motion are assumed, and a solution is obtained based on each assumption. The correct solution is obtained after comparing the various solutions or by observing that those solutions that are not applicable lead to physically impossible conclusions. These ideas are illustrated in the solution of the following two examples.

■ **Example 9.21**

A force P is applied to a crate as shown in Figure E9.21(a). The mass of the crate is 200 kg and the coefficient of static friction $\mu_s = 0.35$. Determine the value for P for impending motion of the crate. Is this motion sliding or tipping? Assume that the weight of the crate acts through its geometric center.

Solution

Because the type of impending motion is not known, the two possible types of motion need to be investigated and the value of the force P corresponding to each type determined. By comparing the two values of P, we can ascertain the type of motion that will actually occur.

Assume Impending Motion Up the Plane

The free body diagram corresponding to this type of motion is shown in Figure E9.21(b). Because sliding is impending, $F = \mu_s N$, and there are only two unknown quantities N and P in the system. Apply two of the three available equilibrium equations using the coordinate system given in Figure E9.21(b). Thus,

$$\sum F_x = 0, \quad 0.35N + 1962 \sin 20° - P = 0, \qquad \text{(a)}$$

and

$$\sum F_y = 0, \quad N - 1962 \cos 20° = 0.$$

Solve Eqs. (a) and (b) simultaneously to obtain

$$P = 1316.3 \text{ N.}$$

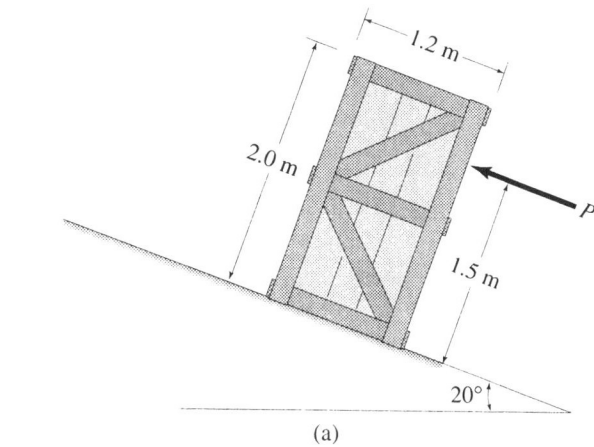

(a)

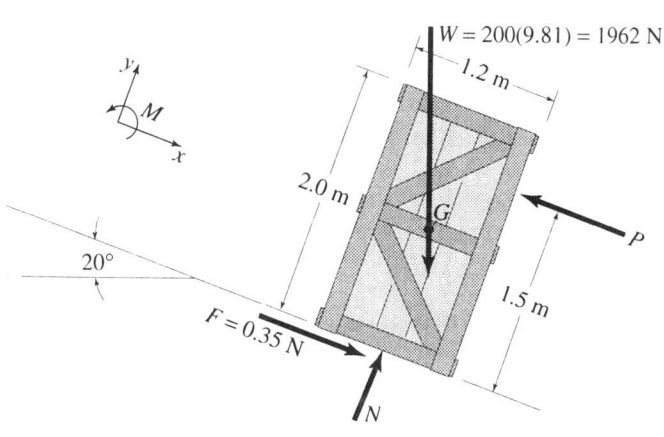

(b)

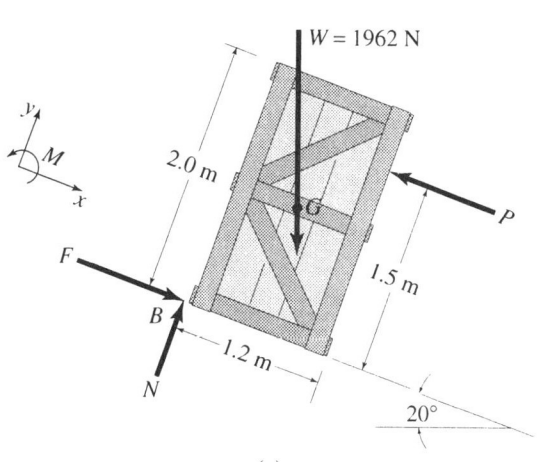

FIGURE E9.21.

(c)

Assume
Impending Tipping
About the Lower
Left-Hand Corner

The free-body diagram corresponding to this type of motion is shown in Figure E9.21(c). In investigating tipping, an infinitesimal ccw angular rotation of the crate is assumed, so that the forces F and N act at point B, as shown. Because we do not have impending sliding, the friction force F is *not* equal to $\mu_s N$, and we have three unknown quantities N, F, and P. Apply the three available equations of equilibrium using the coordinate system given in Figure E9.21(c). Thus

$$\sum F_x = 0, \quad F + 10962 \sin 20° - P = 0, \tag{c}$$

$$\sum F_y = 0, \quad N - 1962 \cos 20° = 0, \tag{d}$$

and

$$\sum M_B = 0, \quad P(1.5) - 1962 \cos 20°(0.6) - 1962 \sin 20°(1.0) = 0. \tag{e}$$

To obtain the value of P, only Eq. (e) is needed. Of course, Eqs. (c) and (d) may be used to obtain the values of N and F if necessary. Thus, from Eq. (e),

$$P = 1185.8 \text{ N}.$$

Therefore, by comparing the value of P needed for impending sliding (1316.3 N) to that needed for impending tipping (1184.8 N), we conclude that the crate would slide and the applied load is

$$P = 1185 \text{ N}. \qquad \text{ANS.}$$

A check on our computations will be to determine the values of F and N from Eqs. (c) and (d) and to compare the determined value of F to the maximum available friction force which is equal to $\mu_s N$. Thus, from Eq. (c) and (d),

$$F = 513.3 \text{ N},$$

$$N = 1843.7 \text{ N},$$

and

$$F_{max} = \mu_s N = 645.3 \text{ N}.$$

Thus, the friction force required for tipping (513.8 N) is smaller than that available in the system (645.3 N), and, therefore, the crate will tip and not slide.

■ Example 9.22

Refer to the system shown in Figure E9.22(a). The coefficient of static friction is 0.4 between block A and its inclined plane and 0.3 between block C and its inclined plane. The weights of the blocks $W_A = 75$ lb and $W_C = 100$ lb. Blocks A and C are connected by weightless links which are pin-jointed at A, B, and C. The vertical load P is applied at

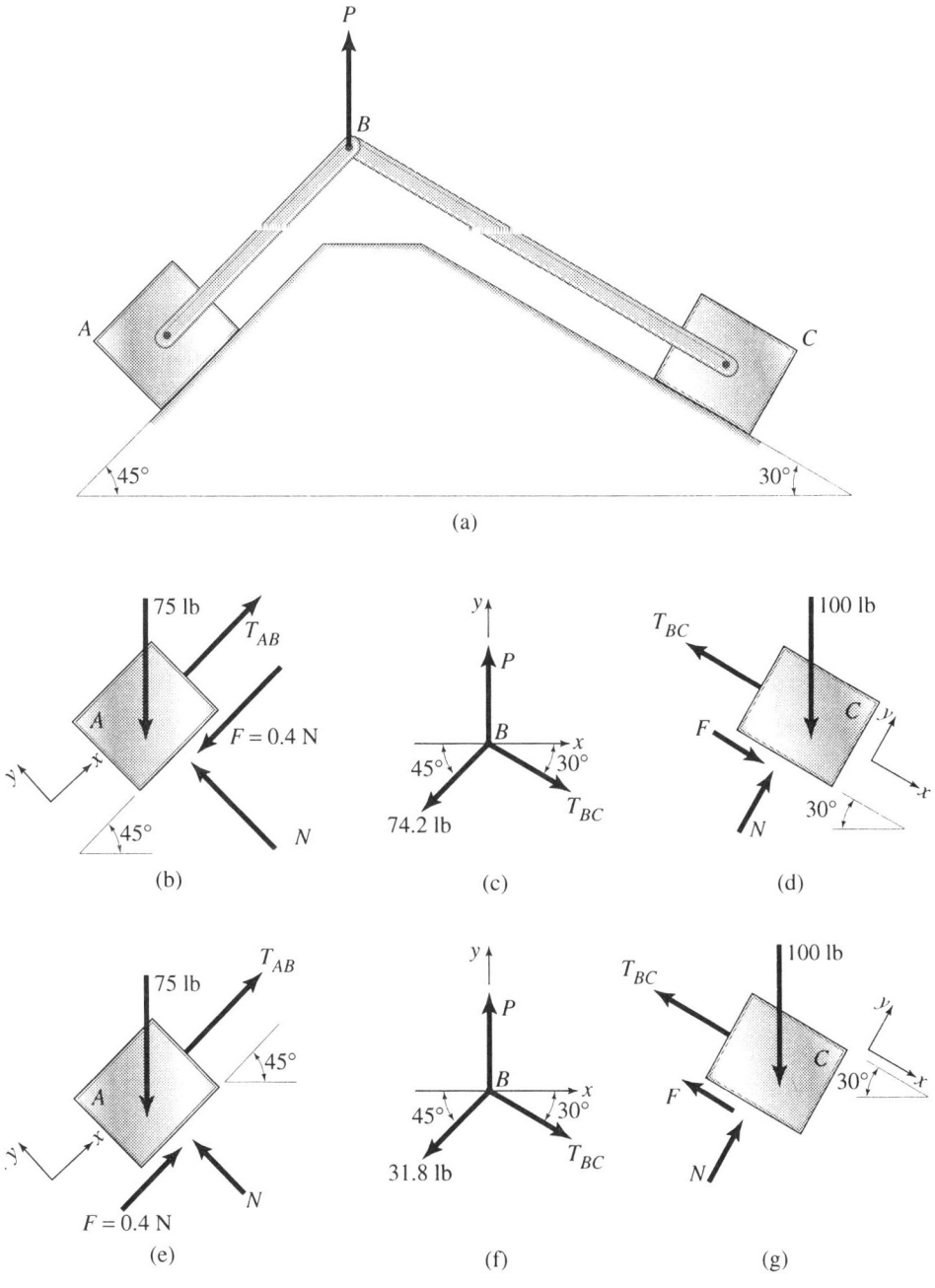

FIGURE E9.22.

pin B. Assume that all pins are frictionless, and determine the range of values of P to insure that neither block moves on its inclined plane.

Solution

Maximum P

Assume that block A is on the verge of moving up its inclined plane, and construct the free-body diagram, as shown in Figure E9.22(b). Because motion is impending, the frictional force $F = \mu_s N$, and there are two unknown quantities N and T_{AB} where T_{AB} is the tension in member AB. Thus,

$$\sum F_x = 0, \quad T_{AB} - 0.4N - 75\sin 45° = 0, \tag{a}$$

and

$$\sum F_y = 0, \quad N - 75\cos 45° = 0. \tag{b}$$

A simultaneous solution of Eqs. (a) and (b) yields

$$N = 530 \text{ lb}$$

and

$$T_{AB} = 74.2 \text{ lb.}$$

Now, consider the free-body diagram of the joint at B, as shown in Figure E9.22(c), and apply the conditions of equilibrium. Thus,

$$\sum F_x = 0, \quad T_{BC}\cos 30° - 74.2\cos 45° = 0, \tag{c}$$

and

$$\sum F_y = 0, \quad P - 74.2\sin 30° - T_{BC}\sin 45° = 0. \tag{c}$$

Solving Eqs. (c) and (d) leads to

$$T_{BC} = 60.6 \text{ lb}$$

and

$$P = 82.8 \text{ lb.}$$

The question that needs to be answered now is whether block C would slide under the action of $P = 82.8$ lb. The free-body diagram of block C is shown in Figure E9.22(d). Note that, in this case, we cannot make the assumption that $F = \mu_s N$. Rather, the required friction force F has to be determined from the conditions of equilibrium and compared to the maximum available friction force. Thus,

$$\sum F_x = 0, \quad F - 60.6 + 100\sin 30° = 0, \tag{e}$$

and

$$\sum F_y = 0, \quad N - 100\cos 30° = 0. \tag{f}$$

Equations (e) and (f) yield

$$F = 10.6 \text{ lb}$$

and

$$N = 86.6 \text{ lb.}$$

The maximum available friction force is given by

$$F_{max} = \mu_s N = 0.3(86.6) = 26.0 \text{ lb.}$$

Therefore, the maximum available friction force (26.0 lb) exceeds the force required for the equilibrium of block C (10.6 lb). Thus, block C does not move, and the maximum value of P is 82.8 lb.

Minimum P

Assume that block A is on the verge of moving down its inclined plane, and construct the free-body diagram, as shown in Figure E9.22(e). Thus,

$$\sum F_x = 0, \quad T_{AB} + 0.4N - 75 \sin 45° = 0, \tag{g}$$

and

$$\sum F_y = 0, \quad N - 75 \cos 45° = 0. \tag{h}$$

Equation (g) and (h) yields

$$N \; 53.0 \text{ lb}$$

and

$$T_{AB} = 31.8 \text{ lb.}$$

The free-body diagram of the joint at B is shown in Figure E9.22(f). Thus,

$$\sum F_x = 0, \quad T_{BC} \cos 30° - 31.8 \cos 45° = 0, \tag{i}$$

and

$$\sum F_y = 0, \quad P - 31.8 \sin 45° - T_{BC} \sin 30° = 0. \tag{j}$$

The solution of Eqs. (i) and (j) leads to

$$T_{BC} = 26.0 \text{ lb}$$

and

$$P = 35.5 \text{ lb.}$$

At this point, we need to check to see whether or not block C moves under the action of $P = 35.5$ lb. The free-body diagram of block C is shown in Figure E9.22(g). Thus,

$$\sum F_x = 0, \quad 100 \sin 30° - 26.0 - F = 0, \tag{k}$$

and

$$\sum F_y = 0, \quad N - 100 \cos 30° = 0. \tag{l}$$

Equations (k) and (l) yield

$$F = 24.0 \text{ lb}$$

and

$$N = 86.6 \text{ lb.}$$

The maximum available friction force is given by

$$F_{max} = \mu_s N = 0.3(86.6) = 26.0 \text{ lb}$$

which is more than the friction force needed for the equilibrium of block C. Therefore, block C does not move, and the minimum value of P is 35.5 lb. Thus, the required range of values of P is

$$35.5 \leq P \leq 82.8 \text{ lb.} \qquad\qquad \text{ANS}$$

■

Problems

9.140 A force P is applied horizontally at the top of the block whose mass is 100 kg as shown in Figure P9.140. If the coefficient of static friction is 0.25, determine whether the block will slide or tip and the value of the force P necessary to initiate the motion. Assume that the weight of the block acts through its geometric center.

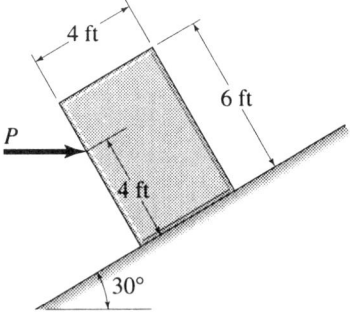

FIGURE P9.141.

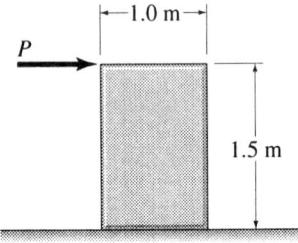

FIGURE P9.140.

9.141 A horizontal force P is applied to a block as shown in Figure P9.141 in an effort to move the block up the inclined

plane. The weight of the block is 750 lb, and the coefficient of static friction is 0.30. Determine the value of P for impending motion of the block. Is this motion sliding or tipping? Assume that the weight of the block acts through its geometric center.

9.142 A force of 5 kN is applied to the crate as shown in Figure P9.142. Determine the range of values of θ ($-90° < \theta < 90°$) for which the crate (a) slides and (b) tips.

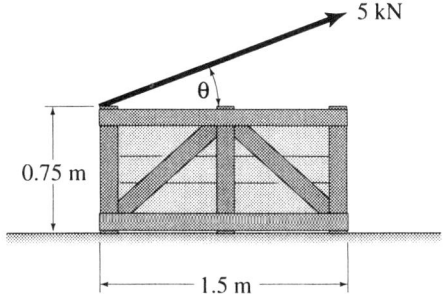

5 kN

θ

0.75 m

1.5 m

FIGURE P9.142.

The crate has a mass of 1000 kg, and the coefficient of static friction is 0.35. Assume that the weight of the crate acts through its geometric center.

9.143 Solve Problem 9.142 for a coefficient of static friction of 0.25.

9.144 Two blocks A and C are connected by the weightless pin-jointed members AB and BC, as shown in Figure P9.144. Let the coefficient of static friction be 0.5 for all contacting surfaces between the blocks and the inclined planes. Determine, for $0° \leq \theta \leq 90°$, the largest value of P and the corresponding angle θ that may be applied at joint B if neither block is to move.

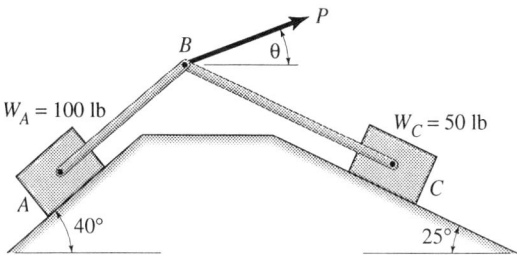

P

B

θ

$W_A = 100$ lb

$W_C = 50$ lb

A

40°

C

25°

FIGURE P9.144.

9.145 Refer to Problem 9.144, and compute the smallest value of P and the corresponding angle θ that may be applied at joint B if neither block is to move.

9.146 Consider the system shown in Figure P9.146. The coefficient of static friction is 0.30 between the sliders at A and B and their guides. The connecting rod has a length L and a weight W which may be assumed to act through its geometric center. In terms of W, determine the range of values of the force P so that the system remains in equilibrium in the position shown.

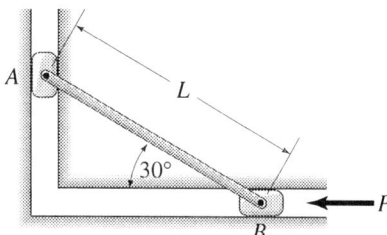

A

L

30°

B

P

FIGURE P9.146.

9.147 A homogeneous rod of length $L = 3.5$ m is placed in the position shown in Figure P9.147. The coefficient of static friction is 0.3, and the mass of the rod is 50 kg. Determine the range of values of the dimension h for which the rod can be placed in equilibrium. Assume that

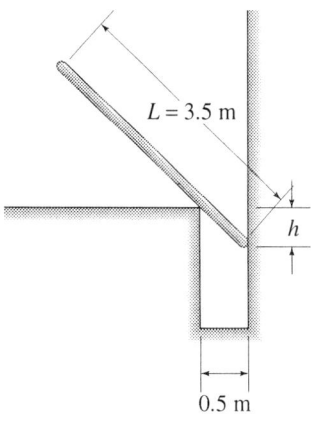

$L = 3.5$ m

h

0.5 m

FIGURE P9.147.

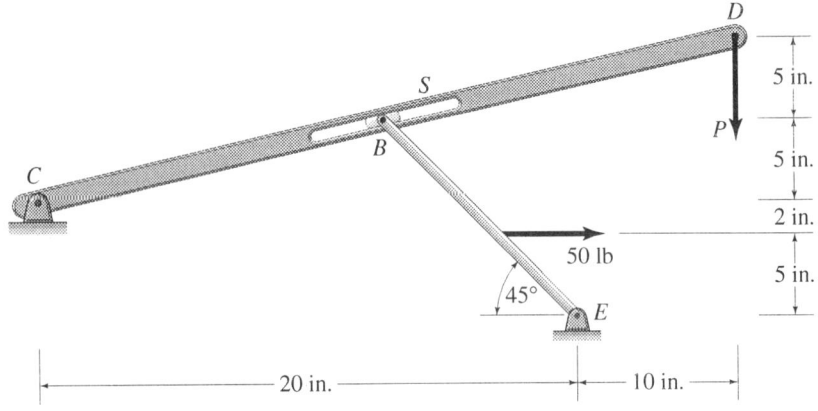

FIGURE P9.148.

the weight of the rod acts through its geometric center.

9.148 In Figure P9.148, member CD is hinged to a fixed support at C and to a slider S at B. The slider is attached to member BE which is hinged to a fixed support at E. A horizontal force of 50 lb is applied to member BE as shown. A vertical force P is applied at D to member CD. Determine the range of values of P for which the system is in equilibrium. Assume that the members are weightless and that the coefficient of static friction is 0.35 between the slider and the slot.

9.149 Blocks A and B are placed on platform CD which is hinged at C, as shown in Figure P9.149. The mass of A is 50 kg

and that of B is 75 kg. The coefficients of static friction are given in the diagram. Determine the angle θ at which motion impends. Express the answer in terms of μ_A and μ_B. Find the angle θ for the case when $\mu_A = 0.30$ and $\mu_B = 0.4$.

9.150 The hinged arm AB rests on the outer surface of a spool whose inner core rests on a horizontal surface, as shown in Figure P9.150. The coefficients of friction are given in the diagram. The weight of the spool $W = 20$ lb which

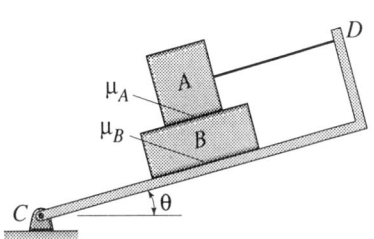

FIGURE P9.149.

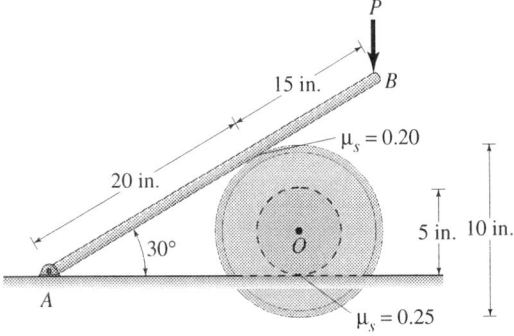

FIGURE P9.150.

may be assumed to act through point O. Determine the largest value of the load P that may be applied without causing impending motion of the spool.

9.151 A beam weighing 800 lb is placed in the position shown in Figure P9.151. A 10° wedge of negligible dimensions is placed under the left end of the beam and a horizontal force P is applied to it, as shown. Assume that the coefficient of static friction is 0.30 for all contacting surfaces. Determine the largest magnitude of the force P that may be applied to cause impending motion of the system. Assume that the weight of the beam acts through its geometric center.

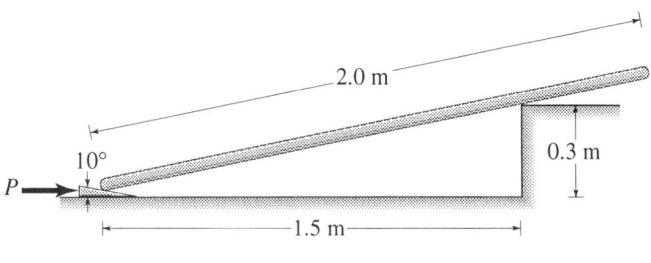

FIGURE P9.151.

Review Problems

9.152 A weightless block is subjected to the weight $W = 300$ lb and the force P, as shown in Figure P9.152. If the coefficient of static friction is 0.25, determine the magnitude of P (a) to produce impending motion up the plane and (b) to keep the block from sliding down the plane.

9.153 A block of weight W is placed on a conveyor belt which moves at constant speed under the action of the force $Q = 200$ N, as shown in Figure P9.153. If the block is not to move with the belt but is to remain where it was placed, determine the weight of the block and the coefficient of kinetic friction between the block and the belt.

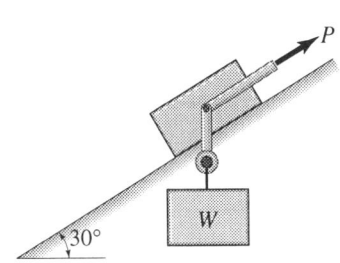

FIGURE P9.152.

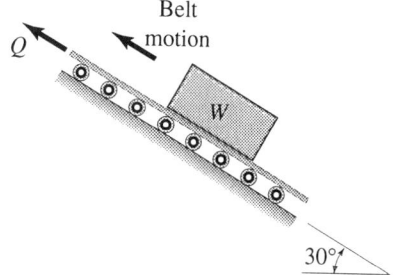

FIGURE P9.153.

9.154 A block of weight W is placed on a conveyor belt which moves at constant speed under the action of the force $Q = 50$ lb, as shown in Figure P9.154. If the block is held in position and prevented from moving with the belt by a cord, as shown, and if the coefficient of kinetic friction is 0.30 between the block and the belt determine its weight and the tension in the cord.

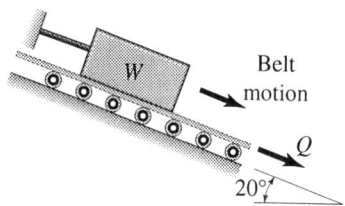

FIGURE P9.154.

9.155 A painter weighing 800 N climbs up a ladder, as shown in Figure P9.155. The

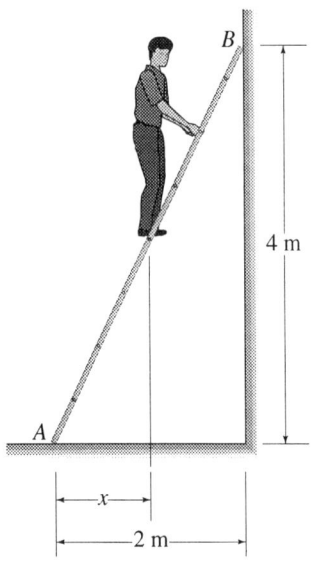

FIGURE P9.155.

coefficient of static friction is 0.15 at B. What must be the minimum coefficient of static friction at A, if the ladder is not to slide when $x = 1.5$ m?

9.156 A sliding door weighing 200 lb is suspended from supports at A and B, as shown in Figure P9.156. If a force $P = 40$ lb, applied as shown, is needed to place the door on the verge of sliding, determine the coefficients of friction at A and B, if assumed equal, and the normal reactions at A and B. Assume that the weight of the door acts through its geometric center.

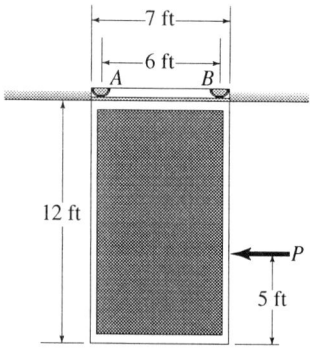

FIGURE P9.156.

9.157 The hydraulic cylinder exerts a force of 1.5 kN downward on lever AB shown in Figure P9.157. If the coefficient of static friction is 0.5 at C, determine the couple that must be applied to the drum to place it on the verge of cw rotation.

9.158 A block of weight W rests on a horizontal plane and is subjected to a force P, as shown in Figure P9.158. Determine the magnitudes of P and y in terms of W, b, and μ_s so that the block is on the verge of tipping and sliding at the same time.

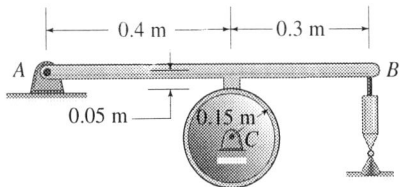

FIGURE P9.157.

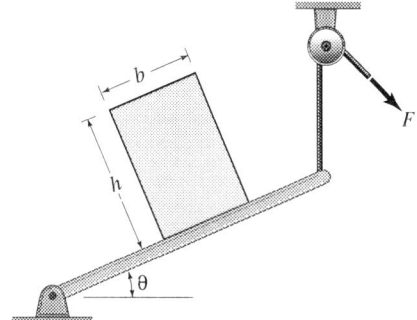

FIGURE P9.159.

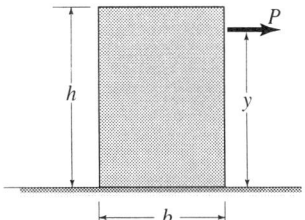

FIGURE P9.158.

Assume that the weight W acts through the block's geometric center.

9.159 A block of weight W, shown in Figure P9.159, is placed on a plane whose angle of inclination θ may be varied at will. If the coefficient of static friction is μ_s, develop expressions in terms of μ_s for the angle θ and for the ratio b/h that would cause the block to be on the verge of sliding and tippping simultaneously. What are the values of θ and b/h for the

case when $\mu_s = 0.3$? Assume that the weight W acts through the block's geometric center.

9.160 The right end of the beam shown in Figure P9.160 is to be raised by means of a $10°$ wedge. If the coefficient of static friction is 0.25 and the beam weighs 50 lb/ft, determine the horizontal force Q needed to bring the wedge to the verge of motion. Assume that the weight of the beam acts through its geometric center.

9.161 A heavy piece of equipment A weighing 8 kN is to be moved to the left by means of two $10°$ wedges, as shown in Figure P9.161. If the coefficient of friction is 0.5 between A and the floor and is 0.3 for all other surfaces, determine the vertical force Q needed to place A on the verge of motion.

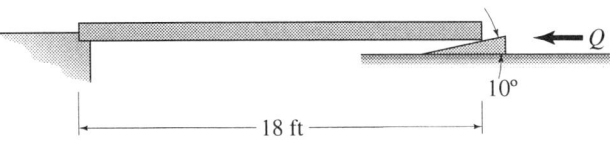

FIGURE P9.160.

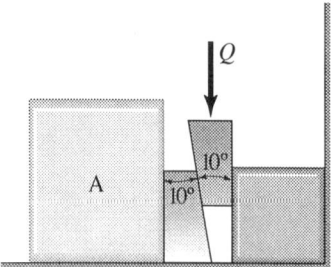

FIGURE P9.161.

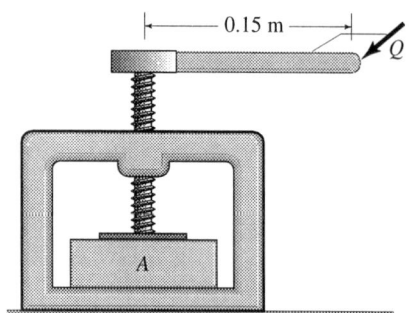

FIGURE P9.163.

9.162 Consider the square-threaded gear system shown in Figure P9.162. The worm gear has a mean radius $r_m = 0.25$ in., is double threaded (i.e., $n = 2$), and its pitch is 0.15 in. If a moment of 10 lb in. is required to turn the worm gear at constant speed and if $\mu_k = 0.10$, determine the couple that must be applied to gear A to accomplish the purpose.

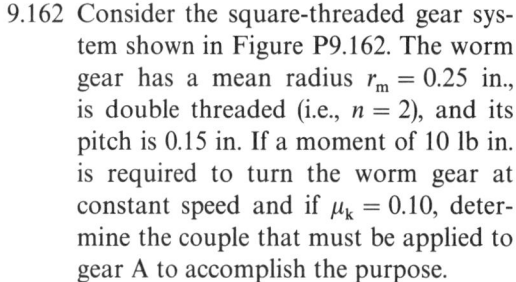

FIGURE P9.162.

9.163 The press, shown schematically in Figure P9.163, is to be designed to apply a compressive force of 0.50 kN on block A when a force $Q = 0.04$ kN is applied, as shown. The screw is to be square, single-threaded with a pitch of 0.007 m. If the coefficient of friction is 0.4, determine the mean radius of the thread. Hint: Re-

fer to Example 9.9, develop an expression containing r_m, and solve by trial and error.

9.164 A V-belt ($\psi = 120°$) is used to transmit a couple from the electric motor at A to the pulley at B, as shown in Figure P9.164. If the couple needed at B is 100 lb·ft and the coefficient of static friction is 0.3 between the belt and pulley B, determine the minimum torque that the motor at A must supply to accomplish the purpose and the minimum coeffi-

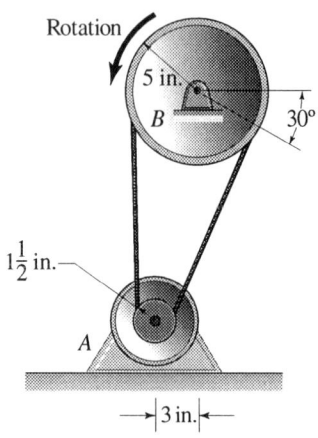

FIGURE P9.164.

cient of friction between the belt and the pulley at A.

9.165 A crate on a 30° inclined plane has a mass of 100 kg and is attached to a rope (flat belt) that passes over a fixed cylindrical peg, as shown in Figure P9.165. If the coefficient of static friction is 0.3, determine (a) the force F needed to keep the crate from sliding down the plane and (b) the force F needed to initiate motion of the crate up the plane.

on the mooring rope. If a boy can exert a force $Q = 50$ lb, as shown, determine the number of turns that he must wrap the mooring rope around the capstan in order to resist the wind. Let $\mu_s = 0.2$.

9.167 Refer to the band brake shown in Figure P9.167. In terms of P, r, and μ_s, determine the couple that needs to be applied to the brake drum to bring it to the verge of rotation (a) in a cw sense and (b) in a ccw sense.

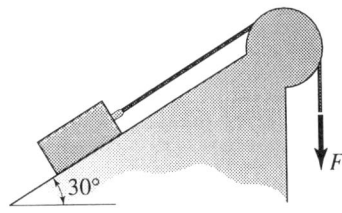

FIGURE P9.165.

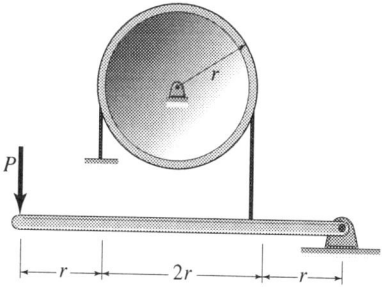

FIGURE P9.167.

9.166 The mooring rope (flat belt) of a sail boat is wrapped around a capstan, as shown in Figure P9.166. The wind on the sail boat produces a force of 3000 lb

9.168 If a force $P = 80$ N is applied to the end of the lever, as shown in Figure P9.168, determine the couple that must be applied to the brake drum to keep it rotat-

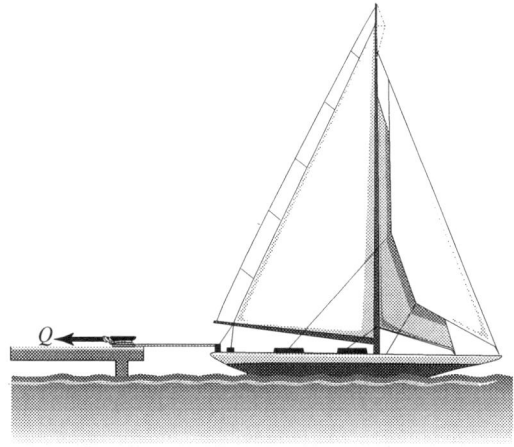

FIGURE P9.166.

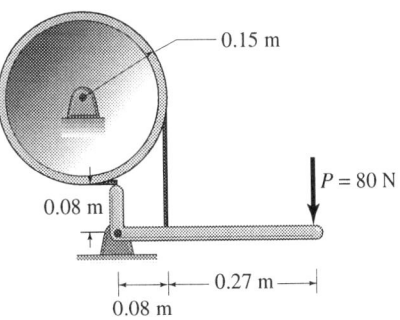

FIGURE P9.168.

ing at constant speed in a ccw sense. Let the coefficient of kinetic friction be 0.20.

9.169 A cylinder of weight W is placed on a 15° inclined plane, as shown in Figure P9.169. If a horizontal force $P = W/2$ is required to roll the cylinder up the plane at constant speed, determine the coefficient of rolling resistance. Assume that the weight W acts through point O.

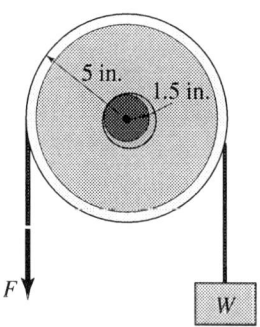

FIGURE P9.170.

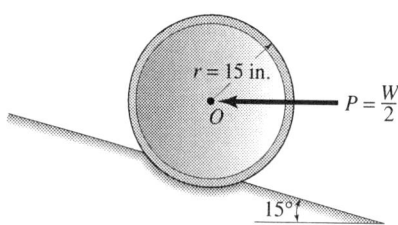

FIGURE P9.169.

9.170 A pulley with a radius of 5 in. and of neglibible weight fits loosely over a shaft with a radius of 1.5 in., as shown in

Figure P9.170. If a force $F = 225$ lb is needed to lift a weight W of 200 lb at constant speed, determine the coefficient of kinetic friction.

9.171 In the pivot bearing shown in Figure P9.171, the pressure p between the shaft and the bearing support varies linearly as shown. In terms of P, D_O, and μ_s, develop an expression for the torque T needed for impending rotation of the shaft. If $P = 1.5$ kN, $D_O = 0.160$ m, and $\mu_s = 0.2$, find T.

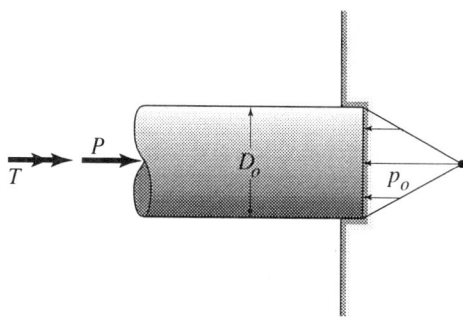

FIGURE P9.171.

10

Centers of Gravity, Centers of Mass, and Centroids

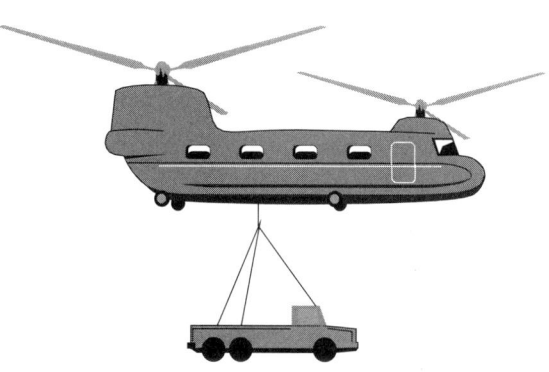

Centers of gravity play important roles in our daily lives. This photograph shows a helicopter carrying a truck so that it is hooked through their centers of gravity.

The analysis and design of many engineering systems require knowledge of the fundamental characteristics of these systems. The location of the center of gravity for a given member of which a system is composed or of the entire system is often required. The center of gravity is critical when dealing with forces and moments acting on a given system. Many man-made and natural systems encountered by engineers have smooth geometries that readily lend themselves to elegant analysis by calculus. For discrete bodies, other mathematical techniques are required. Irrespective of the system being analyzed, the basic theory is the same.

The design of concrete dams, cars, aircrafts, and other systems involves estimating forces and moments exerted on these systems. Whereas a concrete dam may have a smooth body that can be described mathematically, a car is composed of many smooth bodies that result in a rather complex system. In Chapter 5, we defined and applied the concepts of moments of forces with respect to a point and a line or an axis. In this chapter, we extend the moment concept to masses, volumes, lines, and areas. This enables us to locate the mass center or the center of gravity of a body and the centroid of a given volume, area, or line.

Emphasis is placed on integrating to solve problems in Sections 10.1 through 10.4. Symmetry is exploited to conclude that certain integrals will vanish. Even though integration is utilized for a solution, the appropriate way to formulate a solution is to begin with differential elements in pictorial form and write corresponding differential equations. Integration of these equations with appropriate limits will enable us to solve the problems.

Finite summations replace integrations to solve problems of Section 10.5 for composite bodies, areas, and lines. In Section 10.6, we derive and employ two theorems which enable us to conveniently determine surface areas and volumes of solids of revolution.

Challenging problems dealing with fluid statics complete this chapter. Free-body diagrams and two fundamental concepts enable us to solve these problems. These concepts are that pressure increases linearly with depth below the free surface of a liquid at rest and that, at a given depth, the pressure is the same in all directions.

10.1
Centers of Gravity and of Mass

Center of Gravity

A rigid body was defined as an infinite collection of particles whose distances remain fixed with respect to each other. Each of these particles has a differential weight $d\mathbf{W}$ which represents the force of gravity on the particle and, therefore, is directed toward the center of Earth. If we assume that the center of Earth is at a relatively large distance from the body, approaching infinity, then, as shown in Figure 10.1(a), all of the forces $d\mathbf{W}$ will be parallel. In Figure 10.1(a), these forces are shown directed along the negative z axis along which the center of Earth is assumed to lie. Thus, the total weight of the body $\mathbf{W}$ is the resultant of all the forces $d\mathbf{W}$ in the body (i.e., $\mathbf{W} = \int d\mathbf{W}$) and is assumed to act through point G known as the *center of gravity*. Note that the position of the center of gravity in Figure 10.1(a) is given by the coordinates $\bar{x}$, $\bar{y}$, and $\bar{z}$. To determine these coordinates, we use the fact that the moment of the resultant $\mathbf{W}$ about a given coordinate axis is equal to the sum of the moments of all the differential weights $d\mathbf{W}$ about the same axis. Thus, for example, to find $\bar{x}$, we observe that the magnitude of the moment of the resultant $\mathbf{W}$ with respect to the y axis (i.e., $\bar{x}W$), must be equal to the magnitude of the sum of the moments of all forces $d\mathbf{W}$ about the same axis (i.e., $\int x\, dW$). Thus,

$$\bar{x}W = \int x\, dW$$

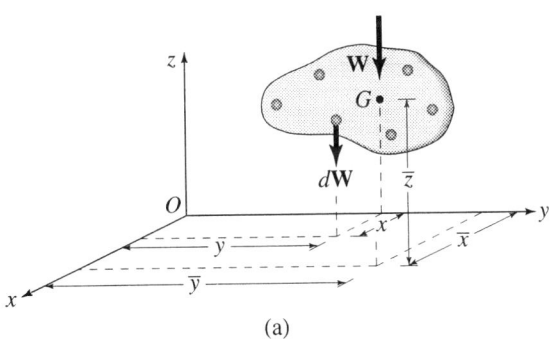

(a)

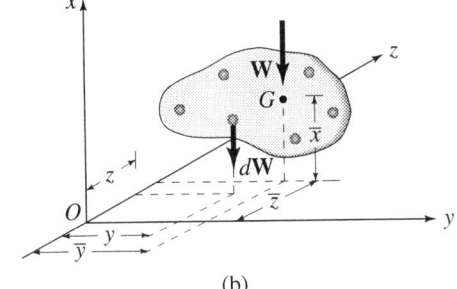

(b)

FIGURE 10.1.

from which

$$\bar{x} = \frac{\displaystyle\int x \, dW}{W} = \frac{\displaystyle\int x \, dW}{\displaystyle\int dW}. \qquad\qquad (10.1\text{a})$$

Similarly, we conclude that

$$\bar{y} = \frac{\displaystyle\int y \, dW}{W} = \frac{\displaystyle\int y \, dW}{\displaystyle\int dW} \qquad\qquad (10.1\text{b})$$

and

$$\bar{z} = \frac{\displaystyle\int z \, dW}{W} = \frac{\displaystyle\int z \, dW}{\displaystyle\int dW}. \qquad\qquad (10.1\text{c})$$

Note that Eq. (10.1c) was derived from Figure 10.1(b) which is obtainable from Figure 10.1(a) by rotating the coordinate system through 90° about the y axis.

Center of Mass

Let us now express the magnitude of the differential weight dW in terms of the differential mass dm (i.e., $dW = g \, dm$). Thus, if we replace dW by the product $g \, dm$ in Eqs. (10.1) and if we assume that the acceleration of gravity g is constant for a given body (a good assumption for bodies of practical interest on the surface of Earth), this constant may be canceled, and, from Eqs. (10.1),

$$\bar{x} = \frac{\displaystyle\int x \, dm}{m} = \frac{\displaystyle\int x \, dm}{\displaystyle\int dm}, \qquad\qquad (10.2\text{a})$$

$$\bar{y} = \frac{\displaystyle\int y \, dm}{m} = \frac{\displaystyle\int y \, dm}{\displaystyle\int dm}, \qquad\qquad (10.2\text{b})$$

and

$$\bar{z} = \frac{\displaystyle\int z \, dm}{m} = \frac{\displaystyle\int z \, dm}{\displaystyle\int dm}. \qquad\qquad (10.2\text{c})$$

Equations (10.2) locate a point in space, inside or outside the body known as the *center of mass* of the body, which may be defined as the point where we may assume that the entire mass of the body is concentrated. Note that the center of mass coincides with the center of gravity in the case of bodies for which the acceleration of gravity g may be assumed constant, the assumption made throughout this textbook. Thus, the same symbol G is used to designate both the center of mass and the center gravity. We observe, here, that the concept of the center of mass becomes significant in studying the dynamics of rigid bodies.

10.2 Centroid of Volume, Area, or Line

The *centroid* of a volume, an area, or a line is a purely geometric property of these objects and may be defined as a point where we may assume that the entire volume, area, or line is concentrated. These quantities are useful in studying certain subjects in mechanics. For example, the centroid of a volume is encountered in the study of *fluid statics* and the centroid of an area in the study of *mechanics of materials*, in particular, in the study of the bending of beams.

Centroid of Volume

Let us now return to Eqs. (10.1) and express the magnitude of the differential weight dW in terms of a differential volume dV and the weight density γ, which may or may not vary from point to point in the rigid body. Thus, if we replace dW by the product $\gamma\,dV$ and assume that γ is constant (i.e., the body is *homogeneous*), this constant may be canceled, and we conclude, from Eqs. (10.1), that

$$\bar{x} = \frac{\displaystyle\int x\,dV}{V} = \frac{\displaystyle\int x\,dV}{\displaystyle\int dV}, \qquad (10.3a)$$

$$\bar{y} = \frac{\displaystyle\int y\,dV}{V} = \frac{\displaystyle\int y\,dV}{\displaystyle\int dV}, \qquad (10.3b)$$

and

$$\bar{z} = \frac{\displaystyle\int z\,dV}{V} = \frac{\displaystyle\int z\,dV}{\displaystyle\int dV}. \qquad (10.3c)$$

Equations (10.3) locate a point in space, inside or outside the volume, known as the *centroid* of the volume which, as stated earlier, is a purely geometric property of the body representing the point at which we may assume that the entire volume is concentrated. To distinguish it from the *center of gravity* or *center of mass* G, the centroid is given the symbol C. Note that in the case of a homogeneous body, the centroid C of its volume coincides with its center of gravity G.

Centroid of Area

Consider the thin plane plate shown in Figure 10.2 which has a constant thickness and a constant weight density β expressed in terms of *weight per unit area*. Thus, a differential element of area dA of this plate would have a differential weight $dW = \beta\, dA$. If the product $\beta\, dA$ is substituted for dW in Eqs. (10.1), we find that the constant density β may be dropped and the resulting equations become

$$\bar{x} = \frac{\displaystyle\int x\, dA}{A} = \frac{\displaystyle\int x\, dA}{\displaystyle\int dA}, \tag{10.4a}$$

$$\bar{y} = \frac{\displaystyle\int y\, dA}{A} = \frac{\displaystyle\int y\, dA}{\displaystyle\int dA}, \tag{10.4b}$$

and

$$\bar{z} = \frac{\displaystyle\int z\, dA}{A} = \frac{\displaystyle\int z\, dA}{\displaystyle\int dA}. \tag{10.4c}$$

Equations (10.4) locate a point in space, inside or outside the area defined by the contour of the thin plate known as the centroid C of this area. As stated above, the centroid of an area is a purely geometric property of the area and may be thought of as the point where we may assume that the entire area is concentrated. Because a plane area represents a two-dimensional entity, it is usually placed in the plane defined by a two-dimensional coordinate system, say the x-y system.

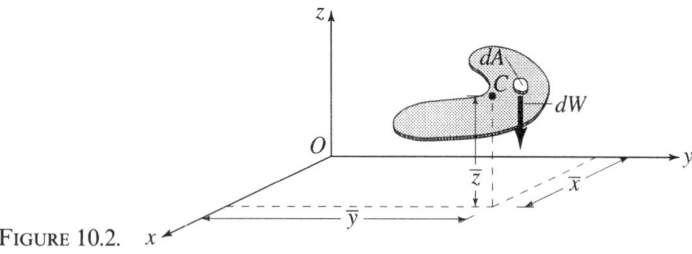

FIGURE 10.2.

Thus, Eq. (10.4b) yields $\bar{z} = 0$ and the centroid coordinates of the area are fully ascertained by finding $\bar{x}$ and $\bar{y}$ from Eqs. (10.4a) and (10.4b).

Because of the mathematical similarity between the moment of a force about the y axis given by $\int x\, dW$ and the integral $\int x\, dA$ in Eq. (10.4a), this latter integral is referred to as the *first moment of area* about the y axis. Also, the integral $\int y\, dA$ is known as the *first moment of area* about the x axis. In Chapter 11, we will deal with integrals of the form $\int x^2\, A$ and $\int y^2\, dA$, known as second moments of areas and called *area moments of inertia*.

Centroid of Lines

Let us now consider a body shaped in the form of a thin wire, as shown in Figure 10.3. We will assume that the cross-sectional area of the wire is constant and that its density α, expressed in terms of *weight per unit length*, is also constant. Thus, a differential length dL of this wire would have a differential weight $dW = \alpha\, dL$. If we substitute the product $\alpha\, dL$ for dW in Eqs. (10.1), we find that the constant density α cancels out and the resulting equations become

$$\bar{x} = \frac{\int x\, dL}{L} = \frac{\int x\, dL}{\int dL}, \tag{10.5a}$$

$$\bar{y} = \frac{\int y\, dL}{L} = \frac{\int y\, dL}{\int dL}, \tag{10.5b}$$

and

$$\bar{z} = \frac{\int z\, dL}{L} = \frac{\int z\, dL}{\int dL}. \tag{10.5c}$$

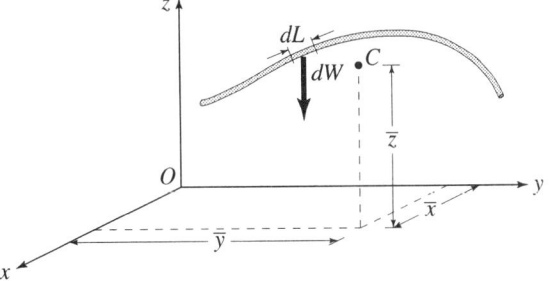

FIGURE 10.3.

Equations (10.5) locate a point in space, on or outside the line, known as the centroid C of the line defined by geometry of the wire. As stated earlier, the *centroid* of a line is a purely geometric property of the line and may be thought of as the point where we may, assume that the entire length of the line is concentrated.

10.3 Composite Objects

A composite object is defined as one consisting of two or more simple geometric objects. Consider, for example, the case of a composite body consisting, as shown in Figure 10.4, of a sphere of weight W_1 attached to a rod of weight W_2 which, in turn, is attached to a cylinder of weight W_3. We may determine the coordinates $\bar{x}$, $\bar{y}$, and $\bar{z}$ by locating its center of gravity from the coordinates of the centers of gravity of the individual components of the entire body. For example, the principle of moments applied with respect to the y axis states that the magnitude of the moment of the resultant $\sum W = W_1 + W_2 + W_3$ about the y axis must be equal to the sum of the magnitude of the moments produced by the individual weights about the same axis. We, therefore, conclude that

$$(\sum W)\bar{x} = W_1 x_1 + W_2 x_2 + W_3 x_3 = \sum Wx$$

from which

$$\bar{x} = \frac{\sum Wx}{\sum W}. \tag{10.6a}$$

Similarly, we may show that

$$\bar{y} = \frac{\sum Wy}{\sum W} \tag{10.6b}$$

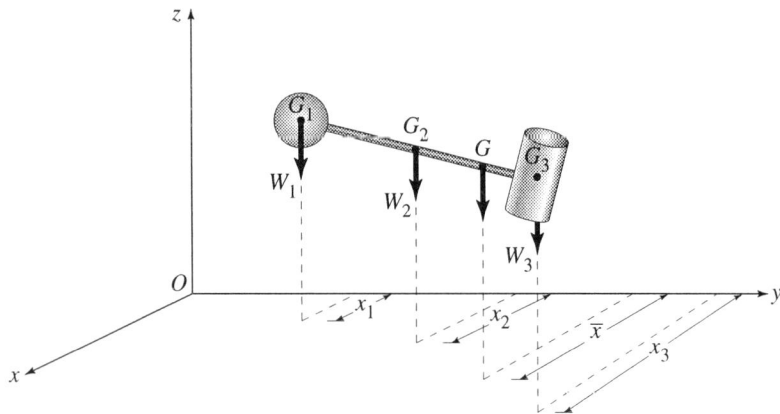

FIGURE 10.4.

and

$$\bar{z} = \frac{\sum Wz}{\sum W}.$$ (10.6c)

Comparing Eqs. (10.6) to Eqs. (10.1), which was derived to determine the center of gravity for an infinite number of particles, we conclude that, in the case of a finite number of parts making up a composite body, the process of integration is replaced by a process of summation that includes *all* of the parts in the system. In the same manner, we replace the integration process by a summation process in Eqs. (10.2) to find the *center of mass* for a composite object. Thus,

$$\bar{x} = \frac{\sum mx}{\sum m},$$ (10.7a)

$$\bar{y} = \frac{\sum my}{\sum m},$$ (10.7b)

and

$$\bar{z} = \frac{\sum mz}{\sum m}.$$ (10.7c)

Similarly, for the *centroid* of a composite volume, Eqs. (10.3) yield

$$\bar{x} = \frac{\sum Vx}{\sum V},$$ (10.8a)

$$\bar{y} = \frac{\sum Vy}{\sum V},$$ (10.8b)

and

$$\bar{z} = \frac{\sum Vz}{\sum V}.$$ (10.8c)

For the *centroid* of a composite area, Eqs. (10.4) lead to

$$\bar{x} = \frac{\sum Ax}{\sum A},$$ (10.9a)

$$\bar{y} = \frac{\sum Ay}{\sum A},$$ (10.9b)

and

$$\bar{z} = \frac{\sum Az}{\sum A}.$$ (10.9c)

Finally, for the *centroid* of a composite line, Eqs. (10.5) yield

$$\bar{x} = \frac{\sum Lx}{\sum L},$$ (10.10a)

$$\bar{y} = \frac{\sum Ly}{\sum L},$$ (10.10b)

and

$$\bar{z} = \frac{\sum Lz}{\sum L}.$$ (10.10c)

Note that, to determine the center of gravity, center of mass, or centroid of a composite object using the above equations, we need to know these properties for the simple individual parts composing the object. A selection of simple parts and their properties is given in Appendices A and B. In Section 10.4, we will learn how to determine such properties using the process of integration.

Occasionally, we encounter composite objects that possess one or more axes or planes of symmetry. In such cases, it is possible to determine, by *inspection*, one or more of the coordinates locating the center of gravity, center of mass, or centroid of the composite object. Consider, for example, the composite area shown in Figure 10.5(a). The x-y

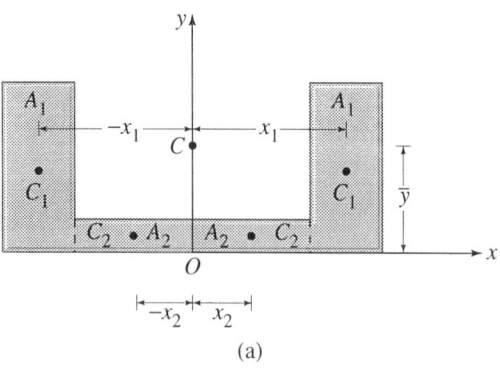

(a)

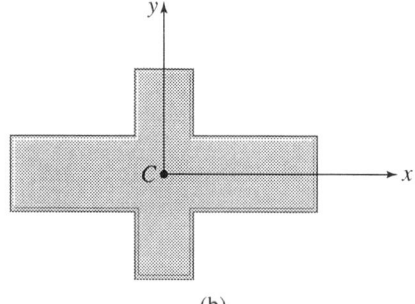

FIGURE 10.5.   (b)

coordinate system was chosen so that the y axis is an axis of symmetry. We may view this composite area as consisting of four rectangles, two with areas denoted by A_1 and two with areas denoted by A_2. Because of the symmetry about the y axis, the x coordinate of the centroid of A_1 on the left of this axis is equal in magnitude but opposite in sign to that on the right of the axis. The same statement applies to the two areas denoted by A_2. Thus, by Eq. (10.9a),

$$\bar{x} = \frac{A_1(x_1) + A_1(-x_1) + A_2(x_2) + A_2(-x_2)}{2A_1 + 2A_2} = 0.$$

Because of symmetry about the y axis, the centroid C of the composite area lies on this axis and the only coordinate needing determination is $\bar{y}$, by Eq. (10.9b). Obviously, we did not need to apply Eq. (10.9a) to show that $\bar{x} = 0$; inspection and observation should lead us to the same conclusion.

Another example where symmetry is used to advantage is shown by the composite area of Figure 10.5(b) in which both the x and y axes are axes of symmetry. Using the same argument as above, we conclude, by inspection, that the centroid of this composite area lies on both axes of symmetry leading us to conclude that the centroid C must be at the intersection of the two axes.

Similarly, it may be shown that, if a composite object possesses a plane of symmetry, its center of gravity, center of mass, or centroid lies in this plane of symmetry.

The following examples illustrate some of the concepts discussed above.

■ **Example 10.1**

Determine the location of the center of gravity for the composite body shown in Figure E10.1(a). Assume that the weight density of the rectangular part is γ (N/m^3) and that of the semicircular part is 2.5γ.

Solution

A coordinate system is established as shown in Figure E10.1(b). Note that the composite body is symmetric with respect to the x-z and y-z planes and, therefore,

$$\bar{x} = \bar{y} = 0. \hspace{3cm} \text{ANS.}$$

Thus, the only coordinate that needs to be determined to locate the center of gravity G is the $\bar{z}$ coordinate.

The composite body is subdivided into three component parts, a semicircular part of weight W_1 with center of gravity G_1, a rectangular part of weight W_2 with center of gravity G_2 and the circular hollow part of weight W_3 with center of gravity G_3. Now, using the information in Appendix B,

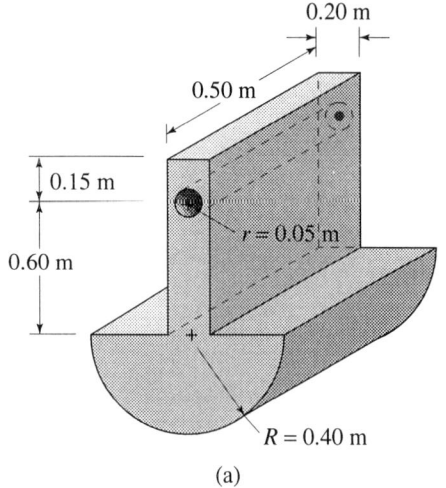

(a)

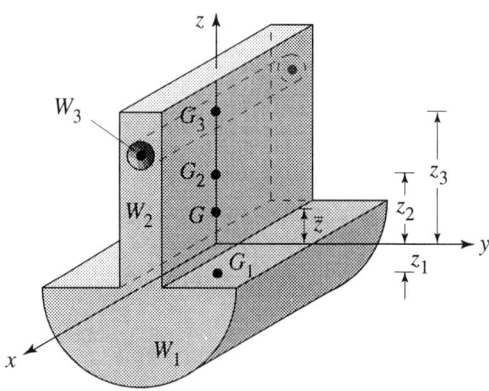

Figure E10.1.   (b)

$$W_1 = V_1 \gamma_1 = \frac{\pi}{2}(0.40^2)(0.5)(2.5\gamma) = 0.31416\gamma \text{ N},$$

$$z_1 = -\frac{4(0.40)}{3\pi} = -0.16977 \text{ m},$$

$$W_2 = V_2 \gamma_2 = (0.20)(0.75)(0.5)(\gamma) = 0.075\gamma \text{ N},$$

$$z_2 = \frac{1}{2}(0.75) = 0.375 \text{ m},$$

and

$$W_3 = V_3 \gamma_3 = \pi(0.05^2)(0.5)(\gamma) = 0.00393\gamma \text{ N},$$

$$z_3 = 0.6 \text{ m}.$$

Thus, using Eq. (10.6c),

$$\bar{z} = \frac{\sum Wz}{\sum W} = \frac{W_1 z_1 + W_2 z_2 - W_3 z_3}{W_1 + W_2 - W_3}$$

$$= \frac{[(0.31416)(-0.16977) + (0.075)(0.375) - (0.00393)(0.6)]\gamma}{0.31416 + 0.075 - 0.00393}$$

$$= -0.0716 \text{ m.} \qquad\qquad\qquad \text{ANS.}$$

The negative sign of the answer indicates that the center of gravity of the composite body is below the x-y plane, not above it, as assumed in Figure E10.1(b). Also, in both numerator and denominator of the above equation, the contribution of W_3 is negative because it represents a part that has been taken away from the composite body.

■ Example 10.2

Locate the centroid for the composite area shown in Figure E10.2(a).

Solution

A coordinate system is chosen as shown in Figure E10.2(b). Note that the composite area may be subdivided into three component parts, a triangular area A_1 with centroid C_1, a rectangular area A_2 with centroid C_2, and a hollow circular area A_3 with centroid C_3. These areas and the corresponding centroid locations are found with the aid of Appendix A. Thus,

$$A_1 = \tfrac{1}{2}(9)(12) = 54 \text{ in.}^2,$$

$$x_1 = \tfrac{1}{3}(9) = 3 \text{ in.},$$

$$y_1 = 10 + \tfrac{1}{3}(12) = 14 \text{ in.},$$

$$A_2 = (25)(10) = 250 \text{ in.}^2,$$

$$x_2 = \tfrac{1}{2}(25) = 12.5 \text{ in.},$$

$$y_2 = \tfrac{1}{2}(10) = 5 \text{ in.},$$

and

$$A_3 = \pi(10^2) = 28.274 \text{ in.}^2,$$

$$x_3 = 17 \text{ in.},$$

$$y_3 = 5 \text{ in.}.$$

By Eq. (10.9a),

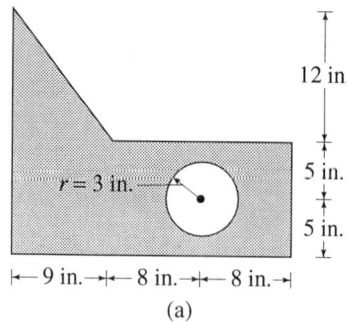

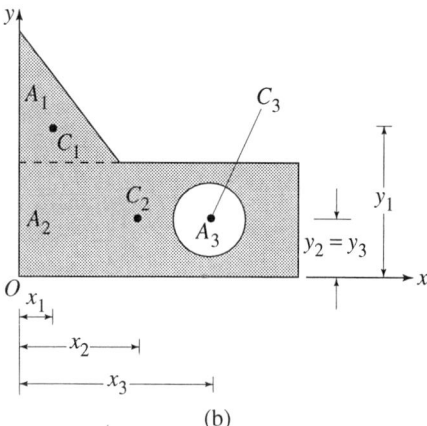

FIGURE E10.2.

$$\bar{x} = \frac{\sum Ax}{\sum A} = \frac{A_1 x_1 + A_2 x_2 - A_3 x_3}{A_1 + A_2 - A_3}$$

$$= \frac{54(3) + 250(12.5) - 28.274(17)}{54 + 250 - 28.274}$$

$$= 10.18 \text{ in.} \qquad\qquad \text{ANS.}$$

Also, by Eq. (10.9b),

$$\bar{y} = \frac{\sum Ay}{\sum A} = \frac{A_1 y_1 + A_2 y_2 - A_3 y_3}{A_1 + A_2 - A_3}$$

$$= \frac{54(14) + 250(5) - 28.274(5)}{54 + 250 - 28.274}$$

$$= 6.76 \text{ in.} \qquad\qquad \text{ANS.}$$

Note that the contribution of W_3 is negative in the equations above because it represents a part that has been taken away from the composite area.

■ **Example 10.3**

Locate the centroid of the homogeneous wire bent into the two-dimensional shape shown in Figure E10.3(a). Use the given x-y coordinate system.

Solution

The wire may be subdivided into three component segments, a straight segment of length L_1 with centroid C_1, a straight segment of length L_2 with centroid C_2 and a semicircular segment of length L_3 with centroid C_3. These lengths and the corresponding centroid locations are found with the aid of Appendix A. Thus,

$$L_1 = \sqrt{16^2 + 10^2} = 18.868 \text{ in.},$$

$$x_1 = -(5 + 8) = -13 \text{ in.},$$

$$y_1 = -5 \text{ in.},$$

$$L_2 = 12 \text{ in.},$$

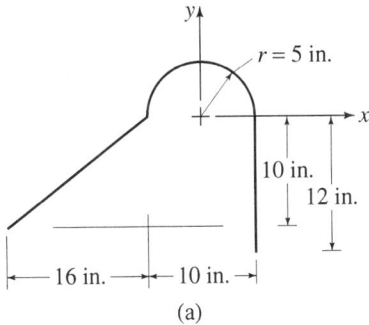

(a)

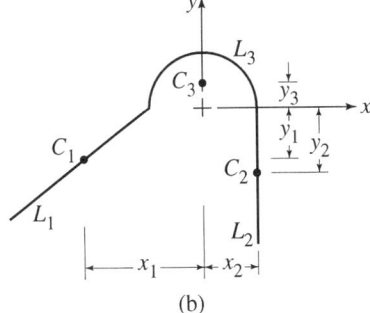

(b)

FIGURE E10.3.

$$x_2 = 5 \text{ in.,}$$

$$y_2 = -6 \text{ in.,}$$

and

$$L_3 = \pi(5) = 15.708 \text{ in.,}$$

$$x_3 = 0 \text{ in.,}$$

$$y_3 = \frac{2}{\pi}(5) = 3.183 \text{ in..}$$

By Eq. (10.10a),

$$\bar{x} = \frac{\sum Lx}{\sum L} = \frac{L_1 x_1 + L_2 x_2 + L_3 x_3}{L_1 + L_2 + L_3}$$

$$= \frac{(18.868)(-13) + (12)(5) + (15.708)(0)}{18.868 + 12 + 15.708}$$

$$= -3.66 \text{ in.} \hspace{3cm} \text{ANS.}$$

By Eq. (10.10b),

$$\bar{y} = \frac{\sum Ly}{\sum L} = \frac{L_1 y_1 + L_2 y_2 + L_3 y_3}{L_1 + L_2 + L_3}$$

$$= \frac{(18.868)(-5) + (12)(-6) + (15.708)(3.183)}{18.868 + 12 + 15.708}$$

$$= -2.50 \text{ in.} \hspace{3cm} \text{ANS.}$$

The negative sign for $\bar{x}$ indicates that the centroid of the composite wire is to the left of the y axis and the negative sign for $\bar{y}$ shows that it is below the x axis.

Problems

10.1 As shown in Figure P10.1, a homogeneous composite body consists of two hemispheres. In terms of R, determine the y coordinate of the centroid of this body.

10.2 Two spheres are fastened to the ends of a cylinder to form a homogeneous composite body, as shown in Figure P10.2.

In terms of R, find the x coordinate of the centroid of this composite body.

10.3 A cylinder and a cone form the homogeneous composite body of Figure P10.3. Determine the y coordinate of its center of mass if the mass density for the cylinder is ρ and that for the cone is 1.5ρ.

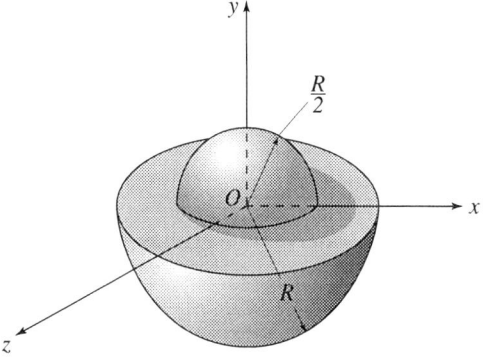

FIGURE P10.1.

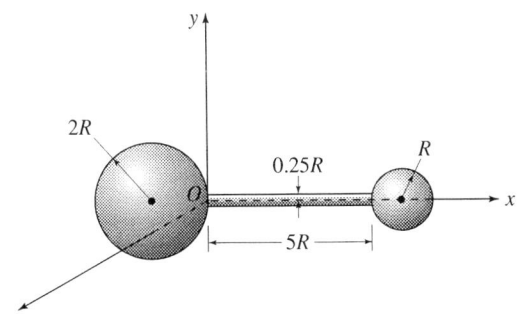

FIGURE P10.2.

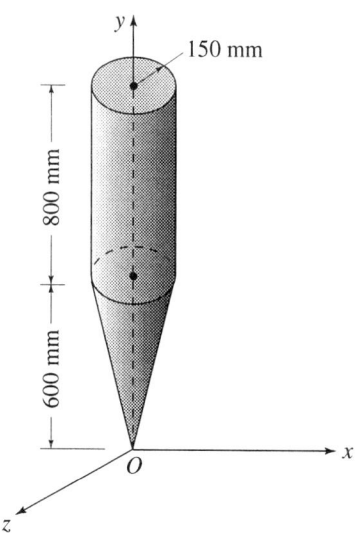

FIGURE P10.3.

10.4 A paraboloid, generated by revolving $y = \frac{1}{9}x^2$ about the x axis is fastened to a hemisphere to form a homogeneous composite body, as shown in Figure P10.4. Find the x coordinate of this composite body.

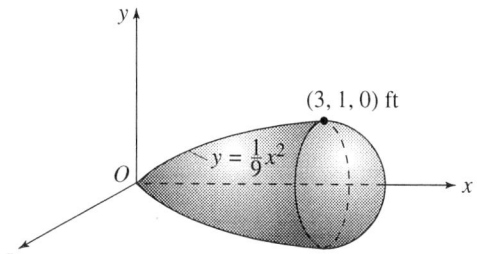

FIGURE P10.4.

10.5 A hemisphere is placed atop a cylinder to form the composite body shown in Figure P10.5. Determine the coordinates of its center of gravity if the weight density of the hemisphere is 0.3 lb/in.^{3} and that for the cylinder is 0.1 lb/in.^{3}

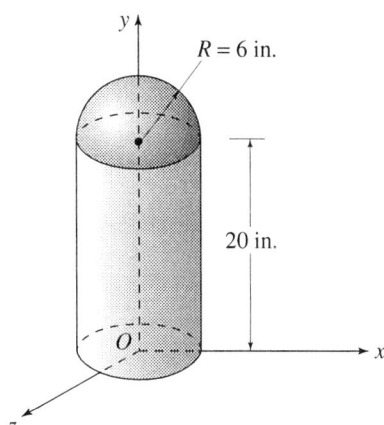

FIGURE P10.5.

10.6 A homogeneous composite body consisting of a hemisphere, a cylinder, and a cone is shown in Figure P10.6. Determine the x coordinate of its centroid.

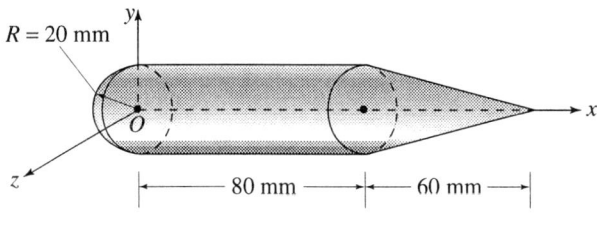

FIGURE P10.6.

10.7 A hemispherical cutout is made in the top of a cylinder, as shown in Figure P10.7. Find the coordinates of the centroid of the resulting body.

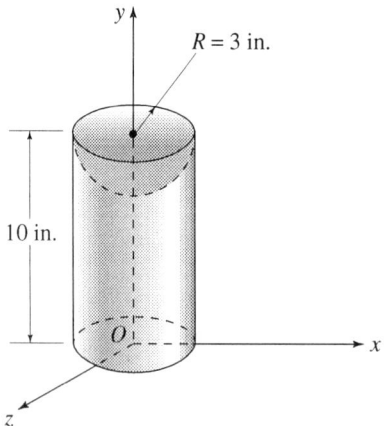

FIGURE P10.7.

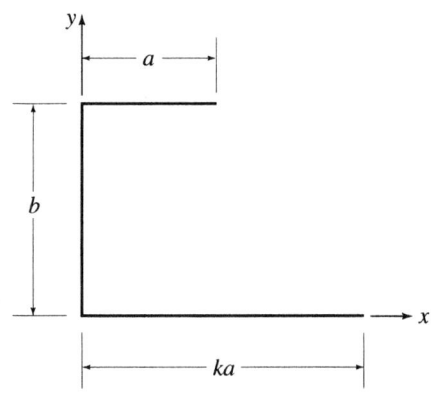

FIGURE P10.8.

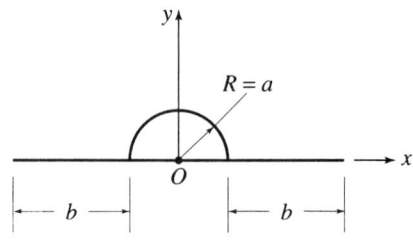

FIGURE P10.9.

10.8 Refer to Figure P10.8, and locate the centroid of this composite line (or thin wire) in terms of a, b, and k, where k is a dimensionless constant. Specialize your results for (a) $k = 0$, and (b) $b = 2a$ and $k = 2$.

10.9 Determine the centroidal coordinates of the composite line shown in Figure P10.9 in terms of a and b. Specialize the results for the case when $a = b/2$.

10.10 In terms of a and b, express the coordinates of the centroid for the composite area shown in Figure P10.10. Specialize your results for (a) $a \to 0$ and (b) $a = b$.

10.11 Locate the x coordinate of the centroid of the composite solid shown in Figure P10.11.

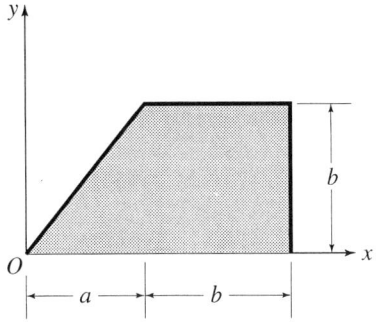

FIGURE P10.10.

10.12 In terms of a, b, and c, determine the coordinates of the centroid for the composite area depicted in Figure P10.12. Specialize your results for the case when $b = 0$ and $c = 2a$.

10.13 Refer to the composite area shown in Figure P10.13, and determine the centroidal coordinates $\bar{x}$ and $\bar{y}$ as functions of a and b.

10.14 A hemisphere is attached to the top of a paraboloid to form the composite solid shown in Figure P10.14. Locate the centroid of this composite body by ex-

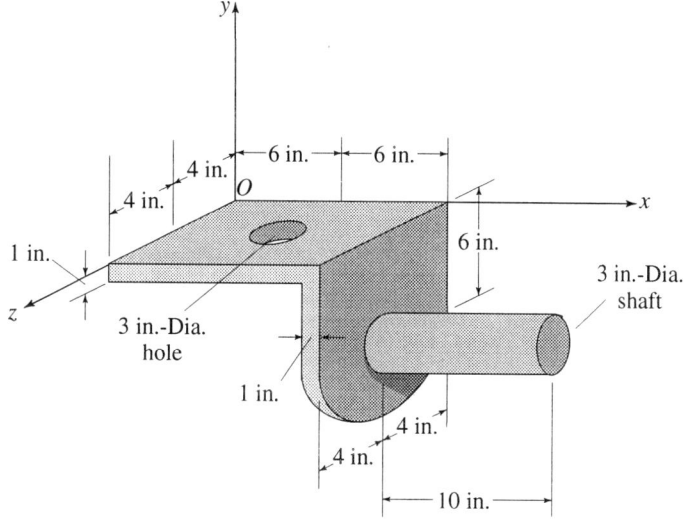

FIGURE P10.11.

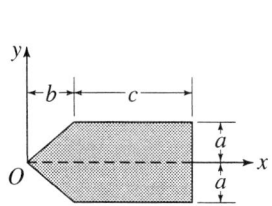

FIGURE P10.12.

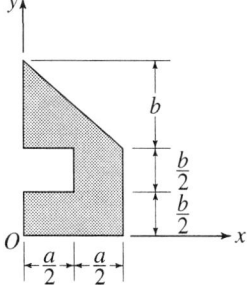

FIGURE P10.13.

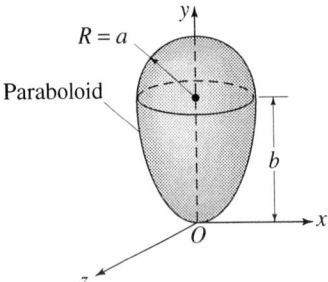

FIGURE P10.14.

pressing $\bar{y}$ as a function of a and b. Then divide through by b to yield $\bar{y}/b$ as a function of a/b. Specialize your results for the case when (a) $\dfrac{a}{b} \to 0$ and (b) $\dfrac{a}{b} \to 100$.

10.15 Consider the three lines of length a, b, and c which lie along the x, y, and z axes, respectively, as shown in Figure P10.15. Determine the three coordinates of the centroid of this composite line. Specialize your general results for the case where (a) $a = 0$ and $b = c$, (b) $a = 0$ and $b = 0$, and (c) $a = b = c$.

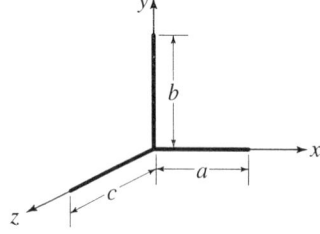

FIGURE P10.15.

10.16 Refer to Figure P10.16, and note that the hole of radius b and the circle of radius a are to have a common horizontal tangent at point A. Determine the y coordinate of the centroid C of the composite area. Divide the equation for $\bar{y}$ by a which will enable you to express $\bar{y}/a$ as a function of b/a.

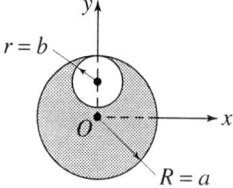

FIGURE P10.16.

10.17 The geometric axes of a right circular cylinder and a conical hole coincide as depicted in Figure P10.17. Express the y coordinate of the centroid of this composite body as a function of a, b, and h. Divide this equation through by a which yields $\bar{y}/a$ as a function of the dimensionless ratios b/a and h/a. Assign (h/a) a value of 2.0, and plot $\bar{y}/a$ vs. b/a as this latter ratio varies from 0 to 1.

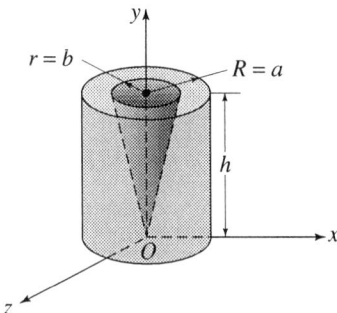

FIGURE P10.17.

10.18 Locate the centroid of the composite line shown in Figure P10.18. Express $\bar{x}/a$ and $\bar{y}/a$ as functions of the dimensionless ratio b/a. Investigate the case $b/a = 0$, and comment on the result.

10.19 Refer to the composite area shown in Figure P10.19, and express the y coordinate of its centroid in terms of a, b, and c. Check your result by considering the special cases (a) $a = 0$ and $c = b$ and (b) $a = b$ and $c = 0$.

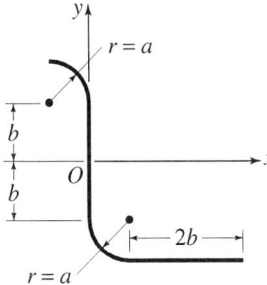

FIGURE P10.18.

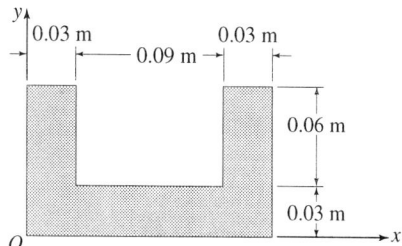

FIGURE P10.24.

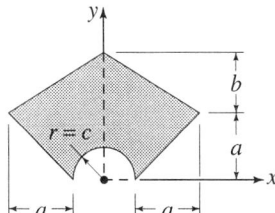

FIGURE P10.19.

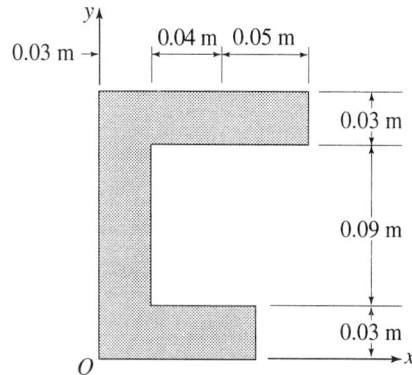

FIGURE P10.25.

10.20 Refer to Figure P10.8, let $a = 0.2$ m, $b = 0.4$ m, and $k = 2$, and compute the x and y coordinates of the centroid of the composite line.

10.21 Refer to Figure P10.16, let $a = 10$ in. and $b = 0.4$ in., and determine the y coordinate of the centroid of the composite area.

10.22 Refer to Figure P10.17, let $a = 0.6$ m, $b = 0.3$ m, and $h = 1.0$ m, and compute the y coordinates of the centroid of the composite volume.

10.23 Refer to Figure P10.11, and determine the y coordinates of the centroid of the composite solid.

10.24 Determine the coordinates for the centroid of the composite area shown in Figure P10.24.

10.25 Compute the coordinates for the centroid of the composite area shown in Figure P10.25.

10.26 Locate the centroid of the composite area shown in Figure P10.26.

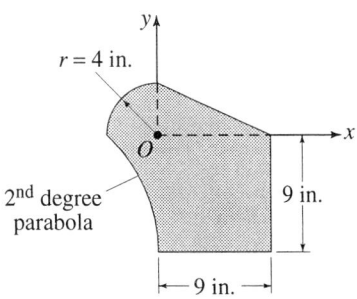

FIGURE P10.26.

10.27 Determine the coordinates of the centroid of the area shown in Figure P10.27.

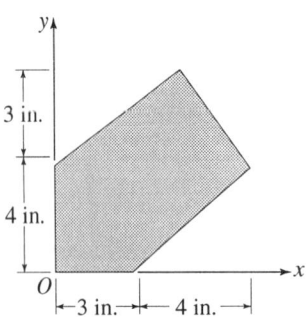

FIGURE P10.27.

10.28 A composite area is shown in Figure P10.28. Find the x and y coordinates of its centroid.

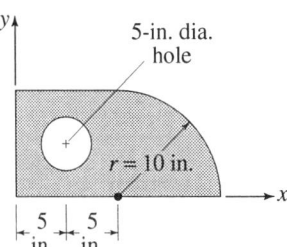

FIGURE P10.28.

10.29 A wire is bent in the shape shown in Figure P10.29. Locate its centroid with respect to the given origin.

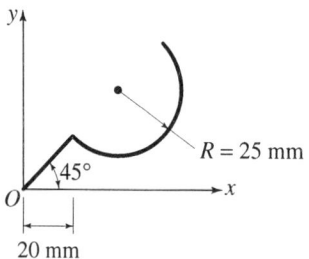

FIGURE P10.29.

10.30 Find the x and y coordinates of the centroid of the line shown in Figure P10.30.

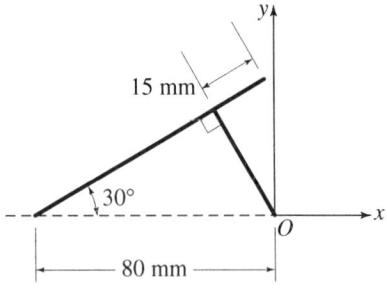

FIGURE P10.30.

10.31 Part of a plastic logo fabricated from a thin rod is shown in Figure P10.31. Assuming that its constant diameter is very small compared to the length given, locate its centroid.

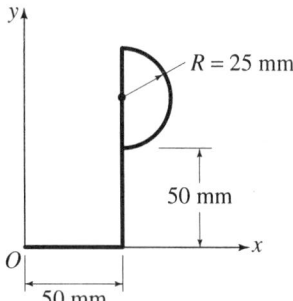

FIGURE P10.31.

10.32 The neon sign shown in Figure P10.32 is made of thin glass tubing. Determine the coordinates of its centroid.

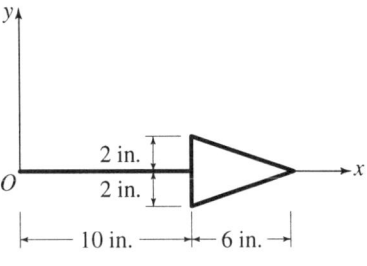

FIGURE P10.32.

10.33 A composite wire is shaped as shown in Figure P10.33. Find the length L so that the centroid of the composite wire is at the intersection of the semicircular and straight segments.

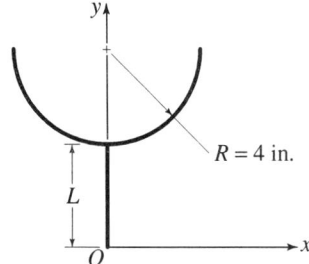

FIGURE P10.33.

10.34 A wire is bent in the shape of an equilateral triangle, as shown in Figure P10.34. Show that the y coordinate of the centroid of this wire is the same as the y coordinate of the centroid of the enclosed triangular area. Do you think this result could be generalized for all triangularly shaped wires?

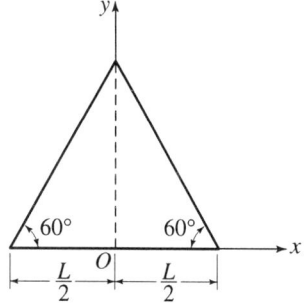

FIGURE P10.34.

10.35 Refer to Figure P10.35, and determine the length L so that the centroid of the composite wire is at $y = 1.5R$.

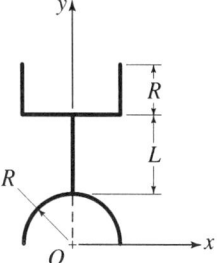

FIGURE P10.35.

10.4
Centroids by
Integration

In Section 10.1, we defined the *center of gravity* and the center of mass and learned that, for the case of a body where the acceleration of gravity g may be assumed constant, these two centers coincide. Also, in Section 10.2, we defined the *centroid of a volume* and discovered that it coincides with the *center of gravity* for the case of a homogeneous body (i.e., a body for which the weight density γ is constant). Throughout this section, we will assume that both g and γ are constants for a given body, and, therefore, we conclude that the *center of gravity, center of mass*, and *centroid* all lie at the same point in space. Thus, our focus in this section will be to learn how to use *integration* to determine the centroids of objects, not only for solids (volumes) but also for lines and plane areas. We will begin with the simplest of these cases and, then, proceed to the more difficult ones. Before we begin, however, we should point out that, as in the case of composite objects discussed in Section 10.3, the centroid of an individual object, possessing a plane or axis of symmetry, lies on this plane or axis.

Area

A plane area is a two-dimensional entity that may be defined in a two-dimensional x-y plane. Thus, only Eqs. (10.4a) and (10.4b) are needed to completely define its centroid. These two equations, slightly modified, have been renumbered and are shown below as Eqs. (10.11).

$$\bar{x} = \frac{\int x_e \, dA}{A} = \frac{\int x_e \, dA}{dA}, \tag{10.11a}$$

and

$$\bar{y} = \frac{\int y_e \, dA}{A} = \frac{\int y_e \, dA}{dA}. \tag{10.11b}$$

where x_e and y_e signify the x and y coordinates, respectively, of the centroid of the element of area dA.

It is possible to avoid double integration if we select an element of area as a thin rectangle either parallel to the x axis, as in Figure 10.6(a), or parallel to the y axis, as in Figure 10.6(b). In Figure 10.6(a), $dA = (x_1 - x_2) dy$, $x_e = x_2 + \left(\dfrac{x_1 - x_2}{2}\right) = \dfrac{x_1 + x_2}{2}$, and $y_e = y$. In Figure 10.6(b), $dA = (y_2 - y_1) dx$, $x_e = x$, and $y_e = y_1 + \left(\dfrac{y_2 - y_1}{2}\right) = \dfrac{y_1 + y_2}{2}$.

Depending upon the functions $f_1(x)$ and $f_2(x)$ and the centroid coordinate being determined, one choice may be more advantageous than the other. Examples 10.4 and 10.5 illustrate the determination of the centroids of areas by integration.

Line

Here, we will consider only the centroid of a line that lies in a plane, say the x-y plane. Thus, only the first two of Eqs. (10.5) will be needed to completely establish its centroid. These two equations, slightly modified, are assigned new numbers, as shown below:

$$\bar{x} = \frac{\int x_e \, dL}{L} = \frac{\int x_e \, dL}{\int dL} \tag{10.12a}$$

and

$$\bar{y} = \frac{\int y_e \, dL}{L} = \frac{\int y_e \, dL}{\int dL} \tag{10.12b}$$

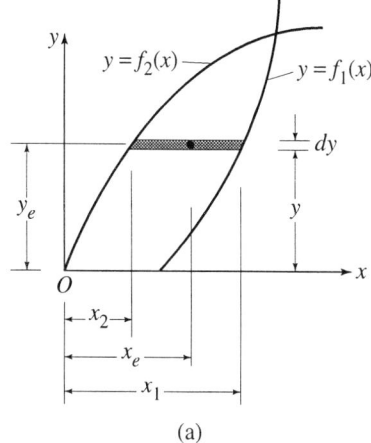

(a)

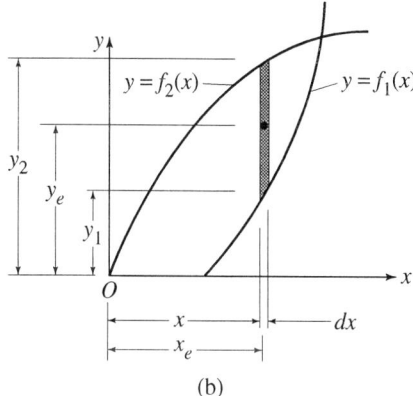

(b)   FIGURE 10.6.

where x_e and y_e represent the x and y coordinates, respectively, of the centroid of the element of length dL. As shown in Figure 10.7, the element of length dL may be expressed in terms of the variables x and y using the Pythagorean theorem. Thus,

$$(dL)^2 = (dx)^2 + (dy)^2,$$

$$dL = \sqrt{(dx)^2 + (dy)^2},$$
(10.13a)

$$dL = \sqrt{1 + \left(\frac{dy}{dx}\right)^2}\, dx.$$

Alternatively,

$$dL = \sqrt{1 + \left(\frac{dx}{dy}\right)^2}\, dy.$$
(10.13b)

Also, $x_e = x$ and $y_e = y$, as shown in Figure 10.7. Examples 10.6 and 10.7 show how the centroid of a line may be located.

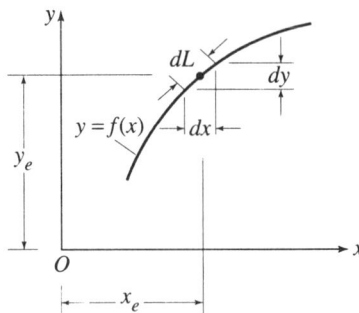

FIGURE 10.7.

Volume

Equations (10.3) provide the means of determining the centroid of a given volume. These equations, slightly modified, have been renumbered and are shown below as Eqs. (10.14).

$$\bar{x} = \frac{\int x_e \, dV}{V} = \frac{\int x_e \, dV}{\int dV}, \qquad (10.14a)$$

$$\bar{y} = \frac{\int y_e \, dV}{V} = \frac{\int y_e \, dV}{\int dV}, \qquad (10.14b)$$

and

$$\bar{z} = \frac{\int z_e \, dV}{V} = \frac{\int z_e \, dV}{\int dV}, \qquad (10.14c)$$

where x_e, y_e, and z_e represent the centroidal coordinates of the element of volume dV.

When dealing with a volume of revolution, as shown in Figure 10.8, which was generated by revolving the shaded area about the y axis, the element of volume is conveniently chosen as a thin circular disk of radius $z = x$ centered at the axis of revolution. In such a case, the element of volume may be found as the product of the area of the disk (πz^2) times its thickness dy, i.e., $dV = \pi z^2 \, dy$. Using the function $[z = f(x)]$, we can express z in terms of y, and the element of volume becomes a function of single variable y. Also, the centroidal coordinates for this element of volume are $x_e = z_e = 0$ and $y_e = y$. Thus, a single integration, based upon Eq. (10.14b), enables us to find $\bar{y}$ for the volume. Note that, because both the x-y and the y-z planes are planes of symmetry (i.e., the volume is symmetric with respect to the y axis of

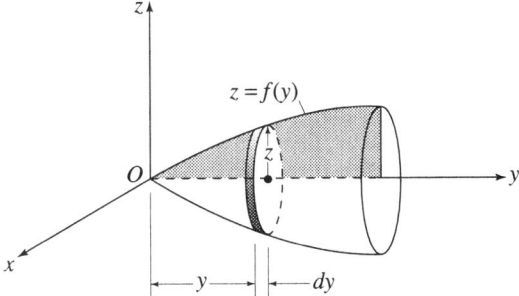

$z = f(y)$

$\bar{z}$

O

y

x

y dy

FIGURE 10.8.

revolution), the centroid of the volume lies on this axis and $\bar{x} = \bar{z} = 0$. On the other hand, when a volume cannot be generated by revolving an area about an axis, the selection of the element of volume may be such that multiple integrations become necessary.

Examples 10.8 and 10.9 illustrate the determination of the centroids of volumes.

■ **Example 10.4**

Locate the x and y coordinates of the centroid of the shaded area shown in Figure E10.4(a). Use a differential element of area parallel to the y axis, and confirm the answers by using a differential element of area parallel to the x axis.

Solution

A differential element of area parallel to the y axis is shown in Figure E10.4(b), and its properties are given by

$$dA = y\, dx = 2x^{1/2}\, dx,$$

$$x_e = x,$$

and

$$y_e = y/2 = \frac{2x^{1/2}}{2} = x^{1/2}.$$

Note that, in the above equations, y was replaced by x using the given function ($y = 2x^{1/2}$), thus, expressing all of the properties in terms of the single variable x. To use Eqs. (10.11), we need to evaluate the following integrals:

$$\int x_e\, dA = \int_0^4 x(2x^{1/2})\, dx$$

$$= \frac{4}{5}(x^{5/2})\Big|_0^4 = 25.6 \text{ m}^3,$$

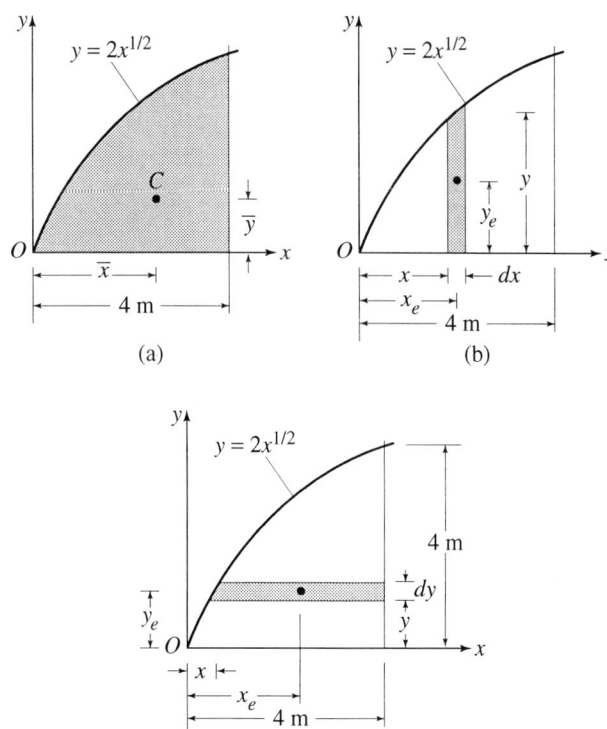

FIGURE E10.4.

$$\int y_e\, dA = \int_0^4 (x^{1/2})(2x^{1/2})\, dx$$

$$= x^2 \big|_0^4 = 16 \text{ m}^3,$$

and

$$A = \int dA = \int_0^4 2x^{1/2}\, dx$$

$$= \frac{4}{3}(x^{3/2}) \bigg|_0^4 = 10.667 \text{ m}^2.$$

Thus, by Eqs. (10.11),

$$\bar{x} = \frac{\displaystyle\int x_e\, dA}{\displaystyle\int dA} = \frac{25.6}{10.667} = 2.40 \text{ m,} \qquad\qquad \text{ANS.}$$

and

$$\bar{y} = \frac{\int y_e\, dA}{\int dA} = \frac{16}{10.667} = 1.500 \text{ m.} \qquad \text{ANS.}$$

Now a differential element of area parallel to the x axis is chosen, as shown in Figure E10.4(c), the properties of which are

$$dA = (4 - x)\, dy = (4 - y^2/4)\, dy,$$

$$x_e = x + (4 - x)/2 = x/2 + 2 = \frac{1}{8}y^2 + 2$$

and

$$y_e = y.$$

Note that, in the above equations, x was replaced by y using the given function $(x = \frac{1}{4}y^2)$ to express all of the properties in terms of the single variable y. Thus, proceeding as before,

$$\int x_e\, dA = \int_0^4 \left(\frac{1}{8}y^2 + 2\right)\left(4 - \frac{1}{4}y^2\right) dy$$

$$= \int_0^4 \left(8 - \frac{1}{32}y^4\right) dy$$

$$= \left(8y - \frac{1}{160}y^5\right)\Big|_0^4 = 25.6 \text{ m}^3,$$

$$\int y_e\, dA = \int_0^4 y\left(4 - \frac{1}{4}y^2\right) dy$$

$$= \left(2y^2 - \frac{1}{16}y^4\right)\Big|_0^4 = 16 \text{ m}^3,$$

and

$$A = \int dA = \int_0^4 \left(4 - \frac{1}{4}y^2\right) dy$$

$$= \left(4y - \frac{1}{12}y^3\right)\Big|_0^4 = 10.667 \text{ m}^2.$$

Note that the upper limit in the above integrals was found from the given function corresponding to $x = 4$ m. Thus, by Eqs. (10.11), we obtain values for x and y locating the centroid C which are identical to the values obtained earlier. The position of the centroid is shown in Figure E10.4(a).

■ **Example 10.5**

Consider the quadrant of the circular area of radius $r = b$ shown in Figure E10.5(a), and determine the location of its centroid.

Solution

Inspection of the given area reveals that a straight line making a 45° angle ccw with the x axis is an axis of symmetry and, therefore, the centroid of the area lies on this line. It follows that $\bar{x} = \bar{y}$, and we need find only one of these coordinates. Also, it is convenient, in this case, to deal in terms of polar coordinates. However, to avoid a double integral, we define the element of area dA, as shown in Figure E10.5(b). Thus, using the properties of a triangular area given in Appendix A,

$$dA = (\tfrac{1}{2}b)(b\,d\theta) = \frac{1}{2}b^2\,d\theta,$$

$$x_e = \frac{2}{3}b\cos\theta,$$

and

$$y_e = \frac{2}{3}b\sin\theta.$$

To use Eqs. (10.11a) to find $\bar{x}$, we need to evaluate the following integrals:

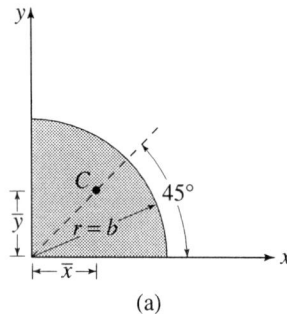

(a)

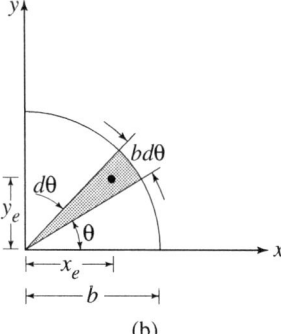

FIGURE E10.5.   (b)

$$\int x_e \, dA = \int_0^{\pi/2} \left(\frac{2}{3}b\cos\theta\right)\left(\frac{1}{2}b^2\,d\theta\right)$$

$$= \frac{b^3}{3}\int_0^{\pi/2}\cos\theta\,d\theta$$

$$= \frac{b^3}{3}(\sin\theta)\Big|_0^{\pi/2} = \frac{b^3}{3},$$

and

$$\int dA = \int_0^{\pi/2}\left(\frac{1}{2}b^2\,d\theta\right)$$

$$= \frac{1}{2}b^2\theta\Big|_0^{\pi/2} = \frac{\pi b^2}{4}.$$

Therefore, by Eq. (10.11a),

$$\bar{x} = \frac{\displaystyle\int x_e\,dA}{\displaystyle\int dA} = \frac{b^3/3}{(\pi b^2/4)} = \left(\frac{4}{3\pi}\right)b \qquad\qquad \text{ANS.}$$

and, as stated above,

$$\bar{y} = \bar{x} = \left(\frac{4}{3\pi}\right)b. \qquad\qquad \text{ANS.}$$

The location of the centroid is shown in Figure E10.5(a).

Note that the same answer would be obtained had we decided to find $\bar{y}$ instead of $\bar{x}$. Thus,

$$\int y_e\,dA = \int_0^{\pi/2}\left(\frac{2}{3}b\sin\theta\right)\left(\frac{1}{2}b^2\,d\theta\right)$$

$$= \frac{b^3}{3}\int_0^{\pi/2}\sin\theta\,d\theta$$

$$= \frac{b^3}{2}(-\cos\theta)\Big|_0^{\pi/2} = \frac{b^3}{3}.$$

Thus, by Eq. (10.11b),

$$\bar{y} = \frac{\displaystyle\int y_e\,dA}{\displaystyle\int dA} = \frac{b^3/3}{(\pi b^2/4)} = \left(\frac{4}{3\pi}\right)b$$

and, of course, $\bar{x}$ would be the same as $\bar{y}$.

■ **Example 10.6**   Locate the centroid of the parabolic line shown in Figure E10.6 which extends from the origin to point (2, 16) in.

Solution   Let us determine the element of length dL in terms of the variable x. Thus, using the given function,

$$\frac{dy}{dx} = 8x.$$

Substituting in Eq. (10.13a),

$$dL = \sqrt{1 + 64x^2}\, dx.$$

Also,

$$x_e = x,$$

and

$$y_e = y = 4x^2.$$

To use Eqs. (10.12), we need to evaluate the following integrals:

$$\int x_e\, dL = \int_0^2 x\sqrt{1 + 64x^2}\, dx$$

$$= \int_0^2 8x\sqrt{\frac{1}{64} + x^2}\, dx$$

$$= \frac{8}{3}\sqrt{\left(\frac{1}{64} + x^2\right)^3}\,\Bigg|_0^2 = 21.458 \text{ in.}^2,$$

and

$$\int y_e\, dL = \int_0^2 4x^2\sqrt{1 + 64x^2}\, dx$$

$$= \int_0^2 32x^2\sqrt{\frac{1}{64} + x^2}\, dx$$

$$= 8x\sqrt{\left(\frac{1}{64} + x^2\right)^3}\,\Bigg|_0^2 = 128.751 \text{ in.}^2.$$

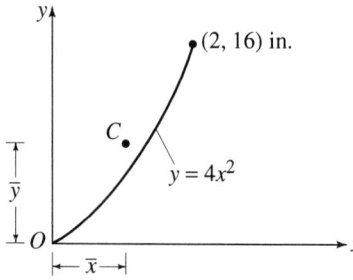

FIGURE E10.6.

Note that the integrals above were evaluated with the aid of Appendix E. Therefore, by Eqs. (10.12),

$$\bar{x} = \frac{\displaystyle\int x_e \, dL}{\displaystyle\int dL} = \frac{21.458}{16.125} = 1.331 \text{ in.,} \qquad \text{ANS.}$$

and

$$\bar{y} = \frac{\displaystyle\int y_e \, dL}{\displaystyle\int dL} = \frac{128.751}{16.125} = 7.98. \qquad \text{ANS.}$$

■ **Example 10.7**

Locate the centroid of the thin wire bent in the shape of a circular arc of radius b, as shown in Figure E10.7(a).

Solution

Inspection of the given line reveals that it is symmetric with respect to an axis making a $45°$ angle ccw with the x axis. Therefore, the centroid of the circular line lies on this axis. It follows, therefore, that $\bar{x} = \bar{y}$, and we need to determine only one of these two coordinates. In this case, it is convenient to deal with polar coordinates.

To determine $\bar{x}$ by using Eq. (10.12a), we need to define an element of length dL and the corresponding centroidal coordinate x_e. Thus, referring to Figure E10.7(b),

$$dL = b \, d\theta$$

and

$$x_e = x = b \cos \theta.$$

The integrals needed for Eq. (10.12a) are evaluated as follows:

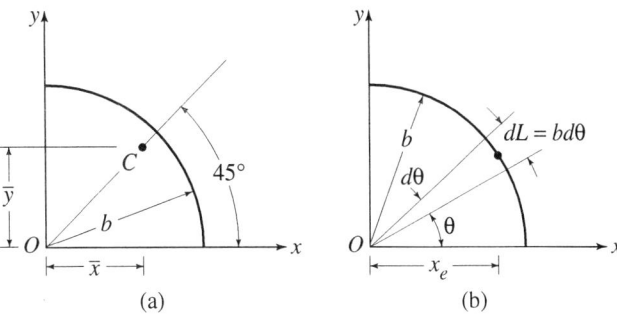

FIGURE E10.7.   (a)   (b)

$$\int x_e \, dL = \int_0^{\pi/2} (b \cos \theta)(b \, d\theta)$$

$$= b^2 (\sin \theta)|_0^{\pi/2} = b^2,$$

and

$$L = \int dL = \int_0^{\pi/2} b \, d\theta$$

$$= b\theta \Big|_0^{\pi/2} = \frac{b\pi}{2}.$$

Therefore, Eq. (10.12a) yields

$$\bar{y} = \bar{x} = \frac{\displaystyle\int x_e \, dL}{\displaystyle\int dL} = \frac{b^2}{(b\pi/2)} = \frac{2b}{\pi}. \qquad \text{ANS.}$$

■ **Example 10.8**

Locate the centroid of the volume shown in Figure E10.8 which was generated by revolving the shaded area one complete revolution about the y axis.

Solution

Note that, because the volume is symmetric with respect to the y axis of revolution, the centroid lies on this axis and

$$\bar{x} = \bar{z} = 0. \qquad \text{ANS.}$$

To determine $\bar{y}$, we use Eq. (10.14b). Thus, as shown in Figure E10.8,

$$dV = (\pi z^2) \, dy = \pi \left(\frac{a^2}{b} \right) y \, dy,$$

$$y_e = y,$$

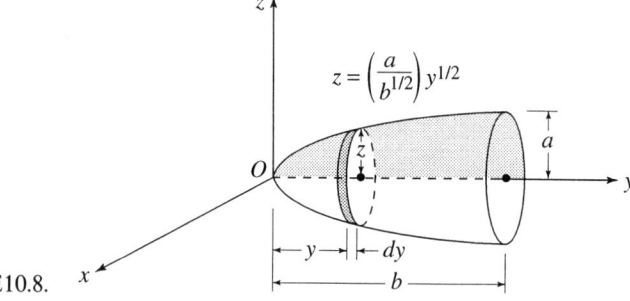

FIGURE E10.8.

$$\int y_e \, dV = \int_0^b (y)(\pi)\left(\frac{a^2}{b}\right) y \, dy$$

$$= \left(\frac{\pi a^2}{3b}\right)(y^3)\Big|_0^b = \frac{\pi a^2 b^2}{3},$$

and

$$V = \int dV = \int_0^b \pi\left(\frac{a^2}{b}\right) y \, dy$$

$$= \left(\frac{\pi a^2}{2b}\right)(y^2)\Big|_0^b = \frac{\pi a^2 b}{2}.$$

Thus, by Eq. (10.14b),

$$\bar{y} = \frac{\displaystyle\int y_e \, dV}{\displaystyle\int dV} = \frac{\dfrac{\pi a^2 b^2}{3}}{\dfrac{\pi a^2 b}{2}} = \frac{2}{3}b. \qquad\qquad \text{ANS.}$$

■ **Example 10.9** Find the y and z coordinates for the centroid of the solid shown in Figure E10.9.

Solution From the geometry given in Figure E10.9, we can show that the equations for lines AC and BC are, respectively, $x = a - \dfrac{a}{b}y$ and $z = c - \dfrac{c}{b}y$. To avoid multiple integrations, a triangular differential element of volume is now selected, as shown, so such that

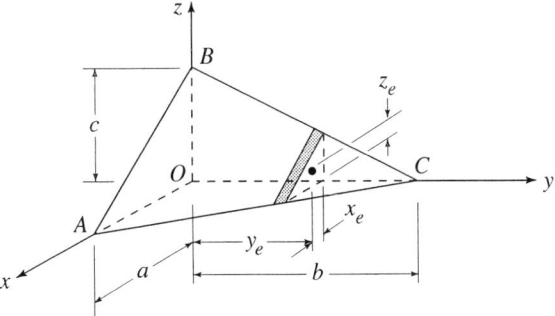

FIGURE E10.9.

$$dV = A\,dy = \left(\frac{1}{2}xy\right)dy$$

$$= \frac{1}{2}\left(a - \frac{a}{b}y\right)\left(c - \frac{c}{b}y\right)dy$$

$$= \frac{1}{2}\left[ac - 2\left(\frac{ac}{b}\right)y + \left(\frac{ac}{b^2}\right)y^2\right]dy$$

where x and z were replaced by the variable y using the equations for lines AC and BC. Also, the y coordinate for the centroid of the element of volume is $y_e = y$ and its z coordinate is $z_e = \frac{1}{3}z = \frac{1}{3}\left(c - \frac{c}{b}y\right)$. This latter centroid location was obtained with the aid of Appendix A.

To determine $\bar{y}$ and $\bar{z}$ locating the centroid of the given solid, we need to evaluate the following:

$$\int y_e\,dV = \int_0^b \frac{1}{2}\left[acy - 2\left(\frac{ac}{b}\right)y^2 + \left(\frac{ac}{b^2}\right)y^3\right]dy$$

$$= \frac{1}{2}\left[\left(\frac{ac}{2}\right)y^2 - \frac{2}{3}\left(\frac{ac}{b}\right)y^3 + \left(\frac{ac}{4b^2}\right)y^4\right]\Bigg|_0^b$$

$$= \frac{acb^2}{4},$$

$$\int z_e\,dV = \int_0^b \frac{1}{3}\left(c - \frac{c}{b}y\right)\left(\frac{1}{2}\right)\left[ac - 2\left(\frac{ac}{b}\right)y + \left(\frac{ac}{b^2}\right)y^2\right]dy$$

$$= \int_0^b \frac{1}{6}\left[ac^2 - 3\left(\frac{ac^2}{b}\right)y + 3\left(\frac{ac^2}{b^2}\right)y^2 - 3\left(\frac{ac^2}{b^3}\right)y^3\right]dy$$

$$= \frac{1}{6}\left[ac^2y - \frac{3}{2}\left(\frac{ac^2}{b}\right)y^2 + \left(\frac{ac^2}{b^2}\right)y^3 - \frac{1}{4}\left(\frac{ac^2}{b^3}\right)y^4\right]\Bigg|_0^b$$

$$= \frac{abc^2}{24},$$

and

$$V = \int dV = \int_0^b \frac{1}{2}\left[ac - 2\left(\frac{ac}{b}\right)y + \left(\frac{ac}{b^2}\right)y^2\right]dy$$

$$= \frac{1}{2}\left[acy - \left(\frac{ac}{b}\right)y^2 + \frac{1}{3}\left(\frac{ac}{b^2}\right)y^3\right]\Bigg|_0^b$$

$$= \frac{abc}{6}.$$

Thus, by Eqs. (10.14b) and (10.14c),

$$\bar{y} = \frac{\displaystyle\int y_e \, dV}{\displaystyle\int dV} = \frac{\dfrac{ab^2c}{24}}{\dfrac{abc}{6}} = \frac{b}{4},$$ ANS.

and

$$\bar{z} = \frac{\displaystyle\int z_e \, dV}{\displaystyle\int dV} = \frac{\dfrac{abc^2}{24}}{\dfrac{abc}{6}} = \frac{c}{4}.$$ ANS.

As an exercise, the student can show that $\bar{x} = a/4$.

■

Problems

10.36 Find the centroid of the area bounded by the three straight lines given by $x = 0$, $y = 0$, and $x + y = 1$.

10.37 Find the x coordinate of the centroid of the area bounded by the parabola and the straight line shown in Figure P10.37.

10.38 Find the coordinates of the centroid of the area between the parabola $y = 1 - x^2$ and the straight line $y = 1 - x$.

10.39 In Figure P10.39, an area is shown which is bounded by the parabola $x = 4y^2$ and the vertical straight line $x = 4$. Determine the coordinates of the centroid of this area.

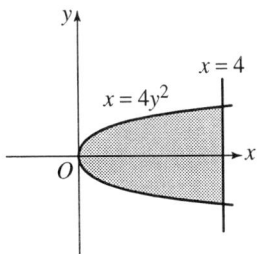

FIGURE P10.39.

10.40 Locate the centroid of the area bounded by $x = 0$, $y = a$, and $y = x^2$. Regard a as a given constant.

10.41 Determine the coordinates of the centroid of the area bounded by $y = x$ and $y = x^{1/2}$.

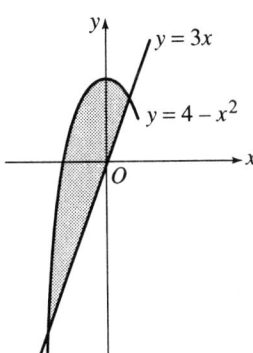

FIGURE P10.37.

10.42 Locate the centroid of the area under the half circle above the x axis. The circle has a center at the origin and a radius a.

10.43 Locate the centroid of the area bounded by the parabola and the straight line shown in Figure P10.43. Coordinates are given in inches.

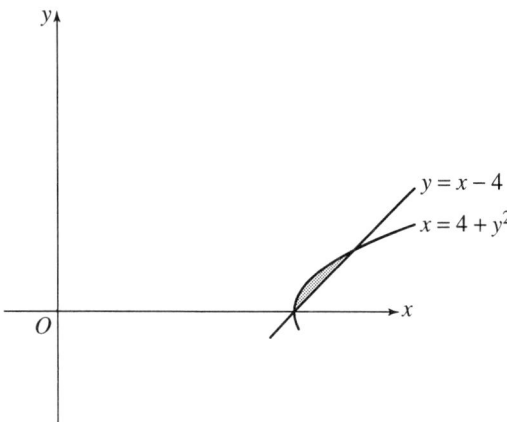

FIGURE P10.43.

10.44 An area is bounded by $y = x^3$, $y = 0$, and $x = 2$. Determine the area, and locate its centroid if the coordinates are given in meters.

10.45 Consider the area under the curved line defined by the cosine function on the interval $(-\pi/2$ to $\pi/2)$ and bounded by $y = 0$. Determine the centroid of this area.

10.46 An area is bounded by $y = x^3$, $x = 0$, and $y = 8$. Determine this area, and locate its centroid if the coordinates are given in meters.

10.47 Determine the coordinates of the centroid of the area shown in Figure P10.47.

10.48 Find the y coordinate of the centroid of the area shown in Figure P10.48.

10.49 Determine the coordinates of the centroid of the area under the half sine wave shown in Figure P10.49. Coordinates are stated in meters.

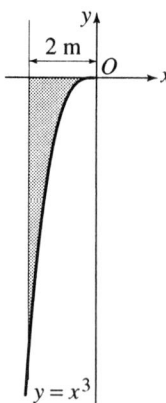

FIGURE P10.47.

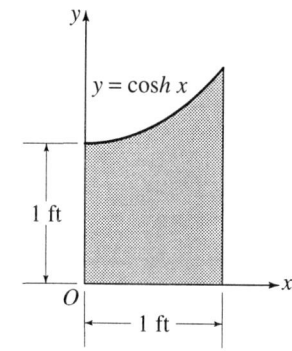

FIGURE P10.48.

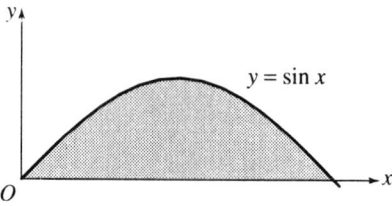

FIGURE P10.49.

10.50 A straight line segment connects point A (2, 4) with point B (5, 8). By the Pythagorean theorem, the length of this line segment is 5, and its centroid lies at the midpoint of the line segment at C

(3.5, 6.0). Write appropriate differential equations, and integrate them to verify the length and centroid coordinates. Coordinates are expressed in inches.

10.51 Find the y coordinate of the centroid of the arc of a parabola shown in Figure P10.51.

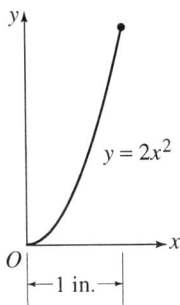

FIGURE P10.51.

10.52 Refer to Problem 10.51 and find the x coordinate of the centroid of the parabolic arc.

10.53 A homogeneous rod is bent in the parabolic shape shown in Figure P10.53. Determine the y coordinate of its centroid.

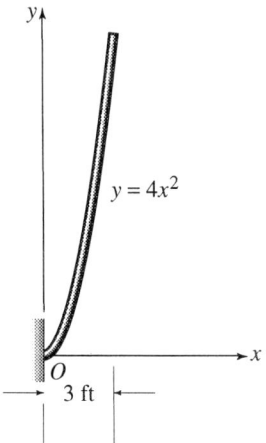

FIGURE P10.53.

10.54 If the rod of Problem 10.53 weighs 2 lb/ft, find the reaction components at the fixed support at O.

10.55 A cone is generated by revolving the line $y = x$ through one complete revolution about the y axis. Consider the portion of this cone bounded by the planes $y = 0$ and $y = 10$, and find its volume and the coordinates of its centroid. Coordinates are expressed in inches.

10.56 A geometric solid is bounded by five planes whose equations are given by $x = 0$, $y = 0$, $z = 0$, $x + y = 2$, and $z = 4$. Determine the volume of this solid and the coordinates of its centroid. Coordinates are stated in meters.

10.57 A body of revolution is generated by rotating the parabola $x = 9y^2$ about the y axis, as shown in Figure P10.57. Determine the coordinates of the centroid of this volume.

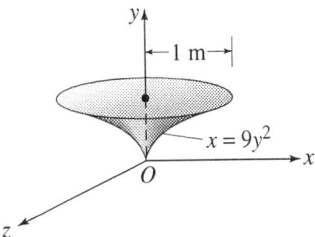

FIGURE P10.57.

10.58 A volume is generated by revolving the parabola $x = 4 + y^2$ about the x axis, as shown in Figure P10.58. Determine the coordinates of the centroid of this volume.

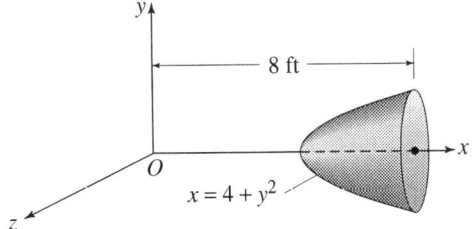

FIGURE P10.58.

10.59 A cone is generated by revolving the straight line $y = \frac{1}{2}x - 2$ about the x axis as shown in Figure P10.59. Determine the coordinates of the centroid of this volume.

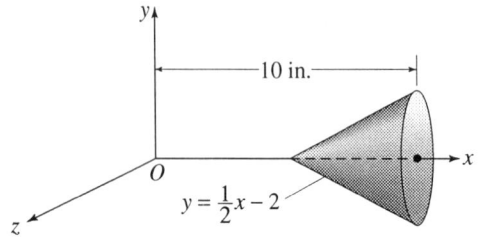

FIGURE P10.59.

10.60 A solid of revolution is shown in Figure P10.60. Determine the coordinates of the centroid of this volume.

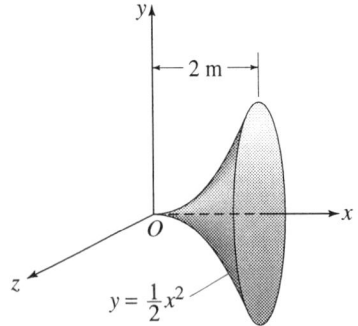

FIGURE P10.60.

10.61 Refer to Figure P10.61, and determine the coordinates of the centroid of the volume shown.

10.62 Consider a sphere with radius a centered at the origin of an x-y-z coordinate system. Locate the centroid of the volume enclosed by the portion of the sphere in the first octant. Hint: Use spherical coordinates.

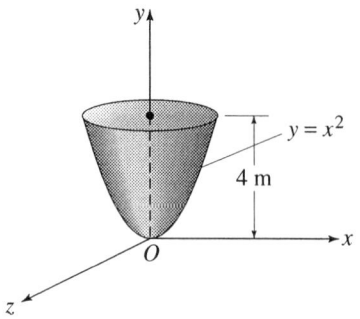

FIGURE P10.61.

10.63 Locate the center of gravity of a hemisphere of constant weight density. Hint: Choose an origin at the center of the whole sphere, and consider only the upper half.

10.64 Refer to Figure P10.64, and determine the coordinates of the mass center of the homogeneous body of revolution shown.

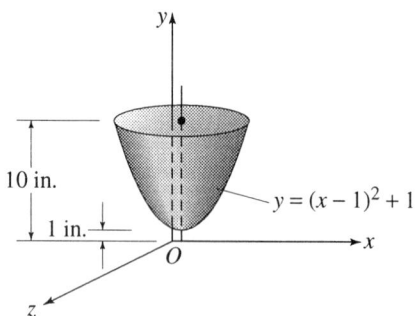

FIGURE P10.64.

10.65 A wedge shaped body of constant density is bounded by five planes whose equations are given by $x = 0$, $y = 0$, $z = 0$, $z = L$ and $x + y = 1$. Locate the center of gravity of this body.

10.66 A body of revolution of uniform density is shown in Figure P10.66. Determine the coordinates of the mass center of this body.

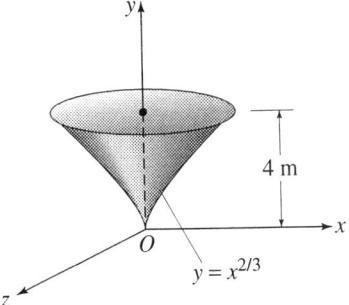

FIGURE P10.66.

10.67 A body of constant mass density is bounded by the curved parabolic surface $by + x^2 - a^2 = 0$ and the following three planes: $z = 0$, $z = L$, and $y = 0$. Determine the coordinates of the center of mass of this body.

10.68 Consider the quarter cylindrical body lying in the first octant which is of constant mass density and is bounded by the curved surface $x^2 + z^2 = a^2$ and the following four planes: $x = 0$, $y = 0$, $z = 0$, and $y = b$. Find the coordinates of the mass center of this body. (Hint: Use cylindrical coordinates.)

10.69 Locate the mass center of the body of revolution shown in Figure P10.69. Assume that the mass density is constant.

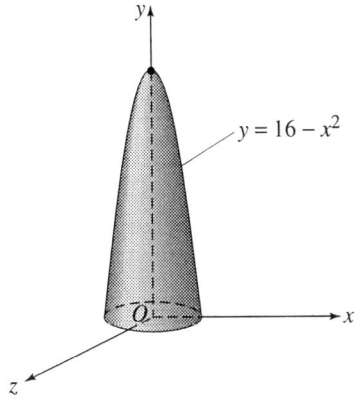

FIGURE P10.69.

10.70 As shown in Figure P10.70, a body of revolution of constant mass density is generated by revolving the shaded area through one complete revolution about the y axis. Locate the center of mass of this body.

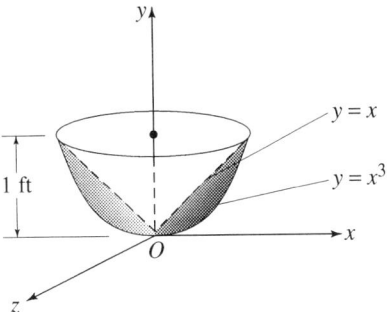

FIGURE P10.70.

10.71 Determine the coordinates of the center of gravity of the homogeneous paraboloid shown in Figure P10.71 which is generated by revolving the parabola $y^2 = (b^2/a)x$ through one complete revolution about the x axis.

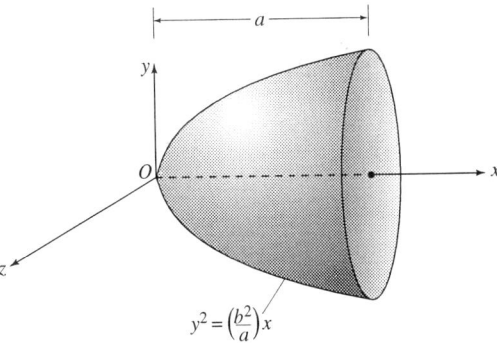

FIGURE P10.71.

10.72 The body of revolution shown in Figure P10.72 has a constant mass density. Locate its mass center.

10.73 Refer to Figure P10.73, and locate the mass center of this body of revolution which has a constant mass density.

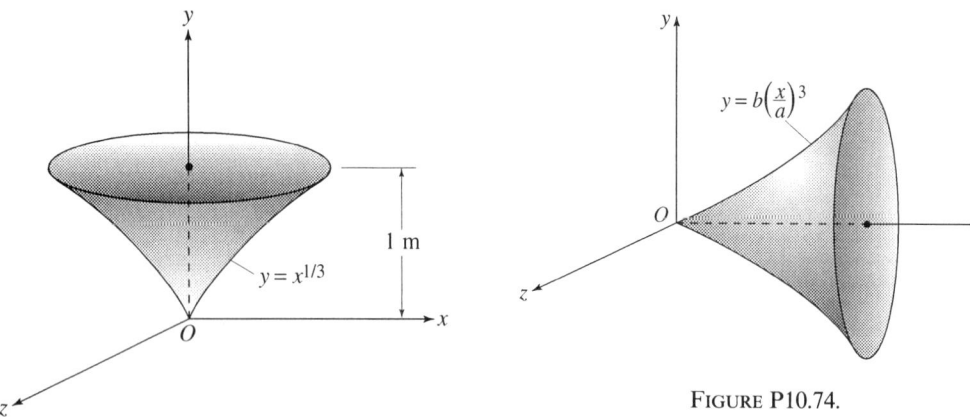

FIGURE P10.72.

FIGURE P10.74.

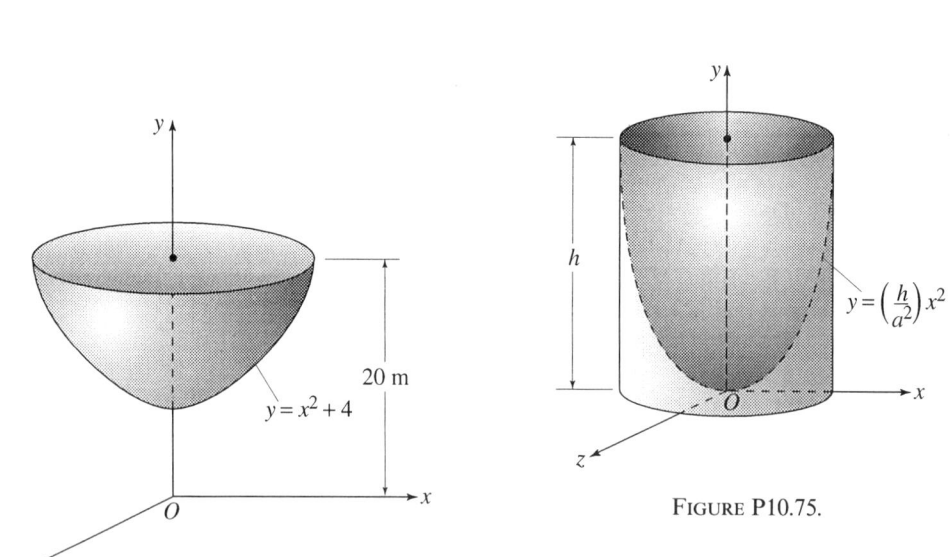

FIGURE P10.73.

FIGURE P10.75.

10.74 A body of constant mass density is generated by revolving the curve $y = b(x/a)^3$ through one complete revolution about the x axis, as shown in Figure P10.74. Determine the coordinates of the mass center of this body.

10.75 The y axis is the axis of symmetry for a cylindrically shaped body of base radius a and height h. A paraboloidally shaped hole is cut from the cylinder by revolving the parabola $y = h(x/a)^2$ about the y axis, as shown in Figure P10.75. Locate its mass center assuming that the mass density of the body is constant.

10.5*
Theorems of Pappus and Guldinus

Pappus, a Greek geometer, is generally credited with the discovery of two useful theorems during the third century A.D. However, the theorems were later reintroduced by the Swiss mathematician Guldinus who lived during the latter part of the sixteenth and the early part of the seventeenth centuries. Whether or not Guldinus had access to the writings of Pappus has not been established. The first of these two theorems deals with the surface area of revolution which is the surface area generated by rotating a plane curve through a complete or partial revolution about a fixed nonintersecting axis. Thus, the straight line AB shown in Figure 10.9(a) would generate the surface area of a truncated cone shown in Figure 10.9(b) if it were rotated through one complete revolution about the fixed y axis. The second theorem deals with the volume of a body of revolution which is the volume of a body developed by rotating a plane area through a complete or partial revolution about a fixed nonintersecting axis. Thus, the shaded area ABCD, shown in Figure 10.9(c), would develop the volume of a truncated cone, shown in Figure 10.9(d), if it were rotated through one complete revolution about the fixed y axis.

Let us now consider a differential length dL of the plane curve BD, shown in Figure 10.10(a), which is to be rotated through one complete revolution about the fixed x axis. The surface area dA developed by dL is expressed by

$$dA = 2\pi y \, dL,$$

and the total area A developed by line BD becomes

$$A = 2\pi \int y \, dL = 2\pi \bar{y} L. \tag{10.15a}$$

Note that, by Eq. (10.5b), the integral $\int y \, dL$ was replaced by the product $\bar{y}L$, where, as shown in Figure 10.10(b), $\bar{y}$ is the y coordinate of the centroid of line BD and L is its length. Generally speaking, however, the line may not experience a complete but only a partial revolution about the x axis defined by the angle β. In such a case, the angle 2π, corresponding to a complete revolution, is replaced by the angle β, and Eq. (10.15a) becomes

$$A = \beta \bar{y} L. \tag{10.15b}$$

The product $2\pi \bar{y}$ in Eq. (10.15a) for one complete revolution and $\beta \bar{y}$ in Eq. (10.15b) for a partial revolution represent the distance moved by the centroid of line BD and L is its length. Therefore, based on Eqs. (10.15) we may state the following theorem known as the *first theorem of Pappus and Guldinus:*

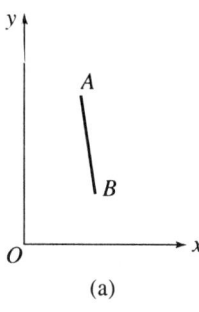

(a)

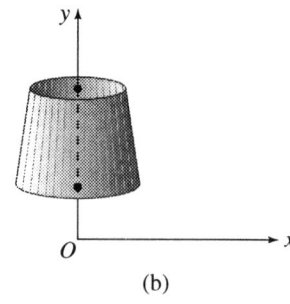

(b)

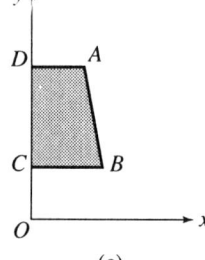

(c)

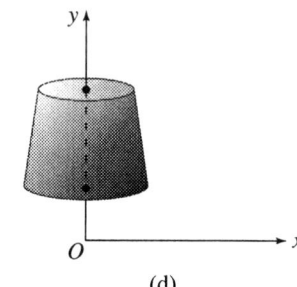

(d)

FIGURE 10.9.

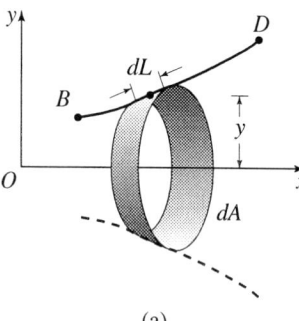

(a)

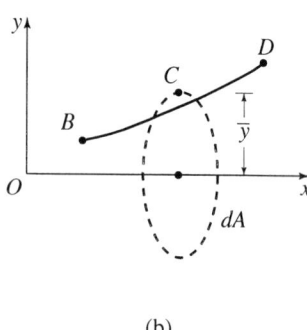

(b)

FIGURE 10.10.

First Theorem

The area of a surface of revolution is equal to the distance moved by the centroid of the generating curve multiplied by its length.

Now, consider a differential area dA of the area BDEF shown in Figure 10.11(a), which is to be revolved through one complete revolution about the fixed x axis. The volume dV developed by dA in one complete revolution is expressed by

$$dV = 2\pi y\, dA,$$

and the total volume V developed by are BDEF becomes

$$V = 2\pi \int y\, dA,$$

$$= 2\pi \bar{y} A. \qquad (10.16a)$$

Note that, by Eq. (10.4b), the integral $\int y\, dA$ was replaced by the product $\bar{y}A$, where, as shown in Figure 10.11(b), $\bar{y}$ is the y coordinate of the centroid of area BDEF and A is its area. Generally speaking, however, the area may not experience a complete but only a partial revolution about the x axis, defined by the angle β. In such a case, the angle 2π, corresponding to a complete revolution, is replaced by the angle β, and Eq. (10.16a) becomes

$$V = \beta \bar{y} A. \qquad (10.16b)$$

The product $2\pi\bar{y}$ in Eq. (10.16a) for one complete revolution and $\beta\bar{y}$ in Eq. (10.16b) for a partial revolution represent the distance moved by the centroid of area BDEF, and A is the magnitude of this area. Therefore, based on Eqs. (10.16), we may state the following theorem known as the *second theorem of Pappus and Guldinus:*

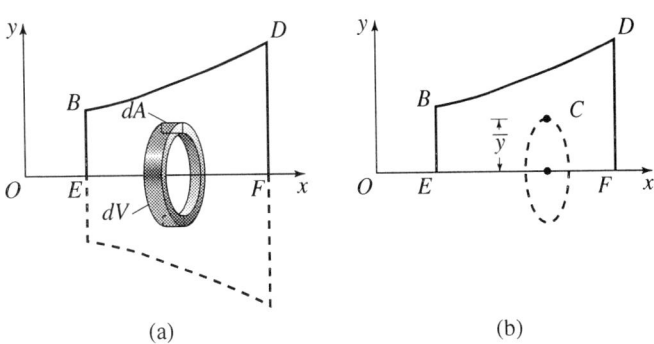

FIGURE 10.11.   (a)   (b)

Second Theorem

The volume of a body of revolution is equal to the distance moved by the centroid of the generating area multiplied by the magnitude of this area.

The following examples illustrate the use of the theorems of Pappus and Guldinus.

■ Example 10.10

(a) Find the surface area generated by revolving lines BD and BED that form the semicircular shape, shown in Figure E10.10(a), through one half of a revolution about the *y* axis.

(b) Find the volume generated by rotating the shaded area, shown in Figure E10.10(b), through one half of a revolution about the *y* axis.

Solution

(a) The centroid locations for lines BD and BED are indicated in Figure E10.10(a) by the symbols x_1 and x_2, respectively. These two quantities are found by the aid of Appendix A. Note that the total surface area generated is the sum of the surface areas developed by each line. Thus, using the first theorem of Pappus and Guildinus,

$$A = A_1 + A_2 = \pi \bar{x}_1 L_1 + \pi \bar{x}_2 L_2$$

$$= \pi(1.0)(1.0) + \pi(1.4775)[\pi(0.5)] = 10.43 \text{ m}^2. \qquad \text{ANS.}$$

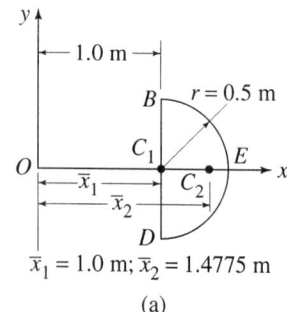

$\bar{x}_1 = 1.0 \text{ m}; \bar{x}_2 = 1.4775 \text{ m}$

(a)

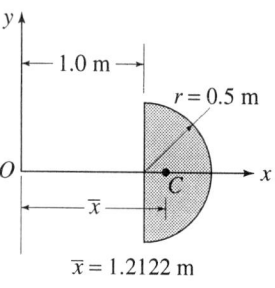

$\bar{x} = 1.2122 \text{ m}$

FIGURE E10.10.   (b)

(b) The centroid location for the given area is indicated in Figure E10.10(b) by the symbol $\bar{x}$ which is determined by the aid of Appendix A. Thus, using the second theorem of Pappus and Guildinus,

$$V = \pi\bar{x}A = \pi(1.2122)\left[\frac{\pi(0.5)^2}{2}\right] = 1.495 \text{ m}^3. \qquad \text{ANS.}$$

■ **Example 10.11**

Use the theorems of Pappus and Guildinus to show that the truncated cone, shown in Figure E10.11(a), has a lateral surface area $A = \pi(r + R)\sqrt{h^2 + (R - r)^2}$ and a volume $V = \pi h[r^2 + \frac{1}{3}(R - r)(R + 2r)]$.

Solution

Consider line BD shown in Figure E10.11(b). If this line is rotated through one complete revolution about the y axis, the required surface area would be generated. Thus, by Eq. (10.15a),

$$A = 2\pi\bar{x}L = 2\pi\left(\frac{r + R}{2}\right)\sqrt{h^2 + (R - r)^2}$$

$$= \pi(r + R)\sqrt{h^2 + (R - r)^2}. \qquad \text{ANS.}$$

Now, consider the area BDOE shown in Figure E10.11(c). If this area is rotated through one complete revolution about the y axis, the required volume would be generated. Note that area BDOE has been subdivided into a rectangular area A_1 and a triangular area A_2. The

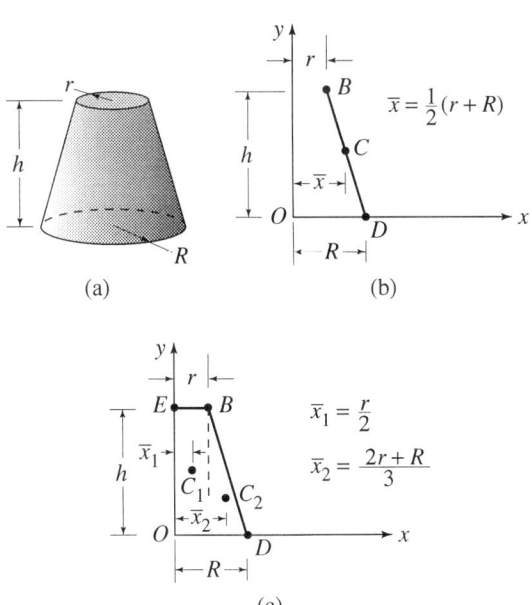

(a)

(b)

$\bar{x} = \frac{1}{2}(r + R)$

(c)

$\bar{x}_1 = \frac{r}{2}$

$\bar{x}_2 = \frac{2r + R}{3}$

FIGURE E10.11.

centroidal distances $\bar{x}_1$ and $\bar{x}_2$, given in Figure E10.11(c), were obtained with the aid of Appendix A. Thus, using Eq. (10.16a),

$$V = 2\pi\bar{x}_1 A_1 + 2\pi\bar{x}_2 A_2$$

$$= 2\pi\left(\frac{r}{2}\right)(rh) + 2\pi\left(\frac{2r + R}{3}\right)\left(\frac{1}{2}\right)(R - r)h$$

$$= \pi r^2 h + \frac{\pi}{3}h(2r + R)(R - r)$$

$$= \pi h\left[r^2 + \frac{1}{3}(R - r)(R + 2r)\right]. \qquad\qquad \text{ANS.}$$

■ **Example 10.12**   Compute the volume of concrete needed to construct the amphitheater-like structure shown in Figure E10.12(a).

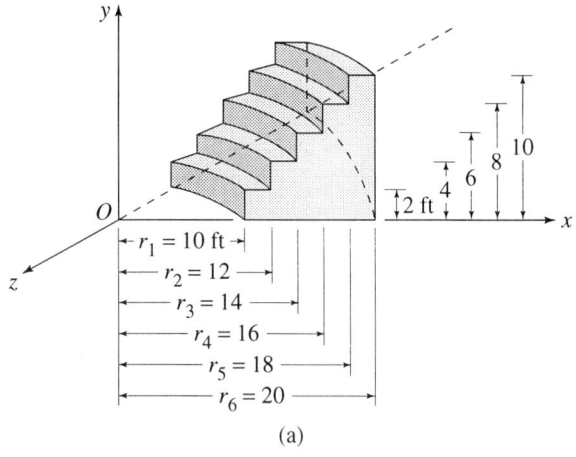

(a)

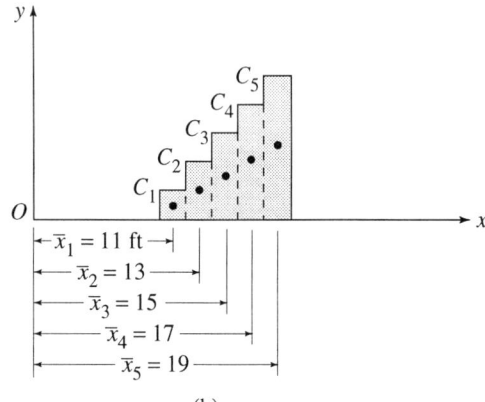

FIGURE E10.12.   (b)

Solution

The volume of concrete needed may be obtained by rotating the cross-sectional area, shown in Figure E10.12(b), through one fourth of a revolution $\left(\dfrac{\pi}{2}\ \text{rad}\right)$ about the y axis. This area is subdivided into five rectangular areas and their centroid locations determined, as shown, with the aid of Appendix A. Thus, using the second theorem of Pappus and Guldinus,

$$V = V_1 + V_2 + \cdots + V_5$$

$$= \frac{\pi}{2}\bar{x}_1 A_1 + \frac{\pi}{2}\bar{x}_2 A_2 + \cdots + \frac{\pi}{2}\bar{x}_5 A_5$$

$$= \frac{\pi}{2}(11)(2 \cdot 2) + \frac{\pi}{2}(13)(2 \cdot 4) + \frac{\pi}{2}(15)(2 \cdot 6) + \frac{\pi}{2}(17)(2 \cdot 8)$$

$$+ \frac{\pi}{2}(19)(2 \cdot 10)$$

$$= 1539\ \text{ft}^3. \hspace{4cm} \text{ANS.}$$

Problems

10.76 The tank depicted in Figure P10.76 is to be coated inside with a sealer to prevent chemical reactions with the fluid to be stored. Determine the surface area, including the top and bottom.

10.77 The cross section of a welded box girder is shown in Figure P10.77. It is to be used as part of a highway bridge, curved in plan, with a radius of curvature of 1000 ft over a central angle of 30°. The radius of curvature is measured horizontally to the vertical y axis of symmetry of the cross section. If steel weighs 490 lb./ft³, find the weight of this curved girder in tons.

10.78 An initial production run of 1000 units is anticipated for the large wastebasket shown in Figure P10.78. Find the volume of plastic material required for 1000 of these wastebaskets. Note that the two diameters are given from mid-thickness to mid-thickness.

10.79 A machine component is to be fabricated from a 0.500-in. diameter rod. It is to be bent in toroidal shape with a radius of 20 in. and a central angle of 60°.

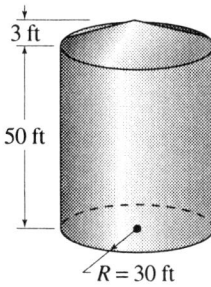

3 ft

50 ft

$R = 30$ ft

FIGURE P10.76.

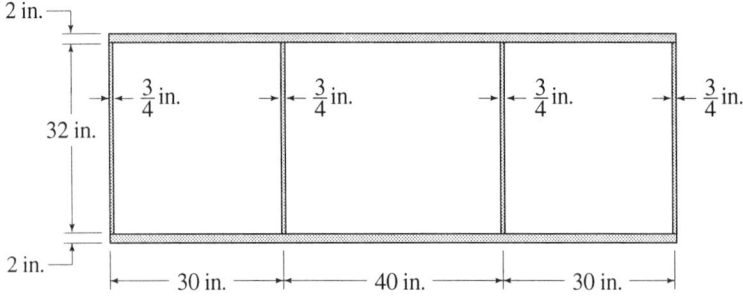

FIGURE P10.77.

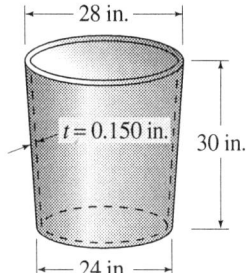

FIGURE P10.78.

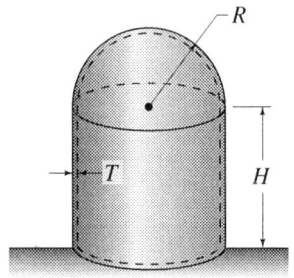

FIGURE P10.80.

The radius of the torus is measured to the center of the rod's circular cross-section. Find the surface area and volume of this component. If it is to be galvanized with a coating which is 0.040 in. thick determine the volume of coating material required. Omit the small circular ends of this component in your calculations.

10.80 A preliminary design of an astronomical observatory is represented in Figure P10.80. It is idealized as a hemisphere atop a cylinder, both with a wall thickness T. Express the following as functions of R, H, and T: (a) the volume of reinforced concrete in the structure, ignoring the floor slab. (b) the enclosed

volume to be air conditioned. Note that the radius R is measured to the wall's mid-thickness.

10.81 Refer to Problem 10.80, and calculate answers for $R = 20$ m, $H = 25$ m, and $T = 0.10$ m.

10.82 Refer to Problem 10.80 and calculate answers for $R = 60$ ft., $H = 75$ ft., and $T = 1.5$ ft.

10.83 A ribbed circular arch highway bridge is shown in Figure P10.83. The primary structure consists of two reinforced concrete ribs, one on each side. Express the volume of reinforced concrete in the two ribs as a function of L, H, W, and D.

10.84 Refer to Problem 10.83 and calculate the number of cubic yards of reinforced

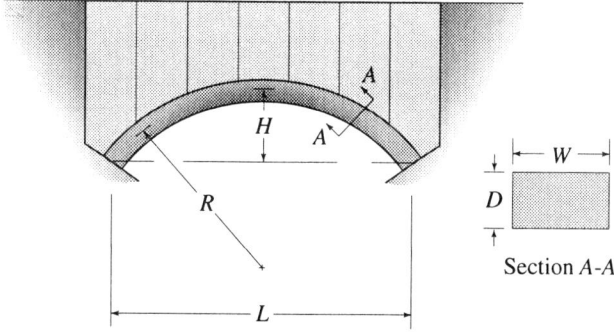

FIGURE P10.83.

concrete required for the two ribs if $L =$ 500 ft., $H = 75$ ft., $D = 18$ in., and $W = 36$ in. If reinforced concrete weighs 150 lb./ft^3, find the tonnage in the two ribs.

10.85 One of the main members of a Ferris wheel has the cross-section shown in Figure P10.85. Find the volume of steel, expressed in ft^3, required for this main member if $R = 100$ ft. and the cross-sectional area $A = 14.4$ in^2.

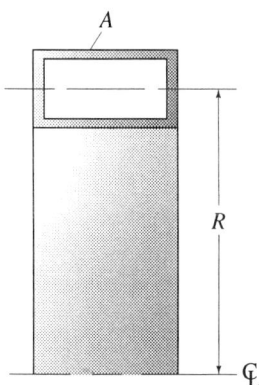

FIGURE P10.85.

10.86 Refer to Figure P10.86 and, using the theorems of Pappus and Guldinus along with direct integration, find the surface area and the volume of the torus gener-

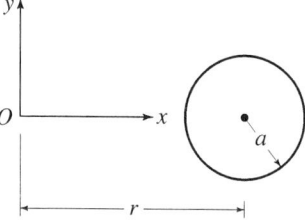

FIGURE P10.86.

ated by rotating the circle of radius a through one complete revolution about the y axis.

10.87 Verify that the surface area and volume equations for a sphere of radius a are $A = 4\pi a^2$ and $V = \frac{4}{3}\pi a^3$, respectively.

10.88 Verify that the lateral surface area and volume equations for a cylinder of radius a and height h are $A = 2\pi ah$ and $V = \pi a^2 h$, respectively. Note that A does not include the area of the cylinder bases which equals $2\pi a^2$.

10.89 Verify that the lateral surface area and volume equations for a right circular cone of base radius a and attitude h are $A = \pi a\sqrt{a^2 + h^2}$ and $V = \frac{1}{3}\pi a^2 h$, respectively. Note that A does not include the area of the cone base.

10.90 Refer to Figure P10.90, and determine the lateral surface area and volume of a solid of revolution generated by revolving the equilateral triangle ABC through an angle of π radians about the y axis. Note that the lateral surface area does not include the area of the two triangular bases of this solid.

10.91 Refer to Figure P10.91, and determine the lateral surface area and volume of a solid of revolution generated by revolving the square ABCD through an angle of 60° about the y axis. Note that the lateral surface area does not include the area of the two square bases of this solid.

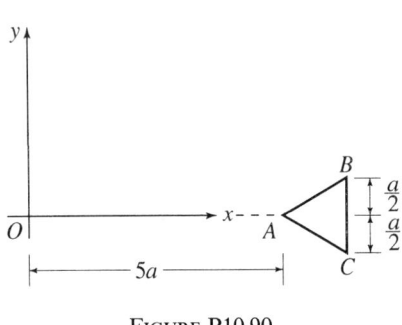

FIGURE P10.90.

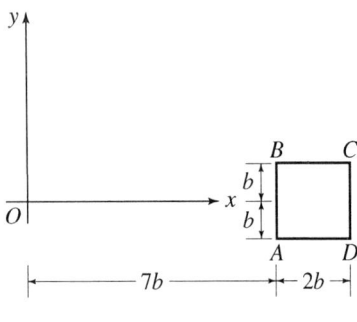

FIGURE P10.91.

10.6*
Fluid Statics

Fluid Pressure

In studying the forces created by an *incompressible** fluid at rest on a submerged body, it is convenient to deal with the term *fluid pressure*, which represents a force per unit area at a given point in the fluid. Thus, such units as lb/in.^{2} in the U.S. Customary system and N/m^2 (known as the *Pascal*) in the SI system are in use. It is also convenient to deal with *gage* rather than *absolute* pressure, because their difference (the atmospheric pressure) is constant and exists at all points in the fluid.

When an object is submerged in an incompressible fluid, it is subjected to pressures at all points on its surface. *The pressure p at a given point is a linear function of the depth of the point below the surface of the fluid, and, according to Pascal's law, this pressure is the same in all directions.* To prove these statements we consider a body of fluid below the x-y plane, as shown in Figure 10.12(a), of weight density $\gamma = g\rho$, where ρ is its mass density. Let us examine the vertical equilibrium of a column of fluid of differential cross-sectional area $dA = dx\,dy$ and depth z, and construct its free-body diagram as shown in Figure

---

* Incompressible fluids are those that are assumed to experience no change in volume when subjected to forces. This classification includes most liquids but excludes all gases.

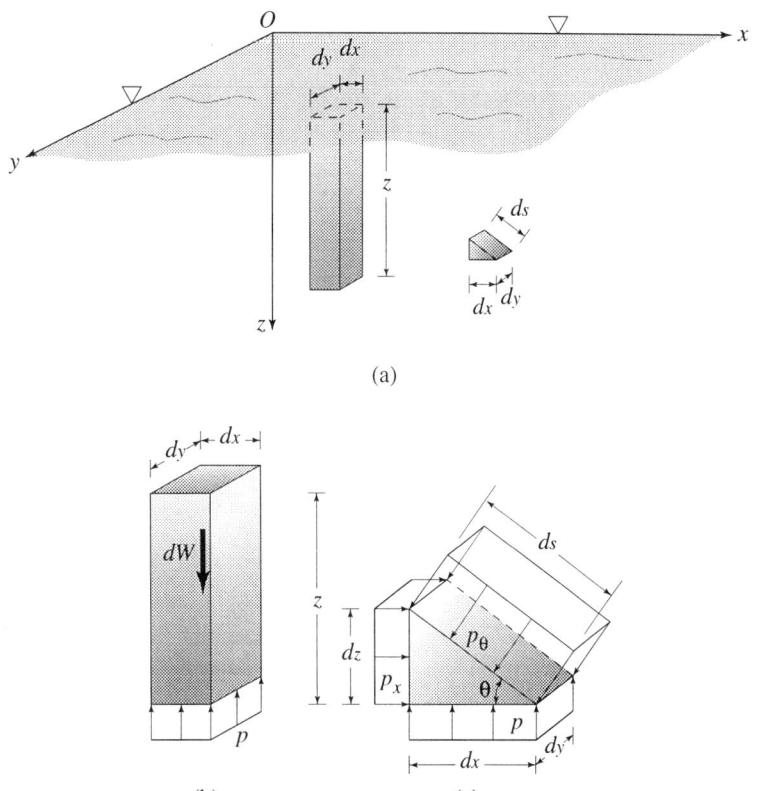

FIGURE 10.12.   (b)   (c)

10.12(b). The top of this column coincides with the free surface of fluid and is subjected to zero gage pressure whereas the bottom at depth z is subjected to gage pressure p. The lateral sides of this column are also subjected to pressures, not shown because only the vertical equilibrium of the column is to be considered. Thus, the column is in vertical equilibrium under the two vertical forces $dW = \gamma \, dV = \gamma z \, dx \, dy$ which represents the weight of the column of fluid and the upward force, $p \, dA = p \, dx \, dy$, produced by the gage pressure at depth z. Therefore,

$$\sum F_z = 0, \qquad \gamma z \, dx \, dy - p \, dx \, dy = 0$$

from which

$$p = \gamma z. \qquad (10.17a)$$

Equation (10.17a) states that the fluid pressure at any point is the product of the weight density (a constant) multiplied by the depth of the point below the free surface of the fluid. Also, because $\gamma = g\rho$, Eq.

(10.17a) may be restated in the form

$$p = g\rho z. \tag{10.17b}$$

The weight and mass densities for some selected fluids are given in Table 10.1.

To prove that the pressure p at any depth is the same in all directions, we consider, at depth z, the equilibrium in the x-z plane, of a triangular differential element of constant width dy whose free-body diagram is shown in Figure 10.12(c). Because of the differential nature of this element, we may assume its weight to be negligible and, ignoring forces in the y direction, conclude that there are three forces acting on it: the upward force on its horizontal area given by $p\,dx\,dy$ where, as in Eq. (10.17), p represents the pressure in the z direction; the horizontal force on its vertical area given by $p_x\,dz\,dy$ where p_x is the pressure in the x direction; and the inclined force on the area defined by the angle θ given by $p_\theta\,ds\,dy$ where p_θ is the pressure on this inclined plane. Thus,

$$\sum F_x = 0, \qquad p_\theta\,ds\,dy\sin\theta - p_x\,dz\,dy = 0.$$

Because $dz = ds\sin\theta$, we conclude that

$$p_\theta = p_x$$

Also,

TABLE 10.1. Weight and Mass Densities for Some Selected Fluids

U.S. Customary Units				
Fluid	Temp °F	Specific Gravity	Weight Density γ (lb/ft^3)	Mass Density ρ (slug/ft^3)
Water	39.2	1.000	62.43	1.94
Water	70	0.998	62.3	1.93
Mercury	32	13.60	847	26.3
Glycerin	68	1.262	78.62	2.44
Olive Oil	59	0.918	57.2	1.78

SI Units				
Fluid	Temp °C	Specific Gravity	Weight Density γ (kN/m^3)	Mass Density ρ (kg/m^3)
Water	4.00	1.000	9.81	1000
Water	21.1	0.998	9.79	998
Mercury	0.0	13.60	133.4	13600
Glycerin	20.0	1.262	12.38	1262
Olive Oil	15.0	0.918	9.00	918

$$\sum F_z = 0, \qquad p_\theta \, ds \, dy \cos \theta - p \, dz \, dy = 0.$$

Because $dx = ds\cos \theta$, it follows that

$$p_\theta = p.$$

We conclude from the above two relationship, therefore, that

$$p_\theta = p_x = p \qquad (10.18)$$

which states Pascal's law that, at a given point in a fluid, the *pressure is the same in all directions.*

Force on Submerged Flat Surfaces

Let us now determine the fluid force developed on a submerged flat surface AB, as shown in Figure 10.13 which has a constant depth D perpendicular to the page. We will define the quantity w as the force per unit of length along AB. Thus, the differential force dR acting on a differential length ds along AB is $dR = w \, ds$. This differential force may also be expressed as the product $p \, dA$ where $dA = D \, ds$. We conclude, therefore, that $w \, ds = pD \, ds$ from which

$$w = pD. \qquad (10.19)$$

Thus, at A in Figure 10.13, $w_A = p_A D = \gamma z_A D$, at B, it is $w_B = p_B D = \gamma z_B D$, and, at any point between A and B, $w = \gamma z D$. Therefore, the force per unit of length w is a linear function of the depth z and, as shown in Figure 10.13, it varies linearly along the length AB of the flat surface.

Based on the results obtained in Section 4.6, we conclude that the magnitude of the resultant R of the fluid pressure on a flat surface is equal to the area under the trapezoidal distribution and, as shown in

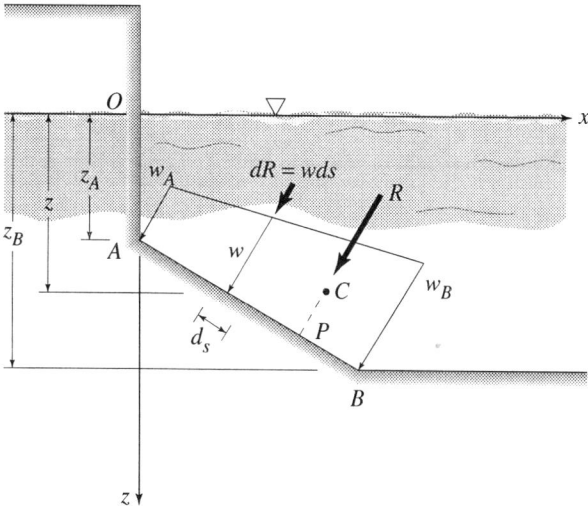

FIGURE 10.13.

Figure 10.13, passes through its centroid C. Note that point P on the flat surface along the line of action of the resultant R is known as the *center of pressure*.

Force on Submerged Curved Surface

The determination of the fluid force on a submerged curved surface requires a little more effort. Consider, for example, the submerged curved surface AB shown in Figure 10.14(a). The resultant and its line of action of the fluid pressure on a submerged curved surface may be determined by, first, finding its components R_x and R_z and their lines of action expressed by the centroidal coordinates $\bar{z}$ and $\bar{x}$, respectively. The magnitude of the resultant is, then, found from $R = \sqrt{R_x^2 + R_z^2}$, its direction from $\theta = \tan^{-1}\left(\dfrac{R_z}{R_x}\right)$ where θ is the angle it makes with the x axis, and, of course, its point of application is fully ascertained once $\bar{x}$ and $\bar{z}$ are found. An outline of such a solution for R_x and R_z and the centroidal coordinates $\bar{z}$ and $\bar{x}$ is given below.

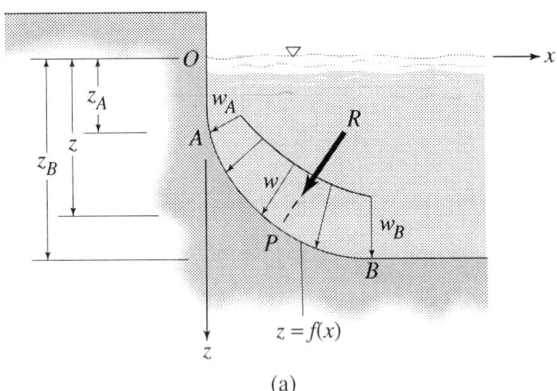

(a)

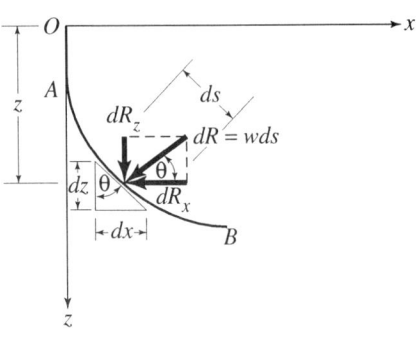

FIGURE 10.14.   (b)

Referring to Figure 10.14(b), we conclude that

$$dR_x = dR \cos \theta = w \, ds \cos \theta$$
$$= pD \, ds \cos \theta = pD \, dz,$$

and

$$R_x = D \int_{z_A}^{z_B} p \, dz. \qquad (10.20\text{a})$$

Also, by the principle of moments, we conclude that

$$\bar{z} R_x = \int z \, dR_x$$

from which

$$\bar{z} = \frac{\int z \, dR_x}{R_x}. \qquad (10.20\text{b})$$

Similarly,

$$dR_z \sim dR \sin \theta = w \, ds \sin \theta$$
$$= pD \, ds \sin \theta = pD \, dx,$$

and

$$R_z = D \int_{x_A}^{x_B} p \, dx. \qquad (10.21\text{a})$$

Also,

$$\bar{x} R_z = \int x \, dR_z,$$

$$\bar{x} = \frac{\int x \, dR_z}{R_z}. \qquad (10.21\text{b})$$

The analysis presented above for a curved surface is general enough to use for analyzing forces acting on any submerged surface. However, as stated earlier, some simplification of the procedure becomes possible when dealing with a straight surface where the variation of the pressure p and, hence, that of w is linear along the length of the surface.

The use of the concepts above will be illustrated in solving the following examples.

■ **Example 10.13**   The hydraulic gate shown in Figure E10.13(a) is proposed for a channel that is 3 m deep into the page. Opening and closing of this gate is

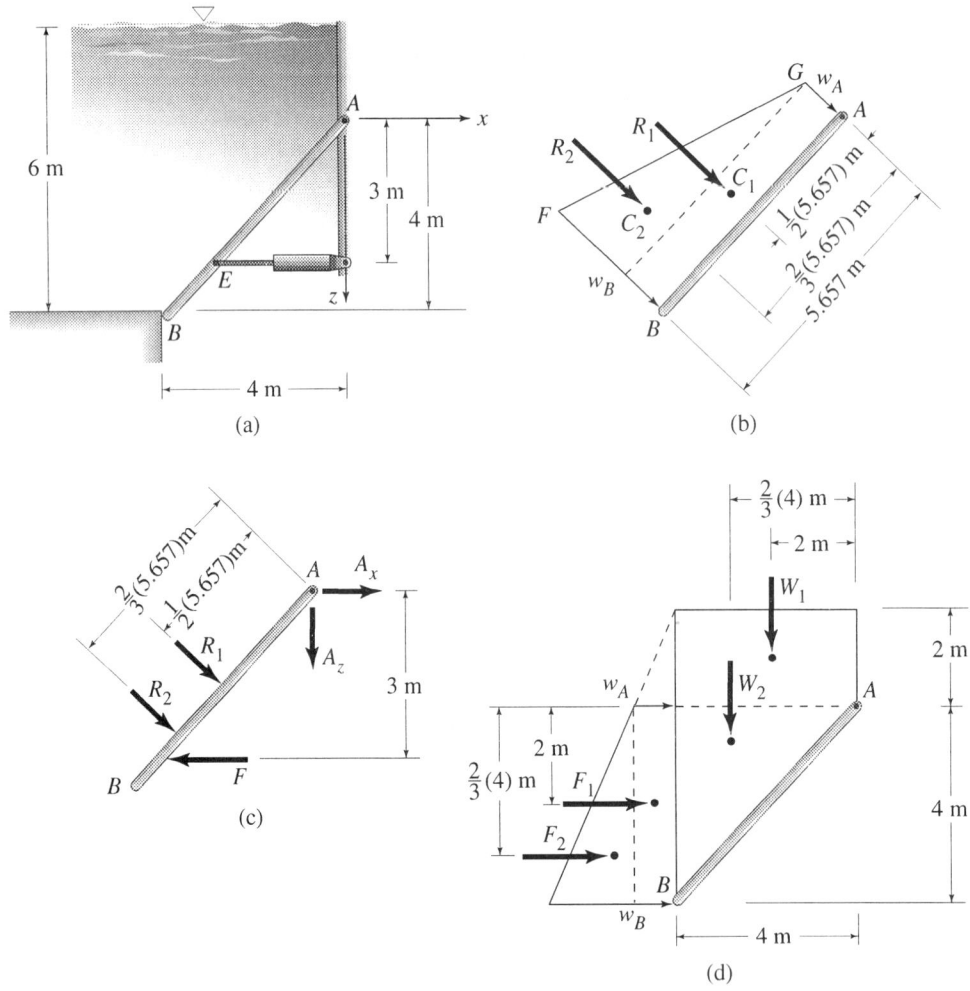

FIGURE E10.13.

controlled by the horizontal hydraulic cylinder at E. Find the force that the hydraulic cylinder must exert on the gate to keep it in the closed position. The mass density for water is $\rho = 1000$ kg/m^3.

Solution

The resultant hydraulic force R on the gate may be found using the method suggested by Eqs. (10.20) and (10.21). However, because the pressure distribution in this case is linear, as shown in Figure E10.13(b), the resultant R and its location may be found by finding the trapezoidal area ABFG and its centroid. Thus, by Eq. (10.19),

$$w_A = p_A D = \gamma z_A D = \gamma \rho z_A D$$

$$= (9.81)(1000)(2)(3) = 58.86 \text{ kN/m},$$

and

$$w_B = p_B D = \gamma z_B D = g \rho z_B D$$

$$= (9.81)(1000)(6)(3) = 176.58 \text{ kN/m}.$$

The resultant force R may be considered in terms of its two components R_1 and R_2, as shown in Figure E10.13(b), where R_1 represents the rectangular portion and R_2 the triangular portion of the trapezoid. Thus,

$$R_1 = (58.86)(5.657) = 332.971 \text{ kN}$$

and acts through C_1 the centroid of the rectangular area. Also,

$$R_2 = \tfrac{1}{2}(176.58 - 58.86)(5.657) = 332.971 \text{ kN}$$

and acts through C_2, the centroid of the triangular area.

The free-body diagram of the gate is shown in Figure E10.13(c) where F represents the force in the hydraulic cylinder. Thus,

$$\sum M_A = 0, \qquad 332.971 \left(\frac{5.657}{2}\right) + 332.971 \left(\frac{2}{3}\right)(5.657) - F(3) = 0,$$

$$F = 733 \text{ kN.} \qquad\qquad\qquad \text{ANS.}$$

An alternate method of finding the effect of the resultant R is to deal with its components W_1, W_2, F_1, and F_2 shown in Figure E10.13(d). The quantities W_1 and W_2 account for the weight of water above the gate, and F_1 and F_2 represent the forces due to the horizontal pressure on the gate. The magnitudes of these quantities are determined below, and their locations relative to the hinge at A are shown in Figure E10.13(d):

$$W_1 = \gamma V_1 = (9.81)(1000)(4)(2)(3) = 235.44 \text{ kN},$$

$$W_2 = \gamma V_2 = (9.81)(1000)(\tfrac{1}{2})(4)(4) = 235.44 \text{ kN},$$

$$F_1 = w_A(4) = 58.86(4) = 235.44 \text{ kN},$$

and

$$F_2 = \tfrac{1}{2}(w_B - w_A)(4) = \tfrac{1}{2}(176.58 - 58.86)(4) = 235.44 \text{ kN}.$$

The fact that all of these quantities have exactly the same numerical value is purely coincidental.

As an exercise, the student can use the above forces in constructing the free-body diagram of the gate and show that the hydraulic cylinder force $F = 733$ kN, identical to that found above.

■ **Example 10.4**

A dam has a parabolic curved surface defined by $z = 60 - 0.6x^2$, shown in Figure E10.14(a). Determine the resultant force R that the water exerts on the dam. Let the weight density of water $\gamma = 62.4 \text{ lb/ft}^3$ and consider a section of the dam which is 1 ft deep into the page.

Solution

For any depth z below the free surface of the water, the pressure $p = \gamma z = 62.4z$. The differential force $dR = p\,dA = 62.4z\,(ds)$. Note that because the depth into the page is 1 ft, $dA = (1)\,ds = ds$. Thus, as

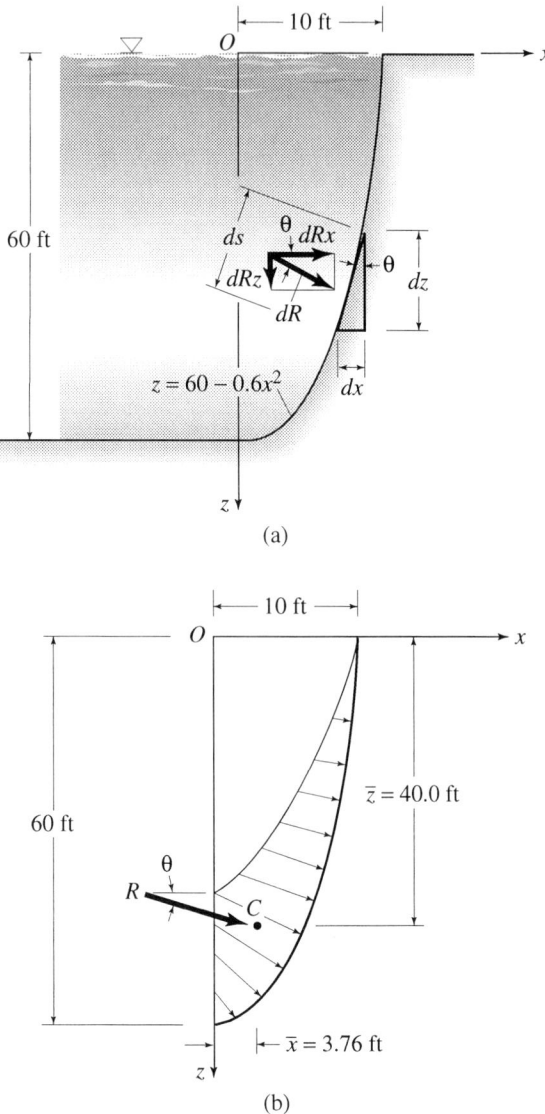

FIGURE E10.14.

shown in Figure E10.14(a),

$$dR_x = dR \cos \theta = 62.4z \, ds \cos \theta$$

$$= 62.4z \, dz$$

and

$$R_x = \int dR_x = 62.4 \int_0^{60} z \, dz$$

$$= 31.2(z^2)|_0^{60} = 112{,}320 \text{ lb}$$

Using the principle of moments with respect to the origin,

$$\bar{z}R_x = \int z \, dR_x = 62.4 \int_0^{60} z^2 \, dz$$

$$= \frac{62.4}{3}(z^3)\bigg|_0^{60} = 4{,}492{,}800 \text{ lb,}$$

$$\bar{z} = \frac{\int z \, dR_x}{R_x} = \frac{4{,}492{,}800}{112{,}320} = 40.0 \text{ ft.,} \qquad \text{ANS.}$$

$$dR_z = dR \sin \theta = 62.4z \, ds \sin \theta$$

$$= 62.4z \, dx$$

$$= 62.4(60 - 0.6x^2) \, dx,$$

and

$$R_z = \int dR_z = 62.4 \int_0^{10} (60 - 0.6x^2) \, dx$$

$$= 62.4(60 - 0.2x^2)|_0^{10} = 24{,}960 \text{ lb.}$$

Also by the principle of moments with respect to the origin,

$$\bar{x}R_z = \int x \, dR_z = 62.4 \int_0^{10} (60x - x^3) \, dx$$

$$= 62.4\left((60x^2 - 0.15x^4)\right)\bigg|_0^{10} = 93{,}600 \text{ lb} \cdot \text{ft.}$$

$$\bar{x} = \frac{\int x \, dR_z}{R_z} = \frac{93{,}600}{24{,}900} = 3.76 \text{ ft.} \qquad \text{ANS.}$$

Therefore,

$$R = \sqrt{R_x^2 + R_x^2} = 115{,}060 = 115{,}100 \text{ lb} \qquad \text{ANS.}$$

where

$$\theta = \tan^{-1}\left(\frac{R_z}{R_x}\right) = \tan^{-1}\left(\frac{24,960}{112,320}\right) = 12.53°. \qquad \text{ANS.}$$

■

Problems

Use $\gamma = 62.4$ lb/ft³ or $\gamma = 9.81$ kN/m³ for the density of water in solving the following problems.

10.92   A triangular gate with a 4 ft base and 3 ft height is used to control the flow of water through a triangular opening in a dam as shown by the front and side views of Figure P10.92. Determine the total fluid force acting on the gate, indicating its direction, sense, and point of application.

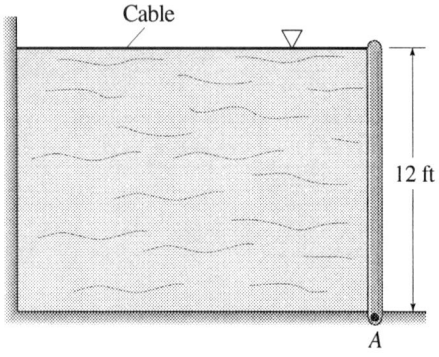

FIGURE P10.93.

at A. Determine the force resisted by the cable and the hinge force.

10.94   Refer to the gate shown in Figure P10.94 which extends 9 ft perpendicular to the page. The rectangular gate

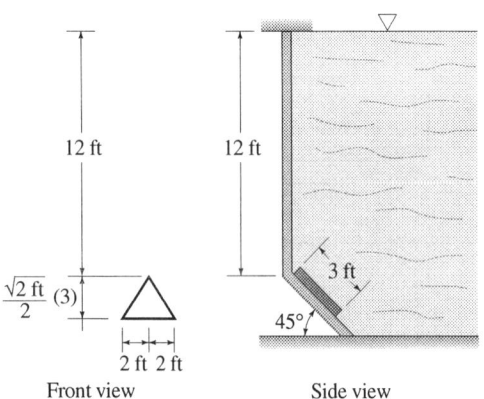

Front view   Side view

FIGURE P10.92.

10.93   The gate shown in Figure P10.93 extends 10 ft perpendicular to the page. It is held against the water pressure by a cable at the water surface and a hinge

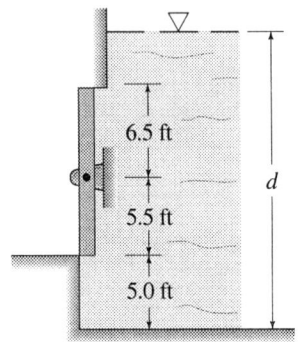

FIGURE P10.94.

will open when the depth d becomes large enough. Determine the minimum depth d which will open the gate.

10.95   A solid cone weighs 6000 lb, has a base diameter of 5 ft, and an altitude of 6 ft. It covers a circular opening at the bottom of the tank containing water 8 ft deep. What vertical force would be required to lift the cone off the bottom?

10.96   Determine the horizontal and vertical force components exerted by the water on the quadrant of a thin walled circular cylinder, 12 ft long into the page, shown in Figure P10.96.

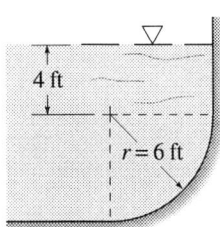

4 ft

$r = 6$ ft

FIGURE P10.96.

10.97   The gate shown in Figure P10.97 is hinged at H and rests against a smooth vertical wall at S. It has a dimension of 2 m perpendicular to the page. De-

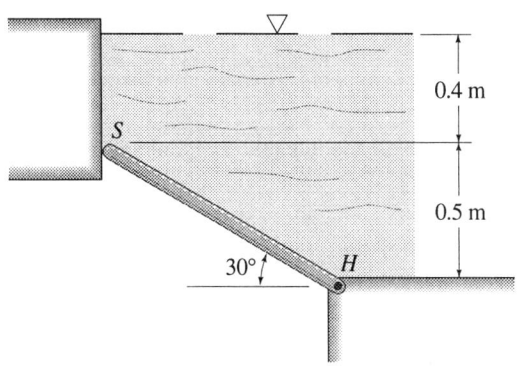

0.4 m

S

0.5 m

30°   H

FIGURE P10.97.

termine the forces exerted by the water on the gate at S and H.

10.98   A preliminary design for an undersea laboratory is shown in Figure P10.98. It is to be anchored with 8 cables which are to be spaced uniformly around the base and attached to ballast on the sea floor. If the laboratory structure and contents may be approximated by a uniform weight throughout its volume of 6 kN/m³, determine the force in each of the cables.

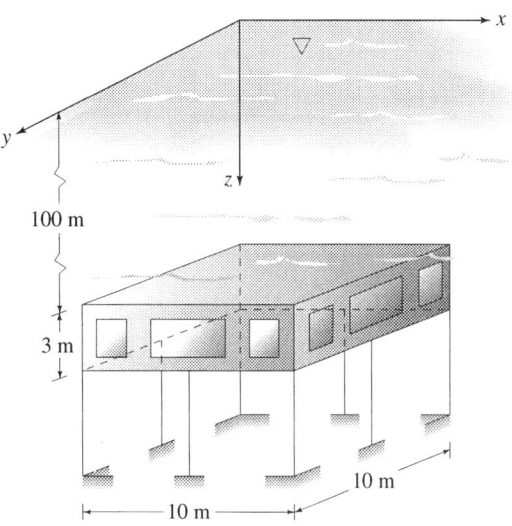

x

y

z

100 m

3 m

10 m

10 m

FIGURE P10.98.

10.99   Water stands on one side of a vertical gate as shown in Figure P10.99. Use integration to determine the total force on the gate and to locate the center of pressure (a) if the gate is a 6 ft × 12 ft rectangle and (b) if the gate is a symmetric 6 ft × 12 ft triangle with its vertex at the top.

10.100 A vertical parabolic gate, shown in Figure P10.100, is submerged in a fluid with a specific gravity of 0.90. Use

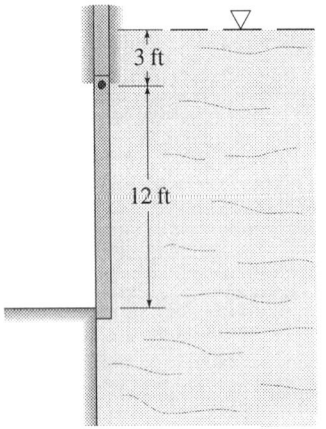

FIGURE P10.99.

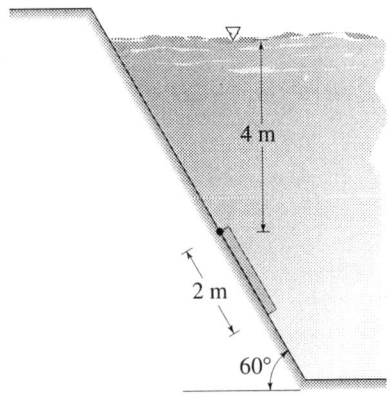

FIGURE P10.101.

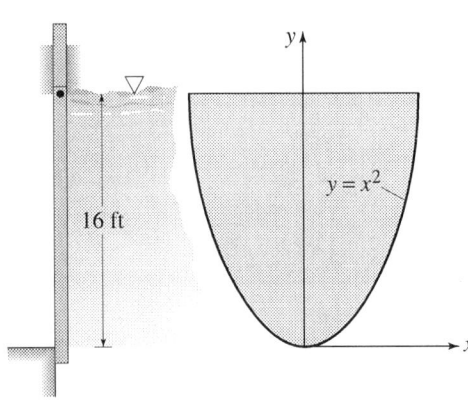

FIGURE P10.100.

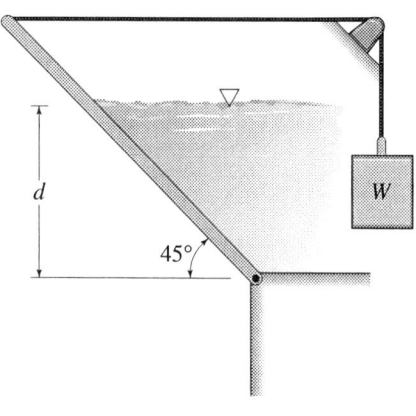

FIGURE P10.102.

integration to find the total fluid force exerted on this gate and to locate the center of pressure.

10.101 A rectangular gate, measuring 2 m × 3 m, is submerged beneath a fluid of specific gravity of 0.95, as shown in Figure P10.101. Determine the total force exerted on this gate by the fluid and the force per unit length exerted on the gate along its bottom edge. Ignore friction.

10.102 What depth of water d will cause the gate shown in Figure P10.102 to begin to fall. Neglect the weight of the gate and assume the peg to be friction-less. The gate is a rectangle 6 m × 2 m with the smaller dimension measured perpendicular to the page. Let $W = 25$ kN.

10.103 A hemispherical dome is attached with rivets to the tank shown in Figure P10.103. The tank and dome are filled with a fluid of specific gravity of 0.85. If

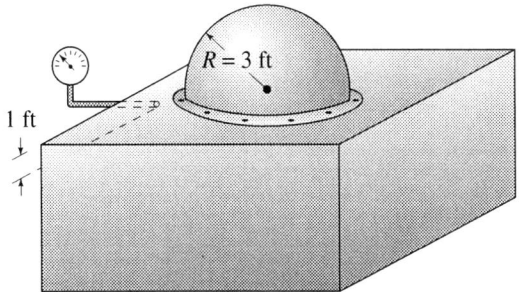

FIGURE P10.103.

the gage pressure is 10 psi, determine the total force which the rivets must resist to hold the dome in place. Gage pressure is the pressure in excess of atmospheric pressure.

10.104 Determine the total force exerted by the water on the dam shown in Figure P10.104. Find the moment of this force with respect to point O.

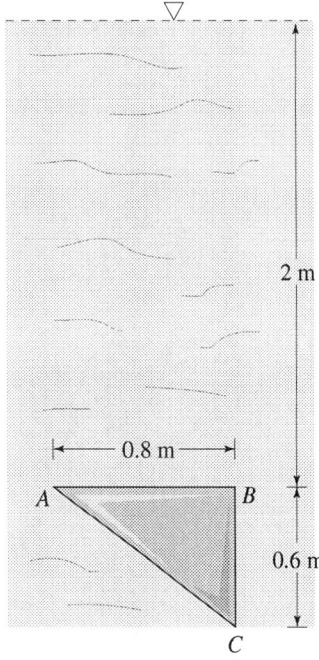

FIGURE P10.105.

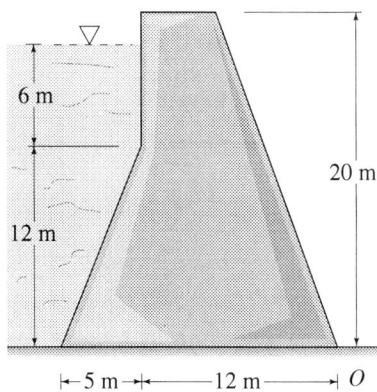

FIGURE P10.104.

10.105 Refer to Figure P10.105, and find the forces acting on each face of the trian-

gular prism ABC which has a dimension of 0.1 m perpendicular to the page. What must be the uniform mass density of this prism for it to be in equilibrium under the action of its own weight and fluid pressure forces?

10.106 A 5-m long taintor gate is shown atop a dam in Figure P10.106. The water level coincides with the center support point O of the gate. Determine the horizontal and vertical force components exerted by the water on the gate.

10.107 A schematic of a hydraulic jack is shown in Figure P10.107. Determine the force P for the following numerical inputs: diameter of large cylinder $D_1 = 0.6$ m; diameter of small cylinder $D_2 = 0.1$ m; $a = 0.2$ m, $b = 1$ m, $W = 40$ kN, $c = 0.05$ m, and the fluid is oil with a specific gravity of 0.85.

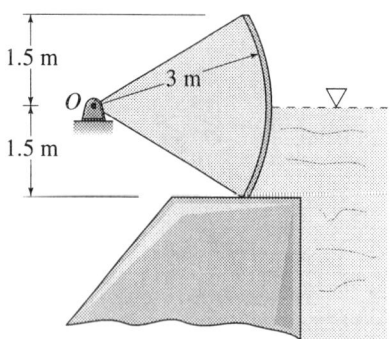

FIGURE P10.106.

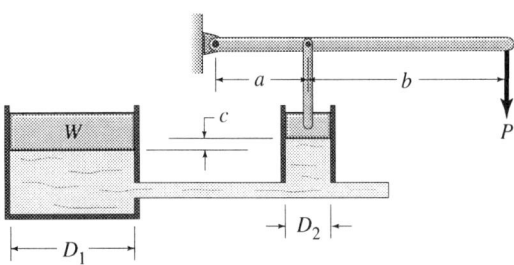

FIGURE P10.107.

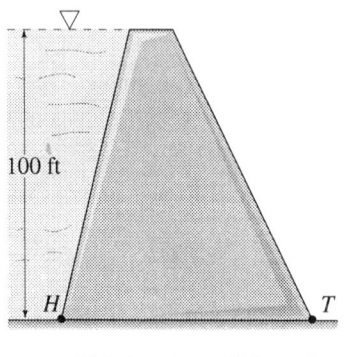

|-25 ft-+|----+|----50 ft----|
            15 ft

FIGURE P10.108.

10.108 Refer to the concrete dam shown in Figure P10.108, and determine the vertical and horizontal components of the force exerted on the base of this dam by the bed rock below it. Locate the action line of this reactive force with respect to point T. Assume a depth into the page of 1 ft. The unit weight of concrete is 144 lb/ft^3.

10.109 The cross section of a concrete dam is shown in Figure P10.109. Find the horizontal and vertical forces exerted on the base of the dam by the bed rock below it. Assume a depth into the page of 1 ft. Locate the point on the base, with respect to point T, where the action line of the vertical resistance force intersects the base of this dam. The density of concrete is 144 lb/ft^3.

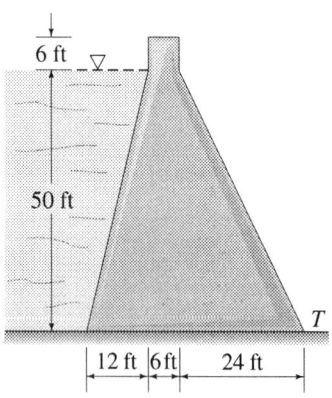

| 12 ft | 6 ft | 24 ft |

FIGURE P10.109.

10.110 The concrete dam shown in Figure P10.110 has a length of 10 m measured perpendicular to the page. Determine the force components exerted on the dam by the bed rock foundation. Locate the action line of the vertical force component with respect to an axis through point O. The density of concrete is 22.64 kN/m^3.

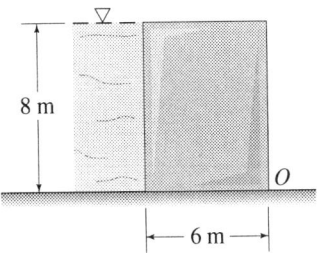

FIGURE P10.110.

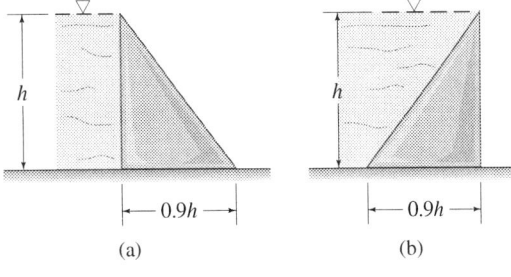

(a)   (b)

FIGURE P10.111.

10.111 A relatively small triangular dam is to be built, and you have been asked to compare designs A and B of Figure P10.111. Base your selection on two criteria: (i) the normal force exerted on the dam by the underlying bed rock must lie within the middle third of the base of the dam, and (ii) the ratio of the horizontal friction force to this normal force must be equal to or less than 0.5. Assume the specific gravity of concrete is 2.31 and a unit depth into the page.

Review Problems

10.112 Determine the coordinates of the centroid of the composite area shown in Figure P10.112.

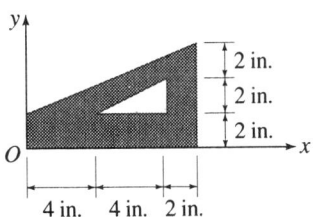

FIGURE P10.112.

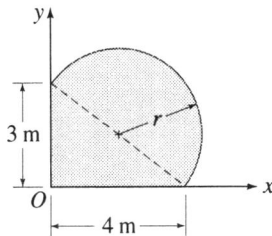

FIGURE P10.113.

10.113 Locate the centroid of the composite area shown in Figure P10.113.

10.114 Refer to Figure P10.114, and find the angle α so that the centroid of the composite wire coincides with the hinge at O.

10.115 Find the radius r of the small cylinder in Figure P10.115 so that the centroid of the composite object lies at the juncture between the two cylinders.

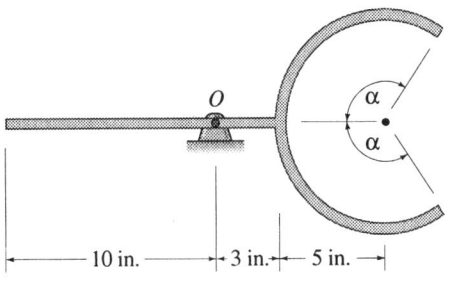

FIGURE P10.114.

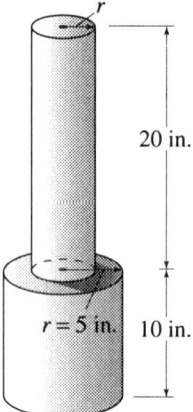

FIGURE P10.115.

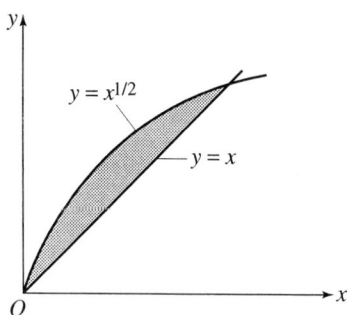

FIGURE P10.117.

10.116 Refer to Figure P10.116 which consists of a solid semisphere on top of a cylinder with a conical cavity. Determine the value of h so that the centroid of the composite object lies at point O.

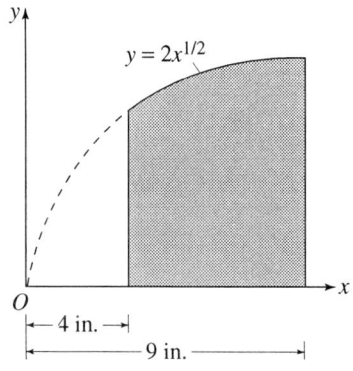

FIGURE P10.118.

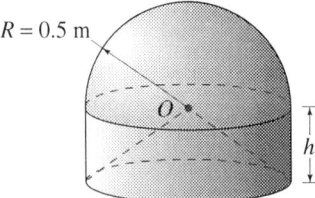

FIGURE P10.116.

10.117 By integration, determine the coordinates of the centroid of the shaded area shown in Figure P10.117. Use a differential element of area parallel to the y axis, and express the answers in terms of meters.

10.118 By integration, determine the coordinates of the centroid of the shaded area shown in Figure P10.118.

10.119 By integration, find the y coordinate of the centroid of the shaded area shown in Figure P10.119. Express the answer in inches.

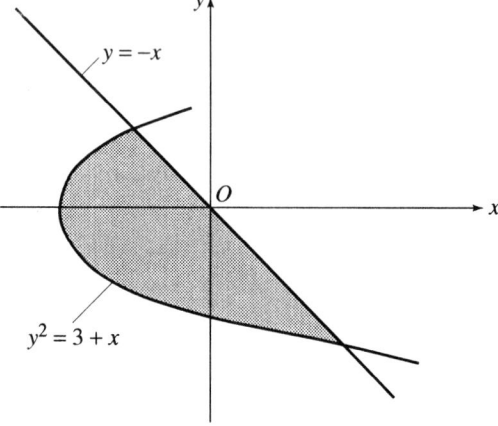

FIGURE P10.119.

10.120 A uniform rod weighing 20 N/m is bent in the shape shown in Figure P10.120. It is held in position by a hinge at O and by a string at B. Determine the tension in the string.

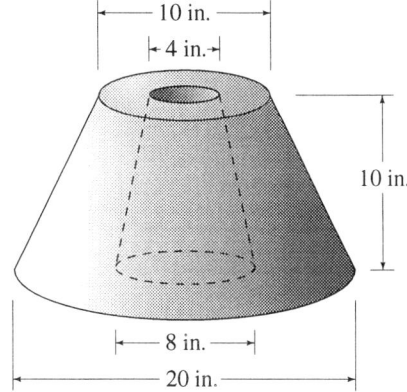

FIGURE P10.122.

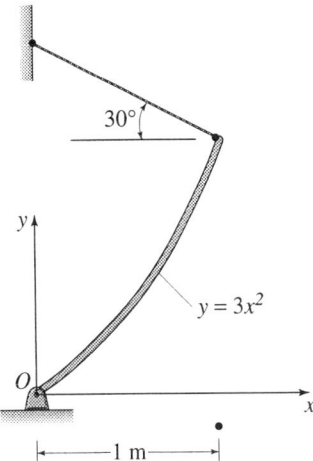

FIGURE P10.120.

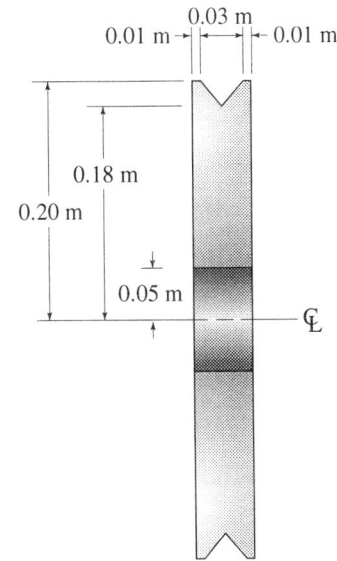

FIGURE P10.123.

10.121 A body of revolution of constant density is generated by rotating the shaded area of Figure P10.117 through one complete revolution about the x axis. Use integration to find the coordinates of the centroid of this body. Express the answer in meters.

10.122 Use integration to find the coordinates of the centroid of the homogeneous hollow truncated cone shown in Figure P10.122.

10.123 The cross sectional area of a cast iron pulley designed for a V-belt is shown in Figure P10.123. If cast iron has a mass density of 7200 kg/m³, determine the weight of the pulley.

10.124 The shape of a proposed water tower is shown in Figure P10.124. The radii are

given in ft at 5-ft intervals for the first 45 ft of the 50-ft height and at a height of 49 ft. Use the theorems of Pappus and Guldinus to find the approximate values of its volume and surface area. Hint: Approximate the curve within each 5-ft segment by a straight line.

10.125 Refer back to the shaded area shown in Figure P10.119. Determine the volume

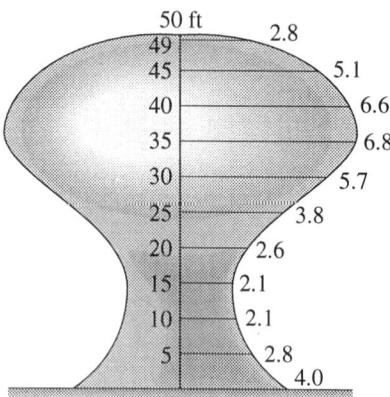

FIGURE P10.124.

generated by rotating this area one complete revolution about an axis located at $y = 2.0$ in. Hint: Use the results obtained in the solution of Problem 10.119.

10.126 A fresh water container 4 m in depth perpendicular to the page has a parabolic shape as shown in Figure P10.126. It is supported at A by a hinge and at B by three cables, one at each end and one at the center of the 4-m depth. Determine the tension in each of the three cables. The weight density for water is $\gamma = 9.81$ kN/m^3.

10.127 A proposed gate for a reservoir has a parabolic shape as shown in Figure P10.127. Determine the resultant hydraulic force acting on the gate and the center of pressure at which this resultant acts. The weight density for water is $\gamma = 62.4$ lb/ft^3.

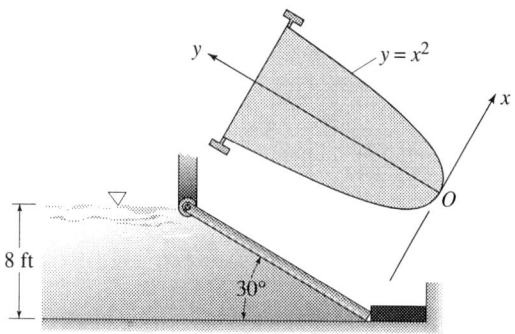

FIGURE P10.127.

10.128 Determine the magnitude, direction, and point of application of the resultant force exerted by the bed rock on the concrete dam shown in Figure P10.128. Concrete has a mass density of 2310 kg/m^3 and water a mass density of 1000 kg/m^3. Consider a section of the dam 1 ft deep into the page.

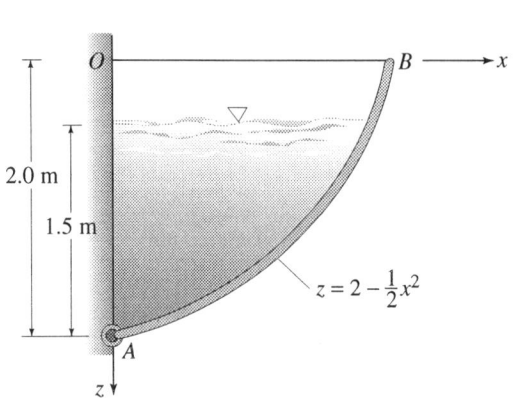

FIGURE P10.126.

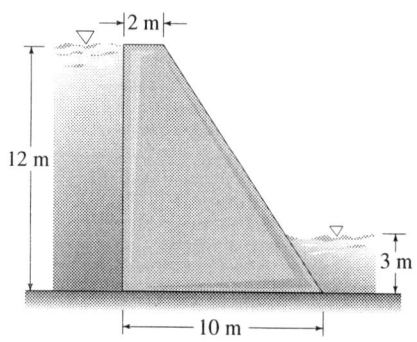

FIGURE P10.128.

11

Moments and Products of Inertia

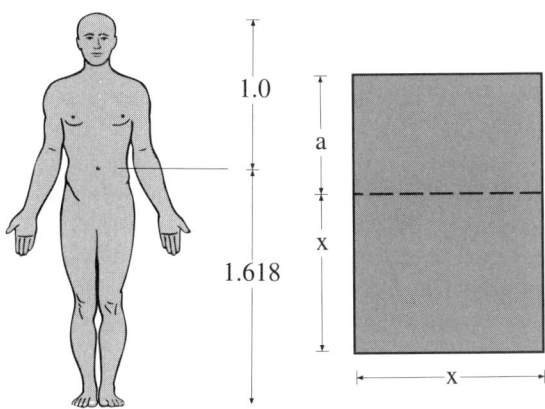

The golden rule as discovered by the Greeks more than 2500 years ago.

The classical Greeks sought to incorporate unifying principles of beauty and perfection in their science, art, and architecture. Even a basic task of dividing objects into smaller parts was influenced by an underlying philosophical concept of perfect proportions. The most fundamental application of the divine proportion was used by the Greeks from about the fifth century B.C., and it is known today as the golden section or golden rule. This rule simply states that the ratio of the small part to the large part is equal to the ratio of the large part to the whole. Therefore, the human body shown above satisfies this rule because $1/1.618 = 1.618/2.618$.

To discover the golden section for the rectangle above, we need to solve for x from the quadratic equation defined by $x/(x + a) = a/x$. This produces a positive root of $x = a(1 + \sqrt{5})/2$. Therefore, the ratio $x/a = 1.618$. The architecture of the ancient Greeks and many modern buildings is designed to fit this golden rule! The dimensions are set so that the ratio of length to width is 1.618. Furthermore, famous paintings and even the perfect human body have measurements that correspond to this rule. Just look at 3×5 pictures. You will see a ratio of 5.0 inch/ 3.0 inch which gives a ratio of 1.60, close to 1.618! The basic message here is that engineers are often called upon to find properties, such as centroids and moments of inertia, of objects and structures of varying shapes and sizes. Some shapes are not as aesthetically pleasing as those developed from the golden rule.

In this chapter, methods will be introduced to compute the moments of inertia for areas and for masses about specified axes. Procedures will be developed to locate the so-called principal axes of inertia for both areas and for masses. These are the axes with respect to which moments of inertia assume their maximum and minimum values. The concepts of principal axes and principal moments of inertia for areas are very significant in the solution of many structural problems and those of principal axes and principal moments of inertia for masses are very important in discussing the accelerated motion of bodies.

11.1
Concepts
and
Definitions

In the next course dealing with dynamics, we encounter an integral which has significance when dealing with the accelerated rotation of bodies. This integral, $\int r^2\, dm$, which defines a physical property of a body, is given the name *mass moment of inertia* and is usually designated by the symbol I. In this integral, dm is a differential element of mass and r represents the perpendicular distance measured from a given axis to the differential element dm. For example, if a rigid body has accelerated rotation about a fixed axis through its mass center G, the governing equation of motion is $\sum M_{\mathrm{G}} = I_{\mathrm{G}}\alpha$, where $\sum M_{\mathrm{G}}$ represents the sum of all the moments about the axis through G, I_{G} the mass moment of inertia with respect to the same axis and α is the angular acceleration of the body. A physical meaning for the mass moment of inertia I_{G} may be obtained by solving for it from the governing equation of motion. Thus, $I_{\mathrm{G}} = \sum M_{\mathrm{G}}/\alpha$ which states that the physical property, known as mass moment of inertia, may be obtained as the ratio of the moments acting on the rigid body to the acceleration they produce on the body. It should be emphasized that the ratio $\sum M_{\mathrm{G}}/\alpha$ remains constant because I_{G} is a physical constant for a given body.

Also, in analyzing the bending of beams and the twisting of shafts in courses dealing with mechanics of materials, integrals are developed which are mathematically similar to that defined as the mass moment of inertia. These integrals, $\int x^2\, dA$ or $\int r^2\, dA$, define certain properties of areas and are named *second moments of areas* to distinguish them from the *first moment of areas* ($\int x\, dA$ or $\int r\, dA$) used in Chapter 10 for determining the centroids of areas. Because these integrals are mathematically similar in form to that defining the mass moment of inertia, they are commonly referred to as area moments of inertia although the name *second moments of areas* would be much more precise. In these integrals, the quantity dA is a differential element of area and the quantities x and r represent perpendicular distances measured from specific axes to the element of area dA. As in the case of the mass moment of inertia, the symbol I is used for the area moment of inertia. This practice does not create any ambiguity because the meaning is always clear from the context. As an example relating to area moment of inertia, we consider the case of a beam bent by a moment M. The equation that yields the magnitude of the *normal stress** σ, known as the *flexure* equation, is $\sigma = \dfrac{My}{I}$, where M is the moment acting on the beam, I is the area moment of inertia about a centroidal axis (known as the *neutral axis*), and y is the perpendicular distance from this axis to

---

*A simple definition of normal stress σ, which suffices for our needs here, is normal force per unit of area. A more precise definition may be found in books on mechanics of materials.

the point in the beam where the stress σ is desired. Therefore, a physical meaning for the area moment of inertia may be obtained by solving the flexure equation for I which yields $I = My/\sigma = M/(\sigma/y)$. This equation states that the physical property, known as area moment of inertia, may be obtained as the ratio of the moment M acting on the beam to the quantity σ/y. It should be noted that the quantity σ/y is a constant because σ varies in direct proportion to its distance y from the neutral axis. Also, because the moment at a given position in a beam is constant, it follows that the area moment of inertia for a specific cross-sectional area is also a constant physical property of this area.

Because of the significance of the properties known as *mass* and *area moment of inertia* in subsequent courses, it is necessary to develop methods to determine them.

Area Moment of Inertia

Consider a differential element of area dA located at the position defined by the coordinates x and y in an area, as shown in Figure 11.1. The *rectangular area moment of inertia* or the *second moment of area* with respect to the x and y axes in the plane of the area are defined, respectively, by

and

$$\left. \begin{aligned} I_x &= \int y^2 \, dA \\ I_y &= \int x^2 \, dA. \end{aligned} \right\} \tag{11.1}$$

Henceforth, the simplified term *moment of inertia of area* will be used instead of the terms *rectangular moment of inertia* of area and *second moment of area*.

If the moment of inertia of area is determined with respect to an axis perpendicular to the area, it is referred to as the polar area moment of inertia and is traditionally designated by the symbol J to distinguish it from the moments of inertia I_x and I_y. Thus, the polar moment of inertia of the area of Figure 11.1 with respect to a z axis through point

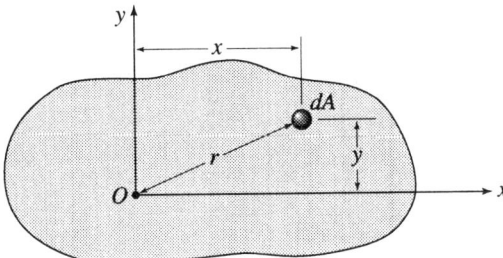

FIGURE 11.1.

O is defined by

$$J_O = \int r^2 \, dA. \tag{11.2}$$

Because $r^2 = x^2 + y^2$, Eq. (11.2) may be expressed in terms of the moments of inertia I_x and I_y as

$$J_O = \int (x^2 + y^2) \, dA = \int y^2 \, dA + \int x^2 \, dA = I_x + I_y \tag{11.3}$$

Thus, the polar moment of inertia of an area, with respect to an out-of-plane axis through point O, is equal to the sum of the moments of inertia with respect to any two orthogonal in-plane axes intersecting at point O.

It is seen from the definition of the area moment of inertia that this quantity has a compound unit equal to a length raised to the fourth power. Thus, in the U.S. Customary (British Gravitational) system such units as in.^{4} and ft^4 are common. In the SI system, the most commonly used units are mm^4 and m^4. Also, from the definition, the area moment of inertia is always a positive quantity.

In solving many types of problems in mechanics, it is convenient to define a quantity known as the *radius of gyration*. Consider, for example, an area A whose moments of inertia with respect to the x and y axes are I_x and I_y, respectively. Let us imagine that the entire area A is concentrated at a single point as shown in Figure 11.2. The perpendicular distances of this point from the coordinates axes, k_x amd k_y, are chosen to satisfy the following relationships:

and

$$\left.\begin{array}{l} I_x = k_x^2 A \quad \text{or} \quad k_x = \sqrt{\dfrac{I_x}{A}}, \\[2mm] I_y = k_y^2 A \quad \text{or} \quad k_y = \sqrt{\dfrac{I_y}{A}}. \end{array}\right\} \tag{11.4}$$

The quantities k_x and k_y in Eq. (11.4) are known as the *radii of gyration of the area* with respect to the x and y axes, respectively. A similar expression may be written for the polar radius of gyration of area with

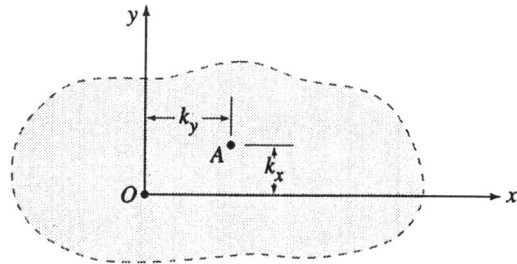

FIGURE 11.2.

respect to any axis perpendicular to the plane of the area. Thus, with respect to an axis through point O in Figure 11.2, for example,

$$k_O = \sqrt{\frac{J_O}{A}}. \tag{11.5}$$

It should be noted from Eqs. (11.4) and (11.5) that the radii of gyration of area have units of length. Thus such units as the in. and the ft are used in the U.S. Customary system and the mm and m in the SI system.

Mass Moment of Inertia

Now, consider a differential element of mass dm located at any position defined by the coordinates x, y, and z in a body, as shown in Figure 11.3. The perpendicular distances from the differential element of mass dm to the x, y, and z coordinate axes are r_x, r_y, and r_z, respectively. By definition, the mass moments of inertia of the body with respect to the x, y, and z axes, respectively, are

$$\left. \begin{array}{l} I_x = \displaystyle\int r_x^2 \, dm, \\[2mm] I_y = \displaystyle\int r_y^2 \, dm, \\[2mm] I_z = \displaystyle\int r_z^2 \, dm. \end{array} \right\} \tag{11.6}$$

and

From the geometry of Figure 11.3, $r_x^2 = y^2 + z^2$, $r_y^2 = x^2 + z^2$, and $r_z^2 = x^2 + y^2$. Therefore, Eqs. (11.6) may be written in the form

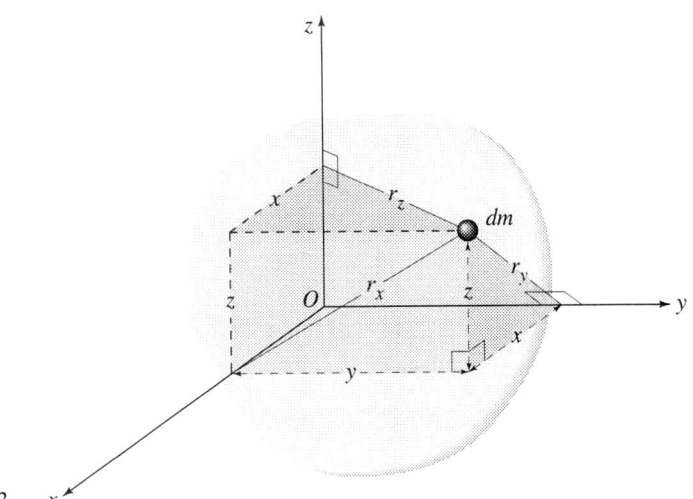

FIGURE 11.3.

$$I_x = \int (y^2 + z^2) \, dm,$$

$$I_y = \int (x^2 + z^2) \, dm, \qquad (11.7)$$

and

$$I_z = \int (x^2 + y^2) \, dm.$$

From the definitions presented in Eq. (11.6), we note that the mass moment of inertia has a compound unit equal to a length squared multiplied by the unit of mass. Thus, such units as slug·ft² and lb·s²·ft are used in U.S. Customary system and kg·m² in the SI system. Also, from the definition, the mass moment of inertia is always a positive quantity.

Consider a body of mass m whose moments of inertia with respect to the x, y, and z axes are I_x, I_y, and I_z, respectively. Assume that the entire mass of this body is concentrated at a single point, as shown in Figure 11.4. The perpendicular distances of this point from the coordinate axes, k_x, k_y, and k_z, are chosen to satisfy the following relationships:

$$I_x = k_x^2 m \qquad \text{or} \qquad k_x = \sqrt{\frac{I_x}{m}},$$

$$I_y = k_y^2 m \qquad \text{or} \qquad k_y = \sqrt{\frac{I_y}{m}}, \qquad (11.8)$$

and

$$I_z = k_z^2 m \qquad \text{or} \qquad k_z = \sqrt{\frac{I_z}{m}}.$$

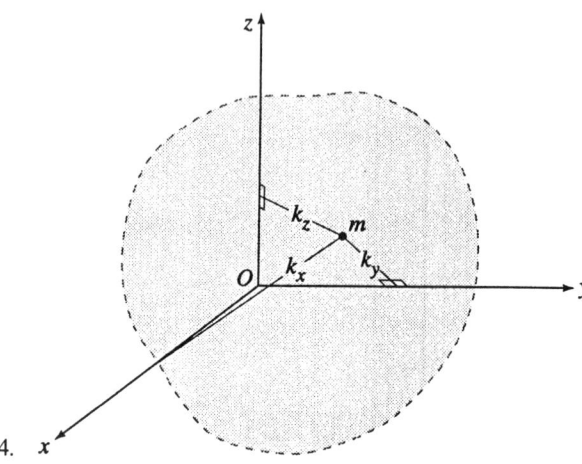

FIGURE 11.4.

The quantities k_x, k_y, and k_z in Eq. (11.8) are known as the *radii of gyration of mass* with respect to the x, y, and z axes, respectively. From the definitions given in Eq. (11.8), note that the radii of gyration of mass have units of length. Thus, such units as the in. and the ft are used in the U.S. Customary system and the mm and m in the SI system.

11.2 Parallel-Axis Theorems

Area

The parallel-axis theorem for area moments of inertia is a useful tool because it relates moments of inertia with respect to two parallel axes, *one of which passes through the centroid of the area.*

Consider the area shown in Figure 11.5. The X-Y coordinate system has its origin at the centroid C of the area. A second coordinate system (x-y) has its origin at point O so that x is at a distance b from, and parallel to X, and y is at a distance a from, and parallel to Y. Thus, $x = X + a$, and $y = Y + b$. By the first of Eqs. (11.1), we can write the moment of inertia of area with respect to the x axis as

$$I_x = \int y^2 \, dA = \int (Y + b)^2 \, dA = \int Y^2 \, dA + b^2 \int dA$$

$$+ \, 2b \int Y \, dA. \tag{11.9}$$

In the last form of Eq. (11.9), the first integral is I_X, the area moment of inertia with respect to the centroidal X axis. The second integral is the area A and, consequently, the second term is Ab^2. By Eq. (10.4), the last integral, $\int Y \, dA = A\overline{Y}$. The quantity $\overline{Y}$, in this case, is zero because it is measured from a centroidal axis. Therefore, it follows that $\int Y \, dA = 0$. Thus, Eq. (11.9) may be written as

$$I_x = I_X + Ab^2. \tag{11.10}$$

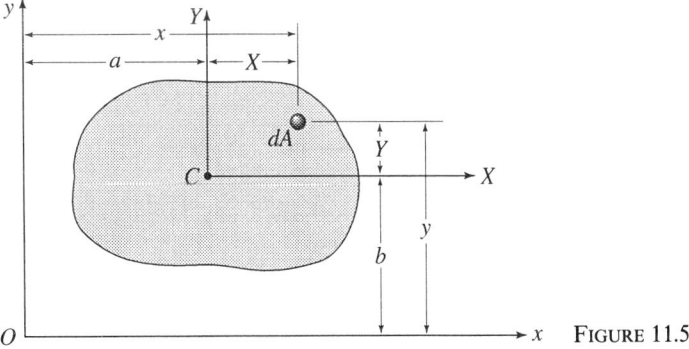

FIGURE 11.5.

Similarly,

$$I_y = I_Y + Aa^2. \qquad (11.11)$$

Either of the two relationships given in Eqs. (11.10) and (11.11) expresses the fact that the area moment of inertia with respect to any axis is equal to the area moment of inertia with respect to a parallel axis through the *centroid* of the area (i.e., the *centroidal moment of inertia*) plus the product of the area and the square of the distance between the two axes. This very important relationship is known as the *parallel-axis theorem* for area moments of inertia. The theorem may be represented in general form by

$$I = I_C + Ad^2 \qquad (11.12)$$

where I is the area moment of inertia with respect to any axis in the plane of the area A, I_C is the area moment of inertia with respect to a parallel centroidal axis, and d is the distance between the two axes.

Using the same steps that led to Eq. (11.12) we may develop a parallel axis theorem for polar moments of inertia of an area. Thus,

$$J = J_C + Ad^2 \qquad (11.13)$$

where J is the polar moment of inertia with respect to any axis perpendicular to the area A, J_C is the polar moment of inertia with respect to a parallel centroidal axis, and d is the distance between the two axes.

It should be noted that Eq. (11.12) leads to the important conclusion that area moments of inertia have a minimum value with respect to centroidal axes. This is significant, for example, in discussing the theory underlying the bending of beams.

A parallel-axis theorem may also be developed for area radii of gyration. For example, substituting in Eqs. (11.12) and (11.13) for the values of I and J, respectively, in terms of the radii of gyration,

$$k^2 = k_C^2 + d^2 \qquad (11.14)$$

where k is the rectangular or polar area radius of gyration with respect to any axis, k_C is the rectangular or polar area radius of gyration with respect to a parallel centroidal axis, and d is the distance between the two axes.

■ Example 11.1

The moment of inertia of the area shown in Figure E11.1, is 3200 in.[4] with respect to the x axis, and 4700 in.[4] with respect to the y axis. Determine the area A and the centroidal moments of inertia I_X and I_Y if $I_X = I_Y$. Also, determine the polar moment of inertia with respect to a centroidal axis and the corresponding radius of gyration.

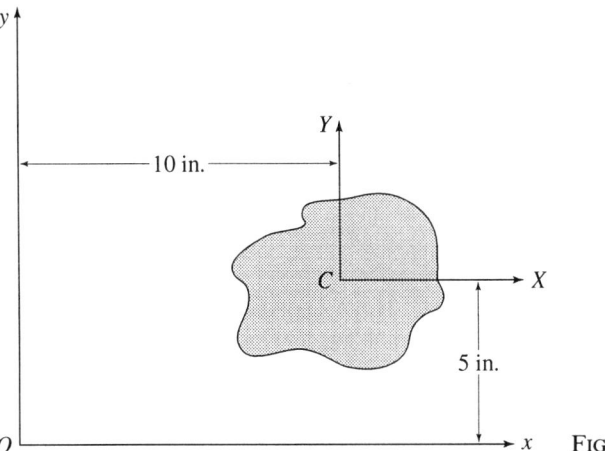

FIGURE E11.1.

Solution

By Eq. (11.12)

$$I_x = 3200 = I_X + A(5)^2 \qquad \text{(a)}$$

and

$$I_y = 4700 = I_Y + A(10)^2. \qquad \text{(b)}$$

Also,

$$I_X = I_Y. \qquad \text{(c)}$$

Solving Eqs. (a), (b) and (c) simultaneously,

$$A = 20.0 \text{ in.}^2 \qquad \text{ANS.}$$

and

$$I_X = I_Y = 2.70 \times 10^3 \text{ in.}^4. \qquad \text{ANS.}$$

By Eq. (11.3),

$$J_C = I_X + I_Y = 5.40 \times 10^3 \text{ in.}^4. \qquad \text{ANS.}$$

By Eq. (11.7)

$$k_C = \sqrt{\frac{J_C}{A}} = 16.43 \text{ in.}^4. \qquad \text{ANS.}$$

∎

Masses

Now consider the body of mass m shown in Figure 11.6. The X-Y-Z coordinate system has its origin at the center of mass G of the body. A second coordinate system (x-y-z) has its origin at point O so that x is at a distance $(b^2 + c^2)^{1/2}$ from and parallel to X, y is at a distance $(a^2 + c^2)^{1/2}$ from and parallel to Y, and z is at a distance $(a^2 + b^2)^{1/2}$

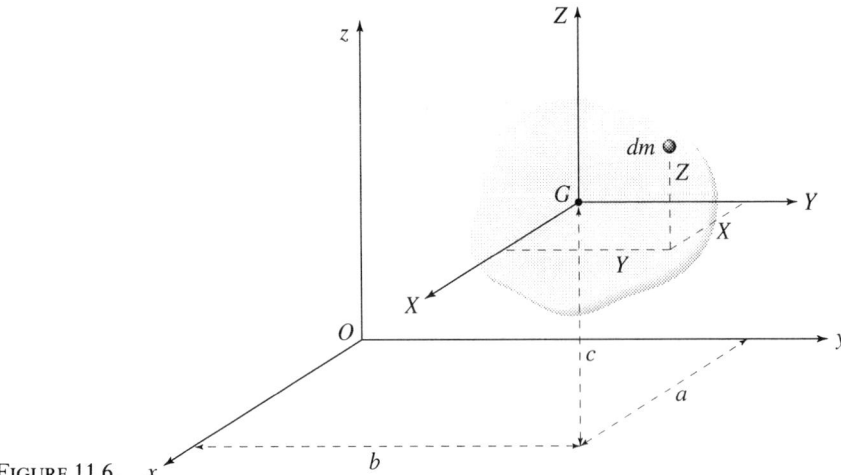

FIGURE 11.6.

from and parallel to Z. Thus, $x = X + a$, $y = Y + b$, and $z = Z + c$. By the first of Eqs. (11.7), we may write the moment of inertia of mass with respect to the x axis as

$$I_x = \int (y^2 + z^2)\,dm = \int [(Y + b)^2 + (Z + c)^2]\,dm$$

$$= \int (Y^2 + Z^2)\,dm + (b^2 + c^2) \int dm + 2b \int Y\,dm$$

$$+ 2c \int Z\,dm. \qquad (11.15)$$

The first integral in Eq. (11.15) represents I_X, the mass moment of inertia with respect to an X axis through the center of mass G. Because the second integral is the mass m, the second term is $(b^2 + c^2)m = d_x^2 m$, where d_x represents the perpendicular distance between axes X and x. As explained earlier in deriving Eq. (11.10), the third and fourth integrals must vanish because they are the first moments of mass with respect to planes passing through the center of mass G of the body. Therefore, Eq. (11.15) reduces to

$$I_x = I_X + md_x^2 \qquad (11.16)$$

Similarly,

$$\left. \begin{aligned} I_y &= I_Y + md_y^2 \\ I_z &= I_Z + md_z^2 \end{aligned} \right\} \qquad (11.17)$$

where d_y and d_z represent the perpendicular distance between the Y and y axes and that between the Z and z axes, respectively.

The relationships given in Eqs. (11.16) and (11.17) express the fact that the mass moment of inertia, with respect to any axis, is equal to the mass moment of inertia with respect to a parallel axis through the *center of mass* (i.e. *centroidal moment of inertia*) plus the product of the mass and the square of the distance between the two axes. This relationship is known as the parallel-axis theorem for mass moments of inertia. The theorem may be represented in general form by

$$I = I_G + md^2 \qquad (11.18)$$

where I is the mass moment of inertia with respect to any axis, I_G is the mass moment of inertia with respect to a parallel centroidal axis, and d is the distance between the two axes.

In a manner similar to that used in deriving Eq. (11.14), we may develop a parallel axes theorem for mass radii of gyration. Thus, if we substitute in Eq. (11.18) for the values of I and I_G in terms of radii of gyration,

$$k^2 = k_G^2 + d^2 \qquad (11.19)$$

where k is the mass radius of gyration with respect to any axis, k_G is the mass radius of gyration with respect to a parallel centroidal axis, and d is the distance between the two axes.

■ **Example 11.2**   The mass moment of inertia of the cylindrical body, shown in Figure E11.2, with respect to the x axis passing through point O is $(871/48)mD^2$

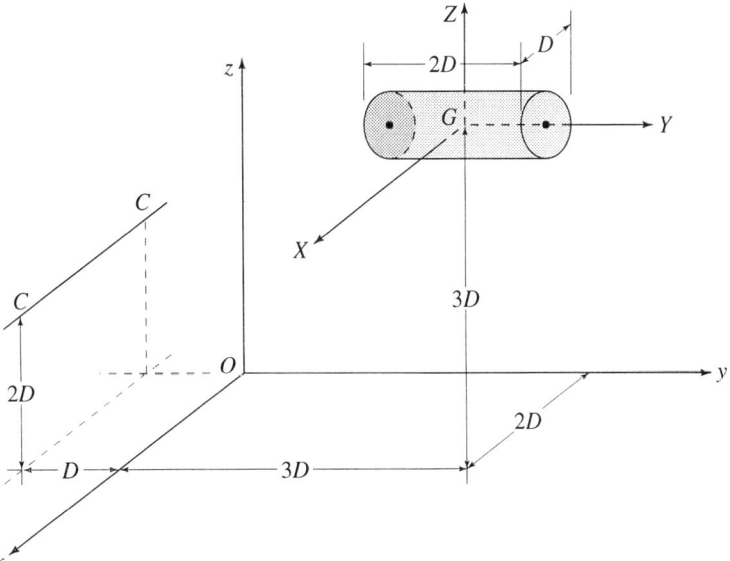

FIGURE E11.2. x

where m is the mass and D is the diameter of the cylindrical body. Determine (a) the mass moment of inertia with respect to the axis CC located as shown and parallel to the x axis, (b) the radius of gyration of mass with respect to the axis CC, and (c) the radius of gyration of mass with respect to the X axis passing through the center of mass.

Solution

(a) By Eq. (11.18),

$$I_{CC} = I_X + md^2$$

and

$$I_X = I_x - md^2 = \left(\frac{871}{48}\right)mD^2 - m[(3D)^2 + (3D)^2] = \left(\frac{7}{48}\right)mD^2.$$

Also,

$$I_{CC} = I_X + md^2 = \left(\frac{7}{48}\right)mD^2 + m[D^2 + (4D)^2] = \left(\frac{823}{48}\right)mD^2 \quad \text{ANS.}$$

(b) By Eq. (11.8),

$$k_{CC} = \sqrt{\frac{I_{CC}}{m}} = \left(\sqrt{\frac{823}{48}}\right)D. \qquad\qquad \text{ANS.}$$

(c) By Eq. (11.8),

$$k_x = \sqrt{\frac{I_x}{m}} = \left(\sqrt{\frac{7}{48}}\right)D. \qquad\qquad \text{ANS.}$$

■

Problems

AREAS: Problems 11.1–11.9; MASSES: Problems 11.10–11.14

11.1   The moment of inertia of the area, shown in Figure P11.1, is 400 in.^{4} with respect to the x axis and 2800 in.^{4} with respect to the y axis. Determine the area A and the centroidal moments of inertia I_X and I_Y if $I_X = \frac{1}{2}I_Y$, $a = 7$ in., and $b = 2$ in. Also, find the polar moments of inertia with respect to axes through points C and O.

11.2   The following information is provided about the area shown in Figure P11.1: $I_X = 18 \times 10^{-6}$ m^4, $I_Y = 12 \times 10^{-6}$ m^4, $A = 4 \times 10^{-4}$ m^2, $a = 2b$, and $I_x = \frac{1}{2}I_y$. Find the distances a and b and the quantities I_x and I_y. What is the centroidal polar radius of gyration for this area?

11.3   The area moments of inertia of the triangle, shown in Figure P11.3, with respect to the x and y axes are $I_x = \frac{1}{12}bh^3$ and $I_y = \frac{1}{12}hb^3$, respectively, where b and

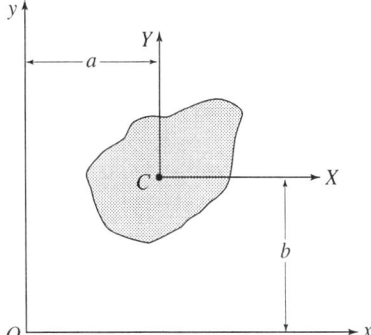

FIGURE P11.1.

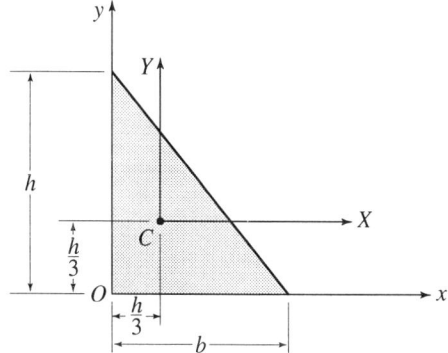

FIGURE P11.3.

h are defined in the sketch. Determine (a) the moments of inertia with respect to the centroidal X and Y axes, (b) the polar moments of inertia with respect to axes through points C and O, (c) the radius of gyration of area with respect to the centroidal X axis, and (d) the polar radius of gyration of area with respect to an axis through point O.

11.4 The following information is given about the area shown in Figure P11.3: $I_X = 200 \times 10^{-6}$ m^4, $J_C = 300 \times 10^{-6}$ m^4, $I_x = 500 \times 10^{-6}$ m^4, $I_y = 200 \times 10^{-6}$ m^4.

Determine (a) the values of *b* and *h* and (b) the radius of gyration k_Y.

11.5 The moment of inertia of the rectangular area shown in Figure P11.5 with respect to the A–A axis is $I_{A-A} = \frac{1}{3}bh^3$. Determine the area moment of inertia with respect to the axes B–B and C–C.

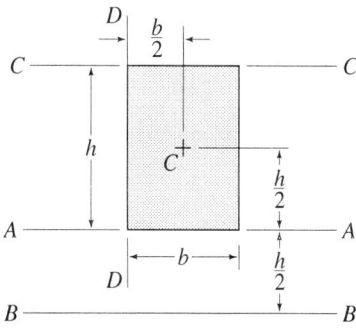

FIGURE P11.5.

11.6 Refer to Problem 11.5. The area moment of inertia of the rectangle is $I_{D-D} = \frac{1}{3}hb^3$ with respect to the D–D axis. Determine the polar radius of gyration with respect to an axis through the centroid C.

11.7 For a diameter $D = 2$ in., the moment of inertia of the circular area shown in Figure P11.7 is $I_{A-A} = 9.25\pi$ in.^{4} with respect to the A–A axis. Determine the polar moment of inertia of the area with

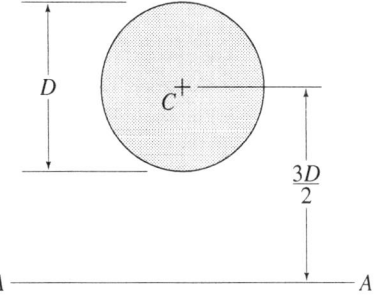

FIGURE P11.7.

respect to an axis through the centroid C.

11.8 Refer to Problem 11.7, and determine the area moment of inertia with respect to any axis tangent to the circumference of the circular area.

11.9 The polar moment of inertia for the area in Figure P11.9 is $J_O = 7900$ in.^{4} with respect to an axis through point O. Determine the radii of gyration k_x and k_y if $A = 20$ in.^{2} and $I_X = I_Y$.

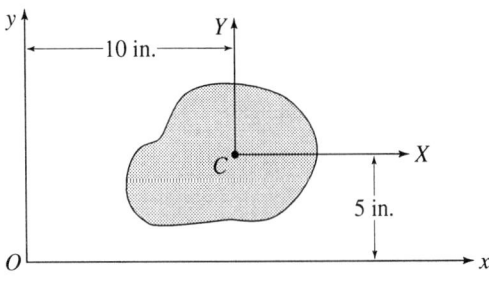

FIGURE P11.9.

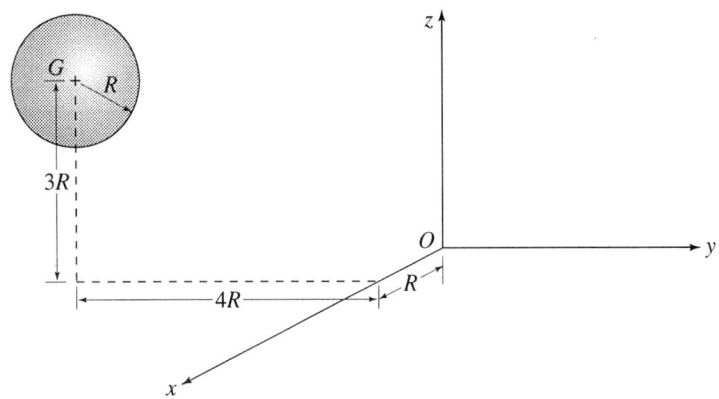

FIGURE P11.10.

11.10 The mass moment of inertia of the sphere, shown in Figure P11.10, with respect to the y axis, is $I_y = (\frac{52}{5})mR^3$ where m is the mass and R is the radius of the sphere. Determine the mass moment of inertia and the mass radius of gyration with respect to the z axis.

11.11 Refer to Problem 11.10 and determine the mass radius of gyration with respect to the x axis.

11.12 The centroidal mass moment of inertia of the sphere, shown in Figure P11.10, is $I_G = 0.08$ kg·m^2. If $R = 0.20$ m and the mass of the sphere $m = 5$ kg, determine the radius of gyration of the sphere with

respect to (a) the x axis and (b) the y axis.

11.13 The mass moment of inertia of the body, shown in Figure P11.13, is $I_x = 5000$ kg·m^2 with respect to the x axis and $I_z = 4000$ kg·m^2 with respect to the z axis. Determine the mass moments of inertia with respect to the X and Z axes passing through the center of mass G if $I_X = 1.5\, I_Z$. Also, determine the mass of the body.

11.14 Refer to Problem 11.13, and determine the mass radius of gyration of the body with respect to the X and z axes.

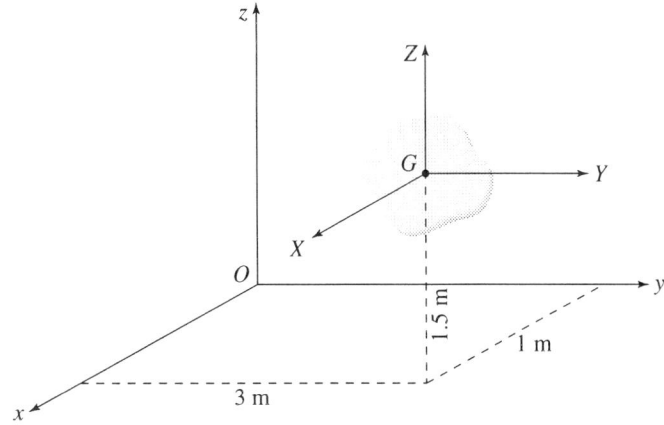

FIGURE P11.13.

11.3
Moments of
Inertia by
Integration

Area

Area moments of inertia for a given geometric shape may be deter-mined by direct integration using the fundamental definition given in Eqs. (11.1). A differential element of area dA is selected and expressed in terms of the chosen coordinate variables. Because a differential element of area is a two-dimensional physical quantity, its mathematical expression could contain as many as two variables which would lead to a double integration process. However, a careful selection of the differential element of area would usually permit determining the area moment of inertia by a single integration.

All of the examples and problems in this section have been selected to permit determining area moments of inertia with a single integral. The following examples illustrate the methods used in solving such problems. Moments of inertia for several commonly used areas have been obtained by these methods and are listed in Appendix A.

■ Example 11.3

Refer to the rectangular area shown in Figure E11.3 and determine (a) the area moment of inertia with respect to the x axis and (b) the area radius of gyration with respect to a centroidal axis parallel to the x axis.

Solution

(a) By Eq. (11.1)

$$I_x = \int y^2 \, dA$$

The differential element of area dA could be chosen in a variety of ways. It is usually most convenient to select it so that it is *parallel* to

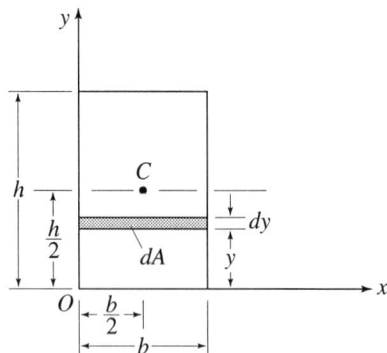

FIGURE E11.3.

the axis with respect to which the area moment of inertia is desired. Thus, all points on the element of area dA are equidistant from the desired axis and the distance y from this axis to the centroid of the element of area dA is equal to the distance y in Eq. (11.1). Such a differential element of area $dA = b\,dy$ is shown in Figure E11.3. Therefore,

$$I_x = \int_0^h y^2(b\,dy) = b\int_0^h y^2\,dy = b\left[\frac{y^3}{3}\right]_0^h = \frac{1}{3}bh^3. \qquad \text{ANS.}$$

(b) The centroid of the rectangular section is located at point C and an axis parallel to the x axis through this point is established as shown in Figure E11.3. Using the parallel-axis theorem, Eq. (11.12),

$$I_C = I_x - Ad^2 = \frac{1}{3}bh^3 - (bh)\left(\frac{h}{2}\right)^2 = \frac{1}{12}bh^3.$$

Therefore, by Eq. (11.4),

$$k_C = \sqrt{\frac{I_C}{A}} = \sqrt{\frac{\frac{1}{12}bh^3}{bh}} = \frac{h}{\sqrt{12}}. \qquad \text{ANS.}$$

■ **Example 11.4**

Refer to the shaded parabolic area shown in Figure E11.4, and determine (a) the moment of inertia of the shaded area with respect to the y axis using a differential element of area *parallel* to the y axis and (b) the moment of inertia of the shaded area with respect to the y axis using a differential element of area *perpendicular* to the y axis.

Solution

(a) A differential element of area dA_1, *parallel* to the y axis, is shown in Figure E11.4 and may be determined as

$$dA_1 = (a - y)\,dx = (a - 4x^2)\,dx.$$

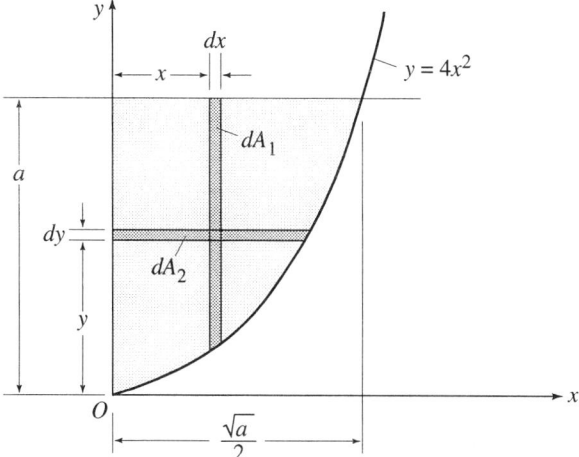

Note that all points of dA_1 are equidistant from the y axis, and, therefore, the distance x from this axis to the centroid of dA_1 is equal to x. Thus, by Eq. (11.1),

$$I_y = \int x^2 \, dA = \int_0^{\sqrt{a}/2} x^2(a - 4x^2) \, dx = \left[\frac{ax^3}{3} - \frac{4x^5}{5}\right]_0^{\sqrt{a}/2}$$

$$= \frac{a^{5/2}}{60}.$$ ANS.

(b) A differential element of area dA_2, perpendicular to the y axis, is also shown in Figure E11.4. The solution to this part of the problem may be obtained by utilizing the results of Example 11.3. In that example, the area moment of inertia of a rectangle with respect to its base was found to be $I = \frac{1}{3}bh^3$. Therefore, if we view dA_2 as a rectangle whose height is x and whose base is dy, then, its differential area moment of inertia with respect to the y axis is given by

$$dI_y = \tfrac{1}{3}(dy)x^3 = \tfrac{1}{3}(dy)(\tfrac{1}{8}y^{3/2}).$$

Therefore,

$$I_y = \int dI_y = \frac{1}{24}\int_0^a y^{3/2} \, dy = \frac{1}{24}\left[\frac{2}{5}y^{5/2}\right]_0^a = \frac{a^{5/2}}{60}.$$ ANS.

■ Example 11.5

Consider the shaded area shown in Figure E11.5, and determine (a) its moment of inertia with respect to the x axis and (b) its polar moment of inertia with respect to an axis through point O.

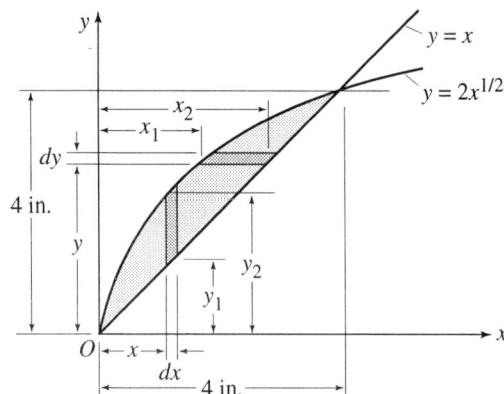

FIGURE E11.5.

Solution

(a) A differential element of area dA is chosen parallel to the x axis, as shown in Figure E11.5. Thus,

$$dA = (x_2 - x_1)\,dy$$

where $x_1 = \frac{1}{4}y^2$ and $x_2 = y$. Therefore,

$$dA = (y - \frac{1}{4}y^2)\,dy,$$

and, by Eq. (11.1),

$$I_x = \int y^2\,dA = \int_0^4 y^2\left(y - \frac{1}{4}y^2\right)dy = \left[\frac{y^4}{4} - \frac{y^5}{20}\right]_0^4$$

$$= 12.8 \text{ in.}^4 \qquad\qquad\qquad\qquad \text{ANS.}$$

(b) The polar moment of inertia with respect to an axis through point O may be found by Eq. (11.3). Thus,

$$J_O = I_x + I_y.$$

The moment of inertia I_y is found in the same manner used to determine I_x. A differential element of area dA is selected parallel to the y axis, as shown in Figure E11.5. Thus,

$$dA = (y_2 - y_1)\,dx$$

where $y_1 = x$ and $y_2 = 2x^{1/2}$. Therefore,

$$dA = (2x^{1/2} - x)\,dx,$$

and

$$I_y = \int x^2\,dA = \int_0^4 x^2(2x^{1/2} - x)\,dx = \left[\frac{4x^{7/2}}{7} - \frac{x^4}{4}\right]_0^4 = 9.14 \text{ in.}^4$$

Therefore,

$$J_O = 12.80 + 9.14 = 21.9 \text{ in.}^4 \qquad\qquad \text{ANS.}$$

■

Mass

As in the case of area moments of inertia, the basic definitions given by Eqs. (11.6) and (11.7) are used to determine the mass moments of inertia for specific masses. A differential element of mass dm is selected and expressed in terms of the chosen coordinate variables and the mass density. Because a differential element of mass is a three-dimensional physical quantity, its mathematical expression could contain as many as three variables which would lead to a triple integration process. However, with a judicious choice of the differential element of mass, it is usually possible to determine the mass moment of inertia of a body by using single or double integrations.

The mass moment of inertia for many bodies of practical interest may be conveniently determined by considering that such bodies consist of an infinite number of thin plates each of which has a constant differential thickness. As shown in the following development, the mass moment of inertia for a differential plate dI may be determined in terms of the moment of inertia of the area defined by the contour of the differential plate. The mass moment of inertia for the entire body is, then, obtained by integrating dI. Using this and other techniques, mass moments of inertia for several commonly used bodies have been determined and are listed in Appendix B.

Consider the homogeneous thin plate of constant thickness t shown in Figure 11.7. The mass moments of inertia of this plate are to be determined with respect to the x, y, and z axes. Note that the plate lies in the y-z plane and that the x axis is perpendicular to the plate.

The mass moment of inertia of the plate with respect to any in-plane axis, such as the y axis, is given by Eq. (11.6). Thus,

$$I_y = \int r_y^2 \, dm$$

where $r_y = z$, $dm = \rho \, dV = \rho(t \, dA)$ and ρ is the constant mass density.

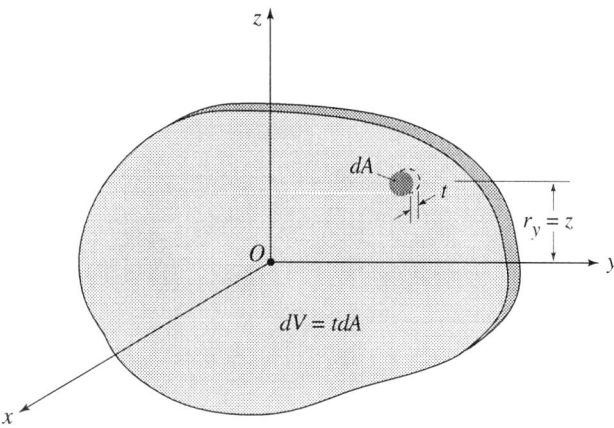

FIGURE 11.7.

Therefore,

$$I_y = \rho t \int z^2 \, dA$$

in which $\int z^2 \, dA$ is the moment of inertia of the cross-sectional area of the thin plate with respect to the y axis. Consequently, the mass moment of inertia of the thin plate with respect to the centroidal y axis may be written in the form

$$(I_{\text{mass}})_y = \rho t (I_{\text{area}})_y. \tag{11.20a}$$

Similarly,

$$(I_{\text{mass}})_z = \rho t (I_{\text{area}})_z, \tag{11.20b}$$

and

$$(I_{\text{mass}})_x = \rho t (J_{\text{area}})_x. \tag{11.20c}$$

Equations (11.20a) to (11.20c) express the fact that *the mass moment of inertia of a thin homogeneous plate, with respect to any axis, is equal to the moment of inertia of the cross-sectional area of the plate, with respect to the same axis, multiplied by the constant thickness of the plate and the constant mass density of the material of the plate*. This important relationship is useful in determining mass moments of inertia of bodies by integration as illustrated in the following examples.

■ **Example 11.6**

The body shown in Figure E11.6 is generated by revolving the area bounded by the x axis, the curve $y = (2x)^{1/2}$, and the line $x = h$ about the x axis through one complete revolution. If the mass density of the

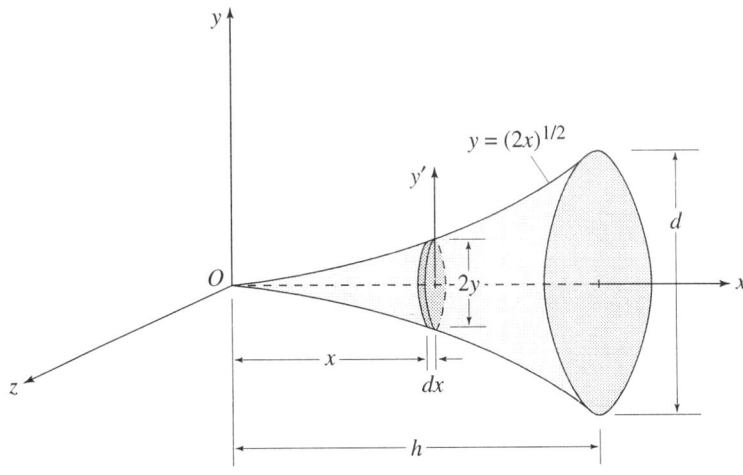

FIGURE E11.6.

body ρ is a constant, determine the mass moment of inertia of the body with respect to (a) the x axis and (b) the y or z axes.

Solution

(a) Select a thin circular plate of thickness dx for the differential element of mass dm. From Appendix A, the polar moment of inertia of a circular area with respect to a centroidal axis is

$$(J_{area})_x = \frac{\pi}{32} D^4.$$

Substituting this value in Eq. (11.20c),

$$(I_{mass})_x = \rho t \left(\frac{\pi}{32}\right) D^4 = \rho t \left(\frac{\pi D^2}{4}\right)\left(\frac{D^2}{8}\right) = \left(\frac{D^2}{8}\right) m$$

where the mass m of the thin plate was substituted for the quantity $\rho t \left(\frac{\pi D^2}{4}\right)$. Therefore, the mass moment of inertia of the differential circular plate with respect to the x axis is given by

$$dI_x = \left(\frac{D^2}{8}\right) dm = \frac{(2y)^2}{8} dm = \left(\frac{y^2}{8}\right) dm = x \, dm$$

where

$$dm = \rho \left(\frac{\pi}{4}\right)(2y)^2 \, dx = \rho \left(\frac{\pi}{4}\right)[2(2x)^{1/2}]^2 \, dx = 2\rho\pi x \, dx$$

and

$$m = \int dm = 2\rho\pi \int_0^h x \, dx = \rho\pi h^2.$$

Also,

$$dI_x = 2\rho\pi x^2 \, dx,$$

and

$$I_x = 2\rho\pi \int_0^h x^2 \, dx = \frac{2\rho\pi h^2}{3} = \frac{2}{3} mh = \frac{1}{12} md^2. \qquad \text{ANS.}$$

(b) Refer to the thin circular plate selected in part (a) for the differential element dm. From Appendix A, the moment of inertia of a circular area with respect to an in-plane diametrical axis is $(I_{area})_y = \left(\frac{\pi}{64}\right) D^4$. Substituting this value in Eq. (11.20a),

$$(I_{mass})_y = \rho t \left(\frac{\pi}{64}\right) D^4 = \rho t \left(\frac{\pi D^2}{4}\right)\left(\frac{D^2}{16}\right) = \left(\frac{D^2}{16}\right) m.$$

Thus, the differential mass moment of inertia of the thin plate with respect to an in-plane y' axis is given by

$$dI_{y'} = \left(\frac{D^2}{16}\right) dm = \frac{(2y)^2}{16} dm = \left(\frac{y^2}{4}\right) dm = \left(\frac{x}{2}\right) dm$$

and, with respect to the coordinate y axis by the parallel-axis theorem, Eq. (11.18), it is given by

$$dI_y = \left(\frac{x}{2}\right) dm + x^2 \, dm = \left(\frac{x}{2} + x^2\right) dm.$$

But, from part (a),

$$dm = 2\rho\pi x \, dx,$$

and

$$m = \rho\pi h^2.$$

Therefore,

$$dI_y = \left(\frac{x}{2} + x^2\right)(2\rho\pi x \, dx),$$

and

$$I_y = I_z = \rho\pi \left(\int_0^h x^2 \, dx + 2\int_0^h x^3 \, dx\right) = \rho\pi h^2 \left(\frac{h}{3} + \frac{h^2}{2}\right)$$

$$= m\left(\frac{h}{3} + \frac{h^2}{2}\right) = m\left(\frac{d^2}{24} + \frac{d^4}{128}\right). \qquad\qquad \text{ANS.}$$

Despite its appearance, this answer is dimensionally homogeneous because the numerical constants in the answer are *not* dimensionless. This is due to the fact that the given relationship, $y = (2x)^{1/2}$, requires that the constant 2 have units of length to make it dimensionally consistent.

■ **Example 11.7**

Determine the mass moment of inertia of the triangular prismatic solid shown in Figure E11.7 with respect to the coordinate y axis. Assume that the mass density of the body, ρ is a constant.

Solution

For a differential element of mass dm, select the thin triangular plate shown in Figure E11.7. From Appendix A, the centroidal moment of inertia of a triangular area is $\frac{1}{36}bh^3$. Therefore, for the differential element of mass in Figure E11.7 with respect to the y' axis,

$$dI_{y'} = \rho \, dx \left(\frac{1}{36} ac^3\right).$$

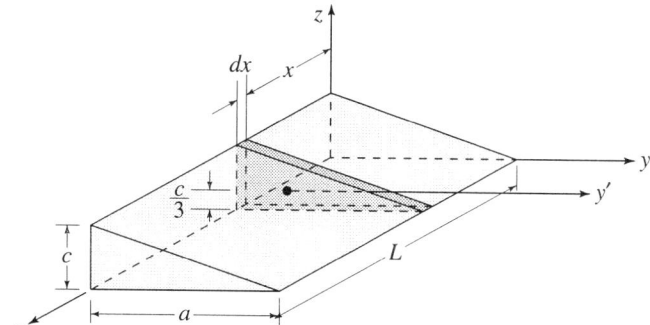

FIGURE E11.7.

Using the parallel axis theorem, the mass moment of inertia of this differential element of mass with respect to the coordinate y axis is given by

$$dI_y = dI_{y'} + \left[x^2 + \left(\frac{c}{3}\right)^2\right]dm = \rho\,dx\left(\frac{1}{36}\right)ac^3 + \left[x^2 + \left(\frac{c}{3}\right)^2\right]dm.$$

Because $dm = \rho\left(\dfrac{ac}{2}\right)dx$, dI_y may be written as

$$dI_y = \frac{1}{36}ac^3\rho\,dx + \frac{1}{2}ac\rho x^2\,dx + \frac{1}{18}ac^3\rho\,dx = \frac{ac^3\rho}{12}dx + \frac{ac\rho}{2}x^2\,dx$$

and

$$I_y = \frac{ac^3\rho}{12}\int_0^L dx + \frac{ac\rho}{2}\int_0^L x^2\,dx = \frac{ac^3\rho}{12}L + \frac{ac\rho}{6}L^3$$

$$= \rho\left(\frac{ac}{2}\right)L\left(\frac{c^2}{6} + \frac{L^2}{3}\right) = m\left(\frac{c^2}{6} + \frac{L^2}{3}\right) \qquad \text{ANS.}$$

where $m = \rho\left(\dfrac{ac}{2}\right)L$ is the mass of the triangular prismatic solid.

■ **Example 11.8**

Derive an exact equation for the mass moment of inertia of the hollow circular cylinder shown in Figure E11.8(a), with respect to the y axis. The cylinder of constant mass density ρ has an outside radius r_o, inside radius r_i, and height h. Then, consider a *thin*-walled hollow cylinder, and derive an approximate equation for its mass moment of inertia with respect to the y axis. This thin-walled cylinder has a thickness $t = r_o - r_i$ and a mean radius $r_m = (r_o + r_i)/2$. Finally, derive an equation for the percentage error in the approximate equation for the thin-walled hollow cylinder as a function of $\beta = t/r_o$. Plot this percentage error function and discuss it.

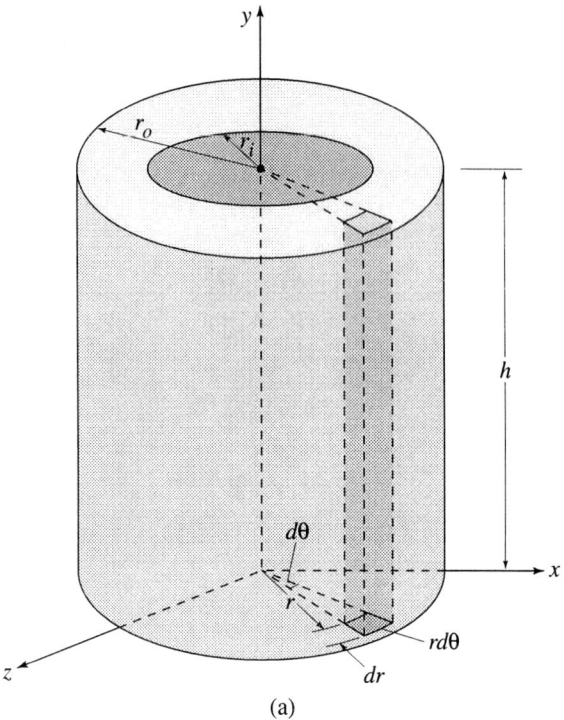

(a)

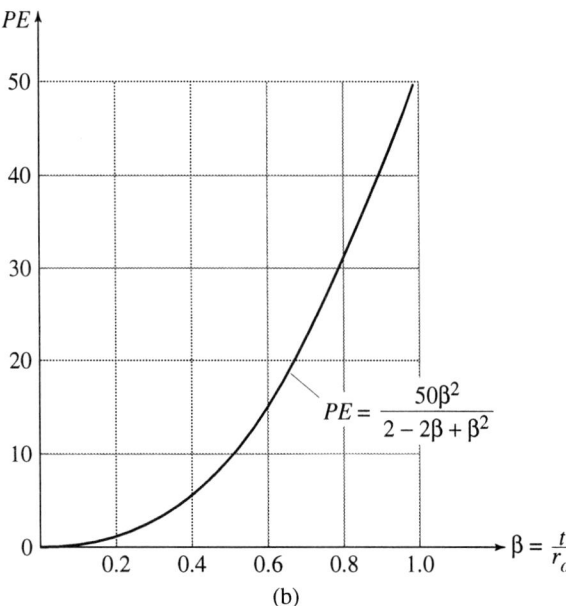

$$PE = \frac{50\beta^2}{2 - 2\beta + \beta^2}$$

(b)

FIGURE E11.8.

Solution

We begin by developing an equation for the mass of the hollow cylinder. As shown in Figure E11.8(a), the equation for the differential volume becomes

$$dV = hr\, dr\, d\theta.$$

The differential mass is the product of the mass density and the differential volume. Thus

$$dm = \rho\, dV = \rho hr\, dr\, d\theta.$$

Therefore,

$$m = \int dm = \rho h \int_{r_i}^{r_o} r\, dr \int_0^{2\pi} d\theta = \pi\rho h(r_o^2 - r_i^2).$$

The differential mass moment of inertia with respect to the y axis is given by

$$dI_y = r^2\, dm.$$

Substitute for dm to give

$$dI_y = r^2 \rho hr\, dr\, d\theta.$$

Therefore,

$$I_y = \rho h \int_{r_i}^{r_o} r^3\, dr \int_0^2 d\theta = \frac{\pi\rho h}{2}(r_o^4 - r_i^4)$$

$$= \frac{\pi\rho h}{2}(r_o^2 - r_i^2)(r_o^2 + r_i^2) = \frac{1}{2}m(r_o^2 + r_i^2). \qquad \text{ANS.}$$

Using primed quantities, for the *thin*-walled circular cylinder, the differential volume becomes

$$dV' = htr_m\, d\theta$$

where t and r_m are defined in the statement of the problem. Hence, as before,

$$dm' = \rho\, dV' = \rho htr_m\, d\theta$$

from which

$$m' = \rho htr_m \int_0^{2\pi} d\theta = 2\pi\rho htr_m.$$

Because the cylinder is thin-walled, we may write

$$dI_y' = r_m^2\, dm'.$$

Substitution for dm' yields

$$dI_y' = r_m^2 \rho htr_m\, d\theta$$

and

$$I'_y = \rho h t r_m^3 \int_0^{2\pi} d\theta = 2\pi \rho h t r_m^3 = 2\pi \rho h t r_m (r_m^2) = m' r_m^2. \qquad \text{ANS.}$$

We note that substitution for t and r_m in terms of r_o and r_i reveals that $m = m'$. This will greatly simplify the percentage error equation because m and m' may be canceled,

By definition the percentage error (PE) is given by

$$\text{PE} = \left(\frac{\text{Exact value} - \text{Approximate value}}{\text{Exact value}} \right) 100$$

$$= \left(\frac{I_y - I'_y}{I_y} \right) 100$$

$$\text{PE} = \left[\frac{\frac{1}{2} m (r_o^2 + r_i^2) - m' r_m^2}{\frac{1}{2} m (r_o^2 + r_i^2)} \right] 100.$$

Cancel m and m', substitute $r_m = \frac{1}{2} m (r_o + r_i)$ and $r_i = r_o - t$ and perform some algebraic steps to determine that,

$$\text{PE} = \frac{50 t^2}{2 r_o^2 - 2 r_o t + t^2}.$$

Divide numerator and denominator by r_o^2, and let $\beta = t/r_o$ to conclude that

$$\text{PE} = \frac{50 \beta^2}{2 - 2\beta + \beta^2}. \qquad \text{ANS.}$$

This percentage error function plotted in Figure E11.8(c), is representative of many such functions which compare exact and approximate solutions. It enables us to define precisely what we mean by *thin-walled* in the context of the mass moment of inertia of a hollow circular cylindrical shape with respect to its geometric axis. For example, if we wish to use the approximate equation for I_y with a maximum error of 10% then we would limit $\beta = t/r_o$ to values between 0 and 0.5 as revealed by the plotted function. In this particular case, we would usually use the exact solution, but there are problems of interest to engineers for which only an approximate solution is far simpler to employ. If only an approximate solution is available, then, comparisons are made with experimental results. Because computers and pocket calculators are widely available, engineers generally prefer to employ exact solutions when they are obtainable.

Problems

11.15 Determine the moment of inertia of the triangular area shown in Figure P11.15 with respect to the x axis. Express your answer in terms of the dimensions b and h. Also, determine the moment of inertia and radius of gyration with respect to a centroidal axis parallel to the x axis.

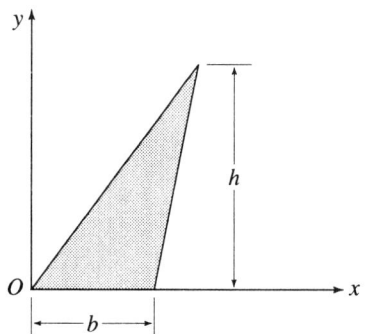

FIGURE P11.15.

11.16 Determine the moment of inertia of the circular area shown in Figure P11.16 with respect to the centroidal Y axis.

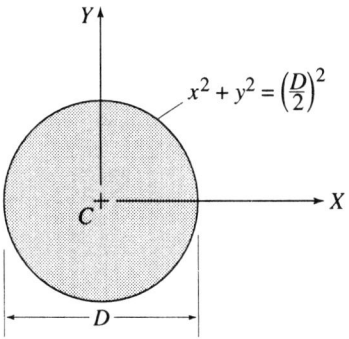

FIGURE P11.16.

What is the polar radius of gyration with respect to an axis through the centroid C?

11.17 Find the moment of inertia I_x for the semielliptical area shown in Figure P11.17. What is the radius of gyration with respect to the x axis?

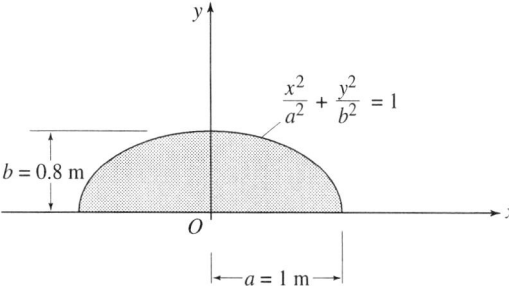

FIGURE P11.17.

11.18 Find the moment of inertia I_y for the semielliptical area shown in Figure P11.17.

11.19 Show that the moment of inertia of the thin rectangular area, shown in Figure P11.19, with respect to the x axis may be written as

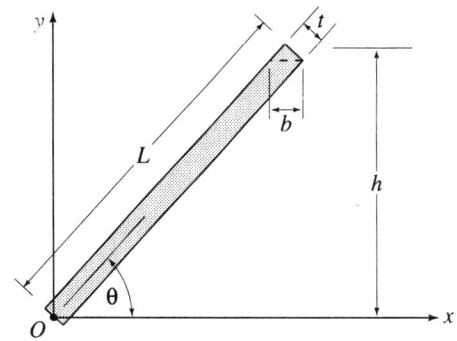

FIGURE P11.19.

$$I_x = \frac{1}{3}bh^3 \text{ where } b = \frac{t}{\sin\theta}$$

and

$$h = L\sin\theta.$$

11.20 Find the moment of inertia I_x of the shaded area shown in Figure P11.20.

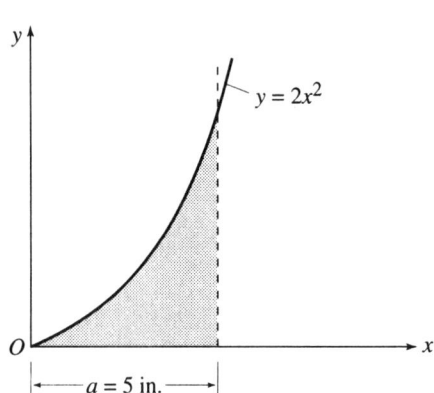

FIGURE P11.20.

11.21 Find the moment of inertia I_y of the shaded area shown in Figure P11.20. What is the moment of inertia with respect to a centroidal y axis?

11.22 Determine the moment of inertia I_x for the shaded area shown in Figure P11.22.

11.23 Determine the moment of inertia I_y for the shaded area shown in Figure P11.22.

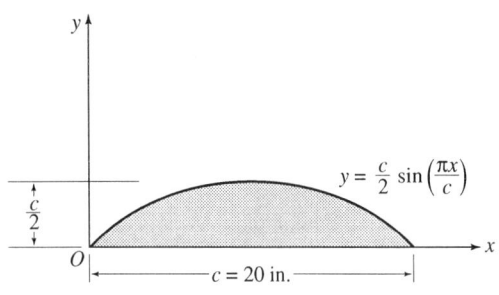

FIGURE P11.22.

What is the radius of gyration with respect to a centroidal y axis?

11.24 Find the moment of inertia I_x for the shaded area shown in Figure P11.24. What is the radius of gyration with respect to the x axis?

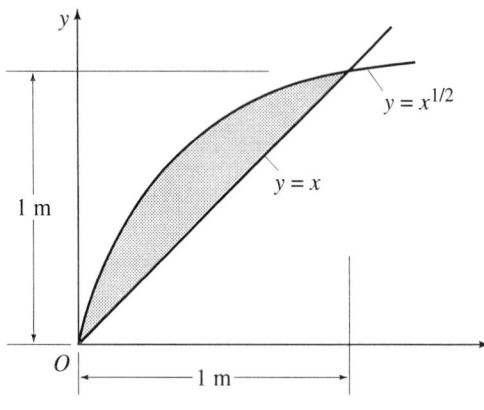

FIGURE P11.24.

11.25 Find the moment of inertia I_y for the shaded area shown in Figure P11.24. What is the radius of gyration with respect to the y axis?

11.26 Find the moment of inertia I_x for the shaded area shown in Figure P11.26.

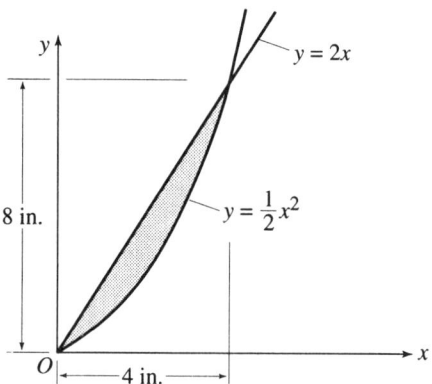

FIGURE P11.26.

11.27 Find the moment of inertia I_y for the shaded area shown in Figure P11.26. What is the polar moment of inertia with respect to an axis through point O?

11.28 Determine the moment of inertia I_x for the shaded area shown in Figure P11.28. What is the radius of gyration with respect to the x axis?

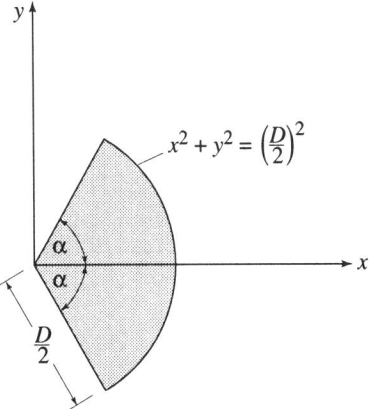

FIGURE P11.30.

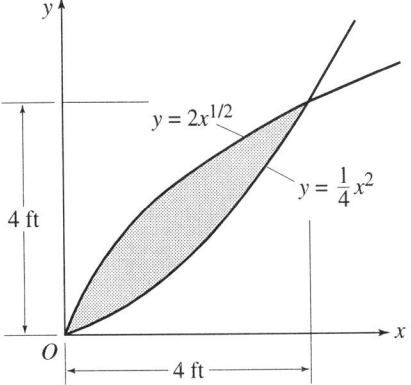

FIGURE P11.28.

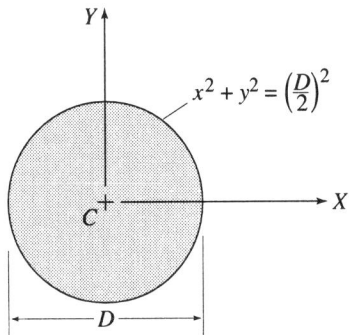

FIGURE P11.31.

11.29 Determine the moment of inertia I_y for the shaded area shown in Figure P11.28. What is the radius of gyration with respect to the y axis?

11.30 Find the moment of inertia of the circular segment shown in Figure P11.30 with respect to the x axis. Specialize the answer to the case of a complete circular area for which $\alpha = \pi$. Note that D is the diameter of the circular segment.

11.31 With respect to an axis through point C perpendicular to the page, determine the mass moment of inertia of a thin plate whose cross-section is the circular area shown in Figure P11.31. Assume that both the thickness of the plate and the mass density of the plate material are constants. Express your answer in terms of m, the mass of the plate.

11.32 With respect to the x axis, determine the mass moment of inertia of a thin plate whose cross-section is the shaded area shown in Figure P11.32. Assume that both the thickness of the plate and the mass density of the plate material are constants. Express your answer in terms of m, the mass of the plate.

11.33 With respect to the y axis, determine the mass moment of inertia of a thin plate whose cross-section is the shaded area shown in Figure P11.33. Assume that both the thickness of the plate and the mass density of the plate material are constants. Express your answer in terms of m, the mass of the plate.

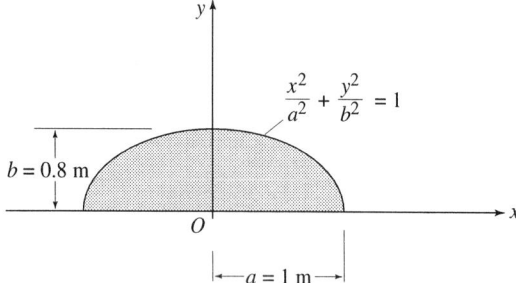

$$\frac{x^2}{a^2} + \frac{y^2}{b^2} = 1$$

$b = 0.8$ m

$a = 1$ m

FIGURE P11.32.

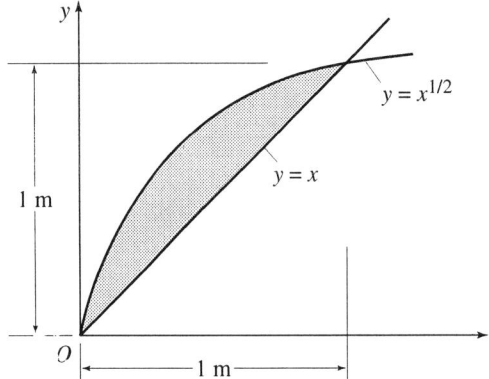

$y = x^{1/2}$

$y = x$

1 m

1 m

FIGURE P11.33.

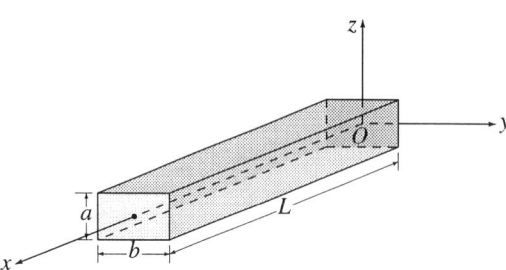

FIGURE P11.34.

11.35 Determine the mass moment of inertia I_y of the prismatic bar shown in Figure P11.34. Assume that the mass density is a constant, and express your answer in terms of the mass m and the dimensions of the bar.

11.36 Find the mass moments of inertia I_X, I_Y, and I_Z of the spherical solid shown in Figure P11.36. Assume that the mass density is a constant, and express your answer in terms of the mass m and the dimensions of the spherical solid.

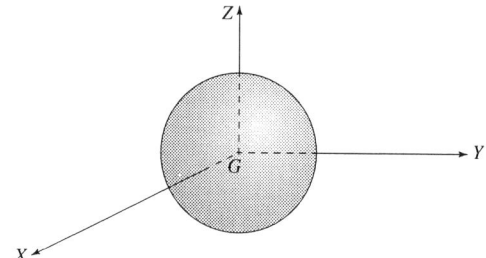

FIGURE P11.36.

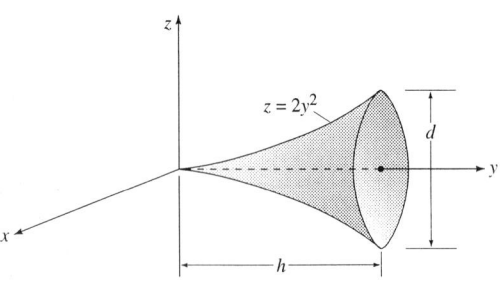

$z = 2y^2$

d

h

FIGURE P11.37.

11.37 With respect to the y axis, determine the mass moment of inertia of the solid shown in Figure P11.37. Assume that the mass density of the solid is a constant, and express your answer in terms of the mass m and the dimensions of the solid.

11.34 With respect to the x axis, compute the mass moment of inertia of the prismatic bar shown in Figure P11.34. Assume that the mass density is a constant, and express your answer in terms of the mass m and the dimensions of the bar.

11.38 With respect to the x axis, determine the mass moment of inertia of the solid shown in Figure P11.37. What is the mass moment of inertia with respect to a centroidal x axis? Assume that the mass density of the solid is a constant, and express your answers in terms of the mass m and the dimensions of the solid.

11.39 With respect to the z axis, find the mass moment of inertia of the solid shown in Figure P11.39. Assume that the mass density of the solid is a constant, and express your answer in terms of the mass m and the dimensions of the solid.

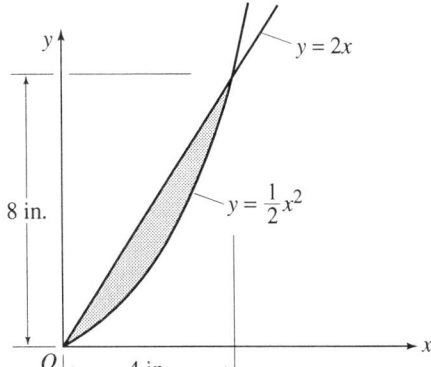

FIGURE P11.41.

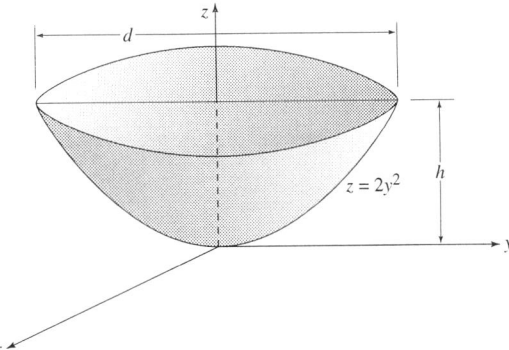

FIGURE P11.39.

11.40 With respect to the y axis, find the mass moment of inertia of the solid shown in Figure P11.39. What is the mass moment of inertia with respect to a centroidal y axis? Assume that the mass density of the solid is a constant, and express your answers in terms of the mass m and the dimensions of the solid.

11.41 A solid is generated by revolving the shaded area shown in Figure P11.41 through one complete revolution about the x axis. Compute its mass moment of inertia with respect to the x axis. Let the mass density $\rho = 0.01$ slug/in^3.

11.42 With respect to the y axis, compute the mass moment of inertia of the solid described in Problem 11.41.

11.43 A solid is generated by revolving the shaded area shown in Figure P11.43 through one complete revolution about

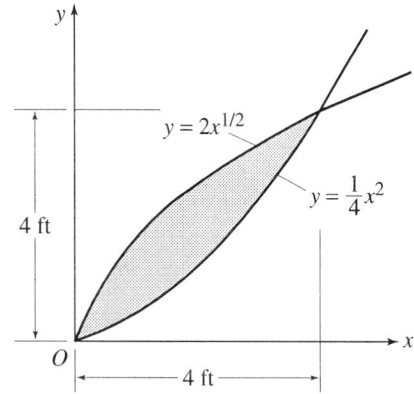

FIGURE P11.43.

the y axis. Compute its mass moment of inertia with respect to the y axis. Let the mass density $\rho = 10$ slug/ft^3.

11.44 Refer to Example 11.8 (p. 721), and determine the mass moment of inertia of the hollow cylinder with respect to the z axis. Assume that the mass density of the cylinder is a constant, and express your answer in terms of the mass m and the height h of the cylinder.

11.45 Refer to Example 11.8 (p. 721), and determine the mass moment of inertia of the hollow cylinder with respect to a Z axis that is parallel to the given z axis and passes through the center of mass of the cylinder. Assume that the mass density of the cylinder is a constant, and express your answer in terms of the mass m and the height h of the cylinder.

11.46 Refer to the equation for the percentage error (PE) as a function of $\beta = t/r_o$, shown in Example 11.8 (p. 721). If you wish to limit the percentage error to a maximum of 4%, determine the permissible range of values of β. Determine this range analytically, and, then, confirm your result by reading from the plot shown in Example 11.8.

11.4
Moments of Inertia of Composite Areas and Masses

A composite area is one that may be decomposed into two or more simple geometric components. Determining the moment of inertia of a composite area with respect to any axis is based upon the fundamental definition given by Eq. (11.1). Consider, for example, the first of Eqs. (11.1), $I_x = \int y^2 \, dA$. The integrand, $y^2 \, dA$, represents the quantity dI_x, the moment of inertia of the differential element of area dA with respect to the x axis. Therefore, $I_x = \int y^2 \, dA = \int dI_x$. For a composite area consisting of a finite number of components, integration may be replaced by summation, and a general relationship may be written as

$$ I = \int dI \Rightarrow \sum_{i=1}^{n} I_i \tag{11.21} $$

Area

In Eq. (11.21), I represents the moment of inertia of the composite area with respect to some axis, i is the summation index, and I_i represents the moment of inertia, with respect to the same axis, of the ith component of the composite area. Therefore, the moment of inertia of a composite area, with respect to some axis, is equal to the algebraic sum of the area moments of inertia of the components with respect to the same axis. In finding the area moment of inertia of the components with respect to the desired axis, it is necessary to use the parallel-axis theorem given by Eq. (11.12). Thus, Eq. (11.21) may be rewritten in the form

$$ I = \sum_{i=1}^{n} (I_C + Ad^2)_i \tag{11.22} $$

In applying Eq. (11.22), it is desirable to have access to values of moments of inertia for simple areas. A set of values of moments of inertia for a selected group of simple areas is given in Appendix A. The following examples illustrate the use of Eq. (11.22) in solving problems.

■ Example 11.9

Determine the moments of inertia of the composite area shown in Figure E11.9(a) with respect to centroidal X and Y axes. Find the radii of gyration with respect to these two axes.

Solution

Decompose the given area into the three rectangular areas A_1, A_2, and A_3 as shown in Figure E11.9(a). The centroid C of the composite area is found by the methods discussed in Chapter 10. The centroid is

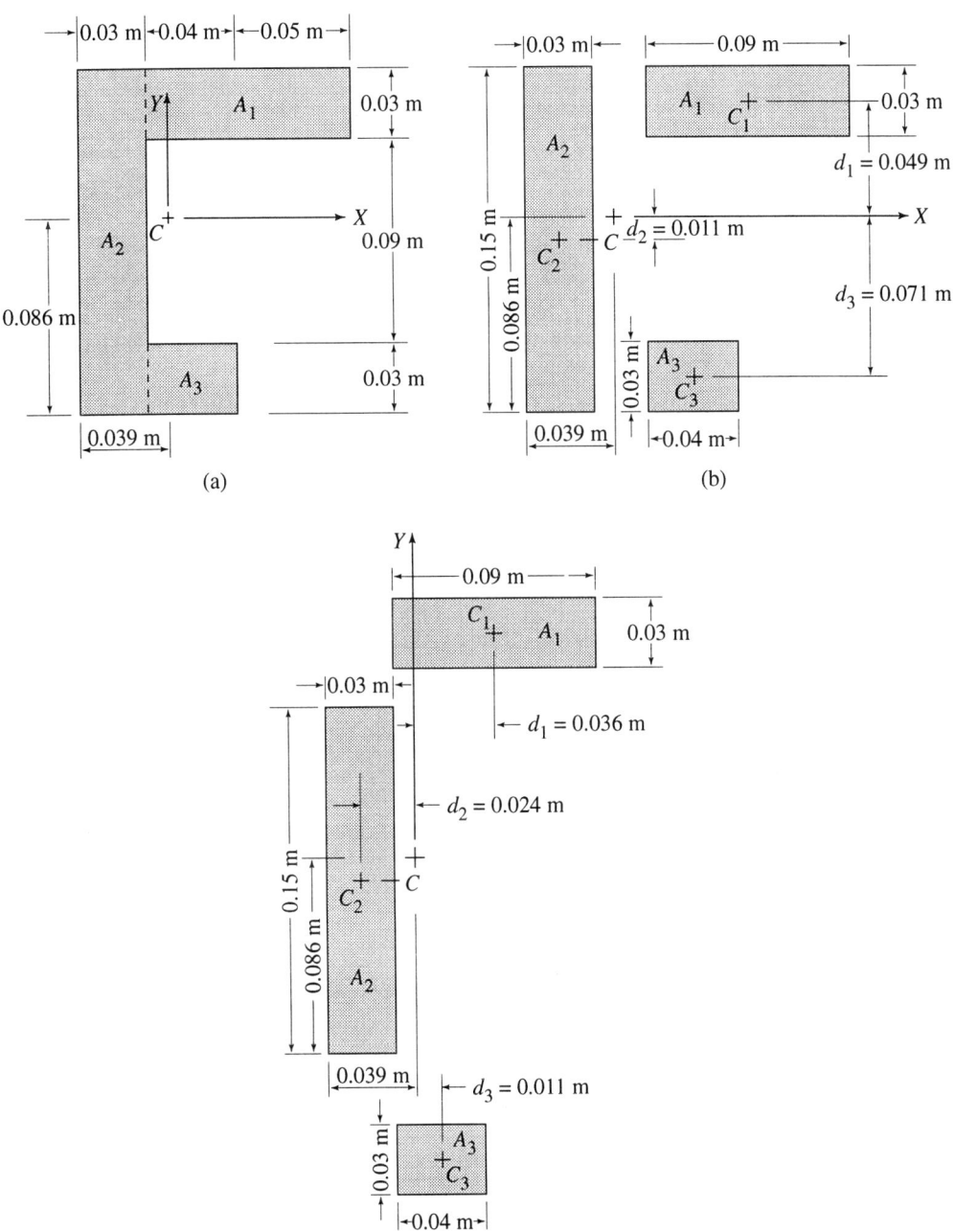

FIGURE E11.9.

located, and centroidal X and Y axes are established, as shown in Figure E11.9(a).

By Eq. (11.22),

$$I_X = \sum_{i=1}^{n} [(I_C + Ad^2)_X]_i$$

$$= [(I_C + Ad^2)_X]_1 + [(I_C + Ad^2)_X]_2 + [(I_C + Ad^2)_X]_3$$

where the subscripts 1, 2, and 3 refer to the three component areas A_1, A_2, and A_3. For convenience, these three areas are properly located with respect to the desired axis (i.e., the X axis), as shown in Figure E11.9(b). Using the values of centroidal moments of inertia given in Appendix A,

$$I_X = \left[\frac{1}{12} bh^3 + Ad^2 \right]_1 + \left[\frac{1}{12} bh^3 + Ad^2 \right]_2 + \left[\frac{1}{12} bh^3 + Ad^2 \right]_3 ,$$

$$I_x = \frac{1}{12}(0.09)(0.03)^3 + (0.09)(0.03)(0.049)^2$$

$$+ \frac{1}{12}(0.03)(0.15)^3 + (0.03)(0.15)(0.011)^2$$

$$+ \frac{1}{12}(0.04)(0.03)^3 + (0.04)(0.03)(0.071)^2$$

$$= 21.8 \times 10^{-6} \text{ m}^4. \qquad \text{ANS.}$$

Similarly, by Eq. (11.22),

$$I_Y = \sum_{i=1}^{n} [(I_C + Ad^2)_Y]_i$$

$$= [(I_C + Ad^2)_Y]_1 + [(I_C + Ad^2)_Y]_2 + [(I_C + Ad^2)_Y]_3.$$

The three component areas A_1, A_2, and A_3 are properly located with respect to the desired axis (i.e., the Y axis) as shown in Figure E11.9(c). Using the values of centroidal moments of inertia given in Appendix A,

$$I_Y = \left[\frac{1}{12} bh^3 + Ad^2 \right]_1 + \left[\frac{1}{12} bh^3 + Ad^2 \right]_2 + \left[\frac{1}{12} bh^3 + Ad^2 \right]_3 ,$$

$$I_Y = \frac{1}{12}(0.03)(0.09)^3 + (0.03)(0.09)(0.036)^2$$

$$+ \frac{1}{12}(0.15)(0.09)^3 + (0.15)(0.03)(0.024)^2$$

$$+ \frac{1}{12}(0.03)(0.04)^3 + (0.03)(0.04)(0.011)^2$$

$$= 8.56 \times 10^{-6} \text{ m}^4. \qquad \text{ANS.}$$

By Eq. (11.4),

$$k_X = \sqrt{\frac{I_X}{A}} = 5.10 \times 10^{-2} \text{ m},$$ ANS.

and

$$k_Y = \sqrt{\frac{I_Y}{A}} = 3.19 \times 10^{-2} \text{ m}.$$ ANS.

■ **Example 11.10**

(a) Determine the moment of inertia of the composite area, shown in Figure E11.10(a), with respect to the y axis.

(b) What is the polar moment of inertia with respect to an axis through point O?

Solution

(a) The composite area of Figure E11.10(a) may be viewed as the algebraic sum of the four areas A_1, A_2, A_3, and A_4 shown in Figure E11.10(b) properly positioned with respect to the desired axis (i.e., the y axis). The composite area of Figure E11.10(a) may be obtained from Figure E11.10(b) by adding areas A_1 and A_2 and, then, subtracting areas A_3 and A_4. Thus, by Eq. (11.22),

$$I_y = \sum_{i=1}^{n} [(I_C + Ad^2)_y]_i$$

$$= [(I_C + Ad^2)_y]_1 + [(I_C + Ad^2)_y]_2$$

$$- [(I_C + Ad^2)_y]_3 - [(I_C + Ad^2)_y]_4$$

where the subscripts 1, 2, 3, and 4 refer to the four component areas A_1, A_2, A_3, and A_4. Using the values given in Appendix A for moments of inertia with respect to centroidal axes,

$$I_Y = \left[\frac{1}{36}bh^3 + Ad^2\right]_1 + \left[\frac{1}{12}bh^3 + Ad^2\right]_2$$

$$- \left[\frac{1}{128}\pi D^4 + Ad^2\right]_3 - \left[\frac{1}{12}bh^3 + Ad^2\right]_4$$

$$= \frac{1}{36}(15)(9)^3 + \frac{1}{2}(15)(9)(6)^2 + \frac{1}{12}(12)(9)^3 + (12)(9)(4.5)^2$$

$$- \frac{1}{128}\pi(3)^3 - \frac{1}{8}\pi(3)^2(3.5)^2 - \frac{1}{12}(4)(3)^2 - (4)(3)(3.5)^2$$

$$= 5.45 \times 10^3 \text{ in.}^4.$$ ANS.

(b) The polar moment of inertia may be found by using Eq. (11.3). This equation requires knowledge of I_x and I_y, which was found in part (a).

As in part (a), we decompose the given area into four areas A_1, A_2, A_3, and A_4, as shown in Figure E11.10(c) properly positioned with respect to the desired axis (i.e., the x axis). Using Eq. (11.22) and values of centroidal moments of inertia from Appendix A,

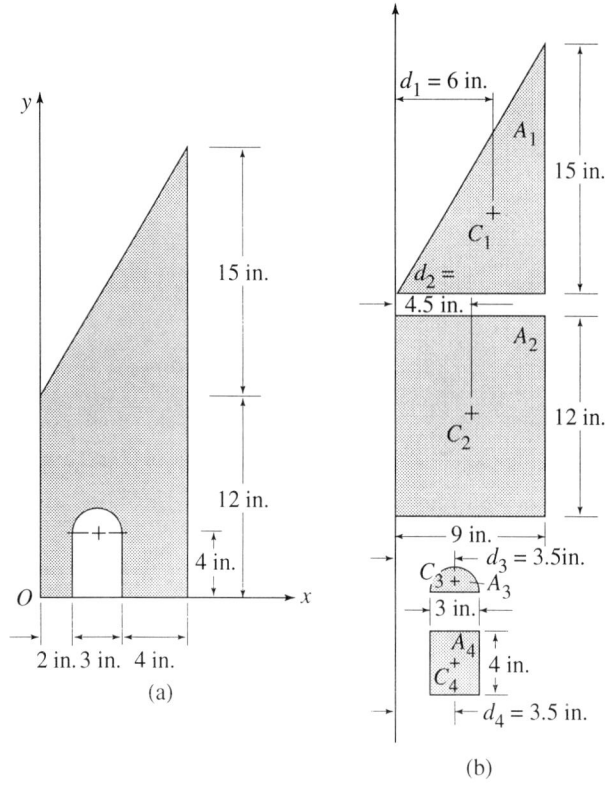

(a)

(b)

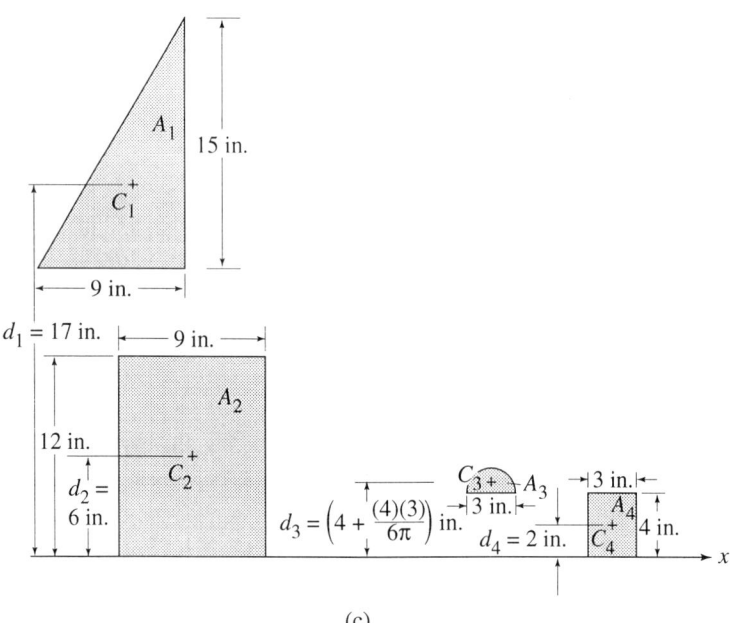

FIGURE E11.10.

(c)

$$I_x = \sum_{i=1}^{n} [(I_C + Ad^2)_x]_i$$

$$= [(I_C + Ad^2)_x]_1 + [(I_C + Ad^2)_x]_2$$

$$- [(I_C + Ad^2)_x]_3 - [(I_C + Ad^2)_x]_4$$

$$I_x = \left[\frac{1}{36}bh^3 + Ad^2\right]_1 + \left[\frac{1}{12}bh^3 + Ad^2\right]_2$$

$$- \left[\frac{9\pi^2 - 64}{1152\pi}D^4 + Ad^2\right]_3 - \left[\frac{1}{12}bh^3 + Ad^2\right]_4$$

$$= \frac{1}{36}(9)(15)^3 + \frac{1}{2}(9)(15)(17)^2 + \frac{1}{12}(9)(12)^3$$

$$+ (9)(12)(6)^2 - \left(\frac{9\pi^2 - 64}{1152\pi}\right)(3)^4$$

$$- \frac{1}{8}\pi(3)^2 \left[4 + \frac{(4)(3)}{6\pi}\right]^2 - \frac{1}{12}(3)(4)^3 - (3)(4)(2)^2$$

$$= 25{,}394{,}71 \text{ in}^4 = 25.39 \times 10^3 \text{ in.}^4.$$

By Eq. (11.3),

$$J_O = I_x + I_y = 3.08 \times 10^4 \text{ in.}^4. \qquad \text{ANS.}$$

Mass

A composite mass is one that may be decomposed into two or more simple geometric components. The same type of development used for the moment of inertia of a composite area may be used to show that the moment of inertia of a composite mass, with respect to any axis, is equal to the algebraic sum of the mass moments of inertia of the components with respect to the same axis. In finding the mass moments of inertia of the component parts with respect to the desired axis, it becomes necessary to use the parallel-axis theorem expressed in Eq. (11.17), and an expression similar to that in Eq. (11.22) may be written for mass moments of inertia. Thus,

$$I = \sum_{i=1}^{n} (I_G + md^2)_i \qquad (11.23)$$

where I represents the moment of inertia of the composite mass with respect to any axis and i is the summation index.

In applying Eq. (11.23), it is desirable to have access to values of mass moments of inertia for simple geometric shapes. A set of values of mass moments of inertia for a selected group of simple geometric shapes is given in Appendix B.

The following examples illustrate the use of Eq. (11.23) in solving problems.

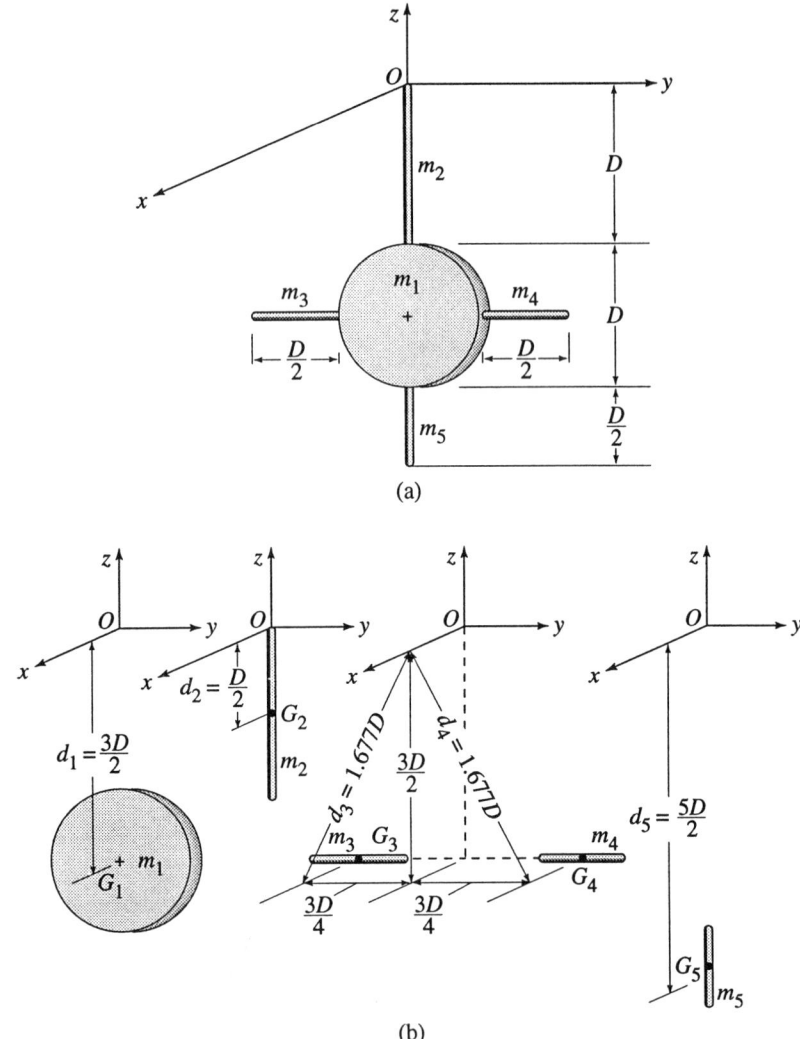

FIGURE E11.11.

(b)

■ **Example 11.11**

Four slender rods are attached to a thin disk, as shown in Figure E11.11(a). Let the mass of the longest rod be m_r and that for the disk be m_d, and determine the moment of inertia of the composite mass with respect to the x axis. Express the answer in terms of m_r, m_d, and D.

Solution

The composite mass of Figure E11.11(a) may be viewed as a composite of the five masses m_1, m_2, m_3, m_4, and m_5. These masses are labeled in Figure E11.11(a) and properly positioned separately in Figure E11.11(b) with respect to the desired axis (i.e., the x axis), to show clearly the distances to this axis from the mass centers of the five component masses. Therefore, the moment of inertia of the composite mass with respect to the x axis may be obtained by adding the moments of inertia, with respect to the same axis, of the five separate masses. Thus, by Eq. (11.23),

$$I_x = \sum_{i=1}^{n} [(I_G + md^2)_x]_1$$
$$= [(I_G + md^2)_x]_1 + [(I_G + md^2)_x]_2 + [(I_G + md^2)_x]_3$$
$$+ [(I_G + md^2)_x]_4 + [(I_G + md^2)_x]_5.$$

Using the values given in Appendix B for the mass moments of inertia with respect to centers of mass and noting that the moments of inertia of m_3 and m_4 are identical,

$$I_x = \left[\frac{1}{8}mD^2 + md^2\right]_1 + \left[\frac{1}{12}mL^2 + md^2\right]_2$$
$$+ 2\left[\frac{1}{12}mL^2 + md^2\right]_3 + \left[\frac{1}{12}mL^2 + md^2\right]_5$$
$$= \frac{1}{8}m_d D^2 + m_d\left(\frac{3D}{2}\right)^2 + \frac{1}{12}m_r D^2 + m_r\left(\frac{D}{2}\right)^2$$
$$+ 2\left[\frac{1}{12}\left(\frac{m_r}{2}\right)\left(\frac{D}{2}\right)^2 + \left(\frac{m_r}{2}\right)(1.677D)^2\right]$$
$$+ \frac{1}{12}\left(\frac{m_r}{2}\right)\left(\frac{D}{2}\right)^2 + \left(\frac{m_r}{2}\right)\left(\frac{5D}{4}\right)^2$$
$$= 2.3750m_d D^2 + 5.7082m_r D^2$$
$$= (2.3750m_d + 5.7082m_r)D^2. \qquad \text{ANS.}$$

■ Example 11.12

Determine the moment of inertia of the composite mass shown in Figure E11.12(a) with respect to the y axis. What is the radius of gyration with respect to this axis? The material has a mass density $\rho = 500 \text{ kg/m}^3$.

Solution

The composite mass may be viewed as a composite of the four masses m_1, m_2, m_3, and m_4 shown in Figure E11.12(b). The composite mass of Figure E11.12(a) may be obtained from the four component masses in Figure E11.12(b) by, first, adding the solid conical body of mass m_1 to the solid cylindrical body of mass m_2 and, then, subtracting the solid conical body of mass m_3 and the solid cylindrical body of mass m_4. All of the component masses in Figure E11.12(b) are positioned properly with respect to the desired axis (i.e., the y axis) to show the distances from this axis to the center of masses of the four component parts. Thus, by Eq. (11.23),

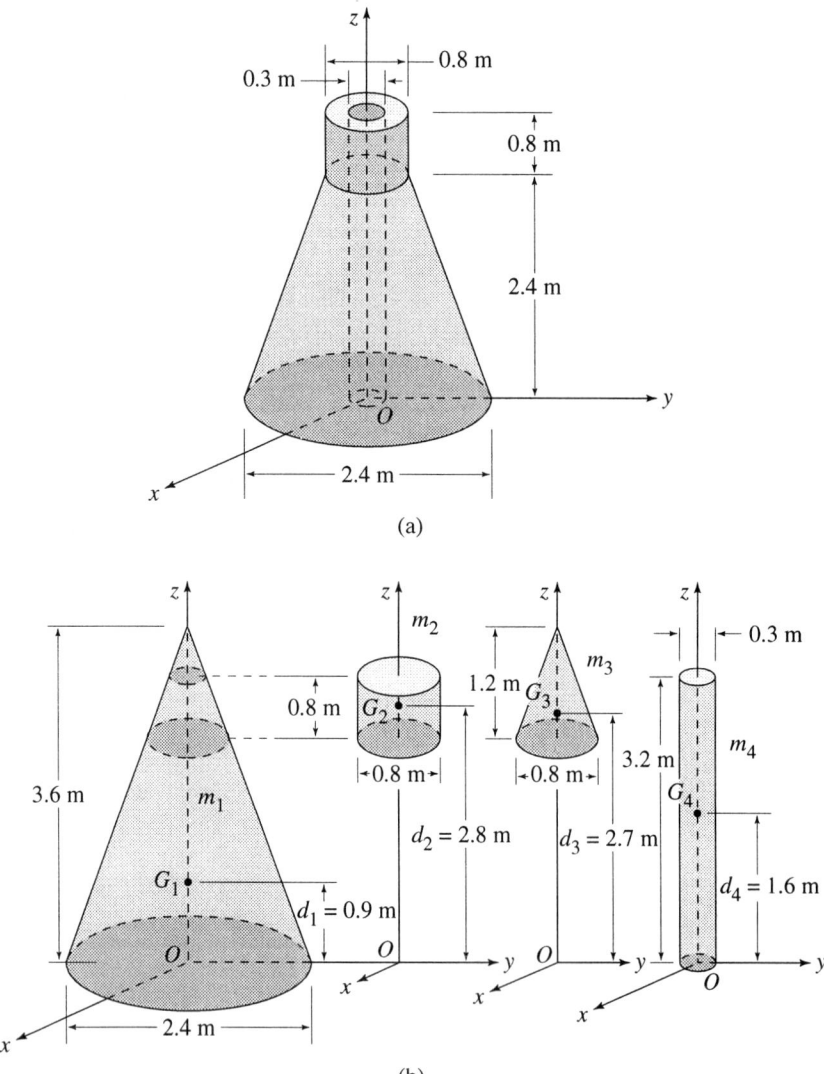

FIGURE E11.12.

$$I_y = \sum_{i=1}^{n} [(I_G + md^2)_y]_i$$

$$= [(I_G + md^2)_y]_1 + [(I_G + md^2)_y]_2$$

$$- [(I_G + md^2)_y]_3 - [(I_G + md^2)_y]_4.$$

Using the values given in Appendix B for mass moments of inertia with respect to centers of mass,

$$I_y = \left[\frac{3}{80}m(D^2 + h^2) + md^2\right]_1 + \left[\frac{1}{12}m\left(\frac{3}{4}D^2 + h^2\right) + md^2\right]_2$$

$$- \left[\frac{3}{80}m(D^2 + h^2) + md^2\right]_3 - \left[\frac{1}{12}m\left(\frac{3}{4}D^2 + h^2\right) + md^2\right]_4$$

$$= \left[\frac{3}{80}m_1(2.4^2 + 3.6^2) + m_1(0.9)^2\right]_1$$

$$+ \left[\frac{1}{12}m_2(0.6^2 + 0.8^2) + m_2(2.8)^2\right]_2$$

$$- \left[\frac{3}{80}m_3(0.8^2 + 1.2^2) + m_3(2.7)^2\right]_3$$

$$- \left[\frac{1}{12}m_4(0.225^2 + 3.2^2) + m_4(1.6)^2\right]_4$$

$$= 1.512m_1 + 7.923m_2 - 7.368m_3 - 3.418m_4.$$

The values of m_1, m_2, m_3, and m_4 are found as follows:

$$m_1 = V_1\rho = \left(\frac{\pi D^2}{12}\right)h\rho = \frac{\pi(2.4)^2}{12}(3.6)(5000) = 27143.36 \text{ kg},$$

$$m_2 = V_2\rho = \left(\frac{\pi D^2}{4}\right)h\rho = \frac{\pi(0.8)^2}{4}(0.8)(5000) = 2010.62 \text{ kg},$$

$$m_3 = V_3\rho = \left(\frac{\pi D^2}{12}\right)h\rho = \frac{\pi(0.8)^2}{12}(1.2)(5000) = 1005.31 \text{ kg},$$

and

$$m_4 = V_4\rho = \left(\frac{\pi D^2}{12}\right)h\rho = \frac{\pi(0.3)^2}{12}(3.2)(5000) = 1130.97 \text{ kg}.$$

Therefore,

$$I_y = 1.512(27143.36) + 7.923(2010.62)$$

$$- 7.368(1005.31) - 3.418(1130.97)$$

$$= 4.57 \times 10^4 \text{ kg·m}^2. \qquad \text{ANS.}$$

By Eq. (11.8), $k_y = \sqrt{\dfrac{I_y}{m}}$ where $m = m_1 + m_2 - m_3 - m_4 = 2.70 \times 10^4$ kg. Therefore,

$$k_y = \sqrt{\frac{4.57 \times 10^4}{2.70 \times 10^4}} = 1.301 \text{ m}. \qquad \text{ANS.}$$

■

Problems

11.47 Determine the moment of inertia I_x and the radius of gyration k_x for the composite area shown in Figure P11.47.

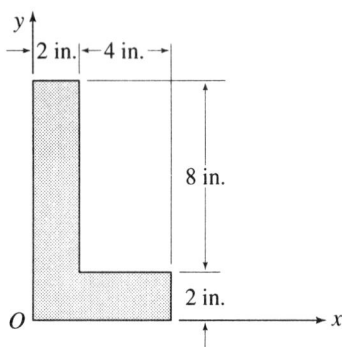

FIGURE P11.47.

11.48 Refer to the composite area shown in Figure P11.47, and determine the moment of inertia I_y and the radius of gyration k_y.

11.49 Determine the moment of inertia I_x and the radius of gyration k_x for the composite area shown in Figure P11.49.

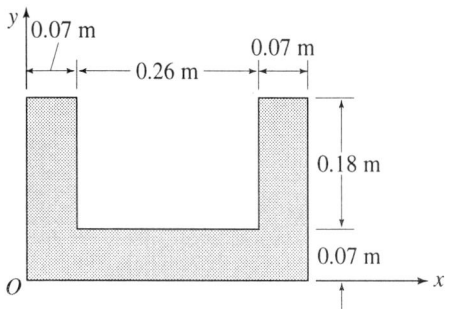

FIGURE P11.49.

11.50 Refer to the composite area shown in Figure P11.49, and compute the moment of inertia I_y and the radius of gyration k_y.

11.51 Compute the moment of inertia I_x and the radius of gyration k_x for the composite area shown in Figure P11.51.

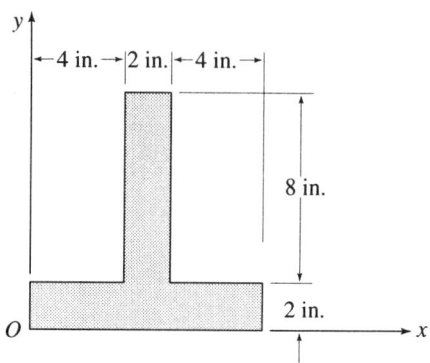

FIGURE P11.51.

11.52 Refer to the composite area shown in Figure P11.51, and find the moment of inertia I_y and the radius of gyration k_y.

11.53 Determine the centroidal moment of inertia I_X and the centroidal radius of gyration k_X for the composite area shown in Figure P11.53.

11.54 Refer to the composite area shown in Figure P11.53, and find the centroidal moment of inertia I_Y and the centroidal radius of gyration k_Y.

11.55 Compute the centroidal moment of inertia I_X and the centroidal radius of gyration k_X for the composite area shown in Figure P11.55.

11.56 Refer to the composite area shown in Figure P11.55, and determine the centroidal moment of inertia I_Y and the centroidal radius of gyration k_Y.

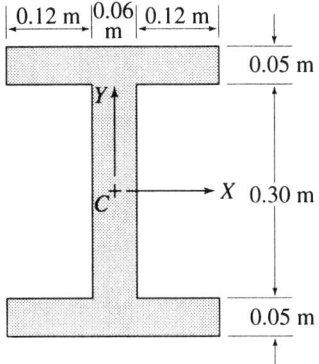

|← 0.12 m →|← 0.06 m →|← 0.12 m →|

0.05 m

Y

C + → X   0.30 m

0.05 m

FIGURE P11.53.

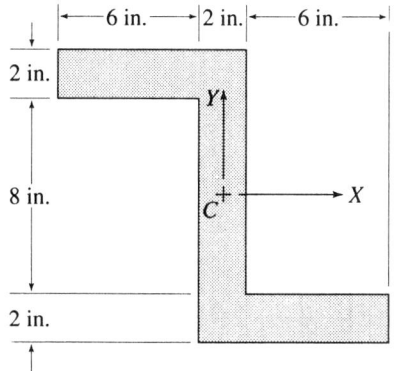

|← 6 in. →|← 2 in. →|← 6 in. →|

2 in.

Y

8 in.

C + → X

2 in.

FIGURE P11.55.

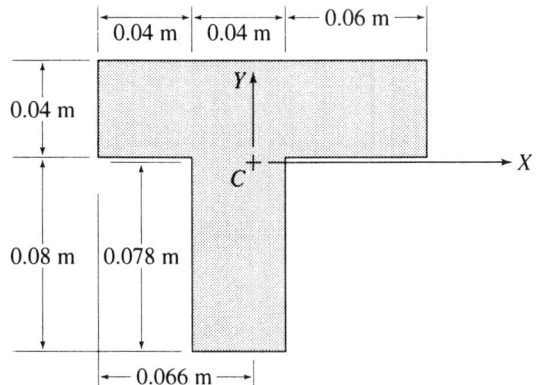

|← 0.04 m →|← 0.04 m →|← 0.06 m →|

0.04 m

Y

C + → X

0.08 m   0.078 m

|← 0.066 m →|

FIGURE P11.57.

11.57 Find the centroidal moment of inertia I_X and the centroidal radius of gyration k_X for the composite area shown in Figure P11.57.

11.58 Refer to the composite area shown in Figure P11.57, and compute the centroidal moment of inertia I_Y and the centroidal radius of gyration k_Y.

11.59 In terms of D_o and D_i, determine the centroidal moments of inertia I_X and I_Y for the cross-section of a hollow cylinder shown in Figure P11.59. What is the polar moment of inertia with respect to an axis through the centroid C?

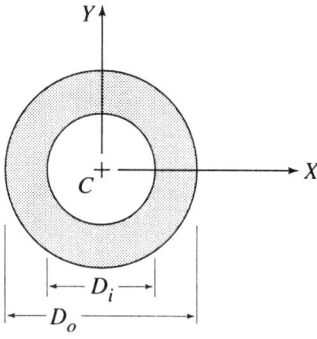

Y

C + → X

|← D_i →|

|← D_o →|

FIGURE P11.59.

11.60 In terms of the dimension a, compute the centroidal moments of inertia I_X and I_Y for the cross-sectional area shown in Figure P11.60.

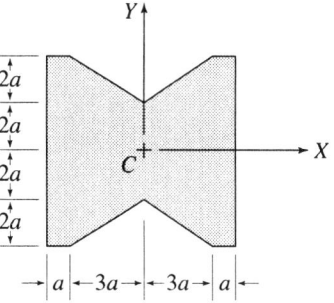

Y

2a
2a
2a
2a

C + → X

|→ a ←|← 3a →|← 3a →|→ a ←|

FIGURE P11.60.

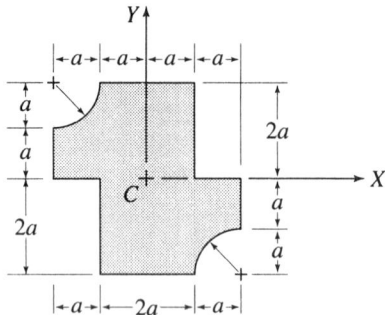

FIGURE P11.61.

11.61 In terms of the dimension a, find the centroidal moments of inertia I_X and I_Y for the cross-sectional area shown in Figure P11.61.

11.62 A column is fabricated by welding four standard L6 × 6 × 1 equal-leg angles, as shown in cross-section in Figure P11.62. From a structural design handbook, the area and centroidal moments of inertia for each angle are $A_a = 11.0$ in^2 and $(I_x)_a = (I_y)_a = 35.5$ in.^{4}, respectively. The centroid C_a of each angle is located as shown in the lower left-hand corner of Figure P11.62. Find the centroidal moments of inertia I_X and I_Y for the cross-sectional area of the column. Neglect

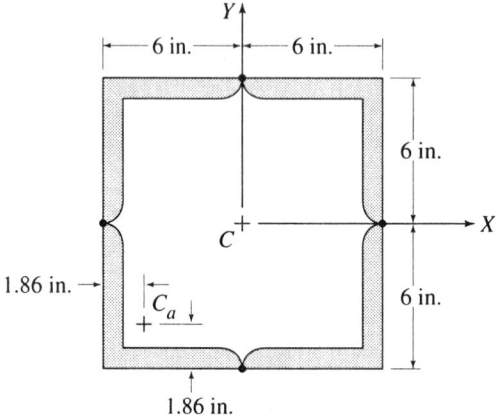

FIGURE P11.62.

the areas of the welds in your computations.

11.63 A beam is fabricated by welding two standard C10 × 30 channels, as shown in cross section in Figure P11.63. From a structural design handbook, the area and centroidal moments of inertia for each channel are $A_c = 8.82$ in.^{2} and $(I_x)_c = 3.94$ in.^{4} and $(I_y)_c = 103.0$ in.^{4}. The centroid C_c of each channel is located as shown in the lower part of Figure P11.63. Compute the centroidal moments of inertia I_X and I_Y for the cross-sectional area of the beam. Neglect the areas of the welds in your computations.

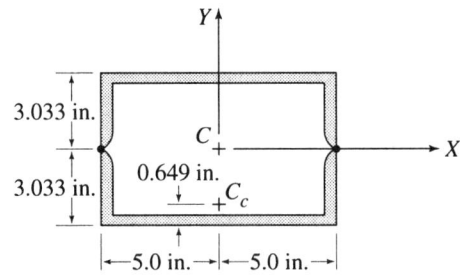

FIGURE P11.63.

11.64 A beam is fabricated by welding together four L5 × 5 × 3/4 angles to a 1 in. × 12 in. plate, as shown in Figure P11.64. From a structural design handbook, the area and centroidal moments of inertia for each angle are $A_a = 6.94$ in.^{2} and $(I_x)_a = (I_y)_a = 15.70$ in.^{4}, respectively. The centroid C_a of each angle is located as shown in the lower right-hand corner of Figure P11.64. Compute the centroidal moment of inertia I_X and the centroidal radius of gyration k_X for the beam's cross-section.

11.65 Refer to the area described in Problem 11.64, and determine the centroidal moment of inertia I_Y and the centroidal radius of gyration k_Y.

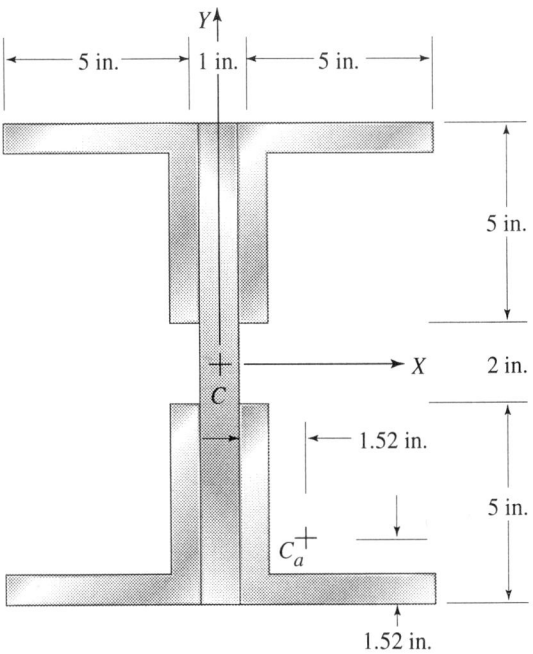

FIGURE P11.64.

11.66 An unsymmetric, composite, cross-sectional area is developed by fastening together a standard L8 × 8 × 11/2 angle and a standard L4 × 4 × 3/4 angle to a 2 in. × 16 in. rectangular plate, as shown in Figure P11.66. The necessary properties of the standard angles were obtained from a structural design handbook and are shown adjacent to these sections in Figure P11.66. Note that the given moments of inertia are with respect to the centroidal axis of the angles which are located in the figure. (a) Locate the centroid of the composite area, and establish a centroidal X-Y coordinate system parallel to the x-y systems shown in Figure P11.66. (b) Compute the moment of inertia and radius of gyration, with respect to the centroidal X axis, for the composite area.

11.67 Refer to the composite area shown in Figure P11.66, and compute the moment of inertia and radius of gyration

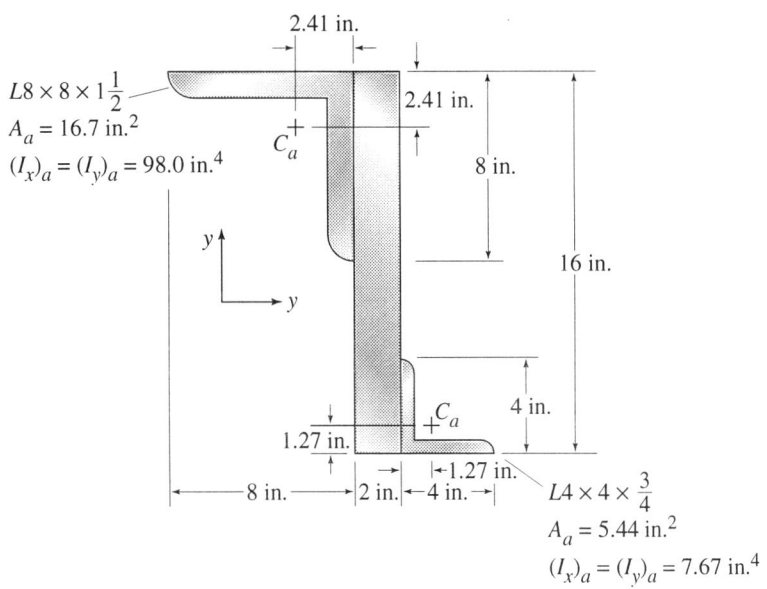

FIGURE P11.66.

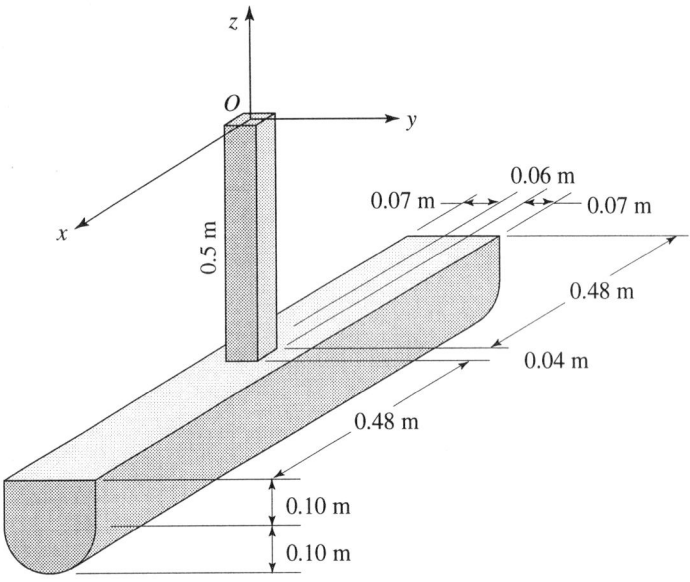

FIGURE P11.68.

with respect to the centroidal Y axis described in part (a) of Problem 11.66.

11.68 Determine the moment of inertia and radius of gyration, with respect to the x axis, for the composite mass shown in Figure P11.68. Express your answers in terms of the constant mass density ρ (kg/m^3).

11.69 Refer to the composite mass of Figure P11.68, and compute the moment of inertia and radius of gyration with respect to the y axis. Express your answers in terms of the constant mass density ρ (kg/m^3).

11.70 Refer to the composite mass of Figure P11.68, and compute the moment of inertia and radius of gyration with respect to the z axis. Express your answers in terms of the constant mass density ρ (kg/m^3).

11.71 Find the moment of inertia and radius of gyration, with respect to the x and y axes, for the composite mass shown in

Figure P11.71. Express your answers in terms of the constant mass density ρ (slug/in.^{3}).

11.72 Refer to the composite mass shown in Figure P11.71, and find the moment of inertia and radius of gyration with respect to the z axis. Express your answers in terms of the constant mass density ρ (slug/in.^{3}).

11.73 Compute the moment of inertia and radius of gyration of the aluminum flywheel, shown in Figure P11.73, with respect to the centroidal Z axis. The mass density for aluminum is approximately 5.2 slug/ft^3.

11.74 Refer to the flywheel shown in Figure P11.73, and compute the moments of inertia and radii of gyration with respect to the centroidal X and Y axes.

11.75 The composite body shown in Figure P11.75 consists of two thin aluminum plates OABC and OCDE and five steel slender roads DG, BG, GF, EF, and

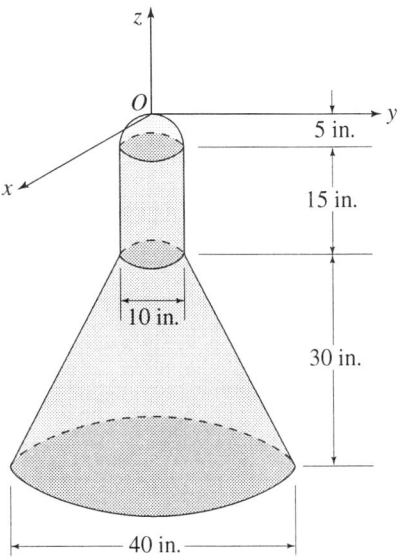

FIGURE P11.71.

AF. The mass density for the aluminum plates is 27 kg/m² and that for the steel rods is 0.6 kg/m. Determine the mass moment of inertia with respect to the x axis.

11.76 Refer to the composite body described in Problem 11.75, and determine the mass moment of inertia and radius of gyration with respect to the y axis. Compute these quantities with respect to the z axis.

11.77 A machine member made of steel is shown in Figure P11.77. The mass density for steel is approximately 15 slug/ft³. Determine the mass moment of inertia and radius of gyration with respect to the x axis.

11.78 Refer to the machine member described in Problem 11.77, and determine the mass moment of inertia and radius of gyration with respect to the y axis.

11.79 Refer to the machine member described in Problem 11.77, and compute the mass moment of inertia and radius of gyration with respect to the z axis.

11.80 A composite member made of aluminum is shown in Figure P11.80. The

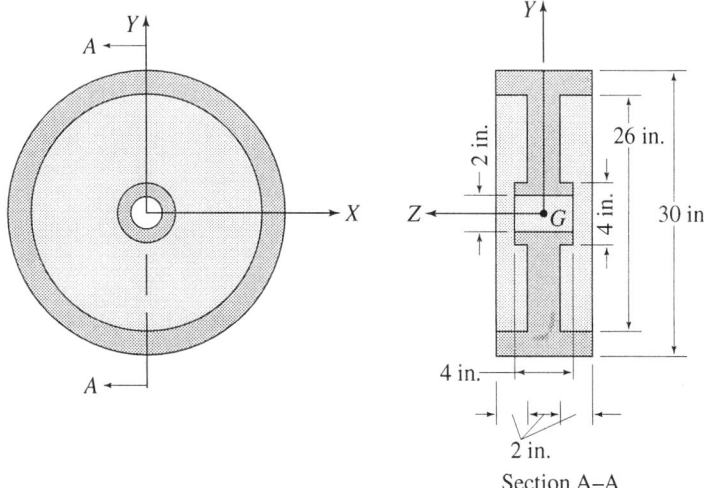

Section A–A

FIGURE P11.73.

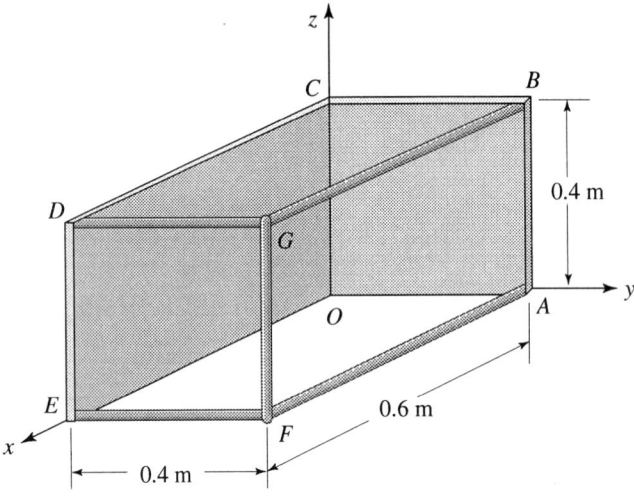

FIGURE P11.75.

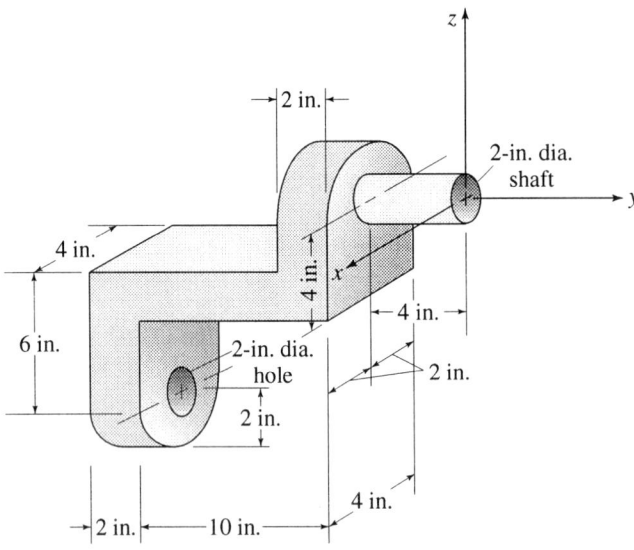

FIGURE P11.77.

mass density for aluminum is approximately 2700 kg/m³. Determine the mass moment of inertia and radius of gyration with respect to the x axis.

11.81 Refer to the composite member described in Problem 11.80, and determine the mass moment of inertial and radius of gyration with respect to the y axis.

11.82 Refer to the composite member described in Problem 11.80, and determine the mass moment of inertia and radius of gyration with respect to the z axis.

11.83 Four slender rods are attached to a thin disk as shown in Figure P11.83. Let the mass of the longest rod be m_r and that for the disk be m_d, and determine the

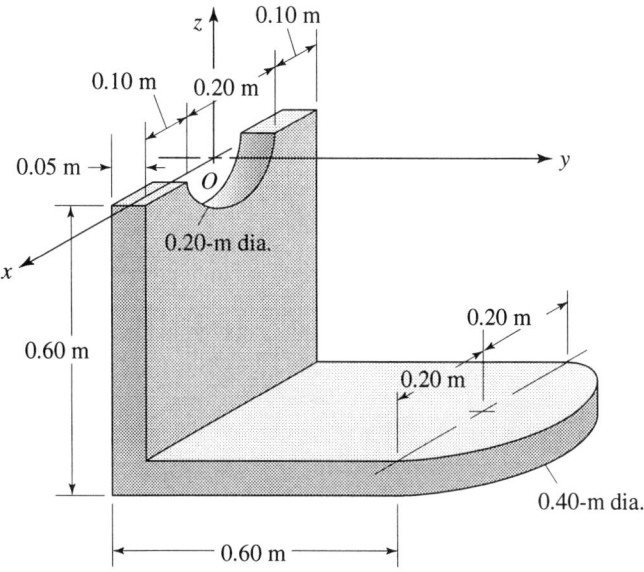

FIGURE P11.80.

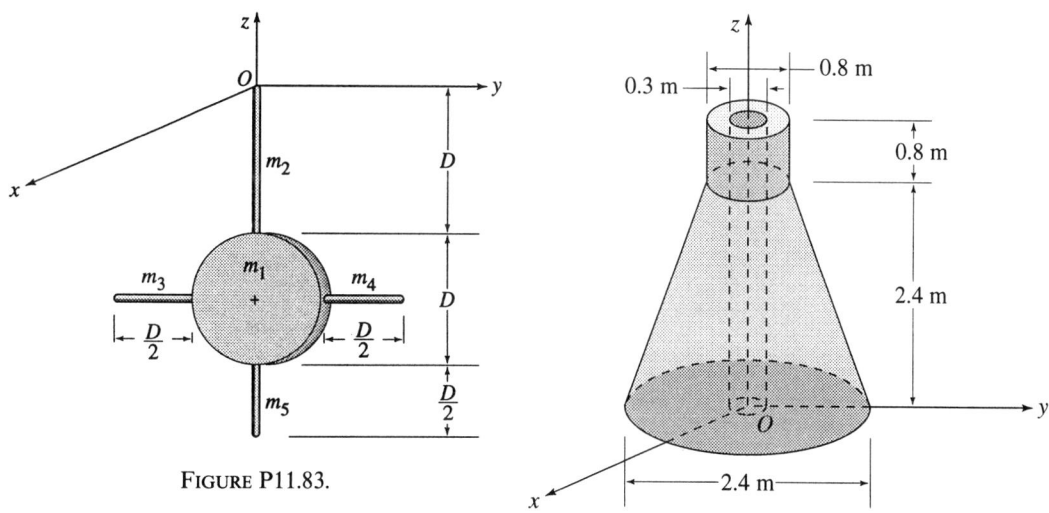

FIGURE P11.83.

FIGURE P11.85.

moment of inertia of the composite mass with respect to the y axis. Express your answer in terms of m_r, m_d, and D.

11.84 Refer to the composite mass described in Problem 11.83, and determine the mass moment of inertia with respect to the z axis. Express your answer in terms of m_r, m_d, and D.

11.85 Determine the moment of inertia of the composite mass shown in Figure P11.85 with respect to the z axis. What is the radius of gyration with respect to this axis? The material has a mass density $\rho = 5000 \text{ kg/m}^3$.

11.5*
Area Product
of Inertia

In solving many structural problems, it becomes necessary to locate the principal axes of inertia for a given cross-sectional area. These are the axes in the plane of the area that define the maximum and minimum moments of inertia. As shown in the next section, the mathematical relationships leading to the determination of the principal axes of inertia contain the integral $\int xy\,dA$ which is given the name *product of inertia* and, in this text, assigned the symbol I_{xy}. Thus, by definition,

$$I_{xy} = \int xy\,dA. \qquad (11.24)$$

The equation represents a mathematical property of an area with respect to a specific set of perpendicular axes, as illustrated in Figure 11.8. Examination of Eq. (11.24) along with Figure 11.8 reveals that the magnitude and sign of the product of inertia for a given area depend upon the location of the perpendicular axes relative to the area. Thus, unlike the moments of inertia I_x and I_y, which can have only positive values, the product of inertia I_{xy} may have positive, zero, or negative values. Also, from the mathematical definition given in Eq. (11.24), the product of inertia, like moments of inertia, has units of length raised to the fourth power. Thus, in the U.S. Customary system such units as in.^{4} and ft^4 are common. In the SI system, the most commonly used units are mm^4 and m^4.

The fundamental definition of the product of inertia I_{xy} given in Eq. (11.24) leads us to conclude that, *if either x or y is an axis of symmetry for the area, the product of inertia I_{xy} is zero*. This conclusion follows from the fact that, for every element of area dA on one side of the axis of symmetry, there is an equal element of area dA on the other side, canceling its effect. As we shall see shortly, this property is very useful in determining products of inertia of composite areas.

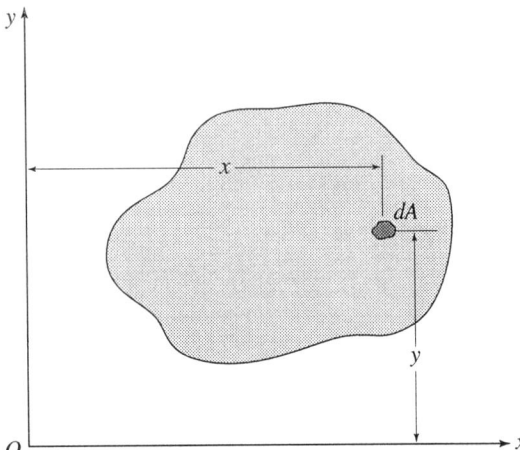

FIGURE 11.8.

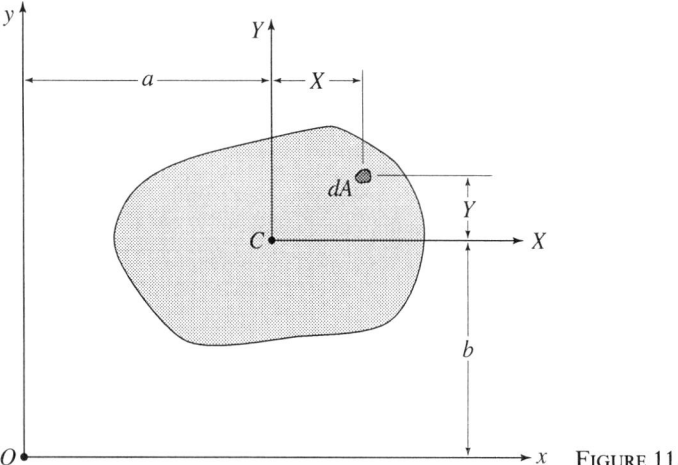

FIGURE 11.9.

As in the case of moments of inertia, a parallel-axis theorem may be developed for products of inertia. Thus, consider the area shown in Figure 11.9 where the X and Y axes are centroidal axes and x and y are any axes parallel to X and Y. Thus, $x = X + a$, $y = Y + b$, and, by Eq. (11.24),

$$I_{xy} = \int xy \, dA = \int (X + a)(Y + b) \, dA$$

$$= \int XY \, dA + b \int X \, dA + a \int Y \, dA + ab \int dA. \quad (11.25)$$

The first integral in Eq. (11.25) is I_{XY}, the second and third integrals represent first moments of the area with respect to centroidal axes and, therefore, must each be zero, and the fourth integral is, of course, the area A. Thus, Eq. (11.25) becomes

$$I_{xy} = I_{XY} + abA \quad (11.26)$$

which represents the parallel axis theorem for products of inertia. In combination with the property of the zero product of inertia for axes of symmetry, this theorem is useful in determining products of inertia of simple and composite areas.

The following examples show how products of inertia may be determined for simple and composite areas.

■ **Example 11.13**   Determine the product of inertia I_{xy} of the shaded area shown in Figure E11.13.

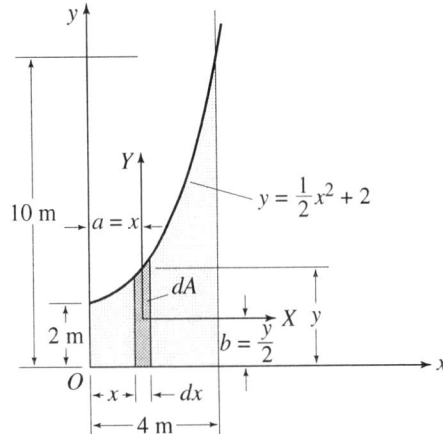

FIGURE E11.13.

Solution

Consider the differential element of area $dA = y\, dx$, shown in Figure E11.13. The product of inertia I_{xy} may be found by summing the differential products of inertia dI_{xy} for all differential elements of area dA in the entire area. Thus,

$$I_{xy} = \int dI_{xy} \qquad\text{(a)}$$

where the differential quantity dI_{xy} may be found by the parallel-axis theorem expressed in Eq. (11.26). Therefore,

$$dI_{xy} = dI_{XY} + ab\, dA \qquad\text{(b)}$$

where $dI_{XY} = 0$ because it is the differential product of inertia of dA with respect to the symmetric, centroidal X and Y axes. As shown in Figure E11.13, the quantities a and b locating the centroid of dA from the x-y coordinate system are equal to x and $y/2$, respectively. Therefore, Eq. (b) becomes

$$dI_{xy} = x\left(\frac{y}{2}\right)dA = \frac{1}{2}xy^2\, dx. \qquad\text{(c)}$$

Because $y = \frac{1}{2}x^2 + 2$ as given in Figure E11.13, Eq. (c) becomes

$$dI_{xy} = \frac{1}{2}x\left(\frac{1}{4}x^4 + 2x^2 + 4\right)dx = \frac{1}{8}x^5 + x^3 + 2x. \qquad\text{(d)}$$

Substituting Eq. (d) in Eq. (a) yields

$$I_{xy} = \int_0^4 \left(\frac{1}{8}x^5 + x^3 + 2x\right)dx = \frac{1}{48}x^6 + \frac{1}{4}x^4 + x^2 \Big]_0^4$$

$$= 165.3 \text{ m}^4. \qquad\qquad\qquad\text{ANS.}$$

■ **Example 11.14**   (a) Compute the product of inertia of the composite area, shown in Figure E11.14(a), with respect to the x-y coordinate system.

(b) Determine the product of inertia of the composite area of Figure E11.14(a) with respect to centroidal X and Y axes parallel, respectively, to the given x and y axes.

Solution   (a) The composite area may be broken down into the two simple rectangular areas $A_1 = 3$ in.^{2} and $A_2 = 9$ in.^{2}, as shown in Figure E11.14(a). Through the centroids C_1 and C_2 of areas A_1 and A_2 establish subscripted X-Y coordinate systems whose axes are parallel to the given x and y axes, as shown in Figure E11.14(a).

The basic definition of the product of inertia may be written in the form

$$I_{xy} = \int dI_{xy} \tag{a}$$

which is applicable for an infinite number of differential areas dA. In the case of a composite area consisting of a finite number of simple areas, the process of integration expressed in Eq. (a) becomes a process

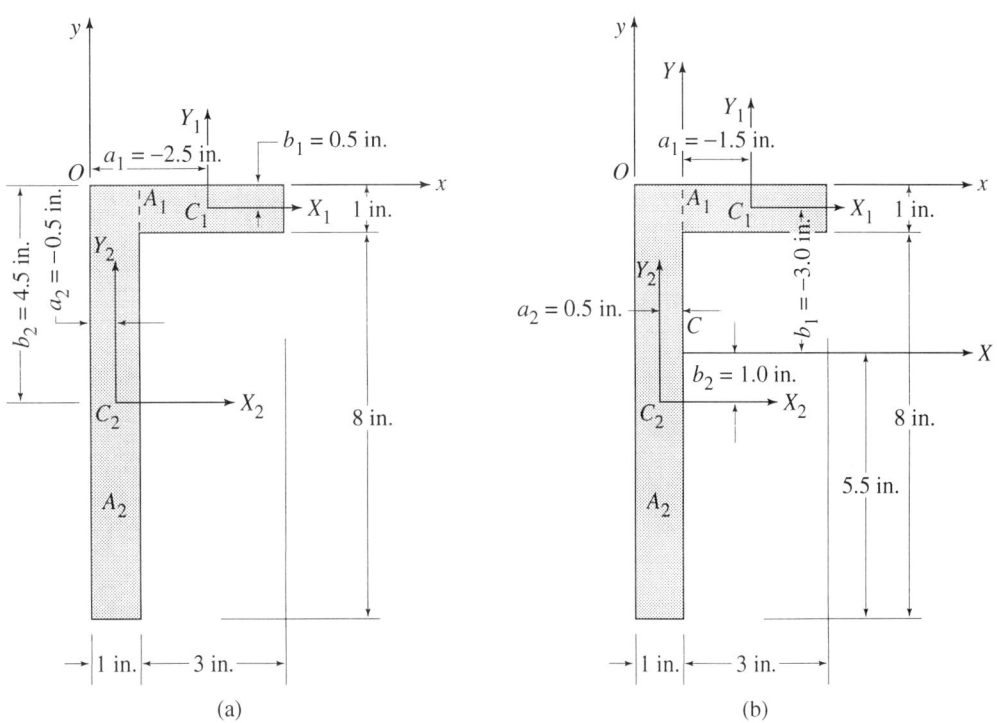

(a)   (b)

FIGURE E11.14.

of summation. Thus, for the composite area of Figure E11.14(a),

$$I_{xy} = \sum_{i=1}^{n} (I_{xy})_i. \tag{b}$$

In this particular case, because $n = 2$, Eq. (b) may be written as

$$I_{xy} = (I_{xy})_1 + (I_{xy})_2 \tag{c}$$

where $(I_{xy})_1$ and $(I_{xy})_2$ are determined by the parallel-axis theorem, Eq. (11.26). Thus,

$$(I_{xy})_1 = I_{X_1 Y_1} + (a_1)(b_1)A_1. \tag{d}$$

Because X_1 and Y_1 are axes of symmetry, $I_{X_1 Y_1} = 0$. Also, the distances a_1 and b_1 locate the centroid C_1 with respect to the given x and y axes, as shown in Figure E11.14(a). Thus Eq. (d) becomes

$$(I_{xy})_1 = 0 + (-2.5)(0.5)(3) = -3.75 \text{ in.}^4.$$

Similarly,

$$(I_{xy})_2 = I_{X_2 Y_2} + (a_2)(b_2)A_2 = 0 + (-0.5)(4.5)(9) = -20.25 \text{ in.}^4.$$

Hence, Eq. (c) yields

$$I_{xy} = -3.75 - 20.25 = -24.0 \text{ in.}^4. \qquad \text{ANS.}$$

(b) The centroid C of the composite area is found and located as shown in Figure E11.14(b). An X-Y coordinate system is established using C as an origin. As in part (a), the composite area is decomposed into the two simple rectangular areas A_1 and A_2 and, as before,

$$I_{XY} = \sum_{i=1}^{n} (I_{XY})_i = (I_{XY})_1 + (I_{XY})_2 \tag{e}$$

where $(I_{XY})_1$ and $(I_{XY})_2$ are found by the parallel-axis theorem. Thus, proceeding as in part (a),

$$(I_{XY})_1 = I_{X_1 Y_1} + (a_1)(b_1)A_1 = 0 + (-1.5)(-3.0)(3) = 13.5 \text{ in.}^4,$$

and

$$(I_{XY})_2 = I_{X_2 Y_2} + (a_2)(b_2)A_2 = 0 + (0.5)(1.0)(9) = 4.5 \text{ in.}^4.$$

Therefore, by Eq. (e),

$$I_{XY} = 13.5 + 4.5 = 18.0 \text{ in.}^4. \qquad \text{ANS.}$$

Alternatively, once the product of inertia I_{xy} for the composite area is determined, the product of inertia I_{XY} for the composite area may be found directly from the parallel-axis theorem. Thus, by Eq. (11.26),

$$I_{XY} = I_{xy} - abA = -24.0 - (1.0)(-3.5)(12) = 18.0 \text{ in.}^4. \quad \text{ANS.}$$

■

Problems

11.86 (a) Compute the product of inertia I_{xy} for the area shown in Figure P11.86. (b) Use both the results of part (a) and the parallel-axis theorem to find the product of inertia I_{XY} with respect to centroidal X and Y axes.

11.87 (a) Compute the product of inertia I_{xy} for the area shown in Figure P11.87. (b)

Use both the results of part (a) and the parallel-axis theorem to find the product of inertia I_{XY} with respect to centroidal X and Y axes. Express your answers in terms of a.

11.88 (a) Compute the product of inertia I_{xy} for the area shown in Figure P11.88. (b) Use both the results of part (a) and the parallel axis theorem to find the product of inertia I_{XY} with respect to centroidal X and Y axes. Express your answers in terms of R.

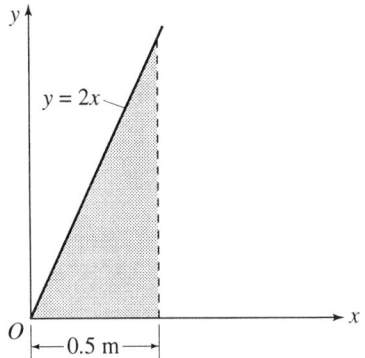

FIGURE P11.86.

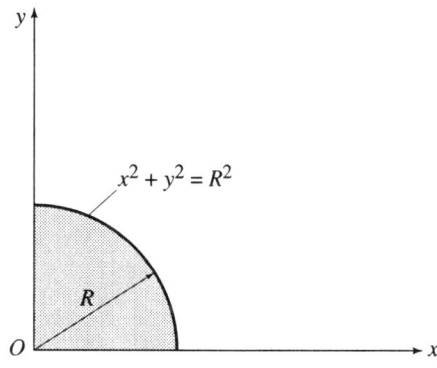

FIGURE P11.88.

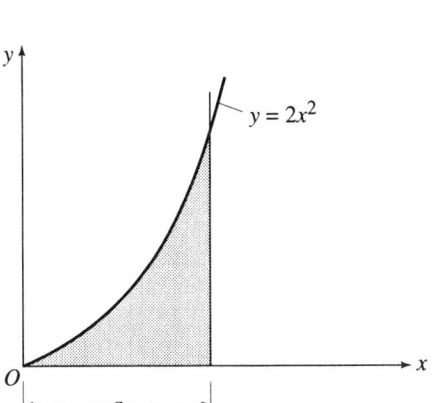

FIGURE P11.87.

11.89 (a) Compute the product of inertia I_{xy} for the area shown in Figure P11.89. (b) Use both the results of part (a) and the parallel-axis theorem to obtain the product of inertia I_{XY} with respect to centroidal X and Y axes.

11.90 (a) Compute the product of inertia I_{xy} for the area shown in Figure P11.90. (b) Use both the results of part (a) and the parallel-axis theorem to obtain the product of inertia I_{XY} with respect to centroidal X and Y axes.

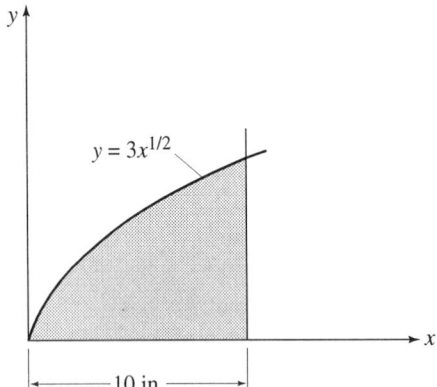

FIGURE P11.89.

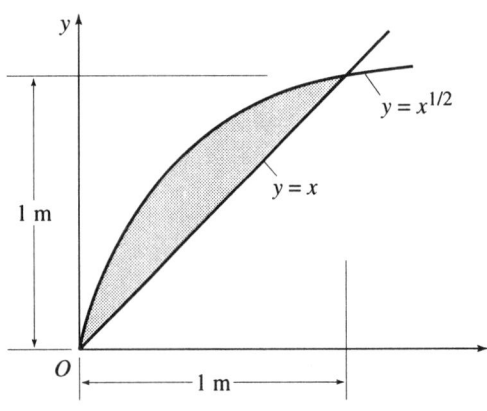

FIGURE P11.90.

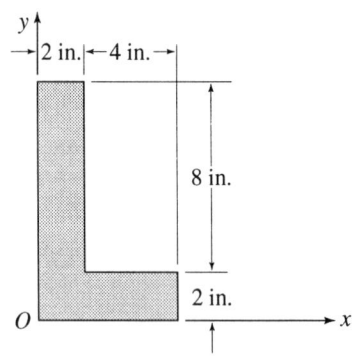

FIGURE P11.91.

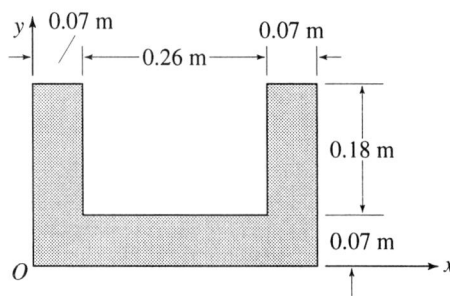

FIGURE P11.92.

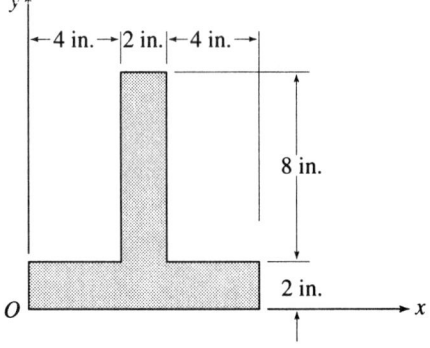

FIGURE P11.93.

11.91 (a) Determine the product of inertia I_{xy} for the composite area shown in Figure P11.91. (b) Use the results of part (a) and the parallel-axis theorem to find the product of inertia I_{XY} with respect to centroidal X and Y axes.

11.92 Compute the product of inertia I_{xy} for the composite area shown in Figure P11.92. What is the product of inertia with respect to centroidal X and Y axes?

11.93 Determine the product of inertia I_{xy} for the composite area shown in Figure P11.93. What is the product of inertia with respect to centroidal X and Y axes?

11.94 Find the product of inertia I_{XY} for the composite area shown in Figure P11.94 with respect to the centroidal X and Y axes.

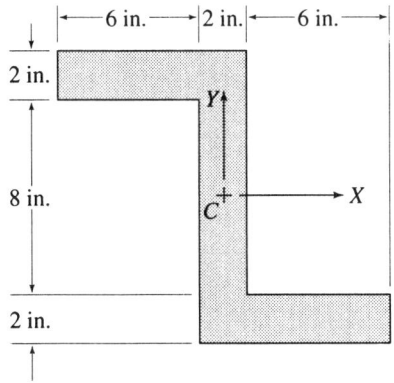

FIGURE P11.94.

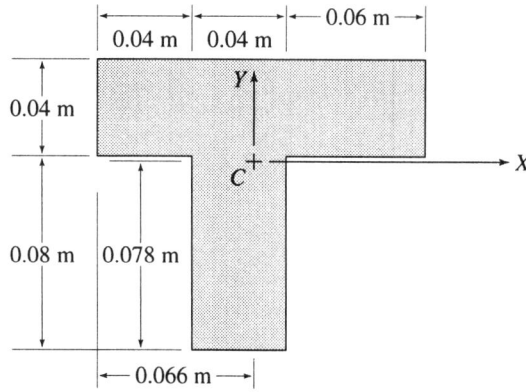

FIGURE P11.95.

11.95 Compute the product of inertia I_{XY} for the composite area shown in Figure P11.95 with respect to the centroidal X and Y axes.

11.6*
Area
Principal
Axes and
Principal
Moments of
Inertia

As stated in a previous section, the solution of many structural problems requires that we determine the *principal axes of inertia* for a given area. These are axes in the plane of the area that define the maximum and minimum moments of inertia, known as *principal moments of inertia*. Determining the principal axes and principal moments of inertia for a given area may be accomplished by expressing the moment of inertia of the area with respect to an arbitrary axis.

Let us consider the moment of inertia of an area with respect to an arbitrary axis n as shown in Figure 11.10. The axis n is inclined to the axis x through the angle α which is positive when measured in the ccw

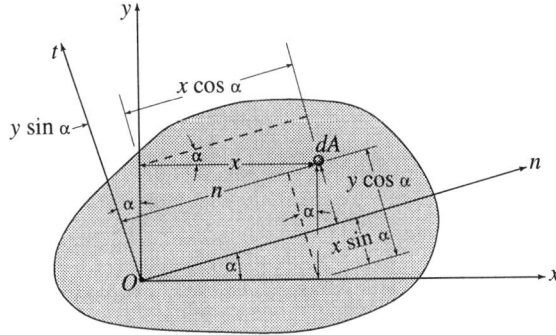

FIGURE 11.10.

direction. Note that the axis t is also inclined to the axis y through the same angle α and, therefore, the n and t axes are perpendicular to each other. From the basic definition of area moment of inertia, Eq. 11.1,

$$I_n = \int t^2 \, dA \tag{11.27}$$

From the geometry of Figure 11.10, we conclude that

and
$$\left. \begin{array}{l} t = y \cos \alpha - x \sin \alpha \\ n = x \cos \alpha + y \sin \alpha. \end{array} \right\} \tag{11.28}$$

Thus,

$$I_n = \int (y \cos \alpha - x \sin \alpha)^2 \, dA$$

$$= \cos^2 \alpha \int y^2 \, dA + \sin^2 \alpha \int x^2 \, dA - 2 \sin \alpha \cos \alpha \int xy \, dA$$

$$= I_x \cos^2 \alpha + I_y \sin^2 \alpha - 2I_{xy} \sin \alpha \cos \alpha \tag{11.29}$$

where the integrals $\int y^2 \, dA$, $\int x^2 \, dA$ and $\int xy \, dA$ were replaced by their equivalents I_x, I_y, and I_{xy}, respectively. Using the trigonometric identities $\sin^2 \alpha = \dfrac{(1 - \cos 2\alpha)}{2}$, $\cos^2 \alpha = \dfrac{(1 + \cos 2\alpha)}{2}$, and $2 \sin \alpha \cos \alpha = \sin 2\alpha$, we rewrite Eq. (11.29) in the form

$$I_n = \tfrac{1}{2}(I_x + I_y) + \tfrac{1}{2}(I_x - I_y) \cos 2\alpha - I_{xy} \sin 2\alpha. \tag{11.30}$$

The area moment of inertia I_t may be obtained similarly or more directly from Eq. (11.30) by replacing α by the quantity $\alpha + 90°$. Thus,

$$I_t = \tfrac{1}{2}(I_x + I_y) - \tfrac{1}{2}(I_x - I_y) \cos 2\alpha + I_{xy} \sin 2\alpha. \tag{11.31}$$

Now let us consider the area product of inertia with respect to the orthogonal n and t axes. From Eq. (11.24) and the geometry expressed in Eqs. (11.28),

$$I_{nt} = \int nt \, dA = \int (x \cos \alpha + y \sin \alpha)(y \cos \alpha - x \sin \alpha) \, dA$$

$$= \sin \alpha \cos \alpha \left[\int y^2 \, dA - \int x^2 \, dA \right]$$

$$+ (\cos^2 \alpha - \sin^2 \alpha) \int xy \, dA$$

$$= (I_x - I_y) \sin \alpha \cos \alpha + I_{xy}(\cos^2 \alpha - \sin^2 \alpha). \tag{11.32}$$

Using the trigonometric relationships $\sin \alpha \cos \alpha = (\sin 2\alpha)/2$ and

$\cos^2 \alpha - \sin^2 \alpha = \cos 2\alpha$, we rewrite Eq. (11.32) as

$$I_{nt} = \tfrac{1}{2}(I_x - I_y)\sin 2\alpha + I_{xy}\cos 2\alpha. \qquad (11.33)$$

Differentiating I_n with respect to α in Eq. (11.30) and setting the result equal to zero, we obtain the values of α for which I_n is either a maximum or a minimum. Thus,

$$2\alpha = \tan^{-1}\left[\frac{-I_{xy}}{\tfrac{1}{2}(I_x - I_y)} \right] \qquad (11.34)$$

which defines two values of 2α differing by $180°$ or two values of α differing by $90°$. Corresponding to one of these two values of α, the moment of inertia I_n is a maximum, and, corresponding to the second, I_n is a minimum. These maximum and minimum moments of inertia, denoted in this text, respectively, as I_u and I_v, are known as the *principal moments of inertia* of area. The corresponding u and v axes are known as the *principal axes of inertia*. Principal moments and principal axes of inertia are significant concepts in beam and column theories.

Without loss of generality, the magnitude of I_x may be assumed larger than that of I_y. In such a case, Eq. (11.34) defines angles whose tangents are negative. Such angles occur in the second and fourth quadrants as shown in Figure 11.11 where, by the Pythagorean theorem,

$$R = \sqrt{[(I_x - I_y)/2]^2 + I_{xy}^2}. \qquad (11.35)$$

Also, from the geometry of Figure 11.11,

and
$$\left.\begin{array}{l} \sin 2\alpha = \pm I_{xy}/R, \\[2mm] \cos 2\alpha = \mp(I_x - I_y)/2R \end{array}\right\} \qquad (11.36)$$

where the upper signs correspond to the angle in the second quadrant and the lower signs to the angle in the fourth quadrant. Substitution of

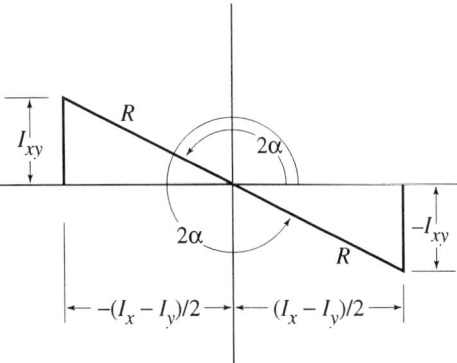

FIGURE 11.11.

these values in Eq. (11.30) yields

$$
\left.
\begin{aligned}
I_u &= I_{max} = (I_x + I_y)/2 + \sqrt{[(I_x - I_y)/2]^2 + I_{xy}^2} \\
\text{and} \quad I_v &= I_{min} = (I_x + I_y)/2 - \sqrt{[(I_x - I_y)/2]^2 + I_{xy}^2}.
\end{aligned}
\right\} \quad (11.37)
$$

Also, substitution of Eqs. (11.36) in Eq. (11.33) yields a zero value for the product of inertia which means that the area product of inertia is zero with respect to the principal axes of inertia i.e., $I_{uv} = 0$. Recalling that the product of inertia is zero with respect to a set of orthogonal axes if at least one of them is an axis of symmetry, we conclude that an *axis of symmetry is a principal axis of inertia*. However, this does not mean that principal axes of inertia are necessarily axes of symmetry.

Because the polar moment of inertia, with respect to an axis through point O, is equal to the sum of the moments of inertia, with respect to any two orthogonal in-plane axes intersecting at point O, we conclude that

$$
J_O = I_x + I_y = I_n + I_t = I_u + I_v. \quad (11.38)
$$

These useful relationships could also have been obtained by separately adding Eq. (11.30) to Eq. (11.31) and, then, adding the two relationships in Eq. (11.37).

■ Example 11.15

Determine the principal centroidal axes and the principal centroidal moments of inertia for the cross-sectional area shown in Figure E11.15.

Solution

The centroid C is located and centroidal X and Y axes are established as shown in Figure E11.15. The moments of inertia I_X and I_Y and the product of inertia I_{XY}, determined by the procedures discussed in preceding sections, are

$$
I_X = 10.944 \times 10^{-5} \text{ m}^4,
$$

$$
I_Y = 1.216 \times 10^{-5} \text{ m}^4,
$$

and

$$
I_{XY} = -1.792 \times 10^{-5} \text{ m}^4.
$$

By Eq. (11.34),

$$
2\alpha = \tan^{-1}\left[\frac{-I_{xy}}{\frac{1}{2}(I_x - I_y)}\right] = \tan^{-1}\left[\frac{-(-1.792 \times 10^{-5})}{\frac{1}{2}(10.944 - 1.216)(10^{-5})}\right]
$$

$$
= \tan^{-1}(0.3684)
$$

Thus, the two sets of values of 2α are

$$
2\alpha_1 = 20.2°, \qquad \alpha_1 = 10.1° \text{ defining the U axis,}
$$

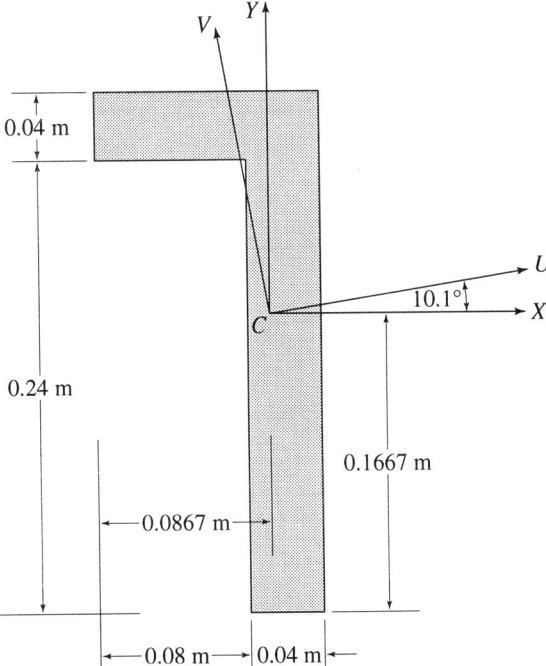

0.04 m

V Y

U

10.1° X

C

0.24 m

0.1667 m

0.0867 m

0.08 m   0.04 m

FIGURE E11.15.

and

$$2\alpha_2 = 200.2°, \qquad \alpha_1 = 100.1° \text{ defining the V axis.}$$

The U and V axes are shown properly oriented with respect to the centroidal X and Y axes. By Eqs. (11.37),

$$I_U = I_{max} = (I_x + I_y)/2 + \sqrt{[(I_x - I_y)/2]^2 + I_{xy}^2}$$
$$= \{(10.944 + 1.216)/2$$
$$+ \sqrt{[(10.944 - 1.216)/2]^2 + (1.792)^2}\} \times 10^{-5}$$
$$= 11.26 \times 10^{-5} \text{ m}^4, \qquad\qquad \text{ANS.}$$

and

$$I_V = I_{min} = (I_x + I_y)/2 - \sqrt{[(I_x - I_y)/2]^2 + I_{xy}^2}$$
$$= \{(10.944 + 1.216)/2$$
$$- \sqrt{[(10.944 - 1.216)/2]^2 + (1.792)^2}\} \times 10^{-5}$$
$$= 0.896 \times 10^{-5} \text{ m}^4. \qquad\qquad \text{ANS.}$$

■ **Example 11.16**   Consider the cross-sectional area in Figure E11.16 which was analyzed in Example 11.15 and where the following properties were determined:

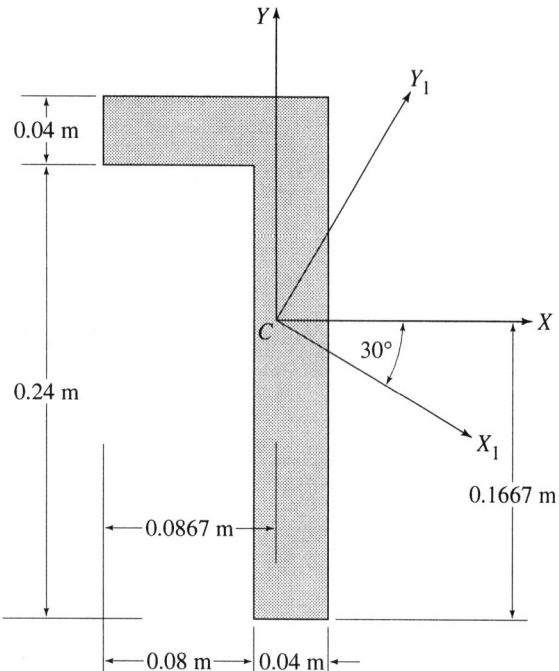

FIGURE E11.16.

$$I_X = 10.944 \times 10^{-5} \text{ m}^4,$$

$$I_Y = 1.216 \times 10^{-5} \text{ m}^4,$$

and

$$I_{XY} = -1.792 \times 10^{-5} \text{ m}^4.$$

Determine (a) the moment of inertia with respect to a centroidal axis oriented 30° cw from the X axis and (b) the product of inertia with respect to a set of orthogonal axes one of which is defined in part (a).

Solution

(a) Locate an X_1 axis 30° cw from the X axis, as shown in Figure E11.16. By Eq. (11.30),

$$I_{X_1} = \tfrac{1}{2}(I_X + I_Y) + \tfrac{1}{2}(I_X - I_Y)\cos 2\alpha - I_{XY}\sin 2\alpha$$

$$I_{X_1} = \{\tfrac{1}{2}(10.944 + 1.216) + \tfrac{1}{2}(10.944 - 1.216)\cos 2(-30°)$$

$$- (-1.792)\sin 2(-30°)\} \times 10^{-5}$$

$$= 6.96 \times 10^{-5} \text{ m}^4. \qquad\qquad \text{ANS.}$$

(b) The centroidal Y_1 axis is located perpendicular to the X_1 axis, as shown in Figure E11.16. By Eq. (11.33).

$$I_{X_1Y_1} = \tfrac{1}{2}(I_X - I_Y)\sin 2\alpha + I_{XY}\cos 2\alpha$$

$$= \{\tfrac{1}{2}(10.944 - 1.216)\sin 2(-30°)$$

$$+ (-1.792)\cos 2(-30°)\} \times 10^{-5}$$

$$= -5.11 \times 10^{-5} \text{ m}^4. \qquad\qquad \text{ANS.}$$

11.7*
Mohr's Circle for Area Moments and Products of Inertia

Equations (11.30) to (11.33) developed in the preceding section may be used to determine the values of I_n, I_t, and I_{nt} for any arbitrary $n\text{-}t$ coordinate system defined by the angle α from the $x\text{-}y$ coordinate system. Also, Eqs. (11.34) to (11.37) may be used to find the principal axes and principal moments of inertia. These determinations may be accomplished by direct substitution, as illustrated in Examples 11.15 and 11.16 or by a semigraphical method, named the Mohr's circle solution, after the German engineer Otto Mohr (1835–1918) who first introduced it. *Mohr's* circle solution is developed in the following paragraphs.

If the first term in the right-hand side of Eq. (11.30) is moved to the left-hand side and the square of the resulting equation is added to the square of Eq. (11.33),

$$[I_n - \tfrac{1}{2}(I_x + I_y)]^2 + [I_{nt} - 0]^2 = [\tfrac{1}{2}(I_x - I_y)]^2 + I_{xy}^2 \quad (11.39)$$

Examination of Eq. (11.39) reveals that if we replace $[I_n - (I_x + I_y)/2]^2$ by $(x - a)^2$, $[I_{nt} - 0]^2$ by $(y - b)^2$ and $(I_x - I_y)/2]^2 + I_{xy}^2$ by R^2, we would obtain the familiar equation $(x - a)^2 + (y - b)^2 = R^2$, which is the equation of a circle in the $x\text{-}y$ plane with a radius R and a center located at (a, b). Thus, if we establish a coordinate system with the horizontal axis labeled I_n and the vertical axis labeled I_{nt}, Eq. (11.39) will be a circle whose radius is given by

$$R = \sqrt{[(I_x - I_y)/2]^2 + I_{xy}^2}. \qquad (11.40)$$

Also, the center of this circle is located at $[(I_x + I_y)/2, 0]$. In other words, the center C of this circle lies on the I_n axis at a distance from the origin O given by

$$OC = (I_x + I_y)/2 \qquad (11.41)$$

After establishing a convenient $I_n - I_{nt}$ coordinate system, Mohr's circle may be constructed by locating the center C from Eq. (11.41) and determining the radius R from Eq. (11.40). A more direct and more useful construction consists of locating two diametrically opposite points on the circumference of the circle. Thus, consider the $x\text{-}y$ coordinate system through any point O for the area shown in Figure 11.12(a).

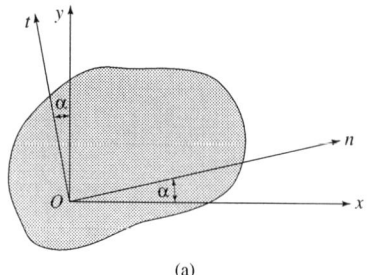

(a)

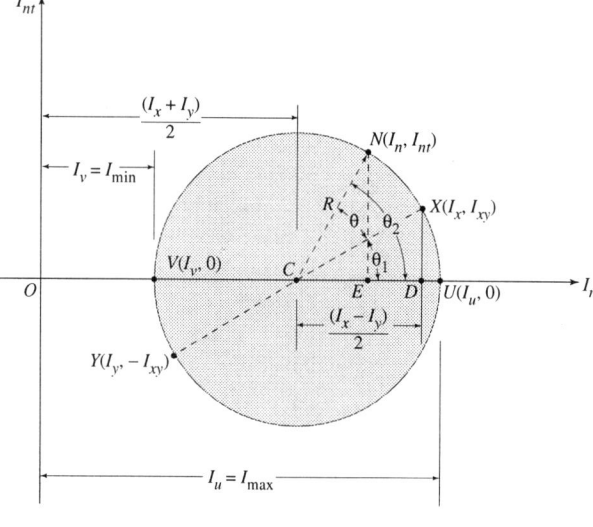

FIGURE 11.12.   (b)

Let us assume that the quantities I_x, I_y, and I_{xy} are known. The arbitrary n-t coordinate system is defined by the angle α as shown. When $\alpha = 0$, the n axis coincides with the x axis and, from Eqs. (11.30) and (11.33), we conclude that $I_n = I_x$ and $I_{nt} = I_{xy}$. These two values (i.e., I_x and I_{xy}) locate point X in Figure 11.12(b). When $\alpha = 90°$, the n axis coincides with the y axis, and, from Eqs. (11.30) and 11.33), we conclude that $I_n = I_y$ and $I_{nt} = -I_{xy}$. These two values (i.e., I_y and $-I_{xy}$) locate point Y in Figure 11.12(b). As shown in Example 11.19, points X and Y are diametrically opposite and their angular separation in Mohr's circle is 180°. Therefore, the line joining them is a diameter of Mohr's circle, and its intersection with the I_n axis locates point C, the center of Mohr's circle. Once the diameter is determined and the center located, Mohr's circle can be constructed. Usually, only a freehand sketch of the circle is sufficient because the required values are obtained *not* by measurement but by trigonometric analysis.

As stated above, points X and Y on the circumference of Mohr's circle are diametrically opposite (i.e., 180° from each other). This implies that the 90° angle between the x and y axes in the given cross-sectional area is doubled when constructing Mohr's circle. It can be shown (see Example 11.19) that *any angle in the given area is represented by twice this angle in Mohr's circle*. Thus, any axis, such as n in Figure 11.12(a) which is located by the ccw angle α from the x axis, is represented by point N located on the circle by rotating from point X through a ccw angle equal to 2α. This concept is stated symbolically by the relationship

$$\theta = 2\alpha \qquad (11.42)$$

where α is the angle in the given area and θ is the corresponding angle in Mohr's circle, as shown in Figure 11.12.

There are basically two objectives in constructing Mohr's circle for moments and products of inertia of area. The first is to locate the principal axes of inertia and to determine the corresponding principal moments of inertia for a given area. These quantities are represented by point U (maximum value) and V (minimum value) in Figure 11.12(b). The second objective is to compute the moments and products of inertia for a given area with respect to inclined axes. The fulfillment of these two objectives requires determining certain geometric properties of Mohr's circle as follows:

$$OC = (I_x + I_y)/2$$

$$R = \sqrt{(CD)^2 + (DX)^2} = \sqrt{[(I_x - I_y)/2]^2 + I_{xy}^2},$$

$$I_u = I_{max} = OC + R,$$

$$I_v = I_{min} = OC - R,$$

$$\theta_1 = \tan^{-1}\left[\frac{I_{xy}}{(I_x - I_y)/2}\right],$$

and

$$\theta_2 = \theta + \theta_1 = \tan^{-1}\left[\frac{I_{nt}}{I_n - OC}\right].$$

Note that all of these relations were obtained directly from the geometry of Mohr's circle in Figure 11.12(b) and that they are the same in essence, if not in form, as those obtained earlier in Section 11.6.

The concepts above are illustrated in the following examples.

■ **Example 11.17**

Consider the shaded area shown in Figure E11.17(a) for which the following properties are given:

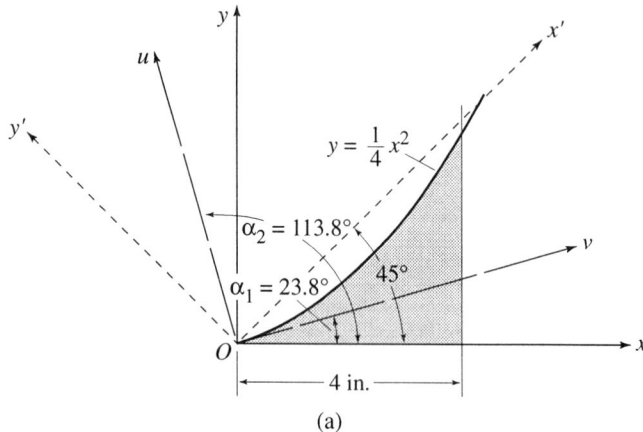

(a)

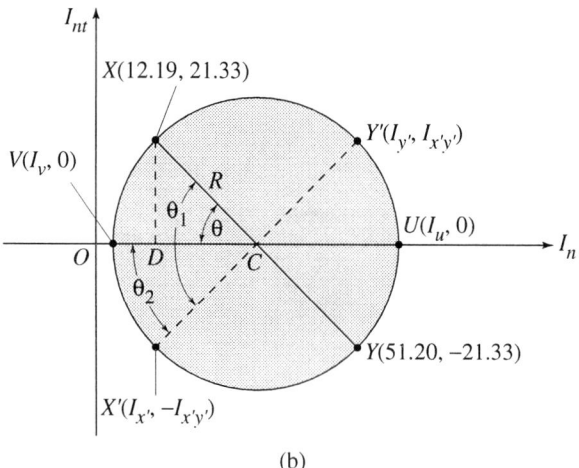

FIGURE E11.17.   (b)

$$I_x = 11.19 \text{ in.}^4,$$

$$I_y = 51.20 \text{ in.}^4,$$

and

$$I_{xy} = 21.33 \text{ in.}^4.$$

Use Mohr's circle to (a) locate the principal axes of inertia u and v passing through point O and the corresponding principal moments of inertia I_u and I_v; (b) determine the moment of inertia $I_{x'}$ and the product of inertia $I_{x'y'}$ where x' and y' are orthogonal axes in which x' is 45° ccw from the x axis as shown.

Solution

(a) Establish an $I_n = I_{nt}$ coordinate system as shown in Figure E11.17(b). Locate point X with coordinates $I_x = 11.19$ and $I_{xy} = 21.33$. Similarly, locate point Y whose coordinates are $I_y = 51.20$ and $I_{xy} = -21.33$. Connect points X and Y, locate the center C on the I_n axis, and sketch Mohr's circle as shown in Figure E11.17(b). The needed computations are carried out as follows:

$$OC = (I_x + I_y)/2 = 31.7,$$

$$CD = OC - OD = 19.51,$$

$$R = \sqrt{(CD)^2 + (DX)^2} = 28.91,$$

$$I_u = I_{max} = OC + R = 60.6 \text{ in.}^4, \qquad \text{ANS.}$$

$$I_v = I_{min} = OC - R = 2.79 \text{ in.}^4, \qquad \text{ANS.}$$

$$2\alpha = \theta = \tan^{-1}\left|\frac{DX}{DC}\right| = 47.6°,$$

$$\alpha_1 = 23.8°, \qquad \text{ANS.}$$

and

$$\alpha_2 = 23.8° + 90° = 113.8°. \qquad \text{ANS.}$$

Because point V on the circle is located ccw from point X, it follows that the v principal axis is oriented at $\alpha_1 = 23.8°$ ccw from the x axis in the given area. The u principal axis is oriented 90° from the v axis or at $\alpha_2 = 113.8°$ ccw from the x axis. These relationship are shown in Figure E11.17(a).

(b) Locate point X' on the circle representing the x' axis at an angle $\theta_1 = 2(45) = 90°$ ccw from point X as shown in Figure E11.17(b). The y' axis would be represented by point Y' which is diametrically opposite point X' on the circle. The calculations are performed as follows:

$$\theta_2 = \theta_1 - \theta = 42.4°,$$

$$I_{x'} = OC - R\cos\theta_2 = 10.35 \text{ in.}^4, \qquad \text{ANS.}$$

$$I_{x'y'} = -R\sin\theta_2 = -19.49 \text{ in.}^4. \qquad \text{ANS.}$$

■ **Example 11.18**

Refer to the composite area shown in Figure E11.18(a). The centroidal moments and product of inertia are given as follows:

$$I_x = 11.541 \times 10^{-6} \text{ m}^4,$$

$$I_y = 10.968 \times 10^{-6} \text{ m}^4,$$

and

$$I_{xy} = -7.818 \times 10^{-6} \text{ m}^4.$$

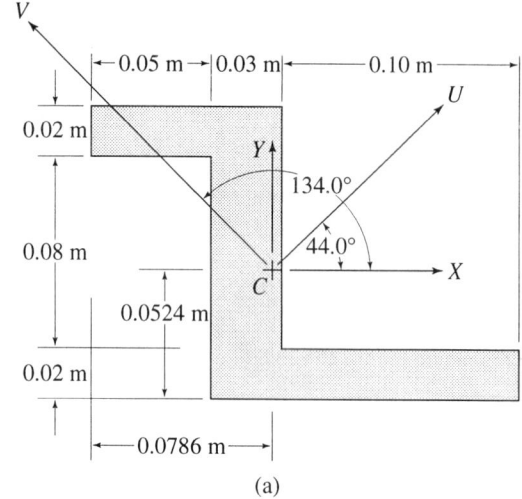

(a)

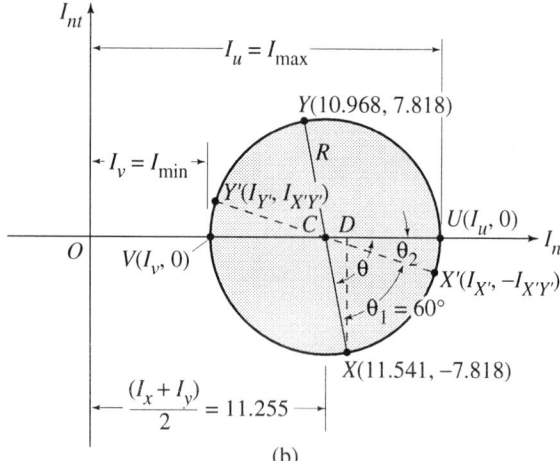

FIGURE E11.18.   (b)

Construct Mohr's circle for moments and products of inertia for this composite area and determine (a) The principal centroidal moments of inertia I_U and I_V. Locate the U and V principal centroidal axes of inertia with respect to the centroidal X and Y axes. (b) The centroidal moment of inertia $I_{X'}$ and the centroidal product of inertia $I_{X'Y'}$ corresponding to an X'-Y' (not shown) coordinate system where the X' axis is inclined $30°$ ccw to the X axis.

Solution

(a) Establish an $I_n - I_{nt}$ coordinate system, as shown in Figure E11.18(b). Locate point X whose coordinates are $I_X = 11.541 \times 10^{-6}$ and $I_{XY} = -7.818 \times 10^{-6}$. Note that the common factor 10^{-6} has been omitted from the construction. Similarly, locate point Y whose coordinates are $I_Y = 10.968 \times 10^{-6}$ and $I_{XY} = 7.818 \times 10^{-6}$, omitting the factor 10^{-6}. Connect points X and Y, locate the center C, and sketch Mohr's circle, as shown. The computations are carried out as follows:

$$OC = (I_X + I_Y)/2 = 11.255,$$

$$CD = OD - OC = 0.286,$$

$$R = \sqrt{(CD)^2 + (DX)^2} = 7.823,$$

$$I_U = I_{\max} = OC + R = 19.08 \times 10^{-6} \text{ m}^4, \qquad \text{ANS.}$$

and

$$I_V = I_{\min} = OC - R = 3.43 \times 10^{-6} \text{ m}^4, \qquad \text{ANS.}$$

$$2\alpha = \theta = \tan^{-1}\left|\frac{DX}{CD}\right| = 87.90°,$$

$$\alpha_1 = 44.0°, \qquad \text{ANS.}$$

and

$$\alpha_2 = 44.0° + 90° = 134.0°. \qquad \text{ANS.}$$

Because point U is located ccw from point X in Mohr's circle, it follows that the U principal axis is oriented at an angle $\alpha_1 = 44.0°$ ccw from the X axis in the composite area, as shown in Figure E11.18(a). Obviously the V principal axis is 90° ccw from the U axis or at $\alpha_2 = 134.0°$ ccw from the X axis.

(b) Because the X' axis is inclined 30° ccw from the X axis, point X' on Mohr's circle is located by rotating ccw from point X through the angle $\theta_1 = 2(30) = 60°$. Note that an axis Y', perpendicular to the X' axis, would be represented by point Y', diametrically opposite point X' in Mohr's circle. The coordinates of point X' provide the values of $I_{X'}$ and $I_{X'Y'}$. Thus

$$\theta_2 = \theta - \theta_1 = 27.90°,$$

$$I_{X'} = OC + R\cos\theta_2 = 18.17 \times 10^{-6} \text{ m}^4, \qquad \text{ANS.}$$

and

$$I_{X'Y'} = -R\sin\theta_2 = -3.66 \times 10^{-6} \text{ m}^4. \qquad \text{ANS.}$$

■ **Example 11.19**   Consider the arbitrary area shown in Figure E11.19(a) in relation to an x-y coordinate system. Assume that I_x, I_y, and I_{xy} are known. The ccw angle α locates an n-t coordinate system from the x-y coordinate sys-

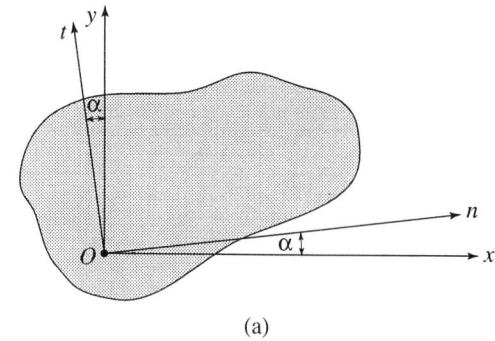

(a)

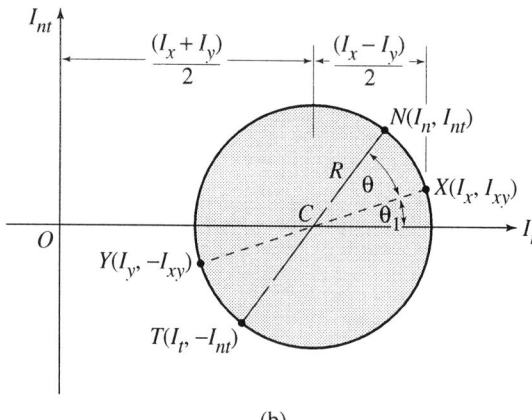

FIGURE E11.19.   (b)

tem. Construct Mohr's circle for moments and products of inertia for this area, and show that the point on Mohr's circle representing the n axis is located at an angle 2α from the point that represents the x axis. In other words, prove that any angle in the given area is represented by twice this angle in Mohr's circle.

Solution

After establishing an $I_n - I_{nt}$ coordinate system, Mohr's circle is constructed by locating point $X(I_x, I_{xy})$ and $Y(I_y, -I_{xy})$ and connecting them to establish the center of the circle, point C, as shown in Figure E11.19(b). Let point $N(I_n, I_{nt})$ be located on the circle at a ccw angle θ from point X. Point $T(I_t, -I_{nt})$ is diametrically opposite point N on the circle. From the geometry in Figure E11.19(b),

$$\sin(\theta_1 + \theta) = I_{nt}/R. \tag{a}$$

Substituting, for I_{nt} in Eq. (a), its value from Eq. (11.33),

$$\sin(\theta_1 + \theta) = \frac{\frac{1}{2}(I_x - I_y)\sin 2\alpha + I_{xy}\cos 2\alpha}{R}. \tag{b}$$

Also, from Figure E11.19(b),

$$\sin \theta_1 = I_{xy}/R \qquad \text{(c)}$$

and

$$\cos \theta_1 = (I_x - I_y)/2R \qquad \text{(d)}$$

From trigonometry,

$$\sin(\theta_1 + \theta) = \cos \theta_1 \sin \theta + \sin \theta_1 \cos \theta. \qquad \text{(e)}$$

Substitution of Eqs. (c) and (d) in Eq. (e) yields

$$\sin(\theta_1 + \theta) = \frac{\frac{1}{2}(I_x - I_y)\sin \theta + I_{xy}\cos \theta}{R} \qquad \text{(f)}$$

Therefore, by comparing Eqs. (b) and (f), we conclude that

$$\sin \theta = \sin 2\alpha$$

and

$$\cos \theta = \cos 2\alpha$$

which shows that $\theta = 2\alpha$. Thus, any angle measured in the actual area is doubled when represented in Mohr's circle. Therefore, because points X and Y in Mohr's circle represent, respectively, the x and y axes which are separated by a 90° angle, it follows that points X and Y must be located on the circle so that the angle between them is $2 \times 90° = 180°$, i.e., points X and Y must be diametrically opposite, as was stated earlier. Furthermore, because the analysis above shows that the sign of θ is the same as the sign of α, it follows that the angle θ in Mohr's circle has the same direction (cw or ccw) as the angle α in the actual area.

Problems

Use Equations to solve the following problems.

11.96 The following information is provided for the shaded area shown in Figure 11.96: $I_x = 9.222 \times 10^{-2}$ m^4 and $I_y = 4.514 \times 10^{-2}$ m^4 (a) Locate the principal axes of inertia through point O, and find the corresponding principal moments of inertia. (b) Determine the moment of inertia $I_{x'}$ and the product of inertia $I_{x'y'}$ where x' and y' are orthogonal axes.

11.97 The following information is provided for the shaded area in Figure 11.97: $I_x = 29,762$ in.^{4} and $I_{xy} = 5,208$ in.^{4} (a) Locate the principal axes of inertia through point O, and find the corresponding principal moments of inertia.

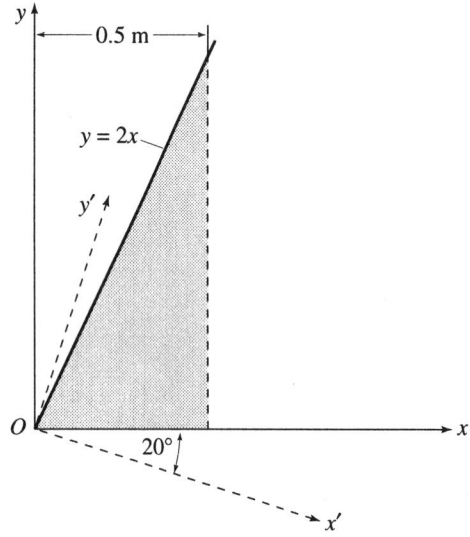

FIGURE P11.96.

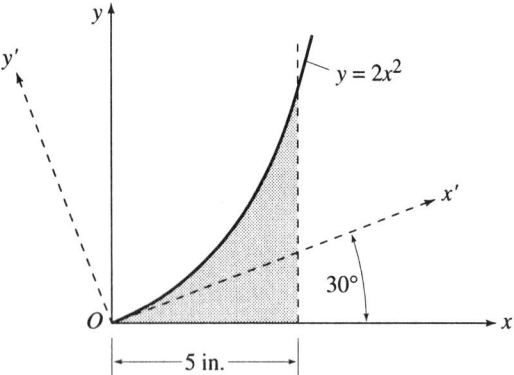

FIGURE P11.97.

(b) Compute the moment of inertia $I_{x'}$ and the product of inertia $I_{x'y'}$ where x' and y' are orthogonal axes.

11.98   Consider the shaded area shown in Figure P11.98. Given $A = 63.25$ in.^{2}, $I_x = 1138.4$ in.^{4}, $I_y = 2710.5$ in.^{4}, and $I_{xy} = 2710.5$ in.^{4} (a) Locate the principal centroidal axes of inertia (through centroid C), and find the correspond-

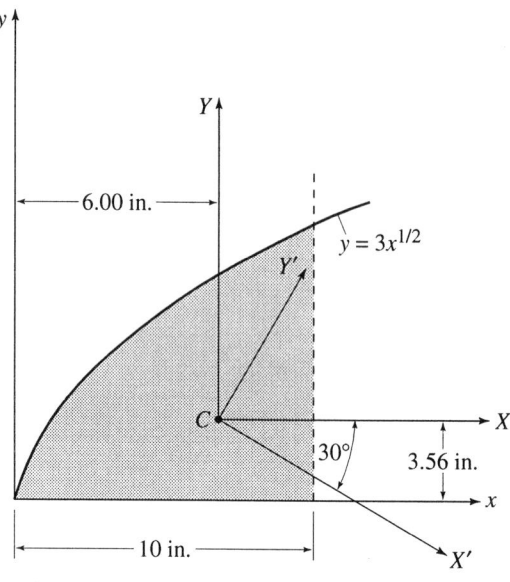

FIGURE P11.98.

ing principal moments of inertia. (b) What is the moment of inertia $I_{X'}$ and the product of inertia $I_{X'Y'}$? The X' and Y' axes are perpendicular to each other.

11.99   The shaded area shown in Figure 11.99 has the following properties: $I_x = 5.00 \times 10^{-2}$ m^4, $I_y = 3.571 \times 10^{-2}$ m^4 and $I_{xy} = 4.167 \times 10^{-2}$ m^4, (a) Locate

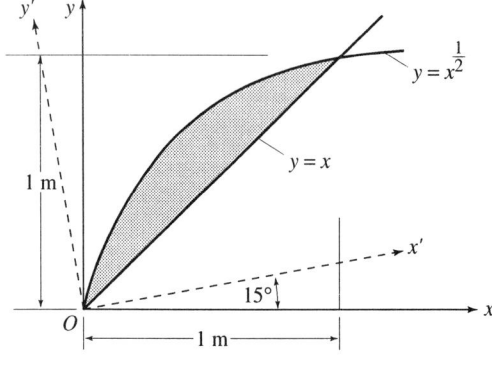

FIGURE P11.99.

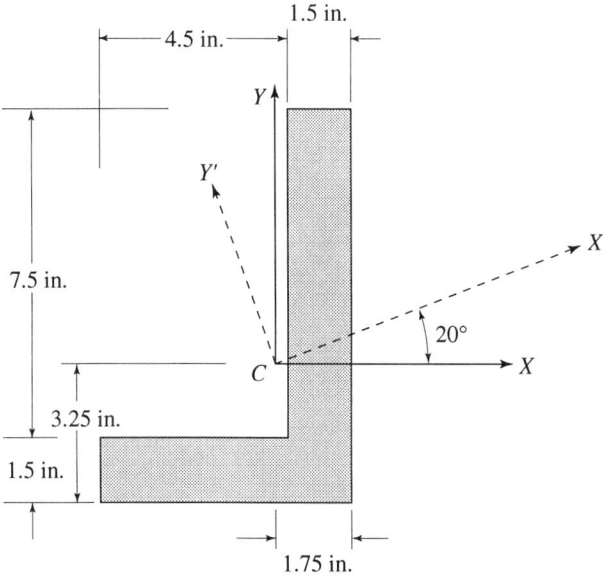

FIGURE P11.100.

the principal axes of inertia through point O, and find the corresponding principal moments of inertia. (b) Determine the moment of inertia $I_{x'}$ and the product of inertia $I_{x'y'}$ where x' and y' are perpendicular axes.

11.100 Consider the cross-sectional area shown in Figure P11.100 for which the following centroidal properties are given: $I_X = 55.67$ in.^{4}, $I_Y = 54.42$ in.^{4}, and $I_{XY} = 50.63$ in.^{4}. (a) Compute the principal centroidal moments of inertia, and locate the principal centroidal axes. (b) Find the moment of inertia $I_{X'}$ and the product of inertia $I_{X'Y'}$ where X' and Y' are orthogonal axes.

11.101 The following properties are provided for the cross-sectional area shown in Figure P11.101: $I_x = 0.759 \times 10^{-3}$ m^4, $I_y = 3.198 \times 10^{-3}$ m^4 and $I_{xy} = 0.998 \times 10^{-3}$ m^4. (a) Locate the principal axes of inertia through point O, and

find the corresponding principal moments of inertia. (b) Locate the centroid of the area, and find the principal centroidal axes and moments of inertia.

11.102 The following properties are given for the cross-sectional area shown in Figure P11.102: $I_x = 688.0$ in.^{4}, $I_y = 1072.0$ in.^{4} and $I_{xy} = 580.0$ in.^{4}. (a) Locate the

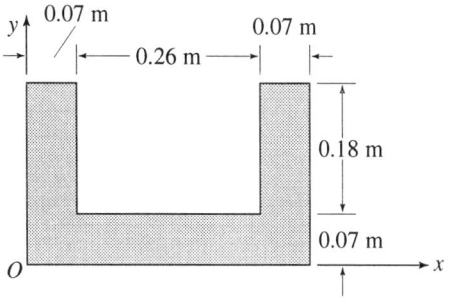

FIGURE P11.101.

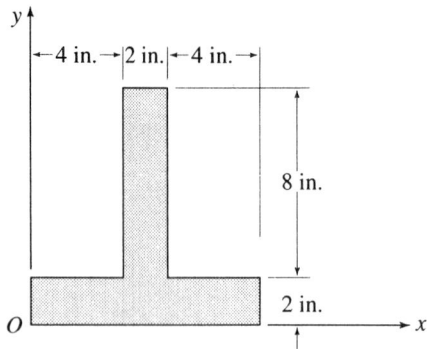

FIGURE P11.102.

principal axes of inertia through point O, and find the corresponding principal moments of inertia. (b) Locate the centroid of the area, and find the principal centroidal axes and moments of inertia.

11.103 Consider the cross-sectional area shown in Figure P11.103 for which the following centroidal properties are given:

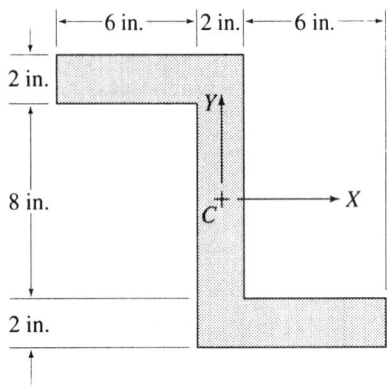

FIGURE P11.103.

$I_X = 896.0$ in.^{4} and $I_Y = 464.0$ in.^{4}. Locate the principal centroidal axes, and find the principal centroidal moments of inertia.

11.104 The following centroidal properties are provided for the cross-sectional area shown in Figure P11.104: $I_Y = 9.778 \times 10^{-6}$ m^4 and $I_{XY} = 1.222 \times 10^{-6}$ m^4. Locate the principal centroidal axes, and find the principal centroidal moments of inertia.

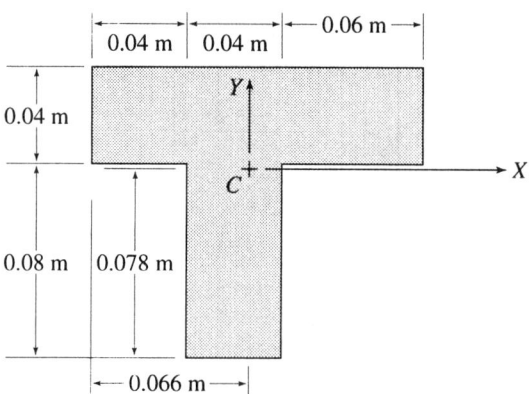

FIGURE P11.104.

Use Mohr's circle to solve the following problems.

11.105 Problem 11.96
11.106 Problem 11.97
11.107 Problem 11.98
11.108 Problem 11.99
11.109 Problem 11.100
11.110 Problem 11.101
11.111 Problem 11.102
11.112 Problem 11.103
11.113 Problem 11.104

11.8*
Mass Principal Axes and Principal Moments of Inertia

In this section, we will consider the moment of inertia of a mass with respect to an arbitrary axis q passing through some point O, as shown in Figure 11.13. From the fundamental definition of mass moment of inertia, we may write the mass moment of inertia with respect to the arbitrary q axis, I_q, as

$$I_q = \int s^2 \, dm \qquad (11.43)$$

where s represents the perpendicular distance from the element of mass dm to the q axis, as shown in Figure 11.13. In view of the three-dimensional nature of this problem, it is very convenient to express the distance s using vector algebra. Thus, denoting, by $\mathbf{r}$, the position vector of the differential mass dm from point O and, by α, the angle between $\mathbf{r}$ and the arbitrary q axis, we conclude that the perpendicular distance $s = r \sin \alpha$ where r is the magnitude of $\mathbf{r}$. By the definition of the vector product of two vectors, we may express the vector distance $\mathbf{s}$ by

$$\mathbf{s} = \boldsymbol{\lambda}_q \times \mathbf{r} \qquad (11.44)$$

where $\boldsymbol{\lambda}_q$ is a unit vector along the q axis. Thus,

$$\mathbf{s} = \begin{vmatrix} \mathbf{i} & \mathbf{j} & \mathbf{k} \\ \lambda_x & \lambda_y & \lambda_z \\ x & y & z \end{vmatrix} = (\lambda_y z - \lambda_z y)\mathbf{i} + (\lambda_z x - \lambda_x z)\mathbf{j}$$
$$+ (\lambda_x y - \lambda_y x)\mathbf{k} \qquad (11.45)$$

where λ_x, λ_y and λ_z are the x, y, and z components of the unit vector $\boldsymbol{\lambda}_q$ (i.e., the directional cosines of the q axis) and x, y, and z are the components of the position vector $\mathbf{r}$ (i.e., the coordinates of the differential

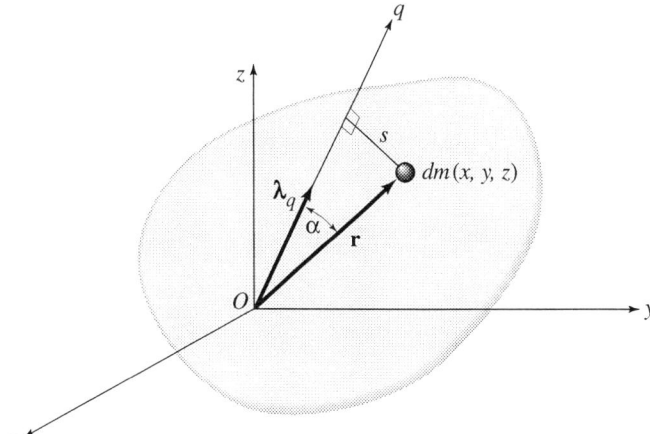

FIGURE 11.13.

mass dm). Therefore,

$$s^2 = \mathbf{s} \cdot \mathbf{s} = (\lambda_y z - \lambda_z y)^2 + (\lambda_z x - \lambda_x z)^2 + (\lambda_x y - \lambda_y x)^2. \quad (11.46)$$

Thus, Eq. (11.43) may be written as

$$I_q = \int [(\lambda_y z - \lambda_z y)^2 + (\lambda_z x - \lambda_x z)^2 + (\lambda_x y - \lambda_y x)^2] \, dm. \quad (11.47)$$

Squaring the quantities in Eq. (11.47) and rearranging terms,

$$
\begin{aligned}
I_q &= \lambda_x^2 \int (y^2 + z^2) \, dm + \lambda_y^2 \int (x^2 + z^2) \, dm + \lambda_z^2 \int (x^2 + y^2) \, dm \\
&\quad - 2\lambda_x \lambda_y \int xy \, dm - 2\lambda_x \lambda_z \int xz \, dm - 2\lambda_y \lambda_z \int yz \, dm \\
&= \lambda_x^2 I_x + \lambda_y^2 I_y + \lambda_z^2 I_z - 2\lambda_x \lambda_y \int xy \, dm \\
&\quad - 2\lambda_z \lambda_x \int xz \, dm - 2\lambda_y \lambda_z \int yz \, dm \quad (11.48)
\end{aligned}
$$

where the integrals $\int (y^2 + z^2) \, dm$, $\int (x^2 + z^2) \, dm$, and $\int (x^2 + y^2) \, dm$ were replaced, respectively, by their equivalent mass moments of inertia I_x, I_y, and I_z. The mixed integrals $\int xy \, dm$, $\int xz \, dm$, and $\int yz \, dm$ in Eq. (11.48) are known as the mass products of inertia with respect to the x-y-z coordinate system. In keeping with the notation for area products of inertia, we define the following three mass products of inertia:

and
$$
\left.
\begin{aligned}
I_{xy} &= \int xy \, dm, \\[1em]
I_{xz} &= \int xz \, dm, \\[1em]
I_{yz} &= \int yz \, dm.
\end{aligned}
\right\} \quad (11.49)
$$

By referring to Figure 11.14, we may interpret $I_{xy} = \int xy \, dm$ as the mass product of inertia with respect to the z-x and y-z planes, $I_{xz} = \int xz \, dm$ as the mass product of inertia with respect to the x-y and y-z planes, and $I_{yz} = \int yz \, dm$ as the mass product of inertia with respect to the x-y and z-x planes. Thus, for example, I_{xy} is zero if either z-x or y-z is a plane of symmetry. Similarly, I_{xz} is zero if either y-x and y-z is a plane of symmetry, and I_{yz} is zero if either x-y or z-x is a plane of symmetry. As in the case of area products of inertia, these conditions of symmetry are useful in determining mass products of inertia. Note that planes of symmetry are not the only planes for which the mass product of inertia is zero. Whether due to symmetry or not, if the mass product of inertia vanishes, the corresponding planes are known as *principal planes*.

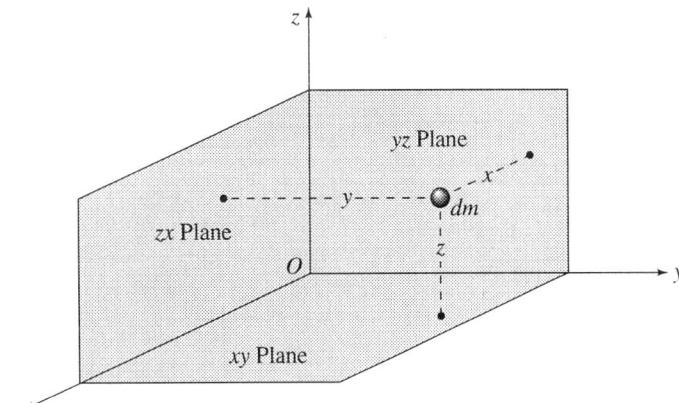

FIGURE 11.14.

Substituting Eq. (11.49) in Eq. (11.48) we obtain

$$I_q = \lambda_x^2 I_x + \lambda_y^2 I_y + \lambda_z^2 I_z - 2\lambda_x\lambda_y I_{xy} - 2\lambda_x\lambda_z I_{xz} - 2\lambda_y\lambda_z I_{yz}. \quad (11.50)$$

As in the case of area products of inertia, a parallel axis theorem may be developed for mass products of inertia. Thus, consider the body of mass m shown in Figure 11.15 where the X-Y-Z coordinate system originates at the mass center G and the x, y, and z axes are parallel, respectively, to the X, Y, and Z axes and have an origin at an arbitrary point O. Because $x = X + a$, $y = Y + b$, and $z = Z + c$,

$$I_{xy} = \int xy \, dm = \int (X + a)(Y + b) \, dm$$

$$= \int XY \, dm + ab \int dm + a \int Y \, dm + b \int X \, dm. \quad (11.51)$$

The first integral is I_{XY}, the mass product of inertia with respect to the centroidal Z-X and Z-Y planes. Because $\int dm = m$, the second term may be written as $(ab)m$. The third and fourth integrals must vanish because they represent the first moments of mass with respect to planes through the center of mass, point G. Thus,

$$I_{xy} = I_{XY} + (ab)m. \quad (11.52)$$

Similarly

$$I_{zx} = I_{ZX} + (ac)m, \quad (11.53)$$

and

$$I_{yz} = I_{YZ} + (bc)m. \quad (11.54)$$

Let us return now to Eq. (11.50). This equation defines the mass moment of inertia I_q with respect to an arbitrary q axis through some

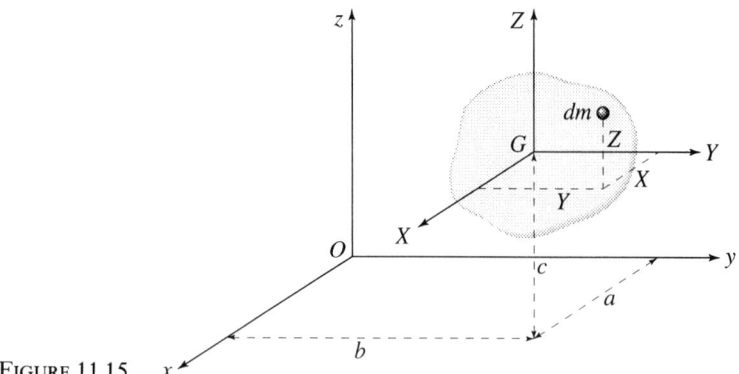

FIGURE 11.15.

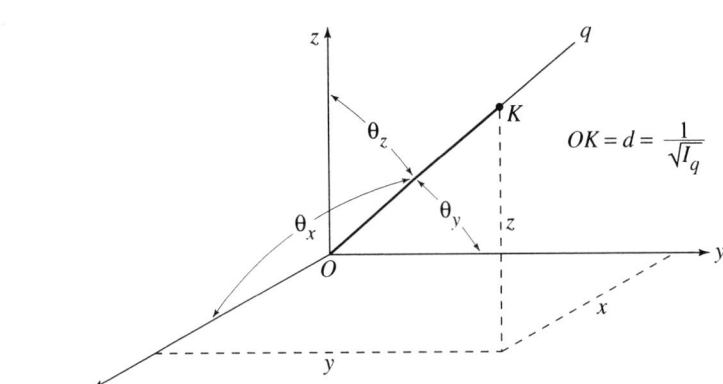

FIGURE 11.16.

point O, in terms of the directional cosines of the axis q, and the mass moments and products of inertia with respect to the x-y-z coordinate system originating at the same point O, as shown in Figure 11.13. By varying the directional cosines, we can obtain the mass moment of inertia I_q with respect to different q axes. As shown in Figure 11.16, if we locate a point K on axis q at a distance $OK = d = 1/\sqrt{I_q}$, we conclude that

$$I_q = 1/(OK)^2 = 1/d^2, \tag{11.55a}$$

$$\left. \begin{aligned} \lambda_x &= \cos \theta_x = x/d \\ \lambda_y &= \cos \theta_y = y/d \\ \lambda_z &= \cos \theta_z = z/d \end{aligned} \right\} \tag{11.55b}$$

Substituting Eqs. (11.55) in Eq. (11.50) and simplifying yields

$$1 = I_x x^2 + I_y y^2 + I_z z^2 - 2I_{xy}xy - 2I_{zx}zx - 2I_{yz}yz. \qquad (11.56)$$

As will be shown below, Eq. (11.56) is a quadric surface which defines an ellipsoid with center at point O, in Figure 11.17(a). Such an ellipsoid is known as the *ellipsoid of inertia* for the body at the specific point O. The ellipsoid of inertia is unique for a specific point O on the body and

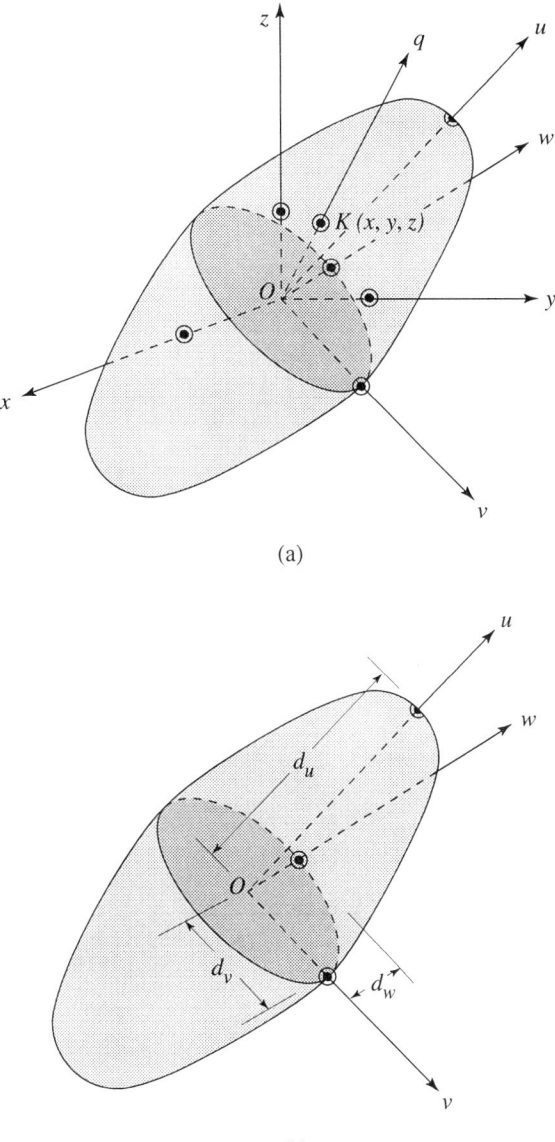

(a)

(b)   FIGURE 11.17.

the coordinates of points on its surface are proportional to the mass moments and mass products of inertia with respect to the coordinate axes originating at point O. If the coordinate axes through point O are changed, the mass moments and mass products of inertia would also change, although the ellipsoid of inertia remains the same. If the x-y-z coordinate axes through point O are rotated to coincide with the principal axes u, v, and w of the ellipsoid, the mass products of inertia I_{uv}, I_{uw}, and I_{vw} vanish because, under these conditions, the planes u-v, u-w and v-w are principal planes. If the rotation is such that x coincides with u, y with v, and z with w, Eq. (11.56) yields

$$1 = I_u u^2 + I_v v^2 + I_w w^2. \tag{11.57a}$$

Because $I_u = 1/d_u^2$, $I_v = 1/d_v^2$ and $I_w = 1/d_w^2$ by Eq. (11.55a) it follows that

$$1 = \frac{u^2}{d_u^2} + \frac{v^2}{d_v^2} + \frac{w^2}{d_w^2} \tag{11.57b}$$

which is the equation of an ellipsoid in term of the principal u, v, and w axes, as shown in Figure 11.17(b). The quantities I_u, I_v, and I_w in Eq. (11.57a) are known as the *mass principal moments of inertia*, and the corresponding u, v, and w axes are the *mass principal axes of inertia*.

Again, if the axes rotation is such that x coincides with u, y with v, and z with w, Eq. (11.50) yields

$$I_q = \lambda_u^2 I_u + \lambda_v^2 I_v + \lambda_w^2 I_w \tag{11.58}$$

where the quantities λ_u, λ_v, and λ_w are the directional cosines of the arbitrary axis q relative to the u, v, and w principal axes of inertia.

The computation of the three principal mass moments of inertia of a body from knowledge of the mass moments and mass products of inertia with respect to an arbitrary x-y-z coordinate system requires solving a cubic equation. It can be shown (see Section 22.2) that the needed equation may be written in the form

$$\begin{vmatrix} (I_x - I) & -I_{xy} & -I_{xz} \\ -I_{yx} & (I_y - I) & -I_{yz} \\ -I_{zx} & -I_{zy} & (I_z - I) \end{vmatrix} = 0 \tag{11.59}$$

which is a determinant expressing a 3×3 matrix known as the *inertial tensor*. Upon expanding this determinant and knowing that $I_{yx} = I_{xy}$, $I_{zx} = I_{xz}$, and $I_{zy} = I_{yz}$, we may rewrite the equation in the form,

$$(I_x - I)(I_y - I)(I_z - I) - I_{xy}^2(I_z - I) - I_{xz}^2(I_y - I)$$

$$- I_{yz}^2(I_x - I) - 2I_{xy}I_{xz}I_{yz} = 0. \tag{11.60}$$

Equation (11.60) is a cubic equation in I whose three roots provide the mass principal moments of inertia I_u, I_v, and I_w. It can also be shown (see Section 22.2) that the directional cosines λ_u, λ_v, and λ_w corresponding, respectively, to the u, v, and w principal axes of inertia may be determined by solving any two of the following three equations plus the relationship $\lambda_x^2 + \lambda_y^2 + \lambda_z^2 = 1$ three times, once for the u axis, once for the v axis, and once for the w axis:

and
$$\left. \begin{aligned} (I_x - I)\lambda_x - I_{xy}\lambda_y - I_{xz}\lambda_z &= 0, \\ -I_{yx}\lambda_x + (I_y - I)\lambda_y - I_{yz}\lambda_z &= 0, \\ -I_{zx}\lambda_x - I_{zy}\lambda_y + (I_z - I)\lambda_z &= 0. \end{aligned} \right\} \qquad (11.61)$$

In Eq. (11.61), the quantity I represents one of the three principal moments of inertia, say I_u, and the quantities λ_u, λ_v, and λ_w will, then, be the directional cosines for the corresponding u axis.

The three principal mass moments of inertia I_u, I_v, and I_w represent the maximum, minimum, and intermediate values of the mass moments of inertia through a point O inside or outside of a body. For the sake of consistency with area principal moments of inertia, we will reserve I_u for the maximum, I_v for the minimum, and I_w for the intermediate value.

The following examples illustrate some of the concepts discussed above.

■ **Example 11.20**

Use integration to determine the mass products of inertia I_{xy}, I_{xz}, and I_{yz} for the body shown in Figure E11.20. Assume that the mass density ρ is a constant. Express the answers in terms of the mass m of the body.

Solution

Select a rectangular element of mass dm as shown in Figure E11.20 where $dm = \rho\, dV = \rho(xz\, dy)$. The width x and the height z of the rect-

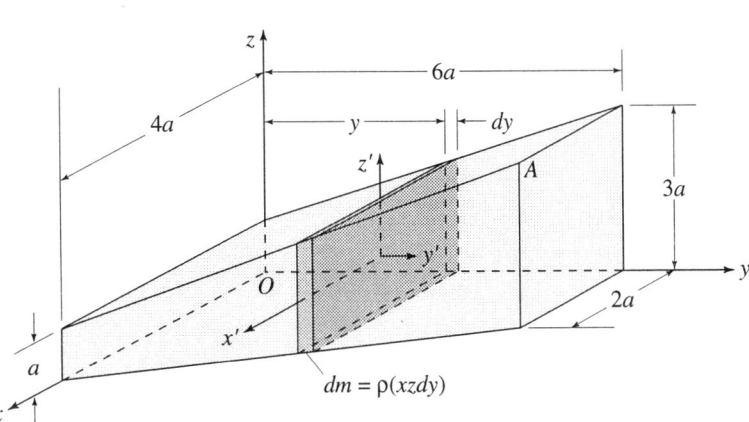

FIGURE E11.20.

angular element vary according to the relationships

$$x = 4a - \tfrac{1}{3}y$$

and

$$z = a + \tfrac{1}{3}y.$$

Thus,

$$dm = \rho(4a - \tfrac{1}{3}y)(a + \tfrac{1}{3}y)\,dy = \rho(4a^2 + ay - \tfrac{1}{9}y^2)\,dy,$$

and

$$m = \rho \int \left(4a^2 + ay - \frac{1}{9}y^2 \right) dy = \rho \left[4a^2 y + \frac{a}{2}y^2 - \frac{1}{27}y^3 \right]_0^{6a} = 34a^3\rho.$$

Observing that the centroidal $z'x'$ and $y'z'$ planes for the element dm are planes of symmetry and using the parallel axis theorem, Eq. (11.52),

$$dI_{xy} = 0 + \left(\frac{x}{2}\right)(y)\,dm = \frac{1}{2}\rho\left(4a - \frac{1}{3}y\right)(y)\left(4a^2 + ay - \frac{1}{9}y^2\right)dy$$

$$= \frac{1}{2}\rho\left(16a^3 y + \frac{8}{3}a^2 y^2 - \frac{7}{9}ay^3 + \frac{1}{27}y^4\right)dy.$$

Therefore,

$$I_{xy} = \frac{1}{2}\rho \int_0^{6a} \left(16a^3 y + \frac{8}{3}a^2 y^2 - \frac{7}{9}ay^3 + \frac{1}{27}y^4\right)dy$$

$$= 142.8\rho a^5 = 4.2a^2(34a^3\rho)$$

$$= 4.200a^2 m. \qquad\qquad \text{ANS.}$$

Similarly, by Eq. (11.53),

$$I_{zx} = \frac{1}{4}\rho \int_0^{6a} \left(16a^4 + 8a^3 y^2 + \frac{1}{9}a^2 y^2 - \frac{2}{9}ay^3 + \frac{1}{81}y^4\right)dy$$

$$= 48.8\rho a^5 = 1.435a^2(34a^3\rho)$$

$$= 1.435a^2 m. \qquad\qquad \text{ANS.}$$

and, by Eq. (11.54),

$$I_{yz} = \frac{1}{2}\rho \int_0^{6a} \left(4a^4 y + \frac{7}{3}a^2 y^2 + \frac{2}{9}ay^3 - \frac{1}{27}y^4\right)dy$$

$$= 127.2\rho a^5 = 3.741a^2(34a^3\rho)$$

$$= 3.741a^2 m. \qquad\qquad \text{ANS.}$$

■ **Example 11.21**

Use integration to find the mass moment of inertia with respect to the line OA for the body shown in Figure E11.21. Express the answer in

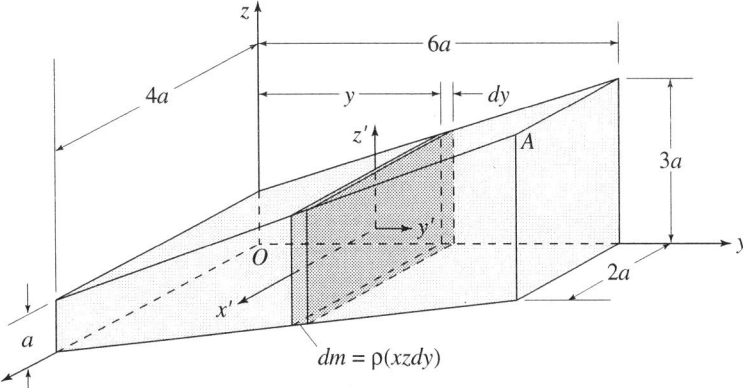

FIGURE E11.21.

terms of the mass m of the body. Assume that the mass density ρ is a constant. Note that this is the same body used in Example 11.20.

Solution

Determination of the mass moment of inertia I_{OA} may be accomplished by using Eq. (11.50). This equation requires knowledge of the mass moments of inertia I_x, I_y, and I_z as well as the directional cosines of line OA. Using the parallel axis theorem for mass moments of inertia, Eq. (11.18), and the information given in Appendix B,

$$dI_x = \frac{1}{12}z^2 \, dm + \left[y^2 + \left(\frac{z}{2} \right)^2 \right] dm = \left(y^2 + \frac{1}{3}z^2 \right) dm.$$

From Example 11.20, $z = a + \frac{1}{3}y$, and $dm = \rho \left(4a^2 + ay - \frac{1}{9}y^2 \right) dy$. Therefore,

$$dI_x = \left[y^2 + \frac{1}{3}\left(a + \frac{1}{3}y \right)^2 \right] \rho \left(4a^2 + ay - \frac{1}{9}y^2 \right) dy$$

$$= \rho \left(\frac{4}{3}a^4 + \frac{11}{9}a^3 y + \frac{117}{27}a^2 y^2 + \frac{82}{81}ay^3 - \frac{28}{243}y^4 \right) dy,$$

and

$$I_x = \int_0^{6a} dI_x = 490.8\rho a^5 = 14.435a^2 (34a^3 \rho) = 14.435a^2 m.$$

Similarly,

$$dI_y = \frac{1}{12}(x^2 + z^2) \, dm + \left[\left(\frac{x}{2} \right)^2 + \left(\frac{z}{2} \right)^2 \right] dm = \left(\frac{1}{3}x^2 + \frac{1}{3}z^2 \right) dm$$

$$= \frac{1}{3}\rho \left(68a^4 + 9a^3 y - 3a^2 y^2 + \frac{4}{9}ay^3 - \frac{2}{81}y^4 \right) dy,$$

and

$$I_y = \int_0^{6a} dI_y = 153.2\rho a^5 = 4.506a^2(34a^3\rho) = 4.506a^2m.$$

Also,

$$dI_z = \frac{1}{12}x^2\,dm + \left[\left(\frac{x}{2}\right)^2 + y^2\right]dm = \left(\frac{1}{3}x^2 + y^2\right)dm$$

$$= \frac{4}{3}\rho\left(16a^4 + \frac{4}{3}a^3y + 2a^2y^2 + \frac{23}{27}ay^3 - \frac{7}{81}y^4\right)dy,$$

and

$$I_z = \int_0^{6a} dI_z = 540.8\rho a^5 = 15.906a^2(34a^3\rho) = 15.906a^2m.$$

Find a unit vector λ_{OA} along line OA. Thus,

$$\lambda_{OA} = \frac{2a\mathbf{i} + 6a\mathbf{j} + 3a\mathbf{k}}{\sqrt{(2a)^2 + (6a)^2 + (3a)^2}} = \left(\frac{2}{7}\right)\mathbf{i} + \left(\frac{6}{7}\right)\mathbf{j} + \left(\frac{3}{7}\right)\mathbf{k},$$

and

$$(\lambda_{OA})_x = \frac{2}{7},$$

$$(\lambda_{OA})_y = \frac{6}{7},$$

and

$$(\lambda_{OA})_z = \frac{3}{7}.$$

Also, from the solution of Example 11.20,

$$I_{xy} = 4.200a^2m,$$
$$I_{xz} = 1.435a^2m,$$

and

$$I_{yz} = 3.741a^2m.$$

Thus, by Eq. (11.50),

$$I_{OA} = \left(\frac{2}{7}\right)^2(14.435a^2m) + \left(\frac{6}{7}\right)^2(4.506a^2m) + \left(\frac{3}{7}\right)^2(15.906a^2m)$$

$$- \left(\frac{2}{7}\right)\left(\frac{6}{7}\right)(4.200a^2m) - \left(\frac{2}{7}\right)\left(\frac{3}{7}\right)(1.435a^2m)$$

$$- 2\left(\frac{6}{7}\right)\left(\frac{3}{7}\right)(3.741a^2m)$$

$$= 2.253a^2m. \qquad\qquad\qquad \text{ANS.}$$

■ **Example 11.22**

Refer to the body shown in Figure E11.22. Determine the mass principal moments of inertia I_u, I_v, and I_w. Also, determine the directional cosines corresponding to each of the three mass principal axes of inertia u, v, and w through point O. The following values are obtained from the solution of Examples 11.20 and 11.21:

$$I_x = 14.435a^2m, \; I_y = 4.506a^2m, \; I_z = 15.906a^2m,$$

and

$$I_{xy} = 4.200a^2m, \; I_{xz} = 1.435a^2m, \; I_{yz} = 3.741a^2m.$$

Solution

Substituting the given mass moments and products of inertia in Eq. (11.60) and simplifying yields

$$I^3 - 34.847I^2 + 332.625I - 497.618 = 0.$$

Any iterative technique, such as Newton's method, may be used to obtain the three roots as

$$\left.\begin{array}{l} I_u = 17.049a^2m, \\ I_w = 15.970a^2m, \\ I_v = 1.828a^2m. \end{array}\right\} \qquad \text{ANS.}$$

and

Substitute the value $I_u = 17.049a^2m$ for I in Eq. (11.61), and simplify to obtain the following relationships:

$$2.614\lambda_x + 4.200\lambda_y + 1.435\lambda_z = 0,$$

$$4.200\lambda_x + 12.543\lambda_y + 3.741\lambda_z = 0,$$

and

$$1.435\lambda_x + 3.741\lambda_y + 1.143\lambda_z = 0.$$

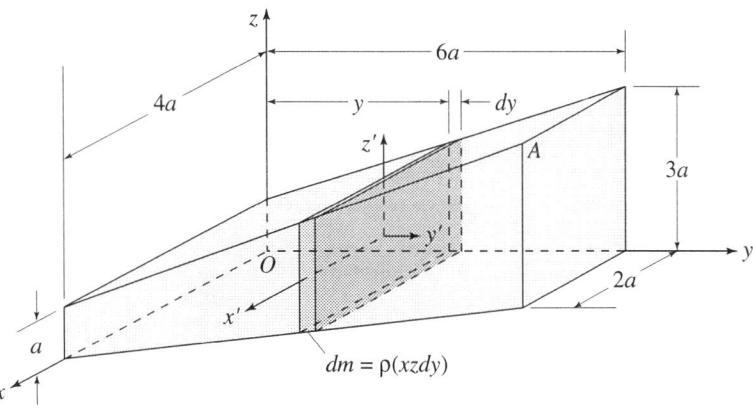

FIGURE E11.22.

Any two of the above three equations plus the relationship $\lambda_x^2 + \lambda_y^2 + \lambda_z^2 = 1$ may be solved simultaneously to obtain the directional cosines for the principal u axis. Thus,

$$\left.\begin{array}{l} \lambda_x = -0.145, \\[4pt] \lambda_y = -0.238, \\[4pt] \lambda_z = 0.960. \end{array}\right\} \qquad \text{ANS.}$$

and

Proceeding in the same manner for I_w, the directional cosines for the principal w axis are

$$\left.\begin{array}{l} \lambda_x = 0.932, \\[4pt] \lambda_y = -0.358, \\[4pt] \lambda_z = 0.052. \end{array}\right\} \qquad \begin{array}{l} \\[4pt] \text{ANS.} \\[4pt] \text{ANS.} \end{array}$$

and

Similarly, the directional cosines for the principal v axis are

$$\left.\begin{array}{l} \lambda_x = 0.332, \\[4pt] \lambda_y = 0.903, \\[4pt] \lambda_z = 0.274. \end{array}\right\} \qquad \begin{array}{l} \\[4pt] \text{ANS.} \\[4pt] \text{ANS.} \end{array}$$

and

∎

Problems

11.114 Use integration to compute the mass products of inertia I_{xy}, I_{xz}, and I_{yz} for the body shown in Figure P11.114. Assume that the mass density ρ is a constant. Express the answers in terms of the mass m of the body.

11.115 Refer to Problem 11.114. Determine the mass moment of inertia with respect to line OA. Express the answer in terms of the mass m of the body.

11.116 Use integration to compute the mass products of inertia I_{xy}, I_{xz}, and I_{yz} for the body shown in Figure P11.116.

The weight density of the material $\rho = 490 \text{ lb/ft}^3$.

11.117 Determine the products of inertia I_{xy}, I_{xz}, and I_{yz} for the homogeneous bent rod shown in Figure P11.117. Express answers in terms of the mass m of the entire rod and the dimension a.

11.118 Refer to Problem 11.117. Compute the mass moment of inertia with respect to line OA. Express the answer in terms of the mass m of the entire rod and the dimension a.

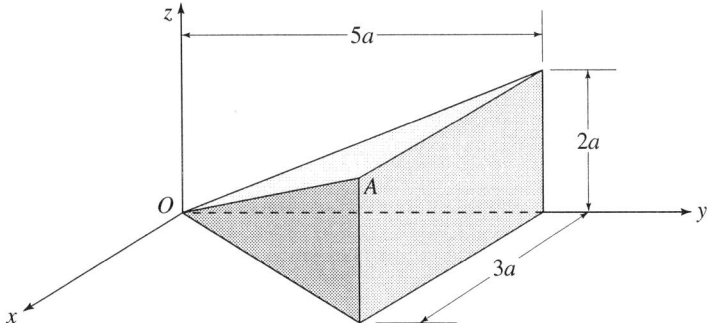

FIGURE P11.114.

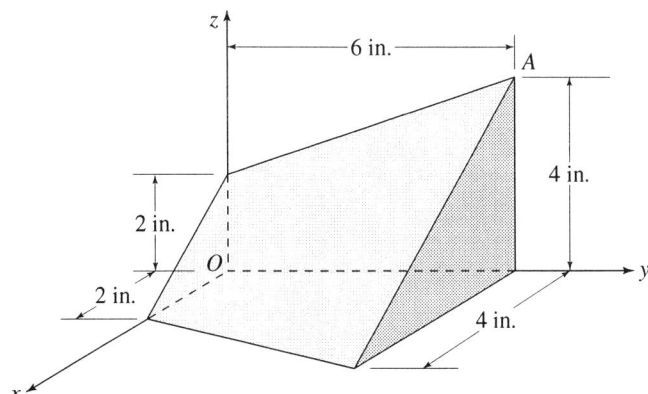

FIGURE P11.116.

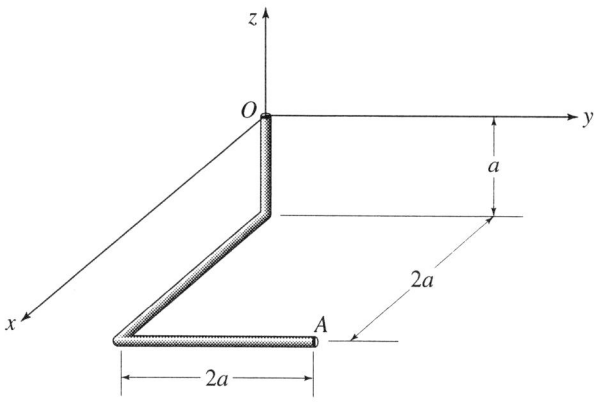

FIGURE P11.117.

11.119 Find the products of inertia I_{xy}, I_{xz}, and I_{yz} for the homogeneous bent plate shown in Figure P11.119. The thickness of the plate is uniform with $t = 0.03$ m. The material is steel whose mass density $\rho = 7850$ kg/m³.

11.120 Refer to Problem 11.119. Determine the mass moment of inertia with respect to line OA.

11.121 Find the products of inertia I_{xy} and I_{xz} for the composite body shown in Figure P11.121. Assume that the mass density ρ is a constant. Express the answer in terms of the mass m of the composite body.

11.122 Use the results of Problems 11.114 and 11.115 to compute the mass principal moments of inertia with respect to axes

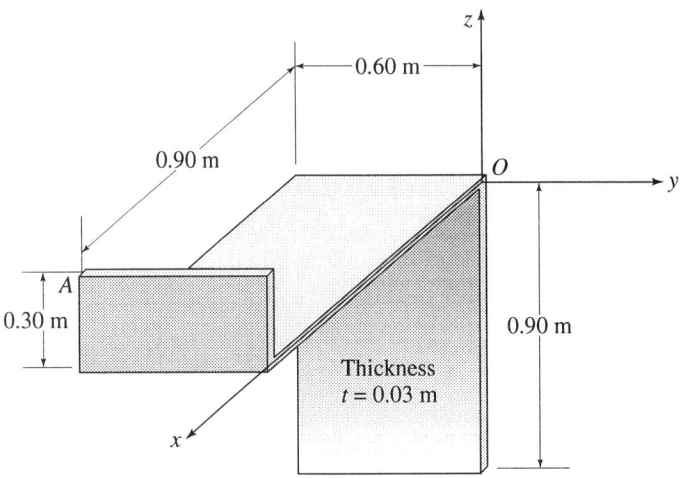

FIGURE P11.119.

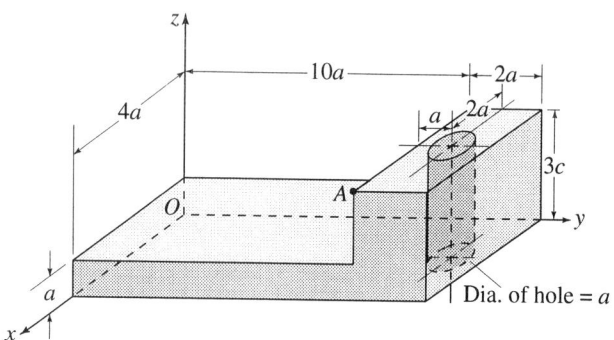

FIGURE P11.121.

through point O in Figure P11.114. Also, compute the directional cosines for the u axis with respect to which the mass moment of inertia is the maximum.

11.123 Use the results of Problems 11.117 and 11.118 to compute the mass principal moments of inertia with respect to axes through point O in Figure P11.117. Also, find the directional cosines for the v axis with respect to which

the mass moment of inertia is the minimum.

11.124 Use the results of Problems 11.119 and 11.120 to determine the mass principal moments of inertia with respect to axes through point O in Figure P11.119. Also, compute the directional cosines for the w axis with respect to which the mass moment of inertia is intermediate between the maximum and the minimum.

Review Problems

11.125 Consider the area shown in Figure P11.125, and find the moment of inertia I_x and the radius of gyration k_x.

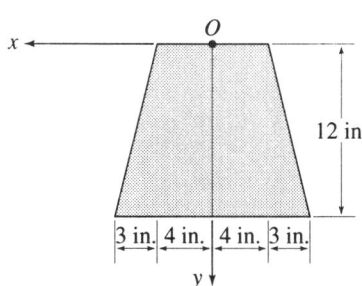

FIGURE P11.125.

11.126 Consider the area shown in Figure P11.125, and find the polar moment of inertia and polar radius of gyration about an axis through point O.

11.127 Determine the moment of inertia I_y and the radius of gyration k_y for the shaded area shown in Figure P11.127.

11.128 Determine the polar moment of inertia and polar radius of gyration about an axis through point O for the shaded area shown in Figure P11.127.

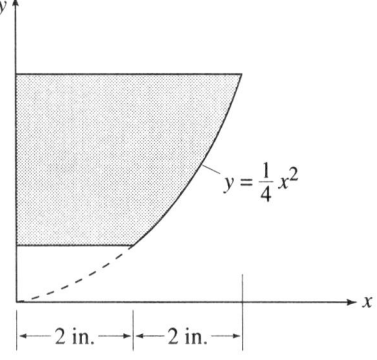

FIGURE P11.127.

11.129 A solid of revolution is developed by rotating the shaded area, shown in Figure P11.129, one complete revolution

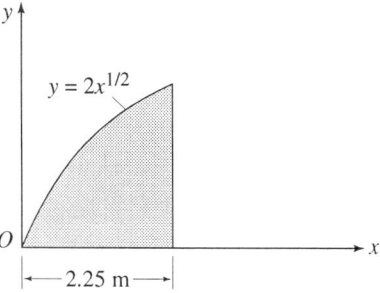

FIGURE P11.129.

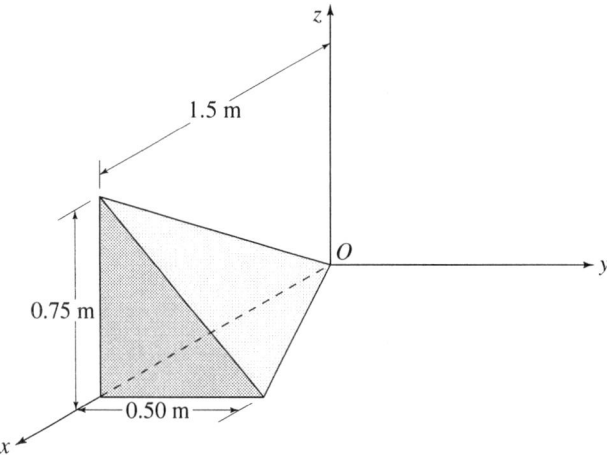

FIGURE P11.130.

about the x axis. Compute its mass moment of inertia about the x axis, expressing it in terms of its constant mass density ρ.

11.130 Determine the mass moment of inertia I_y for the homogeneous prism shown in Figure P11.130. The mass density for the prism material $\rho = 2500$ kg/m^3.

11.131 Determine the moments of inertia with respect to the x and y axes for the composite area shown in Figure P11.131.

11.132 Refer to the composite area shown in Figure P11.131, and determine the moments of inertia with respect to centroidal X and Y axes.

11.133 A beam is fabricated by welding two L6 × 6 × 1 angles to a C12 × 30 channel, as shown in Figure P11.133. From a structural design handbook, the area and centroidal moments of inertia for each angle are $A_a = 11.0$ in.^{2} and $(I_x)_a = (I_y)_a = 35.5$ in.^{4}. Also, the area and centroidal moments of inertia for the channel are $A_c = 8.82$ in.^{2}, $(I_x)_c = 5.14$ in.^{4}, and $(I_y)_c = 162$ in.^{4}. Determine the

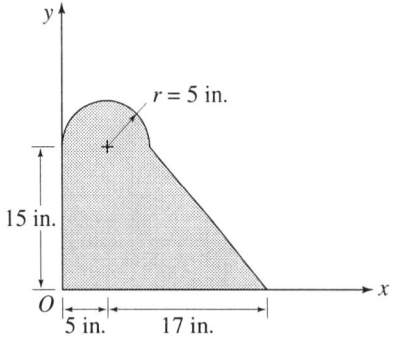

FIGURE P11.131.

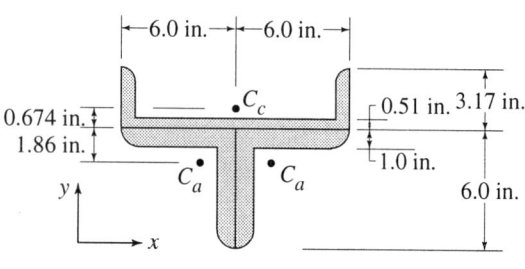

FIGURE P11.133.

moment of inertia and the radius of gyration of the composite cross-section with respect to the *centroidal X* axis. Note that the centroids of the two angles and that of the channel are located in Figure P11.133.

11.134 Three rods, each with a length $L = 0.75$ m and a mass density of 1.5 kg/m, are welded together as shown in Figure P11.134. Determine the mass moments of inertia of the composite mass with respect to the x, y, and z axes.

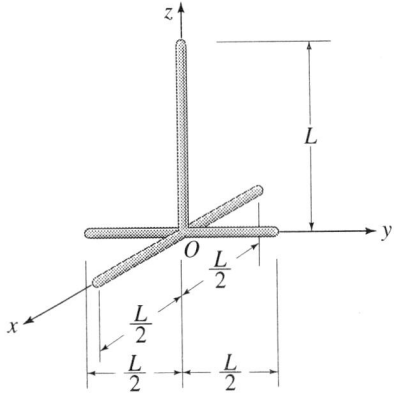

FIGURE P11.134.

11.135 The machine component shown in Figure P11.135 is made from a 0.006-m

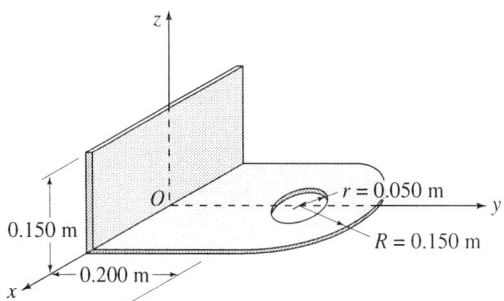

FIGURE P11.135.

aluminum plate for which the mass density is approximately 2700 kg/m³. Determine its mass moment of inertia with respect to the x axis.

11.136 By measuring the period of oscillations when a body is suspended from a point, it is possible to determine its mass radius of gyration with respect to an axis through that point. When the machine element, shown in Figure P11.136, was suspended from point A, it was found that $k_A = 0.41$ m and when suspended from point B, $k_B = 0.28$ m. Determine the dimension b that locates the mass center vertically. Each of the four holes has a radius of 0.05 m.

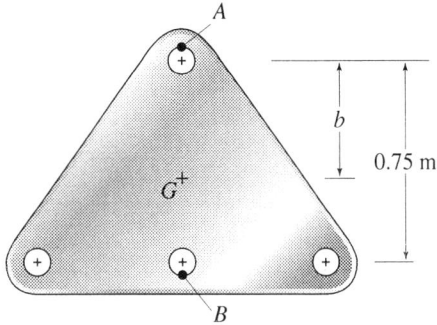

FIGURE P11.136.

11.137 (a) Compute the product of inertia I_{xy} for the shaded area shown in Figure P11.137. (b) Compute the product of inertia I_{xy} of the shaded area shown previously in Figure P11.127.

11.138 Refer to the area shown previously in Figure P11.127. Use equations to find the principal axes through point O and the corresponding principal moments of inertia.

11.139 Solve Problem 11.138 by using Mohr's circle.

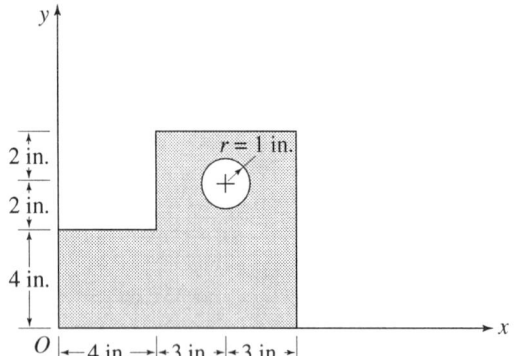

FIGURE P11.137.

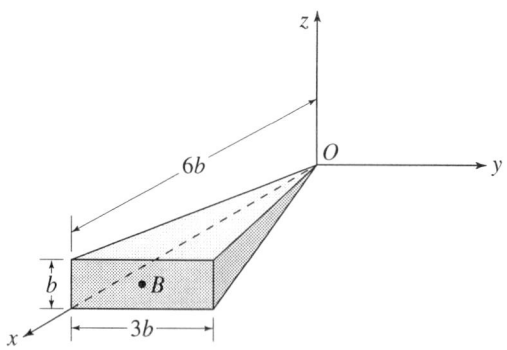

FIGURE P11.141.

11.140 Consider the cross-sectional area shown in Figure P11.140 and determine the principal *centroidal U* and *V* axes and the corresponding principal centroidal moments of inertia I_U and I_V.

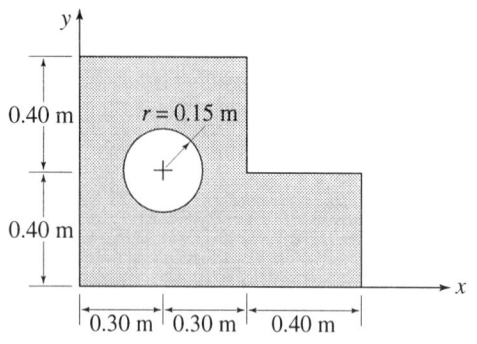

FIGURE P11.140.

11.141 Compute the mass moments of inertia I_x, I_y, and I_z for the solid shown in Figure P11.141. Express the answers in terms of the dimension b and the mass m of the solid. The mass density ρ for the material is constant.

11.142 Refer to the solid shown in Figure P11.141, and compute the mass products of inertia I_{xy}, I_{yz}, and I_{zx}. Express the answers in terms of the dimension b and the mass m of the solid. The mass density ρ for the material is constant. Also, determine the mass moment of inertia with respect to line OB where B has the coordinates: $(6b, 1.5b, 0.5b)$.

11.143 Use the results obtained in Problems 11.141 and 11.142 to determine the mass principal moments of inertia with respect to axes through point O in Figure P11.141. Also, compute the directional cosines for the v axis with respect to which the mass moment of inertia is the minimum.

12

Virtual Work and Stationary Potential Energy

Schematic of an offshore oil rig. This rig is one of many examples where virtual work could be used efficiently in analyzing static problems.

In all of the preceding chapters, equilibrium problems were investigated using the equations of statics, i.e., $\sum \mathbf{F} = \mathbf{0}$ and $\sum \mathbf{M} = \mathbf{0}$. Another method, developed in this chapter and known as the method of virtual work, is based upon considerations of the work done by a force. Although this method is derived from an entirely different premise, it will be shown that it is, nevertheless, equivalent to the equations of statics. However, the method of virtual work has distinct advantages particularly when dealing with the equilibrium of systems of connected rigid bodies, because it is not necessary to dismember the system into separate rigid bodies. Rather, the entire system may be viewed as a single entity in determining the virtual work of all forces acting on it because internal forces of action and reaction at the frictionless connections perform a total amount of work which vanishes.

For the case of conservative systems, the method of virtual work is equivalent to the principle of stationary potential energy. This principle makes it possible to determine not only the equilibrium positions of a conservative system of rigid bodies but also to examine the states of these equilibrium positions and to identify them as stable, unstable, or neutral equilibrium positions.

12.1 Differential Work of a Force

Consider a force $\mathbf{F}$ acting on a particle P which is constrained to move along some path in space, as shown in Figure 12.1. Let us assume that the particle moves from position P_1, where the position vector is $\mathbf{r}$, to an adjacent position P_2 where the position vector is $\mathbf{r} + d\mathbf{r}$. The differential vector quantity $d\mathbf{r}$ is defined as the displacement of the particle as it moves from P_1 to P_2. Note that although the vector $d\mathbf{r}$ is tangent to the path of the particle in position P_1, the motion of the particle actually takes place along the path through the distance ds which represents the magnitude dr of the vector. It should be pointed out that, in the limit when both dr and ds approach zero, they become equal to each other in magnitude and direction.

As the force $\mathbf{F}$ moves the particle through the displacement $d\mathbf{r}$, it performs a differential quantity of *work dU* defined as the dot product of $\mathbf{F}$ and $d\mathbf{r}$. Thus,

$$dU = \mathbf{F} \cdot d\mathbf{r}. \tag{12.1}$$

We recall from an earlier discussion that the dot product of two vector quantities yields a scalar quantity whose magnitude is obtained by multiplying the magnitudes of the two vectors with each other and with the cosine of the angle between them. Thus, if the angle between $\mathbf{F}$ and $d\mathbf{r}$ is β as, shown in Figure 12.1, the work of the force $\mathbf{F}$ during a displacement $d\mathbf{r}$ has a magnitude given by

$$dU = F \, ds \cos \beta. \tag{12.2}$$

From Eq. (12.2), we conclude that the units of work must be those obtained by multiplying units of force by those of length. Thus, in the U.S. customary system, such units as lb·ft and k·in. are used. In the SI system, the most commonly used unit is the N·m, also known as a *joule*. Although the units of work are the same as those for the moment

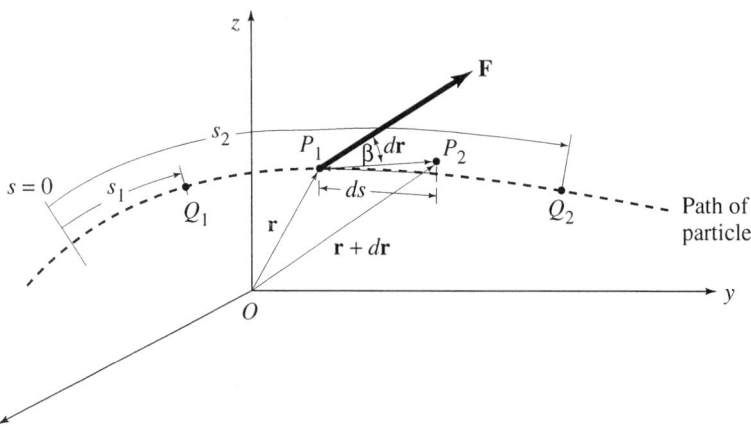

FIGURE 12.1.

of a force, recall that *work* is a scalar and *moment* is a vector and that they represent two entirely different physical quantities.

Equation (12.2) shows that the work of a force depends upon the angle that it makes with the path of the particle. Thus, when $\beta = 90°$ (i.e., $\cos \beta = 0$), the work of the force is zero. In other words, a force that is perpendicular to the direction of motion performs no work. Examples of such forces are shown in Figure 12.2(a) which indicates a block of weight W moving horizontally to the right on a frictionless surface. Because both the weight W and the normal reaction N act in a direction perpendicular to the path of motion as shown in the free-body diagram of Figure 12.2(b), they perform no work. Also, a force performs positive work for acute values of β (because the cosine of angles smaller the $90°$ is positive) and negative work for obtuse values of β (because the cosine of angles larger than $90°$ is negative). Thus, referring again to Figure 12.2, the force F_1 performs positive work and the force F_2 negative work during the horizontal motion of the block to the right.

Equation (12.2) also shows that, unless a displacement ds is experienced, no work is performed by the force. Thus, forces that act at fixed positions perform no work because they undergo no displacement (i.e., $ds = 0$). Examples of such forces are given in Figure 12.3(a) which shows member AB attached to a spring at A and to a frictionless hinge at B. If the member experiences differential rotation $d\theta$ about the hinge at B, the reactive forces B_x and B_y, shown in the free-body diagram of Figure 12.3(b), do no work because they do not displace. However, the applied force Q and the force F in the spring will perform work during any rotation of the member.

Internal forces that occur in pairs of equal magnitude and of opposite sense do a net amount of work equal to zero during any motion of the particle on which they act. Consider, for example, the frictionless and weightless pin-connected mechanism ABC shown in Figure 12.4(a). The free-body diagram of the frictionless pin at B is given in Figure 12.4(b) and shows that the pin is subjected to the two equal and oppo-

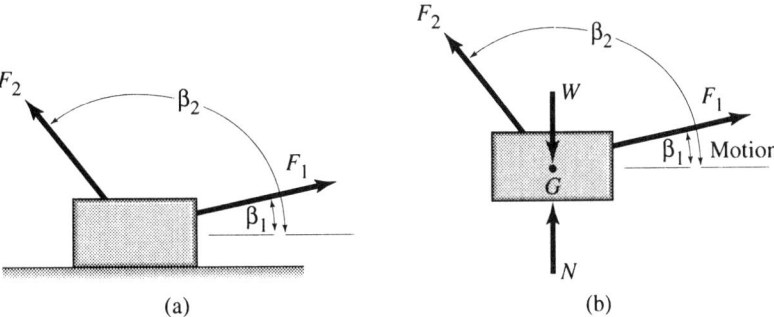

FIGURE 12.2.   (a)   (b)

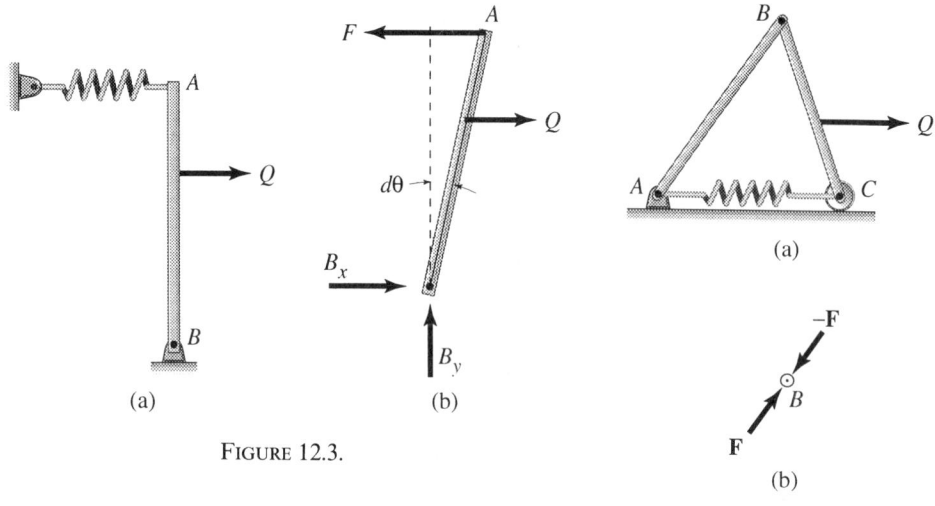

FIGURE 12.3.

FIGURE 12.4.

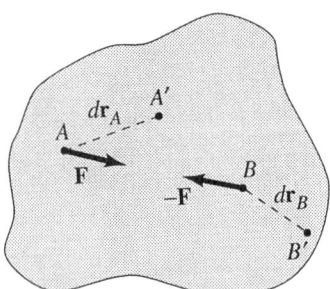

FIGURE 12.5.

site forces **F** and −**F**. When the mechanism is subjected to the force Q, joint B experiences motion during which the two forces **F** and −**F**, acting on the pin perform equal amounts of work. However, while the force **F** performs positive work, the force −**F** performs negative work, leading to a net amount of work which is zero.

Now, consider the case of a rigid body which may be assumed to consist of an infinite number of particles held together in fixed positions relative to one another by equal and opposite forces that they exert on each other. The two particles A and B of a rigid body and the forces of attraction **F** that they exert on each other are shown in Figure 12.5. During a movement of the rigid body, points A and B move through the unequal displacements $d\mathbf{r}_A$ and $d\mathbf{r}_B$, respectively. However, because the body is rigid, the distance from A to B is constant. It follows, therefore, that the components of $d\mathbf{r}_A$ and $d\mathbf{r}_B$ along the line AB must be equal which means that the work performed by **F** must be exactly equal and opposite to that performed by $-\mathbf{F}$. Consequently, the work performed by the internal forces $\mathbf{F}$ and $-\mathbf{F}$ during any movement of the rigid body must be zero. *Thus, only the external forces acting on the rigid body need be considered in determining the work performed during any movement the rigid body may experience.*

12.2 Differential Work of a Couple

The work performed by a couple may be obtained by determining, separately, the work performed by each of the two forces of the couple and, then, algebraically adding the results. Thus, consider the rigid body shown in Figure 12.6 which is constrained to move in the plane of the page. The rigid body is subjected to the two forces $\mathbf{F}$ and $-\mathbf{F}$, also acting in the plane of the page, which give rise to a couple $\mathbf{M}$ pointing out of the paper and have a magnitude of $M = Fc$. The rigid body experiences an infinitesimal displacement from position $A_1 B_1$ to position $A_2 B_2$. This displacement may be viewed as a differential translation $d\mathbf{r}_1$ from $A_1 B_1$ to $A_2 B_1'$ followed by a differential rotation $d\boldsymbol{\theta}$ of magnitude $d\theta$ about an axis perpendicular to the page at A_2. This differential rotation results in the displacement $d\mathbf{r}_2$ that brings the rigid body to its final position $A_2 B_2$. During the translation $d\mathbf{r}_1$, the work done by $\mathbf{F}$ is exactly equal but opposite to that performed by $-\mathbf{F}$. Therefore, during this part of the motion, the work of one force cancels that of the other and the net work is zero. During the differential rotation $d\boldsymbol{\theta}$, however, although the force $-\mathbf{F}$ at A_2 does no work, the force $\mathbf{F}$ at B_1' performs a differential quantity of work $dU = F\,ds_2$. Because $ds_2 = c\,d\theta$, the work performed may be written as $dU = Fc\,d\theta$. Remembering that the product Fc is the magnitude M of the couple, we may express the work of a couple by the relationship

$$dU = M\,d\theta. \tag{12.3}$$

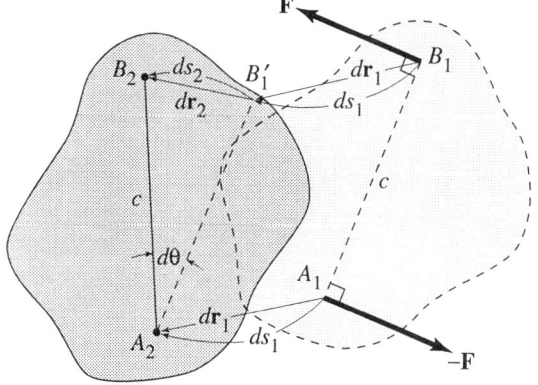

FIGURE 12.6.

In general, the work of a couple may be expressed in vector form by

$$dU = \mathbf{M} \cdot d\boldsymbol{\theta} = M \cos \beta \, d\theta. \tag{12.4}$$

where β represents the angle between vector $\mathbf{M}$ and vector $d\boldsymbol{\theta}$. If the direction of $\mathbf{M}$ coincides with that of $d\boldsymbol{\theta}$, as in our case, the angle β is zero and $dU = M \, d\theta$ as given by Eq. (12.3). Note that because $d\theta$ is expressed in radians, Eq. (12.3) shows that the units of dU are the same as the units of the moment M, namely units of force multiplied by units of length as explained earlier. Note also that the work of a couple is positive if its sense is in the same direction as the rotation and negative if its sense is opposite to the rotation.

12.3
The Concept
of Finite Work

Refer, once again, to Figure 12.1 which is repeated here for convenience. Let us consider the quantity of work performed by the force $\mathbf{F}$ as it moves the particle along its path through a finite displacement from position Q_1 to position Q_2. This amount of work may be viewed as the sum of an infinite number of differential quantities dU performed by $\mathbf{F}$ as it moves the particle through an infinitely large number of differential displacements $d\mathbf{r}$ along the path from Q_1 to Q_2. In other words, the work of a force during a finite displacement of its point of application may be obtained by integrating Eq. (12.1). Thus,

$$U_{1 \to 2} = \int_{U_1}^{U_2} dU = \int_{Q_1}^{Q_2} \mathbf{F} \cdot d\mathbf{r} \tag{12.5}$$

where $U_{1 \to 2}$ represents the work done by the force $\mathbf{F}$ as it moves the particle through the displacement from position Q_1 to position Q_2.

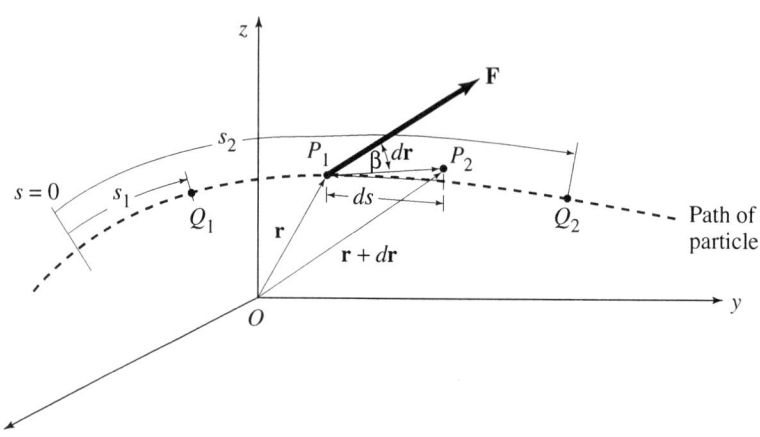

FIGURE 12.1.

Equation (12.5) may also be expressed in the form

$$U_{1 \to 2} = \int_{s_1}^{s_2} (F \cos \beta)\, ds \qquad (12.6)$$

where the quantity s is measured along the path of the particle between position s_1 and position s_2, and the quantity $F \cos \beta$ represents the component of the force $\mathbf{F}$ along the tangent to the path. Note that integration of Eq. (12.6) is possible only under the following two conditions: (a) the force $F \cos \beta$ is constant and may, therefore, be taken out of the integral sign and (b) a relationship exists among the quantities, F, β, and s. Note also, that, physically, the integral of Eq. (12.6) represents the area between s_1 and s_2 under the curve expressing the relationship between $F \cos \beta$ and s as shown in Figure 12.7.

Equation (12.6) may be viewed in one of two ways. In the first, the integrand is written as expressed in Eq. (12.6) where the component of the force in the direction of the displacement (i.e., $F \cos \beta$) moves through the displacement ds. In the second, Eq. (12.6) may be written in the form $U_{1 \to 2} = \int_{s_1}^{s_2} F (ds \cos \beta)$ where the force F moves through the component of the displacement in the direction of the force (i.e., $ds \cos \beta$). Both views are useful but, under certain conditions, one view may have advantages over the other.

Using Eq. (12.3) we may find the work of a couple during a finite rotation from θ_1 to θ_2. Thus,

$$U_{1 \to 2} = \int_{\theta_1}^{\theta_2} M\, d\theta. \qquad (12.7)$$

It should be noted that Eq. (12.7) may be integrated only under conditions similar to those given in the case of Eq. (12.6), namely, that M be a constant or that a relationship is known between M and θ. Note also that the integral of Eq. (12.7) represents the area between θ_1 and θ_2 under the curve expressing the relationship between M and θ, as shown in Figure 12.8.

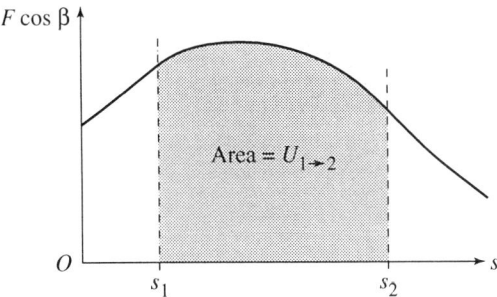

FIGURE 12.7.

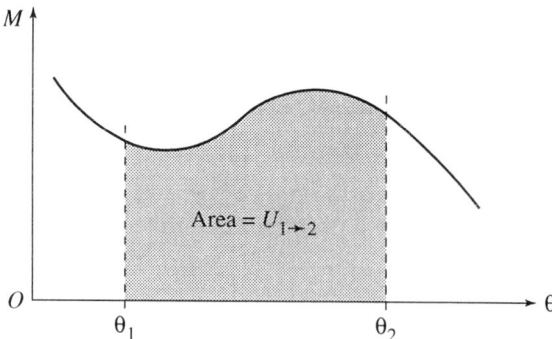

FIGURE 12.8.

θ_1 θ_2

■ **Example 12.1**

A worker needs to move the 300-lb crate from position 1 to position 2 a distance of 40 ft. along a rough inclined plane, as shown in Figure E12.1(a). Compute the total work performed by the worker assuming that $\mu_k = 0.25$ and that the pulley is frictionless. Assume motion at constant velocity.

Solution

The free-body diagram of the crate is shown in Figure E12.1(b).

$$\sum F_y = 0, \quad N - 300\cos 20° = 0,$$

$$N = 281.9 \text{ lb},$$

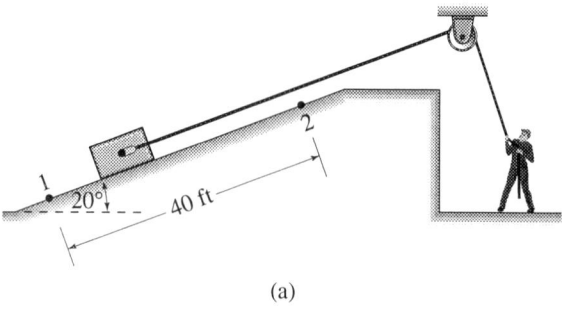

(a)

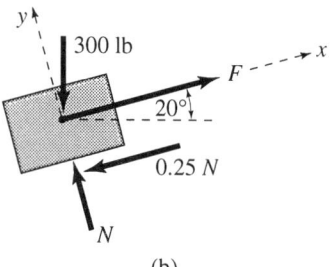

FIGURE E12.1.   (b)

and

$$\sum F_x = 0, \quad F - 300 \sin 20° - 0.25N = 0,$$

$$F = 173.1 \text{ lb}.$$

Because the pulley is frictionless, the force exerted by the worker is equal to F, which is a constant during the moving operation. The work done by the worker can now be determined by Eq. (12.6). Thus,

$$U_{1 \to 2} = \int_{s_1}^{s_2} (F \cos \beta) \, ds.$$

Because F is constant and $\beta = 0°$, it follows that

$$U_{1 \to 2} = F \int_{s_1}^{s_2} ds = F(s_2 - s_1) = F\Delta s$$

where $\Delta s = 40$ ft. Therefore,

$$U_{1 \to 2} = 173.1(40) = 6920 \text{ lb·ft}. \qquad \text{ANS.}$$

■ **Example 12.2**

The mechanism shown in Figure E12.2 is used to raise the weight W by application of the couple M at joint B. The magnitude of the couple needed is found to be $M = \dfrac{1}{2} WL \cos\left(\dfrac{\alpha}{2}\right)$ where α is the angle between member BC and the horizontal. (a) Determine the work done by the couple M as α changes from 20° to 90°. Express the answer in terms of W and L. (b) Let $W = 750$ N, $L = 2$ m, and find a numerical value for the work of M as α changes from 20° to 90°.

Solution

(a) The work of couple M may be obtained by using Eq. (12.7). Thus,

$$U_{1 \to 2} = \int_{20°}^{90°} \frac{1}{2} WL \cos\left(\frac{\alpha}{2}\right) d\alpha = WL \sin\left(\frac{\alpha}{2}\right)\Bigg]_{20°}^{90°}$$

$$= 0.533WL \qquad \text{ANS.}$$

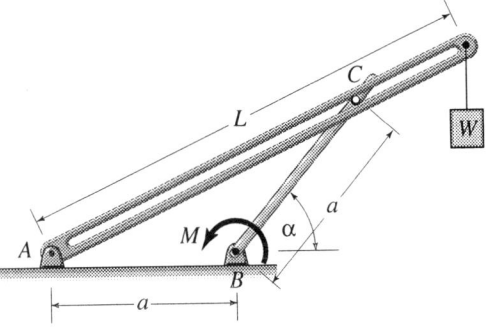

FIGURE E12.2.

(b) Substituting the values given for W and L in the above equation for $U_{1 \to 2}$,

$$U_{1 \to 2} = 0.533(750)(2) = 800 \text{ N·m.}$$ ANS.

■ **Example 12.3**   A force $\mathbf{F} = [(xy + x^2)\mathbf{i} + (2y + x)\mathbf{j}]$ kN acts on a particle that moves from the origin at O to point A as shown in Figure E12.3. Determine the work performed by $\mathbf{F}$ if the path followed is (a) $y = 4x$ and (b) $y = x^2$.

Solution   (a) By Eq. (12.5),

$$U_{1 \to 2} = \int_{Q_1}^{Q_2} \mathbf{F} \cdot d\mathbf{r}.$$

Now, if we substitute the equation of the path $y = 4x$ in the force expression,

$$\mathbf{F} = [x(4x) + x^2]\mathbf{i} + [2(4x) + x]\mathbf{j} = (5x^2)\mathbf{i} + (9x)\mathbf{j}.$$

The position vector from the origin to any point in the x-y plane is given by

$$\mathbf{r} = x\mathbf{i} + y\mathbf{j} = x\mathbf{i} + 4x\mathbf{j}$$

so that

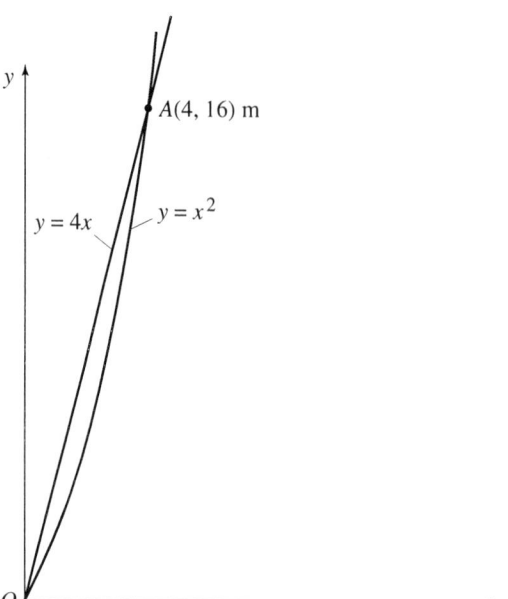

FIGURE E12.3.

$$d\mathbf{r} = (dx)\mathbf{i} + (4dx)\mathbf{j}.$$

Therefore,

$$
\begin{aligned}
U_{1 \to 2} &= \int_0^4 [(5x^2)\mathbf{i} + (9x)\mathbf{j}]\cdot[(dx)\mathbf{i} + (4dx)\mathbf{j}] \\
&= \int_0^4 (5x^2 + 36x)\,dx = \left.\frac{5x^3}{3} + 18x^2\right|_0^4 \\
&= 395 \text{ kN·m.} \qquad\qquad\qquad\qquad\qquad \text{ANS.}
\end{aligned}
$$

(b) Again by, Eq. (12.5),

$$U_{1 \to 2} = \int_{Q_1}^{Q_2} \mathbf{F}\cdot d\mathbf{r}.$$

Now, if we substitute the equation of the path $y = x^2$ in the force expression,

$$\mathbf{F} = [x(x^2) + x^2]\mathbf{i} + [2(x^2) + x]\mathbf{j} = (x^3 + x^2)\mathbf{i} + (2x^2 + x)\mathbf{j}$$

and, as in part (a),

$$
\begin{aligned}
\mathbf{r} &= (x)\mathbf{i} + (y)\mathbf{j} = (x)\mathbf{i} + (x^2)\mathbf{j}, \\
d\mathbf{r} &= (dx)\mathbf{i} + (2x\,dx)\mathbf{j}.
\end{aligned}
$$

Therefore,

$$
\begin{aligned}
U_{1 \to 2} &= \int_0^4 [(x^3 + x^2)\mathbf{i} + (2x^2 + x)\mathbf{j}]\cdot[(dx)\mathbf{i} + (2x\,dx)\mathbf{j}] \\
&= \int_0^4 (5x^3 + 3x^2)\,dx = \left.\frac{5x^4}{4} + x^3\right|_0^4 \\
&= 384 \text{ kN·m.} \qquad\qquad\qquad\qquad\qquad \text{ANS.}
\end{aligned}
$$

Note that the work done by a force depends upon the path traveled by its point of application. This is generally true except for conservative forces which will be discussed in Section 1.5.

■

Problems

12.1 A worker makes use of the system shown in Figure P12.1 to raise the weight W from position 1 to position 2, a total height h. Let $W = 500$ lb, $h = 20$ ft, assume that the circular peg at the top of the system is frictionless, and compute the work performed by (a) the weight W and (b) the worker. Assume motion at constant speed.

12.2   Refer to the system shown in Figure
P12.1. Let $W = 10$ kN, $h = 10$ m, and
the coefficient of friction between the
cable and the circular peg $\mu = 0.3$. Determine (a) the work performed by the
weight W as it rises from position 1 to
position 2 and (b) the work performed
by the worker during this process.

12.3   A disabled sports car weighing 2500 lb.
is being pulled up a steep inclined plane
at constant speed by a tow truck, as
shown in Figure P12.3. Determine (a)
the work done by the tow truck as it
moves the sports car a distance of 40 ft
up the inclined plane and (b) the work
done by the weight of the sports car.
Assume frictionless conditions on the
inclined plane.

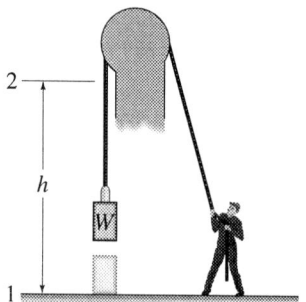

FIGURE P12.1.

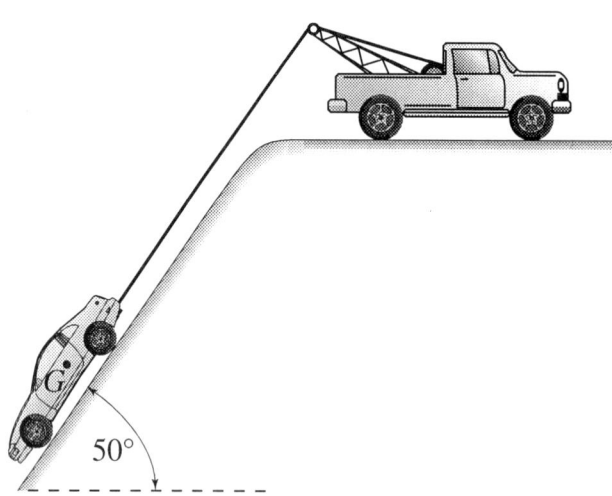

FIGURE P12.3.

12.4   Under the influence of constant forces,
such as Q and F in Figure P12.4, a block
of weight W moves a distance s along a
horizontal plane from position 1 to position 2. If the coefficient of kinetic friction
between the block and the horizontal
plane $\mu_k = 0.15$, $W = 150$ N, $Q = 0$, $F =$

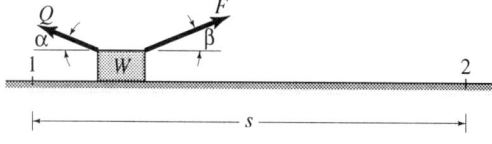

FIGURE P12.4.

100 N, $\beta = 25°$, and $s = 25$ m, determine (a) the work done by F, (b) the work done by the force of friction, and (c) the work done by the weight W.

12.5   Refer to Problem 12.4. Let $\mu_k = 0.10$, $W = 75$ lb, $F = 0$, $Q = 30$ lb, $\alpha = 30°$, and $s = 40$ ft. Determine (a) the work performed by Q as the block moves from 1 to 2, (b) the work done by the force of friction during this interval, and (c) the work done by the weight W.

12.6   Under the influence of several constant forces including F, a block of weight W moves along an inclined plane a distance s from position 1 to position 2, as shown in Figure P12.6. Let $W = 10$ kN, $F = 5$ kN, $\beta = 45°$, $s = 10$ m, and the coefficient of kinetic friction between the block and the inclined plane $\mu_k = 0.20$. Determine (a) the work done by F, (b) the work done by the frictional forces, and (c) the work done by the weight W.

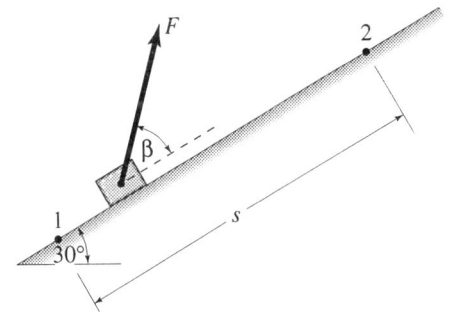

FIGURE P12.6.

12.7   Under the influence of several constant forces including F, a block of weight W moves along an inclined plane a distance s from position 2 to position 1, as shown in Figure P12.6. Let $W = 4$ k, $F = 2$ k, $\beta = 35°$, $s = 25$ ft, and the coefficient of kinetic friction between the block and the incline plane $\mu_k = 0.10$. Determine (a) the work done by F, (b) the work performed by the frictional forces, and (c) the work done by the weight W.

12.8   A screw driver is used to drive a screw into a timber beam, as shown in Figure P12.8. The operation requires the application of a driving force $F = 15$ lb and a turning couple $M = 5$ lb·ft. Determine the total work performed by the screw driver in turning the screw one complete turn. The pitch of the screw (i.e., the advance of the screw per complete turn) is 0.05 in. Assume that F and M remain constant during the operation.

12.9   The jammed drill shaft of an oil drilling rig needs to be retracted a distance of 10 m by application of a pull $F = 5$ kN and a couple $M = 0.5$ kN·m as shown in Figure P12.9. Assume that, during the retraction process, both F and M remain constant and that the drill shaft makes a total of five complete turns in the direction of the torque. If the weight of the shaft $W = 700$ N, compute the total work performed by F, W, and M during the retraction process.

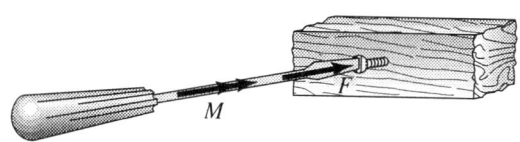

FIGURE P12.8.

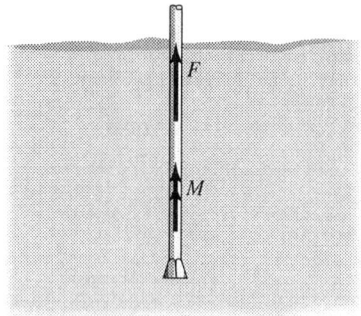

FIGURE P12.9.

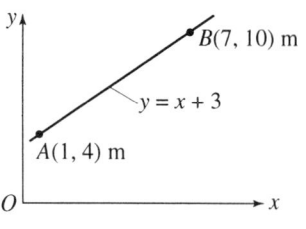

FIGURE P12.11.

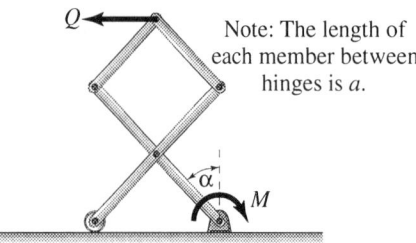

Note: The length of each member between hinges is *a*.

FIGURE P12.10.

A to point B along the straight path given by the equation $y = x + 3$ as shown in Figure P12.11. Determine the work done by **F** during this process.

12.12 A force $\mathbf{F} = [(2x + y)\mathbf{i} + (x^2 + 4y)\mathbf{j}]$ lb acts on a particle that moves from the origin at 0 to point A as shown in Figure P12.12. Determine the work done by **F** if the path followed is (a) $y = x$ and (b) $y = 5x - x^2$.

12.10 Equilibrium of the system shown in Figure P12.10 requires that the couple M be equal to $Qa\cos\alpha$. Determine the work done by the couple M as α changes from $15°$ to $75°$ if $Q = 100$ lb and $a = 10$ in.

12.11 A force $\mathbf{F} = [(x + y)\mathbf{i} + (2x + 3y)\mathbf{j}]$ N acts on a particle that moves from point

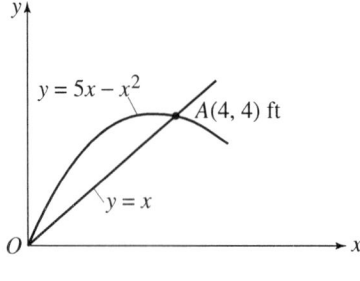

FIGURE P12.12.

12.4
The Concept of Virtual Work

Unlike the real differential displacement $d\mathbf{r}$ discussed in the preceding sections, the term *virtual displacement* is used to signify an assumed or fictitious differential displacement that does not actually take place. To distinguish it from the real displacement $d\mathbf{r}$, the virtual displacement is given the symbol $\delta\mathbf{r}$ and referred to sometimes as a *variation*. However, it should be kept in mind that both of these displacements are first-order differential quantities. In addition to being small, these virtual displacements must be *consistent* with the constraints of a given problem. As will be seen shortly, the concept of virtual displacements is extremely useful in examining the equations of equilibrium, particularly, the equilibrium of systems of connected rigid bodies.

Particle A in Figure 12.9 is subjected to the forces F_1 to F_n. This particle may be or is in equilibrium under the action of these forces, and the displacement from A_1 to A_2 may or may not necessarily take place. If we imagine that a virtual displacement δr occurs, moving the particle to position A_2, a *virtual work* $\delta U = F \cdot \delta r$ is performed by each of the forces acting on the particle. It is assumed that the forces remain *constant* during a differential virtual displacement. The total virtual work performed by all forces acting on the particle during this virtual displacement becomes

$$\delta U = F_1 \cdot \delta r + F_2 \cdot \delta r + F_3 \cdot \delta r + \cdots + F_n \cdot \delta r = (\sum F) \cdot \delta r \quad (12.8)$$

Because the particle is in equilibrium, it follows that $\sum F = 0$ and, therefore, the product $(\sum F) \cdot \delta r$ vanishes, which means that the total virtual work δU must be zero. This conclusion gives rise to a very important principle known as the *principle of virtual work* which was first introduced by John Bernoulli in 1717. It states that:

> *The virtual work performed by all forces acting on a particle*
> *in equilibrium is zero for any virtual displacement.*

This principle is stated mathematically by the equation

$$\delta U = 0. \quad (12.9)$$

Combining Eqs. (12.8) and (12.9), we conclude that

$$\delta U = (\sum F) \cdot \delta r = (\sum F_x)\delta x + (\sum F_y)\delta y + (\sum F_z)\delta z = 0 \quad (12.10)$$

which leads to the conclusion that, when the virtual work is zero,

$$\sum F_x = 0,$$

$$\sum F_y = 0,$$

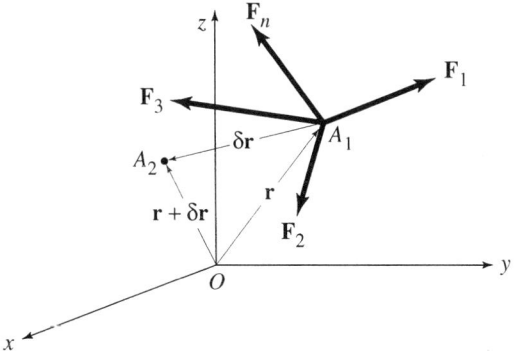

FIGURE 12.9.

and

$$\sum F_z = 0.$$

Therefore, the principle of virtual work is another way of stating the conditions of equilibrium for a particle. Thus, because the conditions of equilibrium are both necessary and sufficient, it follows that the principle of virtual work is also both *necessary* and *sufficient* for the equilibrium of a particle.

As in the case of real displacements and real work discussed in the previous sections, it can be shown that, in dealing with a rigid body, *only* the external forces acting on the body need be considered in determining the virtual work performed during a virtual displacement. Also, the virtual work of a couple M acting on a rigid body may be determined from the relationship

$$\delta U = \mathbf{M} \cdot \delta \boldsymbol{\theta} = (M \cos \beta) \delta \theta \qquad (12.11)$$

where $\delta \boldsymbol{\theta}$ is a virtual displacement (rotation) of the body and β represents the angle between vectors $\mathbf{M}$ and $\delta \boldsymbol{\theta}$. Therefore, the principle of virtual work as it applies to a rigid body may be stated as:

The virtual work performed by all external forces and couples acting on a rigid body in equilibrium is zero for any virtual displacement.

Thus, Eq. (12.9) is also valid in the case of a rigid body in equilibrium if the virtual work δU includes the work of *all* external forces and couples.

The principle of virtual work may be used for solving equilibrium problems relating to individual particles and individual rigid bodies. This approach, however, presents no advantage over the standard method of constructing a free-body diagram and applying the conditions of equilibrium as in earlier chapters. The real advantage of the principle of virtual work is realized when dealing with the equilibrium of a system of connected rigid bodies because there would be no need to dismember the system of connected rigid bodies into two or more free-body diagrams to solve the problem. As will be shown shortly, the use of the method of virtual work makes it possible to solve such a problem using one single free-body diagram, i.e., the free-body diagram of the entire system of connected rigid bodies.

Two observations need to be made about systems of connected rigid bodies. The first is the assumption that the joint between any two adjacent rigid bodies in the system is frictionless. Under these ideal conditions, the net work performed by the two equal and opposite forces at the joint is zero as discussed in Section 12.1. The second has to do with the question of the degree of freedom of the system. By

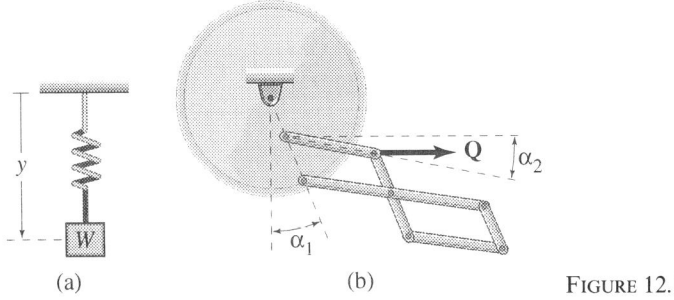

(a)   (b)   FIGURE 12.10.

definition, the *degree of freedom* of a system is the number of independent variables required to fully define the configuration of the system. The number of independent variables in a given case must be *consistent* with the constraining supports of the system. Thus, for example, consider the weight-spring system shown in Figure 12.10(a). The position of this system can be completely described by specifying the one variable y. Thus, this system is referred to as a one-degree-of-freedom system. The mechanism shown in Figure 12.10(b), however, requires two variables α_1 and α_2 for complete definition of its configuration. Therefore, such a mechanism has two degrees of freedom. Systems with more than two degrees of freedom will not be considered in this textbook, although their analysis may be performed similarly to that used for analyzing systems with two degrees of freedom.

The principle of virtual work may now be stated for a system of connected rigid bodies:

> *The virtual work performed by all external forces and couples acting on a system of connected rigid bodies in equilibrium is zero for any independent virtual displacement consistent with the constraints.*

Therefore, Eq. (12.9) $\delta U = 0$ is still valid for the case of a system of connected rigid bodies provided that the work δU is interpreted as the work of all forces and couples applied externally to the system.

The following examples illustrate the application of the principle of virtual work in solving problems dealing with the equilibrium of systems of connected rigid bodies.

■ **Example 12.4**

Consider the system shown in Figure E12.4(a) which is used to transport people from one level to another within the range of values of α between $0°$ and $90°$ using a mechanism (not shown) that applies a couple M at hinge A. The total weight of the passengers and the seat assembly including member CD is W_1 and may be assumed to act at G.

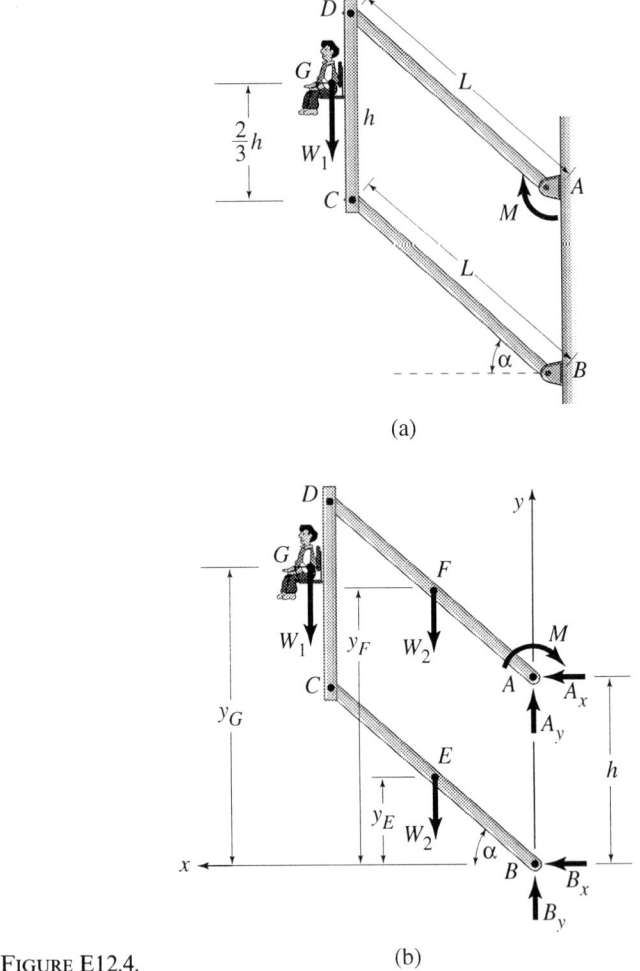

FIGURE E12.4.   (b)

The weight of each of the two uniform members AD and BC is W_2 and may be assumed to act through the geometric centers of these members. (a) Develop an expression that relates M to α. (b) Find the work performed by M as α changes from $0°$ to $90°$.

Solution

(a) The free-body diagram of the entire system for any angular displacement α is shown in Figure E12.4(b) referred to an x-y coordinate system with origin at B. The reactions A_x, A_y at A and B_x, B_y at B perform no work because their points of application do not displace, and the only loads that do perform work are the forces W_1 and W_2 and the couple M. This is a one-degree-of-freedom system because we can fully define its configuration by specifying the one variable α. From the geometry in Figure E12.4(b), the following relationships are established:

$$
\left.
\begin{aligned}
&y_\mathrm{E} = \frac{L}{2}\sin\alpha, && \delta y_\mathrm{E} = \frac{L}{2}\cos\alpha\,\delta\alpha, \\[2mm]
&y_\mathrm{G} = L\sin\alpha + \frac{2}{3}h, && \delta y_\mathrm{G} = L\cos\alpha\,\delta\alpha, \\[2mm]
&y_\mathrm{F} = \frac{L}{2}\sin\alpha + h, && \delta y_\mathrm{F} = \frac{L}{2}\cos\alpha\,\delta\alpha.
\end{aligned}
\right\} \tag{a}
$$

and

If the system is given a cw virtual angular displacement $\delta\alpha$, the load W_1 at G experiences a virtual displacement δy_G, the load W_2 at E a virtual displacement δy_E, and the load W_2 at F a virtual displacement δy_F. Thus, by the principle of virtual work $\delta U = 0$,

$$
-W_1 \delta y_\mathrm{G} - W_2 \delta y_\mathrm{E} - W_2 \delta y_\mathrm{F} + M\delta\alpha = 0. \tag{b}
$$

Note that, whereas the couple M performs positive work because it acts in the same sense as the virtual angular displacement $\delta\alpha$, the forces W_1 and W_2 both perform negative work because they act opposite to their respective virtual displacements. Substituting from Eqs. (a) in Eq. (b) and after simplifying,

$$
(M - W_1 L \cos\alpha - W_2 L \cos\alpha)\delta\alpha = 0.
$$

Because $\delta\alpha \neq 0$, it follows that the quantity in the brackets must vanish, and, therefore,

$$
M = (W_1 + W_2)L\cos\alpha. \qquad \text{ANS.}
$$

(b) The work performed by the couple M as it displaces the system from $\alpha = 0°$ to $\alpha = 90°$ may be determined by Eq. (12.7). Thus,

$$
\begin{aligned}
U_{1 \to 2} &= \int_{\alpha_1}^{\alpha_2} M\, d\alpha = \int_0^{\pi/2} (W_1 + W_2)L\cos\alpha\, d\alpha \\[2mm]
&= (W_1 + W_2)L\sin\alpha \Big|_0^{\pi/2} \\[2mm]
&= (W_1 + W_2)L. \qquad \text{ANS.}
\end{aligned}
$$

■ **Example 12.5**

A special-purpose scale is constructed, as shown in Figure E12.5(a), consisting of two identical parallel-bar mechanisms, only one of which is shown. The spring of constant k is unstretched with a length L_u, when $\theta = 0°$, and all hinged joints may be assumed frictionless. Assume that the weights of all linkages in the system are negligibly small compared to the total weight W on the scale. The weight W is placed on the

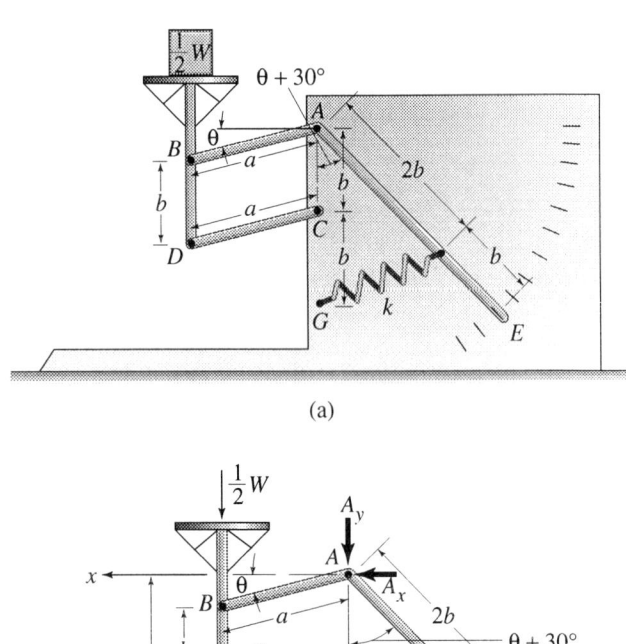

FIGURE E12.5.   (b)

scale so that each mechanism carries $W/2$. The arm AE is integral with member BA of the mechanism, and the pointer at E describes an arc of a circle of radius AE when loads are placed on the scale.

(a) Derive a relationship between the weight W and the angle θ which defines the orientation of member BAE.

(b) Determine the weight W if the equilibrium angle (i) $\theta = 30°$ and (ii) $\theta = 60°$. Let $k = 3000$ lb/ft and $b = a/2 = 4$ ft.

Solution

(a) Consider the free-body diagram of the mechanism referred to an x-y coordinate system with origin at point A, as shown in Figure E12.5(b). The forces A_x, A_y, C_x, and C_y do no work because they do not displace. The only forces that perform work in this case are $\frac{1}{2}W$ and the force F in the spring.

This mechanism has one degree of freedom because we can define completely its configuration by specifying the one variable θ. Thus, from the geometry in Figure E12.5(b), the following relationships are determined:

$$y_B = a \sin \theta, \quad \delta y_B = a \cos \theta \, \delta \theta, \tag{a}$$

and

$$L^2 = (2b)^2 + (2b)^2 - 2(2b)(2b)\cos(\theta + 30°) = 8b^2[1 - \cos(\theta + 30°)]$$

where L represents the stretched length of the spring for any value of θ. Therefore,

$$2L \, \delta L = 8b^2 \sin(\theta + 30°) \, \delta \theta$$

from which

$$\delta L = \frac{4b^2 \sin(\theta + 30°)}{L} \, \delta \theta = \frac{4b \sin(\theta + 30°)}{\sqrt{8[1 - \cos(\theta + 30°)]}} \, \delta \theta. \tag{b}$$

When the system is given a ccw virtual angular displacement $\delta \theta$, the load $W/2$ experiences a virtual displacement δy_B and the spring force F a virtual displacement δL. By the principle of virtual work $\delta U = 0$,

$$\tfrac{1}{2} W \delta y_B - F \delta L = 0. \tag{c}$$

The force $W/2$ performs positive work because it acts in the same sense as its virtual displacement δy_B. The force F performs negative work because it acts opposite to the virtual displacement δL. The spring force is $F = ks$ (see Eq. (2.14)), where $s = L - L_u$ represents the stretch in the spring and L_u is its unstretched length. Thus,

$$s = L - L_u = \sqrt{8b^2[1 - \cos(\theta + 30°)]} - 1.035b$$

$$= b\{\sqrt{8[1 - \cos(\theta + 30°)]} - 1.035\}.$$

Therefore, the force in the spring becomes

$$F = kb\{\sqrt{8[1 - \cos(\theta + 30°)]} - 1.035\}. \tag{d}$$

Substituting from Eqs. (a), (b) and (d) in Eq. (c),

$$\left\{ \frac{1}{2} Wa \cos \theta - \frac{4kb^2[\sqrt{8[1 - \cos(\theta + 30°)]} - 1.035]}{\sqrt{8[1 - \cos(\theta + 30°)]}} \sin(\theta + 60°) \right\} \delta \theta$$

$$= 0.$$

Because $\delta \theta \neq 0$, it follows that the quantity within the large brackets must vanish. Therefore,

$$W = \frac{8kb^2[\sqrt{8[1 - \cos(\theta + 30°)]} - 1.035] \sin(\theta + 60°)}{a \cos \theta \sqrt{8[1 - \cos(\theta + 30°)]}}. \qquad \text{ANS.}$$

(b) (i) Substituting the given data for $\theta = 30°$,

$$W = \frac{8(3000)(16)\left[\sqrt{8[1 - \cos 60°]} - 1.035\right]\sin 60°}{a\cos 30°\sqrt{8[1 - \cos 60°]}}$$

$$= 23{,}200 \text{ lb.} \qquad\qquad\qquad\qquad\qquad\qquad \text{ANS.}$$

(ii) Substituting the given data for $\theta = 60°$,

$$W = \frac{8(3000)(16)\left[\sqrt{8[1 - \cos 90°]} - 1.035\right]\sin 90°}{a\cos 60°\sqrt{8[1 - \cos 90°)}}$$

$$= 60{,}900 \text{ lb.} \qquad\qquad\qquad\qquad\qquad\qquad \text{ANS.}$$

■ **Example 12.6**

Consider the system, shown in Figure E12.6(a), consisting of the two uniform pin-connected members AB and BC each of length $L = 1$ m and the two identical springs with a spring constant k = 300 N/m. The weight of member AB $W_1 = 200$ N and that of BC $W_2 = 300$ N. Determine the angles α_1 and α_2 ($0° \leq \alpha_1 \leq 90°$ and $0° \leq \alpha_2 \leq 90°$) corresponding to the equilibrium of the system. Assume frictionless conditions and that both springs are unstretched when $\alpha_1 = \alpha_2 = 0°$. The frictionless rollers in the horizontal track permit the springs to remain vertical.

Solution

The free-body diagram of the system is shown in Figure E12.6(b) referred to an x-y coordinate system with origin at A. Note that the only

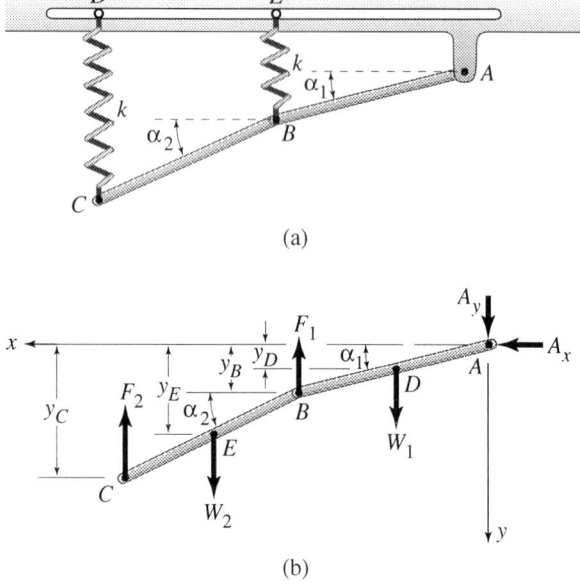

(a)

(b)

FIGURE E12.6.

forces performing work are the spring forces F_1 and F_2 and the member weights W_1 and W_2. Note also that this system has two degrees of freedom because the two independent variables α_1 and α_2 are needed for complete definition of its configuration. The solution of this problem, therefore, requires applying the principle of virtual work $\delta U = 0$, twice; once, using a virtual angular displacement $\delta\alpha_1$ while α_2 is held constant, and the second, a virtual angular displacement $\delta\alpha_2$ while α_1 is kept fixed.

The geometry in Figure E12.6(b) leads to the following relationships:

$$\left.\begin{aligned}
y_D &= \tfrac{1}{2}L\sin\alpha_1, \\
y_B &= L\sin\alpha_1, \\
y_E &= L\sin\alpha_1 + \tfrac{1}{2}L\sin\alpha_2, \\
y_C &= L\sin\alpha_1 + L\sin\alpha_2.
\end{aligned}\right\} \qquad \text{(a)}$$

and

Step 1: *System displaced by $\delta\alpha_2$ while α_1 remains constant*

Because α_1 is constant, the following virtual displacements are obtained from Eqs. (a):

$$\left.\begin{aligned}
\delta y_D &= \delta y_B = 0, \\
\delta y_E &= \tfrac{1}{2}L\cos\alpha_2\,\delta\alpha_2, \\
\delta y_C &= L\cos\alpha_2\,\delta\alpha_2.
\end{aligned}\right\} \qquad \text{(b)}$$

and

Therefore, the principle of virtual work $\delta U = 0$ reduces to

$$W_2\delta y_E - F_2\delta y_C = 0.$$

Replacing F_2 by $ky_C = k\,(L\sin\alpha_1 + L\sin\alpha_2)$ substituting from Eqs. (b) and simplifying,

$$\cos\alpha_2[\tfrac{1}{2}W_2 - kL(\sin\alpha_1 + \sin\alpha_2)] = 0.$$

Therefore,

$$\cos\alpha_2 = 0 \Rightarrow \alpha_2 = 90°,$$

and

$$\tfrac{1}{2}W_2 - kL(\sin\alpha_1 + \sin\alpha_2) = 0$$

from which

$$\sin\alpha_1 + \sin\alpha_2 = \frac{W_2}{2kL}. \qquad \text{(c)}$$

Step 2: *System displaced by $\delta\alpha_1$, while α_2 remains constant.*

Because α_2 is constant, the following virtual displacements are obtained from Eqs. (a):

$$\left.\begin{array}{l} \delta y_D = \tfrac{1}{2} L \cos \alpha_1 \, \delta \alpha_1, \\[4pt] \delta y_B = L \cos \alpha_1 \, \delta \alpha_1, \\[4pt] \delta y_E = L \cos \alpha_1 \, \delta \alpha_1, \\[4pt] \delta y_C = L \cos \alpha_1 \, \delta \alpha_1. \end{array}\right\} \tag{d}$$

and

Thus, the principle of virtual work $\delta U = 0$ leads to

$$W_1 \delta y_D + W_2 \delta y_E - F_1 \delta y_B - F_2 \delta y_C = 0.$$

Replacing F_1 by $k(L \sin \alpha_1)$, F_2 by $k(L \sin \alpha_1 + L \sin \alpha_2)$, substituting from Eqs. (d) and simplifying,

$$\cos \alpha_1 \left[(\tfrac{1}{2} W_1 + W_2) - kL(2 \sin \alpha_1 + \sin \alpha_2) \right] = 0.$$

Therefore,

$$\cos \alpha_1 = 0 \Rightarrow \alpha_1 = 90°,$$

and

$$\tfrac{1}{2} W_1 + W_2 - kL(2 \sin \alpha_1 + \sin \alpha_2) = 0$$

from which

$$2 \sin \alpha_1 + \sin \alpha_2 = \frac{\tfrac{1}{2} W_1 + W_2}{kL}. \tag{e}$$

Solving Eqs. (c) and (e) simultaneously,

$$\left.\begin{array}{l} \alpha_1 = \sin^{-1}\left(\dfrac{W_1 + W_2}{2kL}\right), \\[12pt] \alpha_2 = \sin^{-1}\left(-\dfrac{W_1}{2kL}\right). \end{array}\right\} \tag{f}$$

and

Equations (f) are valid only if $W_1 \leq 2kL$ and $(W_1 + W_2) \leq 2kL$. The minus sign in the equation for α_2 implies that this angle is not ccw as assumed in the diagrams but cw, instead, regardless of the values of W_1, W_2, k, and L.

Therefore, there are two possible equilibrium configurations (positions) defined as follows:

$$\alpha_1 = \alpha_2 = 90°, \qquad \text{ANS.}$$

and

$$\left.\begin{array}{l} \alpha_1 = \sin^{-1}\left(\dfrac{W_1 + W_2}{2kL}\right) = \sin^{-1}\left[\dfrac{200 + 300}{2(300)(1)}\right] = 56.4°, \\[14pt] \alpha_2 = \sin^{-1}\left(-\dfrac{W_1}{2kL}\right) = \sin^{-1}\left[-\dfrac{200}{2(300)(1)}\right] = -19.47°. \end{array}\right\} \quad \text{ANS.}$$

Problems

12.13 Determine the force F developed in the spring for any value of the angle α and of the force Q which is applied vertically at D, as shown in Figure P12.13. Assume a frictionless collar at B. If $Q = 50$ lb, $\alpha = 30°$, and the deformation of the spring is measured to be 1.5 in., find the necessary spring constant. The spring is undeformed when $\alpha = 90°$.

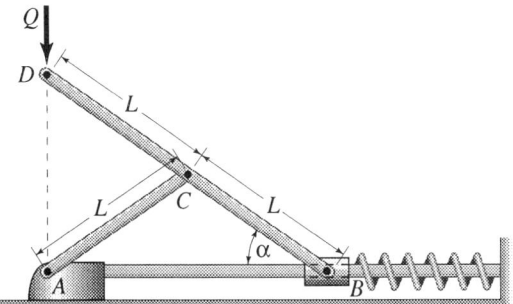

FIGURE P12.13.

12.14 A worker weighing 700 N stands on a 4-m ladder as shown in Figure P12.14. Let the wall be smooth at A and the floor be rough at B with a coefficient of friction μ. The weight of the ladder is 60 N and acts at C. Determine the minimum value of μ needed for equilibrium. Assume that the weight of the worker acts at point D on the ladder.

12.15 A package of weight W is supported by the two parallel links while a force P and a couple $M = PL$ are applied as shown in Figure P12.15. (a) Develop an expression for P as a function of W and α consistent with the equilibrium of the system. (b) If $W = 500$ lb, $P = 50$ lb, and $L = 5a = 2b = 30$ in., determine the angle α $(0° \leq \alpha \leq 90°)$ at equilibrium

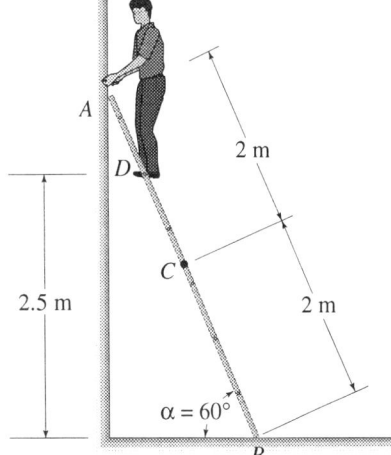

FIGURE P12.14.

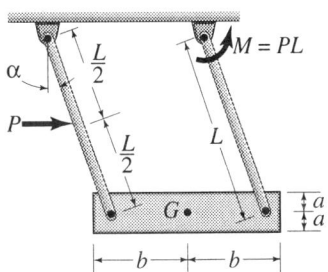

FIGURE P12.15.

(Hint: Use a trial-and-error procedure to solve for α).

12.16 The system shown in Figure P12.16 consists of a weightless rod CB hinged at O and supported by two springs k_1 and k_2. A crate of weight W is suspended at A. Determine the angle β $(0° \leq \beta \leq 90°)$ for equilibrium if $W = 900$ N, $L = 3$ m, $k_1 = 400$ N/m, and $k_2 = 200$ N/m. The frictionless rollers permit the springs to remain vertical. Assume that the two springs are unstretched when $\beta = 0°$.

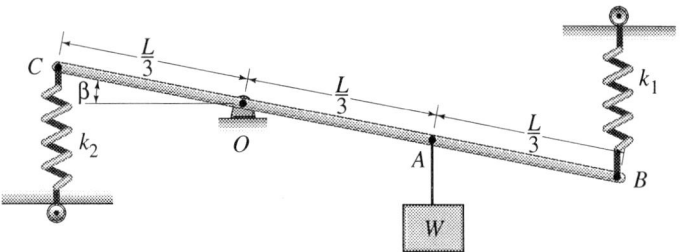

FIGURE P12.16.

12.17 The mechanism, shown in Figure P12.17, consists of two uniform members AB and BC each of length a and weight W hinged together at B, to a fixed hinge at A, and to a roller at C that moves in a horizontal frictionless guide. A linear spring BD of spring constant k is attached to a roller at D that permits the spring to remain horizontal for any position of the mechanism defined by the angle β. When β is zero, the spring is unstretched. (a) Develop a relationship between the force Q and the angle β for the range $0 \leq \beta \leq \pi/2$. (b) What is the force F in the spring for any value of β between 0 and $\pi/2$? (c) Find numerical values for Q and F corresponding to the following data: $k = 30$ lb/in., $a = 2$ ft., $W = 20$ lb, and $\beta = 30°$.

12.18 The pulley arrangement shown in Figure P12.18 is used to lift the weight W

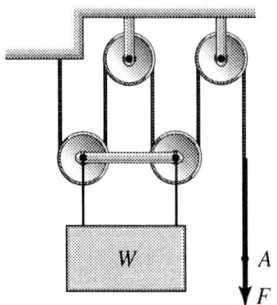

FIGURE P12.18.

by applying a force F at A. Develop a relationship between W and F assuming that all pulleys are frictionless and weightless.

12.19 The pulley arrangement shown in Figure P12.19 is used to pull the weight W along the inclined rough plane for which the coefficient of friction is μ. If all pulleys are assumed weightless and frictionless, (a) determine the relationship between the force F and the weight W and (b) find the value of μ for the case when $W = 750$ lb, $F = 250$ lb, and $\theta = 30°$.

12.20 A weightless missile launcher is shown in Figure P12.20. The launch position is adjusted by changing the angle α through application of the couple M at joint B. The weight of the missile is W and may be assumed to act at point G.

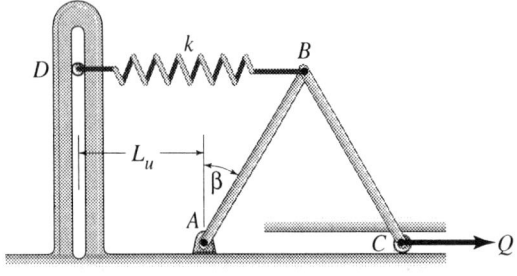

FIGURE P12.17.

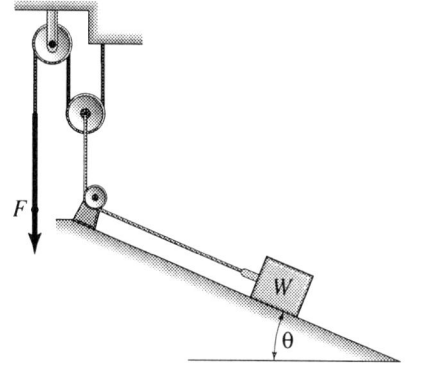

FIGURE P12.19.

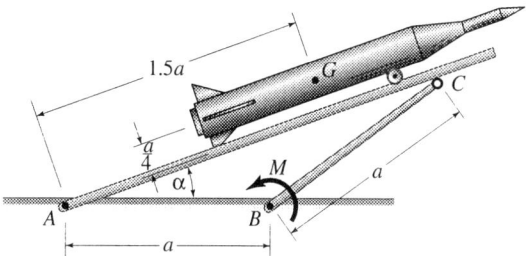

FIGURE P12.20.

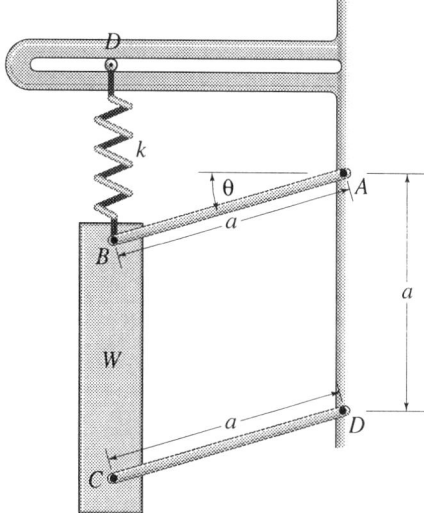

FIGURE P12.21.

(a) Develop an expression for the couple M in terms of the weight W and the angle α. (b) Find the total work performed by the couple M in changing the angle α from zero to $\pi/3$.

12.21 Assume frictionless joints and weightless members AB and CD in Figure P12.21. The spring with a constant $k = 2000$ N/m is unstretched when $\theta = 0°$. Determine the angle θ ($0° \leq \theta \leq 90°$) for equilibrium if $W = 1000$ N and $a = 0.8$ m. The frictionless roller at D allows the spring to remain vertical.

12.22 Refer to Problem 12.21. Remove the linear spring k and replace it with a torsional spring at joint A having a constant K which is measured in terms of N·m/ rad. (a) Compute the value that K

must have if the equilibrium angle θ is to be 30°. (b) Determine the value of the equilibrium angle θ ($0° \leq \theta \leq 90°$) if $K = 500$ N·m/rad.

12.23 The height h of the crate in the delivery truck is varied by applying a couple M to two identical mechanisms only one of which is shown in Figure P12.23. Develop an expression for the height h as a function of the couple M and the weight of the crate W.

12.24 Assume that all hinges of the mechanism, shown in Figure P12.24, are frictionless. What is the weight W that may be sup-

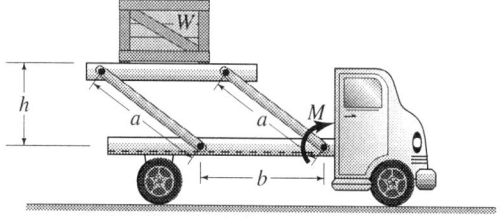

FIGURE P12.23.

ported by a force $F = 50$ lb if $a = 10$ in., $b = 20$ in., and $c = 10$ in?

12.25 Refer to the mechanism shown in Figure P12.25 in which all hinges and the roller are frictionless and all members are weightless. The spring of constant k is unstretched when $\alpha = 0°$. Develop an expression for the force P in terms of the angle α and the spring constant k.

12.26 The position of the work platform in the truck is changed by the application of the couple M to two identical mecha-

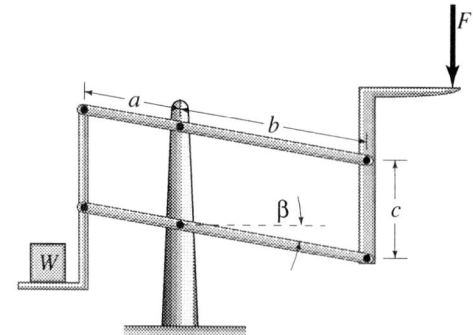

FIGURE P12.24.

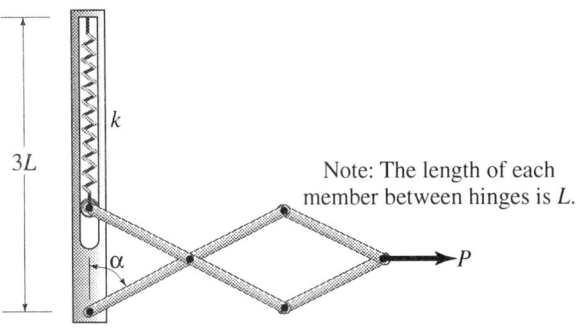

Note: The length of each member between hinges is L.

FIGURE P12.25.

nisms only one of which is shown in Figure P12.26. (a) Derive a relationship expressing the couple M as a function of the height h and the weight W of the platform. (b) What weight W is being supported when $M = 5$ kN·m, $a = 1.5$ m, and $h = 4$ m?

12.27 The force Q developed in the jaws of the clamp shown in Figure P12.27 is controlled by turning the handle H. Each of the two threaded shafts attached to the handle H has a pitch (i.e., the advance of the shaft per complete turn) equal to p. (a) Develop a relation between the force Q and the couple M applied to the handle. (b) If the force Q is to be 50 lb, determine the couple M that needs to be

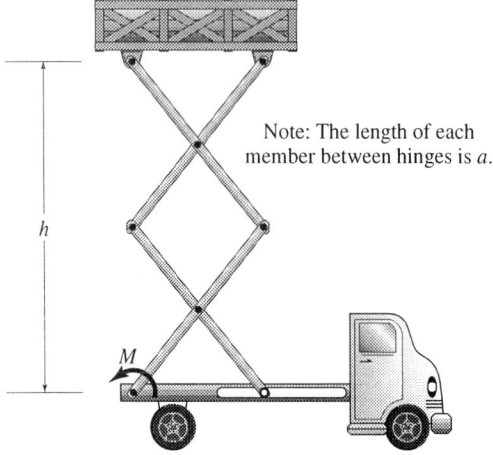

Note: The length of each member between hinges is a.

FIGURE P12.26.

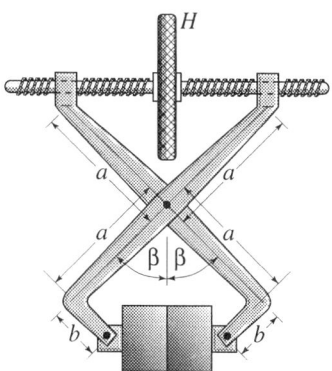

FIGURE P12.27.

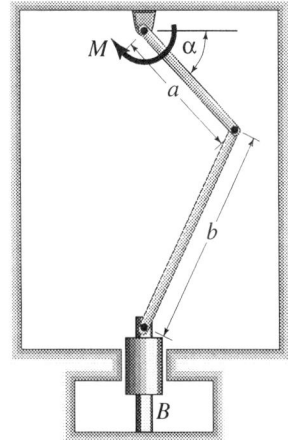

FIGURE P12.29.

applied to the handle for the following conditions: $a = 6$ in., $b = 3$ in., $p = 0.1$ in., and $\beta = 30°$.

12.28 The press shown in Figure P12.28 consists of two identical toggle mechanisms which are operated by two identical hydraulic cylinders. What is the force F developed in each of the two hydraulic cylinders when the press force on block B is Q? Assume frictionless conditions throughout.

12.29 Refer to Figure P12.29. What is the moment M required to produce a compressive force F at B equal to 2 kN when $\alpha = 60°$. Assume frictionless conditions throughout and weightless members. Let $a = b/2 = 0.2$ m.

12.30 A special-purpose lift is constructed as shown in Figure P12.30. It consists of two identical parallel-bar mechanisms only one of which is shown. Assume frictionless conditions and that each of the two mechanisms carries one half of the total load lifted. Derive an expression relating the force F in the hydraulic cylinder AB and the weight W.

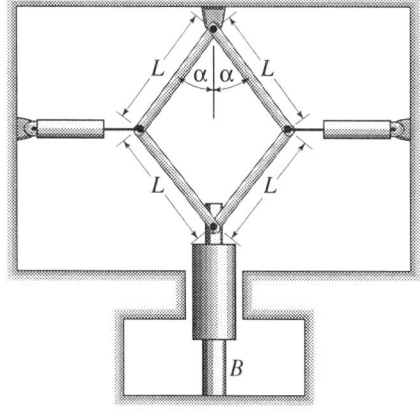

FIGURE P12.28.

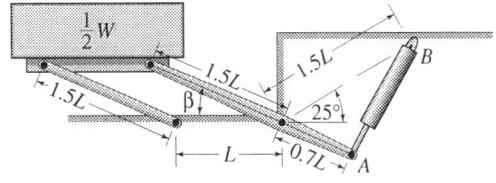

FIGURE P12.30.

12.31 Refer to the mechanism shown in Figure P12.31. Assume frictionless conditions at the hinges and weightless linkages.

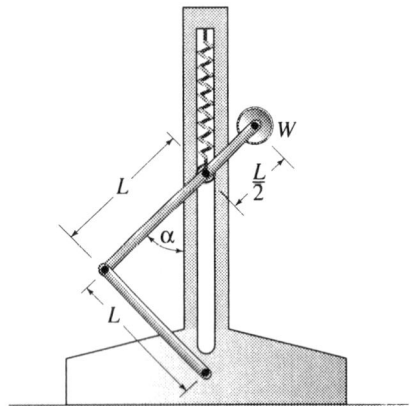

FIGURE P12.31.

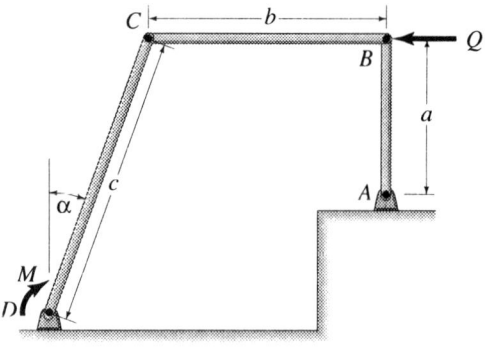

FIGURE P12.33.

The spring is unstretched when $\alpha = 0°$ and its spring constant is $k = 20$ lb/in. If $W = 150$ lb and $L = 18$ in, compute the angle α $(0° \leq \alpha \leq 90°)$ for equilibrium.

12.32 Refer to the mechanism shown in Figure P12.32. Determine the force F developed in the spring for the case when $M = 1.5$ kN·m, $a = b/3 = h/2 = 0.4$ m, and the equilibrium angle $\beta = 60°$. Assume frictionless conditions and weightless members. The spring is unstretched when $\beta = 0°$.

12.33 The mechanism shown in Figure P12.33 is subjected to the simultaneous action of the force Q and the couple M. De-

velop a relationship between Q and M for a specific value of the equilibrium angle α and for the condition shown where AB is vertical and BC is horizontal. What is the force Q for the case when $a = 2$ m, $b = 3$ m, $c = 4$ m, $M = 10$ kN·m, and $\alpha = 45°$? Assume frictionless conditions and weightless members.

12.34 Consider the Scotch-yoke mechanism shown in Figure P12.34. If $M = 75$ lb·ft., $a = 2$ ft., $b = 4$ ft., and $\beta = 60°$, determine the force P for equilibrium. Assume frictionless conditions and weightless members.

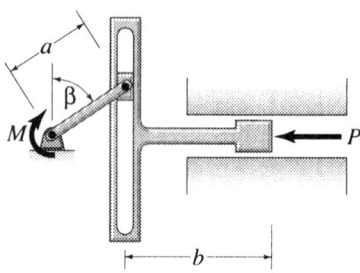

FIGURE P12.34.

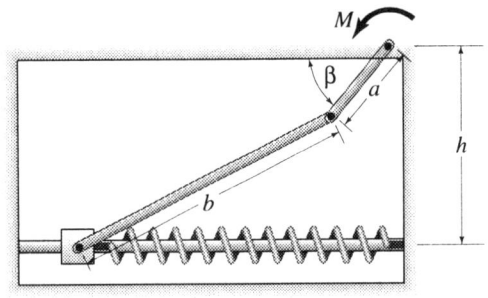

FIGURE P12.32.

12.35 Each of the two identical links AB and BC in Figure P12.35 has a length L and weight W. The spring is unstretched

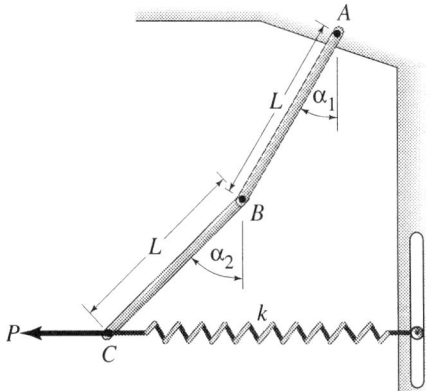

FIGURE P12.35.

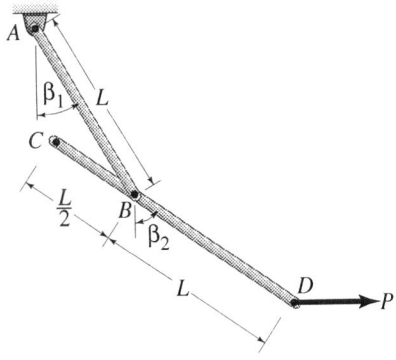

FIGURE P12.36.

when $\alpha_1 = \alpha_2 = 0°$. Assume frictionless conditions, and determine the angles α_1 and α_2 for equilibrium. The frictionless roller at D allows the spring to remain horizontal. Let $W = 10$ lb., $k = 1$ lb/in., $L = 30$ in., and $P = 10$ lb. The angles α_1 and α_2 vary between zero and 90°.

12.36 The mechanism shown in Figure P12.36 consists of two uniform links AB of length L and CD of length $3L/2$ hinged together at B and to a fixed support at A. The weight of AB is W and that of CD is $3W/2$. A horizontal force P is applied at D. Assume frictionless conditions, and determine the equilibrium angles β_1 and β_2. What are these angles for the case when $P = 4W$? The angles β_1 and β_2 vary between zero and 90°.

12.5*
Work of Conservative Forces

The work of a force was defined in Section 12.3 by Eq. (12.6) which is repeated here for convenience. Thus,

$$U_{1\to2} = \int_{s_1}^{s_2} (F \cos \beta)\, ds. \qquad (12.6)$$

The quantity s is the distance measured along the path between position s_1 and s_2 and $F \cos \beta$ represents the components of the force F along the tangent to the path. As stated previously, integration of Eq. (12.6) is possible only if (a) the force $F \cos \beta$ is constant or (b) we can relate the quantities F and β to the path s. In either case, it is seen that, in general, the work $U_{1\to2}$ depends upon the path followed by the point of application of the force. There are special cases of forces, however, for which the work $U_{1\to2}$ is independent of the specific path followed and depends only on the initial and final positions expressed by the symbols s_1 and s_2. Such forces are known as *conservative* forces. Also, a mechanical system acted upon by only conservative forces is

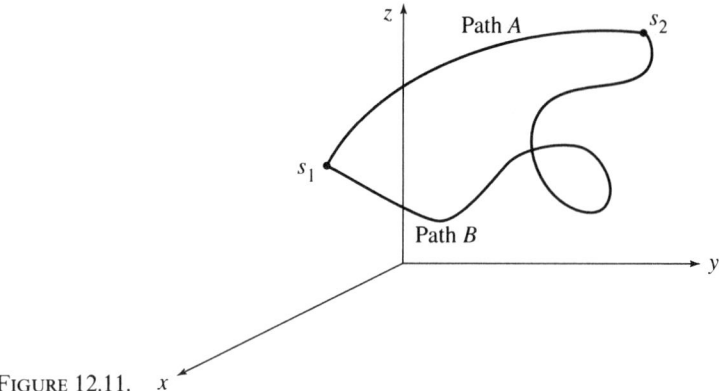

FIGURE 12.11. x

known as a *conservative system*. Thus, a conservative force would perform the same amount of work in moving from position s_1 to position
s_2 whether path A or path B is followed as shown in Figure 12.11. One
consequence of this fact is that a conservative force moving through a
closed loop from s_1 to s_2 and back to s_1, *using any path*, would perform
no work. On the other hand, a nonconservative force performs work
that depends upon the specific path followed. Therefore, the work performed by a nonconservative force in moving from position s_1 to position s_2 would have one value for path A and an entirely different value
for path B and, of course, its work for a closed loop would *not* be zero.*

---

* A more mathematically precise treatment of the above concepts is given in
the footnote:
  Consider the definition of work given by

$$U_{1\to2} = \int_{\varrho_1}^{\varrho_2} \mathbf{F}\cdot d\mathbf{r}.$$

Let $\mathbf{F}$, a conservative force, be the *gradient* of a scalar function $V = f(x, y, z)$. By
definition, the gradient of a scalar function is written symbolically as

$$\nabla V = \left(\frac{\partial V}{\partial x}\right)\mathbf{i} + \left(\frac{\partial V}{\partial y}\right)\mathbf{j} + \left(\frac{\partial V}{\partial z}\right)\mathbf{k}$$

Thus, because $\mathbf{F} = \nabla V$ and $d\mathbf{r} = (dx)\mathbf{i} + (dy)\mathbf{j} + (dz)\mathbf{k}$, it follows that

$$U_{1\to2} = \int_{\varrho_1}^{\varrho_2} \left[\left(\frac{\partial V}{\partial x}\right)\mathbf{i} + \left(\frac{\partial V}{\partial y}\right)\mathbf{j} + \left(\frac{\partial V}{\partial z}\right)\mathbf{k}\right]\cdot[(dx)\mathbf{i} + (dy)\mathbf{j} + (dz)\mathbf{k}]$$

$$= \int_{\varrho_1}^{\varrho_2} \frac{\partial V}{\partial x}dx + \frac{\partial V}{\partial y}dy + \frac{\partial V}{\partial z}dz$$

Because $dV = \dfrac{\partial V}{\partial x}dx + \dfrac{\partial V}{\partial y}dy + \dfrac{\partial V}{\partial z}dz,$

$$U_{1\to2} = \int_{V_1}^{V_2} dV = V_2 - V_1$$

The work of two familiar conservative forces is discussed below.

Work by the Weight of a Body

Consider the weight W which moves along some smooth path from position 1 to position 2, as shown in Figure 12.12. The work of the weight W may be obtained by the use of Eq. (12.5), $U_{1 \rightarrow 2} = \displaystyle\int_{Q_1}^{Q_2} \mathbf{F} \cdot d\mathbf{r}$, where $\mathbf{F} = -W\mathbf{k}$ and $d\mathbf{r} = (dx)\mathbf{i} + (dy)\mathbf{j} + (dz)\mathbf{k}$. If it is assumed that W remains constant as it moves along the path, we may write

$$U_{1 \rightarrow 2} = -\int_1^2 (W\mathbf{k}) \cdot [(dx)\mathbf{i} + (dy)\mathbf{j} + (dz)\mathbf{k}]$$

$$= -W \int_1^2 dz = W(z_1 - z_2). \tag{12.12}$$

Equation (12.12) expresses the fact that the work of the weight W of a body depends *only* upon the initial and final positions without regard to the path followed. Therefore, the weight W is a conservative force. Also, Eq. (12.12) shows that the work of W is a function only of the difference in elevation (i.e., the vertical displacement) between positions 1 and 2 and does *not* depend upon any horizontal displacements that W may experience during its motion. Furthermore, the work of W is positive, if its elevation is decreased, and negative, if it is increased.

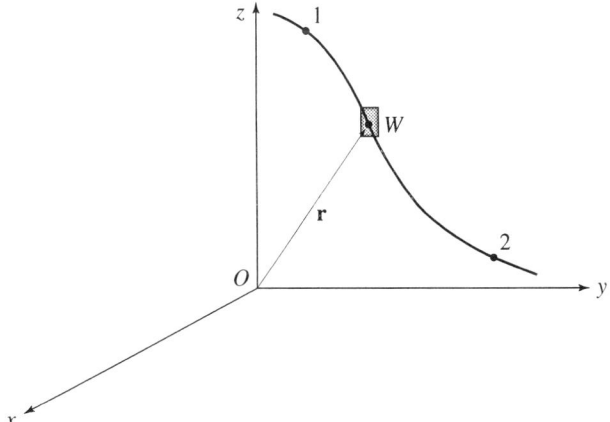

FIGURE 12.12.

where V_1 and V_2 are the values of the scalar function V at positions 1 and 2, respectively. Thus, if the force $\mathbf{F}$ is the gradient of a scalar function, the work done by such a force is a function *only* of its initial and final positions. Also, such a force (known as a conservative force) would perform no work around a closed path.

Work by the Force in a Spring

The characteristics of a linearly elastic spring were discussed in Chapter 2. In summary, however, a linearly elastic spring is one in which an induced force F is directly proportional to the deformation s, according to Eq. (2.14), which is repeated here for convenience. Thus,

$$F = ks. \tag{2.14}$$

where the constant k is the factor of proportionality between F and s, known as the spring constant. The word elastic refers to the fact that, if the deforming force is removed, the spring would return to its initial dimensions and shape.

Consider the linearly elastic spring shown in Figure 12.13. End A of the spring is attached to a fixed support and end B to a weightless rigid body that can slide freely on the frictionless inclined plane. The unstretched length of the spring is L_u, which defines the equilibrium position of the system shown in Figure 12.13.

The work performed by the spring force F on the rigid body, during a displacement from position 1 to position 2, may be determined by Eq. (12.6). Thus, because $\beta = 0°$ and $F = ks$, it follows that

$$U_{1 \to 2} = - \int_{s_1}^{s_2} ks \, ds = -\frac{k}{2}(s_2^2 - s_1^2) \tag{12.13}$$

where the minus sign is introduced to account for the fact that the spring force F is opposite to the direction of the displacement. Thus, when the spring-rigid-body system is displaced *away* from its equilibrium position, the work performed by the spring force on the rigid body is *negative*. If, on the other hand, the spring-rigid-body system is allowed to *return* to its equilibrium position, the work performed by the spring force on the rigid body is *positive* during this displacement. Equation (12.13) also expresses the fact that the work of a spring force on an attached rigid body is a function only of the initial and final positions regardless of the path taken by the rigid body. Therefore, the spring force is a conservative force.

It should be pointed out that the behavior of the spring discussed

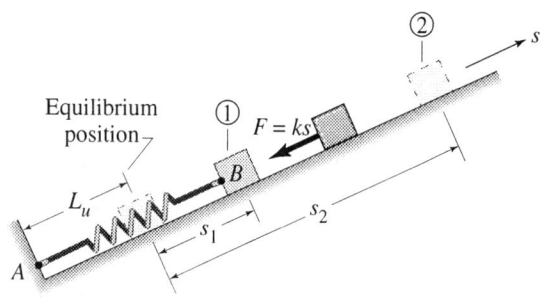

Equilibrium position

FIGURE 12.13.

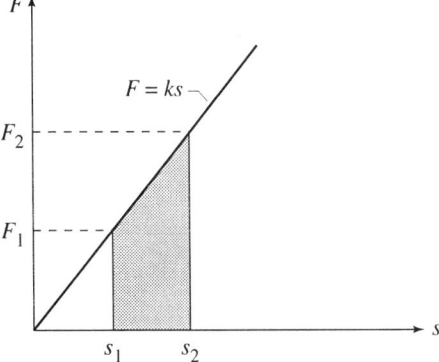

FIGURE 12.14.

above is similar to that of a rod or wire made of a linearly elastic material because the force F developed in the rod (wire) is directly proportional to its deformations, i.e., $F = ks$. However, in the case of a rod or wire, k is usually expressed in terms of the cross-sectional area A, the initial length L, and the modulus of elasticity E by the relation $k = AE/L$ where the modulus of elasticity E is a property that measures the *stiffness* of the material. Thus, once the constant k is known, the treatment of a rod or wire is identical to that of the spring.

The *magnitude* of the work $U_{1\to2}$ given by Eq. (12.13) may be interpreted by reference to Figure 12.14 which is a plot of the equation $F = ks$. By Eq. (12.13),

$$|U_{1\to2}| = \frac{k}{2}s_2^2 - \frac{k}{2}s_1^2 = \frac{1}{2}(ks_2)s_2 - \frac{1}{2}(ks_1)s_1 = \frac{1}{2}F_2s_2 - \frac{1}{2}F_1s_1 \quad (12.14)$$

where F_1 and F_2 are the forces in the spring corresponding to positions 1 and 2, respectively. Thus, Eq. (12.14) shows that the *magnitude* of the work may be obtained by finding the trapezoidal area that is shown shaded in Figure 12.14. The sign of this work is obtained as explained earlier.

■ **Example 12.7**

The conservative system shown in Figure E12.7(a) consists of a uniform member AB of weight $W = 100$ lb attached to a fixed frictionless hinge at A and to a linearly elastic spring at B. The other end of the spring is attached to a fixed support at C. The member is rotated cw about hinge A in the vertical x-y plane by agents not shown in the diagram. Let the unstretched length of the spring $L_u = 5$ ft and its spring constant $k = 150$ lb/ft. Determine the total work performed on member AB as it rotates from (a) $\theta = 0°$ to $\theta = 90°$, (b) $\theta = 0°$ to $\theta = 180°$, and (c) $\theta = 180°$, to $\theta = 270°$.

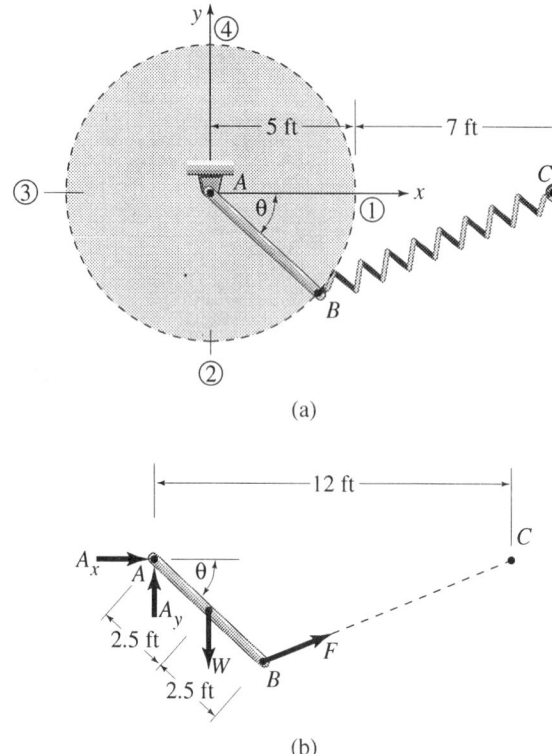

FIGURE E12.7.   (b)

Solution

The free-body diagram of member AB is shown in Figure E12.7(b) for some arbitrary position defined by the angle θ. The reaction components A_x and A_y do no work because they do not displace. Thus, the only forces that perform work are W and F.

(a) The total work performed on member AB is the algebraic sum of the work done by W and F. Thus, by Eq. (12.12),

$$U_{1\to2} = W(y_1 - y_2) = 100[0 - (-2.5)] = 250 \text{ lb·ft},$$

and, by Eq. (12.13),

$$U_{1\to2} = -\tfrac{1}{2}k(s_2^2 - s_1^2)$$

where

$$s_1 = L_1 - L_u = 7 - 5 = 2 \text{ ft}$$

and

$$s_2 = L_2 - L_u = \sqrt{12^2 + 5^2} - 5 = 8 \text{ ft}.$$

Therefore, the spring force performs work $U_{1\to2}$ given by

$$U_{1\to2} = -\frac{150}{2}(8^2 - 2^2) = -4{,}500 \text{ lb·ft.}$$

Thus, the total work done on member AB, when θ changes from zero to 90° is

$$(U_{1\to2})_{\text{Total}} = 250 - 4500 = -4250 \text{ lb·ft.} \qquad \text{ANS.}$$

(b) In a manner similar to that followed in the solution of part (a), for the weight W,

$$U_{1\to3} = W(y_1 - y_3) = 100(0 - 0) = 0,$$

and, for the spring force F,

$$U_{1\to3} = -\tfrac{1}{2}k(s_3^2 - s_1^2)$$

where $s_1 = 2$ ft, as found in part (a) and $s_3 = L_3 - L_u = 17 - 5 = 12$ ft. Therefore,

$$U_{1\to3} = -\tfrac{1}{2}k(12^2 - 2^2) = -10{,}500 \text{ lb·ft.}$$

Thus, the total work performed on member AB as θ changes from zero to 180° is

$$(U_{1\to3})_{\text{Total}} = 0 - 10{,}500 = -10{,}500 \text{ lb·ft.} \qquad \text{ANS.}$$

(c) Similarly to parts (a) and (b), for the weight W,

$$U_{3\to4} = W(y_3 - y_4) = 100(0 - 2.5) = -250 \text{ lb·ft.}$$

For the spring force F,

$$U_{3\to4} = -\tfrac{1}{2}k(s_4^2 - s_3^2).$$

Because $s_3 = 12$ ft and $s_4 = 8$ ft, it follows that

$$U_{3\to4} = -\frac{150}{2}(8^2 - 12^2) = 6{,}000 \text{ lb·ft.}$$

Therefore, as θ changes from 180° to 270°, the total work performed on member AB is

$$(U_{3\to4})_{\text{Total}} = -250 + 6{,}000 = 5{,}750 \text{ lb·ft.} \qquad \text{ANS.}$$

Note that whereas the work of the spring force was negative in parts (a) and (b), it is positive in part (c) because, during the movement from $\theta = 180°$ to $\theta = 270°$, the spring force is in the same direction as the displacement, i.e., the spring system is moving back toward its equilibrium position.

Problems

In the following problems, motion of the conservative systems from one position to another is assumed to occur even though the moving agents may not be indicated. Also, assume that springs in compression are prevented from collapsing sideways.

12.37 A block of weight $W = 200$ lb moves horizontally in a frictionless guide as shown in Figure P12.37. A spring with an unstretched length $L_u = 1$ ft and a spring constant $k = 50$ lb/in. is attached to the block as indicated. Determine the total work done by W and by the spring on the block as it moves from position 1 to position 2.

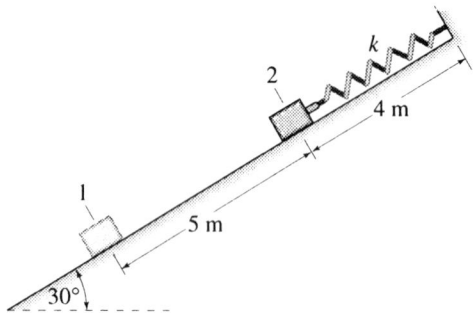

FIGURE P12.38.

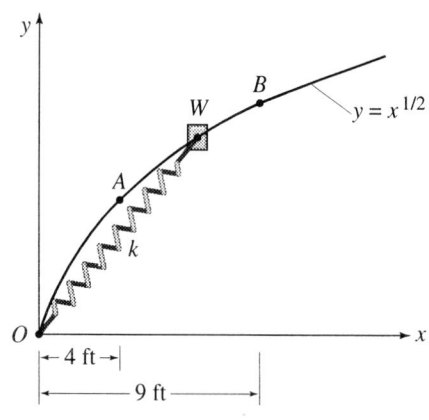

FIGURE P12.39.

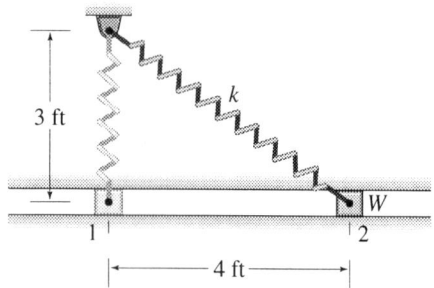

FIGURE P12.37.

12.38 Determine the total work performed by the weight W and by the spring on the block shown in Figure P12.38 as it moves 5 m up the smooth inclined plane. The spring attached to the block has an unstretched length $L_u = 2$ m and a spring constant $k = 10$ kN/m. The weight of the block $W = 1.5$ kN.

12.39 A block of weight $W = 10$ lb is constrained to move in a vertical plane along the frictionless guide defined by

the equation $y = x^{1/2}$ as shown in Figure P12.39. The attached spring has an unstretched length $L_u = 3$ ft and a spring constant $k = 200$ lb/ft. Determine the total work done by the weight W and by the spring on the block as it moves (a) from point A to point B and (b) from point B to the point defining the unstretched length of the spring.

12.40 A block of weight $W = 50$ N moves in a vertical plane along a frictionless track defined by the equation $y = 2x^2$, as

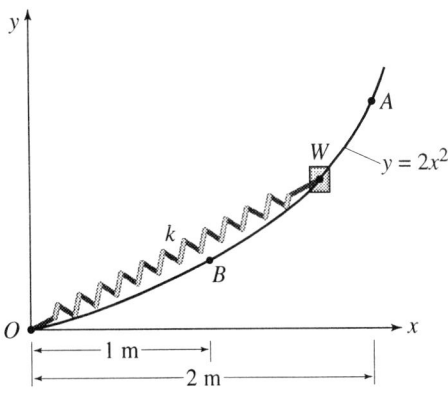

FIGURE P12.40.

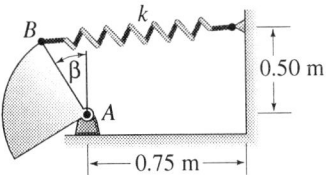

FIGURE P12.42.

shown in Figure P12.40. The attached spring has an unstretched length $L_u = 1$ m and a spring constant $k = 500$ N/m. Compute the total work done by W and by the spring on the block as it moves from point A to point B.

12.41 A uniform member AB of weight $W = 150$ lb in Figure P12.41 is attached to a frictionless hinge at A and to a linear spring at B. The spring has an unstretched length $L_u = 1.5$ ft and a spring constant $k = 600$ lb/ft. Determine the total work done by W and by the spring on member AB as the system moves in a vertical plane from $\alpha = 0°$ to $\alpha = 90°$.

12.42 A homogeneous plate cut in the shape of a quarter circle of radius 0.50 m

and weight $W = 600$ N is attached to a smooth hinge at A and to a linear spring at B, as shown in Figure P12.42. The unstretched length of the spring $L_u = 0.50$ m and its spring constant $k = 700$ N/m. Compute the total work done by W and by the spring on the plate as the system moves in a vertical plane from $\beta = 0°$ to $\beta = 90°$.

12.43 A homogeneous plate cut in the shape of a right-angle triangle has a weight $W = 50$ lb and is attached to a smooth hinge at A and to a linear spring at B, as shown in Figure P12.43. The spring has an unstretched length $L_u = 2.5$ ft and a spring constant $k = 300$ lb/ft. Determine the total work done by W and by the spring on the plate as the system moves in a vertical plane (a) from $\theta = 30°$ to $\theta = 90°$ and (b) from $\theta = 90°$ to $\theta = 0°$.

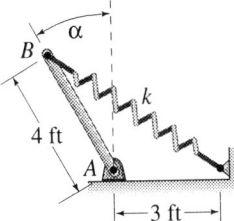

FIGURE P12.41.

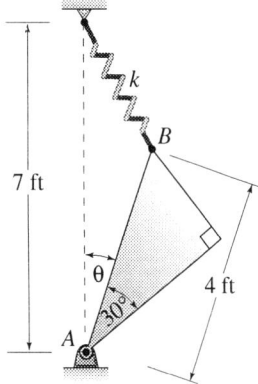

FIGURE P12.43.

12.44 Two identical homogeneous links each of weight W are attached to frictionless hinges at one end and to a common spring at the other, as shown in Figure P12.44. A mechanism (not shown) rotates the two links through identical angles β. When $\beta = 0$, the spring is unstretched. Determine the total work performed on the two links by W and by the spring as β changes from zero to 90° if (a) the system rotates in a vertical plane and (b) the system rotates in a horizontal plane. Express answers in terms of W, k, and b.

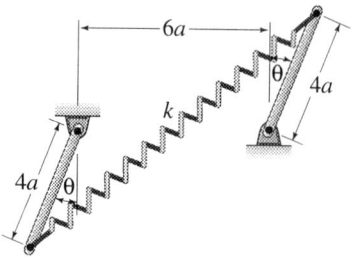

FIGURE P12.45.

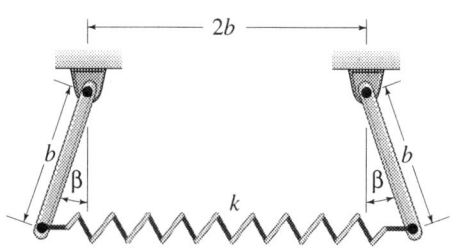

FIGURE P12.44.

12.45 Two identical homogeneous links each of weight W are attached to frictionless hinges at one end and to a common spring at the other, as shown in Figure P12.45. A mechanism (not shown) rotates the two links through identical angles θ. When $\theta = 0°$, the spring is unstretched. Compute the total work performed on the two links by W and by the spring as θ changes from zero to 90° if (a) the system rotates in a vertical plane and (b) the system rotates in horizontal plane. Express answers in terms of W, k, and a.

12.46 A sphere of weight W is attached to one end of a weightless link of length L whose other end is attached to a frictionless hinge at A, as shown in Figure P12.46. One end of a spring, with spring constant k, is attached to the sphere and the other to a fixed support at B. The spring is unstretched when $\alpha = 0°$. If the system moves in a vertical plane from $\alpha = 0°$ to $\alpha = 90°$, determine the value of the spring constant k if the total work performed on the sphere is zero. Express the answer in terms of W and L.

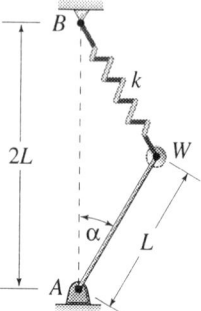

FIGURE P12.46.

12.6*
The Concept of Potential Energy

The term *potential energy* of a mechanical system is used to signify the capacity or potential of a system to do work. For example, the weight W sitting on the table at height h above the floor, as shown in Figure 12.15(a), has the ability to perform a certain amount of work relative to the floor. In other words, if the weight W is released from the table and allowed to drop, it would be able to do Wh units of work assuming the floor is used as a reference or datum plane. Consider also the spring-weight system shown in Figure 12.15(b) where the weight W slides freely on a frictionless surface. In the displaced position shown, the system has a certain amount of energy stored in it as a consequence of the spring deformation. If the system is released and allowed to return to the equilibrium position, it has the capacity to do $\frac{1}{2}ks^2$ units of work on the weight W where s is the displacement of the spring.

In Section 12.5, we concluded that the work of a conservative force is a function only of its initial and final positions and is independent of the path it follows. The work performed by the weight of a body is given by Eq. (12.12) which states that the work is a function only of the difference in elevation from the initial to the final position of the body. Also, if the elevation of the weight W is increased, the work is negative, *but* its capacity to do work becomes greater and we say that its *potential energy due to gravity* V_g increases. On the other hand, if the elevation of the weight W is decreased, the work is positive, *but* its potential energy V_g decreases. Therefore, we may state that the change in potential energy ΔV_g is equal to the *negative* of the work performed as the body changes elevation from one level to another. Thus, using Eq. (12.12),

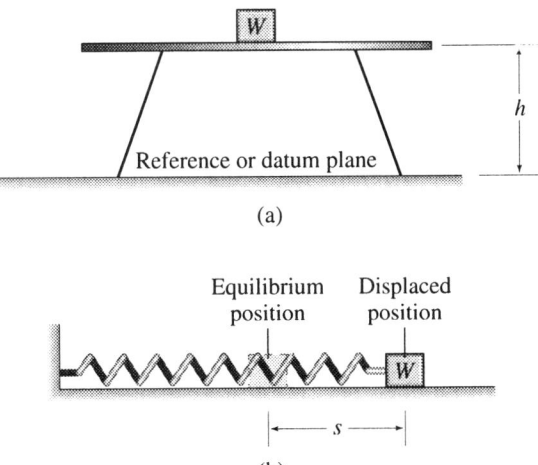

(a)

(b)   FIGURE 12.15.

$$\Delta V_g = -U_{1\to2} = -W(z_1 - z_2) = Wz_2 - Wz_1$$
$$= (V_g)_2 - (V_g)_1 \tag{12.15}$$

where elevation 2 is lower than elevation 1. Equation (12.15) may be written in the form

$$(V_g)_2 = \Delta V_g + (V_g)_1 \tag{12.16}$$

which states that the gravitational potential energy of the body in the final elevation is equal to that in the initial elevation plus the change ΔV_g. However, as will be shown in Section 12.7, only *changes* in potential energy are significant and, therefore, the potential energy in the final elevation may be measured from *any* convenient reference or datum plane. Thus, the potential energy of a body due to gravity may be expressed as

$$V_g = -U_{1\to2} = Wz \tag{12.17}$$

where z is the height of the body measured from any arbitrary position.

As expressed in Eq. (12.13), the work done by a linearly elastic member, such as a spring, on an attached body is a function *only* of its initial and final positions. Also, the work of the spring is negative if it is displaced away from its equilibrium position, but its capacity to do work increases, and we say that its *potential energy due to elastic behavior* V_e increases. On the other hand, the work of the spring is positive if it is allowed to return toward its equilibrium position, but its potential energy V_e decreases. Thus, we may state that the change in potential energy is equal to the *negative* of the work done by the spring. Therefore, by Eq. (12.13),

$$\Delta V_e = -U_{1\to2} = \tfrac{1}{2}k(s_2^2 - s_1^2) = \tfrac{1}{2}ks_2^2 - \tfrac{1}{2}ks_1^2$$
$$= (V_e)_2 - (V_e)_1 \tag{12.18}$$

It follows, therefore, that

$$(V_e)_2 = \Delta V_e + (V_e)_1 \tag{12.19}$$

which states that the elastic potential energy of an elastic system, such as a spring, in its final position is equal to that in the initial position plus the change ΔV_e. However, the elastic potential energy $(V_e)_1$ is zero because it represents the potential energy of the elastic system in its equilibrium position. Thus, the elastic potential energy $(V_e)_2$ becomes equal to the change ΔV_e which *must be measured from the equilibrium position of the elastic system*. Therefore, in the case of a spring, for example,

$$V_e = -U_{1\to2} = \tfrac{1}{2}ks^2 \tag{12.20a}$$

where s is the deformation of the spring measured from its equilibrium position. It may also be shown that the potential energy of a torsional spring is given by

$$V_e = \tfrac{1}{2}K\theta^2. \tag{12.20b}$$

In this equation, K is the torsional spring constant measured in terms of units of torque per unit of rotation, and θ is the rotation of the spring.

Consequently, if a given mechanical system possesses gravitational potential energy V_g and elastic potential energy V_e, its total potential energy V will be given by the algebraic sum of V_g and V_e. Thus,

$$V = V_g + V_e \tag{12.21}$$

where V would be a function of one or more variables depending upon the degree of freedom of the mechanical system under consideration and is known as the *potential energy function*. Also, because the potential energy is equal to the negative of the work done, it follows that

$$V = -U_{1\to 2} \tag{12.22}$$

where $U_{1\to 2}$ is the work performed by conservative forces. As a matter of fact, Eq. (12.22) is valid only for the case of conservative systems.

12.7*
The Principle of Stationary Potential Energy

The principle of virtual work discussed in Section 1.4 may be formulated in terms of potential energy. According to the principle of virtual work, if a mechanical system is in equilibrium, the first variation of the work done δU must vanish. If the work δU is performed by conservative forces, the Eq. (12.22) leads to

$$\delta V = -\delta U = 0 \tag{12.23}$$

from which

$$\delta V = 0. \tag{12.24}$$

Equation (12.24) is an expression of the principle of *stationary potential energy* which states that a necessary and *sufficient condition for the equilibrium of a conservative system is that the first variation of its potential energy function must vanish*. Thus, in the case of a one-degree-of-freedom system, for example, the potential energy function V is expressible in terms of one single variable, say α. In such a case, $V = f(\alpha)$, Also, $\delta V = \left(\dfrac{dV}{d\alpha}\right)\delta\alpha$ and, because $\delta\alpha \neq 0$, the condition expressed in Eq. (12.24) reduces to

$$\frac{dV}{d\alpha} = 0. \tag{12.25}$$

Equation (12.25) expresses the fact that a necessary and sufficient condition for the equilibrium of a conservative system with one degree of freedom is that its potential energy function $V = f(\alpha)$ has a stationary value, i.e., a maximum, a minimum, or a point of inflection. Therefore, to determine the equilibrium positions in terms of the variable α, we would differentiate the potential energy function V with respect to α and set the result equal to zero. The solution of the resulting equation would, then, provide the values(s) of α that define the equilibrium configuration(s) of the system.

In the case of a conservative system with more than one degree of freedom, i.e., one that requires more than one variable, say α_1, α_2, ..., α_n for complete definition, the potential energy function $V = f(\alpha_1, \alpha_2, \ldots, \alpha_n)$. For such a case, the first variation δV is determined by the rules of calculus for differentiating a function of several variables. Using these rules, along with Eq. (12.24),

$$\delta V = \frac{\partial V}{\partial \alpha_1} \delta \alpha_1 + \frac{\partial V}{\partial \alpha_2} \delta \alpha_2 + \cdots + \frac{\partial V}{\partial \alpha_n} \delta \alpha_n = 0. \qquad (12.26)$$

As was discussed in Example 12.6, the virtual displacements are applied, one at a time, independently of each other and, because these virtual displacements are different from zero, it follows that necessary and sufficient conditions for the equilibrium of a conservative system with more than one degree of freedom may be stated by

$$\frac{\partial V}{\partial \alpha_i} = 0 \quad (i = 1, 2, \ldots, n). \qquad (12.27)$$

Equations (12.27) lead to a system of simultaneous equations which may be solved to obtain the values of α_1, α_2, ..., α_n that define the equilibrium configuration of a conservative system. Thus, as in the case of a conservative system with one degree of freedom, *a necessary and sufficient condition for the equilibrium of a conservative system with more than one degree of freedom is that its potential energy function must have stationary values.*

The following examples illustrate the use of the principle of stationary potential energy in solving problems dealing with the equilibrium of conservative systems.

■ **Example 12.8**

Consider the conservative system shown in Figure E12.8(a). It consists of a rigid weightless member AB of length $2b$ attached at A to a frictionless hinge and at B to a flexible and inextensible cable that passes over a small frictionless peg at D. The other end of the cable is connected to a linear spring as shown. The spring has a constant k and is unstretched when α is equal to zero. A horizontal conservative force P is applied to member AB at point C. (a) Develop an expression relating P to the equilibrium angle α. (b) Find the value(s) of the angle α defining the equilibrium of this system for the case when $P = 2kb$.

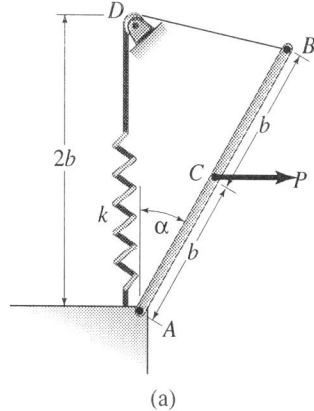

(a)

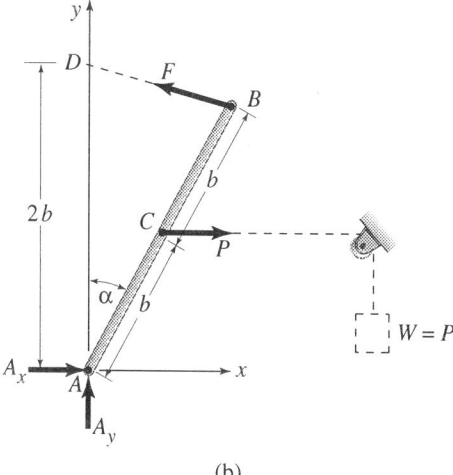

(b)

FIGURE E12.8.

Solution

The free body diagram of member AB is shown in Figure E12.8(b) in reference to an x-y coordinate system. The forces A_x and A_y perform no work because they do not displace. Therefore, the only forces that perform work and contribute to the potential energy of the system are the spring force F and the applied force P. The potential energy developed by the conservative force P is determined by considering the equivalent force system shown to the right of the free-body diagram. This equivalent system considers P as being produced by a dropping weight $W = P$. Therefore, by Eq. (12.21), $V = V_g + V_e$ where

$$V_g = -Px_C = -Pb\sin\alpha$$

and

$$V_e = \tfrac{1}{2}k(DB)^2 = \tfrac{1}{2}k[8b^2(1 - \cos\alpha)].$$

Therefore,

$$V = -Pb \sin \alpha + 4kb^2(1 - \cos \alpha).$$

(a) By the principle of stationary potential energy, Eq. (12.25),

$$\frac{dV}{d\alpha} = -Pb \cos \alpha + 4kb^2 \sin \alpha = 0$$

from which

$$P = 4kb \tan \alpha. \qquad\qquad \text{ANS.}$$

(b) Substituting for P the value $2kb$,

$$2kb = 4kb \tan \alpha,$$

and

$$\tan \alpha = \tfrac{1}{2},$$

from which

$$\alpha = 26.6°. \qquad\qquad \text{ANS.}$$

■ **Example 12.9**   The system shown in Figure E12.9(a) consists of a straight rigid member ABCD attached at A to a frictionless hinge and at B and C to

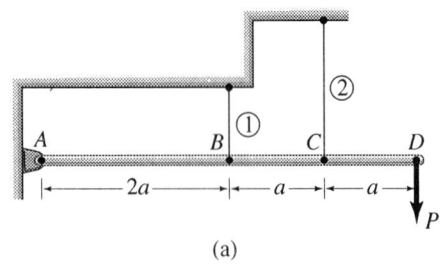

(a)

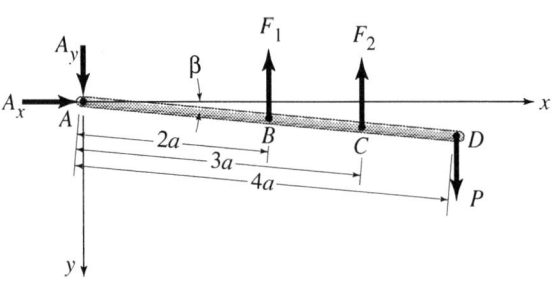

FIGURE E12.9.   (b)

elastic cables 1 and 2 with constants $k_1 = 12 \times 10^5$ lb/in. and $k_2 = 6 \times 10^5$ lb/in., respectively. The member is subjected to a vertical conservative load $P = 10,000$ lb at D. Use the principle of stationary potential energy to find (a) the angle β defining the orientation of member ABCD with the horizontal after the load is applied and (b) the forces F_1 and F_2 developed in the two cables. Let $a = 2$ ft.

Solution

The free-body diagram of member ABCD is shown in Figure E12.9(b). The force system is statically indeterminate and is not solvable by the equations of equilibrium. However, it may be solved by the principle of stationary potential energy. The forces A_x and A_y do no work because they do not displace. The only forces that perform work and contribute to the potential energy function are F_1, F_2, and P. The potential function of the applied force P is found as in Example 12.8. Thus, by Eq. (12.21), $V = V_g + V_e$ where

$$V_g = -P(4a \sin \beta),$$

$$V_e = \frac{1}{2}k_1(2a \sin \beta)^2 + \frac{1}{2}k_2(3a \sin \beta)^2,$$

and

$$V = -4aP \sin \beta + 2a^2 k_1 \sin^2 \beta + \frac{9}{2}a^2 k_2 \sin^2 \beta.$$

In view of the large values of the constants k_1 and k_2 the rotation of member ABCD will be extremely small and the quantity $\sin \beta$ may be approximated by the angle β. Therefore,

$$V = -4aP\beta + 2a^2 k_1 \beta^2 + \frac{9}{2}a^2 k_2 \beta^2.$$

(a) By the principle of stationary potential energy, Eq. (12.25),

$$\frac{dV}{d\beta} = -4aP + 4a^2 k_1 \beta + 9a^2 k_2 \beta = 0$$

from which

$$\beta = \frac{4P}{a(4k_1 + 9k_2)}$$

$$= \frac{4(10,000)}{2(12)[(4)(12)(10^5) + (9)(6)(10^5)]} = 1.634 \times 10^{-4}\,\text{rad}$$

$$= 0.0094°.$$

ANS.

The value obtained for β justifies the assumption that $\sin \beta = \beta$.

(b)

$$F_1 = k_1 s_1 = k_1(2a\beta) = (12 \times 10^2)(2)(2 \times 12)(1.634 \times 10^{-4})$$

$$= 9410 \text{ lb.} \hspace{4cm} \text{ANS.}$$

Similarly,

$$F_2 = k_2 s_2 = k_2(3\alpha\beta) = (6 \times 10^2)(3)(2 \times 12)(1.634 \times 10^{-4})$$

$$= 7060 \text{ lb.} \hspace{4cm} \text{ANS.}$$

■

Problems

Solve the following Problems using the principle of stationary potential energy.

12.47 Problem 12.13 (p. 815). Assume that Q is a conservative force.

12.48 Problem 12.16 (p. 815).

12.49 Problem 12.17 (p. 816). Assume that Q is a conservative force.

12.50 Problem 12.21 (p. 817).

12.51 Problem 12.25 (p. 818). Assume that P is a conservative force.

12.52 Problem 12.31 (p. 819).

12.53 Problem 12.36 (p. 821). Assume that P is a conservative force.

12.54 The vertical rigid member ADB shown in Figure P12.54 is attached at A by a frictionless hinge and at B and D to horizontal elastic cables. A conservative force $P = 20$ kN is applied as shown. If the elastic constant $k_{BC} = 3k_{DE} = 5 \times 10^8$ N/m determine (a) the orientation with respect to the vertical of member ACB after the load P is applied and (b) the forces in the two cables BC and DE.

12.55 The horizontal rigid member ABC is attached to a frictionless hinge at C and to elastic cables at A and B, as shown in Figure P12.55. A conservative vertical load $P = 25$ k is applied as shown. Let

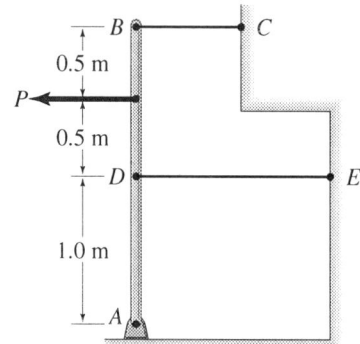

FIGURE P12.54.

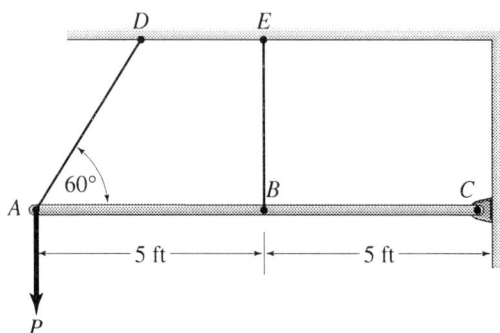

FIGURE P12.55.

the elastic constant $k_{AD} = \frac{3}{4}k_{BE} = 2 \times 10^3$ k/in., and find (a) the orientation with respect to the horizontal of member ABC after the load P is applied and (b) the forces developed in the two cables.

12.56 The horizontal rigid member ABC shown in Figure P12.56 is attached to a frictionless hinge at C and to an elastic cable at B. An elastic compression support, known as a strut, is also used to support the member at A. Assume that the strut does not buckle sideways. A conservative load $P = 50$ kN is applied as shown. Let $k_{BD} = 8 \times 10^8$ N/m and $k_{AE} = 2 \times 10^8$ N/m, and determine (a) the orientation of member ABC with respect to the horizontal after the load P is applied and (b) the forces in cable BD and strut AE.

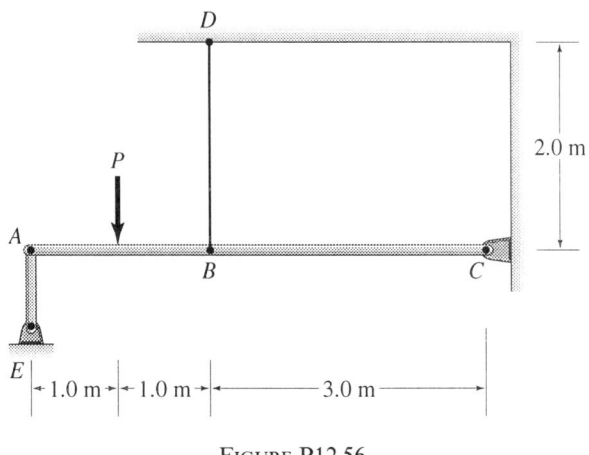

FIGURE P12.56.

12.8*
States of
Equilibrium

In Section 1.7, we developed the principle of stationary potential energy that made it possible to find the equilibrium positions of a conservative system. However, these equilibrium positions may be *stable*, *unstable*, or *neutral*.

Consider, for example, the homogeneous circular disk of weight W which is attached to the frictionless hinge at A, as shown in Figure 12.16(a). This disk is in a state of *stable equilibrium* because, if forced

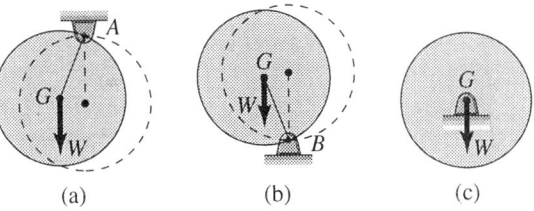

(a)   (b)   (c)   FIGURE 12.16.

away from the equilibrium position shown by the broken lines and, then, released, it will naturally return to it. Note that, when this disk is disturbed from its stable equilibrium position to any other position, its potential energy increases. This, obviously, means that, in the stable equilibrium position, the potential energy of the disk has a *minimum* value. Now, consider the homogeneous disk of weight W attached to the frictionless hinge at B and *carefully* placed in the equilibrium position shown by the broken lines in Figure 12.16(b). This disk is said to be in a state of *unstable equilibrium* because, if disturbed from the equilibrium position and, then, released, it will *not* return to it but will naturally continue its movement away from it. When this disk is displaced from its unstable equilibrium position to *any* other position, its potential energy decreases. Thus, in the unstable equilibrium position, the potential energy of the disk has a *maximum* value. Finally, the homogeneous disk shown in Figure 12.16(c) is attached to a frictionless hinge at its center of gravity G. This disk is in a state of *neutral equilibrium* because, if displaced from the equilibrium position shown to any other position, it will remain in this new position with no tendency to return to or move further from it. In this case, the potential energy of the disk remains *constant* regardless of the position to which it is displaced.

The above concept may be generalized and expressed mathematically for any conservative system. For example, in the case of a system with one degree of freedom represented by the variable α, the potential energy function is $V = f(\alpha)$ and, as shown in Section 12.7, the equilibrium positions are obtained from the condition $\dfrac{dV}{d\alpha} = 0$. Whether these equilibrium positions are stable, unstable, or neutral may be ascertained by examining higher derivatives of the potential energy function. For a stable equilibrium position, the function V must have a *minimum* value and $\dfrac{d^2 V}{d\alpha^2}$ must be *positive*. For an unstable position, the function V must have a *maximum* value and $\dfrac{d^2 V}{d\alpha^2}$ must be *negative*, and, for a neutral equilibrium position, the function V must have a *constant* value and *all* of its derivatives must be zero. These conclusions are depicted in Figure 12.17 and are summarized by the following relationships:

$$\left. \begin{array}{l} \dfrac{dV}{d\alpha} = 0, \\[2ex] \dfrac{d^2 V}{d\alpha^2} > 0; \end{array} \right\} \quad \text{Stable equilibrium} \qquad (12.28)$$

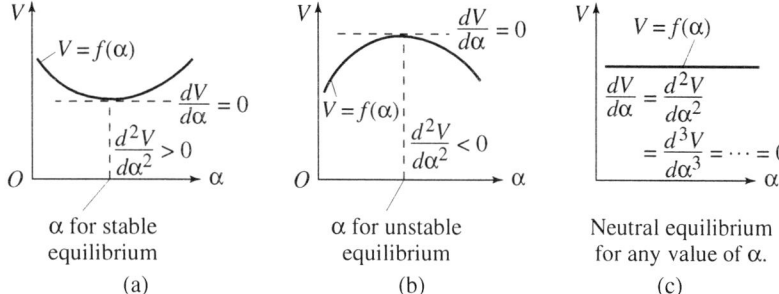

FIGURE 12.17.

(a) α for stable equilibrium

(b) α for unstable equilibrium

(c) Neutral equilibrium for any value of α.

$$\left.\begin{array}{c} \dfrac{dV}{d\alpha} = 0, \\[2mm] \dfrac{d^2V}{d\alpha^2} < 0; \end{array}\right\} \quad \text{Unstable equilibrium} \qquad (12.29)$$

$$\frac{dV}{d\alpha} = \frac{d^2V}{d\alpha^2} = \cdots = \frac{d^nV}{d\alpha^n} = 0. \quad \text{Neutral equilibrium} \qquad (12.30)$$

In the case of a system with more than one degree of freedom, the potential energy function is given by $V = f(\alpha_1, \alpha_2, \ldots, \alpha_n)$, and the equilibrium positions are obtained from the conditions $\dfrac{\partial V}{\partial \alpha_i} = 0$ for $i = 1, 2, \ldots, n$. However, examination of the states of equilibrium becomes more and more complicated as the number of degrees of freedom increases. According to the theory of functions of *two variables*, a system is in a state of *stable equilibrium* if, and only if, *all* of the following conditions are met. If not, the system is unstable.

and

$$\left.\begin{array}{c} \dfrac{\partial V}{\partial \alpha_1} = \dfrac{\partial V}{\partial \alpha_2} = 0, \\[3mm] \left(\dfrac{\partial^2 V}{\partial \alpha_1 \partial \alpha_2}\right)^2 - \left(\dfrac{\partial^2 V}{\partial \alpha_1^2}\right)\left(\dfrac{\partial^2 V}{\partial \alpha_2^2}\right) < 0, \\[3mm] \dfrac{\partial^2 V}{\partial \alpha_1^2} > 0 \quad \text{or} \quad \dfrac{\partial^2 V}{\partial \alpha_2^2} > 0. \end{array}\right\} \qquad (12.31)$$

The following examples illustrate the use of Eqs. (12.28) to (12.31) for determining the positions of equilibrium and examining the stability of these positions for the case of conservative systems.

■ **Example 12.10**

The conservative system shown in Figure E12.10(a) consists of a rigid, weightless, right-angle bent bar ACB attached to a frictionless hinge at

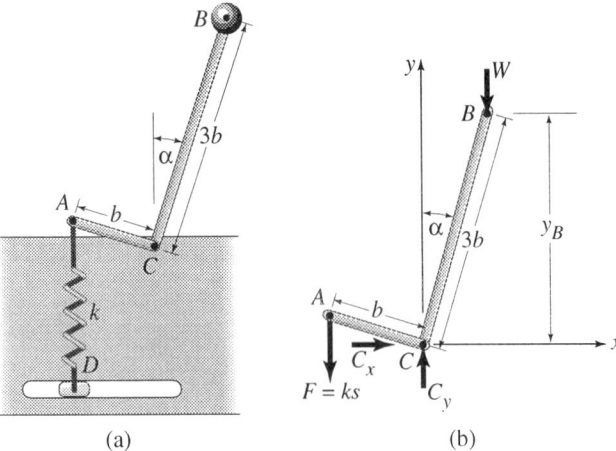

FIGURE E12.10.   (a)   (b)

C, to a sphere of weight W at B, and to a vertical spring at A. The frictionless roller at D allows the spring to remain vertical. The spring has a constant k and is unstretched when $\alpha = 0°$. Determine the value(s) of α $(0° \leq \alpha \leq 90°)$ that define the equilibrium position(s) of the system and indicate if these are stable, unstable, or neutral.

Solution

The free-body diagram of the bent bar is shown in Figure E12.10(b) in reference to an x-y coordinate system. The force components C_x and C_y do no work because they do not displace. The only forces that perform work and contribute to the potential energy of the system are the conservative forces F and W. This is a one-degree-of-freedom system because the variable α is the only one needed to define fully the position of the system. Therefore, by Eq. (12.21), $V = V_g + V_e$ where

$$V_g = Wy_B = W(3b\cos\alpha)$$

and

$$V_e = \frac{1}{2}ks^2 = \frac{1}{2}kb^2\sin^2\alpha.$$

Thus, the potential energy function becomes

$$V = 3bW\cos\alpha + \frac{1}{2}kb^2\sin^2\alpha. \tag{a}$$

Equilibrium

By the principle of stationary potential energy, Eq. (12.25),

$$\frac{dV}{d\alpha} = -3bW\sin\alpha + kb^2\sin\alpha\cos\alpha = 0 \tag{b}$$

from which

$$\sin \alpha (kb^2 \cos \alpha - 3bW) = 0.$$

Therefore, there are two possible solutions

$$\sin \alpha = 0 \Rightarrow \alpha = 0, \tag{c}$$

and

$$kb^2 \cos \alpha - 3bW = 0,$$

$$\alpha = \cos^{-1}\left(\frac{3W}{kb}\right). \tag{d}$$

Stability

The stability of equilibrium is investigated by exploring the second derivative of the potential energy function as indicated by Eqs. (12.28) to (12.30). Thus,

$$\frac{d^2V}{d\alpha^2} = -3bW \cos \alpha + kb^2 \cos^2 \alpha - kb^2 \sin^2 \alpha. \tag{e}$$

For $\alpha = 0$,

$$\frac{d^2V}{d\alpha^2} = -3bW + kb^2. \tag{f}$$

Therefore, the system is in a state of stable equilibrium *only* if $kb^2 > 3bW$ or if $kb > 3W$. Otherwise, the system is unstable for the case when $\alpha = 0$.

For $\alpha = \cos^{-1}\left(\frac{3W}{kb}\right)$,

$$\frac{d^2V}{d\alpha^2} = \frac{9W^2}{k} - kb^2. \tag{g}$$

From Eq. (g), it would appear that the position $\alpha = \cos^{-1}\left(\frac{3W}{kb}\right)$ would be stable only if $\frac{9W^2}{k} - kb^2 > 0$ or if $kb < 3W$. However, if these values are substituted in Eq. (d), we would obtain $\cos \alpha > 1$. This is obviously impossible and leads us to conclude that the position defined by $\alpha = \cos^{-1}\left(\frac{3W}{kb}\right)$ is not a possible equilibrium position (stable or unstable) for the case when $kb < 3W$. However, for $kb < 3W$, the equilibrium position $\alpha = 0$ does exist, but is unstable. On the other hand, if $kb > 3W$, Eq. (g) tells us that the equilibrium position $\alpha = \cos^{-1}\left(\frac{3W}{kb}\right)$ does exist, but is unstable. Also, for $kb > 3W$, as discussed earlier, the equilibrium position $\alpha = 0$ exists and is stable. The following tabulation summarizes the above conclusions

Equilibrium and Stability

α	$kb > 3W$	$kb < 3W$
0	Equilibrium exists and is stable	Equilibrium exists but is unstable
$\cos^{-1}\left(\dfrac{3W}{kb}\right)$	Equilibrium exists but is unstable	Equilibrium does not exist

■ **Example 12.11**

The rigid homogeneous member AB shown in Figure E12.11(a) has a weight $W = 2400$ N and is attached to a frictionless hinge at A. Also, a torsional spring with a spring constant $K = 400$ N·m/rad is placed at A and restrains the rotation of member AB. Let $a = 1.50$ m, and determine the value(s) of the angle θ ($0° \leq \theta \leq 180°$) defining the equilibrium position(s) of this conservative system. Indicate whether these equilibrium positions are stable, unstable, or neutral. The spring is undeformed when $\theta = 0°$.

Solution

The free-body diagram of member AB is shown in Figure E12.11(b) along with an x-y coordinate system. The force components A_x and A_y perform no work because they do not displace. Therefore, the only forces that do work are the weight W of member AB and the couple $M = K\theta$ developed in the torsional spring. The position of this system may be completely specified by giving the angle θ and, therefore, it is a system with one degree of freedom. Thus, by Eq. (12.21), $V = V_g + V_e$ where

$$V_g = W y_G = W\left(\frac{a}{2}\cos\theta\right)$$

and

$$V_e = \frac{1}{2}K\theta^2.$$

Therefore,

$$V = W\left(\frac{a}{2}\right)\cos\theta + \frac{1}{2}K\theta^2.$$

Equilibrium

By the principle of stationary potential energy, Eq. (12.25),

$$\frac{dV}{d\theta} = -W\left(\frac{a}{2}\right)\sin\theta + K\theta = 0.$$

Substituting the given numerical values and simplifying,

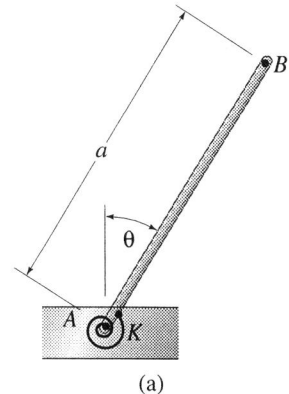

(a)

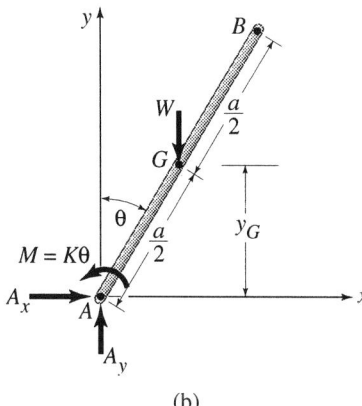

(b)   FIGURE E12.11.

$$\sin \theta - 0.2222\theta = 0.$$

A trial-and-error solution yields the following two values of θ in the range $0° \leq \theta \leq 180°$:

$$\theta = 0°,$$ ANS.

and

$$\theta = 145.6°.$$ ANS.

Stability

The stability of equilibrium may be examined by investigating the second derivative of the potential energy function as indicated in Eqs. (12.28) to (12.30). Thus,

$$\frac{d^2V}{d\theta^2} = -W\left(\frac{a}{2}\right)\cos \theta + K.$$

For $\theta = 0°$,

$$\frac{d^2V}{d\theta^2} = -2400(0.75)(1.0) + 400 = -1400.$$

Because $\frac{d^2V}{d\theta^2}$ is negative, the equilibrium position defined by $\theta = 0°$ is unstable.   ANS.

For $\theta = 145.6°$,

$$\frac{d^2V}{d\theta^2} = -2400(0.75)(-0.825) + 400 = 1885.$$

The second derivative is positive, and the equilibrium position defined by $\theta = 145.6°$ is stable.   ANS.

■ **Example 12.12**

Consider the system shown in Figure E12.12(a), consisting of the two uniform pin-connected members AB and BC each of length $L = 1$ m and the two identical springs with a spring constant $k = 300$ N/m. The weight of member AB $W_1 = 200$ N and that of member BC $W_2 = 300$ N. Determine the angles α_1 and α_2 ($0° \leq \alpha_1 \leq 90°$ and $0° \leq \alpha_2 \leq 90°$) corresponding to the equilibrium of the system. Assume frictionless conditions and that both springs are unstretched when $\alpha_1 = \alpha_2 = 0$.

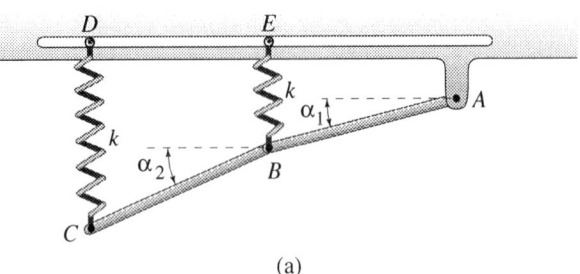

(a)

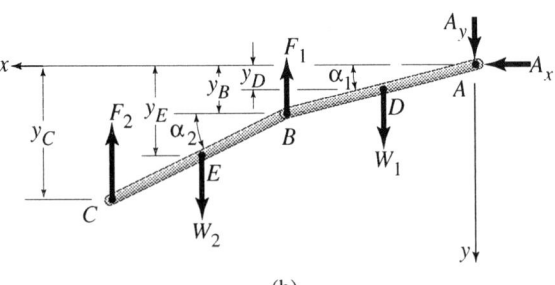

(b)

FIGURE E12.12.

The frictionless rollers in the horizontal track permit the springs to remain vertical. Investigate the stability of the equilibrium positions.

Solution

Note that this system is the same as that which was solved by the method of virtual work in Example 12.6. The free-body diagram of the system is shown in Figure E12.12(b) referred to an x-y coordinate system with origin at A. The only forces that perform work are the spring forces F_1 and F_2 and the member weights W_1 and W_2. The system has two degrees of freedom because it requires the two variables α_1 and α_2 to specify its position fully.

By Eq. (12.21), $V = V_g + V_e$ where

$$V_g = -W_1 y_D - W_2 y_E = -W_1 \left(\frac{L}{2} \sin \alpha_1 \right) - W_2 \left(L \sin \alpha_1 + \frac{L}{2} \sin \alpha_2 \right)$$

and

$$V_e = \frac{1}{2} k s_1^2 + \frac{1}{2} k s_1^2 = \frac{1}{2} k y_B^2 + \frac{1}{2} k y_C^2,$$

$$= \frac{1}{2} k (L \sin \alpha_1)^2 + \frac{1}{2} k (L \sin \alpha_1 + L \sin \alpha_2)^2.$$

Therefore,

$$V = -\frac{W_1 L}{2} \sin \alpha_1 - W_2 L \sin \alpha_1 - \frac{W_2 L}{2} \sin \alpha_2$$

$$+ \frac{k}{2} (L \sin \alpha_2)^2 + \frac{k}{2} (L \sin \alpha_1 + L \sin \alpha_2)^2.$$

Equilibrium

By the principle of stationary potential energy, Eq. (12.27),

$$\frac{\partial V}{\partial \alpha_1} = -\frac{W_1 L}{2} \cos \alpha_1 - W_2 L \cos \alpha_1 + k(L \sin \alpha_1)(L \cos \alpha_1)$$

$$+ k(L \sin \alpha_1 + L \sin \alpha_2)(L \cos \alpha_1) = 0.$$

After simplification, this condition becomes

$$\cos \alpha_1 \left[-\left(\frac{1}{2} W_1 + W_2 \right) + kL(2 \sin \alpha_1 + \sin \alpha_2) \right] = 0.$$

Therefore,

$$\cos \alpha_1 = 0 \Rightarrow \alpha_1 = 90°,$$

and

$$kL(2 \sin \alpha_1 + \sin \alpha_2) - \left(\frac{1}{2} W_1 + W_2 \right) = 0$$

from which

$$2 \sin \alpha_1 + \sin \alpha_2 = \frac{\frac{1}{2} W_1 + W_2}{kL}. \tag{a}$$

Also,

$$\frac{\partial V}{\partial \alpha_2} = -\frac{1}{2} W_2 L \cos \alpha_2 + k(L \sin \alpha_1 + L \sin \alpha_2)(L \cos \alpha_2) = 0.$$

Combining terms,

$$\cos \alpha_2 \left[-\frac{1}{2} W_2 + kL(\sin \alpha_1 + \sin \alpha_2) \right] = 0.$$

Therefore,

$$\cos \alpha_2 = 0 \Rightarrow \alpha_2 = 90°,$$

and

$$kL(\sin \alpha_1 + \sin \alpha_2) - \frac{1}{2} W_2 = 0$$

from which

$$\sin \alpha_1 + \sin \alpha_2 = \frac{\frac{1}{2} W_2}{kL}. \tag{b}$$

Solving equations (a) and (b) simultaneously,

$$\alpha_1 = \sin^{-1} \left(\frac{W_1 + W_2}{2kL} \right) = \sin^{-1} \left[\frac{200 + 300}{2(300)(1)} \right] = 56.4°,$$

and

$$\alpha_2 = \sin^{-1} \left(\frac{-W_1}{2kL} \right) = \sin^{-1} \left[-\frac{200}{2(300)(1)} \right] = -19.5°.$$

The minus sign in the value for α_2 indicates that this angle is not ccw as assumed in the diagram, but cw. Thus, there are two possible equilibrium configurations (positions) defined as follows:

$$\alpha_1 = \alpha_2 = 90.0°, \qquad \text{ANS.}$$

and

$$\alpha_1 = 56.4° \quad \text{and} \quad \alpha_2 = -19.5°. \qquad \text{ANS.}$$

Stability

The stability of the two equilibrium positions is investigated by exploring the quantities in Eq. (12.31).

Substituting the given numerical values in the first derivative with respect to α_1 and simplifying,

$$\frac{\partial V}{\partial \alpha_1} = -400 \cos \alpha_1 + 600 \sin \alpha_1 \cos \alpha_1 + 300 \sin \alpha_2 \cos \alpha_1.$$

Therefore,

$$\frac{\partial^2 V}{\partial \alpha_1^2} = \sin \alpha_1 (400 - 300 \text{ in } \alpha_2) + 600 \cos 2\alpha_1.$$

Similarly,

$$\frac{\partial V}{\partial \alpha_2} = -150 \cos \alpha_2 + 300 \sin \alpha_1 \cos \alpha_2 + 300 \sin \alpha_2 \cos \alpha_2.$$

Thus,

$$\frac{\partial^2 V}{\partial \alpha_2} = \sin \alpha_2 (150 - 300 \sin \alpha_1) + 300 \cos 2\alpha_2.$$

The mixed second partial may be obtained either from $\dfrac{\partial V}{\partial \alpha_1}$ or from $\dfrac{\partial V}{\partial \alpha_2}$, and it reduces to

$$\frac{\partial^2 V}{\partial \alpha_1 \partial \alpha_2} = 300 \cos \alpha_1 \cos \alpha_2.$$

For $\alpha_1 = \alpha_2 = 90°$,

$$\frac{\partial^2 V}{\partial \alpha_1 \partial \alpha_2} = 0,$$

$$\frac{\partial^2 V}{\partial \alpha_1^2} = -500,$$

and

$$\frac{\partial^2 V}{\partial \alpha_2^2} = -450.$$

Therefore

$$\left(\frac{\partial^2 V}{\partial \alpha_1 \partial \alpha_2} \right)^2 - \left(\frac{\partial^2 V}{\partial \alpha_1^2} \right) \left(\frac{\partial^2 V}{\partial \alpha_2^2} \right) = 0 - 225{,}000 < 0.$$

Although the second condition in Eq. (12.31) is satisfied, the third is not, and the equilibrium position defined by $\alpha_1 = \alpha_2 = 90°$ is *unstable*.   ANS.

For $\alpha_1 = 56.4°$ and $\alpha_2 = -19.5°$,

$$\frac{\partial^2 V}{\partial \alpha_1 \partial \alpha_2} = 156.5,$$

$$\frac{\partial^2 V}{\partial \alpha_1^2} = 184.1,$$

and

$$\frac{\partial^2 V}{\partial \alpha_2^2} = 266.5.$$

Therefore,

$$\left(\frac{\partial^2 V}{\partial \alpha_1 \partial \alpha_2}\right)^2 - \left(\frac{\partial^2 V}{\partial \alpha_1^2}\right)\left(\frac{\partial^2 V}{\partial \alpha_2^2}\right) = 24{,}492 - 49{,}062.7 = -24{,}570.4 < 0.$$

Thus, the second and third conditions in Eq. (12.31) are satisfied, and the equilibrium position defined by $\alpha_1 = 56.4°$ and $\alpha_2 = -19.5°$ is *stable*.   ANS.

Problems

12.57 The potential energy function for a conservative system with one degree of freedom represented by the variable μ is given by $V = \mu^3 - 12\mu + 3$. Determine the equilibrium position(s) for this system and state if they are stable, unstable, or neutral. Plot a graph of V vs. μ, and indicate the equilibrium positions on this graph.

12.58 Assume frictionless joints and weightless members AB and CD in Figure P12.58. The spring with a constant $k = 2000$ N/m is unstretched when $\theta = 0°$. The frictionless roller at D allows the spring to remain vertical. If $W = 1000$ N and $a = 0.8$ m, determine the values of θ ($0° \le \theta \le 90°$) that define the equilibrium positions of this system. State if these positions are stable, unstable, or neutral.

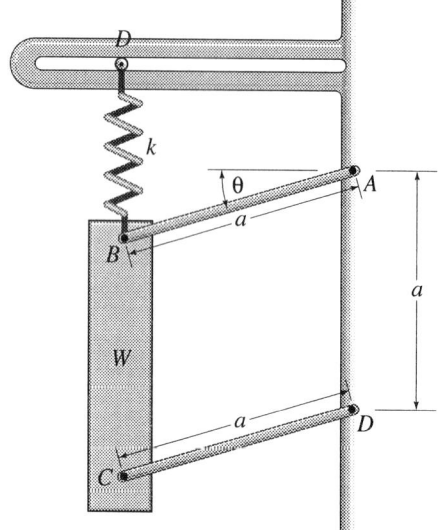

FIGURE P12.58.

12.59 Refer to the mechanism shown in Figure P12.59 in which all hinges and the roller are frictionless. The spring of constant k is unstretched when $\alpha = 0$. Determine the value(s) of α ($0° \leq \alpha \leq 90°$) for equilibrium and state if stable, unstable, or neutral. Let the conservative force $P = kL$.

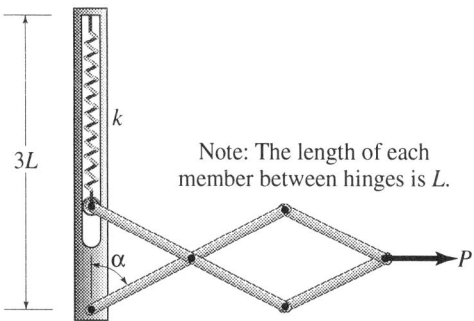

Note: The length of each member between hinges is L.

FIGURE P12.59.

12.60 Assume frictionless conditions at the hinges and weightless linkages for the mechanism in Figure P12.60. The spring is unstretched when $\alpha = 0°$, and its spring constant $k = 20$ lb/in. If $W =$

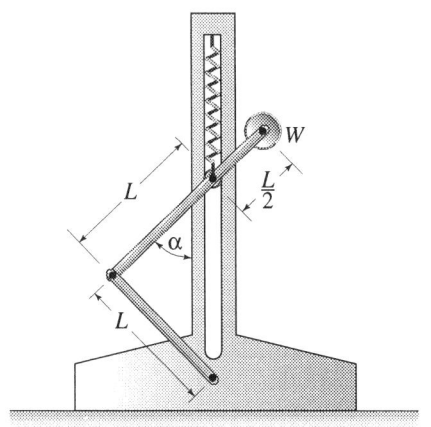

FIGURE P12.60.

150 lb and $L = 18$ in., find the values of α ($0° \leq \alpha \leq 90°$) corresponding to the equilibrium positions of the mechanism. State if these equilibrium positions are stable, unstable, or neutral.

12.61 Show that the conservative system shown in Figure P12.61 has two equilibrium positions defined by $\beta = 90°$ and $\beta = \sin^{-1}\left(\dfrac{W}{2kL}\right)$. State the conditions under which both of these equilibrium positions are stable. Assume frictionless and weightless rollers and that the spring is unstretched when $\beta = 0°$.

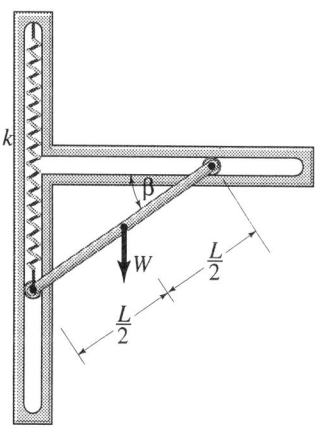

FIGURE P12.61.

12.62 The system shown in Figure P12.62 consists of two identical members each of weight W and length $2b$. The two members are attached to a frictionless hinge at A and to a linear spring at their other ends as shown. The spring has a constant k and an unstretched length when $\alpha = 15°$. Let $W = 2kb$, and determine the value(s) of α ($0° \leq \alpha \leq 90°$) defining the *stable* equilibrium position(s) of the system.

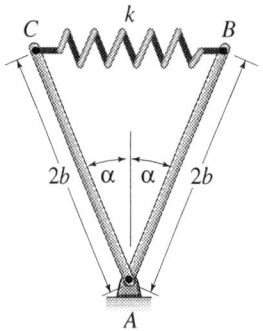

FIGURE P12.62.

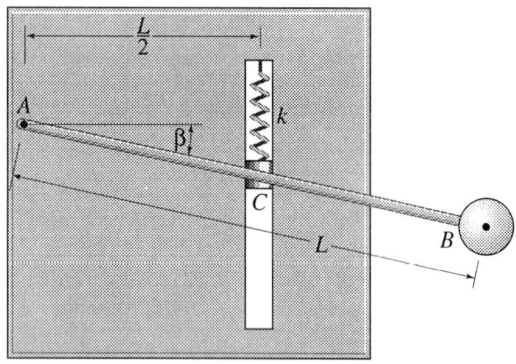

FIGURE P12.64.

12.63 Refer to Figure P12.63. The spring has a constant k and is unstretched when $\theta = 0°$. The rod is rigid and weightless, and frictionless conditions may be assumed at the hinge A and the peg B. In terms of W and L, determine the range of values of k for a stable equilibrium position defined by $\theta = \sin^{-1}\left(\dfrac{2W}{kL}\right)$. Does the system have any other equilibrium positions? If so, are they stable, unstable, or neutral?

the rod but is constrained to move in the vertical track. The spring of constant k is unstretched when $\beta = 0°$. Let $W = \dfrac{kL}{4}$, and determine the equilibrium position(s) $(0° \leq \alpha \leq 90°)$ of the system. Are these positions stable, unstable, or neutral?

12.65 Refer to Figure P12.65. The disk of radius a and weight W rotates about the frictionless hinge at A which is provided with a torsional spring of constant K (not shown) that inhibits the rotation of the disk. Also attached to the top of the disk is a linear spring of constant k. Both springs are undeformed when $\alpha = 0$. Let $K = 2ka^2$ and the weight of the sphere be $W = ka$, and determine the

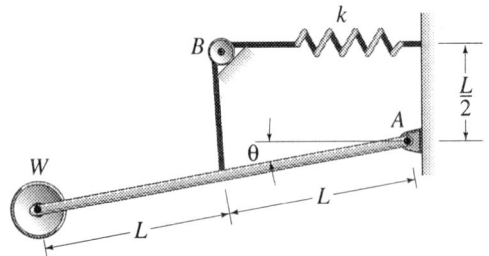

FIGURE P12.63.

12.64 Refer to Figure P12.64. The rod AB is rigid and weightless and is attached to a frictionless hinge at A, and at B, to a sphere of weight W. The frictionless and weightless slider at C can slide freely on

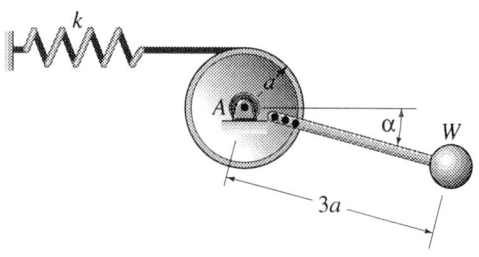

FIGURE P12.65.

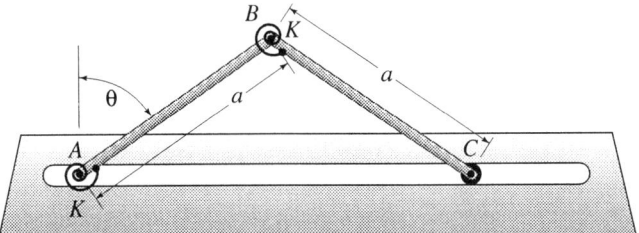

FIGURE P12.66.

value(s) of α $(0° \le \alpha \le 90°)$ defining the equilibrium position(s) of the system. State if these position(s) are stable, unstable, or neutral. The arm connecting the sphere to the disk is rigidly fastened to the disk and may be assumed weightless.

12.66 The mechanism shown in Figure P12.66 consists of two identical homogeneous links AB and BC each of length a and weight W. The roller at C may be assumed frictionless and the frictionless hinges at A and B are provided with identical torsional springs, each with a spring constant K. If the equation defining the equilibrium positions is $\theta - 2\sin\theta = 0$, determine the needed value of the torsional spring constant K in terms of W and a. Assume that the two torsional springs are undeformed when $\theta = 0°$. Find the value(s) of θ $(0° \le \alpha \le 180°)$ for equilibrium and state if stable, unstable, or neutral. Let $W = 5$ kN and $a = 1.0$ m.

12.67 The weightless rigid bar ACB, shown in Figure P12.67, is attached to a frictionless hinge at A which is provided with a torsional spring of constant K. The weightless slider at C can slide freely on rod ACB, but is constrained to move in the horizontal track under the influence of the conservative horizontal force P. If the equation defining the equilibrium

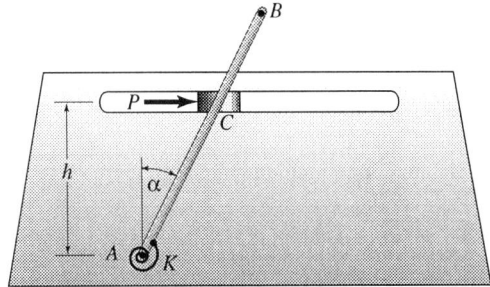

FIGURE P12.67.

position(s) is $\sec^2 \alpha - 3\alpha = 0$, determine the needed value of the torsional spring constant K in terms of P and h. Assume that the torsional spring is undeformed when $\alpha = 0°$. Find the value(s) of α $(0° \le \alpha \le 90°)$ for equilibrium and state if stable, unstable, or neutral. Let $P = 2k$ and $h = 10$ ft.

12.68 Consider the system shown in Figure P12.68 which consists of two identical rigid and weightless links AB and CD that support a weight W. A torsional spring with a spring constant K is placed at A and is undeformed when $\theta = 0°$. Assume frictionless conditions at all hinges, and determine the value of K in terms of W and L so that the system is in a state of stable equilibrium when $\theta = 0°$.

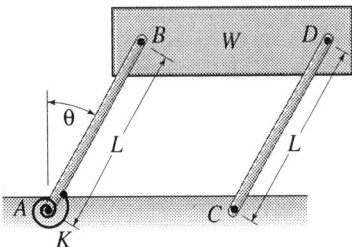

FIGURE P12.68.

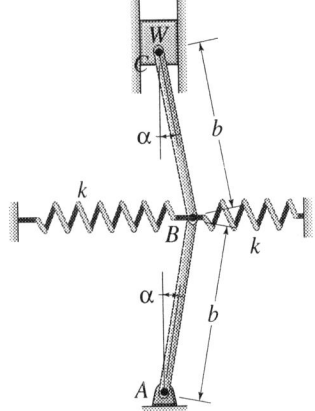

FIGURE P12.69.

12.69 The two identical rigid bars AB and BC shown in Figure P12.69 may be assumed weightless and all hinges may be assumed frictionless. The two identical springs are undeformed when $\alpha = 0°$ and are able to resist compression with-

out buckling sideways. If each spring constant is k, in terms of W and b, determine the range of values of k for which the equilibrium position $\alpha = 0°$ is stable.

12.70 The two identical rigid and weightless members AB and BC are attached to each other by a hinge at B equipped with a torsional spring of constant K, as shown in Figure P12.70. The spring is undeformed when $\beta = 0$. In terms of K and L, determine the range of values of the conservative force P for which the equilibrium position $\beta = 0°$ is stable. Assume frictionless conditions.

12.71 The potential energy function for a conservative system with two degrees of freedom represented by the variables x and y is $V = x^2 + x^3 - 6y$. Determine the equilibrium position(s) for this system and state if they are stable or unstable.

12.72 The two identical rods AB and BC of Figure P12.72 may be assumed rigid and weightless. The frictionless hinges at A and B are equipped with identical torsional springs which are undeformed when $\theta_1 = \theta_2 = 0°$. If each spring has a spring constant K, in terms of K and L, determine the range of values of the conservative force P for which the equilibrium position defined by $\theta_1 = \theta_2 = 0°$ is stable.

12.73 Each of the two identical links AB and BC in Figure P12.73 has a length L and weight W. The spring is unstretched when $\alpha_1 = \alpha_2 = 0°$. Assume frictionless

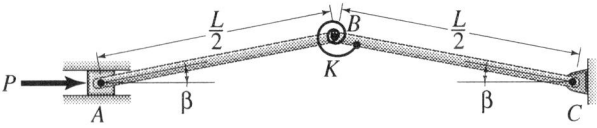

FIGURE P12.70.

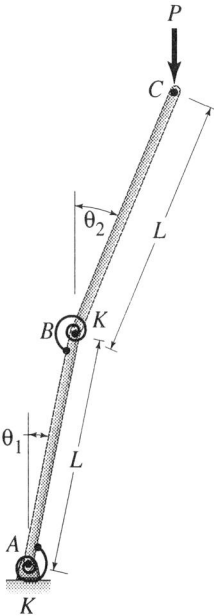

FIGURE P12.72.

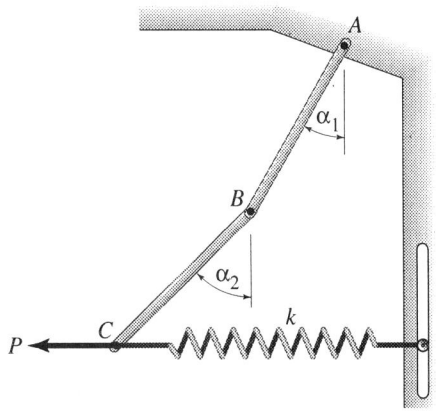

FIGURE P12.73.

conditions and determine the relationship that must exist between the angles α_1 and α_2 for equilibrium of the system, for any values of the quantities P, k, L, and W. The force P is conservative.

12.74 Consider Example 12.10 (p. 841), and allow the angle α to vary in the range $0° \leq \theta \leq 360°$. Are there any new values of α for which equilibrium of the system is possible? Investigate the stability of these new equilibrium positions.

Review Problems

12.75 Member AB, shown in Figure P12.75, has a weight W and a length L. It is supported by an overhead spring of spring constant k. Assume frictionless conditions and use the method of virtual work to find the angle θ consistent with the equilibrium of the member. The two small wheels are weightless. Express the answer in terms of W, k, and L. What is the magnitude of θ when $W = 10$ N, $k = 15$ N/m, and $L = 0.75$ m.

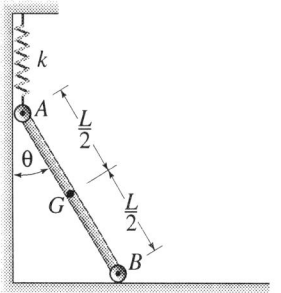

FIGURE P12.75.

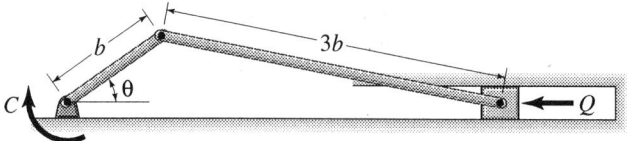

FIGURE P12.76.

12.76 Consider the mechanism shown in Figure P12.76. Use the method of virtual work to develop an expression, consistent with equilibrium, for the couple C in terms of the piston force Q, the dimension b, and the angle θ. Specialize this expression for the case where $Q = 1000$ lb., $b = 4$ in., and $\theta = 50°$.

12.77 The collar at A of weight W_1 is attached to weightless rod ABC, as shown in Figure P12.77. The rod is supported by a peg at B and carries a load W_2 at C. Assume frictionless conditions, and use the method of virtual work to develop an expression, consistent with equilibrium, for W_2 in terms of W_1, L, b, and θ. What condition must be valid between the angle θ and the ratio L/b?

12.78 The system shown in Figure P12.78 consists of two identical uniform links each of length b and weight W. The

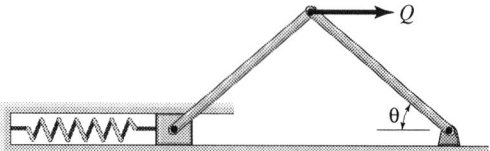

FIGURE P12.78.

spring, of spring constant k, is unstretched when $\theta = 0°$. Assume frictionless conditions, and use the method of virtual work to find an expression, consistent with equilibrium, for the spring constant k in terms of W, Q, b, and θ. What is the magnitude of k if the system is in equilibrium for $\theta = 20°$ when $W = 20$ N, $Q = 75$ N, and $b = 0.80$ m.

12.79 A side view of a garage door consisting of a rectangular plate ABC of weight $W = 100$ lb is shown in Figure P12.79. Two cables, attached to two identical motorized pulleys, one on each side of the door, serve to open and close the

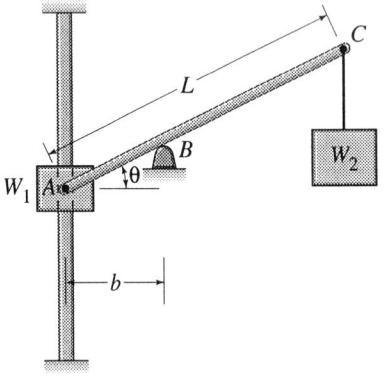

FIGURE P12.77.

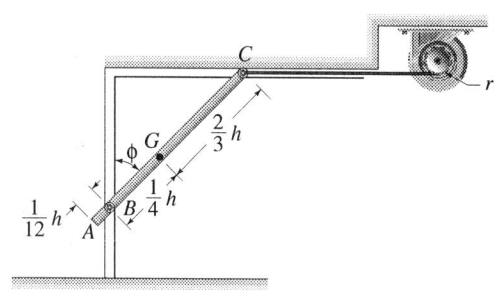

FIGURE P12.79.

door at constant speed. If the angle ϕ is to vary between zero and 80°, determine the maximum couple that each motor must provide to open the door such that $\phi = 80°$. The height h of the door is 12 ft. and the radius r of the motorized pulleys is 3 in. Use the method of virtual work.

12.80 Solve Problem 12.75 by the principle of stationary potential energy.

12.81 Solve Problem 12.77 by the principle of stationary potential energy.

12.82 Solve Problem 12.78 by the principle of stationary potential energy.

12.83 A spool of outer radius $2R$ and inner radius R is shown in Figure P12.83. A linear spring, of spring constant k, is attached to the outer rim of the spool and a torsional spring of spring constant K is attached at the hinge at A (not shown). Both springs are undeformed when $\theta = 0°$. One end of a cord is attached to the inner core of the spool and the other end to a weight W as shown. Assume frictionless conditions, and use the principle of stationary potential energy to find the angle(s) θ consistent with the equilibrium of the system. Is this equi-

librium position(s) stable, unstable, or neutral?

12.84 The system shown in Figure P12.84 consists of two identical rigid and weightless links AB and CD that support a weight W. A linear spring, of spring constant k, connects joints D and A and is unstretched when $\theta = 0°$. Assume frictionless conditions, and in terms of W and b, determine the magnitude of k so that the equilibrium position $\theta = 0°$ is stable.

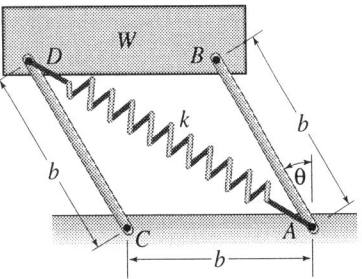

FIGURE P12.84.

12.85 Assume that the two identical rods, each of length b in Figure P12.85, are rigid

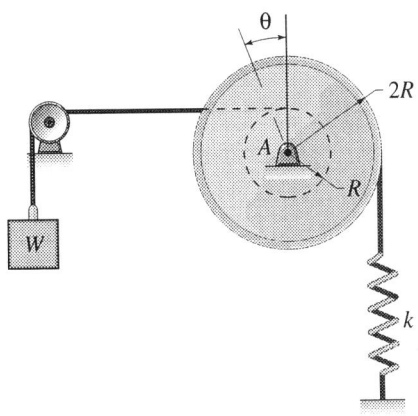

FIGURE P12.83.

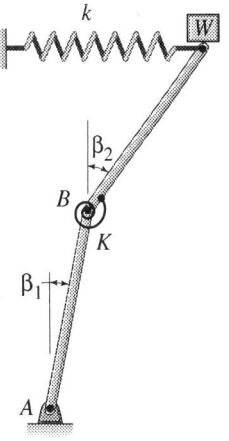

FIGURE P12.85.

and weightless. The linear spring at C has a spring constant k and the torsional spring at B has a spring constant $K = 2kb^2$. Determine the range of values of k for which the equilibrium position defined by $\beta_1 = \beta_2 = 0°$ is stable. Assume frictionless conditions and that both springs are undeformed when $\beta_1 = \beta_2 = 0°$.

Appendix A
Properties of
Selected Lines
and Areas

Shape	Length or Area	Centroid Location	Centroidal Moments of Inertia	Centroidal Radii of Gyration
Arc of a circle 	$L = 2\beta R$	$\bar{x} = 0$ $\bar{y} = \dfrac{R \sin \beta}{\beta}$		
Arc of a quarter circle 	$L = \dfrac{\pi R}{2}$	$\bar{x} = \bar{y} = \dfrac{2R}{\pi}$		

Shape	Length or Area	Centroid Location	Centroidal Moments of Inertia	Centroidal Radii of Gyration
Rectangular Area	$A = bh$	$\bar{x} = \bar{y} = 0$	$I_x = \dfrac{1}{12}bh^3$ $I_y = \dfrac{1}{12}hb^3$	$r_x = \dfrac{h}{\sqrt{12}}$ $r_y = \dfrac{b}{\sqrt{12}}$
Triangular Area	$A = \dfrac{1}{2}bh$	$\bar{y} = \dfrac{1}{3}h$	$I_x = \dfrac{1}{36}bh^2$	$r_x = \dfrac{h}{\sqrt{18}}$
Circular sector area $\alpha = \pi$ leads to a circular area	$A = \alpha R^2$ $A = \pi R^2$	$\bar{x} = \dfrac{2}{3}\dfrac{R\sin\alpha}{\alpha}$ $\bar{y} = 0$ $\bar{x} = 0$ $\bar{y} = 0$	$I_x = \dfrac{R^4}{4}\left(\alpha - \dfrac{1}{2}\sin 2\alpha\right)$ $I_x = I_y = \dfrac{\pi R^4}{4}$ $J_c = I_x + I_y = \dfrac{\pi R^4}{2}$	$r_x = r_y = \dfrac{R}{2}$

Shape	Length or Area	Centroid Location	Centroidal Moments of Inertia	Centroidal Radii of Gyration
Semicircular area	$A = \pi R^2/2$	$\bar{x} = 0$ $\bar{y} = \dfrac{4R}{3\pi}$	$I_x = I_y = \dfrac{\pi R^4}{8}$ $J_O = \dfrac{\pi R^4}{4}$	
Elliptical area	$A = \pi ab$	$\bar{x} = 0$ $\bar{y} = 0$	$I_x = \dfrac{\pi ab^3}{4}$ $I_y = \dfrac{\pi ba^3}{4}$	$r_x = \dfrac{b}{2}$ $r_y = \dfrac{a}{2}$
nth-degree Parabolic quadrant	$A = \dfrac{nab}{n+1}$	$\bar{x} = \dfrac{(n+1)a}{2(n+2)}$ $\bar{y} = \dfrac{(n+1)b}{2n+1}$		
nth-degree Parabolic spandrel	$A = \dfrac{ab}{n+1}$	$\bar{x} = \dfrac{(n+1)a}{n+2}$ $\bar{y} = \dfrac{(n+1)b}{2(2n+1)}$		

Appendix B
Properties of
Selected Masses

Body and dimensions	Volume	Center of mass	Centroidal Moments of Inertia
Solid circular cylinder	$V = \pi R^2 L$	$\bar{x} = 0$ $\bar{y} = 0$ $\bar{z} = 0$	$I_x = I_z = \dfrac{1}{12}m(3R^2 + L^2)$ $I_y = \dfrac{1}{2}mR^2$
Thin cylindrical shell R: Mean radius	$V = 2\pi R t L$	$\bar{x} = 0$ $\bar{y} = 0$ $\bar{z} = 0$	$I_x = I_z = \dfrac{1}{4}m\left(2R^2 + \dfrac{1}{3}L^2\right)$ $I_y = mR^2$

Body and dimensions	Volume	Center of mass	Centroidal Moments of Inertia
Long slender rod $\left(\begin{array}{c}\text{Cross-}\\\text{sectional}\\\text{area A}\end{array}\right)$	$V = AL$	$\bar{x} = 0$ $\bar{y} = 0$ $\bar{z} = 0$	$I_x = I_z = \dfrac{1}{12}mL^2$
Thin rectangular plate 	$V = bht$	$\bar{x} = 0$ $\bar{y} = 0$ $\bar{z} = 0$	$I_x = \dfrac{1}{12}mh^2$ $I_y = \dfrac{1}{12}m(h^2 + b^2)$ $I_z = \dfrac{1}{12}mb^2$
Rectangular prism (or parallelepiped) 	$V = bhL$	$\bar{x} = 0$ $\bar{y} = 0$ $\bar{z} = 0$	$I_x = \dfrac{1}{12}m(L^2 + h^2)$ $I_y = \dfrac{1}{12}m(h^2 + b^2)$ $I_z = \dfrac{1}{12}m(L^2 + b^2)$

Body and dimensions	Volume	Center of mass	Centroidal Moments of Inertia
Solid right circular cone	$V = \dfrac{1}{12}\pi D^2 L$	$\bar{x} = 0$ $\bar{z} = 0$ $\bar{y} = \dfrac{1}{4}L$	$I_x = I_z = \dfrac{3}{80}m\left(4R^2 + \dfrac{1}{3}L^2\right)$ $I_y = \dfrac{3}{30}mR^2$
Thin circular conical shell R: Mean radius	$V = \pi R t L$	$\bar{x} = 0$ $\bar{z} = 0$ $\bar{y} = \dfrac{1}{3}L$	$I_x = I_z = \dfrac{1}{12}m\left(\dfrac{1}{2}R^2 + \dfrac{1}{9}L^2\right)$ $I_y = \dfrac{1}{12}mR^2$
Solid sphere	$V = \dfrac{4}{3}\pi R^3$	$\bar{x} = 0$ $\bar{y} = 0$ $\bar{z} = 0$	$I_x = I_y = I_z = \dfrac{2}{5}mR^2$
Thin spherical shell R: Mean radius	$V = 4\pi R t$	$\bar{x} = 0$ $\bar{y} = 0$ $\bar{z} = 0$	$I_x = I_y = I_z = \dfrac{2}{3}mR^2$

Body and dimensions	Volume	Center of mass	Centroidal Moments of Inertia
Thin circular disk	$V = \pi R^2 t$	$\bar{x} = 0$ $\bar{y} = 0$ $\bar{z} = 0$	$I_x = I_z = \dfrac{1}{4} mR^2$ $I_y = \dfrac{1}{2} mR^2$
Hemisphere	$V = \dfrac{2}{3} \pi R^3$	$\bar{x} = 0$ $\bar{y} = 0$ $\bar{z} = \dfrac{3}{8} R$	$I_x = I_y = \dfrac{83}{640} mR^2$ $I_z = \dfrac{2}{5} mR^2$
Circular paraboloid	$V = \dfrac{1}{2} \pi h a^2$	$\bar{x} = 0$ $\bar{y} = 0$ $\bar{z} = \dfrac{1}{3} h$	$I_x = I_y = \dfrac{1}{15} m(3a^2 + h^2)$ $I_z = \dfrac{1}{3} ma^2$
Semicylinder	$V = \dfrac{1}{2} \pi R^2 L$	$\bar{x} = 0$ $\bar{y} = 0$ $\bar{z} = \dfrac{4R}{3\pi}$	$I_x = I_z = \dfrac{1}{4} m\left(R^2 + \dfrac{L^2}{3} \right)$ $I_y = \dfrac{1}{2} mR^2$

Appendix C
Useful Mathematical Relations

Approximations of Areas

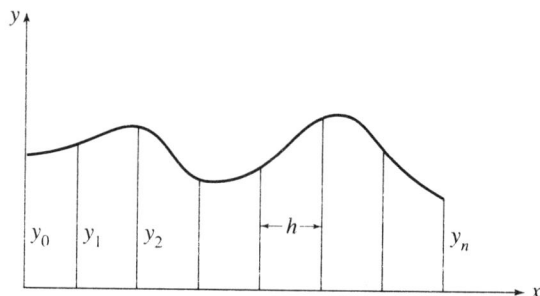

Trapezoidal Rule

$$A = h\left(\frac{1}{2}y_0 + y_1 + y_2 + \cdots + y_{n-1} + \frac{1}{2}y_n\right)$$

Simpson's Rule (Note that n must be even)

$$A = \frac{h}{3}(y_0 + 4y_1 + 2y_2 + 4y_3 + 2y_4 + \cdots + 2y_{n-2} + 4y_{n-1} + y_n)$$

Trigonometric Functions

$$\sin\theta = \frac{b}{c}; \quad \csc\theta = \frac{1}{\sin\theta}$$

$$\cos\theta = \frac{a}{c}; \quad \sec\theta = \frac{1}{\cos\theta}$$

$$\tan\theta = \frac{b}{a}; \quad \cot\theta = \frac{1}{\tan\theta}$$

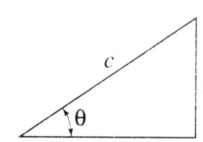

Selected Trigonometric Identities

$\sin(\theta_1 \pm \theta_2) = \sin\theta_1 \cos\theta_2 \pm \cos\theta_1 \sin\theta_2$

$\cos(\theta_1 \pm \theta_2) = \cos\theta_1 \cos\theta_2 \mp \sin\theta_1 \sin\theta_2$

$\sin 2\theta = 2\sin\theta \cos\theta$

$\cos 2\theta = \cos^2\theta - \sin^2\theta$

$\sin^2\theta + \cos^2\theta = 1$

$2\cos^2\theta = 1 + \cos 2\theta$

$2\sin^2\theta = 1 - \cos 2\theta$

$\sec^2\theta = 1 + \tan^2\theta$

$\csc^2\theta = 1 + \cot^2\theta$

Hyperbolic Functions

$\sinh u = \dfrac{e^u - e^{-u}}{2}$

$\cosh u = \dfrac{e^u + e^{-u}}{2}$

$\tanh u = \dfrac{e^u - e^{-u}}{e^u + e^{-u}}$

Series Expansions

$\sin u = u - \dfrac{u^3}{3!} + \dfrac{u^5}{5!} - \dfrac{u^7}{7!} + \cdots .$

$\cos u = 1 - \dfrac{u^2}{2!} + \dfrac{u^4}{4!} - \dfrac{u^6}{6!} + \cdots .$

$\sinh u = u + \dfrac{u^3}{3!} + \dfrac{u^5}{5!} + \dfrac{u^7}{7!} + \cdots .$

$\cosh u = 1 + \dfrac{u^2}{2!} + \dfrac{u^4}{4!} + \dfrac{u^6}{6!} + \cdots .$

Quadratic Equations

$Ax^2 + Bx + C = 0$

$x = \dfrac{-B \pm \sqrt{B^2 - 4AC}}{2A}$

Appendix D
Selected Derivatives

$$\frac{d}{du}(au) = a\frac{du}{dx}$$

$$\frac{d}{dx}(uv) = u\frac{dv}{dx} + v\frac{du}{dx}$$

$$\frac{d}{dx}\left(\frac{u}{v}\right) = \frac{1}{v^2}\left(v\frac{du}{dx} - u\frac{dv}{dx}\right)$$

$$\frac{d}{dx}(u^n) = nu^{n-1}\left(\frac{du}{dx}\right)$$

$$\frac{d}{dx}[f(u)] = \frac{d}{du}[f(u)]\left(\frac{du}{dx}\right)$$

$$\frac{d}{dx}(\ln u) = \frac{1}{u}\left(\frac{du}{dx}\right)$$

$$\frac{d}{dx}(e^u) = e^u\left(\frac{du}{dx}\right)$$

$$\frac{d}{dx}(\sin u) = \cos u\left(\frac{du}{dx}\right)$$

$$\frac{d}{dx}(\cos u) = -\sin u\left(\frac{du}{dx}\right)$$

$$\frac{d}{dx}(\tan u) = \sec^2 u\left(\frac{du}{dx}\right)$$

$$\frac{d}{dx}(\csc) = -(\csc u)(\cot u)\left(\frac{du}{dx}\right)$$

$$\frac{d}{dx}(\sec) = (\sec u)(\tan u)\left(\frac{du}{dx}\right)$$

$$\frac{d}{dx}(\cot u) = -(\csc^2 u)\left(\frac{du}{dx}\right)$$

$$\frac{d}{dx}(\sinh u) = \cosh u\left(\frac{du}{dx}\right)$$

$$\frac{d}{dx}(\cosh u) = \sinh u \left(\frac{du}{dx}\right)$$

$$\frac{d}{dx}(\tanh u) = \operatorname{sech}^2 u \left(\frac{du}{dx}\right)$$

$$\frac{d}{dx}(\operatorname{csch} u) = -(\operatorname{csch} u)(\operatorname{coth} u)\left(\frac{du}{dx}\right)$$

$$\frac{d}{dx}(\operatorname{sech} u) = -(\operatorname{sech} u)(\tanh u)\left(\frac{du}{dx}\right)$$

$$\frac{d}{dx}(\operatorname{coth} u) = -\operatorname{csch}^2 u \left(\frac{du}{dx}\right)$$

Appendix E
Selected Integrals

$$\int \sin x \, dx = -\cos x + c$$

$$\int \cos x \, dx = \sin x + c$$

$$\int \tan x \, dx = \ln \sec x + c$$

$$\int \csc x \, dx = \ln \tan \left(\frac{x}{2}\right) + c$$

$$\int \sec x \, dx = \ln \tan \left(\frac{\pi}{4} + \frac{x}{2}\right) + c$$

$$\int \cot x \, dx = \ln \sin x + c$$

$$\int x \sin(ax) \, dx = \frac{1}{a^2} \sin(ax) - \frac{x}{a} \cos(ax) + c$$

$$\int x^2 \sin(ax) \, dx = \frac{2x}{a^2} \sin(ax) - \left(\frac{a^2 x^2 - 2}{a^3}\right) \cos(ax) + c$$

$$\int x \cos(ax) \, dx = \frac{1}{a^2} \cos(ax) + \frac{x}{a} \sin(ax) + c$$

$$\int x^2 \cos(ax) \, dx = \frac{2x \cos(ax)}{a^2} + \left(\frac{a^2 x^2 - 2}{a^3}\right) \sin(ax) + c$$

$$\int \sinh x \, dx = \cosh x + c$$

$$\int \cosh x \, dx = \sinh x + c$$

$$\int \tanh x \, dx = \ln(\cosh x) + c$$

$$\int \operatorname{csch} x \, dx = \ln \left[\tanh \left(\frac{x}{2}\right) \right] + c$$

$$\int \text{sech}\, x\, dx = \tan^{-1}(\sinh x) + c$$

$$\int \coth x\, dx = \ln(\sinh x) + c$$

$$\int a\, dx = ax + c$$

$$\int u\, dv = uv - \int v\, du + c$$

$$\int x^n\, dx = \frac{x^{n+1}}{n+1} + c \quad \text{except for } n = -1$$

$$\int e^{ax}\, dx = \frac{e^{ax}}{a} + c$$

$$\int \ln x\, dx = x\ln x - x + c$$

$$\int \frac{dx}{x(a + bx)} = -\frac{1}{a}\ln\left(\frac{a + bx}{x}\right) + c$$

$$\int \frac{dx}{a + bx^2} = \frac{1}{\sqrt{ab}}\tan^{-1}\left(x\sqrt{\frac{b}{a}}\right) + c \quad \text{for } a > 0 \text{ and } b > 0$$

$$\int \frac{dx}{a^2 - b^2x^2} = \frac{1}{2ab}\ln\left(\frac{a + bx}{a - bx}\right) + c$$

$$\int \frac{dx}{a + bx} = \frac{1}{b}\ln(a + bx) + c$$

$$\int \frac{x\, dx}{a + bx} = \frac{x}{b} - \frac{a}{b^2}\ln(a + bx) + c$$

$$\int \sqrt{a + bx}\, dx = \frac{2}{3b}\sqrt{(a + bx)^3} + c$$

$$\int x\sqrt{a + bx}\, dx = -\frac{2(2a - 3bx)\sqrt{(a + bx)^3}}{15b^2} + c$$

$$\int x^2\sqrt{a + bx}\, dx = \frac{2(8a^2 - 12abx + 15b^2x^2)\sqrt{(a + bx)^3}}{105b^3} + c$$

$$\int \frac{dx}{\sqrt{a + bx}} = \frac{2\sqrt{(a + bx)}}{b} + c$$

$$\int \frac{x\, dx}{\sqrt{a + bx}} = -\frac{2(2a - bx)\sqrt{(a + bx)}}{3b^2} + c$$

$$\int \frac{x^2\,dx}{\sqrt{a+bx}} = \frac{2(8a^2 - 4abx + 3b^2x^2)\sqrt{(a+bx)}}{15b^3} + c$$

$$\int \frac{dx}{x\sqrt{a+bx}} = \frac{1}{\sqrt{a}}\ln\left(\frac{\sqrt{(a+bx)} - \sqrt{a}}{\sqrt{(a+bx)} + \sqrt{a}}\right) + c \quad a > 0$$

$$\int \frac{dx}{x\sqrt{a+bx}} = \frac{2}{\sqrt{-a}}\tan^{-1}\sqrt{\frac{(a+bx)}{a}} + c \quad a < 0$$

$$\int \frac{dx}{x^2\sqrt{a+bx}} = -\frac{\sqrt{(a+bx)}}{ax} - \frac{b}{2a}\int \frac{dx}{x\sqrt{a+bx}} + c$$

$$\int \sqrt{a^2 - x^2}\,dx = \frac{1}{2}\left(x\sqrt{a^2 - x^2} + a^2\sin^{-1}\frac{x}{|a|}\right) + c$$

$$\int x\sqrt{a^2 - x^2}\,dx = -\frac{1}{3}\sqrt{(a^2 - x^2)^3} + c$$

$$\int x^2\sqrt{a^2 - x^2}\,dx = -\frac{x}{4}\sqrt{(a^2 - x^2)^3}$$

$$+ \frac{a^2}{8}\left(x\sqrt{(a^2 - x^2)} + a^2\sin^{-1}\frac{x}{|a|}\right) + c \quad (a > 0)$$

$$\int \sqrt{a^2 \pm x^2}\,dx = \frac{1}{2}[x\sqrt{x^2 \pm a^2} \pm a^2\ln(x\sqrt{x^2 \pm a^2})] + c$$

$$\int x\sqrt{a^2 \pm x^2}\,dx = \frac{1}{3}\sqrt{(x^2 \pm a^2)^3} + c$$

$$\int x^2\sqrt{a^2 \pm x^2}\,dx = \frac{x}{4}\sqrt{(x^2 \pm a^2)^3} \mp \left(\frac{a^2}{8}\right)x\sqrt{x^2 \pm a^2}$$

$$- \left(\frac{a^4}{8}\right)\ln(x + \sqrt{x^2 \pm a^2}) + c$$

$$\int \frac{dx}{\sqrt{x^2 \pm x^2}} = \ln(x + \sqrt{x^2 \pm a^2}) + c$$

$$\int \frac{x\,dx}{\sqrt{x^2 \pm x^2}} = \sqrt{x^2 \pm a^2} + c$$

$$\int \frac{x^2\,dx}{\sqrt{x^2 \pm a^2}} = \frac{x}{2}\sqrt{x^2 \pm a^2} \mp \left(\frac{a^2}{2}\right)\ln(x + \sqrt{x^2 \pm a^2}) + c$$

$$\int x\ln x\,dx = \left(\frac{x^2}{2}\right)\ln x - \frac{x^2}{4} + c$$

Appendix F
Supports and
Connections

Type of Connection or Support	Reactive Force Components	Special Features
Undeformed length L_u Deformation s W Spring with attached weight 1	$F = ks = W$ W	The force F in a deformed spring is directed along its axis. The sense of this force is such that it is tension if the spring is stretched and compression if it is shortened. Also, $F = ks$ where k is known as the spring constant, equal to the force needed to deform the spring a unit distance and s is the total deformation.
Short link Flexible cable 2	F	One reactive force component F of known direction since it must act along the axis of the link or cable. This force F is unknown only in magnitude.

Type of Connection or Support	Reactive Force Components	Special Features
Flexible cable or belt around frictionless pulley. Flexible cable or belt around frictionless peg.		A frictionless pulley or frictionless peg changes the direction but not the magnitude of the force in the flexible cable or flexible belt. Thus, the force in the flexible cable or flexible belt is a constant along the entire length and must always be tension.
Frictionless hinge or pin.	Member Hinge	A frictionless hinge prevents any translation but allows rotation of the member about the pin axis. The reaction at the hinge is usually expressed in terms of its x and y components. Thus, the reaction at a frictionless hinge contains two unknowns.

Type of Connection or Support	Reactive Force Components	Special Features
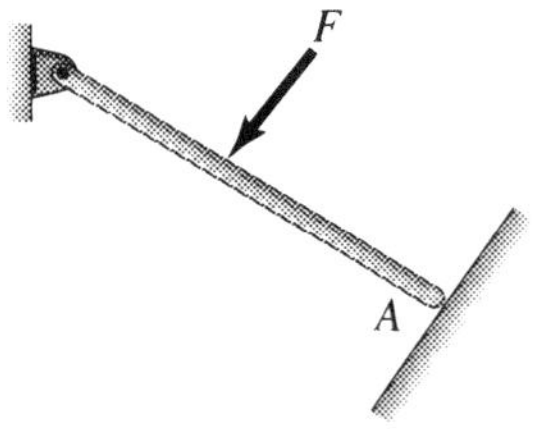 Frictionless plane	Plane / Member	A frictionless plane prevents translation in a direction perpendicular to the plane but allows rotation and translation along the plane. Thus, the support reaction consists of only one component N perpendicular to the frictionless plane.
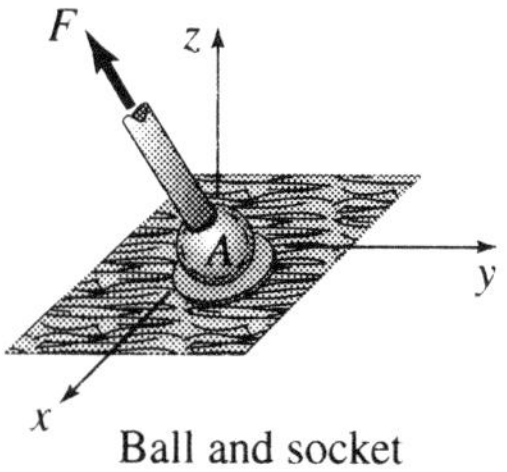 Ball and socket 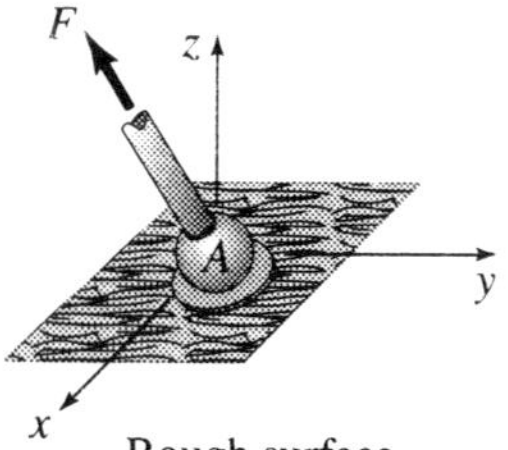 / Rough surface	Member	These supports prevent translation in any direction but permit rotation about any axis. The reaction at the support is generally expressed in terms of its x, y, and z components. Thus, the reaction at the support contains three unknowns.

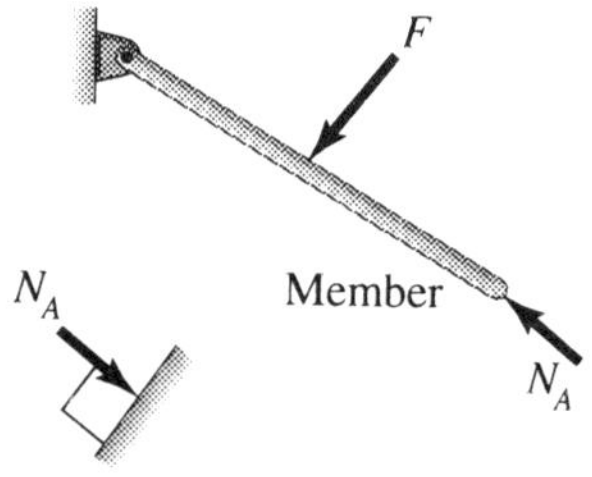

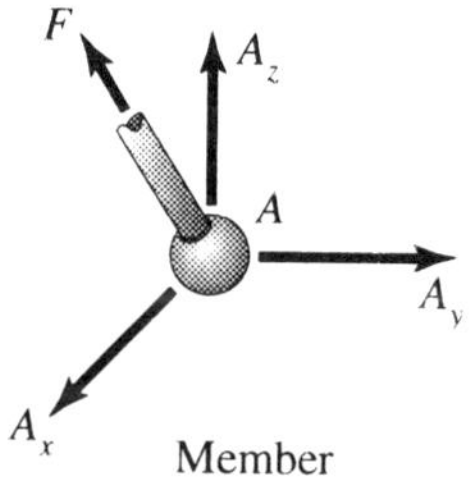

5

6

Type of Connection or Support	Reactive Force Components	Special Features
Roller / Rocker	Member / Plane	A roller or rocker support is capable of resisting a force only in a direction perpendicular to the plane supporting the roller or rocker. Both allow rotation and translation parallel to the plane and cannot transmit a moment or a force parallel to the plane. Thus, the support reaction consists of only one component N perpendicular to the plane.
7		
Collar on rod / Slider in slot	Axis of rod or slot	As in cases 5 and 7, the support reaction consists of only one component N perpendicular to the axis of the rod or the slot.
8		
Fixed support		A fixed support prevents translation in any direction and rotation about a z axis at the support. Thus, the support reaction consists of x and y components as well as a moment about the z axis.
9		

Type of Connection or Support	Reactive Force Components	Special Features
Universal joint 10		This universal joint allows only relative rotations about the y and z axis but no relative translation along any axis. Thus, there is only one reactive moment component and three reactive force components.
Bearing resisting no axial thrust. 11		This bearing allows the shaft to translate only in the x direction and to rotate only about the x axis. Therefore, there are two force components and two moment components.
Bearing resisting axial thrust Three-dimensional hinge 12		The bearing resisting axial thrust and the hinge are identical in action in that they allow only rotation about the axis of the shaft. Thus, there are three force components and two moment components.

Type of Connection or Support	Reactive Force Components	Special Features
 Three-dimensional fixed support		The fixed support prevents translation along and rotation about any axis. Therefore, this support develops three force and three moment reactive components.

13

Answers

Chapter 1

1.1 $F = 24.0$ N $= 5.40$ lb.
1.3 $F = 32.6$ lb $= 145.0$ N.
1.5 $F = 25.0$ lb $= 111.2$ N $\angle 10°$.
1.7 $W = 1177$ N, $F = 834$ N.
1.9 $F = gm$.
1.11 $F = 3.543 \times 10^{16}$ MN.
1.13 $I_A = \frac{2}{3}mr^2 + md^2$ is dimensionally homogeneous.
1.15   (a) 292 kg; (b) 7.62 m; (c) 1779 N; (d) 542 J.
1.17   (a) 1475 ft·lb/s; (b) 10,880 psi; (c) 82.0 ft/s; (d) 4.47×10^4 mph.
1.19   (a) 5; (b) 5; (c) 3, 2, or 1; (d) 3; (e) 4 or 3; (f) 4; (g) 4; (h) 3.
1.21 $W = [R_E/(R_E + h)]^2 W_0$, $W = (\frac{4}{9})W_0$, $W = (\frac{16}{25})W_0$.
1.23   (a) $m = 20.0$ kg $= 1.371$ slug; (b) $W = 74.4$ N $= 16.73$ lb.

Chapter 2

2.1 $R = 6500$ lb $\nwarrow\uparrow\,$ 2.62°.
2.3 $R = 6.18$ k 0.9° .
2.5 $P_x = 650$ lb $\rightarrow$, $P_y = 375$ lb $\uparrow$.
2.7 $P = 5.32$ lb, $P_y = 1.820$ lb.
2.9 $F_1 = 122.0$ lb, $F_2 = 165.6$ lb.
2.11 $\beta = 40.1°$, $F_{AB} = 11.24$ k or 7.12 k.
2.13 $F_{AC} = 3.42$ k, $F_{AB} = 5.00$ k.
2.15 $\mathbf{F} = (7.25\mathbf{i} + 3.38\mathbf{j})$ k.
2.17 $\mathbf{F} = 10.50(0.675\mathbf{i} + 0.738\mathbf{j})$ k.
2.19 $\mathbf{P} = 64.0(0.781\mathbf{i} + 0.625\mathbf{j})$ lb.
2.21 $\mathbf{Q} = 6.35(-0.835\mathbf{i} + 0.551\mathbf{j})$ k.
2.23 $F_x = 10.03$ k, $F_y = 7.03$ k.
2.25 $R = 5.87$ k $\angle 74.7°$.
2.27 $R = 2500$ lb $\angle 86.9°$.
2.29 $\alpha_3 = -33.4°$, $R = 18.44$ k.
2.31 $Q_2 = 3.41$ k, $\alpha = 20.5°$.
2.41 $F_{BA} = 375$ lb, $F_{BC} = 751$ lb.
2.43 $s_{BA} = 1.173$ in., $s_{BC} = 0.1284$ in.

2.45 $F_1 = 65.8$ k $\overset{25°}{\nearrow}$, $F_4 = 62.8$ k $\overset{40°}{\angle}$.

2.47 $F_1 = 12.30$ k, $\alpha_1 = 75.0°$.

2.49 $\alpha = 23.6°$, $N = 458$ lb.

2.51 $k = 73.1$ lb/in., $N = 90.0$ lb.

2.53 $\alpha_B = 55.0°$, $N_B = 172.1$ lb, $N_A = 246$ lb.

2.55 (a) $\beta = 60.9°$; (b) $N = 48.9$ lb.

2.57 $F_{AB} = 5.36$ k, $F_{BC} = 20.7$ k.

2.59 $T = 611$ lb, $\beta = 2.64°$.

2.61 $m = 61.2$ kg.

2.63 $W = 750$ N.

2.65 $N_A/W = 2\sin(150° - \alpha)$, $N_B/W = 2\sin\alpha$.

2.67 $\alpha = 55.8°$, $\beta = 82.8°$.

2.69 (a) $\mathbf{R} = (5.50\mathbf{i} - 7.99\mathbf{j})$ k, (b) $\mathbf{R} = 9.70$ k $\overset{39.5°}{\searrow}$.

2.71 $-17.02 \le F \le 9.32$ kN.

2.73 $m_C = 9.38$ kg, $H = 0.1662$ m.

2.75 (a) $\gamma = 115.0°$; (b) $AB = 1.829$ m, $BC = 0.371$ m; (c) $F = 39.9$ N.

Chapter 3

3.1 $F_x = 247$ lb, $F_y = 210$ lb, $F_z = -131.1$ lb.

3.3 $Q = 3270$ lb, $\alpha = 122.4°$, $\beta = 132.3°$, $\gamma = 59.3°$.

3.5 $\alpha = 48.2°$, $\gamma = 45.6°$, $P_y = 0.97$ k, $P_z = 2.62$ k.

3.7 $\alpha = 140.2°$, $\beta = 59.2°$, $\gamma = 112.6°$.

3.9 $\mathbf{Q} = (3.65\mathbf{i} + 7.82\mathbf{j} + 14.94\mathbf{k})$ k.

3.11 $\mathbf{r}_{A/O} = (-3\mathbf{i} + 5\mathbf{j} + 7\mathbf{k})$ ft, $\alpha = 109.2°$, $\beta = 56.7°$, $\gamma = 39.8°$.

3.13 $\mathbf{r}_{O/B} = (-1.125\mathbf{i} - 1.949\mathbf{j} + 3.900\mathbf{k})$ m, $\alpha = 104.5°$, $\beta = 115.7°$, $\gamma = 29.9°$.

3.15 (a) $F_x = -447$ lb, $F_y = 224$ lb, $F_z = -559$ lb;
      (b) $\alpha = 126.6°$, $\beta = 72.6°$, $\gamma = 138.2°$.

3.17 (a) $T_x = -2.22$ kN, $T_y = -3.80$ kN, $T_z = 0.950$ kN;
      (b) $\alpha = 119.6°$, $\beta = 147.6°$, $\gamma = 77.8°$.

3.19 $x = 2.35$ ft, $y = -10.55$ ft, $z = 3.67$ ft.

3.21 $\theta_0 = 118.0°$.

3.32 $\theta_B = 48.4°$.

3.25 $\theta = 84.9°$.

3.27 $F_{BD} = 45.0$ lb.

3.29 $F_{DC} = 903$ lb.

3.31 $R = 749$ lb, $\alpha = 130.4°$, $\beta = 68.5°$, $\gamma = 131.9°$.

3.33 (a) $\mathbf{R} = (-63.7\mathbf{i} + 856\mathbf{j} \quad 1198\mathbf{k})$ N,
      (b) $\alpha = 92.5°$, $\beta = 54.4°$, $\gamma = 144.5°$.

3.35 $Q_{3x} = -75.0$ N, $Q_{3y} = 650$ N, $Q_{3z} = -525$ N.

3.37 $F_{AB} = 5220$ N, $F_{AD} = 1450$ N, $F_{AC} = 4320$ N.

3.39 $P_1 = 0.1635$ kN, $P_2 = 8.24$ kN, $P_3 = 6.13$ kN.

3.41 $F = 7.07$ kN, $\alpha = 134.5°$, $\beta = 56.4°$, $\gamma = 116.7°$.

3.43 $F_A = 3.26$ kN, $F_B = -8.35$ kN, $F_C = -2.57$ kN.

3.45 $m = 55.1$ kg, $F_{AB} = 179.1$ N, $F_{AC} = 236$ N.
3.47   (a) $F_{BA} = -3400$ N, $F_{BC} = -6190$ N, $F_{BD} = 7720$ N;
         (b) $C_x = 3340$ N, $C_y = 3340$ N, $C_z = 4000$ N.
3.49   (a) $m = 183.5$ kg, $F_{BD} = 2.39$ kN, $F_{BC} = 1.506$ kN;
         (b) $A_x = 2.44$ kN, $A_y = 1.620$ kN, $A_z = 4.06$ kN.
3.51 $k = 412$ kN/m, $F_{AB} = 18.03$ kN, $F_{AC} = 18.03$ kN.
3.53 $W = 1343$ N.
3.55 $m = 960$ kg.
3.57   (a) $T_{AO} = 1.541$ kN; (b) $\theta = 25.1°$.
3.59   (a) $F_{1y} = -4.00$ kN, $F_{2x} = 24.0$ kN, $F_{3z} = 5.00$ kN;
         (b) $F_{1y} = -1.00$ kN, $F_{2x} = 0$, $F_{3z} = -10.00$ kN.
3.61 $F_{OA} = 5.44$ kN, $F_{OB} = 5.44$ kN, $F_{OC} = 9.08$ kN.
3.63 $P_B = 2.75$ kN, $F_{AB} = 4.12$ kN, $A_y = -2.75$ kN, $A_z = 1.372$ kN,
         $B_x = -2.75$ kN, $B_z = -1.372$ kN.

Chapter 4

4.1   (a) $M_O = 888$ N·m ↺; (b) $M_O = 888$ N·m ↺.
4.2   (a) $M_O = 346$ lb·in. ↺; (b) $\theta = 90°$; (c) $r = 17.30$ in.
4.5   (a) $M_A = 51.9$ kN·m ↺; (b) $Q = 10.38$ kN ↓.
4.7   (a) $(M_A)_1 = 77.9$ k·ft ↻, $(M_A)_2 = 67.9$ k·ft ↻,
        $(M_A)_R = 145.8$ k·ft ↻.
4.9 $M_O = 4160$ lb·in. ↺.
4.11 $M_O = 25.4$ N·m ↺.
4.13   (a) $M_O = 4160$ lb·ft ↺; (b) $W = 1107$ lb.
4.15   (a) $M_A = 2.41$ kN·m ↺; (b) $M_A = 2.41$ kN·m ↺.
4.17   (a) $M = 129.0$ lb·ft ↺; (b) $\mathbf{M} = -(129.0\mathbf{k})$ lb·ft.
4.19   (a) $M = 146.5$ kN·m ↻; (b) $\mathbf{M} = (146.5\mathbf{k})$ kN·m.
4.21 $M_R = 3780$ N·m ↺.
4.23 $F = 750$ N.
4.25 $F = 1118$ N.
4.27 $R = 100$ N $60°$ ↗ , $M_R = 47.6$ N·m ↻.
4.29 $R = 500$ N $15°$ ↗ , $M_R = 150.0$ k·in. ↺.
4.31 $F = 37.5$ kN.
4.33 $R = 70.7$ lb $45°$ ↗ , $M_R = 30.0$ lb·ft ↺.
4.35 $R = 14.50$ kN ↓, $M_R = 63.5$ kN·m ↺.
4.37 $R = 25.1$ k $69.6°$ ↙ , $M_R = 114.5$ k·ft ↻.
4.39 $P_1 = 911$ N ↑, $R = 1171$ N $25.5°$ ↗ .
4.41 $M = 149.9$ k·ft ↺, $R = 32.3$ k $60°$ ↙ .
4.43 $R = 19.70$ kN $66°$ ↙ , $\bar{x} = 4.00$ m right of A.
4.45 $a = 3.60$ in., $R = 335$ lb $63.4°$ ↙ .
4.47 $R = 40.0$ k ↓, $M_R = 265$ k·ft ↻.
4.49 $R = 25.0$ k ↓, $M_R = 203$ k·ft ↺.
4.51 $R = 33.0$ k ↓, $M_R = 286$ k·ft ↺.
4.53 $R = 5.33$ kN, $M_R = 1.325$ kN·m ↺.

4.55 $R = 300$ lb   36.90° ◥ , $M_R = 670$ lb·ft ⤵ .

4.57 $a = 12.00$ ft, $w = 0.500$ k/ft.

4.59   (a) $R = 66.7$ k ↓, $\bar{x} = 15.00$ ft right of the free end of the beam;
       (b) same as in part (a).

4.77 $A_y = 1.5wL - 0.25w(x^2/L)$, $(A_y/wL) = 1.5 - 0.25(x/L)^2$,
       $B_y = 1.5wL + wx + 0.25w(x^2/L)$,
       $(B_y/wL) = 1.5 + (x/L) + 0.25(x/L)^2$.

4.79 $A_x = 0$, $A_y = 114.6$ kN ↑, $B_y = 77.4$ kN ↑.

4.81 $x = 3.00$ ft, $A_x = 0$, $A_y = B_y = 13.50$ k ↑.

4.83 $A_x = wL\cos\theta$, $(A_x/wL) = \cos\theta$, $A_y = 0.5wL + wL\sin\theta$,
       $(A_y/wL) = 0.5 + \sin\theta$, $M_A = 0.125wL^2 + wL^2\sin\theta$,
       $(M_A/wL^2) = 0.125 + \sin\theta$.

4.85 $T = 17.50$ kN, $A_x = 15.16$ kN ←, $A_y = 13.75$ kN ↑.

4.87 $k = 20.0$ lb/in., $B_x = 0$, $B_y = 10.00$ lb ↑.

4.89 $A_x = 0$, $A_y = 0$, $F_{BD} = 1.867$ k ◢50° , $T_{DA} = 1.430$ k,
       $N_D = 1.200$ k →.

4.91 $W = 80.9$ lb, $A_x = 18.84$ lb ◥10° , $A_y = 67.7$ lb ⤢10° .

4.93 $N_B = 750$ N →, $C_x = 750$ N ←, $C_y = 0$, $M_C = 750$ N·m ⤵ .

4.95 $E_x = 4.00$ kN ←, $E_y = 7.00$ kN ↑, $B_x = 1.000$ kN →,
       $B_y = 1.000$ kN ↑.

4.97 $C_x = 0$, $C_y = 5.33$ kN ↑, $B_x = 1.000$ kN ←, $B_y = 1.000$ kN ↓.

4.99 $A_x = 0.630$ kN ←, $A_y = 15.00$ kN ↑, $E_x = 15.63$ kN →.

4.101 $C_x = 5.00$ k →, $C_y = 5.00$ k ↓, $B_x = 10.00$ k ←,
       $B_y = 10.00$ k ↑.

4.103 $A_x = 484$ N ←, $A_y = 649$ N ↓, $B_x = 165.5$ N ←,
       $B_y = 433$ N ↑.

4.105 $N_A = 1133$ N →, $N_B = 2270$ N 60° ◥ .

4.107 $N_B = 345$ lb 60° ◢ , $N_C = 500$ lb ◤30° .

4.109 $s_A = \dfrac{3W}{2k}$, $s_B = \dfrac{W}{2k}$, scale A more accurate.

4.111 $h = \dfrac{Wd}{T}$, $L = W$, $D = T$.

4.113 $\alpha_{MIN.} = 6.20°$, $N_A = 2640$ N, $F_A = 1000$ N.

4.115 $\theta = 62.9°$.

4.117 $P = 22.3$ lb.

4.119 (a) $F_A = 15.00$ k ←, $F_B = 15.00$ k →;
       (b) $Q_A = 6.43$ k ↑, $Q_C = 6.43$ k ↓.

4.121 (a) $R = 11.31$ kN ◣45° , $M_R = 26.4$ kN·m ⤵;
       (b) $R = 11.31$ kN ◣45° , $\bar{x} = 3.30$ m.

4.123 $R = 26.8$ kN 63.4° ◣ , $\bar{x} = 5.44$ m right of A.

4.125 $R = 49.7$ kN ↑, $\bar{x} = 11.87$ m left of A.

4.127 $T = 24.3$ kN, $A_x = 15.62$ kN ←, $A_y = 2.09$ kN ↑.

4.129 $F_D = 90.4$ lb.

4.131 $T = 147.2$ N, $A_x = 343$ N $\leftarrow$, $A_y = 0$.

4.133 (a) $F_{AB} = \left(\dfrac{30,000\cos\theta - 9,000\sin\theta}{180\sin\theta}\right)\sqrt{369 - 360\cos\theta}$;

      (b) $\theta = 5°$, $F_{AB} = 5,970$ lb; $\theta = 10°$, $F_{AB} = 3,410$ lb;

         $\theta = 15°$, $F_{AB} = 2,640$ lb; $\theta = 20°$, $F_{AB} = 2,260$ lb;

         $\theta = 25°$, $F_{AB} = 2,010$ lb.

Chapter 5

5.1   (a) $M_O = (428\mathbf{i} + 296\mathbf{j} + 20\mathbf{k})$ lb·ft;

     (b) $M_B = (349\mathbf{i} + 781\mathbf{j} + 395\mathbf{k})$ lb·ft.

5.3   (a) $M_O = -(6.00\mathbf{i} + 22.0\mathbf{j} + 53.0\mathbf{k})$ N·m;

     (b) $M_B = -(10.00\mathbf{i} + 39.6\mathbf{j} + 96.4\mathbf{k})$ N·m.

5.5 $M_O = (-12,130\mathbf{i} + 10,110\mathbf{j})$ lb·ft, $\alpha = 140.2°$, $\beta = 50.2°$,

     $\gamma = 90.0°$.

5.7 $M_A = (29.1\mathbf{i} - 43.7\mathbf{j})$ k·ft, $d_A = 2.63$ ft.

5.9 $M_A = (6.12\mathbf{j} - 12.25\mathbf{k})$ N·m, $\alpha = 90.0°$, $\beta = 63.4°$, $\gamma = 153.5°$.

5.11 $M_B = (56.0\mathbf{i} + 20.0\mathbf{j} + 70.0\mathbf{k})$ k·ft, $d_B = 8.75$ ft.

5.13 (a) $M_O = -(1040\mathbf{i} + 400\mathbf{j})$ N·m;

      (b) $M_O = -(1040\mathbf{i} + 400\mathbf{j})$ N·m.

5.15 $x = -3.52$ m, $y = 4.28$ m, $z = -1.142$ m and

     $x = 2.14$ m, $y = -4.22$ m, $z = 3.11$ m.

5.17 $\mathbf{A}\cdot(\mathbf{B} \times \mathbf{C}) = \mathbf{0}$, where $\mathbf{A}$, $\mathbf{B}$ and $\mathbf{C}$ are coplanar.

5.19 $U_x = -32.1$, $U_z = 44.1$, and $U_x = -54.5$, $U_z = -0.743$.

5.21 (a) $\mathbf{M}_x = (6.64\mathbf{i})$ kN·m, $\mathbf{M}_y = 0$, $\mathbf{M}_z = (-8.42\mathbf{k})$ kN·m;

      (b) $\mathbf{M}_{AB} = (-4.11\mathbf{i} + 3.29\mathbf{k})$ kN·m.

5.23 (a) $\mathbf{M}_{AB} = -(70.1\mathbf{i} + 112.2\mathbf{j})$ kN·m;

      (b) $\mathbf{M}_{BC} = (16.17\mathbf{i} - 25.9\mathbf{j})$ kN·m.

5.25 $Q = 68.7$ kN.

5.27 (a) $\mathbf{M}_x = 0$, $\mathbf{M}_y = (1006\mathbf{j})$ lb·ft, $\mathbf{M}_z = (503\mathbf{k})$ lb·ft;

      (b) $\mathbf{M}_{OB} = (232\mathbf{i} + 123.9\mathbf{j} - 92.9\mathbf{k})$ lb·ft.

5.29 $d = 0.246$ m.

5.31 (a) $\mathbf{M}_{AD} = -(225\mathbf{i})$ k·ft; (b) $\mathbf{M}_{EF} = (-52.4\mathbf{i} + 23.6\mathbf{k})$ k·ft.

5.33 $d = 4.85$ ft.

5.35 $\mathbf{M}_R = 2170(0.447\mathbf{i} + 0.894\mathbf{k})$ lb·ft, $\alpha = 63.4°$, $\beta = 90.0°$,

     $\gamma = 26.6°$.

5.37 $\mathbf{M}_F = \mathbf{M}_P$.

5.39 $\mathbf{M}_x = -(2590\mathbf{i})$ lb·in., $\mathbf{M}_y = 0$, $\mathbf{M}_z = -(1554\mathbf{k})$ lb·in.

5.41 $\mathbf{M}_x = -(24.0\mathbf{i})$ kN·m, $\mathbf{M}_y = -(18.00\mathbf{j})$ kN·m, $\mathbf{M}_z - 0$.

5.43 $\mathbf{M}_x = (590\mathbf{i})$ k·in., $\mathbf{M}_y = (1273\mathbf{j})$ k·in., $\mathbf{M}_z = -(23.6\mathbf{k})$ k·in.

5.45 $\mathbf{R} = -(5.84\mathbf{i} + 3.24\mathbf{j} + 14.21\mathbf{k})$ kN, $\mathbf{M}_R = (48.9\mathbf{i} - 60.9\mathbf{j})$ kN·m.

5.47 $\mathbf{R} = (50.0\mathbf{i} + 40.0\mathbf{j})$ lb, $\mathbf{M}_R = (-120.0\mathbf{i} + 900\mathbf{j} + 500\mathbf{k})$ lb·ft,

     $\mathbf{M}_{OA} = (500\ \mathbf{k})$ lb·ft.

5.49 $\mathbf{R} = (25.0\mathbf{j} - 12.00\mathbf{k})$ kN,

     $\mathbf{M}_R = -(212\mathbf{i} + 60.0\mathbf{j} + 125.0\mathbf{k})$ kN·m,

     $\mathbf{M}_{AC} = -(125.0\mathbf{k})$ kN·m.

5.51 $\mathbf{R} = -(110\mathbf{k})$ k, at $x = 16.36$ ft and $y = 21.8$ ft.

5.53 $F_O = 2.00$ k $\downarrow$, $F_A = 5.50$ k $\downarrow$, $F_B = 2.50$ k $\downarrow$.

5.55   At $x = 13.50$ ft and $y = 30.0$ ft.

5.57 $\mathbf{R} = -(5.48\mathbf{i} + 3.24\mathbf{j} + 14.21\mathbf{k})$ kN,
       $\mathbf{M}_C = (2.10\mathbf{i} + 1.163\mathbf{j} + 5.11\mathbf{k})$ kN·m, at $y = -0.875$ m and
       $z = 10.62$ m.

5.59 $\mathbf{R} = (25.0\mathbf{j} - 12.00\mathbf{k})$ kN, $\mathbf{M}_C = \mathbf{0}$, at $x = 0$, $y = 9.33$ ft.

5.61 $A = 18.53$ k $\uparrow$, $B = 27.5$ k $\uparrow$, $C = 24.7$ k $\uparrow$.

5.63 $d = -0.316$ ft, $T_{AD} = 904$ lb, $T_{CF} = 596$ lb.

5.65 $F_{DC} = 28.9$ k.

5.67 $T_{DC} = 454$ N.

5.69 $B_y = -2730$ lb.

5.71 $C_x = 15.15$ kN, $C_y = -41.0$ kN, $E_x = -20.2$ kN, $E_y = 31.0$ kN,
       $E_z = 27.3$ kN.

5.73 $C_x = 7.50$ k, $C_y = -25.0$ k, $M_{Cy} = 140.0$ k·ft,
       $G_x = -7.50$ k, $G_y = 25.0$ k, $G_z = -10.00$ k.

5.75 $x_G = -0.556$ m, $T_{MIN.} = 9.98$ kN.

5.77 $W = 200$ lb, $A_x = 0$, $A_y = 37.5$ lb, $A_z = 100.0$ lb,
       $B_y = -137.5$ lb, $B_z = 100.0$ lb.

5.79 $A_x = -0.231W$, $A_y = 1.788W$, $A_z = -0.327W$, $T_{BE} = 1.861W$,
       $T_{DF} = 0.496W$.

5.81 $A_x = -4.17$ k, $A_z = 2.50$ k, $B_x = 8.34$ k, $B_y = 3.34$ k, $B_z = 0$,
       $T_{CD} = 5.90$ k.

5.83   Statically indeterminate to first degree.

5.85   Statically indeterminate to second degree.

5.87   Statically indeterminate to second degree.

5.89   Unstable.

5.91   Improperly constrained.

5.93   Statically indeterminate to third degree.

5.95   Improperly constrained.

5.97 $F_x = 5.74$ N and 0.956 N, $F_y = 8.19$ N and 1.808 N,
       $F_z = -0.215$ N and -9.79 N.

5.99 $W_x = 13.18$ and -12.36, $W_y = 1.544$ and -4.84.

5.101 $\mathbf{P} = (35.8\mathbf{i} + 17.89\mathbf{j} - 44.7\mathbf{k})$ lb,
       $\mathbf{M}_O = -(897\mathbf{i} + 447\mathbf{j} + 895\mathbf{k})$ lb·in.

5.103 $F_A = 2.50$ kN $\uparrow$, $F_B = 7.50$ kN $\downarrow$, $F_C = 15.00$ kN $\downarrow$.

5.105 $T_{AB} = 986$ N.

5.107 $T_{CD} = 6.91$ k.

5.109 $T = 5.50$ kN, $A_y = -3.23$ kN, $A_z = 3.81$ kN,
       $B_x = 0$, $B_y = -6.77$ kN, $B_z = 1.690$ kN.

Chapter 6

6.1 $F_{AB} = 0$, $F_{AC} = 10.00$ kN (T), $F_{BC} = 8.00$ kN (C).

6.3 $F_{AB} = F_{BC} = 5.00$ k (T), $F_{AD} = F_{CD} = 6.40$ k (C),
       $F_{BD} = 8.00$ kN (T).

6.5   Partial answers: $F_{AB} = F_{BC} = 2P$ (T), $F_{DE} = F_{EF} = P$ (C),
      $F_{EC} = 2.829P$ (C)

6.7   Partial answers: $F_{AB} = 3.46P$ (T), $F_{CD} = 2.94P$ (C).

6.9 $F_{AC} = F_{CD} = F_{DE} = F_{EB} = 0$, $F_{AB} = 13.34$ k (T),
      $F_{AD} = 14.90$ k (C), $F_{DB} = 18.86$ k (C).

6.11 Partial answers: $F_{AB} = 0.33P$ (T), $F_{CE} = 1.88P$ (C),
      $F_{CE} = 2.0P$ (T).

6.13 $F_{1-3} = F_{3-4} = 0$, $F_{1-2} = 20.0$ kN (C), $F_{2-3} = 25.0$ kN (T),
      $F_{2-4} = 15.00$ kN (C).

6.15 $T_{1-2} = 0$, $T_{1-3} = -Q$.

6.17 $F_{1-2} = 80.0$ k (C), $F_{1-8} = 62.5$ k (T), $F_{2-8} = 20.0$ k (T),
      $F_{8-9} = 62.6$ k (T).

6.19 $F_{1-6} = 30.0$ kN (C), $F_{2-6} = 150.0$ kN (C),
      $F_{6-7} = 120.0$ kN (T).

6.21 $T_{1-2} = 3.35P$ (C), $T_{1-6} = 5.00P$ (T).

6.23 Partial answers: $F_{AB} = 110.0$ k (T), $F_{BC} = 110.0$ k (T),
      $F_{CD} = 14.14$ k (C).

6.25 Partial answers: $F_{1-2} = F_{2-3} = F_{1-4} = F_{2-4} = 0$,
      $F_{3-4} = 14.14$ k (C).

6.27 $F_{BE} = F_{DA} = F_{FE} = F_{FC} = 0$.

6.29 $F_{1-2} = 10.00$ kN (C), $F_{1-4} = 8.00$ kN (T), $F_{2-3} = 10.00$ kN (C).

6.31 $F_{FB} = F_{BG} = F_{CG} = 0$.

6.33 $F_{BF} = F_{GC} = F_{DH} = 0$.

6.35 $F_{5-11} = F_{4-11} = F_{4-10} = F_{3-10} = F_{3-9} = 0$.

6.37 $F_{BC} = 26.6$ k (C), $F_{BF} = 4.67$ k (T), $F_{EF} = 23.3$ k (T).

6.39 $T_{KL} = 4.5P$ (C), $T_{KD} = 0.707P$ (T), $T_{CD} = 4.0P$ (T).

6.41 $F_{FG} = 5.50P$ (C), $F_{FC} = P$ (C), $F_{BC} = 5.63P$ (T).

6.43 $T_{KL} = 11.2P$ (C), $T_{CL} = 0$, $T_{CD} = 10.0P$ (T).

6.45 $F_{HJ} = 2.33P$ (T), $F_{HE} = 3.77P$ (C), $F_{DE} = 3.33P$ (T).

6.47 Partial answers: $F_{AC} = 1.562$ k (T), $F_{BD} = 4.16$ k (C),
      $F_{AD} = 1.046$ k (T).

6.49 $T_{EF} = 7.80Q$ (T), $T_{FA} = 2.83Q$ (C), $T_{AB} = 5.80Q$ (C).

6.51 $T_1 = 180.0$ k (C), $T_2 = 28.3$ k (T), $T_3 = 14.15$ k (C),
      $T_4 = 170.0$ k (T).

6.53 $F_{JK} = 5.87P$ (C), $F_{CK} = 0.40P$ (T), $F_{CD} = 5.62P$ (T).

6.55 $T_{5-6} = (\sqrt{5}/3)(\frac{8}{5}Q - \frac{6}{5}P)$, $T_{6-11} = \sqrt{2}(\frac{11}{15}Q + \frac{1}{5}P)$,
      $T_{11-12} = (1/5)(Q + 3P)$.

6.57 $T_1 = 12.75P$ (C) $= 127.5$ k (C), $T_2 = 2.97P$ (C) $= 29.7$ k (C),
      $T_3 = 16.85P$ (T) $= 168.5$ k (T).

6.59 $F_{3-4} = 1.008$ k (C), $F_{3-9} = 1.965$ k (C), $F_{8-9} = 2.68$ k (T).

6.61 $T_1 = 93.8$ k (C), $T_2 = 14.85$ k (C), $T_3 = 103.4$ k (T).

6.63 $F_1 = 65.6$ k (T), $F_2 = 61.9$ k (T), $F_3 = 109.4$ k (C).

6.65 Partial answers: (a) statically indeterminate to third degree;
      (c) improperly constrained and unstable;
      (e) statically determinate and properly constrained.

6.67 Partial answers: (b) statically indeterminate to first degree;
(d) statically indeterminate to first degree;
(f) improperly constrained and unstable.

6.69 Solution not possible because truss is unstable.

6.71 (a) statically determinate and properly constrained;
(b) improperly constrained and unstable;
(c) improperly constrained and unstable.

6.73 Solution not possible because truss is unstable.

6.75 Partial answers: $T_{1-2} = 30.0$ k (T), $T_{5-12} = 7.07$ k (C).

6.77 $F_{3-4} = 1.5P$ (C), $F_{4-5} = 1.5P$ (C).

6.79 Partial answers: $F_{5-12} = 0.707Q$ (T), $F_{16-17} = 1.50Q$ (T).

6.81 $T_{4-5} = 3.5P$ (C), $T_{5-11} = 0.707P$ (C), $T_{14-15} = 4.0P$ (T).

6.83 Partial answers: $F_{1-3} = 0.75P$ (T), $F_{4-7} = 1.77P$ (C),
$F_{11-12} = 1.25P$ (T).

6.85 Partial answers: Pin 3, $0.581F_{3-6} + 0.707F_{3-4} = 0$,
$0.814F_{3-6} - F_{2-3} - 0.707F_{3-4} = 0$.

6.87 $F_{5-6} = 0.333P$ (C), $F_{6-15} = 0.547P$ (C), $F_{15-16} = 0.833P$ (T).

6.89 $A_x = 0$, $A_z = 54.4$ kN, $B_y = 0$, $B_z = 16.67$ kN, $C_x = 0$,
$C_z = 28.9$ kN.

6.91 Partial answers: $T_{AB} = 0$, $T_{CD} = 12,760$ lb (C),
$T_{CB} = 3,000$ lb (T).

6.93 $A_x = 0$, $A_z = 17.50$ kN, $B_y = 0$, $B_z = 2.50$ kN, $C_x = 0$,
$C_z = 40.0$ kN.

6.95 Partial answers: $B_y = 0$, $C_x = -40.0$ k, $A_x = -20.0$ k.

6.97 $T_{OA} = 3,120$ lb (T), $T_{OB} = 3,500$ lb (C), $T_{OC} = 1,077$ lb (T).

6.99 $T_{OA} = 1,253$ lb (T), $T_{OB} = 1,482$ lb (C), $T_{OC} = 2,010$ lb (T).

6.101 $F_{1-3} = 90.1$ kN (C), $F_{2-3} = 75.0$ kN (T).

6.103 Partial answers: $T_{AB} = 20.9$ kN (C), $T_{CD} = 16.54$ kN (T),
$T_{BD} = 0$.

6.105 $F_{1-2} = F_{1-3} = F_{2-4} = F_{3-4} = F_{4-5} = F_{3-5} = 0$,
$F_{2-3} = 25.0$ kN (C).

6.107 $F_{FG} = 26.7$ k (C), $F_{FB} = 13.36$ k (T).

6.109 $F_{9-10} = 7.50$ k (C), $F_{2-3} = 7.50$ k (T).

6.111 $F_{CE} = 16.50$ kN (C), $F_{CF} = 7.50$ kN (T), $F_{EB} = 12.50$ kN (T).

6.113 $F_{AD} = 9.71$ kN (C), $F_{BE} = 4.91$ kN (T), $F_{CF} = 4.27$ kN (C).

6.115 $T_{BC} = 8.84$ kN (T), $T_{BD} = 22.5$ kN (C), $T_{BA} = 12.12$ kN (T).

Chapter 7

7.1 $T = W/2 \sin \theta$. For $W = 1000$ lb, $\theta = 30°$ and $90°$, $T = 1000$ lb
and 500 lb, respectively.

7.3 $r = 0.804$ ft, $L = 0.100$ ft, $F_{DE} = 114.9$ lb, $A_x = 57.4$ lb $\leftarrow$,
$A_y = 99.5$ lb $\uparrow$.

7.5 $N_A = 300$ N $\angle 30°$, $N_B = 520$ N $60°$.

7.7 $P_{MIN.} = 170.1$ lb, $\phi = 49.1°$.

7.9 $T_A = 157.0$ lb, $T_C = 133.0$ lb.

7.11 $T_{AB} = 22.4$ lb, $T_{CD} = 37.7$ lb, $\phi = 65.1°$.

7.13 $T_{AB} = 48.1$ lb, $T_{EG} = 56.4$ lb, $\phi = 56.7°$.

7.15 $P = 4.66$ k, $F_{CD} = 3.65$ k (C), $F_{EG} = 6.35$ k (C).

7.17 $T_{AB} = 43.7$ lb, $T_{EG} = 26.8$ lb, $\phi = 72.1°$.

7.19 $\theta = 0$, $F_{AB} = \left(\dfrac{a+b}{b}\right)P$, $C = \left(\dfrac{a}{b}\right)P \rightarrow$.

7.21 $N_A = N_D = 57.7$ N, $N_B = N_E = 28.9$ N, $N_C = 170.0$ N,
$N_F = 210$ N.

7.23 $\theta = 69.4°$, $F_{AB} = 99.2$ lb (C), $N_C = 117.1$ lb, $N_D = 58.6$ lb,
$N_E = 101.4$ lb.

7.25   Partial answers: $\dfrac{D_1}{D_2} = \sqrt{2}$, $N_A = N_D = 0.707W_1$,
$N_B = N_F = 1.061W_1$.

7.27 $\dfrac{D_1}{D_2} = \sqrt{2}$, $N_A = N_D = 141.4$ lb, $N_E = N_C = 70.8$ lb,
$N_F = N_B = 212$ lb.

7.29 $\theta = 60.0°$, $T_{AB} = 1.155W$, $F_{CD} = 0.577W$.

7.31 $P = 0.578W$, $T_{AB} = 0.911W$, $T_{CD} = 0.244W$.

7.33 $P = 115.6$ lb, $T_{AB} = 182.2$ lb, $T_{CD} = 48.8$ lb.

7.35 $N_B = 966$ N, $A_x = 337$ N $\rightarrow$, $A_y = 283$ N $\downarrow$.

7.37 $T_{AC} = 1.667$ kN, $B_x = 1.334$ kN $\leftarrow$, $B_y = 1.000$ kN $\uparrow$.

7.39   For member ACE only: $R_C = 134.6$ kN $^{21.8°}\!\!\nearrow$,
$A_x = 53.6$ kN $\rightarrow$, $A_y = 21.4$ kN $\downarrow$, $E_x = 71.4$ kN,
$E_y = 28.6$ kN $\downarrow$.

7.41   For member CBD only: $T_{BA} = 933$ N, $C_x = 400$ N $\rightarrow$,
$C_y = 1012$ N $\downarrow$, $D_x = 400$ N $\rightarrow$, $D_y = 1493$ N $\uparrow$.

7.43   For member ACE only: $R_C = 5.40$ k $\downarrow$, $A_x = 0.508$ k $\leftarrow$,
$A_y = 3.13$ k $\uparrow$, $E_x = 0.508$ k $\rightarrow$, $F_y = 2.27$ k $\uparrow$.

7.45 $B_x = 693$ lb $\leftarrow$, $B_y = 400$ lb $\uparrow$, $T_{AE} = 339$ lb, $C_x = 933$ lb $\leftarrow$,
$C_y = 160.0$ lb $\uparrow$.

7.47 $F_{BC} = \left(\dfrac{b+c}{c\sin\theta}\right)Q$, $A_x = P - \left(\dfrac{b+c}{a}\right)Q \rightarrow$, $A_y = \left(\dfrac{b}{c}\right)Q \uparrow$.

7.49   For member ABC only: $A_x = \left(\dfrac{a+b}{b}\right)Q \rightarrow$, $A_y = \left(\dfrac{a}{b}\right)P \downarrow$,
$B_x = \left(\dfrac{a+b}{b}\right)Q \leftarrow$, $B_y = \left(\dfrac{a+b}{b}\right)P \uparrow$.

7.51   For member ADE only: $F_A = \left(\dfrac{13}{15}\right)P$ $^{12}\!\nearrow^{5}$,
$F_D = \left(\dfrac{5}{6}\right)P$ $^{3}\!\searrow^{4}$, $E_x = \left(\dfrac{1}{3}\right)P \rightarrow$, $E_y = \left(\dfrac{9}{5}\right)P \uparrow$.

7.53   For member BC only: $B_x = 1,000$ lb $\rightarrow$, $B_y = 1,000$ lb $\uparrow$,
$C_x = 1,000$ lb $\leftarrow$, $C_y = 1,000$ lb $\uparrow$.

7.55   For member AB only: $A_x = 0$, $A_y = 0.667$ kN $\downarrow$, $B_x = 0$,
$B_y = 0.667$ kN $\uparrow$.

7.57   For member BC only: $B_x = 250$ lb $\leftarrow$, $B_y = 600$ lb $\downarrow$,
$C_x = 250$ lb $\rightarrow$, $C_y = 600$ lb $\uparrow$.

7.59   For member ABC only: $A_x = 180.0$ lb $\rightarrow$, $A_y = 25.0$ lb $\uparrow$,
$B_x = 180.0$ lb $\leftarrow$, $B_y = 75.0$ lb $\uparrow$.

7.61   For member CED only: $C_x = 1.600$ k $\rightarrow$, $C_y = 1.000$ k $\downarrow$,
$D_x = 0.400$ k $\rightarrow$, $D_y = 1.000$ k $\uparrow$, $T_E = 2$ k.

7.63   For member ACE only: $R_C = 469$ lb $\overset{5}{\underset{6}{\diagup}}$,
$A_x = 389$ lb $\rightarrow$, $A_y = 100.0$ lb $\downarrow$, $E_x = 29.2$ lb $\leftarrow$,
$E_y = 400$ lb $\uparrow$.

7.65   For member AB only: $A_x = 750$ lb $\rightarrow$, $A_y = 563$ lb $\uparrow$,
$B_x = 750$ lb $\leftarrow$, $B_y = 563$ lb $\uparrow$.

7.67   Partial answers: $A_x = \left(\dfrac{W}{2}\right)\left[\left(\dfrac{r}{h}\right)\left(\dfrac{\operatorname{ctn}(\theta/2)}{\sin(\theta/2)}\right) - \tan(\theta/2)\right]$,
$A_y = \dfrac{W}{2}$.

7.69   For member CDF only: $R_D = 1.223wa$ $\overset{1}{\underset{2}{\diagup}}$,
$C_x = 0.547wa$ $\rightarrow$, $C_y = 1.703wa$ $\downarrow$, $F_x = 0.547wa$ $\rightarrow$,
$F_y = 2.247wa$ $\uparrow$.

7.71   For member BCE only: $F_{AB} = 309$ lb $45^\circ \diagup$.
$F_{ED} = 489$ lb $\overset{1}{\underset{2}{\diagup}}$, $C_x = 219$ lb. $\leftarrow$, $C_y = 681$ lb $\uparrow$.

7.73 $A_x = P \rightarrow$, $A_y = \left(\dfrac{5}{4}\right)P \downarrow$, $B_x = P \leftarrow$, $B_y = \left(\dfrac{5}{4}\right)P \uparrow$.

7.75 $F_{AC} = 2.333P\sin\theta \uparrow$, $F_{BC} = 0.754P\sin\theta + 0.849P\cos\theta$ $\overset{1}{\underset{1}{\diagup}}$,
$F_{BD} = 0.961P\sin\theta - 0.721P\cos\theta$ $\overset{3}{\underset{2}{\diagdown}}$,

7.77 $B_x = 0.889wb \rightarrow$, $B_y = 1.278wb \uparrow$, $F_{CA} = 1.693wb$ $\overset{2}{\diagdown}{}^{2.3}$
$F_{CE} = 6.066wb$ $\overset{1}{\underset{4}{\diagup}}$, $D_x = 4.410wb \rightarrow$, $D_y = 1.471wb \uparrow$.

7.79 $B_x = 106.7$ lb $\rightarrow$, $B_y = 153.4$ lb $\uparrow$, $F_{CA} = 203$ lb $\overset{2}{\diagdown}{}^{2.3}$,
$F_{CE} = 728$ lb $\overset{1}{\underset{4}{\diagup}}$, $D_x = 529$ lb $\rightarrow$, $D_y = 176.5$ lb $\uparrow$.

7.81 $F_B = 0.566$ k $45^\circ \diagdown$, $A_x = 0.400$ k $\rightarrow$, $A_y = 0.400$ k $\downarrow$.

7.83   Member ABC only: $A_x = 2Q \leftarrow$, $A_y = 2Q \uparrow$,
$N_B = 1.414Q$ $45^\circ \diagdown$, $C_x = 3Q \leftarrow$, $C_y = 3Q \downarrow$.

7.85   Member DBE only: $D_x = 200$ lb $\leftarrow$, $D_y = 100.0$ lb $\downarrow$,
$N_B = 141.4$ lb $\diagdown 45^\circ$, $E_x = 100.0$ lb $\rightarrow$, $E_y = 200$ lb $\uparrow$.

7.87   Member AB only: $A_x = 1.5P \rightarrow$, $A_y = 0.5P \downarrow$, $B_x = 1.5P \leftarrow$,
$B_y = 1.5P \uparrow$.

7.89 $F_{BC} = \dfrac{2bw}{\cos^2\theta} \uparrow$, $A_x = 2bw\tan\theta \leftarrow$, $A_y = 2bw\tan^2\theta \downarrow$.

7.91 $F_{BC} = 226$ lb $\uparrow$, $A_x = 72.8$ lb $\leftarrow$, $A_y = 26.5$ lb $\downarrow$.

7.93 $T_{CA} = 200$ lb $\downarrow$, $T_{DA} = 245$ lb $\searrow^{45°}$, $T_{DB} = 473$ lb $\uparrow$.

7.95 Member DGI only: $D_x = 0$, $D_y = 1.537W \downarrow$, $G_y = 2.537W \uparrow$.

7.97 Member AEF only ($\theta = 90°$): $A_x = P \leftarrow$, $A_y = 0$,

$$F_{ED} = 2.917P \swarrow^3_4, F_x = 1.333P \rightarrow, F_y = 1.750P \uparrow.$$

7.99 Member ABC only: $A_x = 0.717W \leftarrow$, $A_y = 0.463W \downarrow$,
$T_{CABLE} = W$, $N_B = 0.010W \rightarrow$, $C_y = 0.756W \uparrow$.

7.101 $A_x = \left(\dfrac{a}{a+b}\right)P$, $A_y = 0.577\left(\dfrac{b}{a+b}\right)P$, $F_B = 1.155\left(\dfrac{b}{a+b}\right)P$.

7.103 Member ABC only: $B_x = 2.0P \leftarrow$, $B_y = 2.845P \uparrow$,
$C_x = 2.0P \rightarrow$, $C_y = 0.845P \downarrow$.

7.105 Member BC only: (a) $B_x = \dfrac{M_O}{2a} \leftarrow$, $B_y = \dfrac{M_O}{2a} \downarrow$, $C_x = \dfrac{M_O}{2a} \rightarrow$,
$C_y = \dfrac{M_O}{2a} \uparrow$.

7.107 $A_x = \dfrac{2P}{1+\alpha} \leftarrow$, $A_y = \dfrac{2\alpha P}{1+\alpha} \uparrow$, $B_x = \dfrac{2P}{1+\alpha} \leftarrow$ on AB,
$B_y = \dfrac{(1-\alpha)P}{1+\alpha} \uparrow$ on AB, $C_x = \dfrac{2P}{1+\alpha} \rightarrow$, $C_y = \dfrac{2P}{1+\alpha} \uparrow$.

7.109 $A_x = 0.714$ k $\rightarrow$, $A_y = 6.79$ k $\uparrow$, $B_x = 0.714$ k $\leftarrow$ on AB,
$B_y = 3.21$ k $\uparrow$ on AB, $C_x = 10.71$ k $\leftarrow$, $C_y = 3.21$ k $\uparrow$.

7.111 $F = \left(\dfrac{c+d}{2c}\right)P \operatorname{ctn} \theta$.

7.113 $F = 1423$ lb, $\dfrac{F}{P} = 14.23$.

7.115 (a) $F_{BD} = 19.21$ k, (b) $C_x = 9.61$ k $\rightarrow$, $C_y = 10.64$ k $\downarrow$.

7.117 $T = 7.17$ k, $F_{CD} = 2.14$ k (C), $F_{EF} = 8.00$ k (C).

7.119 $F = 103.3$ lb, $A_x = 0$, $A_y = 133.3$ lb.

7.121 $C = 217$ lb·in. $\circlearrowright$.

7.123 $C = 0.921$ kN·m $\circlearrowright$, $A_x = 2.28$ kN $\rightarrow$, $A_y = 4.00$ kN $\uparrow$.

7.125 $R = 693$ N $\rightarrow$.

7.127 $F_{CD} = 3.97$ kN (C), $A_x = 3.97$ kN $\rightarrow$,
$A_y = 9.40$ kN $\downarrow$ both on AC.

7.129 $M = \left[\dfrac{bc}{(a+b)\tan\theta}\right]P \circlearrowright$.

7.131 $M = 1067$ N·m $\circlearrowright$.

7.133 $Q = 129.2$ lb, $A_x = 0$, $A_y = 210$ lb $\uparrow$, $N_C = 80.8$ lb $\uparrow$,
$B_y = 316$ lb $\downarrow$ all on lower handle.

7.135 $p_0 = 281$ kN/m^2, $L_{MAX.} = 16.03$ m.

7.137 $P = 189.0$ N, $A_x = 67.0$ N $\rightarrow$, $A_y = 439$ N $\uparrow$.

7.139 $C_2 = 4000$ lb·in. $\circlearrowright$.

7.141 $F_{CD} = 1707$ lb (C), $F_{BF} = 754$ lb (C).

7.143 $D_y = 7.50$ tons $\uparrow$, $E_y = 17.50$ tons $\uparrow$, $F_{AB} = 19.00$ tons (C).

7.145 (a) $F_B = 500$ N $\leftarrow$; (b) $N_D = 274$ N $\downarrow$,
$N_E = 274$ N $\uparrow$ both on the rod; (c) $C = 27.4$ N·m $\circlearrowright$;
(d) $A_x = 500$ N $\rightarrow$, $A_y = 0$ both on the disk.

7.147 $C_x = 5.00$ kN $\rightarrow$, $C_y = 6.25$ kN $\uparrow$, $D_x = 5.00$ kN $\leftarrow$,
$D_y = 22.5$ kN $\downarrow$, $E_y = 16.25$ kN $\uparrow$.

7.149 $A_x = 0$, $A_y = 0.830$ kN $\downarrow$, $B_x = 6.67$ kN $\rightarrow$, $B_y = 15.00$ kN $\uparrow$,
$C_x = 6.67$ kN $\leftarrow$, $C_y = 14.17$ kN $\downarrow$.

7.151 $B_x = 60.00$ kN $\rightarrow$, $B_y = 87.4$ kN $\uparrow$, $F_E = 127.3$ kN $45°$,
$F_y = 2.63$ kN $\uparrow$.

7.153 $F_{DC} = 24.9$ kN $\rightarrow$, $F_{EB} = 44.8$ kN $13.39°$,
$F_x = 18.68$ kN $\rightarrow$, $F_y = 10.37$ kN $\uparrow$.

7.155 $A_x = 4.00$ kN $\leftarrow$, $A_y = 39.0$ kN $\uparrow$, $F_{BC} = 5.66$ kN $45°$.

7.157 $A_x = 13.58$ kN $\leftarrow$, $A_y = 11.79$ kN $\uparrow$, $B_x = 8.94$ kN $\rightarrow$,
$B_y = 14.47$ kN $\downarrow$, $C_x = 4.64$ kN $\rightarrow$, $C_y = 2.68$ kN $\uparrow$.

7.159 $P = 12.63$ lb.

7.161 $E_x = 18.09$ k $\leftarrow$, $E_y = 2.32$ k $\downarrow$, $F_D = 29.2$ k $\rightarrow$,
$F_B = 11.35$ k $11.77°$,

7.163 $F_E = 116.6$ N $\begin{smallmatrix}5\\3\end{smallmatrix}$, $C_x = 210$ N $\rightarrow$, $C_y = 0$,
$F_x = 150.0$ N $\leftarrow$, $F_y = 100$ N $\downarrow$.

7.165 $F_E = 1194$ N.

Chapter 8

8.1 $R = 19.00$ kN $\leftarrow$.

8.3 $R = 24.0$ kN $\uparrow$.

8.5 $T = 14.00$ kN·m.

8.7   Uniformly loaded beam only: $V = 0$, $M = \dfrac{wL^2}{8}$.

8.9 $V_1 = \dfrac{P}{3}$, $M_1 = \dfrac{Pb}{3}$.

8.11 $V_1 = -\dfrac{3wL}{2}$, $M_1 = -\dfrac{13wL^2}{8}$.

8.13 $V_1 = 0.180w_0 L$, $M_1 = -0.036w_0 L^2$.

8.15 $V_C = 0$, $M_C = \dfrac{PL}{3}$.

8.17   Part (b) only: $V_B^R = 40.0$ kN, $M_B^R = -60.0$ kN·m.

8.19 $V_1 = 2.50$ k, $M_1 = -25.0$ k·ft.

8.21   Between A and B, $V = 0$, $M = -32.0$ kN·m.

8.23 $V_B^L = -V_B^R = -\dfrac{w_0 L}{6}$, $M_B^L = M_B^R = -\dfrac{w_0 L^2}{6}$.

8.25 $V_1 = V_2 = -wL$, $M_1 = -\dfrac{wL^2}{2}$, $M_2 = -wL^2$.

8.27   Bean (a) only: $V_C = -\dfrac{w_0 L^2}{24}$, $M_C = \dfrac{w_0 L^2}{16}$.

8.29 $V_A = \dfrac{7wL}{4}$, $M_A = -\dfrac{5wL^2}{4}$.

8.31 $V_C = 0$, $M_C = \dfrac{5wL^2}{12}$.

8.33 $V_E^L = -\dfrac{wb}{2}$, $M_E^L = -\dfrac{wb^2}{2}$.

8.35 Segment BC only ($b < x < 2b$): $V = 0$, $M = Pb$.

8.37 Segment AB only ($L < x < 2L$): $V = wL$, $M = wL\left(\dfrac{L}{2} - x\right)$.

8.39 Segment BC only ($L < x < 2L$): $V = w\left(\dfrac{5}{4}L - x\right)$,
$M = \dfrac{w}{2}\left(\dfrac{5}{2}Lx - L^2 - x^2\right)$.

8.41 ($0 < x < L$): $V = \dfrac{w_0}{2}\left(\dfrac{L}{3} - \dfrac{x^2}{L}\right)$, $M = \dfrac{w_0}{6}\left(Lx - \dfrac{x^3}{L}\right)$.

8.43 Segment AB only ($0 < x < L$): $V = 0$, $M = -C_0$.

8.45 Segment BC only ($0 < x < 2L$): $V = \dfrac{3}{2}wL - wx$,
$M = \dfrac{w}{2}(3Lx - x^2 - 2L^2)$.

8.47 Segment AB only ($0 < x_1 < L$): $V = \dfrac{w}{2}(L + 2x_1)$,
$M = -\dfrac{w}{2}(Lx_1 + x_1^2)$.

8.49 Segment BC only ($2b < x < 3b$): $V = P$, $M = P(x - 3b)$.

8.51 Segment AB only ($0 < x < 2b$): $V = P$, $M = Px$.

8.53 Segment BC only ($L < x < 2L$): $V = -\dfrac{w_0 L}{2}$,
$M = -\dfrac{w_0 L}{2}\left(x - \dfrac{2L}{3}\right)$.

8.75 (a) $H = 43.5$ kN, $A_y = 13.91$ kN ↑, $B_y = 26.1$ kN ↑;
(b) $h_C = 3.20$ m.

8.77 $P = 46.4$ k, $H = 65.5$ k, $A_y = 65.5$ k ↑, $B_y = 40.9$ k ↑,
$T_{MAX.} = 92.6$ k.

8.79 $H = 16.53$ kN, $A_y = 34.9$ kN ↑, $B_y = 35.1$ kN ↑,
$T_{MAX.} = 38.8$ kN.

8.81 $H = 20.0$ kN, $A_y = 7.50$ kN ↓, $B_y = 7.50$ kN ↓,
$T_{MAX.} = 22.4$ kN.

8.83 $H = 73.6$ k, $A_y = 19.63$ k ↑, $B_y = 30.4$ k ↑, $T_{MAX.} = 79.6$ k.

8.85 $H = 37.5$ k, $h_C = 16.80$ ft.

8.87 (a) $H = 3750$ k; (b) $h = 0.4x - 8 \times 10^{-4}x^2$; (c) $A_y = 1350$ k ↑,
$B_y = 1650$ k ↑; (d) $F_A = 3990$ k, $F_B = 4100$ k.

8.89 $F_{MAX.} = 1077$ k, $F_{MIN.} = 1000$ k.

8.91 (a) $S_T = 120.4$ ft; (b) $\theta_A = -\theta_B = 7.61°$; (c) $F_{MAX.} = 1818$ lb.

8.93 $y = 300\cosh\left(\dfrac{x}{300}\right)$, $F_{MAX.} = 65.7$ k.

8.95 $y = 400 \cosh\left(\dfrac{x}{400}\right)$, $F_A = F_B = 63.7$ k.

8.97 $F_1 = 0$, $V_1 = -0.5wL$, $M_1 = 0.5wL^2$.

8.99 Section 2 only: $F_2 = P$, $V_2 = P$, $M_2 = -PL$.

8.101 Section 1 only: $F_1 = 0$, $V_1 = -P$, $M_1 = -PL$.

8.103 Section 2 only: $F_2 = 0$, $V_2 = -P$, $M_2 = PL$.

8.105 Section 1 only: $F_1 = 0$, $V_1 = wL$, $M_1 = 1.5wL^2$.

8.107 Section 2 only: $F_2 = P$, $V_2 = 1.5P$, $M_2 = 0.625PL$.

8.109 $F_1 = wR\left(\dfrac{1}{2} + \cos^2\theta\right)$, $V_1 = wR\cos\theta\left(\sin\theta - \dfrac{1}{2}\right)$,

$M_1 = \dfrac{wR^2}{2}(1 - \sin\theta + \cos^2\theta)$.

8.111 $F_1 = 0.5P$, $V_1 = -0.5P$, $M_1 = -0.25PR$.

8.113 $M = 0$.

8.115 $F_1 = -\dfrac{wL}{4}$, $V_1 = \dfrac{wL}{4}$, $M_1 = -\dfrac{wL^2}{8}$.

8.117 $F_1 = wL$, $V_1 = -\dfrac{wL}{2}$, $M_1 = \dfrac{wL^2}{2}$.

8.119 $F_1 = 2P/3$, $V_1 = 2P/3$, $M_1 = -PL/3$.

8.121 $F_1 = -3P$, $V_1 = -3P$, $M_1 = 3PL/2$.

8.135 $V_1 = -4.00$ kN, $M_1 = -20.0$ kN·m.

8.137 Section 2 only: $V_2 = -4.33$ k, $M_2 = 26.5$ k·ft.

8.139 $V_1 = 0.600$ kN, $M_1 = -7.50$ kN·m.

8.141 Segment BC only $(3 < x < 5$ ft): $F = 30 + 10x$, $T = 15x$.

8.145 $F = 21.9$ k.

8.147 $F_{MAX.} = 5540$ k.

8.149 $F_2 = 15.00$ kN, $V_2 = 0$, $M_2 = 0$.

8.151 $F_2 = 17.81$ k, $V_2 = 0$, $M_2 = -150.0$ k·ft.

Chapter 9

9.1 $N = 69.3$ lb, $F = 40.0$ lb, $\mu_s = 0.577$.

9.3 $N = 500$ lb, $F = 100.0$ lb, $\mu_s = 0.200$.

9.5 $Q = 52.2$ lb, $N = 226$ lb.

9.7 $P = 1017$ lb, $N = 1202$ lb.

9.9 (a) $Q = 3040$ lb; (b) $Q = 961$ lb.

9.11 $W_A = 399$ N.

9.13 $W_B = 2080$ N.

9.15 (a) $N = 400$ N, $F = 100.0$ N; (b) No.

9.17 For tipping about the right edge, P must be at very top of block.

9.19 $M = \left(\dfrac{rmg\mu_s}{1 + \mu^2}\right)[(1 + \mu_s)\cos\theta + (1 - \mu_s)\sin\theta]$.

9.21 (a) $T = 532$ N; (b) $\mu_{MIN.} = 0.1550$.

9.23 (b) $\theta = \tan^{-1}\left(\dfrac{2\mu}{1-\mu^2}\right)$.

9.25 $\theta = 50.2°$.

9.27 $P = 71.3$ lb, $Q = 60.1$ lb.

9.29 $C = 947$ N·m.

9.31 $N_A = 87.5$ N, $N_B = 217$ N. For $\mu_A = 0.8$, rod is not in equilibrium.

9.33 $T = 342$ N, $\mu = 0.364$.

9.35 $\mu_A = \mu_B = 0.483$.

9.37 $\mu_{MIN.} = 0.372$.

9.39 $P = 550$ lb.

9.41 $P_{AB} = 297$ lb, $Q = 257$ lb.

9.43 $Q/P = \dfrac{2-\mu}{2}$.

9.45 $P = \left(\dfrac{3}{16}\right)W$.

9.47 $Q_x = 20.8$ lb.

9.49 $P = 8.126W$.

9.51 $N_A = \left(\dfrac{b\cos\theta}{a+b}\right)P$, $N_B = \left(\dfrac{a}{a+b}\right)P$, $F_A = \left(\dfrac{b\sin\theta}{a+b}\right)P$.

9.53 $N_A = 2W$, $N_B = \frac{1}{2}W$, $F_A = \frac{1}{2}W$, μ_s is independent of a/b.

9.55 $F_A = 126.4$ N.

9.57 (a) $P = 6.96$ kN; (b) $P = 2.98$ kN.

9.59 $P = 0.827$ kN.

9.61 $W_{MAX.} = 21.0$ kN; self locking.

9.63 $W_E = 300$ lb, $W_F = 300$ lb.

9.65 $W = 360$ N.

9.67 $P = 761$ N, MA $= 1.314$ up the plane.
     $P = 134.9$ N, MA $= 7.41$ down the plane.

9.69 $P = 127.1$ lb, MA $= 0.787$.

9.71 (a) $P_{MIN.} = 39.8$ lb ▱11.47°; (b) $P_{MIN.} = 95.5$ lb ▱28.5°.

9.73 $\dfrac{W_1}{W_2} = 2(\sin\theta + \mu_s\cos\theta)$.

9.75 $\dfrac{W_2}{W_1} = 4(\sin\theta + \mu_s\cos\theta)$.

9.77 MA $= 4.00$.

9.79 $L = 3.22$ in.

9.81 $F_A = 5770$ N.

9.83 $\theta = \tan^{-1}\mu$.

9.85 $T_2 = 194.7$ lb, $M_O = 47.4$ lb·ft, $O_x = 295$ lb $\leftarrow$, $O_y = 20.0$ lb $\uparrow$.

9.87 (a) $T = 1709$ N; (b) $T = 2340$ N.

9.89 $T_{TOP} = 101.7$ kN, $T_{BOT.} = 1.678$ kN.

9.91 $T = 0.1417$ lb.

9.93 $P = 0.739$ kN.

9.95   (a) $P = 1089$ N; (b) $P = 2270$ N.

9.97 $W = 20.9$ k.

9.99 $T_{TOP} = 619$ lb, $T_{BOT.} = 452$ lb, $A_x = 1009$ lb $\leftarrow$, $A_y = 0$,
       $B_x = 1071$ lb $\rightarrow$, $B_y = 3000$ lb $\uparrow$.

9.101 $P = 0.390W$, $Q = 0.284W$. Scheme (b) is preferred.

9.103 $\dfrac{P}{W} = \dfrac{3 - \mu_s}{18\mu_s}$.

9.105 (a) $\dfrac{Q}{W} = e^{2\mu_s\pi}$, (b) $\dfrac{Q}{W} = \dfrac{1}{e^{2\mu_s\pi}}$.

9.107 $P = 24.0$ lb.

9.109 $C = 10.86Pd$.

9.111 $C = 0.445Wr$.

9.113 $W_2 = \left(\dfrac{W_1}{e^{\mu_s(\pi/2 + \theta)}}\right)(\sin\theta - \mu\cos\theta)$.

9.115 $W_2 = 99.0$ N.

9.117 (a) $C = 282$ lb·ft; (b) $N_A = N_B = 716$ lb.

9.119 $P = 0.123W(3.5e^{\mu_s\pi} + 1)/(e^{\mu_s\pi} - 1)$, $T_{TOP} = 0.8We^{\mu_s\pi}/(e^{\mu_s\pi} - 1)$,
       $T_{BOT.} = 0.8W/(e^{\mu_s\pi} - 1)$.

9.121 $\mu_{MIN.} = 0.441$, $N = 1{,}493$ N.

9.123 $T = 8.55$ lb.

9.125 $P = \left[\dfrac{a + \mu c}{\mu r(a + b)}\right]M_O$, $C_x = \dfrac{M_O}{r} \rightarrow$, $C_y = \left[\dfrac{b - \mu c}{\mu r(a + b)}\right]M_O \uparrow$.

9.127 (a) $D_0 = 6.40$ in., $D_i = 4.00$ in.; (b) $T = 5290$ lb·in.

9.129 (a) $p = \dfrac{16P}{3\pi D^2}$; (b) $T = \dfrac{7\mu_s PD}{9\sqrt{2}}$.

9.131 $T = 28.8$ lb·in.

9.133 $T = 245$ lb·in.

9.135 (a) $F_2 = 691$ lb; (b) $F_2 = 566$ lb.

9.137 (a) $F_2 = 1{,}383$ lb; (b) $F_2 = 1{,}131$ lb.

9.139 $C = 16{,}160$ lb·in.

9.141 $P = 796$ lb, sliding.

9.143 (a) $-47.6° \le \theta \le 75.6°$; (b) $75.6° < \theta < 90°$.

9.145 $P = 26.7$ lb, $\theta = 36.9°$.

9.147 $0.339 \le h \le 0.788$ m.

9.149 $\theta = 40.9°$.

9.151 $P = 247$ lb.

9.153 $W = 400$ N, $\mu_k = 0.577$.

9.155 $\mu_{MIN.} = 0.368$.

9.157 $M = 0.1853$ kN·m.

9.159 $\theta = \tan^{-1}\mu_s$, $\dfrac{b}{h} = \mu_s$.

9.161 $Q = 4.29$ kN.

9.163 $r_m = 0.0267$ m.

9.165 (a) $F = 125.8$ N; (b) $F = 1{,}397$ N.

9.167 (a) $C = 4Pr(e^{\mu_s\pi} - 1)$; (b) $C = 4Pr\left(1 - \dfrac{1}{e^{\mu_s\pi}}\right)$.

9.169 $a = 3.01$ in.

9.171 $T = \dfrac{\mu_s PD_0}{4}$.

Chapter 10

10.1 $\bar{y} = -\left(\dfrac{5}{16}\right)R$.

10.3 $\bar{y} = 850$ mm.

10.5 $\bar{x} = \bar{z} = 0$, $\bar{y} = 13.95$ in.

10.7 $\bar{x} = \bar{z} = 0$, $\bar{y} = 4.03$ in.

10.9 $\bar{x} = 0$, $\bar{y} = \dfrac{a^2}{a(1 + \pi/2) + b}$. For $a = \dfrac{b}{2}$, $\bar{y} = \dfrac{1}{6 + \pi}$.

10.11 $\bar{x} = 11.07$ in.

10..13 $\bar{x} = \dfrac{29a}{60}$, $\bar{y} = \dfrac{47b}{60}$.

10.15 (a) $\bar{x} = 0$, $\bar{y} = \bar{z} = \dfrac{c}{4}$; (b) $\bar{x} = \bar{y} = 0$, $\bar{z} = \dfrac{c}{2}$; (c) $\bar{x} = \bar{y} = \bar{z} = \dfrac{c}{6}$

10.17 For $\dfrac{h}{a} = 2$, $\dfrac{\bar{y}}{a} = \dfrac{3[2 - (b/a)^2]}{2[3 - (b/a)^2]}$.

10.19 (a) $\bar{y} = 0.777$ b; (b) $\bar{y} = b$.

10.21 $\bar{y} = -1.667$ in.

10.23 $\bar{y} = -3.55$ in.

10.25 $\bar{x} = 0.039$ m, $\bar{y} = 0.086$ m.

10.27 $\bar{x} = 3.02$ in., $\bar{y} = 3.08$ in.

10.29 $\bar{x} = 38.6$ mm, $\bar{y} = 22.1$ mm.

10.31 $\bar{x} = 50.0$ mm, $\bar{y} = 47.7$ mm.

10.33 $L = 6.05$ in.

10.35 $L = 0.936\ R$.

10.37 $\bar{x} = -1.500$ units.

10.39 $\bar{x} = 2.40$ units, $\bar{y} = 0$.

10.41 $\bar{x} = \dfrac{2}{5}$ units, $\bar{y} = \dfrac{1}{2}$ units.

10.43 $\bar{x} = 4.40$ in., $\bar{y} = 0.500$ in.

10.45 $\bar{x} = 0$, $\bar{y} = 0.393$ units.

10.47 $\bar{x} = -1.600$ m, $\bar{y} = -2.29$ m.

10.49 $\bar{x} = \dfrac{\pi}{2}$ m, $\bar{y} = \dfrac{\pi}{8}$ m.

10.51 $\bar{y} = 0.894$ in.

10.53 $\bar{y} = 17.93$ ft.

10.55 $V = 1047.2$ in.^{3}, $\bar{x} = \bar{z} = 0$, $\bar{y} = 7.50$ in.

10.57 $\bar{x} = \bar{z} = 0$, $\bar{y} = 0.278$ units.

10.59 $\bar{x} = 8.50$ in., $\bar{y} = \bar{z} = 0.$

10.61 $\bar{x} = \bar{z} = 0, \bar{y} = 2.67$ m.

10.63 $\bar{x} = \bar{z} = 0, \bar{y} = \dfrac{3}{8}r.$

10.65 $\bar{x} = \bar{y} = \dfrac{1}{3}$ units, $\bar{z} = \dfrac{L}{2}.$

10.67 $\bar{x} = 0, \bar{y} = \dfrac{2a^2}{5b}, \bar{z} = \dfrac{L}{2}.$

10.69 $\bar{x} = \bar{z} = 0, \bar{y} = 5.33$ units.

10.71 $\bar{x} = \dfrac{2}{3}a, \bar{y} = \bar{z} = 0.$

10.73 $\bar{x} = \bar{z} = 0, \bar{y} = 14.67$ units.

10.75 $\bar{x} = \bar{z} = 0, \bar{y} = \left(\dfrac{1}{3}\right)h.$

10.77 $W = 442$ tons.

10.79 $A = 32.9$ in.^{2}, $V = 4.11$ in.^{3}, Vol. of coating $= 1.316$ in.^{3}.

10.81   (a) $V = 565$ m^3; (b) $V = 47{,}900$ m^3.

10.83 $V = 2\left[\left(H + \dfrac{L^2}{4H}\right)(WD)\right]\sin^{-1}\left(\dfrac{L}{H + L^2/4H}\right).$

10.85 $V = 62.8$ ft^3.

10.87 $A = 4\pi a^2, V = \dfrac{4}{3}\pi a^3.$

10.89 $A = \pi a\sqrt{a^2 + h^2}, V = \dfrac{1}{3}\pi a^2 h.$

10.91 $A = \dfrac{64}{3}\pi b^2, V = \dfrac{32}{3}\pi b^3.$

10.93 $T = 14{,}908$ lb, $A_x = 29{,}900$ lb $\leftarrow$, $A_y = 0.$

10.95 $Q = 13{,}350$ lb $\uparrow$.

10.97 $S_x = 11.12$ kN $\rightarrow$, $H_x = 4.74$ kN $\leftarrow$, $H_y = 11.05$ kN $\uparrow$.

10.99   (a) $R = 40{,}400$ lb $\leftarrow$, $\bar{y} = 7.30$ ft below the hinge;
        (b) $R = 24{,}700$ lb $\leftarrow$, $\bar{y} = 8.73$ ft below the hinge.

10.101 $R = 272$ kN $30°\!\!\swarrow$, $Q = 48.0$ kN/m.

10.103 $F_R = 36{,}200$ lb.

10.105 $\gamma_P = \gamma.$

10.107 $P = 0.1848$ kN $\downarrow$.

10.109 $R_x = 78{,}000$ lb $\leftarrow$, $R_y = 196{,}720$ lb, $x = 18.14$ ft left of T.

10.111 Design A is acceptable. Design B is not.

10.113 $\bar{x} = 2.14$ m, $\bar{y} = 1.837$ m.

10.115 $r = 2.50$ in.

10.117 $\bar{x} = 0.400$ m, $\bar{y} = 0.500$ m.

10.119 $\bar{y} = -0.500$ in.

10.121 $\bar{x} = 0.500$ m, $\bar{y} = \bar{z} = 0.$

10.123 $W = 390$ N.

10.125 $V = 122.7$ in.^{3}.

10.127 $R = 17,040$ lb $\angle 60°$, $\bar{y} = 6.86$ ft from the origin along the y axis.

Chapter 11

11.1 $A = 48.8$ in.^{2}, $I_X = 205$ in.^{4}, $I_Y = 410$ in.^{4}, $J_C = 615$ in.^{4}, $J_O = 3200$ in.^{4}.

11.3   (a) $I_X = \left(\dfrac{1}{36}\right)bh^3$, $I_Y = \left(\dfrac{1}{36}\right)hb^3$; (c) $k_X = \left(\dfrac{1}{\sqrt{18}}\right)h$.

11.5 $I_{B-B} = \left(\dfrac{13}{12}\right)bh^3$, $I_{C-C} = \left(\dfrac{1}{3}\right)bh^3$.

11.7 $J_C = 0.50\pi$ in.^{4}.

11.9 $k_x = 12.65$ in., $k_y = 15.33$ in.

11.11 $k_x = \left(\dfrac{\sqrt{127}}{5}\right)R$.

11.13 $I_X = 2000$ kg·m^2, $I_Z = 1333$ kg·m^2, $m = 267$ kg.

11.15 $I_x = \left(\dfrac{1}{12}\right)bh^3$, $I_X = \left(\dfrac{1}{36}\right)bh^3$, $k_X = \left(\dfrac{1}{\sqrt{18}}\right)h$.

11.17 $I_x = 0.201$ m^4, $k_x = 0.400$ m.

11.19 $I_x = \left(\dfrac{1}{3}\right)bh^3$, where $b = \dfrac{t}{\sin\theta}$, $h = L\sin\theta$.

11.21 $I_y = 1{,}250$ in.^{4}, $I_Y = 78.1$ in.^{4}.

11.23 $I_y = 15{,}140$ in.^{4}, $I_Y = 2{,}410$ in.^{4}.

11.25 $I_y = 0.036$ m^4, $k_y = 0.465$ m.

11.27 $I_y = 25.6$ in.^{4}, $J_O = 98.7$ in.^{4}.

11.29 $I_y = 21.9$ ft^4, $k_y = 2.03$ ft.

11.31 $I_Z = \left(\dfrac{D^2}{8}\right)(m)$.

11.33 $I_y = 0.216(m)$.

11.35 $I_y = \left(\dfrac{1}{3}\right)\left(L^2 + \dfrac{a^2}{4}\right)(m)$.

11.37 $I_y = \left(\dfrac{10}{9}\right)h^4(m)$.

11.39 $I_z = \left(\dfrac{h}{6}\right)(m)$.

11.41 $I_z = 22.9$ slug·in.^{2}.

11.43 $I_y = 3.57$ slug·in.^{2}.

11.45 $I_Z = \left(\dfrac{h^2}{12}\right)(m)$.

11.47 $I_x = 677$ in.^{4}, $k_x = 4.92$ in.

11.49 $I_x = 759 \times 10^{-6}$ m^4, $k_x = 0.1194$ m.

11.51 $I_x = 688$ in.^{4}, $k_x = 4.37$ in.

11.53 $I_X = 1060 \times 10^{-6} \ m^4$, $k_X = 0.1486$ m.

11.55 $I_X = 896$ in.^{4}, $k_X = 4.32$ in.

11.57 $I_X = 9.78 \times 10^{-6} \ m^4$, $k_X = 3.34 \times 10^{-2}$ m.

11.59 $I_X = I_Y = \dfrac{\pi}{64}(D_0^4 - D_i^4)$, $J_C = \dfrac{\pi}{32}(D_0^4 - D_i^4)$.

11.61 $I_X = 11.50a^4$, $I_Y = 7.50a^4$.

11.63 $I_X = 108.1$ in.^{4}, $I_Y = 206$ in.^{4}.

11.65 $I_Y = 177.1$ in.^{4}, $k_Y = 2.11$ in.

11.67 $I_Y = 302$ in.^{4}, $k_Y = 2.36$ in.

11.69 $I_y = 0.01564\rho$ kg·m^2, $k_y = 0.651$ m.

11.71 $I_x = I_y = 28.955 \times 10^6 \rho$ slug·in^2, $k_x = k_y = 40.2$ in.

11.73 $I_z = 6.22$ slug·ft^2, $k_z = 0.985$ ft.

11.75 $I_x = 1.094$ kg·m^2.

11.77 $I_x = 1.428$ slug·ft^2, $k_x = 0.968$ ft.

11.79 $I_z = 1.229$ slug·ft^2, $k_z = 0.898$ ft.

11.81 $I_y = 17.42$ kg·m^2, $k_y = 0.504$ m.

11.83 $I_y = (2.3125m_d + 5.0834m_r)D^2$.

11.85 $I_z = 11.830$ kg·m^2, $k_z = 0.662$ m.

11.87 (a) $I_{xy} = \left(\dfrac{1}{3}\right)a^6$; (b) $I_{XY} = \left(\dfrac{1}{30}\right)a^6$.

11.89 (a) $I_{xy} = 1{,}500$ in.^{4}; (b) $I_{XY} = 149.8$ in.^{4}.

11.91 (a) $I_{xy} = 132.0$ in.^{4}; (b) $I_{XY} = 0$.

11.93 $I_{xy} = 580$ in.^{4}, $I_{XY} = 0$.

11.95 $I_{XY} = 1.222 \times 10^{-6} \ m^4$.

11.97 (a) $I_u = 30{,}700$ in.^{4}, $I_v = 329$ in.^{4};
       (b) $I_{x'} = 18{,}120$ in.^{4}, $I_{x'y'} = 14{,}950$ in.^{4}.

11.99 (a) $I_u = 8.51 \times 10^{-2} \ m^4$, $I_v = 5.80 \times 10^{-2} \ m^4$,
       (b) $I_{x'} = 2.82 \times 10^{-2} \ m^4$, $I_{x'y'} = 3.97 \times 10^{-2} \ m^4$.

11.101 (a) $I_u = 3.55 \times 10^{-3} \ m^4$, $I_v = 0.403 \times 10^{-3} \ m^4$;
       (b) $I_U = 1.070 \times 10^{-3} \ m^4$, $I_V = 0.287 \times 10^{-3} \ m^4$.

11.103 $I_U = 1{,}206$ in.^{4}, $I_V = 153.6$ in.^{4}.

11.105 (a) $I_u = 10.78 \times 10^{-2} \ m^4$, $I_v = 2.96 \times 10^{-2} \ m^4$;
       (b) $I_{x'} = 10.68 \times 10^{-2} \ m^4$, $I_{x'y'} = 0.881 \times 10^{-2} \ m^4$.

11.107 (a) $I_U = 542$ in.^{4}, $I_V = 229$ in.^{4};
       (b) $I_{X'} = 490$ in.^{4}, $I_{X'Y'} = 116.4$ in.^{4}.

11.109 (a) $I_U = 176.6$ in.^{4}, $I_V = 33.5$ in.^{4};
       (b) $I_{X'} = 111.3$ in.^{4}, $I_{X'Y} = 71.3$ in.^{4}.

11.111 (a) $I_u = 1{,}491$ in.^{4}, $I_v = 269$ in.^{4};
       (b) $I_U - 314$ in.^{4}, $I_V = 172.0$ in.^{4}.

11.113 $I_U = 11.00 \times 10^{-6} \ m^4$, $I_V = 8.56 \times 10^{-6} \ m^4$.

11.115 $I_{OA} = 1.806a^2(m)$.

11.117 $I_{xy} = \left(\dfrac{4}{5}\right)a^2(m)$, $I_{xz} = -\left(\dfrac{6}{5}\right)a^2(m)$, $I_{yz} = -\left(\dfrac{2}{5}\right)a^2(m)$.

11.119 $I_{xy} = -28.6$ kg·m^2, $I_{xz} = 5.72$ kg·m^2, $I_{yz} = 15.26$ kg·m^2.

11.121 $I_{xy} = 14.21a^2(m)$, $I_{xz} = 1.702a^2(m)$.

11.123 $I_u = 3.43a^2(m)$, $I_w = 3.27a^2(m)$, $I_v = 0.374a^2(m)$. For the v axis: $\lambda_x = 0.815$, $\lambda_y = 0.322$, $\lambda_z = -0.482$.

11.125 $I_x = 7{,}200$ in.^{4}, $k_x = 7.39$ in.

11.127 $I_y = 33.1$ in.^{4}, $k_y = 1.882$ in.

11.129 $I_x = 95.426\rho$ kg·m^2.

11.131 $I_x = 26{,}200$ in.^{4}, $I_y = 24{,}600$ in.^{4}.

11.133 $I_X = 116.6$ in.^{4}, $k_X = 1.945$ in.

11.135 $I_x = 0.0643$ kg·m^2.

11.137 (a) $I_{xy} = 1{,}276$ in.^{4}, (b) $I_{xy} = 5{,}210$ in.^{4}.

11.139 $I_u = 99.3$ in.^{4}, $I_v = 6.44$ in.^{4}.

11.141 $I_x = 2b^2(m)$, $I_y = 22.05b^2(m)$, $I_z = 23.85b^2(m)$.

11.143 $I_u = 24.0b^2(m)$, $I_w = 23.4b^2(m)$, $I_v = 0.491b^2(m)$. For the v axis: $\lambda_x = 0.967$, $\lambda_y = 0.244$, $\lambda_z = 0.0791$.

Chapter 12

12.1 (a) $U = -Wh$; (b) $U = Wh$.

12.3 (a) $U = 76{,}600$ lb·ft; (b) $U = -76{,}600$ lb. ft.

12.5 (a) $U = -1039$ lb·ft; (b) $U = -240$ lb·ft; (c) $U = 0$.

12.7 (a) $U = -41.0$ k·ft; (b) $U = -4.57$ k·ft, (c) $U = 50.0$ k·ft.

12.9 $U = 58.7$ kN·m.

12.11 $U = 240$ N·m.

12.13 $k = 57.7$ lb/in.

12.15 (a) $P = \left(\dfrac{2\sin\alpha}{2 + \cos\alpha}\right)W$; (b) $\alpha = 8.60°$.

12.17 (a) $Q = \frac{1}{2}(ka\sin\beta - W\tan\beta)$; (b) $F = ka\sin\beta$; (c) $Q = 174.2$ lb, $F = 360$ lb.

12.19 (a) $F = \frac{1}{2}W(\mu\cos\theta + \sin\theta)$; (b) $\mu = 0.1925$.

12.21 $\theta = 38.7°$.

12.23 $h = \sqrt{a^2 - (M/W)^2}$.

12.25 $P = \left(\dfrac{4}{3}\right)kL(1 - \cos\alpha)\tan\alpha$.

12.27 (a) $M = \left[\dfrac{p(a\cos\beta + b\sin\beta)}{a\cos\beta}\right]Q$; (b) $M = 6.44$ lb·in.

12.29 $M = 0.289$ kN·m.

12.31 $\alpha = 0°$ and $42.3°$.

12.33 $Q = \dfrac{M}{c\cos\alpha}$, $Q = 3.54$ kN.

12.35 $\alpha_1 = 14.36°$, $\alpha_2 = 63.4°$.

12.37 $U = -3{,}600$ lb·ft.

12.39 (a) $U = -4{,}000$ lb·ft; (b) $U = 4{,}220$ lb·ft.

12.41 $U = -5{,}100$ lb·ft.

12.43 (a) $U = -4{,}160$ lb·ft; (b) $U = 4{,}460$ lb·ft.

12.45   (a) $U = -8ka^2$; (b) $U = -8ka^2$.

12.47   See answer to 12.13.

12.49   See answers to 12.17.

12.51   See answer to 12.25.

12.53 $\beta_1 = 63.4°$, $\beta_2 = 84.6°$.

12.55   (a) $\alpha = 1.639 \times 10^{-5}$ rad ccw from the horizontal;
        (b) $F_{AD} = 16.39$ kN, $F_{BE} = 39.3$ kN.

12.57 $\mu = 2$ units, stable; $\mu = -2$ units, unstable.

12.59 $\alpha = 58.0°$, stable.

12.61 $\beta = 90°$, $k < \left(\dfrac{W}{2L}\right)$; $\beta = \sin^{-1}\left(\dfrac{W}{2kL}\right)$, $k > \left(\dfrac{W^2}{4}\right)$.

12.63 $k > \left(\dfrac{2W}{L}\right)$.

12.65 $\alpha = 42.3°$, stable.

12.67 $K = 3Ph$; $\alpha = 22.4°$, stable; $\alpha = 53.2°$, stable.

12.69 $k > \left(\dfrac{W}{b}\right)$.

12.71 $(0, \sqrt{2})$, stable; $(0, -\sqrt{2})$, unstable.

12.73 $\alpha_1 = \left(\dfrac{1}{3}\right)(\tan^{-1}\alpha_2)$.

12.75 $\theta = \cos^{-1}\left(1 - \dfrac{W}{2kL}\right)$, $\theta = 56.3°$.

12.77 $W_2 = W_1 \Big/ \left[\left(\dfrac{L}{b}\right)\cos^3\theta - 1\right]$, $\dfrac{L}{b} > \left(\dfrac{1}{\cos^3\theta}\right)$.

12.79 $T = 51.6$ lb·ft.

12.81   See answers to 12.77.

12.83 $\theta = \dfrac{WR}{4R^2k + K}$, stable.

12.85 $k > \left(\dfrac{W}{2b}\right)$.

Index

NuSolve Installation

The users must provide their own DOS in order to operate the program. The following procedure will enable the users to copy the NuSolve 1.0 into the hard drive C.

Insert the NuSolve disk into the A drive and type **A:** When the prompt A> appears, type **Setup C** and press the return key. This will cause the prompt C:\NUSOLVE> to appear. Now, type **NUSOLVE** and press the return key. This is the final step that will cause the title of the program and authors to appear.

NuSolve 1.0 Limitations and Execution

In all cases, the user can select a particular program by simply typing the number appearing to the left of the menu or by using the arrow keys on the keyboard. In all cases, you can use the arrow keys to move between options. just remember the following:

(1)   Type Q to exist the NuSolve program.
(2)   Use the arrow keys to move between options.
(3)   Press the return key to execute the next step.
(4)   Most programs permit the user to save the data.
(5)   You can read saved data files.
(6)   Matrices or data is displayed using a spreadsheet format.
(7)   Matrices are limited to 25×25 or less.
(8)   Integrals are evaluatcd using thc trapezoidal and Simpson's rule up to 100 data points.
(9)   Roots of polynomials up to 25th degrees including complex roots.
(10)   Solves up to 25 equations by 25 unknowns.

Macintosh Software and the Internet

Currently, engineering education is being impacted by the expanded use of the Internet. The WWW provides a wonderful source of computer programs for solving engineering problems. Consequently, students who wish to use a Macintosh computer may download programs from the WWW. This is not an easy exercise considering the overwhelming number of sites that must be examined. The authors wish to minimize the amount of effort required by suggesting the following two programs:

MathPad *(free noncommercial distribution)*
(`www:http://pubpages.unh.edu/~whd/MathPad/`)

This is a general-purpose graphing scientific calculator. It uses a text window rather than simulating buttons on a hand-held calculator. This live scratched interface allows the user to see and edit the entire calculation. Many examples of engineering problems are provided. These include roots, graphs, systems of equations, and other applications.

Xfunctions *(free noncommercial distribution)*
(`www:http://hws3.hws.edu:9000/eck/index.html`)

This is a general-purpose program that helps students concentrate on learning more about mathematics, rather than learning about computers. It is fun just to use and execute. The program provides animation of a family of functions of the form $f(x,k)$; graphing of derivatives and tangent lines; Riemann sums, with graphical display; graphs of parametrically defined curves; integral curves of vector fields; and 3-D plots of arbitrary functions of the form $z = f(x,y)$.

1. UNITS CONVERTER

This programs permits the user to convert units from US Customary units to SI and vice versa. The user should make a selection of the units to be converted by typing the option number, then using the arrow keys to move between options.

2. MATRIX MULTIPLICATION

This program gives the product of matrix [A] and [B] provided the number of rows in [B] is equal to the number of columns in [A]. The matrices are limited to 25×25 or less.

$$[R] = [A][B]$$

Note that the result can be saved and the resulting matrix multiplied by yet another matrix. This process may be continued indefinitely to multiply several matrices.

3. MATRIX INVERSION

Using any of the two programs in this menu, the user is asked to input a square matrix [A] and its inverse is calculated as the [R] matrix. This implies that

$$[A][R] = [R][A] = [I]$$

where [I] is the identity matrix and [R] = $[A]^{-1}$.

3.1 Nonsymmetrical Matrices
This program calculates the inverse of a square matrix of size 25×25 or less provided its determinant is not equal to zero.

3.2 Symmetrical Matrices
This program calculates the inverse of a square symmetrical matrix of size 25×25 or less provided its determinant is not equal to zero. The main advantage of this program is that it requires less input.

4. LINEAR EQUATIONS

This menu provides a program for solving linear algebraic equations with real coefficients. Hence, given

$$[A]\{x\} = \{b\}$$

The user is as required to input the coefficient matrix [A] and the $\{b\}$ vector. The solution is then displayed as the $\{x\}$ vector.

5. NONLINEAR EQUATIONS

5.1 Newton's Method
This program approximates the roots of a nonlinear algebraic equation using the Newton method of tangents. The user is asked to input the equation using normal arithmetic operations. The program will then analyze the expression and approximate its derivative automatically, thus eliminating one of the most significant drawback associated with Newton's method. The derivative is approximated using a 7-term central finite difference formula. The root is displayed once an initial root, number of iterations (normally 10 iterations is adequate) and an h-value (h is normally taken as 0.01 to 0.001) are specified.

It is recommended that the user first plot the inputted expression so that an appropriate estimate of the root(s) can be made. This can be done using Menu 10.

5.2 Roots of Polynomials
This programs uses the Lin–Bairstow method of quadratic factors which is one of the most efficient techniques for determining both *real* and/or *complex* roots of polynomials with real coefficients. Thus, given

$$f(x) = A_0 + A_1 x + A_2 x^2 + \cdots + A_n x^n = 0$$

The user is asked to input the coefficients A_0 through A_n. This program enables the user to approximate the real and/or complex roots of a given polynomial of order 25 or less.

6. EIGENVALUES

This program approximates the eigenvalues and the corresponding Eigenvectors for a symmetrical matrix of 25×25 size or less. The user is asked to input the upper triangular coefficients along with the minimum allowed off-diagonal coefficient to be used in the Jacobi method transformation. This value may be taken as 0.01. Therefore, given an $n \times n$ symmetrical matrix $[A]$, then

$$[[A] - l[I]] \{x\} = \{0\}$$

The program computes the eigenvalues l_1, l_2, ... , and l_n along with the corresponding Eigenvectors $\{v\}_1$, $\{v\}_2$,..., and $\{v\}_n$. The Eigenvector matrix is displayed as a square matrix $[\{v\}_1, \{v\}_2,..., \text{and } \{v\}_n]$.

7. CURVE FITTING

7.1 Fitting $y = ax + b$
This program calculates the regression coefficient for a linear relationship between a dependent variable y_i and an independent variable x_i. The derived equation $y = ax + b$ represents the line of best fit. Furthermore, the correlation coefficient of determination is also calculated so that the user is able to evaluate the quality of the regression equation.

7.2 Fitting $y = ax^b$
This program calculates the regression coefficient for a linear relationship between a dependent variable y_i and an independent variable x_i. The derived equation $y = ax + b$ represents the line of best fit. Furthermore, the correlation coefficient of determination is also calculated so that the user is able to evaluate the quality of the regression equation.

7.3 Fitting $y = a \ln x + b$
This program calculates the regression coefficient for a linear relationship between a dependent variable y_i and an independent variable x_i. The derived equation $y = ax + b$ represents the line of best fit. Furthermore, the correlation coefficient of determination is also calculated so that the user is able to evaluate the quality of the regression equation.

8. DERIVATIVES

This program approximates the first and second derivative of any continuous function at a given point specified by the user.

8.1 First Derivative
This program approximates the first derivative of any continuous function at a given point. The user is permitted to input the function directly from the screen and to select appropriate interval values (typically taken as 0.01 to 0.001). The forward, central, and backward difference approximations are calculated automatically using 7-term formulas.

8.2 Second Derivative
This program approximates the second derivative of any continuous function at a given point. The user is permitted to input the function directly from the screen and to select appropriate interval values (typically taken as 0.01 to 0.001). The forward, central, and backward difference approximations are calculated automatically using 7-term formulas.

9. INTEGRATION

9.1 Trapezoidal Rule
This program approximates the integral of a given function once the limits of integration and the number of intervals are specified. The user is permitted to input the function from the screen and to assume the number of increments for the inputted function.

$$I = \int_{x_1}^{x_2} f(x)\,dx = \frac{h}{2}[f(x_1) + 2f(x_1 + h)$$
$$+ 2f(x_1 + h) + \cdots + f(x_2)]$$

where h is the increment between two adjacent base points. This program approximate the integral of the function using up to 100 increments.

9.2 Simpson's 1/3 Rule

This program approximates the integral of a given function once the limits of integration and the number of intervals are specified. The user is permitted to input the function from the screen and to assume the number of increments. Note that the number of increment must be even. Thus,

$$I = \int_{x_1}^{x_2} f(x)\,dx = \frac{h}{3}[f(x_1) + 4f(x_1 + h)$$
$$+ 2f(x_1 + h) + \cdots + f(x_2)]$$

where h is the increment between two adjacent base points. This program approximate the integral of the function using up to 100 increments.

10. GRAPHICS

10.1 Plotting Data

This program permits the user to plot any arbitrary data set (x_i, y_i) and connect the points with a series of straight lines. The program is limited to 25 data points and provides the user with several options. These options include showing the grid lines and limits on the x and y axes. The user is asked to input the number of data sets to be plotted and once inputted, the graph will appear on the screen. The user can then change the attributes of the graph using the function keys as specified on the screen. For example pressing F1 will enable the user to change the minimum value of the x-axis.

10.2 Plotting Functions

This program permits the user to plot a function defined by the user. The graph will normally be a smooth curve. The exception is a function that behaves wildly such as the $\tan(x)$. In such cases, you must reduce the limits on the x-axis so that a potion of the curve is plotted. The user is asked to input the number function to be plotted and once inputted, the graph will appear on the screen. The user can then change the attributes of the graph using the function keys as specified on the screen. For example pressing F1 will enable the user to change the minimum value of the x-axis.

NuSolve 1.0
Statics for Engineers
© 1997 Springer-Verlag New York, Inc.